NOTES ON
ACI 318-05

BUILDING CODE REQUIREMENTS FOR STRUCTURAL CONCRETE

with Design Applications

Edited by: Mahmoud E. Kamara
             Basile G. Rabbat

Portland Cement Association
5420 Old Orchard Road
Skokie, Illinois 60077-1083
847.966.6200 Fax 847.966.9781
www.cement.org

An organization of cement companies to improve and extend the uses of portland cement and concrete through market development, engineering, research, education, and public affairs work.

© 2005 Portland Cement Association

Ninth edition, First printing, 2005

Printed in U.S.A.

Library of Congress catalog card number 2005906398

ISBN 0-89312-245-9

Portland Cement Association ("PCA") is a not-for-profit organization and provides this publication solely for the continuing education of qualified professionals. THIS PUBLICATION SHOULD ONLY BE USED BY QUALIFIED PROFESSIONALS who possess all required license(s), who are competent to evaluate the significance and limitations of the information provided herein, and who accept total responsibility for the application of this information. OTHER READERS SHOULD OBTAIN ASSISTANCE FROM A QUALIFIED PROFESSIONAL BEFORE PROCEEDING.

PCA and its members make no express or implied warranty in connection with this publication or any information contained herein. In particular, no warranty is made of merchantability or fitness for a particular purpose. PCA and its members disclaim any product liability (including without limitation any strict liability in tort) in connection with this publication or any information contained herein.

About the building on the cover

Cover photo: Trump International Hotel and Tower. Photo courtesy of SOM
Owner: 401 North Wabash Joint Venture LLC
Renderings: SOM
Architect: SOM
Structural Engineer: SOM

Preface

The first edition of this reference manual was developed to aid users in applying the provisions of the 1971 edition of "Building Code Requirements for Reinforced Concrete (ACI 318-71)." The second through fifth editions updated the material in conformity with provisions of the 1977 code edition, the 1980 code supplement, and the 1983 and 1989 code editions, respectively. The sixth, seventh and eighth editions addressed the 1995, 1999, and 2002 editions of "Building Code Requirements for Structural Concrete (ACI 318-95), (ACI 318-99), and (ACI 318-02)." Through eight editions, much of the initial material has been revised to better emphasize the subject matter, and new chapters added to assist the designer in proper application of the ACI 318 design provisions.

This ninth edition reflects the contents of "Building Code Requirements for Structural Concrete (ACI 318-05)." The text and design examples have been revised to reflect, where possible, comments received from users of the""Notes" who suggested improvements in wording, identified errors, and recommended items for inclusion or deletion.

The primary purpose for publishing this manual is to assist the engineer and architect in the proper application of the ACI 318-05 design standard. The emphasis is placed on "how-to-use" the code. For complete background information on the development of the code provisions, the reader is referred to the""Commentary on Building Code Requirements for Structural Concrete (ACI 318R-05)" which, starting with the 1989 edition, has been published together with the code itself under the same cover.

This manual is also a valuable aid to educators, contractors, materials and products manufacturers, building code authorities, inspectors, and others involved in the design, construction, and regulation of concrete structures.

Although every attempt has been made to impart editorial consistency to the thirty-four chapters, some inconsistencies probably still remain. A few typographical and other errors are probably also to be found. PCA would be grateful to any reader who would bring such errors and inconsistencies to our attention. Other suggestions for improvement are also most sincerely welcome.

Basile G. Rabbat
Engineered Structures Department

Acknowledgments

The following Portland Cement Association staff members have contributed to update of this document:

Iyad (Ed) M. Alsamsam	Daniel Antoniak	Attila Beres
Mahmoud E. Kamara	K. Mark Kluver	Joseph J. Messersmith, Jr
Michael Mota	Basile G. Rabbat	Stephen V. Skalko
Amy Reineke Trygestad	Zhijie (Max) Yan	

The following consultants assisted with update of the 2005 edition:

Kenneth B. Bondy, Consulting Structural Engineer, West Hills, CA, updated Part 26, Prestressed Slab Systems
Ronald A. Cook, University of Florida, Gainesville, FL, developed Part 34, Anchoring to Concrete, for the 2002 edition and updated it for this edition
S.K. Ghosh, S.K. Ghosh Associates Inc., Northbrook, IL, helped update Part 29, Special Provisions for Seismic Design, and reviewed extensively Part 1, General Requirements
David A. Fanella of Graef, Anhalt, Schloemer Associates, Inc. developed Example 29.8.

Dale McFarlane of PCA staff played a crucial role in the production of this publication. He was responsible for the word processing, layout and formatting of this large and complex manuscript. His assistance is very much appreciated.

Finally, sincere gratitude must be expressed to the authors and contributors of various parts of the first through eighth editions of the "Notes." Their initial work is carried over into this edition, although their names are no longer separately identified with the various parts. Robert F. Mast, BERGER/ABAM, Federal Way, WA, updated Parts 5, 6, 7, 8, 24, and 25, of the 2002 edition, all pertaining to application of the Unified Design Provisions. Last, but not least, James Doyle, Consultant, Chaplin, KY, performed a thorough review of Part 10, Deflections, for the 2002 edition.

Basile G. Rabbat

Credits

Parts	Author(s)	Reviewers
1	James J. Messersmith, S.K. Ghosh	Mark Kluver, Stephen V. Skalko
2	K. Mark Kluver	James J. Messersmith, Stephen V. Skalko
3	Stephen V. Skalko	James J. Messersmith, K. Mark Kluver
4	Michael Mota	Basile G. Rabbat
5	Mahmoud E. Kamara	Iyad (Ed) M. Alsamsam
6	Mahmoud E. Kamara	Iyad (Ed) M. Alsamsam
7	Mahmoud E. Kamara	Daniel Antoniak
8	Daniel Antoniak	Basile G. Rabbat
9	Mahmoud E. Kamara	Iyad (Ed) M. Alsamsam
10	Mahmoud E. Kamara	Basile G. Rabbat
11	Mahmoud E. Kamara	Daniel Antoniak
12	Daniel Antoniak	Basile G. Rabbat
13	Daniel Antoniak	Basile G. Rabbat
14	Basile G. Rabbat	Mahmoud E. Kamara
15	Basile G. Rabbat	Mahmoud E. Kamara
16	Zhijie (Max) Yan	Mahmoud E. Kamara
17	Attila Beres	Basile G. Rabbat
18	Mahmoud E. Kamara	Michael Mota
19	Mahmoud E. Kamara	Michael Mota
20	Mahmoud E. Kamara	Michael Mota
21	Zhijie (Max) Yan	Michael Mota
22	Zhijie (Max) Yan	Amy Reineke Trygestad
23	Mahmoud E. Kamara	Basile G. Rabbat
24	Mahmoud E. Kamara	Basile G. Rabbat
25	Mahmoud E. Kamara	Basile G. Rabbat
26	Kenneth B. Bondy	Mahmoud E. Kamara
27	Basile G. Rabbat	Mahmoud E. Kamara
28	Basile G. Rabbat	Mahmoud E. Kamara
29	S.K. Ghosh, David A. Fanella	Basile G. Rabbat
30	James J. Messersmith	K. Mark Kluver, Stephen V. Skalko
31	Mahmoud E. Kamara	Basile G. Rabbat
32	Mahmoud E. Kamara	Basile G. Rabbat
33	Mahmoud E. Kamara	Basile G. Rabbat
34	Ronald A. Cook	Basile G. Rabbat

Contents

1 General Requirements ... 1-1
1.1 SCOPE ... 1-1
1.1.6 Soil-Supported Slabs ... 1-6
1.1.8 Special Provisions for Earthquake Resistance ... 1-6
1.2 DRAWINGS AND SPECIFICATIONS ... 1-11
1.2.1 Items Required to be Shown ... 1-12
1.3 INSPECTION ... 1-12
1.3.4 Records of Inspection ... 1-13
1.3.5 Special Inspections ... 1-13
REFERENCES ... 1-14

2 Materials, Concrete Quality ... 2-1
3—MATERIALS ... 2-1
UPDATE FOR THE '05 CODE ... 2-1
3.1 TESTS OF MATERIALS ... 2-1
3.2 CEMENTS ... 2-1
3.3 AGGREGATES ... 2-2
3.4 WATER ... 2-2
3.5 STEEL REINFORCEMENT ... 2-3
3.5.2 Welding of Reinforcement ... 2-3
3.5.3 Deformed Reinforcement ... 2-3
3.6 ADMIXTURES ... 2-7
3.6.9 Silica Fume ... 2-7
4—DURABILITY REQUIREMENTS ... 2-7
GENERAL CONSIDERATIONS ... 2-7
4.1 WATER-CEMENTITIOUS MATERIALS RATIO ... 2-8
4.2 FREEZING AND THAWING EXPOSURES ... 2-8
4.2.3 Concrete Exposed to Deicing Chemicals ... 2-9
4.3 SULFATE EXPOSURES ... 2-10
4.4 CORROSION PROTECTION OF REINFORCEMENT ... 2-10
5—CONCRETE QUALITY, MIXING, AND PLACING ... 2-12
UPDATE FOR THE '05 CODE ... 2-12
5.1.1 Concrete Proportions for Strength ... 2-12
5.1.3 Test Age for Strength of Concrete ... 2-12
5.2 SELECTION OF CONCRETE PROPORTIONS ... 2-12
5.3 PROPORTIONING ON THE BASIS OF FIELD EXPERIENCE AND/OR TRIAL MIXTURES ... 2-13
5.3.1 Sample Standard Deviation ... 2-13
5.3.2 Required Average Strength ... 2-16
5.3.3 Documentation of Average Compressive Strength ... 2-17
5.4 PROPORTIONING WITHOUT FIELD EXPERIENCE OR TRIAL MIXTURES ... 2-18
5.6 EVALUATION AND ACCEPTANCE OF CONCRETE ... 2-18
5.6.1 Laboratory and Field Technicians ... 2-18
5.6.2 Frequency of Testing ... 2-19
5.6.5 Investigation of Low-Strength Test Results ... 2-20

REFERENCES ... 2-22
Example 2.1—Selection of Water-Cementitious Materials Ratio for
   Strength and Durability ... 2-24
Example 2.2—Strength Test Data Report ... 2-26
Example 2.3—Selection of Concrete Proportions by Trial Mixtures ... 2-29
Example 2.4—Frequency of Testing ... 2-31
Example 2.5—Frequency of Testing ... 2-32
Example 2.6—Acceptance of Concrete ... 2-33
Example 2.7—Acceptance of Concrete ... 2-34

3 Details of Reinforcement ... 3-1
GENERAL CONSIDERATIONS ... 3-1
7.1 STANDARD HOOKS ... 3-1
7.2 MINIMUM BEND DIAMETERS ... 3-2
7.3 BENDING ... 3-2
   7.3.2 Field Bending of Reinforcing Bars ... 3-2
7.5 PLACING REINFORCEMENT ... 3-3
   7.5.1 Support for Reinforcement ... 3-3
   7.5.2 Tolerances in Placing Reinforcement ... 3-5
   7.5.4 "Tack" Welding ... 3-6
7.6 SPACING LIMITS FOR REINFORCEMENT ... 3-7
   7.6.6 Bundled Bars ... 3-7
   7.6.7 Prestressing Steel and Ducts ... 3-8
7.7 CONCRETE PROTECTION FOR REINFORCEMENT ... 3-9
7.8 SPECIAL REINFORCEMENT DETAILS FOR COLUMNS ... 3-9
7.9 CONNECTIONS ... 3-10
7.10 LATERAL REINFORCEMENT FOR COMPRESSION MEMBERS ... 3-10
   7.10.4 Spirals ... 3-10
   7.10.5 Ties ... 3-11
7.11 LATERAL REINFORCEMENT FOR FLEXURAL MEMBERS ... 3-12
   7.11.3 Closed Ties or Stirrups ... 3-13
7.12 SHRINKAGE AND TEMPERATURE REINFORCEMENT ... 3-14
7.13 REQUIREMENTS FOR STRUCTURAL INTEGRITY ... 3-14
   7.13.1 General Structural Integrity ... 3-15
   7.13.2 Cast-in-Place Joists and Beams ... 3-15
   7.13.3 Precast Concrete Construction ... 3-16
   7.13.4 Lift-Slab Construction ... 3-17
REFERENCES ... 3-17
Example 3.1—Placing Tolerance for Rebars ... 3-18

4 Development and Splices of Reinforcement ... 4-1
GENERAL CONSIDERATIONS ... 4-1
12.1 DEVELOPMENT OF REINFORCEMENT—GENERAL ... 4-1
12.2 DEVELOPMENT OF DEFORMED BARS AND DEFORMED WIRE IN TENSION ... 4-1
   12.2.5 Excess Reinforcement ... 4-5
12.3 DEVELOPMENT OF DEFORMED BARS AND DEFORMED WIRE IN
   COMPRESSION ... 4-5
12.4 DEVELOPMENT OF BUNDLED BARS ... 4-6
12.5 DEVELOPMENT OF STANDARD HOOKS IN TENSION ... 4-6

12.5.2 Development Length	4-7
12.5.3 Modification Factors	4-7
12.5.4 Standard Hook at Discontinuous Ends	4-8
12.6 MECHANICAL ANCHORAGE	4-8
12.7 DEVELOPMENT OF WELDED DEFORMED WIRE REINFORCEMENT IN TENSION	4-8
12.8 DEVELOPMENT OF WELDED PLAIN WIRE REINFORCEMENT IN TENSION	4-9
12.9 DEVELOPMENT OF PRESTRESSING STRAND	4-11
12.10 DEVELOPMENT OF FLEXURAL REINFORCEMENT—GENERAL	4-13
12.11 DEVELOPMENT OF POSITIVE MOMENT REINFORCEMENT	4-15
12.12 DEVELOPMENT OF NEGATIVE MOMENT REINFORCEMENT	4-17
12.13 DEVELOPMENT OF WEB REINFORCEMENT	4-17
12.13.4 Anchorage for Bent-Up Bars	4-20
12.13.5 Closed Stirrups or Ties	4-21
12.14 SPLICES OF REINFORCEMENT—GENERAL	4-21
12.14.2 Lap Splices	4-22
12.14.3 Mechanical and Welded Splices	4-22
12.15 SPLICES OF DEFORMED BARS AND DEFORMED WIRE IN TENSION	4-23
12.16 SPLICES OF DEFORMED BARS IN COMPRESSION	4-25
12.16.1 Compression Lap Splices	4-25
12.16.4 End-Bearing Splices	4-25
12.17 SPECIAL SPLICE REQUIREMENTS FOR COLUMNS	4-25
12.17.2 Lap Splices in Columns	4-25
12.17.3 Mechanical or Welded Splices in Columns	4-28
12.17.4 End Bearing Splices in Columns	4-28
12.18 SPLICES OF WELDED DEFORMED WIRE REINFORCEMENT IN TENSION	4-28
12.19 SPLICES OF WELDED PLAIN WIRE REINFORCEMENT IN TENSION	4-29
CLOSING REMARKS	4-29
REFERENCES	4-30
Example 4.1—Development of Bars in Tension	4-31
Example 4.2—Development of Bars in Tension	4-34
Example 4.3—Development of Bars in Tension	4-36
Example 4.4—Development of Flexural Reinforcement	4-39
Example 4.5—Lap Splices in Tension	4-49
Example 4.6—Lap Splices in Compression	4-53
Example 4.7—Lap Splices in Columns	4-55

5 Design Methods and Strength Requirements ... 5-1

UPDATE FOR THE '05 CODE	5-1
8.1 DESIGN METHODS	5-1
8.1.1 Strength Design Method	5-2
8.1.2 Unified Design Provisions	5-3
9.1 STRENGTH AND SERVICEABILITY—GENERAL	5-3
9.1.1 Strength Requirements	5-3
9.1.2 Serviceability Requirements	5-6
9.1.3 Appendix C	5-6
9.2 REQUIRED STRENGTH	5-7
9.3 DESIGN STRENGTH	5-9
9.3.1 Nominal Strength vs. Design Strength	5-9
9.3.2 Strength Reduction Factors	5-9

9.3.3 Development Lengths for Reinforcement	5-10
9.3.5 Structural Plain Concrete	5-10
9.4 DESIGN STRENGTH FOR REINFORCEMENT	5-11
REFERENCES	5-11

6 General Principles of Strength Design ... 6-1

UPDATE FOR THE '05 CODE	6-1
GENERAL CONSIDERATIONS	6-1
INTRODUCTION TO UNIFIED DESIGN PROVISIONS	6-1
10.2 DESIGN ASSUMPTIONS	6-2
10.2.1 Equilibrium of Forces and Compatibility of Strains	6-2
10.2.2 Design Assumption #1	6-4
10.2.3 Design Assumption #2	6-5
10.2.4 Design Assumption #3	6-6
10.2.5 Design Assumption #4	6-7
10.2.6 Design Assumption #5	6-7
10.2.7 Design Assumption #6	6-9
10.3 GENERAL PRINCIPLES AND REQUIREMENTS	6-13
10.3.1 Nominal Flexural Strength	6-13
10.3.2 Balanced Strain Condition	6-16
10.3.3 Compression-Controlled Sections	6-16
10.3.4 Tension-Controlled Sections and Transition	6-17
10.3.5 Maximum Reinforcement for Flexural Members	6-18
10.3.6 Maximum Axial Strength	6-20
10.3.7 Nominal Strength for Combined Flexure and Axial Load	6-21
10.5 MINIMUM REINFORCEMENT OF FLEXURAL MEMBERS	6-24
10.15 TRANSMISSION OF COLUMN LOADS THROUGH FLOOR SYSTEM	6-25
10.17 BEARING STRENGTH ON CONCRETE	6-25
REFERENCES	6-27
Example 6.1—Moment Strength Using Equivalent Rectangular Stress Distribution	6-28
Example 6.2—Design of Beam with Compression Reinforcement	6-30
Example 6.3—Maximum Axial Load Strength vs. Minimum Eccentricity	6-33
Example 6.4—Load-Moment Strength, P_n and M_n, for Given Strain Conditions	6-34

7 Design for Flexure and Axial Load ... 7-1

GENERAL CONSIDERATIONS—FLEXURE	7-1
DESIGN OF RECTANGULAR SECTIONS WITH TENSION REINFORCEMENTS ONLY	7-1
DESIGN PROCEDURE FOR SECTIONS WITH TENSION REINFORCEMENT ONLY	7-4
DESIGN PROCEDURE FOR SECTIONS WITH MULTIPLE LAYERS OF STEEL	7-6
DESIGN PROCEDURE FOR RECTANGULAR SECTIONS WITH COMPRESSION REINFORCEMENT (see Part 6)	7-7
DESIGN PROCEDURE FOR FLANGED SECTIONS WITH TENSION REINFORCEMENT (see Part 6)	7-8
GENERAL CONSIDERATIONS – FLEXURE AND AXIAL LOAD	7-9
GENERAL CONSIDERATIONS – BIAXIAL LOADING	7-10
BIAXIAL INTERACTION STRENGTH	7-10
FAILURE SURFACES	7-12
A. Bresler Reciprocal Load Method	7-12
B. Bresler Load Contour Method	7-13

    C. PCA Load Contour Method 7-15
    MANUAL DESIGN PROCEDURE 7-21
    REFERENCES 7-21
    Example 7.1—Design of Rectangular Beam with Tension Reinforcement Only 7-23
    Example 7.2—Design of One-Way Solid Slab 7-27
    Example 7.3—Design of Rectangular Beam with Compression Reinforcement 7-29
    Example 7.4—Design of Flanged Section with Tension Reinforcement Only 7-33
    Example 7.5—Design of Flanged Section with Tension Reinforcement Only 7-35
    Example 7.6—Design of One-Way Joist 7-38
    Example 7.7—Design of Continuous Beams 7-43
    Example 7.8—Design of a Square Column for Biaxial Loading 7-46

8 Redistribution of Negative Moments in Continuous Flexural Members 8-1
    UPDATE FOR THE '05 CODE 8-1
    BACKGROUND 8-1
    8.4 REDISTRIBUTION OF NEGATIVE MOMENTS IN CONTINUOUS
        FLEXURAL MEMBERS 8-1
    REFERENCE 8-4
    Example 8.1—Moment Redistribution 8-5
    Example 8.2—Moment Redistribution 8-8

9 Distribution of Flexural Reinforcement 9-1
    UPDATE FOR THE '05 CODE 9-1
    GENERAL CONSIDERATIONS 9-1
    10.6 BEAMS AND ONE-WAY SLABS 9-2
        10.6.4 Distribution of Tension Reinforcement 9-2
        10.6.5 Corrosive Environments 9-3
        10.6.6 Distribution of Tension Reinforcement in Flanges of T-Beams 9-4
        10.6.7 Crack Control Reinforcement in Deep Flexural Members 9-4
    13.4 TWO-WAY SLABS 9-5
    REFERENCES 9-5
    Example 9.1—Distribution of Reinforcement for Effective Crack Control 9-6
    Example 9.2—Distribution of Flexural Reinforcement in Deep Flexural Member
        with Flanges 9-7

10 Deflections 10-1
    GENERAL CONSIDERATIONS 10-1
    9.5 CONTROL OF DEFLECTIONS 10-1
        ACI 318 Method 10-6
        Alternate Method 10-7
        9.5.3 Two-Way Construction (Nonprestressed) 10-10
        9.5.4 Prestressed Concrete Construction 10-16
        9.5.5 Composite Construction 10-17
    REFERENCES 10-20
    Example 10.1—Simple-Span Nonprestressed Rectangular Beam 10-22
    Example 10.2—Continuous Nonprestressed T-Beam 10-26
    Example 10.3—Slab System Without Beams (Flat Plate) 10-32
    Example 10.4—Two-Way Beam Supported Slab System 10-41
    Example 10.5—Simple-Span Prestressed Single T-Beam 10-43

Example 10.6—Unshored Nonprestressed Composite Beam 10-49
Example 10.7—Shored Nonprestressed Composite Beam 10-54

11 Design for Slenderness Effects 11-1
UPDATE FOR THE '05 CODE. 11-1
GENERAL CONSIDERATIONS 11-1
CONSIDERATION OF SLENDERNESS EFFECTS 11-3
10.10 SLENDERNESS EFFECTS IN COMPRESSION MEMBERS 11-3
    10.10.1 Second-Order Frame Analysis 11-3
10.11 APPROXIMATE EVALUATION OF SLENDERNESS EFFECTS 11-4
    10.11.1 Section Properties for Frame Analysis 11-4
    10.11.2 Radius of Gyration 11-4
    10.11.3, 10.12.1 Unsupported and Effective Lengths of Compression Members 11-5
    10.11.4 Non-Sway Versus Sway Frames 11-9
    10.11.6 Moment Magnifier d for Biaxial Bending 11-9
    10.12.2, 10.13.2 Consideration of Slenderness Effects 11-9
    10.12.3 Moment Magnification—Nonsway Frames 11-10
    10.13.3 Moment Magnification—Sway Frames 11-12
    10.13.4 Calculation of $\delta_s M_s$ 11-12
    10.13.5 Location of Maximum Moment 11-13
    10.13.6 Structural Stability Under Gravity Loads 11-14
    10.13.7 Moment Magnification for Flexural Members 11-14
SUMMARY OF DESIGN EQUATIONS 11-15
REFERENCES 11-19
Example 11.1—Slenderness Effects for Columns in a Nonsway Frame 11-20
Example 11.2—Slenderness Effects for Columns in a Sway Frame 11-32

12 Shear 12-1
UPDATE FOR THE '05 CODE. 12-1
GENERAL CONSIDERATIONS 12-1
11.1 SHEAR STRENGTH 12-1
    11.1.2 Limit on $\sqrt{f_c'}$ 12-2
    11.1.3 Computation of Maximum Factored Shear Force 12-2
11.2 LIGHTWEIGHT CONCRETE 12-4
11.3 SHEAR STRENGTH PROVIDED BY CONCRETE FOR
    NONPRESTRESSED MEMBERS 12-4
11.5 SHEAR STRENGTH PROVIDED BY SHEAR REINFORCEMENT 12-6
    11.5.1 Types of Shear Reinforcement 12-6
    11.5.4 Anchorage Details for Shear Reinforcement 12-6
    11.5.5 Spacing Limits for Shear Reinforcement 12-6
    11.5.6 Minimum Shear Reinforcement 12-6
    11.5.7 Design of Shear Reinforcement 12-7
Design Procedure for Shear Reinforcement 12-8
17—COMPOSITE CONRETE FLEXURAL MEMBERS 12-10
17.4 VERTICAL SHEAR STRENGTH 12-10
17.5 HORIZONTAL SHEAR STRENGTH 12-10
17.6 TIES FOR HORIZONTAL SHEAR 12-11
REFERENCES 12-12
Example 12.1—Design for Shear - Members Subject to Shear and Flexure Only 12-13

Example 12.2—Design for Shear - with Axial Tension... 12-16
Example 12.3—Design for Shear - with Axial Compression 12-18
Example 12.4—Design for Shear - Concrete Floor Joist... 12-20
Example 12.5—Design for Shear - Shear Strength at Web Openings...................... 12-22
Example 12.6—Design for Horizontal Shear 12-26

13 Torsion 13-1
UPDATE FOR THE '05 CODE.. 13-1
BACKGROUND 13-1
    11.6.1 Threshold Torsion 13-4
    11.6.2 Equilibrium and Compatibility - Factored Torsional Moment T_u 13-5
    11.6.3 Torsional Moment Strength 13-6
    11.6.4 Details of Torsional Reinforcement 13-7
    11.6.5 Minimum Torsion Reinforcement 13-8
    11.6.6 Spacing of Torsion Reinforcement 13-8
    11.6.7 Alternative Design for Torsion 13-8
ZIA-HSU ALTERNATIVE DESIGN PROCEDURE FOR TORSION 13-8
REFERENCES 13-11
Example 13.1—Precast Spandrel Beam Design for Combined Shear and Torsion............ 13-12

14 Shear Friction 14-1
GENERAL CONSIDERATIONS 14-1
11.7 SHEAR-FRICTION 14-1
    11.7.1 Applications 14-2
    11.7.3 Shear-Transfer Design Methods 14-2
    11.7.4 Shear-Friction Design Method 14-2
    11.7.5 Maximum Shear-Transfer Strength 14-5
    11.7.7 Normal Forces 14-5
    11.7.8 — 11.7.10 Additional Requirements 14-5
DESIGN EXAMPLES.. 14-6
Example 14.1—Shear-Friction Design 14-7
Example 14.2—Shear-Friction Design (Inclined Shear Plane) 14-9

15 Brackets, Corbels and Beam Ledges 15-1
GENERAL CONSIDERATIONS 15-1
11.9 LIMITATIONS OF BRACKET AND CORBEL PROVISIONS 15-1
    11.9.1 - 11.9.5 Design Provisions 15-2
BEAM LEDGES 15-3
    11.9.6 Development and Anchorage of Reinforcement 15-7
REFERENCES 15-8
Example 15.1—Corbel Design 15-9
Example 15.2—Corbel Design...Using Lightweight Concrete and Modified
    Shear-Friction Method.. 15-12
Example 15.3—Beam Ledge Design... 15-16

16 Shear in Slabs 16-1
UPDATE FOR THE '05 CODE.. 16-1
11.12 SPECIAL PROVISIONS FOR SLABS AND FOOTINGS.............................. 16-1
    11.12.1 Critical Shear Section 16-1

  11.12.2 Shear Strength Requirement for Two-Way Action 16-3
  11.12.3 Shear Strength Provided by Bars, Wires, and Single or Multiple-Leg Stirrups 16-5
  11.12.4 Shear Strength Provided by Shearheads 16-7
  Other Type of Shear Reinforcement 16-9
  11.12.5 Effect of Openings in Slabs on Shear Strength 16-10
  11.12.6 Moment Transfer at Slab-Column Connections 16-10
 REFERENCES 16-17
 Example 16.1—Shear Strength of Slab at Column Support 16-18
 Example 16.2—Shear Strength for Non-Rectangular Support 16-20
 Example 16.3—Shear Strength of Slab with Shear Reinforcement 16-22
 Example 16.4—Shear Strength of Slab with Transfer of Moment 16-31

17 Strut-And-Tie Models 17-1
 GENERAL 17-1
 A.1 DEFINITIONS 17-1
 A.2 STRUT-AND-TIE MODEL DESIGN PROCEDURE 17-5
 A.3 STRENGTH OF STRUTS 17-5
 A.4 STRENGTH OF TIES 17-7
 A.5 STRENGTH OF NODAL ZONES 17-8
 REFERENCES 17-9
 Example 17.1—Design of Deep Flexural Member by the
  Strut-and-Tie Model 17-10
 Example 17.2—Design of Column Corbel 17-17

18 Two-Way Slab Systems 18-1
 UPDATE FOR THE '05 CODE 18-1
 13.1 SCOPE 18-1
  13.1.4 Deflection Control—Minimum Slab Thickness 18-2
  13.2 DEFINITIONS 18-4
  13.2.1 Design Strip 18-4
  13.2.4 Effective Beam Section 18-4
 13.3 SLAB REINFORCEMENT 18-5
 13.4 OPENINGS IN SLAB SYSTEMS 18-6
 13.5 DESIGN PROCEDURES 18-6
  13.5.4 Shear in Two-Way Slab Systems 18-8
  13.5.3 Transfer of Moment in Slab-Column Connections 18-10
 SEQUEL 18-11

19 Two-Way Slabs — Direct Design Method 19-1
 GENERAL CONSIDERATIONS 19-1
 PRELIMINARY DESIGN 19-1
  13.6.1 Limitations 19-2
  13.6.2 Total Factored Static Moment for a Span 19-2
  13.6.3 Negative and Positive Factored Moments 19-5
  13.6.4 Factored Moments in Column Strips 19-6
  13.6.5 Factored Moments in Beams 19-10
  13.6.6 Factored Moments in Middle Strips 19-10
  13.6.9 Factored Moments in Columns and Walls 19-10
 DESIGN AID — DIRECT DESIGN MOMENT COEFFICIENTS 19-11

    Example 19.1—Two-Way Slab without Beams Analyzed by the Direct Design Method 19-15
    Example 19.2—Two-Way Slab with Beams Analyzed by the Direct Design Method 19-23

20 Two-Way Slabs — Equivalent Frame Method 20-1
    GENERAL CONSIDERATIONS 20-1
    PRELIMINARY DESIGN 20-1
        13.7.2 Equivalent Frame 20-1
        13.7.3 Slab-Beams 20-2
        13.7.4 Columns 20-5
        13.7.5 Torsional Members 20-5
        Equivalent Columns (R13.7.4) 20-8
        13.7.6 Arrangement of Live Load 20-9
        13.7.7 Factored Moments 20-10
    Appendix 20A DESIGN AIDS FOR MOMENT DISTRIBUTION CONSTANTS 20-13
    Example 20.1—Two-Way Slab Without Beams Analyzed by Equivalent Frame Method 20-20
    Example 20.2—Two-Way Slab with Beams Analyzed by Equivalent Frame Method 20-32

21 Walls 21-1
    UPDATE FOR THE '05 CODE 21-1
    14.1 SCOPE 21-1
    14.2 GENERAL 21-1
    14.3 MINIMUM WALL REINFORCEMENT 21-2
    14.4 WALLS DESIGNED AS COMPRESSION MEMBERS 21-3
    14.5 EMPIRICAL DESIGN METHOD 21-4
    14.8 ALTERNATE DESIGN OF SLENDER WALLS 21-6
    11.10 SPECIAL SHEAR PROVISIONS FOR WALLS 21-9
    DESIGN SUMMARY 21-9
    REFERENCES 21-12
    Example 21.1—Design of Tilt-up Wall Panel by Chapter 10 (14.4) 21-13
    Example 21.2—Design of Bearing Wall by Empirical Design Method (14.5) 21-19
    Example 21.3—Design of Precast Panel by the Alternate Design Method (14.8) 21-21
    Example 21.4—Shear Design of Wall 21-29

22 Footings 22-1
    UPDATE FOR THE '05 CODE 22-1
    GENERAL CONSIDERATIONS 22-1
    15.2 LOADS AND REACTIONS 22-1
    15.4 MOMENT IN FOOTINGS 22-2
    15.5 SHEAR IN FOOTINGS 22-3
    15.8 TRANSFER OF FORCE AT BASE OF COLUMN, WALL, OR REINFORCED PEDESTAL 22-5
    PLAIN CONCRETE PEDESTALS AND FOOTINGS 22-6
    REFERENCE 22-6
    Example 22.1—Design for Base Area of Footing 22-7
    Example 22.2—Design for Depth of Footing 22-8
    Example 22.3—Design for Footing Reinforcement 22-10
    Example 22.4—Design for Transfer of Force at Base of Column 22-13
    Example 22.5—Design for Transfer of Force by Reinforcement 22-15
    Example 22.6—Design for Transfer of Horizontal Force at Base of Column 22-17
    Example 22.7—Design for Depth of Footing on Piles 22-20

23 Precast Concrete 23-1
GENERAL CONSIDERATIONS 23-1
16.2 GENERAL 23-1
16.3 DISTRIBUTION OF FORCES AMONG MEMBERS 23-2
16.4 MEMBER DESIGN 23-2
16.5 STRUCTURAL INTEGRITY 23-4
16.6 CONNECTION AND BEARING DESIGN 23-4
16.7 ITEMS EMBEDDED AFTER CONCRETE PLACEMENT 23-4
16.8 MARKING AND IDENTIFICATION 23-4
16.9 HANDLING 23-5
16.10 STRENGTH EVALUATION OF PRECAST CONSTRUCTION 23-5
REFERENCES 23-5
Example 23.1—Load Distribution in Double Tees 23-6

24 Prestressed Concrete — Flexure 24-1
UPDATE FOR THE '05 CODE 24-1
GENERAL CONSIDERATIONS 24-1
PRESTRESSING MATERIALS 24-2
NOTATION AND TERMINOLOGY 24-3
18.2 GENERAL 24-4
18.3 DESIGN ASSUMPTIONS 24-4
18.4 SERVICEABILITY REQUIREMENTS — FLEXURAL MEMBERS 24-5
18.5 PERMISSIBLE STRESSES IN PRESTRESSING TENDONS 24-6
18.6 LOSS OF PRESTRESS 24-6
ESTIMATING PRESTRESS LOSSES 24-7
COMPUTATION OF LOSSES 24-7
    Elastic Shortening of Concrete (ES) 24-7
    Creep of Concrete (CR) 24-7
    Shrinkage of Concrete (SH) 24-8
    Relaxation of Tendons (RE) 24-8
    Friction 24-9
SUMMARY OF NOTATION 24-9
18.7 FLEXURAL STRENGTH 24-10
18.8 LIMITS FOR REINFORCEMENT OF FLEXURAL MEMBERS 24-12
18.9 MINIMUM BONDED REINFORCEMENT 24-15
18.10.4 Redistribution of Negative Moments in Continuous Prestressed Flexural Memebers ... 24-17
18.11 COMPRESSION MEMBERS — COMBINED FLEXURE AND AXIAL LOADS 24-17
REFERENCES 24-17
Example 24.1—Estimating Prestress Losses 24-18
Example 24.2—Investigation of Stresses at Prestress Transfer and at Service Load 24-21
Example 24.3—Flexural Strength of Prestressed Member
    Using Approximate Value for f_{ps} 24-24
Example 24.4—Flexural Strength of Prestressed Member
    Based on Strain Compatibility 24-26
Example 24.5—Tension-Controlled Limit for Prestressed Flexural Member 24-28
Example 24.6—Cracking Moment Strength and Minimum Reinforcement Limit for
    Non-composite Prestressed Member 24-30
Example 24.7—Cracking Moment Strength and Minimum Reinforcement

  Limit for Composite Prestressed Member... 24-32
 Example 24.8—Prestressed Compression Member.. 24-34
 Example 24.9—Cracked Section Design When Tension Exceeds $12\sqrt{f_c'}$............... 24-36

25 Prestressed Concrete Shear.. 25-1
 UPDATE FOR THE '05 CODE.. 25-1
 GENERAL CONSIDERATIONS.. 25-1
  Web Shear.. 25-1
  Flexure-Shear in Prestressed Concrete.. 25-2
 11.1 SHEAR STRENGTH FOR PRESTRESSED MEMBERS.................................... 25-5
  11.1.2 Concrete Strength.. 25-5
  11.1.3 Location for Computing Maximum Factored Shear..................................... 25-5
 11.2 LIGHTWEIGHT CONCRETE... 25-5
 11.4 SHEAR STRENGTH PROVIDED BY CONCRETE
  FOR PRESTRESSED MEMBERS.. 25-6
  11.4.1 NOTATION... 25-6
  11.4.2 Simplified Method... 25-6
  11.4.3 Detailed Method.. 25-7
  11.4.4, 11.4.5 Special Considerations for Pretensioned Members............................ 25-9
 11.5 SHEAR STRENGTH PROVIDED BY SHEAR REINFORCEMENT FOR
  PRESTRESSED MEMBERS... 25-9
 REFERENCE.. 25-9
 Example 25.1—Design for Shear (11.4.1)... 25-10
 Example 25.2—Shear Design Using Fig. 25-4... 25-15
 Example 25.3—Design for Shear (11.4.2)... 25-18

26 Prestressed Slab Systems.. 26-1
 UPDATE FOR THE '05 CODE.. 26-1
 INTRODUCTION.. 26-1
  11.12.2 Shear Strength.. 26-1
  11.12.6 Shear Strength with Moment Transfer.. 26-2
  18.3.3 Permissible Flexural Tensile Stresses
  18.4.2 Permissible Flexural Compressive Stresses.. 26-2
  18.7.2 f_{ps} for Unbonded Tendons... 26-3
 18.12 SLAB SYSTEMS.. 26-3
 REFERENCES.. 26-4
 Example 26.1—Two-Way Prestressed Slab System.. 26-5

27 Shells and Folded Plate Members.. 27-1
 INTRODUCTION.. 27-1
 GENERAL CONSIDERATIONS.. 27-1
 19.2 ANALYSIS AND DESIGN.. 27-2
  19.2.6 Prestressed Shells... 27-2
  19.2.7 Design Method.. 27-2
 19.4 SHELL REINFORCEMENT.. 27-2
  19.4.6 Membrane Reinforcement... 27-2
  19.4.8 Concentration of Reinforcement... 27-3
  19.4.10 Spacing of Reinforcement... 27-3

28 Strength Evaluation of Existing Structures ... 28-1
INTRODUCTION ... 28-1
20.1 STRENGTH EVALUATION - GENERAL ... 28-1
20.2 DETERMINATION OF REQUIRED DIMENSIONS AND MATERIAL PROPERTIES ... 28-2
20.3 LOAD TEST PROCEDURE ... 28-3
20.4 LOADING CRITERIA ... 28-3
20.5 ACCEPTANCE CRITERIA ... 28-4
20.6 PROVISION FOR LOWER LOAD RATING ... 28-5
20.7 SAFETY ... 28-5
REFERENCES ... 28-5

29 Special Provisions for Seismic Design ... 29-1
UPDATE FOR THE '05 CODE ... 29-1
BACKGROUND ... 29-2
GENERAL CONSIDERATIONS ... 29-2
21.2 GENERAL REQUIREMENTS ... 29-4
    21.2.1 Scope ... 29-4
    21.2.2 Analysis and Proportioning of Structural Members ... 29-5
    21.2.3 Strength Reduction Factors ... 29-5
    21.2.4, 21.2.5 Limitations on Materials ... 29-5
    21.2.6 Mechanical Splices ... 29-6
    21.2.7 Welded Splices ... 29-7
    21.2.8 Anchoring to Concrete ... 29-7
21.3 FLEXURAL MEMBERS OF SPECIAL MOMENT FRAMES ... 29-7
    21.3.1 Scope ... 29-7
    21.3.2 Flexural Reinforcement ... 29-10
    21.3.3 Transverse Reinforcement ... 29-10
    21.3.4 Shear Strength Requirements ... 29-11
21.4 SPECIAL MOMENT FRAME MEMBERS SUBJECTED TO BENDING
AND AXIAL LOAD ... 29-13
    21.4.1 Scope ... 29-13
    21.4.2 Minimum Flexural Strength of Columns ... 29-13
    21.4.3 Longitudinal Reinforcement ... 29-16
    21.4.4 Transverse Reinforcement ... 29-16
    21.4.5 Shear Strength Requirements ... 29-19
21.5 JOINTS OF SPECIAL MOMENT FRAMES ... 29-20
    21.5.2 Transverse Reinforcement ... 29-21
    21.5.3 Shear Strength ... 29-23
    21.5.4 Development Length of Bars in Tension ... 29-24
21.6 SPECIAL MOMENT FRAMES CONSTRUCTED USING PRECAST CONCRETE ... 29-25
    21.6.1 Special Moment Frames with Ductile Connections ... 29-25
    21.6.2 Special Moment Frames with Strong Connections ... 29-25
    21.6.3 Non-emulative Design ... 29-28
21.7 SPECIAL REINFORCED CONCRETE STRUCTURAL WALLS AND
COUPLING BEAMS ... 29-30
    21.7.2 Reinforcement ... 29-30
    21.7.3 Design Forces ... 29-33
    21.7.4 Shear Strength ... 29-33
    21.7.5 Design for Flexural and Axial Loads ... 29-34

21.7.6 Boundary Elements of Special Reinforced Concrete Structural Walls 29-34
21.7.7 Coupling Beams 29-36
21.8 SPECIAL STRUCTURAL WALLS CONSTRUCTED USING PRECAST CONCRETE 29-38
21.9 STRUCTURAL DIAPHRAGMS AND TRUSSES 29-38
21.9.5 Reinforcement 29-38
21.9.7 Shear Strength 29-39
21.9.8 Boundary Elements of Structural Diaphragms 29-39
21.10 FOUNDATIONS 29-40
21.10.2 Footings, Foundation Mats, and Pile Caps 29-40
21.10.3 Grade Beams and Slabs on Grade 29-40
21.10.4 Piles, Piers, and Caissons 29-41
21.11 FRAME MEMBERS NOT PROPORTIONED TO RESIST FORCES INDUCED BY
EARTHQUAKE MOTIONS 29-42
21.12 REQUIREMENTS FOR INTERMEDIATE MOMENT FRAMES 29-43
21.13 INTERMEDIATE PRECASAT STRUCTURAL WALLS 29-44
REFERENCES 29-47
Example 29.1—Design of a 12-Story Cast-in-Place
Frame-Shearwall Building and its Components 29-48
Example 29.2—Proportioning and Detailing of Flexural Members
of Building in Example 29.1 29-51
Example 29.3—Proportioning and Detailing of Columns of Building in Example 29.1 29-60
Example 29.4—Proportioning and Detailing of Exterior Beam-Column
Connection of Building in Example 29.1 29-69
Example 29.5—Proportioning and Detailing of Interior Beam-Column
Connection of Building in Example 29.1 29-72
Example 29.6—Proportioning and Detailing of Structural Wall of Building
In Example 29.1 29-76
Example 29.7—Design of 12-Story Precast Frame Building using Strong Connections 29-83
Example 29.8—Design of Slab Column Connections According to 21.11.5 29-97

30 Structural Plain Concrete 30-1
BACKGROUND 30-1
22.1, 22.2 SCOPE AND LIMITATIONS 30-1
22.3 JOINTS 30-1
22.4 DESIGN METHOD 30-2
22.5 STRENGTH DESIGN 30-3
22.6 WALLS 30-4
22.6.5 Empirical Design Method 30-4
22.6.3 Combined Flexure and Axial Load 30-4
Comparison of the Two Methods 30-11
22.7 FOOTINGS 30-12
22.8 PEDESTALS 30-16
22.10 PLAIN CONCRETE IN EARTHQUAKE-RESISTING STRUCTURES 30-16
REFERENCES 30-17
APPENDIX 30A 30-18
Example 30.1—Design of Plain Concrete Footing and Pedestal 30-27
Example 30.2—Design of Plain Concrete Basement Wall 30-31

31 Alternate (Working Stress) Design Method ... 31-1
INTRODUCTION ... 31-1
GENERAL CONSIDERATIONS ... 31-1
COMPARISON OF WORKING STRESS DESIGN WITH STRENGTH DESIGN ... 31-2
SCOPE (A.1 OF '99 CODE) ... 31-4
GENERAL (A.2 OF '99 CODE) ... 31-5
PERMISSIBLE SERVICE LOAD STRESSES (A.3 of '99 CODE) ... 31-5
FLEXURE (A.5 OF '99 CODE) ... 31-5
DESIGN PROCEDURE FOR FLEXURE ... 31-5
SHEAR AND TORSION (A.7 OF '99 CODE) ... 31-6
REFERENCES ... 31-7
Example 31.1—Design of Rectangular Beam with Tension Reinforcement Only ... 31-8

32 Alternative Provisions for Reinforced and Prestressed Concrete Flexural and Compression Members ... 32-1
B.1 SCOPE ... 32-1
B.8.4 REDISTRIBUTION OF NEGATIVE MOMENTS IN CONTINUOUS NONPRESTRESSED FLEXURAL MEMBERS ... 32-1
B.10.3 GENERAL PRINCIPLES AND REQUIREMENTS – NONPRESTRESSED MEMBERS ... 32-4
B.18.1 SCOPE-PRESTRESSED CONCRETE ... 32-5
B.18.8 LIMITS FOR REINFORCEMENT OF PRESTRESSED FLEXURAL MEMBERS ... 32-6
B.18.10.4 REDISTRIBUTION OF NEGATIVE MOMENTS IN CONTINUOUS PRESTRESSED FLEXURAL MEMBERS ... 32-6

33 Alternative Load and Strength Reduction Factors ... 33-1
C.1 GENERAL ... 33-1
C.2 REQUIRED STRENGTH ... 33-1
C.3 DESIGN STRENGTH ... 33-3

34 Anchoring to Concrete ... 34-1
UPDATE FOR THE '05 CODE ... 34-1
INTRODUCTION ... 34-1
HISTORICAL BACKGROUND OF DESIGN METHODS ... 34-2
GENERAL CONSIDERATIONS ... 34-2
DISCUSSION OF DESIGN PROVISIONS ... 34-3
D.1 DEFINITIONS ... 34-3
D.2 Scope ... 34-6
D.3 GENERAL REQUIREMENTS ... 34-6
D.4 GENERAL REQUIREMENTS FOR STRENGTH OF ANCHORAGE ... 34-7
D.5 DESIGN REQUIREMENTS FOR TENSILE LOADING ... 34-7
    D.5.1 Steel Strength of Anchor in Tension ... 34-8
    D.5.2 Concrete Breakout Strength of Anchor in Tension ... 34-8
    D.5.3 Pullout Strength of Anchor in Tension ... 34-10
    D.5.4 Concrete Side-Face Blowout Strength of Headed Anchor in Tension ... 34-11
D.6 DESIGN REQUIREMENTS FOR SHEAR LOADING ... 34-11
    D.6.1 Steel Strength of Anchor in Shear ... 34-11
    D.6.2 Concrete Breakout Strength of Anchor in Shear ... 34-12
    D.6.3 Concrete Pryout Strength of Anchor in Shear ... 34-13

D.7 INTERACTION OF TENSILE AND SHEAR FORCES 34-13
D.8 REQUIRED EDGE DISTANCES, SPACING, AND THICKNESSES
    TO PRECLUDE SPLITTING FAILURE 34-13
D.9 INSTALATION OF ANCHORS 34-14
DESIGN TABLES FOR SINGLE CAST-IN ANCHORS 34-14
NOTES FOR TENSION TABLES 34-5A, B AND C 34-15
NOTES FOR SHEAR TABLES 34-6A, B AND C 34-16
Example 34.1—Single Headed Bolt in Tension Away from Edges 34-29
Example 34.2—Group of Headed Studs in Tension Near an Edge 34-34
Example 34.3—Group of Headed Studs in Tension Near an Edge with Eccentricity 34-39
Example 34.4—Single Headed Bolt in Shear Near an Edge 34-43
Example 34.5—Single Headed Bolt in Tension and Shear Near an Edge 34-50
Example 34.6—Group of L-Bolts in Tension and Shear Near Two Edges 34-56
Example 34.7—Group of Headed Bolts in Moment and Shear Near an
    Edge in a Region of Moderate or High Seismic Risk 34-66
Example 34.8—Single Post-Installed Anchor in Tension and Shear Away from Edges 34-77

1

General Requirements

A significant renaming of the ACI 318 standard took place with the 1995 edition; in the document title, "Reinforced Concrete," was changed to "Structural Concrete" in recognition of the then new Chapter 22 - Structural Plain Concrete. Prior to the '95 code, design and construction requirements for structural members of plain concrete were contained in a separate companion document to ACI 318, designated ACI 318.1. The requirements for structural plain concrete of the former ACI 318.1 code are now incorporated in Chapter 22.

1.1*   SCOPE

As the name implies, *Building Code Requirements for Structural Concrete (ACI 318-05)* is meant to be adopted by reference in a general building code, to regulate the design and construction of buildings and structures of concrete. Section 1.1.1 emphasizes the intent and format of the ACI 318 document and its status as part of a legally adopted general building code. The ACI 318 code has no legal status unless adopted by a state or local jurisdiction having power to regulate building design and construction through a legally appointed building official. It is also recognized that when the ACI code is made part of a legally adopted general building code, that general building code may modify some provisions of ACI 318 to reflect local conditions and requirements. For areas where there is no general building code, there is no law to make ACI 318 the "code." In such cases, the ACI code defines minimum acceptable standards of design and construction practice, even though it has no legal status.

A provision in 1.1.1, new to ACI 318-02 and unchanged in ACI 318-05 requires that the minimum specified compressive strength of concrete be not less than 2500 psi. This provision is also included in 5.1.1. While the commentary does not explain why this provision was added, it was most likely included because an identical requirement was in *The BOCA National Building Code* (NBC), and *Standard Building Code* (SBC) for several editions, and it was also adapted into the 2000 *International Building Code* (IBC) and remains in the 2003 IBC.

Also new to 1.1.1 of ACI 318-02 and unchanged in ACI 318-05 is a statement that "No maximum specified compressive strength (of concrete) shall apply unless restricted by a specific code provision." The impetus for adding this was the fact that some local jurisdictions, most notably in southern California, were in effect, if not formally, imposing maximum limits on strength of concrete used in structures in regions of high seismic risk (UBC Seismic Zone 3 or 4). Committee 318 felt that it was advisable to add the statement to make it known to regulators that possible need for limitations on concrete strength are considered when new code provisions are introduced, and unless concrete strength is specifically limited by other provisions of ACI 318, no maximum upper limit on strength is deemed necessary. The Committee has been making adjustments in the standard on an ongoing basis to account for sometimes differing properties of high-strength concrete.

In the past, most jurisdictions in the United States adopted one of the three following model building codes, now referred to as *legacy codes*, to regulate building design and construction. *The BOCA National Building Code* (NBC), published by the Building Officials and Code Administrators International[1.1], was used primarily in the

---

*Section numbers correspond to those of ACI 318-05.

northeastern states; the *Standard Building Code* (SBC), published by the Southern Building Code Congress International[1.2], was used primarily in the southeastern states; and the *Uniform Building Code* (UBC), published by the International Conference of Building Officials[1.3], was used mainly in the central and western United States. All three of these model codes used the ACI 318 standard to regulate design and construction of structural elements of concrete in buildings or other structures. *The BOCA National Building Code* and the *Standard Building Code* adopted ACI 318 primarily by reference, incorporating only the construction requirements (Chapter 4 through 7) of ACI 318 directly within Chapter 19 of their documents. The *Uniform Building Code* reprinted ACI 318 in its entirety in Chapter 19. It is essential that designers of concrete buildings in jurisdictions still regulated by the UBC refer to Chapter 19, as some ACI 318 provisions were modified and some provisions were added to reflect, in most cases, more stringent seismic design requirements. To clearly distinguish where the UBC differed from ACI 318, the differing portions of UBC Chapter 19 were printed in italics.

Many states and local jurisdictions that formerly adopted one of the three legacy codes, have adopted the *International Building Code* (IBC), developed by the International Code Council[1.A]. The 2000 edition (first edition) of the IBC adopted ACI 318-99 by reference, and the 2003 edition of the IBC[1.A] adopts ACI 318-02 by reference." Portions of Chapters 3 – 7 of ACI 318 have been included in IBC Sections 1903 – 1907. A few modifications have been made to the reproduced ACI 318 provisions and these are indicated by the text printed in italics. Additional modifications to provisions in other Chapters of ACI 318 are contained in IBC Section 1908. Many of these were necessary to coordinate ACI 318 provisions for seismic design (Chapter 21) with the IBC's seismic design provisions.

As this book goes to press, it is anticipated that the 2006 edition of the IBC will adopt ACI 318-05 by reference. In addition, most of the text from ACI 318 had been transcribed into IBC Sections 1903 – 1907 will be removed and replaced with references to the ACI code. IBC Section 1908 will continue to contain modifications to the provisions of ACI 318, most of which are related to seismic design issues.

In the fall of 2002, the National Fire Protection Association (NFPA) issued the first edition (2003) of its *Building Construction and Safety Code NFPA 5000*[1.B] which adopted ACI 318-02 by reference. While there were no modifications to ACI 318 within the first edition of NFPA 5000, it adopted the modifications to ACI 318 contained in Section A.9.9 of ASCE 7-02[1.C]. As this book goes to press, the 2006 edition of NFPA 5000 is nearing completion and it is anticipated that it will adopt ACI 318-05 by reference and the modifications to ACI 318-05 contained in Section 14.2 of ASCE 7-05, including its Supplement Number 1. Only a few jurisdictions scattered throughout the country have adopted NFPA 5000 and it appears that it will not be able to supplant the IBC as the model of choice.

Whichever building code governs the design, be it a model code or locally developed code, the prudent designer should always refer to the governing code to determine the edition of ACI 318 that is adopted and if there are any modifications to it.

Seismic Design Practice — Earthquake design requirements in two of the three legacy codes were based on the 1991 edition of the *NEHRP (National Earthquake Hazards Reduction Program) Recommended Provisions for the Development of Seismic Regulations for New Buildings.*[1.5] The *BOCA National Building Code* (NBC) and the *Standard Building Code* (SBC) incorporated the NEHRP recommended provisions into the codes, with relatively few modifications. The *Uniform Building Code* (UBC), published by the International Conference of Building Officials which traditionally followed the lead of the Structural Engineers Association of California (SEAOC), had its seismic provisions based on the *Recommended Lateral Force Requirements and Commentary*[1.6] (the SEAOC "Blue Book") published by the Seismology Committee of SEAOC. The *SEAOC Blue Book* in its 1996 and 1999 editions, adopted many of the features of the 1994 NEHRP provisions.[1.E]

The designer should be aware that there were important differences in design methodologies between the UBC and the NBC and SBC for earthquake design. Even with the different design methodologies, it is important to note that a building designed under the NBC or SBC earthquake design criteria and the UBC criteria provided a similar level of safety and that the two sets of provisions (NBC and SBC versus UBC) were substantially equivalent.[1.7]

The seismic design provisions of the 2000 edition of the *International Building Code*, were based on the 1997 edition of the *NEHRP Recommended Provisions for Seismic Regulations for New Buildings and Other Structures*.[1.8] Major differences between the 2000 IBC and the NBC and SBC seismic provisions, that were based on the '91 NEHRP Provisions[1.5], included:

1. Seismic ground motion maps of the 1991 edition were replaced with spectral response acceleration maps at periods of 0.2 second and 1.0 second.

2. The 1991 maps gave ground motion parameters that had a 10% probability of exceedance in 50 years (i.e., approximately a 475-year return period). The 1997 maps were based on a maximum considered earthquake (MCE), and for most regions the MCE ground motion was defined with a uniform likelihood of exceedance of 2% in 50 years (return period of about 2500 years).

3. Seismic detailing requirements, were triggered by building use and estimated ground motion on rock in the '91 edition; the trigger was revised to include the amplifying effects of soft soils overlying rock. This might require buildings on soft soils in areas that were traditionally considered to be subject to low or moderate seismic hazard to be detailed for moderate and high seismic risk, respectively.

4. In the '91 edition, the amplifying effects of soft soils were ignored in calculating the design base shear for short period buildings. These effects were now taken into consideration, and resulted in significant increases in base shear for short period buildings on soft soils in areas subject to low seismic hazard.

5. A reliability/redundancy factor was introduced for buildings subject to high seismic risk. This was done to force designers to either add redundancies to the seismic force-resisting system or to pay a penalty in the form of designing for a higher base shear.

6. It became a requirement to design every building for a lateral force at each floor equal to 1% of the effective seismic weight at that level. Seismic design of buildings subject to negligible or very low seismic risk (e.g., located in Seismic Zone 0, or assigned to SPC A) has traditionally not been required by building codes. This new requirement meant that in areas where seismic design had traditionally been ignored (e.g., south Florida, and much of Texas), designers now needed to make sure that these so-called index forces did not control the design of the lateral force-resisting system. These index forces instead of wind are liable to control design of the lateral force-resisting system of larger concrete buildings, such as parking structures, or long narrow buildings, such as hotels/motels.

For a comprehensive comparison of the major differences between the seismic design requirements of the 2000 IBC, and the last editions of the NBC, SBC and UBC, see *Impact of the Seismic Design Provisions of the International Building Code*[1.9].

"The seismic design requirements of the 2003 IBC are based on ASCE 7-02, which in turn is based on the 2000 edition of the *NEHRP (National Earthquake Hazards Reduction Program) Recommended Provisions for Seismic Regulations for New Buildings and Other Strcutures*[1.D]. A comprehensive discussion of changes in the structural provisions from the 2000 to the 2003 IBC has been provided in Ref. 1.F. The 2003 IBC saw the beginning of a philosophical shift from the code containing almost all the seismic design provisions, as was the case with the 2000 IBC, to one in which the code only has the simplified design provisions. For design of buildings requiring other than simplified analysis procedured, the 2003 IBC references ASCE-7-02. It is anticipated that the 2006 IBC will carry this shift to its conclusion and remove virtually all the seismic design provions from the code and reference the provisions of ASCE 7-05, including its Supplement Number 1. It should be pointed out that ASCE 7-05 will be based on the 2003 edition of the *NEHRP (National Earthquake Hazards Reduction Program) Recommended Provisions for Seismic Regulations for New Buildings and Other Strcutures*[1.E]. Supplement Number 1 to ASCE 7-05 updates the seismic design provisions by referencing the latest editions of material design standards, such as ACI 318-05.

For seismic design, the 2003 edition of NFPA 5000 adopts by reference ASCE 7-02. It is anticipated that the 2006 edition of NFPA 5000 will reference ASCE 7-05, including Supplement Number 1."

Differences in Design Methodology — The UBC earthquake design force level was based on the seismic zone, the structural system, and the building use (occupancy). These design considerations were used to determine a design base shear. As the anticipated level of ground shaking increased, the design base shear increased. Similarly, as the need for post-disaster functionality increased, the design base shear was increased.

As with the UBC, the NBC and the SBC provisions increased the design base shear as the level of ground shaking increased. In the NBC and SBC, this was done not through a seismic zone factor Z, but through a coefficient A_v representing effective peak velocity-related acceleration or a coefficient A_a representing effective peak acceleration (for definitions of these terms, see the Commentary to the NEHRP Provisions[1.5]). These two quantities were given on separate contour maps that took the place of the seismic zoning map of the UBC. The NBC and the SBC utilized a "seismic performance category" (SPC) that took into account the level of seismicity and the building occupancy. Based on the SPC of the building, different design criteria such as drift limits and detailing requirements were specified. The IBC provision also increase the design base shear as the level of ground shaking increases. However, in the IBC, the A_a and A_v maps are replaced with spectral response acceleration maps at periods of 0.2 second and 1.0 second, respectively, the IBC replaces the "seismic performance category" of the NBC and the SBC with a "seismic design category" (SDC). This is more than a change of terminology, because in addition to considering the occupancy of the structure and the estimated ground motion on rock, also considered is the modification of ground motion due to the amplifying effects of soft soils overlying rock. Based on the SDC of the building, different design criteria such as drift limits and detailing requirements are specified. As in the UBC, the NBC and the SBC, the IBC earthquake provisions factor into design the effects of site geology and soil characteristics and the type and configuration of the structural framing system.

Another major difference between the provisions of the 1994 and earlier editions of the UBC and those of the IBC, NBC and SBC is in the magnitude of the design base shear. The designer should note that the earthquake design forces of the IBC, NBC and SBC, and the 1994 and earlier editions of the UBC cannot be compared by simply looking at the numbers, since one set of numbers is based on strength design and the other set is based on working or allowable stress design. NBC and SBC design earthquake forces were strength level while pre-1997 UBC forces were service load level. IBC also provides strength level design earthquake forces. The difference shows up in the magnitude of the response modification coefficient, commonly called the "R" factor. In the NBC and SBC provisions, the term was R; in the IBC, the term is R; in the pre-1997 UBC it was R_w, with the "w" subscript signifying "working" load level design forces. The difference also becomes apparent in the load factors to be applied to the earthquake force effects (E). In the NBC and SBC, the load factor for earthquake force effects was 1.0, as it is in the IBC. In the pre-1997 UBC, for reinforced concrete design, a load factor of 1.4 was applied to the earthquake force effects. Thus, for reinforced concrete, when comparing the base shear calculated by the pre-1997 UBC with that calculated by the 2000 or 2003 IBC, or 1993, 1996 or 1999 NBC, or the 1994, 1997 or 1999 SBC, the designer must multiply the UBC base shear by 1.4.

The seismic design force of the 1997 UBC was at strength level, rather than service level. The change was accomplished by changing the former response modification factors, R_w, to strength-based R-factors, similar to those found in the IBC, NBC and SBC. Since the load combinations of Section 9.2 of ACI 318-95, reproduced in Section 1909.2 of the 1997 UBC, were intended to be used with service level loads, the UBC had to adopt strength-based load combinations that were intended to be used with strength level seismic forces. Therefore, the 1997 UBC required that when concrete elements were to be designed for seismic forces or the effects thereof, the strength-based load combinations of UBC Section 1612.2.1 must be used. These load combinations were based on the load combinations of ASCE 7-95[1.10]. The 1997 UBC also required that when concrete elements were being designed for seismic forces or the effects thereof using the UBC load combination, a multiplier of 1.1 must be applied to amplify the required strengths. This was felt to be necessary at the time because of a presumed incompatibility between the strength reduction factors of Section 9.3 of ACI 318 and the strength design load combination of ASCE 7-95 that were incorporated into the 1997 UBC. After actual seismic designs were performed using the 1997 UBC provisions, it was apparent that use of the 1.1 multiplier resulted in overly conservative designs when compared to the 1994 UBC. Based on a study of the appropriateness of using the multiplier, the SEAOC Seismology Committee has gone on record recommending that it not be used. For additional information on this subject, see Ref. 1.11. The multiplier has now been removed from the 2001 California Building Code[1.G], which is based on the 1997 UBC.

The vertical distribution of base shear along the height of a building also differs between the UBC and the IBC, NBC and SBC. For shorter buildings (with a fundamental period less than or equal to 0.7 second), the UBC required that the design base shear be distributed to the different floor levels along the height in proportion to the product of the weights assigned to floor levels and the heights of the floors above the building base (in accordance with the first mode of vibration of the building). For taller buildings (with fundamental period greater than 0.7 second), the design base shear was divided into two parts. The first part was applied as a concentrated force at the top of the building (to account for higher modes of vibration), with the magnitude being in proportion to the fundamental period of the building, this concentrated force was limited to 25% of the design base shear. The remainder of the design base shear was required to be distributed as specified for shorter buildings. In the NBC and SBC, a fraction of the base shear was applied at a floor level in proportion to the product of weight applied to the floor and height (above the base) raised to the power k, where k is a coefficient based on building period. The IBC and SBC specify a k of 1 (linear distribution of V) for $T \leq 0.5$ sec. These loads specified a k of 2 (parabolic distribution of V) for $T \geq 2.5$ sec. For 0.5 sec. $< T <$ 2.5 sec., two choices were available. One might interpolate between a linear and a parabolic distribution by finding a k-value between 1 and 2, depending upon the period; or one might use a parabolic distribution (k = 2), which is always more conservative. The IBC uses the same distribution as the NBC and the SBC.

Lastly, the detailing requirements, also termed ductility or toughness requirements, which are applicable to structures in regions of moderate to high seismic risk, or assigned to intermediate or high seismic performance or design categories, were similar in the three legacy codes. These requirements are essential to impart to buildings the ability to deform beyond the elastic limit and to undergo many cycles of extreme stress reversals. Fortunately, for reinforced concrete structures, all three legacy codes adopted and the IBC now adopts the ACI 318 standard including Chapter 21 — Special Provisions for Seismic Design. However, the designer will need to refer to the governing model code for any modifications to the ACI 318 seismic requirements. Portions of UBC Chapter 19 that differ substantially from the ACI were printed in italics. The NBC and SBC also included some modifications to the ACI document, most notably for prestressed concrete structures assigned to SPC D or E. Likewise, the 2000 IBC included modifications to ACI 318 in Section 1908, most of which recognize precast concrete systems not in Chapter 21 of ACI 318-99 for use in structures assigned to SDC D, E or F. Section 1908 of the 2003 IBC contains fewer modifications partly because design provisions for precast concrete structures in SDC, D, E or F included in ACI 318-02.

Metric in Concrete Construction — Metric is back. In 1988, federal law mandated the metric system as the preferred system of measurement in the United States. In July 1990, by executive order, all federal agencies were required to develop specific timetables for transition to metric. Some federal agencies involved in construction generally agreed to institute the use of metric units in the design of federal construction by January 1994.

The last editions of the three legacy codes featured and the IBC features both inch-pound (U.S. Customary) and SI-metric (Systeme International) units. The "soft" metric equivalents were or are given in the three legacy codes, generally in parentheses after the English units.

It is noteworthy that when metric conversion was first proposed in the 1970s, some of the standards-writing organizations began preparing metric editions of some of their key documents. The American Concrete Institute first published a "hard" metric companion edition to the ACI 318 standard, ACI 318M-83, in 1983. The current ACI 318 standard is available as ACI 318-05 (U.S. Customary units) and ACI 318M-05 (SI-metric units). ACI 318-H05, for the first time, is a soft metric, rather than a hard metric, version of ACI 318-05. Within the same time period, the American Society for Testing and Materials (ASTM) published metric companions to many of its ASTM standards. For example, Standard Specifications A 615M and A 706M for steel bars for concrete reinforcement were developed as metric companions to A 615 and A 706. The older editions of these metric standards were in rounded metric (hard metric) numbers and included ASTM standard metric reinforcing bars. Due to the expense of maintaining two inventories, one for bars in inch-pound units and another for bars in hard metric units, reinforcing bar manufacturers convinced the standards writers to do away with the hard metric standards and develop metric standards based upon soft conversion of ASTM standard inch-pound bars. The latest editions of the ASTM metric reinforcing bar standards reflect this philosophy. Since all federally financed projects have to be designed and constructed in metric, bar manufacturers decided in 1997 that rather than produce the same bars with two different systems of designating size and strength (i.e., inch-pound and metric), they would produce bars with only one system of marking and that would be the system prescribed for the soft metric converted bars. Thus, it is now commonplace to see reinforcing bars with metric size and strength designations on a job that was designed in inch-pound units. It is important to remember that if this occurs on your job, the bars are identical to the inch-pound bars that were specified, except for the markings designating size and strength.

This Ninth edition of the "Notes" is presented in the traditional U.S. Customary units. Largely because of the large volume of this text, unlike in most other PCA publications, no soft metric conversion has been included.

1.1.6 Soil-Supported Slabs

Prior to the 1995 edition of the code, it did not explicitly state whether soil-supported slabs, commonly referred to as slabs-on-grade or slabs-on-ground, were regulated by the code. They were explicitly excluded from the 1995 edition of ACI 318 "…unless the slab transmits vertical loads from other portions of the structure to the soil." The 1999 edition expanded the scope by regulating slabs-on-grade that "… transmit vertical loads or *lateral forces* from other portions of the structure to the soil." Mat foundation slabs and other slabs on ground which help support the structure vertically and/or transfer lateral forces from the supported structure to the soil should be designed according to the applicable provisions of the code, especially Chapter 15 - Footings. The design methodology for typical slabs-on-grade differs from that for building elements, and is addressed in References 1.12 and 1.13. Reference 1.12 describes the design and construction of concrete floors on ground for residential, light industrial, commercial, warehouse, and heavy industrial buildings. Reference 1.13 gives guidelines for slab thickness design for concrete floors on grade subject to loadings suitable for factories and warehouses.

In addition to the modification to 1.1.6, a new Section 21.8, Foundations, was added in Chapter 21 — Special Provisions for Seismic Design in the 1999 edition of ACI 318. Due to sections being added to Chapter 21 in the 2002 edition, these provisions are now in 21.10. Section 21.10.3.4 indicates that "slabs on grade that resist seismic forces from walls or columns that are part of the lateral-force-resisting system shall be designed as structural diaphragms in accordance with 21.9." In this location of Chapter 21, the provisions only apply in regions of high seismic risk, or to structures assigned to high seismic performance or design categories. In regions of low or moderate seismic risk, or for structures assigned to low or intermediate seismic performance or design categories, the provisions of Chapters 1 through 18, or Chapter 22 apply to such slabs, by virtue of the new provision in 1.1.6 (see Table 1-3).

1.1.8 Special Provisions for Earthquake Resistance

Since publication of the 1989 code, the special provisions for seismic design have been located in the main body of the code to ensure adoption of the special seismic design provisions when a jurisdiction adopts the ACI code as part of its general building code. With the continuing high interest nationally in the proper design of buildings for earthquake performance, the code's emphasis on seismic design of concrete buildings continues with this edition. Chapter 21 represents the latest in special seismic detailing of reinforced concrete buildings for earthquake performance.

The landmark volume, *Design of Multistory Concrete Buildings for Earthquake Motions* by Blume, Newmark, and Corning[1.H], published by the Portland Cement Association (PCA) in 1961, gave major impetus to the design and construction of concrete buyildings in regions of high seismicity. In the decades since, significant strides have been made in the earthquake resistant design and construction of reinforced concrete buildings. Significant developments have occurred in the building codes arena as well. However, a comprehensive guide to aid the designer in the detailed seismic design of concrete buildings was not available until PCA published *Design of Concrete Buildings for Earthquake and Wind Forces* by S.K. Ghosh and August W. Domel, Jr. in 1992[1.J].

That design manual illustrated the detailed design of reinforced concrete buildings utilizing the various structural systems recognized in U.S. seismic codes. All designs were according to the provisions of the 1991 edition of the *Uniform Building Code* (UBC), which had adopted, with modifications, the seismic detailing requirements of the 1989 edition of *Building Code Requirements for Reinforced Concrete* (ACI 318-89, Revised 1992). Design of the same building was carriedd out for regions of high, moderate, and low seismicity, and for wind, so that it would be apparent how design and detailing changed with increased seismic risk at the site of the structure.

The above publication was updated to the 1994 edition of the UBC, in which ACI 318-89, Revised 1992, remained the reference standard for concrete design and construction, although a new procedure for the design of reinforced concrete shear walls in combined bending and axial compression was introduced in the UBC itself. The updated publication by S.K. Ghosh, August W. Domel, Jr., and David A. Fanella was issued by PCA in 1995.

Since major changes occurred between the 1994 and 1997 editions of the UBC as discussed above, a new book titled *Design of Concrete Buildings for Earthquake and Wind Forces According to the 1997 Uniform Building Code*[1.15] was developed. It discussed the major differences in the design requirements between the 1994 and the 1997 editions of the UBC. Three different types of concrete structural framing systems were designed and detailed for earthquake forces representing regions of high seismicity (Seismic Zones 3 and 4). Although the design examples focused on regions of high seismicity, one chapter discussed the detailing requirements for structures located in regions of low, moderate, and high seismicity. Design of the basic structural systems for wind was also illustrated. As in this "Notes" publication, the emphasis has placed on "how-to-use" the various seismic design and detailing provisions of the latest and possibly the last UBC.

PCA publication, *Design of Low-Rise Concrete Buildings for Earthquake Forces*[1.16], was a companion document to that described above; however, its focus was on designing concrete buildings under the 1996 and 1997 editions of *The BOCA National Building Code* (NBC) and the *Standard Building Code* (SBC), respectively. As indicated previously, the seismic provisions of the last editions of the NBC and SBC were almost identical, and were based on the 1991 edition of the *NEHRP Recommended Provisions for the Development of Seismic Regulations for New Buildings*.[1.5] With the two exceptions noted below, the book was also applicable to the 1993 and 1999 editions of the NBC, and the 1994 and 1999 editions of the SBC. The only difference between the loading requirements of the 1993 and the 1996 and 1999 NBC was that the load combinations to be used for seismic design under the 1993 edition of the NBC were identical to those that had to be used under all three editions of the SBC. Whereas, the 1996 and 1999 NBC adopted by reference the strength design load combinations of ASCE 7-95[1.9]. The second exception was that different editions of ACI 318 were adopted by the various editions of the codes as illustrated in the table below.

Model Code	Edition	Edition of ACI 318 adopted by Model Code
NBC	1993	1989, Revised 1992
NBC	1996	1995
NBC	1999	1995
SBC	1994	1989
SBC	1997	1995
SBC	1999	1995
IBC	2000	1999
IBC	2003	2002
NFPA 5000	2003	2002

Since designing for seismic forces in areas that had traditionally adopted the NBC or SBC was relatively new, the book provided excellent background information for the structural engineer. Since the overwhelming majority of all buildings constructed in this country are low-rise, that was the focus of this book. For its purpose, low-rise was defined as less than 65 feet in height or having a fundamental period of vibration of less than 0.7 second.

To assist the designer in understanding and using the special detailing requirements of Chapter 21 of the Code, PCA developed a publication titled *Seismic Detailing of Concrete Buildings*[1.17]. Numerous tables and figures illustrated the provisions for buildings located in regions of moderate and high seismic risk – IBC Seismic Design Categories C, D, E and F. While the book was based on the '99 edition of the Code, which was referenced by the 2000 IBC, most of the provisions are applicable to ACI 318-02 and ACI 318-05.

In recent years, the building code situation in this country has changed drastically. The seismic design provisions of the IBC represent revolutionary changes from those of model codes it was developed to replace. This created a need for a new publication similar to the volume first issued by PCA in 1992. To fill that need, PCA and the International Code Council (ICC) published *Seismic and Wind Design of Concrete Buildings*: 2000 IBC, ASCE 7-98, ACI 318-99 by S.K. Ghosh and David A. Fanella in 2003[1.K].

An update of the above publication to the 2003 IBC, *Seismic and Wind Design of Concrete Buildings*: 2003 IBC, ASCE 7-02, ACI 318-02 by S.K. Ghosh, David A. Fanella, and Xuemei Liang[1.L] has recently been published by PCA and ICC. In Chapter 1, an introduction to earthquake-resistant design is provided, along with summaries of the seismic and wind design provisions of the 2003 IBC. Chapter 2 is devoted to an office building utilizing a dual shear wall-frame interactive system in one direction and a moment-resisting frame in the orthogonal direction. Designs for Seismic Design Categories (SDC) A, C, D, and E are illustrated in both directions. Chapter 3 features a residential building, which utilizes a shear-wall frame interactive system in SDC A and B and a building frame system for lateral resistance in SDC C, D, and E. Chapter 4 presents the design of a school building with a moment-resisting frame system in SDC B, C, and D. A residential building utilizing a bearing wall system is treated in Chapter 5. Design is illustrated for SDC A, B, C, D, and E. The final (sixth) chapter is devoted to design of a precast parking structure utilizing the building frame system in SDC B, C, and D. While design is always for the combination of gravity, wind, and seismic forces, wind forces typically govern the design in the low seismic design categories (particularly A), and earthquake forces typically govern in the high seismic design categories (particularly D and above). Detailing requirements depend on the seismic design category, regardless of whether wind or seismic forces govern the design. This publication is designed to provide an appreciation on how design and detailing change with changes in the seismic design category.

1.1.8.1 Structures at Low Seismic Risk — For concrete structures located in regions of low seismic hazard or assigned to low seismic performance or design categories (no or minor risk of damage—Sesimic Design Category A or B), no special design or detailing is required; thus, the general requirements of the code, excluding Chapter 21, apply. Concrete structures proportioned by the general requirements of the code are considered to have a level of toughness adequate for low earthquake intensities.

The designer should be aware that the general requirements of the code include several provisions specifically intended to improve toughness, in order to increase resistance of concrete structures to earthquake and other catastrophic or abnormal loads. For example, when a beam is part of the lateral force-resisting system of a structure, a portion of the positive moment reinforcement must be anchored at supports to develop its yield strength (see 12.11.2). Similarly, hoop reinforcement must be provided in certain types of beam-column connections (see 11.11.2). Other design provisions introduced since publication of the 1971 code, such as those requiring minimum shear reinforcement (see 11.5.5) and improvements in bar anchorage and splicing details (Chapter 12), also increase toughness and the ability of concrete structures to withstand reversing loads due to earthquakes. With publication of the 1989 code, provisions addressing special reinforcement for structural integrity (see 7.13) were added, to enhance the overall integrity of concrete structures in the event of damage to a major supporting element or abnormal loading.

1.1.8.2 Structures at Moderate/Intermediate or High Seismic Risk — For concrete structures located in regions of moderate seismic hazard, or assigned to intermediate seismic performance or design categories (moderate risk of damage—Seismic Design Category C), 21.12 includes certain reinforcing details, in addition to those contained in Chapters 1 through 18, that are applicable to reinforced concrete moment frames (beam-column or slab-column framing systems) required to resist earthquake effects. Reflecting terminology that has been in use in the model codes over at least the past decade, frames detailed in accordance with 21.12 are now referred to as *Intermediate Moment Frames*. These so-called "intermediate" reinforcement details will serve to accommodate an appropriate level of inelastic behavior if the frame is subjected to an earthquake of such magnitude as to require it to perform inelastically. There are no design or detailing requirements in addition to those of Chapters 1 through 18 for other structural components of structures at moderate seismic risk (including structural walls (shearwalls)) regardless of whether they are assumed in design to be part of the seismic-force-resisting system or not. Structural walls proportioned by the general requirements of the code are considered to have sufficient toughness at drift levels anticipated in regions of moderate seismicity.

The type of framing system provided for earthquake resistance in a structure at moderate seismic risk will dictate whether any special reinforcement details need to be incorporated in the structure.

If the lateral force-resisting system consists of moment frames, the details of 21.12 for *Intermediate Moment Frames* must be provided, and 21.2.2.3 shall also apply. Note that even if a load combination including wind load effects (see 9.2.1) governs design versus a load combination including earthquake force effects, the intermediate reinforcement details must still be provided to ensure a limited level of toughness in the moment resisting frames. Whether or not the specified earthquake forces govern design, the frames are the only defense against the effects of an earthquake.

For a combination frame-shearwall structural system, inclusion of the intermediate details will depend on how the earthquake loads are "assigned" to the shearwalls and the frames. If the total earthquake forces are assigned to the shearwalls, the intermediate detailing of 21.12 is not required for the frames. If frame-shearwall interaction is considered in the analysis, with some of the earthquake forces to be resisted by the frames, then the intermediate details of 21.12 are required to toughen up the frame portion of the dual framing system. Model codes have traditionally considered a dual system to be one in which at least 25% of the design lateral forces are capable of being resisted by the moment frames. If structural walls resist total gravity and lateral load effects, no intermediate details are required for the frames; the general requirements of the code apply.

For concrete structures located in regions of high seismic hazard, or assigned to high seismic performance or design categories (major risk of damage—Seismic Design Category D, E, or F), all structural components must satisfy the applicable special proportioning and detailing requirements of Chapter 21 (excluding 21.12 and 21.13). If, for purposes of design, some of the frame members are not considered as part of the lateral force resisting system, special consideration is still required in the proportioning and detailing of these frame members (see 21.11). The special provisions for seismic design of Chapter 21 are intended to provide a monolithic reinforced concrete structure with adequate toughness to respond inelastically under severe earthquake motions.

Unlike previous editions of the Code, the 2002 and 2005 edition specifically addresses precast concrete systems for use in structures in regions of moderate or high seismic hazard, or in structures assigned to intermediate or high seismic performance or design categories. A special moment frame can either be cast-in-place or erected with precast elements. A precast concrete special moment frame must comply with all the requirements for cast-in-place frames (21.2 through 21.5), plus 21.6. In addition, the requirements for an ordinary moment frame must be satisfied (Chapters 1 through 18). Since there is no provision for an intermediate moment frame made of precast elements, by implication such frames erected in structures in regions of moderate seismic hazard, or assigned to intermediate seismic performance or design categories must either be special moment frames, or be qualified under the performance criteria of 21.2.1.5. In the 2002 code, the definition of "ordinary moment frame" was revised to clarify that such a frame can either be cast-in-place or constructed with precast elements, both of which must comply with Chapters 1 through 18.

Two new precast structural walls were added to the 2002 code; an intermediate precast structural wall, and a special precast structural wall. The intermediate precast structural wall must comply with Chapters 1 through 18, plus 21.13. Section 21.13 does not address the wall itself, but covers the connection between individual wall panels, and the connection of wall panels to the foundation. Wherever precast wall panels are used to resist seismic lateral forces in structures in regions of moderate seismic hazard or assigned to intermediate seismic performance or design categories, they must comply with the requirements for an intermediate precast structural wall, or special precast structural wall. By implication, a wall composed of precast elements designed in accordance with Chapters 1 through 18, but not complying with either of these requirements can only be used in structures in regions of low seismic hazard, or in structures assigned to low seismic performance or design categories.

The special precast structural wall must comply with Chapters 1 through 18, plus 21.2, 21.7, 21.13.2 and 21.13.3. Wherever precast wall panels are used to resist seismic lateral forces in structures in regions of high seismic hazard, or assigned to high seismic performance or design categories, they must comply with the requirements for a special precast structural wall.

The ACI 318 proportioning and detailing requirements for lateral force-resisting structural systems of reinforced concrete are summarized in Table 1-1.

Table 1-1 Sections of Code to be Satisfied

Component resisting earthquake effect unless otherwise noted		Level of Seismic Hazard or Assigned Seismic Performance of Design Categories as Defined in Code Section Indicated		
		Low/A,B/A,B 21.2.1.2	Intermediate/C/C 21.2.1.3	High/D,E/D,E,F 21.2.1.4
Frame members	Cast-in-Place	Ch. 1–18	Ch. 1–18, 21.2.2.3, 21.12	Ch. 1–18, 21.2–21.5
	Precast	Ch. 1–18	Note 1	Ch. 1–18, 21.2–21.6
Structural walls and	Cast-in-Place	Ch. 1-18, or Ch. 22	Ch. 1–18	Ch. 1–18. 21.2, 21.7
	Precast	Ch. 1–18, or Ch. 22	Ch. 1–18, 21.13	Ch. 1–18, 21.2, 21.7, 21.8
Structural diaphragms and trusses		Ch. 1–18	Ch. 1–18	Ch. 1–18, 21.2, 21.9
Foundations		Ch. 1–18, or Ch. 22	Ch. 1–18	Ch. 1–18, 21.2, 21.10, 22.10
Frame members assumed not to resist earthquake forces		Ch. 1–18	Ch. 1–18	Ch. 1–18, 21.11

Note 1: There are no provisions for constructing an intermediate moment frame with precast elements. See 21.2.1.5.

1.1.8.3 Seismic Hazard Level Specified in General Building Code — This code has traditionally addressed levels of seismic hazard as "low," "moderate," or "high." Precise definitions of seismic hazard levels are under the jurisdiction of the general building code, and have traditionally been designated by zones (related to intensity of ground shaking). The model codes specify which sections of Chapter 21 must be satisfied, based on the seismic hazard level. As a guide, in the absence of specific requirements in the general building code, seismic hazard levels and seismic zones generally correlate as follows:

Seismic Hazard Level	Seismic Zone
Low	0 and 1
Moderate	2
High	3 and 4

The above correlation of seismic hazard levels and seismic zones refers to the *Uniform Building Code*[1.3].

However, with the adoption of the 1991 NEHRP Provisions into *The BOCA National Building Code* and the *Standard Building Code*, the designer needed to refer to the governing model code to determine appropriate seismic hazard level and corresponding special provisions for earthquake resistance. The NBC, SBC, and '91 NEHRP Provisions, on which the seismic design requirements of the two legacy model codes were based, assigned a building to a Seismic Performance Category (SPC). The SPC expressed hazard in terms of the nature and use of the building and the expected ground shaking on rock at the building site. To determine the SPC of a structure, one had to first determine its Seismic Hazard Exposure Group. Essential facilities were assigned to Seismic Hazard Exposure Group III, assembly buildings and other structures with a large number of occupants were assigned to Group II. Buildings and other structures not assigned to Group II or III, were considered to belong to Group I (see the governing code for more precise definitions of these Seismic Hazard Exposure Groups). The next step was to determine the effective peak velocity-related acceleration coefficient, A_v, given on a contour map that formed part of the NBC and the SBC. With these two items, the structure's SPC could be determined from a table in the governing code that was similar to Table 1-2, which is reproduced from the 1991 NEHRP Provisions.

Table 1-2 Seismic Performance Category[1.5]

Value of A_v	Seismic Hazard Exposure Group		
	I	II	III
$A_v < 0.05$	A	A	A
$0.05 \leq A_v < 0.10$	B	B	C
$0.10 \leq A_v < 0.15$	C	C	C
$0.15 \leq A_v < 0.20$	C	D	D
$0.20 \leq A_v$	D	D	E

"In the 2000 and 2003 editions of the *International Building Code,* the seismic design requirements are based on the 1997 and 2000 edition of the NEHRP Provisions[1.8], respectively. In the IBC the seismic hazard is expressed in a manner that is similar to that of the NBC and the SBC, but with one important difference." The IBC also considers the amplifying effects of softer soils on ground shaking in assigning seismic hazard. The terminology used in the IBC for assigning hazard and prescribing detailing and other requirements is the Seismic Design Category (SDC). The SDC of a building is determined in a manner similar to the SPC in the NBC and the SBC. First the building is assigned to a Seismic Use Group (SUG), which is the same as the Seismic Hazard Exposure Group of the NBC and the SBC. At this point the IBC process becomes more involved. Instead of determining one mapped value of expected ground shaking, two spectral response acceleration values are determined from two different maps; one for a short (0.2 second) period and the other for a period of 1 second. These values are then adjusted for site soil effects and multiplied by two-thirds to arrive at design spectral acceleration values. Knowing the SUG and the design spectral response acceleration values (S_{DS} and S_{D1}), Knowing the SUG and the design spectral response acceleration values,one enters two different tables to determine the SDC based on the two design values. The governing SDC is the higher of the two, if they differ.

"As a guide, for purposes of determining the applicability of special proportioning and detailing requirements of Chapter 21 of the ACI code, Table 1-3 shows the correlation between UBC seismic zones; the Seismic Performance Categories of the NBC, SBC, 1994 (and earlier) NEHRP and ASCE 7-95 (and earlier); and the Seismic Design Categories of the IBC, NFPA 5000, 1997 (and later) NEHRP, and ASCE 7-98 (and later)."

1.2   DRAWINGS AND SPECIFICATIONS

If the design envisioned by the engineer is to be properly implemented in the field, adequate information needs to be included on the drawings or in the specifications, collectively known as the construction documents. The code has for many editions included a list of items that need to be shown on the construction documents.

1.2.1 Items Required to be Shown

The information required to be included as a part of the construction documents remains essentially unchanged from the 1999 code; however, in the 2002 code item "e" was expanded to require that anchors be shown on the drawings. Enough information needs be shown so anchors can be installed with the embedment depth and edge distances the engineer assumed in the design. In addition, where "supplemental reinforcement" (see definition in D.1) was assumed in the design, the location of the reinforcement with respect to the anchors needs to be indicated.

Table 1-3 — Correlation Between Seismic Hazard Levels of ACI 318 and Other Codes and Standards

Code, Standard or Resource Document And Edition	Assigned Seismic Performance or Design Categories and Level of Seismic Risk as Defined in Code Section		
	Low (21.2.1.2)	Moderate/Intermediate (21.2.1.3)	High (21.2.1.4)
BOCA National Building Code 1993, 1996, 1999	SPC^1 A, B	SPC C	SPC D, E
Standard Building Code 1994, 1997, 1999	SPC A, B	SPC C	SPC D, E
Uniform Building Code 1991, 1994, 1997	Seismic Zone 0, 1	Seismic Zone 2	Seismic Zone 3, 4
International Building Code 2000, 2003	SDC^2 A, B	SDC C	SDC D, E, F
NFPA 5000-2003	SDC^2 A, B	SDC C	SDC D, E, F
$ASCE^3$ 7-93, 7-95	SPC^1 A, B	SPC C	SPC D, E
$NEHRP^4$ 1991, 1994	SPC^1 A, B	SPC C	SPC D, E
$ASCE^3$ 7-98, 7-02	SDC^2 A, B	SDC C	SDC D, E, F
$NEHRP^5$ 1997	SDC^2 A, B	SDC C	SDC D, E, F

1. SPC = Seismic Performance Category as defined in building code, standard or resource document
2. SDC = Seismic Design Category as defined in building code, standard or resource document
3. Minimum Design Loads for Buildings and Other Structures
4. NEHRP (National Earthquake Hazards Reduction Program) Recommended Provisions for Seismic Regulations for New Buildings
5. NEHRP (National Earthquake Hazards Reduction Program) Recommended Provisions for Seismic Regulations for New Buildings and Other Structures

1.3 INSPECTION

The ACI code requires that concrete construction be inspected as required by the legally adopted general building code. In the absence of inspection requirements in the general building code or in an area where a building code has not been adopted, the provisions of 1.3 may serve as a guide to providing an acceptable level of inspection for concrete construction. In cases where the building code is silent on this issue or a code has not been adopted, concrete construction, at a minimum; should be inspected by a registered design professional, someone under the supervision of a registered design professional, or a qualified inspector. Individuals professing to be qualified to perform these inspections should be required to demonstrate their competence by becoming certified. Voluntary certification programs for inspectors of concrete construction have been established by the American Concrete Institute (ACI), and International Code Council (ICC). Other similar certification programs may also exist.

The IBC, adopted extensively in the U.S. to regulate building design and construction, and NFPA 5000 require varying degrees of inspection of concrete construction. However, administrative provisions such as these are

---
*Commentary section numbers are preceded by an "R" (e.g., R1.3.5 refers to Comentary Section R1.3.5).

frequently amended when the model code is adopted locally. The engineer should refer to the specific inspection requirements contained in the general building code having jurisdiction over the construction.

In addition to periodic inspections performed by the building official or his representative, inspections of concrete structures by special inspectors may be required; see discussion below on 1.3.5. The engineer should check the local building code or with the local building official to ascertain if special inspection requirements exist within the jurisdiction where the construction will be occurring. Degree of inspection and inspection responsibility should be set forth in the contract documents. However, it should be pointed out that most codes with provisions for special inspections do not permit the contractor to retain the special inspector. Normally they require that the owner enter into a contract with the special inspector. Therefore, if the frequency and type of inspections are shown in the project's construction documents, it should be made clear that the costs for providing these services are not to be included in the bid of the general contractor.

1.3.4 Records of Inspection

Inspectors and inspection agencies will need to be aware of the wording of 1.3.4. Records of inspection must be preserved for two years after completion of a project, or longer if required by the legally adopted general building code. Preservation of inspection records for a minimum two-year period after completion of a project is to ensure that records are available, should disputes or discrepancies arise subsequent to owner acceptance or issuance of a certificate of occupancy, concerning workmanship or any violations of the approved construction documents, or the general building code requirements.

1.3.5 Special Inspections

Continuous inspection is required for placement of all reinforcement and concrete for special moment frames (beam and column framing systems) resisting earthquake-induced forces in structures located in regions of high seismic hazard, or in structures assigned to high seismic performance or design categories. Special moment frames of cast-in-place concrete must comply with the 21.2 - 21.5. Special moment frames constructed with precast concrete elements must comply with the additional requirements of 21.6. For information on how the model building codes in use in the U.S. assign seismic hazard, see Table 1-3. The code stipulates that the inspections must be made by a qualified inspector under the supervision of the engineer responsible for the structural design or under the supervision of an engineer with demonstrated capability for supervising inspection of special moment frames resisting seismic forces in regions of high seismic hazard, or in structures assigned to high seismic performance or design categories. R1.3.5* indicates that qualification of inspectors should be acceptable to the jurisdiction enforcing the general building code.

This inspection requirement is patterned after similar provisions contained in the *The BOCA National Building Code* (NBC), *International Building Code* (IBC), *Standard Building Code* (SBC), and the *Uniform Building Code* (UBC), referred to in those codes as "special inspections." The specially qualified inspector must "demonstrate competence for inspection of the particular type of construction requiring special inspection." See Section 1.3 above for information on voluntary certification programs for concrete special inspectors. Duties and responsibilities of the special inspector are further outlined as follows:

1. Observe the work for conformance with the approved construction documents.
2. Furnish inspection reports to the building official, the engineer or architect of record, and other designated persons.
3. Submit a final inspection report indicating whether the work was in conformance with the approved construction documents and acceptable workmanship.

The requirement for special inspections by a specially qualified special inspector was long a part of the *Uniform Building Code*; however, it was adopted much later in the NBC and the SBC. With the adoption of the NEHRP recommended earthquake provisions by the IBC, NFPA 5000, NBC and the SBC, the need for special inspections came to the forefront. An integral part of the NEHRP provisions is the requirement for special inspections of the seismic-force resisting systems of buildings in intermediate and high seismic performance or design categories.

By definition, special inspection by a special inspector implies continuous inspection of construction. For concrete construction, special inspection is required during placement of all reinforcing steel, during the taking of samples of concrete used for fabricating strength test cylinders, and during concrete placing operations. The special inspector need not be present during the entire time reinforcing steel is being placed, provided final inspection of the in-place reinforcement is performed prior to concrete placement. Generally, special inspections are not required for certain concrete work when the building official determines that the construction is of a minor nature or that no special hazard to public safety exists. Special inspections are also not required for precast concrete elements manufactured under plant control where the plant has been prequalified by the building official to perform such work without special inspections.

Another "inspection" requirement in the IBC, NFPA 5000, and UBC that was not part of the NBC or the SBC is the concept of "structural observation". Under the UBC, structural observation was required for buildings located in high seismic risk areas (Seismic Zone 3 or 4). Under the IBC, it was required for more important structures assigned to seismic design category D, E or F, or sited in an area where the basic wind speed exceeds 110 miles per hour (3-second gust speed). NFPA 5000 has requirements that are similar to those of the IBC. Under the UBC, the owner is required to retain the engineer or architect in responsible charge of the structural design work or another engineer or architect designated by the engineer or architect responsible for the structural design to perform visual observation of the structural framing system at significant stages of construction and upon completion, for general conformance to the approved plans and specifications. Under the IBC and NFPA 5000, any registered design professional qualified to perform the work can be retained for the purpose of making structural observations. At the completion of the project, and prior to issuance of the certificate of occupancy, the engineer or architect is required to submit a statement in writing to the building official indicating that the site visits have been made and noting any deficiencies that have not been corrected.

With ever-increasing interest in inspection of new building construction in the U.S., especially in high seismic risk areasand high wind areas, the designer will need to review the inspection requirements of the governing general building code, and ascertain the role of the engineer in the inspection of the construction phase.

REFERENCES

1.1   *The BOCA National Building Code*, Building Officials and Code Administrators International, Country Club Hills, IL, 1999.

1.2   *Standard Building Code*, Southern Building Code Congress International, Birmingham, AL, 1999.

1.3   *Uniform Building Code*, International Conference of Building Officials, Whittier, CA, 1997.

1.4   *International Building Code*, 2000 Edition, International Code Council, Inc., Falls Church, VA, 2000.

1.5   *NEHRP (National Earthquake Hazards Reduction Program) Recommended Provisions for the Development of Seismic Regulations for New Buildings, Part 1-Provisions, Part 2-Commentary*, Building Seismic Safety Council, Washington, D.C., 1991.

1.6   *Recommended Lateral Force Requirements and Commentary*, Structural Engineers Association of California, San Francisco, CA, 1999.

1.7   *Report and Findings — Uniform Building Code Provisions Compared With the NEHRP Provisions*, prepared for National Institute of Standards and Technology, prepared by International Conference of Building Officials, Whittier, CA, July 1992.

1.8   *NEHRP (National Earthquake Hazards Reduction Program) Recommended Provisions for Seismic Regulations for New Buildings and Other Structures, Part 1-Provisions, Part 2-Commentary*, Building Seismic Safety Council, Washington, D.C., 1997.

1.9   Ghosh, S.K., Impact of the Seismic Design Provisions of the International Building Code, Structures and Code Institute, Northbrook, IL, 2001 (PCA Publication LT254).

1.10   American Society of Civil Engineers (1995), *Minimum Design Loads for Buildings and Other Structures*, ASCE 7-95 Standard, ASCE, New York, NY.

1.11   Ghosh, S.K. (1998), "Design of Reinforced Concrete Buildings Under the 1997 UBC," *Building Standards, May/June 1998*, pp. 20-24.

1.12   *Concrete Floors on Ground*, Publication EB075.03D, Portland Cement Association, Skokie, IL, Revised 2001.

1.13   *Slab Thickness Design for Industrial Concrete Floors on Grade*, Publication IS195.01D, Portland Cement Association, Skokie, IL, Revised 1996.

1.14   Ghosh, S.K., Domel, A.W. Jr., and Fanella, D.A., *Design of Concrete Buildings for Earthquake and Wind Forces*, Publication EB113.02D, Portland Cement Association, Skokie, IL, 1995.

1.15   Fanella, D.A., and Munshi, J. A., *Design of Concrete Buildings for Earthquake and Wind Forces According to the 1997 Uniform Building Code*, Publication EB117.02D, Portland Cement Association, Skokie, IL, 1998.

1.16   Fanella, D.A., and Munshi, J. A., *Design of Low-Rise Concrete Buildings for Earthquake Forces*, Publication EB004.02D, Portland Cement Association, Skokie, IL, 1998.

1.17   Fanella, D.A., Seismic Detailing of Concrete Buildings, Publication SP382, Portland Cement Association, Skokie, IL, 2000.

ADDITIONAL REFERENCES

1.A   *International Building Code*, 2003 Edition, International Code Council, Inc., Falls Church, VA 2002.

1.B   *Building Construction and Safety Code NFPA 5000*, National Fire Protection Association, Quincy, MA, 2002.

1.C   American Society of Civil Engineers (2002), *Minimum Design Loads for Buildings and Other Structures*, ASCE 7-02 Standard, ASCE, Reston, VA.

1.D   *NEHRP (National Earthquake Hazards Reduction Program) Recommended Provisions for Seismic Regulations for New Buildings and Other Structures, Part 1—Provisions, Part 2—Commentary*, Building Seismic Safety Council, Washington, D.C., 2000.

1.E   *NEHRP (National Earthquake Hazards Reduction Program) Recommended Provisions for Seismic Regulations for New Buildings and Other Structures, Part 1—Provisions, Part 2—Commentary*, Building Seismic Safety Council, Washington, D.C., 2003.

1.F   Ghosh, S.K., and Fanella, D.A., and Liang, X., *Seismic and Wind Design of Reinforced Concrete Buildings*: 2003 IBC, ASCE 7-02, ACI 318-02, Portland Cement Assocaition, Skokie, IL, and International Code Council, Falls Church, VA, 2005.

1.G California Building Standards Commission, 2001 California Building Code: California Code of Regulations Title 24, Part 2, Volume 2, Sacramento, CA, 2002. VA 2002.

1.H Blume, J.A., Newmark, N.A., and Corning, L.H., *Design of Multistory Concrete Buildings for Earthquake Motions*, Portland Cement Association, Skokie, IL, 1961.

1.J Ghosh, S.K., and Domel, Jr., A.W., *Design of Concrete Buildings for Earthquake and Wind Forces*, Portland Cement Association, Skokie, IL, and International Conference of Building Officials, Whittier, CA, 1992.

1.K Ghosh, S.K., Fanella, D.A, *Seismic and Wind Design of Reinforced Concrete Buildings:* 2000 IBC, ASCE 7-98, ACI 318-99, Portland Cement Association, Skokie, IL, and International Code Council, Falls Church, VA, 2003.

1.L Ghosh, S.K., Fanella, D.A., and Liang, X., *Seismic and Wind Design of Reinforced Concrete Buildings: 2003 IBC, ASCE 7-02, ACI 318-02*, Portland Cement Association, Skokie, IL, and International Code Council, Falls Church, VA, 2005.

2

Materials, Concrete Quality

CHAPTER 3—MATERIALS

UPDATE FOR THE '05 CODE

New to the '05 Edition of the Code is a change of the term "welded wire fabric" to "welded wire reinforcement" to correct a common misinterpretation that welded wire fabric is not an appropriate alternate to conventional reinforcing bars.

3.1   TESTS OF MATERIALS

Provisions in 3.1.3 (and in 1.3.4) require the inspecting engineer and architect responsible for maintaining availability of complete test records during construction. The provisions of 3.1.3 also require that records of tests of materials and of concrete must be retained by the inspector for two years after completion of a project, or longer if required by the locally adopted building code. Retention of test records for a minimum two-year period after completion of a project is to ensure that records are available should questions arise (subsequent to owner acceptance or issuance of the certificate of occupancy) concerning quality of materials and of concrete, or concerning any violations of the approved plans and specifications or of the building code.

This is required because engineers and architect do not normally inspect concrete, whereas inspectors are typically hired for this purpose. The term "inspector" is defined in 1.3.1. For many portions of the United States, the term "inspector" may be assumed to be the "special inspector", as defined in the legally adopted building codes. When a special inspector is not employed, other arrangements with the code official will be necessary to insure the availability and retention of the test records.

3.2   CEMENTS

Cement used in the work must correspond to that on which the selection of concrete proportions for strength and other properties was based. This may simply mean the same type of cement or it may mean cement from the same source. In the case of a plant that has determined the standard deviation from tests involving cements from several sources, the former would apply. The latter would be the case if the standard deviation of strength tests used in establishing the required target strength was based on one particular type of cement from one particular source.

In ACI 318-02, ASTM C 1157 (Performance Specification for Blended hydraulic Cement) was recognized for the first time. The ASTM C 1157 standard differs from ASTM C 150 and ASTM C 595 in that it does not establish the chemical composition of the different types of cements. However, individual constituents used to manufacture ASTM C 1157 cements must comply with the requirements specified in the standard. The standard also provides for several optional requirements, including one for cement with low reactivity to alkali-reactive aggregates.

Shrinkage-compensating concrete, made using expansive cement conforming to ASTM C 845 (specification for expansive hydraulic cement), minimizes the potential for drying shrinkage cracks. Expansive cement expands slightly during the early hardening period after initial setting. When expansion is restrained by reinforcement, expansive cement concrete can also be used to (1) compensate for volume decrease due to drying shrinkage, (2) induce tensile stress in the reinforcement (post-tensioning), and (3) stabilize the long term dimensions of post-tensioned structures with respect to original design. The major advantage of using expansive cement in concrete is in the control and reduction of drying shrinkage cracks.

The proportions of the concrete mix assume additional importance when expansive cement is used in conjunction with some admixtures. The beneficial effects of using expansive cement may be less or may have the opposite effect when some admixtures are used in concrete containing expansive cement. Section 3.6.8 flags this concern. Trial mixtures should be made with the selected admixtures and other ingredients of expansive cement concrete to observe the effects of the admixtures on the properties of the fresh and the hardened concrete.

Also, when expansive cement concrete is specified, the design professional must consider certain aspects of the design that may be affected. Code sections related to such design considerations include:

- Section 8.2.4 - Effects of forces due to expansion of shrinkage-compensating concrete must be given consideration in addition to all the other effects listed.

- Section 9.2.7 - Structural effects due to expansion of shrinkage-compensating concrete must be included in T which is included in the load combinations of Eq. (9-2).

3.3   AGGREGATES

The nominal maximum aggregate size is limited to (i) one-fifth the narrowest dimension between sides of forms, (ii) one-third the depth of the slab, and (iii) three-quarters the minimum clear spacing between reinforcing bars or prestressing tendons or ducts. The limitations on nominal maximum aggregate size may be waived if the workability and methods of consolidation of the concrete are such that the concrete can be placed without honeycomb or voids. The engineer must decide whether the limitations on maximum size of aggregate may be waived.

3.4   WATER

Over the past numbers of years environmental regulations associated with the disposal of water from concrete production operations have caused larger amounts of non-potable water (i.e., sources not fit for human consumption) to be used as mixing water in hydraulic cement concrete. Use of this water source needs to be limited by the solids content in the water. A new ASTM standard [2.2], which has not as yet been incorporated into ACI 318, provides a test method for this measurement by means of measuring water density.

In addition to limiting the amount of solids in mixing water, maximum concentrations of other materials that impact the quality of concrete must be limited. These include levels of chloride ion, sulfates, and alkalies. Another ASTM standard [2.3], which has also not as yet been incorporated into ACI 318, provides upper limits for these materials, as well as the total solids content in mixing water.

The chief concern over high chloride content is the possible effect of chloride ions on the corrosion of embedded reinforcing steel or prestressing tendons, as well as concrete containing aluminum embedments or which are cast against stay-in-place galvanized metal forms. Limitations placed on the maximum concentration of chloride ion that are contributed by the ingredients including water, aggregates, cement, and admixtures are given in Chapter 4, Table 4.4.1. These limitations that specifically apply to corrosion protection of reinforcement are measured in water soluble chloride ion in concrete, percent by weight of cement. The previously cited ASTM standard limits the chloride ions in ppm (parts per million) and only applies to that contributed by the mixing water.

3.5 STEEL REINFORCEMENT

3.5.2 Welding of Reinforcement

ACI 318-05 references the latest edition of the Structural Welding Code for Reinforcing Steel - ANSI/AWS D1.4-98. All welding of reinforcing bars must be performed in strict compliance with the D1.4 requirements. Recent revisions to D1.4 deserve notice. Most notably, the preheat requirements for A 615 steel bars require consideration if the chemical composition of the bars is not known. See discussion on 12.14.3 in Part 4.

The engineer should especially note the welding restrictions of 21.2.7 for reinforcement in earthquake force resisting structural members in buildings in regions of high seismic risk or in structures assigned to high seismic performance or design categories. Because these structural elements may perform beyond the elastic range of response, under potentially extreme effects of major earthquakes, welding of reinforcing steel, especially welded splices, must be performed in strict adherence with the welding procedures outlined in ANSI/AWS D1.4. These procedures include adequate inspection.

Section R3.5.2 provides guidance on welding to existing reinforcing bars (which lack mill test reports) and on field welding of cold drawn wire and welded wire. Cold drawn wire is used as spiral reinforcement, and wires or welded wire reinforcement may occasionally be field welded. Special attention is necessary when welding cold drawn wire to address possible loss of its yield strength and ductility. Electric resistance welding, as covered by ASTM A 185 and A 497, is an acceptable welding procedure used in the manufacture of welded wire reinforcement. When welded splices are used in lieu of required laps, pull tests of representative samples or other methods should be specified to determine that an acceptable level of specified strength of steel is provided. "Tack" welding (welding of cross bars) of deformed bars or wire reinforcement is not permitted unless authorized by the engineer (see 7.5.4).

The last paragraph of R3.5.2 states that welding of wire is not covered in ANSI/AWS D1.4. Actually, ANSI/AWS D1.4 addresses the welding of all forms of steel reinforcement, but lacks certain critical information for wire or welded wire reinforcemnt (e.g., preheats and electrode selection are not discussed). However, it is recommended that field welding of wire and welded wire reinforcement follow the applicable provisions of ANSI/AWS D1.4, such as certification of welders, inspection procedures, and other applicable welding procedures.

3.5.3 Deformed Reinforcement

Only deformed reinforcement as defined in Chapter 2 may be used for nonprestressed reinforcement, except that plain bars and plain wire may be used for spiral reinforcement. Welded plain wire reinforcement is included under the code definition of deformed reinforcement. Reinforcing bars rolled to ASTM A 615 specifications are the most commonly specified for construction. Rail and axle steels (ASTM A 616 and ASTM A 617, respectively) were deleted from ACI 318-02 and replaced by ASTM A 996 (Specification for Rail-Steel and Axle-Steel Deformed Bars for Concrete Reinforcement). Deformed reinforcement meeting ASTM A 996 is marked with the letter *R* and must meet more restrictive provisions for bend tests than was required by the previous two specification standards that ASTM A996 replaced. Rail steel (ASTM A 996) is not generally available, except in a few areas of the country.

ASTM A 706 covers low-alloy steel deformed bars (Grade 60 only) intended for special applications where welding or bending or both are of importance. Reinforcing bars conforming to A 706 should be specified wherever critical or extensive welding of reinforcement is required, and for use in reinforced concrete structures located in regions of high seismic risk or in structures assigned to high seismic performance or design categories where more bendability and controlled ductility are required. The special provisions of Chapter 21 for seismic design require that reinforcement resisting earthquake-induced flexural and axial forces in frame members and in wall boundary elements forming parts of structures located in regions of high seismic risk or in structures assigned to high seismic performance or design categories comply with ASTM A 706 (see 21.2.5). Grades 40 and 60 ASTM A 615 bars are also permitted in these members if (a) the ratio of actual ultimate tensile strength to the actual tensile yield strength is not less than 1.25, and (b) the actual yield strength based on mill tests does not exceed the specified yield strength by more than 18,000 psi (retests are not permitted to exceed the specified yield strength by more than an additional 3000 psi).

Before specifying A 706 reinforcement, local availability should be investigated. Most rebar producers can make A 706 bars, but generally not in quantities less than one heat of steel for each bar size ordered. A heat of steel varies from 50 to 200 tons, depending on the mill. A 706 in lesser quantities of single bar sizes may not be immediately available from any single producer. Notably, A 706 is being specified more and more for reinforced concrete structures in high seismic risk areas (see Table 1-1 in Part 1). Not only are structural engineers specifying it for use in earthquake-resisting elements of buildings, but also for reinforced concrete bridge structures. Also, A 706 has long been the choice of precast concrete producers because it is easier and more cost effective for welding, especially in the various intricate bearing details for precast elements. This increased usage should impact favorably on the availability of this low-alloy bar.

Section 9.4 permits designs based on a yield strength of reinforcement up to a maximum of 80,000 psi. Currently there is no ASTM specification for a Grade 80 reinforcement. However, deformed reinforcing bars No. 6 through No. 18 with a yield strength of 75,000 psi (Grade 75) are included in the ASTM A 615 specification. "Section 3.5.3.2 requires that the yield strength of deformed bars with a specified yield strength greater than 60,000 psi be taken as the stress corresponding to a strain of 0.35 percent. The 0.35 percent strain limit is to ensure that the elasto-plastic stress-strain curve assumed in 10.2.4 will not result in unconservative values of member strength. Therefore, the designer should be aware that if ASTM A 615, Grade 75 bars are specified, the project specifications need to include a requirement that the yield strength of the bars shall be determined in accordance with Section 9.2.2 of the ASTM A 615 specification." Certified mill test reports should be obtained from the supplier when Grade 75 bars are used. Before specifying Grade 75, local availability should be investigated. The higher yield strength No. 6 through No. 18 bars are intended primarily as column reinforcement. They are used in conjunction with higher strength concrete to reduce the size of columns in high-rise buildings and other applications where high capacity columns are required. Wire used to manufacture both plain and deformed welded wire reinforcement can have a specified yield strength in excess of 60,000 psi. It is permissible to take advantage of the higher yield strength provided the specified yield strength, f_y, used in the design corresponds to the stress at a strain of 0.35 percent.

In recent years manufacturers of reinforcing bars have switched their production entirely to soft metric bars. The physical dimensions (i.e., diameter, and height and spacing of deformations) of the soft-metric bars are no different than the inch-pound bars that were manufactured for many years. The only difference is that the bar size mark that is rolled onto the bar is based on SI metric units. Metric bar sizes and bar marks are based on converting the bar's inch-pound diameter to millimeters and rounding to the nearest millimeter. For example, a No. 4, or 1/2-in. diameter bar, becomes a No. 13 bar since its diameter is 12.7 mm. See Table 2-1 for a complete listing of all 11 ASTM standard reinforcing bar sizes.

ASTM standard specifications A 615, A 706 and A 996 have requirements for bars in both inch-pound and SI metric units; therefore, they have dual designations (e.g., ASTM A 706/A 706M). Each specification provides criteria for one or more grades of steel, which are summarized in Table 2-2.

The minimum required yield strength of the steel used to produce the bars has been changed slightly within ASTM A 615M. The latest edition of the ASTM A 615M bar specifications have a Grade 280, or 280 megapascals (MPa) minimum yield strength, which was previously designated as Grade 300. Soft converting Grade 40 or 40,000 psi yield strength steel will result in a metric yield strength of 275.8 MPa (1,000 psi = 6.895 MPa), which is more closely designated as Grade 280, than the previous Grade 300 designation.

Table 2-1 Inch-Pound and Soft Metric Bar Sizes

Inch-Pound		Metric	
Size No.	Dia. (in.)	Size No.	Dia. (mm)
3	0.375	10	9.5
4	0.500	13	12.7
5	0.625	16	15.9
6	0.750	19	19.1
7	0.875	22	22.2
8	1.000	25	25.4
9	1.128	29	28.7
10	1.270	32	32.3
11	1.410	36	35.8
14	1.693	43	43.0
18	2.257	57	57.3

Table 2-2 ASTM Specifications - Grade and Min. Yield Strength

ASTM Specification	Grade/Minimum Yield Strength	
	Inch-Pound (psi)	Metric (MPa)
A 615 and A 615M	40/40,000 60/60,000 75/75,000	280/280 420/420 520/520
A 996 and A 996M	40/40,000 50/50,000 60/60,000	280/280 350/350 420/420
A 706/A 706M	60/60,000	420/420

When design and construction proceed in accordance with the ACI 318 Code, using customary inch-pound units, the use of soft metric bars will have only a very small effect on the design strength or allowable load-carrying capability of members. For example, where the design strength of a member is a function of the steel's specified yield strength, f_y, the use of soft metric bars increases the strength approximately 1.5% for grade 420 [(420 - 413.7)/413.7].

3.5.3.5 - 3.5.3.6 Welded Plain and Deformed Wire Reinforcement—On occasion, building department plan reviewers have questioned the use of welded wire reinforcement as an alternative to conventional reinforcing bars for structurally reinforced concrete applications. This usually occurs during the construction phase when reinforcing bars shown on the structural drawings are replaced with welded wire reinforcement through a change order. The code officials' concern probably stems from the commonly accepted industry terminology for welded wire reinforcement used as "nonstructural" reinforcement for the control of crack widths for slabs-on-ground.

Wire sizes for welded wire reinforcement range from W1.4 (10 gauge) to W4 (4 gauge). Plain wire is denoted by the letter "W" followed by a number indicating cross-sectional area in hundredths of a square inch. Styles of welded wire reinforcement used to control crack widths in residential and light industrial slabs-on-grade are 6 x 6 W1.4 x W1.4, 6 x 6 W2 x W2, 6 x 6 W2.9 x W2.9 and 6 x 6 W4 x W4. These styles of welded wire reiforcement weigh 0.21 lb, 0.30, 0.42 lb and 0.55 lb per square foot respectively, and are manufactured in rolls, although they are also available in sheets. Smaller wire sizes are not typically used as an alternative to conventional reinforcing bars. Welded wire reinforcement used for structural reinforcement is typically made with a wire size larger than W4. The term "welded wire reinforcement" has replaced the term "welded wire fabric" in this edition of the code to help correct this misinterpretation.

Substitution of welded wire reinforcement for reinforcing bars may be requested for construction or economic considerations. Whatever the reason, both types of reinforcement, either made with welded wire or reinforcing bars are equally recognized and permitted by the code for structural reinforcement. Both welded deformed wire reinforcement and welded plain wire reinforcement are included under the code definition for deformed reinforcement. Welded deformed wire reinforcement utilizes wire deformations plus welded intersections for bond and anchorage. (Deformed wire is denoted by the letter "D" followed by a number indicating cross-sectional area in hundredths of a square inch.) Welded plain wire reinforcement bonds to concrete by positive mechanical anchorage at each wire intersection. This difference in bond and anchorage for plain versus deformed reinforcement is reflected in the development of lap splices provisions of Chapter 12.

3.5.3.7 Coated Reinforcement—Appropriate references to the ASTM specifications for coated reinforcement, A 767 (galvanized) and A 775 (epoxy-coated), are included in the code to reflect increased usage of coated bars. Coated welded wire reinforcement is available with an epoxy coating (ASTM A 884), with wire galvanized before welding (ASTM A 641), and with welded wire galvanized after welding (ASTM A 123). The most common coated bars and welded wire are epoxy-coated reinforcement for corrosion protection. Epoxy-coated reinforcement provides a viable corrosion protection system for reinforced concrete structures. Usage of epoxy-coated reinforcement has become commonplace for many types of reinforced concrete construction such as parking garages (exposed to deicing salts), wastewater treatment plants, marine structures, and other facilities located near coastal areas where the risk of corrosion of reinforcement is higher because of exposure to seawater—particularly if the climate is warm and humid.

Designers specifying epoxy-coated reinforcing bars should clearly outline in the project specifications special hardware and handling methods to minimize damage to the epoxy coating during handling, transporting, and placing coated bars, and placing of concrete.[2.5, 2.6] Special hardware and handling methods include:

1. Using nylon lifting slings, or padded wire rope slings.

2. Using spreader bars for lifting bar bundles, or lifting bundles at the third points with nylon or padded slings. Bundling bands should be made of nylon, or be padded.

3. Storing coated bars on padded or wooden cribbing.

4. Not dragging coated bars over the ground, or over other bars.

5. Minimizing walking on coated bars and dropping tools or other construction materials during or after placing the bars.

6. Using bar supports of an organic material or wire bar supports coated with an organic material such as epoxy or vinyl compatible with concrete.

7. Using epoxy- or plastic-coated tie wire, or nylon-coated tie wire to minimize damage or cutting into the bar coating.

8. Setting up, supporting and moving concrete conveying and placing equipment carefully to minimize damage to the bar coating.

Project specifications should also address field touch-up of the epoxy coating after bar placement. Permissible coating damage and repair are included in the ASTM A 775 and in Ref. 2.5. Reference 2.6 contains suggested project specification provisions for epoxy-coated reinforcing bars.

The designer should be aware that epoxy-coated reinforcement requires increased development and splice lengths for bars in tension (see 12.2.4.3).

3.6 ADMIXTURES

3.6.9 Silica Fume

Silica fume (ASTM C 1240) gets its name because it extracted from the fumes of electric furnaces that produce ferrosilicon or silicon metal. By the time it is collected and prepared as an admixture for concrete it has become a very finely divided solid-microsilica. Silica fume is generally used in concrete for one or more of the following reasons. When used in conjunction with high-range water reducing admixtures, it makes it possible to produce concrete with compressive strengths of 20,000 psi (138 MPa) or higher. It is also used to achieve a very dense cement paste matrix to reduce the permeability of concrete. This provides better corrosion protection to reinforcing steel, particularly when the concrete will be subject to direct or indirect applications of deicing chemicals, such as in bridge decks or in parking garages, respectively.

Mix proportioning, production methods (mixing and handling), and the placing and curing procedures for silica fume concrete require a more concentrated quality control effort than for conventional concretes. It is imperative that the engineer, concrete supplier, and the contractor work as a team to ensure consistently high quality when silica fume concrete is specified.

Note, concrete containing silica fume can be almost black, dark gray, or practically unchanged from the color of cement, depending on the dosage of silica fume. The greatest differences in color will occur in concretes made with cements that are light in color. Mix proportions may also affect variations in color. If color difference is a problem (architectural concrete), the darkest brand of cement available should be used, and different trial mixtures should be tried during the mix design process.

CHAPTER 4—DURABILITY REQUIREMENTS

GENERAL CONSIDERATIONS

The special exposure conditions addressed by the code are located exclusively in Chapter 4 - Durability Requirements, to emphasize the importance of special exposures on concrete durability. The code provisions for concrete proportioning and strength evaluation are located in Chapter 5 - Concrete Quality, Mixing, and Placing. As stated in 5.1.1, selection of concrete proportions must be established to provide for both (a) resistance to special exposures as required by Chapter 4, and (b) conformance with the strength requirements of Chapter 5.

Resistance to special exposures is addressed in 4.2 - Freeze-thaw exposure, 4.3 - Sulfate exposure, and 4.4 - Corrosion protection of reinforcement. Conformance with strength test requirements is addressed in 5.6 - Evaluation and acceptance of concrete. Depending on design and exposure requirements, the lower of the water-cementitious materials ratios required for the structural design requirements and for the concrete exposure conditions must be specified (see Example 2.1)

Unacceptable deterioration of concrete structures in many areas due to severe exposure to freezing and thawing, to deicing salts used for snow and ice removal, to sulfate in soil and water, and to chloride exposure have warranted a stronger code emphasis on the special exposure requirements. Chapter 4 directs special attention to the need for considering concrete durability, in addition to concrete strength.

In the context of the code, durability refers to the ability of concrete to resist deterioration from the environment or the service in which it is placed. Properly designed and constructed concrete should serve its intended function without significant distress throughout its service life. The code, however, does not include provisions for especially severe exposures such as to acids or high temperatures, nor is it concerned with aesthetic considerations such as surface finishes. Items like these, which are beyond the scope of the code, must be covered specifically in the project specifications. Concrete ingredients and proportions must be selected to meet the minimum requirements stated in the code and the additional requirements of the construction documents.

In addition to the proper selection of cement, adequate air entrainment, maximum water-cementitious materials ratio, and limiting chloride ion content of the materials, other requirements essential for durable concrete exposed to adverse environments include: low slump, adequate consolidation, uniformity, adequate cover of reinforcement, and sufficient moist curing to develop the potential properties of the concrete.

4.1   WATER-CEMENTITIOUS MATERIALS RATIO

The traditional "water-cement" ratio was renamed "water-cementitious materials ratio" starting with the 1989 edition of code in recognition of the other cementitious materials permitted by the code to satisfy the code limitations on w/c for concrete durability. The notation, "w/c", is commonly used for "water-cementitious materials" in the term w/c ratio. The definition for "cementitious materials" (see Chapter 2) permits the use of cementitious materials other than portland cement and blended hydraulic cements to satisfy the code's w/c limitation. For the calculation of w/c ratios, as limited by the code, the "cementitious materials" may include:

- portland cement (ASTM C 150)
- blended hydraulic cement (ASTM C 595 and ASTM C 1157)
- expansive hydraulic cement (ASTM C 845)

either by themselves, or in combination with:

- fly ash (ASTM C 618)
- raw or calcinated natural pozzolans (ASTM C 618)
- ground granulated blast-furnace slag (ASTM C 989)
- silica fume (ASTM C 1240)

4.2   FREEZING AND THAWING EXPOSURES

For concrete that will be exposed to freezing and thawing while moist or to deicer salts, air-entrained concrete must be specified with minimum air contents for severe and moderate exposure, as set forth in Table 4.2.1. Project specifications should allow the air content of the delivered concrete to be within (-1.5) and (+1.5) percentage points of Table 4.2.1 target values. Severe exposure is a cold climate where the concrete may be exposed almost continuously to wet freeze-thaw conditions or where deicing salts are used. Examples include pavements, bridge decks, sidewalks, and parking garages. For severe exposure conditions, the code also imposes a maximum limit on the w/c ratio and minimum f'_c (See 4.4.2). A moderate exposure is a cold climate where concrete may be exposed to freezing, but will only occasionally be exposed to moisture prior to freezing, and where no deicing salts are used. Examples are certain exterior walls, beams, girders and slabs not in direct contact with soil.

Intentionally entraining air in concrete significantly improves the resistance of hardened concrete to freezing when exposed to water and deicing salts. Concrete that is dry or contains only a small amount of moisture is essentially not affected by even a large number of cycles of freezing and thawing. Sulfate resistance is also improved by air entrainment.

The entrainment of air in concrete can be accomplished by adding an air-entraining admixture at the mixer, by using an air-entraining cement, or by a combination of both. Air-entraining admixtures, added at the mixer, must conform to ASTM C 260 (3.6.4); air-entraining cements must comply with the specifications in ASTM C 150 and C 595 (3.2.1). Air-entraining cements are sometimes difficult to obtain; and their use has been decreasing as the popularity of air-entraining admixtures has increased. ASTM C 94, Standard Specifications for Ready Mixed Concrete, which is adopted by reference in the ACI code (5.8.2), requires that air content tests be conducted. The frequency of these tests is the same as required for strength evaluation. Samples of concrete must be obtained in accordance with ASTM C 172 and tested in accordance with ASTM C 173 or C 231.

Normal weight concrete that will be exposed to freeze-thaw conditions while wet and exposed to deicing salts must be proportioned so that both a maximum w/c ratio and a minimum compressive strength are provided (Table 4.2.2). Requiring both criteria helps to ensure that the desired durability will actually be obtained in the field. Generally, the required average concrete strength, f'_{cr}, used to develop the mix design will be 500 to 700 psi higher than the specified compressive strength, f'_c. It is also more difficult to accurately determine the w/c ratio of concrete during production then controlling compressive strength. Thus, when selecting an f'_c, it should be reasonable consistent with the w/c ratio required for durability. Using this approach, minimum strengths required for durability provide an effective backup quality control check to the w/c ratio limitation which is more essential to durability.

A minimum strength only is specified for lightweight aggregate concrete, due to the variable absorption characteristics of lightweight aggregates, which makes the calculated w/c ratios meaningless.

Concrete used in water-retaining structures or exposed to severe exposure conditions as described above must be virtually impermeable or watertight. Low permeability not only improves freezing and thawing resistance, especially in the presence of deicing salts, but also improves the resistance of concrete to chloride ion penetration. Concrete that is intended to have low permeability to water must be proportioned so that the specified w/c ratio does not exceed 0.50. If concrete is to be exposed to freezing and thawing in a moist condition, the specified w/c ratio must be no more than 0.45. Also, for corrosion protection of reinforcement in concrete exposed to deicing salts (wet freeze-thaw conditions), and in concrete exposed to seawater (including seawater spray), the concrete must be proportioned so that the specified w/c ratio does not exceed 0.40.

For the above exposure conditions, the corresponding minimum concrete strengths indicated in Table 4.2.2 must also be satisfied for normal- and lightweight, aggregate concretes. Design Example 2.1 illustrates mix proportioning to satisfy both a w/c ratio and a strength requirement for concrete durability.

4.2.3 Concrete Exposed to Deicing Chemicals

Table 4.2.3 limits the type and amount of portland cement replacement permitted in concrete exposed to deicing salts. The amount of fly ash or other pozzolan, or both, is limited to 25 percent of the total weight of cementitious materials. Slag and silica fume are similarly limited to 50 percent and 10 percent, respectively, of the total weight. If fly ash (or other pozzolan) plus slag and silica fume are used as partial cement replacement, the total weight of the combined replacement materials cannot exceed 50 percent of the total weight of cementitious materials, with the maximum percentage of each type of replacement not to exceed the individual percentage limitations. If slag is excluded from the cement replacement combination, the total weight of the combined replacement cannot exceed 35 percent, with the individual percentages of each also not to be exceeded.

As an example: If a reinforced concrete element is to be exposed to deicing salts, Table 4.2.2 limits the w/c ratio to 0.40. If the mix design requires 280 lb of water to produce an air-entrained concrete mix of a given slump, the total weight of cementitious materials cannot be less than 280/0.40 = 700 lbs. The 700 lbs of "cementitious materials" may be all portland cement or a combination of portland cement and fly ash, pozzolan, slag, or silica fume.

If fly ash is used as portland cement replacement, the maximum amount of fly ash is limited to 0.25 (700) = 175 lbs, maintaining the same w/c = 280/(525+175) = 0.40.

If slag is the total replacement, the maximum is limited to 0.50 (700) = 350 lbs, with w/c = 280/(350+350) = 0.40.

If the cement replacement is a combination of fly ash and slag, the maximum amount of the combination is limited to 0.50 (700) = 350 lbs, with the fly ash portion limited to 0.25 (700) = 175 lbs of the total combination, with w/c = 280/(350+175+175) = 0.40.

If the cement replacement is a combination of fly ash and silica fume (a common practice in high performance concrete), the maximum amount of the combination is limited to 0.35 (700) = 245 lbs, and the silica fume portion limited to 0.10 (700) = 70 lbs, with w/c = 280/(385+245+70) = 0.40.

Obviously, other percentages of cement replacement can be used so long as the combined and individual percentages of Table 4.2.3 are not exceeded.

It should be noted that the portland cement replacement limitations apply only to concrete exposed to the potential damaging effects of deicing chemicals. Research has indicated that fly ash, slag, and silica fume can reduce concrete permeability and chloride ingress by providing a more dense and impermeable cement paste. As to the use of fly ash and other pozzolans, and especially silica fume, it is noteworthy that these cement replacement admixtures are commonly used in high performance concrete (HPC) to decrease permeability and increase strength.

4.3 SULFATE EXPOSURES

Sulfate attack of concrete can occur when it is exposed to soil, seawater, or groundwater having a high sulfate content. Measures to reduce sulfate attack include the use of sulfate-resistant cement. The susceptibility to sulfate attack is greater for concrete exposed to moisture, such as in foundations and slabs on ground, and in structures directly exposed to seawater. For concrete that will be exposed to sulfate attack from soil or water, sulfate-resisting cement must be specified. Table 4.3.1 lists the appropriate types of sulfate-resisting cements and maximum water-cementitious materials ratios and corresponding minimum concrete strengths for various exposure conditions. Degree of exposure is based on the amount of water-soluble sulfate concentration in soil or on the amount of sulfate concentration in water. Note that Table 4.3.1 lists seawater under "moderate exposure," even though it generally contains more than 1500 ppm of sulfate concentration. The reason is that the presence of chlorides in seawater inhibits the expansive reaction that is characteristic of sulfate attack.[2.1]

In selecting a cement type for sulfate resistance, the principal consideration is the tricalcium aluminate (C_3A) content. Cements with low percentages of C_3A are especially resistant to soils and waters containing sulfates. Where precaution against moderate sulfate attack is important, as in drainage structures where sulfate concentrations in groundwater are higher than normal, but not necessarily severe (0.10 - 0.20 percent), Type II portland cement (maximum C_3A content of eight percent per ASTM C 150) must be specified.

Type V portland cement must be specified for concrete exposed to severe sulfate attack—principally where soils or groundwaters have a high sulfate content. The high sulfate resistance of Type V cement is attributed to its low tricalcium aluminate content (maximum C_3A content of five percent).

Certain blended cements (C 595) also provide sulfate resistance. Other types of cement produced with low C_3A contents are usable in cases of moderate to severe sulfate exposure. Sulfate resistance also increases with air-entrainment and increasing cement contents (decreasing water-cementitious materials ratios).

Before specifying a sulfate resisting cement, its availability should be checked. Type II cement is usually available, especially in areas where resistance to moderate sulfate attack is needed. Type V cement is available only in particular areas where it is needed to resist severe and very severe sulfate environments. Blended cements may not be available in many areas.

4.4 CORROSION PROTECTION OF REINFORCEMENT

Chlorides can be introduced into concrete through its ingredients: mixing water, aggregates, cement, and admixtures, or through exposure to deicing salts, seawater, or salt-laden air in coastal environments. The chloride ion content limitations of Table 4.4.1 are to be applied to the chlorides contributed by the concrete ingredients, not to chlorides from the environment surrounding the concrete (chloride ion ingress). Chloride ion limits are the responsibility of the concrete production facility which must ensure that the ingredients used in the production of concrete (cement, water, aggregate, and admixtures) result in concrete with chloride ion contents within the limits given for different exposure conditions. When testing is performed to determine chloride ion content of the individual ingredients, or samples of the hardened concrete, test procedures must conform to ASTM C 1218, as indicated in 4.4.1. In addition to a high chloride content, oxygen and moisture must be present to induce the corrosion process. The availability of oxygen and moisture adjacent to embedded steel will vary with the in-service exposure condition, which varies among structures, and between different parts of the same structure.

If significant amounts of chlorides may be introduced into the hardened concrete from the concrete materials to be used, the individual concrete ingredients, including water, aggregates, cement, and any admixtures, must be tested to ensure that the total chloride ion concentration contributed from the ingredients does not exceed the limits of Table 4.4.1. These limits have been established to provide a threshold level to avoid corrosion of the embedded reinforcement prior to service exposure. Chloride limits for corrosion protection also depend upon the type of construction and the environment to which the concrete is exposed during its service life, as indicated in Table 4.4.1.

Chlorides are present in variable amounts in all of the ingredients of concrete. Both water soluble and insoluble chlorides exist; however, only water soluble chlorides induce corrosion. Tests are available for determining either the water soluble chloride content or the total (soluble plus insoluble) chloride content. The test for soluble chloride is more time-consuming and difficult to control, and is therefore more expensive than the test for total chloride. An initial evaluation of chloride content may be obtained by testing the individual concrete ingredients for total (soluble plus insoluble) chloride content. If the total chloride ion content is less than that permitted by Table 4.4.1, water-soluble chloride need not be determined. If the total chloride content exceeds the permitted value, testing of samples of the hardened concrete for water-soluble chloride content will need to be performed for direct comparison with Table 4.4.1 values. Some of the soluble chlorides in the ingredients will react with the cement during hydration and become insoluble, further reducing the soluble chloride ion content, the corrosion-inducing culprit. Of the total chloride ion content in hardened concrete, only about 50 to 85 percent is water soluble; the rest is insoluble. Note that hardened concrete should be at least 28 days of age before sampling.

Chlorides are among the more abundant materials on earth, and are present in variable amounts in all of the ingredients of concrete. Potentially high chloride-inducing materials and conditions include: use of seawater as mixing water or as washwater for aggregates, since seawater contains significant amounts of sulfates and chlorides; use of marine-dredged aggregates, since such aggregates often contain salt from the seawater; use of aggregates that have been contaminated by salt-laden air in coastal areas; use of admixtures containing chloride, such as calcium chloride; and use of deicing salts where salts may be tracked onto parking structures by vehicles. The engineer needs to be cognizant of the potential hazard of chlorides to concrete in marine environments or other exposures to soluble salts. Research has shown that the threshold value for a water soluble chloride content of concrete necessary for corrosion of embedded steel can be as low as 0.15 percent by weight of cement. When chloride content is above this threshold value, corrosion is likely if moisture and oxygen are readily available. If chloride content is below the threshold value, the risk of corrosion is low.

Depending on the type of construction and the environment to which it is exposed during its service life, and the amount and extent of protection provided to limit chloride ion ingress, the chloride level in concrete may increase with age and exposure. Protection against chloride ion ingress from the environment is addressed in 4.4.2, with reference to Table 4.2.2. A maximum water-cementitious materials ratio of 0.40 and a minimum strength of 5000 psi must be provided for corrosion protection of "reinforcement in concrete exposed to chlorides from deicing chemicals, brackish water, seawater or spray from these sources." Resistance to corrosion of embedded steel is also improved with an increase in the thickness of concrete cover. Section R7.7.5 recommends a minimum concrete cover of 2 in. for cast-in-place walls and slabs, and 2-1/2 in. for other members, where concrete will be exposed to external sources of chlorides in service. For plant-produced precast members, the corresponding recommended minimum concrete covers are 1-1/2 in. and 2 in., respectively.

Other methods of reducing environmentally caused corrosion include the use of epoxy-coated reinforcing steel[2.4, 2.5, 2.6], corrosion-inhibiting admixtures, surface treatments, and cathodic protection. Epoxy coating of reinforcement prevents chloride ions from reaching the steel. Corrosion-inhibiting admixtures attempt to chemically arrest the corrosive reaction. Surface treatments attempt to stop or reduce chloride ion penetration at the exposed concrete surface. Cathodic protection methods reverse the corrosion current flow through the concrete and reinforcing steel. It should be noted that, depending on the potential severity of the chloride exposure, and the type and importance of the construction, more than one of the above methods may be combined to provide "added" protection. For example, in prestressed parking deck slabs in cold climates where deicing salts are used for snow and ice removal, all conventional reinforcement and the post-tensioning tendons may be epoxy-coated, with the entire tendon system including the anchorages encapsulated in a watertight protective system especially manufactured for aggressive environments. In addition, special high performance (impermeable) concrete may be used, with the entire deck surface covered with a multi-layer membrane surface treatment. Such extreme protective measures may be cost-effective, considering the alternative. Performance tests for chloride permeability of concrete mixtures may also be used to assure corrosion resistance. ASTM C 1202, which was introduced starting with the 2002 edition of the code, provides a test method for an electrical indication of concrete's ability to resist chloride ion penetration. It is based on AASHTO T 277-83, which was previously referenced in the code.

CHAPTER 5—CONCRETE QUALITY, MIXING, AND PLACING

UPDATE FOR THE '05 CODE

New to the '05 Edition of the Code is the change of the term "standard deviation" to "sample standard deviation" because from a scientific point of view, the Code provides a check of a sample standard deviation, and not a true (absolute) standard deviation for each concrete mixture. The new notation of sample standard deviation is "s_s".

5.1.1   Concrete Proportions for Strength

Concrete mix designs are proportioned for strength based on probabilistic concepts that are intended to ensure that adequate strength will be developed in the concrete. It is emphasized in 5.1.1 that the required average compressive strength, f'_{cr}, of concrete produced must exceed the larger of the value of f'_c specified for the structural design requirements and the minimum strength required for the special exposure conditions set forth in Chapter 4. Concrete proportioned by the code's probabilistic approach may produce strength tests which fall below the specified compressive strength, f'_c. Section 5.1.1 introduces this concept by noting that it is the code's intent to "minimize frequency of strength below f'_c." If a concrete strength test falls below f'_c, the acceptability of this lower strength concrete is provided for in Section 5.6.3.3.

A minimum 2500 psi specified compressive strength, f'_c, is required by Section 5.1.1 of the code. This makes the code consistent with minimum provisions that are contained in several legacy model building codes, and the *International Building Code* (IBC).

5.1.3   Test Age for Strength of Concrete

Section 5.1.3 permits f'_c to be based on tests at ages other than the customary 28 days. If other than 28 days, the test age for f'_c must be indicated on the design drawings or in the specifications. Higher strength concretes, exceeding 6000 psi compression strength, are often used in tall buildings can justifiably have test ages longer than the customary 28 days. For example, in high-rise structures requiring high-strength concrete, the process of construction is such that the columns of the lower floors are not fully loaded until a year or more after commencement of construction. For this reason, specified compressive strengths, f'_c, based on 56- or 90-day test results are commonly specified.

5.2   SELECTION OF CONCRETE PROPORTIONS

Recommendations for proportioning concrete mixtures are given in detail in *Design and Control of Concrete Mixtures*.[2.1] Recommendations for selecting proportions for concrete are also given in detail in "Standard

Practice for Selecting Proportions for Normal, Heavyweight, and Mass Concrete" (ACI 211.1)[2.7] and "Standard Practice for Selecting Proportions for Structural Lightweight Concrete" (ACI 211.2).[2.8]

The use of field experience or laboratory trial batches (see 5.3) is the preferred method for selecting concrete mixture proportions. When no prior experience or trial batch data are available, permission may be granted by the registered design professional to base concrete proportions on "other experience or information" as prescribed in 5.4.

5.3 PROPORTIONING ON THE BASIS OF FIELD EXPERIENCE AND/OR TRIAL MIXTURES

5.3.1 Sample Standard Deviation

For establishing concrete mixture proportions, emphasis is placed on the use of laboratory trial batches or field experience as the basis for selecting the required water-cementitious materials ratio. The code emphasizes a statistical approach to establishing the required average compressive strength of concrete, f'_{cr}, or "target strength" required to ensure attainment of the specified compressive strength, f'_c. If an applicable sample standard deviation, s_s, from strength tests of the concrete is known, the target strength level for which the concrete must be proportioned is established. Otherwise, the proportions must be selected to produce a conservative target strength sufficient to allow for a high degree of variability in strength test results. For background information on statistics as it relates to concrete, see "Recommended Practice for Evaluation of Compression Test Results of Concrete"[2.9] and "Statistical Product Control."[2.10]

Concrete used in background tests to determine sample standard deviation is considered to be "similar" to that specified, if it was made with the same general types of ingredients, under no more restrictive conditions of control over material quality and production methods than are specified to exist on the proposed work, and if its specified strength does not deviate by more than 1000 psi from that f'_c specified. A change in the type of concrete or a significant increase in the strength level may increase the sample standard deviation. Such a situation might occur with a change in the type of aggregate; i.e., from natural aggregate to lightweight aggregate or vice versa, or with a change from non-air-entrained concrete to air-entrained concrete. Also, there may be an increase in sample standard deviation when the average strength level is raised by a significant amount, although the increment in sample standard deviation should be somewhat less than directly proportional to the strength increase. When there is reasonable doubt as to its reliability, any estimated sample standard deviation used to calculate the required average strength should always be on the conservative (high) side.

Sample standard deviations are normally established by at least 30 consecutive tests on representative materials. If less than 30, but at least 15 tests are available, Section 5.3.1.2 provides for a proportional increase in the calculated sample standard deviation as the number of consecutive tests decrease from 29 to 15.

Statistical methods provide valuable tools for assessing the results of strength tests. It is important that concrete technicians understand the basic language of statistics and be capable of effectively utilizing the tool to evaluate strength test results.

Figure 2-1 illustrates several fundamental statistical concepts. Data points represent six (6) strength test results[*] from consecutive tests on a given class of concrete. The horizontal line represents the average of tests that is designated $\bar{X}$. The average is computed by adding all test values and dividing by the number of values summed; i.e., in Fig. 2-1:

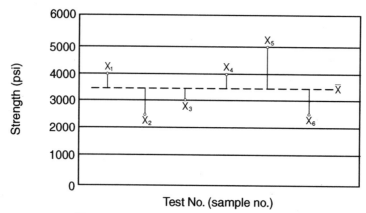

Figure 2-1 Illustration of Statistical Terms

$$\overline{X} = (4000 + 2500 + 3000 + 4000 + 5000 + 2500)/6 = 3500 \text{ psi}$$

The average $\overline{X}$ gives an indication of the overall strength level of the concrete tested.

It would also be informative to have a single number which would represent the variability of the data about the average. The up and down deviations from the average (3500 psi) are given as vertical lines in Fig. 2-1. If one were to accumulate the total length of the vertical lines without regard to whether they are up or down, and divide that total length by the number of tests, the result would be the average length, or the average distance from the average strength:

$$(500 + 1000 + 500 + 500 + 1500 + 1000)/6 = 833 \text{ psi}$$

This is one measure of variability. If concrete test results were quite variable, the vertical lines would be long. On the other hand, if the test results were close, the lines would be short.

In order to emphasize the impact of a few very high or very low test values, statisticians recommend the use of the square of the vertical line lengths. The square root of the sum of the squared lengths divided by one less than the number of tests (some texts use the number of tests) is known as the standard deviation. This measure of variability is commonly designated by the letter s_s. Mathematically, s_s is expressed as:

$$s_s = \sqrt{\frac{\Sigma\left(X - \overline{X}\right)^2}{n - 1}}$$

where

s_s = standard deviation, psi
Σ indicates summation
X = an individual strength test result, psi
$\overline{X}$ = average strength, psi
n = number of tests

For example, for the data in Fig. 2-1, the sample standard deviation would be:

$$s_s = \sqrt{\frac{\left(X_1 - \overline{X}\right)^2 + \left(X_2 - \overline{X}\right)^2 + \left(X_3 - \overline{X}\right)^2 + \left(X_4 - \overline{X}\right)^2 + \left(X_5 - \overline{X}\right)^2 + \left(X_6 - \overline{X}\right)^2}{6 - 1}}$$

---

* A strength test result is the average of the strengths of two cylinders made from the same batch of concrete and tested at the same time.

which is calculated below.

Deviation $(X - \overline{X})$	$(X - \overline{X})^2$
(length of vertical lines)	(length squared)
4000 - 3500 = + 500	+ 250,000
2500 - 3500 = - 1000	+ 1,000,000
3000 - 3500 = - 500	+ 250,000
4000 - 3500 = + 500	+ 250,000
5000 - 3500 = + 1500	+ 2,250,000
2500 - 3500 = - 1000	+ 1,000,000
Total	+ 5,000,000

$$s_s = \sqrt{\frac{5,000,000}{5}} = 1,000 \text{ psi (a very large value)}$$

For concrete strengths in the range of 3000 to 4000 psi, the expected sample standard deviation, representing different levels of quality control, will range as follows:

Sample Standard Deviation	Representing
300 to 400 psi	Excellent Quality Control
400 to 500 psi	Good
500 to 600 psi	Fair
> 600 psi	Poor Quality Control

For the very-high-strength, so called high-performance concrete (HPC), with strengths in excess of 10,000 psi, the expected sample standard deviation will range as follows:

300 to 500 psi	Excellent Quality Control
500 to 700 psi	Good
> 700 psi	Poor Quality Control

Obviously, it would be time consuming to actually calculate s_s in the manner described above. Most hand-held scientific calculators are programmed to calculate sample standard deviation directly. The appropriate mathematical equations are programmed into the calculator with the user simply entering the statistical data (test values), then pressing the appropriate function key to obtain sample standard deviation directly. Example 2.2 illustrates a typical statistical evaluation of strength test results.

The coefficient of variation, V, is simply the standard deviation expressed as a percentage of the average value. The mathematical formula is:

$$V = \frac{s_s}{\overline{X}} \times 100\%$$

For the test results of Fig. 2-1:

$$V = \frac{1000}{3500} \times 100 = 29\%$$

Standard deviation may be computed either from a single group of successive tests of a given class of concrete or from two groups of such tests. In the latter case, a <u>statistical average</u> value of standard deviation is to be used, calculated by usual statistical methods as follows:

$$s_{s3} = \sqrt{\frac{(n_1 - 1)(s_{s1})^2 + (n_2 - 1)(s_{s2})^2}{n_{total} - 2}}$$

where

n_1 = number of samples in group 1

n_2 = number of samples in group 2

$n_{total} = n_1 + n_2$

s_{s1} or s_{s2} is calculated as follows:

$$s_s = \sqrt{\frac{(X_1 - \overline{X})^2 + (X_2 - \overline{X})^2 + \ldots + (X_n - \overline{X})^2}{n - 1}}$$

For ease of computation,

$$s_s = \sqrt{\frac{X_1^2 + X_2^2 + X_3^2 + \ldots + X_n^2 - n\overline{X}^2}{n - 1}}$$

or

$$s_s = \sqrt{\frac{(X_1^2 + X_2^2 + X_3^2 + \ldots + X_n^2) - \frac{(X_1 + X_2 + X_3 + \ldots + X_n)^2}{n}}{n - 1}}$$

where $X_1, X_2, X_3, \ldots X_n$ are the individual strength test results and n is the total number of strength tests.

5.3.2 Required Average Strength

Where the concrete production facility has a record based on at least 30 consecutive strength tests representing materials and conditions similar to those expected (or a record based on 15 to 29 consecutive tests with the calculated sample standard deviation modified by the applicable factor from Table 5.3.1.2), the strength used as the basis for selecting concrete proportions for specified compressive strengths, f'_c, equal to or greater than 5000 psi
must be the larger of:

$f'_{cr} = f'_c + 1.34s_s$ *Table 5.3.2.1, (5-1)*

and $f'_{cr} = f'_c + 2.33s_s - 500$ *Table 5.3.2.1, (5-2)*

For specified compressive strengths, f'_c over 5000, the strength used as the basis for selecting concrete proportions must be the larger of:

$f'_{cr} = f'_c + 1.34s_s$ *Table 5.3.2.1, (5-1)*

and $f'_{cr} = 0.90f'_c + 2.33s_s$ *Table 5.3.2.1, (5-3)*

If the sample standard deviation is unknown, the required average strength f'_{cr} used as the basis for selecting concrete proportions must be determined from Table 5.3.2.2:

For f'_c less than 3000 psi $f'_{cr} = f'_c + 1000$ psi
           between 3000 and 5000 psi $f'_{cr} = f'_c + 1200$ psi
           greater than 5000 psi $f'_{cr} = f'_c + 1.10f'_c + 700$ psi

Formulas for calculating the required target strengths are based on the following criteria:

1. A probability of 1 in 100 that the average of 3 consecutive strength tests will be below the specified strength, f'_c: $f'_{cr} = f'_c + 1.34s_s$, and

2. A probability of 1 in 100 that an individual strength test will be more than 500 psi below the specified strength f'_c,: $f'_{cr} = f'_c + 2.33s_s - 500$ (for concrete strengths not over 5000 psi), and

3. A probability of 1 in 100 that an individual strength test will be more than $0.90f'_c$ below the specified strength f'_c (for concrete strengths in excess of 5000 psi): $f'_{cr} = 0.90f'_c + 2.33s_s$.

Criterion (1) will produce a higher required target strength than Criterion (2) for low to moderate standard deviations, up to 500 psi. For higher standard deviations, Criterion (2) will govern.

The average strength provisions of Section 5.3.2 are intended to provide an acceptable level of assurance that concrete strengths are satisfactory when viewed on the following basis: (1) the average of strength tests over an appreciable time period (three consecutive tests) is equal to or greater than the specified compressive strength, f'_c; or (2) an individual strength test is not more than 500 psi below (for concrete strengths not over 5000 psi); or (3) an individual strength test is not more than $0.90f'_c$ below (for concrete strengths in excess of 5000 psi).

5.3.3 Documentation of Average Compressive Strength

Mix approval procedures are necessary to ensure that the concrete furnished will actually meet the strength requirements. The steps in a mix approval procedure can be outlined as follows:

1. Determine the expected sample standard deviation from past experience.

   a. This is done by examining a record of 30 consecutive tests made on a similar mix.

   b. If it is difficult to find a similar mix on which 30 consecutive tests have been conducted, the sample standard deviation can be computed from two mixes, if the total number of tests equals or exceeds 30. The sample standard deviations are computed separately and then averaged by the statistical averaging method already described.

2. Use the sample standard deviation to select the appropriate target strength from the larger of Table 5.3.2.1 (5-1), (5-2) and (5-3).

   a. For example, if the sample standard deviation is 450 psi, then overdesign must be by the larger of:

   $1.34(450) = 603$ psi
   $2.33(450) - 500 = 549$ psi

   Thus, for a 3000 psi specified strength, the average strength used as a basis for selecting concrete mixture proportions must be 3600 psi.

b. Note that if no acceptable test record is available, the average strength must be 1200 psi greater than f'_c (i.e., 4200 psi average for a specified 3000 psi concrete), see Table 5.3.2.2.

3. Furnish data to document that the mix proposed for use will give the average strength needed. This may consist of:

a. A record of 30 tests of field concrete. This would generally be the same test record that was used to document the sample standard deviation, but it could be a different set of 30 results; or

b. Laboratory strength data obtained from a series of trial batches.

Where the average strength documentation for strengths over 5000 psi are based on laboratory trail mixtures, it is permitted to increase f'_{cr} calculated in Table 5.3.2.2 to allow for a reduction in strength from laboratory trials to actual concrete production.

Section 5.3.3.2(c) permits tolerances on slump and air content when proportioning by laboratory trial batches. The tolerance limits are stated at maximum permitted values, because most specifications, regardless of form, will permit establishing a maximum value for slump or air content. The wording also makes it clear that these tolerances on slump and air content are to be applied only to laboratory trial batches. Selection of concrete proportions by trial mixtures is illustrated in Example 2.3.

5.4 PROPORTIONING WITHOUT FIELD EXPERIENCE OR TRIAL MIXTURES

When no field or trail mixture data are available, "other experience or information" may be used to select a water-cementitious materials ratio. This mixture proportioning option, however, is permitted only when approved by the project engineer/architect. Note that this option must, of necessity, be conservative, requiring a rather high target overstrength (overdesign) of 1200 psi. If, for example, the specified strength is 3000 psi, the strength used as the basis for selecting concrete mixture proportions (water-cementitious materials ratio) must be based on 4200 psi. In the interest of economy of materials, the use of this option for mix proportioning should be limited to relatively small projects where the added cost of obtaining trial mixture data is not warranted. Note also that this alternative applies only for specified compressive strengths of concrete up to 5000 psi; for higher concrete strengths, proportioning by field experience or trial mixture data is required. The '99 Edition of the code limited the maximum strength proportioned without field experience or trial mixtures to 4000 psi.

5.6 EVALUATION AND ACCEPTANCE OF CONCRETE

5.6.1 Laboratory and Field Technicians

The concrete test procedures prescribed in the code require personnel with specific knowledge and skills. Experience has shown that only properly trained field technicians and laboratory personnel who have been certified under nationally recognized programs can consistently meet the standard of control that is necessary to provide meaningful test results. Section 5.6.1 of the code requires that tests performed on fresh concrete at the job site and procedures required to prepare concrete specimens for strength tests must be performed by a "qualified field testing technician". Commonly performed field tests which will require qualified field testing technicians include; unit weight, slump, air content and temperature; and making and curing test specimens. Field technicians in charge of these duties may be qualified through certification in the ACI Concrete Field Testing Technician – Grade I Certification Program.

Section 5.6.1 also requires that "qualified laboratory technicians" must perform all required laboratory tests. Laboratory technicians performing concrete testing may be qualified by receiving certification in accordance with requirements of ACI Concrete Laboratory Testing Technician, Concrete Strength Testing Technician, or the requirements of ASTM C 1007.

The following discussion on Chapter 5 of code addresses the selection of concrete mixture proportions for strength, based on probabilistic concepts.

5.6.2 Frequency of Testing

Proportioning concrete by the probabilistic basis of the code requires that a statistically acceptable number of concrete strength tests be provided. Requiring that strength tests be performed according to a prescribed minimum frequency provides a statistical basis.

The code minimum frequency criterion for taking samples for strength tests** , based on a per day and a per project criterion (the more stringent governs***) for each class of concrete, is summarized below.

5.6.2.1 Minimum Number of Strength Tests Per Day—This number shall be no less than:

- Once per day, nor less than,
- Once for each 150 cu yds of concrete placed, nor less than,
- Once for each 5000 sq ft of surface area of slabs or walls placed.

5.6.2.2 Minimum Number of Strength Tests Per Project —This number shall not be less than:

- Five strength tests from five (5) randomly selected batches or from each batch if fewer than five batches.

If the total quantity of concrete placed on a project is less than 50 cu yds, 5.6.2.3 permits strength tests to be waived by the building official.

According to the ASTM Standard for making concrete test specimens in the field (ASTM C 31), test cylinders should be 6 ×12 in., unless required otherwise by the project specifications.

With the increased use of very-high-strength concretes (in excess of 10,000 psi), the standard 6 × 12 in. cylinder requires very high capacity testing equipment which is not readily available in many testing laboratories. Consequently, most projects that specify very-high-strength concrete specifically permit the use of the smaller 4 × 8 in. cylinders for strength specimens. The 4 × 8 in. cylinder requires about one-half the testing capacity of the 6 × 12 in. specimen. Also, most precast concrete producers use the 4 × 8 in. cylinders for in-house concrete quality control.

It should be noted that the total number of cylinders cast for a project will normally exceed the code minimum number needed to determine acceptance of concrete strength (two cylinders per strength test). A prudent total number for a project may include additional cylinders for information (7-day tests) or to be field cured to check early strength development for form stripping, plus one or two in reserve, should a low cylinder break occur at the 28-day acceptance test age.

Example 2.4 illustrates the above frequency criteria for a large project (5.6.2.1 controls). Example 2.5 illustrates a smaller project (5.6.2.2 controls).

5.6.3.3 Acceptance of Concrete—The strength level of an individual class of concrete is considered satisfactory if both of the following criteria are met:

1. No single test strength (the average of the strengths of at least two cylinders from a batch) shall be more than 500 psi below the specified compressive strength when f'_c is 5000 psi or less (i.e., less than 2500 psi for a

---

** Strength test = average of two cylinder strengths (see 5.6.1.4).
*** On a given project, if total volume of concrete is such that frequency of testing required by 5.6.1.1 would provide less than five tests for a given class of concrete, the per project criterion will govern.

specified 3000 psi concrete strength); or is more than 10 percent below f'_c if over 5000 psi.

2. The average of any three consecutive test strengths must equal or exceed the specified compressive strength f'_c.

Examples 2.6 and 2.7 illustrate "acceptable" and "low strength" strength test results, respectively, based on the above code acceptance criteria.

5.6.5 Investigation of Low-Strength Test Results

If the average of three strength test results in a row is below the specified strength, steps must be taken to increase the strength level of the concrete (see 5.6.3.4). If a single strength test result falls more than 500 psi below the specified strength when f'_c is 5000 or less, or is more than 10 percent below f'_c if over 5000 psi, there may be more serious problems, and an investigation is required according to the procedures outlined in 5.6.5 to ensure structural adequacy.

Note that for acceptance of concrete, a single strength test result (one "test") is always the average strength of 2 cylinders broken at the designated test age, usually 28 days. Due to the many potential variables in the production and handling of concrete, concrete acceptance is <u>never</u> based on a single cylinder break. Two major reasons for low strength test results[3.1] are: (1) improper handling and testing of the cylinders – found to contribute to the majority of low strength investigations; and (2) reduced concrete strength due to an error in production, or the addition of too much water to the concrete at the job site. The latter usually occurs because of delays in placement or requests for a higher slump concrete. High air content due to an over-dosage of air entraining admixture at the batch plant has also contributed to low strength.

If low strength is reported, it is imperative that the investigation follows a logical sequence of possible cause and effect. All test reports should be reviewed and results analyzed before any action is taken. The pattern of strength test results should be studied for any clue to the cause. Is there any indication of actual violation of the specifications? Look at the slump, air content, concrete and ambient temperatures, number of days cylinders were left in the field and under what curing conditions, and any reported cylinder defects.

If the deficiency justifies investigation, testing accuracy should be verified first, and then the structural requirements compared with the measured strength. Of special interest in the early investigation should be the handling and testing of the test cylinders. Minor discrepancies in curing cylinders in mild weather will probably not affect strength much, but if major violations occur, large reductions in strength may result. Almost all deficiencies involving handling and testing of cylinders will lower strength test results. A number of simultaneous violations may contribute to significant reductions. Examples include: extra days in the field; curing over 80°F; frozen cylinders; impact during transportation; delay in moist curing at the lab; improper caps; and insufficient care in breaking cylinders.

For in-place concrete investigation, it is essential to know where in the structure the "tested concrete" is located and which batch (truck) the concrete is from. This information should be part of the data recorded at the time the test cylinders were molded. If test results are found deficient, in-place strength testing may be necessary to ascertain compliance with the code and construction documents. If strength is greater than that actually needed, there is little point in investigating the in-place strength. However, if testing procedures conform to the standards and the test results indicate that concrete strength is lower than required for the member in question, further investigation of the in-place concrete may be required (see 5.6.5).

The laboratory should be held responsible for deficiencies in its procedures. Use of qualified lab personnel is essential. Personnel sampling concrete, making test cylinders and operating lab equipment must be qualified by the ACI certification program or equivalent (see 5.6.1).

If core testing should be required, core drilling from the area in question should be performed according to the procedures outlined in ASTM C 42. The testing of cores requires great care in the operation itself and in the interpretation of the results. Detailed procedures are given in ASTM C 42. The following highlights proper core drilling and testing procedures:

1. Wait 14 days (minimum) before core drilling.
2. Drill 3 cores from the questionable area.
3. Drill cores with a diamond bit.
4. Drill core with a diameter of 2-1/2 in. (minimum) or 2 × maximum aggregate size.
5. Avoid any reinforcing steel in the drilled cores.
6. Drill a minimum core length of 1 × core diameter, but preferably 2 × core diameter.
7. If possible, drill completely through member.
8. Allow 2 in. extra length at the core end to be broken out.
9. Use wooden wedges to remove end portions to be broken out.
10. Saw broken ends to plane surfaces.
11. If concrete is dry under service conditions, air dry the cores for 7 days (60 to 70°F, 60% relative humidity). Test the cores dry.
12. If concrete is wet under service conditions, soak the cores in water (73.4 ± 3°F) for 40 hours. Test the cores wet.
13. Cap the core ends with 1/8 in. thick (or less) capping material.
14. Accurately center the core in the testing machine.
15. Correct the strength for length-to-diameter ratio less than 2, as shown below (interpolate between listed values):

Length-to-Diameter Ratio	Strength Correction Factor
1.94 - 2.10	1.00
1.75	0.98
1.50	0.96
1.25	0.93
1.00	0.87

In addition to the procedures contained in ASTM C 42, the Commentary to 5.6.5 cautions that where a water-cooled bit is used to obtain cores, the coring process causes a moisture gradient between the exterior and interior of the core, which will adversely affect the core's compressive strength. Thus, a restriction on the commencement of core testing is imposed to provide a minimum time for the moisture gradient to dissipate.

There were several significant changes to the '02 Edition of the code that affect the storage and testing of drilled cores. The provisions in 5.6.5.3 have been completely revised to require that immediately after drilling, cores must have any surface water removed by wiping and be placed in watertight bags or containers prior to transportation and storage. The cores must be tested no earlier than 48 hours, nor more than 7 days after coring unless approved by the registered design professional. In prior editions, storage conditions and restrictions on when testing could be performed were different for concrete in structures that would be "dry" or "superficially wet" under service conditions.

In evaluating core test results, the fact that core strengths may not equal the strength specified for molded cylinders should not be a cause for concern. Specified compressive strengths, f'_c, allow a large margin for the unknowns of placement and curing conditions in the field as well as for normal variability. For cores actually taken from the structure, the unknowns have already exerted their effect, and the margin of measured strength above required strength can logically be reduced.

Section 5.6.5.4 states that the concrete will be considered structurally adequate if the average strength of three cores is at least 85 percent of f'_c, with no single core strength less than 75 percent of the specified compressive strength. The concrete can be considered acceptable from the standpoint of strength if the core test results for a given location meet these requirements. The structural engineer should examine cases where core strength values fail to meet the above criteria, to determine if there is cause for concern over structural adequacy. If the results of properly made core tests are so low as to leave structural integrity in doubt, further action may be required.

As a last resort, load tests may be required to check the adequacy of structural members which are seriously in doubt. Generally such tests are suited only for flexural members—floors, beams, and the like—but they may sometimes be applied to other members. In any event, load testing is a highly specialized endeavor that should be performed and interpreted only by an engineer fully qualified in the proper techniques. Load testing procedures and criteria for their interpretation are given in code Chapter 20.

In those rare cases where a structural element fails the load test or where structural analysis of unstable members indicates an inadequacy, appropriate corrective measures must be taken. The alternatives, depending on individual circumstances, are:

- Reducing the load rating to a level consistent with the concrete strength actually obtained.
- Augmenting the construction to bring its load-carrying capacity up to original expectations. This might involve adding new structural members or increasing the size of existing members.
- Replacing the unacceptable concrete.

REFERENCES

2.1 *Design and Control of Concrete Mixtures* by Kosmatka, Steven H.; Kerkhoff, Beatrix; and Panarese, William C; EB001, Fourteenth Edition, Portland Cement Association, Skokie, IL, 2002.

2.2 ASTM C 1603, "Standard Test Method for Measurement of Solids in Water", American Society for Testing and Materials, West Conshohocken, PA, 2004.

2.3 ASTM C 1603, "Standard Specification for Mixing Water Used in the Production of Hydraulic Cement Concrete", American Society for Testing and Materials, West Conshohocken, PA, 2004.

2.4 ASTM D 3963-99, "Standard Specification for Fabrication and Jobsite Handling of Epoxy-Coated Reinforcing Steel Bars", American Society for Testing and Materials, West Conshohocken, PA, 1999.

2.5 ASTM A 775-04, "Standard Specification for Epoxy-Coated Steel Reinforcing Bars", American Society for Testing and Materials, West Conshohocken, PA, 2004.

2.6 ASTM A 934-04, "Standard Specification for Epoxy-Coated Prefabricated Steel Reinforcing Bars," American Society for Testing and Materials, West Conshohocken, PA, 2004.

2.7 ACI Committee 211.1, "Standard Practice for Selecting Proportions for Normal, Heavyweight and Mass Concrete (ACI 211.1-91 (Reapproved 2002)," American Concrete Institute, Farmington Hills, MI, 2002.

2.8 ACI Committee 211.2, "Standard Practice for Selecting Proportions for Structural Lightweight Concrete (ACI 211.2-98)," American Concrete Institute, Farmington Hills, MI, 1998.

2.9 ACI Committee 214, "Recommended Practice for Evaluation of Compression Test Results of Concrete (ACI 214-77, Reapproved 1997)," American Concrete Institute, Farmington Hills, MI, 1997.

2.10 "Guideline Manual to Quality Assurance / Quality Control" NRMCA Publication 190, National Ready Mixed Concrete Association, Silver Springs, MD, 1999.

2.11 "What, Why & How? Low Concrete Cylinder Strength," *Concrete in Practice*, CIP-9, National Ready Mixed Concrete Association, Silver Spring, MD, 2000.

Example 2.1—Selection of Water-Cementitious Materials Ratio for Strength and Durability

Concrete is required for a loading dock slab that will be exposed to moisture in a severe freeze-thaw climate, but not subject to deicers. A specified compressive strength f'_c of 3000 psi is used for structural design. Type I cement with 3/4-in. maximum size normal weight aggregate is specified.

Calculations and Discussion	Code Reference

1. Determine the required minimum strength and maximum w/c ratio for the proposed concrete work to satisfy both design strength and exposure requirements. — *5.2.1*

   For concrete exposed to freezing and thawing in a moist condition, Table 4.2.2 requires a maximum water-cementitious materials ratio of 0.45, and a minimum strength f'_c of 4500 psi. — *4.2.2*

   Since the required strength for the exposure conditions is greater than the required strength for structural design ($f'_c = 3000$ psi), the strength for the exposure requirements ($f'_c = 4500$ psi) governs.

2. Select a w/c ratio to satisfy the governing required strength, $f'_c = 4500$ psi.

   Concrete exposed to freezing and thawing must be air-entrained, with air content indicated in Table 4.2.1. For concrete in a cold climate and exposed to wet freeze-thaw conditions, a target air content of 6% is required for a 3/4-in. maximum size aggregate. — *4.2.1*

   Selection of water-cementitious materials ratio for required strength should be based on trial mixtures or field data made with actual job materials, to determine the relationship between w/c ratio and strength. — *5.3*

   Assume that the strength test data of Example 2.2, with an established sample standard deviation of 353 psi, represent materials and conditions similar to those expected for the proposed concrete work: — *5.3.1.1*

   a. normal weight, air-entrained concrete
   b. specified strength (4000 psi) within 1000 psi of that required for the proposed work (4500 psi)
   c. 30 strength test results.

   For a sample standard deviation of 353 psi, the required average compressive strength f'_{cr} to be used as the basis for selection of concrete proportions must be the larger of — *5.3.2*

   $$f'_{cr} = f'_c + 1.34s_s = 4500 + 1.34(353) = 5000 \text{ psi, or}$$ — *Eq. 5-1*

   $$f'_{cr} = f'_c + 2.33s_s - 500 = 4500 + 2.33(353) - 500 = 4800 \text{ psi}$$ — *Eq. 5-2*

   Therefore, $f'_{cr} = 5000$ psi.

	Code
Example 2.1 (cont'd) **Calculations and Discussion**	**Reference**

Note: The average strength required for the mix design should equal the specified strength plus an allowance to account for variations in materials; variations in methods of mixing, transporting, and placing the concrete; and variations in making, curing, and testing concrete cylinder specimens. For this example, with a sample standard deviation of 353 psi, an allowance of 500 psi for all those variations is made.

Typical trial mixture or field data strength curves are given in Ref. 2.1. Using the field data strength curve, Fig. 9-2 of Ref. 2.1, reproduced in Fig. 2-2, the required water-cementitious materials ratio (w/c) approximately equals 0.38 for an f'_{cr} of 5000 psi. (Use of the typical data curve of Fig. 2-2 is for illustration purposes only; a w/c versus required strength curve that is reflective of local materials and conditions should be used in an actual design situation.)

Since the required w/c ratio of 0.38 for the 4500 psi specified strength is less than the 0.45 required by Table 4.2.2, the 0.38 value must be used to establish the mixture proportions. Note that the specified strength, $f'_c = 4500$ psi, is the strength that is expected to be equaled or exceeded by the average of any set of three consecutive strength tests, with no individual test more than 500 psi below the specified 4500 psi strength.   *5.6.3.3*

As a follow up to this example, the test records of Example 2.2 could probably be used (by the concrete producer) to demonstrate that the concrete mix for which the records were generated will produce the required average strength f'_{cr} of the concrete work for this project. For the purpose of documenting the average strength potential of the concrete mix, the concrete producer need only select 10 consecutive tests from the total of 30 tests that represent a higher average than the required average of 5000 psi. Realistically, the average of the total 30 test results (4835 psi) is close enough to qualify the same concrete mix for the proposed work.   *5.3.3*

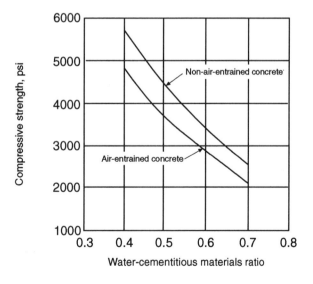

Figure 2-2 Typical Trial Mixture or Field Data Strength Curves

Example 2.2—Strength Test Data Report

Calculate the mean and sample standard deviation for the 30 strength tests results given below, using the formula for sample standard deviation given in R5.3.1. The project specifications call for column concrete to be normal weight, air-entrained, with a specified strength of 4000 psi.

Test No.	Date of Test	28-Day #1	28-Day #2	28-Day Average	28-Day Average (3-Consecutive)
1	05-March-04	4640	4770	4705	
2	06-March-04	4910	5100	5005	
3	10-March-04	4570	4760	4665	4792
4	12-March-04	4800	5000	4900	4857
5	13-March-04	5000	4900	4950	4838
6	17-March-04	4380	4570	4475	4775
7	19-March-04	4630	4820	4725	4717
8	21-March-04	4800	4670	4735	4645
9	25-March-04	5020	4940	4980	4813
10	28-March-04	4740	4900	4820	4845
11	30-March-04	4300	4110	4205	4668
12	02-April-04	4280	3620	3950	4325
13	05-April-04	4740	4880	4810	4322
14	08-April-04	4870	5040	4955	4592
15	09-April-04	4590	4670	4630	4798
16	15-April-04	4420	4690	4555	4713
17	16-April-04	4980	5070	5025	4737
18	19-April-04	4900	4860	4880	4820
19	20-April-04	5690	5570	5630	5178
20	22-April-04	5310	5310	5310	5273
21	24-April-04	5080	4970	5025	5322
22	28-April-04	4640	4440	4540	4958
23	01-May-04	5090	5080	5085	4883
24	03-May-04	5430	5510	5470	5032
25	07-May-04	5290	5360	5325	5293
26	10-May-04	4700	4770	4735	5177
27	11-May-04	4880	5040	4960	5007
28	15-May-04	5000	4890	4945	4880
29	16-May-04	4810	4670	4740	4882
30	18-May-04	4250	4400	4325	4670

Calculations and Discussion	Code Reference

Computation of the mean strength and sample standard deviation is shown in the following table. The sample standard deviation of 353 psi represents excellent quality control for the specified 4000 psi concrete.

Note that the concrete supplied for this concrete work satisfies the acceptance criteria of 5.6.3.3; no single strength test (28-day average of two cylinders) falls below the specified strength (4000 psi) by more than 500 psi (3500 psi), and the average of each set of 3 consecutive strength tests exceeds the specified strength (4000 psi).

			Code
Example 2.2 (cont'd)		**Calculations and Discussion**	**Reference**

Test No.	28-day Strength, X, psi	$X - \bar{X}$, psi	$(X - \bar{X})^2$
1	4705	-130	16,900
2	5005	170	28,900
3	4665	-170	28,900
4	4900	65	4,225
5	4950	115	13,225
6	4475	-360	129,600
7	4725	-110	12,100
8	4735	-100	10,000
9	4980	145	21,025
10	4820	-15	225
11	4205	-630	396,900
12	3950	-885	783,225
13	4810	-25	625
14	4955	100	10,000
15	4630	-205	42,025
16	4555	-280	78,400
17	5025	190	36,100
18	4880	45	2,025
19	5630	795	632,025
20	5310	475	225,625
21	5025	190	36,100
22	4540	-295	87,025
23	5085	250	62,500
24	5470	635	403,225
25	5325	490	240,100
26	4735	-100	10,000
27	4960	125	15,625
28	4945	110	12,100
29	4740	-95	9,025
30	4325	-510	260,100
Σ	145,060		3,607,850

Number of Tests = 30
Maximum Strength = 5630 psi
Minimum Strength = 3950 psi

Mean Strength $= \dfrac{145{,}060}{30} = 4835$ psi

Sample Standard Deviation $= \sqrt{\dfrac{3{,}607{,}850}{29}} = 353$ psi

The single low strength test (3950 psi) results from the very low break for cylinder #2 (3620 psi) of test No. 12. The large disparity between cylinder #2 and cylinder #1 (4280 psi), both from the same batch, would seem to indicate a possible problem with the handling and testing procedures for cylinder #2.

Example 2.2 (cont'd)	Calculations and Discussion	Code Reference

Interestingly, the statistical data from the 30 strength test results can be filed for use on subsequent projects to establish a mix design, where the concrete work calls for normal weight, air-entrained concrete with a specified strength within 1000 psi of the specified 4000 psi value (3000 to 5000 psi). The target strength for mix proportioning would be calculated using the 353 psi sample standard deviation in code Eqs. (5-1) and (5-2). The low sample standard deviation should enable the "ready-mix company" to produce an economical mix for similar concrete work. The strength test data of this example are used to demonstrate that the concrete mix used for this project qualifies for the proposed concrete work of Example 2.1.

Example 2.3—Selection of Concrete Proportions by Trial Mixtures

Establish a water-cementitious materials ratio for a concrete mixture on the basis of the specified compressive strength of the concrete to satisfy the structural design requirements.

Project Specifications:

f'_c = 3000 psi (normal weight) at 28 days

3/4-in. max. size aggregate

5% total air content

4 in. max slump

Kona sand and gravel

Type I Portland Cement

Assume no strength test records are available to establish a target strength for selection of concrete mixture proportions. The water-cementitious materials ratio is to be determined by trial mixtures. See 5.3.3.2.

Calculations and Discussion	Code Reference

1. Without strength test results, use Table 5.3.2.2 to establish a target strength, f'_{cr}.   *5.3.2.2*

    For f'_c = 3000 psi,
    $f'_{cr} = f'_c + 1200 = 3000 + 1200 = 4200$ psi

2. Trial Mixture Procedure   *5.3.3.2*

    Trial mixtures should be based on the same materials as proposed for the concrete work. Three (3) concrete mixtures with three (3) different water-cementitious materials ratios (w/c) should be made to produce a range of strengths that encompass the target strength f'_{cr}. The trial mixtures should have a slump within ± 0.75 in. of the maximum specified (3.25 to 4.75 in.), and a total air content within ± 0.5% of the volume required by the project specifications (4.5 to 5.5%). Three (3) test cylinders per trial mixture should be made and tested at 28 days. The test results are then plotted to produce a strength versus w/c ratio curve to be used to establish an appropriate w/c ratio for the target strength f'_{cr}.

    To illustrate the trial mixture procedure, assume trial mixtures and test data as shown in Table 2-3. Based on the test results plotted in Fig. 2-3 for the three trial mixtures, the maximum w/c ratio to be used as the basis for proportioning the concrete mixture with a target strength, f'_{cr}, of 4200 psi by interpolation, is 0.49.

    Using a water-cementitious materials ratio of 0.49 to produce a concrete with a specified strength of 3000 psi results in a significant overdesign. Referring to Fig. 2-2, Example 2.1, for a w/c ratio of 0.49, a strength level approximating 3800 psi can be expected for air-entrained concrete. The required extent of mix overdesign, when sufficient strength data are not available to establish a sample standard deviation, should be apparent.

| Example 2.3 (cont'd) | Calculations and Discussion | | Code Reference |

Table 2-3 Trial Mixture Data

Trial Mixtures	Batch No. 1	Batch No. 2	Batch No. 3
Selected w/c ratio	0.45	0.55	0.65
Measured slump, in.	3.75	4.25	4.50
Measured air content, %	4.4	5.3	4.8
Test results, psi:			
Cylinder #1	4650	3900	2750
Cylinder #2	4350	3750	2900
Cylinder #3	4520	3650	2850
Average	4510	3770	2830

As strength test data become available during construction, the amount by which the value of f'_{cr} must exceed the specified value of f'_c (1200 psi) may be reduced using a sample standard deviation calculated from the actual job test data, producing a more economical concrete mix.

5.5

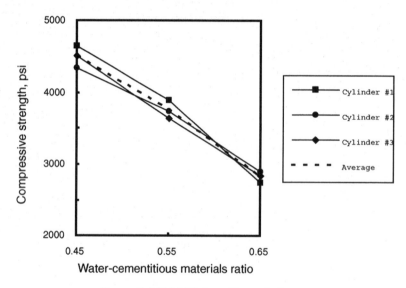

Figure 2-3 Trial Mixture Strength Curve

Example 2.4—Frequency of Testing

Determine the minimum number of test cylinders that must be cast to satisfy the code <u>minimum</u> sampling frequency for strength tests. Concrete placement = 200 cu yd per day for 7 days, transported by 10 cu yd truck mixers. This is a larger project where the minimum number of test cylinders <u>per day</u> of concrete placement (see 5.6.2.1) is greater than the minimum number <u>per project</u> (see 5.6.2.2).

	Code
Calculations and Discussion	**Reference**

1. Total concrete placed on project = 200 (7) = 1400 cu yd

2. Total truck loads (batches) required ≈ 1400/10 ≈ 140

3. Truck loads required to be sampled per day = 200/150 = 1.3   *5.6.2.1*

4. 2 truck loads must be sampled per day

5. Total truck loads required to be sampled for project = 2 (7) = 14

6. Total number of cylinders required to be cast for project = 14 (2 cylinders per test)
   = 28 (minimum)   *5.6.2.4*

It should be noted that the total number of cylinders required to be cast for this project represents a code required minimum number only that is needed for determination of acceptable concrete strength. Addition cylinders should be cast to provide for 7-day breaks, to provide field cured specimens to check early strength development for form removal or for determining when to post-tension prestressing tendons, and to keep one or two in reserve, should a low cylinder break occur at 28-day.

Example 2.5—Frequency of Testing

Determine the minimum number of test cylinders that must be cast to satisfy the code <u>minimum</u> sampling frequency for strength tests. Concrete is to be placed in a 100 ft × 75 ft × 7-1/2 in. slab, and transported by 10 cu yd truck mixers. This is a smaller project where the minimum required number of test cylinders is based on the frequency criteria of 5.6.2.2.

Calculations and Discussion	Code Reference

1. Total surface area placed = 100 × 75 = 7500 sq ft

2. Total concrete placed on project = $7500 \times 7.5 \times \frac{1 \text{ ft}}{12 \text{ in.}} / 27 = 174$ cu yd

3. Total truck loads (batches) required ≈ 174/10 ≈ 18

4. Required truck loads sampled per day = 174/150 = 1.2   5.6.2.1
   = 7500/5000 = 1.5

5. But not less than 5 truck loads (batches) per project   5.6.2.2

6. Total number of cylinders cast for project = 5 (2 cylinders per test)
   = 10 (minimum)   5.6.2.4

It should again be noted that the total number of cylinders cast represents a code required <u>minimum</u> number only for acceptance of concrete strength. A more prudent total number for a project may include additional cylinders.

Example 2.6—Acceptance of Concrete

The following table lists strength test data from 5 truck loads (batches) of concrete delivered to the job site. For each batch, two cylinders were cast and tested at 28 days. The specified strength of the concrete f'_c is 4000 psi. Determine the acceptability of the concrete based on the strength criteria of 5.6.3.3.

Test No.	Cylinder #1	Cylinder #2	Test Average	Average of 3 Consecutive Tests
1	4110	4260	4185	—
2	3840	4080	3960	—
3	4420	4450	4435	4193
4	3670	3820	3745	4047
5	4620	4570	4595	4258

Calculations and Discussion	Code Reference

The average of the two cylinder breaks for each batch represents a single strength test result. Even though the lowest of the five strength test results (3745 psi) is below the specified strength of 4000 psi, the concrete is considered acceptable because it is not below the specified value by more than 500 psi for concrete with an f'_c not over 5000 psi; i.e., not below 3500 psi. The second acceptance criterion, based on the average of three (3) consecutive tests, is also satisfied by the three consecutive strength test averages shown. The procedure to evaluate acceptance based on 3 strength test results in a row is shown in the right column. The 4193 psi value is the average of the first 3 consecutive test results: (4185 + 3960 + 4435)/3 = 4193 psi. The average of the next 3 consecutive tests is calculated as (3960 + 4435 + 3745)/3 = 4047 psi, after the 4185 psi value is dropped from consideration. The average of the next 3 consecutive values is calculated by dropping the 3960 psi value. For any number of strength test results, the consecutive averaging is simply a continuation of the above procedure. Thus, based on the code acceptance criteria for concrete strength, the five strength tests results are acceptable, both on the basis of individual test results and the average of three consecutive test results.

Example 2.7—Acceptance of Concrete

The following table lists strength test data from 5 truck loads (batches) of concrete delivered to the job site. For each batch, two cylinders were cast and tested at 28 days. The specified strength of the concrete f'_c is 4000 psi. Determine the acceptability of the concrete based on the strength criteria of 5.6.3.3.

Test No.	Cylinder #1	Cylinder #2	Test Average	Average of 3 Consecutive Tests
1	3620	3550	3585	—
2	3970	4060	4015	—
3	4080	4000	4040	3880*
4	4860	4700	4780	4278
5	3390	3110	3250**	4023

*Average of 3 consecutive tests below f'c (4000 psi).
**One test more than 500 psi below specified value.

Calculations and Discussion

Code Reference

Investigation of low-strength test results is addressed in 5.6.5. If average of three "tests" in a row dips below the specified strength, steps must be taken to increase the strength of the concrete. If a "test" falls more than 500 psi below the specified strength for concrete with an f'_c not over 5000 psi, there may be more serious problems, requiring an investigation to ensure structural adequacy; and again, steps taken to increase the strength level. For investigations of low strength, it is imperative that the location of the questionable concrete in the structure be known, so that the engineer can make an evaluation of the low strength on the structural adequacy of the member or element.

5.6.5

Based on experience,[2.11] the major reasons for low strength test results are (1) improper sampling and testing, and (2) reduced concrete quality due to an error in production, or the addition of too much water to the concrete at the job site, caused by delays in placement or requests for wet or high slump concrete. High air content can also be a cause of low strength.

The test results for the concrete from Truck 5 are below the specified value, especially the value for cylinder #2, with the average strength being only 3250 psi. (Note that no acceptance decisions are based on the single low cylinder break of 3110 psi. Due to the many variables in the production, sampling and testing of concrete, acceptance or rejection is always based on the average of at least 2 cylinder breaks.)

3

Details of Reinforcement

GENERAL CONSIDERATIONS

Good reinforcement details are vital to satisfactory performance of reinforced concrete structures. Standard practice for reinforcing steel details has evolved gradually. The Building Code Committee (ACI 318) continually collects reports of research and practice related to structural concrete, suggests new research needed, and translates the results into specific code provisions for details of reinforcement.

The ACI Detailing Manual[3.1] provides recommended methods and standards for preparing design drawings, typical details, and drawings for fabrication and placing of reinforcing steel in reinforced concrete structures. Separate sections of the manual define responsibilities of both the engineer and the reinforcing bar detailer. The CRSI Manual of Standard Practice[3.2] provides recommended industry practices for reinforcing steel. As an aid to designers, Recommended Industry Practices for Estimating, Detailing, Fabrication, and Field Erection of Reinforcing Materials are included in Ref. 3.2, for direct reference in project drawings and specifications. The WRI Structural Welded Wire Fabric Detailing Manual[3.3] provides information on detailing welded wire reinforcement systems.

7.1 STANDARD HOOKS

Requirements for standard hooks and minimum finished inside bend diameters for reinforcing bars are illustrated in Tables 3-1 and 3-2. The standard hook details for stirrups and ties apply to No. 8 and smaller bar sizes only.

Table 3-1 Standard Hooks for Primary Reinforcement

Bar size, No.	Min. finished bend dia.[a]
3 through 8	$6d_b$
9, 10, 11	$8d_b$
14 and 18	$10d_b$

[a] Measured on inside of bar.

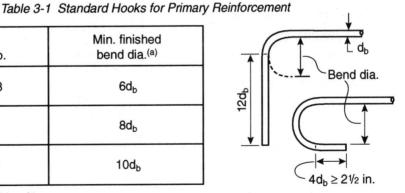

Table 3-2 Standard Hooks for Stirrups and Tie Reinforcement

Bar size, No.	Min. finished bend dia.(b)
3 through 5	$4d_b$
6 through 8	$6d_b$

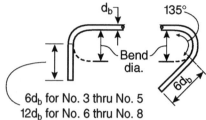

$6d_b$ for No. 3 thru No. 5
$12d_b$ for No. 6 thru No. 8

(b) Measured on inside of bar.

Moment resisting frames used to resist seismic lateral forces in regions of high seismic risk or in structures assigned to high seismic performance or design categories (see Table 1-1), must be designed as special moment frames as defined in 21.1. In special moment frames, detailing of transverse reinforcement in beams and columns must comply with 21.3.3 and 21.4.4, respectively. Except for circular hoops which are required to have seismic hooks with a 90-degree bend on the free ends, the ends of hoops and crossed ties must terminate in seismic hooks with 135-degree bends. These hooks are necessary to effectively anchor the free ends within the confined core so satisfactory performance is achieved in areas of members where inelastic behavior may occur. See Part 29 of this publication for discussion and illustrations of this special detailing requirement.

7.2 MINIMUM BEND DIAMETERS

Minimum bend diameter for a reinforcing bar is defined as "the diameter of bend measured on the *inside* of the bar." Minimum bend diameters, expressed as multiples of bar diameters, are dependent on bar size; for No. 3 to No. 8 bars, the minimum bend diameter is 6 bar diameters; for No. 9 to No. 11 bars, the minimum bend diameter is 8 bar diameters; and for No. 14 and No. 18 bars, the minimum bend diameter is 10 bar diameters. Exceptions to these provisions are:

1. For stirrups and ties in sizes No. 5 and smaller, the minimum bend diameter is 4 bar diameters. For No. 6 through No. 8 stirrups and ties, the minimum bend diameter is 6 bar diameters.

2. For welded wire reinforcement used for stirrups and ties, the inside diameter of the bend must not be less than four wire diameters for deformed wire larger than D6 and two wire diameters for all other wire. Welded intersections must be at least four wire diameters away from bends with inside diameters of less than eight wire diameters.

7.3 BENDING

All reinforcement must be bent cold unless otherwise permitted by the engineer. For unusual bends, special fabrication including heating may be required and the engineer must give approval to the techniques used.

7.3.2 Field Bending of Reinforcing Bars

Reinforcing bars partially embedded in concrete are frequently subjected to bending and straightening in the field. Protruding bars often must be bent to provide clearance for construction operations. Field bending and straightening may also be required because of incorrect fabrication or accidental bending. According to 7.3.2, bars partially embedded in concrete must not be field bent without authorization of the engineer unless shown on the plans. Test results[3,4] provide guidelines for field bending and straightening, and heating if necessary, of bars partially embedded in concrete. As an aid to the engineer on proper procedure, the recommendations of Ref. 3.4 are stated below. ASTM A 615 Grade 60 deformed bars were used in the experimental work on which the recommendations are based.

1. Field bending/straightening should be limited to bar sizes No. 11 and smaller. Heat should be applied for bending/straightening bar sizes No. 6 through No. 11, or for bending/straightening bar sizes No. 5 and smaller when those bars have been previously bent. Previously unbent bars of sizes No. 5 and smaller may be bent/straightened without heating.

2. A bending tool with bending diameter as shown in Table 3-3(a) should be used. Any bend should be limited to 90 degrees.

3. In applying heat for field bending/straightening, the steel temperature should be at or above the minimum temperature shown in Table 3-3(b) at the end of the heating operation, and should not exceed the maximum temperature shown during the heating operation.

4. In applying heat for field bending/straightening, the entire length of the portion of the bar to be bent (or the entire length of the bend to be straightened) should be heated plus an additional 2 in. at each end. For bars larger than No. 9, two heat tips should be used simultaneously at opposite sides of the bar to assure a uniform temperature throughout the thickness of the bar.

5. Before field bending/straightening, the significance of possible reductions in the mechanical properties of bent/straightened bars, as indicated in Table 3-3(c), should be evaluated.

7.5 PLACING REINFORCEMENT

7.5.1 Support for Reinforcement

Support for reinforcement, including tendons and post-tensioning ducts, is required to adequately secure the reinforcement against displacement during concrete placement. The CRSI Manual of Standard Practice[3.2] gives an in-depth treatise on types and typical sizes of supports for reinforcement. Types and typical sizes of wire bar supports are illustrated in Table 3-4. In addition to wire bar supports, bar supports are also available in precast concrete, cementitious fiber-reinforced and plastic materials. If the concrete surface will be exposed during service, consideration must be given to the importance of the appearance of the concrete surface and the environment to which it will be exposed. For example, if the concrete surface will be exposed directly to the weather or to a humid environment, it is likely that rust spots or stains will eventually show if unprotected bright steel barsupports are used. As outlined in the CRSI manual, bar supports are available in four classes of protection, depending on their expected exposure and the amount of corrosion protection required. Based on current industry practice, the available classes of protection are:

Class 1　Maximum Protection
　　　　　Plastic protected bar supports intended for use in situations of moderate to severe exposure and/or situations requiring light grinding (1/16 in. maximum) or sandblasting of the concrete surface.

Class 1A　Maximum Protection (For Use With Epoxy-Coated Reinforcement Bars) Epoxy-, vinyl-, or plastic coated bright basic wire bar supports intended for use in situations of moderate to maximum exposure where no grinding or sandblasting of the concrete surface is required. Generally, they are used when epoxy-coated reinforcing bars are required.

Class 2　Moderate Protection
　　　　　Stainless steel protected steel wire bar supports intended for use in situations of moderate exposure and/or situations requiring light grinding (1/16 in. maximum) or sandblasting of the concrete surface. The bottom of each leg is protected with a stainless steel tip.

Class 3　No Protection
　　　　　Bright basic wire bar supports with no protection against rusting. Unprotected wire bar supports are intended for use in situations where surface blemishes can be tolerated, or where supports do not come into contact with a concrete surface which is exposed.

Table 3-3 Field Bending and Straightening of Reinforcing Bars[3.4]

(a) Ratio of Bend Diameter to Bar Diameter

Bar Size, No.	Bend inside diameter/bar diameter	
	Not Heated	Heated
3, 4, 5	8	8
6, 7, 8, 9	Not permitted	8
10, 11	Not permitted	10

(b) Temperature Limits for Heating Bars

Bar Size, No.	Minimum Temperature (°F)	Maximum Temperature (°F)
3, 4	1200	1250
5, 6	1350	1400
7, 8, 9	1400	1450
10, 11	1450	1500

(c) Percent Reduction in Mechanical Properties of Bent and Straightened Bars

Bending Condition	Bar Size, No.	% Yield Strength Reduction	% Ultimate Tensile Strength Reduction	% Elongation Reduction
Cold	3, 4	—	—	20
Cold	5	5	—	30
Hot	All sizes	10	10	20

The engineer will need to specify the proper class of protection in the project specifications. It should be noted that the support system for reinforcement is usually detailed on the reinforcement placing drawings prepared by the "rebar" fabricator. The support system, including the proper class of protection, should be reviewed by the engineer, noting that the bar support size also dictates the cover provided for the reinforcement.

Use of epoxy-coated reinforcing bars will require bar supports made of a dielectrical material, or wire supports coated with a dielectrical material such as epoxy or vinyl, which is compatible with concrete. See discussion on 3.5.3.7, in Part 2 of this document, concerning special hardware and handling to minimize damage to the epoxy coating during handling, transporting, and placing epoxy-coated bars.

Commentary R7.5.1 emphasizes the importance of rigidly supporting the beam stirrups, in addition to the main flexural reinforcement, directly on the formwork. If not supported directly, foot traffic during concrete placement can push the web reinforcement down onto the forms, resulting in loss of cover and potential corrosion problems. It should be noted that the CRSI Manual of Standard Practice[3.2], often referenced in the design documents for placing reinforcing bars, does not specifically address this need for direct web reinforcement support. The placing drawings, usually prepared by the bar fabricator, should show a typical section or detail, so that this support requirement is clear and not overlooked by the ironworkers.

A word of caution on reinforcement displacement during concrete placing operations. If concrete placement is by pumping, it is imperative that the pipelines and the pipeline support system be supported above and independently of the chaired reinforcement by "chain-chairs" or other means. There must be no contact, direct or indirect, with the chaired reinforcement; otherwise, the surging action of the pipeline during pumping operations can, and most assuredly will, completely dislodge the reinforcement. This potential problem is especially acute in relatively thin slab members, especially those containing tendons, where the vertical placement of the reinforcement is most critical. The project specifications should specifically address this potential concrete placement problem.

7.5.2 Tolerances in Placing Reinforcement

The code provides tolerances applied simultaneously to concrete cover and member effective depth, d. With dimension "d" being the most structurally important dimension, any deviation in this dimension, especially for members of lesser depth, can have an adverse effect on the strength provided in the completed construction. The permitted variation from the effective depth d takes this strength reduction into account, with a smaller permitted variation for shallower members. The permitted tolerances are also established to reflect common construction techniques and practices. The critical dimensional tolerances for locating the longitudinal reinforcement are illustrated in Table 3-5, with two exceptions:

1. Tolerance for clear distance to formed soffits must not exceed minus 1/4 in.

2. Tolerance for cover must not exceed minus one-third the minimum concrete cover required in the design drawings and specifications. See Example 3.1

Table 3-4 Types and Sizes of Wire Bar Supports[3.2]

SYMBOL	BAR SUPPORT ILLUSTRATION	BAR SUPPORT ILLUSTRATION PLASTIC CAPPED OR DIPPED	TYPE OF SUPPORT	TYPICAL SIZES
SB	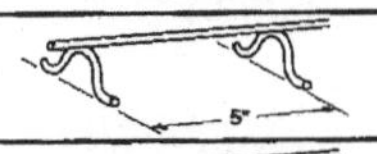	CAPPED 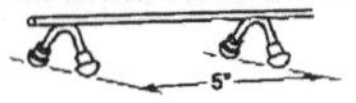	Slab Bolster	¾, 1, 1½, and 2 in. heights in 5 ft and 10 ft lengths
SBU*	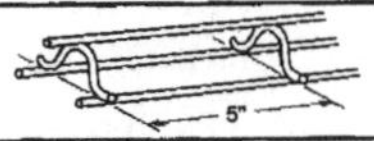		Slab Bolster Upper	Same as SB
BB	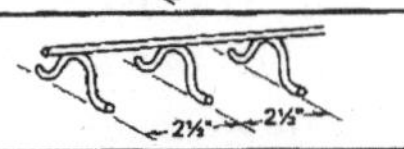	CAPPED 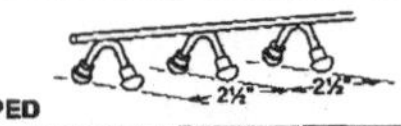	Beam Bolster	1, 1½, 2 to 5 in. heights in increments of ¼ in. In lengths of 5 ft
BBU*	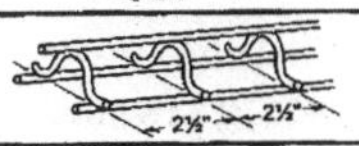		Beam Bolster Upper	Same as BB
BC	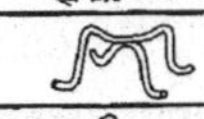	DIPPED 	Individual Bar Chair	¾, 1, 1½, and 1¾ in. heights
JC	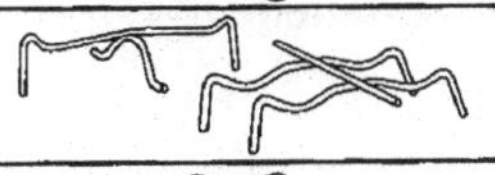	DIPPED 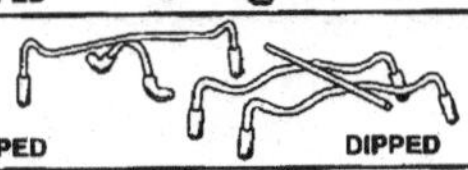 DIPPED	Joist Chair	4, 5, and 6 in. widths and ¾, 1 and 1½ in. heights
HC	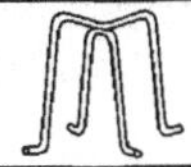	CAPPED 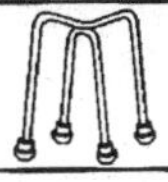	Individual High Chair	2 to 15 in. heights in increments of ¼ in.
HCM*			High Chair for Metal Deck	2 to 15 in. heights in increments of ¼ in.
CHC	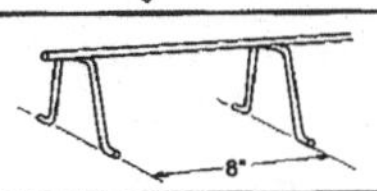	CAPPED 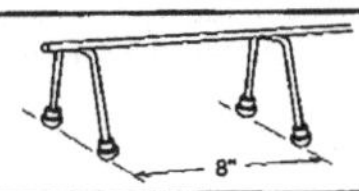	Continuous High Chair	Same as HC in 5 ft and 10 ft lengths
CHCU*	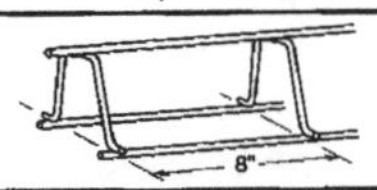		Continuous High Chair Upper	Same as CHC
CHCM*	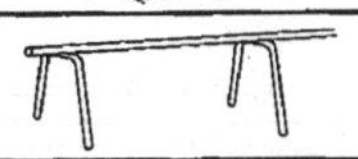		Continuous High Chair for Metal Deck	Up to 5 in. heights in increments of ¼ in.
JCU**	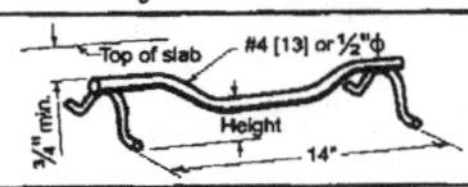	DIPPED 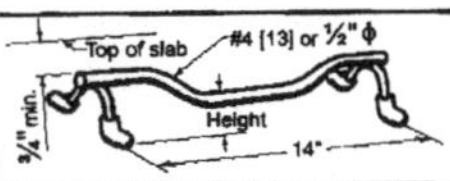	Joist Chair Upper	14 in. span; heights –1 in. thru +3½ in. vary in ¼ in. increments
CS	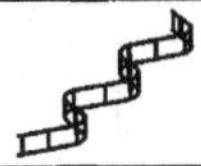		Continuous Support	1½ to 12 in. in increments of ¼ in. In lengths of 6'-8"
SBC			Single Bar Centralizer (Friction)	6 in. to 24 in. diameter

*Usually available in Class 3 only, except on special order.
**Usually available in Class 3 only, with upturned or end-bearing legs.

1 in. = 25.4 mm
1 ft = 304.8 mm

Table 3-5 Critical Dimensional Tolerances for Placing Reinforcement

Effective Depth d	Tolerance on d	Tolerance on Min. Cover
d ≤ 8 in.	± 3/8 in.	- 3/8 in.
d > 8 in.	± 1/2 in.	- 1/2 in.

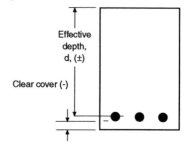

For ends of bars and longitudinal location of bends, the tolerance is ± 2 in., except at discontinuous ends of corbels and brackets where the tolerance is ± 1/2 in. At the discontinuous ends of other members the tolerance is permitted to be ± 1 in. The tolerance for minimum cover in 7.5.2.1 shall also apply. These tolerances are illustrated in Fig. 3-1.

Note that a plus (+) tolerance increases the dimension and a minus (-) tolerance decreases the dimension. Where only a minus tolerance is indicated on minimum cover, there is no limit in the other direction. Quality control during construction should be based on the more restrictive of related tolerances.

In addition to the code prescribed rebar placing tolerances, the engineer should be familiar with ACI Standard 117, *Standard Tolerances for Concrete Construction and Materials*.[3.5] ACI 117 includes tolerances for all measured dimensions, quantities and concrete properties used in concrete construction. The ACI 117 document is intended to be used by direct reference in the project specifications; therefore it is written in a specification format.

The designer must specify and clearly identify cover tolerances as the needs of the project dictate. For example, if concrete is to be exposed to a very aggressive environment, such as deicing chemicals, where the amount of concrete cover to the reinforcement may be a critical durability consideration, the engineer may want to indicate closer tolerances on concrete cover than those permitted by the code, or alternatively, specify a larger cover in recognition of expected variation in the placing of the reinforcement.

7.5.4 "Tack" Welding

Note that welding of crossing bars (tack welding) for assembly of reinforcement is prohibited except as specifically authorized by the engineer. By definition, a tack weld is a small spotweld to facilitate fabrication or field installation of reinforcement, and is not intended as a structural weld. Tack welding can lead to local embrittlement of the steel, and should never be done on reinforcement required by design. As noted in 3.5.2, all welding of reinforcement must conform to controlled welding procedures specified in AWS D1.4, including proper preheat (if required), and welding with electrodes meeting requirements of final welds.

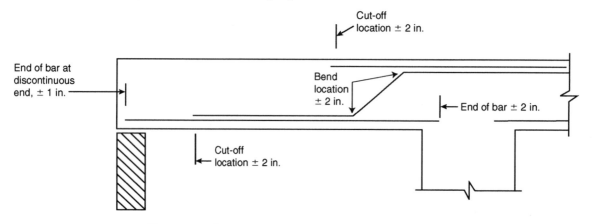

Figure 3-1 Tolerances for Bar Bend and Cutoff Locations

7.6 SPACING LIMITS FOR REINFORCEMENT

Spacing (clear distance) between bars must be as follows:

Minimum Spacing

For members with parallel bars in a layer, the clear spacing between bars must be at least one bar diameter but not less than 1 in.; and for reinforcement in two or more layers, bars in the upper layers must be directly above bars in the bottom layer, with at least 1 in. clear vertically between layers. For spirally reinforced and tied reinforced compression members, the clear distance between longitudinal bars must be at least 1-1/2 bar diameters, but not less than 1-1/2 in. These spacing requirements also apply to clear distance between contact-lap-spliced single or bundled bars and adjacent splices or bars. Section 3.3.2, which contains spacing requirements based on maximum nominal aggregate size, may also be applicable. Clear distances between bars are illustrated in Table 3-6.

Maximum Spacing

In walls and slabs other than concrete joists, primary flexural reinforcement must not be spaced greater than 3 times the wall or slab thickness nor 18 in.

Table 3-6 Minimum Clear Distances Between Bars, Bundles, or Tendons

Reinforcement Type	Type Member	Clear Distance
Deformed bars	Flexural members	$d_b \geq 1$ in.
	Compression members, tied or spirally reinforced	$1.5 d_b \geq 1.5$ in.
Pretensioning tendons	Wires	$4d_b$
	Strands*	$3d_b$

*When $f'_{ci} \geq 4000$ psi, center to center = 1-3/4 in. for 1/2 in. strands
= 2 in. for 0.6 in. strands

7.6.6 Bundled Bars

For isolated situations requiring heavy concentration of reinforcement, bundles of standard bar sizes can save space and reduce congestion for placement and consolidation of concrete. In those situations, bundling of bars in columns is a means to better locating and orienting the reinforcement for increased column capacity; also, fewer ties are required if column bars are bundled.

Bundling of bars (parallel reinforcing bars in contact, assumed to act as a unit) is permitted, but only if such bundles are enclosed by ties or stirrups. Some limitations are placed on the use of bundled bars as follows:

1. No. 14 and No. 18 bars cannot be bundled in beams.

2. If individual bars in a bundle are cut off within the span of beams, such cutoff points must be staggered at least 40 bar diameters.

3. A maximum of two bundled bars in any one plane is implied (three or four adjacent bars in one plane are not considered as bundled bars).

4. For spacing and concrete cover based on bar diameter, d_b, a unit of bundled bars must be treated as a single bar with diameter derived from the total area of all bars in the bundle. Equivalent diameters of bundled bars are given in Table 3-7.

5. A maximum of four bars may be bundled (See Fig. 3-2).

6. Bundled bars must be enclosed within stirrups or ties.

Table 3-7 Equivalent Diameters of Bundled Bars, in.

Bar Size, No.	Bar Diameter	2-Bar Bundle	3-Bar Bundle	4-Bar Bundle
6	0.750	1.06	1.30	1.50
7	0.875	1.24	1.51	1.75
8	1.000	1.42	1.74	2.01
9	1.128	1.60	1.95	2.26
10	1.270	1.80	2.20	2.54
11	1.410	1.99	2.44	2.82
14	1.693	2.39	2.93	3.39

Figure 3-2 Possible Reinforcing Bar Bundling Schemes

7.6.7   Prestressing Steel and Ducts

Prior to the '99 code, distances between prestressed steel were specified in terms of minimum clear distances. The '99 and subsequent codes specifies distances between prestressed steel in terms of minimum center-to-center spacing and requires $4d_b$ for strands and $5d_b$ for wire. When the compressive strength of the concrete at the time of prestress transfer, f'_{ci} is 4000 psi or greater, the minimum center-to-center spacing can be reduced to 1-3/4 in. for strands 1/2-in. nominal diameter or smaller and 2 in. for strands 0.6-in. nominal diameter. These changes were made as a result of research sponsored by the Federal Highway Administration. Center-to-center spacing is now specified because that is the way it was measured in the research. In addition, converting to clear spacing is awkward and unnecessary, and templates used by precast manufacturers have always been fabricated based on center-to-center dimensions. Closer vertical spacing and bundling of prestressed steel is permitted in the middle portion of the span if special care in design and fabrication is employed. Post-tensioning ducts may be bundled if concrete can be satisfactorily placed and provision is made to prevent the tendons from breaking through the duct when tensioned.

7.7 CONCRETE PROTECTION FOR REINFORCEMENT

Concrete cover or protection requirements are specified for members cast against earth, in contact with earth or weather, and for interior members not exposed to weather. Starting with the '02 code, the location of the cover requirements for cast-in-place concrete (prestressed) was reorganized. Cast-in-place concrete (prestressed) immediately follows cast-in-place (nonprestressed). They are then followed by the cover requirements for precast concrete manufactured under plant control conditions. In some cases slightly reduced cover or protection is permitted under the conditions for cast-in-place (prestressed) and precast concrete manufactured under plant control control conditions than permitted for cast-in-place concrete (non-prestressed). The term "manufactured under plant controlled conditions" does not necessarily mean that precast members must be manufactured in a plant. Structural elements precast at the job site (e.g., tilt-up concrete walls) will also qualify for the lesser cover if the control of form dimensions, placing of reinforcement, quality of concrete, and curing procedure are equivalent to those normally expected in a plant operation. Larger diameter bars, bundled bars, and prestressed tendons require greater cover. Corrosive environments or fire protection may also warrant increased cover. Section 18.3.3, which was introduced in the '02 code, requires that prestressed flexural members be classified as Class U (uncracked), Class C (cracked), or Class T (transition between uncracked and cracked). Section 7.7.5.1, also new to the '02 code, requires the cover of 7.7.2 be increased 50% for prestressed members classified as Class C or T where the members are exposed to corrosive environments or other severe exposure conditions. The requirement to increase the cover by 50% may be waived if the precompressed zone is not in tension under sustained load. The designer should take special note of the commentary recommendations (R7.7.5) for increased cover where concrete will be exposed to external sources of chlorides in service, such as deicing salts and seawater. As noted in R7.7, alternative methods of protecting the reinforcement from weather may be used if they provide protection equivalent to the additional concrete cover required in 7.7.1(b), 7.7.2(b), and 7.7.3(a), as compared to 7.7.1(c), 7.7.2(c), and 7.7.3(b), respectively.

7.8 SPECIAL REINFORCEMENT DETAILS FOR COLUMNS

Section 7.8 covers the special detailing requirements for offset bent longitudinal bars and steel cores of composite columns.

Where column offsets of less than 3 in. are necessary, longitudinal bars may be bent, subject to the following limitations:

1. Slope of the inclined portion of an offset bar with respect to the axis of column must not exceed 1 in 6 (see Fig. 3-3).

2. Portions of bar above and below an offset must be parallel to axis of column.

3. Horizontal support at offset bends must be provided by lateral ties, spirals, or parts of the floor construction. Ties or spirals, if used, shall be placed not more than 6 in. from points of bend (see Fig. 3-3). Horizontal support provided must be designed to resist 1-1/2 times the horizontal component of the computed force in the inclined portion of an offset bar.

4. Offset bars must be bent before placement in the forms.

When a column face is offset 3 in. or more, longitudinal column bars parallel to and near that face must not be offset bent. Separate dowels, lap spliced with the longitudinal bars adjacent to the offset column faces, must be provided (see Fig. 3-3). In some cases, a column might be offset 3 in. or more on some faces, and less than 3 in. on the remaining faces, which could possibly result in some offset bent longitudinal column bars and some separate dowels being used in the same column.

Steel cores in composite columns may be detailed to allow transfer of up to 50 percent of the compressive load in the core by direct bearing. The remainder of the load must be transferred by welds, dowels, splice plates, etc. This should ensure a minimum tensile capacity similar to that of a more common reinforced concrete column.

7.9 CONNECTIONS

Enclosures must be provided for splices of continuing reinforcement, and for end anchorage of reinforcement terminating at beam and column connections. This confinement may be provided by the surrounding concrete or internal closed ties, spirals, or stirrups.

7.10 LATERAL REINFORCEMENT FOR COMPRESSION MEMBERS

7.10.4 Spirals

Minimum diameter of spiral reinforcement in cast-in-place construction is 3/8 in. and the clear spacing must be between the limits of 1 in. and 3 in. This requirement does not preclude the use of a smaller minimum diameter

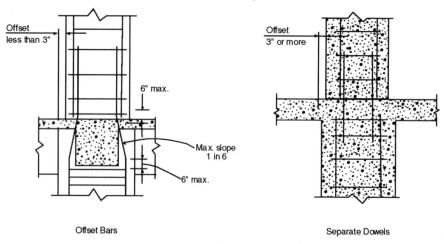

Figure 3-3 Special Column Details

for precast units. Beginning with the '99 code, full mechanical splices complying with 12.14.3 are allowed. Previously, only lap splices and full welded splices were permitted. Editions of the code prior to the '99 required lap splices to be 48 bar or wire diameters, regardless of whether the bar or wire was plain or deformed, or uncoated or epoxy-coated. The '99 code was revised to require that lap splices of plain uncoated and epoxy-coated deformed bar or wire be 72 bar or wire diameters. The required lap splice length for plain uncoated and epoxy-coated deformed bar or wire is permitted to be reduced to 48 bar or wire diameters provided the ends of the lapped bars or wires terminate in a standard 90 degree hook as required for stirrups and ties (7.1.3). The lap splice length for deformed uncoated bar or wire remains unchanged at 48 bar or wire diameters, as does the requirement that the minimum lap splice length be not less than 12 in. Anchorage of spiral reinforcement must be provided by 1-1/2 extra turns at each end of a spiral unit.

Spiral reinforcement must extend from the top of footing or slab in any story to the level of the lowest horizontal reinforcement in slabs, drop panels, or beams above. If beams or brackets do not frame into all sides of the column, ties must extend above the top of the spiral to the bottom of the slab or drop panel (see Fig. 3-4). In columns with capitals, spirals must extend to a level where the diameter or width of capital is twice that of the column.

Spirals must be held firmly in place, at proper pitch and alignment, to prevent displacement during concrete placement. Prior to ACI 318-89, the code specifically required spacers for installation of column spirals. Section 7.10.4.9 now simply states that "spirals shall be held firmly in place and true to line." This performance provision permits alternative methods, such as tying, to hold the fabricated cage in place during construction, which is current practice in most areas where spirals are used. The original spacer requirements were moved to the commentary to provide guidance where spacers are used for spiral installation. Note that the project specifications should cover the spacer requirements (if used) or the tying of the spiral reinforcement.

7.10.5 Ties

In tied reinforced concrete columns, ties must be located at no more than half a tie spacing above the floor slab or footing and at no more than half a tie spacing below the lowest horizontal reinforcement in the slab or drop panel above. If beams or brackets frame from four directions into a column, ties may be terminated not more than 3 in. below the lowest horizontal reinforcement in the shallowest of such beams or brackets (see Fig. 3-5). Minimum size of lateral ties in tied reinforced columns is related to the size of the longitudinal bars. Minimum tie sizes are No. 3 for non-prestressed longitudinal bars No. 10 and smaller, and No. 4 for No. 11 longitudinal bars and larger and for bundled bars. The following restrictions also apply: spacing must not exceed 16 longitudinal bar diameters, 48 tie bar diameters, or the least dimension of the column; every corner bar and alternate bar must have lateral support provided by the corner of a tie or crosstie with an included angle of not more than 135 degree. No unsupported bar shall be farther than 6 in. from a supported bar (see Fig. 3-6). Note that the 6-in. clear distance is measured along the tie.

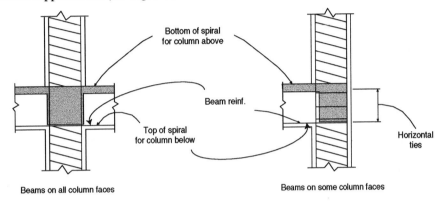

Figure 3-4 Termination of Spirals

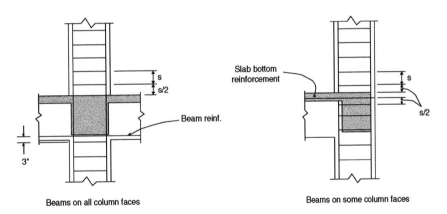

Figure 3-5 Termination of Column Ties

Welded wire reinforcement and continuously wound bars or deformed wire reinforcement of equivalent area may be used for ties. Where main reinforcement is arranged in a circular pattern, it is permissible to use complete circular ties at the specified spacing. This provision allows the use of circular ties at a spacing greater than that specified for spirals in spirally reinforced columns. Anchorage at the end of a continuously wound bar or wire reinforcement should be by a standard hook or by one additional turn of the tie pattern.

Where anchor bolts are provided in the tops of columns or pedestals to attach other structural members, the code requires that these bolts be confined by lateral reinforcement that is also surrounding at least four of the vertical bars in the column or pedestal for continuity of the load transfer at the connection. The lateral ties are required to be a minimum of two No. 4 or three No. 3 bars and must be distributed within the top 5 in. of the column or pedestal.

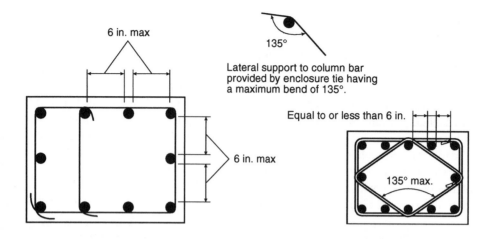

Figure 3-6 Lateral Support of Column Bars by Ties

7.11   LATERAL REINFORCEMENT FOR FLEXURAL MEMBERS

Where compression reinforcement is used to increase the flexural strength of a member (10.3.5.1), or to control long-term deflection [Eq. (9-11)], 7.11.1 requires that such reinforcement be enclosed by ties or stirrups. The purpose of the ties or stirrups is to prevent buckling of the compression reinforcement. Requirements for size and spacing of the ties or stirrups are the same as for ties in tied columns. Welded wire reinforcement of equivalent area may be used. The ties or stirrups must extend throughout the distance where the compression reinforcement is required for flexural strength or deflection control. Section 7.11.1 is interpreted not to apply to reinforcement located in a compression zone to help assemble the reinforcing cage or hold the web reinforcement in place during concrete placement.

Enclosing reinforcement required by 7.11.1 is illustrated by the U-shaped stirrup in Fig. 3-7, for a continuous beam, in the negative moment region; the continuous bottom portion of the stirrup satisfies the enclosure intent of 7.11.1 for the two bottom bars shown. A completely closed stirrup is ordinarily not necessary, except in cases of high moment reversal, where reversal conditions require that both top and bottom longitudinal reinforcement be designed as compression reinforcement.

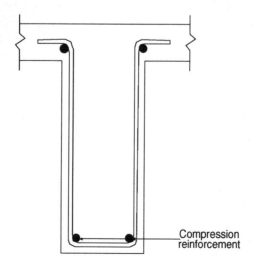

Figure 3-7 Enclosed Compression Reinforcement in Negatve Moment Region

Torsion reinforcement, where required, must consist of completely closed stirrups, closed ties, spirals, or closed cages of welded wire reinforcement as required by 11.6.4.

7.11.3 Closed Ties or Stirrups

According to 7.11.3, a closed tie or stirrup is formed either in one piece with overlapping 90-degree or 135-degree end hooks around a longitudinal bar, or in one or two pieces with a Class B lap splice, as illustrated in Fig. 3-8. The one-piece closed stirrup with overlapping end hooks is not practical for placement. Neither of the closed stirrups shown in Fig. 3-8 is considered effective for members subject to high torsion. Tests have shown that, with high torsion, loss of concrete cover and subsequent loss of anchorage result if the 90-degree hook and lap splice details are used where confinement by external concrete is limited. See Fig. 3-9. The ACI Detailing Manual[3.1] recommends the details illustrated in Fig. 3-10 for closed stirrups used as torsional reinforcement.

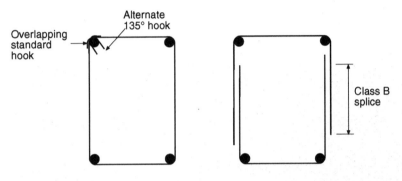

Figure 3-8 Code Definition of Closed Tie or Stirrup

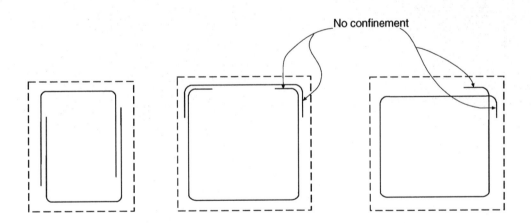

Figure 3-9 Closed Stirrup Details Not Recommended for Members Subject to High Torsion

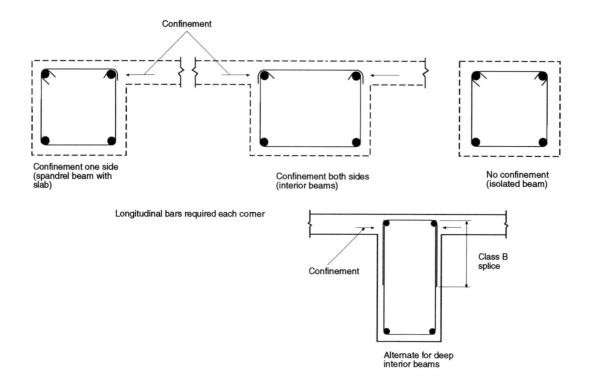

Figure 3-10 Two-Piece Closed Stirrup Details [3.1]
Recommended for Members Subject to High Torsion

7.12   SHRINKAGE AND TEMPERATURE REINFORCEMENT

Minimum shrinkage and temperature reinforcement normal to primary flexural reinforcement is required for structural floor and roof slabs (not slabs on ground) where the flexural reinforcement extends in one direction only. Minimum steel ratios, based on the gross concrete area, are:

1. 0.0020 for Grades 40 and 50 deformed bars;

2. 0.0018 for Grade 60 deformed bars or welded wire reinforcement;

3. $0.0018 \times 60,000/f_y$ for reinforcement with a yield strength greater than 60,000 psi;
   but not less than 0.0014.

Spacing of shrinkage and temperature reinforcement must not exceed 5 times the slab thickness nor 18 in. Splices and end anchorages of such reinforcement must be designed for the full specified yield strength. The minimum steel ratios cited above do not apply where prestressed steel is used.

Bonded or unbonded prestressing tendons may be used for shrinkage and temperature reinforcement in structural slabs (7.12.3). The tendons must provide a minimum average compressive stress of 100 psi on the gross concrete area, based on effective prestress after losses. Spacing of tendons must not exceed 6 ft. Where the spacing is greater than 54 in., additional bonded reinforcement must be provided at slab edges.

7.13   REQUIREMENTS FOR STRUCTURAL INTEGRITY

Structures capable of safely supporting all conventional design loads may suffer local damage from severe local abnormal loads, such as explosions due to gas or industrial liquids; vehicle impact; impact of falling objects; and local effects of very high winds such as tornadoes. Generally, such abnormal loads or events are not design considerations. The overall integrity of a reinforced concrete structure to withstand such abnormal loads can be substantially enhanced by providing relatively minor changes in the detailing of the reinforcement. The intent of 7.13 is to improve the redundancy and ductility of structures. This is achieved by providing, as a minimum, some continuity reinforcement or tie between horizontal framing members. In the event of damage to a major supporting element or an abnormal loading event, the integrity reinforcement is intended to confine any resulting damage to a relatively small area, thus improving overall stability.

It is not the intent of 7.13 that a structure be designed to resist general collapse caused by gross misuse or to resist severe abnormal loads acting directly on a large portion of the structure. General collapse of a structure as the result of abnormal events such as wartime or terrorist bombing, and landslides, are beyond the scope of any practical design.

7.13.1 General Structural Integrity

Since accidents and misuse are normally unforseeable events, they cannot be defined precisely; likewise, providing general structural integrity to a structure is a requirement that cannot be stated in simple terms. The performance provision..."members of a structure shall be effectively tied together to improve integrity of the overall structure," will require a level of judgment on the part of the design engineer, and will generate differing opinions among engineers as to how to effectively provide a general structural integrity solution for a particular framing system. It is obvious that all conditions that might be encountered in design cannot be specified in the code. The code, however, does set forth specific examples of certain reinforcing details for cast-in-place joists, beams, and two-way slab construction.

With damage to a support, top reinforcement which is continuous over the support, but not confined by stirrups, will tend to tear out of the concrete and will not provide the catenary action needed to bridge the damaged support. By making a portion of the bottom reinforcement in beams continuous over supports, some catenary action can be provided. By providing some continuous top and bottom reinforcement in edge or perimeter beams, an entire structure can be tied together; also, the continuous tie provided to perimeter beams of a structure will toughen the exterior portion of a structure, should an exterior column be severely damaged. Other examples of ways to detail for required integrity of a framing system to carry loads around a severely damaged member can be cited. The design engineer will need to evaluate his particular design for specific ways of handling the problem. The concept of providing general structural integrity is discussed in the Commentary of ASCE 7, *Minimum Design Loads for Buildings and Other Structures*.[3.6] The reader is referred to that document for further discussion of design concepts and details for providing general structural integrity.

7.13.2 Cast-in-Place Joists and Beams

Since 1989, the code requires continuous reinforcement in beams around the perimeter of the structure for structural integrity. The required amount is a minimum of one-sixth the tension reinforcement for negative moment at the support and one-fourth of the tension reinforcement for positive moment at the midspan. In either case the code requires a minimum of two bars and, mechanical and welded splices are explicitly permitted for splicing of continuous reinforcement in cast-in-place joists and beams. Figures 3-11 through 3-13 illustrate the required reinforcing details for the general case of cast-in-place joists and beams.

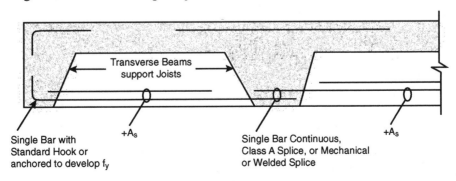

Figure 3-11 Continuity Reinforcement for Joist Construction

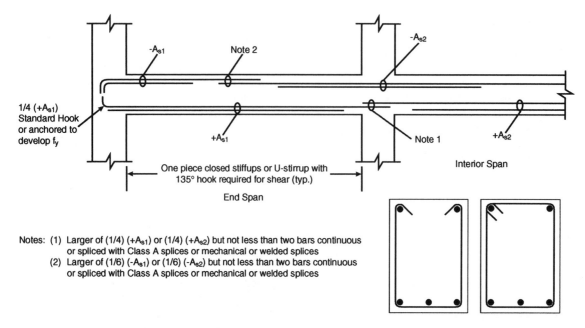

Figure 3-12 Continuity Reinforcement for Perimeter Beams

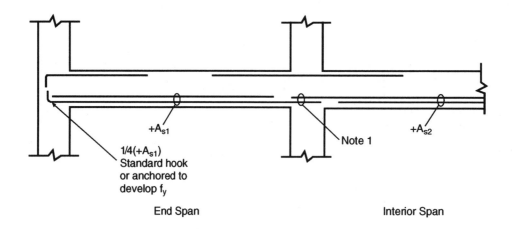

Note: (1) Larger of $(1/4)(+A_{s1})$ or $(1/4)(+A_{s2})$ but not less than two bars continuous or spliced with Class A splices or mechanical or welded splices

Figure 3-13 Continuity Reinforcement for Other Beams without Closed Stirrups

7.13.3 Precast Concrete Construction

While the requirements for structural integrity introduced in ACI 318-89 were prescriptive for cast-in-place construction, the '89 code provided only performance requirements for precast construction. This approach was made necessary because precast structures can be built in a lot of different ways. The code requires tension ties for precast concrete buildings of all heights. Connections that rely solely on friction due to gravity forces are not permitted.

The general requirement for structural integrity (7.13.1) states that "...members of a structure shall be effectively tied together...". The '89 commentary cautioned that for precast concrete construction, connection details should be arranged so as to minimize the potential for cracking due to restrained creep, shrinkage, and temperature movements. Ref. 3.7 contains information on industry practice for connections and detailing requirements. Prescriptive requirements recommended by the PCI for precast concrete bearing wall buildings are given in Ref. 3.8. Prescriptive structural integrity requirements for precast concrete structures were introduced for the first time in Chapter 16 of ACI 318-95 (see discussion in Part 23 of this publication).

7.13.4   Lift-Slab Construction

Section 7.13.4 refers the code user to 13.3.8.6 and 18.12.6 for lift-slab construction.

REFERENCES

3.1   *ACI Detailing Manual — 2004*, Publication SP-66(04), American Concrete Institute, Farmington Hills, MI, 2004.

3.2   *Manual of Standard Practice*, 27th edition, Concrete Reinforcing Steel Institute, Schaumburg, IL, 2001.

3.3   *Structural Welded Wire Fabric Detailing Manual*, WWR-600, Wire Reinforcement Institute, McLean, VA, 1994.

3.4   Babaei, K., and Hawkins, N.M., "Field Bending and Straightening of Reinforcing Steel," *Concrete International: Design and Construction*, V. 14, No. 1, January 1992.

3.5   *Standard Specification for Tolerances for Concrete Construction and Materials*, ACI 117-90, (reapproved 2002), American Concrete Institute, Farmington Hills, MI, 2002.

3.6   *Minimum Design Loads for Buildings and Other Structures*, (ASCE 7-02), American Society of Civil Engineers, Reston, VA, 2002.

3.7   *Design and Typical Details of Connections for Precast and Prestressed Concrete*, Publication MNL-123-88, Precast/Prestressed Concrete Institute, Chicago, IL, 1988.

3.8   PCI Building Code Committee, "Proposed Design Requirements for Precast Concrete," *PCI Journal*, V. 31, No. 6, Nov.-Dec. 1986, pp. 32-47.

Example 3.1—Placing Tolerance for Rebars

For the wall section shown below, with specified clear concrete cover indicated, determine the minimum cover permitted in construction, including the code tolerances on concrete cover.

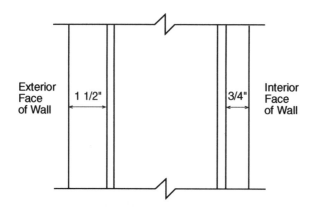

Calculations and Discussion	Code Reference

Tolerance on concrete cover is minus 1/2 in., but in no case may the tolerance be more than 1/3 the specified concrete cover.   7.5.2.1

1. For the exterior face, a measured 1 in. cover (1-1/2 - 1/2) is permitted. Actual bar placement may be within 1 in. of the side forms.

2. For the interior face, a measured 1/2 in. cover (3/4 - 1/4) is permitted. For the 3/4 in. specified cover, the tolerance limit is (1/3)(3/4) = 1/4 in. < 1/2 in.

As noted in the ACI 117 Standard[3.5], tolerances are a means to establish permissible variation in dimension and location, giving both the designer and the contractor parameters within which the work is to be performed. They are the means by which the designer conveys to the contractor the performance expectations.

4

Development and Splices of Reinforcement

GENERAL CONSIDERATIONS

The development length concept for anchorage of deformed bars and deformed wire in tension, is based on the attainable average bond stress over the length of embedment of the reinforcement. This concept requires the specified minimum lengths or extensions of reinforcement beyond all locations of peak stress in the reinforcement. Such peak stresses generally occur in flexural members at the locations of maximum stress and where adjacent reinforcement terminates or is bent.

The strength reduction factor ϕ is not used in Chapter 12 of the code since the specified development lengths already include an allowance for understrength.

12.1   DEVELOPMENT OF REINFORCEMENT—GENERAL

Development length or anchorage of reinforcement is required on both sides of a location of peak stress at each section of a reinforced concrete member. In continuous members, for example, reinforcement typically continues for a considerable distance on one side of a critical stress location so that detailed calculations are usually required only for the side where the reinforcement is terminated.

Until further research is completed and to ensure ductility and safety of structures built with high strength concrete, starting with the 1989 code, the term $\sqrt{f'_c}$ has been limited to 100 psi. Existing design equations for development of straight bars in tension and compression, and standard hooks in tension, are all a function of $\sqrt{f'_c}$. These equations were developed from results of tests on reinforcing steel embedded in concrete with compressive strengths of 3000 to 6000 psi. ACI Committee 318 was prudent in limiting $\sqrt{f'_c}$ at 100 psi pending completion of tests to verify applicability of current design equations to bars in high strength concrete.

12.2   DEVELOPMENT OF DEFORMED BARS AND DEFORMED WIRE IN TENSION

The provisions of 12.2 are based on the work of Orangun, Jirsa, and Breen[4.1], and Sozen and Moehle.[4.2] Development length of straight deformed bars and wires in tension, expressed in terms of bar or wire diameter, is given in 12.2.3 by the general equation:

$$\ell_d = \left(\frac{3}{40} \frac{f_y}{\sqrt{f'_c}} \frac{\psi_t \psi_e \psi_s \lambda}{\left(\frac{c_b + K_{tr}}{d_b} \right)} \right) d_b \qquad \text{Eq. (12-1)}$$

where

ℓ_d = development length, in.

d_b = nominal diameter of bar or wire, in.

f_y = specified yield strength of nonprestressed bar or wire, psi

f'_c = specified compressive strength of concrete, psi

ψ_t = **reinforcement location factor**
 = 1.3 for horizontal reinforcement placed such that more than 12 in. of fresh concrete is cast below the development length or splice
 = 1.0 for other reinforcement

ψ_e = **coating factor**
 = 1.5 for epoxy-coated bars or wires with cover less than $3d_b$ or clear spacing less than $6d_b$
 = 1.2 for all other epoxy-coated bars or wires
 = 1.0 for uncoated reinforcement

The product of ψ_t and ψ_e need not be taken greater than 1.7.

ψ_s = **reinforcement size factor**
 = 0.8 for No. 6 and smaller bars and deformed wires
 = 1.0 for No. 7 and larger bars

λ = **lightweight aggregate concrete factor**
 = 1.3 when lightweight aggregate concrete is used, or
 = $6.7\sqrt{f'_c}/f_{ct}$, but not less than 1.0, when f_{ct} is specified
 = 1.0 for normal weight concrete

c_b = **spacing or cover dimension**, in.
 = the smaller of (1) distance from center of bar or wire being developed to the nearest concrete surface, and (2) one-half the center-to-center spacing of bars or wires being developed

K_{tr} = **transverse reinforcement index**

$$= \frac{A_{tr} f_{yt}}{1500sn}$$

where

A_{tr} = total cross-sectional area of all transverse reinforcement which is within the spacing s and which crosses the potential plane of splitting through the reinforcement being developed, in.^{2}

f_{yt} = specified yield strength of transverse reinforcement, psi

s = maximum spacing of transverse reinforcement within ℓ_d, center-to-center, in.

n = number of bars or wires being developed along the plane of splitting

Note that the term $\left(\dfrac{c_b + K_{tr}}{d_b}\right)$ cannot be taken greater than 2.5 (12.2.3) to safeguard against pullout type failures. In the 1989 and earlier editions of the code, the expression $0.03 d_b f_y / \sqrt{f'_c}$ was specified to prevent pullout type failures.

As a design simplification, it is conservative to assume $K_{tr} = 0$, even if transverse reinforcement is present. If a clear cover of $2d_b$ and a clear spacing between bars being developed of $4d_b$ is provided, variable "c" would equal $2.5d_b$. For the preceding conditions, even if $K_{tr} = 0$, the term $\left(\dfrac{c_b + K_{tr}}{d_b}\right)$ would equal 2.5.

The term $\left(\dfrac{c_b + K_{tr}}{d_b}\right)$ in the denominator of Eq. (12-1) accounts for the effects of small cover, close bar spacing, and confinement provided by transverse reinforcement. To further simplify computation of ℓ_d, preselected values for term $\left(\dfrac{c_b + K_{tr}}{d_b}\right)$ were chosen starting with the 1995 code. As a result, Equation (12-1) can take the simplified forms specified in 12.2.2, and shown below in Table 4-1. For discussion purposes only, the four equations are identified in this table as Equations A through D. Note that these identifiers do not appear in the code.

In Eqs. A and B, the term $\left(\dfrac{c_b + K_{tr}}{d_b}\right) = 1.5$, while in Eqs. C and D, $\left(\dfrac{c_b + K_{tr}}{d_b}\right) = 1.0$. Equations A and C include a reinforcement size factor $\psi_s = 0.8$. The 20 percent reduction is based on comparisons with past provisions and numerous test results.

Equations A and B can only be applied if one of the following two different sets of conditions is satisfied:

Table 4-1 Development Lengths ℓ_d Specified in 12.2.2

	No. 6 and smaller bars and deformed wires	No. 7 and larger bars
Clear spacing of bars or wires being developed or spliced not less than d_b, clear cover not less than d_b, and stirrups or ties throughout ℓ_d not less than the code minimum or Clear spacing of bars or wires being developed or spliced not less than $2d_b$ and clear cover not less than d_b	(Eq. A) $\left(\dfrac{f_y \psi_t \psi_e \lambda}{25 \sqrt{f'_c}}\right) d_b$	(Eq. B) $\left(\dfrac{f_y \psi_t \psi_e \lambda}{20 \sqrt{f'_c}}\right) d_b$
Other cases	(Eq. C) $\left(\dfrac{3 f_y \psi_t \psi_e \lambda}{50 \sqrt{f'_c}}\right) d_b$	(Eq. D) $\left(\dfrac{3 f_y \psi_t \psi_e \lambda}{40 \sqrt{f'_c}}\right) d_b$

Set #1

The following three conditions must simultaneously be satisfied:

1. The clear spacing of reinforcement being developed or spliced should not be less than the diameter of reinforcement being developed, d_b,
2. The clear cover for reinforcement being developed should not be less than d_b, and
3. Minimum amount of stirrups or ties throughout ℓ_d should not be less than the minimum values specified in 11.5.5.3 for beams or 7.10.5 for columns.

Set #2

The following two conditions must simultaneously be satisfied:

1. The clear spacing of reinforcement being developed or spliced should not be less than $2d_b$, and
2. The clear cover should not be less than d_b.

If all the conditions of Set #1 or of Set #2 cannot be satisfied, then Eqs. C or D must be used. Note that Eq. D is identical to Eq. (12-1) with $\left(\dfrac{c_b + K_{tr}}{d_b}\right) = 1.0$ and reinforcement size factor $\gamma = 1.0$.

Although Eqs. A through D are easier to use than Eq. (12-1), the term $\left(\dfrac{c_b + K_{tr}}{d_b}\right)$ can only assume the value of 1.0 (Eqs. C and D) or 1.5 (Eqs. A and B). On the other hand, Eq. (12-1) may require a little extra effort, but the value of expression $\left(\dfrac{c_b + K_{tr}}{d_b}\right)$ can be as high as 2.5. Therefore, the development lengths ℓ_d computed by Eq. (12-1) could be substantially shorter than development lengths computed from the simplified equations of 12.2.2.

The development lengths of Table 4-1 can be further simplified for specific conditions. For example, for Grade 60 reinforcement ($f_y = 60,000$ psi) and different concrete compressive strengths, assuming normal weight concrete ($\lambda = 1.0$) and uncoated ($\psi_e = 1.0$) bottom bars or wires ($\psi_t = 1.0$), values of ℓ_d as a function of d_b can be determined as shown in Table 4-2.

Table 4-2 Development Length ℓ_d for Grade 60, Uncoated, Bottom Reinforcement in Normal Weight Concrete

	f'_c psi	No. 6 and smaller bars and deformed wires	No. 7 and larger bars
Clear spacing of bars being developed or spliced not less than d_b, clear cover not less than d_b, and beam stirrups or column ties throughout ℓ_d not less than the code minimum or Clear spacing of bars being developed or spliced not less than $2d_b$ and clear cover not less than d_b	3000	$44d_b$	$55d_b$
	4000	$38d_b$	$47d_b$
	5000	$34d_b$	$42d_b$
	6000	$31d_b$	$39d_b$
	8000	$27d_b$	$34d_b$
	10,000	$24d_b$	$30d_b$
Other cases	3000	$66d_b$	$82d_b$
	4000	$57d_b$	$71d_b$
	5000	$51d_b$	$64d_b$
	6000	$46d_b$	$58d_b$
	8000	$40d_b$	$50d_b$
	10,000	$36d_b$	$45d_b$

As in previous editions of the code, development length of straight deformed bars or wires, including all modification factors must not be less than 12 in.

12.2.5 Excess Reinforcement

Reduction in ℓ_d may be permitted by the ratio [(A_s required)/(A_s provided)] when excess reinforcement is provided in a flexural member. Note that this reduction does not apply when the full f_y development is required, as for tension lap splices in 7.13, 12.15.1, and 13.3.8.5, development of positive moment reinforcement at supports in 12.11.2, and for development of shrinkage and temperature reinforcement according to 7.12.2.3. Note also that this reduction in development length is not permitted for reinforcement in structures located in regions of high seismic risk or for structures assigned to high seismic performance or design categories (see 21.2.1.4).

Reduced ℓ_d computed after applying the excess reinforcement according to 12.2.5 must not be less than 12 in.

12.3 DEVELOPMENT OF DEFORMED BARS AND DEFORMED WIRE IN COMPRESSION

Shorter development lengths are required for bars in compression than in tension since the weakening effect of flexural tension cracks in the concrete is not present. The development length for deformed bars or deformed wire in compression is $\ell_{dc} = 0.02d_b f_y / \sqrt{f'_c}$, but not less than $0.0003d_b f_y$ or 8 in. Note that ℓ_{dc} may be reduced where excess reinforcement is provided (12.3.3(a)) and where "confining" ties or spirals are provided around the reinforcement (12.3.3(b)). Note that the tie and spiral requirements to permit the 25 percent reduction in development length are somewhat more restrictive than those required for "regular" column ties in 7.10.5 and less restrictive than those required for spirals in 7.10.4. For reference, compression development lengths for Grade 60 bars are given in Table 4-3.

Table 4-3 Compression Development Length ℓ_{dc} (inches) for Grade 60 Bars

Bar Size No.	f'_c (Normal Weight Concrete), psi		
	3000	4000	≥ 4444 *
3	8.2	7.1**	6.8**
4	11.0	9.5	9.0
5	13.7	11.9	11.3
6	16.4	14.2	13.5
7	19.2	16.6	15.8
8	21.9	19.0	18.0
9	24.7	21.4	20.3
10	27.8	24.1	22.9
11	30.9	26.8	25.4
14	37.1	32.1	30.5
18	49.4	42.8	40.6

* For $f'_c \geq$ 4444 psi, minimum basic development length $0.0003d_b f_y$ governs; for Grade 60 bars, $\ell_{dc} = 18d_b$.

** Development length ℓ_{dc} (including applicable modification factors) must not be less than 8 in.

12.4   DEVELOPMENT OF BUNDLED BARS

Increased development length for individual bars within a bundle, whether in tension or compression, is required when 3 or 4 bars are bundled together. The additional length is needed because the grouping makes it more difficult to mobilize resistance to slippage from the "core" between the bars. The modification factor is 1.2 for a 3-bar bundle, and 1.33 for a 4-bar bundle. Other pertinent requirements include 7.6.6.4 concerning cut-off points of individual bars within a bundle, and 12.14.2.2 relating to lap splices of bundled bars.

Where the factors of 12.2 are based on bar diameter d_b, a unit of bundled bars must be treated as a single bar of a diameter derived from the total equivalent area. See Table 3-7 in Part 3 of this document.

12.5   DEVELOPMENT OF STANDARD HOOKS IN TENSION

The current provisions for hooked bar development were first introduced in the 1983 code. They represented a major departure from the hooked-bar anchorage provisions of earlier codes in that they uncoupled hooked-bar anchorages from straight bar development and gave total hooked-bar embedment length directly. The current provisions not only simplify calculations for hook anchorage lengths but also result in a required embedment length considerably less, especially for the larger bar sizes, than that required by earlier codes. Provisions are given in 12.5 for determining the development length of deformed bars with standard end hooks. End hooks can only be considered effective in developing bars in tension, and not in compression (see 12.1.1 and 12.5.5). Only "standard" end hooks (see 7.1) are considered; anchorage capacity of end hooks with larger end diameters cannot be determined by the provisions of 12.5.

In applying the hook development provisions, the first step is to calculate the development length of the hooked bar, ℓ_{dh} from 12.5.2. This length is then multiplied by the applicable modification factor or factors of 12.5.3. Development length ℓ_{dh} is measured from the critical section to the outside end of the standard hook, i.e., the straight embedment length between the critical section and the start of the hook, plus the radius of bend of the hook, plus one-bar diameter. For reference, Fig. 4-1 shows ℓ_{dh} and the standard hook details (see 7.1) for all standard bar sizes. For 180 degree hooks normal to exposed surfaces, the embedment length should provide for a minimum distance of 2 in. beyond the tail of the hook.

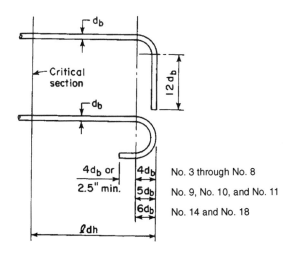

Figure 4-1 Development Length ℓ_{dh} of Standard Hooks

12.5.2 Development Length ℓ_{dh}

The development length, ℓ_{dh}, for standard hooks in tension is given in 12.5.2 as:

$$\ell_{dh} = \left(\frac{0.02\psi_e \lambda f_y}{\sqrt{f'_c}} \right) d_b$$

where $\psi_e = 1.2$ for epoxy-coated reinforcement[4.3] and $\lambda = 1.3$ for lightweight aggregate concrete. For other cases, ψ_e and λ are equal to 1.0.

Table 4-4 lists the development length of hooked bars embedded in normal weight concrete with different specified compressive strengths and uncoated Grade 60 reinforcing bars.

Table 4-4 Development Length ℓ_{dh} (inches) of Standard Hooks for Uncoated Grade 60 Bars*

Bar Size No.	f'_c (Normal Weight Concrete), psi					
	3000	4000	5000	6000	8000	10,000
3	8.2	7.1	6.4	5.8	5.0	4.5
4	11.0	9.5	8.5	7.7	6.7	6.0
5	13.7	11.9	10.6	9.7	8.4	7.5
6	16.4	14.2	12.7	11.6	10.1	9.0
7	19.2	16.6	14.8	13.6	11.7	10.5
8	21.9	19.0	17.0	15.5	13.4	12.0
9	24.7	21.4	19.1	17.5	15.1	13.5
10	27.8	24.1	21.6	19.7	17.0	15.2
11	30.9	26.8	23.9	21.8	18.9	16.9
14	37.1	32.1	28.7	26.2	22.7	20.3
18	49.5	42.8	38.3	35.0	30.3	27.1

* Development length ℓ_{dh} (including modification factors) must not be less than the larger of $8d_b$ or 6 in.

12.5.3 Modification Factors

The ℓ_{dh} modification factors listed in 12.5.3 account for:

- Favorable confinement conditions provided by increased cover (12.5.3(a))
- Favorable confinement provided by transverse ties or stirrups to resist splitting of the concrete (12.5.3(b) and (c))

- More reinforcement provided than required by analysis (12.5.3(d))

The side cover (normal to plane of hook), and the cover on bar extension beyond 90 degree hook referred to in 12.5.3(a) are illustrated in Fig. 4-2.

Note that requirements for 90-degree and 180-degree hooks are clarified in 12.5.3 of the 2002 code. Figures R12.5.3 (a) and R12.5.3 (b) illustrate the cases where the modification factor of 12.5.3 (b) may be used.

After multiplying the development length ℓ_{dh} by the applicable modification factor or factors, the resulting development length ℓ_{dh} must not be less than the larger of $8d_b$ or 6 in.

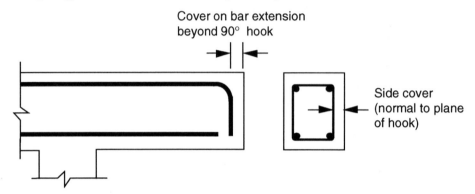

Figure 4-2 Concrete Covers Referenced in 12.5.3(a)

12.5.4   Standard Hook at Discontinuous Ends

Section 12.5.4 is a special provision for hooked bars terminating at discontinuous ends of members, such as at the ends of simply-supported beams, at free ends of cantilevers, and at ends of members framing into a joint where the member does not extend beyond the joint. If the full strength of a hooked bar must be developed, and both side cover and top (or bottom) cover over the hook are less than 2.5 in., 12.5.4 requires the hook to be enclosed within ties or stirrup for the full development length, ℓ_{dh}. Spacing of the ties or stirrup must not exceed $3d_b$, where d_b is the diameter of the hooked bar. In addition, the modification factor of 0.8 for confinement provided by ties or stirrups (12.5.3(b) and (c)) does not apply to the special condition covered by 12.5.4. At discontinuous ends of slabs with concrete confinement provided by the slab continuous on both sides normal to the plane of the hook, the provisions of 12.5.4 do not apply.

12.6   MECHANICAL ANCHORAGE

Section 12.6 permits the use of mechanical devices for development of reinforcement, provided their adequacy without damaging the concrete has been confirmed by tests. Section 12.6.3 reflects the concept that development of reinforcement may consist of a combination of mechanical anchorage plus additional embedment length of the reinforcement. For example, when a mechanical device cannot develop the design strength of a bar, additional embedment length must be provided between the mechanical device and the critical section.

12.7   DEVELOPMENT OF WELDED DEFORMED WIRE REINFORCEMENT IN TENSION

For welded deformed wire reinforcement, development length is measured from the critical section to the end of the wire. As specified in 12.7.1, development of welded deformed wire is computed as the product of ℓ_d from 12.2.2 or 12.2.3 times a wire reinforcement factor from 12.7.2 or 12.7.3. Where provided reinforcement is more than required, development length can be reduced by 12.2.5. In applying 12.2.2 or 12.2.3 to epoxy-coated deformed wire reinforcement, a coating factor $\psi_e = 1.0$ can be used. The resulting development length ℓ_d

cannot be less than 8 in., except in computation of lap splice lengths (see 12.18) and development of web reinforcement (see 12.13). Figure 4-3 shows the development length requirements for welded deformed wire reinforcement.

To apply the wire reinforcement factor of 12.7.2 to the development length of deformed wire reinforcement requires at least one cross wire located within the development length at a distance no less than 2 in. from the critical section. The wire reinforcement factor given in 12.7.2 is the greater of $(f_y - 35,000)/f_y$ or $5d_b/s$, but need not be taken greater than 1.0. s is the spacing between the wires to be developed.

If there is no cross wire within the development length, or the cross wire is less than 2 in. from the critical section, the development length of welded deformed wire reinforcement must be computed from 12.2.2 or 12.2.3. For this condition, the wire reinforcement factor must be taken equal to 1.0 (see 12.7.3).

According to ASTM A497, welded deformed steel wire reinforcement may consist solely of deformed steel wire (ASTM A496), or welded deformed steel wire reinforcement (ASTM A496) in one direction in combination with plain steel wire (ASTM A82) in the orthogonal direction. In the latter case, the reinforcement must be developed according to 12.8 for plain wire reinforcement.

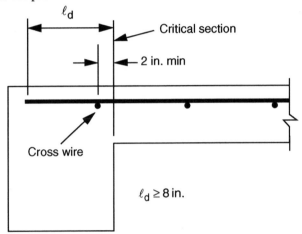

Figure 4-3 Development of Welded Deformed Wire Reinforcement

12.8   DEVELOPMENT OF WELDED PLAIN WIRE REINFORCEMENT IN TENSION

For welded plain wire reinforcement, the development length is measured from the point of critical section to the outermost cross wire. Full development of plain reinforcement ($A_w f_y$) is achieved by embedment of at least two cross wires beyond the critical section, with the closer cross wire located not less than 2 in. from the critical section. Section 12.8 further requires that the length of embedment from critical section to outermost cross wire not be less than $\ell_d = 0.27(A_b/s)(f_y/\sqrt{f'_c})\lambda$, nor less than 6 in. If more reinforcement is provided than that required by analysis, the development length ℓ_d may be reduced by the ratio of (A_s required)/(A_s provided). The 6 in. minimum development length does not apply to computation of lap splice lengths (see 12.19). Figure 4-4 shows the development length requirements for welded plain wire reinforcement.

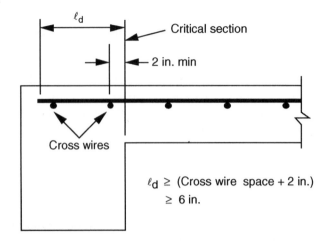

Figure 4-4 Development of Welded Plain Wire Reinforcement

For fabrics made with smaller wires, embedment of two cross wires, with the closer cross wire not less than 2 in. from the critical section, is usually adequate to develop the full yield strength of the anchored wires. Fabrics made with larger (closely spaced) wires will require a longer embedment ℓ_d.

For example, check fabric 6 × 6-W4 × W4 with f'_c = 3000 psi, f_y = 60,000 psi, and normal weight concrete (λ = 1.0).

$$\ell_d = 0.27 \times (A_b/s_s) \times \left(f_y/\sqrt{f'_c}\right) \times \lambda$$

$$= 0.27 \times (0.04/6) \times (60,000/\sqrt{3000}) \times 1.0 = 1.97 \text{ in.}$$

< 6 in.

$<$ (1 space + 2 in.)   governs

Two cross wire embedment plus 2 in. is satisfactory (see Fig. 4-5).

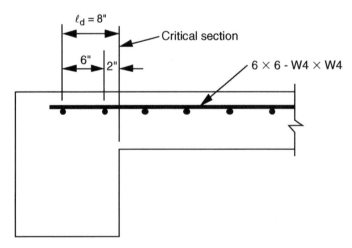

Figure 4-5 Development of 6 × 6-W4 × W4 Welded Wire Reinforcement

Check fabric 6 × 6-W20 × W20:

$$\ell_d = 0.27 \times (0.20/6) \times \left(60{,}000/\sqrt{3000}\right) \times 1.0 = 9.9 \text{ in.}$$

> 6 in.

> (1 space + 2 in.)

As shown in Fig. 4-6, an additional 2 in. beyond the two cross wires plus 2 in. embedment is required to fully develop the W20 fabric. If the longitudinal spacing is reduced to 4 in. (4 × 6-W20 × W20), a minimum ℓ_d of 15 in. is required for full development, i.e. 3 cross wires plus 3 in. embedment.

References 4.4 and 4.5 provide design aids for welded wire reinforcement, including development length tables for both deformed and plain welded wire reinforcement.

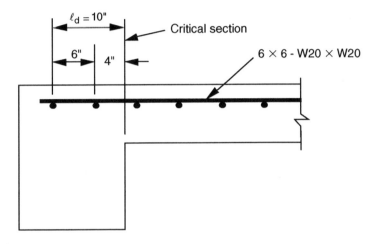

Note: If end support is not wide enough for straight embedment, the development length ℓ_d may be bent down (hooked) into support.

Figure 4-6 Development of 6 × 6-W20 × W20 Fabric

12.9 DEVELOPMENT OF PRESTRESSING STRAND

Prestressed concrete members may be either pretensioned or post-tensioned. In post-tensioned applications, development of tendons is accomplished through mechanical anchorage. Tendons may include strands, wires or high-strength bars.

In pretensioned members, tendons typically consist of seven-wire strands. Development length ℓ_d (in inches) of strands is specified in 12.9.1 and is computed from Eq. (12-2), which was formerly in R12.9:

$$\ell_d = \left(\frac{f_{se}}{3000}\right)d_b + \left(\frac{f_{ps} - f_{se}}{1000}\right)d_b \qquad \text{Eq. (12-2)}$$

where f_{ps} = stress in prestressed reinforcement at nominal strength, psi
f_{se} = effective stress in prestressed reinforcement after all prestress losses, psi
d_b = nominal diameter of strand, in.

The expressions in parentheses are dimensionless.

The term $\left(\dfrac{f_{se}}{3000}\right)d_b$ represents the transfer length of the strand (ℓ_t), i.e., the distance over which the strand should be bonded to the concrete to develop f_{se} in the strand. The second term, $\left[\left(f_{ps}-f_{se}\right)/1000\right]d_b$, represents the flexural bond length, i.e., the additional length over which the strand should be bonded so that a stress f_{ps} may develop in the strand at nominal strength of the member.

Where bonding of one or more strands does not extend to the end of the member, critical sections may be at locations other than those where full design strength is required (see 12.9.2). In such cases, a more detailed analysis may be required. Similarly, where heavy concentrated loads occur within the strand development length, critical sections may occur away from the section that is required to develop full design strength.

Note that two times the development length specified in 12.9.1 is required for "debonded" strands (12.9.3) when the member is designed allowing tension in the precompressed tensile zone under service load conditions.

In some pretensioned applications, total member length may be shorter than two times the development length. This condition may be encountered in very short precast, prestressed concrete members. In such cases, the strands will not be able to develop f_{ps}. Maximum usable stress in underdeveloped strands can be derived as illustrated in Fig. 4-7. The maximum strand stress, f_{max}, at distance ℓ_x from girder end can be determined for the condition of $\ell_t < \ell_x < \ell_d$ as follows:

$$f_{max} = f_{se} + \Delta f$$

$$= f_{se} + \frac{(f_{ps} - f_{se})}{(f_{ps} - f_{se})d_b}\left(\ell_x - \frac{f_{se}}{3000}d_b\right)$$

$$= f_{se} + \frac{\ell_x}{d_b} - \frac{f_{se}}{3000}$$

Therefore,

$$f_{max} = \frac{\ell_x}{d_b} + \frac{2}{3000}f_{se}$$

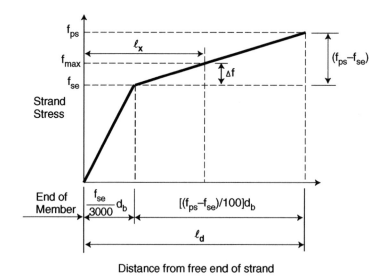

Figure 4-7 Strand Transfer and Development Lengths

12.10 DEVELOPMENT OF FLEXURAL REINFORCEMENT—GENERAL

Section 12.10 gives the basic requirements for providing development length of reinforcement from the points of maximum or critical stress. Figures 4-8(a) and (b) illustrate typical critical sections and code requirements for development and termination of flexural reinforcement in a continuous beam. Points of maximum positive and negative moments $\left(M_u^+ \text{ and } M_u^-\right)$ are critical sections, from which adequate anchorage ℓ_d must be provided. Critical sections are also at points within the span where adjacent reinforcement is terminated; continuing bars must have adequate anchorage ℓ_d from the theoretical cut-off points of terminated bars (see 12.10.4). Note also that terminated bars must be extended beyond the theoretical cut-off points in accordance with 12.10.3. This extension requirement is to guard against possible shifting of the moment diagram due to load variation, settlement of supports, and other unforeseen changes in the moment conditions. Development lengths ℓ_d are determined from 12.2.

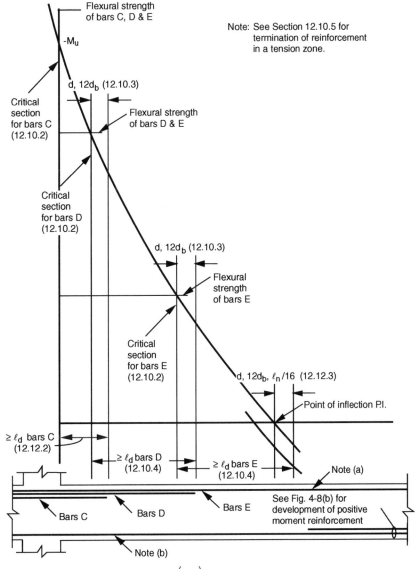

Note (a): Portion of total negative reinforcement $\left(A_s^-\right)$ must be continuous (or spliced with a Class A splice or a mechanical or welded splice satisfying 12.14.3) along full length of perimeter beams (7.13.2.2).

(a) Negative Moment Reinforcement

Figure 4-8 Development of Positive and Negative Moment Reinforcement

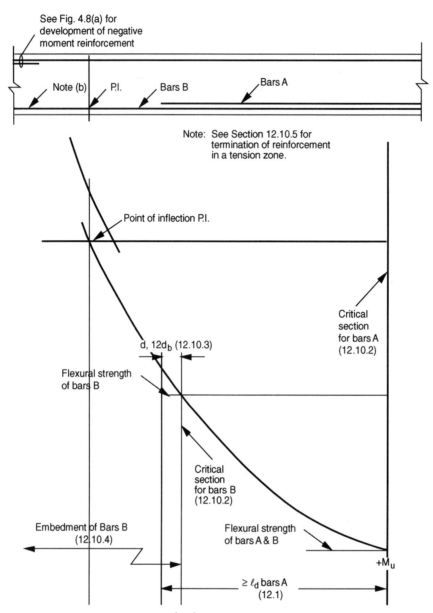

Note (b): Portion of total positive reinforcement $\left(A_s^+\right)$ must be continuous (or spliced with a Class A splice or a mechanical or welded splice satisfying 12.14.3) along full length of perimeter beams and of beams without closed stirrups (7.13.2.2). See also 7.13.2.4.

(b) Positive Moment Reinforcement

Figure 4-8 Development of Positive and Negative Moment Reinforcement
— continued —

Sections 12.10.1 and 12.10.5 address the option of anchoring tension reinforcement in a compression zone. Research has confirmed the need for restrictions on terminating bars in a tension zone. When flexural bars are cut off in a tension zone, flexural cracks tend to open early. If the shear stress in the area of bar cut-off and tensile stress in the remaining bars at the cut-off location are near the permissible limits, diagonal tension cracking tends to develop from the flexural cracks. One of the three alternatives of 12.10.5 must be satisfied to reduce the possible occurrence of diagonal tension cracking near bar cut-offs in a tension zone. Section 12.10.5.2 requires excess stirrup area over that required for shear and torsion. Requirements of 12.10.5 are not intended to apply to tension splices.

Section 12.10.6 is for end anchorage of tension bars in special flexural members such as brackets, members of variable depth, and others where bar stress, f_s, does not decrease linearly in proportion to a decreasing moment. In Fig. 4-9, the development length ℓ_d into the support is probably less critical than the required development length. In such a case, safety depends primarily on the outer end anchorage provided. A welded cross bar of equal diameter should provide an effective end anchorage. A standard end hook in the vertical plane may not be effective because an essentially plain concrete corner might exist near the load and could cause localized failure. Where brackets are wide and loads are not applied too close to the corners, U-shaped bars in a horizontal plane provide effective end hooks.

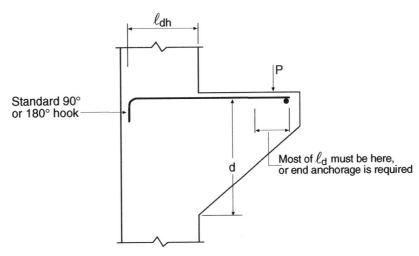

Figure 4-9 Special Member Largely Dependent on End Anchorage

12.11   DEVELOPMENT OF POSITIVE MOMENT REINFORCEMENT

To further guard against possible shifting of moments due to various causes, 12.11.1 requires specific amounts of positive moment reinforcement to be extended along the same face of the member into the support, and for beams, to be embedded into the support at least 6 in. The specified amounts are one-third for simple members and one-fourth for continuous members. In Fig. 4-8(b), for example, the area of Bars "B" would have to be at least one-fourth of the area of reinforcement required at the point of maximum positive moment M_u^+.

Section 12.11.2 is intended to assure ductility in the structure under severe overload, as might be experienced in a strong wind or earthquake. In a lateral load resisting system, full anchorage of the reinforcement extended into the support provides for possible stress reversal under such overload. Anchorage must be provided to develop the full yield strength in tension at the face of the support. The provision will require such members to have bottom bars lapped at interior supports or hooked at exterior supports. The full anchorage requirement does not apply to any excess reinforcement provided at the support.

Section 12.11.3 limits bar sizes for the positive moment reinforcement at simple supports and at points of inflection. In effect, this places a design restraint on flexural bond stress in areas of small moment and large shear. Such a condition could exist in a heavily loaded beam of short span, thus requiring large size bars to be developed within a short distance. Bars should be limited to a diameter such that the development length ℓ_d computed for f_y according to 12.2 does not exceed $(M_n/V_u) + \ell_a$ (12.11.3). The limit on bar size at simple supports is waived if the bars have standard end hooks or mechanical anchorages terminating beyond the centerline of the support. Mechanical anchorages must be equivalent to standard hooks.

The length (M_n/V_u) corresponds to the development length of the maximum size bar permitted by the previously used flexural bond equation. The length (M_n/V_u) may be increased 30% when the ends of the bars are confined by a compressive reaction, such as provided by a column below, but not when a beam frames into a girder.

For the simply-supported beam shown in Fig. 4-10, the maximum permissible ℓ_d for Bars "a" is $1.3 M_n/V_u + \ell_a$. This has the effect of limiting the size of bar to satisfy flexural bond. Even though the total embedment length from the critical section for Bars "a" is greater than $1.3 M_n/V_u + \ell_a$, the size of Bars "a" must be limited so that $\ell_d \leq 1.3 M_n/V_u + \ell_a$. Note that M_n is the nominal flexural strength of the cross-section (without the ϕ factor). As noted previously, larger bar sizes can be accommodated by providing a standard hook or mechanical anchorage at the end of the bar within the support. At a point of inflection (see Fig. 4-11), the positive moment reinforcement must have a development length ℓ_d, as computed by 12.2, not to exceed the value of $(M_n/V_u) + \ell_a$, with ℓ_a not greater than d or $12d_b$, whichever is greater.

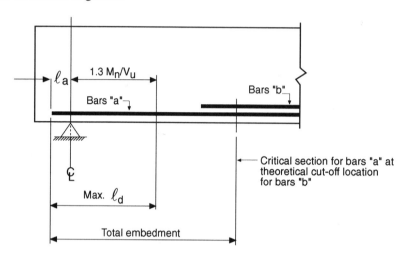

Figure 4-10 Development Length Requirements at Simple Support (straight bars)

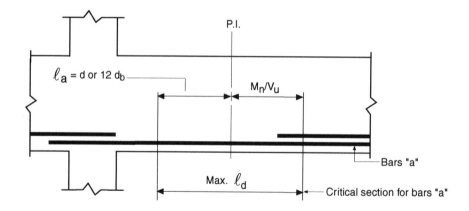

Figure 4-11 Concept for Determining Maximum Size of Bars "a" at Point of Inflection (12.11.3)

Sections 12.11.4 and 12.12.4 address development of positive and negative moment reinforcement in deep flexural members. The provisions specify that at simple supports of deep beams, positive moment tension reinforcement should be anchored to develop its specified yield strength f_y in tension at the face of the support. However, if the design is carried out using the strut-and-tie method of Appendix A, this reinforcement shall be anchored in accordance with A.4.3. At interior supports of deep beams, both positive and negative moment tension reinforcement shall be continuous or be spliced with that of the adjacent spans.

12.12 DEVELOPMENT OF NEGATIVE MOMENT REINFORCEMENT

The requirements in 12.12.3 guard against possible shifting of the moment diagram at points of inflection. At least one-third of the negative moment reinforcement provided at a support must be extended a specified embedment length beyond a point of inflection. The embedment length must be the effective depth of the member d, $12d_b$, or 1/16 the clear span, whichever is greater, as shown in Figs. 4-8 and 4-12. The area of Bars "E" in Fig. 4-8(a) must be at least one-third the area of reinforcement provided for $-M_u$ at the face of the support. Anchorage of top reinforcement in tension beyond interior support of continuous members usually becomes part of the adjacent span top reinforcement, as shown in Fig. 4-12.

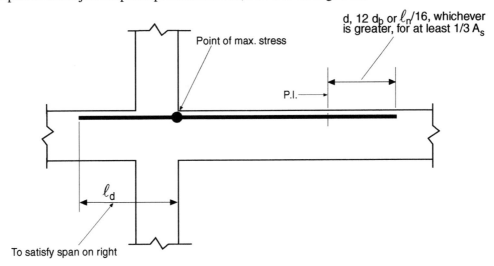

(Usually such anchorage becomes part of adjacent beam reinforcement)

Figure 4-12 Anchorage into Adjacent Beam

Standard end hooks are an effective means of developing top bars in tension at exterior supports as shown in Fig. 4-13. Code requirements for development of standard hooks are discussed above in 12.5.

12.13 DEVELOPMENT OF WEB REINFORCEMENT

Stirrups must be properly anchored so that the full tensile force in the stirrup can be developed at or near mid-depth of the member. To function properly, stirrups must be extended as close to the compression and tension surfaces of the member as cover requirements and proximity of other reinforcement permit (12.13.1). It is equally important for stirrups to be anchored as close to the compression face of the member as possible because flexural tension cracks initiate at the tension face and extend towards the compression zone as member strength is approached.

The ACI code anchorage details for stirrups have evolved over many editions of the code and are based primarily on past experience and performance in laboratory tests. For No. 5 bar and smaller, stirrup anchorage is provided by a standard stirrup hook (90 degree bend plus $6d_b$ extension at free end of bar)* around a longitudinal bar (12.13.2.1). The same anchorage detail is permitted for the larger stirrup bar sizes, No. 6, No. 7, and No. 8, in Grade 40. Note that for the larger bar sizes, the 90 degree hook detail requires a $12d_b$ extension at the free end of the bar (7.1.3(b)). Fig. 4-14 illustrates the anchorage requirement for U-stirrups fabricated from deformed bars and deformed wire.

---

* For structures located in regions of high seismic risk, stirrups required to be hoops must be anchored with a 135-degree bend plus $6d_b$ (but not less than 3 in.) extension. See definition of seismic hook in 21.1.

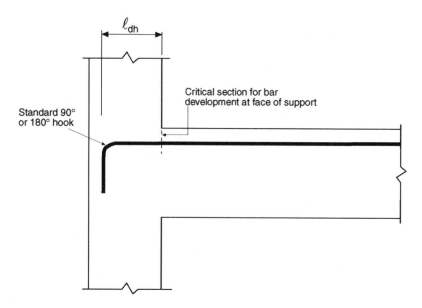

Figure 4-13 Anchorage into Exterior Support with Standard Hook

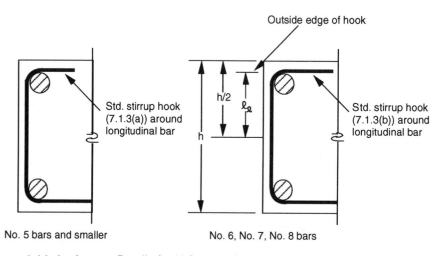

Figure 4-14 Anchorage Details for U-Stirrups (Deformed Bars and Deformed Wires)

For the larger stirrup bar sizes (No. 6, No. 7, or No. 8) in Grade 60, in addition to a standard stirrup hook, an embedment of $0.014 d_b f_{yt} / \sqrt{f'_c}$ between midheight of member and outside end of hook is required. The available embedment length, denoted ℓ_e, must be checked to ensure adequate anchorage at the higher bar force (see 12.13.2.2). The embedment length required is illustrated in Fig. 4-14 and listed in Table 4-5. Minimum depth of member required to accommodate No. 6, No. 7, or No. 8 stirrups fabricated in Grade 60 is also shown in Table 4-6. For practical size of beams where the loads are of such magnitude to require No. 6, No. 7, or No. 8 bar sizes for shear reinforcement, the embedment length required should be easily satisfied, and the designer need only be concerned with providing a standard stirrup hook around a longitudinal bar for proper stirrup end anchorage.

Provisions of 12.13.2.3 covering the use of welded plain wire reinforcement as simple U-stirrups are shown in Fig. 4-15. Requirements for stirrup anchorage (12.13.2.4) detail for straight single leg stirrups formed with welded plain or deformed wire reinforcement is shown in Fig. 4-16. Anchorage of the single leg is provided primarily by the longitudinal wires. Use of welded wire reinforcement for shear reinforcement has become commonplace in the precast, prestressed concrete industry.

Table 4-5 Embedment Length ℓ_ℓ (in.) for Grade 60 Stirrups

Bar Size No.	Concrete Compressive Strength f'_c, psi					
	3000	4000	5000	6000	8000	10,000
6	11.5	10.0	8.9	8.1	7.0	6.3
7	13.4	11.6	10.4	9.5	8.2	7.4
8	15.3	13.3	11.9	10.8	9.4	8.4

Table 4-6 Minimum Depth of Member (in.) to Accommodate Grade 60 No. 6, No. 7, and No. 8 Stirrups

Clear cover to stirrup (in.)	Bar Size No.	Concrete Compressive Strength f'_c, psi					
		3000	4000	5000	6000	8000	10,000
1-1/2	6	26	23	21	20	17	16
	7	30	27	24	22	20	18
	8	34	30	27	25	22	20
2	6	27	24	22	21	18	17
	7	31	28	25	23	21	19
	8	35	31	28	26	23	21

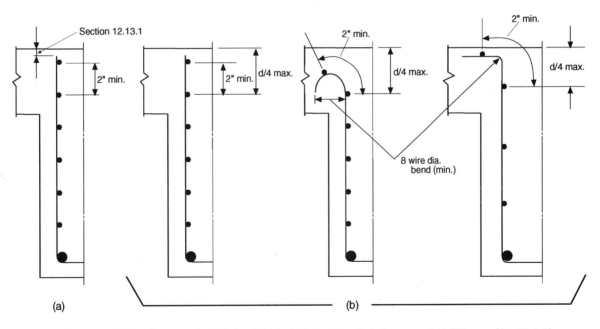

Figure 4-15 Anchorage Details for Welded Plain Wire Reinforcement U-Stirrups (12.13.2.3)

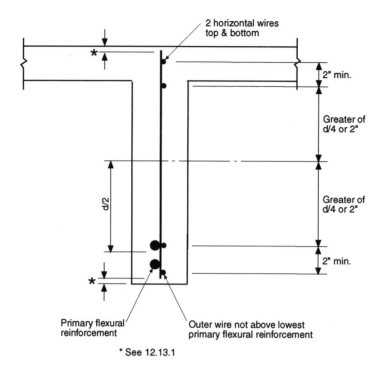

Figure 4-16 Anchorage Details for Welded Wire Reinforcement Single Leg Stirrups (12.13.2.4)

Note that 12.13.3 requires that each bend in the continuous portion of U-stirrups must enclose a longitudinal bar. This requirement is usually satisfied for simple U-stirrups, but requires special attention in bar detailing when multiple U-stirrups are used.

Clarifications of anchorage of web reinforcement made in the 1989 code eliminated the possibility of anchoring web reinforcement without hooking the stirrup around a longitudinal bar. Inquiries have shown that some designers routinely use small bars in joists without hooking them around a longitudinal bar, particularly a continuously bent single leg stirrup called a W-stirrup, accordion stirrup, or snake. To recognize this practice, 12.13.2.5 was introduced starting with the 1995 code.

12.13.4 Anchorage for Bent-Up Bars

Section 12.13.4 gives anchorage requirements for longitudinal (flexural) bars bent up to resist shear. If the bent-up bars are extended into a tension region, the bent-up bars must be continuous with the longitudinal reinforcement. If the bent-up bars are extended into a compression region, the required anchorage length beyond mid-depth of the member (d/2) must be based on that part of f_{yt} required to satisfy Eq. (11-17). For example, if f_{yt} = 60,000 psi and calculations indicate that 30,000 psi is required to satisfy Eq. (11-17), the required anchorage length $\ell'_d = (30,000/60,000)\ell_d$, where ℓ_d is the tension development length for full f_y per 12.2. Fig. 4-17 shows the required anchorage length ℓ'_d.

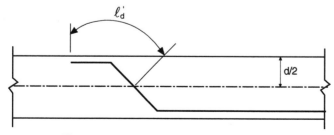

Figure 4-17 Anchorage for Bent-Up Bars

12.13.5 Closed Stirrups or Ties

Section 12.13.5 gives requirements for lap splicing double U-stirrups or ties (without hooks) to form a closed stirrup. Legs are considered properly spliced when the laps are $1.3\ell_d$ as shown in Fig. 4-18, where ℓ_d is determined from 12.2.

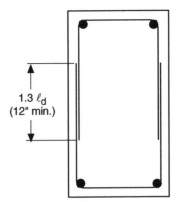

Figure 4-18 Overlapping U-Stirrups to Form Closed Unit

Alternatively, if a lap splice of $1.3\ell_d$ cannot fit within the depth of shallow members, provided that depth of members is at least 18 in., double U-stirrups may be used if each leg extends the full available depth of the member and the force in each leg does not exceed 9000 lb ($A_b f_{yt} \leq 9000$ lb.; see Fig. 4-19).

If stirrups are designed for the full yield strength f_y, No. 3 and 4 stirrups of Grade 40 and only No. 3 of Grade 60 satisfy the 9000 lb limitation.

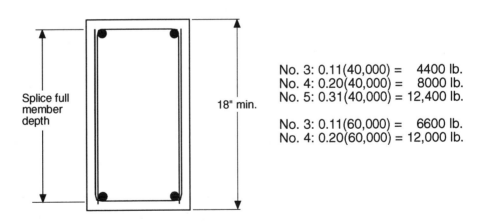

Figure 4-19 Lap Splice Alternative for U-Stirrups

12.14 SPLICES OF REINFORCEMENT—GENERAL

The splice provisions require the engineer to show clear and complete splice details in the contract documents. The structural drawings, notes and specifications should clearly show or describe all splice locations, types permitted or required, and for lap splices, length of lap required. The engineer cannot simply state that all splices shall be in accordance with the ACI 318 code. This is because many factors affect splices of reinforcement, such as the following for tension lap splices of deformed bars:

- bar size
- bar yield strength
- concrete compressive strength
- bar location (top bars or other bars)
- normal weight or lightweight aggregate concrete
- spacing and cover of bars being developed
- enclosing transverse reinforcement
- epoxy coating
- number of bars spliced at one location

It is virtually impossible for a reinforcing bar detailer to know what splices are required at a given location in a structure, unless the engineer explicitly illustrates or defines the splice requirements. Section 12.14.1 states: "Splices of reinforcement shall be made only as required or permitted on the design drawings, or in specifications, or as authorized by the engineer."

Two industry publications are suggested as design reference material for proper splicing of reinforcement. Reference 4.4 provides design aid data in the use of welded wire reinforcement, including development length and splice length tables for both deformed and plain wire reinforcement. Reference 4.5 provides accepted practices in splicing reinforcement; use of lap, mechanical, and welded splices are described, including simplified design data for lap splice lengths.

12.14.2 Lap Splices

Lap splices are not permitted for bars larger than No. 11, either in tension or compression, except:

- No. 14 and No. 18 bars in compression only may be lap spliced to No. 11 and smaller bars (12.16.2), and
- No. 14 and No. 18 bars in compression only may be lap spliced to smaller size footing dowels (15.8.2.3).

Section 12.14.2.2 gives the provisions for lap splicing of bars in a bundle (tension or compression). The lap lengths required for individual bars within a bundle must be increased by 20 percent and 33 percent for 3- and 4-bar bundles, respectively. Overlapping of individual bar splices within a bundle is not permitted. Two bundles must not be lap-spliced as individual bars.

Bars in flexural members may be spliced by noncontact lap splices. To prevent a possible unreinforced section in a spaced (noncontact) lap splice, 12.14.2.3 limits the maximum distance between bars in a splice to one-fifth the lap length, or 6 in. whichever is less. Contact lap splices are preferred for the practical reason that when the bars are wired together, they are more easily secured against displacement during concrete placement.

12.14.3 Mechanical and Welded Splices

Section 12.14.3 permits the use of mechanical or welded splices. A full mechanical splice must develop, in tension or compression, at least 125 percent of the specified yield strength of the bar (12.14.3.2). In a full welded splice, the bars must develop in tension at least 125 percent of the specified yield strength of the bar (12.4.3.4). ANSI/AWS D1.4 allows indirect welds where the bars are not butted. Note that ANSI/AWS D1.4 indicates that wherever practical, direct butt splices are preferable for No. 7 and larger bars. Use of mechanical or welded splices having less than 125 percent of the specified yield strength of the bar is limited to No. 5 and smaller bars (12.14.3.5) in regions of low computed stress. Mechanical and welded splices not meeting 12.14.3.2 and 12.14.3.4 are limited to No. 5 and smaller bars due to the potentially brittle nature of failure at these welds.

Section 12.14.3.3 requires all welding of reinforcement to conform to *Structural Welding Code-Reinforcing Steel* (ANSI/AWS D1.4). Section 3.5.2 requires that the reinforcement to be welded must be indicated on the drawings, and the welding procedure to be used must be specified. To carry out these code requirements properly, the engineer should be familiar with provisions in ANSI/AWS D1.4 and the ASTM specifications for reinforcing bars.

The standard rebar specifications ASTM A615, A616 and A617 do not address weldability of the steel. No limits are given in these specifications on the chemical elements that affect weldability of the steels. A key item in ANSI/AWS D1.4 is carbon equivalent (C.E.). The minimum preheat and interpass temperatures specified in ANSI/AWS D1.4 are based on C.E. and bar size. Thus, as indicated in 3.5.2 and R3.5.2, when welding is required, the ASTM A615, A616 and A617 rebar specifications must be supplemented to require a report of the chemical composition to assure that the welding procedure specified is compatible with the chemistry of the bars.

ASTM A706 reinforcing bars are intended for welding. The A706 specification contains restrictions on chemical composition, including carbon, and C.E. is limited to 0.55 percent. The chemical composition and C.E. must be reported. By limiting C.E. to 0.55 percent, little or no preheat is required by ANSI/AWS D1.4. Thus, the engineer does not need to supplement the A706 specification when the bars are to be welded. However, before specifying ASTM A706 reinforcing bars, local availability should be investigated.

Reference 4.5 contains a detailed discussion of welded splices. Included in the discussion are requirements for other important items such as field inspection, supervision, and quality control.

The ANSI/AWS D1.4 document covers the welding of reinforcing bars only. For welding of wire to wire, and of wire or welded wire reinforcement to reinforcing bars or structural steels, such welding should conform to applicable provisions of ANSI/AWS D1.4 and to supplementary requirements specified by the engineer. Also, the engineer should be aware that there is a potential loss of yield strength and ductility of low carbon cold-drawn wire if wire is welded by a process other than controlled resistance welding used in the manufacture of welded wire reinforcement.

In the discussion of 7.5 in Part 3 of this document, it was noted that welding of crossing bars (tack welding) is not permitted for assembly of reinforcement unless authorized by the engineer. An example of tack welding would be a column cage where the ties are secured to the longitudinal bars by small arc welds. Such welding can cause a metallurgical notch in the longitudinal bars, which may affect the strength of the bars. Tack welding seems to be particularly detrimental to ductility (impact resistance) and fatigue resistance. Reference 4.5 recommends: "Never permit field welding of crossing bars ('tack' welding, 'spot' welding, etc.). Tie wire will do the job without harm to the bars."

12.15   SPLICES OF DEFORMED BARS AND DEFORMED WIRE IN TENSION

Tension lap splices of deformed bars and deformed wire are designated as Class A and B with the length of lap being a multiple of the tensile development length ℓ_d. The two-level splice classification (Class A & B) is intended to encourage designers to splice bars at points of minimum stress and to stagger lap splices along the length of the bars to improve behavior of critical details.

The development length ℓ_d (12.2) used in the calculation of lap length must be that for the full f_y because the splice classifications already reflect any excess reinforcement at the splice location (factor of 12.2.5 for excess A_s must not be used). The minimum length of lap is 12 in.

For lap splices of slab and wall reinforcement, effective clear spacing of bars being spliced at the same location is taken as the clear spacing between the spliced bars (R12.15.1). This clear spacing criterion is illustrated in Fig. 4-20(a). Spacing for noncontact lap splices (spacing between lapped bars not greater than (1/5) lap length nor 6 in.) should be considered the same as for contact lap splices. For lap splices of column and beam bars, effective clear spacing between bars being spliced will depend on the orientation of the lapped bars; see Fig. 4-20(b) and (c), respectively.

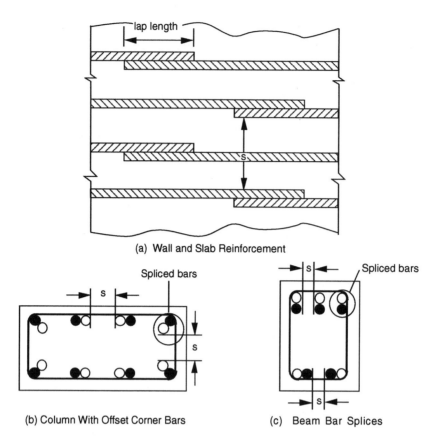

Figure 4-20 Effective Clear Spacing of Spliced Bars

The designer must specify the class of tension lap splice to be used. The class of splice depends on the magnitude of tensile stress in the reinforcement and the percentage of total reinforcement to be lap spliced within any given splice length as shown in Table 4-7. If the area of tensile reinforcement provided at the splice location is more than twice that required for strength (low tensile stress) and 1/2 or less of the total steel area is lap spliced within the required splice length, a Class A splice may be used. Both splice conditions must be satisfied, otherwise, a Class B splice must be used. In other words, if the area of reinforcement provided at the splice location is less than twice that required for strength (high tensile stress) and/or more than 1/2 of the total area is to be spliced within the lap length, a Class B splice must be used.

Table 4-7 Tension Lap Splice Conditions (at splice location)

CLASS A...$1.0\ell_d$	CLASS B...$1.3\ell_d$
(A_s provided) $\geq$ 2 (A_s required) and percent A_s spliced $\leq$ 50	All other conditions

Mechanical or welded splices conforming to 12.14.3 may be used in lieu of tension lap splices. Section R12.15.3 clarifies that such splices need not be staggered although such staggering is encouraged where the area of reinforcement provided is less than twice that required by analysis.

Section 12.15.4 emphasizes that mechanical and welded splices not meeting the requirements of 12.14.3.2 and 12.14.3.4, respectively, are only allowed for No. 5 bars and smaller, and only if certain conditions are met (see 12.15.4.1 and 12.15.4.2).

Splices in tension tie members are required to be made with a full mechanical or welded splice with a 30 in. stagger between adjacent bar splices. See definition of "tension tie member" in R12.15.5.

12.16   SPLICES OF DEFORMED BARS IN COMPRESSION

Since bond behavior of reinforcing bars in compression is not complicated by the potential problem of transverse tension cracking in the concrete, compression lap splices do not require such strict provisions as those specified for tension lap splices. Tests have shown that the strength of compression lap splices depends primarily on end bearing of the bars on the concrete, without a proportional increase in strength even when the lap length is doubled. Thus, the code requires significant longer lap length for bars with a yield strength greater than 60,000 psi.

12.16.1   Compression Lap Splices

Calculation of compression lap splices was simplified starting with the '89 code by removing the redundant calculation for development length in compression. For compression lap splices, 12.16.1 requires the minimum lap length to be simply $0.0005 d_b f_y$ for $f_y = 60,000$ psi or less, but not less than 12 in. For reinforcing bars with a yield strength greater than 60,000 psi, a minimum lap length of $(0.0009 f_y - 24) d_b$ but not less than 12 in. is specified. Lap splice lengths must be increased by one-third for concrete with a specified compressive strength less than 3000 psi.

As noted in the discussion of 12.14.2, No. 14 and No. 18 bars may be lap spliced, in compression only, to No. 11 and smaller bars or to smaller size footing dowels. Section 12.16.2 requires that when bars of a different size are lap spliced in compression, the length of lap must be the compression development length of the larger bar, or the compression lap splice length of the smaller bar, whichever is the longer length.

12.16.4   End-Bearing Splices

Section 12.16.4 specifies the requirements for end-bearing compression splices. End-bearing splices are only permitted in members containing closed ties, closed stirrups or spirals (12.16.4.3). Section R12.16.4.1 cautions the engineer in the use of end-bearing splices for bars inclined from the vertical. End-bearing splices for compression bars have been used almost exclusively in columns and the intent is to limit use to essentially vertical bars because of the field difficulty of getting adequate end bearing on horizontal bars or bars significantly inclined from the vertical. Mechanical or welded splices are also permitted for compression splices and must meet the requirements of 12.14.3.2 or 12.14.3.4, respectively.

12.17   SPECIAL SPLICE REQUIREMENTS FOR COLUMNS

The special splice requirements for columns were significantly simplified in the '89 code. The column splice requirements simplify the amount of calculations that are required compared to previous provisions by assuming that a compression lap splice (12.17.2.1) has a tensile capacity of at least one-fourth f_y (R12.17).

The column splice provisions are based on the concept of providing some tensile resistance at all column splice locations even if analysis indicates compression only at a splice location. In essence, 12.17 establishes the required tensile strength of spliced longitudinal bars in columns. Lap splices, butt-welded splices, mechanical or end-bearing splices may be used.

12.17.2   Lap Splices in Columns

Lap splices are permitted in column bars subject to compression or tension. Type of lap splice to be used will depend on the bar stress at the splice location, compression or tension, and magnitude if tension, due to all factored load combinations considered in the design of the column. Type of lap splice to be used will be governed by the load combination producing the greatest amount of tension in the bars being spliced. The design requirements for lap splices in column bars can be illustrated by a typical column load-moment strength interaction as shown in Fig. 4-21.

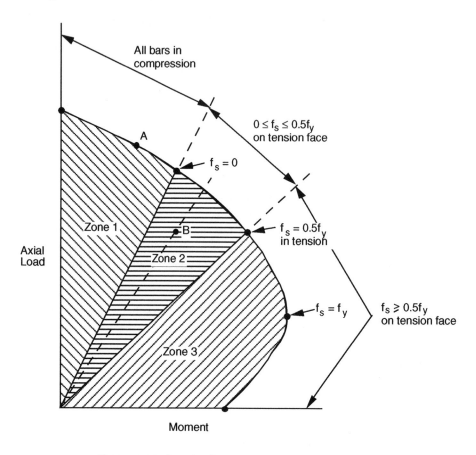

Figure 4-21 Special Splice Requirements for Columns

Bar stress at various locations along the strength interaction curve define segments of the strength curve where the different types of lap splices may be used. For factored load combinations along the strength curve, bar stress can be readily calculated to determine type of lap splice required. However, a design dilemma exists for load combinations that do not fall exactly on the strength curve (below the strength curve) as there is no simple exact method to calculate bar stress for this condition.

A seemingly rational approach is to consider factored load combinations below the strength curve as producing bar stress of the same type, compression or tension, and of the same approximate magnitude as that produced along the segment of the strength curve intersected by radial lines (lines of equal eccentricity) through the load combination point. This assumption becomes more exact as the factored load combinations being investigated fall nearer to the actual strength interaction curve of the column. Using this approach, zones of "bar stress" can be established as shown in Fig. 4-21.

For factored load combinations in Zone 1 of Fig. 4-21, all column bars are considered to be in compression. For load combinations in Zone 2 of the figure, bar stress on the tension face of the column is considered to vary from zero to $0.5f_y$ in tension. For load combinations in Zone 3, bar stress on the tension face is considered to be greater than $0.5f_y$ in tension. Type of lap splice to be used will then depend on which zone, or zones, all factored load combinations considered in the design of the column are located. The designer need only locate the factored load combinations on the load-moment strength diagram for the column and bars selected in the design to determine type of lap splice required. Use of load-moment design charts in this manner will greatly facilitate the design of column bar splices. For example, if factored gravity load combination governed design of the column, say Point A in Fig. 4-21, where all bars are in compression, but a load combination including wind, say Point B in Fig. 4-21, produces some tension in the bars, the lap splice must be designed for a Zone 2 condition (bar stress is tensile but does not exceed $0.5f_y$ in tension).

The design requirements for lap splices in columns are summarized in Table 4-8. Note that the compression lap splice permitted when all bars are in compression (see 12.17.2.1) considers a compression lap length adequate as a minimum tensile strength requirement. See Example 4.8 for design application of the lap splice requirements for columns.

Table 4-8 Lap Splices in Columns

12.17.2.1—Bar stress in compression (Zone 1)*	Use compression lap splice (12.16) modified by factor of 0.83 for ties (12.17.2.4) or 0.75 for spirals (12.17.2.5).
12.17.2.2—Bar stress ≤ $0.5f_y$ in tension (Zone 2)*	Use Class B tension lap splice (12.15) if more than 1/2 of total column bars spliced at same location. or Use Class A tension lap splice (12.15) if not more than 1/2 of total column bars spliced at same location. Stagger alternate splices by ℓ_d.
12.17.2.3—Bar stress > $0.5f_y$ in tension (Zone 3)*	Use Class B tension lap splice (12.15).

* For Zones 1, 2, and 3, see Fig. 4-21.

Sections 12.17.2.4 and 12.17.2.5 provide reduction factors for the compression lap splice when the splice is enclosed throughout its length by ties (0.83 reduction factor) or by a spiral (0.75 reduction factor). Spirals must meet the requirements of 7.10.4 and 10.9.3. When ties are used to reduce the lap splice length, the ties must have a minimum effective area of 0.0015hs. The tie legs in both directions must provide the minimum effective area to permit the 0.83 modification factor. See Fig. 4-22. The 12 in. minimum lap length also applies to these permitted reductions.

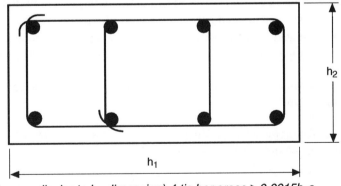

(perpendicular to h_1 dimension) 4 tie bar areas ≥ $0.0015h_1s$
(perpendicular to h_2 dimension) 2 tie bar areas ≥ $0.0015h_2s$

Figure 4-22 Application of 12.17.2.4

With the "basic" lap length for compression lap splices a function of bar diameter d_b and bar yield strength f_y, and three modification factors for ties and spirals and for lower concrete strength, it is convenient to establish compression lap splices simply as a multiple of bar diameter.

> For Grade 60 bars.. $30d_b$
>> enclosed within ties .... $25d_b$
>> enclosed within spirals .... $22.5d_b$
> For Grade 75 bars.. $43.5d_b$
>> enclosed within ties .... $36d_b$
>> enclosed within spirals .... $33d_b$

but not less than 12 in. For f'_c less than 3000 psi, multiply by a factor of 1.33. Compression lap splice tables for the standard bar sizes can be readily developed using the above values.

12.17.3 Mechanical or Welded Splices in Columns

Mechanical or welded splices are permitted in column bars where bar stress is either compressive or tensile for all factored load combinations (Zones 1, 2, and 3 in Fig. 4-21). "Full" mechanical or "full" welded splices must be used; that is, the mechanical or welded splice must develop at least 125 percent of the bar yield strength, $1.25A_bf_y$. Use of mechanical or welded splices of lesser strength is permitted for splicing bars No. 5 and smaller in tension, in accordance with 12.15.4.

12.17.4 End Bearing Splices in Columns

End bearing splices are permitted for column bars stressed in compression for all factored load combinations (Zone 1 in Fig. 4-21). Even though there is no calculated tension, a minimum tensile strength of the continuing (unspliced) bars must be maintained when end bearing splices are used. Continuing bars on each face of the column must provide a tensile strength of $A_sf_y/4$, where A_s is the total area of bars on that face of the column. Thus, not more than 3/4 of the bars can be spliced on each face of the column at any one location. End bearing splices must be staggered or additional bars must be added at the splice location if more than 3/4 of the bars are to be spliced.

12.18 SPLICES OF WELDED DEFORMED WIRE REINFORCEMENT IN TENSION

For tension lap splices of deformed wire reinforcement, the code requires a minimum lap length of $1.3\ell_d$, but not less than 8 in. Lap length is measured between the ends of each reinforcement sheet. The development length ℓ_d is the value calculated by the provisions in 12.7. The code also requires that the overlap measured between the outermost cross wires be at least 2 in. Figure 4-23 shows the lap length requirements.

If there are no cross wires within the splice length, the provisions in 12.15 for deformed wire must be used to determine the length of the lap.

Section 12.18.3 provides requirements for splicing welded wire reinforcement, including deformed wires in one direction and plain wires in the orthogonal direction.

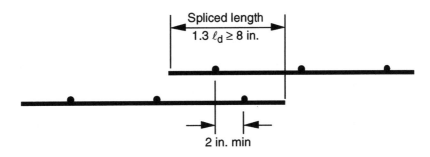

Figure 4-23 Lap Splice Length for Deformed Wire Fabric

12.19 SPLICES OF WELDED PLAIN WIRE REINFORCEMENT IN TENSION

The minimum length of lap for tension lap splices of plain wire reinforcement is dependent upon the ratio of the area of reinforcement provided to that required by analysis. Lap length is measured between the outermost cross wires of each reinforcement sheet. The required lap lengths are shown in Fig. 4-24.

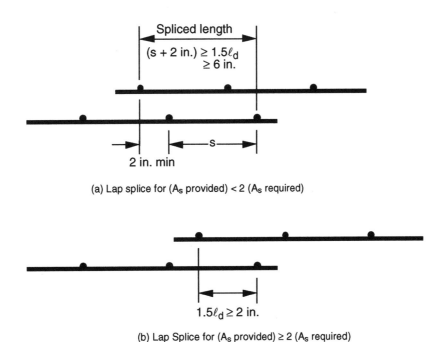

Figure 4-24 Lap Splice Length for Plain Wire Reinforcement

CLOSING REMARKS

One additional comment concerning splicing of temperature and shrinkage reinforcement at the exposed surfaces of walls or slabs: one must assume all temperature and shrinkage reinforcement to be stressed to the full specified yield strength f_y. The purpose of this reinforcement is to prevent excess cracking. At some point in the member, it is likely that cracking will occur, thus fully stressing the temperature and shrinkage reinforcement. Therefore, all splices in temperature and shrinkage reinforcement must be assumed to be those required for development of yield tensile strength. A Class B tension lap splice must be provided for this steel.

REFERENCES

4.1 Orangun, C. O., Jirsa, J. O., and Breen, J. E., "A Reevaluation of Test Data on Development Length and Splices," *ACI Journal, Proceedings* V. 74, Mar. 1977, pp. 114-122.

4.2 Sozen, M.A., and Moehle, J.P., "Selection of Development and Lap-Splice Lengths of Deformed Reinforcing Bars in Concrete Subjected to Static Loads," Report to PCA and CRSI, PCA R&D Serial No. 1868, March 1990.

4.3 Hamad, B. S., Jirsa, J. O., and D'Abreu, N. I., "Effect of Epoxy Coating on Bond and Anchorage of Reinforcement in Concrete Structures," Research Report 1181-1F, Center for Transportation Research, University of Texas at Austin, Dec. 1990, 242 pp.

4.4 *Manual of Standard Practice, Stuctural Welded Wire Reinforcement*, WWR-500, 5th Edition, Wire Reinforcement Institute, Findlay, OH, 1999, 27 pp.

4.5 *Reinforcement Anchorages and Splices*, 4th Edition, Concrete Reinforcing Steel Institute, Schaumburg, IL, 1997.

Example 4.1—Development of Bars in Tension

A beam at the perimeter of the structure has 7-No. 9 top bars over the support. Structural integrity provisions require that at least one-sixth of the tension reinforcement be made continuous, but not less than 2 bars (7.13.2.2). Bars are to be spliced with a Class A splice at midspan. Determine required length of Class A lap splice for the following two cases:

Case A - Development computed from 12.2.2
Case B - Development computed from 12.2.3

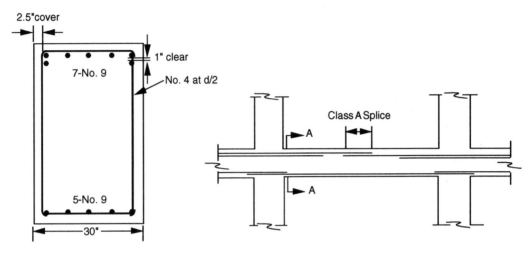

Assume:
   Lightweight concrete
   2.5 in. clear cover to stirrups
   Epoxy-coated bars
   $f_c' = 4000$ psi
   $f_y = 60,000$ psi
   $b = 30$ in. (with bar arrangement as shown)

Calculations and Discussion	Code Reference
It is assumed that development of negative moment reinforcement has been satisfied and, therefore, top bars are stopped away from midspan.	12.12.3
Minimum number of top bars to be made continuous for structural integrity is 1/6 of 7 bars provided, i.e., 7/6 bars or, a minimum of 2 bars. Two corner bars will be spliced at midspan.	7.13.2.2
Class A lap splice requires a $1.0\ell_d$ length of bar lap	12.15.1

Nominal diameter of No. 9 bar = 1.128 in.

CASE A - Section 12.2.2

Refer to Table 4-1. For bars No. 7 and larger, either Eq. B or Eq. D apply. To determine if Eq. B or Eq. D governs, determine clear cover and clear spacing for bars being developed.

Example 4.1 (cont'd)   **Calculations and Discussion**   **Code Reference**

Clear spacing between spliced bars (corner bars)
= [30 - 2 (cover) - 2 (No. 4 stirrup) - 2 (No. 9 bar)]
= [30 - 2 (2.5) - 2 (0.5) - 2(1.128)]
= 21.7 in.
= 19.3d_b

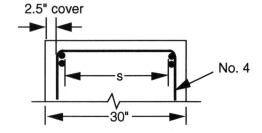

Clear cover to spliced bar = 2.5 + 0.5 = 3.0 in. = 2.7d_b

As clear spacing > 2d_b and clear cover > d_b, Eq. B applies.

$$\ell_d = \left(\frac{f_y \psi_t \psi_e \lambda}{20\sqrt{f'_c}}\right) d_b$$
     12.2.2

ψ_t = 1.3 for top bar   12.2.4

ψ_e = 1.5 for epoxy-coated bar with cover less than 3d_b 12.2.4

$\psi_t \psi_e$ = 1.3 × 1.5 = 1.95; however, product of α and β need not be taken greater than 1.7.   12.2.4

λ = 1.3 for lightweight aggregate concrete   12.2.4

$$\ell_d = \frac{60{,}000\ (1.7)\ (1.3)}{20\sqrt{4000}} (1.128)$$

= 118.3 in.

Class A splice = 1.0ℓ_d = 118.3 in.

CASE B - Section 12.2.3

Application of Eq. (12-1) requires a little more computations, but can result in smaller development lengths.

$$\ell_d = \left(\frac{3}{40} \frac{f_y}{\sqrt{f'_c}} \frac{\psi_t \psi_e \psi_s \lambda}{\left(\frac{c_b + K_{tr}}{d_b}\right)}\right) d_b$$
     Eq. (12-1)

Parameter "c_b" is the smaller of (1) distance from center of bar being developed to the nearest concrete surface, and (2) one-half the center-to-center spacing of bars being developed. Also, note that the term $\left(\frac{c_b + K_{tr}}{d_b}\right)$ cannot exceed 2.5.

Distance from center of bar or wire being developed to the nearest concrete surface
= clear cover to spliced bar + 1/2 bar diameter

	Code
Example 4.1 (cont'd) **Calculations and Discussion**	**Reference**

$= 2.7d_b + 0.5d_b = 3.2d_b$

Center-to-center spacing = clear spacing + $1.0d_b$ = $19.3d_b + 1.0d_b = 20.3d_b$

Therefore, c is the smaller of $3.2d_b$ and $0.5(20.3d_b)$, i.e. $3.2d_b$

No need to compute K_{tr} as c/d_b is greater than 2.5

$\gamma = 1.0$ for No. 7 bar and larger

$$\ell_d = \frac{3\,(60{,}000)\,(1.7)\,(1.0)\,(1.3)}{40\sqrt{4000}\,(2.5)}\,(1.128)$$

$= 71.0$ in.

Class A splice = $1.0\ell_d = 71.0$ in.

The extra computations required to satisfy the general Eq. (12-1) of 12.2.3 can lead to substantial reductions in tension development or splice lengths compared to values computed from the simplified procedure of 12.2.2.

Example 4.2—Development of Bars in Tension

Calculate required tension development length for the No. 8 bars (alternate short bars) in the "sand-lightweight" one-way slab shown below. Use $f'_c = 4000$ psi and $f_y = 60,000$ psi, and uncoated bars.

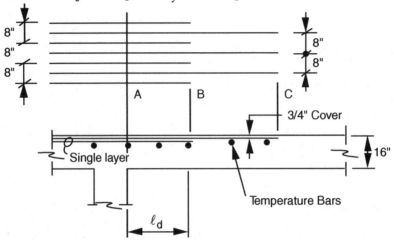

Calculations and Discussion	Code Reference

Calculations for this example will be performed using 12.2.2 and 12.2.3.

Assume short bars are developed within distance AB while long bars are developed within BC.

Nominal diameter of No. 8 bar is 1.00 in.

A. Development length by 12.2.2

Center-to-center spacing of bars being developed = 8 in. = $8d_b$

Clear cover = 0.75 in. = $0.75d_b$

As clear cover is less than d_b, and bar size is larger than No. 7, Eq. D of Table 4-1 applies.

$$\ell_d = \left(\frac{3f_y \psi_t \psi_e \lambda}{40\sqrt{f'_c}}\right) d_b \qquad \text{12.2.2}$$

$\psi_t = 1.3$ for top bar

$\psi_e = 1.0$ for uncoated bars

$\lambda = 1.3$ for lightweight concrete

$$\ell_d = \frac{3\,(60,000)\,(1.3)\,(1.0)\,(1.3)}{40\sqrt{4000}}\,(1.0) = 120.3 \text{ in.}$$

Example 4.2 (cont'd)   **Calculations and Discussion**   **Code Reference**

B. Development length by 12.2.3

$$\ell_d = \left(\frac{3}{40} \frac{f_y}{\sqrt{f'_c}} \frac{\psi_t \psi_e \psi_s \lambda}{\left(\frac{c_b + K_{tr}}{d_b} \right)} \right) d_b$$

Eq. (12-1)

ψ_t = 1.3 for top bar

ψ_e = 1.0 for uncoated bars

ψ_s = 1.0 for No. 7 and larger bars

λ = 1.3 for lightweight concrete

Center-to-center spacing of bars being developed = 8 in. = $8d_b$
Clear spacing between bars being developed = 8 - 1 = 7 in. = $7d_b$

Clear cover = 0.75 in. = $0.75d_b$
Distance "c" from center of bar to concrete surface = 0.75 + 0.5 = 1.25 in. = $1.25d_b$ (governs)

$\qquad\qquad\qquad\qquad\qquad\qquad\qquad\qquad$ = $8d_b/2 = 4d_b$ (center-to-center spacing/2)

c_b = $1.25d_b$ (computed above)

$K_{tr} = \dfrac{A_{tr} f_{yt}}{1500sn} = 0$ (no transverse reinforcement)

$\ell_d = \dfrac{3 \, (60,000) \, (1.3) \, (1.0) \, (1.0) \, (1.3)}{40\sqrt{4000} \, (1.25)} (1.0) = 96.2$ in.

Example 4.3—Development of Bars in Tension

Calculate required development length for the inner 2 No. 8 bars in the beam shown below. The 2 No. 8 outer bars are to be made continuous along full length of beam. Use $f'_c = 4000$ psi (normal weight concrete) and $f_y = 60,000$ psi, and uncoated bars. Stirrups provided satisfy the minimum code requirements for beam shear reinforcement.

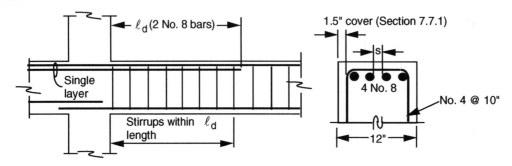

Calculations and Discussion	Code Reference

Calculations for this example will be performed using 12.2.2 and 12.2.3.

Nominal diameter of No. 8 bar = 1.00 in.

A. Development length by 12.2.2

Clear spacing = [12 - 2 (cover) - 2 (No. 4 stirrups) - 4 (No. 8 bars)]/3 spaces
= [12 - 2 (1.5) - 2 (0.50) - 4 (1.00)]/3
= 1.33 in.
= $1.33d_b$

Clear cover = 1.5 + 0.5 = 2.0 in. = $2d_b$

Refer to Table 4-1. Clear spacing between bars being developed more than d_b, clear cover more than d_b, and minimum stirrups provided. Eq. B of Table 4-1 applies.

$$\ell_d = \left(\frac{f_y \psi_t \psi_e \lambda}{20\sqrt{f'_c}}\right) d_b \qquad\qquad 12.2.2$$

ψ_t = 1.3 for top bar

ψ_e = 1.0 for uncoated bars

λ = 1.0 for normal weight concrete

$$\ell_d = \frac{(60,000)\,(1.3)\,(1.0)\,(1.0)}{20\sqrt{4000}}\,(1.0) = 61.7 \text{ in.}$$

Example 4.3 (cont'd)	Calculations and Discussion	Code Reference

B. Development length by 12.2.3

$$\ell_d = \left(\frac{3}{40} \frac{f_y}{\sqrt{f'_c}} \frac{\psi_t \psi_e \psi_s \lambda}{\left(\frac{c_b + K_{tr}}{d_b} \right)} \right) d_b \qquad \text{Eq. (12-1)}$$

ψ_t = 1.3 for top bar

ψ_e = 1.0 for uncoated bars

ψ_s = 1.0 for No. 7 and larger bars

λ = 1.0 for normal weight concrete

Clear spacing = $1.33 d_b$
Center-to-center spacing of bars being developed = 1.33 + 1.0 in. = 2.33 in. = $2.33 d_b$

Clear cover = 1.50 + 0.5 = 2.0 in. = $2 d_b$
Distance from center of bar to concrete surface = 1.5 + 0.5 + 0.5 = 2.5 in. = $2.5 d_b$

c_b = the smaller of (1) distance from center of bar being developed to the nearest concrete surface ($2.5 d_b$), and of (2) one-half the center-to-center spacing of bars being developed ($2.33 d_b / 2 = 1.17 d_b$)

$c_b = 1.17 d_b$

$$K_{tr} = \frac{A_{tr} f_{yt}}{1500 sn}$$

A_{tr} (2-No. 4) = 2 × 0.2 = 0.4 in.^{2}

s = 10 in. spacing of stirrups

n = 2 bars being developed

$$K_{tr} = \frac{0.4 (60{,}000)}{1500 (10)(2)} = 0.80 \text{ in.} = 0.80 d_b$$

$$\left(\frac{c_b + K_{tr}}{d_b} \right) = \frac{1.17 + 0.80}{1.0} = 1.97 < 2.5 \quad \text{O.K.} \qquad 12.2.3$$

Example 4.3 (cont'd)	Calculations and Discussion	Code Reference

$$\ell_d = \frac{3\,(60{,}000)\,(1.3)\,(1.0)\,(1.0)\,(1.0)}{40\,\sqrt{4000}\,(1.97)}\,(1.0) = 47.0 \text{ in.}$$

Example 4.4—Development of Flexural Reinforcement

Determine lengths of top and bottom bars for the exterior span of the continuous beam shown below. Concrete is normal weight and bars are Grade 60. Total uniformly distributed factored gravity load on beam is w_u = 6.0 kips/ft (including weight of beam).

f'_c = 4000 psi
f_y = 60,000 psi
b = 16 in.
h = 22 in.
Concrete cover = 1 1/2 in.

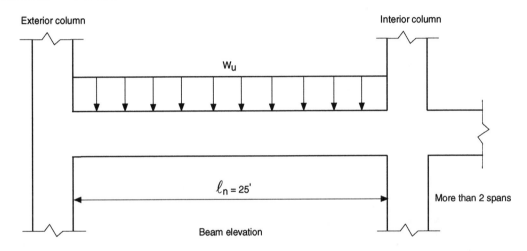

Beam elevation

	Code
Calculations and Discussion	**Reference**

1. Preliminary design for moment and shear reinforcement

    a. Use approximate analysis for moment and shear *8.3.3*

Location	Factored moments & shears
Interior face of exterior support	$-M_u = w_u \ell_n^2/16 = 6\,(25^2)/16 = -234.4$ ft - kips
End span positive	$+M_u = w_u \ell_n^2/14 = 6\,(25^2)/14 = 267.9$ ft - kips
Exterior face of first interior support	$-M_u = w_u \ell_n^2/10 = 6\,(25^2)/10 = -375.0$ ft - kips
Exterior face of first interior support	$V_u = 1.15 w_u \ell_n/2 = 1.15\,(6)\,(25)/2 = 86.3$ kips

    b. Determine required flexural reinforcement using procedures of Part 7 of this publication. With 1.5 in. cover, No. 4 bar stirrups, and No. 9 or No. 10 flexural bars, d ≈ 19.4 in.

Example 4.4 (cont'd) — Calculations and Discussion

Code Reference

M_u	A_s required	Bars	A_s provided
-234.4 ft-kips	2.93 in.²	4 No. 8	3.16 in.²
+267.9 ft-kips	3.40 in.²	2 No. 8 2 No. 9	3.58 in.²
-375.0 ft-kips	5.01 in.²	4 No. 10	5.08 in.²

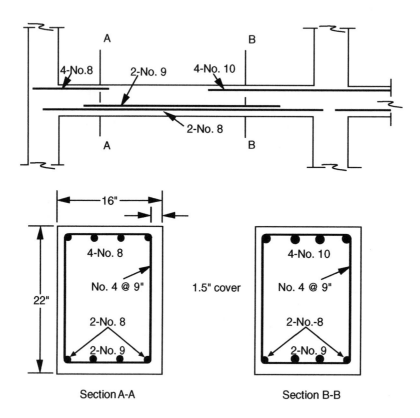

Section A-A   Section B-B

c. Determine required shear reinforcement

    V_u at "d" distance from face of support: *11.1.3.1*

    $V_u = 86.3 - 6(19.4/12) = 76.6$ kips

    $\phi V_c = \phi\left(2\sqrt{f'_c}\, b_w d\right) = 0.75 \times 2\sqrt{4,000} \times 16 \times 19.4/1,000 = 29.5$ kips   *11.1.3.1*

    Try No. 4 U-stirrups @ 7 in. spacing $< s_{max} = \dfrac{d}{2} = 9.7$ in.   *11.5.4.1*

    $\phi V_s = \dfrac{\phi A_v f_y d}{s} = 0.75\,(0.40)\,(60)\,(19.4)/7 = 49.9$ kips   *11.5.6.2*

    $\phi V_n = \phi V_c + \phi V_s = 29.5 + 49.9 = 79.4$ kips > 76.6 kips   O.K.

| Example 4.4 (cont'd) | Calculations and Discussion | Code Reference |

Distance from support where stirrups not required:

$$V_u < \frac{\phi V_c}{2} = \frac{29.5}{2} = 14.8 \text{ kips}$$

11.5.5.1

$V_u = 86.3 - 6x = 14.8$ kips

$x = 11.9$ ft $\approx$ 1/2 span

Use No. 4 U-stirrups @ 7 in. (entire span)

2. Bar lengths for bottom reinforcement

   a. Required number of bars to be extended into supports.

12.11.1

   One-fourth of $(+A_s)$ must be extended at least 6 in. into the supports. With a longitudinal bar required at each corner of the stirrups (12.13.3), at least 2 bars should be extended full length. Extend the 2-No. 8 bars full span length (plus 6 in. into the supports) and cut off the 2-No. 9 bars within the span.

   b. Determine cut-off locations for the 2 No. 9 bars and check other development requirements.

   Shear and moment diagrams for loading condition causing maximum factored positive moment are shown below.

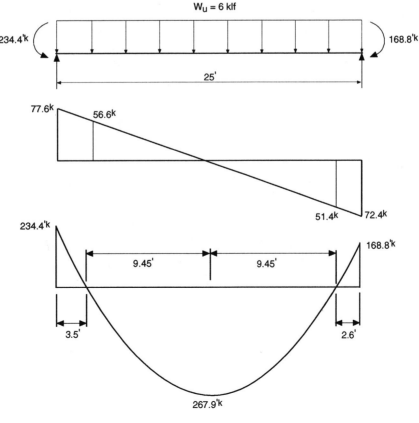

		Code
Example 4.4 (cont'd)	**Calculations and Discussion**	**Reference**

The positive moment portion of the M_u diagram is shown below at a larger scale, including the design moment strengths ϕM_n for the total positive A_s (2-No. 8 and 2-No. 9) and for 2-No. 8 bars separately. For 2-No. 8 and 2-No. 9, ϕM_n = 280.7 ft-kips. For 2-No. 8, ϕM_n = 131.8 ft-kips.

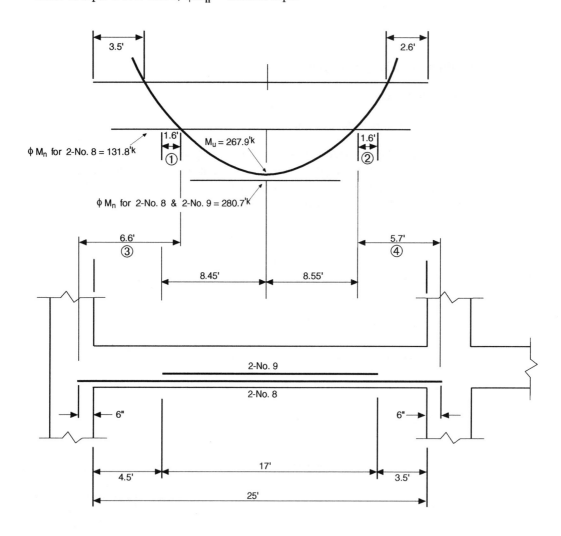

As shown, the 2-No. 8 bars extend full span length plus 6 in. into the supports. The 2-No. 9 bars are cut off tentatively at 4.5 ft and 3.5 ft from the exterior and interior supports, respectively. These tentative cutoff locations are determined as follows:

Dimensions (1) and (2) must be the larger of d or $12d_b$. 12.10.3

d = 19.4 in. = 1.6 ft (governs)

$12d_b$ = 12 (1.128) = 13.5 in.

4-42

| Example 4.4 (cont'd) | Calculations and Discussion | Code Reference |

Within the development length ℓ_d, only 2-No. 8 bars are being developed (2-No. 9 bars are already developed in length 8.45 ft)

Development for No. 8 corner bars, see Table 4-2.

$\ell_d = 47d_b = 47(1.0) = 47$ in. $= 3.9$ ft

Dimension (3): 6.6 ft > 3.9 ft O.K.
Dimension (4): 5.7 ft > 3.9 ft O.K.

Check required development length ℓ_d for 2-No. 9 bars. Note that 2-No. 8 bars are already developed in length 4 ft from bar end.

Clear spacing between 2-No. 9 bars

$[16 - 2(1.5) - 2(0.5) - 2(1.0) - 2(1.128)]/3 = 2.58$ in. $= 2.29d_b > 2d_b$

For No. 9 bar, $\ell_d = 47d_b$ *Table 4-2*

$= 47(1.128) = 53$ in. $= 4.4$ ft < 8.45 ft   O.K.

For No. 8 bars, check development requirements at points of inflection (PI):   *12.11.3*

$$\ell_d \leq \frac{M_n}{V_u} + \ell_a$$ *Eq. (12-3)*

For 2-No. 8 bars, $M_n = 131.8/0.9 = 146.4$ ft-kips

At left PI, $V_u = 77.6 - 6(3.5) = 56.6$ kips

ℓ_a = larger of $12d_b = 12(1.0) = 12$ in. or $d = 19.4$ in. (governs)

$$\ell_d \leq \frac{146.4 \times 12}{56.6} + 19.4 = 50.5 \text{ in.}$$

For No. 8 bars, $\ell_d = 47$ in. < 50.5 in.   O.K.

At right PI, $V_u = 56.8$ kips; by inspection, the development requirements for the No. 8 bars are O.K.

With both tentative cutoff points located in a zone of flexural tension, one of the three conditions of 12.10.5 must be satisfied.

At left cutoff point (4.5 ft from support):

$V_u = 77.6 - (4.5 \times 6) = 50.6$ kips

$\phi V_n = 79.4$ kips (No. 4 U-stirrups @ 7 in.)

$2/3(79.4) = 52.9$ kips > 50.6 kips   O.K.   *12.10.5.1*

| Example 4.4 (cont'd) | Calculations and Discussion | Code Reference |

For illustrative purposes, determine if the condition of 12.10.5.3 is also satisfied:

M_u = 54.1 ft-kips at 4.5 ft from support

A_s required = 0.63 in.^{2}

For 2-No. 8 bars, A_s provided = 1.58 in.^{2}

1.58 in.^{2} > 2 (0.63) = 1.26 in.^{2} O.K.   *12.10.5.3*

3/4 (79.4) = 59.6 kips > 50.6 kips   O.K.   *12.10.5.3*

Therefore, 12.10.5.3 is also satisfied at cutoff location.

At right cutoff point (3.5 ft from support):

V_u = 72.4 - (3.5 × 6) = 51.4 kips

2/3 (ϕV_n) = 52.9 kips > 51.4 kips   O.K.   *12.10.5.1*

Summary: The tentative cutoff locations for the bottom reinforcement meet all code development requirements. The 2-No. 9 bars × 17 ft would have to be placed unsymmetrically within the span. To assure proper placing of the No. 9 bars, it would be prudent to specify a 18 ft length for symmetrical bar placement within the span, i.e., 3.5 ft from each support. The ends of the cut off bars would then be at or close to the points of inflection, thus, eliminating the need to satisfy the conditions of 12.10.5 when bars are terminated in a tension zone. The recommended bar arrangement is shown at the end of the example.

3. Bar lengths for top reinforcement

Shear and moment diagrams for loading condition causing maximum factored negative moments are shown below.

The negative moment portions of the M_u diagram are also shown below at a larger scale, including the design moment strengths ϕM_n for the total negative A_s at each support (4-No. 8 at exterior support and 4-No. 10 at interior support) and for 2-No. 10 bars at the interior support. For 4-No. 8, ϕM_n = 251.1 ft-kips. For 4-No. 10, ϕM_n = 379.5 ft-kips. For 2-No. 10, ϕM_n = 194.3 ft-kips.

Example 4.4 (cont'd)	Calculations and Discussion	Code Reference

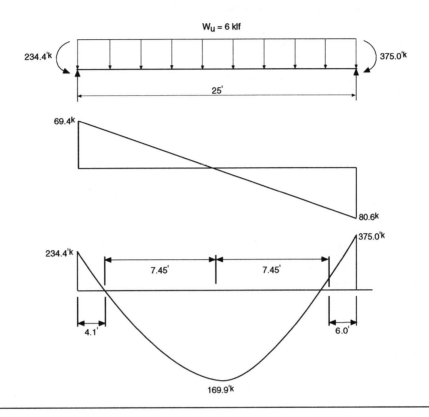

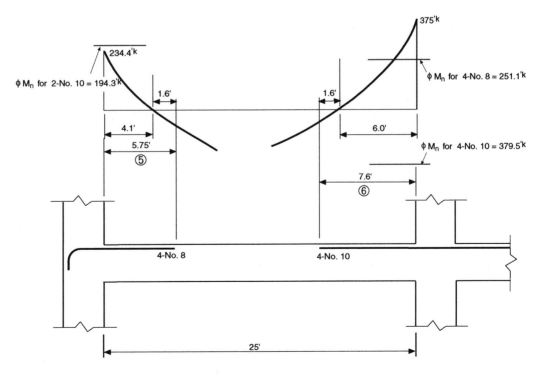

Example 4.4 (cont'd)	Calculations and Discussion	Code Reference

4. Development requirements for 4-No. 8 bars at exterior support

   a. Required number of bars to be extended.

   One-third of ($-A_s$) provided at supports must be extended beyond the point of inflection a distance equal to the greater of d, $12d_b$, or $\ell_n/16$.   **12.12.3**

   d = 19.4 in. = 1.6 ft (governs)

   $12d_b = 12 (1.0) = 12.0$ in.

   $\ell_n/16 = 25 \times 12/16 = 18.75$ in.

   Since the inflection point is located only 4.1 ft from the support, total length of the No. 8 bars will be relatively short even with the required 1.6 ft extension beyond the point of inflection. Check required development length ℓ_d for a cutoff location at 5.75 ft from face of support.

   Dimension (5) must be at least equal to ℓ_d **12.12.2**

   For No. 8 bars, $\ell_d = 47d_b = 47 (1.0) = 47$ in.   *Table 4-2*

   With 4-No. 8 bars being developed at same location (face of support):

   Including top bar effect, $\ell_d = 1.3 (47) = 61.1$ in.

   For No. 8 top bars, $\ell_d = 61.1$ in. = 5.1 ft < 5.75 ft   O.K.

   b. Anchorage into exterior column.

   The No. 8 bars can be anchored into the column with a standard end hook. From Table 4-4, $\ell_{dh} = 19.0$ in. The required ℓ_{dh} for the hook could be reduced if excess reinforcement is considered:

   $$\frac{(A_s \text{ required})}{(A_s \text{ provided})} = \frac{2.93}{3.16} = 0.93$$ **12.5.3(c)**

   $\ell_{dh} = 19 \times 0.93 = 17.7$ in.

   Overall depth of column required would be 17.7 + 2 = 19.7 in.

5. Development requirements for 4-No. 10 bars at interior column

   a. Required extension for one-third of ($-A_s$)   **12.12.3**

	Code
Example 4.4 (cont'd) **Calculations and Discussion**	**Reference**

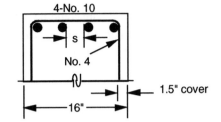

$d = 19.4$ in. $= 1.6$ ft   (governs)

$12d_b = 12 (1.27) = 15.24$ in.

$\ell_n/16 = 18.75$ in.

For No. 10 bars, clear spacing $= [16 - 2 (1.5) - 2 (0.5) - 4 (1.27)]/3$
$= 2.31$ in. $= 1.82d_b > d_b$
Center-to-center spacing $= 2.82d_b$

Cover $= 1.5 + 0.5 = 2.0$ in. $= 1.57d_b > d_b$

Distance from center of bar to concrete surface $= 1.57d_b + 0.5d_b = 2.07d_b$

With minimum shear reinforcement provided and including top bar effect

$\ell_d = 1.3 (47d_b)$ Table 4-2

$= 1.3 (47) (1.27) = 77.6$ in.

Dimension (6) $= 6.0$ ft $+ 1.6$ ft $= 7.6$ ft $> \ell_d = 77.6$ in $= 6.5$ ft   O.K.

6. Summary: Selected bar lengths for the top and bottom reinforcement shown below.

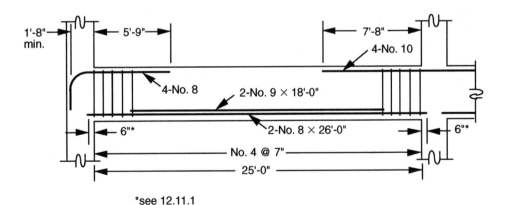

*see 12.11.1

7. Supplementary Requirements

   If the beam were part of a primary lateral load resisting system, the 2-No. 8 bottom bars   12.11.2
   extending into the supports would have to be anchored to develop the bar yield strength
   at the face of supports. At the exterior column, anchorage can be provided by a standard
   end hook. Minimum width of support (overall column depth) required for anchorage of
   the No. 8 bar with a standard hook is a function of the development length ℓ_{dh} from
   Table 4-4, and the appropriate modification factors (12.5.3).

Example 4.4 (cont'd)	Calculations and Discussion	Code Reference

At the interior column, the 2-No. 8 bars could be extended ℓ_d distance beyond the face of support into the adjacent span or lap spliced with extended bars from the adjacent span. Consider a Class A lap splice adequate to satisfy the intent of 12.11.2.

Example 4.5—Lap Splices in Tension

Design the tension lap splices for the grade beam shown below.

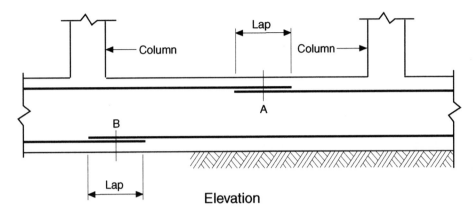

Elevation

$f'_c = 4000$ psi
$f_y = 60,000$ psi, uncoated bars
$b = 16$ in.
$h = 30$ in.
Bar cover = 3.0 in.
4-No. 9 bars top and bottom (continuous)
No. 4 stirrups @ 14 in. (entire span)
$+M_u$ @B = 340 ft-kips
$-M_u$ @A = 120 ft-kips

Preferably, splices should be located away from zones of high tension. For a typical grade beam, top bars should be spliced under the columns, and bottom bars about midway between columns. Even though in this example the splice at A is not a preferred location, the moment at A is relatively small. Assume for illustration that the splices must be located as shown.

Calculations and Discussion	Code Reference

Calculations for this example will be performed using 12.2.3.

Nominal diameter of No. 9 bar = 1.128 in.

Assuming all bars are spliced at the same location

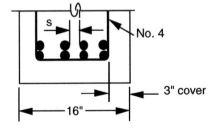

Clear spacing = [16 - 2 (cover) - 2 (No. 4 stirrups) - 4 (No. 9 bars)]/3 spaces
= [16 - 2 (3.0) - 2 (0.50) - 4 (1.128)]/3
= 1.50 in.
= $1.33d_b$

Center-to-center spacing of bars being developed = 1.50 + 1.128 = 2.63 in. = $2.33d_b$

Clear cover = 3.0 + 0.5 = 3.5 in. = $3.1d_b$

Example 4.5 (cont'd)	Calculations and Discussion	Code Reference

Distance from center of bar to concrete surface = $3.0 + 0.5 + (1.128/2) = 4.1$ in. = $3.6d_b$

c = the smaller of (1) distance from center of bar being developed to the nearest concrete surface and (2) one-half the center-to-center spacing of bars being developed

c = $3.6d_b$

 = $2.33d_b/2 = 1.17d_b$ (governs)

Lap Splice of Bottom Reinforcement at Section B

$$\ell_d = \left(\frac{3}{40} \frac{f_y}{\sqrt{f'_c}} \frac{\psi_t \psi_e \psi_s \lambda}{\left(\frac{c_b + K_{tr}}{d_b} \right)} \right) d_b \qquad \text{Eq. (12-1)}$$

ψ_t = 1.0 for bottom bar 12.2.4

ψ_e = 1.0 for uncoated bars

ψ_s = 1.0 for No. 7 and larger bars

λ = 1.0 for normal weight concrete

c_b = $1.17d_b$ (computed above)

$$K_{tr} = \frac{A_{tr} f_{yt}}{1500 sn}$$

A_{tr} = area of 2-No. 4 stirrups = $2(0.2) = 0.4$ in.^{2}
s = 14 in. spacing
n = 4 bars being developed

$$K_{tr} = \frac{0.4 (60,000)}{1500 (14)(4)} = 0.29 \text{ in.} = 0.26d_b$$

$\left(\frac{c_b + K_{tr}}{d_b} \right) = 1.17 + 0.26 = 1.43 < 2.5$ O.K. 12.2.3

$$\ell_d = \frac{3(60,000)(1.0)(1.0)(1.0)(1.0)}{40\sqrt{4,000}(1.43)} (1.128) = 56.1 \text{ in.}$$

		Code
Example 4.5 (cont'd)	**Calculations and Discussion**	**Reference**

A_s required (+M_u @ B = 340 ft-kips) = 3.11 in.^{2}

A_s provided (4 No. 9 bars) = 4.00 in.^{2}

$$\frac{A_s \text{ provided}}{A_s \text{ required}} = \frac{4.00}{3.11} = 1.29 < 2$$

Class B splice required = $1.3\ell_d$

12.15.1

12.15.2

Note: Even if lap splices were staggered (A_s spliced = 50%), a Class B splice must be used with (A_s provided/A_s required) < 2

Class B Splice = $1.3\ell_d$ = 1.3 (56.1) = 72.9 in. = 6.1 ft

It is better practice to stagger alternate lap splices. As a result, the clear spacing between spliced bars will be increased with a potential reduction of development length.

Clear spacing = 2 (1.50) + 1.128 = 4.13 in. = $3.66d_b$

Center-to-center spacing of bars being developed = $3.66d_b + d_b = 4.66d_b$

Distance from center of bar to concrete surface = $3.6d_b$

Thus, $c = \dfrac{4.66d_b}{2} = 2.33d_b$

$$K_{tr} = \frac{2\,(0.2)\,(60{,}000)}{(1500)\,(14)\,(2)} = 0.57 \text{ in.} = 0.51d_b$$

Therefore, $\left(\dfrac{c_b + K_{tr}}{d_b}\right) = 2.33 + 0.51 = 2.84 > 2.5$ Use 2.5.

12.2.3

$$\ell_d = \frac{3\,(60{,}000)\,(1.0)\,(1.0)\,(1.0)\,(1.0)}{40\sqrt{4000}\,(2.5)}\,(1.128) = 32.1 \text{ in.}$$

Class B splice = 1.3 (32.1) = 41.7 in. = 3.5 ft

Use 3 ft-6 in. lap splice @ B and stagger alternate lap splices.

Lap Splice of Top Reinforcement at Section A

As size of top and bottom reinforcement is the same, computed development and splice lengths for top bars will be equal to that of the bottom bars increased by the 1.3 multiplier for top bars. In addition, because positive and negative factored moments are different, the ratio of provided to required reinforcement may affect the type of splice as demonstrated below.

Example 4.5 (cont'd)	Calculations and Discussion	Code Reference

A_s required (+M_u @ A = 120 ft-kips) = 1.05 in.^{2}

A_s provided/A_s required = 4.00/1.05 = 3.81 > 2

If alternate lap splices are staggered at least a lap length (A_s spliced = 50%):

Class A splice may be used = $1.0\ell_d$ 12.15.2

If all bars are lap spliced at the same location (within req'd lap length):

Class B splice must be used = $1.3\ell_d$

Assuming splices are staggered, the top bar multiplier will be 1.3.

Class A splice = 1.3 (1.0) (32.1) = 41.7 in. = 3.5 ft

Use 3 ft-6 in. lap splice @ A also, and stagger alternate lap splices.

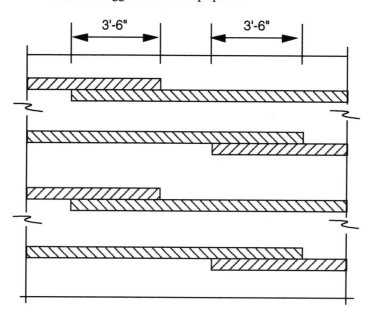

Alternate lap splice stagger arrangement
(Note: bar laps are positioned vertically)

Example 4.6—Lap Splices in Compression

The following two examples illustrate typical calculations for compression lap splices in tied and spirally reinforced columns.

Calculations and Discussion	Code Reference

1. Design a compression lap splice for the tied column shown below. Assume all bars in compression for factored load combinations considered in design (Zone 1 in Fig. 4-21). See also Table 4-8.

   $b = 16$ in.
   $h = 16$ in.
   $f'_c = 4000$ psi
   $f_y = 60,000$ psi
   8-No. 9 bars

   a. Determine lap splice length: *12.16.1*

   For $f_y = 60,000$ psi:

   Length of lap = $0.0005 f_y d_b$, but not less than 12 in.

   $\qquad = 0.0005 (60,000) 1.128 = 34$ in.

   b. Determine column tie requirements to allow an 0.83 reduced lap length: *12.17.2.4*

   Required column ties: No. 3 @ 16 in. o.c.   *7.10.5.2*

   Required spacing of No. 3 ties for reduced lap length:

   effective area of ties $\geq 0.0015 hs$

   $(2 \times 0.11) \geq 0.0015 \times 16s$

   $s = 9.2$ in.

   Spacing of the No. 3 ties must be reduced to 9 in. o.c. throughout the lap splice length to allow a lap length of 0.83 (34 in.) = 28 in.

2. Determine compression lap splice for spiral column shown.

   $f'_c = 4000$ psi
   $f_y = 60,000$ psi
   8-No. 9 bars
   No. 3 spirals

Example 4.6 (cont'd) Calculations and Discussion	**Code Reference**

a. Determine lap splice length   *12.16.1*

    For bars enclosed within spirals, "basic" lap splice length may be multiplied by a factor of 0.75.   *12.17.2.5*

    For $f_y = 60,000$ psi:

    lap = 0.75(34) = 26 in.

    Note: End bearing, welded, or mechanical connections may also be used.   *12.16.3*
    12.16.4

Example 4.7—Lap Splices in Columns

Design the lap splice for the tied column detail shown.

- Continuing bars from column above (4-No. 8 bars)
- Offset bars from column below (4-No. 8 bars)

f'_c = 4000 psi (normal weight)
f_y = 60,000 psi
b = h = 16 in.
4-No. 8 bars (above and below floor level)
No. 3 ties @ 16 in.
Cover = 1.5 in.

Lap splice to be designed for the following factored load combinations:

1. P_u = 465 kips
   M_u = 20 ft-kips

2. P_u = 360 kips
   M_u = 120 ft-kips

3. P_u = 220 kips
   M_u = 100 ft-kips

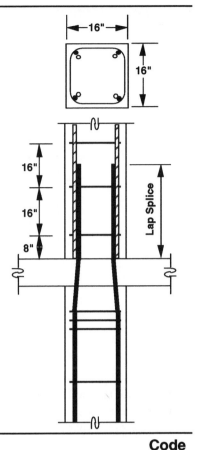

Calculations and Discussion	Code Reference
1. Determine type of lap splice required.	12.17.2

Type of lap splice to be used depends on the bar stress at the splice location due to all factored load combinations considered in the design of the column. For design purposes, type of lap splice will be based on which zone, or zones, of bar stress all factored load combinations are located on the column load-moment strength diagram. See discussion for 12.17.2, and Fig. 4-21. The load-moment strength diagram (column design chart) for the 16 × 16 column with 4-No. 8 bars is shown below, with the three factored load combinations considered in the design of the column located on the interaction strength diagram.

Note that load combination (2) governed the design of the column (selection of 4-No. 8 bars).

Table 4-8

For load combination (1), all bars are in compression (Zone 1), and a compression lap splice could be used. For load combination (2), bar stress is not greater than $0.5f_y$ (Zone 2), so a Class B tension lap splice is required; or, a Class A splice may be used if alternate lap splices are staggered. For load combination (3), bar stress is greater than $0.5f_y$ (Zone 3), and a Class B splice must be used.

	Code
Example 4.7 (cont'd) **Calculations and Discussion**	**Reference**

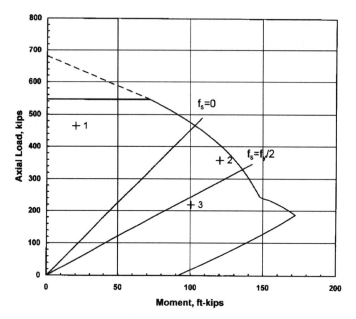

Interaction Diagram for 16 in. × 16 in. Column

Lap splice required for the 4-No. 8 bars must be based on the load combination producing the greatest amount of tension in the bars; for this example, load combination (3) governs the type of lap splice to be used.

Class B splice required = $\ell_d = 1.3\ell_{db}$ 12.15.1

2. Determine lap splice length

Determine tension development length by 12.2.3.

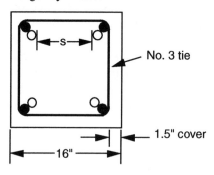

Nominal diameter of No. 8 bar = 1.00 in.

Clear spacing between bars being developed is large and will not govern.

Clear cover = 1.5 + 0.375 = 1.875 in. = $1.875d_b$

Distance from center of bar to concrete surface = 1.875 + 0.5 = 2.375 in. = $2.375d_b$

4-56

		Code
Example 4.7 (cont'd)	**Calculations and Discussion**	**Reference**

c = the smaller of (1) distance from center of bar being developed to the nearest concrete surface, and of (2) one-half the center-to-center spacing of bars being developed

$c = 2.375 d_b$

$$\ell_d = \left(\frac{3}{40} \frac{f_y}{\sqrt{f'_c}} \frac{\psi_t \psi_e \psi_s \lambda}{\left(\frac{c_b + K_{tr}}{d_b} \right)} \right) d_b \qquad \qquad Eq.\ (12\text{-}1)$$

$\psi_t = 1.0$ for vertical bar   *12.2.4*

$\psi_e = 1.0$ for uncoated bars

$\psi_s = 1.0$ for No. 7 and larger bars

$\lambda = 1.0$ for normal weight concrete

$c_b = 2.375 d_b$

$K_{tr} = \dfrac{A_{tr} f_{yt}}{1500 sn}$

A_{tr} = area of 2-No. 3 ties

s = 16 in. spacing

n = 2 bars being developed on one column face

$K_{tr} = \dfrac{2\,(0.11)\,(60{,}000)}{1500\,(16)\,(2)} = 0.275$ in. $= 0.275\,d_b$

$\left(\dfrac{c_b + K_{tr}}{d_b} \right) = 2.375 + 0.275 = 2.65 > 2.5$ Use 2.5   *12.2.3*

$\ell_d = \dfrac{3\,(60{,}000)\,(1.0)\,(1.0)\,(1.0)\,(1.0)}{40\sqrt{4000}\,(2.5)}\,(1.00) = 28.5$ in.

Class B splice = 1.3(28.5) = 37 in.

Use 37 in. lap splice for the 4 No. 8 bars at the floor level indicated.

Blank

5

Design Methods and Strength Requirements

UPDATE FOR THE '05 CODE

Expressions to determine strength reduction factor ϕ within the transition zone were revised in Figure R.9.3.2. The expressions were modified to resolve the inaccuracy reported from some users of the 2002 code.

In pretensioned members, 9.3.2.7 was revised to allow the linear increase of the ϕ factor from 0.75 to 0.9 for sections in flexural members located between the end of the transfer length and the end of the development length. This revision was introduced to remedy a discontinuity in the calculated flexural strength along the length of prentntioned members.

Spiral transverse reinforcement (10.9.3) has been excluded from the upper limit of 80,000 psi yield strength (9.4). Research shows that 100,000 psi yield strength reinforcement can be used for confinement. This change will help reduce congestion, and allow easier concrete consolidation.

8.1 DESIGN METHODS

Two philosophies of design for reinforced concrete have long been prevalent. Working Stress Design was the principal method used from the early 1900s until the early 1960s. Since publication of the 1963 edition of the ACI code, there has been a rapid transition to Ultimate Strength Design, largely because of its more rational approach. Ultimate strength design, referred to in the code as the Strength Design Method (SDM) is conceptually more realistic in its approach to structural safety and reliability at the strength limit state.

The 1956 ACI code (ACI 318-56) was the first code edition which officially recognized and permitted the ultimate strength method of design. Recommendations for the design of reinforced concrete structures by ultimate strength theories were included in an appendix.

The 1963 ACI code (ACI 318-63) treated the working stress and the ultimate strength methods on an equal basis. However, a major portion of the working stress method was modified to reflect ultimate strength behavior. The working stress provisions of the 1963 code, relating to bond, shear and diagonal tension, and combined axial compression and bending, had their basis in ultimate strength.

The 1971 ACI code (ACI 318-71) was based entirely on "ultimate strength design" for proportioning reinforced concrete members, except for section (8.10) devoted to what was called the Alternate Design Method (ADM). The ADM was not applicable to the design of prestressed concrete members. Even in that section, the service load capacities (except for flexure) were given as various percentages of the ultimate strength capacities of other parts of the code. The transition to ultimate strength methods for reinforced concrete design was essentially complete in the 1971 ACI code, with ultimate strength design definitely established as being preferred.

In the 1977 ACI code (ACI 318-77) the ADM was relegated to Appendix B. The appendix location served to separate and clarify the two methods of design, with the main body of the code devoted exclusively to the SDM. The ADM was retained in all editions of the code from 1977 to the 1999 edition, where it was found in Appendix A. In 2002, the code underwent the most significant revisions since 1963. The ADM method was deleted from the 2002 code (ACI 318-02). It is still referenced in Commentary Section R1.1 of the 2002 code. The general serviceability requirements of the main body of the code, such as the provisions for deflection and crack control, must always be satisfied.

A modification to the SDM, referred to as the Unified Design Provisions, was added to the '95 edition of the code. In keeping with tradition, the method was added as Appendix B. The provisions apply to the design of nonprestressed and prestressed members subject to flexure and axial loads. The Unified Design Provisions were incorporated into the body starting with the 2002 code. See 8.1.2 below.

8.1.1 Strength Design Method

The Strength Design Method requires that the design strength of a member at any section should equal or exceed the required strength calculated by the code-specified factored load combinations. In general,

> Design Strength ≥ Required Strength (U)

where

> Design Strength = Strength Reduction Factor (ϕ) x Nominal Strength
>
> ϕ = Strength reduction factor that accounts for (1) the probability of understrength of a member due to variations in material strengths and dimensions, (2) inaccuracies in the design equations, (3) the degree of ductility and required reliability of the loaded member, and (4) the importance of the member in the structure (see 9.3.2).
>
> Nominal Strength = Strength of a member or cross-section calculated using assumptions and strength equations of the Strength Design Method before application of any strength reduction factors.
>
> Required Strength (U) = Load factors × Service load effects. The required strength is computed in accordance with the load combinations in 9.2.
>
> Load Factor = Overload factor due to probable variation of service loads.
>
> Service Load = Load specified by general building code (unfactored).

Notation

Required strength:

> M_u = factored moment (required flexural strength)
> P_u = factored axial force (required axial load strength) at given eccentricity
> V_u = factored shear force (required shear strength)
> T_u = factored torsional moment (required torsional strength)

Nominal strength:

> M_n = nominal flexural strength
> M_b = nominal flexural moment strength at balanced strain conditions
> P_n = nominal axial strength at given eccentricity

P_o = nominal axial strength at zero eccentricity
P_b = nominal axial strength at balanced strain conditions
V_n = nominal shear strength
V_c = nominal shear strength provided by concrete
V_s = nominal shear strength provided by shear reinforcement
T_n = nominal torsional moment strength

Design Strength:

ϕM_n = design flexural strength
ϕP_n = design axial strength at given eccentricity
ϕV_n = design shear strength = $\phi (V_c + V_s)$
ϕT_n = design torsional moment strength

Section R2.2 gives an in-depth discussion on many of the concepts in the Strength Design Method.

8.1.2 Unified Design Provisions

A modification to the Strength Design Method for nonprestressed and prestressed concrete flexural and compression members was introduced in 1995 in Appendix B. This appendix introduced substantial changes in the design for flexure and axial loads. Reinforcement limits, strength reduction factors ϕ, and moment redistribution were affected.

The Unified Design method is similar to the Strength Design Method in that it uses factored loads and strength reduction factors to proportion the members. The main difference is that in the Unified Design Provisions, a concrete section is defined as either compression-controlled or tension-controlled, depending on the magnitude of the net tensile strain in the reinforcement closest to the tension face of a member. The ϕ factor is then determined by the strain conditions at a section at nominal strength. Prior to these provisions, the ϕ factors were specified for cases of axial load or flexure or both in terms of the type of loading.

It is important to note that the Unified Design Provisions do not alter nominal strength calculations. The major differences occur in checking reinforcement limits for flexural members, determining the ϕ factor for columns with small axial load, and computing redistributed moments. Most other applicable provisions in the body of the 1999 code apply to design using the current code.

The code sections displaced by the Unified Design Provisions are now located in Appendix B. These former provisions are still permitted to be used.

In general, the Unified Design Provisions provide consistent means for designing nonprestressed and prestressed flexural and compression members, and produce results similar to those obtained from the Strength Design Method. The examples in Part 6 and Ref. 5.1 illustrate the use of this new design method.

9.1 STRENGTH AND SERVICEABILITY—GENERAL

9.1.1 Strength Requirements

The basic criterion for strength design as indicated in 9.1.1 is as follows:

Design Strength ≥ Required Strength

Strength Reduction Factor (ϕ) × Nominal Strength ≥ Load Factor × Service Load Effects

All structural members and sections must be proportioned to meet the above criterion under the most critical load combination for all possible actions (flexure, axial load, shear, etc.):

$$\phi P_n \geq P_u$$

$$\phi M_n \geq M_u$$

$$\phi V_n \geq V_u$$

$$\phi T_n \geq T_u$$

The above criterion provides for the margin of structural safety in two ways:

1. It decreases the strength by multiplying the nominal strength with the appropriate strength reduction factor ϕ, which is always less than 1. The nominal strength is computed by the code procedures assuming that the member or the section will have the exact dimensions and material properties assumed in the computations. For example, the nominal flexural strength for the singly reinforced section shown in Fig. 5-1 is:

$$M_n = A_s f_y (d - a/2)$$

and the design flexural moment strength is

$$\phi M_n = \phi [A_s f_y (d - a/2)]$$

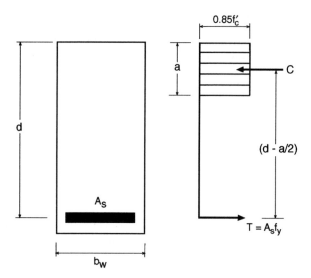

Figure 5-1 Singly Reinforced Section

2. It increases the required strength by using factored loads or the factored internal moments and forces. Factored loads are defined in 2.2 as service loads multiplied by the appropriate load factors. The loads to be used are described in 8.2. Thus, the required flexural strength of the section shown in Fig. 5-1 for dead and live loads is:

$$M_u = 1.2 M_d + 1.6 M_\ell \geq 1.4 M_d \qquad \text{Eqs. (9-1) \& (9-2)}$$

where M_d and M_ℓ are the moments due to service dead and live loads, respectively.

Thus, the design strength requirement for this section becomes:

$$\phi[A_s f_y (d - a/2)] \geq 1.2 M_d + 1.6 M_\ell \geq 1.4 M_d$$

Similarly, for shear acting on the section, the criterion for strength design can be stated as:

$$\phi V_n = \phi(V_c + V_s) \geq V_u$$

$$\phi \left[2\sqrt{f'_c}\, b_w d + \frac{A_v f_y d}{s} \right] \geq 1.2 V_d + 1.6 V_\ell \geq 1.4 V_d$$

The following are the reasons for requiring strength reduction factors and load factors in strength design: [5.2]

1. The strength reduction of materials or elements is required because:

    a. Material strengths may differ from those assumed in design because of:

        - Variability in material strengths—the compression strength of concrete as well as the yield strength and ultimate tensile strength of reinforcement are variable.
        - Effect of testing speed—the strengths of both concrete and steel are affected by the rate of loading.
        - In situ strength vs. specimen strength—the strength of concrete in a structure is somewhat different from the strength of the same concrete in a control specimen.
        - Effect of variability of shrinkage stresses or residual stresses—the variability of the residual stresses due to shrinkage may affect the cracking load of a member, and is significant where cracking is the critical limit state. Similarly, the transfer of compression loading from concrete to steel due to creep and shrinkage in columns may lead to premature yielding of the compression steel, possibly resulting in instability failures of slender columns with small amounts of reinforcement.

    b. Member dimensions may vary from those assumed, due to construction/fabrication tolerances. The following are significant:

        - Formwork tolerances affecting final member dimensions.
        - Rolling and fabrication tolerances in reinforcing bars.
        - Geometric tolerances in cross-section and reinforcement placement tolerances.

    c. Assumptions and simplifications in design equations, such as use of the rectangular stress block and the maximum usable strain of concrete equal to 0.003, introduce both systematic and random inaccuracies.

    d. The use of discrete bar sizes leads to variations in the actual capacity of members. Calculated area of reinforcement has to be rounded up to match the area of an integer number of reinforcing bars.

2. The load factors are required for possible overloading because:

    a. Magnitudes of loads may vary from those determined from building codes. Dead loads may vary because of:

        - Variations in member sizes.
        - Variations in material density.
        - Structural and nonstructural alterations.

        Live loads can vary considerably from time to time and from building to building.

b. Uncertainties exist in the calculation of load effects—the assumptions of stiffnesses, span lengths, etc., and the inaccuracies involved in modeling three-dimensional structures for structural analysis lead to differences between the stresses which actually occur in a building and those estimated in the designer's analysis.

3. Strength reduction and load increase are also required because the consequences of failure may be severe. A number of factors should be considered:

   a. The type of failure, warning of failure, and existence of alternative load paths.
   b. Potential loss of life.
   c. Costs to society in lost time, lost revenue, or indirect loss of life or property due to failure.
   d. The importance of the structural element in the structure.
   e. Cost of replacing the structure.

   By way of background to the numerical values of load factors and strength reduction factors specified in the code, it may be worthwhile reproducing the following paragraph from Ref. 5.2:

   > "The ACI ... design requirements ... are based on an underlying assumption that if the probability of understrength members is roughly 1 in 100 and the probability of overload is roughly 1 in 1000, the probability of overload on an understrength structure is about 1 in 100,000. Load factors were derived to achieve this probability of overload. Based on values of concrete and steel strength corresponding to probability of 1 in 100 of understrength, the strengths of a number of typical sections were computed. The ratio of the strength based on these values to the strength based on nominal strengths of a number of typical sections were arbitrarily adjusted to allow for the consequences of failure and the mode of failure of a particular type of member, and for a number of other sources of variation in strength."

An Appendix to Ref. 5.2 traces the history of development of the current ACI load and strength reduction factors.

9.1.2 Serviceability Requirements

The provisions for adequate strength do not necessarily ensure acceptable behavior of the member at service load levels. Therefore, the code includes additional requirements to provide satisfactory service load performance.

There is not always a clear separation between the provisions for strength and those for serviceability. For actions other than flexure, the detailing provisions in conjunction with the strength requirements are meant to ensure adequate performance at service loads. For flexural action, there are special serviceability requirements concerning short and long term deflections, distribution of reinforcement, crack control, and permissible stresses in prestressed concrete. A consideration of service load deflections is particularly important in view of the extended use of high-strength materials and more accurate methods of design which result in increasingly slender reinforced concrete members.

9.1.3 Appendix C

Starting with thw 2002 code, the load factors and strength reduction factors used in the 1999 and earlier codes were placed in Appendix C. Use of Appendix C is permitted by 9.1.3. However, it is mandatory that both the load combinations and strength reduction factors of Appendix C are used together.

9.2 REQUIRED STRENGTH

As previously stated, the required strength U is expressed in terms of factored loads, or their related internal moments and forces. Factored loads are the service-level loads specified in the general building code, multiplied by appropriate load factors in 9.2. It is important to recognize that earthquake forces computed in accordance with the latest editions of the model buildings codes in use in the U. S. are strength-level forces. Specifically, seismic forces calculated under the 1993 and later editions of *The BOCA National Building Code*, the 1994 and later editions of the *Standard Building Code*, and the 1997 *Uniform Building Code* are strength-level forces. In addition, the 2000 and 2003 *International Building Code* (IBC) developed by the International Code Council have seismic provisions that are strength-level forces.

This development has created confusion within the structural engineering profession since when designing in concrete one must use some load combinations from ACI 318 and others from the governing building code. To assist the structural engineer in understanding the various load combinations and their proper application to design of concrete structural elements governed by one of these codes, a publication was developed by PCA in 1998. *Strength Design Load Combinations for Concrete Elements*[5.3] provides background on the use of the ACI 318 factored load combinations. In addition, it cites the load combinations in the model codes, including the IBC, that must be used for seismic design.

Section 9.2 prescribes load factors for specific combinations of loads. A list of these combinations is shown below. The numerical value of the load factor assigned to each type of load is influenced by the degree of accuracy with which the load can usually be assessed, the variation which may be expected in the load during the lifetime of a structure and the probability of simultaneous occurrence of different load types. Hence, dead loads, because they can usually be more accurately determined and are less variable, are assigned a lower load factor (1.2) as compared to live loads (1.6). Also, weight and pressure of liquids with well-defined densities and controllable maximum heights are assigned a reduced load factor of 1.2 due the lesser probability of overloading. A higher load factor of 1.6 is required for earth and groundwater pressures due to considerable uncertainty of their magnitude and recurrence. Note that while most usual combinations of loads are included, it should not be assumed that all cases are covered. Section 9.2 contains load combination as follows:

$$U = 1.4(D + F) \qquad Eq.\ (9\text{-}1)$$

$$U = 1.2(D + F + T) + 1.6(L + H) + 0.5(L_r \text{ or } S \text{ or } R) \qquad Eq.\ (9\text{-}2)$$

$$U = 1.2D + 1.6(L_r \text{ or } S \text{ or } R) + (1.0L \text{ or } 0.8W) \qquad Eq.\ (9\text{-}3)$$

$$U = 1.2D + 1.6W + 1.0L + 0.5(L_r \text{ or } S \text{ or } R) \qquad Eq.\ (9\text{-}4)$$

$$U = 1.2D + 1.0E + 1.0L + 0.2S \qquad Eq.\ (9\text{-}5)$$

$$U = 0.9D + 1.6W + 1.6H \qquad Eq.\ (9\text{-}6)$$

$$U = 0.9D + 1.0E + 1.6H \qquad Eq.\ (9\text{-}7)$$

where:

- $D =$ dead loads, or related internal moments and forces
- $E =$ load effects of seismic forces, or related internal moments and forces
- $F =$ loads due to weight and pressures of fluids with well-defined densities and controllable maximum heights, or related internal moments and forces
- $H =$ loads due to weight and pressure of soil, water in soil, or other materials, or related internal moments and forces

L =	live loads, or related internal moments and forces
L_r =	roof live load, or related internal moments and forces
R =	rain load, or related internal moments and forces
S =	snow load, or related internal moments and forces
T =	cumulative effect of temperature, creep, shrinkage, differential settlement, and shrinkage-compensating concrete
U =	required strength to resist factored loads or related internal moments and forces
W =	wind load, or related internal moments and forces

Note that in Eqs. (9-1) through (9-7), the effect of one or more loads not acting simultaneously must also be investigated.

Exceptions to the load combination are as follows:

1. The load factor on L in Eq. (9-3), (9-4), and (9-5) shall be permitted to be reduced to 0.5 except for garages, areas occupied as places of public assembly, and all areas where the live load L is greater than 100 lb/ft^2.
2. Where wind load W has not been reduced by a directionality factor, it shall be permitted to use 1.3W in place of 1.6W in Eq. (9-4) and (9-6). Note that the wind load equation in ASCE 7-98 and IBC 2000 includes a factor for wind directionality that is equal to 0.85 for buildings. The corresponding load factor for wind in the load combination equations was increased accordingly (1.3/0.85 = 1.53 rounded up to 1.6). The code allows use of the previous wind load factor of 1.3 when the design wind load is obtained from other sources that do not include the wind directionality factor.
3. Where earthquake load E is based on service-level seismic forces, 1.4E shall be used in place of 1.0E in Eq. (9-5) and (9-7).
4. The load factor on H shall be set equal to zero in Eq. (9-6) and (9-7) if the structural action due to H counteracts that due to W or E. Where lateral earth pressure provides resistance to structural actions from other forces, it shall not be included in H but shall be included in the design resistance.

Other consideration related to load combination are as follows:

1. Resistance to impact effects, where applicable, shall be included with live load (9.2.2).
2. Differential settlement, creep, shrinkage, expansion of shrinkage-compensating concrete, or temperature change shall be based on a realistic assessment of such effects occurring in service (9.2.3).
3. For a structure in a flood zone, the flood load and load combinations of ASCE 7 shall be used (9.2.4).
4. For post-tensioned anchorage zone design, a load factor of 1.2 shall be applied to the maximum prestressing steel jacking force (9.2.5).

For many members, the loads considered are dead, live, wind, and earthquake. Where the F, H, R, S, and T loads are not considered, the seven equations simplify to those given in Table 5-1 below.

Table 5-1 Required Strength for Simplified Load Combinations

Loads	Required Strength	Code Eq. No.
Dead (D) and Live (L)	1.4D	9-1
	1.2D + 1.6L + 0.5L_r	9-2
Dead, Live, and Wind (W)	1.2D + 1.6L_r + 1.0L	9-3
	1.2D + 1.6L_r + 0.8W	9-3
	1.2D + 1.6W + 1.0L + 0.5L_r	9-4
	0.9D + 1.6W	9-6
Dead, Live, and Earthquake (E)	1.2D + 1.0L + 1.0E	9-5
	0.9D + 1.0E	9-7

While considering gravity loads (dead and live), a designer using the code moment coefficients (same coefficients for dead and live loads—8.3.3) has three choices: (1) multiplying the loads by the appropriate load factors, adding them into the total factored load, and then computing the forces and moments due to the total load, (2) computing the effects of factored dead and live loads separately, and then superimposing the effects, or (3) computing the effects of unfactored dead and live loads separately, multiplying the effects by the appropriate load factors, and then superimposing them. Under the principle of superposition, all three procedures yield the same answer. For designers performing a more exact analysis using different coefficients for dead and live loads (pattern loading for live loads), choice (1) does not exist. While considering gravity as well as lateral loads, load effects (due to factored or unfactored loads), of course, have to be computed separately before any superposition can be made.

In determining the required strength for combinations of loads, due regard must be given to the proper sign (positive or negative), since one type of loading may produce effects that either add to or counteract the effect of another load. Even though Eqs. (9-6) and (9-7) have a positive sign preceding the wind (W) or earthquake (E) load, the combinations are to be used when wind or earthquake forces or effects counteract those due to dead loads. When the effects of gravity loads and wind (W) or earthquake (E) loads are additive, Eqs. (9-4), (9-5), and (9-6) must be used.

Consideration must be given to various combinations of loads in determining the most critical design combination. This is of particular importance when strength is dependent on more than one load effect, such as strength under combined moment and axial load, or the shear strength of members carrying axial load.

9.3 DESIGN STRENGTH

9.3.1 Nominal Strength vs. Design Strength

The design strength provided by a member, its connections to other members, and its cross-section, in terms of flexure, axial load, shear, and torsion, is equal to the nominal strength calculated in accordance with the provisions and assumptions stipulated in the code, multiplied by a strength reduction factor ϕ, which is less than unity. The rules for computing the nominal strength are based generally on conservatively chosen limit states of stress, strain, cracking or crushing, and conform to research data for each type of structural action. An understanding of all aspects of the strengths computed for various actions can only be obtained by reviewing the background to the code provisions.

9.3.2 Strength Reduction Factors

The ϕ factors prescribed for structural concrete in 9.3.2 are listed in Table 5-2. The reasons for use of strength reduction factors have been given in earlier sections.

Table 5-2 Strength Reduction Factors ϕ in the Strength Design Method

Tension-controlled sections	0.90

Compression-controlled sections
    Members with spiral reinforcement conforming to 10.9.3
    Other reinforced members
	
	0.70
	0.65

Shear and torsion	0.75
Bearing on concrete (except for post-tensioned anchorage zones)	0.65
Post-tensioned anchorage zones	0.85
Struts, ties, nodal zones and shearing areas in strut-and-tie models (Appendix A)	0.75

Note that a lower ϕ factor is used for compression-controlled (e.g. columns) sections than for tension-controlled (e.g. beams) sections. This is because compression-controlled sections generally have less ductility and are more sensitive to variations in concrete strength. Additionally, the consequences of failure of a column would generally be more severe than those for failure of a beam. Furthermore, columns with spiral reinforcement are assigned a higher ϕ factor than tied columns because the former have greater toughness and ductility.

Tension-controlled sections and compression-controlled sections are defined in 10.3.3. See Part 6 for detailed discussion.

The code permits a linear transition in ϕ between the limits for tension-controlled and compression-controlled sections. For sections in which the net tensile strain in the extreme tension steel at nominal strength is between the limits for compression-controlled and tension-controlled section, ϕ is permitted to be linearly increased from that for compression-controlled sections to 0.90 as the net tensile strain in the extreme tension steel at nominal strength increases from the compression-controlled strain limit to 0.005. This is best illustrated by Figure 5-2.

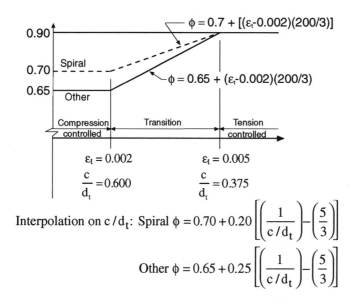

Figure 5-2 Variation of ϕ with Net Tensile ε_t and c/d_t for Grade 60 Reinforcement and for Prestressing Steel

For members subject to flexure and axial load, the design strengths are determined by multiplying both P_n and M_n by the appropriate single value of ϕ.

9.3.3 Development Lengths for Reinforcement

Development lengths for reinforcement, as specified in Chapter 12, do not require a strength reduction modification. Likewise, ϕ factors are not required for splice lengths, since these are expressed in multiples of development lengths.

9.3.5 Structural Plain Concrete

This section specifies that the strength reduction factor $\phi = 0.55$ be used for the nominal strength in flexure, compression, shear, and bearing of plain concrete in Chapter 22 of the code. This is because both the flexural tension strength and the shear strength of plain concrete depend on the tensile strength characteristics of concrete having no reserve strength or ductility in the absence of steel reinforcement.

9.4　DESIGN STRENGTH FOR REINFORCEMENT

An upper limit of 80,000 psi is placed on the yield strength of reinforcing steels other than prestressing steel and spiral transverse reinforcement in 10.9.3. A steel strength above 80,000 psi is not recommended because the yield strain of 80,000 psi steel is about equal to the maximum usable strain of concrete in compression. Currently there is no ASTM specification for Grade 80 reinforcement. However, No. 11, No. 14, and No. 18 deformed reinforcing bars with a yield strength of 75,000 psi (Grade 75) are included in ASTM A615.

In accordance with 3.5.3.2, use of reinforcing bars with a specified yield strength f_y exceeding 60,000 psi requires that f_y be the stress corresponding to a strain of 0.35 percent. ASTM A615 for Grade 75 bars includes the same requirement. The 0.35 percent strain requirement also applies to welded wire reinforcement with wire having a specified yield strength greater than 60,000 psi. Higher-yield-strength wire is available and a value of f_y greater than 60,000 psi can be used in design, provided compliance with the 0.35 percent strain requirement is certified.

There are limitations on the yield strength of reinforcement in other sections of the code:

1. Sections 11.5.2, 11.6.3.4, and 11.7.6: The maximum f_y that may be used in design for shear, combined shear and torsion, and shear friction is 60,000 psi, except that f_y up to 80,000 psi may be used only for shear reinforcement consisting of welded deformed wire reinforcement meeting the requirements of ASTM A497.

2. Sections 19.3.2 and 21.2.5: The maximum specified f_y is 60,000 psi in shells, folded plates and structures governed by the special seismic provisions of Chapter 21.

In addition, the deflection provisions of 9.5 and the limitations on distribution of flexural reinforcement of 10.6 will become increasingly critical as f_y increases.

REFERENCES

5.1　Mast, R. F., "Unified Design Provisions for Reinforced and Prestressed Concrete Flexural and Compression Members," *ACI Structural Journal*, Vol. 89, No. 2, March-April 1992, pp. 185-199.

5.2．MacGregor, J. G., "Safety and Limit States Design for Reinforced Concrete," *Canadian Journal of Civil Engineering*, Vol. 3, No. 4, December 1976, pp. 484-513.

5.3　*Strength Design Load Combinations for Concrete Elements*, Publication IS521, Portland Cement Association, Skokie, Illinois, 1998.

Blank

6

General Principles of Strength Design

UPDATE TO THE '05 CODE

A minor editorial change was made in 2005 in 10.3.5 to clarify that the axial load limit of $0.10 f'_c A_g$ corresponds to the "factored axial load" for nonprestressed flexural members.

GENERAL CONSIDERATIONS

Historically, ultimate strength was the earliest method used in design, since the ultimate load could be measured by test without a knowledge of the magnitude or distribution of internal stresses. Since the early 1900s, experimental and analytical investigations have been conducted to develop ultimate strength design theories that would predict the ultimate load measured by test. Some of the early theories that resulted from the experimental and analytical investigations are reviewed in Fig. 6-1.

Structural concrete and reinforcing steel both behave inelastically as ultimate strength is approached. In theories dealing with the ultimate strength of reinforced concrete, the inelastic behavior of both materials must be considered and must be expressed in mathematical terms. For reinforcing steel with a distinct yield point, the inelastic behavior may be expressed by a bilinear stress-strain relationship (Fig. 6-2). For concrete, the inelastic stress distribution is more difficult to measure experimentally and to express in mathematical terms.

Studies of inelastic concrete stress distribution have resulted in numerous proposed stress distributions as outlined in Fig. 6-1. The development of our present ultimate strength design procedures has its basis in these early experimental and analytical studies. Ultimate strength of reinforced concrete in American design specifications is based primarily on the 1912 and 1932 theories (Fig. 6-1).

INTRODUCTION TO UNIFIED DESIGN PROVISIONS

The Unified Design Provisions introduced in the main body of the code in 2002 do not alter nominal strengths. The nominal strength of a section subject to flexure, axial load, or combinations thereof is the same as it was in previous codes. However, the Unified Design Provisions do alter the calculations of design strengths, which are reduced from nominal strengths by the strength reduction factor ϕ.

The following definitions are related to the Unified Design Provisions, and are given in Chapter 2 of the code. These definitions are briefly explained here, with further detailed discussion under the relevant code sections.

1. Net tensile strain: The tensile strain at nominal strength exclusive of strains due to effective prestress, creep, shrinkage, and temperature. The phrase "at nominal strength" in the definition means at the time the concrete in compression reaches its assumed strain limit of 0.003 (10.2.3). The "net tensile strain" is the strain caused by bending moments and axial loads, exclusive of strain caused by prestressing and by volume changes. The net tensile strain is that normally calculated in nominal strength calculations.

2. Extreme tension steel: The reinforcement (prestressed or nonprestressed) that is the farthest from the extreme compression fiber. The symbol d_t denotes the depth from the extreme compression fiber to the extreme tensile steel. The net tensile strain in the extreme tension steel is simply the maximum tensile steel strain due to external loads.

3. Compression-controlled strain limit: The net tensile strain at balanced strain conditions; see 10.3.2. The definition of balanced strain conditions in 10.3.2 is unchanged from previous editions of the code. Thus, the concrete reaches a strain of 0.003 as the tension steel reaches yield strain. However, 10.3.3 permits the compression-controlled strain limit for Grade 60 reinforcement and for prestressed reinforcement to be set equal to 0.002.

4. Compression-controlled section: A cross-section in which the net tensile strain in the extreme tension steel at nominal strength is less than or equal to the compression-controlled strain limit. The strength reduction factor ϕ for compression-controlled sections is set at 0.65 or 0.7 in 9.3.2.2.

5. Tension-controlled section: A cross-section in which the net tensile strain in the extreme tension steel at nominal strength is greater than or equal to 0.005. The strength reduction factor ϕ for tension-controlled sections is set at 0.9 in 9.3.2.1. However, ACI 318-99 and earlier editions of the code permitted a ϕ of 0.9 to be used for flexural members with reinforcement ratios not exceeding 0.75 of the balanced reinforcement ration ρ_b. For rectangutions, with one layer of tension reinforcement, $0.75\rho_b$ corresponds to a net tensile strain ε_t of 0.00376. The use of ϕ of 0.9 is now permitted only for less heavily reinforced sections with $\varepsilon_t \geq 0.005$.

The use of these definitions is described under 8.4, 9.2, 10.3, and 18.8.

10.2   DESIGN ASSUMPTIONS

10.2.1   Equilibrium of Forces and Compatibility of Strains

Computation of the strength of a member or cross-section by the Strength Design Method requires that two basic conditions be satisfied: (1) static equilibrium and (2) compatibility of strains.

The first condition requires that the compressive and tensile forces acting on the cross-section at "ultimate" strength be in equilibrium, and the second condition requires that compatibility between the strains in the concrete and the reinforcement at "ultimate" conditions must also be satisfied within the design assumptions permitted by the code (see 10.2).

The term "ultimate" is used frequently in reference to the Strength Design Method; however, it should be realized that the "nominal" strength computed under the provisions of the code may not necessarily be the actual ultimate value. Within the design assumptions permitted, certain properties of the materials are neglected and other conservative limits are established for practical design. These contribute to a possible lower "ultimate strength" than that obtained by test. The computed nominal strength should be considered a code-defined strength only. Accordingly, the term "ultimate" is not used when defining the computed strength of a member. The term "nominal" strength is used instead.

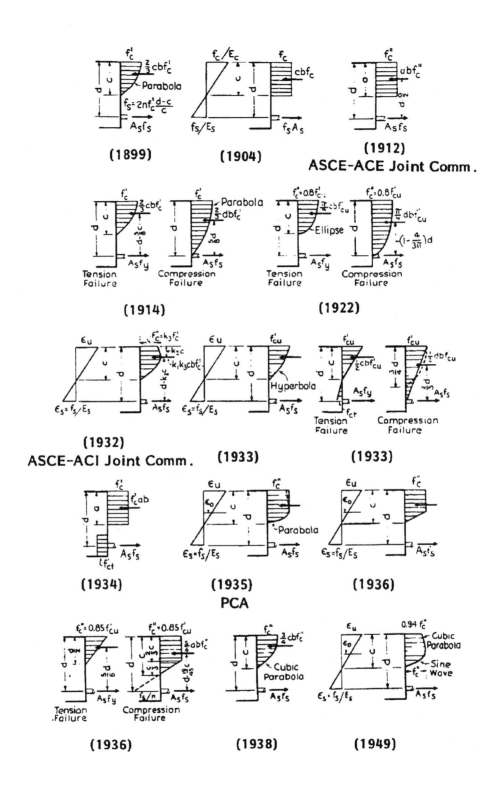

Figure 6-1 Development of Ultimate Strength Theories of Flexure

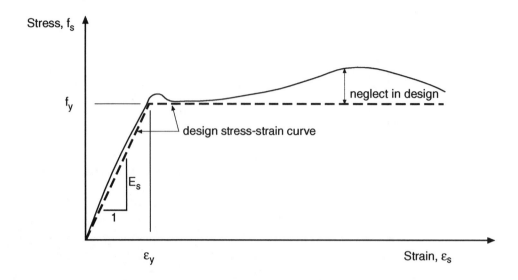

Figure 6-2 Stress-Strain Relationship for Reinforcement

Furthermore, in discussing the strength method of design for reinforced concrete structures, attention must be called to the difference between loads on the structure as a whole and load effects on the cross-sections of individual members. Elastic methods of structural analysis are used first to compute service load effects on the individual members due to the action of service loads on the entire structure. Only then are the load factors applied to the service load effects acting on the individual cross-sections. Inelastic (or limit) methods of structural analysis, in which design load effects on the individual members are determined directly from the ultimate test loads acting on the whole structure, are not considered. Section 8.4, however, does permit a limited redistribution of negative moments in continuous members. The provisions of 8.4 recognize the inelastic behavior of concrete structures and constitute a move toward "limit design." This subject is presented in Part 8.

The computed "nominal strength" of a member must satisfy the design assumptions given in 10.2.

10.2.2   Design Assumption #1

Strain in reinforcement and concrete shall be assumed directly proportional to the distance from the neutral axis.

In other words, plane sections normal to the axis of bending are assumed to remain plane after bending. Many tests have confirmed that the distribution of strain is essentially linear across a reinforced concrete cross-section, even near ultimate strength. This assumption has been verified by numerous tests to failure of eccentrically loaded compression members and members subjected to bending only.

The assumed strain conditions at ultimate strength of a rectangular and circular section are illustrated in Fig. 6-3. Both the strain in the reinforcement and in the concrete are directly proportional to the distance from the neutral axis. This assumption is valid over the full range of loading—zero to ultimate. As shown in Fig. 6-3, this assumption is of primary importance in design for determining the strain (and the corresponding stress) in the reinforcement.

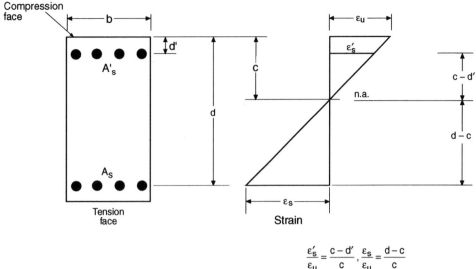

$$\frac{\varepsilon'_s}{\varepsilon_u} = \frac{c-d'}{c}, \frac{\varepsilon_s}{\varepsilon_u} = \frac{d-c}{c}$$

(a) Flexure

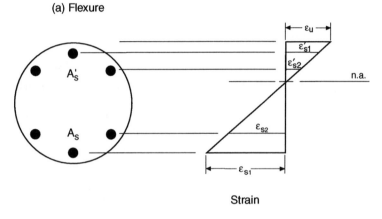

(b) Flexure and Axial Load

Figure 6-3 Assumed Strain Variation

10.2.3 Design Assumption #2

Maximum usable strain at extreme concrete compression fiber shall be assumed equal to $\varepsilon_u = 0.003$.

The maximum concrete compressive strain at crushing of the concrete has been measured in many tests of both plain and reinforced concrete members. The test results from a series of reinforced concrete beam and column specimens, shown in Fig. 6-4, indicate that the maximum concrete compressive strain varies from 0.003 to as high as 0.008. However, the maximum strain for practical cases is 0.003 to 0.004; see stress-strain curves in Fig. 6-5. Though the maximum strain decreases with increasing compressive strength of concrete, the 0.003 value allowed for design is reasonably conservative. The codes of some countries specify a value of 0.0035 for design, which makes little difference in the computed strength of a member.

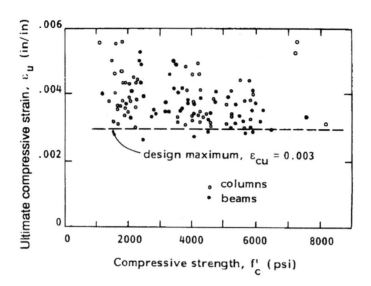

Figure 6-4 Maximum Concrete Compressive Strain, ε_u from Tests of Reinforced Concrete Members

10.2.4 Design Assumption #3

Stress in reinforcement f_s below the yield strength f_y shall be taken as E_s times the steel strain ε_s. For strains greater than f_y/E_s, stress in reinforcement shall be considered independent of strain and equal to f_y.

For deformed reinforcement, it is reasonably accurate to assume that below the yield stress, the stress in the reinforcement is proportional to strain. For practical design, the increase in strength due to the effect of strain hardening of the reinforcement is neglected for strength computations; see actual vs. design stress-strain relationship of steel in Fig. 6-2.

The force developed in the tensile or compressive reinforcement is a function of the strain in the reinforcement ε_s, such that:

when $\varepsilon_s \leq \varepsilon_y$ (yield strain):

$f_s = E_s \varepsilon_s$

$A_s f_s = A_s E_s \varepsilon_s$

when $\varepsilon_s \geq \varepsilon_y$:

$f_s = E_s \varepsilon_y = f_y$

$A_s f_s = A_s f_y$

where ε_s is the value from the strain diagram at the location of the reinforcement; see Fig. 6-3. For design, the modulus of elasticity of steel reinforcement, E_s, is taken as 29,000,000 psi (see 8.5.2).

10.2.5 Design Assumption #4

Tensile strength of concrete shall be neglected in flexural calculations of reinforced concrete.

The tensile strength of concrete in flexure, known as the modulus of rupture, is a more variable property than the compressive strength, and is about 8% to 12% of the compressive strength. The generally accepted value for design is $7.5\sqrt{f'_c}$ (9.5.2.3) for normal-weight concrete. This tensile strength in flexure is neglected in strength design. For practical percentages of reinforcement, the resulting computed strengths are in good agreement with test results. For very small percentages of reinforcement, neglecting the tensile strength of concrete is conservative. It should be realized, however, that the strength of concrete in tension is important in cracking and deflection (serviceability) considerations.

10.2.6 Design Assumption #5

Relationship between concrete compressive stress distribution and concrete strain shall be assumed to be rectangular, trapezoidal, parabolic, or any other shape that results in prediction of strength in substantial agreement with results of comprehensive tests.

This assumption recognizes the inelastic stress distribution in concrete at high stresses. As maximum stress is approached, the stress-strain relationship of concrete is not a straight line (stress is not proportional to strain). The general stress-strain behavior of concrete is shown in Fig. 6-5. The shape of the curves is primarily a function of concrete strength and consists of a rising curve from zero stress to a maximum at a compressive strain between 0.0015 and 0.002, followed by a descending curve to an ultimate strain (corresponding to crushing of the concrete) varying from 0.003 to as high as 0.008. As discussed under Design Assumption #2, the code sets the maximum usable strain at 0.003 for design. The curves show that the stress-strain behavior for concrete becomes notably nonlinear at stress levels exceeding $0.5 f'_c$.

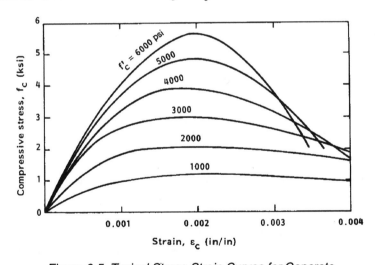

Figure 6-5 Typical Stress-Strain Curves for Concrete

The actual distribution of concrete compressive stress in a practical case is complex and usually not known. However, research has shown that the important properties of the concrete stress distribution can be approximated closely using any one of several different forms of stress distributions (see Fig. 6-1). The three most common stress distributions are the parabolic, the trapezoidal, and the rectangular, each giving reasonable results. At the theoretical ultimate strength of a member in flexure (nominal strength), the compressive stress distribution should conform closely to the actual variation of stress, as shown in Fig. 6-6. In this figure, the maximum stress is indicated by $k_3 f'_c$, the average stress is indicated by $k_1 k_3 f'_c$, and the depth of the centroid of the approximate parabolic distribution from the extreme compression fiber by $k_2 c$, where c is the neutral axis depth.

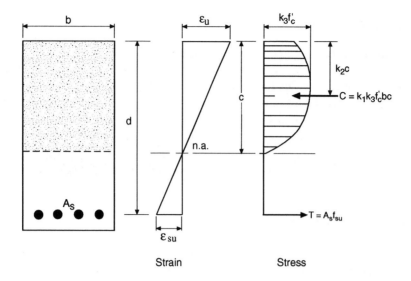

Figure 6-6 Actual Stress-Strain Conditions at Nominal Strength in Flexure

For the stress conditions at ultimate, the nominal moment strength, M_n, may be computed by equilibrium of forces and moments in the following manner:

From force equilibrium (Fig. 6-6):

$$C = T$$

or, $\quad k_1 k_3 f'_c bc = A_s f_{su}$

so that $\quad c = \dfrac{A_s f_{su}}{k_1 k_3 f'_c b}$

From moment equilibrium:

$$M_n = (C \text{ or } T)(d - k_2 c) = A_s f_{su}\left(d - \frac{k_2}{k_1 k_3} \frac{A_s f_{su}}{f'_c b}\right) \quad (1)$$

The maximum strength is assumed to be reached when the strain in the extreme compression fiber is equal to the crushing strain of the concrete, ε_u. When crushing occurs, the strain in the tension reinforcement, ε_{su}, may be either larger or smaller than the yield strain, $\varepsilon_y = f_y/E_s$, depending on the relative proportion of reinforcement to concrete. If the reinforcement amount is low enough, yielding of the steel will occur prior to crushing of the concrete (ductile failure condition). With a very large quantity of reinforcement, crushing of the concrete will occur first, allowing the steel to remain elastic (brittle failure condition). The code has provisions which are intended to ensure a ductile mode of failure by limiting the amount of tension reinforcement. For the ductile failure condition, f_{su} equals f_y, and Eq. (1) becomes:

$$M_n = A_s f_y \left(d - \frac{k_2}{k_1 k_3} \frac{A_s f_y}{f'_c b}\right) \quad (2)$$

If the quantity $k_2/(k_1 k_3)$ is known, the moment strength can be computed directly from Eq. (2). It is not necessary to know the values of k_1, k_2, and k_3 individually. Values for the combined term, as well as the individual k_1 and

k_2 values, have been established from tests and are shown in Fig. 6-7. As shown in the figure, $k_2/(k_1k_3)$ varies from about 0.55 to 0.63. Computation of the flexural strength based on the approximate parabolic stress distribution of Fig. 6-6 may be done using Eq. (2) with given values of $k_2/(k_1k_3)$. However, for practical design purposes, a method based on simple static equilibrium is desirable.

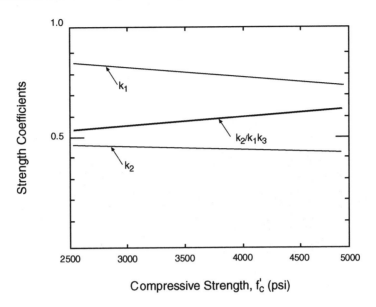

Figure 6-7 Stress-Block Parameters

During the last century, the Portland Cement Association adopted the parabolic stress-strain relationship shown in Fig. 6-8 for much of its experimental and analytical research work. "More exact" stress distributions such as this one have their greatest application with computers and are not recommended for longhand calculations. Recent PCA publications and computer software related to structural concrete design are based entirely on the rectangular stress block.

10.2.7   Design Assumption #6

Requirements of 10.2.6 may be considered satisfied by an equivalent rectangular concrete stress distribution defined as follows: A concrete stress of $0.85 f'_c$ shall be assumed uniformly distributed over an equivalent compression zone bounded by edges of the cross-section and a straight line located parallel to the neutral axis at a distance $a = \beta_1 c$ from the fiber of maximum compressive strain. Distance c from the fiber of maximum compressive strain to the neutral axis shall be measured in a direction perpendicular to that axis. Fraction β_1 shall be taken as 0.85 for strengths f'_c up to 4000 psi and shall be reduced continuously at a rate of 0.05 for each 1000 psi of strength in excess of 4000 psi, but β_1 shall not be taken less than 0.65.

The code allows the use of a rectangular compressive stress block to replace the more exact stress distributions of Fig. 6-6 (or Fig. 6-8). The equivalent rectangular stress block, shown in Fig. 6-9, assumes a uniform stress of $0.85 f'_c$ over a depth $a = \beta_1 c$, determined so that $a/2 = k_2 c$. The constant β_1 is equal to 0.85 for concrete with $f'_c \leq 4000$ psi and reduces by 0.05 for each additional 1000 psi of f'_c in excess of 4000 psi. For high-strength concretes, above 8000 psi, a lower limit of 0.65 is placed on the β_1 factor. Variation in β_1 vs. concrete strength f'_c is shown in Fig. 6-10.

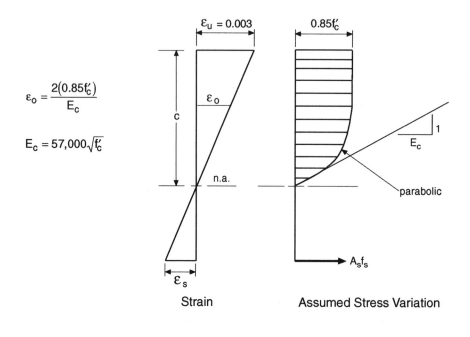

Figure 6-8 PCA Stress-Strain Relationship

The need for a β_1 factor is caused by the variation in shape of the stress-strain curve for different concrete strengths, as shown in Fig. 6-5. For concrete strengths up to 4,000 psi, the shape and centroid of the actual concrete stress block can reasonably be approximated by a rectangular stress block with a uniform stress of 0.85 f'_c and a depth of 0.85 times the depth to the neutral axis. That is to say, with a β_1 of 0.85.

Higher strength concretes have a more linear shape, with less inelastic behavior. For a good approximation of the stress block for concretes with strengths above 4,000 psi, the ratio β_1 of rectangular stress block depth to neutral axis depth needs to be reduced. Thus, the 1963 code required that β_1 "shall be reduced continuously at a rate of 0.05 for each 1,000 psi of strength in excess of 4,000 psi.

As time went by and much higher concrete strengths came into use, it was realized that this reduction in β_1 should not go on indefinitely. Very high strengths have a stress block that approaches a triangular shape. This almost-triangular stress block is best approximated by a rectangular stress block with $\beta_1 = 0.65$. Thus, in the 1977 and later codes, β_1 was set at 0.65 for concrete strengths of 8,000 psi and above.

Using the equivalent rectangular stress distribution (Fig. 6-9), and assuming that the reinforcement yields prior to crushing of the concrete ($\varepsilon_s > \varepsilon_y$), the nominal moment strength M_n may be computed by equilibrium of forces and moments.

From force equilibrium:

$$C = T$$

or, $\quad 0.85 f'_c ba = A_s f_y$

so that $\quad a = \dfrac{A_s f_y}{0.85 f'_c b}$

From moment equilibrium:

$$M_n = (C \text{ or } T)\left(d - \frac{a}{2}\right) = A_s f_y \left(d - \frac{a}{2}\right)$$

Substituting a from force equilibrium,

$$M_n = A_s f_y \left(d - 0.59 \frac{A_s f_y}{f'_c b}\right) \qquad (3)$$

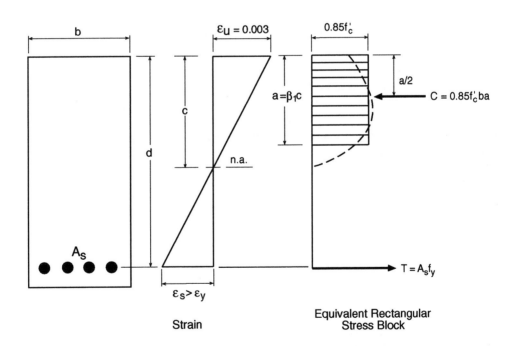

Figure 6-9 Equivalent Rectangular Concrete Stress Distribution (ACI)

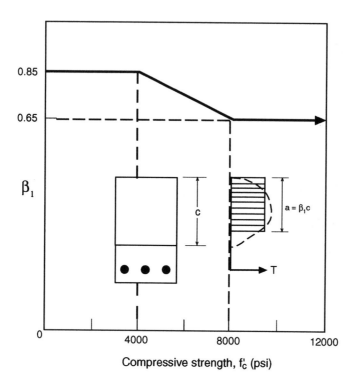

Figure 6-10 Strength Factor β_1

Note that the 0.59 value corresponds to $k_2/(k_1 k_3)$ of Eq. (2). Substituting $A_s = \rho bd$, Eq. (3) may be written in the following nondimensional form:

$$\text{let } \omega = \rho \frac{f_y}{f'_c}$$

$$\frac{M_n}{bd^2 f'_c} = \rho \frac{f_y}{f'_c}\left(1 - 0.59\, \rho\, \frac{f_y}{f'_c}\right) \tag{4}$$

$$= \omega(1 - 0.59\omega)$$

As shown in Fig. 6-11, Eq. (4) is "in substantial agreement with the results of comprehensive tests." However, it must be realized that the rectangular stress block does not represent the actual stress distribution in the compression zone at ultimate, but does provide essentially the same strength results as those obtained in tests. Computation of moment strength using the equivalent rectangular stress distribution and static equilibrium is illustrated in Example 6.1.

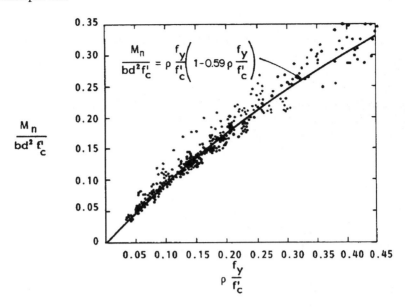

Figure 6-11 Tests of 364 Beams Controlled by Tension ($\varepsilon_s > \varepsilon_y$)

10.3 GENERAL PRINCIPLES AND REQUIREMENTS

10.3.1 Nominal Flexural Strength

Nominal strength of a member or cross-section subject to flexure (or to combined flexure and axial load) must be based on equilibrium and strain compatibility using the design assumptions of 10.2. Nominal strength of a cross-section of any shape, containing any amount and arrangement of reinforcement, is computed by applying the force and moment equilibrium and strain compatibility conditions in a manner similar to that used to develop the nominal moment strength of the rectangular section with tension reinforcement only, as illustrated in Fig. 6-9. Using the equivalent rectangular concrete stress distribution, expressions for nominal moment strength of rectangular and flanged sections (typical sections used in concrete construction) are summarized as follows:

a. Rectangular section with tension reinforcement only (see Fig. 6-9):

Expressions are given above under Design Assumption #6 (10.2.7).

b. Flanged section with tension reinforcement only:

When the compression flange thickness is equal to or greater than the depth of the equivalent rectangular stress block a, moment strength M_n is calculated by Eq. (3), just as for a rectangular section with width equal to the flange width. When the compression flange thickness h_f is less than a, the nominal moment strength M_n is (see Fig. 6-12):

$$M_n = (A_s - A_{sf}) f_y \left(d - \frac{a}{2}\right) + A_{sf} f_y \left(d - \frac{h_f}{2}\right) \tag{5}$$

where

A_{sf} = area of reinforcement required to equilibrate compressive strength of overhanging flanges

= $0.85 f'_c (b - b_w) h_f / f_y$

$a = (A_s - A_{sf}) f_y / 0.85 f'_c b_w$

b = width of effective flange (see 8.10)

b_w = width of web

h_f = thickness of flange

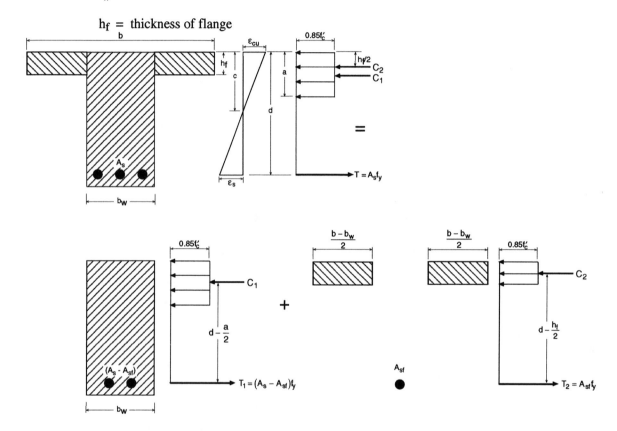

Figure 6-12 Strain and Equivalent Stress Distribution for Flanged Section

c. Rectangular section with compression reinforcement:

For a doubly reinforced section with compression reinforcement A'_s, two possible situations can occur (see Fig. 6-13):

i. Compression reinforcement A'_s yields:

$f'_s = f_y$

$$a = \frac{(A_s - A'_s) f_y}{0.85 f'_c b} \tag{6}$$

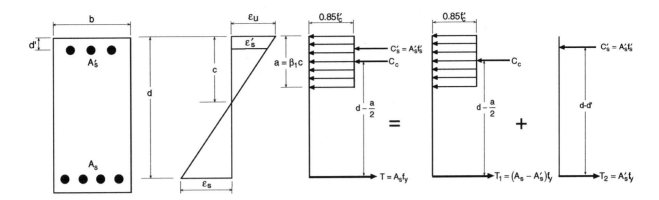

Figure 6-13 Strain and Equivalent Stress Distribution of Doubly Reinforced Rectangular Section

The nominal moment strength is:

$$M_n = (A_s - A'_s) f_y (d - \frac{a}{2}) + A'_s f_y (d - d') \quad (7)$$

Note that A'_s yields when the following (for Grade 60 reinforcement, with $\varepsilon_y = 0.00207$) is satisfied:

$$d'/c \leq 0.31$$

where $c = \dfrac{a}{\beta_1}$

ii. Compression reinforcement does not yield:

$$f'_s = E_s \varepsilon'_s = E_s \varepsilon_u \left(\frac{c - d'}{c}\right) < f_y \quad (8)$$

The neutral axis depth c can be determined from the following quadratic equation:

$$c^2 - \frac{(A_s f_y - 87 A'_s) c}{0.85 \beta_1 f'_c b} - \frac{87 A'_s d'}{0.85 \beta_1 f'_c b} = 0$$

where f'_c and f_y have the units of ksi. The nominal moment strength is:

$$M_n = 0.85 f'_c a b (d - \frac{a}{2}) + A'_s f'_s (d - d') \quad (9)$$

where

$$a = \beta_1 c$$

Alternatively, the contribution of compression reinforcement may be neglected and the moment strength calculated by Eq. (3), just as for a rectangular section with tension reinforcement only.

d. For other cross-sections, the nominal moment strength M_n is calculated by a general analysis based on equilibrium and strain compatibility using the design assumptions of 10.2.

e. Nominal flexural strength M_n of a cross-section of a composite flexural member consisting of cast-in-place and precast concrete is computed in a manner similar as that for a regular reinforced concrete section. Since the "ultimate" strength is unrelated to the sequence of loading, no distinction is made between shored and unshored members in strength computations (see 17.2.4).

10.3.2   Balanced Strain Condition

A balanced strain condition exists at a cross-section when the maximum strain at the extreme compression fiber just reaches $\varepsilon_u = 0.003$ simultaneously with the first yield strain of $\varepsilon_s = \varepsilon_y = f_y/E_s$ in the tension reinforcement. This balanced strain condition is shown in Fig. 6-14.

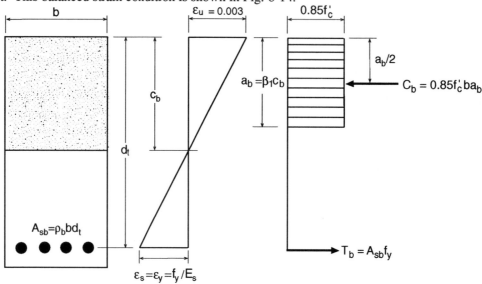

Figure 6-14 Balanced Strain Condition in Flexure

The ratio of neutral axis depth c_b to extreme depth d_t to produce a balanced strain condition in a section with tension reinforcement only may be obtained by applying strain compatibility conditions. Referring to Fig. 6-14, for the linear strain condition:

$$\frac{c_b}{d_t} = \frac{\varepsilon_u}{\varepsilon_u + \varepsilon_y}$$

$$= \frac{0.003}{0.003 + f_y/29,000,000} = \frac{0.003}{0.003 + \varepsilon_y}$$

Note that for Grade 60 steel, 10.3.3 permits the steel strain ε_y to be rounded to 0.002. Substituting into the above equation, the ratio $c_b/d_t = 0.6$. This value applies to all sections with Grade 60 steel, not just to rectangular sections.

10.3.3   Compression-Controlled Sections

Sections are compression-controlled when the net tensile strain in the extreme tension steel is equal to or less than the compression-controlled strain limit at the time the concrete in compression reaches its assumed strain limit of 0.003. The compression-controlled strain limit is the net tensile strain in the reinforcement at balanced strain conditions. For Grade 60 reinforcement, and for all prestressed reinforcement, it is permitted to set the compression-controlled strain limit equal to 0.002.

Note that when other grades of reinforcement are used, the compression-controlled strain limit is not 0.002. This changes the compression-controlled strain limit, and that changes the "transition" equations for the strength reduction factor given in Fig. 5-2 in Part 5.

10.3.4 Tension-Controlled Sections and Transition

Sections are tension-controlled when the net tensile strain in the extreme tension steel is equal to or greater than 0.005 just as the concrete in compression reaches its assumed strain limit of 0.003. Sections with net tensile strain in the extreme tension steel between the compression-controlled strain limit and 0.005 constitute a transition region between compression-controlled and tension-controlled sections.

Figure 6-15 shows the stress and strain conditions at the limit for tension-controlled sections. This limit is important because it is the limit for the use of $\phi = 0.9$ (9.3.2.1). Critical parameters at this limit are given a subscript t. Referring to Fig. 6-15, by similar triangles:

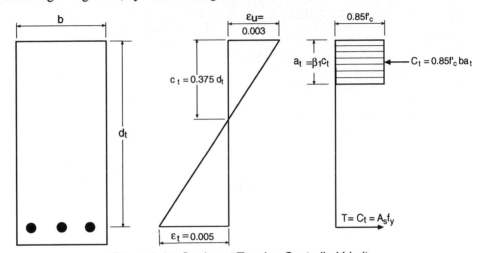

Figure 6-15 Strains at Tension-Controlled Limit

$c_t = 0.375 d_t$

$a_t = \beta_1 c_t = 0.375 \beta_1 d_t$

$C_t = 0.85 f'_c b a_t = 0.319 \beta_1 f'_c b d_t$

$T = A_s f_y = C_t$

$A_s = 0.319 \beta_1 f'_c b d_t / f_y$

$$\rho_t = A_s / (b d_t) = 0.319 \beta_1 f'_c / f_y \tag{10}$$

$$\omega_t = \frac{\rho_t f_y}{f'_c} = 0.319 \beta_1 \tag{11}$$

$$M_{nt} = \omega_t (1 - 0.59 \omega_t) f'_c b d_t^2 \qquad \text{from Eq. (4)}$$

$$R_{nt} = \frac{M_{nt}}{b d_t^2} = \omega_t (1 - 0.59 \omega_t) f'_c \tag{12}$$

Values for ρ_t, ω_t, and R_{nt} are given in Table 6-1.

Table 6-1 Design Parameters at Strain Limit of 0.005 for Tension-Controlled Sections

		$f'_c = 3000$ $\beta_1 = 0.85$	$f'_c = 4000$ $\beta_1 = 0.85$	$f'_c = 5000$ $\beta_1 = 0.80$	$f'_c = 6000$ $\beta_1 = 0.75$	$f'_c = 8000$ $\beta_1 = 0.65$	$f'_c = 10{,}000$ $\beta_1 = 0.65$
R_{nt}		683	911	1084	1233	1455	1819
ϕR_{nt}		615	820	975	1109	1310	1637
ω_t		0.2709	0.2709	0.2550	0.2391	0.2072	0.2072
ρ_t Grade	40	0.02032	0.02709	0.03187	0.03586	0.04144	0.05180
	60	0.01355	0.01806	0.02125	0.02391	0.02762	0.03453
	75	0.01084	0.01445	0.01700	0.01912	0.02210	0.02762

10.3.5 Maximum Reinforcement for Flexural Members

Since 2002, the body of the code defines reinforcement limits in terms of net tensile strain, ε_t, instead of the balanced ratio ρ/ρ_b that was used formerly. For rectangular sections with one layer of Grade 60 steel, a simple relationship between ε_t and ρ/ρ_b exists (see Fig. 6-16):

$$c = \frac{0.003 d_t}{\varepsilon_t + 0.003}$$

$$a = \beta_1 c = \frac{0.003 \beta_1 d_t}{\varepsilon_t + 0.003}$$

At balanced:

$$a_b = \frac{0.003 \beta_1 d_t}{(60/29{,}000) + 0.003} = 0.592\,\beta_1 d_t$$

$$\frac{\rho}{\rho_b} = \frac{a}{a_b} = \frac{0.00507}{\varepsilon_t + 0.003}$$

or,

$$\varepsilon_t = \frac{0.00507}{\rho/\rho_b} - 0.003$$

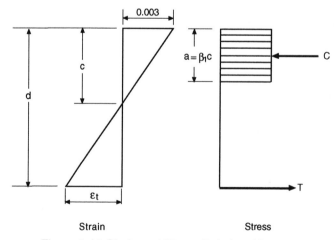

Figure 6-16 Strain and Stress Relationship

This relationship is shown graphically in Fig. 6-17.

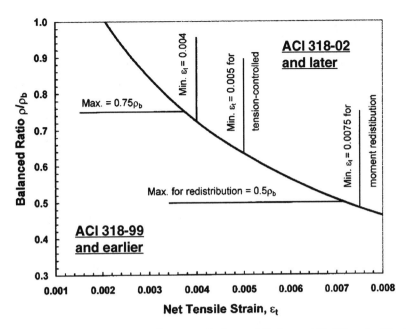

Figure 6-17 Relationship Between Balanced Ratio and Net Tensile Strain

Since 2002, the code limits the maximum reinforcement in a flexural member (with axial load less than $0.1\,f'_c A_g$) to that which would result in a net tensile strain ε_t at nominal strength not less than 0.004. This compares to the former code limit of $0.75\,\rho_b$, which results in an ε_t of 0.00376. Furthermore, at the net tensile strain limit of 0.004, the ϕ factor is reduced to 0.812. For heavily reinforced members, the overall safety margin (load factor/ϕ) is about the same as by 318-99, despite the reduced load factors. See Fig. 6-18.

The strength of tension-controlled sections is clearly controlled by steel strength, which is less variable than concrete strength and this offers greater reliability. For tension-controlled flexural members, since 2002, the ACI code permits a ϕ of 0.9 to be used, despite the reduced load factors introduced in 2002. As Fig. 6-18 shows, the new code reduces the strength requirement by about 10 percent for tension-controlled sections.

As discussed in Part 7, it is almost always advantageous to limit the net tensile strain in flexural members to a minimum of 0.005, even though the code permits higher amounts of reinforcement producing lower net tensile strains. Where member size is limited and extra strength is needed, it is best to use compression reinforcement to limit the net tensile strain so that the section is tension-controlled.

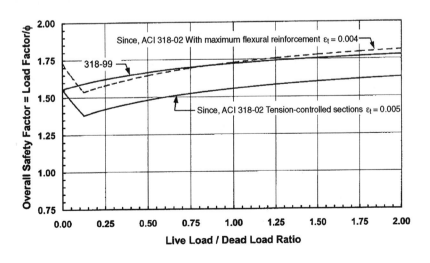

Figure 6-18 Overall Safety Factor for Flexural Members

10.3.6 Maximum Axial Strength

The strength of a member in pure compression (zero eccentricity) is computed by:

$$P_o = 0.85 f'_c A_g + f_y A_{st}$$

where A_{st} is the total area of reinforcement and A_g is the gross area of the concrete section. Refinement in concrete area can be considered by subtracting the area of concrete displaced by the steel:

$$P_o = 0.85 f'_c (A_g - A_{st}) + f_y A_{st} \qquad (13)$$

Pure compression strength P_o represents a hypothetical loading condition. Prior to the 1977 ACI code, all compression members were required to be designed for a minimum eccentricity of 0.05h for spirally reinforced members or 0.10h for tied reinforced members (h = overall thickness of member). The specified minimum eccentricities were originally intended to serve as a means of reducing the axial design load strength of a section in pure compression and were included to: (1) account for accidental eccentricities, not considered in the analysis, that may exist in a compression member, and (2) recognize that concrete strength is less than f'_c at sustained high loads.

Since the primary purpose of the minimum eccentricity requirement was to limit the axial strength for design of compression members with small or zero computed end moments, the 1977 code was revised to accomplish this directly by limiting the axial strength to 85% and 80% of the axial strength at zero eccentricity (P_o), for spiral and tied reinforcement columns, respectively.

For spirally reinforced members,

$$P_{n(max)} = 0.85 P_o = 0.85 [0.85 f'_c (A_g - A_{st}) + f_y A_{st}] \qquad (14)$$

For tied reinforced members,

$$P_{n(max)} = 0.80 P_o = 0.80 [0.85 f'_c (A_g - A_{st}) + f_y A_{st}] \qquad (15)$$

The maximum axial strength, $P_{n(max)}$, is illustrated in Fig. 6-19. In essence, design within the cross-hatched portion of the load-moment interaction diagram is not permitted. The 85% and 80% values approximate the axial strengths at e/h ratios of 0.05 and 0.10 specified in the 1971 code for spirally reinforced and tied reinforced members, respectively (see Example 6.3). The designer should note that R10.3.6 and R10.3.7 state that "Design aids and computer programs based on the minimum eccentricity requirement of the 1963 and 1971 ACI Building Codes may be considered equally applicable for usage."

The current provisions for maximum axial strength also eliminate the concerns expressed by engineers about the excessively high minimum design moments required for large column sections, and the often asked question as to whether the minimum moments were required to be transferred to other interconnecting members (beams, footings, etc.).

Note that a minimum moment (minimum eccentricity requirement) for slender compression members in a braced frame is given in 10.12.3.2. If factored column moments are very small or zero, the design of these columns must be based on the minimum moment $P_u (0.6 + 0.03h)$.

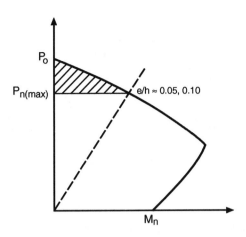

Figure 6-19 Maximum Axial Strength (10.3.6)

10.3.7 Nominal Strength for Combined Flexure and Axial Load

The strength of a member or cross-section subject to combined flexure and axial load, M_n and P_n, must satisfy the same two conditions as required for a member subject to flexure only: (1) static equilibrium and (2) compatibility of strains. Equilibrium between the compressive and tensile forces includes the axial load P_n acting on the cross-section. The general condition of the stress and the strain in concrete and steel at nominal strength of a member under combined flexure and axial compression is shown in Fig. 6-20. The tensile or compressive force developed in the reinforcement is determined from the strain condition at the location of the reinforcement.

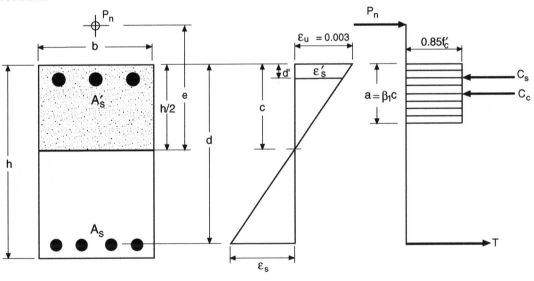

$\varepsilon_s \leq \varepsilon_y$ (Compression Controls)

$\varepsilon_s = \varepsilon_y$ (Balanced Condition)

$0.005 > \varepsilon_s > \varepsilon_y$ (Transition)

$\varepsilon_s \geq 0.005\varepsilon_y$ (Tension Controls)

Figure 6-20 Strain and Equivalent Stress Distribution for Section Subject to Combined Flexure and Axial Load

Referring to Fig. 6-20,

$$T = A_s f_s = A_s (E_s \varepsilon_s) \quad \text{when } \varepsilon_s < \varepsilon_y$$

or $\quad T = A_s f_y \quad$ when $\varepsilon_s \geq \varepsilon_y$

$$C_s = A_s' f_s' = A_s' (E_s \varepsilon_s') \quad \text{when } \varepsilon_s' < \varepsilon_y$$

or $\quad C_s = A_s' f_y \quad$ when $\varepsilon_s' \geq \varepsilon_y$

$$C_c = 0.85 f_c' b a$$

The combined load-moment strength (P_n and M_n) may be computed by equilibrium of forces and moments.

From force equilibrium:

$$P_n = C_c + C_s - T \qquad (16)$$

From moment equilibrium about the mid-depth of the section:

$$M_n = P_n e = C_c \left(\frac{h}{2} - \frac{a}{2}\right) + C_s \left(\frac{h}{2} - d'\right) + T\left(d - \frac{h}{2}\right) \qquad (17)$$

For a known strain condition, the corresponding load-moment strength, P_n and M_n, can be computed directly. Assume the strain in the extreme tension steel, A_s, is at first yield ($\varepsilon_s = \varepsilon_y$). This strain condition with simultaneous strain of 0.003 in the extreme compression fiber defines the "balanced" load-moment strength, P_b and M_b, for the cross-section.

For the linear strain condition:

$$\frac{c_b}{d} = \frac{\varepsilon_u}{\varepsilon_u + \varepsilon_y} = \frac{0.003}{0.003 + f_y/29{,}000{,}000} = \frac{87{,}000}{87{,}000 + f_y}$$

so that $\quad a_b = \beta_1 c_b = \left(\dfrac{87{,}000}{87{,}000 + f_y}\right)\beta_1 d$

Also $\quad \dfrac{c_b}{c_b - d'} = \dfrac{\varepsilon_u}{\varepsilon_s'}$

so that $\quad \varepsilon_s' = 0.003\left(1 - \dfrac{d'}{c_b}\right) = 0.003\left[1 - \dfrac{d'}{d}\left(\dfrac{87{,}000 + f_y}{87{,}000}\right)\right]$

and $\quad f_{sb}' = E_s \varepsilon_s' = 87{,}000\left[1 - \dfrac{d'}{d}\left(\dfrac{87{,}000 + f_y}{87{,}000}\right)\right] \quad$ but not greater than f_y

From force equilibrium:

$$P_b = 0.85 f_c' b a_b + A_s' f_{sb}' - A_s f_y \qquad (18)$$

6-22

From moment equilibrium:

$$M_b = P_b e_b = 0.85 f'_c b a_b \left(\frac{h}{2} - \frac{a}{2}\right) + A'_s f'_{sb} \left(\frac{h}{2} - d'\right) + A_s f_y \left(\frac{d}{2} - h\right) \quad (19)$$

The "balanced" load-moment strength defines only one of many load-moment combinations possible over the full range of the load-moment interaction relationship of a cross-section subject to combined flexure and axial load. The general form of a strength interaction diagram is shown in Fig. 6-21. The load-moment combination may be such that compression exists over most or all of the section, so that the compressive strain in the concrete reaches 0.003 before the tension steel yields ($\varepsilon_s < \varepsilon_y$) (compression-controlled segment); or the load combination may be such that tension exists over a large portion of the section, so that the strain in the tension steel is greater than the yield strain ($\varepsilon_s > \varepsilon_y$) when the compressive strain in the concrete reaches 0.003 (transition or tension-controlled segment). The "balanced" strain condition ($\varepsilon_s = \varepsilon_y$) divides these two segments of the strength curve. The linear strain variation for the full range of the load-moment interaction relationship is illustrated in Fig. 6-21.

Under pure compression, the strain is uniform over the entire cross-section and equal to 0.003. With increasing load eccentricity (moment), the compressive strain at the "tension face" gradually decreases to zero, then becomes tensile, reaching the yield strain ($\varepsilon_s = \varepsilon_y$) at the balanced strain condition. For this range of strain variations, the strength of the section is governed by compression ($\varepsilon_s = -0.003$ to ε_y). Beyond the balanced strain condition, the steel strain gradually increases up to the state of pure flexure corresponding to an infinite load eccentricity ($e = \infty$). For this range of strain variations, strength is governed by tension ($\varepsilon_s > \varepsilon_y$). With increasing eccentricity, more and more tension exists over the cross-section. Each of the many possible strain conditions illustrated in Fig. 6-22 describes a point, P_n and M_n, on the load-moment curve (Fig. 6-21). Calculation of P_n and M_n for four different strain conditions along the load-moment strength curve is illustrated in Example 6.4.

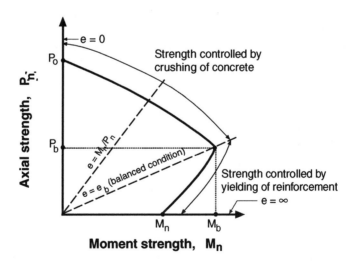

Figure 6-21 Axial Load-Moment Interaction Diagram

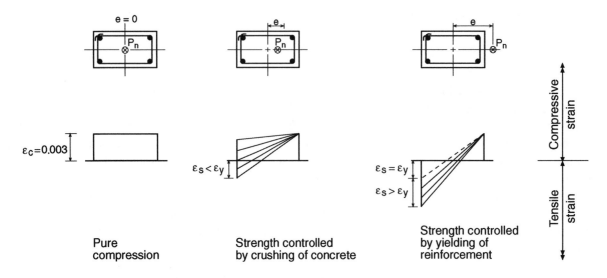

Figure 6-22 Strain Variation for Full Range of Load-Moment Interaction

10.5   MINIMUM REINFORCEMENT OF FLEXURAL MEMBERS

Members with cross-sections much larger than required for strength, for architectural or other reasons, could fail suddenly because of small amounts of tensile reinforcement. The computed moment strength of such sections, assuming reinforced concrete behavior and using cracked section analyses, could become less than that of a corresponding unreinforced concrete section computed from its modulus of rupture. To prevent failure in such situations, a minimum amount of tensile reinforcement is specified in 10.5.

The minimum reinforcement ratio $\rho_{min} = 200/f_y$ was originally derived to provide the same 0.5% minimum (for mild steel grade) as required in earlier versions of the ACI code. This minimum reinforcement is adequate for concrete strengths of about 4000 psi and less. The '95 version of the code recognizes that $\rho_{min} = 200/f_y$ may not be sufficient for f'_c greater than about 5000 psi. The code has accordingly revised 10.5.1 and 10.5.2 to specify the following minimum amounts of steel:

At every section of flexural members where tensile reinforcement is required,

$$A_{s,min} = \frac{3\sqrt{f'_c}}{f_y}b_w d \geq \frac{200}{f_y}b_w d \qquad \text{Eq. (10-3)}$$

Note that $3\sqrt{f'_c}$ and 200 are equal when $f'_c = 4444$ psi. Thus, $3\sqrt{f'_c}b_w d/f_y$ controls when $f'_c > 4444$ psi; otherwise, $200 b_w d/f_y$ controls.

Equation (10-4) of ACI 318-99 was removed and replaced with the following statement, which says the same thing as the former Eq. (10-4):

> 10.5.2 For statically determinate members with a flange in tension, the area $A_{s,min}$ shall be equal to or greater than the value given by Eq. (10-3) with b_w replaced by either $2b_w$ or the width of the flange, whichever is smaller.

Note that the requirements of 10.5.1 and 10.5.2 need not be applied if at every section the area of tensile reinforcement provided is at least one-third greater than that required by analysis (see 10.5.3). For structural slabs and footings (10.5.4), the flexural reinforcement cannot be less than that required for temperature and shrinkage (7.12).

10.15 TRANSMISSION OF COLUMN LOADS THROUGH FLOOR SYSTEM

When the column concrete strength does not exceed the floor concrete strength by more than 40 percent, no special precautions need be taken in computing the column strength (10.15). For higher column concrete strengths, ACI provisions limit the assumed column strength unless concrete puddling is used in the slab at, and around the column (10.15.1), see Figure 6-23. For columns laterally supported on four sides by beams of approximately equal depth or by slabs, the code permits the strength of the column to be based on an assumed concrete strength in the column joint equal to 75 percent of column concrete strength plus 35 percent of floor concrete strength (10.15.3). In the application of 10.15.3, the ratio of column concrete strength to slab concrete strength was limited to 2.5 for design. This effectively limits the assumed column strength to a maximum of 2.225 times the floor concrete strength.

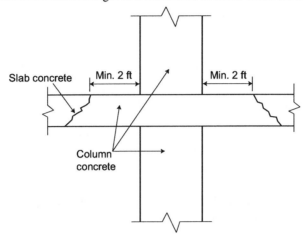

Figure 6-23 Puddling at Slab-Column

Puddling is an intricate procedure that requires coordination between the engineer and the contractor. Special attention should be paid to avoid cold joints and to ensure that the specified column concrete is placed where it is intended. Current industry practice indicates that it is frequently avoided in mainstream high-rise construction since it requires additional time to properly execute in the field. When utilized, a procedure for proper placing and blending of the two concrete types should be clearly called out in the project documents.

10.17 BEARING STRENGTH ON CONCRETE

Code-defined bearing strength (P_{nb}) of concrete is expressed in terms of an average bearing stress of $0.85 f_c'$ over a bearing area (loaded area) A_1. When the supporting concrete area is wider than the loaded area on all sides, the surrounding concrete acts to confine the loaded area, resulting in an increase in the bearing strength of the supporting concrete. With confining concrete, the bearing strength may be increased by the factor $\sqrt{A_2/A_1}$, but not greater than 2, where $\sqrt{A_2/A_1}$ is a measure of the confining effect of the surrounding concrete. Evaluation of the strength increase factor $\sqrt{A_2/A_1}$ is illustrated in Fig. 6-24.

For the usual case of a supporting concrete area considerably greater than the loaded area $\left(\sqrt{A_2/A_1} > 2\right)$, the nominal bearing stress is $2(0.85 f_c')$.

Referring to Fig. 6-25,

    a.   For the supported surface (column):

$$P_{nb} = 0.85 f_c' A_1$$

         where f_c' is the specified strength of the column concrete.

b. For supporting surface (footing):

$$P_{nb} = 0.85 f'_c A_1 \sqrt{\frac{A_2}{A_1}} \text{ and } \sqrt{\frac{A_2}{A_1}} \leq 2.0$$

where f'_c is the specified strength of the footing concrete.

The design bearing strength is ϕP_{nb}, where, for bearing on concrete, $\phi = 0.65$. When the bearing strength is exceeded, reinforcement must be provided to transfer the excess load.

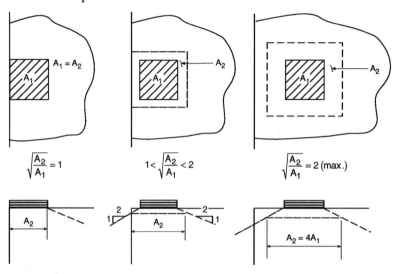

Figure 6-24 Measure of Confinement $\sqrt{A_2/A_1} \leq 2$ Provided by Surrounding Concrete

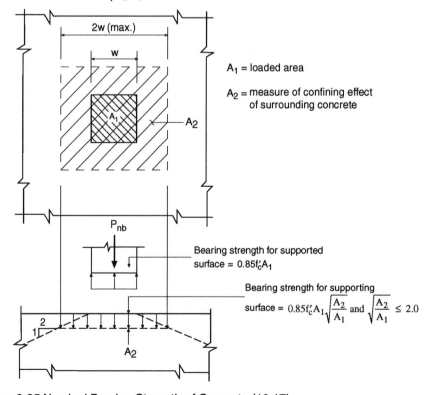

Figure 6-25 Nominal Bearing Strength of Concrete (10.17)

REFERENCES

6.1 Hognestad, E., Hanson, N.W., and McHenry, D., "Concrete Stress Distribution in Ultimate Strength Design," *ACI Journal, Proceedings* Vol. 52, December 1955, pp. 455-479; also *PCA Development Department Bulletin D6*.

6.2 Hognestad, E., "Ultimate Strength of Reinforced Concrete in American Design Practice," Proceedings of a Symposium on the Strength of Concrete Structures, London, England, May 1955; also *PCA Development Department Bulletin D12*.

6.3 Hognestad, E., "Confirmation of Inelastic Stress Distribution in Concrete," *Journal of the Structural Division, Proceedings ASCE*, Vol. 83, No. ST2, March 1957; pp. 1189-1—1189-17 also *PCA Development Department Bulletin D15*.

6.4 Mattock, A.H., Kriz, L.B., and Hognestad, E., "Rectangular Concrete Stress Distribution in Ultimate Strength Design," *ACI Journal, Proceedings*, Vol. 57, February 1961, pp. 875-928; also *PCA Development Department Bulletin D49*.

6.5 Wang, C.K., and Salmon, C.G., *Reinforced Concrete Design*, Fourth Edition, Harper & Row Publishers, New York, N.Y. 1985.

Example 6.1—Moment Strength Using Equivalent Rectangular Stress Distribution

For the beam section shown, calculate moment strength based on static equilibrium using the equivalent rectangular stress distribution shown in Fig. 6-9. Assume $f'_c = 4000$ psi and $f_y = 60,000$ psi. For simplicity, neglect hanger bars.

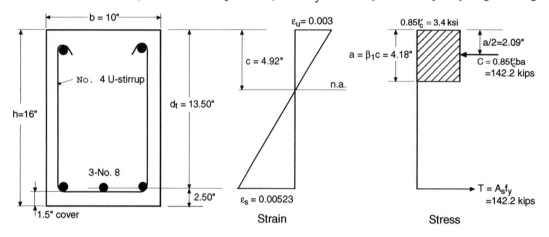

Calculations and Discussion	Code Reference

1. Define rectangular concrete stress distribution.
   10.2.7

   $d = d_t = 16 - 2.5 = 13.50$ in.
   10.0

   $A_s = 3 \times 0.79 = 2.37$ in.^{2}

   Assuming $\varepsilon_s > \varepsilon_y$,

   $T = A_s f_y = 2.37 \times 60 = 142.2$ kips
   10.2.4

   $$a = \frac{A_s f_y}{0.85 f'_c b} = \frac{142.2}{0.85 \times 4 \times 10} = 4.18 \text{ in.}$$

2. Determine net tensile strain ε_s and ϕ

   $$c = \frac{a}{\beta_1} = \frac{4.18}{0.85} = 4.92 \text{ in.}$$

   $$\varepsilon_s = \left(\frac{d_t - c}{c}\right)0.003 = \left(\frac{13.50 - 4.92}{4.92}\right)0.003 = 0.00523 > 0.005$$

   Therefore, section is tension-controlled
   10.3.4

   $\phi = 0.9$
   9.3.2.1

   $\varepsilon_s = 0.00523 > 0.004$ which is minimum for flexural members
   10.3.5

   This also confirms that $\varepsilon_s > \varepsilon_y$ at nominal strength.

Example 6.1 (cont'd)	Calculations and Discussion	Code Reference

3. Determine nominal moment strength, M_n, and design moment strength, ϕM_n.

$$M_n = A_s f_y \left(d - \frac{a}{2}\right) = 142.2 \, (13.50 - 2.09) = 1{,}622.5 \text{ in.-kips} = 135.2 \text{ ft-kips}$$

$$\phi M_n = 0.9(135.2) = 121.7 \text{ ft-kips}$$

9.3.2.1

4. Minimum reinforcement.

$$A_{s,min} = \frac{3\sqrt{f'_c}}{f_y} b_w d \geq \frac{200 b_w d}{f_y}$$

Eq. (10-3)

Since $f'_c < 4444$ psi, $200 b_w d / f_y$ governs:

$$\frac{200 b_w d}{f_y} = \frac{200 \times 10 \times 13.50}{60{,}000} = 0.45 \text{ in.}^2$$

A_s (provided) = 2.37 in.^{2} > $A_{s,min}$ = 0.45 in.^{2} O.K.

Example 6.2—Design of Beam with Compression Reinforcement

A beam cross-section is limited to the size shown. Determine the required area of reinforcement for a factored moment $M_u = 516$ ft-kips. $f'_c = 4000$ psi, $f_y = 60,000$ psi.

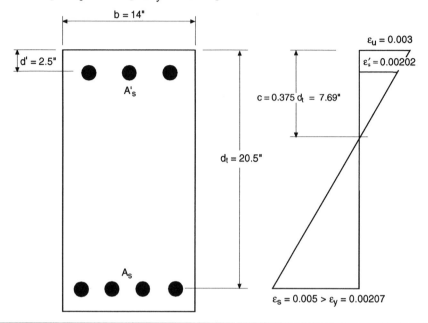

Calculations and Discussion

Code Reference

1. Check if compression reinforcement is required, using $\phi = 0.9$

   $M_n = M_u / \phi = 516 / 0.9 = 573$ ft-kips

   $R_n = \dfrac{M_n}{bd_t^2} = \dfrac{573 \times 12 \times 1000}{214 \times 20.5^2} = 1169$

   This exceeds the maximum R_{nt} of 911 for tension-controlled sections of 4000 psi concrete. (see Table 6-1.) Also, it appears likely that two layers of tension reinforcement will be necessary. But, for simplicity, assume that $d_t = d$.

2. Find the nominal strength moment M_{nt} resisted by the concrete section, without compression reinforcement, and M'_n to be resisted by the compression reinforcement.

   $M_{nt} = R_n bd^2 = 911 \times 14 \times 20.5^2 / (1000 \times 12) = 447$ ft-kips

   $M'_n = M_n - M_{nt} = 573 - 447 = 126$ ft-kips

3. Determine the required compression steel

   The strain in the compression steel at nominal strength is just below yield strain, as shown in the strain diagram above.

		Code
Example 6.2 (cont'd)	**Calculations and Discussion**	**Reference**

$f'_s = E_s \varepsilon_s = 29,000 \times 0.00202 = 58.7$ psi $= 58.7$ ksi

$$A'_s = \frac{M'_n}{f'_s(d-d')} = \frac{126 \times 12}{58.7 \, (20.5-2.5)} = 1.43 \text{ in.}^2$$

4. Determine the required tension steel

   $A_s = \rho_t(bd) + A'_s(f'_s/f_y)$

   From Table 6-1, $\rho_t = 0.01806$, so that

   $A_s = 0.01806(14)(20.5) + 1.43(58.7/60)$

   $= 5.18 + 1.40 = 6.58$ in.^{2}

5. Alternative solution

   Required nominal strength

   $$M_n = \frac{M_u}{\phi} = \frac{516}{0.9} = 573 \text{ ft-kips}$$

   a. Determine maximum moment without compression reinforcement M_{nt}, using $\phi = 0.9$:

      $c = 0.375 d_t = 0.375 \times 20.5 = 7.69$ in.

      $a = \beta_1 c = 0.85 \times 7.69 = 6.54$ in.

      $C = T = 3.4 \times 6.54 \times 14 = 311.3$ kips

      $M_{nt} = T\left(d_t - \frac{a}{2}\right) = 311.3\left(20.5 - \frac{6.54}{2}\right) = 5363.7$ kip-in. $= 447.0$ kip-ft

   b. Required area of tension steel to develop M_{nt}:

      $$A_{s,nt} = \frac{311.3}{60} = 5.19 \text{ in.}^2$$

   c. Additional moment (573−447 = 126 ft-kips) must be developed in T-C couple between tension steel and compression steel.

      Additional tension steel required: $\Delta A_s = \dfrac{126 \times 12}{(20.5-2.5) \times 60} = 1.40$ in.^{2}

      Total tension steel required: $A_s = 5.19 + 1.40 = 6.59$ in.^{2}

	Code
Example 6.2 (cont'd) Calculations and Discussion	**Reference**

Compression steel required: $\quad A'_s = \dfrac{126 \times 12}{(20.5-2.5) \times 58.7} = 1.43 \text{ in.}^2$

6. Comparison to Example 6.2 of *Notes on ACI 318-99* designed by ACI 318-99:

Example 6.2 of *Notes on ACI 318-99* was designed by the 1999 code for an M_u of 580 ft-kips. By current code, assuming a live-to-dead load ratio of 0.5 for this beam, the beam could be designed as a tension-controlled section for an M_u of 516 ft-kips. The results for the required reinforcement are

	by 318-99	Since 2002
Compression reinforcement A'_s	1.49 in.²	1.43 in.²
Tension reinforcement A_s	7.63 in.²	6.58 in.²

The reduction in tension reinforcement is a result of the lower load factors in the current code. However, the compression reinforcement requirement is about the same. This is caused by the need for ductility in order to use the ϕ of 0.9 for flexure.

Example 6.3—Maximum Axial Load Strength vs. Minimum Eccentricity

For the tied reinforced concrete column section shown below, compare the nominal axial load strength P_n equal to $0.80P_o$ with P_n at $0.1h$ eccentricity. $f'_c = 5000$ psi, $f_y = 60,000$ psi.

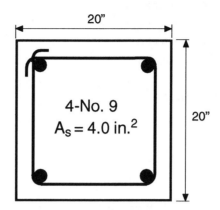

Calculations and Discussion	Code Reference

Prior to ACI 318-77, columns were required to be designed for a minimum eccentricity of $0.1h$ (tied) or $0.05h$ (spiral). This required tedious computations to find the axial load strength at these minimum eccentricities. With the 1977 ACI code, the minimum eccentricity provision was replaced with a maximum axial load strength: $0.80P_o$ (tied) or $0.85P_o$ (spiral). The 80% and 85% values were chosen to approximate the axial load strengths at e/h ratios of 0.1 and 0.05, respectively.

1. In accordance with the minimum eccentricity criterion:

    At $e/h = 0.10$: $P_n = 1543$ kips (computer solution)

2. In accordance with maximum axial load strength criterion: 10.3.5.2

    $P_{n(max)} = 0.80P_o = 0.80 [0.85 f'_c (A_g - A_{st}) + f_y A_{st}]$ Eq. (10-2)

    $= 0.80 [0.85 \times 5 (400 - 4.0) + (60 \times 4.0)] = 1538$ kips

Depending on material strengths, size, and amount of reinforcement, the comparison will vary slightly. Both solutions are considered equally acceptable.

Example 6.4—Load-Moment Strength, P_n and M_n, for Given Strain Conditions

For the column section shown, calculate the load-moment strength, P_n and M_n, for four strain conditions:

1. Bar stress near tension face of member equal to zero, $f_s = 0$
2. Bar stress near tension face of member equal to $0.5f_y$ ($f_s = 0.5f_y$)
3. At limit for compression-controlled section ($\varepsilon_t = 0.002$)
4. At limit for tension-controlled sections ($\varepsilon_t = 0.005$).

Use $f'_c = 4000$ psi, and $f_y = 60,000$ psi.

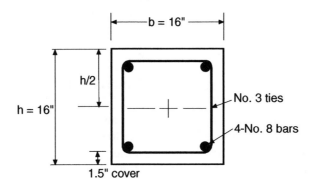

Calculations and Discussion	Code Reference

1. Load-moment strength, P_n and M_n, for strain condition 1: $\varepsilon_s = 0$

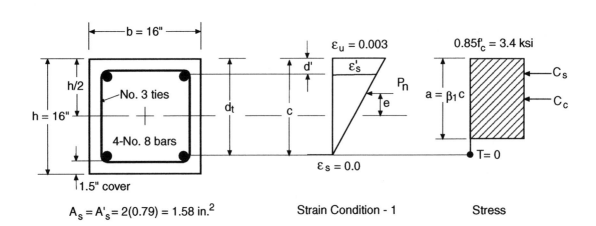

$A_s = A'_s = 2(0.79) = 1.58$ in.^{2} Strain Condition - 1   Stress

		Code
Example 6.4 (cont'd)	**Calculations and Discussion**	**Reference**

a. Define stress distribution and determine force values.   10.2.7

$d' = \text{Cover} + \text{No. 3 tie dia.} + \dfrac{d_b}{2} = 1.5 + 0.375 + 0.5 = 2.38$ in.

$d_t = 16 - 2.38 = 13.62$ in.

Since $\varepsilon_s = 0$, $c = d_t = 13.62$ in.   10.2.7.2

$a = \beta_1 c = 0.85(13.62) = 11.58$ in.   10.2.7.1

where $\beta_1 = 0.85$ for $f'_c = 4000$ psi   10.2.7.3

$C_c = 0.85 f'_c ba = 0.85 \times 4 \times 16 \times 11.58 = 630.0$ kips   10.2.7

$\varepsilon_y = \dfrac{f_y}{E_s} = \dfrac{60}{29,000} = 0.00207$ 10.2.4

From strain compatibility:

$\varepsilon'_s = \varepsilon_u \left(\dfrac{c - d'}{c} \right) = 0.003 \left(\dfrac{13.62 - 2.38}{13.62} \right) = 0.00248 > \varepsilon_y = 0.00207$ 10.2.2

Compression steel has yielded.

$C_s = A'_s f_y = 1.58(60) = 94.8$ kips

b. Determine P_n and M_n from static equilibrium.

$P_n = C_c + C_s = 630.0 + 94.8 = 724.8$ kips   *Eq. (16)*

$M_n = P_n e = C_c \left(\dfrac{h}{2} - \dfrac{a}{2} \right) + C_s \left(\dfrac{h}{2} - d' \right)$ *Eq. (17)*

$= 630(8.0 - 5.79) + 94.8(8.0 - 2.38) = 1925.1$ in.-kips $= 160.4$ ft-kips

$e = \dfrac{M_n}{P_n} = \dfrac{1925.1}{724.8} = 2.66$ in.

Therefore, for strain condition $\varepsilon_s = 0$:

Design axial load strength, $\phi P_n = 0.65(724.8) = 471.1$ kips   9.3.2.2

Design moment strength, $\phi M_n = 0.65(160.4) = 104.3$ ft-kip

Example 6.4 (cont'd)	Calculations and Discussion	Code Reference

2. Load-moment strength, P_n and M_n, for strain condition 2: $\varepsilon_s = 0.5\varepsilon_y$

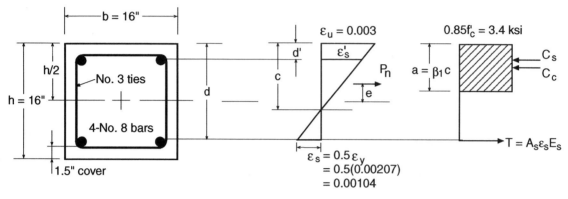

Strain Condition - 2   Stress

a. Define stress distribution and determine force values.
   10.2.7

$d' = 2.38$ in., $d_t = 13.62$ in.

From strain compatibility:

$$\frac{c}{0.003} = \frac{d_t - c}{0.5\varepsilon_y}$$

$$c = \frac{0.003 d_t}{0.5\varepsilon_y + 0.003} = \frac{0.003 \times 13.62}{0.00104 + 0.003} = 10.13 \text{ in.}$$

Strain in compression reinforcement:

$$\varepsilon'_s = \varepsilon_u \left(\frac{c - d'}{c}\right) = 0.003 \left(\frac{10.13 - 2.38}{10.13}\right) = 0.00230 > \varepsilon_y = 0.00207$$

Compression steel has yielded.

$a = \beta_1 c = 0.85 (10.13) = 8.61$ in.
10.2.7.1

$C_c = 0.85 f'_c b a = 0.85 \times 4 \times 16 \times 8.61 = 468.4$ kips
10.2.7

$C_s = A'_s f_y = 1.58 (60) = 94.8$ kips

$T = A_s f_s = A_s (0.5 f_y) = 1.58 (30) = 47.4$ kips

Example 6.4 (cont'd)	Calculations and Discussion	Code Reference

b. Determine P_n and M_n from static equilibrium.

$P_n = C_c + C_s - T = 468.4 + 94.8 - 47.4 = 515.8$ kips   *Eq. (16)*

$M_n = P_n e = C_c \left(\dfrac{h}{2} - \dfrac{a}{2} \right) + C_s \left(\dfrac{h}{2} + d' \right) + T \left(d - \dfrac{h}{2} \right)$ *Eq. (17)*

$= 468.4 (8.0 - 4.31) + 94.8 (8.0 - 2.38) + 47.4 (13.62 - 8.0)$

$= 2527.6$ in.-kips $= 210.6$ ft-kips

$e = \dfrac{M_n}{P_n} = \dfrac{2527.6}{515.8} = 4.90$ in.

Therefore, for strain condition $\varepsilon_s = 0.5 \varepsilon_y$:

Design axial load strength, $\phi P_n = 0.65 (515.8) = 335.3$ kips   *9.3.2.2*

Design moment strength, $\phi M_n = 0.65 (210.6) = 136.9$ ft-kips

3. Load-moment strength, P_n and M_n, for strain condition 3: $\varepsilon_s = \varepsilon_y$

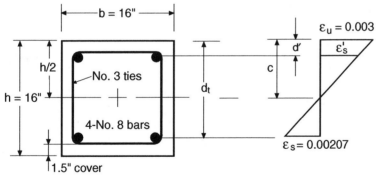
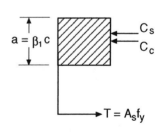

Strain Condition - 3

a. Define stress distribution and determine force values.   *10.2.7*

$d' = 2.38$ in., $d_t = 13.62$ in.

From strain compatibility:

$\dfrac{c}{0.003} = \dfrac{d_t - c}{\varepsilon_y}$

6-37

Example 6.4 (cont'd)	Calculations and Discussion	Code Reference

$$c = \frac{0.003 d_t}{\varepsilon_y + 0.003} = \frac{0.003 \times 13.62}{0.00207 + 0.003} = 8.06 \text{ in.}$$

Note: The code permits the use of 0.002 as the strain limit for compression-controlled sections with Grade 60 steel. It is slightly conservative, and more consistent, to use the yield strain of 0.00207.

Strain in compression reinforcement:

$$\varepsilon'_s = \varepsilon_u \left(\frac{c - d'}{c} \right) = 0.003 \left(\frac{8.06 - 2.38}{8.06} \right) = 0.00211 > \varepsilon_y = 0.00207$$

Compression steel has yielded.

$a = \beta_1 c = 0.85 (8.06) = 6.85$ in.   10.2.7.1

$C_c = 0.85 f'_c b a = 0.85 \times 4 \times 16 \times 6.85 = 372.7$ kips   10.2.7

$C_s = A'_s f_y = 1.58 (60) = 94.8$ kips

$T = A_s f_s = A_s f_y = 1.58 (60) = 94.8$ kips

b. Determine P_n and M_n from static equilibrium.

$P_n = C_c + C_s - T = 372.7 + 94.8 - 94.8 = 372.7$ kips   Eq. (16)

$$M_n = P_n e = C_c \left(\frac{h}{2} - \frac{a}{2} \right) + C_s \left(\frac{h}{2} + d' \right) + T \left(d - \frac{h}{2} \right) \quad \text{Eq. (17)}$$

$ = 372.7 (8.0 - 3.43) + 94.8 (8.0 - 2.38) + 94.8 (13.62 - 8.0)$

$ = 2770.5$ in.-kips $= 230.9$ ft-kips

$$e = \frac{M_n}{P_n} = \frac{2770.5}{372.7} = 7.43 \text{ in.}$$

Therefore, for strain condition $\varepsilon_s = \varepsilon_y$:

Design axial load strength, $\phi P_n = 0.65 (372.7) = 242.3$ kips   9.3.2.2

Design moment strength, $\phi M_n = 0.65 (230.9) = 150.1$ ft-kips

| Example 6.4 (cont'd) | Calculations and Discussion | Code Reference |

4. Load-moment strength, P_n and M_n, for strain condition 4: $\varepsilon_s = 0.005$

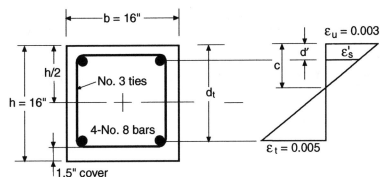
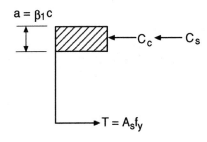

Strain Condition - 4

a. Define stress distribution and determine force values.   *10.2.7*

$d' = 2.38$ in., $d_t = 13.62$ in.

From strain compatibility:

$$\frac{c}{0.003} = \frac{d-c}{0.005}$$

$$c = \frac{0.003d}{0.005 + 0.003} = \frac{0.003 \times 13.62}{0.005 + 0.003} = 5.11 \text{ in.}$$

Strain in compression reinforcement:

$$\varepsilon'_s = \varepsilon_u \left(\frac{c-d'}{c}\right) = 0.003 \left(\frac{5.11 - 2.38}{5.11}\right) = 0.00160 < \varepsilon_y = 0.00207$$

Compression steel has not yielded.

$$f'_s = \varepsilon'_s E'_s = 0.00160 (29,000) = 46.5 \text{ ksi}$$

$a = \beta_1 c = 0.85 (5.11) = 4.34$ in.   *10.2.7.1*

$C_c = 0.85 f'_c ba = 0.85 \times 4 \times 16 \times 4.34 = 236.2$ kips   *10.2.7*

$C_s = A'_s f_y = 1.58 (46.5) = 73.5$ kips

$T = A_s f_s = A_s (f_y) = 1.58 (60) = 94.8$ kips

		Code
Example 6.4 (cont'd)	**Calculations and Discussion**	**Reference**

b. Determine P_n and M_n from static equilibrium.

$P_n = C_c + C_s - T = 236.2 + 73.5 - 94.8 = 214.9$ kips   *Eq. (16)*

$M_n = P_n e = C_c \left(\dfrac{h}{2} - \dfrac{a}{2}\right) + C_s \left(\dfrac{h}{2} + d'\right) + T\left(d - \dfrac{h}{2}\right)$ *Eq. (17)*

$\quad = 236.2 (8.0 - 2.17) + 73.5 (8.0 - 2.38) + 94.8 (13.62 - 8.0)$

$\quad = 2322.9$ in.-kips $= 193.6$ ft-kips

$e = \dfrac{M_n}{P_n} = \dfrac{2322.9}{214.9} = 10.81$ in.

Therefore, for strain condition $\varepsilon_s = 0.005$:

Design axial load strength, $\phi P_n = 0.9 (214.9) = 193.4$ kips   *9.3.2.2*

Design moment strength, $\phi M_n = 0.9 (193.6) = 174.2$ ft-kips

A complete interaction diagram for this column is shown in Fig. 6-25. In addition, Fig. 6-26 shows the interaction diagram created using the Portland Cement Association computer program pcaColumn.

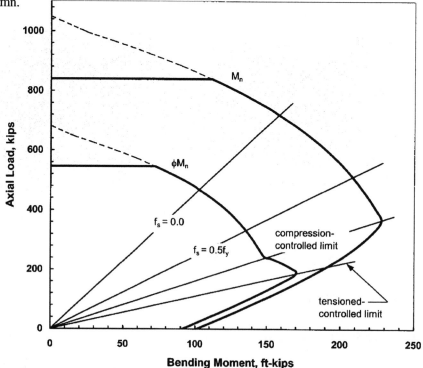

Figure 6-25 Interaction Diagram

Example 6.4 (cont'd)	Calculations and Discussion	Code Reference

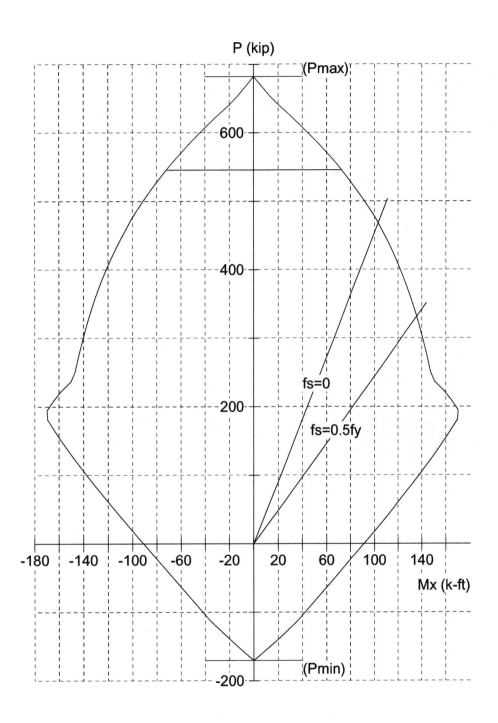

Figure 6-26 Interaction Diagram from pcaColumn

Blank

7

Design for Flexure and Axial Load

GENERAL CONSIDERATIONS—FLEXURE

For design or investigation of members subjected to flexure (beams and slabs), the nominal strength of the member cross-section (M_n) must be reduced by the strength reduction factor ϕ to obtain the design strength (ϕM_n) of the section. The design strength (ϕM_n) must be equal to or greater than the required strength (M_u). In addition, the serviceability requirements for deflection control (9.5) and distribution of reinforcement for crack control (10.6) must also be satisfied.

Examples 7.1 through 7.7 illustrate proper application of the various code provisions that govern design of members subject to flexure. The design examples are prefaced by step-by-step procedures for design of rectangular sections with tension reinforcement only, rectangular sections with multiple layers of steel, rectangular sections with compression reinforcement, and flanged sections with tension reinforcement only.

DESIGN OF RECTANGULAR SECTIONS WITH TENSION REINFORCEMENT ONLY[7.1]

In the design of rectangular sections with tension reinforcement only (Fig. 7-1), the conditions of equilibrium are:

1. Force equilibrium:

$$C = T \qquad (1)$$

$$0.85 f'_c ba = A_s f_y = \rho b d f_y$$

$$a = \frac{A_s f_y}{0.85 f'_c b} = \frac{\rho d f_y}{0.85 f'_c}$$

2. Moment equilibrium:

$$M_n = (C \text{ or } T)\left(d - \frac{a}{2}\right)$$

$$M_n = \rho b d f_y \left[d - \frac{0.5 \rho d}{0.85} \frac{f_y}{f'_c}\right] \qquad (2)$$

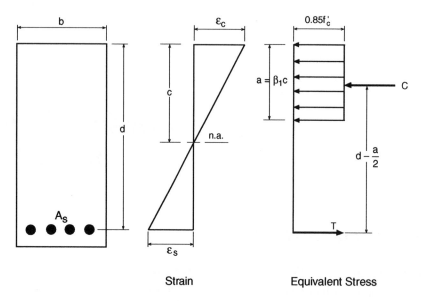

Figure 7-1 Strain and Equivalent Stress Distribution in Rectangular Section

A nominal strength coefficient of resistance R_n is obtained when both sides of Eq. (2) are divided by bd^2:

$$R_n = \frac{M_n}{bd^2} = \rho f_y \left(1 - \frac{0.5\rho f_y}{0.85 f'_c}\right) \qquad (3)$$

When b and d are preset, ρ is obtained by solving the quadratic equation for R_n:

$$\rho = \frac{0.85 f'_c}{f_y}\left(1 - \sqrt{1 - \frac{2R_n}{0.85 f'_c}}\right) \qquad (4)$$

The relationship between ρ and R_n for Grade 60 reinforcement and various values of f'_c is shown in Fig. 7-2.

Equation (3) can be used to determine the steel ratio ρ given M_u or vice-versa if the section properties b and d are known. Substituting $M_n = M_u/\phi$ into Eq. (3) and dividing each side by f'_c:

$$\frac{M_u}{\phi f'_c bd^2} = \frac{\rho f_y}{f'_c}\left(1 - \frac{0.5\rho f_y}{0.85 f'_c}\right)$$

Define $\omega = \dfrac{\rho f_y}{f'_c}$

Substituting ω into the above equation:

$$\frac{M_u}{\phi f'_c bd^2} = \omega(1 - 0.59\omega) \qquad (5)$$

Table 7-1, based on Eq. (5), has been developed in order to serve as a design aid for either design or investigation of sections having tension reinforcement only where b and d are known.

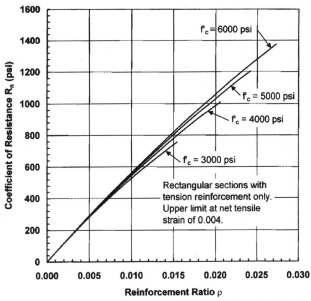

Figure 7-2 Strength Curves (R_n vs. ρ) for Grade 60 Reinforcement

Table 7-1 Flexural Strength $M_u / \phi f'_c bd^2$ or $M_n / f'_c bd^2$ of Rectangular Sections with Tension Reinforcement Only

ω	.000	.001	.002	.003	.004	.005	.006	.007	.008	.009
0.00	0	.0010	.0020	.0030	.0040	.0050	.0060	.0070	.0080	0090
0.01	.0099	.0109	.0119	.0129	.0139	.0149	.0159	.0168	.0178	.0188
0.02	.0197	.0207	.0217	.0226	.0236	.0246	.0256	.0266	.0275	.0285
0.03	.0295	.0304	.0314	.0324	.0333	.0343	.0352	.0362	.0372	.0381
0.04	.0391	.0400	.0410	.0420	.0429	.0438	.0448	.0457	.0467	.0476
0.05	.0485	.0495	.0504	.0513	.0523	.0532	.0541	.0551	.0560	.0569
0.06	.0579	.0588	.0597	.0607	.0616	.0626	.0634	.0643	.0653	.0662
0.07	.0671	.0680	.0689	.0699	.0708	.0717	.0726	.0735	.0744	.0753
0.08	.0762	.0771	.0780	.0789	.0798	.0807	.0816	.0825	.0834	.0843
0.09	.0852	.0861	.0870	.0879	.0888	.0897	.0906	.0915	.0923	.0932
0.10	.0941	.0950	.0959	.0967	.0976	.0985	.0994	.1002	.1001	.1020
0.11	.1029	.1037	.1046	.1055	.1063	.1072	.1081	.1089	.1098	.1106
0.12	.1115	.1124	.1133	.1141	.1149	.1158	.1166	.1175	.1183	.1192
0.13	.1200	.1209	.1217	.1226	.1234	.1243	.1251	.1259	.1268	.1276
0.14	.1284	.1293	.1301	.1309	.1318	.1326	.1334	.1342	.1351	.1359
0.15	.1367	.1375	.1384	.1392	.1400	.1408	.1416	.1425	.1433	.1441
0.16	.1449	.1457	.1465	.1473	.1481	.1489	.1497	.1506	.1514	.1522
0.17	.1529	.1537	.1545	.1553	.1561	.1569	.1577	.1585	.1593	.1601
0.18	.1609	.1617	.1624	.1632	.1640	.1648	.1656	.1664	.1671	.1679
0.19	.1687	.1695	.1703	.1710	.1718	.1726	.1733	.1741	.1749	.1756
0.20	.1764	.1772	.1779	.1787	.1794	.1802	.1810	.1817	.1825	.1832
0.21	.1840	.1847	.1855	.1862	.1870	.1877	.1885	.1892	.1900	.1907
0.22	.1914	.1922	.1929	.1937	.1944	.1951	.1959	.1966	.1973	.1981
0.23	.1988	.1995	.2002	.2010	.2017	.2024	.2031	.2039	.2046	.2053
0.24	.2060	.2067	.2075	.2082	.2089	.2096	.2103	.2110	.2117	.2124
0.25	.2131	.2138	.2145	.2152	.2159	.2166	.2173	.2180	.2187	.2194
0.26	.2201	.2208	.2215	.2222	.2229	.2236	.2243	.2249	.2256	.2263
0.27	.2270	.2277	.2284	.2290	.2297	.2304	.2311	.2317	.2324	.2331
0.28	.2337	.2344	.2351	.2357	.2364	.2371	.2377	.2384	.2391	.2397
0.29	.2404	.2410	.2417	.2423	.2430	.2437	.2443	.2450	.2456	.2463
0.30	.2469	.2475	.2482	.2488	.2495	.2501	.2508	.2514	.2520	.2527

$M_n / f'_c bd^2 = \omega(1 - 0.59\omega)$, where $\omega = \rho f_y / f'_c$

For design: Using factored moment M_u, enter table with $M_u/\phi f'_c bd^2$; find ω and compute steel percentage $\rho = \omega f'_c / f_y$

For investigation: Enter table with $\omega = \rho f_y / f'_c$; find value of $M_n / f'_c bd^2$ and solve for nominal strength, M_n.

Figure 7-3 shows the effect of the strength reduction factor ϕ. In particular, it shows what happens when the limit for tension-controlled sections with a ϕ of 0.9 is passed. As can be seen from Fig. 7-3, there is no benefit in designing a flexural member that is below the tension-controlled strain limit of 0.005. Any gain in strength with higher reinforcement ratios is offset by the reduction in the strength reduction factor ϕ at higher reinforcement ratios. Therefore, flexural members are more economical when designed as tension-controlled sections.

One might wonder "why even permit higher amounts of reinforcement and lower net tensile strains if there is no advantage?" In many cases, the provided steel is above the optimum at the limit for tension-controlled sections. The "flat" portion of the curve in Fig. 7-3 allows the designer to provide excess reinforcement above that required (considering discrete bar sizes) without being penalized for "being above a code limit."

Although flexural members should almost always be designed as tension-controlled sections with $\varepsilon_t \geq 0.005$, it often happens that columns with small axial load and large bending moments are in the "transition region" with ε_t between 0.002 and 0.005, and ϕ is somewhere between that for compression-controlled sections and that for tension-controlled sections.

Columns are normally designed using interaction charts or tables. The "breakpoint" for ε_t of 0.005 and $\phi = 0.9$ may fall above or below the zero axial load line on the interaction diagrams.

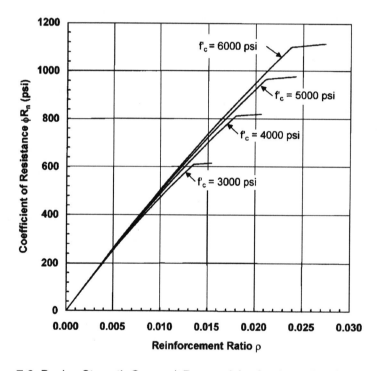

Figure 7-3 Design Strength Curves (ϕR_n vs. ρ) for Grade 60 Reinforcement

DESIGN PROCEDURE FOR SECTIONS WITH TENSION REINFORCEMENT ONLY

Step 1: Select an approximate value of tension reinforcement ratio ρ equal to or less than ρ_t, but greater than the minimum (10.5.1), where the reinforcement ratio ρ_t is given by:

$$\rho_t = \frac{0.319\beta_1 f'_c}{f_y}$$

where $\beta_1 = 0.85$ for $f'_c \leq 4000$ psi

$\qquad = 0.85 - 0.05 \left(\dfrac{f'_c - 4000}{1000} \right)$ for 4000 psi $< f'_c < 8000$ psi

$\qquad = 0.65$ for $f'_c \geq 8000$ psi

Values of ρ_t are given in Table 6-1.

Step 2: With ρ preset ($\rho_{min} \leq \rho \leq \rho_t$) compute bd^2 required:

$$bd^2 \text{ (required)} = \dfrac{M_u}{\phi R_n}$$

where $R_n = \rho f_y \left(1 - \dfrac{0.5 \rho f_y}{0.85 f'_c} \right)$, $\phi = 0.90$ for flexure with $\rho \leq \rho_t$, and $M_u =$ applied factored moment (required flexural strength)

Step 3: Size the member so that the value of bd^2 provided is greater than or equal to the value of bd^2 required.

Step 4: Based on the provided bd^2, compute a revised value of ρ by one of the following methods:

1. By Eq. (4) where $R_n = M_u / \phi bd^2$ (exact method)

2. By strength curves such as those shown in Fig. 7-2 and Fig. 7-3. Values of ρ are given in terms of $R_n = M_u / \phi bd^2$ for Grade 60 reinforcement.

3. By moment strength tables such as Table 7-1. Values of $\omega = \rho f_y / f'_c$ are given in terms of moment strength $M_u / \phi f'_c bd^2$.

4. By approximate proportion

$$\rho \approx \text{(original } \rho\text{)} \dfrac{\text{(revised } R_n\text{)}}{\text{(original } R_n\text{)}}$$

Note from Fig. 7-2 that the relationship between R_n and ρ is approximately linear.

Step 5: Compute required A_s:

$A_s =$ (revised ρ) (bd provided)

When b and d are preset, the required A_s is computed directly from:

$A_s = \rho$ (bd provided)

where ρ is computed using one of the methods outlined in Step 4.

DESIGN PROCEDURE FOR SECTIONS WITH MULTIPLE LAYERS OF STEEL

The simple and conservative way to design a beam with two layers of tension steel is to take d_t equal to d, the depth to the centroid of all the tension steel. However, the code does permit the designer to take advantage of the fact that d_t, measured to the center of the layer farthest from the compression face, is greater than d. The only time this would be necessary is when designing at or very close to the strain limit of 0.005 for tension-controlled sections.

Figure 7-4 shows strain and stress diagrams for a section with multiple layers of steel with the extreme steel layer at the tension-controlled strain limit of 0.005. Let ρ_2 stand for the maximum ρ (based on d) for this section.

$$\rho_2 = \frac{C}{f_y bd}$$

However,

$$\rho_t = \frac{C}{f_y bd_t}$$

Therefore,

$$\frac{\rho_2}{\rho_t} = \frac{d_t}{d}$$

$$\rho_2 = \rho_t \left(\frac{d_t}{d}\right) \tag{6}$$

Additional information can be found in the strain diagram of Fig. 7-4. The yield strain of Grade 60 reinforcement is 0.00207. By similar triangles, any Grade 60 steel that is within $0.366 d_t$ of the bottom layer will be at yield. This is almost always the case, unless steel is distributed on the side faces. Also, compression steel will be at yield if it is within $0.116 d_t$ (or, $0.31c$) of the compression face.

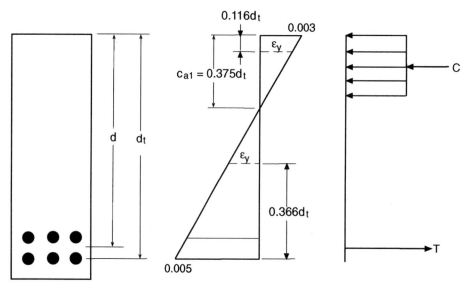

Figure 7-4 Multiple Layers of Reinforcement

DESIGN PROCEDURE FOR RECTANGULAR SECTIONS WITH COMPRESSION REINFORCEMENT (see Part 6)

Steps are summarized for the design of rectangular beams (with b and d preset) requiring compression reinforcement (see Example 7.3).

Step 1. Check to see if compression reinforcement is needed. Compute

$$R_n = \frac{M_n}{bd^2}$$

Compare this to the maximum R_n for tension-controlled sections given in Table 6-1. If R_n exceeds this, use compression reinforcement.

If compression reinforcement is needed, it is likely that two layers of tension reinforcement will be needed. Estimate d_t/d ratio.

Step 2. Find the nominal moment strength resisted by a section without compression reinforcement, and the additional moment strength M'_n to be resisted by the compression reinforcement and by added tension reinforcement.

From Table 6-1, find ρ_t. Then, using Eq. (6);

$$\rho = \rho_t \left(\frac{d_t}{d}\right)$$

$$\omega = \rho \frac{f_y}{f'_c}$$

Determine M_{nt} from Table 7-1.

Compute moment strength to be resisted by compression reinforcement:

$$M'_n = M_n - M_{nt}$$

Step 3. Check yielding of compression reinforcement

If $d'/c < 0.31$, compressive reinforcement has yielded and $f'_s = f_y$

See Part 6 to determine f'_s when the compression reinforcement does not yield.

Step 4. Determine the total required reinforcement, A'_s and A_s

$$A'_s = \frac{M'_n}{(d-d')f'_s}$$

Step 5: Check moment capacity

$$\phi M_n = \phi\left[(A_s - A'_s)f_y\left(d - \frac{a}{2}\right) + A'_s f_y (d - d')\right] \geq M_u$$

where

$$a = \frac{(A_s - A'_s)f_y}{0.85 f'_c b}$$

DESIGN PROCEDURE FOR FLANGED SECTIONS WITH TENSION REINFORCEMENT (see Part 6)

Steps are summarized for the design of flanged sections with tension reinforcement only (see Examples 7.4 and 7.5).

Step 1: Determine effective flange width b according to 8.10.

Using Table 7-1, determine the depth of the equivalent stress block a, assuming rectangular section behavior with b equal to the flange width (i.e., $a \leq h_f$):

$$a = \frac{A_s f_y}{0.85 f'_c b} = \frac{\rho d f_y}{0.85 f'_c} = 1.18 \omega d$$

where ω is obtained from Table 7-1 for $M_u/\phi f'_c bd^2$. Assume tension-controlled section with $\phi = 0.9$.

Step 2: If $a \leq h_f$, determine the reinforcement as for a rectangular section with tension reinforcement only. If $a > h_f$, go to step 3.

Step 3: If $a > h_f$, compute the required reinforcement A_{sf} and the moment strength ϕM_{nf} corresponding to the overhanging beam flange in compression:

$$A_{sf} = \frac{C_f}{f_y} = \frac{0.85 f'_c (b - b_w) h_f}{f_y}$$

$$\phi M_{nf} = \phi\left[A_{sf} f_y \left(d - \frac{h_f}{2}\right)\right]$$

Step 4: Compute the required moment strength to be carried by the beam web:

$$M_{uw} = M_u - \phi M_{nf}$$

Step 5: Using Table 7-1, compute the reinforcement A_{sw} required to develop the moment strength to be carried by the web:

$$A_{sw} = \frac{0.85 f'_c b_w a_w}{f_y}$$

where $a_w = 1.18\omega_w d$ with ω_w obtained from Table 7-1 for $M_{uw}/\phi f'_c b_w d^2$.

Alternatively, obtain A_{sw} from the following:

$$A_{sw} = \frac{\omega_w f'_c b_w d}{f_y}$$

Step 6: Determine the total required reinforcement:

$$A_s = A_{sf} + A_{sw}$$

Step 7: Check to see if section is tension-controlled, with $\phi = 0.9$.

$$c = a_w / \beta_1$$

If $c/d_t \leq 0.375$, section is tension-controlled
If $c/d_t > 0.375$, add compression reinforcement

Step 8: Check moment capacity:

$$\phi M_n = \phi\left[\left(A_s - A_{sf}\right) f_y \left(d - \frac{a_w}{2}\right) + A_{sf} f_y \left(d - \frac{h_f}{2}\right)\right] \geq M_u$$

where $A_{sf} = \dfrac{0.85 f'_c (b - b_w) h_f}{f_y}$

$a_w = \dfrac{(A_s - A_{sf}) f_y}{0.85 f'_c b_w}$

GENERAL CONSIDERATIONS—FLEXURE AND AXIAL LOAD

Design or investigation of a short compression member (without slenderess effect) is based primarily on the strength of its cross-section. Strength of a cross-section under combined flexure and axial load must satisfy both force equilibrium and strain compatibility (see Part 6). The combined nominal axial load and moment strength (P_n, M_n) is then multiplied by the appropriate strength reduction factor ϕ to obtain the design strength (ϕP_n, ϕM_n) of the section. The design strength must be equal to or greater than the required strength:

$$(\phi P_n, \phi M_n) \geq (P_u, M_u)$$

All members subjected to combined flexure and axial load must be designed to satisfy this basic criterion. Note that the required strength (P_u, M_u) represents the structural effects of the various combinations of loads and forces to which a structure may be subjected; see Part 5 for discussion on 9.2.

A "strength interaction diagram" can be generated by plotting the design axial load strength ϕP_n against the corresponding design moment strength ϕM_n; this diagram defines the "usable" strength of a section at different eccentricities of the load. A typical design load-moment strength interaction diagram is shown in Fig. 7-5, illustrating the various segments of the strength curve permitted for design. The "flat-top" segment of the design strength curve defines the limiting axial load strength $P_{n,max}$; see Part 5 for a discussion on 10.3.6. As the design axial load strength ϕP_n decreases, a transition occurs between the compression-controlled limit and the tension-controlled limit, as shown in the figure. Example 6.4 illustrates the construction of an interaction diagram.

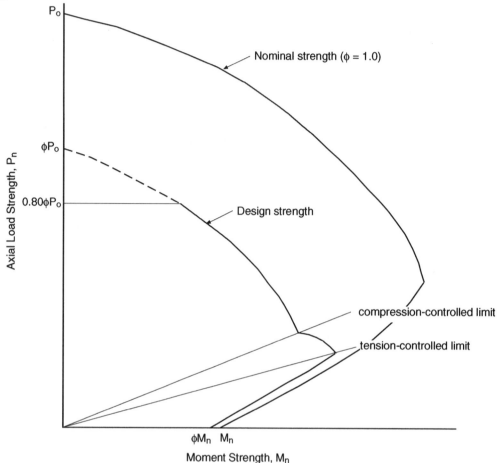

Figure 7-5 Design Load-Moment Strength Diagram (tied column)

GENERAL CONSIDERATIONS—BIAXIAL LOADING

Biaxial bending of columns occurs when the loading causes bending simultaneously about both principal axes. The commonly encountered case of such loading occurs in corner columns. Design for biaxial bending and axial load is mentioned in R10.3.6 and R10.3.7. Section 10.11.6 addresses moment magnifiers for slenderness consideration of compression members under biaxial loading. Section R10.3.6 states that "corner and other columns exposed to known moments about each axis simultaneously should be designed for biaxial bending and axial load." Two methods are recommended for combined biaxial bending and axial load design: the Reciprocal Load Method and the Load Contour Method. Both methods, and an extension of the Load Contour Method (PCA Load Contour Method), are presented below.

BIAXIAL INTERACTION STRENGTH

A uniaxial interaction diagram defines the load-moment strength along a single plane of a section under an axial load P and a uniaxial moment M. The biaxial bending resistance of an axially loaded column can be represented schematically as a surface formed by a series of uniaxial interaction curves drawn radially from the P axis (see

Fig. 7-6). Data for these intermediate curves are obtained by varying the angle of the neutral axis (for assumed strain configurations) with respect to the major axes (see Fig. 7-7).

The difficulty associated with the determination of the strength of reinforced columns subject to combined axial load and biaxial bending is primarily an arithmetic one. The bending resistance of an axially loaded column about a particular skewed axis is determined through iterations involving simple but lengthy calculations. These extensive calculations are compounded when optimization of the reinforcement or cross-section is sought.

For uniaxial bending, it is customary to utilize design aids in the form of interaction curves or tables. However, for biaxial bending, because of the voluminous nature of the data and the difficulty in multiple interpolations, the development of interaction curves or tables for the various ratios of bending moments about each axis is impractical. Instead, several approaches (based on acceptable approximations) have been developed that relate the response of a column in biaxial bending to its uniaxial resistance about each major axis.

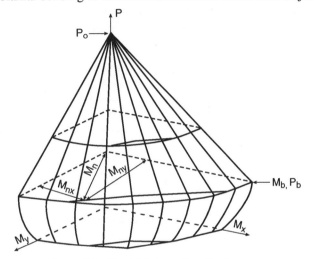

Figure 7-6 Biaxial Interaction Surface

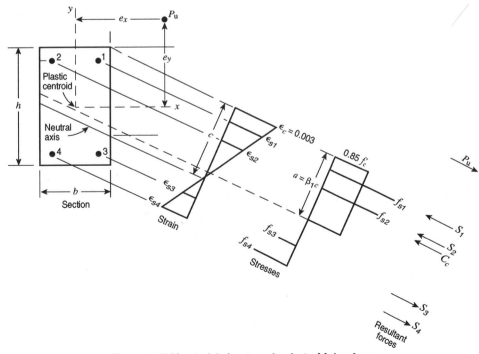

Figure 7-7 Neutral Axis at an Angle to Major Axes

FAILURE SURFACES

The nominal strength of a section under biaxial bending and compression is a function of three variables P_n, M_{nx} and M_{ny} which may be expressed in terms of an axial load acting at eccentricities $e_x = M_{ny}/P_n$ and $e_y = M_{nx}/P_n$ as shown in Fig. 7-8. A failure surface may be described as a surface produced by plotting the failure load P_n as a function of its eccentricities e_x and e_y, or of its associated bending moments M_{ny} and M_{nx}. Three types of failure surfaces have been defined.[7.4, 7.5, 7.6] The basic surface S_1 is defined by a function which is dependent upon the variables P_n, e_x and e_y, as shown in Fig. 7-9(a). A reciprocal surface can be derived from S_1 in which the reciprocal of the nominal axial load P_n is employed to produce the surface S_2 ($1/P_n$, e_x, e_y) as illustrated in Fig. 7-9(b). The third type of failure surface, shown in Fig. 7-9(c), is obtained by relating the nominal axial load P_n to the moments M_{nx} and M_{ny} to produce surface S_3 (P_n, M_{nx}, M_{ny}). Failure surface S_3 is the three-dimensional extension of the uniaxial interaction diagram previously described.

A number of investigators have made approximations for both the S_2 and S_3 failure surfaces for use in design and analysis.[7.6 - 7.10] An explanation of these methods used in current practice, along with design examples, is given below.

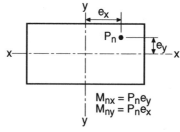

Reinforcing bars not shown

Figure 7-8 Notation for Biaxial Loading

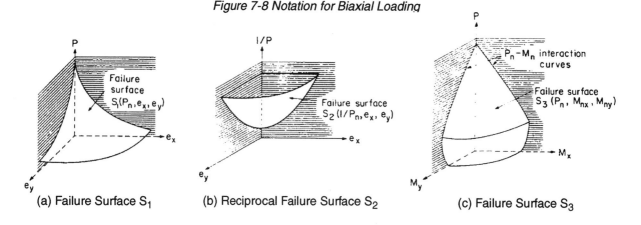

(a) Failure Surface S_1 (b) Reciprocal Failure Surface S_2 (c) Failure Surface S_3

Figure 7-9 Failure Surfaces

A. Bresler Reciprocal Load Method

This method approximates the ordinate $1/P_n$ on the surface S_2 ($1/P_n$, e_x, e_y) by a corresponding ordinate $1/P'_n$ on the plane S'_2 ($1/P'_n$, e_x, e_y), which is defined by the characteristic points A, B and C, as indicated in Fig. 7-10. For any particular cross-section, the value P_o (corresponding to point C) is the load strength under pure axial compression; P_{ox} (corresponding to point B) and P_{oy} (corresponding to point A) are the load strengths under uniaxial eccentricities e_y and e_x, respectively. Each point on the true surface is approximated by a different plane; therefore, the entire surface is approximated using an infinite number of planes.

The general expression for axial load strength for any values of e_x and e_y is as follows:[7.6]

$$\frac{1}{P_n} \approx \frac{1}{P'_n} = \frac{1}{P_{ox}} + \frac{1}{P_{oy}} - \frac{1}{P_o}$$

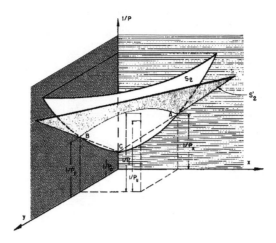

Figure 7-10 Reciprocal Load Method

Rearranging variables yields:

$$P_n \approx \frac{1}{\frac{1}{P_{ox}} + \frac{1}{P_{oy}} - \frac{1}{P_o}} \qquad (7)$$

where
- P_{ox} = Maximum uniaxial load strength of the column with a moment of $M_{nx} = P_n e_y$
- P_{oy} = Maximum uniaxial load strength of the column with a moment of $M_{ny} = P_n e_x$
- P_o = Maximum axial load strength with no applied moments

This equation is simple in form and the variables are easily determined. Axial load strengths P_o, P_{ox}, and P_{oy} are determined using any of the methods presented above for uniaxial bending with axial load. Experimental results have shown the above equation to be reasonably accurate when flexure does not govern design. The equation should only be used when:

$$P_n \geq 0.1 f'_c A_g \qquad (8)$$

B. Bresler Load Contour Method

In this method, the surface S_3 (P_n, M_{nx}, M_{ny}) is approximated by a family of curves corresponding to constant values of P_n. These curves, as illustrated in Fig. 7-11, may be regarded as "load contours."

The general expression for these curves can be approximated[7.6] by a nondimensional interaction equation of the form

$$\left(\frac{M_{nx}}{M_{nox}}\right)^\alpha + \left(\frac{M_{ny}}{M_{noy}}\right)^\beta = 1.0 \qquad (9)$$

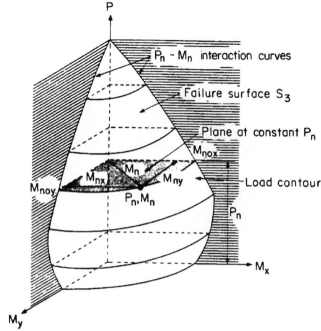

Figure 7-11 Bresler Load Contours for Constant P_n on Failure Surface S_3

where M_{nx} and M_{ny} are the nominal biaxial moment strengths in the direction of the x and y axes, respectively. Note that these moments are the vectorial equivalent of the nominal uniaxial moment M_n. The moment M_{nox} is the nominal uniaxial moment strength about the x-axis, and M_{noy} is the nominal uniaxial moment strength about the y-axis. The values of the exponents α and β are a function of the amount, distribution and location of reinforcement, the dimensions of the column, and the strength and elastic properties of the steel and concrete. Bresler[7.6] indicates that it is reasonably accurate to assume that $\alpha = \beta$; therefore, Eq. (9) becomes

$$\left(\frac{M_{nx}}{M_{nox}}\right)^\alpha + \left(\frac{M_{ny}}{M_{noy}}\right)^\alpha = 1.0 \qquad (10)$$

which is shown graphically in Fig. 7-12.

When using Eq. (10) or Fig. 7-12, it is still necessary to determine the α value for the cross-section being designed. Bresler indicated that, typically, α varied from 1.15 to 1.55, with a value of 1.5 being reasonably accurate for most square and rectangular sections having uniformly distributed reinforcement.

With α set at unity, the interaction equation becomes linear:

$$\frac{M_{nx}}{M_{nox}} + \frac{M_{ny}}{M_{noy}} = 1.0 \qquad (11)$$

Equation (11), as shown in Fig. 7-12, would always yield conservative results since it underestimates the column capacity, especially for high axial loads or low percentages of reinforcement. It should only be used when

$$P_n < 0.1 f'_c A_g \qquad (12)$$

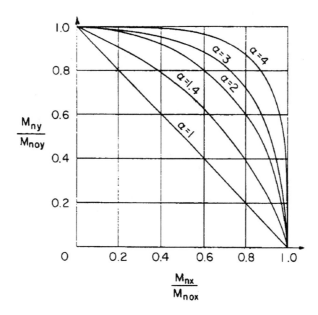

Figure 7-12 Interaction Curves for Bresler Load Contour Method (Eq.(9))

C. PCA Load Contour Method

The PCA approach described below was developed as an extension of the Bresler Load Contour Method. The Bresler interaction equation [Eq. (10)] was chosen as the most viable method in terms of accuracy, practicality, and simplification potential.

A typical Bresler load contour for a certain P_n is shown in Fig. 7-13(a). In the PCA method,[7.11] point B is defined such that the nominal biaxial moment strengths M_{nx} and M_{ny} at this point are in the same ratio as the uniaxial moment strengths M_{nox} and M_{noy}. Therefore, at point B

$$\frac{M_{nx}}{M_{ny}} = \frac{M_{nox}}{M_{noy}} \tag{13}$$

When the load contour of Fig. 7-13(a) is nondimensionalized, it takes the form shown in Fig. 7-13(b), and the point B will have x and y coordinates of β. When the bending resistance is plotted in terms of the dimensionless parameters P_n/P_o, M_{nx}/M_{nox}, M_{ny}/M_{noy} (the latter two designated as the relative moments), the generated failure surface S_4 (P_n/P_o, M_{nx}/M_{nox}, M_{ny}/M_{noy}) assumes the typical shape shown in Fig. 7-13(c). The advantage of expressing the behavior in relative terms is that the contours of the surface (Fig. 7-13(b))—i.e., the intersection formed by planes of constant P_n/P_o and the surface—can be considered for design purposes to be symmetrical about the vertical plane bisecting the two coordinate planes. Even for sections that are rectangular or have unequal reinforcement on the two adjacent faces, this approximation yields values sufficiently accurate for design.

The relationship between α from Eq. (10) and β is obtained by substituting the coordinates of point B from Fig. 7-13(a) into Eq. (10), and solving for α in terms of β. This yields:

$$\alpha = \frac{\log 0.5}{\log \beta}$$

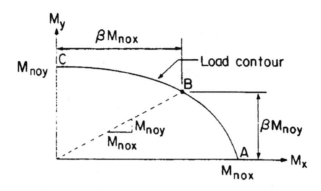

Figure 7-13(a) Load Contour of Failure Surface s_3 along Plane of Constant P_n

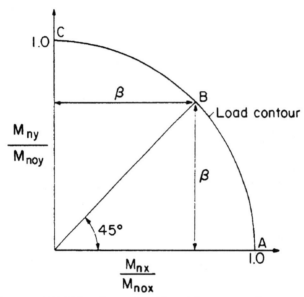

Figure 7-13(b) Nondimensional Load Contour at Constant P_n

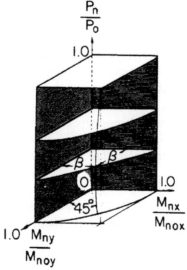

Figure 7-13(c) Failure Surface $S_4 \left(\dfrac{P_n}{P_o}, \dfrac{M_{nx}}{M_{nox}}, \dfrac{M_{ny}}{M_{noy}} \right)$

7-16

Thus, Eq. (10) may be written as:

$$\left(\frac{M_{nx}}{M_{nox}}\right)^{\left(\frac{\log 0.5}{\log \beta}\right)} + \left(\frac{M_{ny}}{M_{noy}}\right)^{\left(\frac{\log 0.5}{\log \beta}\right)} = 1.0 \qquad (14)$$

For design convenience, a plot of the curves generated by Eq. (14) for nine values of β are given in Fig. 7-14. Note that when $\beta = 0.5$, its lower limit, Eq. (14) is a straight line joining the points at which the relative moments equal 1.0 along the coordinate planes. When $\beta = 1.0$, its upper limit, Eq. (14) is two lines, each of which is parallel to one of the coordinate planes.

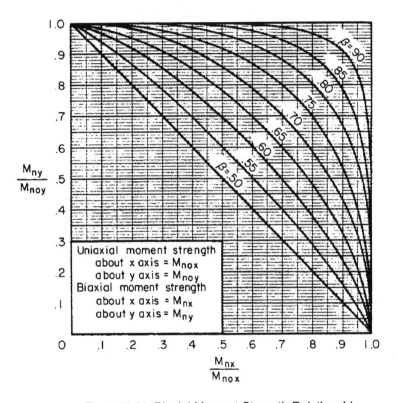

Figure 7-14 Biaxial Moment Strength Relationship

Values of β were computed on the basis of 10.2, utilizing a rectangular stress block and the basic principles of equilibrium. It was found that the parameters γ, b/h, and f'_c had minor effect on the β values. The maximum difference in β was about 5% for values of P_n/P_o ranging from 0.1 to 0.9. The majority of the β values, especially in the most frequently used range of P_n/P_o, did not differ by more than 3%. In view of these small differences, only envelopes of the lowest β values were developed for two values of f_y and different bar arrangements, as shown in Figs. 7-15 and 7-16.

As can be seen from Figs. 7-15 and 7-16, β is dependent primarily on the ratio P_n/P_o and to a lesser, though still significant extent, on the bar arrangement, the reinforcement index ω and the strength of the reinforcement.

Figure 7-14, in combination with Figs. 7-15 and 7-16, furnish a convenient and direct means of determining the biaxial moment strength of a given cross-section subject to an axial load, since the values P_o, M_{nox}, and M_{noy} can be readily obtained by methods described above.

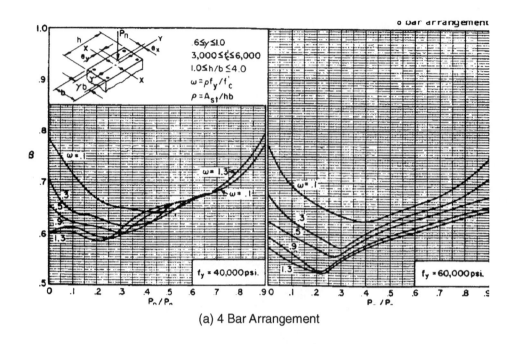

(a) 4 Bar Arrangement

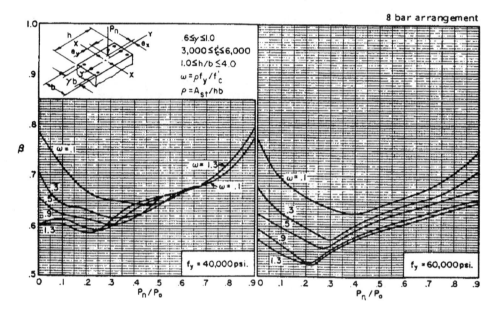

(b) 8 Bar Arrangement

Figure 7-15 Biaxial Design Constants

While investigation of a given section has been simplified, the determination of a section which will satisfy the strength requirements imposed by a load eccentric about both axes can only be achieved by successive analyses of assumed sections. Rapid and easy convergence to a satisfactory section can be achieved by approximating the curves in Fig. 7-14 by two straight lines intersecting at the 45 degree line, as shown in Fig. 7-17.

7-18

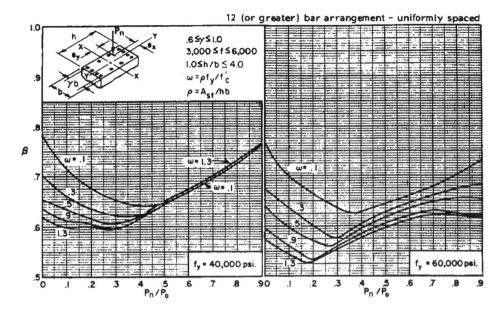

(a) 6, 8, and 10 Bar Arrangement

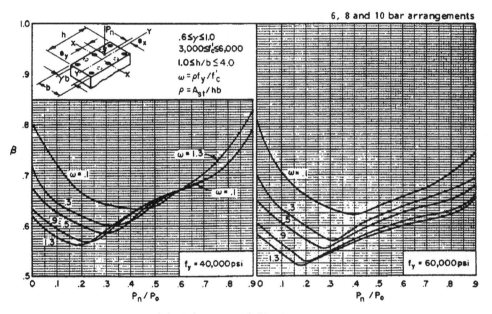

(b) 12 (or greater) Bar Arrangement

Figure 7-16 Biaxial Design Constants

By simple geometry, it can be shown that the equation of the upper lines is:

$$\frac{M_{nx}}{M_{nox}} \left(\frac{1-\beta}{\beta} \right) + \frac{M_{ny}}{M_{noy}} = 1 \text{ for } \frac{M_{ny}}{M_{nx}} > \frac{M_{noy}}{M_{nox}} \tag{15}$$

7-19

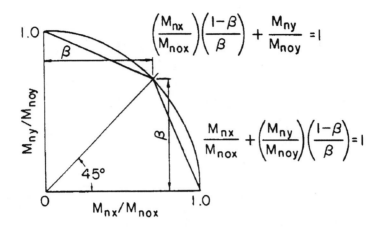

Figure 7-17 Bilinear Approximation of Nondimensionalized Load Contour (Fig. 7-13(b))

which can be restated for design convenience as follows:

$$M_{nx}\left(\frac{M_{noy}}{M_{nox}}\right)\left(\frac{1-\beta}{\beta}\right) + M_{ny} = M_{noy} \quad (16)$$

For rectangular sections with reinforcement equally distributed on all faces, Eq. (16) can be approximated by:

$$M_{nx}\frac{b}{h_a}\left(\frac{1-\beta}{\beta}\right) + M_{ny} \approx M_{noy} \quad (17)$$

The equation of the lower line of Fig. 7-17 is:

$$\frac{M_{nx}}{M_{nox}} + \frac{M_{ny}}{M_{noy}}\left(\frac{1-\beta}{\beta}\right) = 1 \text{ for } \frac{M_{ny}}{M_{nx}} < \frac{M_{noy}}{M_{nox}} \quad (18)$$

or

$$M_{nx} + M_{ny}\left(\frac{M_{nox}}{M_{noy}}\right)\left(\frac{1-\beta}{\beta}\right) = M_{nox} \quad (19)$$

For rectangular sections with reinforcement equally distributed on all faces,

$$M_{nx} + M_{ny}\frac{h_a}{b}\left(\frac{1-\beta}{\beta}\right) \approx M_{nox} \quad (20)$$

In design Eqs. (17) and (20), the ratio b/h_a or h_a/b must be chosen and the value of β must be assumed. For lightly loaded columns, β will generally vary from 0.55 to about 0.70. Hence, a value of 0.65 for β is generally a good initial choice in a biaxial bending analysis.

MANUAL DESIGN PROCEDURE

To aid the engineer in designing columns for biaxial bending, a procedure for manual design is outlined below:

1. Choose the value of β at 0.65 or use Figs. 7-15 and 7-16 to make an estimate.

2. If M_{ny}/M_{nx} is greater than b/h, use Eq. (17) to calculate an approximate equivalent uniaxial moment strength M_{noy}. If M_{ny}/M_{nx} is less than b/h_a, use Eq. (20) to calculate an approximate equivalent uniaxial moment strength M_{nox}.

3. Design the section using any of the methods presented above for uniaxial bending with axial load to provide an axial load strength P_n and an equivalent uniaxial moment strength M_{noy} or M_{nox}.

4. Verify the section chosen by any one of the following three methods:

    a. <u>Bresler Reciprocal Load Method</u>:

    $$P_n \leq \frac{1}{\frac{1}{P_{ox}} + \frac{1}{P_{oy}} - \frac{1}{P_o}} \tag{7}$$

    b. <u>Bresler Load Contour Method</u>:

    $$\frac{M_{nx}}{M_{nox}} + \frac{M_{ny}}{M_{noy}} \leq 1.0 \tag{11}$$

    c. <u>PCA Load Contour Method</u>: Use Eq. (14) or,

    $$\frac{M_{nx}}{M_{nox}}\left(\frac{1-\beta}{\beta}\right) + \frac{M_{ny}}{M_{noy}} \leq 1.0 \quad \text{for} \quad \frac{M_{ny}}{M_{nx}} > \frac{M_{noy}}{M_{nox}} \tag{15}$$

    $$\frac{M_{nx}}{M_{nox}} + \frac{M_{ny}}{M_{noy}}\left(\frac{1-\beta}{\beta}\right) \leq 1.0 \quad \text{for} \quad \frac{M_{ny}}{M_{nx}} < \frac{M_{noy}}{M_{nox}} \tag{18}$$

REFERENCES

7.1 Wang, C.K. and Salmon, C.G., *Reinforced Concrete Design*, Fourth Edition, Harper & Row Publishers, New York, N.Y., 1985.

7.2 Mast, R.F., (1992), "Unified Design Provisions for Reinforced and Prestressed Concrete Flexural and Compression Members", *ACI Structural Journal*, V.89, pp. 185-199.

7.3 Munshi, J.A., (1998), "Design of Reinforced Concrete Flexural Sections by Unified Design Approach," *ACI Structural Journal*, V.95, pp. 618-625.

7.4 Pannell, F. N., "The Design of Biaxially Loaded Columns by Ultimate Load Methods," *Magazine of Concrete Research*, London, July 1960, pp. 103-104.

7.5 Pannell, F. N., "Failure Surfaces for Members in Compression and Biaxial Bending," *ACI Journal, Proceedings* Vol. 60, January 1963, pp. 129-140.

7.6 Bresler, Boris, "Design Criteria for Reinforced Columns under Axial Load and Biaxial Bending," *ACI Journal, Proceedings* Vol. 57, November 1960, pp. 481-490, discussion pp. 1621-1638.

7.7 Furlong, Richard W., "Ultimate Strength of Square Columns under Biaxially Eccentric Loads," *ACI Journal, Proceedings* Vol. 58, March 1961, pp. 1129-1140.

7.8 Meek, J. L., "Ultimate Strength of Columns with Biaxially Eccentric Loads," *ACI Journal, Proceedings* Vol., 60, August 1963, pp. 1053-1064.

7.9 Aas-Jakobsen, A., "Biaxial Eccentricities in Ultimate Load Design," *ACI Journal, Proceedings* Vol. 61, March 1964, pp. 293-315.

7.10 Ramamurthy, L. N., "Investigation of the Ultimate Strength of Square and Rectangular Columns under Biaxially Eccentric Loads," Symposium on Reinforced Concrete Columns, American Concrete Institute, Detroit, MI, 1966, pp. 263-298.

7.11 *Capacity of Reinforced Rectangular Columns Subject to Biaxial Bending*, Publication EB011D, Portland Cement Association, Skokie, IL, 1966.

7.12 *Biaxial and Uniaxial Capacity of Rectangular Columns*, Publication EB031D, Portland Cement Association, Skokie, IL, 1967.

Example 7.1—Design of Rectangular Beam with Tension Reinforcement Only

Select a rectangular beam size and required reinforcement A_s to carry service load moments $M_D = 56$ ft-kips and $M_L = 35$ ft-kips. Select reinforcement to control flexural cracking.

$f'_c = 4000$ psi
$f_y = 60{,}000$ psi

Calculations and Discussion	Code Reference

1. To illustrate a complete design procedure for rectangular sections with tension reinforcement only, a minimum beam depth will be computed using the maximum reinforcement permitted for tension-controlled flexural members, ρ_t. The design procedure will follow the method outlined on the preceding pages.   *10.3.4*

   Step 1. Determine maximum tension-controlled reinforcement ratio for material strengths $f'_c = 4000$ psi and $f_y = 60{,}000$ psi.

   $\rho_t = 0.01806$ from Table 6-1

   Step 2. Compute bd^2 required.

   Required moment strength:

   $M_u = (1.2 \times 56) + (1.6 \times 35) = 123.2$ ft-kips   *Eq. (9-2)*

   $$R_n = \rho f_y \left(1 - \frac{0.5 \rho f_y}{0.85 f'_c}\right)$$

   $$= (0.01806 \times 60{,}000)\left(1 - \frac{0.5 \times 0.01806 \times 60{,}000}{0.85 \times 4000}\right) = 911 \text{ psi}$$

   $$bd^2 \text{ (required)} = \frac{M_u}{\phi R_n} = \frac{123.2 \times 12 \times 1000}{0.90 \times 911} = 1803 \text{ in.}^3$$

   Step 3. Size member so that bd^2 provided $\geq bd^2$ required.

   Set $b = 10$ in. (column width)

   $$d = \sqrt{\frac{1803}{10}} = 13.4 \text{ in.}$$

   Minimum beam depth $\approx 13.4 + 2.5 = 15.9$ in.

| Example 7.1 (cont'd) | Calculations and Discussion | Code Reference |

For moment strength, a 10 × 16 in. beam size is adequate. However, deflection is an essential consideration in designing beams by the Strength Design Method. Control of deflection is discussed in Part 10.

Step 4. Using the 16 in. beam depth, compute a revised value of ρ. For illustration, ρ will be computed by all four methods outlined earlier.

$d = 16 - 2.5 = 13.5$ in.

1. By Eq. (4) (exact method):

$$R_n = \frac{M_u}{\phi(bd^2 \text{ provided})} = \frac{123.2 \times 12 \times 1000}{0.90(10 \times 13.5^2)} = 901 \text{ psi}$$

$$\rho = \frac{0.85 f'_c}{f_y}\left(1 - \sqrt{1 - \frac{2R_n}{0.85 f'_c}}\right)$$

$$= \frac{0.85 \times 4}{60}\left(1 - \sqrt{1 - \frac{2 \times 901}{0.85 \times 4000}}\right) = 0.0178$$

2. By strength curves such as shown in Fig. 7-2:

for $R_n = 901$ psi, $\rho \approx 0.0178$

3. By moment strength tables such as Table 7-1:

$$\frac{M_u}{\phi f'_c bd^2} = \frac{123.2 \times 12 \times 1000}{0.90 \times 4000 \times 10 \times 13.5^2} = 0.2253$$

$\omega \approx 0.2676$

$$\rho = \frac{\omega f'_c}{f_y} = 0.2676 \times \frac{4}{60} = 0.0178$$

4. By approximate proportion:

$$\rho \approx (\text{original } \rho) \frac{(\text{revised } R_n)}{(\text{original } R_n)}$$

$$\rho = 0.01806 \times \frac{901}{911} = 0.0179$$

Example 7.1 (cont'd) — **Calculations and Discussion** — **Code Reference**

Step 5. Compute A_s required.

$$A_s = (\text{revised } \rho)(\text{bd provided})$$

$$= 0.0178 \times 10 \times 13.5 = 2.40 \text{ in.}^2$$

2. A review of the correctness of the computations can be made by considering statics.

$$T = A_s f_y = 2.40 \times 60 = 144.0 \text{ kips}$$

$$a = \frac{A_s f_y}{0.85 f'_c b} = \frac{144.0}{0.85 \times 4 \times 10} = 4.24 \text{ in.}$$

Design moment strength:

$$\phi M_n = \phi\left[A_s f_y\left(d - \frac{a}{2}\right)\right] = 0.9\left[144.0\left(13.5 - \frac{4.24}{2}\right)\right]$$

$$= 1{,}475 \text{ in.-kips} = 122.9 \text{ ft-kips} \approx \text{required } M_u = 123.2 \text{ ft-kips} \quad \text{O.K.}$$

3. Select reinforcement to satisfy distribution of flexural reinforcement requirements of 10.6. *10.6*

   A_s required = 2.40 in.^{2}

   For illustrative purposes, select 1-No. 9 and 2-No. 8 bars ($A_s = 2.40$ in.^{2}). For practical design and detailing, one bar size for total A_s is preferable.

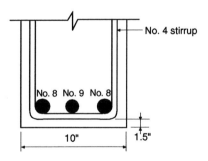

$c_c = 1.5 + 0.5 = 2.0$ in.

Maximum spacing allowed,

Example 7.1 (cont'd)   **Calculations and Discussion**   **Code Reference**

$$s = 15\left(\frac{40,000}{f_s}\right) - 2.5c_c \leq 12\left(\frac{40,000}{f_s}\right)$$ *Eq. (10-4)*

Use $f_s = \frac{2}{3} f_y = 40$ ksi

$$s = 15\left(\frac{40,000}{40,000}\right) - 2.5 \times 2 = 10 \text{ in. (governs)}$$

or, $s = 12\left(\dfrac{40,000}{40,000}\right) = 12$ in.

or, refer to Table 9-1: for $f_s = 40$ ksi and $c_c = 2$, $s = 10$ in.

$$\text{Spacing provided} = \frac{1}{2}\left\{10 - 2\left(1.5 + 0.5 + \frac{1.0}{2}\right)\right\}$$

$$= 2.50 \text{ in.} < 10 \text{ in.} \quad \text{O.K.}$$

Example 7.2—Design of One-Way Solid Slab

Determine required thickness and reinforcement for a one-way slab continuous over two or more equal spans. Clear span $\ell_n = 18$ ft.

f'_c = 4000 psi
f_y = 60,000 psi
Service loads: w_d = 75 psf (assume 6-in. slab), w_ℓ = 50 psf

Calculations and Discussion	Code Reference

1. Compute required moment strengths using approximate moment analysis permitted by 8.3.3. Design will be based on end span.

   Factored load $q_u = (1.2 \times 75) + (1.6 \times 50) = 170$ psf *Eq. (9-2)*

   Positive moment at discontinuous end integral with support:

   $+M_u = q_u \ell_n^2 / 14 = 0.170 \times 18^2/14 = 3.93$ ft-kips/ft *8.3.3*

   Negative moment at exterior face of first interior support:

   $-M_u = q_u \ell_n^2 / 14 = 0.170 \times 18^2/10 = 5.51$ ft-kips/ft *8.3.3*

2. Determine required slab thickness. *10.3.3*

   Choose a reinforcement percentage ρ equal to about $0.5\rho_t$, or one-half the maximum permitted for tension-controlled sections, to have reasonable deflection control.

   From Table 6-1, for $f'_c = 4000$ psi and $f_y = 60,000$ psi: $\rho_t = 0.01806$

   Set $\rho = 0.5 (0.01806) = 0.00903$

   Design procedure will follow method outlined earlier:

   $$R_n = \rho f_y \left(1 - \frac{0.5\rho f_y}{0.85 f'_c}\right)$$

   $$= (0.00903 \times 60,000) \left(1 - \frac{0.5 \times 0.00903 \times 60,000}{0.85 \times 4000}\right) = 499 \text{ psi}$$

   Required $d = \sqrt{\dfrac{M_u}{\phi R_n b}} = \sqrt{\dfrac{5.51 \times 12,000}{0.90 \times 499 \times 12}} = 3.50$ in.

   Assuming No. 5 bars, required $h_a = 3.50 + 0.31/2 + 0.75 = 4.41$ in.

   The above calculations indicate a slab thickness of 4.5 in. is adequate. However, Table 9-5(a) indicates a minimum thickness of $\ell/24 \geq 9$ in., unless deflections are computed. Also note that Table 9-5(a) is applicable only to "members in one-way construction not supporting or attached

| Example 7.2 (cont'd) | Calculations and Discussion | Code Reference |

to partitions or other construction likely to be damaged by large deflections." Otherwise deflections must be computed.

For purposes of illustration, the required reinforcement will be computed for $h_a = 4.5$ in., $d = 3.59$ in.

3. Compute required negative moment reinforcement.

$$R_n = \frac{M_u}{\phi bd^2} = \frac{5.51 \times 12 \times 1000}{0.9 \times 12 \times 3.59^2} = 475$$

$$\rho \approx 0.00903 \left(\frac{475}{499}\right) = 0.00860$$

$-A_s$ (required) $= \rho bd = 0.00860 \times 12 \times 3.59 = 0.37$ in.^{2}/ft

Use No. 5 @ 10 in. ($A_s = 0.37$ in.^{2}/ft)

4. For positive moment, use Table 7-1:

$$\frac{M_u}{\phi f'_c bd^2} = \frac{3.93 \times 12,000}{0.9 \times 4000 \times 12 \times 3.59^2} = 0.0847$$

From Table 7-1, $\omega \approx 0.090$

$$\rho = \frac{\omega f'_c}{f_y} = 0.090 \times \frac{4}{60} = 0.006$$

$+A_s$ (required) $= \rho bd = 0.006 \times 12 \times 3.59 = 0.258$ in.^{2}/ft

Use No. 4 @ 9 in. ($A_s = 0.27$ in.^{2}/ft) or No. 5 @ 12 in. ($A_s = 0.31$ in.^{2}/ft)

Example 7.3—Design of Rectangular Beam with Compression Reinforcement

A beam cross-section is limited to the size shown. Determine the required area of reinforcement for service load moments M_D = 430 ft-kips and M_L = 175 ft-kips. Check crack control requirements of 10.6.

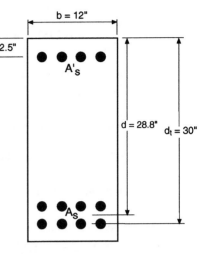

f'_c = 4000 psi
f_y = 60,000 psi

Calculations and Discussion	Code Reference

1. Determine required reinforcement.

    Step 1. Determine if compression reinforcement is needed.

    $$M_u = 1.2M_D + 1.6M_L = 796 \text{ ft-kips}$$ Eq. (9-2)
    $$M_n = M_u / \phi = 796/0.9 = 884 \text{ ft-kips}$$

    $$R_n = \frac{M_n}{bd^2} = \frac{884 \times 12 \times 1000}{12 \times 30^2} = 982$$

    This exceeds the maximum R_n of 911 for tension-controlled sections of 4000 psi concrete, without compression reinforcement. (see Table 6-1.) Also, it appears likely that two layers of tension reinforcement will be necessary. Estimate $d = d_t - 1.2$ in. $= 28.8$ in.

    Step 2. Find the nominal strength moment resisted by the concrete section, without compression reinforcement.

    $$\rho_t = 0.01806 \text{ from Table 6-1}$$

    $$\rho = \rho_t \left(\frac{d_t}{d}\right) = 0.01806 \left(\frac{30}{28.8}\right) = 0.01881$$ (6)

    $$\omega = \rho \frac{f_y}{f'_c} = 0.01881 \times \frac{60}{4} = 0.282$$

7-29

Example 7.3 (cont'd)   **Calculations and Discussion**   **Code Reference**

$$\frac{M_{nt}}{f'_c b d^2} = 0.2351 \text{ from Table 7-1}$$

$M_{nt} = 0.2351 \times 4 \times 12 \times 28.8^2 = 9{,}360 \text{ in.-kips} = 780 \text{ ft-kips}$
resisted by the concrete

Required moment strength to be resisted by the compression reinforcement:

$M'_n = 884 - 780 = 104 \text{ ft-kips}$

Step 3. Determine the compression steel stress f'_s.

Check yielding of compression reinforcement. Since the section was designed at the tension-controlled net tensile strain limit $\varepsilon_t = 0.005$, $c_{a1}/d_t = 0.375$

$c_{a1} = 0.375 d_t = 0.375 \times 30 = 11.25 \text{ in.}$

$d'/c_{a1} = 2.5/11.25 = 0.22 < 0.31$

Compression reinforcement yields at the nominal strength ($f'_s = f_y$)

Step 4. Determine the total required reinforcement:

$$A'_s = \frac{M'_n}{f_y(d-d')}$$

$$= \frac{104 \times 12 \times 1000}{60{,}000 \,(28.8 - 2.5)} = 0.79 \text{ in.}^2$$

$A_s = 0.79 + \rho b d$

$ = 0.79 + (0.01881 \times 12 \times 28.8) = 7.29 \text{ in.}^2$

Step 5. Check moment capacity.

When the compression reinforcement yields:

$$a = \frac{(A_s - A'_s)f_y}{0.85 f'_c b} = \frac{6.50 \times 60}{0.85 \times 4 \times 12} = 9.56 \text{ in.}$$

$$\phi M_n = \phi \left[(A_s - A'_s) f_y \left(d - \frac{a}{2} \right) + A'_s f_y (d - d') \right]$$

$$= 0.9 \left[6.50 \times 60 \left(28.8 - \frac{9.56}{2} \right) + (0.79 \times 60)(28.8 - 2.5) \right] / 12$$

Example 7.3 (cont'd) — **Calculations and Discussion** — **Code Reference**

$ = 796$ ft-kips $= M_u = 796$ ft-kips   O.K.

2. Select reinforcement to satisfy control of flexural cracking criteria of 10.6.

   Compression reinforcement:

   Select 2-No. 6 bars ($A'_s = 0.88$ in.$^2 > 0.79$ in.^{2})

   Tension reinforcement:

   Select 6-No. 10 bars in two layers ($A_s = 7.62$ in.$^2 > 7.29$ in.^{2})

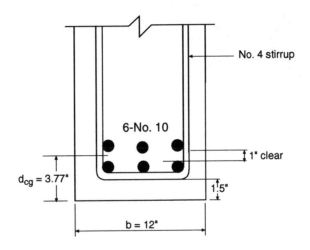

   Maximum spacing allowed,

   $$s = 15\left(\frac{40{,}000}{f_s}\right) - 2.5c_c \leq 12\left(\frac{40{,}000}{f_s}\right) \qquad \textit{Eq. (10-4)}$$

   $c_c = 1.5 + 0.5 = 2.0$ in.

   Use $f_s = \frac{2}{3}f_y = 40$ ksi

   $$s = 15\left(\frac{40{,}000}{40{,}000}\right) - (2.5 \times 2) = 10 \text{ in. (governs)}$$

   $$\text{or,} = 12\left(\frac{40{,}000}{40{,}000}\right) = 12 \text{ in.}$$

   $$\text{Spacing provided} = \frac{1}{2}\left\{12 - 2\left(1.5 + 0.5 + \frac{1.27}{2}\right)\right\}$$

   $$= 4.68 \text{ in.} < 10 \text{ in.}   \text{O.K.}$$

7-31

Example 7.3 (cont'd)	Calculations and Discussion	Code Reference

4. Stirrups or ties are required throughout distance where compression reinforcement is required for strength.
       7.11.1

   Max. spacing = 16 × long. bar dia. = 16 × 0.75 = 12 in. (governs)   7.10.5.2

                = 48 × tie bar dia. = 48 × 0.5 = 24 in.

                = least dimension of member = 12 in.

Use s_{max} = 12 in. for No. 4 stirrups

Using the simplified assumption of $d = d_t$, the extra steel is only 1.2 percent (calculations are not shown).

Example 7.4—Design of Flanged Section with Tension Reinforcement Only

Select reinforcement for the T-section shown, to carry service dead and live load moments of M_D = 72 ft-kips and M_L = 88 ft-kips.

f'_c = 4000 psi
f_y = 60,000 psi

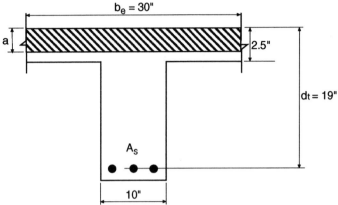

	Calculations and Discussion	Code Reference

1. Determine required flexural strength.

   $M_u = (1.2 \times 72) + (1.6 \times 88) = 227$ ft-kips   *Eq. (9-2)*

2. Using Table 7-1, determine depth of equivalent stress block a, as for a rectangular section. Assume $\phi = 0.9$.

   $$\frac{M_u}{\phi f'_c b d^2} = \frac{227 \times 12}{0.9 \times 4 \times 30 \times 19^2} = 0.0699$$

   From Table 7-1, $\omega \approx 0.073$

   $$a = \frac{A_s f_y}{0.85 f'_c b} = \frac{\rho d f_y}{0.85 f'_c} = 1.18 \omega d = 1.18 \times 0.073 \times 19 = 1.64 \text{ in.} < 2.5 \text{ in.}$$

   With $a < h_f$, determine A_s as for a rectangular section (see Ex. 7.5 for the case when $a > h_f$).

   Check ϕ:

   $c_{a1} = a/\beta_1 = 1.64/0.85 = 1.93$ in.

   $c_{a1}/d_t = 1.93/19 = 0.102 < 0.375$

   Section is tension-controlled, and $\phi = 0.9$.

7-33

| Example 7.4 (cont'd) | Calculations and Discussion | Code Reference |

3. Compute A_s required.

$$A_s f_y = 0.85 f'_c b a$$

$$A_s = \frac{0.85 \times 4 \times 30 \times 1.64}{60} = 2.78 \text{ in.}^2$$

Alternatively,

$$A_s = \rho b d = \omega \frac{f'_c}{f_y} b d$$

$$= 0.073 \times \frac{4}{60} \times 30 \times 19 = 2.77 \text{ in.}^2$$

Try 3-No. 9 bars ($A_s = 3.0$ in.^{2}).

4. Check minimum required reinforcement.   10.5

For $f'_c < 4444$ psi,   Eq. (10-3)

$$\rho_{min} = \frac{200}{f_y} = \frac{200}{60,000} = 0.0033$$

$$\frac{A_s}{b_w d} = \frac{3.0}{10 \times 19} = 0.0158 > 0.0033 \quad \text{O.K.}$$

5. Check distribution of reinforcement.   10.6

Maximum spacing allowed,

$$s = 15\left(\frac{40,000}{f_s}\right) - 2.5 c_c \leq 12\left(\frac{40,000}{f_s}\right)$$

     Eq. (10-4)

$c_c = 1.5 + 0.5 = 2.0$ in.

Use $f_s = \frac{2}{3} f_y = 40$ ksi

$$s = 15\left(\frac{40,000}{40,000}\right) - (2.5 \times 2) = 10 \text{ in. (governs)}$$

$$s = 12\left(\frac{40,000}{40,000}\right) = 12 \text{ in.}$$

$$\text{Spacing provided} = \frac{1}{2}\left\{10 - 2\left(1.5 + 0.5 + \frac{1.128}{2}\right)\right\}$$

$$= 2.44 \text{ in.} < 10 \text{ in.} \quad \text{O.K.}$$

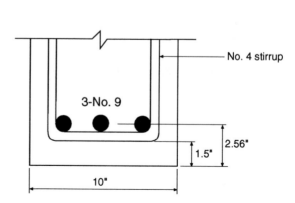

Example 7.5—Design of Flanged Section with Tension Reinforcement Only

Select reinforcement for the T-section shown, to carry a factored moment of M_u = 400 ft-kips.

f'_c = 4000 psi
f_y = 60,000 psi

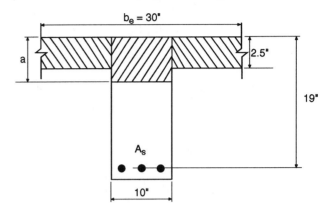

Calculations and Discussion	Code Reference

1. Determine required reinforcement.

    Step 1. Using Table 7-1, determine depth of equivalent stress block a, as for a rectangular section.

    Assume tension-controlled section, $\phi = 0.9$.

    $M_n = M_u / \phi = 400/0.9 = 444$ ft-kips

    Assume a < 2.5 in.

    $$\frac{M_n}{f'_c b d^2} = \frac{444 \times 12}{4 \times 30 \times 19^2} = 0.123$$

    From Table 7-1, $\omega \approx 0.134$

    $$a = \frac{A_s f_y}{0.85 f'_c b} = 1.18 \omega d$$

    $= 1.18 \times 0.134 \times 19 = 3.0$ in. > 2.5 in.

    Step 2. Since the value of a as a rectangular section exceeds the flange thickness, the equivalent stress block extends in the web, and the design must be based on T-section behavior. See Example 7.4 when a is less than the flange depth.

    Step 3. Compute required reinforcement A_{sf} and nominal moment strength M_{nf} corresponding to the overhanging beam flange in compression (see Part 6).

    Compressive strength of flange

Example 7.5 (cont'd)	Calculations and Discussion	Code Reference

$$C_f = 0.85 f'_c (b - b_w) h_f$$

$$= 0.85 \times 4 (30 - 10) 2.5 = 170 \text{ kips}$$

Required A_{sf} to equilibrate C_f:

$$A_{sf} = \frac{C_f}{f_y} = \frac{170}{60} = 2.83 \text{ in.}^2$$

Nominal moment strength of flange:

$$M_{nf} = \left[A_{sf} f_y \left(d - \frac{h_f}{2} \right) \right]$$

$$= [2.83 \times 60 (19 - 1.25)]/12 = 251 \text{ ft-kips}$$

Step 4. Required nominal moment strength to be carried by beam web:

$$M_{nw} = M_n - M_{nf} = 444 - 251 = 193 \text{ ft-kips}$$

Step 5. Using Table 7-1, compute reinforcement A_{sw} required to develop moment strength to be carried by the web.

$$\frac{M_{nw}}{f'_c b d^2} = \frac{193 \times 12}{4 \times 10 \times 19^2} = 0.1604$$

From Table 7-1, $\omega_w \approx 0.179$

$$\rho_w = 0.179 \times \frac{4}{60} = 0.01193$$

Step 6. Check to see if section is tension-controlled, with $\phi = 0.9$:

$$\rho_t = 0.01806 \text{ from Table 6-1}$$

Therefore, $\rho_w < \rho_t$ and section is tension-controlled ($\phi = 0.9$)

$$A_{sw} = \rho_w bd = 0.01193 \times 10 \times 19 = 2.27 \text{ in.}^2$$

Step 7. Total reinforcement required to carry factored moment $M_u = 400$ ft-kips:

$$A_s = A_{sf} + A_{sw} = 2.83 + 2.27 = 5.10 \text{ in.}^2$$

Step 8. Check moment capacity.

$$\phi M_n = \phi \left[(A_s - A_{sf}) f_y \left(d - \frac{a_w}{2} \right) + A_{sf} f_y \left(d - \frac{h_f}{2} \right) \right]$$

7-36

| Example 7.5 (cont'd) | Calculations and Discussion | Code Reference |

$$a_w = \frac{(A_s - A_{sf})f_y}{0.85 f'_c b_w}$$

$$= \frac{(5.10 - 2.83) \times 60}{0.85 \times 4 \times 10} = 4.01 \text{ in.}$$

$$\phi M_n = 0.9\left[(5.10 - 2.83)\,60\left(19 - \frac{4.01}{2}\right) + (2.83 \times 60)\left(19 - \frac{2.5}{2}\right)\right]/12$$

$$= 400 \text{ ft-kips} = M_u = 400 \text{ ft-kips} \quad \text{O.K.}$$

2. Select reinforcement to satisfy crack control criteria. 10.6

   Try 5-No. 9 bars in two layers ($A_s = 5.00$ in.^{2}) (2% less than required, assumed sufficient)

   Maximum spacing allowed,

   $$s = 15\left(\frac{40,000}{f_s}\right) - 2.5 c_c \leq 12\left(\frac{40,000}{f_s}\right) \quad \text{Eq. (10-4)}$$

   $c_c = 1.5 + 0.5 = 2.0$ in.

   Use $f_s = \frac{2}{3} f_y = 40$ ksi

   $$s = 15\left(\frac{40,000}{40,000}\right) - (2.5 \times 2) = 10 \text{ in. (governs)}$$

   $$s = 12\left(\frac{40,000}{40,000}\right) = 12 \text{ in.}$$

   $$\text{Spacing provided} = \frac{1}{2}\left\{10 - 2\left(1.5 + 0.5 + \frac{1.128}{2}\right)\right\}$$

   $$= 2.44 \text{ in.} < 10 \text{ in.} \quad \text{O.K.}$$

Note: Two layers of reinforcement are required, which may not have been recognized when d was assumed to be 19 in. Also, the provided steel is slightly less than required. Therefore, the overall height should be a little more than $d + d_{cg} = 22.41$ in., or the steel should be increased.

Example 7.6—Design of One-Way Joist

Determine the required depth and reinforcement for the one-way joist system shown below. The joists are 6 in. wide and are spaced 36 in. o.c. The slab is 3.5 in. thick.

f'_c = 4000 psi
f_y = 60,000 psi
Service DL = 130 psf (assumed total for joists and beams plus superimposed dead loads)
Service LL = 60 psf

Width of spandrel beams = 20 in.
Width of interior beams = 36 in.

Columns: interior = 18 × 18 in.
exterior = 16 × 16 in.
Story height (typ.) = 13 ft

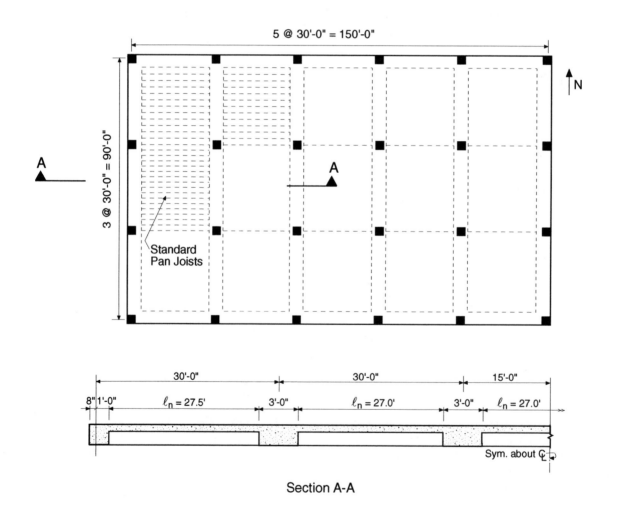

Section A-A

		Code
Example 7.6 (cont'd)	**Calculations and Discussion**	**Reference**

1. Compute the factored moments at the faces of the supports and determine the depth of the joists.

   $w_u = [(1.2 \times 0.13) + (1.6 \times 0.06)] \times 3 = 0.756$ kips/ft   *Eq. (9-2)*

   Using the approximate coefficients, the factored moments along the span are summarized in the table below.   8.3.3

Location	M_u (ft-kips)
End span	
Ext. neg.	$w_u \ell_n^2/24 = 0.756 \times 27.5^2/24 = 23.8$
Pos.	$w_u \ell_n^2/14 = 0.756 \times 27.5^2/14 = 40.8$
Int. neg.	$w_u \ell_n^2/10 = 0.756 \times 27.25^2/10 = 56.1$
Interior span	
Pos.	$w_u \ell_n^2/16 = 0.756 \times 27^2/16 = 34.4$
Neg.	$w_u \ell_n^2/11 = 0.756 \times 27^2/11 = 50.1$

   For reasonable deflection control, choose a reinforcement ratio ρ equal to about one-half ρ_t. From Table 6-1, $\rho_t = 0.01806$.

   Set $\rho = 0.5 \times 0.01806 = 0.00903$

   Determine the required depth of the joist based on $M_u = 56.1$ ft-kips:

   $$\omega = \frac{\rho f_y}{f'_c} = \frac{0.00903 \times 60}{4} = 0.1355$$

   From Table 7-1, $M_u/\phi f'_c b d^2 = 0.1247$

   $$d = \sqrt{\frac{M_u}{\phi f'_c b_w (0.1247)}} = \sqrt{\frac{56.1 \times 12}{0.9 \times 4 \times 6 \times 0.1247}} = 15.8 \text{ in.}$$

   $h_a \approx 15.8 + 1.25 = 17.1$ in.

   From Table 9-5(a), the minimum required thickness of the joist is

   $$h_{min} = \frac{\ell}{18.5} = \frac{30 \times 12}{18.5} = 19.5 \text{ in.}$$

   Use a 19.5-in. deep joist (16 + 3.5).

Example 7.6 (cont'd)   **Calculations and Discussion**   **Code Reference**

2. Compute required reinforcement.

   a. End span, exterior negative

   $$\frac{M_u}{\phi f'_c b d^2} = \frac{23.8 \times 12}{0.9 \times 4 \times 6 \times 18.25^2} = 0.0397$$

   From Table 7-1, $\omega \approx 0.041$

   $$A_s = \frac{\omega b d f'_c}{f_y} = \frac{0.041 \times 6 \times 18.25 \times 4}{60} = 0.30 \text{ in.}^2$$

   For $f'_c < 4444$ psi, use

   $$A_{s,\,min} = \frac{200 b_w d}{f_y} = \frac{200 \times 6 \times 18.25}{60,000} = 0.37 \text{ in.}^2 > A_s \qquad \text{Eq. (10-3)}$$

   Distribute bars uniformly in top slab:

   $$A_s = \frac{0.37}{3} = 0.123 \text{ in.}^2/\text{ft}$$

   Use No. 3 @ 10 in. ($A_s = 0.13$ in.^{2}/ft)

   b. End span, positive

   $$\frac{M_u}{\phi f'_c b d^2} = \frac{40.8 \times 12}{0.9 \times 4 \times 36 \times 18.25^2} = 0.0113$$

   From Table 7-1, $\omega \approx 0.012$

   $$A_s = \frac{\omega b d f'_c}{f_y} = \frac{0.012 \times 36 \times 18.25 \times 4}{60} = 0.53 \text{ in.}^2$$

   Check rectangular section behavior:

   $$a = \frac{A_s f_y}{0.85 f'_c b} = \frac{0.53 \times 60}{0.85 \times 4 \times 36} = 0.26 \text{ in.} < 3.5 \text{ in.} \quad \text{O.K.}$$

   Use 2-No. 5 bars ($A_s = 0.62$ in.^{2})

   c. End span, interior negative

   $$\frac{M_u}{\phi f'_c b d^2} = \frac{56.1 \times 12}{0.9 \times 4 \times 6 \times 18.25^2} = 0.0936$$

	Code
Example 7.6 (cont'd) Calculations and Discussion	Reference

From Table 7-1, $\omega \approx 0.100$

$$A_s = \frac{\omega b d f'_c}{f_y} = \frac{0.100 \times 6 \times 18.25 \times 4}{60} = 0.73 \text{ in.}^2$$

Distribute reinforcement uniformly in slab:

$$A_s = \frac{0.73}{3} = 0.24 \text{ in.}^2/\text{ft}$$

Use No. 5 @ 12 in. for crack control considerations in slabs (see Table 9-1).

d. The reinforcement for the other sections is obtained in a similar fashion. The following table summarizes the results. Note that at all sections, the requirements in 10.6 for crack control are satisfied.

Location	M_u	A_s	Reinforcement
End span	(ft-kips)	(in.^{2})	
Ext. neg.	23.8	0.37	No. 3@10 in.
Pos.	40.8	0.53	2-No. 5
Int. neg.	56.1	0.73	No. 5@12 in.*
Interior span			
Pos.	34.4	0.42	2-No. 5
Neg.	50.1	0.65	No. 5@12 in.*

*Maximum 12 in. spacing required for crack control in slab.

e. The slab reinforcement normal to the ribs is often located at mid-depth of the slab to resist both positive and negative moments.

$$\text{Use } M_u = \frac{w_u \ell_n^2}{12} = \frac{0.185 \times 2.5^2}{12} = 0.096 \text{ ft-kips}$$

where $w_u = 1.2 (44 + 30) + 1.6 (60)$

$ = 185 \text{ psf} = 0.185 \text{ kips/ft}^2$

$$\frac{M_u}{\phi f'_c b d^2} = \frac{0.096 \times 12}{0.9 \times 4 \times 12 \times 1.75^2} = 0.0087$$

From Table 7-1, $\omega \approx 0.0087$

$$A_s = \frac{\omega b d f'_c}{f_y} = \frac{0.0087 \times 12 \times 1.75 \times 4}{60} = 0.01 \text{ in.}^2/\text{ft}$$

Example 7.6 (cont'd)	Calculations and Discussion	Code Reference

For slabs, minimum reinforcement is governed by the provisions in 7.12.2.1:

$A_{s,min} = 0.0018 \times 12 \times 3.5 = 0.08$ in.^{2}/ft

$s_{max} = 5h = 5 \times 3.5 = 17.5$ in. (governs)   7.12.2.2

$\phantom{s_{max}} = 18$ in.

Use No. 3 @ 16 in. ($A_s = 0.08$ in.^{2}/ft)

3. Shear at supports must be checked. Since the joists meet the requirements in 8.11, the contribution of the concrete to shear strength V_c is permitted to be 10% more than that specified in Chapter 11.   8.11.8

Example 7.7—Design of Continuous Beams

Determine the required depth and reinforcement for the support beams along the interior column line in Example 7.6. The width of the beams is 36 in.

f'_c = 4000 psi
f_y = 60,000 psi
Service DL = 130 psf (assumed total for joists and beams plus superimposed dead loads)
Service LL = 60 psf

Columns: interior = 18 × 18 in.
exterior = 16 × 16 in.
Story height (typ.) = 13 ft

Calculations and Discussion	Code Reference

1. Compute the factored moments at the faces of the supports and determine the depth of the beam.

   $w_u = [(1.2 \times 0.13) + (1.6 \times 0.06)] \times 30 = 7.56$ kips/ft *Eq. (9-2)*

   Using the approximate coefficients, the factored moments along the span are summarized in the table below. *8.3.3*

Location	M_u (ft-kips)
End span	
Ext. neg.	$w_u \ell_n^2 / 16 = 7.56 \times 28.58^2/16 = 385.9$
Pos.	$w_u \ell_n^2 / 14 = 7.56 \times 28.58^2/14 = 441.1$
Int. neg.	$w_u \ell_n^2 / 10 = 7.56 \times 28.54^2/10 = 615.8$
Interior span	
Pos.	$w_u \ell_n^2 / 16 = 7.56 \times 28.50^2/16 = 383.8$

For overall economy, choose a beam depth equal to the joist depth used in Example 7.6.

Check the 19.5-in. depth for $M_u = 615.8$ ft-kips:

From Table 6-2,

$$\phi R_{nt} = 820 = \frac{M_{ut}}{bd^2}$$

$M_{ut} = 820 \times 36 \times 17^2 / 1000 = 8531$ in.-kips = 711 ft-kips

		Code
Example 7.7 (cont'd)	**Calculations and Discussion**	**Reference**

$M_u < M_{ut}$

Section will be tension-controlled without compresion reinforcement.

Check beam depth based on deflection criteria in Table 9.5(a):

$$h_{min} = \frac{\ell}{18.5} = \frac{30 \times 12}{18.5} = 19.5 \text{ in.} \quad \text{O.K.}$$

Use a 36 × 19.5 in. beam.

2. Compute required reinforcement:

   a. End span, exterior negative

   $$\frac{M_u}{\phi f'_c b d^2} = \frac{385.9 \times 12}{0.9 \times 4 \times 36 \times 17^2} = 0.1236$$

   From Table 7-1, $\omega \approx 0.134$

   $$A_s = \frac{\omega b d f'_c}{f_y} = \frac{0.134 \times 36 \times 17 \times 4}{60} = 5.47 \text{ in.}^2$$

   For $f'_c < 4444$ psi, use

   $$A_{s,min} = \frac{200 b_w d}{f_y} = \frac{200 \times 36 \times 17}{60,000} = 2.04 \text{ in.}^2 \qquad \text{Eq. (10-3)}$$

   Use 7-No. 8 bars ($A_s = 5.53$ in.^{2})

   Check distribution of flexural reinforcement requirements of 10.6.

   Maximum spacing allowed,

   $$s = 15\left(\frac{40,000}{f_s}\right) - 2.5 c_c \leq 12\left(\frac{40,000}{f_s}\right) \qquad \text{Eq. (10-4)}$$

   $c_c = 1.5 + 0.5 = 2.0$ in.

   Use $f_s = \frac{2}{3} f_y = 40$ ksi

   $$s = 15\left(\frac{40,000}{40,000}\right) - 2.5 \times 2 = 10 \text{ in. (governs)}$$

   $$s = 12\left(\frac{40,000}{40,000}\right) = 12 \text{ in.}$$

Example 7.7 (cont'd) — Calculations and Discussion — Code Reference

$$\text{Spacing provided} = \frac{1}{6}\left\{36 - 2\left(1.5 + 0.5 + \frac{1.0}{2}\right)\right\}$$

$$= 5.17 \text{ in.} < 10 \text{ in.} \quad \text{O.K.}$$

b. End span, positive

$$\frac{M_u}{\phi f'_c bd^2} = \frac{441.1 \times 12}{0.9 \times 4 \times 36 \times 17^2} = 0.1413$$

From Table 7-1, $\omega \approx 0.156$

$$A_s = \frac{\omega bd f'_c}{f_y} = \frac{0.156 \times 36 \times 17 \times 4}{60} = 6.37 \text{ in.}^2$$

Use 11-No. 7 bars ($A_s = 6.60$ in.²)

Note that this reinforcement satisfies the cracking requirements in 10.6.4, and fits adequately within the beam width. It can also conservatively be used at the midspan section of the interior span.

c. End span, interior negative

$$\frac{M_u}{\phi f'_c bd^2} = \frac{615.8 \times 12}{0.9 \times 4 \times 36 \times 17^2} = 0.1973$$

From Table 7-1, $\omega \approx 0.228$

$$A_s = \frac{\omega bd f'_c}{f_y} = \frac{0.228 \times 36 \times 17 \times 4}{60} = 9.30 \text{ in.}^2$$

Use 10-No. 9 bars ($A_s = 10.0$ in.²)

This reinforcement is adequate for cracking and spacing requirements as well.

Example 7.8—Design of a Square Column for Biaxial Loading

Determine the required square tied column size and reinforcement for the factored load and moments given. Assume the reinforcement is equally distributed on all faces.

P_u = 1200 kips, M_{ux} = 300 ft-kips, M_{uy} = 125 ft-kips

f'_c = 5000 psi, f_y = 60,000 psi

Calculations and Discussion	Code Reference

1. Determine required nominal strengths, assuming compression-controlled behavior: 9.3.2.2(b)

$$P_n = \frac{P_u}{\phi} = \frac{1200}{0.65} = 1846 \text{ kips}$$

$$M_{nx} = \frac{M_{ux}}{\phi} = \frac{300}{0.65} = 461.5 \text{ ft-kips}$$

$$M_{ny} = \frac{M_{uy}}{\phi} = \frac{125}{0.65} = 192.3 \text{ ft-kips}$$

2. Assume β = 0.65

3. Determine an equivalent uniaxial moment strength M_{nox} or M_{noy}.

$$\frac{M_{ny}}{M_{nx}} = \frac{192.3}{465.1} = 0.42 \text{ is less than } \frac{b}{h_a} = 1.0 \text{ (square column)}$$

Therefore, using Eq. (20)

$$M_{nox} \approx M_{nx} + M_{ny} \frac{h_a}{b} \left(\frac{1-\beta}{\beta} \right)$$

$$= 461.5 + \left[192.3 \times (1.0) \left(\frac{1-0.65}{0.65} \right) \right] = 565.1 \text{ ft-kips}$$

4. Assuming a 24 in. square column, determine the reinforcement required to provide an axial load strength P_n = 1846 kips and an equivalent uniaxial moment strength M_{nox} = 565.1 ft-kips

The figure below is an interaction diagram for this column with 4-No. 11 bars. The section is adequate with this reinforcement for (P_n, M_{nox}).

		Code
Example 7.8 (cont'd)	Calculations and Discussion	Reference

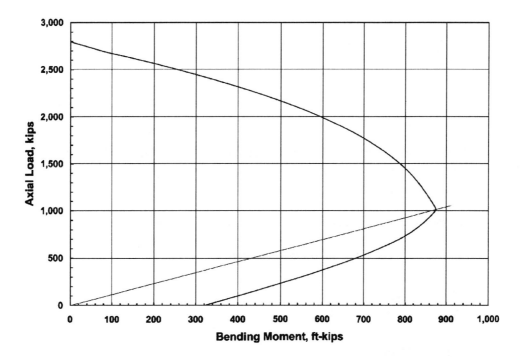

5. Selected section will now be checked for biaxial strength by each of the three methods presented in the discussion.

   a. <u>Bresler Reciprocal Load Method</u>

   Check $P_n \geq 0.1 f'_c A_g$ (8)

   1714 kips > 0.1 (5) (576) = 288 kips   O.K.

   To employ this method, P_o, P_{ox}, and P_{oy} must be determined.

   $P_o = 0.85 f'_c (A_g - A_{st}) + A_{st} f_y$

        = 0.85 (5) (576 − 6.24) + 6.24 (60) = 2796 kips

   P_{ox} is the uniaxial load strength when only M_{nx} acts on the column. From the interaction diagram, P_{ox} = 2225 kips when M_{nx} = 461.5 ft-kips.

   Similarly, P_{oy} = 2575 kips when M_{ny} = 192.3 ft-kips. Note that both P_{ox} and P_{oy} are greater than the balanced axial force, so that the section is compression-controlled.

   Using the above values, Eq. (7) can now be evaluated:

   $$P_n = 1846 \text{ kips} \leq \frac{1}{\frac{1}{P_{ox}} + \frac{1}{P_{oy}} - \frac{1}{P_o}}$$

		Code
Example 7.8 (cont'd)	**Calculations and Discussion**	**Reference**

$$< \frac{1}{\frac{1}{2225} + \frac{1}{2575} - \frac{1}{2796}} = 2083 \text{ kips O.K}$$

b. <u>Bresler Load Contour Method</u>

Due to a lack of available data, a conservative α value of 1.0 is chosen. Although $P_u > 0.1 f'_c A_g$, the necessary calculations will be carried out for example purposes. Since the section is symmetrical, M_{nox} is equal to M_{noy}.

From the interaction diagram, $M_{nox} = 680$ ft-kips for $P_n = 1846$ kips.

Using the above value, Eq. (11) can now be evaluated:

$$\frac{M_{nx}}{M_{nox}} + \frac{M_{ny}}{M_{noy}} = \frac{461.5}{680} + \frac{192.3}{680} = 0.68 + 0.28 = 0.96 < 1.0 \text{ O.K.}$$

c. <u>PCA Load Contour Method</u>

To employ this method, P_o, M_{nox}, M_{noy} and the true value of β must first be found.

$$P_o = 0.85 f'_c (A_g - A_{st}) + A_{st} f_y$$

$$= 0.85 (5) (576 - 6.24) + 6.24 (60) = 2796 \text{ kips}$$

Since the section is symmetrical, M_{nox} and M_{noy} are equal.

From the interaction diagram, $M_{nox} = 680$ ft-kips for $P_n = 1846$ kips.

Having found P_o and using ρ_g (actual), the true β value is determined as follows:

$$\frac{P_n}{P_o} = \frac{1846}{2796} = 0.66, \quad \omega = \frac{\rho_g f_y}{f'_c} = \frac{(6.24/24^2)}{5} = 0.13$$

From Fig. 7-15(a), read $\beta = 0.66$

Using the above values, Eq. (13) can now be evaluated:

$$\left(\frac{M_{nx}}{M_{nox}}\right)^{\left(\frac{\log 0.5}{\log \beta}\right)} + \left(\frac{M_{ny}}{M_{noy}}\right)^{\left(\frac{\log 0.5}{\log \beta}\right)} \leq 1.0$$

Example 7.8 (cont'd)   **Calculations and Discussion**   **Code Reference**

$\log 0.5 = -0.3$

$\log \beta = \log 0.66 = -0.181$

$\dfrac{\log 0.5}{\log \beta} = 1.66$

$\left(\dfrac{461.5}{680}\right)^{1.66} + \left(\dfrac{192.3}{680}\right)^{1.66} = 0.53 + 0.12 = 0.65 < 1.0$ O.K.

This section can also be checked using the bilinear approximation.

Since $\dfrac{M_{ny}}{M_{nx}} < \dfrac{M_{noy}}{M_{nox}}$, Eq. (17) should be used.

$\dfrac{M_{nx}}{M_{nox}} + \dfrac{M_{ny}}{M_{noy}}\left(\dfrac{1-\beta}{\beta}\right) = \dfrac{461.5}{680} + \dfrac{192.3}{680}\left(\dfrac{1-0.66}{0.66}\right)$

$= 0.68 + 0.15 = 0.83 < 1.0$ O.K.

Blank

8

Redistribution of Negative Moments in Continuous Flexural Members

UPDATE FOR THE '05 CODE

BACKGROUND

The behavior of a concrete member is affected by its reinforcement layout. For example, consider a three span reinforced concrete beam built monolithically with reinforcement provided only at the bottom of the beam. Prior to cracking, the beam behaves as three continuous spans. After cracking over the interior supports, the three-span beam will behave as three simply supported spans. Therefore, after cracking, redistribution of internal forces occurs in the system. However, the cracks over the interior supports may become large and unacceptable from a serviceability point of view. Section 8.4 sets rules for redistribution of negative moments in continuous beams provided they have sufficient ductility. The redistribution provisions allow for adequate serviceability.

The provisions of 8.4 are beneficial when evaluating existing structures or during the design of new structures. The procedure recognizes that the moment envelop is the result of different transient load patterns (8.9). For example, when considering the pattern that produces the largest negative moment, the designer can reduce that negative moment. This reduction, however, will cause an increase of the concurrent positive moment is the midspan. Similarly, increasing the negative moment over supports will reduce the positive moment in the midpsan. By increasing and decreasing negative moments over supports of continuous members, the negative and positive moments can be optimized and the required amount of flexural reinforcement can be economized. This procedure is illustrated in Examples 8.1 and 8.2.

8.4 REDISTRIBUTION OF NEGATIVE MOMENTS IN CONTINUOUS FLEXURAL MEMBERS

Section 8.4 permits a redistribution of negative moments in continuous flexural members if the net tensile strain exceeds a specified amount. This provision recognizes the inelastic behavior of concrete structures and constitutes a move toward "limit design." Application of moment redistribution, in many cases, results in substantial decrease in total required reinforcement, which allows avoiding reinforcement congestion or reduction of concrete dimensions.

A maximum 10 percent adjustment of negative moments was first permitted in the 1963 ACI Code. Experience with the use of that provision, though satisfactory, was still conservative. The 1971 code increased the maximum adjustment percentage up to 20 percent depending on the reinforcement indices. The increase was justified by additional knowledge of ultimate and service load behavior obtained from tests and analytical studies. Moment redistribution was allowed for both nonprestressed and prestressed members but different specifications were used for each type of member. Starting with the 2002 revision of the code, 8.4 specified the negative moment redistribution factor in terms of the net tensile strain, ε_t. This unified provision applies equally to both nonprestressed and prestressed members. Former provisions involving reinforcement indices may still be used as prescribed in B.8.4 and B.18.10.4.

According to 8.9, continuous members must be designed to resist more than one configuration of live loads. An elastic analysis is performed for each loading configuration, and an envelope moment value is obtained for the design of each section. Thus, for any of the loading conditions considered, certain sections in a given span will reach the ultimate moment while others will have reserve capacity. Tests have shown that a structure can continue to carry additional loads if the sections that reached their moment capacities continue to rotate as plastic hinges and redistribute the moments to other sections until a collapse mechanism forms.

Recognition of this additional load capacity beyond the intended original design suggests the possibility of redesign with resulting savings in material. Section 8.4 allows a redesign by decreasing or increasing the elastic negative moments for each loading condition (with the corresponding changes in positive moment required by statics). These moment changes may be such as to reduce both the maximum positive and negative moments in the final moment envelope. In order to ensure proper rotation capacity, the net tensile strain in the sections at the support must conform to 8.4. Example 8.1 illustrates this requirement.

Limits of applicability of 8.4 may be summarized as follows:

1. Provisions apply to continuous nonprestressed and prestressed flexural members.
2. Provisions do not apply to members designed by the approximate moments of 8.3.3, or to slab systems designed by the Direct Design Method (13.6.1.7).
3. Bending moments must be determined by analytical methods, such as moment distribution, slope deflection, etc. Redistribution is not allowed for moments determined through approximate methods.
4. Redistribution is only permitted when the net tensile strain is not less than 0.0075 (8.4.3).
5. Maximum allowable percentage increase or decrease of negative moment is equal to $1000\,\varepsilon_t$, but not more than 20 percent (8.4.1).
6. Adjustment of negative moments is made for each loading configuration considered. Members are then proportioned for the maximum adjusted moments resulting from all loading conditions.
7. Adjustment of negative support moments for any span requires adjustment of positive moments in the same span (8.4.2). A decrease of a negative support moment requires a corresponding increase in the positive span moment for equilibrium.
8. Static equilibrium must be maintained at all joints before and after moment redistribution.
9. In the case of unequal negative moments on the two sides of a fixed support (i.e., where adjacent spans are unequal), the difference between these two moments is taken into the support. Should either or both of these negative moments be adjusted, the resulting difference between the adjusted moments is taken into the support.
10. Moment redistribution may be carried out for additional cycles. After each cycle of redistribution, a new allowable percentage increase or decrease in negative moment is calculated. After the first iteration, the reduction is typically 15 percent off its final value, which is usually reached after three cycles.

The permissible percentage redistribution is defined in terms of the net tensile strain ε_t. In general, the design procedures outlined in Part 7 of the Notes can be used to determine the location of the neutral axis, c, which allows calculating ε_t from the expression

$$\varepsilon_t = 0.003\left(\frac{d_t}{c}-1\right) \qquad (1)$$

However, in the case of a section with a rectangular compression block and one layer of tension reinforcement only ($d_t = d$), an explicit relation between the net tensile strain, ε_t, and the nondimensional coefficient of resistance,

$$R_n / f'_c = M_n / (f'_c bd^2) = M_u / (\phi f'_c bd^2) \quad (2)$$

can be derived as follows (see Fig. 8-1).

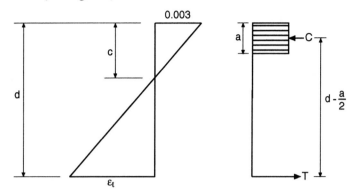

Figure 8-1 Strains and Stresses

Setting $r = c/d_t$, the depth of the concrete stress block, a, and the concrete stress block resultant, C, can respectively be expressed as:

$$a = \beta_1 c = \beta_1 r d \quad (3)$$

$$C = 0.85 f'_c ba = 0.85 f'_c b \beta_1 r d \quad (4)$$

Substituting Eq. (3) and Eq. (4) into the equilibrium condition for internal and external moments:

$$M_n = C\left(d - \frac{a}{2}\right) \quad (5)$$

results in:

$$\frac{M_n}{f'_c bd^2} = 0.85 \beta_1 r \left(1 - \frac{\beta_1 r}{2}\right) \quad (6)$$

with the nondimensional coefficient of resistance [see Eq. (2)] on the left hand side. Solving Eq. (6) with respect to r yields:

$$r = \frac{1 - \sqrt{1 - \frac{40}{17} \frac{R_n}{f'_c}}}{\beta_1} \quad (7)$$

Substituting r into Eq. (1) gives

$$\varepsilon_t = 0.003 \left(\frac{\beta_1}{1 - \sqrt{1 - \frac{40}{17} \frac{R_n}{f'_c}}} - 1 \right) \quad (8)$$

Note that Eq. (8) does not involve steel strength and is valid for use with all types of steel, including prestressing steel. Figure 8-2 shows the relationship between permissible redistribution, net tensile strain, and coefficient of resistance.

The following procedure may be utilized to determine the permissible moment redistribution.

1. Determine factored bending moments at supports by analytical elastic methods. Compute coefficients of resistance using Eq (2). Use $\phi = 0.90$ because the assumption $\varepsilon_t \geq 0.0075$ implies a tension-controlled section.
2. Use Eq. (8) to calculate ε_t, and if it satisfies $\varepsilon_t \geq 0.0075$ then determine the corresponding permissible percent redistribution $1000\, \varepsilon_t \leq 20\%$.

    Alternatively enter Fig. 8-2 with value of R_n / f_c'. Move up to intersect the appropriate curve, and move left to find the permissible percent redistribution. Interpolate between curves if needed.

3. Adjust support moments, and corresponding positive moments to satisfy equilibrium.

It usually happens that the steel provided using discrete bar sizes is somewhat more than that required. This reduces ε_t and the permissible percent redistribution slightly. However, the excess steel increases the strength far more than the change in percent redistribution. For example, referring to Fig. 8-2, the curve for 4,000 psi concrete shows a coefficient of resistance of 0.112 when $\varepsilon_t = 0.015$ and a 15 percent redistribution. If so much extra steel were provided that ε_t was reduced to 0.010, with a permissible redistribution of 10 percent, the coefficient of resistance increases from 0.112 to 0.150. Thus, a 5 percent reduction in permissible redistribution is accompanied by a 34 percent increase in strength. Consequently, it is not necessary to calculate the slight reduction in permissible redistribution, because it is offset by a far greater increase in strength.

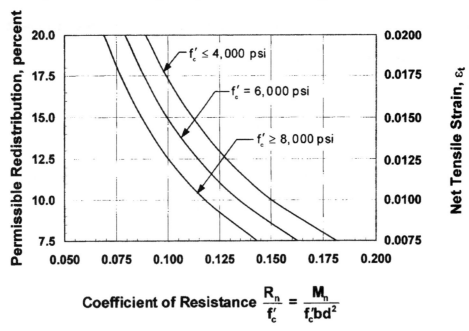

Figure 8-2 Permissible Moment Redistribution

REFERENCE

8.1 Mast, R.F., (1992), "Unified Design Provisions for Reinforced and Prestressed Concrete Flexural and Compression Members,"" *ACI Structural Journal*, V. 89, pp. 185-199.

Example 8.1—Moment Redistribution

Determine required reinforcement for the one-way joist floor shown, using moment redistribution to reduce total reinforcement.

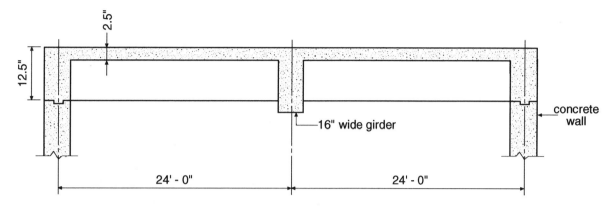

Joist-slab: 10 + 2.5 × 5 + 25 (10-in. deep form + 2.5-in. slab, 5-in. wide form spaced @ 25 in. o.c.)

f'_c = 4000 psi
f_y = 60,000 psi
DL = 80 psf
LL = 100 psf

For simplicity, continuity at concrete walls is not considered.

Calculations and Discussion	Code Reference
1. Determine factored loads.	
$\quad$ U = 1.2D + 1.6L	Eq. (9-2)
$\quad w_d = 1.2 \times 0.08 \times 25/12 = 0.200$ kips/ft	
$\quad w_\ell = 1.6 \times 0.10 \times 25/12 = 0.333$ kips/ft	
$\quad w_u = 0.533$ kips/ft per joist	
2. Obtain moment diagrams by elastic analysis.	
$\quad$ Consider three possible load patterns:	
$\quad$ Load pattern I: Factored DL and LL on both spans.	
$\quad$ Load pattern II: Factored DL and LL on one span and factored DL only on the other span.	
$\quad$ Load pattern III: Reverse of load pattern II.	
$\quad$ The elastic moment diagrams for these load cases are shown in the figure (moments shown in ft-kips).	

| Example 8.1 (cont'd) | Calculations and Discussion | Code Reference |

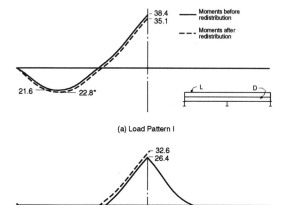

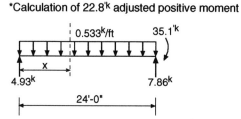

*Calculation of 22.8'k adjusted positive moment

$V = 0$ @ $x = 9.25$ ft

M @ $x = 9.25$ ft $= (4.93 \times 9.25) - 0.533 \times \dfrac{9.25^2}{2}$

$= 22.8^{'k}$

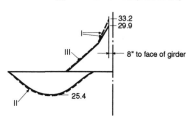

(c) Factored Moment Envelope

Figure 8-3 Redistribution of Moments for Example 8.1

3. Redistribution of negative moments.

    a. Load pattern I:

    The intent is to decrease the negative moment at the support to obtain a new moment envelope.

    From load pattern I: $M_u = -33.2$ ft-kips at face of girder.

    For $b = 5$ in., and $d = 11.5$ in.:

    $$\dfrac{R_n}{f_n} = \dfrac{33.2 \times 12}{0.9 \times 4 \times 5 \times (11.5)^2} = 0.167 \text{ and the permissible reduction}$$

    $$1000\,\varepsilon_t = 3\left(\dfrac{0.85}{1-\sqrt{\dfrac{40}{17}} \times 0.167} - 1\right) = 8.5\%$$

    Decreasing the negative moment $M_u = -38.4$ ft-kips in Fig. 8-3(a) by 8.5%, redistributed moment diagrams are obtained as shown by the dashed lines in Fig. 8-3(a).

    The maximum span moment correspondingly increases to 22.8 ft-kips by equilibrium (see calculation in figure).

| | Example 8.1 (cont'd) | Calculations and Discussion | Code Reference |

b. Load pattern II:

The elastic moment diagram of Load Pattern II is compared with the redistributed moment diagram of Load Pattern I. For savings in span positive moment reinforcement, it is desirable to reduce the span positive moment of 26.3 ft-kips. This can be achieved through redistribution of the negative moment at the support by increasing it by 8.5%, to $26.4 \times 1.085 = 28.6$ ft-kips. As a result the positive moment is reduced from 26.3 to 25.4 ft-kips.

4. Design factored moments.

From the redistributed moment envelope, factored moments and required reinforcement are determined as shown in the following table.

Table 8-1 Summary of Final Design

Section	Load Pattern		Required Steel		Provided Steel		Redistribution, percent
	I	I	A_s (in.²)	ρ	A_s (in.²)	ρ^{**}	
Support Moment* (ft-kips)	-29.9	—	0.52	0.0092	2-No.5	0.0103 (b = 5 in.)	-8.5
Midspan Moment (ft-kips)	—	25.4	0.50	0.0017	2-No. 5	0.0021*** (b = 25 in.)	+8.5 at support

*Calculated at face of support.

**Check $\rho_{min} = \dfrac{3\sqrt{f'_c}}{f_y} = 0.0032$

$\rho_{min} = \dfrac{200}{60,000} = 0.0033$ *(governs)*

*** $\rho = \dfrac{A_s}{b_w d} = \dfrac{0.51}{5 \times 11.5} = 0.0089 > \rho_{min}$

Final note: Moment redistribution has permitted a reduction of 8.5% in the negative moment. Similarly, the positive span moment has been reduced through redistribution of the negative support moment.

Example 8.2—Moment Redistribution

Determine the required reinforcement areas for the spandrel beam at an intermediate floor level as shown, using moment redistribution to reduce total reinforcement required.

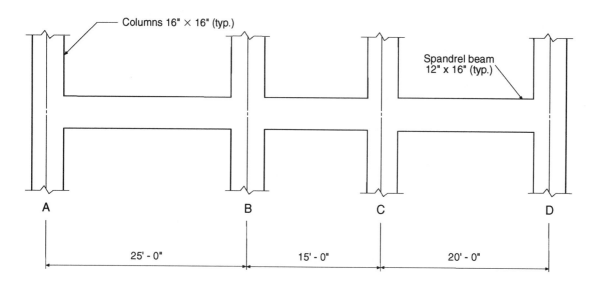

Columns = 16 × 16 in.
Story height = 10 ft
Spandrel beam = 12 × 16 in.
f'_c = 4000 psi
f_y = 60,000 psi
DL = 1167 lb/ft
LL = 450 lb/ft

Calculations and Discussion	Code Reference

1. Determine factored loads.

    U = 1.2D + 1.6L *Eq. (9-2)*

    w_d = 1.2 × 1.167 = 1.4 kips/ft

    w_ℓ = 1.6 × 0.45 = 0.72 kips/ft

    w_u = 2.12 kips/ft

2. Determine the elastic bending moment diagrams for the five load patterns shown in Figs. 8-4 (a) to (e) and the maximum moment envelope values for all load patterns. *8.9.2*

    Maximum negative moments at column counterlines and column faces, and positive midspan moments were determined by computer analysis using pcaBeam program for each of the five loading configurations. Adjusted moments after redistribution are also shown by dashed lines. The values of the adjusted moments are given in parentheses.

8-8

| Example 8.2 (cont'd) | Calculations and Discussion | Code Reference |

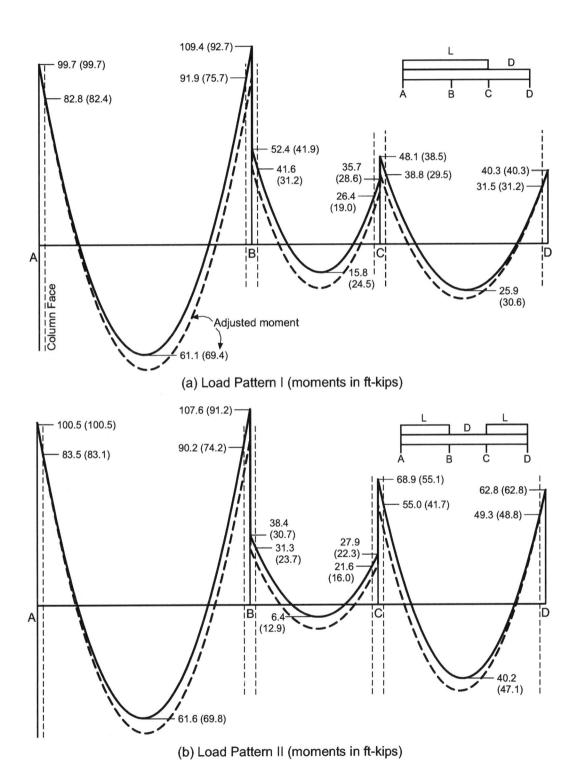

Figure 8-4 Redistribution of Moments for Example 8.2

Example 8.2 (cont'd) | **Calculations and Discussion** | **Code Reference**

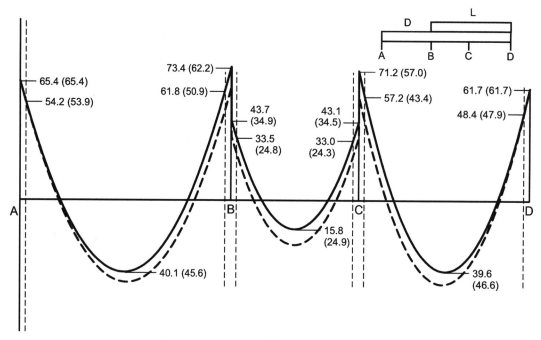

(c) Load Pattern III (moments in ft-kips)

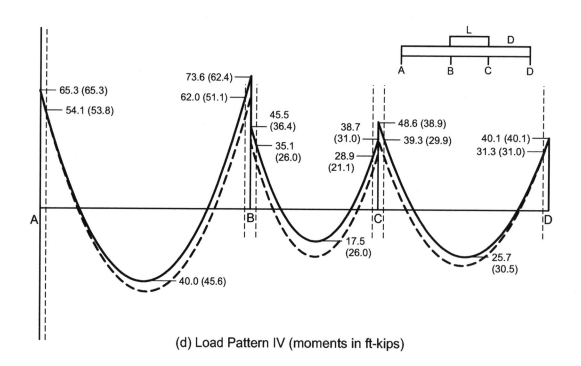

(d) Load Pattern IV (moments in ft-kips)

Figure 8-4 (continued) Redistribution of Moments for Example 8.2

Example 8.2 (cont'd)	Calculations and Discussion	Code Reference

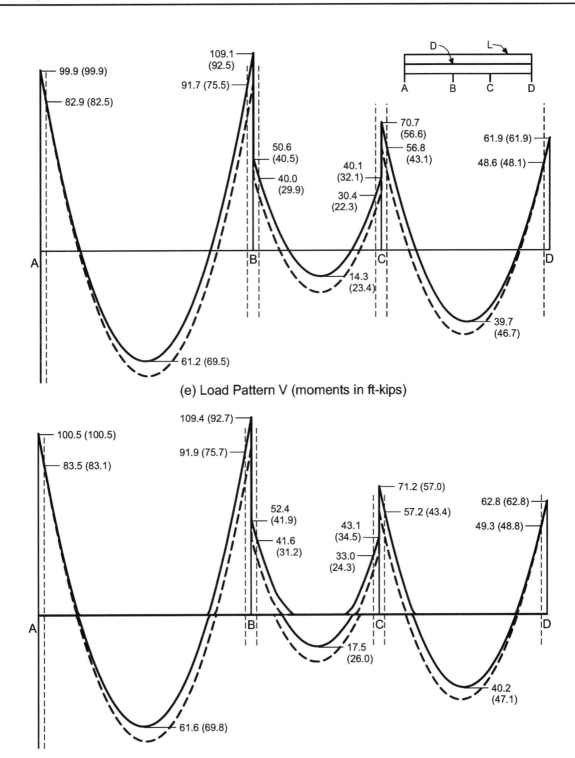

(e) Load Pattern V (moments in ft-kips)

(f) Maximum Moment Envelopes for Pattern Loading
(moments in ft-kips)

Figure 8-4 (continued) Redistribution of Moments for Example 8.2

		Code Reference
Example 8.2 (cont'd)	**Calculations and Discussion**	

3. Determine maximum allowable percentage increase or decrease in negative moments:

   use d = 14.0 in.; cover = 1.5 in.   7.7.1

   Calculate $\dfrac{R_n}{f'_c} = \dfrac{M_u}{\phi f'_c b d^2}$ and corresponding $\varepsilon_t = 0.003 \left(\dfrac{\beta_1}{1 - \sqrt{1 - \dfrac{40}{17}\dfrac{R_n}{f'_c}}} - 1 \right)$.

   For M_u use envelope value at support face. Based on ε_t calculate the adjustment. Iterate until the adjusted moments converge (starts repeating). See Table 8-2.

Table 8-2 Moment Adjustments at Supports

		Support					
		A	B		C		D
		Right	Left	Right	Left	Right	Left
Iteration 1	M_u (ft-kips)	83.5	91.9	41.6	33.0	57.2	49.3
	R_n/f'_c	0.1184	0.1303	0.0589	0.0467	0.0811	0.0699
	ε_t	0.0139	0.0122	0.0325	0.0421	0.0224	0.0267
	Adjustment (%)	13.9	12.2	20.0	20.0	20.0	20.0
Iteration 2	M_u (ft-kips)	71.9	80.7	33.3	26.4	45.8	39.4
	R_n/f'_c	0.1019	0.1143	0.0471	0.0374	0.0649	0.0559
	ε_t	0.0169	0.0146	0.0417	0.0537	0.0291	0.0345
	Adjustment (%)	16.9	14.6	20.0	20.0	20.0	20.0
Iteration 3	M_u (ft-kips)	69.4	78.5				
	R_n/f'_c	0.0984	0.1113				
	ε_t	0.0177	0.0151				
	Adjustment (%)	17.7	15.1				
Iteration 4	M_u (ft-kips)	68.8	78.0				
	R_n/f'_c	0.0975	0.1106				
	ε_t	0.0179	0.0152				
	Adjustment (%)	17.9	15.2				
Iteration 5	M_u (ft-kips)	68.6	77.9				
	R_n/f'_c	0.0972	0.1104				
	ε_t	0.0179	0.0153				
	Adjustment (%)	17.9	15.3				
Iteration 6	M_u (ft-kips)		77.9				
	R_n/f'_c		0.1104				
	ε_t		0.0153				
	Adjustment (%)		15.3				
Final Allowable Adjustment (%)		17.9	15.3	20.0	20.0	20.0	20.0

	Code
Example 8.2 (cont'd) **Calculations and Discussion**	**Reference**

4. Adjustment of moments.

   Note: Adjustment of negative moments, either increase or decrease, is a decision to be made by the engineer. In this example, it was decided to reduce the negative moments on both sides of supports B and C and accept the increase in the corresponding positive moments, and not to adjust the negative moments at the exterior supports A and D.

   Referring to Figs. 8-4(a) through (e), the following adjustments in moments are made.

   Load Pattern I — Fig. (a)
   $M_{B,Left}$ = 109.4 ft-kips (adjustment = 15.3%)
   Reduction to $M_{B,Left}$= –109.44 × 0.153 = 16.7 ft-kips
   AAdjusted $M_{B,Left}$ = –109.4–(–16.7) = –92.7 ft-kips

   Increase in positive moment in span A-B
   M_A = –99.7 ft-kips
   Adjusted $M_{B,Left}$ = –92.7 ft-kips

   Mid-span ordinate on line M_A to $M_{B,Left}$ = $\dfrac{-99.7+(-92.7)}{2}$ = –96.2 ft–kips

   Moment due to uniform load = $w_u \ell^2 / 8$ = 2.12 × 25^2/8 = 165.6 ft-kips

   Adjusted positive moment at mid-span = –96.2 + 165.6 = 69.4 ft-kips

   Decrease in negative moment at the left face of support B
   Ordinate on line M_A to $M_{B,Left}$ = $-99.7+\dfrac{-92.7-(-99.7)}{25.0}$ × 24.33 = 92.9 ft-kips

   Moment due to uniform load = $\dfrac{1}{2} w_u x(\ell - x)$ = $\dfrac{1}{2}$ × 2.12 × 24.33 × (25.0–24.33) = –17.2 ft-kips

   Adjusted negative moment at the left face of support B = –92.9 + 17.2 = –75.7 ft-kips

   Similar calculations are made to determine the adjusted moment at other locations and for other load patterns. Results of the additional calculations are shown in Table 8-3.

5. After the adjusted moments have been determined analytically, the adjusted bending moment diagrams for each loading pattern can be determined. The adjusted moment curves were determined graphically and are indicated by the dashed lines in Figs. 8-4 (a) to (e).

6. An adjusted maximum moment envelope can now be obtained from the adjusted moment curves as shown in Fig. 8-4 (f) by dashed lines.

7. Final steel ratios ρ can now be obtained on the basis of the adjusted moments.

   From the redistributed moment envelopes of Fig. 8-4 (f), the design factored moments and the required reinforcement area are obtained as shown in Table 8-4.

Example 8.2 (cont'd) — Calculations and Discussion

Table 8-3 Moments Before and After Redistribution (moments in ft-kips)

Location	Load Pattern I M_u	Load Pattern I M_{adj}	Load Pattern II M_u	Load Pattern II M_{adj}	Load Pattern III M_u	Load Pattern III M_{adj}	Load Pattern IV M_u	Load Pattern IV M_{adj}	Load Pattern IV M_u	Load Pattern IV M_{adj}
A	-99.7	-99.7	-100.5	-100.5	-65.4	-65.4	-65.3	-65.3	-99.9	-99.9
A Right Face	-82.8	-82.4	-83.5	-83.1	-54.2	-53.9	-54.1	-53.8	-82.9	-82.5
Mid-Span A-B	+61.1	+69.4	+61.6	+69.8	+40.1	+45.6	+40.0	+45.6	+61.2	+69.5
B Left Face	-91.9	-75.7	-90.2	-74.2	-61.8	-50.9	-62.0	-51.1	-91.7	-75.5
B Left Center	-109.4	-92.7	-107.6	-91.2	-73.4	-62.2	-73.6	-62.4	-109.1	-92.5
B Right Center	-52.4	-41.9	-38.4	-30.7	-43.7	-34.9	-45.5	-36.4	-50.6	-40.5
B Right Face	-41.6	-31.2	-31.3	-23.7	-33.5	-24.8	-35.1	-26.0	-40.0	-29.9
Mid-Span B-C	+15.8	+24.5	+6.4	+12.9	+15.8	+24.9	+17.5	+26.0	+14.3	+23.4
C Left Face	-26.4	-19.0	-21.6	-16.0	-33.0	-24.3	-28.9	-21.1	-30.4	-22.3
C Left Center	-35.7	-28.6	-27.9	-22.3	-43.1	-34.5	-38.7	-31.0	-40.1	-32.1
C Right Center	-48.1	-38.5	-68.9	-55.1	-71.2	-57.0	-48.6	-38.9	-70.7	-56.6
C Right Face	-38.8	-29.5	-55.0	-41.7	-57.2	-43.4	-39.3	-29.9	-56.8	-43.1
Mid-Span C-D	+25.9	+30.6	+40.2	+47.1	+39.6	+46.6	+25.7	+30.5	+39.7	+46.7
D Left Face	-31.5	-31.2	-49.3	-48.8	-48.4	-47.9	-31.3	-31.0	-48.6	-48.1
D	-40.3	-40.3	-62.8	-62.8	-61.7	-61.7	-40.1	-40.1	-61.9	-61.9

▓ Final design moments after redistribution

Table 8-4 Summary of Finasl Design

Location		Moment (ft-kips)	Load Case	Required A_s (in^2)	ρ
Support A	Right Face	-83.1	II	1.43	0.0085
Midspan A-B		69.8	II	1.18	0.0070
Support B	Left Face	-75.7	I	1.29	0.0077
Support B	Right Face	-31.2	I	0.51	0.0030
Midspan B-C		26	IV	0.42	0.0025
Support C	Left Face	-24.3	III	0.39	0.0023
Support C	Right Face	-43.4	III	0.72	0.0043
Midspan C-D		47.1	II	0.78	0.0046
Support D	Left Face	-48.8	II	0.81	0.0048

▓ Use $A_{s,min} = 200 \dfrac{b_w d}{f_y} = 200 \times \dfrac{12 \times 14}{60{,}000} = 0.56 \text{ in.}^2$

9

Distribution of Flexural Reinforcement

UPDATE FOR THE '05 CODE

Equation 10-4, for maximum bar spacing to control cracking, was modified to provide results consistent with previous editions of the code while maintaining similar level of crack control. The default steel stress at service load in the equation was increased from $0.6f_y$ to $(2/3)f_y$. The revised equation is intended to recognize the increase in service load stress level in flexural reinforcement resulting from the use of the load combinations introduced in the 2002 code.

The provisions for skin reinforcement in Section 10.6.7 were simplified and made consistent with the requirement for flexural tension reinforcement in 10.6.4. Research [Ref. 9.3] has shown that control of side face cracking can be achieved through proper spacing of the skin reinforcement for selected cover dimension. The research also confirmed that the reinforcement spacing requirements in Section 10.6.4 are sufficient to control side face cracking. To eliminate the confusion regarding the definition of effective depth for multi-layer reinforced members, 10.6.7 is simplified to require skin reinforcement based on the overall depth of the member instead of the effective depth.

GENERAL CONSIDERATIONS

Provisions of 10.6 require proper distribution of tension reinforcement in beams and one-way slabs to control flexural cracking. Structures built in the past using Working Stress Design methods and reinforcement with a yield strength of 40,000 psi or less had low tensile stresses in the reinforcement at service loads. Laboratory investigations have shown that cracking is generally in proportion to the steel tensile stress. Thus, with low tensile stresses in the reinforcement at service loads, these structures exhibited few flexural cracking problems.

With the advent of high-strength steels having yield stresses of 60,000 psi and higher, and with the use of Strength Design methods which allow higher stresses in the reinforcement, control of flexural cracking has assumed more importance. For example, if a beam were designed using Working Stress Design and a steel yield strength of 40,000 psi, the stress in the reinforcement at service loads would be about 20,000 psi. Using Strength Design and a steel yield strength of 60,000 psi, the stress at service loads could be as high as 40,000 psi. If flexural cracking is indeed proportional to steel tensile stress, then it is quite evident that the criteria for crack control must be included in the design process.

Early investigations of crack width in beams and members subject to axial tension indicated that crack width was proportional to steel stress and bar diameter, but was inversely proportional to reinforcement percentage. More recent research using deformed bars has confirmed that crack width is proportional to steel stress. However, other variables such as the quality of concrete and concrete cover were also found to be important. It should be kept in mind that there are large variations in crack widths, even in careful laboratory-controlled work.

For this reason, only a simple crack control expression, designed to give reasonable reinforcing details that are in accord with laboratory work and practical experience, is presented in the code.

10.6 BEAMS AND ONE-WAY SLABS

10.6.4 Distribution of Tension Reinforcement

There are three perceived reasons that were identified early on for limiting the crack widths in concrete. These are appearance, corrosion, and water tightness. The three seldom apply simultaneously in a particular structure. Appearance is important for concrete exposed to view such as wall panels. Corrosion is important for concrete exposed to aggressive environments. Water tightness may be required for marine/sanitary structures. Appearance requires limiting of crack widths on the surface. This can be ensured by locating the reinforcement as close as possible to the surface (by using small cover) to prevent cracks from widening. Corrosion control, on the other hand, is obtained by using better quality concrete and by increasing the thickness of concrete cover. Water tightness requires severe limits on crack widths, applicable only to specialty structures. Thus, it should be recognized that a single provision, such as Eq. (10-4) of this code, may not be sufficient to address the control of cracking for all the three different reasons of appearance, corrosion, and water tightness.

There is a strong correlation between surface crack width and cover d_c, as shown in Fig. 9-1. For a particular magnitude of strain in the steel, the larger the cover, the larger will be the surface crack width affecting the appearance. From 1971 through 1995, the code specified limiting of z-factors based on the concept that the width of surface cracks needs to be limited. The specified values of z = 175 and 145 kips/in. for interior and exterior exposures, respectively, corresponded to the limiting crack widths of 0.016 and 0.013 in. It was assumed that by limiting the crack width to these values, one would achieve corrosion protection. But in order to comply with the specified z-value limits, the method essentially encouraged reduction of the reinforcement cover, which could be detrimental to corrosion protection. Furthermore, the method severely penalized structures with covers more than 2 in. by either reducing the spacing or the service load stress of the reinforcement.

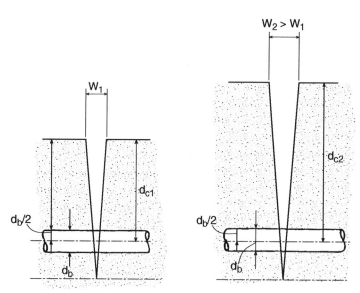

Figure 9-1 Crack Width for Different Cover Thicknesses.

The role of cracks in the corrosion of reinforcement has been found to be controversial. Research [9.1 & 9.2] shows that corrosion is not clearly correlated with surface crack widths in the range normally found with reinforcement stresses at service load level. In fact, it is weakly related to the earlier codes' surface crack width limits of 0.013 to 0.016 in. Further, it has been found that actual crack widths in structures are highly variable. A scatter of the order of ± 50% is observed. This prompted investigation of alternatives to the z factor limits for exterior and interior exposure, as given in the 1995 and earlier editions of the code.

Addressing some of the limitations of the previous approach, a simple and more practical equation has been adopted starting with the 1999 code, which directly limits the maximum reinforcement spacing. The new method is intended to control surface cracks to a width that is generally acceptable in practice but may vary widely in a given structure. The new method, for this reason, does not purport to predict crack widths in the field. According to the new method, the spacing of reinforcement closest to a tension surface shall not exceed that given by

$$s = 15\left(\frac{40,000}{f_s}\right) - 2.5c_c \qquad \text{Eq. (10-4)}$$

but not greater than $12(40,000/f_s)$

where
- s = center-to-center spacing of flexural tension reinforcement nearest to the extreme tension face, in. (where there is only one bar or wire nearest to the extreme tension face, s is the width of the extreme tension face).
- f_s = calculated stress (psi) in reinforcement at service load computed as the unfactored moment divided by the product of steel area and internal moment arm. It is permitted to take f_s as $2/3\, f_y$.
- c_c = clear cover from the nearest surface in tension to the surface of flexural tension reinforcement, in.

Note, in the 1999 and 2002, codes, the default steel stress at service load was $0.6f_y$. To recognize the increase in service load stress level in the flexural reinforcement resulting from the use of the load combinations introduced in the 2002 code, the default steel stess used in (Eq. 10-4) was adjusted in 2005 by increasing it from $0.6f_y$, to $(2/3)f_y$. Note also that contrary to the 1995 provision, this spacing is independent of the exposure condition.

For the usual case of beams with Grade 60 reinforcement with 2 in. clear cover to the tension face and assuming $f_s = 2/3(60,000) = 40,000$ psi, the maximum bar spacing is 10 in. Using the upper limit of Eq. (10-4), the maximum spacing allowed, irrespective of the cover, is 12 in. for $f_s = 40,000$ psi. The spacing limitation is independent of the bar size used. Thus for a required amount of flexural reinforcement, this approach would encourage use of smaller bar sizes to satisfy the spacing criteria of Eq. (10-4).

Although Eq. (10-4) is easy to solve, it is convenient to have a table showing maximum spacing of reinforcement for various amounts of clear cover and different service level steel stress f_s. (see Table 9-1 below).

Table 9-1 Maximum Spacing of Reinforcement

Steel Stress, f_s, (psi)	Clear Cover (in.)							
	3/4	1	1-1/4	1-1/2	1-3/4	2	2-1/2	3
30,000	16	16	16	16	15.63	15	13.75	12.5
40,000	12	12	11.88	11.25	10.63	10	8.75	7.5

* Note, maximum reinforcement spacing is 18 in. (7.6.5, 7.12.2.2, 10.5.4, 14.3.5)

10.6.5 Corrosive Environments

As described under 10.6.4, data are not available regarding crack width beyond which a danger of corrosion exists. Exposure tests indicate that concrete quality, adequate compaction, and ample cover may be of greater importance for corrosion protection than crack width at the concrete surface. The requirements of 10.6.4 do not apply to structures subject to very aggressive exposure or designed to be watertight. Special precautions are required and must be investigated for such cases.

10.6.6   Distribution of Tension Reinforcement in Flanges of T-Beams

For control of flexural cracking in the flanges of T-beams, the flexural tension reinforcement must be distributed over a flange width not exceeding the effective flange width (8.10) or 1/10 of the span, whichever is smaller. If the effective flange width is greater than 1/10 the span, some additional longitudinal reinforcement, as illustrated in Fig. 9-2, must be provided in the outer portions of the flange (see Example 9.2).

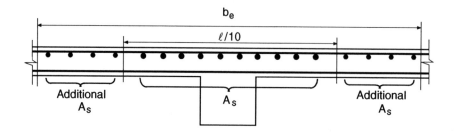

Figure 9-2 Negative Moment Reinforcement for Flanged Floor Beams

10.6.7   Crack Control Reinforcement in Deep Flexural Members

In the past, several cases of wide cracks developing on side faces of deep beams between the main reinforcement and neutral axis (Fig. 9-3(a)) have been observed. These cracks are attributed to the absence of any skin reinforcement, as a result of which cracks in the web widen more as compared to the cracks at the level of flexural tension reinforcement (Fig. 9-3(a)). For flexural members with overall height h exceeding 36 in, the code requires that additional longitudinal skin reinforcement for crack control must be distributed along the side faces of the member. The skin reinforcement must be extended for a distance h/2 from the tension face of the member. The vertical spacing s of the skin reinforcement is computed from 10.6.4 (Eq. 10-4). The code does not specify the size of the skin reinforcement. Research [Ref. 9.3] has shown that control of side face cracking can be achieved through proper spacing of the skin reinforcement for selected cover dimension. The research also confirmed that the reinforcement spacing requirements in Section 10.6.4 are sufficient to control side face cracking. Research has shown that the spacing rather than bar size is of primary importance [Ref. 9.3]. Typically No. 3 to No. 5 bars (or welded wire reinforcement with minimum area of 0.1 in.^{2} per foot of depth) is provided.

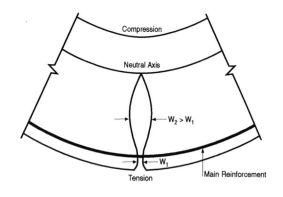 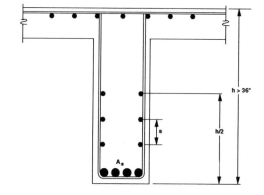

a) Side Face Cracking (Exaggerated)   b) Crack Control "Skin Reinforcement for Deep Beams

Figure 9-3 Skin Reinforcement

Note that the provisions of 10.6 do not directly apply to prestressed concrete members, as the behavior of a prestressed member is considerably different from that of a nonprestressed member. Requirements for proper distribution of reinforcement in prestressed members are given in Chapter 18 of the code and Part 24 of this book.

13.4 TWO-WAY SLABS

Control of flexural cracking in two-way slabs, including flat plates and flat slabs, is usually not a problem, and is not specifically covered in the code. However, 13.3.2 restricts spacing of slab reinforcement at critical moment sections to 2 times the slab thickness, and the area of reinforcement in each direction for two-way slab systems must not be less than that required for shrinkage and temperature (7.12). These limitations are intended in part to control cracking. Also, the minimum thickness requirements for two-way construction for deflection control (9.5.3) indirectly serve as a control on excessive cracking.

REFERENCES

9.1 Darwin, David et al, "Debate: Crack Width, Cover and Corrosion," *Concrete International,* Vol. 7, No. 5, May 1985, American Concrete Institute, Farmington Hills, MI, pp. 20-35.

9.2 Oesterle, R.G., "The Role of Concrete Cover in Crack Control Crieria and Corrosion Protection," RD Serial No. 2054, Portland Cement Association, Skokie, IL, 1997.

9.3 Frosch, R.J., "Modeling and Control of Side Face Beam Cracking,," *ACI Structural Journal,* Vol 99, No. 3, May–June 2002, pp. 376-385.

Example 9.1—Distribution of Reinforcement for Effective Crack Control

Assume a 16 in. wide beam with A_s (required) = 3.00 in.^{2}, and f_y = 60,000 psi. Select various bar arrangements to satisfy Eq. (10-4) for control of flexural cracking.

Calculations and Discussion	Code Reference

1. For 2-No. 11 bars (A_s = 3.12 in.^{2})

    c_c = 1.5 + 0.5 = 2.0 in. (No. 4 stirrup)

    use f_s = 2/3 f_y = 40 ksi

    Maximum spacing allowed,

    $$s = 15\left(\frac{40,000}{40,000}\right) - (2.5 \times 2.0) = 10 \text{ in. Eq. (10-4)}$$

    12(40,000/40,000) = 12 in. > 10 in.

    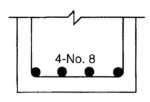

    $$\text{spacing provided} = 16 - 2\left(1.5 + 0.5 + \frac{1.41}{2}\right)$$

    $$= 10.6 \text{ in.} > 10 \text{ in.} \quad \text{N.G.}$$

2. For 4-No. 8 bars (A_s = 3.16 in.^{2})

    c_c = 2.0 in. (No. 4 stirrup)

    Maximum spacing allowed,

    s = 10 in. [Eq. (10-4)]

    $$\text{spacing provided} = \frac{1}{3}\left[16 - 2\left(1.5 + 0.5 + \frac{1.0}{2}\right)\right]$$

    $$= 3.7 \text{ in.} < 10 \text{ in.} \quad \text{O.K.}$$

Example 9.2—Distribution of Reinforcement in Deep Flexural Member with Flanges

Select reinforcement for the T-section shown below.

Span: 50 ft continuous $f'_c = 4000$ psi
                           $f_y = 60,000$ psi

Service load moments:

Positive Moment	Negative Moment
$M_d = +265$ ft-kips	$M_d = -280$ ft-kips
$M_\ell = +680$ ft-kips	$M_\ell = -750$ ft-kips

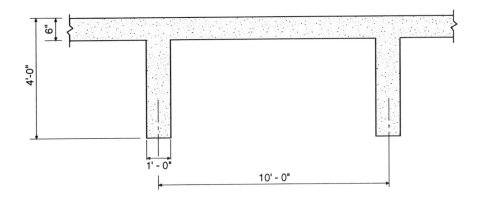

	Code
Calculations and Discussion	**Reference**

1. Distribution of positive moment reinforcement

    a. $M_u = 1.2(265) + 1.6(680) = 1406$ ft-kips   Eq. (9-2)

    Assuming 2 layers of No. 11 bars with 1.5 in. clear cover and No. 4 stirrups,

    $$d_{cg} = \frac{(3 \times 1.56)(2.71) + (2 \times 1.56)(5.12)}{(5 \times 1.56)} = 3.67 \text{ in.}$$

    $d = 48 - 3.67 = 44.3$ in.

    Effective width = 108 in.   8.10.2

    A_s required = 7.18 in.^{2}

    Try 5-No. 11 ($A_s = 7.80$ in.^{2})

| Example 9.2 (cont'd) | Calculations and Discussion | Code Reference |

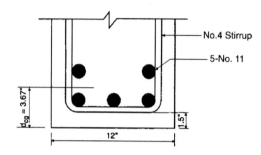

b. Clear cover to the tension reinforcement

$c_c = 1.5 + 0.5 = 2.0$ in. 10.0

Stress in reinforcement at service load: 10.6.4

$$f_s = \frac{+M}{jdA_s} = \frac{(265+680)12}{0.87 \times 44.3 \times 7.80} = 37.7 \text{ ksi}$$

$$s = 15\left(\frac{40,000}{37,700}\right) - 2.5\, c_c$$ Eq. (10-4)

$$= \frac{540}{37.7} - (2.5 \times 2) = 10.9 \text{ in.}$$

$$12\left(\frac{40}{f_s}\right) = 12\left(\frac{40}{37.7}\right)$$

$$= 12.7 \text{ in.} > 10.9 \text{ in.} \quad \text{O.K.}$$

$$\text{Spacing provided} = \frac{1}{2}\left[12 - 2\left(1.5 + 0.5 + \frac{1.41}{2}\right)\right]$$

$$= 3.3 \text{ in.} < 10.9 \text{ in.} \quad \text{O.K.}$$

2. Distribution of negative moment reinforcement

   a. $M_u = 1.2\,(280) + 1.6\,(750) = 1536$ ft-kips

   A_s required $= 8.76$ in.^{2}

   Effective width for tension reinforcement $= 1/10 \times 50 \times 12 = 60$ in. < 108 in.   10.6.6

   Try 9-No. 9 bars @ ≈ 10 in. ($A_s = 9.0$ in.^{2})

| Example 9.2 (cont'd) | Calculations and Discussion | Code Reference |

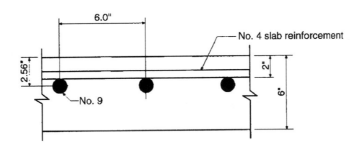

b. $c_c = 2.0$ in.

In lieu of computations for f_s at service load, use $f_s = 2/3 f_y$ as permitted in 10.6.4

Maximum spacing allowed,

$$s = 15\left(\frac{40,000}{40,000}\right) - (2.5 \times 2.0) = 10 \text{ in.} = 10 \text{ in.} \quad \text{O.K.}$$ Eq. (10-4)

c. Longitudinal reinforcement in slab outside 60-in. width.   10.6.6

For crack control outside the 60-in. width, use shrinkage and temperature reinforcement according to 7.12.   7.12

For Grade 60 reinforcement, $A_s = 0.0018 \times 12 \times 6 = 0.130$ in.^{2}/ft

Use No. 4 bars @ 18 in. ($A_s = 0.133$ in.^{2}/ft)

3. Skin reinforcement (h > 36 in.)   10.6.7

The spacing of the skin reinforcement is provided according to equation 10-4. The clear cover of the skin reinforcement is the same as the tension reinforcement; therefore the maximum allowed spacing of the skin reinforcement is 10 in.

Use 3-No. 3 bars uniformly spaced along each face of the beam extending a distance > h/2 beyond the bottom surface of the beam.

Spacing of the skin reinforcement:

$s = (24 - 1.5 - 0.5 - 1.41 - 1 - 1.41/2)/3 = 6.3$ in. < 10 in. OK

Use skin reinforcement at a spacing of 6.0 in.

Similarly, provide No. 3 @ 6.0 in. in the upper half of the depth in the negative moment region.

| Example 9.2 (cont'd) | Calculations and Discussion | Code Reference |

4. Detail section as shown below.

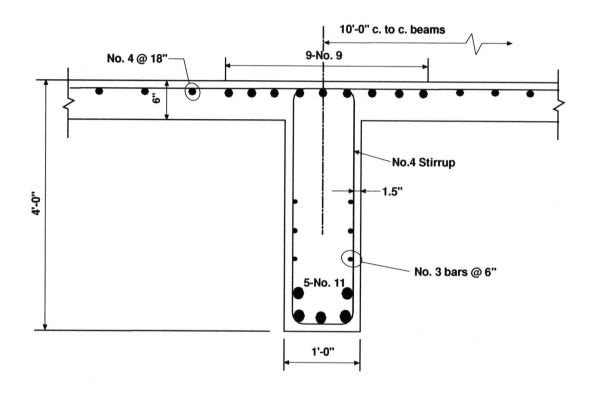

10

Deflections

GENERAL CONSIDERATIONS

The ACI code provisions for control of deflections are concerned only with deflections that occur at service load levels under static conditions and may not apply to loads with strong dynamic characteristics such as those due to earthquakes, transient winds, and vibration of machinery. Because of the variability of concrete structural deformations, designers must not place undue reliance on computed estimates of deflections. In most cases, the use of relatively simple procedures for estimating deflections is justified. In-depth treatments of the subject of deflection control, including more refined methods for computing deformations, may be found in Refs. 10.1 and 10.2.

9.5   CONTROL OF DEFLECTIONS

Two methods are given in the code for controlling deflections of one-way and two-way flexural members. Deflections may be controlled directly by limiting computed deflections [see Table 9.5(b)] or indirectly by means of minimum thickness [Table 9.5(a) for one-way systems, and Table 9.5(c) and Eqs. (9-12) and (9-13) for two-way systems.]

9.5.2.1 Minimum Thickness for Beams and One-Way Slabs (Nonprestressed)—Deflections of beams and one-way slabs supporting loads commonly experienced in buildings will normally be satisfactory when the minimum thickness from Table 9.5(a) (reproduced in Table 10-1) are met or exceeded.

The designer should especially note that this minimum thickness requirement is intended only for members **not** supporting or attached to partitions or other construction likely to be damaged by large deflections. For all other members, deflections need to be computed.

9.5.2.2 Immediate Deflection of Beams and One-Way Slabs (Nonprestressed)—Initial or short-term deflections of beams and one-way slabs occur immediately on the application of load to a structural member. The principal factors that affect the immediate deflection (see Ref. 10.3) of a member are:

    a.   magnitude and distribution of load,
    b.   span and restraint condition,
    c.   section properties and steel percentage,
    d.   material properties, and
    e.   amount and extent of flexural cracking.

Table 8-1 Minimum Thickness for Nonprestressed Beams and One-Way Slabs
(Grade 60 Reinforcement and Normal Weight Concrete)

Member	Minimum Thickness, h			Cantilever
	Simply Supported	One End Continuous	Both Ends Continuous	
One-Way Slabs	ℓ/20	ℓ/24	ℓ/28	ℓ/10
Beams	ℓ/16	ℓ/18.5	ℓ/21	ℓ/8

(1) For f_y other than 60,000 psi, multiply by tabulated values by $(0.4 + f_y/100,000)$ e.g., for grade 40 reinforcement, multiply values by 0.80
(2) For structural lightweight concrete, multiply tabulated values by $(1.65 - 0.005w_c)$ but not less than 1.09, where w_c is the unit weight in lb per cu ft.

The following concrete properties strongly influence the behavior of reinforced flexural members under short-time loads: compressive strength (f'_c), modulus of elasticity (E_c) and modulus of rupture (f_r). The modulus of elasticity particularly shows more variation with concrete quality, concrete age, stress level, and rate or duration of load.

The idealized short-term deflection of a typical reinforced concrete beam is shown in Fig. 10-1. There are two distinct phases of behavior: (i) uncracked behavior, when the applied moment (M_a) is less than the cracking moment (M_{cr}); and (ii) cracked behavior, when the applied moment (M_a) is greater than the cracking moment (M_{cr}). Two different values for the moment of inertia would therefore be used for calculating the deflections: the gross moment of inertia (I_g) for the uncracked section, and the reduced moment of inertia for the cracked section (I_{cr}).

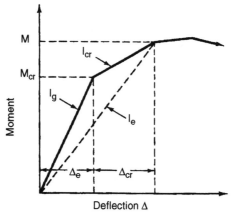

Figure 10-1 Bilinear Moment-Deflection Relationship[10.4]

For the uncracked rectangular beam shown in Fig. 10-2, the gross moment of inertia is used ($I_g = bh^3/12$). The moment of inertia of a cracked beam with tension reinforcement (I_{cr}) is computed in the following manner:

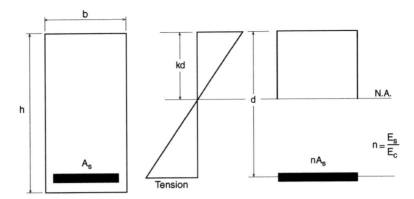

Figure 10-2 Cracked Transformed Section of Singly Reinforced Beam

Taking moment of areas about the neutral axis

$$b \times kd \times \frac{kd}{2} = nA_s(d - kd)$$

use $\quad B = \dfrac{b}{nA_s}$

$$kd = \frac{\sqrt{2Bd + 1} - 1}{B}$$

Moment of inertia of cracked section about neutral axis,

$$I_{cr} = \frac{b(kd)^3}{3} + nA_s(d - kd)^2$$

Expressions for computing the cracked moment of inertia for sections with compression reinforcement and flanged sections, which are determined in a similar manner, are given in Table 10-2.

9.5.2.3, 9.5.2.4 Effective Moment of Inertia for Beams and One-Way Slabs (Nonprestressed)—The flexural rigidity EI of a beam may not be constant along its length because of varying amounts of steel and cracking at different sections along the beam. This, and other material related sources of variability, makes the exact prediction of deflection difficult in practice.

The effective moment of inertia of cantilevers, simple beams, and continuous beams between inflection points is given by

$$I_e = (M_{cr}/M_a)^3 I_g + [1 - (M_{cr}/M_a)^3] I_{cr} \le I_g \qquad \text{Eq. (9-8)}$$

where $\quad M_{cr} = f_r I_g / y_t \qquad$ Eq. (9-9)

M_a = maximum service load moment (unfactored) at the stage for which deflections are being considered

$\quad f_r = 7.5\sqrt{f'_c}$ for normal weight concrete $\qquad$ Eq. (9-10)

For lightweight concrete, f_r is modified according to 9.5.2.3.

The effective moment of inertia I_e provides a transition between the well-defined upper and lower bounds of I_g and

I_{cr} as a function of the level of cracking represented by M_a/M_{cr}. The equation empirically accounts for the effect of tension stiffening—the contribution of uncracked concrete between cracks in regions of low tensile stress.

For each load combination being considered, such as dead load or dead plus live load, deflections should be calculated using an effective moment of inertia [Eq. (9-8)] computed with the appropriate service load moment, M_a. The incremental deflection caused by the addition of load, such as live load, is then computed as the difference between deflections computed for any two load combinations.

Table 10-2 Gross and Cracked Moment of Inertia of Rectangular and Flanged Section

Gross Section	Cracked Transformed Section	Gross and Cracked Moment of Inertia
Rectangular section with A_s	Without compression steel	$n = \dfrac{E_s}{E_c}$ $B = \dfrac{b}{(nA_s)}$ $I_g = \dfrac{bh^3}{12}$ **Without compression steel** $kd = \left(\sqrt{2dB+1}-1\right)/B$ $I_{cr} = b(kd)^3/3 + nA_s(d-kd)^2$
Rectangular section with A'_s and A_s	With compression steel, $(n-1)A'_s$	**With compression steel** $r = (n-1)A'_s/(nA_s)$ $kd = \left[\sqrt{2dB(1+rd'/d)+(1+r)^2}-(1+r)\right]/B$ $I_{cr} = b(kd)^3/3 + nA_s(d-kd)^2 + (n-1)A'_s(kd-d')^2$
Flanged section (T) with A_s	Without compression steel	$n = \dfrac{E_s}{E_c}$ $C = b_w/(nA_s)$, $f = h_f(b-b_w)/(nA_s)$ $y_t = h - \tfrac{1}{2}\left[(b-b_w)h_f^2 + b_w h^2\right]/\left[(b-b_w)h_f + b_w h\right]$ $I_g = (b-b_w)h_f^3/12 + b_w h^3/12 + (b-b_w)h_f(h-h_f/2-y_t)^2 + b_w h(y_t - h/2)^2$ **Without compression steel** $kd = \left[\sqrt{C(2d+h_f f)+(1+f)^2}-(1+f)\right]/C$ $I_{cr} = (b-b_w)h_f^3/12 + b_w(kd)^3/3 + (b-b_w)h_f(kd-h_f/2)^2 + nA_s(d-kd)^2$
Flanged section (T) with A'_s and A_s	With compression steel, $(n-1)A'_s$	**With compression steel** $kd = \left[\sqrt{C(2d+h_f f+2rd')+(f+r+1)^2}-(f+r+1)\right]/C$ $I_{cr} = (b-b_w)h_f^3/12 + b_w(kd)^3/3 + (b-b_w)h_f(kd-h_f/2)^2 + nA_s(d-kd)^2 + (n-1)A'_s(kd-d')^2$

For prismatic members (including T-beams with different cracked sections in positive and negative moment regions), I_e may be determined at the support section for cantilevers and at the midspan section for simple and continuous spans. The use of the midspan section properties for continuous prismatic members is considered satisfactory in approximate calculations primarily because the midspan rigidity has the dominant effect on deflections. Alternatively, for continuous prismatic and nonprismatic members, 9.5.2.4 suggests using the average I_e at the critical positive and negative moment sections. The '83 commentary on 9.5.2.4 suggested the following approach to obtain improved results:

Beams with one end continuous:

$$\text{Avg. } I_e = 0.85 I_m + 0.15 (I_{cont.end}) \quad (1)$$

Beams with both ends continuous:

$$\text{Avg. } I_e = 0.70 I_m + 0.15 (I_{e1} + I_{e2}) \quad (2)$$

where I_m refers to I_e at the midspan section

I_{e1} and I_{e2} refer to I_e at the respective beam ends.

Moment envelopes based on the approximate moment coefficients of 8.3.3 are accurate enough to be used in computing both positive and negative values of I_e (see Example 10.2). For a single heavy concentrated load, only the midspan I_e should be used.

The initial or short-term deflection (Δ_i) for cantilevers and simple and continuous beams may be computed using the following elastic equation given in the '83 commentary on 9.5.2.4. For continuous beams, the midspan deflection may usually be used as an approximation of the maximum deflection.

$$\Delta_i = K (5/48) M_a \ell^2 / E_c I_e \quad (3)$$

where M_a is the support moment for cantilevers and the midspan moment (when K is so defined) for simple and continuous beams

ℓ is the span length as defined in 8.7.

For uniformly distributed loading w, the theoretical values of the deflection coefficient K are shown in Table 10-3.

Since deflections are logically computed for a given continuous span based on the same loading pattern as for maximum positive moment, Eq. (3) is thought to be the most convenient form for a deflection equation.

9.5.2.5 Long-Term Deflection of Beams and One-Way Slabs (Nonprestressed)—Beams and one-way slabs subjected to sustained loads experience long-term deflections. These deflections may be two to three times as large as the immediate elastic deflection that occurs when the sustained load is applied. The long-term deflection is caused by the effects of shrinkage and creep, the formation of new cracks and the widening of earlier cracks. The principal factors that affect long-term deflections (see Ref. 10.3) are:

a. stresses in concrete
b. amount of tensile and compressive reinforcement
c. member size
d. curing conditions
e. temperature
f. relative humidity
g. age of concrete at the time of loading
h. duration of loading

Table 10-3 Deflection Coefficient K

		K
1.	Cantilevers (deflection due to rotation at supports not included)	2.40
2.	Simple beams	1.0
3.	Continuous beams	$1.2 - 0.2 M_o/M_a$
4.	Fixed-hinged beams (midspan deflection)	0.80
5.	Fixed-hinged beams (maximum deflection using maximum moment)	0.74
6.	Fixed-fixed beams	0.60
For other types of loading, K values are given in Ref. 8.2. M_o = Simple span moment at midspan $\left(\dfrac{w\ell^2}{8}\right)$ M_a = Net midspan moment.		

The effects of shrinkage and creep must be approximated because the strain and stress distribution varies across the depth and along the span of the beam. The concrete properties (strength, modulus of elasticity, shrinkage and creep) also vary with mix composition, curing conditions and time. Two approximate methods for estimating long-term deflection appear below.

ACI 318 Method

According to 9.5.2.5, additional long-term deflections due to the combined effects of shrinkage and creep from sustained loads $\Delta_{(cp+sh)}$ may be estimated by multiplying the immediate deflection caused by the sustained load $(\Delta_i)_{sus}$ by the factor λ_Δ; i.e.

$$\Delta_{(cp+sh)} = \lambda(\Delta_i)_{sus} \tag{4}$$

where $\quad \lambda_\Delta = \dfrac{\xi}{1+50\rho'} \qquad\qquad$ *Eq. (9-11)*

Values for ξ are given in Table 10-4 for different durations of sustained load. Figure R9.5.2.5 in the commentary to the code shows the variation of ξ for periods up to 5 years. The compression steel $\rho' = A_s'/bd$ is computed at the support section for cantilevers and the midspan section for simple and continuous spans. Note that sustained loads include dead load and that portion of live load that is sustained. See R9.5.1.

Table 10-4 Time-Dependent Factor ξ (9.5.2.5)

Sustained Load Duration	ξ
5 years and more	2.0
12 months	1.4
6 months	1.2
3 months	1.0

Alternate Method

Alternatively, creep and shrinkage deflections may be computed separately using the following expressions from Refs. 10.2, 10.5, and 10.6. The procedure is summarized in Section 2.6.2 of Ref.10.4.

$$\Delta_{cp} = \lambda_{cp}(\Delta_i)_{sus} \qquad (5)$$

$$\Delta_{sh} = K_{sh}\phi_{sh}\ell^2 \qquad (6)$$

where

$$\lambda_{cp} = k_r C_t;$$

$$k_r = 0.85/(1 + 50\rho')$$

C_t = time dependent creep coefficient (Table 2.1 or Eq. 2.7 of Ref. 10.4)

K_{sh} = shrinkage deflection constant (Table 10-5)

$$\phi_{sh} = A_{sh}(\varepsilon_{sh})_t/h$$

A_{sh} = shrinkage deflection multiplier (Figure 10-3 or Eq. 6.1 below)

$(\varepsilon_{sh})_t$ = time dependent shrinkage strain (Table 2.1 or Eq. 2.8 & 2.9 of Ref. 10.4)

ℓ = beam span length

h = beam depth

The ultimate value of the creep coefficient C_t, denoted as C_u, is dependent on the factors a through h listed above. Likewise, the ultimate value of the time dependent shrinkage strain depends on the varying conditions and is designated $(\varepsilon_{sh})_u$. Typical values for the two properties are discussed in Section 2.3.4 of ACI 435 (Ref. 10.4).

In Ref. 10.4, the ultimate creep coefficient is dependent on six factors:

a. relative humidity
b. age of concrete at load application
c. minimum member dimension
d. concrete consistency
e. fine aggregate content
f. air content

Standard conditions for these six variables are 40% R.H., 3 days (steam cured) or 7 days (moist cured), 6 in. least dimension, 3 in. slump, 50% fine aggregate and 6% air content. For the case of standard conditions, C_u is equal to 2.35. Correction factors are presented in Fig. 2.1 of Ref. 10.4, to adjust the value of C_u for non-standard conditions.

Two variations from standard conditions that might be encountered in normal construction are for relative humidity of 70% and load application taking place at an age of 20 days. The correction factor for the relative humidity is given by the following:

$$K_h^c = 1.27 - 0.0067H$$

where H is the relative humidity in percent. For the case of 70% relative humidity,

$$K_h^c = 1.27 - 0.0067(70) = 0.80$$

Correction for the time of load application is given in the following two expressions for steam or moist curing conditions:

$$K_{to}^c = 1.13(t^{-0.095}) \quad \text{(Steam Cured)}$$

$$K_{to}^c = 1.25(t^{-0.118}) \quad \text{(Moist Cured)}$$

where t is the age of load application in days. For t = 20 days the two equations give 0.85 and 0.88 respectively. The average is 0.865.

If it is assumed that all other conditions remain constant the ultimate creep coefficient for the condition of 70% relative humidity and load application at 20 days becomes, according to the methodology indicated:

$$C_u = (0.80)(0.865)(2.35) = 1.63$$

By comparison, the value for C_u suggested in the 1978 edition of ACI 435, based on relative humidity of 70%, age at load application of 20 days and minimum dimension of 6 in. (the standard case) was $C_u = 1.60$.

An evaluation of ultimate creep strain can also be made. In Ref. 10.4 it is stated that $(\varepsilon_{sh})_u$ is dependent on a set of factors similar to those that affect the ultimate creep coefficient. In particular, the five conditions, and their standard values, are as follows:

a. relative humidity – 40%
b. minimum member dimension – 6 in.
c. fine aggregate content – 50%
d. cement content – 1200 kg/m³
e. air content – 6%

For standard conditions, the ultimate shrinkage strain is 780 x 10⁻⁶. Keeping all applicable conditions the same as used in evaluation of the ultimate creep and use of a cement factor of 6 bags per cubic yard (335 kg/m³), calculation of the appropriate correction factors yields:

$$K_h^s = 1.4 - 0.01H = 1.4 - (0.01)(70) = 0.70 \quad \text{(relative humidity)}$$

$$K_b^s = 0.75 + 0.000214B = 0.75 + (0.000214)(335) = 0.82 \quad \text{(cement content)}$$

Application of the product of the two corrections to the standard value gives:

$$(\varepsilon_{sh})_u = (0.70)(0.82)(780 \times 10^{-6}) = 448 \times 10^{-6}$$

This value compares with 400 x 10⁻⁶ suggested in the 1978 edition of ACI 435.

In summary, an estimate of the values of C_u and $(\varepsilon_{sh})_u$ can be obtained for non-prestressed flexural members using the methodology presented in Section 2.3.4 of Ref. 10.4.

Once the ultimate values for creep and shrinkage are determined, the relationships between these ultimate values and the values at earlier times can be estimated by Eqs. 2.7, 2.8 and 2.9 of ACI 435R[10.4]. The expressions are reproduced below:

$$C_t = \left(\frac{t^{0.6}}{10 + t^{0.6}}\right) C_u \qquad \text{Eq. (2.7) of ACI 435R}$$

Where t represents time, in days, after application of load.

For moist cured concrete, the shrinkage relationship is:

$$(\varepsilon_{sh})_t = \left(\frac{t}{35 + t}\right)(\varepsilon_{sh})_u \qquad \text{Eq. (2.8) of ACI 435R}$$

(t is in days minus 7 after placement)

and for steam cured concrete:

$$(\varepsilon_{sh})_t = \left(\frac{t}{55 + t}\right)(\varepsilon_{sh})_u \qquad \text{Eq. (2.9) of ACI 435R}$$

(t is in days minus 3 after placement)

Comparison of the values for the time dependent creep coefficients and shrinkage strains given in Table 2.1 of ACI 435R and those that result from Eqs. 2.7, 2.8 and 2.9 shows that the values obtained by the two methods vary slightly, particularly for the lower values of time, t. Since the calculation of deflections in concrete structures involves considerable approximation, the use of the time dependent quantities obtained either from the table or from the equations is considered acceptable.

A_{sh} may be taken directly from Fig. 10-3 or computed by the following set of equations which are given in Section 2.6.2 of ACI 435:

$$A_{sh} = 0.7 \cdot (\rho - \rho')^{\frac{1}{3}} \cdot \left(\frac{\rho - \rho'}{\rho}\right)^{\frac{1}{2}} \qquad \text{for } \rho - \rho' \leq 3.0$$

$$= 0.7 \cdot \rho^{\frac{1}{3}} \qquad \text{for } \rho' = 0 \qquad (6.1)$$

$$= 1.0 \qquad \text{for } \rho - \rho' \geq 3.0$$

In the above equations, both ρ and ρ' are expressed in <u>percent</u>, not in decimal fraction as is usual. The ratios are also expressed in <u>percent</u> for determination of A_{sh} from Figure 10-3.

Values for the shrinkage deflection coefficient K_{sh} are given in Table 10-5, assuming equal positive and negative shrinkage curvatures with an inflection point at the quarter-point of continuous spans, which is generally satisfactory for deflection computation.

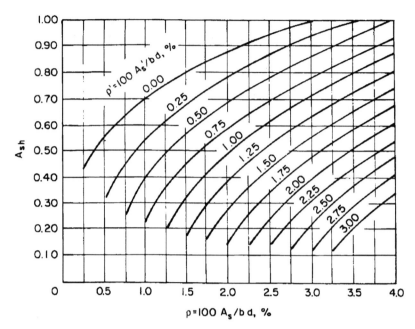

Figure 10-3 Values of A_{sh} for Calculating Shrinkage Deflection

Table 10-5 Shrinkage Deflection Coefficient K_{sh}

	K_{sh}
Cantilevers	0.50
Simple Spans	0.13
Spans with One End Continuous—Multi-Span Beams	0.09
Spans with One End Continuous—Two-Span Beams	0.08
Spans With Both Ends Continuous	0.07

The reinforcement ratios ρ and ρ' used in determining A_{sh} from Fig. 10-3, refer to the support section of cantilevers and the midspan section of simple and continuous beams. For T-beams, use $\rho = 100\ (\rho + \rho_w)/2$ and a similar calculation for any compression steel ρ' in determining A_{sh}, where $\rho_w = A_s/b_w d$. See Example 10.2.

As to the choice of computing creep and shrinkage deflections by Eq. (9-11) or separately by Eqs. (5) and (6), the combined ACI calculation is simpler but provides only a rough approximation, since shrinkage deflections are only indirectly related to the loading (primarily by means of the steel content). One case in which the separate calculation of creep and shrinkage deflections may be preferable is when part of the live load is considered as a sustained load.

All procedures and properties for computing creep and shrinkage deflections apply equally to normal weight and lightweight concrete.

9.5.2.6 Deflection Limits—Deflections computed using the preceding methods are compared to the limits given in Table 9.5(b). The commentary gives information for the correct application of these limits, including consideration of deflections occurring prior to installation of partitions.

9.5.3   Two-Way Construction (Nonprestressed)

Deflections of two-way slab systems with and without beams, drop panels, and column capitals need not be computed when the minimum thickness requirements of 9.5.3 are met. The minimum thickness requirements include the effects of panel location (interior or exterior), panel shape, span ratios, beams on panel edges, supporting columns and capitals, drop panels, and the yield strength of the reinforcing steel.

Table 10-6 Minimum Thickness of Slabs without Interior Beams (Table 9.5(c))

Yield strength, f_y psi*	Without drop panels†		Interior panels	With drop panels†		Interior panels
	Exterior panels			Exterior panels		
	Without edge beams	With edge beams††		Without edge beams	With edge beams††	
40,000	$\dfrac{\ell_n}{33}$ **	$\dfrac{\ell_n}{36}$	$\dfrac{\ell_n}{36}$	$\dfrac{\ell_n}{36}$	$\dfrac{\ell_n}{40}$	$\dfrac{\ell_n}{40}$
60,000	$\dfrac{\ell_n}{30}$	$\dfrac{\ell_n}{33}$	$\dfrac{\ell_n}{33}$	$\dfrac{\ell_n}{33}$	$\dfrac{\ell_n}{36}$	$\dfrac{\ell_n}{36}$
75,000	$\dfrac{\ell_n}{28}$	$\dfrac{\ell_n}{31}$	$\dfrac{\ell_n}{31}$	$\dfrac{\ell_n}{31}$	$\dfrac{\ell_n}{34}$	$\dfrac{\ell_n}{34}$

* For f_y between the values given in the table, minimum thickness shall be determined by linear interpolation.
** For two-way construction, ℓ_n is the length of clear span in the long direction, measured face-to-face of supports in slabs without beams and face-to-face of beams or other supports in other cases.
† Drop panel is defined in 13.2.5.
†† Slabs with beams between columns along exterior edges. The value of α_f for the edge beam shall not be less than 0.8.

Section 9.5.3.2 provides minimum thickness requirements for two-way slab systems without beams between interior columns (flat plates and flat slabs). The minimum thickness is determined directly as a function of span length using Table 9.5(c). The section also provides minimum values for slabs with and without drop panels. The values given in Table 9.5(c) represent the upper limit of slab thicknesses given by Eqs. (9-12) and (9-13). The minimum thickness requirements of 9.5.3.2 are illustrated in Fig. 10-4.

Section 9.5.3.3 provides minimum thickness requirements for two-way slab systems with beams supporting all sides of a panel. It should be noted that these provisions are intended to apply only to two-way systems, that is, systems in which the ratio of long to short span is not greater than 2. For slabs that do not satisfy this limitation, Eqs. (9-12) and (9-13) may give unreasonable results. For such cases, 9.5.2 should be used.

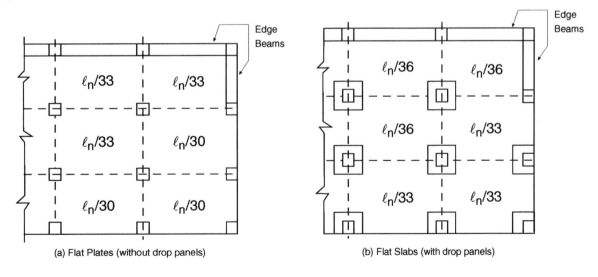

Figure 10-4 Minimum Thickness of Slabs without Interior Beams (Grade 60 Reinforcement)

Figure 10-5 may be used to simplify minimum thickness calculations for two-way slabs. It should be noted in Fig. 10-5 that the difference between the controlling minimum thickness for square panels and rectangular panels having a 2-to-1 panel side ratio is not large.

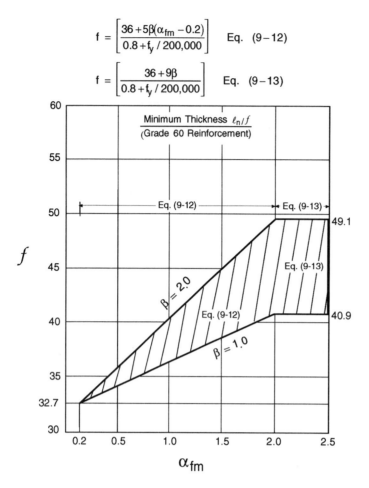

Figure 10-5 Minimum Thickness for Two-Way Beam Supported Slabs

9.5.3.4 Deflection of Nonprestressed Two-Way Slab Systems—

Initial or Short-Term Deflection: An approximate procedure[10.2, 10.7] that is compatible with the Direct Design and Equivalent Frame Methods of code Chapter 13 may be used to compute the initial or short-term deflection of two-way slab systems. The procedure is essentially the same for flat plates, flat slabs, and two-way beam-supported slabs, after the appropriate stiffnesses are computed. The midpanel deflection is computed as the sum of the deflection at midspan of the column strip or column line in one direction, Δ_{cx} or Δ_{cy}, and deflection at midspan of the middle strip in the orthogonal direction, Δ_{mx} or Δ_{my} (see Fig. 10-6). The column strip is the width on each side of column center line equal to 1/4 of the smaller panel dimension. The middle strip is the central portion of the panel which is bounded by two column strips.

For square panels,

$$\Delta = \Delta_{cx} + \Delta_{my} = \Delta_{cy} + \Delta_{mx} \tag{7}$$

For rectangular panels, or for panels that have different properties in the two directions, the average Δ of the two directions is used:

$$\Delta = \left[\left(\Delta_{cx} + \Delta_{my}\right) + \left(\Delta_{cy} + \Delta_{mx}\right)\right]/2 \tag{8}$$

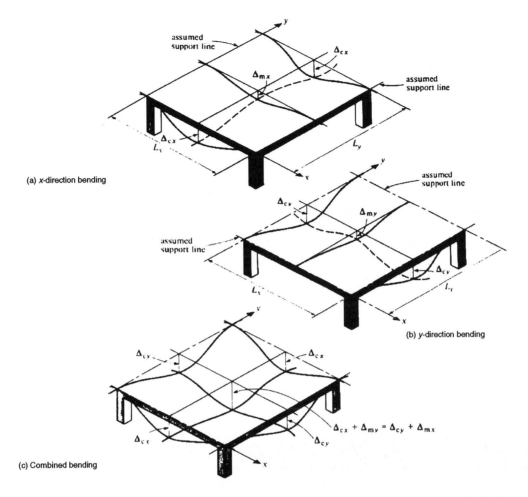

Figure 10-6 Basis for Equivalent Frame Method of Deflection Analysis of Two-Way Slab Systems, with or without Beams

The midspan deflection of the column strip or middle strip in an equivalent frame is computed as the sum of three parts: deflection of panel assumed fixed at both ends, plus deflection of panel due to the rotation at the two support lines. In the x direction, the deflections would be computed using the following expressions:

$$\Delta_{cx} = \text{Fixed } \Delta_{cx} + (\Delta\theta_1)_{cx} + (\Delta\theta_2)_{cx} \qquad \text{for column strip} \qquad (9)$$

$$\Delta_{mx} = \text{Fixed } \Delta_{mx} + (\Delta\theta_1)_{mx} + (\Delta\theta_2)_{mx} \qquad \text{for middle strip}$$

While these equations and the following discussion address only the computation of deflections in the x direction, similar computations to determine Δ_{cy} and Δ_{my} would be necessary to compute deflections in the y direction.

The first step in the process of computing Fixed Δ_{cx} and Fixed Δ_{mx} is to compute the midspan fixed-end deflection of the full-width equivalent frame under uniform loading, given by

$$\text{Fixed } \Delta_{\text{frame}} = \frac{w\ell^4}{384 \, E_c I_{\text{frame}}} \qquad (10)$$

where   w = load per unit area × full width

The effect of different stiffnesses in positive and negative moment regions [primarily when using drop panels and/or I_e in Eq. (9-8)] can be included by using an average moment of inertia as given by Eqs. (1) and (2).

The midspan fixed-end deflection of the column and middle strips is then computed by multiplying Fixed Δ_{frame} (Eq. (10)) by the M/EI ratio of the strips (column or middle) to the full-width frame.

$$\text{Fixed } \Delta_{c,m} = (LDF)_{c,m} \text{ Fixed } \Delta_{frame} \frac{(EI)_{frame}}{(EI)_{c,m}} \quad \text{for column or middle strip} \qquad (11)$$

where $(LDF)_{c,m} = \dfrac{M_{c,m}}{M_{frame}} = $ lateral distribution factor

The distribution of the total factored static moment, M_o, to the column and middle strips is prescribed in 13.6.3 and 13.6.4. In particular, 13.6.4.1, 13.6.4.2 and 13.6.4.4 provide tables which allocate fractions of M_o to the interior and exterior negative moment regions and the positive moment region, respectively, for <u>column</u> strips. The percent of the total not designated for the column strips is allocated to the <u>middle</u> strips. That is, for example, if 75 percent of M_o is designated for the interior negative moment of a column strip, the corresponding moment in the middle strip will be required to sustain 25 percent of M_o. The following expressions provide linear interpolation between the tabulated values given in 13.6.4.1, 13.6.4.2 and 13.6.4.4. Note that all expressions are given as percentages of M_o:

$M^-_{ext} = 100 - 10\beta_t + 12\beta_t (\alpha_{f1}\ell_2/\ell_1)(1 - \ell_2/\ell_1)$ (Exterior negative moment, % M_o)

$M^-_{int} = 75 + 30(\alpha_{f1}\ell_2/\ell_1)(1 - \ell_2/\ell_1)$ (Interior negative moment, % M_o)

$M^+ = 60 + 30(\alpha_{f1}\ell_2/\ell_1)(1.5 - \ell_2/\ell_1)$ (Positive moment, % M_o)

In application of the above expressions, if the actual value of $\alpha_{f1}\ell_2/\ell_1$ exceeds 1.0, the value 1.0 is used. Similarly, if β_t exceeds 2.5, the value 2.5 is used.

In order to calculate the lateral distribution factors (LDF), three cases should be considered:

    a. strips for interior panels
    b. strips in edge panels parallel to the edge
    c. strips in edge panels perpendicular to the edge

Note that in corner panels, *Case c* is used for strips in either direction as there is an exterior negative moment at each outer panel edge. In all cases, the <u>strip</u> moment, used in determination of the LDFs, is taken as the average of the positive and negative moment. Thus, the following formulas are obtained for the three cases:

Case a: LDF = $\frac{1}{2}(M^-_{int} + M^+)$
Case b: LDF = $\frac{1}{2}(M^-_{int} + M^+)$
Case c: LDF = $\frac{1}{2}[\frac{1}{2}(M^-_{int} + M^-_{ext}) + M^+]$

These lateral distribution factors apply to column strips and are expressed in percentages of the total panel moment M_o. The corresponding factors for the middle strips are determined, in general, as follows:

$$LDF_{mid} = 100 - LDF_{col}$$

The remaining terms in Eq. (9), the midspan deflection of column strip or middle strip caused by rotations at the ends $\left((\Delta\theta_1)_{cx}, (\Delta\theta_1)_{mx}, \text{etc.}\right)$, must now be computed. If the ends of the column at the floor above and below are assumed fixed (usual case for an equivalent frame analysis) or ideally pinned, the rotation of the column at the floor in question is equal to the net applied moment divided by the stiffness of the equivalent column.

$$\theta_{frame} = \theta_c = \theta_m = \frac{(M_{net})_{frame}}{K_{ec}} \qquad (12)$$

where K_{ec} = equivalent column stiffness (see 13.7.4)

The midspan deflection of the column strip subjected to a rotation of θ_1 radians at one end with the opposite end fixed is

$$(\Delta\theta_1)_c = \frac{\theta_1 \ell}{8} \qquad (13)$$

The additional deflection terms for the column and middle strips would be computed similarly.

Because θ in Eq. (12) is based on gross section properties, while the deflection calculations are based on I_e, Eq. (14) may be used instead of Eq. (13) for consistency:

$$(\Delta\theta_1)_c = \theta_1 \left(\frac{\ell}{8}\right) \left(\frac{I_g}{I_e}\right)_{frame} \qquad (14)$$

Direct Design Method: The deflection computation procedure described above has been expressed in terms of the equivalent frame method for moment analysis. However, it is equally suited for use with the direct design method in which coefficients are used to calculate moments at critical sections instead of using elastic frame analysis as in case of the equivalent frame method. In the direct design method, design moments are computed using clear spans. When determining deflections due to rotations at the ends of a member, these moments should theoretically be corrected to obtain moments at the center of the columns. However, this difference is generally small and may be neglected. In the case of flat plates and flat slabs, the span measured between the column centerlines is thought to be more appropriate than the clear span for deflection computations.

If all spans are equal and are identically loaded, the direct design method will give no unbalanced moments and rotations except at an exterior column. Therefore, in these cases, rotations need be considered only at the exterior columns. When live load is large compared to the dead load (not usually the case), end rotations may be computed by a simple moment-area procedure in which the effect of pattern loading may be included.

Effective Moment of Inertia: The effective moment of inertia given by Eq. (9-8) is recommended for computing deflections of partially cracked two-way construction. An average I_e of the positive and negative regions in accordance with Eqs. (1) and (2) may also be used.

For the typical cracking locations found empirically, the following moment of inertia values have been shown to be applicable in most cases.

	Case	Inertia
a.	Slabs without beams (flat plates, flat slabs)	
	(i) All dead load deflections—	I_g
	(ii) Dead-plus-live load deflections:	
	For the column strips in both directions—	I_e
	For the middle strips in both directions—	I_g
b.	Slab with beams (two-way beam-supported slabs)	
	(i) All dead load deflections—	I_g
	(ii) Dead-plus-live load deflections:	
	For the column strips in both directions—	I_g
	For the middle strips in both directions—	I_e

The I_e of the equivalent frame in each direction is taken as the sum of the column and middle strip I_e values.

Long-Term Deflection: Since the available data on long-term deflections of two-way construction is too limited to justify more elaborate procedures, the same procedures as those used for one-way members are recommended. Equation (9-11) may be used with $\xi = 2.5$ for sustained loading of five years or longer duration.

9.5.4 Prestressed Concrete Construction

Typical span-depth ratios for general use in design of prestressed members are given in the PCI Design Handbook[10.8] and summarized in Ref. 10.2 from several sources. Starting with the 2002 edition of ACI 318, the Building Code classifies prestressed concrete flexural members, in 18.3.3, as Class U (uncracked), Class T (transition), or Class C (cracked.) For Class U flexural members, deflections must be calculated based on the moment of inertia of the gross section I_g. For Classes T and C, deflections must be computed based on a cracked transformed section analysis or on a bilinear moment-deflection relationship. Reference 10.9 provides a procedure to compute deflection of cracked prestressed concrete members.

Deflection of Noncomposite Prestressed Members—The ultimate (in time) camber and deflection of prestressed members may be computed based on a procedure described in Ref. 10.2. The procedure includes the use of I_e for partially prestressed members (Ref. 10.8) as a suggested method of satisfying 9.5.4.2 for deflection analysis when the computed tensile stress exceeds the modulus of rupture, but does not exceed $12\sqrt{f'_c}$. For detailed information on the deflection of cracked prestressed beams and on the deflection of composite prestressed beams, see Refs. 10.2 and 10.9.

The ultimate deflection of noncomposite prestressed members is obtained as (Refs. 10.2 and 10.10):

$$\underset{(1)}{\Delta_u} = \underset{(1)}{-\Delta_{po}} + \underset{(2)}{\Delta_o} - \underset{(3)}{\left[-\frac{\Delta P_u}{P_o} + (k_r C_u)\left(1 - \frac{\Delta P_u}{2P_o}\right)\right] \Delta_{po}} + \underset{(4)}{(k_r C_u)\Delta_o} + \underset{(5)}{\Delta_s}$$

$$+ \underset{(6)}{(\beta_s k_r C_u) \Delta_s} + \underset{(7)}{\Delta_\ell} + \underset{(8)}{(\Delta_{cp})_\ell} \qquad (15)$$

Term (1) is the initial camber due to the initial prestressing moment after elastic loss, $P_o e$. For example, $\Delta_{po} = P_o e \ell^2 / 8 E_{ci} I_g$ for a straight tendon.

Term (2) is the initial deflection due to self-weight of the beam. $\Delta_o = 5 M_o \ell^2 / 48 E_{ci} I_g$ for a simple beam, where M_o = midspan self-weight moment.

Term (3) is the creep (time-dependent) camber of the beam due to the prestressing moment. This term includes the effects of creep and loss of prestress; that is, the creep effect under variable stress. Average values of the prestress loss ratio after transfer (excluding elastic loss), $(P_o - P_e)/P_e$, are about 0.18, 0.21, and 0.23 for normal, sand, and all-lightweight concretes, respectively. An average value of $C_u = 2.0$ might be reasonable for the creep factor due to ultimate prestress force and self-weight. The k_r factor takes into account the effect of any nonprestressed tension steel in reducing time-dependent camber, using Eq. (16). It is also used in the PCI Design Handbook[8.8] in a slightly different form.

$$k_r = 1/\left[1 + (A_s/A_{ps})\right] \quad \text{for } A_s/A_{ps} < 2 \qquad (16)$$

When $k_r = 1$, Terms (1) + (3) can be combined as:

$$-\Delta_{po} - \left[-\Delta_{po} + \Delta_{pe} + C_u\left(\frac{\Delta_{po} + \Delta_{pe}}{2}\right)\right] = -\Delta_{pe} - C_u\left(\frac{\Delta_{po} + \Delta_{pe}}{2}\right)$$

Term (4) is the creep deflection due to self-weight of the beam. Use the same value of C_u as in Term (3). Since creep due to prestress and self-weight takes place under the combined stresses caused by them, the effect of any nonprestressed tension steel in reducing the creep deformation is included in both the camber Term (3) and the deflection Term (4).

Term (5) is the initial deflection of the beam under a superimposed dead load. $\Delta_s = 5M_s\ell^2/48E_cI_g$ for a simple beam, where M_s = midspan moment due to superimposed dead load (uniformly distributed).

Term (6) is the creep deflection of the beam caused by a superimposed dead load. k_r is the same as in Terms (3) and (4), and is included in this deflection term for the same reason as in Term (4). An average value of $C_u = 1.6$ is recommended, as in Eq. (7) for nonprestressed members, assuming load application at 20 days after placement. β_s is the creep correction factor for the age of the beam concrete when the superimposed dead load is applied at ages other than 20 days (same values apply for normal as well as lightweight concrete): $\beta_s = 1.0$ for age 3 weeks, 0.96 for age 1 month, 0.89 for age 2 months, 0.85 for age 3 months, and 0.83 for age 4 months.

Term (7) is the initial live load deflection of the beam. $\Delta_\ell = 5M_\ell\ell^2/48E_cI_g$ for a simple beam under uniformly distributed live load, where M_ℓ = midspan live load moment. For uncracked members, $I_e = I_g$. For partially cracked noncomposite and composite members, see Refs. 10.2 and 10.3. See also Example 8.5 for a partially cracked case.

Term (8) is the live load creep deflection of the beam. This deflection increment may be computed as $(\Delta_{cp})_\ell = (M_s/M_\ell) C_u \Delta_\ell$, where M_s is the sustained portion of the live load moment and $C_u = 1.6$, for load application at 20 days or multiplied by the appropriate β_s, as in Term (6).

An alternate method of calculation of long-term camber and deflection is the so-called *PCI Multiplier Method* which is presented in both Ref. 10.4 and Ref. 10.8. In that procedure the various instantaneous components of camber or deflection are simply multiplied by the appropriate tabulated coefficients to obtain the additional contributions due to long term effects. The coefficients are given in Table 3.4 of Ref. 10.4 or Table 4.8.2 of Ref. 10.8.

9.5.5   Composite Construction

The ultimate (in time) deflection of unshored and shored composite flexural members may be computed by methods discussed in Refs. 10.2 and 10.10. The methods are reproduced in the following section for both unshored and shored construction. Subscripts 1 and 2 are used to refer to the slab (or effect of the slab, such as under slab dead load) and the precast beam, respectively. Examples 10.6 and 10.7 demonstrate the beneficial effect of shoring in reducing deflections.

9.5.5.1 Shored Construction—For shored composite members, where the dead and live load is resisted by the full composite section, the minimum thicknesses of Table 9.5(a) apply as for monolithic structural members.

The calculation of deflections for shored composite beams is essentially the same as for monolithic beams, except for the deflection due to shrinkage warping of the precast beam, which is resisted by the composite section after the slab has hardened, and the deflection due to differential shrinkage and creep of the composite beam. These effects are represented by Terms (3) and (4) in Eq. (17).

$$\overset{(1)\qquad\quad (2)\qquad\quad (3)\qquad (4)\quad (5)\quad\ (6)}{\Delta_u \;=\; (\Delta_i)_{1+2} \;+\; 1.80k_r(\Delta_i)_{1+2} \;+\; \Delta_{sh}\frac{I_2}{I_c} \;+\; \Delta_{ds} \;+\; (\Delta_i)_\ell \;+\; (\Delta_{cp})_\ell} \qquad (17)$$

When $k_r = 0.85$ (neglecting any effect of slab compression steel) and Δ_{ds} is assumed to be equal to $(\Delta_i)_{1+2}$, Eq. (17) reduces to Eq. (18).

$$\Delta_u = \overset{(1+2+4)}{3.53\,(\Delta_i)_{1+2}} + \overset{(3)}{\Delta_{sh}\frac{I_2}{I_c}} + \overset{(5)}{(\Delta_i)_\ell} + \overset{(6)}{(\Delta_{cp})_\ell} \qquad (18)$$

Term (1) is the initial or short-term deflection of the composite beam due to slab plus precast beam dead load (plus partitions, roofing, etc.), using Eq. (3), with $M_a = M_1 + M_2$ = midspan moment due to slab plus precast beam dead load. For computing $(I_e)_{1+2}$ in Eq. (1), M_a refers to the moment $M_1 + M_2$, and M_{cr}, I_g, and I_{cr} to the composite beam section at midspan.

Term (2) is the creep deflection of the composite beam due to the dead load in Term (1), using Eq. (5). The value of C_u to be used must be a combination of that for the slab and that for the beam. In the case of the slab, an adjusted value of $C_u = 1.74$, based on the shores being removed at 10 days of age for a moist-cured slab, may be used. The beam may be older than 20 days (the standard condition) when the loads are applied, however $C_u = 1.60$ may be used conservatively. An average of the two values may be used as an approximation. For other times of load application, the adjustments can be made in similar fashion using the correction factors, β_s, listed previously in the description of Term (6) of Eq. (15). Index ρ' refers to any compression steel in the slab at midspan when computing k_r.

Term (3) is the shrinkage deflection of the composite beam after the shores are removed, due to the shrinkage of the precast beam concrete, but not including the effect of differential shrinkage and creep which is given by Term (4). Equation (6) may be used to compute Δ_{sh}. Assuming the slab is cast at a precast (steam-cured) beam concrete age of 2 months and that shores are removed about 10 days later. At that time, the shrinkage in the beam is approximately 36% of the ultimate, according to Table 2.1 of ACI 435. The shrinkage strain subsequent to that time will be $(\varepsilon_{sh})_{rem} = (1 - 0.36)(\varepsilon_{sh})_u$. That value should be used in Eq. (6) to calculate the deflection component in this Term.

Term (4) is the deflection due to differential shrinkage and creep. As an approximation, $\Delta_{ds} = (\Delta_i)_{1+2}$ may be used.

Term (5) is the initial or short-term live load deflection of the composite beam, using Eq. (3). The calculation of the incremental live load deflection follows the same procedure as that for a monolithic beam. This is the same as in the method described in connection with Term (9) of Eq. (19) discussed below.

Term (6) is the creep deflection due to any sustained live load, using Eq. (5). In computing this component of deflection, use of an ultimate creep coefficient, $C_u = 1.6$ is conservative. The creep coefficient may be reduced by the factor β_s defined in Term (6) of Eq. (15).

These procedures suggest using midspan values only, which may normally be satisfactory for both simple composite beams and those with a continuous slab as well. See Ref. 10.10 for an example of a continuous slab in composite construction.

9.5.5.2 Unshored Construction—For unshored composite construction, if the thickness of a nonprestressed precast member meets the minimum thickness requirements, deflections need not be computed. Section 9.5.5.2 also states that, if the thickness of an unshored nonprestressed composite member meets the minimum thickness requirements, deflections occurring after the member becomes composite need not be computed, but the long-term deflection of the precast member should be investigated for the magnitude and duration of load prior to beginning of effective composite action.

$$\Delta_u = \overset{(1)}{\underline{(\Delta_i)_2}} + \overset{(2)}{\underline{0.77k_r(\Delta_i)_2}} + \overset{(3)}{\underline{0.83k_r(\Delta_i)_2\frac{I_2}{I_c}}} + \overset{(4)}{\underline{0.36\Delta_{sh}}} + \overset{(5)}{\underline{0.64\Delta_{sh}\frac{I_2}{I_c}}}$$

$$+ \overset{(6)}{\underline{(\Delta_i)_1}} + \overset{(7)}{\underline{1.22k_r(\Delta_i)_1\frac{I_2}{I_c}}} + \overset{(8)}{\underline{\Delta_{ds}}} + \overset{(9)}{\underline{(\Delta_i)_\ell}} + \overset{(10)}{\underline{(\Delta_{cp})_\ell}} \qquad (19)$$

With $k_r = 0.85$ (no compression steel in the precast beam) and Δ_{ds} assumed to be equal to $0.50(\Delta_i)_1$, Eq. (19) reduces to Eq. (20).

$$\Delta_u = \overset{(1+2+3)}{\left(1.65 + 0.71\frac{I_2}{I_c}\right)(\Delta_i)_2} + \overset{(4+5)}{\left(0.36 + 0.64\frac{I_2}{I_c}\right)\Delta_{sh}}$$

$$+ \overset{(6+7+8)}{\left(1.50 + 1.04\frac{I_2}{I_c}\right)(\Delta_i)_1} + \overset{(9)}{(\Delta_i)_\ell} + \overset{(10)}{(\Delta_{cp})_\ell} \qquad (20)$$

In Eqs. (19) and (20), the parts of the total creep and shrinkage occurring before and after slab casting are based on the assumption of a precast beam age of 20 days when its dead load is applied and of 2 months when the composite slab is cast.

Term (1) is the initial or short-term dead load deflection of the precast beam, using Eq. (3), with $M_a = M_2$ = midspan moment due to the precast beam dead load. For computing $(I_e)_2$ in Eq. (9-9), M_a refers to the precast beam dead load, and M_{cr}, I_g, and I_{cr} to the precast beam section at midspan.

Term (2) is the dead load creep deflection of the precast beam up to the time of slab casting, using Eq. (5), with $C_t = 0.48 \times 1.60 = 0.77$ (for 20 days to 2 months; Table 2.1 of ACI 435; for slabs cast at other than 60 days, the appropriate values from Table 2.1 should be used), and the ρ' refers to the compression steel in the precast beam at midspan when computing k_r.

Term (3) is the creep deflection of the composite beam following slab casting, due to the precast beam dead load, using Eq. (5), with the long term creep being the balance after the slab is cast, $C_t = 1.60 - 0.77 = 0.83$. As indicated in Term (3), if the slab is cast at time other than 2 months, C_t will be as determined from Table 2.1 of ACI 435 and the value of C_t to be used for this term will be found as the difference between 1.60 and the value used for Term (2). ρ' is the same as in Term (2). The ratio I_2/I_c modifies the initial stress (strain) and accounts for the effect of the composite section in restraining additional creep curvature (strain) after the composite section becomes effective. As a simple approximation, $I_2/I_c = [(I_2/I_c)_g + (I_2/I_c)_{cr}]/2$ may be used.

Term (4) is the deflection due to shrinkage warping of the precast beam up to the time of slab casting, using Eq. (6), with $(\varepsilon_{sh})_t = 0.36(\varepsilon_{sh})_u$ at age 2 months for steam cured concrete (assumed to be the usual case for precast beams) The multiplier 0.36 is obtained from Table 2.1 of Ref. 10.4. As in the previous two terms, if the slab is cast at time different from 2 months after beam manufacture, the percentage of the ultimate shrinkage strain should be adjusted to reflect the appropriate value from Table 2.1 of ACI 435. $(\varepsilon_{sh})_u = 400 \times 10^{-6}$ in./in.

Term (5) is the shrinkage deflection of the composite beam following slab casting, due to the shrinkage of the precast beam concrete, using Eq.(6), with $\varepsilon_{sh} = 0.64(\varepsilon_{sh})_u$. This term does not include the effect of differential shrinkage and creep, which is given by Term (8). I_2/I_c is the same as in Term (3).

Term (6) is the initial or short-term deflection of the precast beam under slab dead load, using Eq. (3), with the incremental deflection computed as follows: $(\Delta_i)_1 = (\Delta_i)_{1+2} - (\Delta_i)_2$, where $(\Delta_i)_2$ is the same as in Term (1). For computing $(I_e)_{1+2}$ and $(\Delta_i)_{1+2}$ in Eqs. (9-8) and (3), $M_a = M_1 + M_2$ due to the precast beam plus slab dead load at midspan, and M_{cr}, I_g, and I_{cr} refer to the precast beam section at midspan. When partitions, roofing, etc., are placed at the same time as the slab, or soon thereafter, their dead load should be included in M_1 and M_a.

Term (7) is the creep deflection of the composite beam due to slab dead load using Eq. (5), with $C_u = \beta_s \times 1.60$. For loading age of 2 months, $\beta_s = 0.89$ is the appropriate correction factor as noted in Term(6) of Eq. (15). For loading at other times, the appropriate value of β_s should be used. In this term, the initial strains, curvatures and deflections under slab dead load were based on the precast section only. Hence the creep curvatures and deflections refer to the precast beam concrete, although the composite section is restraining the creep curvatures and deflections, as mentioned in connection with Term (3). k_r is the same as in Term (2), and I_2/I_c is the same as in Term (3).

Term (8) is the deflection due to differential shrinkage and creep. As an approximation, $\Delta_{ds} = 0.50 \, (\Delta_i)_1$, may be used.

Term (9) is the initial or short-term deflection due to live load (and other loads applied to the composite beam and not included in Term (6)) of the composite beam, using Eq. (4), with the incremental deflection estimated as follows: $(\Delta_i)_\ell = (\Delta_i)_{d+\ell} - (\Delta_i)_d$, based on the composite section. This is thought to be a conservative approximation, since the computed $(\Delta_i)_d$ is on the low side and thus the computed $(\Delta_i)_\ell$ is on the high side, even though the incremental loads are actually resisted by different sections (members). This method is the same as for Term (5) of Eq. (17), and the same as for a monolithic beam. Alternatively, Eq. (3) may be used with $M_a = M_1$ and $I_e = (I_c)_{cr}$ as a simple rough approximation. The first method is illustrated in Example 8.7 and the alternative method in Example 8.6.

Term (10) is the creep deflection due to any sustained live load applied to the composite beam, using Eq. (5), with $C_u = \beta_s \times 1.60$. As in the other cases, β_s is given for various load application times in the explanation of Term (6) of Eq. (15). ρ' refers to any compression steel in the slab at midspan when computing k_r. This Term corresponds to Term (6) in Eqs. (17) and (18).

REFERENCES

10.1 *Deflections of Concrete Structures*, Special Publication SP 43, American Concrete Institute, Farmington Hills, MI, 1974.

10.2 Branson, D.E., *Deformation of Concrete Structures*, McGraw-Hill Book Co., New York, N.Y., 1977.

10.3 ACI Committee 435, "Deflections of Reinforced Concrete Flexural Members", ACI 435.2R-66, (Reapproved 1989).

10.4 *Control of Deflection in Concrete Structures*, ACI 435R-95, American Concrete Institute, Farmington Hills, MI, 1995, (Reapproved 2000).

10.5 *Designing for Creep and Shrinkage in Concrete Structures*, Special Publication SP 76, American Concrete Institute, Farmington Hills, MI, 1982.

10.6 *Designing for Effects of Creep, Shrinkage, and Temperature in Concrete Structures*, Special Publication SP 27, American Concrete Institute, Farmington Hills, MI, 1971.

10.7 Nilson, A. H., and Walters, D. B., "Deflection of Two-Way Floor Systems by the Equivalent Frame Method," *ACI Journal, Proceedings* Vol. 72, No. 5, May 1975, pp. 210-218.

10.8 *PCI Design Handbook—Precast and Prestressed Concrete*, MNL-120-04, 6th Ed., Prestressed Concrete Institute, Chicago, IL, 2004, 750 pp.

10.9 Mast, R.F., "Analysis of Cracked Prestressed Concrete Sections: A Practical Approach," *PCI Journal*, V. 43, No. 4, July-Aug., 1998, pp. 80-91.

10.10 Branson, D. E., "Reinforced Concrete Composite Flexural Members," Chapter 4, pp. 97-174, and "Prestressed Concrete Composite Flexural Members," Chapter 5, pp. 148-210, *Handbook of Composite Construction Engineering*, G.M. Sabanis, Editor, Van Nostrand Reinhold Co., New York, N.Y., 1979.

Example 10.1—Simple-Span Nonprestressed Rectangular Beam

Required: Analysis of short-term deflections, and long-term deflections at ages 3 months and 5 years (ultimate value)

Data:

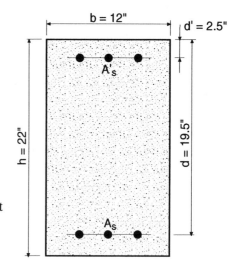

f'_c = 3000 psi (normal weight concrete)
f_y = 40,000 psi
A_s = 3-No. 7 = 1.80 in.^{2}
E_s = 29,000,000 psi
$\rho = A_s/bd = 0.0077$
A'_s = 3-No. 4 = 0.60 in.^{2}
$\rho' = A'_s/bd = 0.0026$

(A'_s not required for strength)
Superimposed dead load (not including beam weight) = 120 lb/ft
Live load = 300 lb/ft (50% sustained)
Span = 25 ft

Calculations and Discussion	Code Reference

1. Minimum beam thickness, for members not supporting or attached to partitions or other construction likely to be damaged by large deflections:

$$h_{min} = \left(\frac{\ell}{16}\right)$$ *Table 9.5(a)*

multiply by 0.8 for f_y = 40,000 psi steel

$$h_{min} = \frac{25 \times 12}{16} \times 0.8 = 15 \text{ in.} < 22 \text{ in.} \quad \text{O.K.}$$

2. Moments:

$$w_d = 0.120 + (12)(22)(0.150)/144 = 0.395 \text{ kips/ft}$$

$$M_d = \frac{w_d \ell^2}{8} = \frac{(0.395)(25)^2}{8} = 30.9 \text{ ft-kips}$$

$$M_\ell = \frac{w_\ell \ell^2}{8} = \frac{(0.300)(25)^2}{8} = 23.4 \text{ ft-kips}$$

$$M_{d+\ell} = 54.3 \text{ ft-kips}$$

$$M_{sus} = M_d + 0.50 M_\ell = 30.9 + (0.50)(23.4) = 42.6 \text{ ft-kips}$$

| Example 10.1 (cont'd) | Calculations and Discussion | Code Reference |

3. Modulus of rupture, modulus of elasticity, modular ratio:

$$f_r = 7.5\sqrt{f'_c} = 7.5\sqrt{3000} = 411 \text{ psi}$$

Eq. (9-10)

$$E_c = w_c^{1.5} 33\sqrt{f'_c} = (150)^{1.5} 33\sqrt{3000} = 3.32 \times 10^6 \text{ psi}$$

8.5.1

$$n_s = \frac{E_s}{E_c} = \frac{29 \times 10^6}{3.32 \times 10^6} = 8.7$$

4. Gross and cracked section moments of inertia, using Table 10-2:

$$I_g = \frac{bh^3}{12} = \frac{(12)(22)^3}{12} = 10{,}650 \text{ in.}^4$$

$$B = \frac{b}{(nA_s)} = \frac{12}{(8.7)(1.80)} = 0.766 \text{ in.}$$

$$r = \frac{(n-1)A'_s}{(nA_s)} = \frac{(7.7)(0.60)}{(8.7)(1.80)} = 0.295$$

$$kd = \left[\sqrt{2dB(1 + rd'/d) + (1+r)^2} - (1+r)\right]/B$$

$$= \left[\sqrt{(2)(19.5)(0.766)\left\{1 + \frac{0.295 \times 2.5}{19.5}\right\} + (1.295)^2} - 1.295\right]/0.766 = 5.77 \text{ in.}$$

$$I_{cr} = \frac{bk^3d^3}{3} + nA_s(d-kd)^2 + (n-1)A'_s(kd-d')^2$$

$$= \frac{(12)(5.77)^3}{3} + (8.7)(1.80)(19.5 - 5.77)^2 + (7.7)(0.60)(5.77 - 2.5)^2$$

$$= 3770 \text{ in.}^4$$

$$\frac{I_g}{I_{cr}} = 2.8$$

5. Effective moments of inertia, using Eq. (9-8):

$$M_{cr} = \frac{f_r I_g}{y_t} = [(411)(10{,}650)/(11)]/(12{,}000) = 33.2 \text{ ft-kips}$$

Eq. (9-9)

a. Under dead load only

		Code
Example 10.1 (cont'd)	**Calculations and Discussion**	**Reference**

$\dfrac{M_{cr}}{M_d} = \dfrac{33.2}{30.9} > 1$. Hence $(I_e)_d = I_g = 10{,}650$ in.^{4}

b. Under sustained load

$$\left(\dfrac{M_{cr}}{M_{sus}}\right)^3 = \left(\dfrac{33.2}{42.6}\right)^3 = 0.473$$

$(I_e)_{sus} = (M_{cr}/M_a)^3 I_g + [1 - (M_{cr}/M_a)^3] I_{cr} \leq I_g$ *Eq. (9-8)*

$= (0.473)(10{,}650) + (1 - 0.473)(3770)$

$= 7025$ in.^{4}

c. Under dead + live load

$$\left(\dfrac{M_{cr}}{M_{d+\ell}}\right)^3 = \left(\dfrac{33.2}{54.3}\right)^3 = 0.229$$

$(I_e)_{d+\ell} = (0.229)(10{,}650) + (1 - 0.229)(3770)$

$= 5345$ in.^{4}

6. Initial or short-time deflections, using Eq. (3): *9.5.2.2*
   9.5.2.3

$$(\Delta_i)_d = \dfrac{K(5/48)M_d \ell^2}{E_c (I_e)_d} = \dfrac{(1)(5/48)(30.9)(25)^2(12)^3}{(3320)(10{,}650)} = 0.098 \text{ in.}$$

$K = 1$ for simple spans (see Table 8-3)

$$(\Delta_i)_{sus} = \dfrac{K(5/48)M_{sus} \ell^2}{E_c (I_e)_{sus}} = \dfrac{(1)(5/48)(42.6)(25)^2(12)^3}{(3320)(7025)} = 0.205 \text{ in.}$$

$$(\Delta_i)_{d+\ell} = \dfrac{K(5/48)M_{d+\ell} \ell^2}{E_c (I_e)_{d+\ell}} = \dfrac{(1)(5/48)(54.3)(25)^2(12)^3}{(3320)(5345)} = 0.344 \text{ in.}$$

$(\Delta_i)_\ell = (\Delta_i)_{d+\ell} - (\Delta_i)_d = 0.344 - 0.098 = 0.246$ in.

<u>Allowable Deflections (Table 9.5(b)):</u>

Flat roofs not supporting and not attached to nonstructural elements likely to be damaged by large deflections—

$(\Delta_i)_\ell \leq \dfrac{\ell}{180} = \dfrac{300}{180} = 1.67$ in. > 0.246 in.   O.K.

		Code
Example 10.1 (cont'd)	**Calculations and Discussion**	**Reference**

Floors not supporting and not attached to nonstructural elements likely to be damaged by large deflections—

$$(\Delta_i)_\ell \leq \frac{\ell}{360} = \frac{300}{360} = 0.83 \text{ in.} > 0.246 \text{ in.} \quad \text{O.K.}$$

7. Additional long-term deflections at ages 3 mos. and 5 yrs. (ultimate value):

   Combined creep and shrinkage deflections, using Eqs. (9-11) and (4):

Duration	ξ	$\lambda = \frac{\xi}{1 + 50\rho'}$	$(\Delta_i)_{sus}$ in.	$(\Delta_i)_\ell$ in.	$\Delta_{cp} + \Delta_{sh} = \lambda(\Delta_i)_{sus}$ in.	$\Delta_{cp} + \Delta_{sh} + (\Delta_i)_\ell$ in.
5-years	2.0	1.77	0.205	0.246	0.363	0.61
3-months	1.0	0.89	0.205	0.246	0.182	0.43

Separate creep and shrinkage deflections, using Eqs. (5) and (6):

For $\rho = 0.0077$; $\rho' = 0.0026$

For $\rho = 100\rho = 0.77$ and $\rho' = 100\rho' = 0.26$, read $A_{sh} = 0.455$ (Fig. 10-3) and $K_{sh} = 0.125$ for simple spans (Table 10-5).

Duration	C_t	$\lambda_{cp} = \frac{0.85 C_t}{1+50\rho'}$	$\Delta_{cp} = \lambda_{cp}(\Delta_i)_{sus}$ in.	ε_{sh} in./in.	$\phi_{sh} = \frac{A_{sh}\varepsilon_{sh}}{h}$ 1/in.	$\Delta_{sh} = K_{sh}\phi_{sh}\ell^2$ in.	$\Delta_{cp} + \Delta_{sh} + (\Delta_i)_\ell$ in.
5-years	1.6 (ultimate)	1.20	0.246	400×10^{-6}	$\frac{0.455 \times 400 \times 10^{-6}}{22}$ $= 8.27 \times 10^{-6}$	$\frac{1}{8} \times 8.27 \times 10^{-6}$ $\times (25 \times 12)^2$ $= 0.093$	$0.246+0.093+0.246$ $= 0.59$
3-months	0.56×1.6 $= 0.9$	0.68	0.14	$0.6 \times 400 \times 10^{-6}$ $= 240 \times 10^{-6}$	4.96×10^{-6}	$= 0.0558$	$0.14+0.056+0.246$ $= 0.44$

<u>Allowable Deflection Table 9.5(b)</u>:

Roof or floor construction supporting or attached to nonstructural elements likely to be damaged by large deflections (very stringent limitation).

$$\Delta_{cp} + \Delta_{sh} + (\Delta_i)_\ell \leq \frac{\ell}{480} = \frac{300}{480} = 0.63 \text{ in.} \quad \text{O.K. by both methods}$$

Roof or floor construction supporting or attached to nonstructural elements not likely to be damaged by large deflections.

$$\Delta_{cp} + \Delta_{sh} + (\Delta_i)_\ell \leq \frac{\ell}{240} = \frac{300}{240} = 1.25 \text{ in.} \quad \text{O.K. by both methods}$$

Example 10.2—Continuous Nonprestressed T-Beam

Required: Analysis of short-term and ultimate long-term deflections of end-span of multi-span beam shown below.

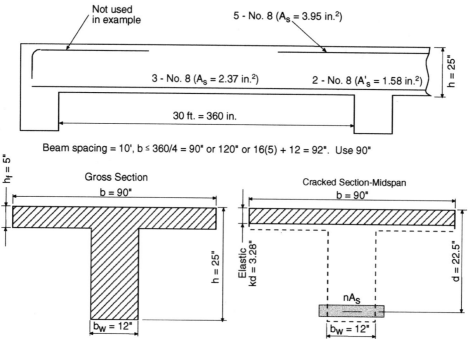

$nA_s = (11.3)(2.37) = 26.8$ in.^{2}
$\rho = 2.37/(90)(22.5) = 0.00117$
$\rho_w = 2.37/(12)(22.5) = 0.00878$
$\rho' = 0$

Data:

f'_c = 4000 psi (sand-lightweight concrete)
f_y = 50,000 psi
w_c = 120 pcf

Beam spacing = 10 ft
Superimposed Dead Load (not including beam weight) = 20 psf
Live Load = 100 psf (30% sustained)

(A'_s is not required for strength)

Beam will be assumed to be continuous at one end only for h_{min} in Table 9.5(a), for Avg. I_e in Eq. (1), and for K_{sh} in Eq. (6), since the exterior end is supported by a spandrel beam. The end span might be assumed to be continuous at both ends when supported by an exterior column.

Calculations and Discussion	Code Reference

1. Minimum thickness, for members not supporting or attached to partitions or other construction likely to be damaged by large deflections:

| Example 10.2 (cont'd) | Calculations and Discussion | Code Reference |

$h_{min} = \dfrac{\ell}{18.5}$ *Table 9.5(a)*

Modifying factors = 1.09 for w_c = 120 pcf [footnote (a) Table 9.5(a)]

= 0.9 for f_y = 50,000 psi [footnote (b) Table 9.5(a)]

$h_{min} = \left(\dfrac{360}{18.5}\right)(0.90)(1.09) = 19.1$ in. $< h = 25$ in.   O.K.

2. Loads and moments:

$w_d = (20 \times 10) + (120)(12 \times 20 + 120 \times 5)/144 = 900$ lb/ft

$w_\ell = (100 \times 10) = 1000$ lb/ft

In lieu of a moment analysis, the ACI approximate moment coefficients may be used as follows: Pos. $M = w\ell_n^2/14$ for positive I_e and maximum deflection, Neg. $M = w\ell_n^2/10$ for negative I_e.   *8.3.3*

a. Positive Moments

Pos. $M_d = \dfrac{w_d \ell_n^2}{14} = \dfrac{(0.900)(30)^2}{14} = 57.9$ ft-kips

Pos. $M_\ell = \dfrac{(1.000)(30)^2}{14} = 64.3$ ft-kips

Pos. $M_{d+\ell} = 57.9 + 64.3 = 122.2$ ft-kips

Pos. $M_{sus} = M_d + 0.30 M_\ell = 57.9 + (0.30)(64.3) = 77.2$ ft-kips

b. Negative Moments

Neg. $M_d = \dfrac{w_d \ell_n^2}{10} = \dfrac{(0.900)(30)^2}{10} = 81.0$ ft-kips

Neg. $M_\ell = \dfrac{(1.000)(30)^2}{10} = 90.0$ ft-kips

Neg. $M_{d+\ell} = 81.0 + 90.0 = 171.0$ ft-kips

Neg. $M_{sus} = M_d + 0.30 M_\ell = 81.0 + (0.30)(90.0) = 108.0$ ft-kips

3. Modulus of rupture, modulus of elasticity, modular ratio:

$f_r = (0.85)(7.5)\sqrt{f'_c} = 6.38\sqrt{4000} = 404$ psi (0.85 for sand lightweight concrete)   *Eq. (9-10)*

9.5.2.3(b)

		Code
Example 10.2 (cont'd)	**Calculations and Discussion**	**Reference**

$E_c = w_c^{1.5} 33\sqrt{f'_c} = (120)^{1.5} 33\sqrt{4000} = 2.74 \times 10^6$ psi *8.5.1*

$n = \dfrac{E_s}{E_c} = \dfrac{29 \times 10^6}{2.74 \times 10^6} = 10.6$

4. Gross and cracked section moments of inertia:

   a. Positive moment section

   $y_t = h - (1/2) [(b - b_w) h_f^2 + b_w h^2]/[(b - b_w) h_f + b_w h]$

   $= 25 - (1/2) [(78) (5)^2 + (12)(25)^2]/[(78) (5) + (12) (25)]$

   $= 18.15$ in.

   $I_g = (b - b_w) h_f^3/12 + b_w h^3/12 + (b - b_w) h_f (h - h_f/2 - y_t)^2 + b_w h (y_t - h/2)^2$

   $= (78) (5)^3/12 + (12) (25)^3/12 + (78) (5) (25 - 2.5 - 18.15)^2$

   $+ (12) (25) (18.15 - 12.5)^2 = 33{,}390$ in.^{4}

   $B = \dfrac{b}{nA_s} = \dfrac{90}{(10.6)(2.37)} = 3.58/\text{in.}$ (Table 10-2)

   $kd = \dfrac{\sqrt{2dB + 1} - 1}{B} = \dfrac{\sqrt{(2)(22.5)(3.58) + 1} - 1}{3.58}$

   $= 3.28$ in. $< h_f = 5$ in.

   Hence, treat as a rectangular compression area.

   $I_{cr} = bk^3d^3/3 + nA_s (d - kd)^2 = (90) (3.28)^3/3 + (10.6) (2.37) (22.5 - 3.28)^2$

   $= 10{,}340$ in.^{4}

   b. Negative moment section

   $I_g = \dfrac{12 \times 25^3}{12} = 15{,}625$ in.^{4}

   $I_{cr} = 11{,}185$ in.^{4} (similar to Example 10.1, for $b = 12$ in., $d = 22.5$ in., $d' = 2.5$ in., $A_s = 3.95$ in.^{2}, $A'_s = 1.58$ in.^{2})

5. Effective moments of inertia, using Eqs. (9-8) and (1):

   a. Positive moment section:

   $M_{cr} = f_r I_g/y_t = [(404)(33{,}390)/(18.15)]/12{,}000 = 61.9$ ft-kips *Eq. (9-9)*

10-28

		Code
Example 10.2 (cont'd)	**Calculations and Discussion**	**Reference**

$M_{cr}/M_d = 61.9/57.9 > 1$. Hence $(I_e)_d = I_g = 33{,}390$ in.^{4}

$(M_{cr}/M_{sus})^3 = (61.9/77.2)^3 = 0.515$

$(I_e)_{sus} = (M_{cr}/M_a)^3 I_g + [1 - (M_{cr}/M_a)^3] I_{cr} \leq I_g$ Eq. (9-8)

$\qquad = (0.515)(33{,}390) + (1 - 0.515)(10{,}340) = 22{,}222$ in.^{4}

$\left(M_{cr}/M_{d+\ell}\right)^3 = (61.9/122.2)^3 = 0.130$

$(I_e)_{d+\ell} = (0.130)(33{,}390) + (1 - 0.130)(10{,}340) = 13{,}336$ in.^{4}

   b. Negative moment section:

$M_{cr} = [(404)(15{,}625)/(12.5)]/12{,}000 = 42.1$ ft-kips   Eq. (9-9)

$(M_{cr}/M_d)^3 = (42.1/81.0)^3 = 0.14$

$(I_e)_d = (0.14)(15{,}625) + (1 - 0.14)(11{,}185) = 11{,}808$ in.^{4} Eq. (9-8)

$(M_{cr}/M_{sus})^3 = (42.1/108.0)^3 = 0.06$

$(I_e)_{sus} = (0.06)(15{,}625) + (1 - 0.06)(11{,}185) = 11{,}448$ in.^{4} Eq. (9-8)

$\left(M_{cr}/M_{d+\ell}\right)^3 = (42.1/171.0)^3 = 0.015$

$(I_e)_{d+\ell} = (0.015)(15{,}625) + (1 - 0.015)(11{,}185) = 11{,}251$ in.^{4} Eq. (9-8)

   c. Average inertia values:

Avg. $(I_e) = 0.85 I_m + 0.15 (I_{cont.\ end})$ Eq. (1)

Avg. $(I_e)_d = (0.85)(33{,}390) + (0.15)(11{,}808) = 30{,}153$ in.^{4}

Avg. $(I_e)_{sus} = (0.85)(22{,}222) + (0.15)(11{,}448) = 20{,}606$ in.^{4}

Avg. $(I_e)_{d+\ell} = (0.85)(13{,}336) + (0.15)(11{,}251) = 13{,}023$ in.^{4}

6. Initial or short-time deflections, with midspan I_e and with avg. I_e: 9.5.2.4

$$(\Delta_i) = K\left(\frac{5}{48}\right)\frac{M_a \ell^2}{E_c I_e}$$ Eq. (3)

$K = 1.20 - 0.20 M_o/M_a = 1.20 - (0.20)\left(w\ell_n^2/8\right)/\left(w\ell_n^2/14\right) = 0.850$ (Table 10-3)

$$(\Delta_i)_d = \frac{K(5/48) M_d \ell^2}{E_c (I_e)_d} = \frac{(0.85)(5/48)(57.9)(30)^2 (12)^3}{(2740)(33{,}390)} = 0.087 \text{ in.}$$

10-29

		Code
Example 10.2 (cont'd)	**Calculations and Discussion**	**Reference**

$= 0.096$ in., using avg. $I_e = 30,149$ in.^{4}

$$(\Delta_i)_{sus} = \frac{K(5/48)M_{sus}\ell^2}{E_c(I_e)_{sus}} = \frac{(0.85)(5/48)(77.2)(30)^2(12)^3}{(2740)(22,222)} = 0.175 \text{ in.}$$

$= 0.188$ in., using avg. $I_e = 20,594$ in.^{4}

$$(\Delta_i)_{d+\ell} = \frac{K(5/48)M_{d+\ell}\ell^2}{E_c(I_e)_{d+\ell}} = \frac{(0.85)(5/48)(122.2)(30)^2(12)^3}{(2740)(13,336)} = 0.460 \text{ in.}$$

$= 0.472$ in., using avg. $I_e = 13,023$ in.^{4}

$(\Delta_i)_\ell = (\Delta_i)_{d+\ell} - (\Delta_i)_d = 0.460 - 0.087 = 0.373$ in.

$= 0.472 - 0.096 = 0.376$ in., using avg. I_e from Eq. (1)

Allowable deflections Table 9.5(b):

For flat roofs not supporting and not attached to nonstructural elements likely to be damaged by large deflections — $(\Delta_i)_\ell \leq \ell/180 = 2.00$ in. > 0.376 in. O.K

For floors not supporting and not attached to nonstructural elements likely to be damaged by large deflections — $(\Delta_i)_\ell \leq \ell/360 = 360/360 = 1.00$ in. O.K.

7. Ultimate long-term deflections:

Using ACI Method with combined creep and shrinkage effects:

$$\lambda = \frac{\xi}{1 + 50\rho'} = \frac{2.0 \text{ (ultimate value)}}{1 + 0} = 2.0 \qquad \text{Eq. (9-11)}$$

$$\Delta_{(cp+sh)} = \lambda(\Delta_i)_{sus} = (2.0)(0.175) = 0.350 \text{ in.} \qquad \text{Eq. (4)}$$

$\Delta_{(cp+sh)} + (\Delta_i)\ell = 0.350 + 0.373 = 0.723$ in.

$= [2(0.188) + 0.376] = 0.752$ using avg. I_e from Eq. (1).

Using Alternate Method with separate creep and shrinkage deflections:

$$\lambda_{cp} = \frac{0.85C_u}{1 + 50\rho'} = \frac{(0.85)(1.60)}{1 + 0} = 1.36$$

$$\Delta_{cp} = \lambda_{cp}(\Delta_i)_{sus} = (1.36)0.175 = 0.238 \text{ in.} \qquad \text{Eq. (5)}$$

		Code
Example 10.2 (cont'd)	**Calculations and Discussion**	**Reference**

$= 1.36(0.188) = 0.256$ in., using avg. I_e Eq. (1).

$$\rho = 100\left(\frac{\rho + \rho_w}{2}\right) = 100\left(\frac{2.37}{90 \times 22.5} + \frac{2.37}{12 \times 22.5}\right)/2$$

$= 100 \, (0.00117 + 0.00878)/2 = 0.498\%$

A_{sh} (from Fig. 10-3) $= 0.555$

$$\phi_{sh} = A_{sh}\frac{(\varepsilon_{sh})_u}{h} = \frac{(0.555)(400 \times 10^{-6})}{25} = 8.88 \times 10^{-6}/\text{in.}$$

$$\Delta_{sh} = K_{sh}\phi_{sh}\ell^2 = (0.090)\left(8.88 \times 10^{-6}\right)(30)^2(12)^2 = 0.104 \text{ in.} \qquad \textit{Eq. (6)}$$

$\Delta_{cp} + \Delta_{sh} + (\Delta_i)_\ell = 0.238 + 0.104 + 0.373 = 0.715$ in.

$\phantom{\Delta_{cp} + \Delta_{sh} + (\Delta_i)_\ell} = (0.256 + 0.104 + 0.376) = 0.736.$, using avg. I_e from Eq. (1).

<u>Allowable deflections Table 9.5(b):</u>

For roof or floor construction supporting or attached to nonstructural elements likely to be damaged by large deflections (very stringent limitation) —

$$\Delta_{cp} + \Delta_{sh} + (\Delta_i)_\ell \leq \frac{\ell}{480} = \frac{360}{480} = 0.75 \text{ in.}$$

All results O.K.

For roof or floor construction supporting or attached to nonstructural elements not likely to be damaged by large deflections —

$$\Delta_{cp} + \Delta_{sh} + (\Delta_i)_\ell \leq \frac{\ell}{240} = \frac{360}{240} = 1.50 \text{ in.} \quad \text{All results O.K.}$$

Example 10.3—Slab System Without Beams (Flat Plate)

Required: Analysis of short-term and ultimate long-term deflections of a corner panel.

Data:

Flat plate with no edge beams, designed by Direct Design Method
Slab f'_c = 3000 psi, Column f'_c = 5000 psi, (normal weight concrete)
f_y = 40,000 psi
Square panels—15 × 15 ft center-to-center of columns
Square columns—14 × 14 in., Clear span, ℓ_n = 15 - 1.17 = 13.83 ft
Story height = 10 ft., Slab thickness, h = 6 in.
The reinforcement in the column strip negative moment regions consists of No. 5 bars at 7.5 in. spacing. Therefore, the total area of steel in a 90-in. strip (half the panel length) is given by:

A_s = (90/7.5)(0.31) = 3.72 sq. in.

The distance from the compressive side of the slab to the center of the steel is:

d = 4.62 in

Middle Strip reinforcement and d values are not required for deflection computations, since the slab remains uncracked in the middle strips.
Superimposed Dead Load = 10 psf
Live Load = 50 psf
Check for 0% and 40% Sustained Live Load

Calculations and Discussion	Code Reference

1. Minimum thickness: 9.5.3.2

   From Table 10-6, with Grade 40 steel:

   Interior panel $h_{min} = \dfrac{\ell_n}{36}$ = (13.83 × 12)/36 = 4.61 in.

   Exterior panel $h_{min} = \dfrac{\ell_n}{33}$ = (13.83 × 12)/33 = 5.03 in.

   Since the actual slab thickness is 6 in., deflection calculations are not required; however, as an illustration, deflections will be checked for a corner panel, to make sure that all allowable deflections per Table 9.5(b) are satisfied.

2. Comment on trial design with regard to deflections:

   Based on the minimum thickness limitations versus the actual slab thickness, it appears likely that computed deflections will meet most or all of the code deflection limitations. It turns out that all are met.

		Code
Example 10.3 (cont'd)	**Calculations and Discussion**	**Reference**

3. Modulus of rupture, modulus of elasticity, modular ratio:

$$f_r = 7.5\sqrt{f'_c} = 7.5\sqrt{3000} = 411 \text{ psi}$$ Eq. (9-10)

$$E_{cs} = w_c^{1.5} 33\sqrt{f'_c} = (150)^{1.5} 33\sqrt{3000} = 3.32 \times 10^6 \text{ psi}$$ 8.5.1

$$E_{cc} = (150)^{1.5} 33\sqrt{5000} = 4.29 \times 10^6 \text{ psi}$$

$$n = \frac{E_s}{E_{cs}} = \frac{29}{3.32} = 8.73$$

4. Service load moments and cracking moment:

$$w_d = 10 + (150)(6.0)/12 = 85.0 \text{ psf}$$

$$(M_o)_d = w_d \ell_2 \ell_n^2/8 = (85.0)(15)(13.83)^2/8000 = 30.48 \text{ ft-kips}$$

$$(M_o)_{d+\ell} = w_d \ell_2 \ell_n^2/8 = (85.0 + 50.0)(15)(13.83)^2/8000 = 48.41 \text{ ft-kips}$$

$$(M_o)_{sus} = (85 + 0.4 \times 50)(15)(13.83)^2/8000 = 37.65 \text{ ft-kips}$$

The moments are distributed to the ends and centers of the column and middle strips according to the coefficients in the tables of Sections 13.6.3.3, 13.6.4.1, 13.6.4.2 and 13.6.4.4. In this case, the span ratio, ℓ_2/ℓ_1, is equal to 1.0. The multipliers of the panel moment, M_o, that are used to make the distribution in an end span are given in the following table:

	Ext. Negative	Positive	Int. Negative
Total Panel	0.26	0.52	0.70
Col. Strip	(1.0)(0.26)	(0.60)(0.52)	(0.75)(0.70)
Mid. Strip	(1.0-1.0)(0.26)	(1.0-0.60)(0.52)	(1.0-0.75)(0.70)

The resulting moments applied to the external and internal ends and to the center span of the column and middle strips are given in the following tables:

Dead Load Moments, ft-kips

	Ext. Negative	Positive	Int. Negative
Total Panel	7.93	15.85	21.34
Col. Strip	7.93	9.51	16.00
Mid. Strip	0	6.34	5.34

10-33

Example 10.3 (cont'd)	Calculations and Discussion	Code Reference

Dead Load + Live Load Moments, ft-kips

	Ext. Negative	Positive	Int. Negative
Total Panel	12.59	25.18	33.89
Col. Strip	12.59	15.10	25.41
Mid. Strip	0	10.07	8.47

Sustained Load Moments, ft-kips (Dead Load + 40% Live Load)

	Ext. Negative	Positive	Int. Negative
Total Panel	9.79	19.58	26.36
Col. Strip	9.79	11.75	19.77
Mid. Strip	0	7.83	6.59

The gross moment of inertia of a panel, referred to as the total equivalent frame moment of inertia is:

$$I_{frame} = \ell_s h^3/12 = (15 \times 12)(6)^3/12 = 3{,}240 \text{ in.}^4$$

For this case, the moment of inertia of a column strip or a middle strip is equal to half of the moment of inertia of the total equivalent frame:

$$I_g = \tfrac{1}{2}(3240) = 1{,}620 \text{ in.}^4$$

The cracking moment for either a column strip or a middle strip is obtained from the standard flexure formula based on the uncracked section as follows:

$$(M_{cr})_{c/2} = (M_{cr})_{m/2} = f_r I_g/y_t = (411)(15 \times 12)(6.0)^3/(4)(12)(3.0)(12{,}000)$$

$$= 9.25 \text{ ft-kips}$$

5. Effective moments of inertia:

A comparison of the tabulated applied moments with the cracking moment shows that the apportioned moment at all locations, except at the interior support of the column strips for the live load and sustained load cases, is less than the cracking moment under the imposed loads. The cracked section moment of inertia is, therefore, only required for the <u>column strips</u> in the negative moment zones. Formulas for computation of the cracked section moment of inertia are obtained from Table 10-2:

$$B = \frac{b}{nA_s} = \frac{\tfrac{1}{2}(15 \times 12)}{8.73 \times 3.72} = 2.77 \left(\frac{1}{\text{in.}}\right)$$

$$kd = \frac{\sqrt{2dB+1}-1}{B} = \frac{\sqrt{2 \times 4.62 \times 2.77 + 1}-1}{2.77} = 1.50 \text{ in.}$$

		Code
Example 10.3 (cont'd)	**Calculations and Discussion**	**Reference**

$$I_{cr} = \frac{b(kd)^3}{3} + nA_s(d-kd)^2 = \frac{90 \times 1.50^3}{3} + 8.73 \times 3.72(4.62 - 1.50)^2 = 417 \text{ in.}^4$$

To obtain an equivalent moment of inertia for the cracked location, apply the Branson modification to the moments of inertia for cracked and uncracked sections. The approximate moment of inertia in the cracked sections is given by the general formula in Equation (9-8) of ACI 318. From the tables developed in Section 4 above, the ratios of the dead load plus live load and sustained load moments to the cracking moment are found as follows:

For live dead load plus live load:

$$\frac{M_{cr}}{M_a} = \frac{18.50}{25.41} = 0.728$$

$$\left(\frac{M_{cr}}{M_a}\right)^3 = 0.386$$

and for the sustained load case (dead load plus 40% live load):

$$\frac{M_{cr}}{M_a} = \frac{18.50}{19.77} = 0.936$$

$$\left(\frac{M_{cr}}{M_a}\right)^3 = 0.819$$

The equivalent moment of inertia for the two cases are now computed by Eq. (9-8) of ACI 318:

For dead load plus live load:

$$I_e = (0.386)1620 + (1-0.386)(417) = 881 \text{ in.}^4$$

For sustained load (dead load + 40% live load):

$$I_e = (0.819)1620 + (1-0.819)(417) = 1402 \text{ in.}^4$$

Finally, the equivalent moment of inertia for the uncracked sections is just the moment of inertia of the gross section, I_g.

To obtain an average moment of inertia for calculation of deflection, the "end" and "midspan" values are then combined according to Equation (1):

For dead load plus live load:

$$\text{Avg. } I_e = 0.85(1620) + 0.15(881)] = 1509 \text{ in.}^4$$

| Example 10.3 (cont'd) | Calculations and Discussion | Code Reference |

For sustained load (dead load + 40% live load):

$$\text{Avg. } I_e = 0.85(1620) + 0.15(1402)] = 1587 \text{ in.}^4$$

To obtain the equivalent moment of inertia for the "equivalent frame", which consists of a column and a middle strip, add the average moments of inertia for the respective strips. For the middle strips, the moment of inertia is that of the gross section, I_g, and for the column strips, the average values computed above are used:

For dead load only:

$$(I_e)_{frame} = 1620 + 1620 = 3240 \text{ in.}^4$$

For dead load plus live load:

$$(I_e)_{frame} = 1620 + 1509 = 3129 \text{ in.}^4$$

For dead load plus 40% live load:

$$(I_e)_{frame} = 1620 + 1587 = 3207 \text{ in.}^4$$

Note: In this case, where a corner panel is considered, there is only half of a column strip along the two outer edges. However, the section properties for half a strip are equal to half of those for a full strip; also, the applied moments to the edge strip are half those applied to an interior strip. Consequently, deflections calculated for either a half or for a full column strip are the same. Strictly, these relationships only apply because all panels are of equal dimensions in both directions. If the panels are not square or if adjacent panels are of differing dimensions, additional calculations would be necessary.

6. Flexural stiffness (K_{ec}) of an exterior equivalent column:

$K_b = 0$ (no beams)   R13.7.4

The stiffness of the equivalent exterior column is determined by combining the stiffness of the upper and lower columns at the outer boundary of the floor with the torsional stiffness offered by a strip of the floor slab, parallel to the edge normal to the direction of the equivalent frame and extending the full panel length between columns. In the case of a corner column, the length is, of course, only half the panel length. The width of the strip is equal to the column dimension normal to the direction of the equivalent frame (ACI 318, R13.7.5).

The column stiffness is computed on the basis of the rotation resulting from application of a moment to the simply supported end of a propped cantilever, $M = 4EI/L$. In this case the result is:

$$K_c = 4E_{cc}I_c/\ell_c = 4E_{cc}[(14)^4/(12)]/[(10)(12)] = 106.7 E_{cc}$$

Since the columns above and below the slab are equal in dimension, the total stiffness of the columns is twice that of a single column:

$$\Sigma K_c = 2K_c = (2)(106.7 E_{cc}) = 213.4 E_{cc}$$

		Code
Example 10.3 (cont'd)	**Calculations and Discussion**	**Reference**

The torsional stiffness of the slab strip is calculated according to the methodology set out in R13.7.5 of ACI 318, $K_t = \Sigma 9 E_{cs}/L_2(1-c_2/L_2)3$. The cross-sectional torsional constant, C, is defined in Section 13.0 of ACI 318.

$$C = (1 - 0.63\ x/y)(x^3 y/3) = (1 - 0.636 \times 6.0/14)(6.0^3 \times 14/3) = 735.8 \text{ in.}^4$$

$$K_t = \frac{\Sigma 9 E_{cs} C}{\ell_2 (1 - c_2/\ell_2)^3} = \frac{(2)(9) E_{cs} (735.8)}{(15)(12)\left(1 - \frac{14}{15 \times 12}\right)^3} = 93.9 E_{cs}$$

For Ext. Frame, $K_t = 93.9 E_{cs}/2 = 47 E_{cs}$, $E_{cc} = (4.29/3.32) E_{cs} = 1.292 E_{cs}$

The equivalent column stiffness is obtained by treating the column stiffness and the torsional member stiffness as springs in series:

$$K_{ec} = \frac{1}{\frac{1}{\Sigma K_c} + \frac{1}{K_t}} = \frac{E_{cs}}{\left(\frac{1}{213.4 \times 1.292}\right) + \left(\frac{1}{93.9}\right)} = 70 E_{cs} = 19{,}370 \text{ ft-kips/rad}$$

For Ext. Frame, $K_{ec} = \dfrac{E_{cs}}{\left(\dfrac{1}{213.4 \times 1.292}\right) + \left(\dfrac{1}{47.0}\right)} = 40.1 E_{cs} = 11{,}090 \text{ ft-kips/rad}$

7. Deflections, using Eqs. (7) to (14):

   Fixed $\Delta_{frame} = w\ell_2 \ell^4 / 384 E_{cs} I_{frame}$ *Eq. (10)*

   $$(\text{Fixed } \Delta_{frame})_{d,d+\ell} = \frac{(85.0 \text{ or } 135.0 \text{ or } 105.0)(15)^5 (12)^3}{(384)\left(3.32 \times 10^6\right)(3240 \text{ or } 3129 \text{ or } 3207)}$$

   $$= 0.027 \text{ in.}, 0.044 \text{ in.}; 0.034$$

   Fixed $\Delta_{c,m} = (LDF)_{c,m} (\text{Fixed } \Delta_{frame})(I_{frame}/I_{c,m})$ *Eq. (11)*

These deflections are distributed to the column and middle strips in the ratio of the total applied moment to the beam stiffness (M/EI) of the respective strips to that of the complete frame. As shown in Step 4 above, the fraction of bending moment apportioned to the column or middle strips varies between the ends and the midspan. Therefore, in approximating the deflections by this method, the average moment allocation fraction (Lateral Distribution Factor - LDF) is used. In addition, since the equivalent moment of inertia changes whenever the cracking moment is exceeded, an average moment of inertia is utilized. This average moment of inertia is computed on the basis of Equation (9-8) from ACI 318 and Eq. (1) of this chapter. Finally, since the modulus of elasticity is constant throughout the slab, the term E occurs in both the numerator and the denominator and is, therefore, omitted. The LDFs are calculated as follows:

	Code
Example 10.3 (cont'd) **Calculations and Discussion**	**Reference**

For the column strip:

$$LDF_c = \tfrac{1}{2}[\tfrac{1}{2}(M^-_{int} + M^-_{ext}) + M^+] = \tfrac{1}{2}[\tfrac{1}{2}(0.75 + 1.00) + 0.60] = 0.738$$

For the middle strip:

$$LDF_m = 1 - LDF_c = 0.262$$

$(\text{Fixed } \Delta_c)_d = (0.738)(0.027)(2) = 0.040 \text{ in.}$

$(\text{Fixed } \Delta_c)_{d+\ell} = (0.738)(0.044)(3129/1509) = 0.067 \text{ in.}$

$(\text{Fixed } \Delta_c)_\ell = 0.067 - 0.040 = 0.027 \text{ in.}$

$(\text{Fixed } \Delta_c)_{sus} = (0.738)(0.034)(3207/1587) = 0.051 \text{ in}$

$(\text{Fixed } \Delta_m)_d = (0.262)(0.027)(2) = 0.014 \text{ in.}$

$(\text{Fixed } \Delta_m)_{d+\ell} = (0.262)(0.044)(3129/1620) = 0.022 \text{ in.}$

$(\text{Fixed } \Delta_m)_\ell = 0.022 - 0.014 = 0.008 \text{ in.}$

$(\text{Fixed } \Delta_m)_{sus} = (0.0.262)(0.034)(3207/1587) = 0.018 \text{ in}$

In addition to the fixed end displacement found above, an increment of deflection must be added to each due to the actual rotation that occurs at the supports. The magnitude of the increment is equal to qL/8. The rotations, q, are determined as the net moments at the column locations divided by the effective column stiffnesses. In this case, the column strip moment at the corner column of the floor is equal to half of 100% of 0.26 x M_o (ACI 318, Sec. 13.6.3.313 and Sec.13.6.4.2). Because the column strip at the edge of the floor is only half as wide as an interior column strip, only half of the apportioned moment acts. The net moments at other columns are either quite small or zero. Therefore they are neglected. The net moments on a corner column for the three loading cases are:

$(M_{net})_d = \tfrac{1}{2} \times 0.26 \times 1.00 \times (M_o)_d = \tfrac{1}{2}[(0.26)(1.00)](30.48) = 3.96 \text{ ft-kips}$

$(M_{net})_{d+\ell} = \tfrac{1}{2} \times 0.26 \times 1.00 \times (M_o)_{d+\ell} = \tfrac{1}{2}[(0.26)(1.00)](48.41) = 6.29 \text{ ft-kips}$

$(M_{net})_{sus} = \tfrac{1}{2} \times 0.26 \times 1.00 \times (M_o)_{sus} = \tfrac{1}{2}[(0.26)(1.00)](37.65) = 4.89 \text{ ft-kips}$

For both column and middle strips,

End $\theta_d = (M_{net})_d / \text{avg. } K_{ec} = 3.96/11,090 = 0.000357 \text{ rad}$ Eq. (12)

End $\theta_{d+\ell} = 6.29/11,090 = 0.000567 \text{ rad}$

		Code
Example 10.3 (cont'd)	**Calculations and Discussion**	**Reference**

End θ_{sus} = 4.89/11090 = 0.000441 rad

$\Delta\theta$ = (End θ) ($\ell/8$) $(I_g/I_e)_{frame}$ Eq. (14)

$(\Delta\theta)_d$ = (0.000357) (15) (12) (1)/8 = 0.008 in.

$(\Delta\theta)_{d+\ell}$ = (0.000567) (15) (12) (1620/1509)/8 = 0.014 in.

$(\Delta\theta)_\ell$ = 0.014 - 0.008 = 0.006 in.

$(\Delta\theta)_{sus}$ = (0.000441)(15)(12)/8 = 0.010 in.

The deflections due to rotation calculated above are for column strips. The deflections due to end rotations for the middle strips will be assumed to be equal to that in the column strips. Therefore, the strip deflections are calculated by the general relationship:

$\Delta_{c,m}$ = Fixed $\Delta_{c,m}$ + $(\Delta\theta)$ Eq.(9)

$(\Delta_c)_d$ = 0.040 + 0.008 = 0.048 in.

$(\Delta_m)_d$ = 0.014 + 0.008 = 0.022 in.

$(\Delta_c)_\ell$ = 0.027 + 0.006 = 0.033 in.

$(\Delta_m)_\ell$ = 0.008 + 0.006 = 0.014 in.

$(\Delta_c)_{sus}$ = 0.051 + (0.010) = 0.061 in.

$(\Delta_m)_{sus}$ = 0.018 + (0.010) = 0.028 in.

Δ = Δ_{cx} + Δ_{my} = midpanel deflection of corner panel   Eq. (7)

$(\Delta_i)_d$ = 0.048 + 0.022 = 0.070 in.

$(\Delta_i)_\ell$ = 0.033 + 0.014 = 0.047 in.

$(\Delta_i)_{sus}$ = 0.061 + 0.028 = 0.089 in.

The long term deflections may be calculated using Eq. (9-11) of ACI 318 (Note: $\rho' = 0$):

For dead load only:

$(\Delta_{cp+sh})_d = 2.0 \times (\Delta_i)_d$ = (2)(0.070) = 0.140 in.

Example 10.3 (cont'd)	Calculations and Discussion	Code Reference

For sustained load (dead load + 40% live load)

$(\Delta_{cp+sh})_{sus} = 2.0 \times (\Delta_i)_{sus} = (2)(0.109) = 0.218$ in.

The long term deflection due to sustained load plus live load is calculated as:

$(\Delta_{cp+sh})_{sus} + (\Delta_i)_\ell = 0.218 + 0.047 = 0.265$ in.

These computed deflections are compared with the code allowable deflections in Table 9.5(b) as follows:

Flat roofs not supporting and not attached to nonstructural elements likely to be damaged by large deflections—

$(\Delta_i)_\ell \le (\ell_n \text{ or } \ell)/180 = (13.83 \text{ or } 15)(12)/180 = 0.92$ in. or 1.00 in., versus 0.047 in. O.K.

Floors not supporting and not attached to nonstructural elements likely to be damaged by large deflections—

$(\Delta_i)_\ell \le (\ell_n \text{ or } \ell)/360 = 0.46$ in. or 0.50 in., versus 0.047 in. O.K.

Roof or floor construction supporting or attached to non-structural elements likely to be damaged by large deflections—

$\Delta_{(cp+sh)} + (\Delta_i)_\ell \le (\ell_n \text{ or } \ell)/480 = 0.35$ in. or 0.38 in., versus 0.265 in. O.K.

Roof or floor construction supporting or attached to non-structural elements not likely to be damaged by large deflections—

$\Delta_{(cp+sh)} + (\Delta_i)_\ell \le (\ell_n \text{ or } \ell)/240 = 0.69$ in. or 0.75 in., versus 0.265 in. O.K.

All computed deflections are found to be satisfactory in all four categories.

Example 10.4—Two-Way Beam Supported Slab System

Required: Minimum thickness for deflection control

Data:

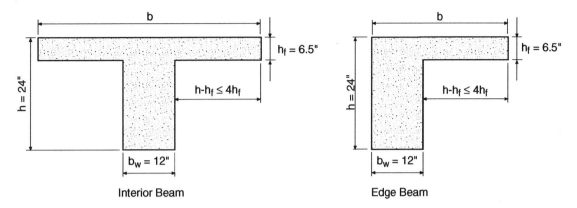

Interior Beam   Edge Beam

f_y = 60,000 psi, slab thickness h_f = 6.5 in.
Square panels—22 × 22 ft center-to-center of columns
All beams—b_w = 12 in. and h = 24 in. ℓ_n = 22 - 1 = 21 ft
It is noted that f'_c and the loading are not required in this analysis.

	Code
Calculations and Discussion	**Reference**

1. Effective width b and section properties, using Table 10-2:

   a. Interior Beam

   I_s = (22) (12) (6.5)³/12 = 6040 in.⁴

   h - h_f = 24 - 6.5 = 17.5 in. ≤ 4h_f = (4) (6.5) = 26 in. O.K.

   Hence, b = 12 + (2) (17.5) = 47 in.

   y_t = h - (1/2) [(b - b_w) h_f^2 + $b_w h^2$]/[(b - b_w) h_f + b_wh]

   = 24 - (1/2) [(35) (6.5)² + (12) (24)²]/[(35) (6.5) + (12) (24)]

   = 15.86 in.

   I_b = (b - b_w) h_f^3/12 + $b_w h^3$/12 + (b - b_w) h_f (h - h_f/2 - y_t)² + b_wh (y_t - h/2)²

   = (35) (6.5)³/12 + (12) (24)³/12 + (35) (6.5) (24 - 3.25 - 15.86)² +

   (12) (24) (15.86 - 12)² = 24,360 in.⁴

   α_f = $E_{cb}I_b/E_{cs}I_s$ = I_b/I_s = 24,360/6040 = 4.03

	Code
Example 10.4 (cont'd) Calculations and Discussion	**Reference**

b. Edge Beam

$I_s = (11)(12)(6.5)^3/12 = 3020$ in.^{4}

$b = 12 + (24 - 6.5) = 29.5$ in.

$y_t = 24 - (1/2)[(17.5)(6.5)^2 + (12)(24)^2]/[(17.5)(6.5) + (12)(24)] = 14.48$ in.

$I_b = (17.5)(6.5)^3/12 + (12)(24)^3/12 + (17.5)(6.5)(24 - 3.25 - 14.48)^2 +$

$\quad (12)(24)(14.48 - 12)^2 = 20,470$ in.^{4}

$\alpha_f = I_b/I_s = 20,470/3020 = 6.78$

α_{fm} and β values:

α_{fm} (average value of α_f for all beams on the edges of a panel):

Interior panel — $\alpha_{fm} = 4.03$

Side panel — $\alpha_{fm} = [(3)(4.03) + 6.78]/4 = 4.72$

Corner panel — $\alpha_{fm} = [(2)(4.03) + (2)(6.78)]/4 = 5.41$

For square panels, β = ratio of clear spans in the two directions = 1

2. Minimum thickness: 9.5.3.3

Since $\alpha_{fm} > 2.0$ for all panels, Eq. (9-13) applies.

$$h_{min} = \frac{\ell_n \left(0.8 + \dfrac{f_y}{200,000}\right)}{36 + 9\beta}$$ Eq. (9-13)

$$= \frac{(21 \times 12)\left(0.8 + \dfrac{60,000}{200,000}\right)}{36 + 9(1)} = 6.16 \text{ in.} \quad \text{(all panels)}$$

Hence, the slab thickness of 6.5 in. > 6.16 in. is satisfactory for all panels, and deflections need not be checked.

Example 10.5—Simple-Span Prestressed Single T-Beam

Required: Analysis of short-term and ultimate long-term camber and deflection.

Data:

8ST36 (Design details from PCI Handbook 3rd Edition, 1985)
Span = 80 ft, beam is partially cracked
f'_{ci} = 3500 psi, f'_c = 5000 psi (normal weight concrete)
f_{pu} = 270,000 psi
E_p = 27,000,000 psi
14 - 1/2 in. dia. depressed (1 Pt.) strands
4 - 1/2 in. dia. nonprestressed strands
(Assume same centroid when computing I_{cr})
P_i = (0.7) (14) (0.153) (270) = 404.8 kips
P_o = (0.90) (404.8) = 364 kips
P_e = (0.78) (404.8) = 316 kips
e_e = 11.15 in., e_c = 22.51 in.
y_t = 26.01 in., A_g = 570 in.^{2}, I_g = 68,920 in.^{4}
Self weight, w_o = 594 lb/ft

Superimposed DL, w_s = (8)(10 psf) = 80 lb/ft is applied at age 2 mos (β_s = 0.76 in Term (6) of Eq. (15))

Live load, w_ℓ = (8)(51 psf) = 408 lb/ft
Capacity is governed by flexural strength

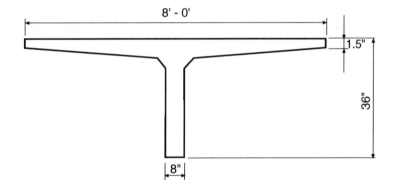

Calculations and Discussion — Code Reference

1. Span-depth ratios (using PCI Handbook):

    Typical span-depth ratios for single T beams are 25 to 35 for floors and 35 to 40 for roofs, versus (80)(12)/36 = 27, which indicates a relatively deep beam. It turns out that all allowable deflections in Table 9.5(b) are satisfied.

2. Moments for computing deflections:

$$M_o = \frac{w_o \ell^2}{8} = \frac{(0.594)(80)^2}{8} = 475 \text{ ft-kips}$$

Example 10.5 (cont'd)   **Calculations and Discussion**   **Code Reference**

($\times\, 0.96 = 456$ ft-kips at 0.4ℓ for computing stresses and I_e— tendons depressed at one point)

$$M_s = \frac{w_s \ell^2}{8} = \frac{(0.080)(80)^2}{8} = 64 \text{ ft-kips} \ (61 \text{ ft-kips at } 0.4\ell)$$

$$M_\ell = \frac{w_\ell \ell^2}{8} = \frac{(0.408)(80)^2}{8} = 326 \text{ ft-kips} \ (313 \text{ ft-kips at } 0.4\ell)$$

3. Modulus of rupture, modulus of elasticity, moment of inertia:

$$f_r = 7.5\sqrt{f'_c} = 7.5\sqrt{5000} = 530 \text{ psi} \qquad \text{Eq. (9-10)}$$

$$E_{ci} = w_c^{1.5}\, 33\sqrt{f'_{ci}} = (150)^{1.5}\, 33\sqrt{3500} = 3.59 \times 10^6 \text{ psi} \qquad 8.5.1$$

$$E_c = w_c^{1.5}\, 33\sqrt{f'_c} = (150)^{1.5}\, 33\sqrt{5000} = 4.29 \times 10^6 \text{ psi}$$

$$n = \frac{E_p}{E_c} = \frac{27 \times 10^6}{4.29 \times 10^6} = 6.3$$

The moment of inertia of the cracked section, at 0.4ℓ, can be obtained by the approximate formula given in Eq. 4.8.2 of the PCI Handbook:

$$I_{cr} = nA_{st}d^2\left(1 - 1.6\sqrt{n\rho}\right) = (6.3)(18 \times 0.153)(30.23)^2 \left(1 - 1.6\sqrt{6.3 \times 0.000949}\right)$$

$$= 13{,}890 \text{ in}^4 \ (\text{at } 0.4\ell)$$

It may be shown that the cracked section moment of inertia calculated by the formulas given in Table 10.2 is very close to the value obtained by the approximate method shown above. The results differ by approximately 1%; therefore either method is suitable for this case.

4. Determination of classification of beams

In order to classify the beam according to the requirements of ACI Section 18.3.3, the maximum flexural stress is calculated and compared to the modulus rupture to determine its classification. The classifications are defined as follows:

         Class U: $f_t \leq 7.5\sqrt{f'_c}$

         Class T: $7.5\sqrt{f'_c} < f_t \leq 12\sqrt{f'_c}$

         Class C: $f_t > 12\sqrt{f'_c}$

Example 10.5 (cont'd)	Calculations and Discussion	Code Reference

The three classes refer to uncracked (U), transition (T) and cracked (C) behavior.

The maximum tensile stress due to service loads and prestressing forces are calculated by the standard formula for beams subject to bending moments and axial loads. It may be shown that the maximum bending stresses in a prestressed beam occur at approximately 0.4ℓ. In the following, the bending moments are those that occur at 0.4ℓ. The eccentricity of the prestressing force at 0.4ℓ, e = 20.24 in, is obtained by linear interpolation between the end eccentricity, $e_e = 11.15$ in and that at the center, $e_e = 22.51$. The calculation proceeds as follow:

$$M_{tot} = M_d + M_\ell$$

$$f_t = \frac{M_{tot} y_t}{I_g} - \frac{P_e e \cdot y_t}{I_g} - \frac{P_e}{A_g}$$

$$= [(456 + 61 + 313)(12) - (316)(20.24)] [(26.01)/68,920] - 316/570$$

$$f_t = 791 \text{ psi}$$

Check the ratio of calculated tensile stress to the square root of f'_c:

$$\frac{f_t}{\sqrt{f'_c}} = \frac{791}{\sqrt{5000}} = 11.2$$

The ratio is between 7.5 and 12, therefore according to the definitions of Section 18.3.3 of ACI 318, the beam classification is T, transition region. Table R18.3.3 requires that deflections for this classification be based on the cracked section properties assuming bi-linear moment deflection behavior; Section 9.5.4.2 of the code allows either a bi-linear moment-deflection approach or calculation of deflections on the basis of an equivalent moment of inertia determined according to Eq. (9-8).

5. Determine live load moment that causes first cracking:

Check the tensile stress due to dead load and prestressing forces only. As noted previously, the maximum tensile stresses occur at approximately 0.4ℓ:

$$f_t = [(456 + 61)(12) - (316)(20.24)] [(26.01)/68920] - 316/570 = -627 \text{ psi}$$

Since the sign is negative, compressive stress is indicated. Therefore, the section is uncracked under the dead load plus prestressing forces and dead load deflections can be based on the moment of inertia of the gross concrete cross section. It was shown above that the maximum tensile stress due to combined dead load plus live load equals 791 psi, which exceeds the modulus of rupture, $f_r = 530$ psi

Therefore, the live load deflections must be computed on the basis of a cracked section analysis because the behavior is inelastic after the addition of full live load. In particular, Table R18.3.3 of ACI 318 requires that bilinear behavior be utilized to determine deflections in such cases, however. Section 9.5.4.2 permits deflections to be computed either on the basis of bilinear behavior or on the basis of an effective moment of inertia.

		Code
Example 10.5 (cont'd)	**Calculations and Discussion**	**Reference**

In order to calculate the deflection assuming bilinear behavior, it is first necessary to determine the fraction of the total live load that causes first cracking. That is, to find the portion of live load that will just produce a maximum tensile stress equal to f_r. The desired value of live load moment can be obtained by re-arranging the equation used above to determine the total tensile stress (for classification), and setting the tensile stress equal to f_r. The moment value is obtained as follows (Note: Quantities calculated at 0.4ℓ):

$$\text{Live Load Cracking Moment} = \frac{f_r I_g}{y_t} + P_e e + \frac{P_e I_g}{A_g y_t} - M_d$$

$= (530)68920)/(12000)(26.01) + 316(20.24/12) + [(316/570)(68920/26.01)]/12 - 517$

$= 117 + 533 + 122 - 517$

$= 255$ ft-kips

The fraction of the live load cracking moment to the total live load moment is:

$255/313 = 0.815$

6. Camber and Deflection, using Eq. (15):

$$\text{Term (1)} \text{---} \Delta_{po} = \frac{P_o(e_c - e_e)\ell^2}{12 E_{ci} I_g} + \frac{P_o e_e \ell^2}{8 E_{ci} I_g} \quad \text{(from PCI Handbook for single point depressed strands)}$$

$$= \frac{(364)(22.51 - 11.15)(80)^2(12)^2}{(12)(3590)(68,920)} + \frac{(364)(11.15)(80)^2(12)^2}{(8)(3590)(68,920)}$$

$= 3.17$ in.

$$\text{Term (2)} \text{---} \Delta_o = \frac{5 M_o \ell^2}{48 E_{ci} I_g} = \frac{(5)(475)(80)^2(12)^3}{(48)(3590)(68,590)} = 2.21 \text{ in.}$$

Term (3) — $k_r = 1/[1 + (A_s/A_{ps})] = 1/[1 + (4/14)] = 0.78$

$$\left[-\frac{\Delta P_u}{P_o} + (k_r C_u)\left(1 - \frac{\Delta P_u}{2 P_o}\right) \right] \Delta_{po}$$

The increment in prestressing force is:

$\Delta P_u = P_o - P_e = 364 - 316 = 48$ kips

It follows that:

$\Delta P_u / P_o = 48 / 364 = 0.13$

		Code
Example 10.5 (cont'd)	**Calculations and Discussion**	**Reference**

Therefore, the deflection is:

$$= [-0.13 + (0.78 \times 2.0)(1 - 0.065)](3.17) = 4.21 \text{ in.}$$

Term (4) — $(k_r C_u) \Delta_o = (0.78)(2.0)(2.21) = 3.45$ in.

Term (5) — $\Delta_s = \dfrac{5 M_s \ell^2}{48 E_c I_g} = \dfrac{(5)(64)(80)^2 (12)^3}{(48)(4290)(68,920)} = 0.25$ in.

Term (6) — $(\beta_s k_r C_u) \Delta_s = (0.76)(0.78)(1.6)(0.25) = 0.24$ in.

Term (7) – Initial deflection due to live load.

The ratio of live load cracking moment to total live load moment was found previously. To calculate the deflection according to bi-linear behavior, the deflection due to the portion of the live load below the cracking value is based on the gross moment of inertia; the deflection due to the remainder of the live load is based on the cracked section moment of inertia. Also, the deflections are based on moments at the center of the span even though the moment that caused initial cracking was evaluated at 0.4ℓ.

The deflection formula used is the standard expression:

$$\Delta = \frac{5 \, ML^2}{48 \, EI}$$

For the portion of the live load applied below the cracking moment load, the value of M is the value calculated above, 255 ft-kips and the moment of inertia is that of the gross section:

$$\Delta_{\ell 1} = 5(255)(80)^2(12)^3/48(3590)(68590) = 1.19 \text{ in}$$

Deflection due to the remainder of the live load is calculated similarly, with a moment of 313-255 = 58 ft-kips and the cracked section moment of inertia, 13,890 in⁴:

$$\Delta_{\ell 2} = 5(58)(80)^2(12)^3/48(3590)(13,890) = 1.34 \text{ in}$$

The total live load deflection is the sum of the previous two components:

$$\Delta_\ell = 1.19 + 1.34 = 2.53 \text{ in.}$$

It can be verified by a separate calculation that the deflection based on the full live load moment and the effective moment of inertia, calculated by Eq. 9-8 of ACI 318, is slightly less than that calculated here on the basis of a bi-linear moment-deflection relationship.
    Combined results and comparisons with code limitations

Example 10.5 (cont'd)	Calculations and Discussion	Code Reference

$$\begin{array}{ccccccc}(1)&(2)&(3)&(4)&(5)&(6)&(7)\\\overline{}&\overline{}&\overline{}&\overline{}&\overline{}&\overline{}&\overline{}\end{array}$$

$\Delta_u = -3.17 + 2.21 - 4.21 + 3.45 + 0.25 + 0.24 + 2.53 = 1.30$ in. $\downarrow$ *Eq. (15)*

Initial Camber $= \Delta_{po} - \Delta_o = 3.17 - 2.21 = 0.96$ in. $\uparrow$ versus 1.6 in. at erection in PCI Handbook

Residual Camber $= \Delta_\ell - \Delta_u = 2.53 - 0.87 = 1.66$ in. $\uparrow$ versus 1.1 in.

Time-Dependent plus Superimposed Dead Load and Live Load Deflection

$= -4.21 + 3.45 + 0.25 + 0.24 + 2.53 = 2.26$ in. or

$= \Delta_u - (\Delta_o - \Delta_{po}) = 0.87 - (-0.96) = 2.26$ in. $\downarrow$

These computed deflections are compared with the allowable deflections in Table 9.5(b) as follows:

$\ell/180 = (80)(12)/180 = 5.33$ in. versus $\Delta_\ell = 2.53$ in.   O.K.

$\ell/360 = (80)(12)/360 = 2.67$ in. versus $\Delta_\ell = 2.53$ in.   O.K.

$\ell/480 = (80)(12)/480 = 2.00$ in. versus Time-Dep. etc. $= 2.26$ in.   O.K.

Note that the long term deflection occurring after attachment of non-structural elements (2.26 in) exceeds the L/480 limit. It actually meets L/425. Since the L/480 limit only applies in case of *nonstructural elements likely to be damaged by large deflections*, the particular use of the beam would have to be considered in order to make a judgment on the acceptability of the computed deflections. Refer to the footnotes following Table 9.5(b) of ACI 318.

Example 10.6—Unshored Nonprestressed Composite Beam

Required: Analysis of short-term and ultimate long-term deflections.

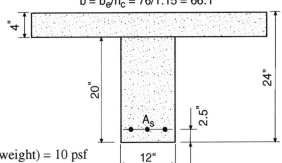

Data:

Normal weight concrete
Slab $f'_c = 3000$ psi
Precast beam $f'_c = 4000$ psi
$f_y = 40,000$ psi
$A_s =$ 3-No. 9 = 3.00 in.^{2}
$E_s = 29,000,000$ psi
Superimposed Dead Load (not including beam and slab weight) = 10 psf
Live Load = 75 psf (20% sustained)
Simple span = 26 ft = 312 in., spacing = 8 ft = 96 in.
$b_e = 312/4 = 78.0$ in., or spacing = 96.0 in., or 16(4) + 12 = 76.0 in.

Calculations and Discussion	Code Reference

1. Minimum thickness for members not supporting or attached to partitions or other construction likely to be damaged by large deflections:

$$h_{min} = \left(\frac{\ell}{16}\right)(0.80 \text{ for } f_y) = \left(\frac{312}{16}\right)(0.80) = 15.6 \text{ in.} < h = 20 \text{ in. or } 24 \text{ in.}$$ *Table 9.5(a)*

2. Loads and moments:

$w_1 = (10 \text{ psf}) (8) + (150 \text{ pcf}) (96) (4)/144 = 480$ lb/ft

$w_2 = (150 \text{ pcf}) (12) (20)/144 = 250$ lb/ft

$w_\ell = (75 \text{ psf}) (8) = 600$ lb/ft

$M_1 = w_1 \ell^2/8 = (0.480) (26)^2/8 = 40.6$ ft-kips

$M_2 = w_2 \ell^2/8 = (0.250) (26)^2/8 = 21.1$ ft-kips

$M_\ell = w_\ell \ell^2/8 = (0.600) (26)^2/8 = 50.7$ ft-kips

3. Modulus of rupture, modulus of elasticity, modular ratio:

$(E_c)_1 = w_c^{1.5} \, 33\sqrt{f'_c} = (150)^{1.5} \, 33\sqrt{3000} = 3.32 \times 10^6$ psi   *8.5.1*

$(f_r)_2 = 7.5\sqrt{f'_c} = 7.5\sqrt{4000} = 474$ psi   *Eq. (9-10)*

$(E_c)_2 = (150)^{1.5} \, 33\sqrt{4000} = 3.83 \times 10^6$ psi   *8.5.1*

		Code
Example 10.6 (cont'd)	**Calculations and Discussion**	**Reference**

$$n_c = \frac{(E_c)_2}{(E_c)_1} = \frac{3.83}{3.32} = 1.15$$

$$n = \frac{E_s}{(E_c)_2} = \frac{29}{3.83} = 7.56$$

4. Gross and cracked section moments of inertia, using Table 10-2:

Precast Section

$I_g = (12)(20)^3/12 = 8000$ in.^{4}

$B = b/(nA_s) = 12/(7.56)(3.00) = 0.529$/in.

$kd = (\sqrt{2dB + 1} - 1)/B = [\sqrt{(2)(17.5)(0.529) + 1} - 1]/0.529 = 6.46$ in.

$I_{cr} = bk^3d^3/3 + nA_s(d - kd)^2 = (12)(6.46)^3/3 + (7.56)(3.00)(17.5 - 6.46)^2 = 3840$ in.^{4}

Composite Section

$y_t = h - (1/2)[(b - b_w)h_f^2 + b_wh^2]/[(b - b_w)h_f + b_wh]$

$= 24 - (1/2)[(54.1)(4)^2 + (12)(24)^2]/[(54.1)(4) + (12)(24)] = 16.29$ in.

$I_g = (b - b_w)h_f^3/12 + b_wh^3/12 + (b - b_w)h_f(h - h_f/2 - y_t)^2 + b_wh(y_t - h/2)^2$

$= (54.1)(4)^3/12 + (12)(24)^3/12 + (54.1)(4)(24 - 2 - 16.29)^2$

$+ (12)(24)(16.29 - 12)^2 = 26,470$ in.^{4}

$B = b/(nA_s) = 66.1/(7.56)(3.00) = 2.914$

$kd = (\sqrt{2dB + 1} - 1)/B = [\sqrt{(2)(21.5)(2.914) + 1} - 1]/2.914$

$= 3.51$ in. $< h_f = 4$ in. Hence, treat as a rectangular compression area.

$I_{cr} = bk^3d^3/3 + nA_s(d - kd)^2 = (66.1)(3.51)^3/3 + (7.56)(3.00)(21.5 - 3.51)^2$

$= 8295$ in.^{4}

$I_2/I_c = [(I_2/I_c)_g + (I_2/I_c)_{cr}]/2 = [(8000/26,470) + (3840/8295)]/2 = 0.383$

5. Effective moments of inertia, using Eq. (9-8):

For Term (1), Eq. (19)—Precast Section,

$M_{cr} = f_rI_g/y_t = (474)(8000)/(10)(12,000) = 31.6$ ft-kips   Eq. (9-9)

$M_{cr}/M_2 = 31.6/21.1 > 1$. Hence $(I_e)_2 = I_g = 8000$ in.^{4}

		Code
Example 10.6 (cont'd)	**Calculations and Discussion**	**Reference**

For Term (6), Eq. (19)—Precast Section,

$[M_{cr}/(M_1 + M_2)]^3 = [31.6/(40.6 + 21.1)]^3 = 0.134$

$(I_e)_{1+2} = (M_{cr}/M_a)^3 I_g + [1 - (M_{cr}/M_a)^3] I_{cr} \leq I_g$ *Eq. (9-8)*

$\qquad = (0.134)(8000) + (1 - 0.134)(3840) = 4400 \text{ in.}^4$

6. Deflection, using Eq. (19):

Term (1) — $(\Delta_i)_2 = \dfrac{K(5/48) M_2 \ell^2}{(E_c)_2 (I_e)_2} = \dfrac{(1)(5/48)(21.1)(26)^2 (12)^3}{(3830)(8000)} = 0.084 \text{ in.}$

Term (2) — $k_r = 0.85$ (no compression steel in precast beam).

$0.77 k_r (\Delta_i)_2 = (0.77)(0.85)(0.084) = 0.055 \text{ in.}$

Term (3) — $0.83 k_r (\Delta_i)_2 \dfrac{I_2}{I_c} = (0.83)(0.85)(0.084)(0.383) = 0.023 \text{ in.}$

Term (4) — $K_{sh} = 1/8$. Precast Section: $\rho = (100)(3.00)/(12)(17.5) = 1.43\%$

From Fig. 8-3, $A_{sh} = 0.789$

$\phi_{sh} = A_{sh} (\varepsilon_{sh})_u / h = (0.789)(400 \times 10^{-6})/20 = 15.78 \times 10^{-6}/\text{in.}$

$\Delta_{sh} = K_{sh} \phi_{sh} \ell^2 = (1/8)(15.78 \times 10^{-6})(26)^2 (12)^2 = 0.192 \text{ in.}$

The ratio of shrinkage strain at 2 months to the ultimate is 0.36 per Table 2.1 of Ref. 10.4 Therefore the shrinkage deflection of the precast beam at 2 months is:

$0.36 \Delta_{sh} = (0.36)(0.192) = 0.069 \text{ in.}$

Term (5) — $0.64 \Delta_{sh} \dfrac{I_2}{I_c} = (0.64)(0.192)(0.383) = 0.047 \text{ in.}$

Term (6) — $(\Delta_i)_1 = \dfrac{K(5/48)(M_1 + M_2) \ell^2}{(E_c)_2 (I_e)_{1+2}} - (\Delta_i)_2$

$\qquad = \dfrac{(1)(5/48)(40.6 + 21.1)(26)^2 (12)^3}{(3830)(4400)} - 0.088 = 0.358 \text{ in.}$

10-51

Example 10.6 (cont'd) — Calculations and Discussion

Term (7) —Creep deflection of the composite beam due to slab dead load. The slab is cast at 2 months. Therefore, the fraction of the creep coefficient, C_u, is obtained my multiplying the value under standard conditions of 1.60 by a b_s value of 0.89 (See explanation of Term (6) in Eq. (15)). The total creep of the beam is reduced by the ratio of the moment of inertia of the beam to the moment of inertia of the composite section. k_r is, as before, taken as 0.85:

$$(0.89)(1.60)k_r(\Delta_i)_1 \frac{I_2}{I_c} = (0.89)(1.60)(0.85)(0.358)(0.383) = 0.166 \text{ in.}$$

Term (8) —Due to the fact that the beam and the slab were cast at different times, there will be some contribution to the total deflection due to the tendency of the two parts to creep and shrink at different rates. It is noted in Table 2.1 of ACI 435R-95 (Ref. 10.4) that the creep and shrinkage at a time of 2 months is almost half of the total. Consequently, behavior of the composite section will be affected by this different age. The proper calculation of the resulting deflection is very complex. In this example, the deflection due to differing age concrete is approximated as one-half of the dead load deflection of the beam due to the slab dead load. Readers are cautioned that this procedure results in only a rough estimate. Half of the dead load deflection is

$$\Delta_{ds} = 0.50(\Delta_i)_1 = (0.50)(0.358) = 0.179 \text{ in. (rough estimate)}$$

Term (9) — Using the alternative method

$$(\Delta_i)_\ell = \frac{K(5/48) M_\ell \ell^2}{(E_c)_2 (I_c)_{cr}} = \frac{(1)(5/48)(50.7)(26)^2 (12)^3}{(3830)(8295)} = 0.194 \text{ in.}$$

Term (10) — $k_r = 0.85$ (neglecting the effect of any compression steel in slab)

$$(\Delta_{cp})_\ell = k_r C_u \left[0.20 (\Delta_i)_\ell \right]$$

$$= (0.85)(1.60)(0.20 \times 0.194) = 0.053 \text{ in.}$$

In Eq. (19), $\Delta_u = 0.084 + 0.055 + 0.023 + 0.069 + 0.047 + 0.358 + 0.166 + 0.179 +$
$0.194 + 0.053$
$= 1.23 \text{ in.}$

Checking Eq. (20) (same solution),

$$\Delta_u = \left(1.65 + 0.71 \frac{I_2}{I_c}\right)(\Delta_i)_2 + \left(0.36 + 0.64 \frac{I_2}{I_c}\right)\Delta_{sh} +$$

Example 10.6 (cont'd)   **Calculations and Discussion**   Code Reference

$$\left(1.05 + 1.21\frac{I_2}{I_c}\right)(\Delta_i)_1 + (\Delta_i)_\ell + (\Delta_{cp})_\ell$$

$= (1.65 + 0.71 \times 0.383)(0.084) + (0.36 + 0.64 \times 0.383)(0.192)$
$\quad + (1.50 + 1.21 \times 0.383)(0.358) + 0.194 + 0.053$
$= 1.23$ in. (same as above) Assuming nonstructural elements are installed after the composite slab has hardened,

$$\Delta_{cp} + \Delta_{sh} + (\Delta_i)_\ell = \text{Terms (3) + (5) + (7) + (8) + (9) + (10)}$$

$$= 0.023 + 0.047 + 0.166 + 0.179 + 0.194 + 0.053 = 0.66 \text{ in.}$$

Comparisons with the allowable deflections in Table 9.5(b) are shown at the end of Design Example 10.7.

Example 10.7—Shored Nonprestressed Composite Beam

Required: Analysis of short-term and ultimate long-term deflections, to show the beneficial effect of shoring in reducing deflections.

Data: Same as in Example 10.6, except that shored construction is used.

Calculations and Discussion	Code Reference

1. Effective moments of inertia for composite section, using Eq. (9-8):

   $M_{cr} = f_r I_g / y_t = (474)(26{,}470)/(16.29)(12{,}000) = 64.2$ ft-kips   Eq. (9-9)

   $M_{cr}/(M_1 + M_2) = [64.2/(40.6 + 21.1)] = 1.04 > 1$

   Hence $(I_e)_{1+2} = I_g = 26{,}470$ in.^{4}

   In Term (5), Eq. (17)—Composite Section,

   $[M_{cr}/(M_1 + M_2 + M_\ell)]^3 = [64.2/(40.6 + 21.1 + 50.7)]^3 = 0.186$

   $(I_e)_{d+\ell} = (M_{cr}/M_a)^3 I_g + [1 - (M_{cr}/M_a)^3] I_{cr} \leq I_g$ Eq. (9-8)

   $= (0.186)(26{,}470) + (1 - 0.186)(8295) = 11{,}675$ in.^{4}

   versus the alternative method of Example 8.6 where $I_e = (I_c)_{cr} = 8295$ in.^{4} was used with the live load moment directly.

2. Deflections, using Eqs. (17) and (18):

   Term (1) — $(\Delta_i)_{1+2} = \dfrac{K(5/48)(M_1 + M_2)\ell^2}{(E_c)_2 (I_e)_{1+2}}$

   $= \dfrac{(1)(5/48)(40.6 + 21.1)(26)^2(12)^3}{(3830)(26{,}470)} = 0.074$ in.

   Term (2) — Creep deflection due to total dead load of beam and slab. The value of C_u for the beam is taken to be 1.60. Consider the value of C_u for the slab to be slightly higher. For shores removed at 10 days, it may be shown by comparison of the correction factors, K_{to}^c for 10 and 20 day load applications (Section 2.3.4, ACI 435, Ref. 10.4) that the ultimate creep coefficient for the slab is approximately 1.74. k_r is conservatively assumed to have a value of 0.85.

   The average creep coefficient for the composite section is:

   Avg. $C_u = \frac{1}{2}(1.60 + 1.74) = 1.67$

   $1.67 k_r (\Delta_i)_{1+2} = (1.67)(0.85)(0.074) = 0.105$ in.

Example 10.7 (cont'd) — Calculations and Discussion

Term (3) — Shrinkage deflection of the precast beam after shores are removed. As indicated in Term 4 of Example 8.6, the fraction of shrinkage of the precast beam at 2 months is 0.36. The shores are assumed to be removed about 10 days after the 2-month point. Therefore, consider the remaining fraction of shrinkage is $1 - 0.36 = 0.64$. Recall that the ultimate shrinkage, $(\varepsilon_{sh})_u = 400 \times 10^{-6}$. Utilize the result found for Δ_{sh} in Term (4) of Example 10.6:

Remaining $(e_{sh}) = (0.64)(400 \times 10^{-6}) = 256 \times 10^{-6}$

$\Delta_{sh} \dfrac{I_2}{I_c} = (256/400)(0.192)(0.383) = 0.047$ in.

Term (4) — Deflection due to differences in shrinkage and creep in the beam and slab. This is a complex issue. For this example, assume that the magnitude of this component is approximated by the initial dead load deflection of the composite section.

$\Delta_{ds} = (\Delta_i)_{1+2} = 0.074$ in. (rough estimate)

Term (5) — $(\Delta_i)_\ell = \dfrac{K(5/48)(M_1 + M_2 + M_\ell)\ell^2}{(E_c)_2 (I_e)_{d+\ell}} - (\Delta_i)_{1+2}$

$= \dfrac{(1)(5/48)(40.6 + 21.1 + 50.7)(26)^2 (12)^3}{(3830)(11{,}675)} - 0.074$ in. $= 0.232$ in.

Term (6) — $k_r = 0.85$ (neglecting the effect of any compression steel in slab),

$(\Delta_{cp})_\ell = k_r C_u [0.20 (\Delta_i)_\ell] = (0.85)(1.60)(0.20 \times 0.232) = 0.063$ in.

In Eq. (17), $\Delta_u = 0.074 + 0.105 + 0.047 + 0.074 + 0.232 + 0.063 = 0.60$ in.

versus 1.23 in. with unshored construction.

This shows the beneficial effect of shoring in reducing the total deflection.

Checking by Eq. (18) (same solution),

$\Delta_u = 3.42 (\Delta_i)_{1+2} + \Delta_{sh} \dfrac{I_2}{I_c} + (\Delta_i)_\ell + (\Delta_{cp})_\ell$

$= (3.42)(0.074) + 0.046 + 0.232 + 0.063 = 0.60$ in. (same as above)

Assuming that nonstructural elements are installed after shores are removed,

$\Delta_{cp} + \Delta_{sh} + (\Delta_i)_\ell = \Delta_u - (\Delta_i)_{1+2} = 0.60 - 0.07 = 0.53$ in.

| Example 10.7 (cont'd) | Calculations and Discussion | Code Reference |

Comparison of Results of Examples 10.6 and 10.7

The computed deflections of $(\Delta_i)_\ell = 0.19$ in. in Example 10.6 and 0.23 in. in Example 10.7; and $\Delta_{cp} + \Delta_{sh} + (\Delta_i)_\ell = 0.66$ in. in Example 10.6 and 0.53 in. in Example 10.7 are compared with the allowable deflections in Table 9.5(b) as follows:

Flat roofs not supporting and not attached to nonstructural elements likely to be damaged by large deflections—

$(\Delta_i)_\ell \leq \ell/180 = 312/180 = 1.73$ in.   O.K.

Floors not supporting and not attached to nonstructural elements likely to be damaged by large deflections—

$(\Delta_i)_\ell \leq \ell/360 = 312/360 = 0.87$ in.   O.K.

Roof or floor construction supporting or attached to nonstructural elements likely to be damaged by large deflections (very stringent limitation)—

$\Delta_{cp} + \Delta_{sh} + (\Delta_i)_\ell \leq \ell/480 = 312/480 = 0.65$ in.

Note that the long term deflection occurring after attachment of non-structural elements (0.66 in) exceeds the L/480 limit. It actually meets L/473. Since the L/480 limit only applies in case of *nonstructural elements likely to be damaged by large deflections*, the particular use of the beam would have to be considered in order to make a judgment on the acceptability of the computed deflections. Refer to the footnotes following Table 9.5(b) of ACI 318

Roof or floor construction supporting or attached to nonstructural elements not likely to be damaged by large deflections—

$\Delta_{cp} + \Delta_{sh} + (\Delta_i)_\ell \leq \ell/240 = 312/240 = 1.30$ in.   O.K.

11

Design for Slenderness Effects

UPDATE FOR THE '05 CODE

Section 10.13.6 was modified to clarify the load factors to be used for investigating the strength and stability of the structure as a whole under gravity loads in 10.13.6(a) and (b).

GENERAL CONSIDERATIONS

Design of columns consists essentially of selecting an adequate column cross-section with reinforcement to support required combinations of factored axial loads P_u and factored (primary) moments M_u, including consideration of column slenderness (secondary moments).

Column slenderness is expressed in terms of its slenderness ratio $k\ell_u/r$, where k is an effective length factor (dependent on rotational and lateral restraints at the ends of the column), ℓ_u is the unsupported column length, and r is the radius of gyration of the column cross-section. In general, a column is slender if its applicable cross-sectional dimension is small in comparison to its length.

For design purposes, the term "short column" is used to denote a column that has a strength equal to that computed for its cross-section, using the forces and moments obtained from an analysis for combined bending and axial load. A "slender column" is defined as a column whose strength is reduced by second-order deformations (secondary moments). By these definitions, a column with a given slenderness ratio may be considered a short column for design under one set of restraints, and a long column under another set. With the use of higher strength concrete and reinforcement, and with more accurate analysis and design methods, it is possible to design smaller cross-sections, resulting in members that are more slender. The need for reliable and rational design procedures for slender columns thus becomes a more important consideration in column design.

A short column may fail due to a combination of moment and axial load that exceeds the strength of the cross-section. This type of a failure is known as "material failure." As an illustration, consider the column shown in Fig. 11-1. Due to loading, the column has a deflection Δ which will cause an additional (secondary) moment in the column. From the free body diagram, it can be seen that the maximum moment in the column occurs at section A-A, and is equal to the applied moment plus the moment due to member deflection, which is $M = P(e + \Delta)$.

Failure of a short column can occur at any point along the strength interaction curve, depending on the combination of applied moment and axial load. As discussed above, some deflection will occur and a "material failure" will result when a particular combination of load P and moment $M = P(e + \Delta)$ intersects the strength interaction curve.

If a column is very slender, it may reach a deflection due to axial load P and a moment Pe such that deflections will increase indefinitely with an increase in the load P. This type of failure is known as a "stability failure," as shown on the strength interaction curve.

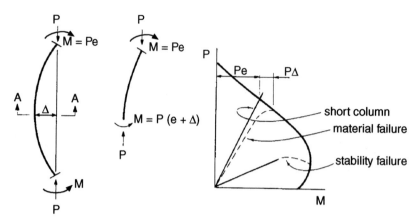

Figure 11-1 Strength Interaction for Slender Columns

The basic concept on the behavior of straight, concentrically loaded, slender columns was originally developed by Euler more than 200 years ago. It states that a member will fail by buckling at the critical load $P_c = \pi^2 EI/(\ell_e)^2$, where EI is the flexural stiffness of the member cross-section, and ℓ_e is the effective length, which is equal to $k\ell_u$. For a "stocky" short column, the value of the buckling load will exceed the direct crushing strength (corresponding to material failure). In members that are more slender (i.e., members with larger $k\ell_u/r$ values), failure may occur by buckling (stability failure), with the buckling load decreasing with increasing slenderness (see Fig. 11-2).

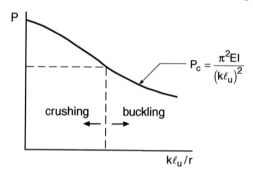

Figure 11-2 Failure Load as a Function of Column Slenderness

As shown above, it is possible to depict slenderness effects and amplified moments on a typical strength interaction curve. Hence, a "family" of strength interaction diagrams for slender columns with varying slenderness ratios can be developed, as shown in Fig. 11-3. The strength interaction diagram for $k\ell_u/r = 0$ corresponds to the combinations of moment and axial load where strength is not affected by member slenderness (short column strength).

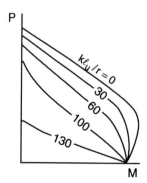

Figure 11-3 Strength Interaction Diagrams for Slender Columns

CONSIDERATION OF SLENDERNESS EFFECTS

Slenderness limits are prescribed for both nonsway and sway frames, including design methods permitted for each slenderness range. Lower-bound slenderness limits are given, below which secondary moments may be disregarded and only axial load and primary moment need be considered to select a column cross-section and reinforcement (short column design). It should be noted that for ordinary beam and column sizes and typical story heights of concrete framing systems, effects of slenderness may be neglected for more than 90 percent of columns in nonsway frames and around 40 percent of columns in sway frames. For moderate slenderness ratios, an approximate analysis of slenderness effects based on a moment magnifier (see 10.12 and 10.13) is permitted. For columns with high slenderness ratios, a more exact second-order analysis is required (see 10.11.5), taking into account material nonlinearity and cracking, as well as the effects of member curvature and lateral drift, duration of the loads, shrinkage and creep, and interaction with the supporting foundation. No upper limits for column slenderness are prescribed. The slenderness ratio limits in 10.12.2 for nonsway frames and 10.13.2 for sway frames, and design methods permitted for consideration of column slenderness, are summarized in Fig. 11-4.

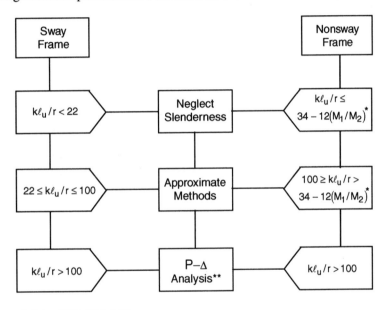

* $34 - 12(M_1/M_2) \leq 40$
** Permitted for any slenderness ratio

Figure 11-4 Consideration of Column Slenderness

10.10 SLENDERNESS EFFECTS IN COMPRESSION MEMBERS

10.10.1 Second-Order Frame Analysis

The code encourages the use of a second-order frame analysis or P-Δ analysis for consideration of slenderness effects in compression members. Generally, the results of a second-order analysis give more realistic values of the moments than those obtained from an approximate analysis by 10.12 or 10.13. For sway frames, the use of second-order analyses will generally result in a more economical design. Procedures for carrying out a second-order analysis are given in Commentary Refs. 10.31-10.32. The reader is referred to R10.10.1, which discusses minimum requirements for an adequate second-order analysis under 10.10.1.

If more exact analyses are not feasible or practical, 10.10.2 permits an approximate moment magnifier method to account for column slenderness. Note, however, that for all compression members with a column slenderness ratio ($k\ell_u/r$) greater than 100 (see Fig. 11-4), a more exact analysis as defined in 10.10.1 must be used for consideration of slenderness effects.

10.11 APPROXIMATE EVALUATION OF SLENDERNESS EFFECTS

The moment magnification factor δ is used to magnify the primary moments to account for increased moments due to member curvature and lateral drift. The moment magnifier δ is a function of the ratio of the applied axial load to the critical or buckling load of the column, the ratio of the applied moments at the ends of the column, and the deflected shape of the column.

10.11.1 Section Properties for Frame Analysis

According to 10.11.1, the factored axial loads P_u, the factored moments at the column ends M_1 and M_2, and the relative lateral story deflections Δ_o shall be computed using an elastic first-order frame analysis taking into account cracked regions along the length of the members. It is usually not economically feasible to perform such calculations even for small structures. Thus, the section properties given in 10.11.1 and summarized in Table 11-1 may be used in the analysis to account for cracking. The values of E, I, and A have been chosen from the results of frame tests and analyses as outlined in code Reference 10.33. It is important to note that for service load analysis of the structure, it is satisfactory to multiply the moments of inertia given in Table 11-1 by 1/0.70 = 1.43 (R10.11.1). Also, the moments of inertia must be divided by $(1 + \beta_d)$ in the case when sustained lateral loads act on the structure (for example, lateral loads resulting from earth pressure) or when the gravity load stability check made in accordance with 10.13.6 is performed.

Table 11-1 Section Properties for Frame Analysis

	Modulus of Elasticity	Moment of Inertia†	Area
Beams		$0.35 I_g$	
Columns		$0.70 I_g$	
Walls - uncracked	E_c from 8.5.1	$0.70 I_g$	$1.0 A_g$
Walls - cracked		$0.35 I_g$	
Flat plates and flat slabs		$0.25 I_g$	

†Divide by $(1 + \beta_d)$ when sustained lateral loads act or for stability checks made in accordance with 10.13.6. For service load analyses, multiply by 1/0.70 = 1.43.

10.11.2 Radius of Gyration

In general, the radius of gyration, r, is $\sqrt{I_g/A_g}$. In particular, r may be taken as 0.30 times the dimension in the direction of analysis for a rectangular section and 0.25 times the diameter of a circular section, as shown in Fig. 11-5.

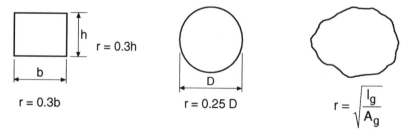

Figure 11-5 Radius of Gyration, r

10.11.3, 10.12.1 Unsupported and Effective Lengths of Compression Members

The unsupported length ℓ_u of a column, defined in 10.11.3, is the clear distance between lateral supports as shown in Fig. 11-6. Note that the length ℓ_u may be different for buckling about each of the principal axes of the column cross-section. The basic Euler equation for critical buckling load can be expressed as $P_c = \pi^2 EI/(\ell_e)^2$, where ℓ_e is the effective length $k\ell_u$. The basic equations for the design of slender columns were derived for hinged ends, and thus, must be modified to account for the effects of end restraint. Effective column length $k\ell_u$, as contrasted to actual unbraced length ℓ_u, is the term used in estimating slender column strength, and considers end restraints as well as nonsway and sway conditions.

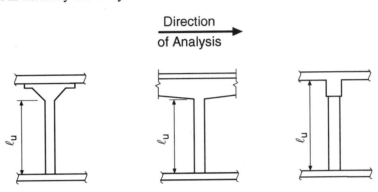

Figure 11-6 Unsupported Length, ℓ_u

At the critical load defined by the Euler equation, an originally straight member buckles into a half-sine wave as shown in Fig. 11-7(a). In this configuration, an additional moment P-Δ acts at every section, where Δ is the lateral deflection at the specific location under consideration along the length of the member. This deflection continues to increase until the bending stress caused by the increasing moment (P-Δ), plus the original compression stress caused by the applied loading, exceeds the compressive strength of concrete and the member fails. The effective length ℓ_e (= $k\ell_u$) is the length between pinned ends, between zero moments or between inflection points. For the pinned condition illustrated in Fig. 11-7(a), the effective length is equal to the unsupported length ℓ_u. If the member is fixed against rotation at both ends, it will buckle in the shape depicted in Fig. 11-7(b); inflection points will occur at the locations shown, and the effective length ℓ_e will be one-half of the unsupported length. The critical buckling load P_c for the fixed-end condition is four times that for a pin-end condition. Rarely are columns in actual structures either hinged or fixed; they are partially restrained against rotation by members framing into the column, and thus the effective length is between $\ell_u/2$ and ℓ_u, as shown in Fig. 11-7(c) as long as the lateral displacement of one end of the column with respect to the other end is prevented. The actual value of the effective length depends on the rigidity of the members framing into the top and bottom ends of the column.

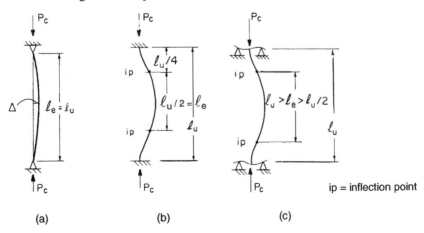

Figure 11-7 Effective Length, ℓ_e (Nonsway Condition)

A column that is fixed at one end and entirely free at the other end (cantilever) will buckle as shown in Fig. 11-8(a). The upper end will deflect laterally relative to the lower end; this is known as sidesway. The deflected shape of such a member is similar to one-half of the sinusoidal deflected shape of the pin-ended member illustrated in Fig. 11-7(a). The effective length is equal to twice the actual length. If the column is fixed against rotation at both ends but one end can move laterally with respect to the other, it will buckle as shown in Fig. 11-8(b). The effective length ℓ_e will be equal to the actual length ℓ_u, with an inflection point (ip) occurring as shown. The buckling load of the column in Fig. 11-8(b), where sidesway is not prevented, is one-quarter that of the column in Fig. 11-7(b), where sidesway is prevented. As noted above, the ends of columns are rarely either completely hinged or completely fixed, but rather are partially restrained against rotation by members framing into the ends of the columns. Thus, the effective length will vary between ℓ_u and infinity, as shown in Fig. 11-8(c). If restraining members (beams or slab) are very rigid as compared to the column, the buckling in Fig. 11-8(b) is approached. If, however, the restraining members are quite flexible, a hinged condition is approached at both ends and the column(s), and possibly the structure as a whole, approaches instability. In general, the effective length ℓ_e depends on the degree of rotational restraint at the ends of the column, in this case $\ell_u < \ell_e < \infty$.

In typical reinforced concrete structures, the designer is rarely concerned with single members, but rather with rigid framing systems consisting of beam-column and slab-column assemblies. The buckling behavior of a frame that is not braced against sidesway can be illustrated by the simple portal frame shown in Fig. 11-9. Without lateral restraint at the upper end, the entire (unbraced) frame is free to move sideways. The bottom end may be pinned or partially restrained against rotation.

In summary, the following comments can be made:

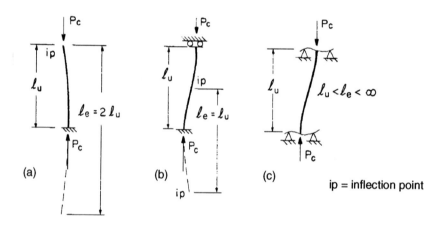

Figure 11-8 Effective Length, ℓ_e (Sway Condition)

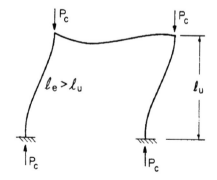

Figure 11-9 Rigid Frame (Sway Condition)

1. For compression members in a nonsway frame, the effective length ℓ_e falls between $\ell_u/2$ and ℓ_u, where ℓ_u is the actual unsupported length of the column.

2. For compression members in a sway frame, the effective length ℓ_e is always greater than the actual length of the column ℓ_u, and may be $2\ell_u$ and higher. In this case, a value of k less than 1.2 normally would not be realistic.

3. Use of the alignment charts shown in Figs. 11-10 and 11-11 (also given in Fig. R10.12.1) allows graphical determination of the effective length factors for compression members in nonsway and sway frames, respectively. If both ends of a column in a nonsway frame have minimal rotational stiffness, or approach $\psi = \infty$, then k = 1.0. If both ends have or approach full fixity, $\psi = 0$, and k = 0.5. If both ends of a column in a sway frame have minimal rotational stiffness, or approach $\psi = \infty$, then k = ∞. If both ends have or approach full fixity, $\psi = 0$, then k = 1.0.

An alternative method for computing the effective length factors for compression members in nonsway and sway frames is contained in R10.12.1. For compression members in a nonsway frame, an upper bound to the effective length factor may be taken as the smaller of the values given by the following two expressions, which are given in the 1972 British Standard Code of Practice (ACI Refs. 10.38 and 10.39):

$$k = 0.7 + 0.05(\psi_A + \psi_B) \leq 1.0$$

$$k = 0.85 + 0.05\psi_{min} \leq 1.0$$

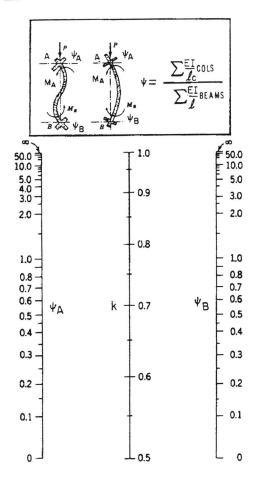

Figure 11-10 Effective Length Factors for Compression Members in a Nonsway Frame

where ψ_A and ψ_B are the values of ψ at the ends of the column and ψ_{min} is the smaller of the two values.

For compression members in a sway frame restrained at both ends, the effective length factor may be taken as (ACI Ref. 10.25):

$$\text{For } \psi_m < 2, \quad k = \frac{20 - \psi_m}{20}\sqrt{1 + \psi_m}$$

$$\text{For } \psi_m \geq 2, \quad k = 0.9\sqrt{1 + \psi_m}$$

where ψ_m is the average of the ψ values at the two ends of the column.

For compression members in a sway frame hinged at one end, the effective length factor may be taken as (ACI Refs. 10.38 and 10.39):

$$k = 2.0 + 0.3\psi$$

where ψ is the column-to-beam stiffness ratio at the restrained end.

In determining the effective length factor k from Figs. 11-10 and 11-11, or from the Commentary equations, the rigidity (EI) of the beams (or slabs) and columns shall be calculated based on the values given in 10.11.1.

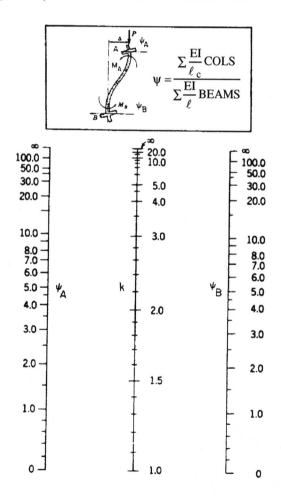

Figure 11-11 Effective Length Factors for Compression Members in a Sway Frame

10.11.4 Nonsway Versus Sway Frames

In actual structures, there is rarely a completely nonsway or sway condition. If it is not readily apparent by inspection, 10.11.4.1 and 10.11.4.2 give two possible ways of determining if a frame is nonsway or not. According to 10.11.4.1, a column in a structure can be considered nonsway if the column end moments due to second-order effects do not exceed 5 percent of the first-order end moments. According to 10.11.4.2, it is also permitted to assume a story within a structure is nonsway if:

$$Q = \frac{\Sigma P_u \Delta_o}{V_{us} \ell_c} \le 0.05 \qquad\qquad Eq.\ (10\text{-}6)$$

where

Q = stability index for a story
ΣP_u = total factored vertical load in the story corresponding to the lateral loading case for which ΣP_u is greatest (R10.11.4)
V_{us} = factored horizontal shear in the story
Δ_o = first-order relative deflection between the top and bottom of the story due to V_u
ℓ_c = column length, measured from center-to-center of the joints in the frame

Note that Eq. (10-6) is not applicable when $V_u = 0$.

10.11.6 Moment Magnifier δ for Biaxial Bending

When biaxial bending occurs in a column, the computed moments about each of the principal axes must be magnified. The magnification factors δ are computed considering the buckling load P_c about each axis separately, based on the appropriate effective lengths and the related stiffness ratios of columns to beams in each direction. Thus, different buckling capacities about the two axes are reflected in different magnification factors. The moments about each of the two axes are magnified separately, and the cross-section is then proportioned for an axial load P_u and magnified biaxial moments.

10.12.2, 10.13.2 Consideration of Slenderness Effects

For compression members in a nonsway frame, effects of slenderness may be neglected when $k\ell_u/r$ is less than or equal to 34 - 12 (M_1/M_2), where M_2 is the larger end moment and M_1 is the smaller end moment. The ratio M_1/M_2 is positive if the column is bent in single curvature, negative if bent in double curvature. Note that M_1 and M_2 are factored end moments obtained by an elastic frame analysis and that the term [34-12M_1/M_2] shall not be taken greater than 40. For compression members in a sway frame, effects of slenderness may be neglected when $k\ell_u/r$ is less than 22 (10.13.2). The moment magnifier method may be used for columns with slenderness ratios exceeding these lower limits.

The upper slenderness limit for columns that may be designed by the approximate moment magnifier method is $k\ell_u/r$ equal to 100 (10.11.5). When $k\ell_u/r$ is greater than 100, an analysis as defined in 10.10.1 must be used, taking into account the influence of axial loads and variable moment of inertia on member stiffness and fixed-end moments, the effect of deflections on the moments and forces, and the effects of duration of loading (sustained load effects). Criteria for consideration of column slenderness are summarized in Fig. 11-4.

The lower slenderness ratio limits will allow a large number of columns to be exempt from slenderness consideration. Considering the slenderness ratio $k\ell_u/r$ in terms of ℓ_u/h for rectangular columns, the effects of slenderness may be neglected in design when ℓ_u/h is less than 10 for compression members in a nonsway frame and with zero moments at both ends. This lower limit increases to 15 for a column in double curvature with equal end moments and a column-to-beam stiffness ratio equal to one at each end. For columns with minimal or zero restraint at both ends, a value of k equal to 1.0 should be used. For stocky columns restrained by flat slab floors, k ranges from about 0.95 to 1.0 and can be conservatively estimated as 1.0. For columns in beam-column frames, k ranges from about 0.75 to 0.90, and can be conservatively estimated as 0.90. If the initial computation of the slenderness ratio based on estimated values of k indicates that effects of slenderness must be considered in the design, a more accurate value of k should be calculated and slenderness re-evaluated. For a compression member in a sway frame with a column-to-beam stiffness ratio equal to 1.0 at both ends, effects of slenderness may be neglected when ℓ_u/h is less than 5. This value reduces to 3 if the beam stiffness is reduced to one-fifth of the column stiffness at each end of the column. Thus, beam stiffnesses at the top and bottom of a column of a high-rise structure where sidesway is not prevented by structural walls or other means will have a significant effect on the degree of slenderness of the column.

The upper limit on the slenderness ratio of $k\ell_u/r$ equal to 100 corresponds to an ℓ_u/h equal to 30 for a compression member in a nonsway frame with zero restraint at both ends. This ℓ_u/h limit increases to 39 for a column-to-beam stiffness ratio of 1.0 at each end.

10.12.3　Moment Magnification—Nonsway Frames

The approximate slender column design equations contained in 10.12.3 for nonsway frames are based on the concept of a moment magnifier δ_{ns} which amplifies the larger factored end moment M_2 on a compression member. The column is then designed for the factored axial load P_u and the amplified moment M_c where M_c is given by:

$$M_c = \delta_{ns} M_2 \qquad \text{Eq. (10-8)}$$

where

$$\delta_{ns} = \frac{C_m}{1 - \dfrac{P_u}{0.75 P_c}} \geq 1.0 \qquad \text{Eq. (10-9)}$$

$$P_c = \frac{\pi^2 EI}{(k\ell_u)^2} \qquad \text{Eq. (10-10)}$$

The critical load P_c is computed for a nonsway condition using an effective length factor k of 1.0 or less. When k is determined from the alignment charts or the equations in R10.12, the values of E and I from 10.11.1 must be used in the computations of ψ_A and ψ_B. Note that the 0.75 factor in Eq. (10-9) is a stiffness reduction factor (see R10.12.3).

In defining the critical column load P_c, the difficult problem is the choice of a stiffness parameter EI which reasonably approximates the stiffness variations due to cracking, creep, and the nonlinearity of the concrete stress-strain curve. In lieu of a more exact analysis, EI shall be taken as:

$$EI = \frac{(0.2 E_c I_g + E_s I_{se})}{1 + \beta_d} \qquad \text{Eq. (10-11)}$$

or

$$EI = \frac{0.4 E_c I_g}{1 + \beta_d} \qquad \text{Eq. (10-12)}$$

The second of these two equations is a simplified approximation to the first. Both equations approximate the lower limits of EI for practical cross-sections and, thus, are conservative. The approximate nature of the EI equations is shown in Fig. 11-12 where they are compared with values derived from moment-curvature diagrams for the case when there is no sustained load ($\beta_d = 0$).

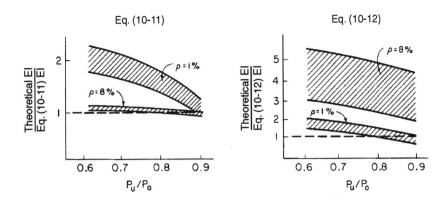

Figure 11-12 Comparison of Equations for EI with EI Values from Moment-Curvature Diagrams

Equation (10-11) represents the lower limit of the practical range of stiffness values. This is especially true for heavily reinforced columns. As noted above, Eq. (10-12) is simpler to use but greatly underestimates the effect of reinforcement in heavily reinforced columns (see Fig. 11-12).

Both EI equations were derived for small e/h values and high P_u/P_o values, where the effect of axial load is most pronounced. The term P_o is the nominal axial load strength at zero eccentricity.

For reinforced concrete columns subjected to sustained loads, creep of concrete transfers some of the load from the concrete to the steel, thus increasing steel stresses. For lightly reinforced columns, this load transfer may cause compression steel to yield prematurely, resulting in a loss in the effective value of EI. This is taken into account by dividing EI by $(1 + \beta_d)$. For nonsway frames, β_d is defined as follows (see 10.11.1):

$$\beta_d = \frac{\text{Maximum factored axial sustained load}}{\text{Maximum factored axial load associated with the same load combination}}$$

For composite columns in which a structural steel shape makes up a large percentage of the total column cross-section, load transfer due to creep is not significant. Accordingly, only the EI of the concrete portion should be reduced by $(1 + \beta_d)$ to account for sustained load effects.

The term C_m is an equivalent moment correction factor. For members without transverse loads between supports, C_m is (10.12.3.1):

$$C_m = 0.6 + 0.4 \left(\frac{M_1}{M_2}\right) \geq 0.4 \qquad \text{Eq. (10-13)}$$

For members with transverse loads between supports, it is possible that the maximum moment will occur at a section away from the ends of a member. In this case, the largest calculated moment occurring anywhere along the length of the member should be magnified by δ_{ns}, and C_m must be taken as 1.0. Figure 11-13 shows some values of C_m, which are a function of the end moments.

If the computed column moment M_2 in Eq. (10-8) is small or zero, design of a nonsway column must be based on the minimum moment $M_{2,min}$ (10.12.3.2):

$$M_{2,min} = P_u (0.6 + 0.03h) \qquad \text{Eq. (10-14)}$$

For members where $M_{2,min} > M_2$, the value of C_m shall either be taken equal to 1.0, or shall be computed by Eq. (10-13) using the ratio of the actual computed end moments M_1 and M_2.

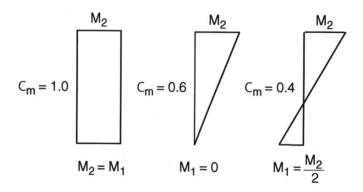

Figure 11-13 Moment Factor C_m

10.13.3 Moment Magnification—Sway Frames

The design of sway frames for slenderness consists essentially of three steps:

1. The magnified sway moments $\delta_s M_s$ are computed in one of three ways:

    a. A second-order elastic frame analysis (10.13.4.1)

    b. An approximate second-order analysis (10.13.4.2)

    c. An approximate magnifier method given in earlier ACI codes (10.13.4.3)

2. The magnified sway moments $\delta_s M_s$ are added to the unmagnified nonsway moments M_{ns} at each end of the column (10.13.3):

$$M_1 = M_{1ns} + \delta_s M_{1s} \qquad \text{Eq. (10-15)}$$

$$M_2 = M_{2ns} + \delta_s M_{2s} \qquad \text{Eq. (10-16)}$$

    The nonsway moments M_{1ns} and M_{2ns} are computed using a first-order elastic analysis.

3. If the column is slender and subjected to high axial loads, it must be checked to see whether moments at points between the column ends are larger than those at the ends. According to 10.13.5, this check is performed using the nonsway magnifier δ_{ns} with P_c computed assuming $k = 1.0$ or less.

10.13.4 Calculation of $\delta_s M_s$

As noted above, there are three different ways to compute the magnified sway moments $\delta_s M_s$. If a second-order elastic analysis is used to compute $\delta_s M_s$, the deflections must be representative of the stage immediately prior to the ultimate load. Thus, the values of EI given in 10.11.1 must be used in the second-order analysis. Note that I must be divided by $(1 + \beta_d)$ where for sway frames, β_d is defined as follows (see 10.11.1):

$$\beta_d = \frac{\text{Maximum factored sustained shear within a story}}{\text{Maximum factored shear in that story}}$$

For wind or earthquake loads, $\beta_d = 0$. An example of a non-zero β_d may occur when members are subjected to earth pressure.

Section 10.13.4.2 allows an approximate second-order analysis to determine $\delta_s M_s$. In this case, the solution of the infinite series that represents the iterative P-Δ analysis for second-order moments is given as follows:

$$\delta_s M_s = \frac{M_s}{1 - Q} \geq M_s \qquad \text{Eq. (10-17)}$$

where

Q = stability index for a story

$$= \frac{\Sigma P_u \Delta_o}{V_{us} \ell_c} \qquad \text{Eq. (10-6)}$$

Note that Eq. (10-17) closely predicts the second-order moments in a sway frame until δ_s exceeds 1.5. For the case when $\delta_s > 1.5$, $\delta_s M_s$ must be computed using 10.13.4.1 or 10.13.4.3.

The code also allows $\delta_s M_s$ to be determined using the magnified moment procedure that was given in previous ACI codes (10.13.4.3):

$$\delta_s M_s = \frac{M_s}{1 - \frac{\Sigma P_u}{0.75 \Sigma P_c}} \geq M_s \qquad \text{Eq. (10-18)}$$

where

ΣP_u = summation of all the factored vertical loads in a story

ΣP_c = summation of the critical buckling loads for all sway-resisting columns in a story

It is important to note that the moment magnification in the columns farthest from the center of twist in a building subjected to significant torsional displacement may be underestimated by the moment magnifier procedure. A three-dimensional second-order analysis should be considered in such cases.

10.13.5 Location of Maximum Moment

When the unmagnified nonsway moments at the ends of the column are added to the magnified sway moments at the same points, one of the resulting total end moments is usually the maximum moment in the column. However, for slender columns with high axial loads, the maximum moment may occur between the ends of the column. A simple way of determining if this situation occurs or not is given in 10.13.5: if an individual compression member has

$$\frac{\ell_u}{r} > \frac{35}{\sqrt{\frac{P_u}{f'_c A_g}}} \qquad \text{Eq. (10-19)}$$

the maximum moment will occur at a point between the ends of the column. In this case, M_2, which is defined in Eq. (10-16), must be magnified by the nonsway moment magnifier given in Eq. (10-9). The column is then designed for the factored axial load P_u and the moment M_c, where M_c is computed from the following:

$$M_c = \delta_{ns} M_2 = \delta_{ns}(M_{2ns} + \delta_s M_{2s}) \qquad \text{Eq. (10-8)}$$

Where:

$$\delta_{ns} = \frac{C_m}{1 - \frac{P_u}{0.75P_c}}$$

Note that to calculate δ_{ns} k is determined according to 10.12.1 and $\delta_{ns} \geq 1.0$, and M_1 and M_2 given by Eqs. (10-15), (10-16) are used to calculate C_m according to Eq. (10-13).

10.13.6 Structural Stability Under Gravity Loads

For sway frames, the possibility of sidesway instability of the structure as a whole under factored gravity loads must be investigated. This is checked in three different ways, depending on the method that is used in determining $\delta_s M_s$:

1. When $\delta_s M_s$ is computed by a second-order analysis (10.13.4.1), the following expression must be satisfied:

$$\frac{\text{Second-order lateral deflections}}{\text{First-order lateral deflections}} \leq 2.5$$

Note that these deflections are based on the applied loading of $1.2P_D$ and $1.6P_L$ plus factored lateral load. The frame should be analyzed twice for this set of applied loads: the first analysis should be a first-order analysis and the second should be a second-order analysis. The lateral load may be the actual lateral loads used in design or it may be a single lateral load applied to the top of the frame. In any case, the lateral load(s) should be large enough to give deflections that can be compared accurately.

2. When $\delta_s M_s$ is computed by the approximate second-order analysis (10.13.4.2), then

$$Q = \frac{\Sigma P_u \Delta_o}{V_{us} \ell_c} \leq 0.60$$

where the value of Q is evaluated using $1.2P_D$ and $1.6P_L$. Note that the above expression is equivalent to $\delta_s = 2.5$. The values of V_{us} and Δ_o may be determined using the actual or any arbitrary set of lateral loads. The above stability check is satisfied if the value of Q computed in 10.11.4.2 is less than or equal to 0.2.

3. When $\delta_s M_s$ is computed using the expressions from previous ACI codes (10.13.4.3), the stability check is satisfied when

$$0 < \delta_s \leq 2.5$$

In this case, ΣP_u and ΣP_c correspond to the factored dead and live loads.

It is important to note that in each of the three cases above, β_d shall be taken as the following:

$$\beta_d = \frac{\text{Maximum factored sustained axial load}}{\text{Maximum factored axial load}}$$

10.13.7 Moment Magnification for Flexural Members

The strength of a laterally unbraced frame is governed by the stability of the columns and by the degree of end restraint provided by the beams in the frame. If plastic hinges form in the restraining beams, the structure approaches a mechanism and its axial load capacity is drastically reduced. Section 10.13.7 requires that the restraining flexural members (beams or slabs) have the capacity to resist the magnified column moments. The ability of the moment magnifier method to provide a good approximation of the actual magnified moments at the member ends in sway frame is a significant improvement over the reduction factor method for long columns prescribed in earlier ACI codes to account for member slenderness in design.

SUMMARY OF DESIGN EQUATIONS

A summary of the equations for the design of slender columns subjected to dead, live and lateral loads, in both nonsway and sway frames is presented in this section. Examples 11.1 and 11.2 illustrate the application of these equations for the design of columns in nonsway and sway frames, respectively.

- **Nonsway Frames**

1. Determine the factored load combinations per 9.2.

    It is assumed in the examples that follow that the load factor for live load is 0.5 (i.e. condition 9.2.1(a) applies) and that the wind load has been reduced by a directionality factor (9.2.1(b)).

    Note that the factored moments $M_{u,top}$ and $M_{u,bot}$ at the top and bottom of the column, respectively, are to be determined using a first-order frame analysis, based on the cracked section properties of the members.

2. For each load combination, determine M_c, where M_c is the largest factored column end moment, including slenderness effects (if required). Note that M_c may be determined by one of the following methods:

    a. Second-order (P-Δ) analysis (10.10.1)

    b. Magnified moment method (only if $k\ell_u/r \leq 100$; see 10.12 and step (3) below)

    Determine the required column reinforcement for the critical load combination determined in step (1) above. Each load combination consists of P_u and M_c.

3. Magnified moment method (10.12):

    Slenderness effects can be neglected when

    $$\frac{k\ell_u}{r} \leq 34 - 12\left(\frac{M_1}{M_2}\right) \qquad \text{Eq. (10-7)}$$

    where $[34-12 M_1/M_2] \leq 40$. The term M_1/M_2 is positive if the column is bent in single curvature, negative if bent in double curvature. If $M_1 = M_2 = 0$, assume $M_2 = M_{2,\,min}$. In this case $k\ell_u/r = 34.0$.

    When slenderness effects need to be considered, determine M_c for each load combination:

    $$M_c = \delta_{ns} M_2 \qquad \text{Eq. (10-8)}$$

    where

    M_2 = larger of $M_{u,bot}$ and $M_{u,top}$

    $\qquad \geq P_u(0.6 + 0.03h) \qquad$ 10.12.3.2

    $$\delta_{ns} = \frac{C_m}{1 - \dfrac{P_u}{0.75 P_c}} \geq 1.0 \qquad \text{Eq. (10-9)}$$

$$P_c = \frac{\pi^2 EI}{(k\ell_u)^2} \qquad \text{Eq. (10-10)}$$

$$EI = \frac{(0.2E_c I_g + E_s I_{se})}{1 + \beta_d} \qquad \text{Eq. (10-11)}$$

or

$$EI = \frac{0.4 E_c I_g}{1 + \beta_d} \qquad \text{Eq. (10-12)}$$

$$\beta_d = \frac{\text{Maximum factored axial sustained load}}{\text{Maximum factored axial load associated with the same load combination}} \qquad 10.11.1$$

$$C_m = 0.6 + 0.4 \left(\frac{M_1}{M_2}\right) \geq 0.4 \quad \text{(for columns without transverse loads)} \qquad \text{Eq. (10-13)}$$

$$= 1.0 \quad \text{(for columns with transverse loads)}$$

The effective length factor k shall be taken as 1.0, or may be determined from analysis (i.e., alignment chart or equations given in R10.12.1). In the latter case, k shall be based on the E and I values determined according to 10.11.1 (see 10.12.1).

- **Sway Frames**

  1. Determine the factored load combinations per 9.2.

     a. Gravity (dead and live) loads

        The moments $(M_{u,bot})_{ns}$ and $(M_{u,top})_{ns}$ at the bottom and top of column, respectively, are to be determined using an elastic first-order frame analysis, based on the cracked section properties of the members.

        The moments M_1 and M_2 are the smaller and the larger of the moments $(M_{u,bot})_{ns}$ and $(M_{u,top})_{ns}$, respectively. The moments M_{1ns} and M_{2ns} are the factored end moments at the ends at which M_1 and M_2 act, respectively.

     b. Gravity (dead and live) plus lateral loads

        The total moments at the top and bottom of the column are $M_{u,top} = (M_{u,top})_{ns} + (M_{u,top})_s$ and $M_{u,bot} = (M_{u,bot})_{ns} + (M_{u,bot})_s$, respectively. The moments M_1 and M_2 are the smaller and the larger of the moments $M_{u,top}$ and $M_{u,bot}$, respectively. Note that at this stage, M_1 and M_2 do not include slenderness effects. The moments M_{1ns} and M_{1s} are the factored nonsway and sway moments, respectively, at the end of the column at which M_1 acts, while M_{2ns} and M_{2s} are the factored nonsway and sway moments, respectively, at the end of the column at which M_2 acts.

     c. Gravity (dead) plus lateral loads

        The definitions for the moments in this load combination are the same as given above for part 1(b).

     d. The effects due to lateral forces acting equal and opposite to the ones in the initial direction of analysis must also be considered in the load combinations given in parts 1(b) and 1(c) above.

2. Determine the required column reinforcement for the critical load combination determined in step (1) above. Each load combination consists of P_u, M_1, and M_2, where now M_1 and M_2 are the total factored end moments, including slenderness effects. Note that if the critical load P_c is computed using EI from Eq. (10-11), it is necessary to estimate first the column reinforcement. Moments M_1 and M_2 are determined by one of the following methods:

   a. Second-order (P-Δ) analysis (10.10.1)

   b. Magnified moment method (only if $k\ell_u/r \leq 100$; see 10.13 and step 3 below)

3. Magnified moment method (see 10.13):

   Slenderness effects can be neglected when

   $$\frac{k\ell_u}{r} < 22 \qquad\qquad 10.13.2$$

   When slenderness effects need to be considered:

   $$M_1 = M_{1ns} + \delta_s M_{1s} \qquad\qquad Eq.\ (10\text{-}15)$$

   $$M_2 = M_{2ns} + \delta_s M_{2s} \qquad\qquad Eq.\ (10\text{-}16)$$

   The moments $\delta_s M_{1s}$ and $\delta_s M_{2s}$ are to be computed by one of the following methods (10.13.4):

   a. Second-order elastic analysis (see 10.13.4.1)

   b. Approximate second-order analysis (10.13.4.2)

   $$\delta_s M_s = \frac{M_s}{1-Q} \geq M_s, \quad 1.0 \leq \delta_s \leq 1.5 \qquad\qquad Eq.\ (10\text{-}17)$$

   where

   $$Q = \frac{\Sigma P_u \Delta_o}{V_{us}\ell_c} \qquad\qquad Eq.\ (10\text{-}6)$$

   c. Approximate magnifier method given in ACI code (see 10.13.4.3):

   $$\delta_s M_s = \frac{M_s}{1 - \dfrac{\Sigma P_u}{0.75\Sigma P_c}} \geq M_s \qquad\qquad Eq.\ (10\text{-}18)$$

   where

   $$P_c = \frac{\pi^2 EI}{(k\ell_u)^2} \qquad\qquad Eq.\ (10\text{-}10)$$

   $$EI = \frac{(0.2 E_c I_g + E_s I_{se})}{1 + \beta_d} \qquad\qquad Eq.\ (10\text{-}11)$$

   or

$$EI = \frac{0.4E_c I_g}{1 + \beta_d} \quad \text{Eq. (10-12)}$$

The effective length factor k must be greater than 1.0 and shall be based on the E and I values determined according to 10.11.1 (see 10.13.1).

4. Check if the maximum moment occurs at the ends of the column or between the ends of the column (10.13.5). If

$$\frac{\ell_u}{r} > \frac{35}{\sqrt{\frac{P_u}{f'_c A_g}}} \quad \text{Eq. (10-19)}$$

the column must be designed for the factored axial load P_u and the moment M_c, where

$$M_c = \delta_{ns} M_2 = \delta_{ns}(M_{2ns} + \delta_s M_{2s})$$

where

$$\delta_{ns} = \frac{C_m}{1 - \frac{P_u}{0.75 P_c}}$$

To calculate δ_{ns} k is determined according to the provisions in 10.12.1 and $\delta_{ns} \geq 1.0$ and M_1 and M_2 given by Eqs. (10-15), (10-16) are used to calculate C_m according to Eq. (10-13).

5. Check the possibility of sidesway instability under gravity loads (10.13.6):

   a. When $\delta_s M_s$ is computed from 10.13.4.1:

   $$\frac{\text{Second-order lateral deflections}}{\text{First-order lateral deflections}} \leq 2.5$$

   based on factored dead and live loads plus factored lateral oad.

   b. When $\delta_s M_s$ is computed from 10.13.4.2:

   $$Q = \frac{\Sigma P_u \Delta_o}{V_u \ell_c} \leq 0.60$$

   based on factored dead and live loads.

   c. When $\delta_s M_s$ is computed from 10.13.4.3:

   $$0 < \delta_s \leq 2.5$$

   where δ_s is computed using ΣP_u and ΣP_c corresponding to the factored dead and live loads.

   In all three cases, β_d shall be taken as:

   $$\beta_d = \frac{\text{Maximum factored sustained axial load}}{\text{Maximum factored axial load}}$$

Reference 11.1 gives the derivation of the design equations for the slenderness provisions outlined above.

REFERENCES

11.1 MacGregor, J. G., "Design of Slender Concrete Columns—Revisited," *ACI Structural Journal,* V. 90, No. 3, May-June 1993, pp. 302-309.

11.2 pcaColumn—Design and Investigation of Reinforced Concrete Column Sections, Portland Cement Association, Skokie, IL 2005.

Example 11.1—Slenderness Effects for Columns in a Nonsway Frame

Design columns A3 and C3 in the first story of the 10-story office building shown below. The clear height of the first story is 21 ft-4 in., and is 11 ft-4 in. for all of the other stories. Assume that the lateral load effects on the building are caused by wind, and that the dead loads are the only sustained loads. Other pertinent design data for the building are as follows:

Material properties:

    Concrete:
        Floors: $f'_c = 4,000$ psi, $w_c = 150$ pcf
        Columns and walls: $f'_c = 6,000$ psi, $w_c = 150$ pcf
        Reinforcement: $f_y = 60$ ksi

Beams: 24×20 in.
Exterior columns: 20×20 in.
Interior columns: 24×24 in.
Shearwalls: 12 in.

Weight of floor joists = 86 psf
Superimposed dead load = 32 psf
Roof live load = 30 psf
Floor live load = 50 psf
Wind loads computed according to ASCE 7.

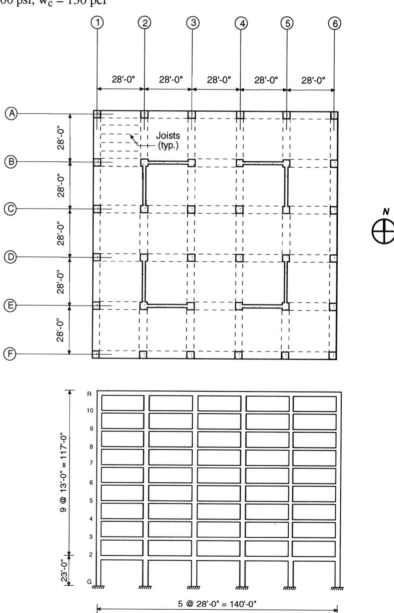

					Code
					Reference

Example 11.1 (cont'd)   **Calculations and Discussion**

1. Factored axial loads and bending moments for columns A3 and C3 in the first story

Column A3

Load Case			Axial Load (kips)	Bending Moment (ft-kips)	
				Top	Bottom
		Dead (D)	718.0	79.0	40.0
		Live (L)*	80.0	30.3	15.3
		Roof live load (L_r)	12.0	0.0	0.0
		Wind (W)	±8.0	±1.1	±4.3
Eq.	No.	Load Combination			
9-1	1	1.4D	1,005.2	110.6	56.0
9-2	2	1.2D + 1.6L + 0.5L_r	995.6	143.3	72.5
9-3	3	1.2D + 0.5L + 1.6L_r	920.8	110.0	55.7
	4	1.2D + 1.6L_r + 0.8W	887.2	95.7	51.4
	5	1.2D + 1.6L_r - 0.8W	874.4	93.9	44.6
9-4	6	1.2D + 0.5L + 0.5L_r + 1.6W	920.4	111.7	62.5
	7	1.2D + 0.5L + 0.5L_r - 1.6W	894.8	108.2	48.8
9-6	8	0.9D + 1.6W	659.0	72.9	42.9
	9	0.9D - 1.6W	633.4	69.3	29.1

*includes live load reduction per ASCE 7

Column C3

Load Case			Axial Load (kips)	Bending Moment (ft-kips)	
				Top	Bottom
		Dead (D)	1,269.0	1.0	0.7
		Live (L)*	147.0	32.4	16.3
		Roof live load (L_r)	24.0	0.0	0.0
		Wind (W)	±3.0	±2.5	±7.7
Eq.	No.	Load Combination			
9-1	1	1.4D	1,776.6	1.4	1.0
9-2	2	1.2D + 1.6L + 0.5L_r	1,770.0	53.0	26.9
9-3	3	1.2D + 0.5L + 1.6L_r	1,634.7	17.4	9.0
	4	1.2D + 1.6L_r + 0.8W	1,563.6	3.2	7.0
	5	1.2D + 1.6L_r - 0.8W	1,558.8	-0.8	-5.3
9-4	6	1.2D + 0.5L + 0.5L_r + 1.6W	1,613.1	21.4	21.3
	7	1.2D + 0.5L + 0.5L_r - 1.6W	1,603.5	13.4	-3.3
9-6	8	0.9D + 1.6W	1,146.9	4.9	13.0
	9	0.9D - 1.6W	1,137.3	-3.1	-11.7

*includes live load reduction per ASCE 7

Note that Columns A3 and C3 are bent in double curvature with the exception of Load Case 7 for Column C3.

2. Determine if the frame at the first story is nonsway or sway

The results from an elastic first-order analysis using the section properties prescribed in 10.11.1 are as follows:

ΣP_u = total vertical load in the first story corresponding to the lateral loading case for which ΣP_u is greatest

Example 11.1 (cont'd)   **Calculations and Discussion**   **Code Reference**

The total building loads are: D = 37,371 kips, L = 3609 kips, and L_r = 605 kips.
The maximum ΣP_u is determined from Eq. (9-4):

$$\Sigma P_u = (1.2 \times 37{,}371) + (0.5 \times 3609) + (0.5 \times 605) + 0 = 46{,}952 \text{ kips}$$

V_{us} = factored story shear in the first story corresponding to the wind loads
     = 1.6 × 324.3 = 518.9 kips   *Eq. (9-4), (9-6)*

Δ_o = first-order relative lateral deflection between the top and bottom of the first story due to V_{us}
     = 1.6 × (0.03 - 0) = 0.05 in.

Stability index $Q = \dfrac{\Sigma P_u \Delta_o}{V_{us} \ell_c} = \dfrac{46{,}952 \times 0.05}{518.9 \times [(23 \times 12) - (20/2)]} = 0.02 < 0.05$ *Eq. (10-6)*

Since Q < 0.05, the frame at the first story level is considered nonsway.   *10.11.4.2*

3. Design of column C3

     Determine if slenderness effects must be considered.

     Using an effective length factor k = 1.0,   *10.12.1*

$$\frac{k\ell_u}{r} = \frac{1.0 \times 21.33 \times 12}{0.3 \times 24} = 35.6$$

The following table contains the slenderness limit for each load case:

Eq.	No.	Axial loads (kips) P_u	Bending Moment (ft-kips) M_{top}	M_{bot}	Curvature	M_1 (ft-kips)	M_2 (ft-kips)	M_1/M_2	Slenderness limit
9-1	1	1776.6	1.4	1.0	Double	1.0	1.4	0.70	40.00
9-2	2	1770.0	53.0	26.9	Double	26.9	53.0	0.51	40.00
9-3	3	1634.7	17.4	9.0	Double	9.0	17.4	0.52	40.00
9-3	4	1564.2	3.7	8.5	Double	3.7	8.5	0.43	39.20
9-3	5	1558.2	-1.3	-6.9	Double	1.3	6.9	0.19	36.27
9-4	6	1613.1	21.4	21.3	Double	21.3	21.4	1.00	40.00
9-4	7	1603.5	13.4	-3.3	Single	3.3	13.4	-0.25	31.02
9-6	8	1146.9	4.9	13.0	Double	4.9	13.0	0.38	38.54
9-6	9	1137.3	-3.1	-11.7	Double	3.1	11.7	0.27	37.18

The least value of $34 - 12\left(\dfrac{M_1}{M_2}\right)$ is obtained from load combination no. 7:

$$34 - 12\left(\frac{M_1}{M_2}\right) = 34 - 12\left(\frac{3.3}{13.4}\right) = 31.02 < 40$$

	Code
Example 11.1 (cont'd) Calculations and Discussion	Reference

Slenderness effects need to be considered for column C3 since $k\ell_u/r > 34 - 12\,(M_1/M_2)$. 10.12.2

The following calculations illustrate the magnified moment calculations for load combination no. 7:

$$M_c = \delta_{ns} M_2$$

where

$$\delta_{ns} = \frac{C_m}{1 - \dfrac{P_u}{0.75 P_c}} \geq 1$$

$$C_m = 0.6 + 0.4\left(\frac{M_1}{M_2}\right) \geq 0.40$$

$$= 0.6 + 0.4\left(\frac{3.3}{13.4}\right) = 0.70$$

$$P_c = \frac{\pi^2 EI}{(k\ell_u)^2}$$

$$EI = \frac{0.2 E_c I_g + E_s I_e}{1 + \beta_d}$$

$$E_c = 57{,}000\,\frac{\sqrt{6000}}{1000} = 4415 \text{ ksi}$$

$$I_g = \frac{24^4}{12} = 27{,}648 \text{ in.}^4$$

$$E_s = 29{,}000 \text{ ksi}$$

Assuming 16-No. 7 bars with 1.5 cover to No. 3 ties as shown in the figure.

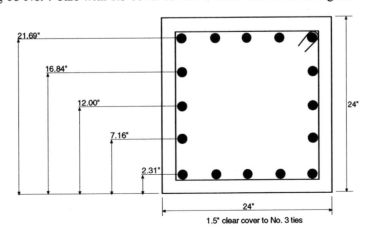

Example 11.1 (cont'd)   **Calculations and Discussion**   **Code Reference**

$$I_{se} = 2\left[(5 \times 0.6)(21.69-12)^2 + (2 \times 0.6)(16.84-12)^2\right]$$

$$= 619.6 \text{ in.}^4$$

Since the dead load is the only sustained load,

$$\beta_d = \frac{1.2P_D}{1.2P_D + 0.5P_L + 0.5P_{Lr} - 1.6W}$$

$$= \frac{1.2 \times 1269}{(1.2 \times 1269) + (0.5 \times 147) + (0.5 \times 24) - (1.6 \times 3)}$$

$$= 0.95$$

$$EI = \frac{(0.2 \times 4415 \times 27{,}648) + (29{,}000 \times 619.6)}{1 + 0.95} = 21.73 \times 10^6 \text{ kip-in.}^2$$

$$P_c = \frac{\pi^2 \times 21.73 \times 10^6}{(1 \times 21.33 \times 12)^2} = 3274 \text{ kips}$$

$$\delta_{ns} = \frac{0.7}{1 - \frac{1603.5}{0.75 \times 3274}} = 2.02$$

Check miminum moment requirement:

$$M_{2,\,min} = P_n(0.6 + 0.03h)$$

$$= 1603.5[0.6 + (0.03 \times 24)]/12$$

$$= 176.4 \text{ ft-kip} > M_2$$

$$M_c = 2.02 \times 176.4 = 356.3 \text{ ft-kip}$$

The following table contains results from a strain compatibility analysis, where compressive strains are taken as positive (see Part 6 and 7).

Therefore, since $fM_n > M_u$ for all $fP_n = P_u$, use a 24 3 24 in. column with 16-No. 7 bars ($r_g = 1.7\%$).

Example 11.1 (cont'd)	Calculations and Discussion	Code Reference

No.	P_u (kips)	M_u (ft-kips)	c (in)	ε_t	ϕ	ϕP_n (kips)	ϕM_n (ft-kips)
1	1776.6	1.4	25.92	0.00049	0.65	1776.6	367.2
2	1770.0	53.0	25.83	0.00048	0.65	1770.0	371.0
3	1634.7	17.4	23.86	0.00027	0.65	1634.7	447.0
4	1563.6	7.0	22.85	0.00015	0.65	1563.6	480.9
5	1558.8	5.3	22.78	0.00014	0.65	1558.8	483.2
6	1613.1	21.4	23.55	0.00024	0.65	1613.1	457.8
7	1603.5	356.3	23.41	0.00022	0.65	1603.5	462.5
8	1146.9	13.0	17.25	-0.00077	0.65	1146.9	609.9
9	1137.3	11.7	17.13	-0.00080	0.65	1137.3	611.7

Design for P_u = and M_c = can be performed manually, by creating an interaction diagram as shown in example 6.4. For this example, Figure 11-14 shows the design srength interaction diagram for Column C3 obtained from the computer program pcaColumn. The figure also shows the axial load and moments for all load combinations.

4. Design of column A3

   a. Determine if slenderness effects must be considered.

   Determine k from the alignment chart of Fig. 11-10 or from Fig. R10.12.1:

   $$I_{col} = 0.7 \left(\frac{20^4}{12}\right) = 9{,}333 \text{ in.}^4 \qquad \textit{10.11.1(b)}$$

   $$E_c = 57{,}000 \; \frac{\sqrt{6{,}000}}{1{,}000} = 4{,}415 \text{ ksi} \qquad \textit{8.5.1}$$

   For the column below level 2:

   $$\left(\frac{E_c I}{\ell_c}\right) = \frac{4{,}415 \times 9{,}333}{[(23 \times 12) - (20/2)]} = 155 \times 10^3 \text{ in.-kips}$$

   For the column above level 2:

   $$\left(\frac{E_c I}{\ell_c}\right) = \frac{4{,}415 \times 9{,}333}{13 \times 12} = 264 \times 10^3 \text{ in.-kips}$$

   $$I_{beam} = 0.35 \left(\frac{24 \times 20^3}{12}\right) = 5{,}600 \text{ in.}^4 \qquad \textit{10.11.1(b)}$$

   $$\frac{EI}{\ell} = \frac{57\sqrt{4{,}000} \times 5{,}600}{28 \times 12} = 60 \times 10^3 \text{ in.-kips}$$

   $$\psi_A = \frac{\Sigma E_c I / \ell_c}{\Sigma E_c I / \ell} = \frac{155 + 264}{60} = 7.0$$

| Example 11.1 (cont'd) | Calculations and Discussion | Code Reference |

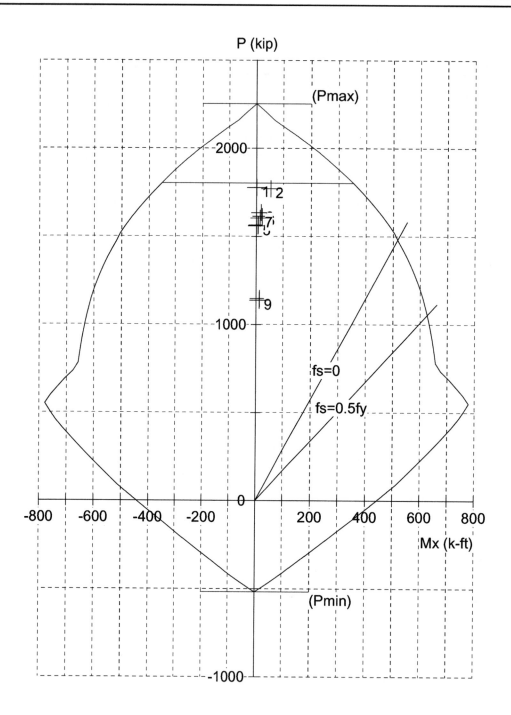

Figure 11-14 Interaction Diagram for Column C3

Example 11.1 (cont'd)   **Calculations and Discussion**   **Code Reference**

Assume $\psi_B = 1.0$ (column essentially fixed at base)

From Fig. R10.12.1(a), k = 0.86.

Therefore, for column A3 bent in double curvature, the least $34 - 12\left(\dfrac{M_1}{M_2}\right)$ is obtained from load combination no. 9:

$$34 - 12\left(\dfrac{-29.1}{69.3}\right) = 39.0$$

$$\dfrac{k\ell_u}{r} = \dfrac{0.86 \times 21.33 \times 12}{0.3 \times 20} = 36.7 < 39.0$$

For column A3 bent in single curvature, the least $34 - 12\left(\dfrac{M_1}{M_2}\right)$ is obtained from load combination no. 8:

$$\dfrac{k\ell_u}{r} = 36.7 > 34 - 12\left(\dfrac{42.9}{72.9}\right) = 26.9$$

Therefore, column slenderness need not be considered for column A3 if bent in double curvature. However, to illustrate the design procedure including slenderness effects for nonsway columns, assume single curvature bending.

b. Determine total moment M_c (including slenderness effects) for each load combination.

$$M_c = \delta_{ns} M_2 \qquad \text{Eq. (10-8)}$$

where

$$\delta_{ns} = \dfrac{C_m}{1 - \dfrac{P_u}{0.75 P_c}} \geq 1.0 \qquad \text{Eq. (10-9)}$$

The following table summarizes magnified moment computations for column A3 for all load combinations, followed by detailed calculations for combination no. 6 to illustrate the procedure.

No.	P_u (kips)	M_2 (ft-kips)	b	$EI \times 10^6$ (kip-in^2)	P_c kips	C_m	d_{ns}	$M_{2,min}$ (ft-kips)	M_c (ft-kips)
1	1005.2	110.6	1.00	9.88	2013	0.80	2.40	100.5	265.6
2	995.6	143.3	0.87	10.60	2158	0.80	2.08	99.6	298.6
3	920.8	110.0	0.94	10.21	2080	0.80	1.96	92.1	215.3
4	887.2	95.7	0.97	10.03	2042	0.82	1.94	88.7	185.3
5	874.4	93.9	0.99	9.96	2028	0.79	1.86	87.4	174.5
6	920.4	111.7	0.94	10.21	2079	0.82	2.01	92.0	224.6
7	894.8	108.2	0.96	10.07	2051	0.78	1.87	89.5	201.8
8	659.0	72.9	0.98	9.98	2033	0.84	1.47	65.9	107.2
9	633.4	69.3	1.00	9.89	2014	0.77	1.32	63.3	91.7

| Example 11.1 (cont'd) | Calculations and Discussion | Code Reference |

Load combination no. 6:

$$U = 1.2D + 0.5L + 0.5L_r + 1.6W$$

$$C_m = 0.6 + 0.4\left(\frac{M_1}{M_2}\right) \geq 0.4 \qquad \text{Eq. (10-13)}$$

$$= 0.6 + 0.4\left(\frac{62.5}{111.7}\right) = 0.82$$

$$P_c = \frac{\pi^2 EI}{(k\ell_u)^2} \qquad \text{Eq. (10-10)}$$

$$EI = \frac{(0.2 E_c I_g + E_s I_{se})}{1 + \beta_d} \qquad \text{Eq. (10-11)}$$

$$E_c = 57,000 \frac{\sqrt{6,000}}{1,000} = 4,415 \text{ ksi} \qquad 8.5.1$$

$$I_g = \frac{20^4}{12} = 13,333 \text{ in.}^4$$

$$E_s = 29,000 \text{ ksi} \qquad 8.5.2$$

Assuming 8-No. 8 bars with 1.5 in. cover to No. 3 ties:

$$I_{se} = 2\left[(3 \times 0.79)\left(\frac{20}{2} - 1.5 - 0.375 - \frac{1.00}{2}\right)^2\right] = 276 \text{ in.}^4$$

Since the dead load is the only sustained load,

$$\beta_d = \frac{1.2 P_D}{1.2 P_D + 0.5 P_L + 0.5 P_{L_r} + 1.6 P_w}$$

$$= \frac{1.2 \times 718}{(1.2 \times 718) + (0.5 \times 80) + (0.5 \times 12) + (1.6 \times 8)} = 0.94$$

$$EI = \frac{(0.2 \times 4,415 \times 13,333) + (29,000 \times 276)}{1 + 0.94} = 10.21 \times 10^6 \text{ kip-in.}^2$$

| Example 11.1 (cont'd) | Calculations and Discussion | Code Reference |

From Eq. (10-12):

$$EI = \frac{0.4E_c I_g}{1+\beta_d}$$

$$= \frac{0.4 \times 4{,}415 \times 13{,}333}{1+0.94} = 12.14 \times 10^6 \text{ kip-in.}^2$$

Using EI from Eq. (10-10), the critical load P_c is:

$$P_c = \frac{\pi^2 \times 10.21 \times 10^6}{(0.86 \times 21.33 \times 12)^2} = 2{,}079 \text{ kips}$$

Therefore, the moment magnification factor δ_{ns} is:

$$\delta_{ns} = \frac{0.82}{1 - \frac{920.4}{0.75 \times 2{,}079}} = 2.01$$

Check minimum moment requirement:

$M_{2,min} = P_u(0.6 + 0.03h)$ *Eq. (10-14)*

$= 920.4\,[0.6 + (0.03 \times 20)]/12$

$= 92.0$ ft-kips $< M_2 = 111.7$ ft-kips

Therefore,

$M_c = 2.01 \times 111.7 = 224.6$ ft-kips

c. Determine required reinforcement.

For the 20 x 20 in. column, try 8-No. 8 bars.

Determine maximum allowable axial compressive force, $\phi P_{n,max}$:

$\phi P_{n,max} = 0.80\phi \left[0.85 f'_c (A_g - A_{st}) + f_y A_{st}\right]$ *Eq. (10-2)*

$= (0.80 \times 0.65)\left[(0.85 \times 6)(20^2 - 6.32) + (60 \times 6.32)\right]$

$= 1{,}241.2$ kips $>$ maximum $P_u = 1{,}005.2$ kips O.K.

Example 11.1 (cont'd)	Calculations and Discussion	Code Reference

The following table contains results from a strain compatibility analysis, where compressive strains are taken as positive (see Parts 6 and 7).

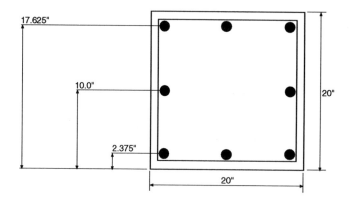

No.	P_u (kips)	M_u (ft-kips)	c (in.)	ε_t	ϕ	ϕP_n (kips)	ϕM_n (ft-kips)
1	1,005.2	265.6	17.81	0.00003	0.65	1,005.2	298.0
2	995.6	298.6	17.64	0.00000	0.65	995.6	301.1
3	920.8	215.3	16.42	-0.00022	0.65	920.8	321.4
4	887.2	185.3	15.88	-0.00033	0.65	887.2	329.3
5	874.4	174.5	15.67	-0.00037	0.65	874.4	332.1
6	920.4	224.6	16.41	-0.00022	0.65	920.4	321.6
7	894.8	201.8	16.00	-0.00030	0.65	894.8	327.6
8	659.0	107.2	12.36	-0.00128	0.65	659.0	364.8
9	633.4	92.4	12.00	-0.00141	0.65	633.4	367.2

Therefore, since $\phi M_n > M_u$ for all $\phi P_n = P_u$, use a 20 × 20 in. column with 8-No. 8 bars ($\rho_g = 1.6\%$). Figure 11-15 obtained from pcaColumn [11,2], contains the design strength interaction diagram for Column A3 with the factored axial loads and magnified moments for all load combinations.

Example 11.1 (cont'd)	Calculations and Discussion	Code Reference

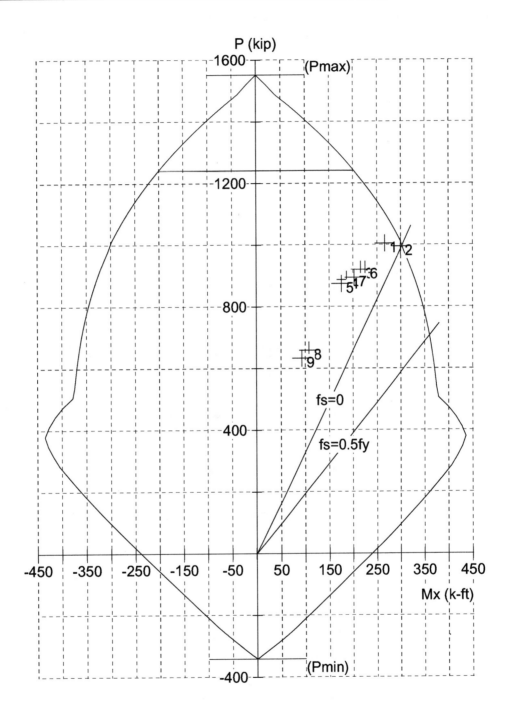

Figure 11-15 Design Strength Interaction Diagram for Column A3

Example 11.2—Slenderness Effects for Columns in a Sway Frame

Design columns C1 and C2 in the first story of the 12-story office building shown below. The clear height of the first story is 13 ft-4 in., and is 10 ft-4 in. for all of the other stories. Assume that the lateral load effects on the building are caused by wind, and that the dead loads are the only sustained loads. Other pertinent design data for the building are as follows:

Material properties:

    Concrete: = 6,000 psi for columns in the bottom two stories (w_c = 150 pcf)
                     = 4,000 psi elsewhere (w_c = 150 pcf)
    Reinforcement: f_y = 60 ksi

Beams: 24 × 20 in.
Exterior columns: 22 × 22 in.
Interior columns: 24 × 24 in.

Superimposed dead load = 30 psf
Roof live load = 30 psf
Floor live load = 50 psf
Wind loads computed according to ASCE 7

				Code Reference
Example 11.2 (cont'd)		**Calculations and Discussion**		

1. Factored axial loads and bending moments for columns C1 and C2 in the first story

   Since this is a symmetrical frame, the gravity loads will not cause appreciable sidesway.

   Column C1

		Load Case	Axial Load (kips)	Bending moment (ft-kips)							
				Top	Bottom						
		Dead (D)	622.4	34.8	17.6						
		Live (L)*	73.9	15.4	7.7						
		Roof live load (L_r)	8.6	0.0	0.0						
		Wind (W) (N-S)	-48.3	17.1	138.0						
		Wind (W) (S-N)	48.3	-17.1	-138.0						
	No.	Load Combination				M_1	M_2	M_{1ns}	M_{2ns}	M_{1s}	M_{2s}
9-1	1	1.4D	871.4	48.7	24.6	24.6	48.7	24.6	48.7	---	---
9-2	2	1.2D + 1.6L + 0.5L_r	869.4	66.4	33.4	33.4	66.4	33.4	66.4	---	---
9-3	3	1.2D + 0.5L + 1.6L_r	797.6	49.5	25.0	25.0	49.5	25.0	49.5	---	---
	4	1.2D + 1.6L_r + 0.8W	722.0	55.4	131.5	55.4	131.5	41.8	21.1	13.7	110.4
	5	1.2D + 1.6L_r - 0.8W	799.3	28.1	-89.3	28.1	-89.3	41.8	21.1	-13.7	-110.4
9-4	6	1.2D + 0.5L + 0.5L_r + 1.6W	710.9	76.8	245.8	76.8	245.8	49.5	25.0	27.4	220.8
	7	1.2D + 0.5L + 0.5L_r - 1.6W	865.4	22.1	-195.8	22.1	-195.8	49.5	25.0	-27.4	-220.8
9-6	8	0.9D + 1.6W	482.9	58.7	236.6	58.7	236.6	31.3	15.8	27.4	220.8
	9	0.9D - 1.6W	637.4	4.0	-205.0	4.0	-205.0	31.3	15.8	-27.4	-220.8

   *includes live load reduction per ASCE 7

   Column C2

		Load Case	Axial Load (kips)	Bending moment (ft-kips)							
				Top	Bottom						
		Dead (D)	1,087.6	-2.0	-1.0						
		Live (L)*	134.5	-15.6	-7.8						
		Roof live load (L_r)	17.3	0.0	0.0						
		Wind (W) (N-S)	-0.3	43.5	205.0						
		Wind (W) (S-N)	0.3	-43.5	-205.0						
	No.	Load Combination				M_1	M_2	M_{1ns}	M_{2ns}	M_{1s}	M_{2s}
9-1	1	1.4D	1,522.6	-2.8	-1.4	-1.4	-2.8	-1.4	-2.8	---	---
9-2	2	1.2D + 1.6L + 0.5L_r	1,529.4	-27.4	-13.7	-13.7	-27.4	-13.7	-27.4	---	---
9-3	3	1.2D + 0.5L + 1.6L_r	1,400.1	-10.2	-5.1	-5.1	-10.2	-5.1	-10.2	---	0.0
	4	1.2D + 1.6L_r + 0.8W	1,332.6	32.4	162.8	32.4	162.8	-2.4	-1.2	34.8	164.0
	5	1.2D + 1.6L_r - 0.8W	1,333.0	-37.2	-165.2	-37.2	-165.2	-2.4	-1.2	-34.8	-164.0
9-4	6	1.2D + 0.5L + 0.5L_r + 1.6W	1,380.5	59.4	322.9	59.4	322.9	-10.2	-5.1	69.6	328.0
	7	1.2D + 0.5L + 0.5L_r - 1.6W	1,381.5	-79.8	-333.1	-79.8	-333.1	-10.2	-5.1	-69.6	-328.0
9-6	8	0.9D + 1.6W	978.4	67.8	327.1	67.8	327.1	-1.8	-0.9	69.6	328.0
	9	0.9D - 1.6W	979.3	-71.4	-328.9	-71.4	-328.9	-1.8	-0.9	-69.6	-328.0

   *includes live load reduction per ASCE 7

2. Determine if the frame at the first story is nonsway or sway

   The results from an elastic first-order analysis using the section properties prescribed in 10.11.1 are as follows:

   ΣP_u = total vertical load in the first story corresponding to the lateral loading case for which ΣP_u is greatest

   The total building loads are: D = 17,895 kips, L = 1,991 kips, L_r = 270 kips. The maximum ΣP_u is from Eq. (9-4):

   $\Sigma P_u = (1.2 \times 17,895) + (0.5 \times 1,991) + (0.5 \times 270) + 0 = 22,605$ kips

		Code
Example 11.2 (cont'd)	**Calculations and Discussion**	**Reference**

V_{us} = factored story shear in the first story corresponding to the wind loads

$= 1.6 \times 302.6 = 484.2$ kips   *Eq. (9-4), (9-6)*

Δ_o = first-order relative deflection between the top and bottom of the first story due to V_u
$= 1.6 \times (0.28 - 0) = 0.45$ in.

Stability index $Q = \dfrac{\Sigma P_u \Delta_o}{V_{us} \ell_c} = \dfrac{22{,}605 \times 0.45}{484.2 \times [(15 \times 12) - (20/2)]} = 0.12 > 0.05$ *Eq. (10-6)*

Since $Q > 0.05$, the frame at the first story level is considered sway.   *10.11.4.2*

3. Design of column C1

    a. Determine if slenderness effects must be considered.

    Determine k from alignment chart in R10.12.1.

    $I_{col} = 0.7 \left(\dfrac{22^4}{12} \right) = 13{,}665$ in.^{4} *10.11.1*

    $E_c = 57{,}000 \, \dfrac{\sqrt{6{,}000}}{1{,}000} = 4{,}415$ ksi   *8.5.1*

    For the column below level 2:

    $\dfrac{E_c I}{\ell_c} = \dfrac{4{,}415 \times 13{,}665}{(15 \times 12) - 10} = 355 \times 10^3$ in.-kips

    For the column above level 2:

    $\dfrac{E_c I}{\ell_c} = \dfrac{4{,}415 \times 13{,}665}{12 \times 12} = 419 \times 10^3$ in.-kips

    $I_{beam} = 0.35 \left(\dfrac{24 \times 20^3}{12} \right) = 5{,}600$ in.^{4} *10.11.1*

    For the beam: $\dfrac{E_c I}{\ell_c} = \dfrac{57\sqrt{4{,}000} \times 5{,}600}{24 \times 12} = 70 \times 10^3$ in.-kips

    $\psi_A = \dfrac{\Sigma E_c I / \ell_c}{\Sigma E_c I / \ell} = \dfrac{355 + 419}{70} = 11.1$

    Assume $\psi_B = 1.0$ (column essentially fixed at base)

				Code
Example 11.2 (cont'd)		**Calculations and Discussion**		**Reference**

From the alignment chart (Fig. R10.12.1(b)), k = 1.9.

$$\frac{k\ell_u}{r} = \frac{1.9 \times 13.33 \times 12}{0.3 \times 22} = 46 > 22$$ 10.13.2

Thus, slenderness effects must be considered.

b. Determine total moment M_2 (including slenderness effects) and the design load combinations, using the approximate analysis of 10.13.4.2.

The following table summarizes magnified moment computations for column C1 for all load combinations, followed by detailed calculations for combinations no. 4 and 5 to illustrate the procedure.

No.	Load Combination	ΣP_u (kips)	Δ_o (in)	V_{us} (kips)	Q	δ_s	M_{2ns} (ft-kips)	M_{2s} (ft-kips)	M_2 (ft-kips)
1	1.4D	25,053					48.7		48.7
2	1.2D+1.6L+0.5Lr	24,795					66.4		66.4
3	1.2D+0.5L+1.6Lr	22,903					49.5		49.5
4	1.2D+1.6Lr+0.8W	21,908	0.28	302.6	0.12	1.14	21.1	110.4	147.0
5	1.2D+1.6Lr-0.8W	21,908	0.28	302.6	0.12	1.14	21.1	-110.4	-104.8
6	1.2D+0.5L+0.5Lr+1.6W	22,605	0.28	484.2	0.08	1.08	25.0	220.8	264.2
7	1.2D+0.5L+0.5Lr-1.6W	22,605	0.45	484.2	0.12	1.14	25.0	-220.8	-226.8
8	0.9D+1.6W	16,106	0.45	484.2	0.09	1.10	15.8	220.8	257.9
9	0.9D-1.6W	16,106	0.45	484.2	0.09	1.10	15.8	-220.8	-226.2

$$M_2 = M_{2ns} + M_{2s}$$ Eq. (10-16)

$$\delta_s M_{2s} = \frac{M_{2s}}{1-Q} \geq M_{2s}$$ Eq. (10-17)

For load combinations no. 4 and 5:

$$U = 1.2D + 1.6L_r \pm 0.8W$$

$$\Sigma P_u = (1.2 \times 17,895) + (1.6 \times 270) \pm 0 = 21,906 \text{ kips}$$

$$\Delta_o = 0.8 \times (0.28 - 0) = 0.22 \text{ in.}$$

$$V_{us} = 0.8 \times 302.6 = 240.1 \text{ kips}$$

$$\ell_c = (15 \times 12) - (20/2) = 170 \text{ in.}$$

$$Q = \frac{\Sigma P_u \Delta_o}{V_{us}\ell_c} = \frac{21,906 \times 0.22}{240.1 \times 170} = 0.12$$

| Example 11.2 (cont'd) | Calculations and Discussion | Code Reference |

$$\delta = \frac{1}{1-Q} = \frac{1}{1-0.12} = 1.14$$

- For sidesway from north to south (load combination no. 4):

$$\delta_s M_{2s} = 1.14 \times 110.4 = 125.9 \text{ ft-kips}$$

$$M_2 = M_{2ns} + \delta_s M_{2s} = 21.1 + 125.9 = 147.0 \text{ ft-kips}$$

$$P_u = 722.0 \text{ kips}$$

- For sidesway from south to north (load combination no. 5):

$$M_{2s} = 0.8 \times 138.0 = 110.4 \text{ ft-kips}$$

$$M_{2su} = 1.2 \times 17.6 + 1.6 \times 0 = 21.1 \text{ ft-kips}$$

$$\delta_s M_{2s} = 1.14 \times (-110.4) = -125.9 \text{ ft-kips}$$

$$M_2 = 21.1 - 125.9 = -104.8 \text{ ft-kips}$$

$$P_u = 799.3 \text{ kips}$$

c. For comparison purposes, recompute $\delta_s M_{2s}$ using the magnified moment method outlined in 10.13.4.3

$$\delta_s M_{2s} = \frac{M_{2s}}{1 - \frac{\Sigma P_u}{0.75 \Sigma P_c}} \geq M_{2s} \qquad \text{Eq. (10-18)}$$

The critical load P_c is calculated from Eq. (10-10) using k from 10.13.1 and EI from Eq. (10-11) or (10-12). Since the reinforcement is not known as of yet, use Eq. (10-12) to determine EI.

For each of the 12 exterior columns along column lines 1 and 4 (i.e., the columns with one beam framing into them in the direction of analysis), k was determined in part 3(a) above to be 1.9.

$$EI = \frac{0.4 E_c I}{1 + \beta_d} = \frac{0.4 \times 4{,}415 \times 22^4}{12(1 + 0)} = 34.5 \times 10^6 \text{ in.}^2\text{-kips} \qquad \text{Eq. (10-12)}$$

$$P_c = \frac{\pi^2 EI}{(k\ell_u)^2} = \frac{\pi^2 \times 34.5 \times 10^6}{(1.9 \times 13.33 \times 12)^2} = 3{,}686 \text{ kips} \qquad \text{Eq. (10-10)}$$

Example 11.2 (cont'd)	Calculations and Discussion	Code Reference

For each of the exterior columns A2, A3, F2, and F3, (i.e., the columns with two beams framing into them in the direction of analysis):

$$\psi_A = \frac{355 + 419}{2 \times 70} = 5.5$$

$$\psi_B = 1.0$$

From the alignment chart, k = 1.75.

$$P_c = \frac{\pi^2 \times 34.5 \times 10^6}{(1.75 \times 13.33 \times 12)^2} = 4{,}345 \text{ kips}$$

Eq. (10-10)

For each of the 8 interior columns:

$$I_{col} = 0.7 \left(\frac{24^4}{12}\right) = 19{,}354 \text{ in.}^4$$

10.11.1

For the column below level 2:

$$\frac{E_c I}{\ell_c} = \frac{4{,}415 \times 19{,}354}{(15 \times 12) - 10} = 503 \times 10^3 \text{ in.-kips}$$

For the column above level 2:

$$\frac{E_c I}{\ell_c} = \frac{4{,}415 \times 19{,}354}{12 \times 12} = 593 \times 10^3 \text{ in.-kips}$$

$$\psi_A = \frac{503 + 593}{2 \times 70} = 7.8$$

$$\psi_B = 1.0$$

From the alignment chart, k = 1.82.

$$EI = 0.4 \times 4{,}415 \times \frac{24^2}{12} = 48.8 \times 10^6 \text{ in.-kips}$$

Eq. (10-12)

$$P_c = \frac{\pi^2 EI}{(k\ell_u)^2} = \frac{\pi^2 \times 48.8 \times 10^6}{(1.82 \times 13.33 \times 12)^2} = 5{,}683 \text{ kips}$$

		Code
Example 11.2 (cont'd)	Calculations and Discussion	Reference

Therefore,

$$\Sigma P_c = 12(3,686) + 4(4,345) + 8(5,683) = 107,076 \text{ kips}$$

The following table summarizes magnified moment computations for column C1 using 10.13.4.3 for all load conditions. The table is followed by detailed calculations for combinations no. 4 and 5 to illustrate the procedure.

No.	Load Combination	ΣP_u (kips)	δ_s (in.)	M_{2ns} (ft-kips)	M_{2s} (ft-kips)	M_2 (ft-kips)
1	1.4D	25,053	---	48.7	---	48.7
2	1.2D + 1.6L + 1.6L$_r$	24,795	---	66.4	---	66.4
3	1.2D + 0.5L + 1.6L$_r$	22,903	---	49.5	---	49.5
4	1.2D + 1.6L$_r$ + 0.8W	21,908	1.38	21.1	110.4	173.5
5	1.2D + 1.6L$_r$ - 0.8W	21,908	1.38	21.1	-110.4	-131.3
6	1.2D + 0.5L + 0.5L$_r$ + 1.6W	22,605	1.39	25.0	220.8	331.9
7	1.2D + 0.5L + 0.5L$_r$ - 1.6W	22,605	1.39	25.0	-220.8	-281.9
8	0.9D + 1.6W	16,106	1.25	15.8	220.8	292.0
9	0.9D - 1.6W	16,106	1.25	15.8	-220.8	-260.3

For load combinations No. 4 and 5:

$$U = 1.2D + 1.6L_r \pm 0.8W$$

$$\delta_s = \frac{1}{1 - \frac{\Sigma P_u}{0.75 \Sigma P_c}} = \frac{1}{1 - \frac{21,908}{0.75 \times 107,076}} = 1.38$$

- For sidesway from north to south (load combination no. 4):

    $\delta_s M_{2s} = 1.38 \times 110.4 = 152.4$ ft-kips

    $M_2 = 21.1 + 152.4 = 173.5$ ft-kips

    $P_u = 722.0$ kips

- For sidesway from south to north (load combination no. 5):

    $\delta_s M_{2s} = 1.38 \times (-110.4) = -152.4$ ft-kips

    $M_2 = 21.1 - 152.4 = -131.3$ ft-kips

    $P_u = 799.3$ kips

Example 11.2 (cont'd) — Calculations and Discussion | Code Reference

A summary of the magnified moments for column C1 for all load combinations is provided in the following table.

No.	Load Combination	P_u (kips)	10.13.4.2 δ_s	10.13.4.2 M_2 (ft-kips)	10.13.4.3 δ_s	10.13.4.3 M_2 (ft-kips)
1	1.4D	871.4	---	48.7	---	48.7
2	1.2D + 1.6L + 0.5L$_r$	869.4	---	66.4	---	66.4
3	1.2D + 0.5L + 1.6L$_r$	797.6	---	49.5	---	49.5
4	1.2D + 1.6L$_r$ + 0.8W	722.0	1.14	147.0	1.38	173.5
5	1.2D + 1.6L$_r$ - 0.8W	799.3	1.14	-104.8	1.38	-131.3
6	1.2D + 0.5L + 0.5L$_r$ + 1.6W	710.9	1.14	276.7	1.39	331.9
7	1.2D + 0.5L + 0.5L$_r$ - 1.6W	865.4	1.14	-226.8	1.39	-281.9
8	0.9D + 1.6W	482.9	1.10	257.9	1.25	292.0
9	0.9D - 1.6W	637.4	1.10	-226.2	1.25	-260.3

d. Determine required reinforcement.

For the 22 × 22 in. column, try 8-No. 8 bars. Determine maximum allowable axial compressive force, $\phi P_{n,max}$:

$$\phi P_{n,max} = 0.80\phi\left[0.85 f'_c \left(A_g - A_{st}\right) + f_y A_{st}\right]$$ Eq. (10-2)

$$= (0.80 \times 0.65)[(0.85 \times 6)(22^2 - 6.32) + (60 \times 6.32)]$$

$$= 1{,}464.0 \text{ kips} > \text{maximum } P_u = 871.4 \text{ kips} \quad \text{O.K.}$$

The following table contains results from a strain compatibility analysis, where compressive strains are taken as positive (see Parts 6 and 7). Use $M_u = M_2$ from the approximate $P - \Delta$ method in 10.13.4.2.

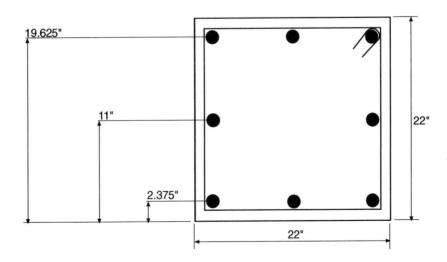

11-39

			Code
Example 11.2 (cont'd)		Calculations and Discussion	Reference

No.	P_u (kips)	M_u (ft-kips)	c (in.)	ε_t	ϕ	ϕP_n (kips)	ϕM_n (ft-kips)
1	871.4	48.7	14.85	-0.00096	0.65	871.4	459.4
2	869.4	66.4	14.82	-0.00097	0.65	869.4	459.7
3	797.6	49.5	13.75	-0.00128	0.65	797.6	468.2
4	722.0	147.0	12.75	-0.00162	0.65	722.0	474.1
5	799.3	-104.8	13.78	-0.00127	0.65	799.3	468.0
6	710.9	276.7	12.61	-0.00167	0.65	710.9	474.8
7	865.4	-226.8	14.76	-0.00099	0.65	865.4	460.2
8	482.9	257.9	7.36	-0.00500	0.90	482.9	557.2
9	637.4	-226.2	11.68	-0.00204	0.65	637.4	478.8

Therefore, since $\phi M_n > M_u$ for all $\phi P_n = P_u$, use a 22 × 22 in. column with 8-No. 8 bars ($\rho_g = 1.3\%$). The same reinforcement is also adequate for the load combinations from the magnified moment method of 10.13.4.3.

4. Design of column C2

   a. Determine if slenderness effects must be considered.

   In part 3(c), k was determined to be 1.82 for the interior columns. Therefore,

   $$\frac{k\ell_u}{r} = \frac{1.82 \times 13.33 \times 12}{0.3 \times 24} = 40.4 > 22$$ 10.13.2

   Slenderness effects must be considered.

   b. Determine total moment M_2 (including slenderness effects) and the design load combinations, using the approximate analysis of 10.13.4.2.

   The following table summarizes magnified moment computation for column C2 for all load combinations, followed by detailed calculations for combinations no. 4 and 5 to illustrate the procedure.

No.	Load Combination	ΣP_u (kips)	Δ_o (in)	V_{us} (kips)	Q	δ_s	M_{2ns} (ft-kips)	M_{2s} (ft-kips)	M_2 (ft-kips)
1	1.4D	25,053	-	-	-	-	2.8	-	2.8
2	1.2D+1.6L+0.5Lr	24,795	-	-	-	-	27.4	-	27.4
3	1.2D+0.5L+1.6Lr	22,903	-	-	-	-	10.2	-	10.2
4	1.2D+1.6Lr+0.8W	21,908	0.28	302.6	0.12	1.14	-1.2	164.0	185.0
5	1.2D+1.6Lr-0.8W	21,908	0.28	302.6	0.12	1.14	-1.2	-164.0	-187.4
6	1.2D+0.5L+0.5Lr+1.6W	22,605	0.45	484.2	0.12	1.14	-5.1	328.0	368.9
7	1.2D+0.5L+0.5Lr-1.6W	22,605	0.45	484.2	0.12	1.14	-5.1	-328.0	-379.1
8	0.9D+1.6W	16,106	0.45	484.2	0.09	1.10	-0.9	328.0	358.6
9	0.9D-1.6W	16,106	0.45	484.2	0.09	1.10	-0.9	-328.0	-360.4

$M_2 = M_{2ns} + M_{2s}$ Eq. (10-16)

$\delta_s M_{2s} = \dfrac{M_{2s}}{1-Q} \geq M_{2s}$ Eq. (10-17)

				Code
Example 11.2 (cont'd)		**Calculations and Discussion**		**Reference**

For load combinations no. 4 and 5:

$$U = 1.2D + 1.6L_r \pm 0.8W$$

From part 3(b), δ_s was determined to be 1.14.

- For sidesway from north to south (load combination no. 4):

$$M_{2s} = 0.8 \times 205.0 = 164.0 \text{ ft-kips}$$

$$M_{2ns} = 1.2(-1.0) + 1.6 \times 0 = 1.2 \text{ ft-kips}$$

$\delta_s M_{2s} = 1.14 \times 164 = 187.0$ ft-kips

$M_2 = M_{2ns} + \delta_s M_{2s} = -1.2 + 187.0 = 185.8$ ft-kips

$P_u = 1{,}332.6$ kips

- For sidesway from south to north (load combination no. 5):

$\delta_s M_{2s} = 1.14 \times (-164) = -187.0$ ft-kips

$M_2 = -1.2 - 187.0 = -188.2$ ft-kips

$P_u = 1{,}333.0$ kips

c. For comparison purposes, recompute $\delta_s M_{2s}$ using the magnified moment method outlined in 10.13.4.3. Use the values of ΣP_u, ΣP_c, and δ_s computed in part 3(c).

No.	Load Combination	ΣP_u (kips)	δ_s (in.)	M_{2ns} (ft-kips)	M_{2s} (ft-kips)	M_2 (ft-kips)
1	1.4D	25,053	---	-2.8	---	-2.8
2	1.2D + 1.6L + 0.5L_r	24,795	---	-27.4	---	-27.4
3	1.2D + 0.5L + 1.6L_r	22,903	---	-10.2	---	-10.2
4	1.2D + 1.6L_r + 0.8W	21,908	1.38	-1.2	164.0	225.1
5	1.2D + 1.6L_r - 0.8W	21,908	1.38	-1.2	-164.0	-227.5
6	1.2D + 0.5L + 0.5L_r + 1.6W	22,605	1.39	-5.1	328.0	451.4
7	1.2D + 0.5L + 0.5L_r - 1.6W	22,605	1.39	-5.1	-328.0	-461.6
8	0.9D + 1.6W	16,106	1.25	-0.9	328.0	409.4
9	0.9D - 1.6W	16,106	1.25	-0.9	-328.0	-411.2

$$U = 1.2D + 1.6L_r \pm 0.8W$$

$\delta_s = 1.38$ from part 3(c)

- For sidesway from north to south (load combination no. 4):

11-41

		Code Reference
Example 11.2 (cont'd)	**Calculations and Discussion**	

$\delta_s M_{2s} = 1.38 \times 164.0 = 226.3$ ft-kips

$M_2 = -1.2 + 226.3 = 225.1$ ft-kips

$P_u = 1,332.6$ kips

- For sidesway from south to north (load combination no. 5):

$\delta_s M_{2s} = 1.38 \times (-164.0) = -226.3$ ft-kips

$M_2 = -1.2 - 226.3 = -227.5$ ft-kips

$P_u = 1,333.0$ kips

A summary of the magnified moments for column C2 under all load combinations is provided in the following table.

No.	Load Combination	P_u (kips)	10.13.4.2		10.13.4.3	
			δ_s	M_2 (ft-kips)	δ_s	M_2 (ft-kips)
1	1.4D	1,522.6	---	-2.8	---	-2.8
2	1.2D + 1.6L + 0.5L$_r$	1,529.0	---	-27.4	---	-27.4
3	1.2D + 0.5L + 1.6L$_r$	1,400.1	---	-10.2	---	-10.2
4	1.2D + 1.6L$_r$ + 0.8W	1,332.6	1.14	185.8	1.38	225.1
5	1.2D + 1.6L$_r$ - 0.8W	1,333.0	1.14	-188.2	1.38	-227.5
6	1.2D + 0.5L + 0.5L$_r$ + 1.6W	1,380.5	1.14	368.8	1.39	451.4
7	1.2D + 0.5L + 0.5L$_r$ - 1.6W	1,381.5	1.14	-379.0	1.39	-461.6
8	0.9D + 1.6W	978.4	1.10	358.6	1.25	409.4
9	0.9D - 1.6W	979.3	1.10	-360.4	1.25	-411.2

d. Determine required reinforcement.

For the 24×24 in. column, try 8-No. 8 bars. Determine maximum allowable axial compressive force, $\phi P_{n,max}$:

$\phi P_{n,max} = 0.80\phi \left[0.85 f'_c \left(A_g - A_{st} \right) + f_y A_{st} \right]$ Eq. (10-2)

$= (0.80 \times 0.65)[(0.85 \times 6)(24^2 - 6.32) + (60 \times 6.32)]$

$= 1,708$ kips $>$ maximum $P_u = 1,529.0$ kips O.K.

The following table contains results from a strain compatibility analysis, where compressive strains are taken as positive (see Parts 6 and 7). Use $M_u = M_2$ from the approximate P−Δ method in 10.13.4.2.

			Code
Example 11.2 (cont'd)		**Calculations and Discussion**	**Reference**

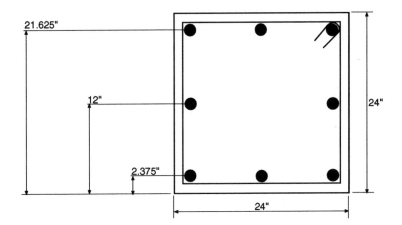

No.	P_u (kips)	M_u (ft-kips)	c (in.)	ε_t	ϕ	ϕP_n (kips)	ϕM_n (ft-kips)
1	1,522.6	-2.8	23.30	0.00022	0.65	1,522.6	438.1
2	1,529.0	-27.4	23.39	0.00023	0.65	1,529.0	435.3
3	1,400.1	-10.2	21.49	-0.00002	0.65	1,400.1	489.7
4	1,332.6	185.8	20.50	-0.00016	0.65	1,332.6	513.3
5	1,333.0	-188.2	20.51	-0.00016	0.65	1,333.0	513.1
6	1,380.5	368.8	21.20	-0.00006	0.65	1,380.5	496.9
7	1,381.5	-379.0	21.22	-0.00005	0.65	1,381.5	496.4
8	978.4	358.6	15.52	-0.00118	0.65	978.4	587.1
9	979.3	-360.4	15.46	-0.00120	0.65	979.3	587.5

Therefore, since $\phi M_n > M_u$ for all $\phi P_n = P_u$, use a 24 × 24 in. column with 8-No. 8 bars ($\rho_g = 1.1\%$). The same reinforcement is also adequate for the load combinations from the magnified moment method of 10.13.4.3. Figure 11-16 obtained from pcaColumn[11.2] shows the design strength diagram for Column C2.

5. Determine if the maximum moment occurs at the end of the member. *10.13.5*

   For column C1:

   $$\frac{\ell_u}{r} = \frac{13.33 \times 12}{0.3 \times 22} = 24.2 < \frac{35}{\sqrt{\frac{871.4}{6 \times 22^2}}} = 63.9 \qquad \text{Eq. (10-19)}$$

   For column C2:

   $$\frac{\ell_u}{r} = \frac{13.33 \times 12}{0.3 \times 24} = 22.2 < \frac{35}{\sqrt{\frac{1,529.0}{6 \times 24^2}}} = 52.6$$

| Example 11.2 (cont'd) | Calculations and Discussion | Code Reference |

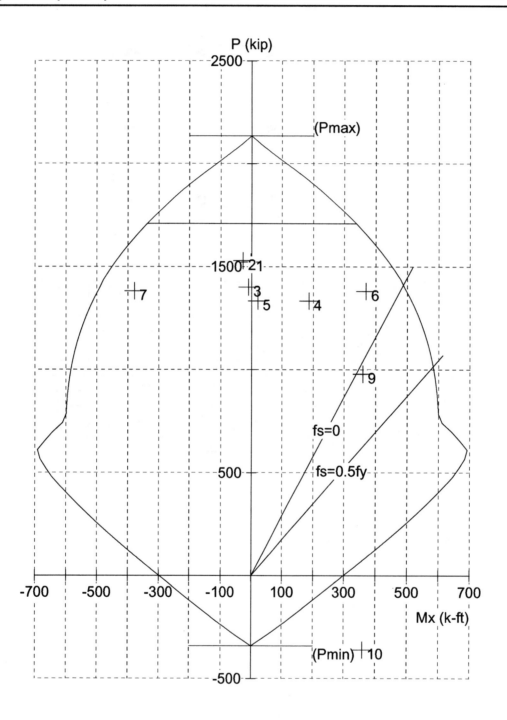

Figure 11-16 Design Strength Interaction Diagram for Column C2

Example 11.2 (cont'd)	Calculations and Discussion	Code Reference

Therefore, for columns C1 and C2, the maximum moment occurs at one of the ends, and the total moment M_2 does not have to be further magnified by δ_{ns}.

6. Check for sidesway instability of the structure.   10.13.6

   a. When using 10.13.4.2 to compute $\delta_s M_s$, the value of Q evaluated using factored gravity loads shall not exceed 0.60. Note that for stability checks, all moments of inertia must be divided by $(1 + \beta_d)$ 10.11.1 where, for this story

   $$\beta_d = \frac{\text{Maximum factored sustained axial load}}{\text{Maximum factored axial load}}$$

   $$= \frac{1.4 P_D}{P_u} = \frac{1.4 \times 17{,}895}{1.4 \times 17{,}895} = 1.0$$

   $$1 + \beta_d = 2.0$$

   Dividing all of the moments of inertia by $(1 + \beta_d)$ is equivalent to increasing the deflections, and consequently Q, by $(1 + \beta_d)$. Thus, at the second floor level,

   $$Q = 2 \times 0.12 = 0.24 < 0.60$$

   Therefore, the structure is stable at this level. In fact, computing the modified Q at each floor level shows that the entire structure is stable.

   b. When using 10.13.4.3 to compute $\delta_s M_s$, the value of δ_s computed using ΣP_u and ΣP_c corresponding to the factored dead and live loads shall be positive and shall not exceed 2.5. For the stability check, the values of EI must be divided by $(1 + \beta_d)$. Thus, the values of P_c must be recomputed considering the effects of β_d.

   $$\Sigma P_c = \frac{107{,}076}{1 + 1} = 53{,}532 \text{ kips}$$

   and $\delta_s = \dfrac{1}{1 - \dfrac{(1.4 \times 17{,}895) + (1.7 \times 2{,}261)}{0.75 \times 53{,}532}} = 2.66 > 2.5$

   The structure is unstable when the magnified moment method of 10.13.4.3 is used.

Blank

12

Shear

UPDATE FOR THE '05 CODE

In the 2005 edition of the code, changes regarding shear are mostly editorial and notation related. Also, 11.5.3 and 17.5.2 have been inserted to clarify the definition of d for prestressed members.

GENERAL CONSIDERATIONS

The relatively abrupt nature of a failure in shear, as compared to a ductile flexural failure, makes it desirable to design members so that strength in shear is relatively equal to, or greater than, strength in flexure. To ensure that a ductile flexural failure precedes a shear failure, the code (1) limits the minimum and maximum amount of longitudinal reinforcement and (2) requires a minimum amount of shear reinforcement in all flexural members if the factored shear force V_u exceeds one-half of the shear strength provided by the concrete, ($V_u > 0.5\phi V_c$), except for certain types of construction (11.5.5.1), (3) specifies a lower strength reduction for shear ($\phi = 0.75$) than for tension-controlled section under flexure ($\phi = 0.90$).

The determination of the amount of shear reinforcement is based on a modified form of the truss analogy. The truss analogy assumes that shear reinforcement resists the total transverse shear. Considerable research has indicated that shear strength provided by concrete V_c can be assumed equal to the shear causing inclined cracking; therefore, shear reinforcement need be designed to carry only the excess shear.

Only shear design for nonprestressed members with clear-span-to-overall-depth ratios greater than 4 is considered in Part 12. Also included is horizontal shear design in composite concrete flexural members, which is covered separately in the second half of Part 12. Shear design for deep flexural members, which have clear-span-to-overall-depth ratios less than 4, is presented in Part 17. Shear design of prestressed members is discussed in Part 25. The alternate shear design method of Appendix A, Strut-and-Tie Models, is discussed in Part 32.

11.1 SHEAR STRENGTH

Design provisions for shear are presented in terms of shear forces (rather than stresses) to be compatible with the other design conditions for the strength design method, which are expressed in terms of loads, moments, and forces.

Accordingly, shear is expressed in terms of the factored shear force V_u, using the basic shear strength requirement:
Design shear strength ≥ Required shear strength

$$\phi V_n \geq V_u \qquad \text{Eq. (11-1)}$$

The nominal shear strength V_n is computed by:

$$V_n = V_c + V_s \qquad \text{Eq. (11-2)}$$

where V_c is the nominal shear strength provided by concrete and V_s is the nominal shear strength provided by shear reinforcement.

Equation (11-2) can be substituted into Eq. (11-1) to obtain:

$$\phi V_c + \phi V_s \geq V_u$$

The required shear strength at any section is computed using Eqs. (11-1) and (11-2), where the factored shear force V_u is obtained by applying the load factors specified in 9.2. The strength reduction factor, $\phi = 0.75$, is specified in 9.3.2.3.

11.1.1.1 Web Openings

Often it is necessary to modify structural components of buildings to accommodate necessary mechanical and electrical service systems. Passing these services through openings in the webs of floor beams within the floor-ceiling sandwich eliminates a significant amount of dead space and results in a more economical design. However, the effect of the openings on the shear strength of the floor beams must be considered, especially when such openings are located in regions of high shear near supports. In 11.1.1.1, the code requires the designer to consider the effect of openings on the shear strength of members. Because of the many variables such as opening shape, size, and location along the span, specific design rules are not stated. However, references are given for design guidance in R11.1.1.1. Generally, it is desirable to provide additional vertical stirrups adjacent to both sides of a web opening, except for small isolated openings. The additional shear reinforcement can be proportioned to carry the total shear force at the section where an opening is located. Example 12.5 illustrates application of a design method recommended in Ref. 12.1.

11.1.2 Limit on $\sqrt{f'_c}$

Concrete shear strength equations presented in Chapter 11 of the Code are a function of $\sqrt{f'_c}$, and had been verified experimentally for members with concrete compressive strength up to 10,000 psi. Due to a lack of test data for members with $f'_c > 10,000$ psi, 11.1.2 limits the value of $\sqrt{f'_c}$ to 100 psi, except as allowed in 11.1.2.1.

Section 11.1.2 does not prohibit the use of concrete with $f'_c > 10,000$ psi; it merely directs the engineer not to count on any strength in excess of 10,000 psi when computing V_c, unless minimum shear reinforcement is provided in accordance with 11.1.2.1.

It should be noted that prior to the 2002 Code, minimum area of transverse reinforcement was independent of the concrete strength. However, tests indicated that an increase in the minimum amount of transverse reinforcement is required for members with high-strength concrete to prevent sudden shear failures when inclined cracking occurs. Thus, to account for this, minimum transverse reinforcement requirements are a function of $\sqrt{f'_c}$.

11.1.3 Computation of Maximum Factored Shear Force

Section 11.1.3 describes three conditions that shall be satisfied in order to compute the maximum factored shear force V_u in accordance with 11.1.3.1 for nonprestressed members:

1. Support reaction, in direction of applied shear force, introduces compression into the end regions of the member.
2. Loads are applied at or near the top of the member.
3. No concentrated load occurs between the face of the support and the location of the critical section, which is a distance d from the face of the support (11.1.3.1).

When the conditions of 11.1.3 are satisfied, sections along the length of the member located less than a distance d from the face of the support are permitted to be designed for the shear force V_u computed at a distance d from the face of the support. See Fig. 12-1 (a), (b), and (c) for examples of support conditions where 11.1.3 would be applicable.

Conditions where 11.1.3 cannot be applied include: (1) members framing into a supporting member in tension, see Fig. 12-1 (d); (2) members loaded near the bottom, see Fig. 12-1 (e); and (3) members subjected to an abrupt change in shear force between the face of the support and a distance d from the face of the support, see Fig. 12-1 (f). In all of these cases, the critical section for shear must be taken at the face of the support. Additionally, in the case of Fig. 12-1 (d), the shear within the connection must be investigated and special corner reinforcement should be provided.

One other support condition is noteworthy. For brackets and corbels, the shear at the face of the support V_u must be considered, as shown in Fig. 12-2. However, these elements are more appropriately designed for shear using the shear-friction provisions of 11.7. See Part 15 for design of brackets and corbels.

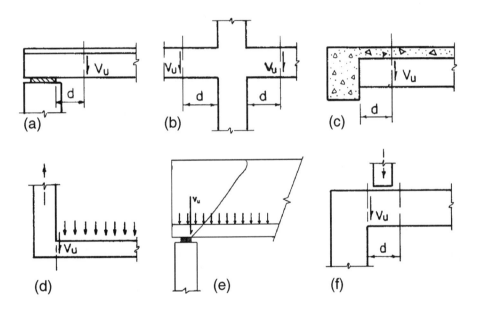

Figure 12-1 Typical Support Conditions for Locating Factored Shear Force V_u

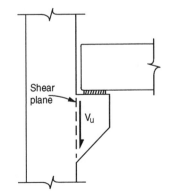

Figure 12-2 Critical Shear Plane for Brackets and Corbels

11.2 LIGHTWEIGHT CONCRETE

Since the shear strength of lightweight aggregate concrete may be less than that of normal weight concrete with equal compressive strength, adjustments in the value of V_c, as computed for normal weight concrete, are necessary.

Except for 11.5.5.3, 11.5.7.9, 11.6.3.1, 11.12.3.2, and 11.12.4.8, when average splitting tensile strength f_{ct} is specified, $f_{ct}/6.7$ is substituted for $\sqrt{f'_c}$ in all equations of Chapter 11. However, the value of $f_{ct}/6.7$ cannot be taken greater than $\sqrt{f'_c}$. When f_{ct} is not specified, $\sqrt{f'_c}$ is reduced using a multiplier of 0.75 for all-lightweight concrete or 0.85 for sand-lightweight concrete. Linear interpolation between these multipliers is allowed when partial sand replacement is used. Section 11.7.4.3 specifies the same multipliers for lightweight concrete.

11.3 SHEAR STRENGTH PROVIDED BY CONCRETE FOR NONPRESTRESSED MEMBERS

When computing the shear strength provided by concrete for members subject to shear and flexure only, designers have the option of using either the simplified equation, $V_c = 2\sqrt{f'_c}\,b_w d$ [Eq. (11-3)], or the more elaborate expression given by Eq. (11-5). In computing V_c from Eq. (11-5), it should be noted that V_u and M_u are the values which occur simultaneously at the section considered. A maximum value of 1.0 is prescribed for the ratio $V_u d/M_u$ to limit V_c near points of inflection where M_u is zero or very small.

For members subject to shear and flexure with axial compression, a simplified V_c expression is given in 11.3.1.2, with an optional more elaborate expression for V_c available in 11.3.2.2. For members subject to shear, flexure and significant axial tension, 11.3.1.3 requires that shear reinforcement must be provided to resist the total shear unless the more detailed analysis of 11.3.2.3 is performed. Note that N_u represents a tension force in Eq. (11-8) and is therefore taken to be negative.

No precise definition is given for "significant axial tension." If there is uncertainty about the magnitude of axial tension, it may be desirable to carry all applied shear by shear reinforcement.

Figure 12-3 shows the variation of shear strength provided by concrete, V_c as function of $\sqrt{f'_c}$, $V_u d/M_u$, and reinforcement ratio ρ_w.

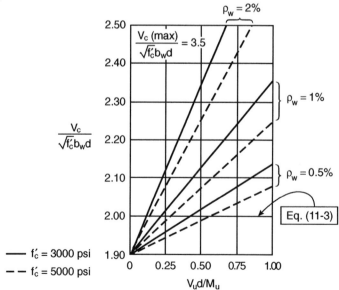

Figure 12-3 Variation of $V_c/\sqrt{f'_c}\,b_w d$ with f'_c, ρ_w, and $V_u d/M_u$ using Eq. (11-5)

Figure 12-4 shows the approximate range of values of V_c for sections under axial compression, as obtained from Eqs. (11-5) and (11-6). Values correspond to a 6×12 in. beam section with an effective depth of 10.8 in. The curves corresponding to the alternate expressions for V_c given by Eqs. (11-4) and (11-7), as well as that corresponding to Eq. (11-8) for members subject to axial tension, are also indicated.

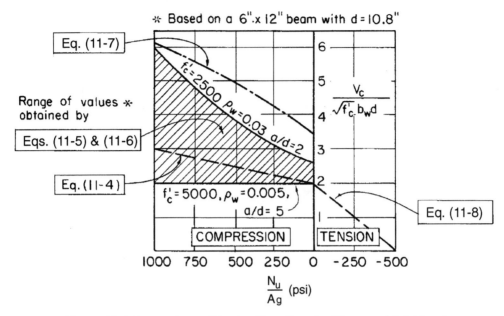

Figure 12-4 Comparison of Design Equations for Shear and Axial Load

Figure 12-5 shows the variation of V_c with N_u/A_g and f'_c for sections subject to axial compression, based on Eq. (11-4). For the range of N_u/A_g values shown, V_c varies from about 49% to 57% of the value of V_c as defined by Eq. (11-7).

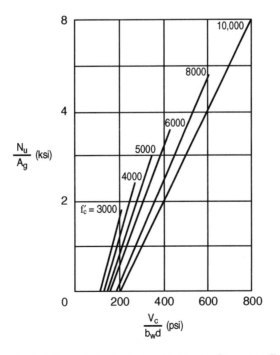

Figure 12-5 Variation of V_c/b_wd with f'_c and N_u/A_g using Eq. (11-4)

11.5 SHEAR STRENGTH PROVIDED BY SHEAR REINFORCEMENT

11.5.1 Types of Shear Reinforcement

Several types and arrangements of shear reinforcement permitted by 11.5.1.1 and 11.5.1.2 are illustrated in Fig. 12-6. Spirals, circular ties, or hoops are explicitly recognized as types of shear reinforcement starting with the 1999 code. Vertical stirrups are the most common type of shear reinforcement. Inclined stirrups and longitudinal bent bars are rarely used as they require special care during placement in the field.

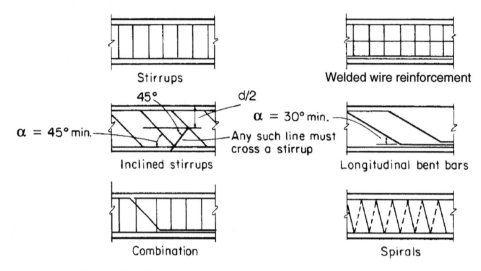

Figure 12-6 Types and Arrangements of Shear Reinforcement

11.5.4 Anchorage Details for Shear Reinforcement

To be fully effective, shear reinforcement must extend as close to full member depth as cover requirements and proximity of other reinforcement permit (12.13.1), and be anchored at both ends to develop the design yield strength of the shear reinforcement. The anchorage details prescribed in 12.13 are presumed to satisfy this development requirement.

11.5.5 Spacing Limits for Shear Reinforcement

Spacing of stirrups and welded wire reinforcement, placed perpendicular to axis of member, must not exceed one-half the effective depth of the member (d/2), nor 24 in. When the quantity $\phi V_s = (V_u - \phi V_c)$ exceeds $\phi 4\sqrt{f'_c}\, b_w d$, maximum spacing must be reduced by one-half to (d/4) or 12 in. Note also that the value of (ϕV_s) shall not exceed $\phi 8\sqrt{f'_c}\, b_w d$ (11.5.7.9). For situations where the required shear strength exceeds this limit, the member size or the strength of the concrete may be increased to provide additional shear strength provided by concrete.

11.5.6 Minimum Shear Reinforcement

When the factored shear force V_u exceeds one-half the shear strength provided by concrete ($V_u > \phi V_c/2$), a minimum amount of shear reinforcement must be provided in concrete flexural members, except for slabs and footings, joists defined by 8.11, and wide, shallow beams (11.5.5.1). When required, the minimum shear reinforcement for nonprestressed members is

$$A_{v,min} = 0.75\, \sqrt{f'_c}\, \frac{b_w s}{f_{yt}} \qquad \text{Eq. (11-13)}$$

but not less than $\dfrac{50 b_w s}{f_{yt}}$

Minimum shear reinforcement is a function of the concrete compressive strength starting with the 2002 Code. Equation (11-13) provides a gradual increase in the minimum required $A_{v,min}$, while maintaining the previous minimum value of $50 b_w s / f_{yt}$.

Note that spacing of minimum shear reinforcement must not exceed $d/2$ or 24 in.

11.5.7 Design of Shear Reinforcement

When the factored shear force V_u exceeds the shear strength provided by concrete, ϕV_c, shear reinforcement must be provided to carry the excess shear. The code provides an equation that defines the required shear strength V_s provided by reinforcement in terms of its area A_v, yield strength f_{yt}, and spacing s. [Eq. (11-15)]. The equation is based on a truss model with the inclination angle of compression diagonals equal to 45 degree.

To assure correct application of the strength reduction factor, ϕ, equations for directly computing required shear reinforcement A_v are developed below. For shear reinforcement placed perpendicular to the member axis, the following method may be used to determine the required area of shear reinforcement A_v, spaced at a distance s:

$$\phi V_n \geq V_u \qquad \text{Eq. (11-1)}$$

where

$$V_n = V_c + V_s \qquad \text{Eq. (11-2)}$$

and

$$V_s = \dfrac{A_v f_{yt} d}{s} \qquad \text{Eq. (11-15)}$$

Substituting V_s into Eq. (11-2) and V_n into Eq. (11-1), the following equation is obtained:

$$\phi V_c + \dfrac{\phi A_v f_{yt} d}{s} \geq V_u$$

Solving for A_v,

$$A_v = \dfrac{(V_u - \phi V_c) s}{\phi f_{yt} d}$$

Similarly, when inclined stirrups are used as shear reinforcement,

$$A_v = \dfrac{(V_u - \phi V_c) s}{\phi f_{yt} (\sin\alpha + \cos\alpha) d}$$

where α is the angle between the inclined stirrup and longitudinal axis of member (see Fig. 12-8).

When shear reinforcement consists of a single bar or group of parallel bars, all bent-up at the same distance from the support,

$$A_v = \dfrac{(V_u - \phi V_c)}{f_y \sin\alpha}$$

where α is the angle between the bent-up portion and longitudinal axis of member, but not less than 30 degree (see Fig. 12-6). For this case, the quantity $(V_u - \phi V_c)$ must not exceed $\phi 3\sqrt{f'_c} b_w d$.

Design Procedure for Shear Reinforcement

Design of a nonprestressed concrete beam for shear involves the following steps:

1. Determine maximum factored shear force V_u at critical sections of the member per 11.1.3 (see Fig. 12-1).
2. Determine shear strength provided by the concrete ϕV_c per Eq. (11-3): $\phi V_c = \phi 2\sqrt{f'_c} b_w d$
   where $\phi = 0.75$ (9.3.2.3).
3. Compute $V_u - \phi V_c$ at the critical section. If $V_u - \phi V_c > \phi 8\sqrt{f'_c} b_w d$, increase the size of the section or the concrete compressive strength.
4. Compute the distance from the support beyond which minimum shear reinforcement is required (i.e., where $(V_u = \phi V_c)$, and the distance from the support beyond which the concrete can carry the total shear force (i.e., where $V_u = \phi V_c/2$).
5. Use Table 12-1 to determine the required area of vertical stirrups A_v or stirrup space s at a few controlling sections along the length of the member, which includes the critical sections.

Where stirrups are required, it is usually more expedient to select a bar size and type (e.g., No. 3 U-stirrups (2 legs)) and determine the required spacing. Larger stirrup sizes at wider spacings are usually more cost effective than smaller stirrup sizes at closer spacings because the latter requires disproportionately high costs for fabrication and placement. Changing the stirrup spacing as few times as possible over the required length also results in cost savings. If possible, no more than three different stirrup spacings should be specified, with the first stirrup located 2 in. from the face of the support.

Table 12-1 Provisions for Shear Design

		$V_u \leq \phi V_c/2$	$\phi V_c/2 < V_u \leq \phi V_c$	$\phi V_c < V_u$
Required area of stirrups, A_v		none	$0.75\sqrt{f'_c}\dfrac{b_w s}{f_{yt}} \geq \dfrac{50 b_w s}{f_{yt}}$	$\dfrac{(V_u - \phi V_c)s}{\phi f_{yt}}$
Stirrup spacing, s	Required	—	$\dfrac{A_v f_{yt}}{0.75\sqrt{f'_c}\, b_w} \leq \dfrac{A_v f_{yt}}{50 b_w}$	$\dfrac{\phi A_v f_{yt}}{V_u - \phi V_c}$
	Maximum	—	$d/2 \leq 24$ in.	$d/2 \leq 24$ in. for $(V_u - \phi V_c) \leq \phi 4\sqrt{f'_c} b_w d$ $d/2 \leq 12$ in. for $\phi 4\sqrt{f'_c} b_w d < (V_u - \phi V_c) \leq \phi 8\sqrt{f'_c} b_w d$

The shear strength requirements are illustrated in Fig. 12-7.

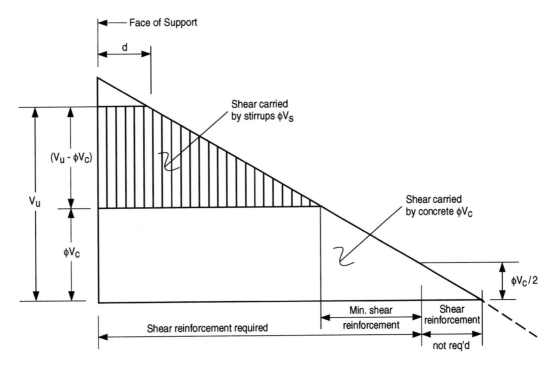

Figure 12-7 Shear Strength Requirements

The expression for shear strength provided by shear reinforcement ϕV_s can be assigned specific force values for a given stirrup size and strength of reinforcement. The selection and spacing of stirrups can be simplified if the spacing is expressed as a function of the effective depth d instead of numerical values. Practical limits of stirrup spacing generally vary from $s = d/2$ to $s = d/4$, since spacing closer than $d/4$ is not economical. With one intermediate spacing at $d/3$, a specific value of ϕV_s can be derived for each stirrup size and spacing as follows:

For vertical stirrups:

$$\phi V_s = \frac{\phi A_v f_{yt} d}{s} \qquad \text{Eq. (11-7)}$$

Substituting d/n for s, where $n = 2, 3$, and 4

$$\phi V_s = \phi A_v f_{yt} n$$

Thus, for No. 3 U-stirrups @ $s = d/2$, $f_{yt} = 60$ ksi and $\phi = 0.75$

$$\phi V_s = 0.75 (2 \times 0.11) 60 \times 2 = 19.8 \text{ kips, say 19 kips}$$

Values of ϕV_s given in Table 12-2 may be used to select shear reinforcement. Note that the ϕV_s values are independent of member size and concrete strength. Selection and spacing of stirrups using the design values for $\phi V_s = (V_u - \phi V_c)$ can be easily solved by numerical calculation or graphically. See Example 12.1.

Table 12-2 Shear Strength ϕV_s for Given Bar Sizes and Spacings

Spacing	Shear Strength ϕV_s (kips)					
	No. 3 U-Stirrups*		No. 4 U-Stirrups*		No. 5 U-Stirrups*	
	Grade 40	Grade 60	Grade 40	Grade 60	Grade 40	Grade 60
d/2	13	19	24	36	37	55
d/3	19	29	36	54	55	83
d/4	26	39	48	72	74	111

* Stirrups with 2 legs (double values for 4 legs, etc.)

CHAPTER 17 — COMPOSITE CONCRETE FLEXURAL MEMBERS

17.4 VERTICAL SHEAR STRENGTH

Section 17.4.1 of the Code permits the use of the entire composite flexural member to resist the design vertical shear as if the member were monolithically cast. Therefore, the requirements of Code Chapter 11 apply.

Section 17.4.3 permits the use of vertical shear reinforcement to serve as ties for horizontal shear reinforcement, provided that the vertical shear reinforcement is extended and anchored in accordance with applicable provisions.

17.5 HORIZONTAL SHEAR STRENGTH

In composite flexural members, horizontal shear forces are caused by the moment gradient resulting from vertical shear force. These horizontal shear forces act over the interface of interconnected elements that form the composite member.

Section 17.5.1 requires full transfer of the horizontal shear forces by friction at the contact surface, properly anchored ties, or both. Unless calculated in accordance with 17.5.4, the factored applied horizontal shear force $V_u \leq \phi V_{nh}$, where ϕV_{nh} is the horizontal shear strength (17.5.3).

The horizontal shear strength is $\phi V_{nh} = 80 b_v d$ for intentionally roughened contact surfaces without the use of ties (friction only), and for surfaces that are not intentionally roughened with the use of minimum ties provided in accordance with 17.6 (17.5.3.1 and 17.5.3.2). When ties per 17.6 are provided, and the contact surface is intentionally roughened to a full amplitude of approximately 1/4 in., the horizontal shear strength is:

$$\phi V_{nh} = (260 + 0.6 \rho_v f_{yt}) \lambda b_v d \leq 500 b_v d \quad (17.5.3.3).$$

The expression for V_{nh} in 17.5.3.3 accounts for the effect of the quantity of reinforcement crossing the interface by including ρ_v, which is the ratio of tie reinforcement area to area of contact surface, or $\rho_v = A_v / b_v s$. It also incorporates the correction factor λ to account for lightweight aggregate concrete per 11.7.4.3. It should also be noted that for concrete compressive strength $f'_c \leq 4444$ psi, the minimum tie reinforcement per Eq. (11-13) is $\rho_v f_{yt} = 50$ psi; substituting this into the above expression, $V_{nh} = 290 \lambda b_v d$. The upper limit of $500 b_v d$ corresponds to $\rho_v f_{yt} = 400$ psi in the case of normal weight concrete (i.e., $\lambda = 1$).

When in computing the horizontal shear strength of a composite flexural member, the following apply:

1. When $V_u > \phi(500 b_v d)$, the shear friction method of 11.7.4 must be used (17.5.2.4). Refer to Part 14 for further details on the application of 11.7.4.
2. No distinction shall be made between shored or unshored members (17.2.4). Tests have indicated that the strength of a composite member is the same whether or not the first element cast is shored or not.
3. Composite members must meet the appropriate requirements for deflection control per 9.5.6.
4. The contact surface shall be clean and free of laitance. Intentionally roughened surface may be achieved by scoring the surface with a stiff bristled broom. Heavy raking or grooving of the surface may be sufficient to achieve "full ¼ in. amplitude."
5. The effective depth d is defined as the distance from the extreme compression fiber for the entire composite section to the centroid of the tension reinforcement. For prestressed member, the effective depth shall not be taken less than 0.80 h (17.5.2).

The code also presents an alternative method for horizontal shear design in 17.5.4. The horizontal shear force that must be transferred across the interface between parts of a composite member is taken to be the change in internal compressive or tensile force, parallel to the interface, in any segment of a member. When this method is used, the limits of 17.5.3.1 through 17.5.3.4 apply, with the contact area A_c substituted for the quantity $b_v d$ in the expressions. Section 17.5.4.1 also requires that the reinforcement be distributed to approximately reflect the variation in shear force along the member. This requirement emphasizes the difference between the design of composite members on concrete and on steel. Slip between the steel beam and composite concrete slab at maximum strength is large, which permits redistribution of the shear force along the member. In concrete members with a composite slab, the slip at maximum strength is small and redistribution of shear resistance along the member is limited. Therefore, distribution of horizontal shear reinforcement must be based on the computed distribution of factored horizontal shear in concrete composite flexural members.

17.6 TIES FOR HORIZONTAL SHEAR

According to 17.6.3, ties are required to be "fully anchored" into interconnected elements "in accordance with 12.13." Figure 12-8 shows some tie details that have been used successfully in testing and design practice. Figure 12-8(a) shows an extended stirrup detail used in tests of Ref. 12.3. Use of an embedded "hairpin" tie, as illustrated in Fig. 12-8(b), is common practice in the precast, prestressed concrete industry. Many precast products are manufactured in such a way that it is difficult to position tie reinforcement for horizontal shear before concrete is placed. Accordingly, the ties are embedded in the plastic concrete as permitted by 16.7.1.

Shear reinforcement that extends from previously-cast concrete and is adequately anchored into the composite portion of a member (Fig. 12-8(c)) may be used as reinforcement (ties) to resist horizontal shear (17.4.3). Therefore, this reinforcement may be used to satisfy requirements for both vertical and horizontal shear.

Example 12.6 illustrates design for horizontal shear.

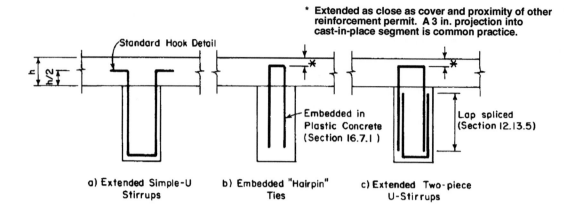

Figure 12-8 Ties for Horizontal Shear

REFERENCES

12.1 Barney, G.B.; Corley, W.G.; Hanson, J.M.; and Parmelee, R.A., "Behavior and Design of Prestressed Concrete Beams with Large Web Openings," *PCI Journal,* V. 22, No. 6, November-December 1977, pp. 32-61. Also, *Research and Development Bulletin* RD054D, Portland Cement Association, Skokie, IL.

12.2 Hanson, N.W., *Precast-Prestressed Concrete Bridges 2. Horizontal Shear Connections*, Development Department Bulletin D35, Portland Cement Association, Skokie, IL, 1960.

12.3 Roller, J.J., and Russell, H.G., "Shear Strength of High Strength Concrete Beams with Web Reinforcement," *ACI Structural Journal*, Vol. 87, No. 2, March-April 1990, pp. 191-198.

Example 12.1—Design for Shear - Members Subject to Shear and Flexure Only

Determine required size and spacing of vertical U-stirrups for a 30-foot span, simply supported beam.

b_w = 13 in.
d = 20 in.
f'_c = 3000 psi
f_{yt} = 40,000 psi
w_u = 4.5 kips/ft

Calculations and Discussion	Code Reference

For the purpose of this example, the live load will be assumed to be present on the full span, so that design shear at centerline of span is zero. (A design shear greater than zero at midspan is obtained by considering partial live loading of the span.) Using design procedure for shear reinforcement outlined in this part:

1. Determine factored shear forces

    @ support: V_u = 4.5 (15) = 67.5 kips

    @ distance d from support:

    V_u = 67.5 - 4.5 (20/12) = 60 kips   11.1.3.1

2. Determine shear strength provided by concrete

    $\phi V_c = \phi 2\sqrt{f'_c}\, b_w d$ Eq. (11-3)

    ϕ = 0.75   9.3.2.3

    ϕV_c = 0.75 (2) $\sqrt{3000} \times 13 \times 20 / 1000$ = 21.4 kips

    V_u = 60 kips > ϕV_c = 21.4 kips

    Therefore, shear reinforcement is required.   11.1.1

3. Compute $V_u - \phi V_c$ at critical section.

    $V_u - \phi V_c$ = 60 − 21.4 = 38.6 kips < $\phi 8\sqrt{f'_c}\, b_w d$ = 85.4 kips   O.K.   11.5.7.9

| Example 12.1 (cont'd) | Calculations and Discussion | Code Reference |

4. Determine distance x_c from support beyond which minimum shear reinforcement is required ($V_u = \phi V_c$):

$$x_c = \frac{V_u \text{ @ support} - \phi V_c}{w_u} = \frac{67.5 - 21.4}{4.5} = 10.2 \text{ ft}$$

Determine distance x_m from support beyond which concrete can carry total shear force ($V_u = \phi V_c / 2$):

$$x_m = \frac{V_u \text{ @ support} - (\phi V_c / 2)}{w_u} = \frac{67.5 - (21.4/2)}{4.5} = 12.6 \text{ ft}$$

5. Use Table 12-1 to determine required spacing of vertical U-stirrups.

    At critical section, $V_u = 60$ kips $> \phi V_c = 21.4$ kips

    $$s \text{ (req'd)} = \frac{\phi A_v f_{yt} d}{V_u - \phi V_c}$$ Eq. (11-15)

    Assuming No. 4 U-stirrups ($A_v = 0.40$ in.²),

    $$s \text{ (req'd)} = \frac{0.75 \times 0.40 \times 40 \times 20}{38.6} = 6.2 \text{ in.}$$

    Check maximum permissible spacing of stirrups:

    $s \text{ (max)} \leq d/2 = 20/2 = 10$ in. (governs)   11.5.5.1

    ≤ 24 in. since $V_u - \phi V_c = 38.6$ kips $< \phi 4\sqrt{f'_c}\, b_w d = 42.7$ kips

    Maximum stirrup spacing based on minimum shear reinforcement:

    $$s \text{ (max)} \leq \frac{A_v f_{yt}}{0.75\sqrt{f'_c}\, b_w} = \frac{0.4 \times 40,000}{0.75\sqrt{3,000}(13)} = 30 \text{ in.}$$ 11.5.6.3

    $$\leq \frac{A_v f_{yt}}{50 b_w} = \frac{0.4 \times 40,000}{50 \times 13} = 24.6 \text{ in.}$$

    Determine distance x from support beyond which 10 in. stirrup spacing may be used:

    $$10 = \frac{0.75 \times 0.4 \times 40 \times 20}{V_u - 21.4}$$

    $V_u - 21.4 = 24$ kips or $V_u = 24 + 21.4 = 45.4$ kips

    $$x = \frac{67.5 - 45.4}{4.5} = 4.9 \text{ ft}$$

		Code
Example 12.1 (cont'd)	**Calculations and Discussion**	**Reference**

Stirrup spacing using No. 4 U-stirrups:

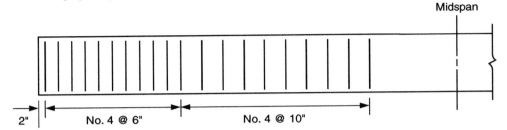

6. As an alternate procedure, use simplified method presented in Table 12-2 to determine stirrup size and spacing.

   At critical section,

   $\phi V_s = V_u - \phi V_c = 60 - 21.4 = 38.6$ kips

   From Table 12-2 for Grade 40 stirrups:

   No. 4 U-stirrups @ d/4 provides $\phi V_s = 48$ kips

   No. 4 U-stirrups @ d/3 provides $\phi V_s = 36$ kips

   By interpolation, No. 4 U-stirrups @ d/3.22 = 38.6 kips

   Stirrup spacing = d/3.22 = 20/3.22 = 6.2 in.

   Stirrup spacing along length of beam is determined as shown previously.

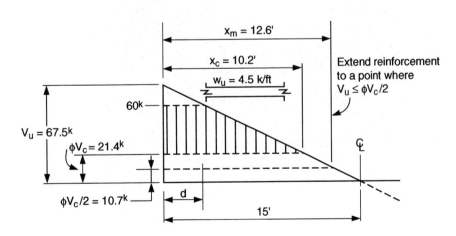

12-15

Example 12.2—Design for Shear - with Axial Tension

Determine required spacing of vertical U-stirrups for a beam subject to axial tension.

f'_c = 3600 psi (sand-lightweight concrete, f_{ct} not specified)
f_{yt} = 40,000 psi
M_d = 43.5 ft-kips
M_ℓ = 32.0 ft-kips
V_d = 12.8 kips
V_ℓ = 9.0 kips
N_d = -2.0 kips (tension)
N_ℓ = -15.2 kips (tension)

Calculations and Discussion	Code Reference

1. Determine factored loads

    9.2.1

    M_u = 1.2 (43.5) + 1.6 (32.0) = 103.4 ft-kips

    Eq. (9-2)

    V_u = 1.2 (12.8) + 1.6 (9.0) = 29.8 kips

    N_u = 1.2 (-2.0) + 1.6 (-15.2) = -26.7 kips (tension)

2. Determine shear strength provided by concrete

    Since average splitting tensile strength f_{ct} is not specified, $\sqrt{f'_c}$ is reduced by a factor of 0.85 (sand-lightweight concrete)

    11.2.1.2

    $$\phi V_c = \phi 2 \left[1 + \frac{N_u}{500 A_g} \right] 0.85 \sqrt{f'_c}\, b_w d$$

    Eq. (11-8)

    ϕ = 0.75

    9.3.2.3

    $$\phi V_c = (0.75)\, 2 \left[1 + \frac{(-26,700)}{500\,(18 \times 10.5)} \right] 0.85\sqrt{3600}\,(10.5)\,16/1000 = 9.2 \text{ kips}$$

3. Check adequacy of cross-section.

    $(V_u - \phi V_c) \leq \phi 8\,(0.85)\sqrt{f'_c}\, b_w d$

    11.5.7.9

    Note: 0.85 is a factor for lightweight concrete per 11.2.1.2

    $(V_u - \phi V_c)$ = 29.8 - 9.2 = 20.6 kips

    $\phi 8(0.85)\sqrt{f'_c}\,b_w d = 0.75 \times 8 \times 0.85\sqrt{3600} \times 10.5 \times 16/1000 = 51.4$ kips > 20.6 kips   O.K.

		Code
Example 12.2 (cont'd)	**Calculations and Discussion**	**Reference**

4. Determine required spacing of U-stirrups

    Assuming No. 3 U-stirrups ($A_v = 0.22$ in.^{2}),

    $$s \text{ (req'd)} = \frac{\phi A_v f_{yt} d}{(V_u - \phi V_c)}$$

    $$= \frac{0.75 \times 0.22 \times 40 \times 16}{20.6} = 5.1 \text{ in.}$$

5. Determine maximum permissible spacing of stirrups

    $V_u - \phi V_c = 20.6$ kips

    $\phi 4(0.85)\sqrt{f'_c} b_w d = 25.7$ kips > 20.6 kips   11.2.1.2

    Therefore, provisions of 11.5.5.1 apply.   11.5.5.3

    s (max) of vertical stirrups $\leq d/2 = 8$ in. (governs)   11.5.5.1
    or ≤ 24 in.

    s (max) of No. 3 U-stirrups corresponding to minimum reinforcement area requirements:

    $$s \text{ (max)} = \frac{A_v f_{yt}}{0.75(0.85)\sqrt{f'_c} b_w} = \frac{0.22 \times 40,000}{0.75 \times 0.85 \times \sqrt{3600} \times 10.5} = 21.9 \text{ in.}$$ 11.5.6.3

    $$s \text{ (max)} = \frac{A_v f_{yt}}{50 b_w} = \frac{0.22 (40,000)}{50 (10.5)} = 16.8 \text{ in.}$$

    s (max) = 8 in. (governs)

    Summary:

    Use No. 3 vertical stirrups @ 5.0 in. spacing.

Example 12.3—Design for Shear - with Axial Compression

A tied compression member has been designed for the given load conditions. However, the original design did not take into account the fact that under a reversal in the direction of lateral load (wind), the axial load, due to the combined effects of gravity and lateral loads, becomes P_u = 10 kips, with essentially no change in the values of M_u and V_u. Check shear reinforcement requirements for the column under (1) original design loads and (2) reduced axial load.

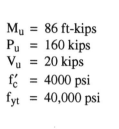

M_u = 86 ft-kips
P_u = 160 kips
V_u = 20 kips
f'_c = 4000 psi
f_{yt} = 40,000 psi

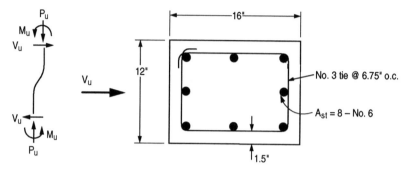

Calculations and Discussion	Code Reference

Condition 1: $P_u = N_u = 160$ kips

1. Determine shear strength provided by concrete

    d = 16 - [1.5 + 0.375 + (0.750/2)] = 13.75 in.

    $$\phi V_c = \phi 2 \left[1 + \frac{N_u}{2000 A_g} \right] \sqrt{f'_c} \, b_w d$$ Eq. (11-4)

    $\phi = 0.75$ 9.3.2.3

    $\phi V_c = 0.75 \, (2) \left[1 + \frac{160,000}{2000 \, (16 \times 12)} \right] \sqrt{4000} \, (12)(13.75)/1000 = 22.2$ kips

    $\phi V_c = 22.2$ kips $> V_u = 20$ kips

2. Since $V_u = 20$ kips $> \phi V_c/2 = 11.1$ kips, minimum shear reinforcement requirements must be satisfied. 11.5.6.1

    No. 3 stirrups (A_v = 0.22 in.^{2})

    $s \, (max) = \dfrac{A_v f_{yt}}{0.75 \sqrt{f'_c} b_w} = \dfrac{0.22 \, (40,000)}{0.75 \sqrt{4000} \, (12)} = 15.5$ in. Eq. (11-13)

    $s \, (max) = \dfrac{A_v f_{yt}}{50 b_w} = \dfrac{0.22 \, (40,000)}{50 \, (12)} = 14.7$ in.

    $s \, (max) = d/2 = 13.75/2 = 6.9$ in. (governs) 11.5.5.1

    Therefore, use of s = 6.75 in. is satisfactory.

		Code
Example 12.3 (cont'd)	**Calculations and Discussion**	**Reference**

Condition 2: $P_u = N_u = 10$ kips

1. Determine shear strength provided by concrete.

$$\phi V_c = 0.75\,(2)\left[1 + \frac{(10,000)}{2000\,(16\times 12)}\right] \times \sqrt{4000}\,(12)(13.75)/1000 = 16.1 \text{ kips}$$

Eq. (11-4)

$\phi V_c = 16.1$ kips $< V_u = 20$ kips

Shear reinforcement must be provided to carry excess shear.

2. Determine maximum permissible spacing of No. 3 ties

$$s\,(\text{max}) = \frac{d}{2} = \frac{13.75}{2} = 6.9 \text{ in.}$$

11.5.4.1

Maximum spacing, d/2, governs for Conditions 1 and 2.

3. Check total shear strength with No. 3 @ 6.75 in.

$$\phi V_s = \phi A_v f_{yt} \frac{d}{s} = \frac{0.75\,(0.22)\,(40)\,(13.75)}{6.75} = 13.4 \text{ kips}$$

Eq. (11-15)

$\phi V_c + \phi V_s = 16.1 + 13.4 = 29.5$ kips $> V_u = 20$ kips   O.K.

Example 12.4—Design for Shear - Concrete Floor Joist

Check shear requirements in the uniformly loaded floor joist shown below.

f'_c = 4000 psi
f_{yt} = 40,000 psi
w_d = 77 psf
w_ℓ = 120 psf

Assumed longitudinal reinforcement:

Bottom bars: 2 – No. 5
Top bars: No. 5 @ 9 in.

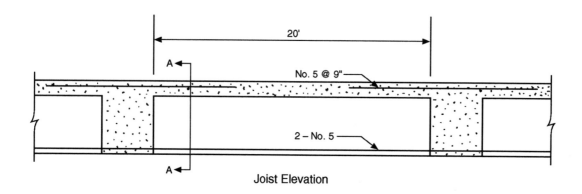

Joist Elevation

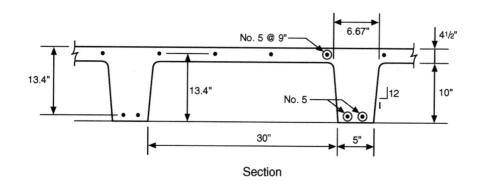

Section

Calculations and Discussion	Code Reference

1. Determine factored load.

   w_u = [1.2 (77) + 1.6 (120)] 35/12 = 830 lb/ft   Eq. (9-2)

Example 12.4 (cont'd)	Calculations and Discussion	Code Reference

2. Determine factored shear force.

    @ distance d from support: 11.1.3.1

    $V_u = 0.83(10) - 0.83(13.4/12) = 7.4$ kips   8.3.3

3. Determine shear strength provided by concrete.

    According to 8.11.8, V_c may be increased by 10 percent.

    Average width of joist web $b_w = (6.67 + 5)/2 = 5.83$ in.

    $\phi V_c = 1.1 \phi 2\sqrt{f'_c}\, b_w d$ 8.11.8 Eq. (11-3)

    $\phi = 0.75$ 9.3.2.3

    $\phi V_c = 1.1(0.75)\, 2\sqrt{4000}\,(5.83)(13.4)/1000 = 8.2$ kips

    $\phi V_c = 8.2$ kips $> V_u = 7.4$ kips   O.K.

    Note that minimum shear reinforcement is not required for joist construction defined by 8.11.   11.5.6.1(b)

    Alternatively, calculate V_c using Eq. (11-5)

    Compute ρ_w and $V_u d/M_u$ at distance d from support:

    $\rho_w = \dfrac{A_s}{b_w d} = \dfrac{(2 \times 0.31)}{(5.83)(13.4)} = 0.0079$

    M_u @ face of support $= \dfrac{w_u \ell_n^2}{11} = \dfrac{0.83(20)^2}{11} = 30.2$ ft-kips   8.3.3

    M_u @ d $= \dfrac{w_u \ell_n^2}{11} + \dfrac{w_u d^2}{2} - \dfrac{w_u \ell_n d}{2}$

    $= 30.2 + \dfrac{(0.83)(13.4/12)^2}{2} - \dfrac{(0.83)(20)(13.4/12)}{2} = 21.5$ ft-kips

    $\dfrac{V_u d}{M_u} = \dfrac{7.4(13.4/12)}{21.5} = 0.38 < 1.0$ O.K.   11.3.2.1

    $\phi V_c = \phi\, 1.1 \left(1.9\sqrt{f'_c} + 2500 \rho_w \dfrac{V_u d}{M_u}\right) b_w d \leq \phi(1.1)\, 3.5\sqrt{f'_c}\, b_w d$

    $= 0.75(1.1)\left[1.9\sqrt{4{,}000} + 2500(0.0079)(0.38)\right](5.83)(13.4)/1{,}000$

    $= 8.2$ kips $< 0.75(1.1)(3.5)\sqrt{4{,}000}\,(5.83)(13.4)/1{,}000 = 14.3$ kips   O.K.

    $\phi V_c = 8.2$ kips $> V_u = 7.4$ kips   O.K.

Example 12.5—Design for Shear - Shear Strength at Web Openings

The simply supported prestressed double tee beam shown below has been designed without web openings to carry a factored load w_u = 1520 lb/ft. Two 10-in.-deep by 36-in.-long web openings are required for passage of mechanical and electrical services. Investigate the shear strength of the beam at web opening A.

This design example is based on an experimental and analytical investigation reported in Ref. 12.1.

Beam f'_c = 6000 psi
Topping f'_c = 3000 psi
f_{pu} = 270,000 psi
f_{yt} = 60,000 psi

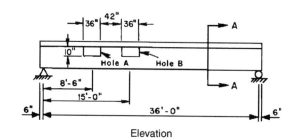

Elevation

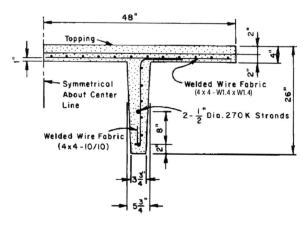

Section A-A
(one-half of the double tee section)

	Code
Calculations and Discussion	**Reference**

This example treats only the shear strength considerations for the web opening. Other strength considerations need to be investigated, such as: to avoid slip of the prestressing strand, openings must be located outside the required strand development length, and strength of the struts to resist flexure and axial loads must be checked. The reader is referred to the complete design example in Ref. 12.1 for such calculations. The design example in Ref. 12.1 also illustrates procedures for checking service load stresses and deflections around the openings.

1. Determine factored moment and shear at center of opening A. Since double tee is symmetric about centerline, consider one-half of double tee section.

    $w_u = \dfrac{1520}{2}$ = 760 lb/ft per tee

    M_u = 0.760 (36/2) (8.5) - 0.760 (8.5)²/2 = 88.8 ft-kips

    V_u = 0.760 (36/2) - 0.760 (8.5) = 7.2 kips

Example 12.5 (cont'd)   **Calculations and Discussion**   **Code Reference**

2. Determine required shear reinforcement adjacent to opening. Vertical stirrups must be provided adjacent to both sides of web opening. The stirrups should be proportioned to carry the total shear force at the opening.

$$A_v = \frac{V_u}{\phi f_{yt}} = \frac{7200}{0.75 \times 60,000} = 0.16 \text{ in.}^2$$

Use No. 3 U-stirrup, one on each side of opening ($A_v = 0.22$ in.^{2})

3. Using a simplified analytical procedure developed in Ref. 12.1, the axial and shear forces acting on the "struts" above and below opening A are calculated. Results are shown in the figure below. The reader is referred to the complete design example in Ref. 12.3 for the actual force calculations. Axial forces should be accounted for in the shear design of the struts.

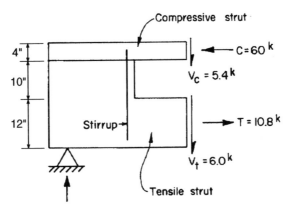

4. Investigate shear strength for tensile strut.

   $V_u = 6.0$ kips

   $N_u = -10.8$ kips

   $d = 0.8h = 0.8(12) = 9.6$ in.   11.5.3

   b_w = average width of tensile strut = $[3.75 + (3.75 + 2 \times 12/22)]/2 = 4.3$ in.

   $$V_c = 2\left(1 + \frac{N_u}{500 A_g}\right)\sqrt{f'_c}\, b_w d \quad\quad\quad\quad Eq.\ (11\text{-}8)$$

   $$= 2\left(1 - \frac{10,800}{500 \times 4.3 \times 12}\right)\sqrt{6000}\,(4.3)(9.6)/1000 = 3.72 \text{ kips}$$

Example 12.5 (cont'd)	Calculations and Discussion	Code Reference

$\phi V_c = 0.75\ (3.72) = 2.8$ kips

$V_u = 6.0$ kips $> \phi V_c = 2.8$ kips

Therefore, shear reinforcement is required in tensile strut.

$$A_v = \frac{(V_u - \phi V_c)\ s}{\phi f_{yt} d}$$

$$= \frac{(6.0 - 2.8)\ 9}{0.75 \times 60 \times 9.6} = 0.07\ \text{in.}^2$$

where $s = 0.75h = 0.75 \times 12 = 9$ in.   11.5.5.1

Use No. 3 single leg stirrups at 9-in. centers in tensile strut, ($A_v = 0.11$ in.^{2}). Anchor stirrups around prestressing strands with 180 degree bend at each end.

5. Investigate shear strength for compressive strut.

    $V_u = 5.4$ kips

    $N_u = 60$ kips

    $d = 0.8h = 0.8\ (4) = 3.2$ in.

    $b_w = 48$ in.

$$V_c = 2\left(1 + \frac{N_u}{2000 A_g}\right) \sqrt{f'_c}\ b_w d \qquad \text{Eq. (11-4)}$$

$$= 2\left(1 + \frac{60{,}000}{2000 \times 48 \times 4}\right) \sqrt{3000}\ (48)\ (3.2)/1000 = 19.5\ \text{kips}$$

$\phi V_c = 0.75\ (19.5) = 14.6$ kips

$V_u = 5.4$ kips $< \phi V_c = 14.6$ kips

Therefore, shear reinforcement is not required in compressive strut.

6. Design Summary - See reinforcement details below.

    a. Use U-shaped No. 3 stirrup adjacent to both edges of opening to contain cracking within the struts.

Example 12.5 (cont'd) | **Calculations and Discussion** | **Code Reference**

b. Use single-leg No. 3 stirrups at 9-in. centers as additional reinforcement in the tensile strut.

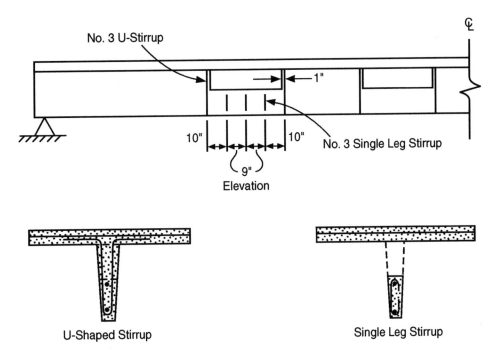

Details of Additional Reinforcement

A similar design procedure is required for opening B.

Example 12.6—Design for Horizontal Shear

For the composite slab and precast beam construction shown, design for transfer of horizontal shear at contact surface of beam and slab for the three cases given below. Assume the beam is simply supported with a span of 30 feet.

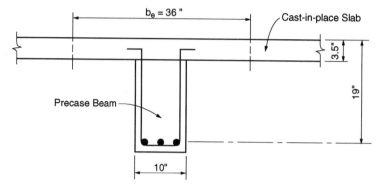

f'_c = 3000 psi (normal weight concrete)
f_{yt} = 60,000 psi

	Calculations and Discussion	Code Reference
Case I:	Service dead load = 315 lb/ft Service live load = 235 lb/ft Factored load = 1.2(315) + 1.6 (235) = 754 lb/ft	Eq. (9-2)

1. Determine factored shear force V_u at a distance d from face of support:.

   V_u = (0.754 × 30/2) – (0.754 × 19/12) = 10.1 kips   *11.1.3.1*

2. Determine horizontal shear strength.   *17.5.3*

   $V_u \leq \phi V_{nh}$ *Eq. (17-1)*

   $\phi V_{nh} = \phi (80 b_v d)$ *17.5.3.1 & 17.5.3.2*

   = 0.75 (80 × 10 × 19)/1000 = 11.4 kips

   V_u = 10.1 kips $\leq \phi V_{nh}$ = 11.4 kips

	Code
Example 12.6 (cont'd) Calculations and Discussion	**Reference**

Therefore, design in accordance with either 17.5.3.1 or 17.5.3.2:

Note: For either condition, top surface of precast beam must be cleaned and free of laitance prior to placing slab concrete.

If top surface of precast beam is intentionally roughened, no ties are required.   *17.5.3.1*

If top surface of precast beam is not intentionally roughened, minimum ties are required in accordance with 17.6.   *17.5.3.2*

3. Determine required minimum area of ties.   *17.6*

$$A_v = \frac{0.75\sqrt{f'_c}\,b_w d}{f_{yt}} \geq \frac{50 b_w s}{f_{yt}}$$ *11.5.5.3*

where s (max) = 4(3.5) = 14 in. < 24 in.   *17.6.1*

$$A_v = \frac{0.75\sqrt{3000}\,(10)(14)}{60,000} = 0.096 \text{ in.}^2 \text{ at 14 in. o.c.}$$

$$\text{Min. } A_v = \frac{50 \times 10 \times 14}{60,000} = 0.117 \text{ in.}^2 \text{ at 14 in. o.c.}$$

$$\text{or } 0.10 \text{ in.}^2/\text{ft}$$

Case II: Service dead load = 315 lb/ft
           Service live load = 1000 lb/ft
           Factored load = 1.2(315) + 1.6(1000) = 1978 lb/ft   *9.2*

1. Determine factored shear force V_u at a distance d from face of support.

$$V_u = (1.98 \times 30/2) - (1.98 \times 19/12) = 26.6 \text{ kips}$$ *11.1.3.1*

2. Determine horizontal shear strength.   *17.5.3*

$$V_u = 26.6 \text{ kips} > \phi V_{nh} = \phi(80 b_v d) = 11.4 \text{ kips}$$

Therefore, 17.5.2.3 must be satisfied. Minimum ties are required as computed above ($A_v = 0.10$ in.^{2}/ft).

$$V_{nh} = \phi(260 + 0.6 \rho_v f_{yt}) \lambda b_v d$$ *17.5.3.3*

Example 12.6 (cont'd)	Calculations and Discussion	Code Reference

where $\rho_v = \dfrac{A_v}{b_v s} = \dfrac{0.10 \text{ in.}^2}{10 \text{ in. (12 in.)}}$

$= 0.00083$

$\lambda = 1.0$ (normal weight concrete)   11.7.4.3

$\phi V_{nh} = 0.75 (260 + 0.6 \times 0.00083 \times 60{,}000)(1.0 \times 10 \times 19)$

$= 0.75 (290) \, 190 = 41.3$ kips

$\phi V_{nh} = 41.3$ kips $< \phi (500 b_v d)/1000 = 71.3$ kips   O.K.   17.5.3.3

$V_u = 26.6$ kips $< \phi V_{nh} = 41.3$ kips

Therefore, design in accordance with 17.5.2.3:

Contact surface must be intentionally roughened to "a full amplitude of approximately 1/4-in.," and minimum ties provided in accordance with 17.6.

3. Compare tie requirements with required vertical shear reinforcement at distance d from face of support.

$V_u = 26.6$ kips

$V_c = 2\sqrt{f'_c}\, b_w d = 2\sqrt{3000} \times 10 \times 19/1000 = 20.8$ kips   Eq. (11-3)

$V_u \leq \phi (V_c + V_s) = \phi V_c + \phi A_v f_{yt} \dfrac{d}{s}$ Eq. (11-15)

Solving for A_v/s:

$\dfrac{A_v}{s} = \dfrac{V_u - \phi V_c}{\phi f_{yt} d} = \dfrac{26.6 - (0.75 \times 20.8)}{0.75 \times 60 \times 19} = 0.013$ in.^{2}/in.

$s_{max} = \dfrac{19}{2} = 9.5$ in. < 24 in.   11.5.5.1

$A_v = 0.013 \times 9.5 = 0.12$ in.^{2}

Provide No. 3 U-stirrups @ 9.5 in. o.c. ($A_v = 0.28$ in.^{2}/ft). This exceeds the minimum ties required for horizontal shear ($A_v = 0.10$ in.^{2}/ft) so the No. 3 U-stirrups @ 9.5 in. o.c. are adequate to satisfy both vertical and horizontal shear reinforcement requirements. Ties must be adequately anchored into the slab by embedment or hooks. See Fig. 12-8.

		Code
Example 12.6 (cont'd)	**Calculations and Discussion**	**Reference**

Case III: Service dead load = 315 lb/ft
            Service live load = 3370 lb/ft
            Factored load = 1.2(315) + 1.6 (3370) = 5770 lb/ft   9.2

1. Determine factored shear force V_u at distance d from support.

   $V_u = (5.77 \times 30/2) - (5.77 \times 19/12) = 77.4$ kips   11.1.3.1

   $V_u = 77.4$ kips $> \phi(500 b_v d) = 0.75(500 \times 10 \times 19)/1000 = 71.3$ kips   17.5.3.4

Since V_u exceeds $\phi(500 b_v d)$, design for horizontal shear must be in accordance with 11.7.4 - Shear-Friction. Shear along the contact surface between beam and slab is resisted by shear-friction reinforcement across and perpendicular to the contact surface.

As required by 17.5.3.1, a varied tie spacing must be used, based on the actual shape of the horizontal shear distribution. The following method seems reasonable and has been used in the past:

Converting the factored shear force to a unit stress, the factored horizontal shear stress at a distance d from span end is:

$$v_{uh} = \frac{V_u}{b_v d} = \frac{77.4}{10 \times 19} = 0.407 \text{ ksi}$$

The shear "stress block" diagram may be shown as follows:

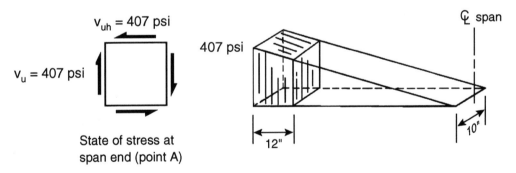

Assume that the horizontal shear is uniform per foot of length, then the shear transfer force for the first foot is:

$V_{uh} = 0.407 \times 10 \times 12 = 48.9$ kips

Example 12.6 (cont'd)	Calculations and Discussion	Code Reference

Required area of shear-friction reinforcement is computed by Eqs. (11-1) and (11-25):

$$V_{uh} \leq \phi V_n = \phi A_{vf} f_{yt} \mu$$ Eq. (11-25)

$$A_{vf} = \frac{V_{uh}}{\phi f_{yt} \mu}$$

If top surface of precast beam is intentionally roughened to approximately 1/4 in., $\mu = 1.0$.   11.7.4.3

$$A_{vf} = \frac{48.9}{0.75 \times 60 \times 1.0} = 1.09 \text{ in.}^2/\text{ft}$$

With No. 5 double leg stirrups, $A_{vf} = 0.62$ in.^{2}

$$s = \frac{0.62 \times 12}{1.09} = 6.8 \text{ in.}$$

Use No. 5 U-stirrups @ 6.5 in. o.c. for a minimum distance of d + 12 in. from span end.

If top surface of precast beam is not intentionally roughened, $\mu = 0.6$.   11.7.4.3

$$A_{vf} = \frac{48.9}{0.75 \times 60 \times 0.6} = 1.81 \text{ in.}^2/\text{ft}$$

$$s = \frac{0.62 \times 12}{1.81} = 4.1 \text{ in.}$$

Use No. 5 U-Stirrups @ 4 in. o.c. for a minimum distance of d + 12 in. from span end.

This method can be used to determine the tie spacing for each successive one-foot length. The shear force will vary at each one-foot increment and the tie spacing can vary accordingly to a maximum of 14 in. toward the center of the span.

Note: Final tie details are governed by vertical shear requirements.

13

Torsion

UPDATE FOR THE '05 CODE

In the 2005 code, the provisions of for torsion design remain essentially unchanged. However, a new section (11.6.7) now permits using alternative procedures for torsion design of solid sections with an aspect ratio, h/b_t, of three or more. Moreover, in addition to standard hooks, 11.6.4.2 allows using seismic hooks to anchor transverse torsional reinforcement.

BACKGROUND

The 1963 code included one sentence concerning torsion detailing. It prescribed use of closed stirrups in edge and spandrel beams and one longitudinal bar in each corner of those closed stirrups. Comprehensive design provisions for torsion were first introduced in the 1971 code. With the exception of a change in format in the 1977 document, the requirements have remained essentially unchanged through the 1992 code. These first generation provisions applied only to reinforced, nonprestressed concrete members. The design procedure for torsion was analogous to that for shear. Torsional strength consisted of a contribution from concrete (T_c) and a contribution from stirrups and longitudinal reinforcement, based on the skew bending theory.

The design provisions for torsion were completely revised in the 1995 code and remain essentially unchanged since then. The new procedure, for solid and hollow members, is based on a thin-walled tube, space truss analogy. This unified approach applies equally to reinforced and prestressed concrete members. Background of the torsion provisions has been summarized by MacGregor and Ghoneim.[13.1] Design aids and design examples for structural concrete members subject to torsion are presented in Ref. 13.2.

For design purposes, the center portion of a solid beam can conservatively be neglected. This assumption is supported by test results reported in Ref. 13.1. Therefore, the beam is idealized as a tube. Torsion is resisted through a constant shear flow q (force per unit length of wall centerline) acting around the centerline of the tube as shown in Fig. 13-1(a). From equilibrium of external torque T and internal stresses:

$$T = 2A_o q = 2A_o \tau t \qquad (1)$$

Rearranging Eq. (1)

$$q = \tau t = \frac{T}{2A_o} \qquad (2)$$

where τ = shear stress, assumed uniform, across wall thickness

   t = wall thickness

   T = applied torque

A_o = area enclosed within the tube centerline [see Fig. 13-1(b)]

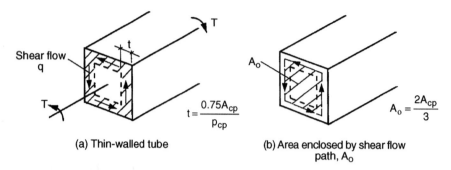

Figure 13-1 Thin-Wall Tube Analogy

When a concrete beam is subjected to a torsional moment causing principal tension larger than $4\sqrt{f'_c}$, diagonal cracks spiral around the beam. After cracking, the tube is idealized as a space truss as shown in Fig. 13-2. In this truss, diagonal members are inclined at an angle θ. Inclination of the diagonals in all tube walls is the same. Note that this angle is not necessarily 45 degree. The resultant of the shear flow in each tube wall induces forces in the truss members. A basic concept for structural concrete design is that concrete is strong in compression, while steel is strong in tension. Therefore, in the truss analogy, truss members that are in tension consist of steel reinforcement or "tension ties." Truss diagonals and other members that are in compression consist of concrete "compression struts." Forces in the truss members can be determined from equilibrium conditions. These forces are used to proportion and detail the reinforcement.

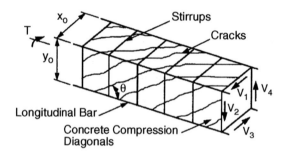

Figure 13-2 Space Truss Analogy

Figure 13-3 depicts a free body extracted from the front vertical wall of the truss of Fig. 13-2. Shear force V_2 is equal to the shear flow q (force per unit length) times the height of the wall y_o. Stirrups are designed to yield when the maximum torque is reached. The number of stirrups intersected is a function of the stirrup spacing s and the horizontal projection $y_o \cot\theta$ of the inclined surface. From vertical equilibrium:

$$V_2 = \frac{A_t f_{yt}}{s} y_o \cot\theta \qquad (3)$$

As the shear flow (force per unit length) is constant over the height of the wall,

$$V_2 = q y_o = \frac{T}{2A_o} y_o \qquad (4)$$

Substituting for V_2 in Eqs. (3) and (4),

$$T = \frac{2A_o A_t f_{yt}}{s} \cot\theta \qquad (5)$$

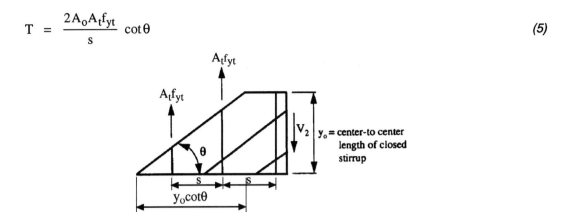

Figure 13-3 Free Body Diagram for Vertical Equilibrium

A free body diagram for horizontal equilibrium is shown in Fig. 13-4. The vertical shear force V_i in wall "i" is equal to the product of the shear flow q times the length of the wall y_i. Vector V_i can be resolved into two components: a diagonal component with an inclination θ equal to the angle of the truss diagonals, and a horizontal component equal to:

$$N_i = V_i \cot\theta$$

Force N_i is centered at the midheight of the wall since q is constant along the side of the element. Top and bottom chords of the free body of Fig. 13-4 are subject to a force $N_i/2$ each. Internally, it is assumed that the longitudinal steel yields when the maximum torque is reached. Summing the internal and external forces in the chords of all the space truss walls results in:

$$\Sigma A_{\ell i} f_y = A_\ell f_y = \Sigma N_i = \Sigma V_i \cot\theta = \Sigma q y_i \cot\theta = \Sigma \frac{T}{2A_o} y_i \cot\theta = \frac{T}{2A_o} \cot\theta \Sigma y_i$$

where $A_\ell f_y$ is the yield force in all longitudinal reinforcement required for torsion.

Rearranging the above equation,

$$T = \frac{2A_o A_\ell f_{y\ell}}{2(x_o + y_o)\cot\theta} \qquad (6)$$

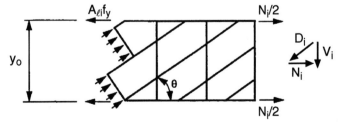

Figure 13-4 Free Body Diagram for Horizontal Equilibrium

11.6.1 Threshold Torsion

Torsion can be neglected if the factored torque T_u is less than $\phi T_{cr}/4$, where T_{cr} is the cracking torque. The cracking torque corresponds to a principal tensile stress of $4\sqrt{f'_c}$. Prior to cracking, thickness of the tube wall "t" and the area enclosed by the wall centerline "A_o" are related to the uncracked section geometry based on the following assumptions:

$$t = \frac{3A_{cp}}{4p_{cp}} \tag{7}$$

$$A_o = \frac{2A_{cp}}{3} \quad \text{(before cracking)} \tag{8}$$

where A_{cp} = area enclosed by outside perimeter of concrete cross-section, in.²

p_{cp} = outside perimeter of concrete cross-section, in.

A_o = area within centerline of the thin-wall tube, in.²

Equations (7) and (8) apply to the uncracked section. For spandrel beams and other members cast monolithically with a slab, parts of the slab overhangs contribute to torsional resistance. Size of effective portion of slab to be considered with the beam is illustrated in Fig. R13.2.4.

Substituting for t from Eq. (7), A_o from Eq. (8), and taking $\tau = 4\sqrt{f'_c}$ in Eq. (1), the cracking torque for nonprestressed members can be derived:

$$T_{cr} = 4\sqrt{f'_c}\left(\frac{A_{cp}^2}{p_{cp}}\right) \tag{9}$$

For prestressed concrete members, based on a Mohr's Circle analysis, the principal tensile stress of $4\sqrt{f'_c}$ is reached at $\sqrt{1 + \frac{f_{pc}}{4\sqrt{f'_c}}}$ times the corresponding torque for nonprestressed members. Therefore, the cracking torque for prestressed concrete members is computed as:

$$T_{cr} = 4\sqrt{f'_c}\left(\frac{A_{cp}^2}{p_{cp}}\right)\sqrt{1 + \frac{f_{pc}}{4\sqrt{f'_c}}} \tag{10}$$

where f_{pc} = compressive stress in concrete, due to prestress, at centroid of section (also see 2.1)

Similarly, for nonprestressed members subjected to an applied axial force, the principal tensile stress of $4\sqrt{f'_c}$ is reached at $\sqrt{1 + \frac{N_u}{4A_g\sqrt{f'_c}}}$ times the corresponding torque, so that the cracking torque is:

$$T_{cr} = 4\sqrt{f'_c}\left(\frac{A_{cp}^2}{p_{cp}}\right)\sqrt{1 + \frac{N_u}{4A_g\sqrt{f'_c}}} \tag{11}$$

where N_u = factored axial force normal to the cross-section (positive for compression)

A_g = gross area of section. For a hollow section, A_g is the area of the concrete only and does not include the area of the void(s) (see 11.6.1).

According to 11.6.1, design for torsion can be neglected if $T_u < \frac{\phi T_{cr}}{4}$, i.e.:

For nonprestressed members:

$$T_u < \phi\sqrt{f'_c}\left(\frac{A_{cp}^2}{p_{cp}}\right) \tag{12}$$

For prestressed members:

$$T_u < \phi\sqrt{f'_c}\left(\frac{A_{cp}^2}{p_{cp}}\right)\sqrt{1 + \frac{f_{pc}}{4\sqrt{f'_c}}} \tag{13}$$

For nonprestressed members subjected to an axial tensile or compressive force:

$$T_u < \phi\sqrt{f'_c}\left(\frac{A_{cp}^2}{p_{cp}}\right)\sqrt{1 + \frac{N_u}{4A_g\sqrt{f'_c}}} \tag{14}$$

It is important to note that A_g is to be used in place of A_{cp} in Eqs. (12) through (14) for hollow sections, where for torsion, a hollow section is defined as having one or more longitudinal voids such that $A_g/A_{cp} < 0.95$ (see R11.6.1). The quantity A_g in this case is the area of the concrete only (i.e., the area of the void(s) are not included), based on the outer boundaries prescribed in 13.2.4. The threshold torsion provisions of 11.6.1 were modified in the 2002 code to apply to hollow sections, since results of tests in code Ref. 11.29 indicate that the cracking torque of a hollow section is approximately (A_g/A_{cp}) times the cracking torque of a solid section with the same outside dimensions. Multiplying the cracking torque by (A_g/A_{cp}) a second time reflects the transition from the circular interaction between the inclined cracking loads in shear and torsion for solid members, to the approximately linear interaction for thin-walled hollow sections.

11.6.2 Equilibrium and Compatibility - Factored Torsional Moment T_u

Whether a reinforced concrete member is subject to torsion only, or to flexure combined with shear, the stiffness of that member will decrease after cracking. The reduction in torsional stiffness after cracking is much larger than the reduction in flexural stiffness after cracking. If the torsional moment T_u in a member cannot be reduced by redistribution of internal forces in the structure, that member must be designed for the full torsional moment T_u (11.6.2.1). This is referred to as "equilibrium torsion." See Fig. R11.6.2.1. If redistribution of internal forces can occur, as in indeterminate structures, the design torque can be reduced. This type of torque is referred to as "compatibility torsion." See Fig. R11.6.2.2. Members subject to compatibility torsion need not be designed for a torque larger than the product of the cracking torque times the strength reduction factor ϕ (0.75 for torsion, see 9.3.2.3). For cases of compatibility torsion where $T_u > \phi T_{cr}$ the member can be designed for ϕT_{cr} only, provided redistribution of internal forces is accounted for in the design of the other members of the structure (11.6.2.2). Cracking torque T_{cr} is computed by Eq. (9) for nonprestressed members, by Eq. (10) for prestressed members, and by Eq. (11) for nonprestressed members subjected to an axial tensile or compressive force. For hollow sections, A_{cp} shall not be replaced with A_g in these equations (11.6.2.2).

11.6.2.4-11.6.2.5 Critical Section—In nonprestressed members, the critical section for torsion design is at distance "d" (effective depth) from the face of support. Sections located at a distance less than d from the face of support must be designed for the torque at distance d from the support. Where a cross beam frames into a girder at a distance less than d from the support, a concentrated torque occurs in the girder within distance d. In such cases, the design torque must be taken at the face of support. The same rule applies to prestressed members, except that h/2 replaces distance d, where h is the overall height of member. In composite members, h is the overall height of the composite section.

11.6.3 Torsional Moment Strength

The design torsional strength should be equal to or greater than the required torsional strength:

$$\phi T_n \geq T_u \qquad \text{Eq. (11-20)}$$

The nominal torsional moment strength in terms of stirrup yield strength was derived above [see Eq.(5)]:

$$T_n = \frac{2 A_o A_t f_{yt}}{s} \cot\theta \qquad \text{Eq. (11-21)}$$

where $A_o = 0.85 A_{oh}$ (this is an assumption for simplicity, see 11.6.3.6)

A_{oh} = area enclosed by centerline of the outermost closed transverse torsional reinforcement as illustrated in Fig. 13-5

θ = angle of compression diagonal, ranges between 30 and 60 degree. It is suggested in 11.6.3.6 to use 45 degree for nonprestressed members and 37.5 degree for prestressed members with prestress force greater than 40 percent of tensile strength of the longitudinal reinforcement.

Note that the definition of A_o used in Eq. (8) was for the uncracked section. Also note that nominal torsional strength T_n is reached after cracking and after the concrete member has undergone considerable twisting rotation. Under these large deformations, part of the concrete cover may have spalled. For this reason, when computing area A_o corresponding to T_n, the concrete cover is ignored. Thus, parameter A_o is related to A_{oh}, the area enclosed by centerline of the outermost closed transverse torsional reinforcement. Area A_o can be determined through rigorous analysis (Ref. 13.3) or simply assumed equal to $0.85 A_{oh}$ (see 11.6.3.6).

Substituting for T from Eq. (5) into Eq. (6) and replacing $2(x_o + y_o)$ with p_h (perimeter of centerline of outermost closed transverse torsional reinforcement), the longitudinal reinforcement required to resist torsion is computed as a function of the transverse reinforcement:

Figure 13-5 Definition of A_{oh}

$$A_\ell = \left(\frac{A_t}{s}\right) p_h \left(\frac{f_{yt}}{f_y}\right) \cot^2\theta \qquad \text{Eq. (11-22)}$$

Note that term (A_t/s) used in Eq. (11-22) is that due to torsion only, and is computed from Eq. (11-21). In members subject to torsion combined with shear, flexure or axial force, the amount of longitudinal and transverse reinforcement required to resist all actions must be determined using the principle of superposition (see 11.6.3.8 and R11.6.3.8). In members subject to flexure, area of longitudinal torsion reinforcement in the flexural compression zone may be reduced to account for the compression due to flexure (11.6.3.9). In prestressed members, the longitudinal reinforcement required for torsion may consist of tendons with a tensile strength $A_{ps}f_{ps}$ equivalent to the yield force of mild reinforcement, $A_\ell f_{y\ell}$, computed by Eq. (11-22).

To reduce unsightly cracking and prevent crushing of the concrete compression struts, 11.6.3.1 prescribes an upper limit for the maximum stress due to shear and torsion, analogous to that due to shear only. In solid sections, stresses due to shear act over the full width of the section, while stresses due to torsion are assumed resisted by a thin-walled tube. See Fig. R11.6.3.1(b). Thus, 11.6.3.1 specifies an elliptical interaction between stresses due to shear and those due to torsion for solid sections as follows:

$$\sqrt{\left(\frac{V_u}{b_w d}\right)^2 + \left(\frac{T_u p_h}{1.7 A_{oh}^2}\right)^2} \le \phi \left(\frac{V_c}{b_w d} + 8\sqrt{f'_c}\right) \qquad \text{Eq. (11-18)}$$

For hollow sections, the stresses due to shear and torsion are directly additive on one side wall [see Fig. R11.6.3.1(a)]. Thus, the following linear interaction is specified:

$$\left(\frac{V_u}{b_w d}\right) + \left(\frac{T_u p_h}{1.7 A_{oh}^2}\right) \le \phi \left(\frac{V_c}{b_w d} + 8\sqrt{f'_c}\right) \qquad \text{Eq. (11-19)}$$

In Eqs. (11-18) and (11-19), V_c is the contribution of concrete to shear strength of nonprestressed (see 11.3) or prestressed (see 11.4) concrete members. Further, the 2005 code clarifies in 11.5.3 that for prestressed members d should be taken as the distance from extreme compression fiber to centroid of the prestressed and nonprestressed longitudinal tension reinforcement, if any, but need not be taken less than 0.8h.

When applying Eq. (11-19) to a hollow section, if the actual wall thickness t is less than A_{oh}/p_h, the actual wall thickness should be used instead of A_{oh}/p_h (11.6.3.3).

11.6.4   Details of Torsional Reinforcement

Longitudinal and transverse reinforcement are required to resist torsion. Longitudinal reinforcement may consist of mild reinforcement or prestressing tendons. Transverse reinforcement may consist of stirrups, welded wire reinforcement, or spiral reinforcement. To control widths of diagonal cracks, the design yield strength of longitudinal and transverse torsional reinforcement must not exceed 60,000 psi (11.6.3.4).

In the truss analogy illustrated in Fig. 13-2, the diagonal compression strut forces bear against the longitudinal corner reinforcement. In each wall, the component of the diagonal struts, perpendicular to the longitudinal reinforcement is transferred from the longitudinal reinforcement to the transverse reinforcement. It has been observed in torsional tests of beams loaded to destruction that as the maximum torque is reached, the concrete cover spalls.[13.3] The forces in the compression struts outside the stirrups, i.e. within the concrete cover, push out the concrete shell. Based on this observation, 11.6.4.2 specifies that the stirrups should be closed, with 135 degree hooks or seismic hooks as defined in 21.1. Stirrups with 90 degree hooks become ineffective when the concrete cover spalls. Similarly, lapped U-shaped stirrups have been found to be inadequate for resisting torsion due to lack of support when the concrete cover spalls. For hollow sections, the distance from the centerline of the transverse torsional reinforcement to the inside face of the wall of the hollow section must not be less than $0.5 A_{oh}/p_h$ (11.6.4.4).

11.6.5 Minimum Torsion Reinforcement

In general, to ensure ductility of nonprestressed and prestressed concrete members, minimum reinforcement is specified for flexure (10.5) and for shear (11.5.6). Similarly, minimum transverse and longitudinal reinforcement is specified in 11.6.5 whenever $T_u > \phi T_{cr}/4$. Usually, a member subject to torsion will also be simultaneously subjected to shear. The minimum area of stirrups for shear and torsion is computed from:

$$(A_v + 2A_t) = 0.75\sqrt{f'_c}\,\frac{b_w s}{f_{yt}} \geq \frac{50 b_w s}{f_{yt}} \qquad \text{Eq. (11-23)}$$

which now accounts for higher strength concretes (see 11.6.5.2).

The minimum area of longitudinal reinforcement is computed from:

$$A_{\ell,\min} = \frac{5\sqrt{f'_c}\,A_{cp}}{f_y} - \left(\frac{A_t}{s}\right)p_h \frac{f_{yt}}{f_y} \qquad \text{Eq. (11-24)}$$

but A_t/s (due to torsion only) must not be taken less than $25 b_w/f_{yt}$.

11.6.6 Spacing of Torsion Reinforcement

Spacing of stirrups must not exceed the smaller of $p_h/8$ and 12 in. For a square beam subject to torsion, this maximum spacing is analogous to a spacing of about d/2 in a beam subject to shear (11.6.6.1).

The longitudinal reinforcement required for torsion must be distributed around the perimeter of the closed stirrups, at a maximum spacing of 12 in. In the truss analogy, the compression struts push against the longitudinal reinforcement which transfers the transverse forces to the stirrups. Thus, the longitudinal bars should be inside the stirrups. There should be at least one longitudinal bar or tendon in each corner of the stirrups to help transmit the forces from the compression struts to the transverse reinforcement. To avoid buckling of the longitudinal reinforcement due to the transverse component of the compression struts, the longitudinal reinforcement must have a diameter not less than 1/24 of the stirrup spacing, but not less than 3/8 in. (11.6.6.2).

11.6.7 Alternative Design for Torsion

Section 11.6.7 introduced in the 2005 code allows using alternative torsion design procedures for solid sections with h/b_t ratio of three or more. According to 2.1, h is defined as overall thickness of height of members, and b_t is width of that part of cross section containing the closed stirrup resisting torsion. This criterion would be easy to apply to rectangular sections. For other cross sections see discussion below.

An alternative procedure can only be used if its adequacy has been proven by comprehensive tests. Commentary R11.6.7 suggests an alternative procedure, which has been described in detail by Zia and Hsu in Ref 13.4. This procedure is briefly outlined below and its application is also illustrated in Example 13.1.

ZIA-HSU ALTERNATIVE DESIGN PROCEDURE FOR TORSION

Zia-Hsu method for torsion design applies to solid rectangular, box, and flanged sections of prestressed and nonprestressed members. In this procedure L-, T-, inverted T-, and I-shaped sections are subdivided into rectangles, provided that these rectangles include closed stirrups and longitudinal reinforcement required for torsion. Equally important is that the stirrups must overlap adjacent rectangles. This alternative method is most appropriate for precast spandrel beams with a tall stem and a small ledge at the bottom of the stem. In this case, the h/b_t ratio is checked for the vertical stem.

The following steps summarize the procedure:

1. Determine the factored shear force V_u and the factored torsional moment T_u

2. Calculate the shear and torsional constant

$$C_t = \frac{b_w d}{\Sigma x^2 y} \quad (15)$$

where b_w is the web width and d is the distance from extreme compression fiber to centroid of longitudinal prestressed and nonprestressed tension reinforcement, if any, but need not be less than 0.80h for prestressed members. The section has to be divided into rectangular components of dimensions x and y ($x < y$) in such a way that the sum of x^2y terms is maximum. For overhanging flanges, however, the width shall not be taken more than three times the flange thickness (i.e. height).

3. Check the threshold (minimum) torsional moment

$$T_{min} = \phi \, 0.5 \sqrt{f'_c} \, \gamma \Sigma x^2 y \quad (16)$$

where $\gamma = \sqrt{1 + \dfrac{10 f_{pc}}{f'_c}}$ is a prestressing factor and f_{pc} is the average prestressing force in the member after losses.

If $T_u \leq T_{min}$, then torsion design is not required. Otherwise proceed to Step 4.

4. Check the maximum permissible torsional moment

$$T_{max} = \frac{\frac{1}{3} C \gamma \sqrt{f'_c} \Sigma x^2 y}{\sqrt{1 + \left(\dfrac{C \gamma V_u}{30 C_t T_u}\right)^2}} \quad (17)$$

where $C = 12 - 10 \dfrac{f_{pc}}{f'_c}$. If $T_u > T_{max}$, then the section is not adequate and needs to be redesigned.

Options are to use a larger cross section, or increase f'_c or f_{pc}.

5. Calculate nominal torsional moment strength provided by concrete under pure torsion

$$T'_c = 0.8 \sqrt{f'_c} \, \Sigma x^2 y (2.5\gamma - 1.5) \quad (18)$$

6. Calculate the nominal shear strength provided by concrete without torsion $V'_c = 2\sqrt{f'_c} \, b_w d$ for nonprestressed members and the smaller of V_{ci} and V_{cw} for prestressed members, where V_{ci} and V_{cw} are defined by Eqs (11-10) and (11-12), respectively.

7. Calculate the nominal torsional moment strength provided by concrete under combined loading

$$T_c = \frac{T_c'}{\sqrt{1+\left(\frac{T_c'}{V_c'}\frac{V_u}{T_u}\right)^2}} \qquad (19)$$

8. Calculate the nominal shear strength provided by concrete under combined loading

$$V_c = \frac{V_c'}{\sqrt{1+\left(\frac{V_c'}{T_c'}\frac{T_u}{V_u}\right)^2}} \qquad (20)$$

9. Compute transverse reinforcement for torsion

   If $T_u > \phi T_c$, then the area of transverse torsional reinforcement required over distance s equals

$$\frac{A_t}{s} = \frac{T_s}{\alpha_t x_1 y_1 f_{yt}} \qquad (21)$$

where:
A_t = area of one leg of a closed stirrup resisting torsion

$$T_s = \frac{T_u}{\phi} - T_c$$

$$\alpha_t = 0.66 + 0.33\left(\frac{y_1}{x_1}\right), \text{ but no more than 1.5}$$

x_1 = shorter center-to-center dimension of a closed stirrup
y_1 = longer center-to-center dimension of a closed stirrup

10. Compute transverse reinforcement for shear

    If $V_u > \phi V_c$, then the area of transverse shear reinforcement required over distance s equals

$$\frac{A_v}{s} = \frac{V_s}{d\, f_{yt}} \qquad (22)$$

where:
A_v = the area of a stirrup (all legs) in section,

$$V_s = \frac{V_u}{\phi} - V_c$$

11. Calculate the total transverse reinforcement

    The total transverse reinforcement required for shear and torsion is equal to

$$\frac{A_v}{s} + 2\frac{A_t}{s}$$

but should not be taken less than $\left(\dfrac{A_v}{s}+2\dfrac{A_t}{s}\right)_{min}$, which is equal to the smaller of

$$50\left(1+12\dfrac{f_{pc}}{f_c}\right)\dfrac{b_w}{f_{yt}} \text{ and } 200\dfrac{b_w}{f_{yt}}.$$

12. Calculate longitudinal torsional reinforcement

    The area of longitudinal torsional reinforcement required is equal to the larger of

    $$A_\ell = 2A_t\left(\dfrac{x_1+y_1}{s}\right) \tag{23}$$

    and

    $$A_\ell = \left[\dfrac{400xs}{f_y}\left(\dfrac{T_u}{T_u+\dfrac{V_u}{3C_t}}\right)-2A_t\right]\left[\dfrac{x_1+y_1}{s}\right] \tag{24}$$

    However, the value calculated from Eq (24) need not exceed the value obtained when the smaller of

    $$50\left(1+12\dfrac{f_{pc}}{f_c}\right)\dfrac{b_w s}{f_{yt}} \text{ and } 200\dfrac{b_w s}{f_{yt}} \text{ is substituted for } 2A_t.$$

Application of the ACI procedure (11.6) and the Zia-Hsu procedure (Ref. 13.4) is illustrated in Example 13.1

REFERENCES

13.1 MacGregor, J.G., and Ghoneim, M.G., "Design for Torsion," *ACI Structural Journal*, March-April 1995, pp. 211-218.

13.2 Fanella, D. A. and Rabbat, B.G., *Design of Concrete Beams for Torsion*, Engineering Bulletin EB106.02D, Portland Cement Association, Skokie, IL 1997.

13.3 Collins, M.P. and Mitchell, D., *Prestressed Concrete Structures*, Prentice Hall, Englewood Cliffs, NJ, 1991, 766 pp.

13.4 Zia, P. and Hsu T.T.C., "Design for Torsion and Shear in Prestressed Concrete Flexural Members", *PCI Journal,* Vol. 25, No. 3, May-June 2004, pp. 34-42.

13.5 pcaBeam-Analysis, design, and investigation of reinforced concrete beams and one-way slab systems, Portland Cement Association, Skokie, IL, 2005.

Example 13.1—Precast Spandrel Beam Design for Combined Shear and Torsion

Design a precast, nonprestressed concrete spandrel beam for combined shear and torsion. Roof members are simply supported on spandrel ledge. Spandrel beams are connected to columns to transfer torsion. Continuity between spandrel beams is not provided.

Compare torsional reinforcement requirements using ACI 318-05 provisions, Zia-Hsu alternative design for torsion, and pcaBeam (Ref 13.5) software.

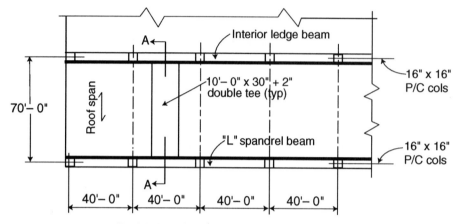

Partial plan of precast roof system

Design Criteria:
Live load = 30 lb/ft^2
Dead load = 90 lb/ft^2 (double tee + topping + insulation + roofing)
f'_c = 5000 psi (w_c = 150 pcf)
f_y = 60,000 psi

Roof members are 10 ft wide double tee units, 30 in. deep with 2 in. topping. Design of these units is not included in this design example. For lateral support, alternate ends of roof members are fixed to supporting beams.

Calculations and Discussion	Code Reference

A. ACI 318 Procedure (11.6)

1. The load from double tee roof members is transferred to the spandrel beam as concentrated forces and torques. For simplicity assume double tee loading on spandrel beam as uniform. Calculate factored loading M_u, V_u, T_u for spandrel beam.

Example 13.1 (cont'd)   **Calculations and Discussion**   **Code Reference**

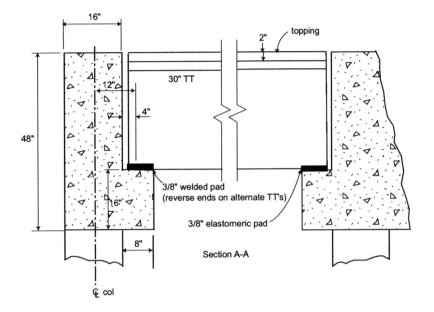

Section A-A

Dead load:
Superimposed = (0.090) (70)/2 = 3.15
Spandrel = [(1.33) (4.00) + (1.33) (0.67)] 0.150 = 0.94
                                                     Total = 4.08 kips/ft

Live load = (0.030) (70)/2 = 1.05 kips/ft

Factored load = (1.2) (4.08) + (1.6) (1.05) = 6.58 kips/ft   *9.2.1*

At center of span, $M_u = \dfrac{6.58 \times 40^2}{8} = 1316$ ft-kips

End shear $V_u = (6.58)(40)/2 = 131.6$ kips

Torsional factored load = $1.2\,(3.15) + 1.2\left(\dfrac{16}{12} \times \dfrac{8}{12} \times 0.150\right) + 1.6\,(1.05) = 5.62$ kips/ft

Eccentricity of double tee reactions relative to centerline of spandrel beam = 8 + 4 = 12 in.

End torsional moment $T_u = 5.62 \left(\dfrac{40}{2}\right)\left(\dfrac{12}{12}\right) = 112.4$ ft-kips

Assumed = 45.5 in.

Critical section for torsion is at the face of the support because of concentrated torques applied by the double tee stems at a distance less than d from the face of the support.   *11.6.2.4*

Example 13.1 (cont'd)	Calculations and Discussion	Code Reference

Critical section for shear is also at the face of support because the load on the spandrel beam is not applied close to the top of the member and because the concentrated forces transferred by the double tee stems are at a distance less than d from the face of the support.

11.1.3.(b)
11.1.3.(c)

Therefore, critical section is 8 in. from column centerline.

At critical section: [20.0 - (8.0/12) = 19.33 ft from midspan]

V_u = 131.6 (19.33/20.0) = 127.20 kips

T_u = 112.4 (19.33/20.0) = 108.6 ft-kips

The spandrel beam must be designed for the full factored torsional moment since it is required to maintain equilibrium.

11.6.2.1

2. Check if torsion may be neglected

11.6.1

Torsion may be neglected if $T_u < \dfrac{\phi T_{cr}}{4}$

$\phi = 0.75$

9.3.2.3

$$T_{cr} = 4\sqrt{f'_c}\left(\dfrac{A_{cp}^2}{P_{cp}}\right)$$

Eq. (9)

A_{cp} = area enclosed by outside perimeter of spandrel beam, including the ledge
= (16)(48) + (16)(8) = 768 + 128 = 896 in.²

p_{cp} = outside perimeter of spandrel beam
= 2(16 + 48) + 2(8) = 144 in.

The limiting value to ignore torsion is:

$$\phi\sqrt{f'_c}\left(\dfrac{A_{cp}^2}{P_{cp}}\right) = 0.75\sqrt{5000}\left(\dfrac{896^2}{144}\right)\dfrac{1}{12,000} = 24.6 \text{ ft-kips} < 108.6 \text{ ft-kips}$$

Eq.(12)

Torsion must be considered.

3. Determine required area of stirrups for torsion

Design torsional strength must be equal to or greater than the required torsional strength:

$\phi T_n \geq T_u$

Eq. (11-20)

	Code
Example 13.1 (cont'd) **Calculations and Discussion**	**Reference**

where

$$T_n = \frac{2A_o A_t f_{yv}}{s} \cot\theta \qquad \text{Eq. (11-21)}$$

$A_o = 0.85 A_{oh}$

A_{oh} = area enclosed by centerline of the outermost closed transverse torsional reinforcement

Assuming 1.25 in. cover (precast concrete exposed to weather) and No. 4 stirrup   *7.7.3(a)*

$A_{oh} = (13)(45) + (8)(13) = 689 \text{ in.}^2$

$A_o = 0.85(689) = 585.6 \text{ in.}^2$

For nonprestressed member, use $\theta = 45$ degree   *11.6.3.6(a)*

Substituting in Eqs. (11-20) and (11-21)

$$\frac{A_t}{s} = \frac{T_u}{2\phi A_o f_{yv} \cot\theta}$$

$$\frac{A_t}{s} = \frac{(108.6)(12,000)}{2(0.75)(586.6)(60,000)(1.0)} = 0.025 \text{ in.}^2/\text{in./leg}$$

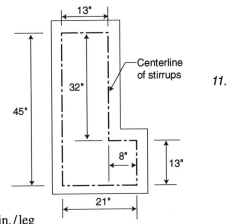

4. Calculate required area of stirrups for shear

$$V_c = 2\sqrt{f'_c}\, b_w d \qquad \text{Eq. (11-3)}$$

$$= 2\sqrt{5000}\,(16)(45.5)/1000$$

$$= 102.95 \text{ kips}$$

From Eqs. (11-1) and (11-2)

$$V_s = \frac{V_u}{\phi} - V_c = \frac{127.2}{0.75} - 102.95 = 66.65 \text{ kips}$$

$$\frac{A_v}{s} = \frac{V_s}{f_{yv} d} = \frac{66.65}{60(45.5)} = 0.024 \text{ in.}^2/\text{in.}$$

5. Determine combined shear and torsion stirrup requirements   *11.6.3.8*

$$\frac{A_t}{s} + \frac{A_v}{2s} = 0.025 + \frac{0.024}{2} = 0.037 \text{ in.}^2/\text{in./leg}$$

		Code
Example 13.1 (cont'd)	**Calculations and Discussion**	**Reference**

Try No. 4 bar, $A_b = 0.20$ in.^{2}

$$s = \frac{0.20}{0.038} = 5.26 \text{ in.} \quad \text{Use 5 in. minimum spacing.}$$

6. Check maximum stirrup spacing

    For torsion spacing must not exceed $p_h/8$ or 12 in.:   11.6.6

    $p_h = 2(13 + 45) + 2(6) = 128$ in.

    $$\frac{p_h}{8} = \frac{128}{8} = 16 \text{ in.}$$

    For shear, spacing must not exceed $d/2$ or 24 in. $\left(V_s = 66.65 \text{ kips} < 4\sqrt{f'_c}b_wd = 205.9 \text{ kips}\right)$:   11.5.5.1, 11.5.5.3

    $$\frac{d}{2} = \frac{45.5}{2} = 22.75 \text{ in.}$$

    Use 5 in. minimum and 12 in. maximum spacing.

7. Check minimum stirrup area

    $$(A_v + 2A_t) = 0.75\sqrt{f'_c}\frac{b_ws}{f_{yt}} = 0.75\sqrt{5,000}\frac{(16)(12)}{60,000} = 0.17 \text{ in.}^2$$

    $$> \frac{50b_ws}{f_{yv}} = \frac{50(16)(12)}{60,000} = 0.16 \text{ in.}^2 \qquad \text{Eq. (11-23)}$$

    Area provided = $2(0.20) = 0.40$ in.^{2} > 0.17 in.^{2} O.K.

8. Determine stirrup layout

    Since both shear and torsion are zero at the center of span, and are assumed to vary linearly to the maximum value at the critical section, the start of maximum stirrup spacing can be determined by simple proportion.

    $$\frac{s \text{ (critical)}}{s \text{ (maximum)}}(19.33) = \frac{5}{12}(19.33) = 8.05 \text{ ft, say 8 ft from midspan.}$$

9. Check for crushing of the concrete compression struts   11.6.3.1

    $$\sqrt{\left(\frac{V_u}{b_wd}\right)^2 + \left(\frac{T_up_h}{1.7A_{oh}^2}\right)^2} \leq \phi\left(\frac{V_c}{b_wd} + 8\sqrt{f'_c}\right) \qquad \text{Eq. (11-18)}$$

Example 13.1 (cont'd)   **Calculations and Discussion**   **Code Reference**

$$\sqrt{\left(\frac{127,200}{(16)(45.5)}\right)^2 + \left(\frac{(108,600 \times 12)(128)}{1.7(689)^2}\right)^2} = 270.64 \text{ psi} < 10\phi\sqrt{f'_c} = 530 \text{ psi O.K.}$$

10. Calculate longitudinal torsion reinforcement   11.6.3.7

$$A_\ell = \left(\frac{A_t}{s}\right) p_h \left(\frac{f_{yt}}{f_y}\right) \cot^2\theta$$ Eq. (11-22)

$$A_\ell = (0.025)(128)\left(\frac{60}{60}\right)(1.0) = 3.20 \text{ in.}^2$$

Check minimum area of longitudinal reinforcement

$$A_{\ell,\min} = \frac{5\sqrt{f'_c}\, A_{cp}}{f_y} - \left(\frac{A_t}{s}\right) p_h \frac{f_{yt}}{f_y}$$ Eq. (11-24)

$\left(\dfrac{A_t}{s}\right)$ must not be less than $\dfrac{25 b_w}{f_{yt}} = \dfrac{25(16)}{60,000} = 0.007 \text{ in.}^2/\text{in.}$ 11.6.5.3

$$A_{\ell,\min} = \frac{5\sqrt{5000}\,(896)}{60,000} - (0.025)(122) = 2.23 \text{ in.}^2 < 3.20 \text{ in.}^2$$

The longitudinal reinforcement required for torsion must be distributed around the perimeter of the closed stirrups, at a maximum spacing of 12 in. The longitudinal bars should be inside the stirrups. There should be at least one longitudinal bar in each corner of the stirrups. Select 12 bars.   11.6.6.2

Area of each longitudinal bar = $\dfrac{3.17}{12}$ = 0.264 in.²   Use No. 5 bars

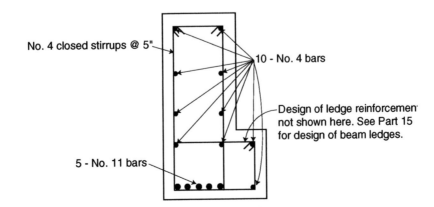

		Code
Example 13.1 (cont'd)	**Calculations and Discussion**	**Reference**

11. Size combined longitudinal reinforcement

    Use No. 5 bars in sides and top corners of spandrel beam. Note that two of the twelve longitudinal bars (bars at the bottom of the web) required for torsion are to be combined with the ledge reinforcement. Design of the ledge reinforcement is not shown here. See Part 15 of this document for design of beam ledges.

    Determine required flexural reinforcement, assuming tension-controlled behavior.

    $\phi = 0.90$ 9.3.2
    From Part 7,

    $$R_n = \frac{M_u}{\phi b d^2} = \frac{1316 \times 12,000}{0.9 \times 16 \times 45.5^2} = 529.73 \text{ psi}$$

    $$\rho = \frac{0.85 f'_c}{f_y}\left(1 - \sqrt{1 - \frac{2R_n}{0.85 f'_c}}\right)$$

    $$= \frac{0.85 \times 5}{60}\left(1 - \sqrt{1 - \frac{2 \times 529.73}{0.85 \times 5000}}\right) = 0.0095$$

    $$A_s = \rho b d = 0.0095 \times 16 \times 45.5 = 6.92 \text{ in.}^2$$

    As bottom reinforcement at midspan, provide (2/12) of the longitudinal torsion reinforcement in addition to the flexural reinforcement.

    $$\left(\frac{2}{12}\right)(3.20) + 6.92 = 7.45 \text{ in.}^2$$

    As bottom reinforcement at end of span, provide (2/12) of the longitudinal torsion   *12.11.1*
    reinforcement plus at least (1/3) the positive reinforcement for flexure:

    $$\left(\frac{2}{12}\right)(3.20) + \left(\frac{6.92}{3}\right) = 2.84 \text{ in.}^2$$

    Use 5-No. 11 bars ($A_s = 7.80 \text{ in.}^2 > 7.45 \text{ in.}^2$)

    Check if section is tension-controlled, based on provided reinforcement.

    From a strain compatibility analysis, conservatively assuming that the section is subjected to flexure only (see Eq. (8) in Part 8),

Example 13.1 (cont'd)   **Calculations and Discussion**   **Code Reference**

$$\varepsilon_t = 0.003 \left(\frac{\beta_1}{1 - \sqrt{1 - \frac{40}{17} \frac{R_n}{f'_c}}} - 1 \right) = 0.003 \left(\frac{0.80}{1 - \sqrt{1 - \frac{40}{17} \frac{529.73}{5000}}} - 1 \right) = 0.015 > 0.005$$

Therefore, section is tension-controlled, and $\phi = 0.90$.   *10.3.4*

Note that for strain compatability analysis including the effects of torsion, see Ref. 13.3.

Extend 2-No. 11 bars to end of girder ($A_s = 3.12$ in.^{2} > 2.84 in.^{2})

Note that the longitudinal torsion reinforcement must be adequately anchored.

B. Zia-Hsu Alternative Torsion Design (Ref. 13-4)

For comparison torsional reinforcement requirements will be determined according to Zia-Hsu alternative design procedure for torsion design. Since a non-prestressed member is considered then $f_{pc} = 0$ will be used.

1. Determine the factored shear force V_u and the factored torsional moment T_u

   Based on calculations in A (ACI 318 Procedure):
   $V_u = 127.2$ kips
   $T_u = 108.6$ ft-kips = 1303.2 in.-kips

2. Calculate the shear and torsional constant

   Compute the largest $\sum x^2 y$ value. Consider Options A and B.

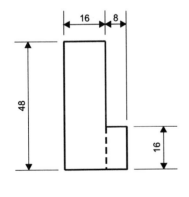
Option A

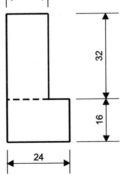
Option B

For Option A:

$$\sum x^2 y = \left(16^2 \times 48\right) + \left(8^2 \times 16\right) = 13{,}312 \text{ in.}^3$$

Example 13.1 (cont'd)	Calculations and Discussion	Code Reference

For Option B:

$$\Sigma x^2 y = \left(16^2 \times 32\right) + \left(16^2 \times 24\right) = 14{,}336 \text{ in.}^3$$

$$C_t = \frac{b_w d}{\Sigma x^2 y} = \frac{16 \times 45.5}{14{,}336} = 0.05078 \frac{1}{\text{in.}} \qquad \textit{Eq. (15)}$$

3. Check the minimum torsional moment

$$T_{min} = \phi 0.5 \sqrt{f_c'}\, \gamma \Sigma x^2 y$$
$$= 0.75 \times 0.5 \times \sqrt{5000} \times 1.0 \times 14{,}336/12{,}000 = 31.76 \text{ ft-kips} \qquad \textit{Eq. (16)}$$

where $\gamma = \sqrt{1 + \frac{10 f_{pc}}{f_c'}} = \sqrt{1 + \frac{10 \times 0}{5000}} = 1.0$

Since $T_u > T_{min}$ torsion design is required.

4. Check the maximum torsional moment

$$T_{max} = \frac{\frac{1}{3} C \gamma \sqrt{f_c'}\, \Sigma x^2 y}{\sqrt{1 + \left(\frac{C \gamma V_u}{30 C_t T_u}\right)^2}}$$

$$= \frac{\frac{1}{3} \times 12.0 \times 1.0 \times \sqrt{5000} \times 14{,}336}{\sqrt{1 + \left(\frac{12.0 \times 1.0 \times 127.2}{30 \times 0.05078 \times 1303.2}\right)^2}} \times \frac{1}{12{,}000} = 267.88 \text{ ft-kips} \qquad \textit{Eq. (17)}$$

where $C = 12 - 10 \frac{f_{pc}}{f_c'} = 12 - 10 \frac{0}{f_c'} = 12.0$

This section is adequate for torsion as $T_u < T_{min}$.

5. Calculate nominal torsional moment strength provided by concrete under pure torsion $\textit{Eq. (18)}$

$$T_c' = 0.8 \sqrt{f_c'}\, \Sigma x^2 y (2.5\gamma - 1.5)$$
$$= \frac{0.8 \times \sqrt{5000} \times 14{,}336 \times (2.5 \times 1.0 - 1.5)}{12{,}000} = 67.58 \text{ ft-kips}$$

		Code
Example 13.1 (cont'd)	**Calculations and Discussion**	**Reference**

6. Calculate the nominal shear strength provided by concrete without torsion

$$V_c' = 2\sqrt{f_c'}\,b_w d = 2 \times \sqrt{5000} \times 16 \times 45.5/1000 = 102.95 \text{ kips}$$

7. Calculate the nominal torsional moment strengths under combined loading

$$T_c = \frac{T_c'}{\sqrt{1+\left(\frac{T_c'}{V_c'}\frac{V_u}{T_u}\right)^2}} = \frac{67.58}{\sqrt{1+\left(\frac{67.58}{102.95}\frac{127.2}{108.6}\right)^2}} = 53.58 \text{ ft-kips} \qquad \text{Eq. (19)}$$

8. Calculate the nominal shear strengths under combined loading

$$V_c = \frac{V_c'}{\sqrt{1+\left(\frac{V_c'}{T_c'}\frac{T_u}{V_u}\right)^2}} = \frac{102.95}{\sqrt{1+\left(\frac{102.95}{67.58}\frac{108.6}{127.2}\right)^2}} = 62.75 \text{ kips}$$

Eq. (20)

9. Compute transverse reinforcement for torsion

$$T_u = 108.6 \text{ ft-kips} > \phi T_c = 0.75 \times 53.32 = 39.99 \text{ ft-kips}.$$

Area of transverse torsional reinforcement required over distance s equals

$$\frac{A_t}{s} = \frac{T_s}{\alpha_t x_1 y_1 f_{yt}} = \frac{1097.8}{1.50 \times 13 \times 45 \times 60} = 0.0208 \frac{\text{in.}^2/\text{in.}}{\text{leg}} \qquad \text{Eq. (21)}$$

where:

$$T_s = \frac{T_u}{\phi} - T_c = \frac{108.6}{0.75} - 53.58 = 91.22 \text{ ft-kips} = 1094.7 \text{ in.-kips}$$

$$\alpha_t = 0.66 + 0.33\left(\frac{y_1}{x_1}\right) = 0.66 + 0.33\left(\frac{45}{13}\right) = 1.80 > 1.50, \text{ use } 1.50$$

$x_1 = 13$ (shorter center-to-center dimension of a closed stirrup),
$y_1 = 45$ (longer center-to-center dimension of a closed stirrup).

10. Compute transverse reinforcement for shear

Example 13.1 (cont'd) — Calculations and Discussion — Code Reference

$V_u = 127.2$ kips $> \phi V_c = 0.75 \times 62.75 = 47.06$ kips

Area of transverse shear reinforcement required over distance s equals

Eq. (22)

$$\frac{A_v}{s} = \frac{V_s}{d\,f_{yt}} = \frac{106.85}{45.5 \times 60} = 0.0391 \frac{in.^2}{in.}$$

where:

$$V_s = \frac{V_u}{\phi} - V_c = \frac{127.2}{0.75} - 62.75 = 106.85 \text{ ft-kips}$$

11. Calculate the total transverse reinforcement

The total transverse reinforcement required for shear and torsion is equal to

$$\frac{A_v}{s} + 2\frac{A_t}{s} = 0.0391 + 2 \times 0.0208 = 0.0807 \frac{in.^2}{in.}$$

which is more than the required minimum of

$$\left(\frac{A_v}{s} + 2\frac{A_t}{s}\right)_{min} = 50\left(1 + 12\frac{f_{pc}}{f_c'}\right)\frac{b_w}{f_y} = 50\left(1 + 12\frac{0}{f_c'}\right)\frac{16}{60,000} = 0.0133 \frac{in.^2}{in.}$$

Assuming a two leg stirrup, the area of one leg should be

$$\frac{A_v}{2s} + \frac{A_t}{s} = 0.0391/2 + 0.0208 = 0.0404 \frac{in.^2/in.}{leg}$$

12. Calculate longitudinal torsional reinforcement

The area of longitudinal torsional reinforcement required is equal to

$$A_\ell = 2A_t\left(\frac{x_1 + y_1}{s}\right) = 2\frac{A_t}{s}(x_1 + y_1) = 2 \times 0.0208 \times (13 + 45) = 2.41 \text{ in.}^2$$

Eq. (23)

which is greater than the smaller of the following two values

Example 13.1 (cont'd)   **Calculations and Discussion**   **Code Reference**

$$A_\ell = \left[\frac{400x}{f_y}\left(\frac{T_u}{T_u + \frac{V_u}{3C_t}}\right) - \frac{2A_t}{s}\right](x_1 + y_1)$$

$$= \left[\frac{400 \times 16}{60,000}\left(\frac{1303.2}{1303.2 + \frac{127.2}{3 \times 0.05078}}\right) - 2 \times 0.0208\right](13+45) = 1.33 \text{ in.}^2$$

$$A_\ell = \left[\frac{400x}{f_y}\left(\frac{T_u}{T_u + \frac{V_u}{3C_t}}\right) - 50\frac{b_w}{f_y}\right](x_1 + y_1) \quad\quad Eq.\ (24)$$

$$= \left[\frac{400 \times 16}{60,000}\left(\frac{1303.2}{1303.2 + \frac{127.2}{3 \times 0.05078}}\right) - 0.0133\right](13+45) = 3.00 \text{ in.}^2$$

Example 13.1 (cont'd)   **Calculations and Discussion**   **Code Reference**

C. pcaBeam Solution

Torsional reinforcement requirements obtained from pcaBeam program are presented graphically in Fig. 13-6. The diagram represents combined shear and torsion capacity in terms of required and provided reinforcement area. The upper part of the diagram is related to the transverse reinforcement and shows that at the face of the support the required reinforcement is

$$\frac{A_v}{s} + 2\frac{A_t}{s} = 0.076 \frac{\text{in.}^2}{\text{in.}}$$

The lower part of the diagram is related to the torsional longitudinal reinforcement and shows that $A_\ell = 3.40$ in.^{2} is required for torsional reinforcement at the face of the support. As shown in Fig. (13-6), close to the supports, Eq. (11-22) governs the required amount of longitudinal torsional reinforcement. As expected, as T_u decreases, so does A_ℓ. However, where Eq. (11-24) for $A_{\ell,\text{min}}$ starts to govern, the amount of longitudinal reinforcement increases, although T_u decreases toward the midspan. This anomaly occurs where the minimum required transverse reinforcement governs.

Torsional reinforcement requirements are compared in Table 13-1. Transverse reinforcement requirements are in good agreement. Higher differences are observed for longitudinal reinforcement. The small discrepancy between ACI 318-05 and pcaBeam program (also based on ACI) can be attributed to numerical round-off errors and to fixed 1.5 in. side cover assumed in pcaBeam.

Table 13-1 Comparison of required torsional reinforcement

Required reinforcement	ACI 318-05	Zia-Hsu	pcaBeam
$\left(\frac{A_v}{s} + 2\frac{A_t}{s}\right)\left[\frac{\text{in.}^2}{\text{in.}}\right]$	0.074	0.081	0.076
$A_\ell \left(\text{in.}^2\right)$	3.20	2.41	3.40

Example 13.1 (cont'd)	Calculations and Discussion	Code Reference

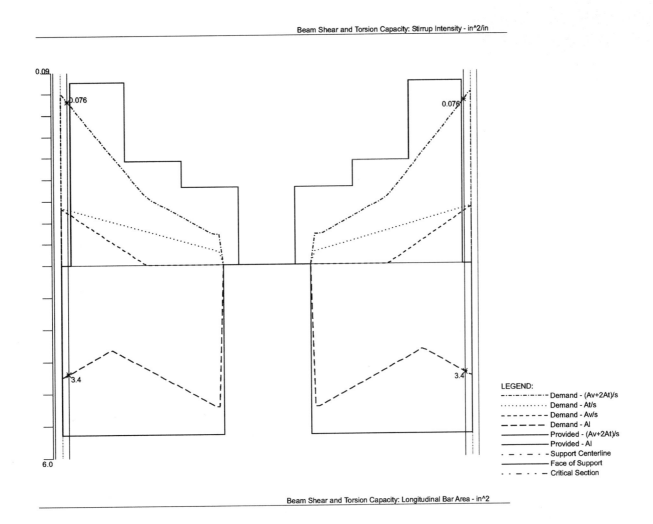

Figure 13-6 Torsional Reinforcement Requirements Obtained from pcaBeam Program

Blank

14

Shear Friction

GENERAL CONSIDERATIONS

Provisions for shear friction were introduced in ACI 318-71. With the publication of ACI 318-83, 11.7 was completely rewritten to expand the shear-friction concept to include applications (1) where the shear-friction reinforcement is placed at an angle other than 90 degrees to the shear plane, (2) where concrete is cast against concrete not intentionally roughened, and (3) with lightweight concrete. In addition, a performance statement was added to allow "any other shear-transfer design methods" substantiated by tests. It is noteworthy that 11.9 refers to 11.7 for the direct shear-transfer in brackets and corbels; see Part 15.

11.7   SHEAR-FRICTION

The shear-friction concept provides a convenient tool for the design of members for direct shear where it is inappropriate to design for diagonal tension, as in precast connections, and in brackets and corbels. The concept is simple to apply and allows the designer to visualize the structural action within the member or joint. The approach is to assume that a crack has formed at an expected location, as illustrated in Fig. 14-1. As slip begins to occur along the crack, the roughness of the crack surface forces the opposing faces of the crack to separate. This separation is resisted by reinforcement (A_{vf}) across the assumed crack. The tensile force ($A_{vf}f_y$) developed in the reinforcement by this strain induces an equal and opposite normal clamping force, which in turn generates a frictional force ($\mu A_{vf}f_y$) parallel to the crack to resist further slip.

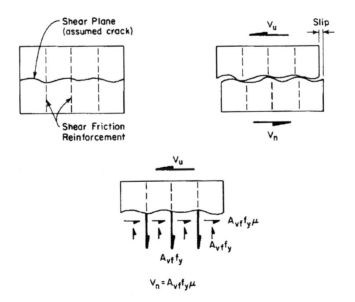

Figure 14-1 Idealization of the Shear-Friction Concept

11.7.1 Applications

Shear-friction design is to be used where direct shear is being transferred across a given plane. Situations where shear-friction design is appropriate include the interface between concretes cast at different times, an interface between concrete and steel, and connections of precast constructions, etc. Example locations of direct shear transfer and potential cracks for application of the shear-friction concept are shown in Fig. 14-2 for several types of members. Successful application of the concept depends on proper selection of location of the assumed slip or crack. In typical end or edge bearing applications, the crack tends to occur at an angle of about 20 degrees to the vertical (see Example 14.2).

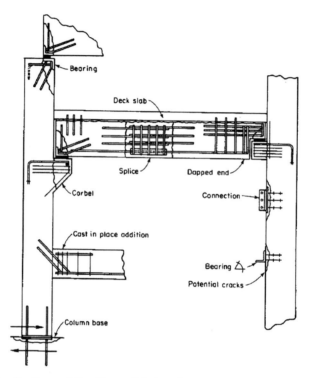

Figure 14-2 Applications of the Shear-Friction Concept and Potential Crack Locations

11.7.3 Shear-Transfer Design Methods

The shear-friction design method presented in 11.7.4 is based on the simplest model of shear-transfer behavior, resulting in a conservative prediction of shear-transfer strength. Other more comprehensive shear-transfer relationships provide closer predictions of shear-transfer strength. The performance statement of 11.7.3 "…any other shear-transfer design methods…" includes the other methods within the scope and intent of 11.7. However, it should be noted that the provisions of 11.7.5 through 11.7.10 apply to whatever shear-transfer method is used. One of the more comprehensive methods is outlined in R11.7.3. Application of the "Modified Shear-Friction Method" is illustrated in Part 15, Example 15.2. The 1992 edition of the code introduced in 17.5.2.3 a modified shear-friction equation. It applies to the interface shear between precast concrete and cast-in-place concrete.

11.7.4 Shear-Friction Design Method

As with the other shear design applications, the code provisions for shear-friction are presented in terms of the nominal shear-transfer strength V_n for direct application in the basic shear strength relation:

Design shear-transfer strength ≥ Required shear-transfer strength

$$\phi V_n \geq V_u \qquad \qquad Eq.\ (11\text{-}1)$$

Note that ϕ is 0.75 for shear and torsion (9.3.2.3). Furthermore, it is recommended that $\phi = 0.75$ be used for all design calculations involving shear-friction, where shear effects predominate. For example, 11.9.3.1 specifies the use of $\phi = 0.75$ for all design calculations in accordance with 11.9 (brackets and corbels). The nominal shear strength V_n is computed as:

$$V_n = A_{vf} f_y \mu \qquad \text{Eq. (11-25)}$$

Combining Eqs. (11-1) and (11-25), the required shear-transfer strength for shear-friction reinforcement perpendicular to the shear plane is:

$$V_u \le \phi A_{vf} f_y \mu$$

The required area of shear-friction reinforcement, A_{vf}, can be computed directly from:

$$A_{vf} = \frac{V_u}{\phi f_y \mu}$$

The condition where shear-friction reinforcement crosses the shear-plane at an angle α other than 90 degrees is illustrated in Fig. 14-3. The tensile force $A_{vf} f_y$ is inclined to the crack and must be resolved into two components: (1) a clamping component $A_{vf} f_y \sin \alpha$ with an associated frictional force $\mu A_{vf} f_y \sin \alpha$, and (2) a component parallel to the crack that directly resists slip equal to $A_{vf} f_y \cos \alpha$. Adding the two components resisting slip, the nominal shear-transfer strength becomes:

$$V_n = \mu A_{vf} f_y \sin \alpha + A_{vf} f_y \cos \alpha$$
$$= A_{vf} f_y (\mu \sin \alpha + \cos \alpha) \qquad \text{Eq. (11-26)}$$

Substituting this into Eq. (11-1):

$$V_u \le \phi [\mu A_{vf} f_y \sin \alpha + A_{vf} f_y \cos \alpha]$$

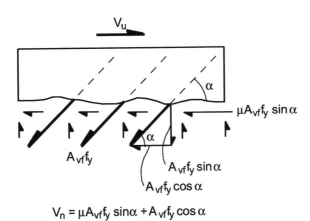

Figure 14-3 Idealization of Inclined Shear-Friction Reinforcement

For shear reinforcement inclined to the crack, the required area of shear-friction reinforcement, A_{vf}, can be computed directly from:

$$A_{vf} = \frac{V_u}{\phi f_y (\mu \sin \alpha + \cos \alpha)}$$

Note that Eq. (11-26) applies only when the shear force V_u produces tension in the shear-friction reinforcement.

The shear-friction method assumes that all shear resistance is provided by friction between crack faces. The actual mechanics of resistance to direct shear are more complex, since dowel action and the apparent cohesive strength of the concrete both contribute to direct shear strength. It is, therefore, necessary to use artificially high values of the coefficient of friction μ in the direct shear-friction equations so that the calculated shear strength will be in reasonable agreement with test results. Use of these high coefficients gives predicted strengths that are a conservative lower bound to test data, as shown in Fig. 14-4. The modified shear-friction design method given in R11.7.3 is one of several more comprehensive methods which provide closer estimates of the shear-transfer strength.

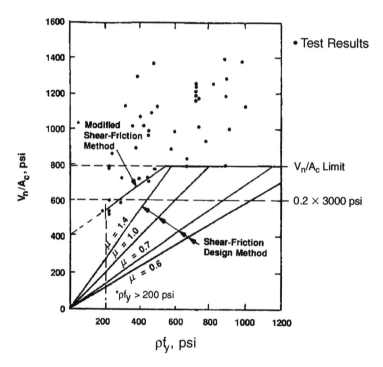

Figure 14-4 Effect of Shear-Friction Reinforcement on Shear Transfer Strength

11.7.4.3   **Coefficient of Friction**—The "effective" coefficients of friction, μ, for the various interface conditions include a parameter λ which accounts for the somewhat lower shear strength of all-lightweight and sand-lightweight concretes. For example, the μ value for all lightweight concrete ($\lambda = 0.75$) placed against hardened concrete not intentionally roughened is 0.6 (0.75) = 0.45. The coefficient of friction for different interface conditions is as follows:

    Concrete placed monolithically 1.4λ

    Concrete placed against hardened concrete with
    surface intentionally roughened as specified in 11.7.9 .... 1.0λ

    Concrete placed against hardened concrete
    not intentionally roughened 0.6λ

    Concrete anchored to as-rolled structural steel by
    headed studs or by reinforcing bars (see 11.7.10) 0.7λ

where $\lambda = 1.0$ for normal weight concrete, 0.85 for sand-lightweight concrete, and 0.75 for "all light weight" concrete.

11.7.5 Maximum Shear-Transfer Strength

The shear-transfer strength V_n cannot be taken greater than $0.2 f'_c$, nor 800 psi times the area of concrete section resisting shear transfer. This upper limit on V_n effectively limits the maximum reinforcement, as shown in Fig. 14-4. Also, for lightweight concretes, 11.9.3.2.2 limits the shear-transfer strength V_n along the shear plane for design applications with low shear span-to-depth ratios a_v/d, such as brackets and corbels. This further restriction on lightweight concrete is illustrated in Example 14.1.

11.7.7 Normal Forces

Equations (11-25) and (11-26) assume that there are no forces other than shear acting on the shear plane. A certain amount of moment is almost always present in brackets, corbels, and other connections due to eccentricity of loads or applied moments at connections. In case of moments acting on a shear plane, the flexural tension stresses and flexural compression stresses are in equilibrium. There is no change in the resultant compression $A_{vf}f_y$ acting across the shear plane and the shear-transfer strength is not changed. It is therefore not necessary to provide additional reinforcement to resist the flexural tension stresses, unless the required flexural tension reinforcement exceeds the amount of shear-transfer reinforcement provided in the flexural tension zone.

Joints may also carry a significant amount of tension due to restrained shrinkage or thermal shortening of the connected members. Friction of bearing pads, for example, can cause appreciable tensile forces on a corbel supporting a member subject to shortening. Therefore, it is recommended, although not generally required, that the member be designed for a minimum direct tensile force of at least $0.2V_u$ in addition to the shear. This minimum force is required for design of connections such as brackets or corbels (see 11.9.3.4), unless the actual force is accurately known. Reinforcement must be provided for direct tension according to 11.7.7, using $A_s = N_{uc}/\phi f_y$, where N_{uc} is the factored tensile force.

Since direct tension perpendicular to the assumed crack (shear plane) detracts from the shear-transfer strength, it follows that compression will add to the strength. Section 11.7.7 acknowledges this condition by allowing a "permanent net compression" to be added to the shear-friction clamping force, $A_{vf}f_y$. It is recommended, although not required, to use a reduction factor of 0.9 for strength contribution from such compressive loads.

11.7.8 — 11.7.10 Additional Requirements

Section 11.7.8 requires that the shear-friction reinforcement be "appropriately placed" along the shear plane. Where no moment acts on the shear plane, uniform distribution of the bars is proper. Where a moment exists, the reinforcement should be distributed in the flexural tension zone.

Reinforcement should be adequately embedded on both sides of the shear plane to develop the full yield strength of the bars. Since space is limited in thin walls, corbels, and brackets, it is often necessary to use special anchorage details such as welded plates, angles, or cross bars. Reinforcement should be anchored in confined concrete. Confinement may be provided by beam or column ties, "external" concrete, or special added reinforcement.

In 11.7.9, if coefficient of friction μ is taken equal to 1.0λ, concrete at the interface must be roughened to a full amplitude of approximately 1/4 in. This can be accomplished by raking the plastic concrete or by bushhammering or chiseling hardened concrete surfaces.

A final requirement of 11.7.10, often overlooked, is that structural steel interfaces must be clean and free of paint. This requirement is based on tests to evaluate the friction coefficient for concrete anchored to unpainted structural steel by studs or reinforcing steel ($\mu = 0.7$). Data are not available for painted surfaces. If painted surfaces are to be used, a lower value of μ would be appropriate.

DESIGN EXAMPLES

In addition to Examples 14.1 and 14.2 of this part, shear-friction design is also illustrated for direct shear-transfer in brackets and corbels (see Part 15), horizontal shear transfer between composite members (see Part 12) and at column/footing connections (see Part 22).

Example 14.1—Shear-Friction Design

A tilt-up wall panel is subject to the factored seismic shear forces shown below. Design the shear anchors assuming lightweight concrete, $w_c = 95$ pcf. $f'_c = 4000$ psi and $f_y = 60,000$ psi.

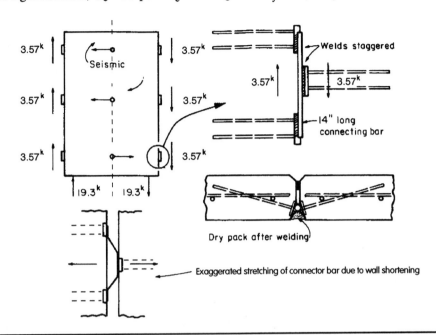

	Code
Calculations and Discussion	**Reference**

1. Design anchor steel using shear-friction method.

   Center plate is most heavily loaded. Try 2 in. × 4 in. × 1/4 in. plate.

   $V_u = 3570$ lb

   $V_u \leq \phi V_n$ *Eq. (11-1)*

   $V_u \leq \phi(A_{vf} f_y \mu)$ *Eq. (11-25)*

   For unpainted steel in contact with all lightweight concrete (95 pcf):

   $\mu = 0.7\lambda = 0.7 \times 0.75 = 0.525$ 11.7.4.3

   $\phi = 0.75$ 9.3.2.3

   Solving for $A_{vf} = \dfrac{V_u}{\phi f_y \mu} = \dfrac{3570}{0.75(60,000)(0.525)} = 0.15$ in.^{2}

   Use 2-No. 3 bars per plate   ($A_{vf} = 0.22$ in.^{2})

   Note: Weld bars to plates to develop full f_y. Length of bar must be adequate to fully develop bar.

| Example 14.1 (cont'd) | Calculations and Discussion | Code Reference |

Check maximum shear-transfer strength permitted for connection. For lightweight aggregate concrete: 11.9.3.2.2

$$V_{n(max)} = \left[0.2 - 0.07\left(\frac{a_v}{d}\right)\right] f'_c b_w d \text{ or } \left[800 - 280\left(\frac{a_v}{d}\right)\right] b_w d$$

For the purposes of the above equations, assume a_v = thickness of plate = 0.25 in., and d = distance from edge of plate to center of farthest attached rebar = 2.5 in.:

$$\frac{a_v}{d} = \frac{0.25}{2.5} = 0.1$$

Assume, for the purposes of the above equations, that $b_w d = A_c$ = contact area of plate:

$b_w d = A_c = 2 \times 4 = 8 \text{ in.}^2$

$V_{n(max)} = [0.2 - 0.07(0.1)](4000)(8) = 6176 \text{ lb}$

or $V_n = [800 - (280 \times 0.1)](8) = 6176 \text{ lb}$

$\phi V_{n(max)} = 0.75(6176) = 4632 \text{ lb}$

$V_u = 3570 \text{ lb} \leq \phi V_{n(max)} = 4632 \text{ lb}$ O.K.   Eq. (11-1)

Use 2 in. $\times$ 4 in. $\times \dfrac{1}{4}$ in. plates, with 2-No. 3 bars.

Example 14.2—Shear-Friction Design (Inclined Shear Plane)

For the pilaster beam support shown, design for shear transfer across the potential crack plane. Assume a crack at an angle of about 20 degrees to the vertical, as shown below. Beam reactions are D = 25 kips, L = 30 kips. Use T = 20 kips as an estimate of shrinkage and temperature change effects. f'_c = 3500 psi and f_y = 60,000 psi.

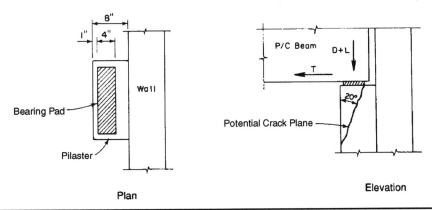

Calculations and Discussion	Code Reference

1. Factored loads to be considered:

   Beam reaction R_u = 1.2D + 1.6L = 1.2 (25) + 1.6 (30) = 30 + 48 = 78 kips   *Eq. (9-2)*

   Shrinkage and temperature effects T_u = 1.6 (20) = 32 kips (governs)   *11.9.3.4*
   but not less than 0.2 (R_u) = 0.2 (78) = 15.6 kips

   Note that the live load factor of 1.6 is used with T, due to the low confidence level in determining shrinkage and temperature effects occurring in service. Also, a minimum value of 20 percent of the beam reaction is considered (see 11.9.3.4 for corbel design).

2. Evaluate force conditions along potential crack plane.

   Direct shear transfer force along shear plane:

   V_u = $R_u \sin\alpha$ + $T_u \cos\alpha$ = 78 (sin70°) + 32 (cos70°)

       = 73.3 + 11.0 = 84.3 kips

   Net tension (or compression) across shear plane:

   N_u = $T_u \sin\alpha$ - $R_u \cos\alpha$ = 32 (sin70°) - 78 (cos70°)

       = 30.1 - 26.7 = 3.4 kips (net tension)

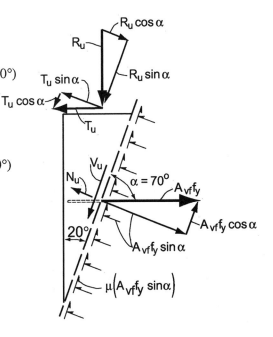

| Example 14.2 (cont'd) | Calculations and Discussion | Code Reference |

If the load conditions were such as to result in net compression across the shear plane, it still should not have been used to reduce the required A_{vf}, because of the uncertainty in evaluating the shrinkage and temperature effects. Also, 11.7.7 permits a reduction in A_{vf} only for "permanent" net compression.

3. Shear-friction reinforcement to resist direct shear transfer. Use μ for concrete placed monolithically.

$$A_{vf} = \frac{V_u}{\phi f_y (\mu \sin\alpha + \cos\alpha)}$$ Eq. (11-26)

$\mu = 1.4\lambda = 1.4 \times 1.0 = 1.4$ 11.7.4.3

$$A_{vf} = \frac{84.3}{0.75 \times 60 \,(1.4 \sin 70° + \cos 70°)} = 1.13 \text{ in.}^2$$ [μ from 11.7.4.3]

4. Reinforcement to resist net tension.

$$A_n = \frac{N_u}{\phi f_y (\sin\alpha)} = \frac{3.4}{0.75 \times 60 \,(\sin 70°)} = 0.08 \text{ in.}^2$$

Since failure is primarily controlled by shear, use $\phi = 0.75$ (see 11.9.3.1 for corbel design).

5. Add A_{vf} and A_n for total area of required reinforcement. Distribute reinforcement uniformly along the potential crack plane.

$A_s = 1.13 + 0.08 = 1.21 \text{ in.}^2$

Use No. 3 closed ties (2 legs per tie)

Number required = $1.21 / [2 \,(0.11)] = 5.5$, say 6.0 ties

Ties should be distributed along length of potential crack plane; approximate length = $5/(\tan 20°) \approx 14$ in.

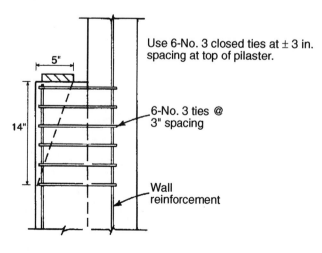

14-10

Example 14.2 (cont'd)	Calculations and Discussion	Code Reference

6. Check reinforcement requirements for dead load only plus shrinkage and temperature effects. Use 0.9 load factor for dead load to maximize net tension across shear plane.

$R_u = 0.9D = 0.9(25) = 22.5$ kips, $T_u = 32$ kips

$V_u = 22.5(\sin 70°) + 32(\cos 70°) = 21.1 + 11.0 = 32.1$ kips

$N_u = 32(\sin 70°) - 22.5(\cos 70°) = 30.1 - 7.7 = 22.4$ kips (net tension)

$$A_{vf} = \frac{32.1}{0.75 \times 60(1.4\sin 70° + \cos 70°)} = 0.43 \text{ in.}^2$$

$$A_n = \frac{22.4}{0.75 \times 60 \times \sin 70°} = 0.53 \text{ in.}^2$$

$A_s = 0.43 + 0.53 = 0.96$ in.^{2} < 1.21 in.^{2}

Therefore, original design for full dead load + live load governs.

7. Check maximum shear-transfer strength permitted

$V_{n(max)} = [0.2f'_c A_c]$ or $[800 A_c]$ *11.7.5*

Taking the width of the pilaster to be 16 in.:

$$A_c = \left(\frac{5}{\sin 20°}\right) \times 16 = 234 \text{ in.}^2$$

$V_{n(max)} = 0.2(3500)(234)/1000 = 164$ kips   (governs)

or $V_{n(max)} = 800(234)/1000 = 187$ kips

$\phi V_{n(max)} = 0.75(164) = 123$ kips

$V_u = 84.3$ kips $\leq \phi V_{n(max)} = 123$ kips   O.K.   *Eq. (11-1)*

Blank

15

Brackets, Corbels and Beam Ledges

GENERAL CONSIDERATIONS

Provisions for the design of brackets and corbels were introduced in ACI 318-71. These provisions were derived based on extensive test results. The 1977 edition of the code permitted design of brackets and corbels based on shear friction, but maintained the original design equations. The provisions were completely revised in ACI 318-83, eliminating the empirical equations of the 1971 and 1977 codes, and simplifying design by using the shear-friction method exclusively for nominal shear-transfer strength V_n. From 1971 through 1999 code, the provisions were strictly limited to shear span-to-depth ratio a_v/d less than or equal to 1.0. Since 2002, the code allows the use of the provisions of Appendix A, Strut-and-tie models, to design brackets and corbels with a_v/d ratios less than 2.0, while the provisions of 11.9 continue to apply only for a_v/d ratios less than or equal to 1.0.

11.9 LIMITATIONS OF BRACKET AND CORBEL PROVISIONS

The design procedure for brackets and corbels recognizes the deep beam or simple truss action of these short-shear-span members, as illustrated in Fig. 15-1. Four potential failure modes shown in Fig. 15-1 shall be prevented: (1) Direct shear failure at the interface between bracket or corbel and supporting member; (2) Yielding of the tension tie due to moment and direct tension; (3) Crushing of the internal compression "strut;" and (4) Localized bearing or shear failure under the loaded area.

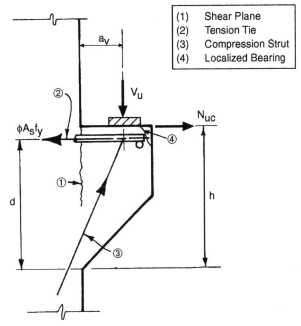

Figure 15-1 Structural Action of Corbel

For brackets and corbels with a shear span-to-depth ratio a_v/d less than 2, the provision of Appendix A may be used for design. The provisions of 11.9.3 and 11.9.4 are permitted with $a_v/d \leq 1$ and the horizontal force $N_{uc} \leq V_n$.

Regardless which design method is used, the provisions of 11.9.2, 11.9.3.2.1, 11.9.3.2.2, 11.9.5, 11.9.6, and 11.9.7 must be satisfied.

When a_v/d is greater than 2.0, brackets and corbels shall be designed as cantilevers subjected to the applicable provisions of flexure and shear.

11.9.1 - 11.9.5   Design Provisions

The critical section for design of brackets and corbels is taken at the face of the support. This section should be designed to resist simultaneously a shear V_u, a moment $M_u = V_u a_v + N_{uc}(h - d)$, and a horizontal tensile force N_{uc} (11.9.3). The value of N_{uc} must be not less than $0.2V_u$, unless special provisions are made to avoid tensile forces (11.9.3.4). This minimum value of N_{uc} is established to account for the uncertain behavior of a slip joint and/or flexible bearings. Also, the tension force N_{uc} typically is due to indeterminate causes such as restrained shrinkage or temperature stresses. In any case it shall be treated as a live load with load factor of 1.6 (11.9.3.4). Since corbel and bracket design is predominantly controlled by shear, 11.9.3.1 specifies that the strength reduction factor ϕ shall be taken equal to 0.75 for all design conditions.

For normal weight concrete, shear strength V_n is limited to the smaller of $0.2 f'_c b_w d$ and $800 b_w d$ (11.9.3.2). For lightweight concrete, V_n is limited by the provisions of 11.9.3.2.2, which are somewhat more restrictive than those for normal weight concrete. Tests show that for lightweight concrete, V_n is a function of f'_c and a_v/d.

For brackets and corbels, the required reinforcement is:

A_{vf} = area of shear-friction reinforcement to resist direct shear V_u, computed in accordance with 11.7 (11.9.3.2).

A_f = area of flexural reinforcement to resist moment $M_u = V_u a_v + N_{uc}(h - d)$, computed in accordance with 10.2 and 10.3 (11.9.3.3).

A_n = area of tensile reinforcement to resist direct tensile force N_{uc}, computed in accordance with 11.9.3.4.

Actual reinforcement is to be provided as shown in Fig. 15-2 and includes:

A_{sc} = primary tension reinforcement

A_h = shear reinforcement (closed stirrups or ties)

This reinforcement is provided such that total amount of reinforcement $A_{sc} + A_h$ crossing the face of support is the greater of (a) $A_{vf} + A_n$, and (b) $3A_f/2 + A_n$ to satisfy criteria based on test results.[15.1]

If case (a) controls (i.e., $A_{vf} > 3A_f/2$):

$$A_{sc} = A_{vf} + A_n - A_h$$
$$= A_{vf} + A_n - 0.5(A_{sc} - A_n) \hspace{2cm} 11.9.4$$

or $A_{sc} = 2A_{vf}/3 + A_n$ (primary tension reinforcement)

and $A_h = (0.5)(A_{sc} - A_n) = A_{vf}/3$ (closed stirrups or ties)   11.9.4

If case (b) controls (i.e., $3A_f/2 > A_{vf}$):

$$A_{sc} = 3A_f/2 + A_n - A_h$$

$$= 3A_f/2 + A_n - 0.5(A_{sc} - A_n)$$

or $\quad A_{sc} = A_f + A_n \quad$ (primary tension reinforcement)

and $\quad A_h = (0.5)(A_{sc} - A_n) = A_f/2 \quad$ (closed stirrups or ties)

In both cases (a) and (b), $A_h = (0.5)(A_{sc} - A_n)$ determines the amount of shear reinforcement to be provided as closed stirrups parallel to A_{sc} and uniformly distributed within $(2/3)d$ adjacent to A_{sc} per 11.9.4.

A minimum ratio of primary tension reinforcement $\rho_{min} = 0.04 f'_c/f_y$ is required to ensure ductile behavior after cracking under moment and direct tensile force (11.9.5).

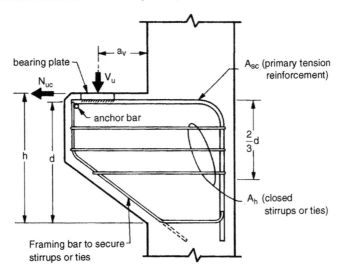

Figure 15-2 Corbel Reinforcement

BEAM LEDGES

Beam with ledges shall be designed for the overall member effects of flexure, shear, axial forces, and torsion, as well as for local effects in the vicinity of the ledge (Refs. 15.2-15.6). The design of beam ledges is not specifically addressed by the code. This section addresses only local failure modes and reinforcement requirements to prevent such failure.

Design of beam ledges is somewhat similar to that of a bracket or corbel with respect to loading conditions. Additional design considerations and reinforcement details need to be considered in beam ledges. Accordingly, even though not specifically addressed by the code, special design of beam ledges is included in this Part. Some failure modes discussed above for brackets and corbels are also shown for beam ledges in Fig. 15-3. However, with beam ledges, two additional failure modes shall be considered (see Fig. 15-3): (5) separation between ledge and beam web near the top of the ledge in the vicinity of the ledge load and (6) punching shear. The vertical load applied to the ledge is resisted by a compression strut. In turn, the vertical component of the inclined compression strut must be picked up by the web stirrups (stirrup legs A_v adjacent to the side face of the web) acting as "hanger" reinforcement to carry the ledge load to the top of beam. At the reentrant corner of the ledge to web intersection, a diagonal crack would extend to the stirrup and run downward next to the stirrup. Accordingly, a slightly larger shear span, a_f, is used to compute the moment due to V_u. Therefore, the critical section for moment is taken at center of beam stirrups, not at face of beam. Also, for beam ledges, the internal moment arm should not be taken greater than $0.8h$ for flexural strength.

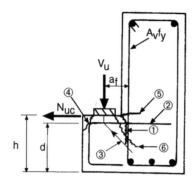

Figure 15-3 Structural Action of Beam Ledge

The design procedure described in this section is based on investigations performed by Mirza and Furlong (Refs. 15.3 to 15.5). The key information needed by the designer is establishing the effective width of ledge for each of the potential failure modes. These effective widths were determined by Mirza and Furlong through analytical investigations, with results verified by large scale testing. Design of beam ledges can also be performed by the strut-and-tie procedure (refer to Part 17 for discussion).

Design to prevent local failure modes requires consideration of the following actions:

1. Shear V_u

2. Horizontal tensile force N_{uc} greater or equal to $0.2V_u$, but not greater than V_u.

3. Moment $M_u = V_u a_f + N_{uc}(h-d)$

Reinforcement for the different failure modes is determined based on the effective widths or critical sections discussed below. In all cases, the required strengths (V_u, M_u, or N_u) should be less than or equal to the design strengths (ϕV_n, ϕM_n, or ϕN_n). The strength reduction factor ϕ is taken equal to 0.75 for all actions, as for brackets and corbels. The strength requirements for different failure modes are shown below for normal weight concrete. When lightweight aggregate concrete is used, modifications should be made per 11.2.

a. Shear Friction

Parameters affecting determination of the shear friction reinforcement are illustrated in Figure 15-4.

$$V_u \leq 0.2\phi f'_c (W + 4a_v)d \tag{1}$$

$$\leq \phi \mu A_{vf} f_y$$

where
d = effective depth of ledge from centroid of top layer of ledge transverse reinforcement to the bottom of the ledge (see Fig. 15-4)

μ = coefficient of friction per 11.7.4.3.

Note that per 11.7.5, $0.2 f'_c \leq 800$ psi.

If $(W + 4a_v) > S$, then $V_u \leq 0.2\phi f'_c Sd$, where S is the distance between center of adjacent bearings on the same ledge.

At ledge ends, $V_u \leq 0.2\phi f'_c (2c)d$, where c is the distance from center of end bearing to the end of the ledge. However, 2c should be less than or equal to the smaller of $(W + 4a_v)$ and S.

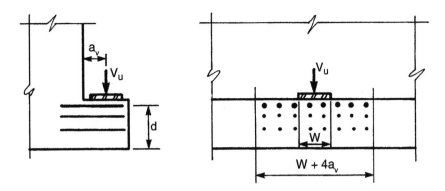

Figure 15-4 Shear Friction

b. Flexure

Conditions for flexure and direct tension are shown in Figure 15-5.

$$V_u a_f + N_{uc}(h-d) \leq \phi A_f f_y (jd) \tag{2}$$

$$N_{uc} \leq \phi A_n f_y$$

The primary tension reinforcement A_{sc} should equal the greater of $(A_f + A_n)$ or $(2A_{vf}/3 + A_n)$. If $(W + 5a_f)$ > S, reinforcement should be placed over distance S. At ledge ends, reinforcement should be placed over distance (2c), where c is the distance from the center of the end bearing to the end of the ledge, but not more than $1/2 (W + 5a_f)$. Reference 15.5 recommends taking $jd = 0.8d$.

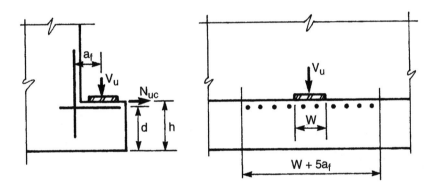

Figure 15-5 Flexure and Direct Tension

c. Punching Shear

Critical perimeter for punching shear design is illustrated in Fig. 15-6.

$$V_u \leq 4\phi\sqrt{f'_c} (W + 2L + 2d_f) d_f \tag{3}$$

where d_f = effective depth of ledge from top of ledge to center of bottom transverse reinforcement (see Fig. 15-6)

Truncated pyramids from adjacent bearings should not overlap. At ledge ends,

$$V_u \leq 4\phi\sqrt{f'_c} (W + L + d_f) d_f$$

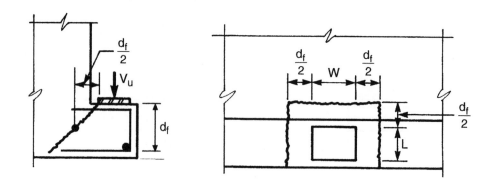

Figure 15-6 Punching Shear

d. Hanger Reinforcement

Hanger reinforcement should be proportioned to satisfy strength. Furthermore, serviceability criteria should be considered when the ledge is subjected to a large number of live load repetitions, as in parking garages and bridges. As shown in Figure 15-7, strength is governed by

$$V_u \leq \phi \frac{A_v f_y}{s} S \qquad (4)$$

where A_v = area of one leg of hanger reinforcement

S = distance between ledge loads

s = spacing of hanger reinforcement

Serviceability is governed by

$$V \leq \frac{A_v (0.5 f_y)}{s} (W + 3av) \qquad (5)$$

where V is the reaction due to service dead load and live load.

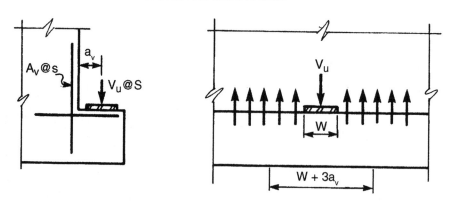

Figure 15-7 Hanger Reinforcement to Prevent Separation of Ledge from Stem

In addition, hanger reinforcement in inverted tees is governed by consideration of the shear failure mode depicted in Figure 15-8:

$$2V_u \leq 2\left[2\phi\sqrt{f'_c}b_f d'_f\right] + \phi \frac{A_v f_y}{s}(W + 2d'_f) \tag{6}$$

where d'_f = flange depth from top of ledge to center of bottom longitudinal reinforcement (see Fig. 15-8)

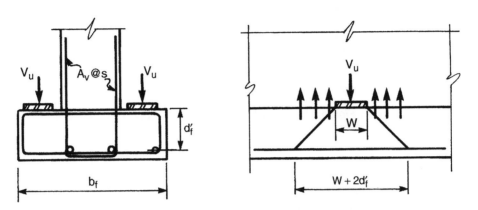

Figure 15-8 Hanger Reinforcement to Prevent Partial Separation of Ledge from Stem and Shear of the Ledge

11.9.6 Development and Anchorage of Reinforcement

All reinforcement must be fully developed on both sides of the critical section. Anchorage within the support is usually accomplished by embedment or hooks. Within the bracket or corbel, the distance between load and support face is usually too short, so that special anchorage must be provided at the outer ends of both primary reinforcement A_{sc} and shear reinforcement A_h. Anchorage of A_{sc} is normally provided by welding an anchor bar of equal size across the ends of A_{sc} (Fig. 15-9(a)) or welding to an armor angle. In the former case, the anchor bar must be located beyond the edge of the loaded area. Where anchorage is provided by a hook or a loop in A_{sc}, the load must not project beyond the straight portion of the hook or loop (Fig. 15-9(b)). In beam ledges, anchorage may be provided by a hook or loop, with the same limitation on the load location (Fig. 15-10). Where a corbel or beam ledge is designed to resist specific horizontal forces, the bearing plate should be welded to A_{sc}.

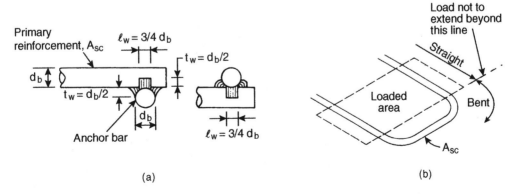

Figure 15-9 Anchorage Details Using (a) Cross-Bar Weld and (b) Loop Bar Detail

The closed stirrups or ties used for A_h must be similarly anchored, usually by engaging a "framing bar" of the same diameter as the closed stirrups or ties (see Fig. 15-2).

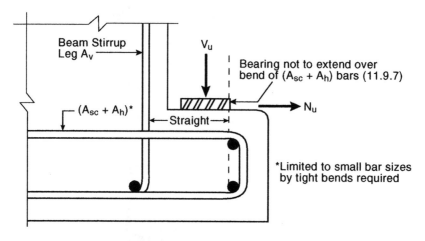

Figure 15-10 Bar Details for Beam Ledge

REFERENCES

15.1 Mattock, Alan H., Chen, K.C., and Soongswang, K., "The Behavior of Reinforced Concrete Corbels," *PCI Journal*, Prestressed Concrete Institute, V. 21, No. 2, Mar.-Apr. 1976, pp. 52-77.

15.2 Klein, G. J., "Design of Spandrel Beams," *PCI Journal*, Vol. 31, No. 5, Sept.-Oct. 1986, pp. 76-124.

15.3 Mirza, Sher Ali, and Furlong, Richard W., "Strength Criteria for Concrete Inverted T-Girder," *Journal of Structural Engineering*, V. 109, No. 8, Aug. 1983, pp. 1836-1853.

15.4 Mirza, Sher Ali, and Furlong, Richard W., "Serviceability Behavior and Failure Mechanisms of Concrete Inverted T-Beam Bridge Bent Caps," *ACI Journal, Proceedings*, V. 80, No. 4, July-Aug. 1983, pp. 294-304.

15.5 Mirza, S. A., and Furlong, R. W., "Design of Reinforced and Prestressed Concrete Inverted T-Beams for Bridge Structures," *PCI Journal*, Vol. 30, No. 4, July-Aug. 1985, pp. 112-136.

15.6 "Design of Concrete Beams for Torsion," Portland Cement Association, Skokie, Illinois, 1999.

Example 15.1—Corbel Design

Design a corbel with minimum dimensions to support a beam as shown below. The corbel is to project from a 14-in. square column. Restrained creep and shrinkage create a horizontal force of 20 kips at the welded bearing.

f'_c = 5000 psi (normal weight)

f_y = 60,000 psi

Beam reactions:

DL = 24 kips
LL = 37.5 kips
T = 20 kips

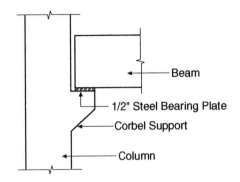

Calculations and Discussion	Code Reference

1. Size bearing plate based on bearing strength on concrete according to 10.17. Width of bearing plate = 14 in.

   $V_u = 1.2(24) + 1.6(37.5) = 88.8$ kips *Eq. (9-2)*

   $V_u \leq \phi P_{nb} = \phi(0.85 f'_c A_1)$ *10.17.1*

   $\phi = 0.65$ *9.3.2.4*

   $88.8 = 0.65(0.85 \times 5 \times A_1) = 2.763 A_1$

   $A_1 = \dfrac{88.8}{2.763} = 32.14$ in.^{2}

   Bearing length = $\dfrac{32.14}{14} = 2.30$ in.

   Use 2.5 in. × 14 in. bearing plate.

2. Determine shear span 'a_v' with 1 in. max. clearance at beam end. Beam reaction is assumed at third point of bearing plate to simulate rotation of supported girder and triangular distribution of stress under bearing pad.

   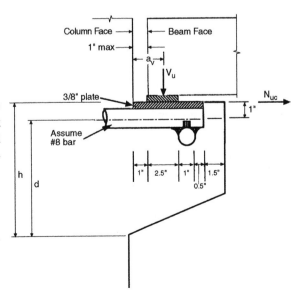

   $a_v = \dfrac{2}{3}(2.5) + 1.0 = 2.67$ in.

   Use $a_v = 3$ in. maximum.

   Detail cross bar just outside outer bearing edge.

Example 15.1 (cont'd)	Calculations and Discussion	Code Reference

3. Determine total depth of corbel based on limiting shear-transfer strength V_n.

    V_n is the least of $V_n = 800 b_w d$ (governs)   **11.9.3.2.1**

    or $V_n = 0.2 f'_c b_w d = (0.2 \times 5000) b_w d = 1000 b_w d$

    Thus, $V_u \leq \phi V_n = \phi(800 b_w d)$

    Required $d = \dfrac{88,800}{0.75 (800 \times 14)} = 10.57$ in.

    Assuming No. 8 bar, 3/8 in. steel plate, plus tolerance,

    $h = 10.57 + 1.0 = 11.57$ in.   Use $h = 12$ in.

    For design, $d = 12.0 - 1.0 = 11.0$ in.

    $\dfrac{a_v}{d} = 0.27 < 1$ O.K.   **11.9.1**

    Also, $N_{uc} = 1.6 \times 20 = 32.0$ kips (treat as live load)

    $N_{uc} < V_u = 88.8$ kips   O.K.

4. Determine shear-friction reinforcement A_{vf}.   **11.9.3.2**

    $A_{vf} = \dfrac{V_u}{\phi f_y \mu} = \dfrac{88.8}{0.75 (60)(1.4 \times 1)} = 1.41$ in.^{2} **11.7.4.1**

5. Determine direct tension reinforcement A_n.

    $A_n = \dfrac{N_{uc}}{\phi f_y} = \dfrac{32.0}{0.75 \times 60} = 0.71$ in.^{2} **11.9.3.1**

6. Determine flexural reinforcement A_f.   **11.7.4.3**

    $M_u = V_u a_v + N_{uc}(h - d) = 88.8(3) + 32(12 - 11) = 298.4$ in.-kips   **11.9.3.3**

    Find A_f using conventional flexural design methods or conservatively use $j_u d = 0.9 d$.

    $A_f = \dfrac{298.4}{0.75(60)(0.9 \times 11)} = 0.67$ in.^{2}

    Note that for all design calculations, $\phi = 0.75$ **11.9.3.1**

| Example 15.1 (cont'd) | Calculations and Discussion | Code Reference |

7. Determine primary tension reinforcement A_s. 11.9.3.5

   $\frac{2}{3}A_{vf} = \frac{2}{3}(1.41) = 0.94 \text{ in.}^2 > A_f = 0.67 \text{ in.}^2$; Therefore, $\frac{2}{3}A_{vf}$ controls design

   $A_{sc} = \frac{2}{3}A_{vf} + A_n = 0.94 + 0.71 = 1.65 \text{ in.}^2$

   Use 2-No. 9 bars, $A_{sc} = 2.0 \text{ in.}^2$

   Check minimum reinforcement: 11.9.5

   $\rho_{min} = 0.04\left(\frac{f'_c}{f_y}\right) = 0.04\left(\frac{5}{60}\right) = 0.0033$

   $A_{sc(min)} = 0.0033\,(14)\,(11) = 0.51 \text{ in.}^2 < A_{sc} = 2.0 \text{ in.}^2$ O. K.

8. Determine shear reinforcement A_h 11.9.4

   $A_h = 0.5\,(A_{sc} - A_n) = 0.5\,(2.0 - 0.71) = 0.65 \text{ in.}^2$

   Use 3-No. 3 stirrups, $A_h = 0.66 \text{ in.}^2$

   Distribute stirrups in two-thirds of effective corbel depth adjacent to A_{sc}.

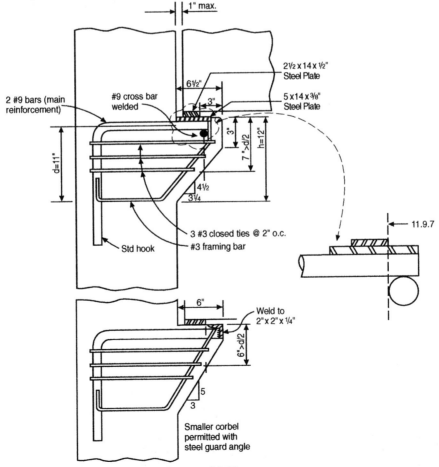

Example 15.2—Corbel Design . . . Using Lightweight Concrete and Modified Shear-Friction Method

Design a corbel to project from a 14-in.-square column to support the following beam reactions:

Dead load = 32 kips

Live load = 30 kips

Horizontal force = 24 kips

f'_c = 4000 psi (all lightweight)

f_y = 60,000 psi

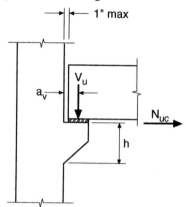

Calculations and Discussion	Code Reference

1. Size bearing plate

   $V_u = 1.2(32) + 1.6(30) = 86.4$ kips   Eq. (9-2)

   $V_u \leq \phi P_{nb} = \phi(0.85 f'_c A_1)$ 10.17.1

   $\phi = 0.65$ 9.3.2.4

   $86.4 = 0.65(0.85 \times 4 \times A_1)$

   Solving, $A_1 = 39.1$ in.²

   Length of bearing required = $\dfrac{39.1}{14} = 2.8$ in.

   Use 14 in. × 3 in. bearing plate.

2. Determine a_v.

   Assume beam reaction to act at outer third point of bearing plate, and 1 in. gap between back edge of bearing plate and column face. Therefore:

   $a_v = 1 + \dfrac{2}{3}(3) = 3$ in.

3. Determine total depth of corbel based on limiting shear-transfer strength V_n. For easier placement of reinforcement and concrete, try h = 15 in. Assuming No. 8 bar:

   $d = 15 - 0.5 - 0.375 = 14.13$ in., say 14 in.

   $\dfrac{a_v}{d} = \dfrac{3}{14} = 0.21 < 1.0$ 11.9.1

   $N_{uc} = 1.6 \times 24 = 38.4$ kips $< V_u = 86.4$ kips   O.K.

		Code
Example 15.2 (cont'd)	**Calculations and Discussion**	**Reference**

For lightweight concrete and $f'_c = 4000$ psi, V_n is the least of: 11.9.3.2.2

$$V_n = \left(800 - 280\frac{a_v}{d}\right)b_w d = [800 - (280 \times 0.21)] \, 14 \times \frac{14}{1000} = 145.3 \text{ kips}$$

$$V_n = \left(0.2 - 0.07\frac{a_v}{d}\right)f'_c b_w d = [0.2 - 0.07(0.21)](4,000)(14)\frac{14}{1000} = 145.3 \text{ kips}$$

$\phi V_n = 0.75(145.3) = 109.0$ kips $> V_u = 86.4$ kips   O.K.

4. Determine shear-friction reinforcement A_{vf}.   11.9.3.2

    Using a Modified Shear-Friction Method as permitted by 11.7.3 (see R11.7.3):

    $$V_n = 0.8 A_{vf} f_y + K_1 b_w d, \text{ with } \frac{A_{vf} f_y}{b_w d} \text{ not less than 200 psi}$$

    For all lightweight concrete, $K_1 = 200$ psi   R11.7.3

    $$V_u \leq \phi V_n = \phi(0.8 A_{vf} f_y + 0.2 b_w d)$$

    Solving for A_{vf}:

    $$A_{vf} = \frac{V_u - \phi(0.2 b_w d)}{\phi(0.8 f_y)}, \text{ but not less than } 0.2 \times \frac{b_w d}{f_y}$$

    $$= \frac{86.4 - (0.75 \times 0.2 \times 14 \times 14)}{0.75(0.8 \times 60)} = 1.58 \text{ in.}^2 \text{ (governs)}$$

    but not less than $0.2 \times \dfrac{b_w d}{f_y} = 0.2 \times \dfrac{14 \times 14}{60} = 0.65$ in.^{2}

    For comparison, compute A_{vf} by Eq. (11-25):   11.7.4.3

    For lightweight concrete,

    $$\mu = 1.4\lambda = 1.4(0.75) = 1.05$$

    $$A_{vf} = \frac{V_u}{\phi f_y \mu} = \frac{86.4}{0.75 \times 60 \times 1.05} = 1.83 \text{ in.}^2 > 1.58 \text{ in.}^2$$

    Note: Modified shear-friction method presented in R11.7.3 would give a closer estimate of shear-transfer strength than the conservative shear-friction method in 11.7.4.1.

5. Determine flexural reinforcement A_f.   11.9.3.3

    $$M_u = V_u a_v + N_{uc}(h - d) = 86.4(3) + 38.4(15 - 14.0) = 297.6 \text{ in.-kips}$$

    Find A_f using conventional flexural design methods, or conservatively use $j_u d = 0.9d$

| Example 15.2 (cont'd) | Calculations and Discussion | Code Reference |

$$A_f = \frac{M_u}{\phi f_y j u d} = \frac{297.6}{0.75 \times 60 \times 0.9 \times 14} = 0.53 \text{ in.}^2$$

Note that for all design calculations, $\phi = 0.75$ 11.9.3.1

6. Determine direct tension reinforcement A_n.   11.9.3.4

$$A_n = \frac{N_{uc}}{\phi f_y} = \frac{38.4}{0.75 \times 60} = 0.85 \text{ in.}^2$$

7. Determine primary tension reinforcement A_{sc}.   11.9.3.5

$$\left(\frac{2}{3}\right) A_{vf} = \left(\frac{2}{3}\right) 1.83 = 1.22 \text{ in.}^2 > A_f = 0.53 \text{ in.}^2; \text{ Therefore, } \left(\frac{2}{3}\right) A_{vf} \text{ controls design.}$$

$$A_{sc} = \left(\frac{2}{3}\right) A_{vf} + A_n = 1.22 + 0.85 = 2.07 \text{ in.}^2$$

Use 3-No. 8 bars, $A_{sc} = 2.37 \text{ in.}^2$

Check $A_{sc(min)} = 0.04 \left(\frac{4}{60}\right) 14 \times 14 = 0.52 \text{ in.}^2 < A_{sc} = 2.37 \text{ in.}^2$ O. K.   11.9.5

8. Determine shear reinforcement A_h.   11.9.4

$A_h = 0.5 (A_{sc} - A_n) = 0.5 (2.37 - 0.85) = 0.76 \text{ in.}^2$

Use 4-No. 3 stirrups, $A_h = 0.88 \text{ in.}^2$

The shear reinforcement is to be placed within two-thirds of the effective corbel depth adjacent to A_{sc}.

$$s_{max} = \left(\frac{2}{3}\right)\frac{14}{4} = 2.33 \text{ in.} \quad \text{Use } 2\frac{1}{4} \text{ in. o.c. stirrup spacing.}$$

9. Corbel details

Corbel will project $(1 + 3 + 2) = 6$ in. from column face.

Use 6-in. depth at outer face of corbel, then depth at outer edge of bearing plate will be

$6 + 3 = 9 \text{ in.} > \frac{14}{2} = 7.0 \text{ in.}$ O.K.   11.9.2

A_{sc} to be anchored at front face of corbel by welding a No. 8 bar transversely across ends of A_{sc} bars.   11.9.6

A_{sc} must be anchored within column by standard hook.

Example 15.2 (cont'd)	Calculations and Discussion	Code Reference

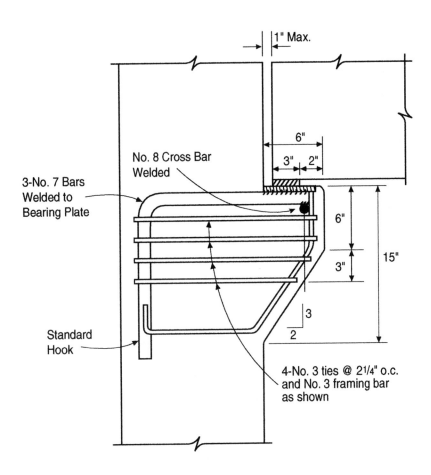

Example 15.3—Beam Ledge Design

f'_c = 5000 psi (normal weight)

f_y = 60,000 psi

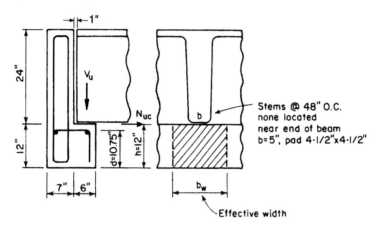
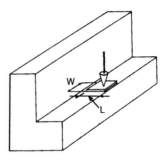

The L-beam shown is to support a double-tee parking deck spanning 64 ft. Maximum service loads per stem are: DL = 11.1 kips; LL = 6.4 kips; total load = 17.5 kips. The loads may occur at any location on the L-beam ledge except near beam ends. The stems of the double-tees rest on 4.5 in. × 4.5 in. × 1/4 in. neoprene bearing pads (1000 psi maximum service load).

Design in accordance with the code provisions for brackets and corbels may require a wider ledge than the 6 in. shown. To maintain the 6-in. width, one of the following may be necessary: (1) Use of a higher strength bearing pad (up to 2000 psi); or (2) Anchoring primary ledge reinforcement A_{sc} to an armor angle.

This example will be based on the 6-in. ledge with 4.5-in.-square bearing pad. At the end of the example an alternative design will be shown.

Note: This example illustrates design to prevent potential local failure modes. In addition, ledge beams should be designed for global effects, not considered in this example. For more details see References 15.2 to 15.6.

Calculations and Discussion	Code Reference

1. Check 4.5 × 4.5 in. bearing pad size (1000 psi maximum service load).

   Capacity = 4.5 × 4.5 × 1.0 = 20.3 kips > 17.5 kips   O.K.

2. Determine shear spans and effective widths for both shear and flexure [Ref. 15.3 to 15.5]. The reaction is considered at outer third point of the bearing pad.

   a. For shear friction

   $$a_v = 4.5 \left[\frac{2}{3}\right] + 1.0 = 4 \text{ in.}$$

   Effective width = $W + 4a_v$ = 4.5 + 4 (4) = 20.5 in.

	Code
Example 15.3 (cont'd) **Calculations and Discussion**	**Reference**

    b. For flexure, critical section is at center of the hanger reinforcement (A_v)

       Assume 1 in. cover and No. 4 bar stirrups

       $a_f = 4 + 1 + 0.25 = 5.25$ in.

       Effective width = $W + 5a_f = 4.5 + 5(5.25) = 30.75$ in.

3. Check concrete bearing strength.

    $V_u = 1.2(11.1) + 1.6(6.4) = 23.6$ kips   *Eq. (9-2)*

    $\phi P_{nb} = \phi(0.85 f'_c A_1)$ *10.17.1*

    $\phi = 0.65$ *9.3.2.4*

    $\phi P_{nb} = 0.65(0.85 \times 5 \times 4.5 \times 4.5) = 55.9$ kips > 23.6 kips   O.K.

4. Check effective ledge section for maximum nominal shear-transfer strength V_n.   *11.9.3.2.1*

    For $f'_c = 5000$ psi: $V_n(\max) = 800 b_w d$, where $b_w = (W + 4a_v) = 20.5$ in.

    $V_n = \dfrac{800(20.5)(10.75)}{1000} = 176.3$ kips

    $\phi = 0.75$ *11.9.3.1*

    $\phi V_n = 0.75(176.3) = 132.2$ kips > 23.6 kips   O.K.

5. Determine shear-friction reinforcement A_{vf}.   *11.9.3.2*

    $A_{vf} = \dfrac{V_u}{\phi f_y \mu} = \dfrac{23.6}{0.75(60)1.4} = 0.37$ in.²/per effective width of 20.5 in.   *11.7.4.1*

    where $\mu = 1.4$ *11.7.4.3*

6. Check for punching shear (Eq. (3))

    $V_u \leq 4\phi\sqrt{f'_c}\,(W + 2L + 2d_f)\,d_f$

    $W = L = 4.5$ in.

    $d_f \approx 10$ in. (assumed)

    $4\phi\sqrt{f'_c}\,(3W + 2d_f)\,d_f = 4 \times 0.75 \times \sqrt{5000}\,[(3 \times 4.5) + (2 \times 10)] \times 10/1000$

          $= 71.1$ kips > 23.6 kips

| Example 15.3 (cont'd) | Calculations and Discussion | Code Reference |

7. Determine reinforcement to resist direct tension A_n. Unless special provisions are made to reduce direct tension, N_u should be taken not less than $0.2V_u$ to account for unexpected forces due to restrained long-time deformation of the supported member, or other causes. When the beam ledge is designed to resist specific horizontal forces, the bearing plate should be welded to the tension reinforcement A_{sc}.

11.9.3.4

$$N_u = 0.2V_u = 0.2(23.6) = 4.7 \text{ kips}$$

$$A_n = \frac{N_u}{\phi f_y} = \frac{4.7}{0.75(60)} = 0.10 \text{ in.}^2/\text{per effective width of 30.75 in. } (0.003 \text{ in.}^2/\text{in.})$$

8. Determine flexural reinforcement A_f.

$$M_u = V_u a_f + N_u(h - d) = 23.6(5.25) + 4.7(12 - 10.75) = 129.8 \text{ in.-kips}$$

Find A_f using conventional flexural design methods. For beam ledges, Ref. 15.5 recommends to use $j_u d = 0.8d$.

11.9.3.3

$$\phi = 0.75$$

11.9.3.1

$$A_f = \frac{129.8}{0.75(60)(0.8 \times 10.75)} = 0.34 \text{ in.}^2/\text{per 30.75 in. width} = 0.011 \text{ in.}^2/\text{in.}$$

9. Determine primary tension reinforcement A_{sc}.

11.9.3.5

$$\left(\frac{2}{3}\right)A_{vf} = \left(\frac{2}{3}\right)0.37 = 0.25 \text{ in.}^2/\text{per 20.5 in. width} = 0.012 \text{ in.}^2/\text{in.}$$

$$A_{sc} = \left(\frac{2}{3}\right)A_{vf} + A_n = 0.012 + 0.003 = 0.015 \text{ in.}^2/\text{in. (governs)}$$

$$A_{sc} = A_f + A_n = 0.011 + 0.003 = 0.014 \text{ in.}^2/\text{in.}$$

Check $A_{sc(min)} = 0.04\left(\frac{f'_c}{f_y}\right)d$ per in. width

11.9.5

$$= 0.04\left(\frac{5}{60}\right)10.75 = 0.036 \text{ in.}^2/\text{in.} > 0.015 \text{ in.}^2/\text{in.}$$

For typical shallow ledge members, minimum A_{sc} by 11.9.5 will almost always govern.

10. Determine shear reinforcement A_h.

$$A_h = 0.5(A_{sc} - A_n) = 0.5(0.036 - 0.003) = 0.017 \text{ in.}^2/\text{in.}$$

11.9.4

		Code
Example 15.3 (cont'd)	**Calculations and Discussion**	**Reference**

11. Determine final size and spacing of ledge reinforcement.

    For $A_{sc} = 0.036$ in.^{2}/in.:

    Try No. 5 bars ($A = 0.31$ in.^{2})

    $$s_{max} = \frac{0.31}{0.036} = 8.6 \text{ in.}$$

    Use No. 5 @ 8 in.

    $A_h = 0.017$ in.^{2}/in. For ease of constructability, provide reinforcement A_h at same spacing of 8 in.

    Provide No. 4 ($A = 0.2$ in.^{2}) @ 8 in. within 2/3d adjacent to A_{sc}.

12. Check required area of hanger reinforcement.

    For strength (Eq. (4)):

    $$A_v = \frac{V_u s}{\phi f_y S}$$

    For $s = 8$ in. and $S = 48$ in.

    $$A_v = \frac{23.6 \times 8}{0.75 \times 60 \times 48} = 0.09 \text{ in.}^2$$

    For serviceability (Eq. (5)):

    $$A_v = \frac{V}{0.5 f_y} \times \frac{s}{(W + 3a_v)}$$

    $V = 11.1 + 6.4 = 17.5$ kips

    $W + 3a_v = 4.5 + (3 \times 4) = 16.5$ in.

    $$A_v = \frac{17.5}{0.5 \times 60} \times \frac{8}{16.5} = 0.28 \text{ in.}^2 \text{ (governs)}$$

    No. 5 hanger bars @ 8 in. are required

    Sufficient stirrups for combined shear and torsion must be provided for global effects in the ledge beam. (See Refs. 15.5 and 15.6)

13. Reinforcement Details

    In accordance with 11.9.7, bearing area (4.5 in. pad) must not extend beyond straight portion of beam ledge reinforcement, nor beyond inside edge of transverse anchor bar. With a 4.5 in. bearing pad, this requires that the width of ledge be increased to 9 in. as shown below. Alternately, a 6 in. ledge with a 3 in. medium strength pad (1500 psi) and the ledge reinforcement welded to an armor angle would satisfy the intent of 11.9.7.

Example 15.3 (cont'd)	Calculations and Discussion	Code Reference

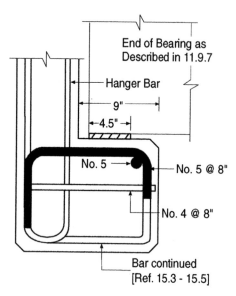

9 in. Ledge Detail

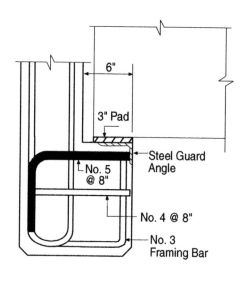

6 in. Ledge Detail
(Alternate)

16

Shear in Slabs

UPDATE FOR THE '05 CODE

The expression $2\sqrt{f_c'}$ is replaced by $\phi\left(2\sqrt{f_c'}\right)$ in 11.12.6.2 to correct a typographical error.

11.12　SPECIAL PROVISIONS FOR SLABS AND FOOTINGS

The provisions of 11.12 must be satisfied for shear design in slabs and footings. Included are requirements for critical shear sections, nominal shear strength of concrete, and shear reinforcement.

11.12.1　Critical Shear Section

In slabs and footing, shear strength in the vicinity of columns, concentrated loads, or reactions is governed by the more severe of two conditions:

- Wide-beam action, or one-way shear, as evaluated by provisions 11.1 through 11.5.
- Two-way action, as evaluated by 11.12.2 through 11.12.6.

Analysis for wide-beam action considers the slab to act as a wide beam spanning between columns. The critical section extends in a plane across the entire width of the slab and is taken at a distance d from the face of the support (11.12.1.1); see Fig. 16-1. In this case, the provisions of 11.1 through 11.5 must be satisfied. Except for long, narrow slabs, this type of shear is seldom a critical factor in design, as the shear force is usually well below the shear capacity of the concrete. However, it must be checked to ensure that shear strength is not exceeded.

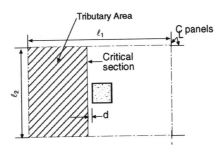

Fig. 16-1 Tributary Area and Critical Section for Wide-Beam Shear

Two-way or "punching" shear is generally the more critical of the two types of shear in slab systems supported directly on columns. Depending on the location of the column, concentrated load, or reaction, failure can occur along two, three, or four sides of a truncated cone or pyramid. The perimeter of the critical section b_o is located in such a manner that it is a minimum, but need not approach closer than a distance $d/2$ from edges or corners of columns, concentrated loads, or reactions, or from changes in slab thickness such as edges of capitals or drop panels (11.12.1.2); see Fig. 16-2. In this case the provisions of 11.12.2 through 11.12.6 must be satisfied. It is important to note that it is permissable to use a rectangular perimeter b_o to define the critical section for square or rectangular columns, concentrated loads, or reaction areas (11.12.1.3).

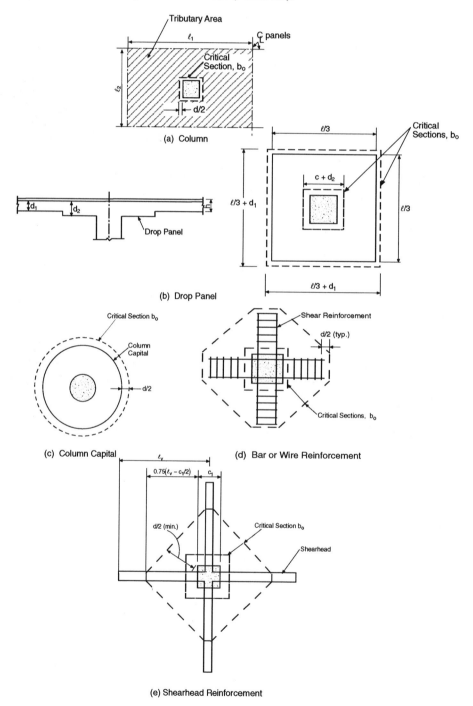

Fig. 16-2 Tributary Areas and Critical Sections for Two-Way Shear

11.12.2 Shear Strength Requirement for Two-Way Action

In general, the factored shear force V_u at the critical shear section shall be less than or equal to the shear strength ϕV_n:

$$\phi V_n \geq V_u \qquad \text{Eq. (11-1)}$$

where the nominal shear strength V_n is:

$$V_n = V_c + V_s \qquad \text{Eq. (11-2)}$$

and

V_c = nominal shear strength provided by concrete, computed in accordance with 11.12.2.1 if shear reinforcement is not used or 11.12.3.1 if shear reinforcement is used.

V_s = nominal shear strength provided by reinforcement, if required, computed in accordance with 11.12.3 if bars, wires, or stirrups are used, or 11.12.4 if shearheads are used. Where moment is transferred between the slab and the column in addition to direct shear, 11.12.6 shall apply.

11.12.2.1 Nominal shear strength provided by concrete V_c for slabs without shear reinforcement

The shear stress provided by concrete at a section v_c is a function of the concrete compressive stress f'_c, and is limited to $4\sqrt{f'_c}$ for square columns. The nominal shear strength provided by concrete V_c is obtained by multiplying v_c by the area of concrete section resisting shear transfer, which is equal to the perimeter of the critical shear section b_o multiplied by the effective depth of the slab d:

$$V_c = 4\sqrt{f'_c}\, b_o d \qquad \text{Eq. (11-35)}$$

Tests have indicated that the value of $4\sqrt{f'_c}$ is unconservative when the ratio of the long and short sides of a rectangular column or loaded area β_c is larger than 2.0. In such cases, the shear stress on the critical section varies as shown in Fig. 16-3. Equation (11-33) accounts for the effect of β_c on the concrete shear strength:

$$V_c = \left(2 + \frac{4}{\beta}\right)\sqrt{f'_c}\, b_o d \qquad \text{Eq. (11-33)}$$

From Fig. 16-3, it can be seen that for $\beta \leq 2.0$ (i.e., square or nearly square column or loaded area), two-way shear action governs, and the maximum concrete shear stress v_c is $4\sqrt{f'_c}$. For values of β value larger than 2.0, the concrete stress decreases linearly to a minimum $2\sqrt{f'_c}$, which is equivalent to shear stress for one-way shear.

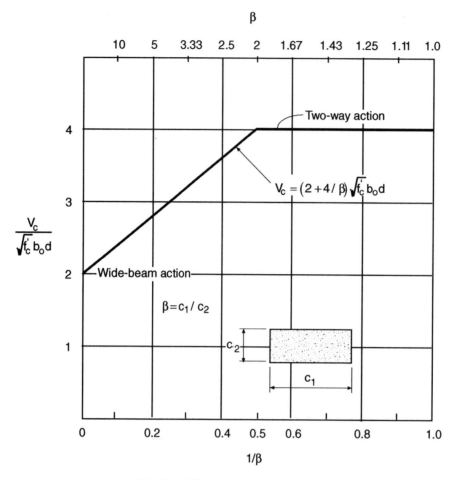

Fig. 16-3 Effect of β on Concrete Shear Strength

Other tests have indicated that v_c decreases as the ratio b_o/d increases. Equation (11-34) accounts for the effect of b_o/d on the concrete shear strength:

$$V_c = \left(\frac{\alpha_s d}{b_o} + 2\right)\sqrt{f'_c}\, b_o d \qquad \qquad Eq.\ (11\text{-}34)$$

Figure 16-4 illustrates the effect of b_o/d for interior, edge, and corner columns, where α_s equals 40, 30, and 20, respectively. For an interior column with $b_o/d \leq 2.0$, the maximum permissible shear stress is $4\sqrt{f'_c}$; see Fig. 16-4. Once $b_o/d > 2.0$, the shear stress decreases linearly to $2\sqrt{f'_c}$ at b_o/d equal to infinity.

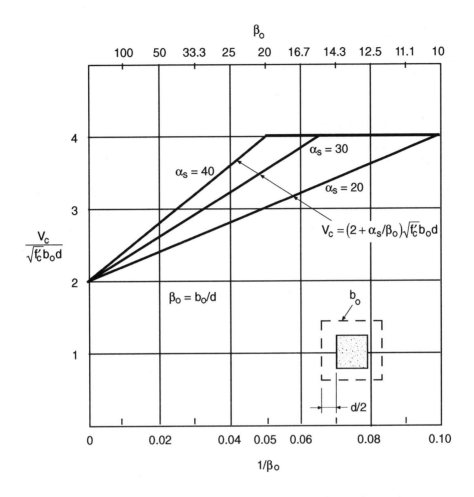

Fig. 16-4 Effect of b_o/d on Concrete Shear Strength

Note that reference to interior, edge, and corner column does not suggest column location in a building, but rather refers to the number of sides of the critical section available to resist the shear stress. For example, a column that is located in the interior of a building, with one side at the edge of an opening, shall be evaluated as an edge column.

The concrete nominal shear strength for two-way shear action of slabs without shear reinforcement is the least of Eqs. (11-33), (11-34), and (11-35) (11.12.2.1). Note that if lightweight concrete is used, 11.2 shall apply.

11.12.3 Shear Strength Provided by Bars, Wires, and Single or Multiple-Leg Stirrups

The use of bars, wires, or single or multiple-leg stirrups as shear reinforcement in slabs is permitted provided that the effective depth of the slab is greater than or equal to 6 inches, but not less than 16 times the shear reinforcement bar diameter (11.12.3). Suggested rebar shear reinforcement consist of properly anchored single-leg, multiple-leg, or closed stirrups that are engaging longitudinal reinforcement at both the top and bottom of the slab (11.12.3.4); see Fig. R11.12.3 (a), (b), (c).

With the use of shear reinforcement, the nominal shear strength provided by concrete V_c shall not be taken greater than $2\sqrt{f'_c}b_o d$ (11.12.3.1), and nominal shear strength V_n is limited to $6\sqrt{f'_c}b_o d$ (11.12.3.2). Thus, V_s must not be greater than $4\sqrt{f'_c}b_o d$.

The area of shear reinforcement A_v is computed from Eq. (11-15), and is equal to the cross-sectional area of all legs of reinforcement on one peripheral line that is geometrically similar to the perimeter of the column section (11.12.3.1):

$$A_v = \frac{V_s s}{f_y d}$$
Eq. (11-15)

The spacing limits of 11.12.3.3 correspond to slab shear reinforcement details that have been shown to be effective. These limits are as follows (see Fig. 16-5):

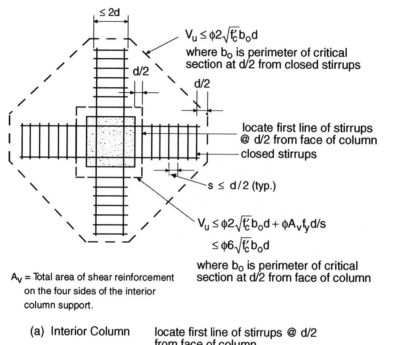
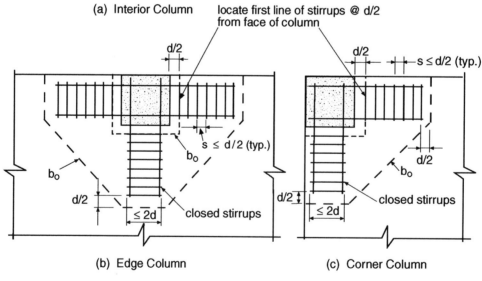

Fig. 16-5 Design and Detailing Criteria for Slabs with Stirrups

1. The first line of stirrups surrounding the column shall be placed at distance not exceeding d/2 from the column face.
2. The spacing between adjacent legs in the first line of shear reinforcement shall not exceed 2d.
3. The spacing between successive lines of shear reinforcement that surround the column shall not exceed d / 2.
4. The shear reinforcement can be terminated when $V_u \leq \phi 2\sqrt{f'_c} b_o d$ (11.12.3.1).

Proper anchorage of the shear reinforcement is achieved by satisfying the provisions of 12.13 (11.12.3.4). Refer to Fig. R11.12.3 and Part 4 for additional details on stirrup anchorage. It should be noted that anchorage requirements of 12.13 may be difficult for slabs thinner than 10 inches. Application of shear reinforcement design using bars or stirrups is illustrated in Example 16.3.

Where moment transfer is significant between the column and the slab, it is recommended to use closed stirrups in a pattern as symmetrical as possible around the column (R11.12.3).

11.12.4 Shear Strength Provided by Shearheads

The provisions of 11.12.4 permit the use of structural steel sections such as I- or channel-shaped sections (shearheads) as shear reinforcement in slabs, provided the following criteria are satisfied:

1. Each arm of the shearhead shall be welded to an identical perpendicular arm with full penetration welds and each arm must be continuous within the column section (11.12.4.1); see Fig. 16-6 (a).
2. Shearhead depth shall not exceed 70 times the web thickness of the steel shape (11.12.4.2); see Fig. 16-6 (b).
3. Ends of each shearhead arm is permitted to be cut at angles not less than 30 deg with the horizontal, provided the tapered section is adequate to resist the shear force at that location (11.12.4.3); see Fig. 16-6 (b).
4. All compression flanges of steel shapes shall be located within 0.3d of compression surface of slab, which in the case of direct shear, is the distance measured from the bottom of the slab (11.12.4.4); see Fig. 16-6 (b).
5. The ratio α_v of the flexural stiffness of the steel shape to surrounding composite cracked slab section of width $c_2 + d$ shall not be less than 0.15 (11.12.4.5); see Fig. 16-6 (c).
6. The required plastic moment strength M_p is computed from the following equation (11.12.4.6):

$$\phi M_p = \frac{V_u}{2n}\left[h_v + \alpha_v(\ell_v - 0.5c_1)\right] \qquad \text{Eq. (11-37)}$$

where:
M_p = plastic moment strength for each shearhead arm required to ensure that the ultimate shear is attained as the moment strength of the shearhead is reached.
ϕ = strength reduction factor for tension-controlled members, equal to 0.9 per 9.3.2.3.
n = number of shearhead arms; see Fig. R11.12.4.7.
ℓ_v = minimum required length of shearhead arm per 11.12.4.7 and 11.12.4.8; see Fig. R11.12.4.7.
h_v = depth of shearhead cross-section; see Fig. 16-6 (b).

7. The critical section for shear shall be perpendicular to the plane of the slab and shall cross each shearhead arm at three-quarters of the distance $(\ell_v - 0.5c_1)$ from the column face to the end of the shearhead arm. The critical section shall be located per 11.12.1.2(a) (11.12.4.7); see Fig. R11.12.4.7.

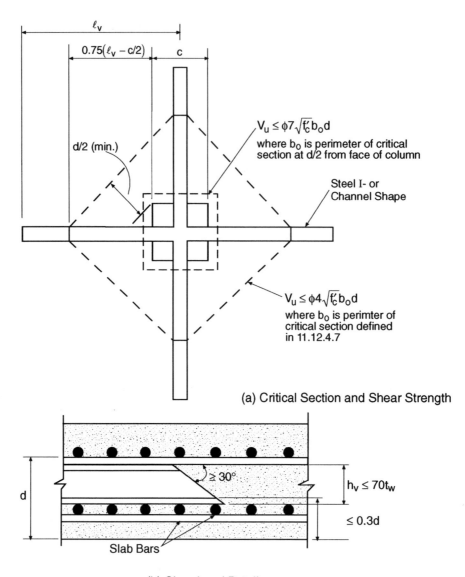

(a) Critical Section and Shear Strength

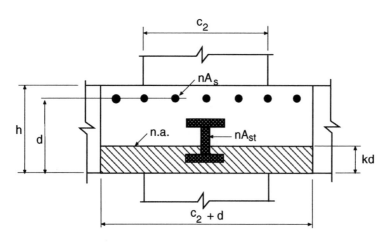

(b) Shearhead Details

(c) Properties of Cracked Composite Slab Section

Fig. 16-6 Design and Detailing Criteria for Slabs with Shearhead Reinforcement

8. The nominal shear strength V_n shall be less than or equal to $4\sqrt{f'_c}b_o d$ on the critical section defined by 11.12.4.7, and $7\sqrt{f'_c}b_o d$ at $d/2$ distance from the column face (11.12.4.8); see Fig. 16-6 (a).

9. Section 11.12.4.9 permits the shearheads to contribute in resisting the slab design moment in the column strip. The moment resistance M_v contributed to each column strip shall be the minimum of:

   a. $\dfrac{\phi\alpha_v V_u}{2n}(\ell_v - 0.5c_1)$ Eq.(11-38)

   b. $0.30M_u$ of the total factored moment in each column strip
   c. the change in column strip moment over the length ℓ_v
   d. the value of M_p computed by Eq. (11-37).

When direct shear and moment are transferred between slab and column, the provisions of 11.12.6 must be satisfied in addition to the above criteria. In slabs with shearheads, integrity steel shall be provided in accordance with 13.3.8.6. Application on the design of shearheads as shear reinforcement is illustrated in Example 16.3.

Other Type of Shear Reinforcement

Slab shear reinforcement consisting of vertical bars mechanically anchored at each end by a plate or a head capable of developing the yield strength of the bars have been used successfully (R11.12.3); see Fig. 16-7. This type of shear reinforcement for slabs can be advantageous when considering their ease of installation and the cost of placement, compared to other types of slab shear reinforcement. Extensive tests, methods of design, and worked-out design examples are presented in Refs. 16.1 through 16.4.

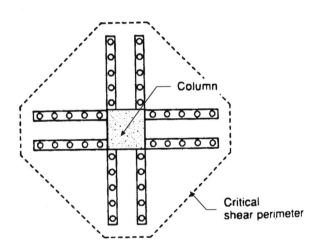

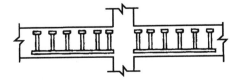

Fig. 16-7 Shear Reinforcement by Headed Studs

11.12.5 Effect of Openings in Slabs on Shear Strength

The effect of openings in slabs on concrete shear strength shall be considered when the opening is located: (1) anywhere within a column strip of a flat slab system and (2) within 10 times the slab thickness from a concentrated load or reaction area. Slab opening effect is evaluated by reducing the perimeter of the critical section b_o by a length equal to the projection of the opening enclosed by two-lines extending from the centroid of the column and tangent to the opening; see Fig 16-8 (a). For slabs with shear reinforcement, the ineffective portion of the perimeter b_o is one-half of that without shear reinforcement; see Fig. 16-8 (b). The one-half factor is interpreted to apply equally to shearhead reinforcement and bar or wire reinforcement as well. Effect of opening in slabs on flexural strength is discussed in Part 18.

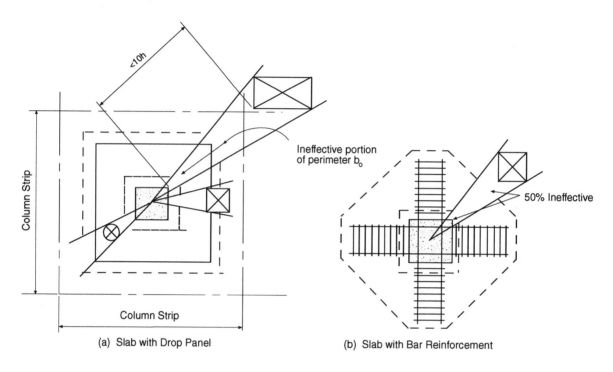

Fig. 16-8 Effect of Openings in Slabs on Shear Strength

11.12.6 Moment Transfer at Slab-Column Connections

For various loading conditions, unbalanced moment M_u can occur at the slab-column connections. For slabs without beams between supports, the transfer of unbalanced moment is one of the most critical design conditions for two-way slab systems. Shear strength at an exterior slab-column connection (without spandrel beam) is especially critical, because the total exterior negative moment must be transferred to the column, which is in addition to the direct shear due to gravity loads; see Fig. 16-9. The designer should not take this aspect of two-way slab design lightly. Two-way slab systems usually are fairly "forgiving" in the event of an error in the amount and or distribution of flexural reinforcement; however, little or no forgiveness is to be expected if shear strength provisions are not fully satisfied.

Note that the provisions of 11.12.6 (or 13.5.3) do not apply to slab systems with beams framing into the column support. When beams are present, load transfer from the slab through the beams to the columns is considerably less critical. Shear strength in slab systems with beams is covered in 13.6.8.

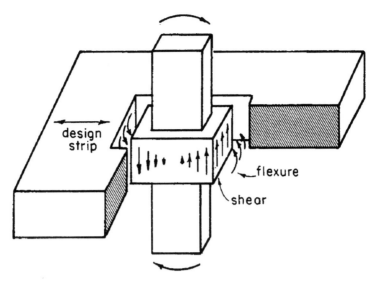

Fig. 16-9 Direct Shear and Moment Transfer

11.12.6.1 Distribution of Unbalanced Moment

The code specifies that the unbalanced moment at a slab-column connection must be transferred from the slab (without beams) to the column by eccentricity of shear in accordance with 11.12.6 and by flexure in accordance with 13.5.3 (11.12.6.1). Studies (Ref. 16.7) of moment transfer between slabs and square columns found that $0.6M_u$ is transferred by flexure across the perimeter of the critical section b_o defined by 11.12.1.2, and $0.4M_u$ by eccentricity of shear about the centroid of the critical section. For a rectangular column, the portion of moment transferred by flexure $\gamma_f M_u$ increases as the dimension of the column that is parallel to the applied moment increases. The fraction of unbalanced moment transferred by flexure γ_f is:

$$\gamma_f = \frac{1}{1+\left(\frac{2}{3}\right)\sqrt{\frac{b_1}{b_2}}} \qquad \text{Eq. (13-1)}$$

and the fraction of unbalanced moment transferred by eccentricity of shear is:

$$\gamma_v = 1 - \gamma_f \qquad \text{Eq. (11-39)}$$

where b_1 and b_2 are the dimensions of the perimeter of the critical section, with b_1 parallel to the direction of analysis; see Fig. 16-10. The relationship of the parameters presented into Eqs. (13-1) and (11-39) is graphically illustrated in Fig. 16-11. Modification or adjustment of γ_f and thus γ_v, is permitted in accordance with 13.5.3.3 for any two-way slab system, except for prestressed slabs. The following modifications are applicable, provided that the reinforcement ratio in the slab within the effective width defined in 13.5.3.2 does not exceed $0.375\rho_b$:

- For unbalanced moments about an axis parallel to the slab edge at exterior supports (i.e., bending perpendicular to the edge), it is permitted to take $\gamma_f = 1.0$ provided that $V_u \leq 0.75\phi V_c$ at an edge column or $V_u \leq 0.5\phi V_c$ at a corner column.
- For unbalanced moments at interior supports and for unbalanced moments about an axis transverse to the edge of exterior supports (i.e., bending parallel to the edge), it is permitted to increase γ_f by up to 25%, provided that $V_u \leq 0.4\phi V_c$.

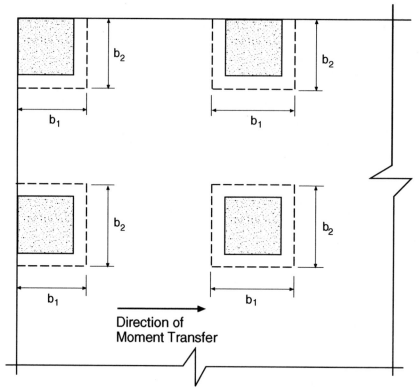

Fig. 16-10 Parameters b_1 and b_2 for Eqs. (11-39) and (13-1)

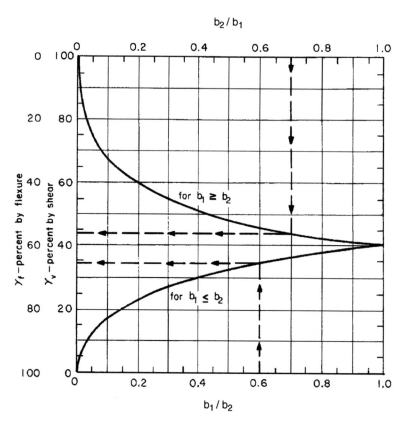

Fig. 16-11 Graphical Solution of Eqs. (13-1) and (11-39)

16-12

The unbalanced moment transferred by eccentricity of shear is $\gamma_v M_u$, where M_u is the unbalanced moment at the centroid of the critical section. The unbalanced moment M_u at an exterior support of an end span will generally not be computed at the centroid of the critical transfer section in the frame analysis. When the Direct Design Method of Chapter 13 is utilized, moments are computed at the face of the support. Considering the approximate nature of the procedure to evaluate the stress distribution due to moment-shear transfer, it seems unwarranted to consider a change in moment to the transfer centroid; use of the moment values from frame analysis (centerline of support) or from 13.6.3.3 (face of support) is accurate enough.

Unbalanced moment transfer between an edge column and a slab without edge beams requires special consideration when slabs are analyzed for gravity loads using the moment coefficients of the Direct Design Method. In this case, unbalanced moment M_u must be set equal to $0.3M_o$ (13.6.3.6), where M_o is the total factored static moment in the span. Therefore, the fraction of unbalanced moment transferred by shear is $\gamma_v M_u = \gamma_v(0.3M_o)$. See Part 19 for further discussion of that special shear strength requirement and its application in Example 19.1. If the Equivalent Frame Method is used, the unbalanced moment is equal to the computed frame moment.

11.12.6.2 Shear Stresses and Strength Computation

Assuming that shear stress resulting from moment transfer by eccentricity of shear varies linearly about the centroid of the critical section defined in 11.12.1.2, the factored shear stresses at the faces of the critical section due to the direct shear V_u and the unbalanced moment transferred by eccentricity of shear $\gamma_v M_u$ are (see Fig. 16-12, and R11.12.6.2):

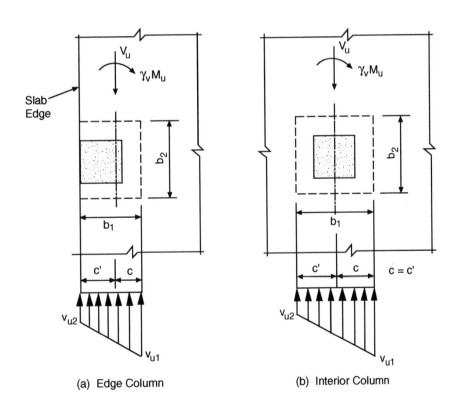

(a) Edge Column (b) Interior Column

Fig. 16-12 Shear Stress Distribution due to Moment-Shear Transfer at Slab-Column Connection

$$v_{u1} = \frac{V_u}{A_c} + \frac{\gamma_v M_u c}{J} \qquad \text{Eq. (1)}$$

$$v_{u2} = \frac{V_u}{A_c} - \frac{\gamma_v M_u c'}{J} \qquad \text{Eq. (2)}$$

where: A_c = area of concrete section resisting shear transfer, equal to the perimeter b_o multiplied by the effective depth d

J = property of critical section analogous to polar moment of inertia of segments forming area A_c.

c and c' = distances from centroidal axis of critical section to the perimeter of the critical section in the direction of analysis

Expressions for $A_c, c, c', J/c$, and J/c', are contained in Fig. 16-13 for rectangular columns and Fig. 16-14 for circular interior columns.

Where biaxial moment transfer occurs, research has shown that the method for evaluating shear stresses due to moment transfer between slabs and column in R.11.12.6.2 is still applicable (Ref. 16.8). There is no need to superimpose the shear stresses due to moments transfer in two directions.

The maximum shear stress v_{u1} computed from Eq. (1) shall not exceed ϕv_n, where ϕv_n is determined from the following (11.12.6.2):

a. For slabs without shear reinforcement: $\phi v_n = \phi v_c$, where ϕv_n is the minimum of:

$$\phi v_c = \phi \left(2 + \frac{4}{\beta}\right)\sqrt{f'_c} \qquad \text{Eq. (11-33)}$$

$$\phi v_c = \phi \left(2 + \frac{\alpha_s d}{b_o}\right)\sqrt{f'_c} \qquad \text{Eq. (11-34)}$$

$$\phi v_c = \phi 4\sqrt{f'_c} \qquad \text{Eq. (11-35)}$$

b. For slabs with shear reinforcement other than shearheads, ϕv_n is computed from (11.12.3):

$$\phi v_n = \phi\left(2\sqrt{f'_c} + \frac{A_v f_y}{b_o s}\right) \le \phi 6\sqrt{f'_c} \qquad \text{Eqs. (11-15), (11.12.3.1), and (11.12.3.2)}$$

where A_v is the total area of shear reinforcement provided on the column sides and b_o is the perimeter of the critical section located at d / 2 distance away from the column perimeter, as defined by 11.12.1.2 (a). Due to the variation in shear stresses, as illustrated in Fig. 16-12, the computed area of shear reinforcement, if required, may be different from one column side to the other. The required area of shear reinforcement due to shear stress v_{u1} at its respective column side is:

$$A_v = (v_{u1} - \phi v_c)\frac{(c + d)s}{\phi f_y} \qquad \text{Eq. (3)}$$

where $(c+d)$ is an effective "beam" width and $v_c = 2\sqrt{f'_c}$. However, R11.12.3 recommends symmetrical placement if shear reinforcement on all column sides. Thus, with symmetrical shear reinforcement assumed on all sides of the column, the required area A_v may be computed from:

$$A_v = (v_{u1} - \phi v_c)\frac{b_o s}{\phi f_y} \qquad \text{Eq. (4)}$$

where A_v is the total area of required shear reinforcement to be extended from the sides of the column, and b_o is the perimeter of the critical section located at d/2 from the column perimeter. With symmetrical reinforcement on all column sides, the reinforcement extending from the column sides with lesscomputed shear stress provides torsional resistance in the strip of slab perpendicular to the direction of analysis.

c. For slabs with shearheads as shear reinforcement, ϕv_n is computed from:

$$\phi v_n = \phi 4\sqrt{f'_c} \geq v_{u1} \qquad \text{11.12.6.3}$$

$$v_{u1} = \frac{V_u}{b_o d} + \frac{\gamma_v M_u c}{J} \leq \phi 4\sqrt{f'_c} \qquad \text{Eq. (1)}$$

where b_o is the perimeter of the critical section defined in 11.12.4.7, c and J are section properties of the critical section located at d/2 from the column perimeter (11.12.6.3), V_u is the direct shear force acting on the critical section defined in 11.12.4.7, and $\gamma_v M_u$ is the unbalanced moment transferred by eccentricity of shear acting about the centroid of the critical section defined in 11.12.1.2(a). Note that this seemingly inconsistent summation of shear stresses occurring at two different critical shear sections is conservative and justified by tests (see R11.12.6.3). At the critical section located d/2 from the column perimeter, v_u shall not exceed $\phi 7\sqrt{f'_c}$ (11.12.4.8); see Fig. 16-5.

Case A: Edge Column (Bending parallel to edge)

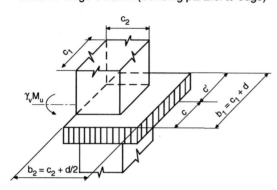

Case B: Interior Column

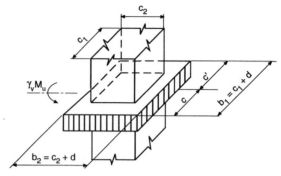

Case C: Edge Column (Bending perpendicular to edge)

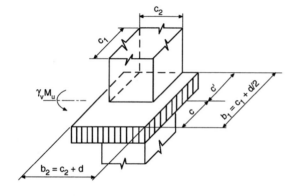

Case D: Corner Column

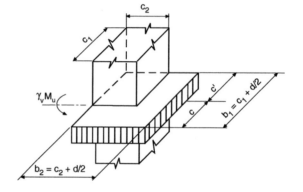

Case	Area of critical section, A_c	Modulus of critical section		c	c'
		J/c	J/c'		
A	$(b_1+2b_2)d$	$\dfrac{b_1 d(b_1+6b_2)+d^3}{6}$	$\dfrac{b_1 d(b_1+6b_2)+d^3}{6}$	$\dfrac{b_1}{2}$	$\dfrac{b_1}{2}$
B	$2(b_1+b_2)d$	$\dfrac{b_1 d(b_1+3b_2)+d^3}{3}$	$\dfrac{b_1 d(b_1+3b_2)+d^3}{3}$	$\dfrac{b_1}{2}$	$\dfrac{b_1}{2}$
C	$(2b_1+b_2)d$	$\dfrac{2b_1^2 d(b_1+2b_2)+d^3(2b_1+b_2)}{6b_1}$	$\dfrac{2b_1^2 d(b_1+2b_2)+d^3(2b_1+b_2)}{6(b_1+b_2)}$	$\dfrac{b_1^2}{2b_1+b_2}$	$\dfrac{b_1(b_1+b_2)}{2b_1+b_2}$
D	$(b_1+b_2)d$	$\dfrac{b_1^2 d(b_1+4b_2)+d^3(b_1+b_2)}{6b_1}$	$\dfrac{b_1^2 d(b_1+4b_2)+d^3(b_1+b_2)}{6(b_1+2b_2)}$	$\dfrac{b_1^2}{2(b_1+b_2)}$	$\dfrac{b_1(b_1+2b_2)}{2(b_1+b_2)}$

Fig. 16-13 Section Properties for Shear Stress Computations – Rectangular Columns

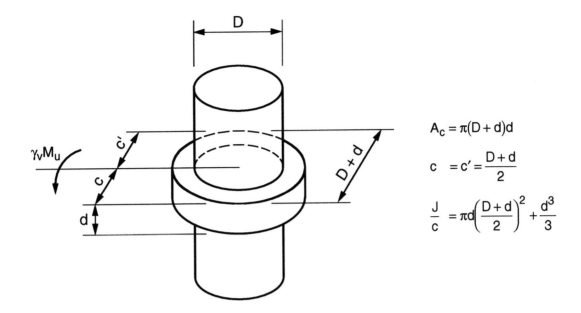

Fig. 16-14 Section Properties for Shear Stress Computations – Circular Interior Column

REFERENCES

16.1 ACI-ASCE Committee 421, "Shear Reinforcement for Slabs", American Concrete Institute, Farmington Hills, MI, 1999

16.2 Dilger, W.H. and Ghali, A., "Shear Reinforcement for Concrete Slabs," *Proceedings ASCE, Journal of the Structural Division*, Vol. 107, No. ST12, December 1981, pp. 2403-2420.

16.3 Elgabry, A.E. and Ghali, A., "Design of Stud-Shear Reinforcement for Slabs," *ACI Structural Journal*, Vol. 87, No. 3, May-June 1990, pp. 350-361.

16.4 Hammill, N. and Ghali, A., "Punching Shear Resistance of Corner Slab-Column Connections," *ACI Structural Journal*, Vol. 91, No. 6, November-December 1994, pp. 697-707.

16.5 Ghosh, S.K., "Aspects of Design of Reinforced Concrete Flat Plate Slab Systems," *Analysis and Design of High-Rise Concrete Buildings*, SP-97, American Concrete Institute, Detroit, MI, 1985, pp. 139-157.

16.6 *Simplified Design - Reinforced Concrete Buildings of Moderate Size and Height*, 3rd edition, Portland Cement Association, Skokie, IL, 2004.

16.7 Hanson, N.W., and Hanson, J.M., "Shear and Moment Transfer Between Concrete Slabs and Columns," Journal, PCA Research and Development Labs, V-10, No. 1, Jan. 1968, pp. 2-16.

16.8 ACI-ASCE Committee 426, "The Shear Strength of Reinforced Concrete Members-Slabs." *Journal of Structural Division*, ASCE, Vol. 100 No. ST8, August 1973, pp. 1543-1591.

Example 16.1—Shear Strength of Slab at Column Support

Determine two-way action shear strength at an interior column support of a flat plate slab system for the following design conditions.

Column dimensions = 48 in. × 8 in.

Slab effective depth d = 6.5 in.

Specified concrete strength f'_c = 4,000 psi

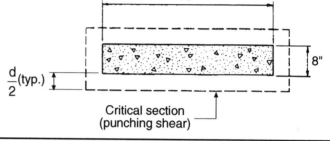

Calculations and Discussion	Code Reference

1. Two-way action shear (punching shear) without shear reinforcement:

$$V_u \leq \phi V_n \quad \text{Eq. (11-1)}$$

$$\leq \phi V_c \quad 11.12.2$$

2. Effect of loaded area aspect ratio β_c:

$$\phi V_c = \phi \left(2 + \frac{4}{\beta}\right) \sqrt{f'_c}\, b_o d \quad \text{Eq. (11-33)}$$

where $\beta = \dfrac{48}{8} = 6$ 11.12.2.1

$b_o = 2(48 + 6.5 + 8 + 6.5) = 138$ in.   11.12.1.2

$\phi = 0.75$ 9.3.2.3

$\phi V_c = 0.75 \times 138 \times 6.5/1{,}000 = 113.5$ kips

3. Effect of perimeter area aspect ratio β_o:

$$\phi V_c = \phi \left(2 + \frac{\alpha_s}{\beta_o}\right) \sqrt{f'_c}\, b_o d \quad \text{Eq. (11-34)}$$

where $\alpha_s = 40$ for interior column support   11.12.2.1

$\beta_o = \dfrac{b_o}{d} = \dfrac{138}{6.5} = 21.2$

$\phi V_c = 0.75 \times 138 \times 6.5/1{,}000 = 165.4$ kips

		Code
Example 16.1 (cont'd)	**Calculations and Discussion**	**Reference**

4. Excluding effect of β and β_o:

    $\phi V_c = \phi 4 \sqrt{f'_c}\, b_o d$ *Eq. (11-35*

    $= 0.75 \times 4 \times \sqrt{4,000} \times 138 \times 6.5 / 1,000 = 170.2$ kips

5. The shear strength ϕV_n is the smallest of the values computed above, i.e.,
   $\phi V_n = 113.5$ kips.

Example 16.2—Shear Strength for Non-Rectangular Support

For the L-shaped interior column support shown, check punching shear strength for a factored shear force of V_u = 125 kips. Use f'_c = 4,000 psi. Effective slab depth = 5.5 in.

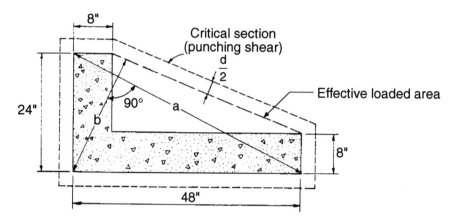

Calculations and Discussion	Code Reference

1. For shapes other than rectangular, R11.12.2.1 recommends that β be taken as the ratio of the longest overall dimension of the effective loaded area a to the largest overall dimension of the effective loaded area b, measured perpendicular to a:

   R11.12.2.1

$$\beta = \frac{a}{b} = \frac{54}{25} = 2.16$$

For the critical section shown, b_o = 141 in.

11.12.1.2

Scaled dimensions of the drawings are used, and should be accurate enough

2. Two-way action shear (punching shear) without shear reinforcement:

$$V_u \leq \phi V_n$$

Eq. (11-1)

$$\leq \phi V_c$$

11.12.2

where the nominal shear strength V_c without shear reinforcement is the lesser of values given by Eqs. (11-33) and (11-34), but not greater than $4\sqrt{f'_c}\, b_o d$:

$$V_c = \left(2 + \frac{4}{\beta}\right)\sqrt{f'_c}\, b_o d$$

Eq. (11-33)

$$= \left(2 + \frac{4}{2.16}\right)\sqrt{4{,}000} \times 141 \times 5.5 / 1{,}000 = 188.9 \text{ kips}$$

$$V_c = \left(2 + \frac{\alpha_s}{\beta_o}\right)\sqrt{f'_c}\, b_o d$$

Eq. (11-34)

16-20

| Example 16.2 (cont'd) | Calculations and Discussion | Code Reference |

where α_s = 40 for interior column support   *11.12.2.1*

$$\beta_o = \frac{b_o}{d} = \frac{141}{5.5} = 25.6$$

$$V_c = \left(2 + \frac{40}{25.6}\right)\sqrt{4,000} \times 141 \times 5.5 / 1,000 = 174.7 \text{ kips}$$

$$V_c = 4\sqrt{f'_c}\, b_o d \quad\quad\quad\quad\quad\quad\quad\quad\quad\quad\quad\quad\quad\quad\quad\quad\quad\quad\quad\quad\quad\quad\quad\quad\quad\quad\quad\quad\quad\quad\quad\quad\quad\quad\quad\quad\quad\quad\quad\quad\quad\quad\quad\quad\quad\quad\quad\quad\quad\quad\quad\quad\quad\quad\quad\quad\quad \textit{Eq. (11-35)}$$

$$= 4\sqrt{4,000} \times 141 \times 5.5 / 1,000 = 196.2 \text{ kips}$$

$\phi V_c = 0.75\,(174.7) = 131$ kips

$V_u = 125$ kips $< \phi V_c = 131$ kips   O.K.

Example 16.3—Shear Strength of Slab with Shear Reinforcement

Consider an interior panel of a flat plate slab system supported by a 12-in. square column. Panel size $\ell_1 = \ell_2 = 21$ ft. Determine shear strength of slab at column support, and if not adequate, increase the shear strength by shear reinforcement. Overall slab thickness h = 7.5 in. (d = 6 in.).

f'_c = 4,000 psi

f_y = 60,000 psi (bar reinforcement)

f_y = 36,000 psi (structural steel)

Superimposed factored load = 160 psf

Column strip negative moment M_u = 175 ft-kips

Calculations and Discussion	Code Reference
1. Wide-beam action shear and two-way action shear (punching shear) without shear reinforcement:	11.12.2
$V_u \leq \phi V_n$	Eq. (11-1)
$\leq \phi V_c$	11.12.2

   a. Since there are no shear forces at the center lines of adjacent panels, tributary areas and critical sections for slab shear are as shown below.

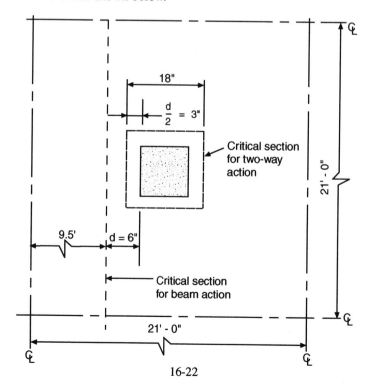

		Code
Example 16.3 (cont'd)	**Calculations and Discussion**	**Reference**

For 7.5-in. slab, factored dead load $q_{Du} = 1.2 \times \dfrac{7.5}{12} \times 150 = 113$ psf   9.2.1

$q_u = 113 + 160 = 273$ psf

a. Wide-Beam Action Shear.

Investigation of wide-beam action shear strength is made at the critical section at a distance d from face of column support.   11.1.3.1

$V_u = 0.273 (9.5 \times 21) = 54.5$ kips

$V_c = 2\sqrt{f'_c} b_w d = 2\sqrt{4,000} (21 \times 21) \times 6/1,000 = 191.3$ kips   Eq. (11-3)

$\phi = 0.75$ 9.3.2.3

$\phi V_c = 0.75 (191.3) = 143.5$ kips $> V_u = 54.5$ kips   O.K.

Wide-beam action will rarely control the shear strength of two-way slab systems.

b. Two-Way Action Shear.

Investigation of two-way action shear strength is made at the critical section b_o located at d/2 from the column perimeter. Total factored shear force to be transferred from slab to column: 11.12.1.2(a)

$V_u = 0.273 (21^2 - 1.5^2) = 119.8$ kips

Shear strength V_c without shear reinforcement: 11.12.2.1

$b_o = 4 (18) = 72$ in.   11.12.1.2(a)

$\beta = \dfrac{12}{12} = 1.0 < 2$

$\beta_o = \dfrac{b_o}{d} = \dfrac{72}{6} = 12 < 20$

$V_c = 4\sqrt{f'_c} b_o d = 4\sqrt{4,000} \times 72 \times 6/1,000 = 109.3$ kips

$\phi = 0.75$ 9.3.2.3

$\phi V_c = 0.75 (109.3) = 82$ kips $< V_u = 119.8$ kips   N.G.

Shear strength of slab is not adequate to transfer the factored shear force $V_u = 119.8$ kips from slab to column support. Shear strength may be increased by:

  i. increasing concrete strength f'_c
 ii. increasing slab thickness at column support, i.e., using a drop panel
iii. providing shear reinforcement (bars, wires, or steel I- or channel-shapes)

The following parts of the example will address all methods to increase shear strength.

	Code
Example 16.3 (cont'd) **Calculations and Discussion**	**Reference**

2. Increase shear strength by increasing strength of slab concrete:

   $V_u \leq \phi V_n$ Eq. (11-1)

   $119,800 \leq 0.75 \left(4\sqrt{f'_c} \times 72 \times 6\right)$

   Solving, $f'_c = 8,545$ psi

3. Increase shear strength by increasing slab thickness at column support with drop panel:

   Provide drop panel in accordance with 13.2.5 (see Fig. 18-18). Minimum overall slab thickness at drop panel = 1.25 (7.5) = 9.375-in. Try a 9.75 in. slab thickness (2.25-in. projection below slab*; d ≈ 8.25 in.). Minimum distance from centerline of column to edge of drop panel = 21/6 = 3.5 ft. Try 7 × 7 ft drop panel.

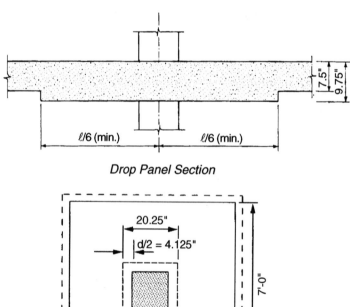

Drop Panel Section

a. Investigate shear strength at critical section b_o located at d/2 from column perimeter.

   Total factored shear force to be transferred —

---

* See Chapter 9 (Design Considerations for Economical Formwork) in Ref. 16.6.

| Example 16.3 (cont'd) | Calculations and Discussion | Code Reference |

For 2.25-in. drop panel projection, $q_{Du} = 1.2 \times \dfrac{2.25}{12} \times 150 = 34$ psf

$V_u = 0.273 (21^2 - 1.69^2) + 0.034 (7^2 - 1.69^2) = 119.6 + 1.6 = 121.2$ kips

$b_o = 4 (12 + 8.25) = 81$ in. **11.12.1.2(a)**

$\beta = 1.0 < 2$

$\beta_o = \dfrac{b_o}{d} = \dfrac{81}{8.25} = 9.8 < 20$

$\phi V_c = \phi 4\sqrt{f'_c}\, b_o d$ **Eq. (11-35)**

$\quad = 0.75 \times 4\sqrt{4,000} \times 81 \times 8.25 > V_u = 121.2$ kips O.K.

b. Investigate shear strength at critical section b_o located at d/2 from edge of drop panel.

Total factored shear force to be transferred —

$V_u = 0.273 (21^2 - 7.5^2) = 105.0$ kips

$b_o = 4 (84 + 6) = 360$ in. **11.12.1.2(b)**

$\beta = \dfrac{84}{84} = 1.0 < 2$ **Eq. (11-35)**

$\beta_o = \dfrac{b_o}{d} = \dfrac{360}{6} = 60 > 20$

$\phi V_c = \phi \left(2 + \dfrac{\alpha_s}{\beta_o} \right)\sqrt{f'_c}\, b_o d = \phi \left(2 + \dfrac{40}{60} \right)\sqrt{f'_c}\, b_o d = \phi 2.67\sqrt{f'_c}\, b_o d$ **Eq. (11-36)**

$\quad = 0.75 \times 2.67\sqrt{4,000} \times 360 \times 6 / 1,000 = 273.2$ kips $> V_u = 105.0$ kips O.K.

Note the significant decrease in potential shear strength at edge of drop panel due to large β_o.

A 7 × 7 ft drop panel with a 2.25-in. projection below the slab will provide adequate shear strength for the superimposed factored loads of 160 psf.

Example 16.3 (cont'd)	Calculations and Discussion	Code Reference

4. Increase shear strength by bar reinforcement (see Figs. R11.12.3(a) and 16-5):

   a. Check effective depth d   *11.12.3*

   Assuming No. 3 stirrups ($d_b = 0.375$ in.),

   $$d = 6 \text{ in.} \geq \begin{cases} 6 \text{ in. O.K.} \\ 16 \times 0.375 = 6 \text{ in. O.K.} \end{cases}$$

   b. Check maximum shear strength permitted with bars.   *11.12.3.2*

   $$V_u \leq \phi V_n \quad \quad \textit{Eq. (11-1)}$$

   $$\phi V_n = \phi\left(6\sqrt{f'_c} b_o d\right) = 0.75\left(6\sqrt{4,000} \times 72 \times 6\right)/1,000 = 123.0 \text{ kips}$$

   $$V_u = 0.273 (21^2 - 1.5^2) = 119.8 \text{ kips} < \phi V_n = 123.0 \text{ kips} \quad \text{O.K.}$$

   c. Determine shear strength provided by concrete with bar shear reinforcement.   *11.12.3.1*

   $$V_c = 2\sqrt{f'_c} b_o d = 2\sqrt{4,000} \times 72 \times 6/1,000 = 54.6 \text{ kips}$$

   $$\phi V_c = 0.75 (54.6) = 41.0 \text{ kips}$$

   d. Design shear reinforcement in accordance with 11.5.

   Required area of shear reinforcement A_v is computed by

   $$A_v = \frac{(V_u - \phi V_c) s}{\phi f_y d}$$

   Assume s = 3 in. (maximum spacing permitted = d/2)   *11.5.4.1*

   $$A_v = \frac{(119.8 - 41.0) \times 3}{0.75 \times 6.0 \times 6} = 0.88 \text{ in.}^2$$

   where A_v is total area of shear reinforcement required on the four sides of the column (see Fig. 16-5).

   $$A_v \text{ (per side)} = \frac{0.88}{4} = 0.22 \text{ in.}^2$$

| Example 16.3 (cont'd) | Calculations and Discussion | Code Reference |

e. Determine distance from sides of column where stirrups may be terminated (see Fig. 16-5).

$$V_u \leq \phi V_c$$ Eq. (11-1)

$$\leq \phi 2\sqrt{f'_c}\, b_o d$$

For square column (see sketch below),

$$b_o = 4\left(12 + a\sqrt{2}\right)$$

$$119,800 \leq 0.75 \times 2\sqrt{4000} \times 4\left(12 + a\sqrt{2}\right) \times 6$$

Solving, $a = 28.7$ in.

Note that the above is a conservative estimate, since V_u at the perimeter of the critical section shown below is considerably lower than 119.8 kips.

Stirrups may be terminated at $d/2 = 3$ in. inside the critical perimeter b_o.

Use 9-No. 3 closed stirrups @ 3 in. spacing ($A_v = 0.22$ in.^{2}) along each column line as shown below.

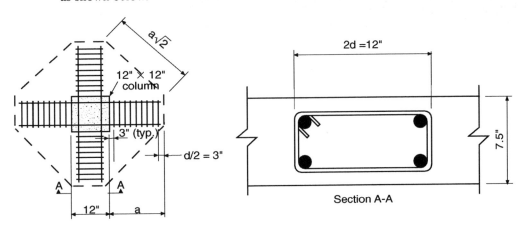

Section A-A

5. Increase shear strength by steel I shapes (shearheads):   11.12.4

   a. Check maximum shear strength permitted with steel shapes (see Fig. 18-8).   11.12.4.8

   $$V_u = 0.273\,(21^2 - 1.5^2) = 119.8 \text{ kips}$$

   $$V_u \leq \phi V_n$$ Eq. (11-1)

   $$\phi V_n = \phi\left(7\sqrt{f'_c}\, b_o d\right)$$ 11.12.4.8

   $$\leq 0.75\left(7\sqrt{4,000} \times 72 \times 6\right)/1,000 = 143.4 \text{ kips} > V_u = 119.8 \text{ kips O.K.}$$

		Code
Example 16.3 (cont'd)	**Calculations and Discussion**	**Reference**

b. Determine minimum required perimeter b_o of a critical section at shearhead ends with shear strength limited to $V_n = 4\sqrt{f'_c}\, b_o d$ (see Fig. 16-6 (b)).

$V_u \leq \phi V_n$ Eq. (11-1)

$119{,}800 \leq 0.75\left(4\sqrt{4{,}000} \times b_o \times 6\right)$

Solving, $b_o = 105.2$ in.   11.12.4.7

c. Determine required length of shearhead arm ℓ_v to satisfy $b_o = 105.2$ in. at $0.75(\ell_v - c_1/2)$.   11.12.4.7

$b_o \approx 4\sqrt{2}\left[\dfrac{c_1}{2} + \dfrac{3}{4}\left(\ell_v - \dfrac{c_1}{2}\right)\right]$ (see Fig. 16-6 (b))

With $b_o = 105.2$ in. and $c_1 = 12$ in., solving, $\ell_v = 22.8$ in.

Note that the above is a conservative estimate, since V_u at the perimeter of the critical section considered is considerably lower than 119.8 kips.

d. To ensure that premature flexural failure of shearhead does not occur before shear strength of slab is reached, determine required plastic moment strength M_p of each shearhead arm.

$\phi M_p = \dfrac{V_u}{2n}\left[h_v + \alpha_v\left(\ell_v - \dfrac{c_1}{2}\right)\right]$ Eq. (11-37)

For a four (identical) arm shearhead, $n = 4$; assuming $h_v = 4$ in. and $\alpha_v = 0.25$:   11.12.4.5

$\phi M_p = \dfrac{119.8}{2(4)}\left[4 + 0.25\left(23.6 - \dfrac{12}{2}\right)\right] = 125.8$ in.-kips

$\phi = 0.9$ (tension-controlled member)   9.3.2.1

Required $M_p = \dfrac{125.8}{0.9} = 139.8$ in.-kips

Try W4 × 13 (plastic modulus $Z_x = 6.28$ in.^{3}) A36 steel shearhead

$M_p = Z_x f_y = 6.28(36) = 226.1$ in.-kips > 139.8 in.-kips   O.K.

e. Check depth limitation of W4 × 13 shearhead.   11.12.4.2

$70 t_w = 70(0.280) = 19.6$ in. > $h_v = 4.16$ in.   O.K.

f. Determine location of compression flange of steel shape with respect to compression surface of slab, assuming 3/4-in. cover and 2 layers of No. 5 bars.   11.12.4.4

$0.3d = 0.3(6) = 1.8$ in. < $0.75 + 2(0.625) = 2$ in.   N.G.

| Example 16.3 (cont'd) | Calculations and Discussion | Code Reference |

Therefore, both layers of the No. 5 bars in the bottom of the slab must be cut.

g. Determine relative stiffness ratio α_v.

For the W4 × 13 shape:

A_{st} = 3.83 in.²

I_s = 11.3 in.⁴

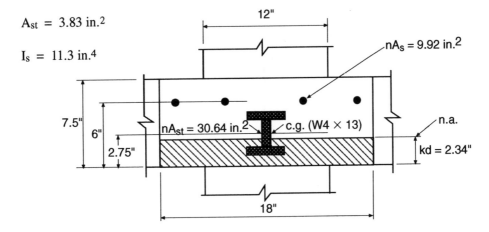

A_s provided for M_u = 175 ft-kips is No. 5 @ 5 in.

c.g. of W4 × 13 from compression face = 0.75 + 2 = 2.75 in.

Effective slab width = c_2 + d = 12 + 6 = 18 in.

Transformed section properties:

For f'_c = 4,000 psi, use $\dfrac{E_s}{E_c} = \dfrac{29,000}{3605} = 8$

Steel transformed to equivalent concrete:

$\dfrac{E_s}{E_c} A_s = 8 \,(4 \times 0.31) = 9.92$ in.²

$\dfrac{E_s}{E_c} A_{st} = 8 \,(3.83) = 30.64$ in.²

Neutral axis of composite cracked slab section may be obtained by equating the static moments of the transformed areas.

$$\dfrac{18 \,(kd)^2}{2} = 30.64 \,(2.75 - kd) + 9.92 \,(6 - kd)$$

where kd is the depth of the neutral axis for the transformed area

Solving, kd = 2.34 in.

Example 16.3 (cont'd)	Calculations and Discussion	Code Reference

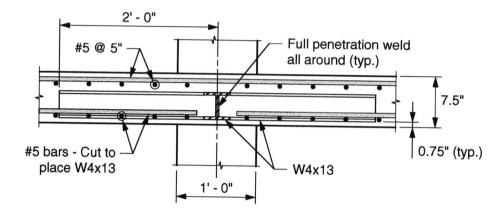

Final Details of Shearhead Reinforcement

$$\text{Composite I} = \frac{18(2.34)^3}{3} + \frac{E_s}{E_c}(I_s \text{ steel shape}) + 9.92(3.66)^2 + 30.64(0.41)^2$$

$$= 76.9 + 8(11.3) + 132.9 + 5.2 = 305.4 \text{ in.}^4$$

$$\alpha_v = \frac{E_s/E_c I_s}{I_{composite}} = \frac{8 \times 11.3}{305.4} = 0.30 > 0.15 \quad \text{O.K.}$$

Therefore, W4 × 13 section satisfies all code requirements for shearhead reinforcement.

h. Determine contribution of shearhead to negative moment strength of column strip.
   11.12.4.9

$$M_v = \frac{\phi \alpha_v V_u}{2n}\left(\ell_v - \frac{c_1}{2}\right) \qquad \text{Eq. (11-38)}$$

$$= \frac{0.9 \times 0.30 \times 119.8}{2 \times 4}(25-6) = 76.8 \text{ in.-kips} = 6.4 \text{ ft-kips}$$

However, M_v must not exceed either M_p = 139.8 in.-kips or 0.3 × 175 × 12 = 630 in.-kips, or the change in column strip moment over the length ℓ_v. For this design, approximately 4% of the column strip negative moment may be considered resisted by the shearhead reinforcement.

Example 16.4—Shear Strength of Slab with Transfer of Moment

Consider an exterior (edge) panel of a flat plate slab system supported by a 16-in. square column. Determine shear strength for transfer of direct shear and moment between slab and column support. Overall slab thickness h = 7.25 in. (d ≈ 6.0 in.). Assume that the Direct Design Method is used for analysis of the slab. Consider two loading conditions:

1. Total factored shear force V_u = 30 kips

   Total factored static moment M_o in the end span = 96 ft-kips

2. V_u = 60 kips

   M_o = 170 ft-kips

   f'_c = 4,000 psi

   f_y = 60,000 psi

	Code
Calculations and Discussion	**Reference**

---

1. Section properties for shear stress computations:

   Referring to Fig. 16-13, edge column bending perpendicular to edge (Case C),

   $$b_1 = c_1 + \frac{d}{2} = 16 + \frac{6}{2} = 19.0 \text{ in.}$$

   $$b_2 = c_2 + d = 16 + 6 = 22.0 \text{ in.}$$

   $$b_o = 2(19.0) + 22 = 60.0 \text{ in.}$$

   $$c = \frac{b_1^2}{2b_1 + b_2}$$

   $$= \frac{19.0^2}{(2 \times 19.0) + 22.0} = 6.02 \text{ in.}$$

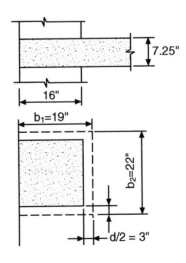

   $$A_c = (2b_1 + b_2)d = 360 \text{ in.}^2$$

   $$\frac{J}{c} = \frac{\left[2b_1^2 d(b_1 + 2b_2) + d^3(2b_1 + b_2)\right]}{6b_1} = 2{,}508 \text{ in.}^3$$

   $$c' = b_1 - c = 19 - 6.02 = 12.98 \text{ in.}$$

   $$\frac{J}{c'} = \left(\frac{J}{c}\right)\left(\frac{c}{c'}\right) = 2{,}508\left(\frac{6.02}{12.98}\right) = 1{,}163 \text{ in.}^3$$

Example 16.4 (cont'd)	Calculations and Discussion	Code Reference

2. Loading condition (1), V_u = 30 kips, M_o = 96 ft-kips:

   a. Portion of unbalanced moment to be transferred by eccentricity of shear.   11.12.6.1

   $\gamma_v = 1 - \gamma_f$ Eq. (11-39)

   For unbalanced moments about an axis parallel to the edge at exterior supports, the value of γ_f can be taken equal to 1.0 provided that $V_u \leq 0.75\phi V_c$.   13.5.3.3

   $V_c = 4\sqrt{f'_c}\, b_o d$ Eq. (11-35)

   $= 4\sqrt{4,000} \times 60 \times 6.0/1,000 = 91.1$ kips

   $\phi = 0.75$ 9.3.2.3

   $0.75\phi V_c = 0.75 \times 0.75 \times 91.1 = 51.2$ kips $> V_u = 30$ kips

   Therefore, all of the unbalanced moment at the support may be considered transferred by flexure (i.e., $\gamma_f = 1.0$ and $\gamma_v = 0$). Note that γ_f can be taken as 1.0 provided that ρ within the effective slab width $3h + c_2 = 21.75 + 16 = 37.75$ in. is not greater than $0.375\rho_b$.

   b. Check shear strength of slab without shear reinforcement.

   Combined shear stress along inside face of critical transfer section.

   $v_{u1} = \dfrac{V_u}{A_c} + \dfrac{\gamma_v M_u c}{J} = \dfrac{30,000}{360} + 0 = 83.3$ psi

   Permissible shear stress:

   $\phi v_n = \phi 4\sqrt{f'_c} = 0.75\left(4\sqrt{4,000}\right) = 189.7$ psi $> v_{u1} = 83.3$ psi O.K.

   Slab shear strength is adequate for the required shear and moment transfer between slab and column.

   Design for the portion of unbalanced moment transferred by flexure $\gamma_f M_u$ must also be considered. See Example 19.1 when using the Direct Design Method. See Example 20.1 for the Equivalent Frame Method.   13.5.3.2

   For the Direct Design Method, $\gamma_f M_u = 1.0 \times (0.26\, M_o) = 25$ ft-kips to be transferred over the effective width of 37.75 in., provided that ρ within the 37.75-in. width $\leq 0.375\, \rho_b$.   13.6.3.3
   13.5.3.3

	Code
Example 16.4 (cont'd) **Calculations and Discussion**	**Reference**

3. Loading condition (2), V_u = 60 kips, M_o = 170 ft-kips:

   a. Check shear strength of slab without shear reinforcement.

   Portion of unbalanced moment to be transferred by eccentricity of shear.   *11.12.6.1*

   $0.75\phi V_c = 51.2$ kips $< V_u = 60$ kips   *13.5.3.3*

   Therefore, $\gamma_v = 1 - \gamma_f$ *Eq. (11-39)*

   $$\gamma_f = \frac{1}{1+\frac{2}{3}\sqrt{\frac{b_1}{b_2}}} = \frac{1}{1+\frac{2}{3}\sqrt{\frac{19.0}{22.0}}} = 0.62$$ *Eq. (13-1)*

   $\gamma_v = 1 - 0.62 = 0.38$

   For the Direct Design Method, the unbalanced moment M_u to be used in the shear stress computation for the edge column = $0.3M_o = 0.3 \times 170 = 51.0$ ft-kips.   *13.6.3.6*

   Combined shear stress along inside face of critical transfer section,

   $$v_{u1} = \frac{V_u}{A_c} + \frac{\gamma_v M_u c}{J}$$

   $$= \frac{60,000}{360} + \frac{0.38 \times 51.0 \times 12,000}{2,508}$$

   $= 166.7 + 92.7 = 259.4$ psi

   $\phi v_n = 189.7$ psi $< v_{u1} = 259.4$ psi   N.G.

   Shear reinforcement must be provided to carry excess shear stress; provide either bar reinforcement or steel I- or channel-shapes (shearheads).

   Increase slab shear strength by bar reinforcement.

Example 16.4 (cont'd)	Calculations and Discussion	Code Reference

b. Check maximum shear stress permitted with bar reinforcement. 11.12.3.2

Check effective depth, d 11.12.3

Assuming No. 3 stirrups ($d_b = 0.375$ in.),

$$d = 6 \text{ in.} \geq \begin{cases} 6 \text{ in. O.K.} \\ 16 \times 0.375 = 6 \text{ in. O.K.} \end{cases}$$

$$v_{u1} \leq \phi 6\sqrt{f'_c}$$

$$\phi v_n = 0.75(6\sqrt{4,000}) = 284.6 \text{ psi}$$

$$v_{u1} = 259.4 \text{ psi} < \phi v_n = 284.6 \text{ psi O.K.}$$

c. Determine shear stress carried by concrete with bar reinforcement. 11.12.3.1

$$\phi V_c = \phi 2\sqrt{f'_c} = 0.75(2\sqrt{4,000}) = 94.9 \text{ psi}$$

d. With symmetrical shear reinforcement on all sides of column, required A_v is

$$A_v = \frac{(v_{u1} - \phi v_c) b_o s}{\phi f_y}$$ Eq. (4)

where b_o is perimeter of critical section located at d/2 from column perimeter

$$b_o = 2(19) + 22 = 60 \text{ in. and}$$

$$s = \frac{d}{2} = 3.0 \text{ in.}$$

$$A_v = \frac{(259.4 - 94.9) \times 60 \times 3.0}{0.75 \times 60,000} = 0.66 \text{ in.}^2$$

A_v is total area of shear reinforcement required on the three sides of the column.

$$A_v \text{ (per side)} = \frac{0.66}{3} = 0.22 \text{ in.}^2$$

Use No. 3 closed stirrups @ 3.0 in. spacing ($A_v = 0.22$ in.²)

A check on the calculations can easily be made; for No. 3 closed stirrups @ 3.0 in.:

$$\phi(v_c + v_s) = \phi\left(2\sqrt{f'_c} + \frac{A_v f_y}{b_o s}\right)$$ Eq. (11-41)

		Code
Example 16.4 (cont'd)	**Calculations and Discussion**	**Reference**

$$= 0.75\left[2\sqrt{4,000} + \frac{(3 \times 0.22) \times 60,000}{60 \times 3.0}\right]$$

$= 0.75\ (126.5 + 220.0) = 259.9\ \text{psi} > v_{u1} = 259.4\ \text{psi}$ O.K.

e. Determine distance from sides of column where stirrups may be terminated.

$V_u \le \phi V_c$ 11.12.3.1

$\phi V_c = \phi 2\sqrt{f'_c} b_o d$

where $b_o = 2a\sqrt{2} + (3 \times 16)$

$60,000 \le 0.75 \times 2\sqrt{4,000}\left(2a\sqrt{2} + 48\right)6.0$

Solving, a = 20.3 in.

Note that the above is a conservative estimate, since V_u at the perimeter of the critical section considered is considerably lower than 60 kips.

No. of stirrups required = (20.3 - d/2)/3.0 = 5.8
(Stirrups may be terminated at d/2 = 3.0 in. inside perimeter b_o)

Use 6-No. 3 closed stirrups @ 3.0 in. spacing along the three sides of the column.
Use similar stirrup detail as for Example 16.3.

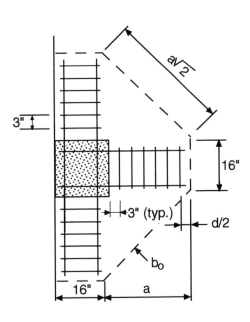

Blank

17

Strut-and-Tie Models

GENERAL

The strut-and-tie model is essentially a truss analogy. It is based on the fact that concrete is strong in compression, and that steel is strong in tension. Truss members that are in compression are made up of concrete, while truss members that are in tension consist of steel reinforcement.

Appendix A, Strut-and-Tie Models, was introduced in ACI 318-02. It provides a design approach, applicable to an array of design problems that do not have an explicit design solution in the body of the code. This method requires the designer to consciously select a realistic load path within the structural member in the form of an idealized truss. Rational detailing of the truss elements and compliance with equilibrium assures the safe transfer of loads to the supports or to other regions designed by conventional procedures. While solutions provided with this powerful design and analysis tool are not unique, they represent a conservative lower bound approach. As opposed to some of the prescriptive formulations in the body of ACI 318, the very visual, rational strut-and-tie model of Appendix A gives insight into detailing needs of irregular regions of concrete structures and promotes ductility.

The design methodology presented in Appendix A is largely based on the seminal articles on the subject by Schlaich et al.[17.1], Collins and Mitchell[17.2], and Marti[17.3]. Since publication of these papers, the strut-and-tie method has received increased attention by researchers and textbook writers (Collins and Mitchell[17.4], MacGregor and Wight[17.5]). MacGregor described the background of provisions incorporated in Appendix A in ACI Special Publication SP-208[17.6]. The present form of Appendix A does not include explicit serviceability provisions (such as deflection control).

A.1 DEFINITIONS

The strut-and-tie design procedure calls for the distinction of two types of zones in a concrete component depending on the characteristics of stress fields at each location. Thus, structural members are divided into B-regions and D-regions.

B-regions represent portions of a member in which the "plane section" assumptions of the classical beam theory can be applied with a sectional design approach.
D-regions are all the zones outside the B-regions where cross-sectional planes do not remain plane upon loading. D-regions are typically assumed at portions of a member where **discontinuities** (or disturbances) of stress distribution occur due to concentrated forces (loads or reactions) or abrupt changes of geometry. Based on St. Venant's Principle, the normal stresses (due to axial load and bending) approach quasi-linear distribution at a distance approximately equal to the larger of the overall height (h) and width of the member, away from the

location of the concentrated force or geometric irregularity. Figure 17-1 illustrates typical discontinuities, D-Regions (cross-hatched areas), and B-Regions.

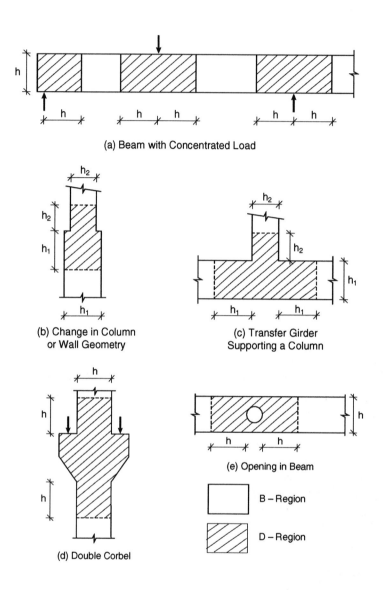

Figure 17-1 Load and Geometric Discontinuities

While B-regions can be designed with the traditional methods (ACI 318 Chapters 10 and 11), the **strut-and-tie model** was primarily introduced to facilitate the design of D-regions, and can be extended to the B-regions as well. The strut-and-tie model depicts the D-region of the structural member with a truss system consisting of compression struts and tension ties connected at nodes as shown in Fig. 17-2. This truss system is designed to transfer the factored loads to the supports or to adjacent B-regions. At the same time, forces in the truss members should maintain equilibrium with the applied loads and reactions.

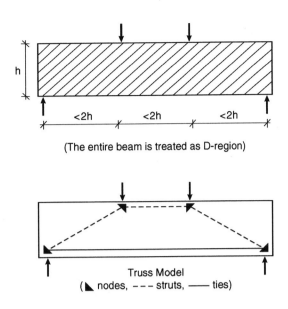

Figure 17-2 Strut-and-Tie Model

Struts are the compression elements of the strut-and-tie model representing the resultants of a compression field. Both parallel and fan shaped compression fields can be modeled by their resultant compression struts as shown in Fig. 17-3.

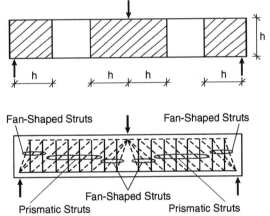

Figure 17-3 Prismatic and Fan-Shaped Struts

Typically compression struts would take a bottle-shape wherever the strut can spread laterally at mid-length. As a design simplification, prismatic compression members commonly idealize struts, however, other shapes are also possible. If the effective concrete compressive strength (f_{ce}) is different at the opposite ends of a strut, a linearly tapering compression member is suggested. This condition may occur, if at the two ends of the strut the nodal zones have different strengths or different bearing lengths. Should the compression stress be high in the strut, reinforcement may be necessary to prevent splitting due to transverse tension. (The splitting crack that develops in a cylinder supported on edge, and loaded in compression is a good example of the internal lateral spread of the compressive stress trajectories).

Ties consist of conventional deformed steel, or prestressing steel, or both, plus a portion of the surrounding concrete that is concentric with the axis of the tie. The surrounding concrete is not considered to resist axial force in the model. However, it reduces the elongation of the tie (tension stiffening), in particular, under service loads. It also defines the zone in which the forces in the struts and ties are to be anchored.

Nodes are the intersection points of the axes of the struts, ties and concentrated forces, representing the joints of a strut-and-tie model. To maintain equilibrium, at least three forces should act on a given node of the model. Nodes are classified depending on the sign of the forces acting upon them (e.g., a C-C-C node resists three compression forces, a C-T-T node resists one compression forces and two tensile forces, etc.) as shown in Fig. 17-4

Figure 17-4 Classification of Nodes

A nodal zone is the volume of concrete that is assumed to transfer strut-and tie forces through the node. The early strut-and-tie models used <u>hydrostatic nodal zones</u>, which were lately superseded by <u>extended nodal zones</u>.

The faces of a **hydrostatic nodal zone** are perpendicular to the axes of the struts and ties acting on the node, as depicted in Fig. 17-5. The term hydrostatic refers to the fact that the in-plane stresses are the same in all directions. (Note that in a true hydrostatic stress state the out-of plane stresses should be also equal). Assuming identical stresses on all faces of a C-C-C nodal zone with three struts implies that the ratios of the lengths of the sides of the nodal zones ($w_{n1} : w_{n2} : w_{n3}$) are proportional to the magnitude of the strut forces ($C_1 : C_2 : C_3$).

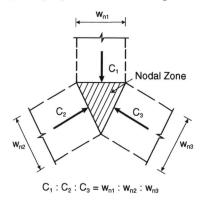

Figure 17-5 Hydrostatic Nodal Zone

The extended nodal zone is a portion of a member bounded by the intersection of the effective strut width, w_s, and the effective tie width, w_t. This is illustrated in Fig. 17-6.

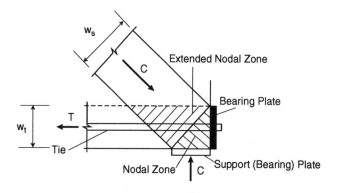

Figure 17-6 Extended Nodal Zone

A.2 STRUT-AND-TIE MODEL DESIGN PROCEDURE

A design with the strut-and-tie model typically involves the following steps:
1. Define and isolate D-regions.
2. Compute resultant forces on each D-region boundary.
3. Devise a truss model to transfer the resultant forces across the D-region. The axes of the struts and ties, respectively, are oriented to approximately coincide with the axes of the compression and tension stress fields.
4. Calculate forces in the truss members.
5. Determine the effective widths of the struts and nodal zones considering the forces from the previous steps and the effective concrete strengths (defined in A.3.2 and A.5.2). Strength checks are based on

$$\phi F_n \geq F_u \quad \quad Eq.\ (A\text{-}1)$$

where F_u is the largest factored force obtained from the applicable load combinations, F_n is the nominal strength of the strut, tie, or node, and the ϕ factor is listed in 9.3.2.6 as 0.75 for ties, strut, nodal zones and bearing areas of strut-and-tie models.
6. Provide reinforcement for the ties considering the steel strengths defined in A.4.1. The reinforcement must be detailed to provide proper anchorage in the nodal zones

In addition to the strength limit states, represented by the strut-and-tie model, structural members should be checked for serviceability requirements. Traditional elastic analysis can be used for deflection checks. Crack control can be verified using provisions of 10.6.4, assuming that the tie is encased in a prism of concrete corresponding to the area of tie (RA.4.2).

There are usually several strut-and-tie models that can be devised for a given structural member and loading condition. Models that satisfy the serviceability requirements the best, have struts and ties that follow the compressive and tensile stress trajectories, respectively. Certain construction rules of strut-and-tie models, e.g., "the angle, θ, between the axes of any strut and any tie entering a single node shall not be taken as less than 25 degree (A.2.5) are imposed to mitigate potential cracking problems and to avoid incompatibilities due to shortening of the struts and lengthening of the ties in almost the same direction.

A.3 STRENGTH OF STRUTS

The nominal compressive strength of a strut without longitudinal reinforcement shall be taken as

$$F_{ns} = f_{ce} A_{cs} \quad \quad Eq.\ (A\text{-}2)$$

to be calculated at the weaker end of the compression member. A_{cs} is the cross-sectional area at the end of the strut. In typical two-dimensional members, the width of the strut (w_s) can be taken as the width of the member. The effective compressive strength of the concrete (f_{ce}) for this purpose shall be taken as the lesser of the concrete strengths at the two sides of the nodal zone/strut interface. Section A.3.2 specifies the calculation of f_{ce} for the strut (detailed below), while A.5.2 provides for the same in the nodal zone (discussed later).

The effective compressive strength of the concrete in a strut is calculated, similarly to basic strength equations, as:

$$f_{ce} = 0.85 \beta_s f_c' \quad \quad Eq.\ (A\text{-}3)$$

The β_s factor accounts for the effect of cracking and possible presence of transverse reinforcement. The strength of the concrete in a strut can be computed with $\beta_s = 1.0$ for struts that have uniform cross sectional area over their length. This is quasi-equivalent to the rectangular stress block in the compression zone of a beam or column. For bottle-shaped struts (Fig. 17-7) with reinforcement placed to resist the splitting forces (satisfying A3.3) $\beta_s = 0.75$ or without adequate confinement to resist splitting forces $\beta_s = 0.6\lambda$ (where λ is a correction factor (11.7.4.3) for lightweight concrete.)

For struts intersecting cracks in a tensile zone, β_s is reduced to 0.4. Examples include strut-and-tie models used to design the longitudinal and transverse reinforcement of the tension flanges of beams, box-girders and walls. For all other cases (e.g., in beam webs where struts are likely to be crossed by inclined cracks), the β_s factor can be conservatively taken as 0.6.

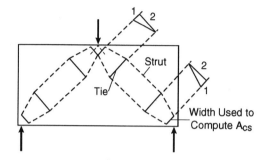

Figure 17-7 Bottle Shaped Compression Strut

Section A.3.3 addresses cases where transverse reinforcement is provided to cross the bottle-shaped struts. The compression forces in the strut may be assumed to spread at a slope 2:1. The rebars are intended to resist the transverse tensile forces resulting from the compression force spreading in the strut. They may be placed in one layer (when the γ angle between the rebar and the axis of the strut is at least 40 degree) or in two orthogonal layers.

To allow for $\beta_s = 0.75$, for concrete strength not exceeding 6000 psi, the reinforcement ratio needed to cross the strut is:

$$\sum \frac{A_{si}}{b_s s_i} \sin \gamma_i \geq 0.003 \qquad \text{Eq. (A-4)}$$

where A_{si} is the total area of reinforcement at spacing s_i in a layer of reinforcement with bars at an angle γ_i to the axis of the strut (shown in Fig. 17-8), and b_s is the width of the strut. Often, this reinforcement ratio cannot be provided due to space limitations. In those cases $\beta_s = 0.6\lambda$ shall be used.

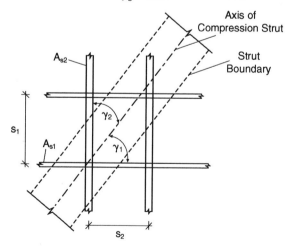

Figure 17-8 Layers of Reinforcement to Restrain Splitting Cracks of Struts

If substantiated by test and analyses, increased effective compressive strength of a strut due to confining reinforcement may be used (e.g., at anchorage zones of prestressing tendons). This topic is discussed in detail in Refs. 17.7 and 17.8.

Additional strength can be provided to the struts by including compression reinforcement parallel to the axis of the strut. These bars must be properly anchored and enclosed by ties or spirals per 7.10. The compressive strength of these longitudinally reinforced struts can be calculated as:

$$F_{ns} = f_{ce}A_{cs} + A_s'f_s' \qquad \text{Eq. (A-5)}$$

where f_s' is the stress in the longitudinal strut reinforcement at nominal strength. It can be either obtained from strain analyses at the time the strut crushes or taken as $f_s' = f_y$ for Grade 40 and 60 rebars.

A.4 STRENGTH OF TIES

The nominal strength of a tie is calculated as the sum of yield strength of the conventional reinforcement plus the force in the prestressing steel:

$$F_{nt} = A_{ts}f_y + A_{tp}(f_{se} + \Delta f_p) \qquad \text{Eq. (A-6)}$$

Note, that A_{tp} is zero if there is no prestressing present in the tie. The actual prestressing stress $(f_{se} + \Delta f_p)$ should not exceed the yield stress f_{py} of the prestressing steel. Also, if not calculated, the code allows to estimate the increase in prestressing steel stress due to factored loads Δf_p, as 60,000 psi for bonded prestressed reinforcement, or 10,000 psi for unbonded prestressed reinforcement.

Since the intent of having a tie is to provide for a tension element in a truss, the axis of the reinforcement centroid shall coincide with the axis of the tie assumed in the model. Depending on the distribution of the tie reinforcement, the <u>effective tie width</u> (w_t) may vary between the following limits:
- The minimum width for configurations where only one layer of reinforcement provided in a tie, w_t can be taken as the diameter of the bars in the tie plus twice the concrete cover to the surface of the ties. Should the tie be wider than this, the reinforcement shall be distributed evenly over the width.
- The upper limit is established as the width corresponding to the width in a hydrostatic nodal zone, calculated as

$$w_{t,max} = F_{nt}/f_{ce}$$

where f_{ce} is the applicable effective compression strength of a nodal zone discussed below.

Nodes shall be able to develop the difference between the forces of truss members connecting to them. Thus, besides providing adequate amount of tie reinforcement, special attention shall be paid to proper anchorage. Anchorage can be achieved using mechanical devices, post-tensioning anchorage devices, standard hooks, or straight bar embedment. The reinforcement in a tie should be anchored before it leaves the extended nodal zone, i.e., at the point defined by the intersection of the centroid of the bars in the tie and the extensions of the outlines of either the strut or the bearing area as shown on Fig.17-9. For truss layouts where more than one tie intersect at a node, each tie force shall be developed at the point where the centroid of the reinforcement in the tie leaves the extended nodal zone. (Note, that transverse reinforcement required by A3.3 shall be anchored according to the provisions of 12.13).

In many cases the structural configuration does not allow to provide for the straight development length for a tie.

For such cases, anchorage is provided through mechanical devices, hooks, or splicing with several layers of smaller bars. These options often require a wider structural member and/or additional confinement reinforcement (e.g., to avoid cracking along the outside of the hooks).

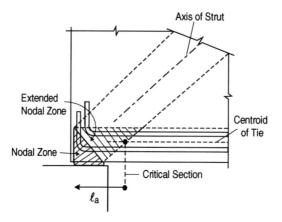

Figure 17-9 Anchorage of Tie Reinforcement

A.5 STRENGTH OF NODAL ZONES

The nominal compression strength at the face of a nodal zone or at any section through the nodal zone shall be:

$$F_{nn} = f_{ce}A_{nz} \qquad \text{Eq. (A-7)}$$

where A_{nz} is taken as the area of the face of the nodal zone that the strut force F_u acts on, if the face is perpendicular to the line of action of F_u. If the nodal zone is limited by some other criteria, the node-to-strut interface may not be perpendicular to the axis of the strut, therefore, the axial stresses in the compression-only strut will generate both shear and normal stresses acting on the interface. In those cases, the A_{nz} parameter shall be the area of a section, taken through the nodal zone perpendicular to the strut axis.

The strut-and-tie model is applicable to three-dimensional situations as well. In order to keep calculations simple, A5.3 allows the area of the nodal faces to be less than that described above. The shape of each face of the nodal zones must be similar to the shape of the projection of the end of the struts onto the corresponding faces of the nodal zones.

The effective compressive strength of the concrete in the nodal zone (f_{ce}) is calculated as:

$$f_{ce} = 0.85\beta_n f_c' \qquad \text{Eq. (A-8)}$$

and must not exceed the effective concrete compressive strength on the face of a nodal zone due to the strut-and-tie model forces, unless confining reinforcement is provided within the nodal zone and its effect is evidenced by tests and analysis. The sign of forces acting on the node influences the capacity at the nodal zones as reflected by the β_n value. The presence of tensile stresses due to ties decreases the nodal zone concrete strength.

$\beta_n = 1.0$ in nodal zones bounded by struts or bearing areas (e.g., C-C-C nodes)
$\beta_n = 0.8$ in nodal zones anchoring one tie (e.g., C-C-T nodes)
$\beta_n = 0.6$ in nodal zones anchoring two or more ties (e.g., C-T-T or T-T-T nodes).

REFERENCES

17.1. Schlaich, J.; Schafer, K. and Jennewein, M, "Toward a Consistent Design of Structural Concrete," PCI Journal, V. 32, No. 3, May-June 1987, pp. 74-150.

17.2. Collins, M.P. and Mitchell, D., "A Rational Approach to Shear Design—The 1984 Canadian Code Provisions," ACI Journal, Vol. 83, No. 6, November-December 1986, pp. 925-933.

17.3. Marti P., "Truss Models in Detailing," Concrete International, ACI, Dec. 1985, pp. 66-73.

17.4 Collins, M.P. and Mitchell, D., Prestressed Concrete Structures, Prentice Hall Inc. Englewood Cliffs, N.J., 1991, 766 pp.

17.5 MacGregor, J. G. and Wight, J.K., Reinforced Concrete Mechanics and Design, 4th Edition, Prentice Hall, Upper Saddle River, N.J., 2005, 1132 pp.

17.6 "Examples for the Design of Structural Concrete with Strut-and-Tie Models," SP-208, American Concrete Institute, Farmington Hills, MI, 2002, 242 pp.

17.7 FIP Commission 3, FIP Recommendations 1996 Practical Design of Structural Concrete, FIP Congress Amsterdam 1996, Federation Internationale de la Precontraint, Lausanne, 1999.

17.8 Bergmeister, K.; Breen, J.E.: and Jirsa, J.O., "Dimensioning of the Nodes and Development of Reinforcement," IABSE Colloquim Stuttgart 1991, International Association for Bridge and Structural Engineering, Zurich, 1991, pp. 551-556.

Example 17.1—Design of Deep Flexural Member by the Strut-and-Tie Model

Determine the required reinforcement for the simply supported transfer girder shown in Fig. 17-10. The single column at midspan subjects the girder to 180 kips dead load and 250 kips live load.

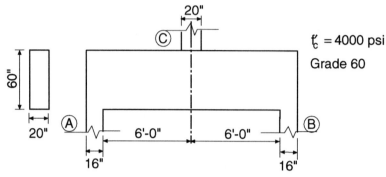

Figure 17-10 Transfer Girder

	Code
Calculations and Discussion	**Reference**

1. Calculate factored load and reactions

   The transfer girder dead load is conservatively lumped to the column load at midspan. Transfer girder dead load is:

   $5(20/12)\,[6 + 6 + (32/12)]\,0.15 = 18.5$ kips

   $P_u = 1.2D + 1.6L = 1.2 \times (18.5 + 180) + 1.6 \times 250 = 640$ kips   *Eq. (9-2)*

   $R_A = R_B = 640/2 = 320$ kips

2. Determine if this beam satisfies the definition of a "deep beam"   *10.7.1*
   11.8.1

   Overall girder height $h = 5$ ft

   Clear span $\ell_n = 12$ ft

   $\dfrac{\ell_n}{h} = \dfrac{12}{5} = 2.4 < 4$

   Member is a "deep beam" and will be designed using Appendix A.

3. Check the maximum shear capacity of the cross section

   $V_u = 320$ kips

   Maximum $\phi V_n = \phi\left(10\sqrt{f'_c}\,b_w d\right)$ *11.8.3*

Example 17.1 (cont'd)   **Calculations and Discussion**   **Code Reference**

$= 0.75(10\sqrt{4000} \times 20 \times 54)/1000 = 512 \text{ kips} > V_u$ O.K.

4. Establish truss model

   Assume that the nodes coincide with the centerline of the columns (supports), and are located 5 in. from the upper or lower edge of the beam as shown in Fig. 17-11. The strut-and-tie model consists of two struts (A-C and B-C), one tie (A-B), and three nodes (A, B, and C). In addition, columns at A and B act as struts representing reactions. The vertical strut at the top of Node C represents the applied load.

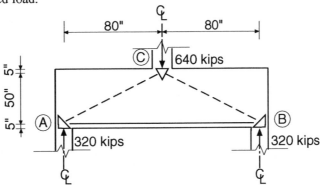

Figure 17-11 Preliminary Truss Layout

The length of the diagonal struts $= \sqrt{50^2 + 80^2} = 94.3$ in.

The force in the diagonal struts $= 320 \dfrac{94.3}{50} = 603$ kips

The force in the horizontal tie $= 320 \dfrac{80}{50} = 512$ kips

Verify the angle between axis of strut and tie entering Node A.

The angle between the diagonal struts and the horizontal tie $= \tan^{-1}(50/80) = 32° > 25°$ O.K.   A.2.5

5. Calculate the effective concrete strength (f_{ce}) for the struts assuming that reinforcement is provided per A.3.3 to resist splitting forces. (See Step 9)

   For the "bottle-shaped" Struts A-C & B-C

   $f_{ce} = 0.85 \beta_s f'_c = 0.85 \times 0.75 \times 4000 = 2550$ psi   *Eq. (A-3)*

   where $\beta_s = 0.75$ per A.3.2.2(a)

   Note, this effective compressive strength cannot exceed the strength of the nodes at both ends of the strut. See A.3.1.

Example 17.1 (cont'd)	Calculations and Discussion	Code Reference

The vertical struts at A, B, and C have uniform cross-sectional area throughout their length.

$\beta_s = 1.0$ *A.3.2.1*

$f_{ce} = 0.85 \times 1.0 \times 4000 = 3400$ psi

6. Calculate the effective concrete strength (f_{ce}) for Nodal Zones A, B, and C

   Nodal Zone C is bounded by three struts. So this is a C-C-C nodal zone
   with $\beta_n = 1.0$ *A.5.2.1*

   $f_{ce} = 0.85\beta_n f'_c = 0.85 \times 0.80 \times 4000 = 2720$ psi   *Eq. (A-8)*

   Nodal Zones A and B are bounded by two struts and a tie. For a C-C-T node:

   $\beta_n = 0.80$

   $f_{ce} = 0.85\beta_n f'_c = 0.85 \times 0.80 \times 4000 = 2720$ psi

7. Check strength at Node C

   Assume that a hydrostatic nodal zone is formed at Node C. This means that the faces of the nodal zone are perpendicular to the axis of the respective struts, and that the stresses are identical on all faces.

   To satisfy the strength criteria for all three struts and the node, the minimum nodal face dimension is determined based on the least strength value of $f_{ce} = 2550$ psi. The same strength value will be used for Nodes A and B as well.

   The strength checks for all components of the strut and tie model are based on
   $\phi F_n \geq F_u$ *Eq. (A-1)*

   where $\phi = 0.75$ for struts, ties, and nodes.   *9.3.2.6*

   The length of the horizontal face of Nodal Zone C is calculated as

   $$\frac{640,000}{0.75 \times 2550 \times 20} = 16.7 \text{ in. (less than column width of 20 in.)}$$

   The length of the other faces, perpendicular to the diagonal struts, can be obtained from proportionality:

   $16.7 \times \dfrac{603}{640} = 15.7$ in.

Example 17.1 (cont'd) — Calculations and Discussion | Code Reference

8. Check the truss geometry.

   At Node C

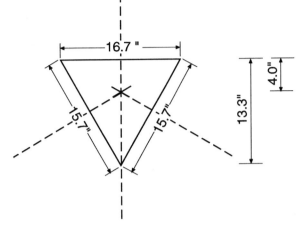

Figure 17-12 Geometry of Node C

The center of the nodal zone is at 4.0 in. from the top of the beam, which is very close to the assumed 5 in.

At Node A

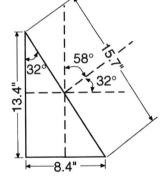

Figure 17-13 Geometry of Node A

The horizontal tie should exert a force on this node to create a stress of 2550 psi. Thus size of the vertical face of the nodal zone is

$$\frac{512{,}000}{0.75 \times 2550 \times 20} = 13.4 \text{ in.}$$

The center of the tie is located $13.4/2 = 6.7$ in. from the bottom of the beam.
This is reasonably close to the 5 in. originally assumed, so no further iteration is warranted.

Width of node at Support A

$$\frac{320{,}000}{0.75 \times 2550 \times 20} = 8.4 \text{ in.}$$

		Code
Example 17.1 (cont'd)	**Calculations and Discussion**	**Reference**

9. Provide vertical and horizontal reinforcement to resist splitting of diagonal struts.

   The angle between the vertical ties and the struts is 90°- 32° = 58° (sin 58° = 0.85)

   Try two overlapping No. 4 ties @ 12 in. O.C. (to accomodate the longitudinal tie reinforcement designed in Step 10, below.

   $$\frac{A_{si}}{bs}\sin\gamma_i = \frac{4 \times 0.20}{20 \times 12} \times 0.85 = 0.00283$$ *Eq. (A-4)*

   and No. 5 horizontal bars @ 12 in. O.C. on each side face (sin 32° = 0.53)

   $$\frac{2 \times 0.31}{20 \times 12} \times 0.53 = 0.00137$$

   $$\sum\frac{A_{si}}{bs}\sin\gamma_i = 0.00283 + 0.00137 = 0.0042 > 0.003 \text{ O.K.}$$ *Eq. (A-4)*

10. Provide horizontal reinforcing steel for the tie

    $$A_{s,req} = \frac{F_u}{\phi f_y} = \frac{512}{0.75 \times 60} = 11.4 \text{ in.}^2$$

    Select 16 - No. 8 $A_s = 12.64$ in.^{2}

    These bars must be properly anchored. The anchorage is to be measured from the point where the tie exits the extended nodal zone as shown in Fig. 17-14.

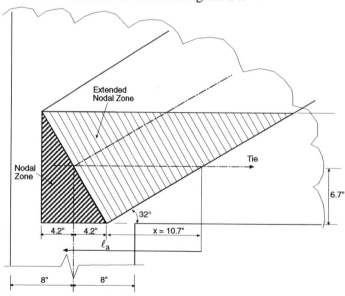

Figure 17-14 Development of Tie Reinforcement Within the Extended Nodal Zone

Example 17.1 (cont'd)	Calculations and Discussion	Code Reference

Distance $x = 6.7/\tan 32 = 10.7$ in.

Available space for a straight bar embedment

$10.7 + 4.2 + 8 - 2.0$ (cover) $= 20.9$ in.

This length is inadequate to develop a straight No. 8 bar.

Development length for a No. 8 bar with a standard 90 deg. hook

$\ell_{dh} = (0.02\psi_e \lambda f_y / \sqrt{f'_c}) d_b$ 12.5.2

$\phantom{\ell_{dh}} = \left(0.02(1.0)(1.0)60{,}000 / \sqrt{4000}\right) 1.0$

$\phantom{\ell_{dh}} = 19.0 < 20.9$ in. O.K.

Note: the 90 degree hooks will be enclosed within the column reinforcement that extends in the transfer girder. (Fig. 17-15) By providing adequate cover and transverse confinement, the development length of the standard hook could be reduced by the modifiers of 12.5.3.

Less congested reinforcement schemes can be devised with reinforcing steel welded to bearing plates, or with the use of prestressing steel.

Comments:

The discrepancy in the vertical location of the nodes results in a negligible (about 1.5 percent) difference in the truss forces. Thus, another iteration is not warranted.

There are several alternative strut-and-tie models that could have been selected for this problem. An alternative truss layout is illustrated in Fig. 17-16. It has the advantage that the force in the bottom chord varies between nodes, instead of being constant between supports. Further, the truss posts carry truss forces, instead of providing vertical reinforcement just for crack control (A.3.3.1). Finally, the diagonals are steeper, therefore the diagonal compression and the bottom chord forces are reduced. The optimum idealized truss is one that requires the least amount of reinforcement.

Example 17.1 (cont'd)	Calculations and Discussion	Code Reference

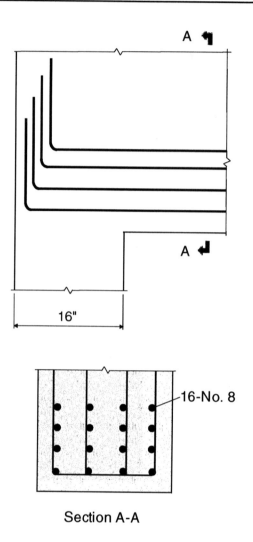

Figure 17-15 Detail of Tie Reinforcement

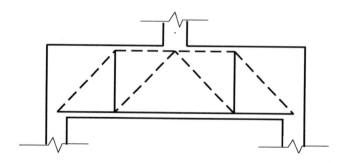

Figure 17-16 Alternative Strut-and-Tie Model

Example 17.2—Design of Column Corbel

Design the single corbel of the 16 in. × 16 in. reinforced concrete column for a vertical force $V_u = 60$ kips and horizontal force $N_u = 12$ kips. Assume $f'_c = 5000$ psi, and Grade 60 reinforcing steel.

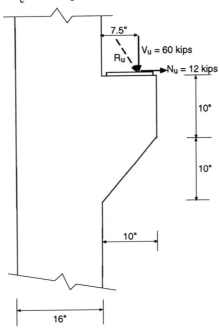

Figure 17-17 Design of Corbel

Calculations and Discussion	Code Reference

1. Establish the geometry of a trial truss and calculate force demand in members.

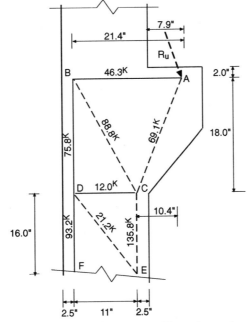

Figure 17-18 Truss Layout

17-17

Example 17.2 (cont'd)	Calculations and Discussion	Code Reference

2. Provide reinforcement for ties

    Use $\phi = 0.75$ 9.3.2.6

    The nominal strength of ties is to be taken as:

    $$F_{nt} = A_{ts}f_y + A_{tp}(f_{se} + \Delta f_p)$$ Eq. (A-6)

    where the last term can be ignored for nonprestressed reinforcement

    <u>Tie AB</u> $F_u = 46.3$ kips

    $$A_{ts} = \frac{F_u}{\phi f_y} = \frac{46.3}{0.75 \times 60} = 1.03 \text{ in.}^2 \text{ Provide 4 - No. 5 } A_{ts} = 1.24 \text{ in.}^2$$

    <u>Tie CD</u> $F_u = 12.0$ kips

    $$A_{ts} = \frac{12.0}{0.75 \times 60} = 0.27 \text{ in.}^2 \text{ Provide No. 4 tie (2 legs) } A_{ts} = 0.40 \text{ in.}^2$$

    <u>Tie BD & DF</u> $P_u = 93.2$ kips

    $$A_{ts} = \frac{93.2}{0.75 \times 60} = 2.07 \text{ in.}^2 \text{ Provide steel in addition of the vertical column reinforcement}$$

    This reinforcement may be added longitudinal bar or a rebar bent at Node A, that is used as Tie AB as well.

3. Calculate strut widths

    It is assumed that transverse reinforcement will be provided in compliance with A.3.3, so a $\beta_s = 0.75$ can be used in calculating the strut length

    $$f_{ce} = 0.85 \beta_s f'_c = 0.85 \times 0.75 \times 5000 = 3187 \text{ kips}$$ Eq. (A-3)

    $$\phi f_{ce} = 0.75 \times 3187 = 2390 \text{ psi}$$

    Calculate the width of struts required

    <u>Strut AC</u> $P_u = 69.1$ kips

    $$w = \frac{69{,}100}{16 \times 2390} = 1.81 \text{ in.}$$

Example 17.2 (cont'd) — Calculations and Discussion — Code Reference

Strut BC

$$w = \frac{88,800}{16 \times 2390} = 2.32 \text{ in.}$$

Strut CE

$$w = \frac{135,800}{16 \times 2390} = 3.55 \text{ in.}$$

Strut DE

$$w = \frac{21.2}{16 \times 2390} = 0.55 \text{ in.}$$

The width of the struts will fit within the concrete column with the corbel.

Provide confinement reinforcement for the struts per A.3.3 in the form of horizontal ties
The angle of the diagonal struts to the horizontal hoops is 58 degree. Provide No. 4 hoops at 4.5 in. on center.

$$\frac{A_s}{bs} \sin\gamma = \frac{2 \times 0.20}{24 \times 4.5} \sin 58° = 0.0031 > 0.003 \quad \text{O.K.}$$

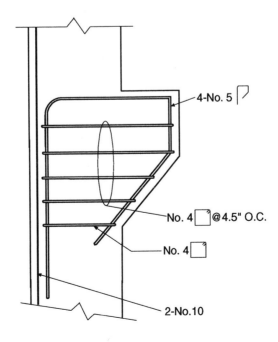

Figure 17-19 Reinforcement Details

Blank

18

Two-Way Slab Systems

UPDATE FOR THE '05 CODE

The primary drop panel definition was moved to Chapter 2. A new Section 13.2.5 was added to give additional dimensional requirements for drop panel if the drop panel is intended to reduce the amount of negative reinforcement over a column or minimum required slab thickness. The new Section 13.2.5 replaces the deleted Sections 13.3.7.1, 13.3.7.2, and 13.3.7.3.

13.1   SCOPE

Figure 18-1 shows the various types of two-way reinforced concrete slab systems in use at the present time that may be designed according to Chapter 13.

A solid slab supported on beams on all four sides (Fig. 18-1(a)) was the original slab system in reinforced concrete. With this system, if the ratio of the long to the short side of a slab panel is two or more, load transfer is predominantly by bending in the short direction and the panel essentially acts as a one-way slab. As the ratio of the sides of a slab panel approaches unity (or as panel approaches a square shape), significant load is transferred by bending in both orthogonal directions, and the panel should be treated as a two-way rather than a one-way slab.

As time progressed and technology evolved, the column-line beams gradually began to disappear. The resulting slab system consisting of solid slabs supported directly on columns is called the flat plate (Fig. 18-1(b)). The two-way flat plate is very efficient and economical and is currently the most widely used slab system for multistory construction, such as motels, hotels, dormitories, apartment buildings, and hospitals. In comparison to other concrete floor/roof systems, flat plates can be constructed in less time and with minimum labor costs because the system utilizes the simplest possible formwork and reinforcing steel layout. The use of flat plate construction also has other significant economic advantages. For instance, because of the shallow thickness of the floor system, story heights are automatically reduced, resulting in smaller overall height of exterior walls and utility shafts; shorter floor-to-ceiling partitions; reductions in plumbing, sprinkler, and duct risers; and a multitude of other items of construction. In cities like Washington, D.C., where the maximum height of buildings is restricted, the thin flat plate permits the construction of the maximum number of stories in a given height. Flat plates also provide for the most flexibility in the layout of columns, partitions, small openings, etc. An additional advantage of flat plate slabs that should not be overlooked is their inherent fire resistance. Slab thickness required for structural purposes will, in most cases, provide the fire resistance required by the general building code, without having to apply spray-on fire proofing, or install a suspended ceiling. This is of particular importance where job conditions allow direct application of the ceiling finish to the flat plate soffit, eliminating the need for suspended ceilings. Additional cost and construction time savings are then possible as compared to other structural systems.

The principal limitation on the use of flat plate construction is imposed by shear around the columns (13.5.4). For heavy loads or long spans, the flat plate is often thickened locally around the columns creating what are known as drop panels. When a flat plate incorporates drop panels, it is called a flat slab (Fig. 18-1(c)). Also for reasons of shear around the columns, the column tops are sometimes flared, creating column capitals. For purposes of design, a column capital is part of the column, whereas a drop panel is part of the slab (13.7.3 and 13.7.4).

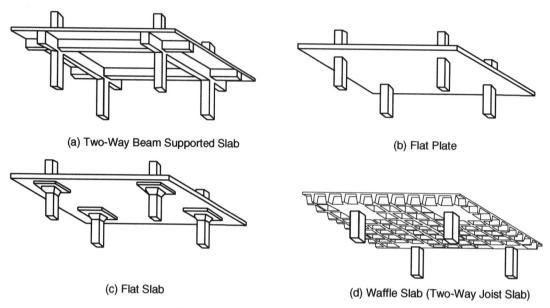

Figure 18-1 Types of Two-Way Slab Systems

Waffle slab construction (Fig. 18-1(d)) consists of rows of concrete joists at right angles to each other with solid heads at the column (needed for shear strength). The joists are commonly formed by using standard square "dome" forms. The domes are omitted around the columns to form the solid heads. For design purposes, waffle slabs are considered as flat slabs with the solid heads acting as drop panels (13.1.3). Waffle slab construction allows a considerable reduction in dead load as compared to conventional flat slab construction since the slab thickness can be minimized due to the short span between the joists. Thus, it is particularly advantageous where the use of long span and/or heavy loads is desired without the use of deepened drop panels or support beams. The geometric shape formed by the joist ribs is often architecturally desirable.

13.1.4   Deflection Control—Minimum Slab Thickness

Minimum thickness/span ratios enable the designer to avoid extremely complex deflection calculations in routine designs. Deflections of two-way slab systems need not be computed if the overall slab thickness meets the minimum requirements specified in 9.5.3. Minimum slab thicknesses for flat plates, flat slabs, and waffle slabs based on Table 9.5(c), and two-way beam-supported slabs based on Eqs. (9-12) and (9-13) are summarized in Table 18-1, where ℓ_n is the clear span length in the long direction of a two-way slab panel. The tabulated values are the controlling minimum thicknesses governed by interior, side, or corner panels assuming a constant slab thickness for all panels making up a slab system. Practical edge beam sizes will usually provide beam-to-slab stiffness ratios α greater than the minimum specified value of 0.8. A "standard" size drop panel that would allow a 10% reduction in the minimum required thickness of a flat slab floor system is illustrated in Fig. 18-2. Note that a drop of larger size and depth may be used if required for shear strength; however, a corresponding lesser slab thickness is not permitted unless deflections are computed.

For design convenience, minimum thicknesses for the six types of two-way slab systems listed in Table 18-1 are plotted in Fig. 18-3.

Refer to Part 10 for a general discussion on control of deflections for two-way slab systems, including design examples of deflection calculations for two-way slabs.

Table 18-1 Minimum Thickness for Two-Way Slab Systems (Grade 60 Reinforcement)

Two-Way Slab System	α_m	β	Minimum h
Flat Plate	—	≤ 2	$\ell_n/30$
Flat Plate with Spandrel Beams[1] [Min. h = 5 in.]	—	≤ 2	$\ell_n/33$
Flat Slab[2]	—	≤ 2	$\ell_n/33$
Flat Slab[2] with Spandrel beams[1] [Min. h = 4 in.]	—	≤ 2	$\ell_n/36$
Two-Way Beam-Supported Slab[3]	≤ 0.2	≤ 2	$\ell_n/30$
	1.0	1	$\ell_n/33$
		2	$\ell_n/36$
	≥ 2.0	1	$\ell_n/37$
		2	$\ell_n/44$
Two-Way Beam-Supported Slab[1,3]	≤ 0.2	≤ 2	$\ell_n/33$
	1.0	1	$\ell_n/36$
		2	$\ell_n/40$
	≥ 2.0	1	$\ell_n/41$
		2	$\ell_n/49$

[1]*Spandrel beam-to-slab stiffness ratio $\alpha \geq 0.8$ (9.5.3.3)*
[2]*Drop panel length $\geq \ell/3$, depth $\geq 1.25h$ (13.3.7)*
[3]*Min. h = 5 in. for $\alpha_m \leq 2.0$; min. h = 3.5 in. for $\alpha_m > 2.0$ (9.5.3.3)*

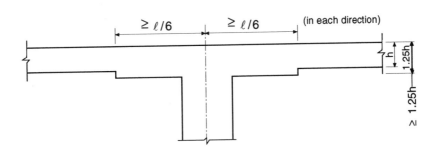

Figure 18-2 Drop Panel Details (13.2.5)

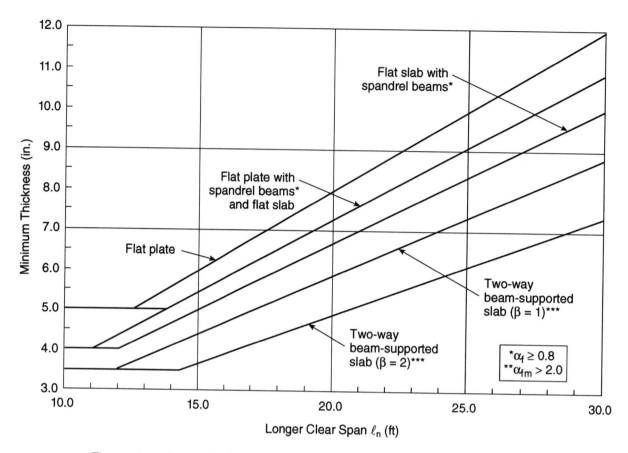

Figure 18-3 Minimum Slab Thickness for Two-Way Slab Systems (see Table 18-1)

13.2   DEFINITIONS

13.2.1   Design Strip

For analysis of a two-way slab system by either the Direct Design Method (13.6) or the Equivalent Frame Method (13.7), the slab system is divided into design strips consisting of a column strip and half middle strip(s) as defined in 13.2.1 and 13.2.2, and as illustrated in Fig. 18-4. The column strip is defined as having a width equal to one-half the transverse or longitudinal span, whichever is smaller. The middle strip is bounded by two column strips. Some judgment is required in applying the definitions given in 13.2.1 for column strips with varying span lengths along the design strip.

The reason for specifying that the column strip width be based on the shorter of ℓ_1 or ℓ_2 is to account for the tendency for moment to concentrate about the column line when the span length of the design strip is less than its width.

13.2.4   Effective Beam Section

For slab systems with beams between supports, the beams include portions of the slab as flanges, as shown in Fig. 18-5. Design constants and stiffness parameters used with the Direct Design and Equivalent Frame analysis methods are based on the effective beam sections shown.

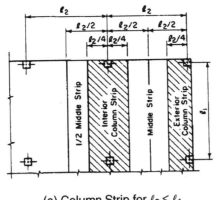

(a) Column Strip for $\ell_2 \leq \ell_1$

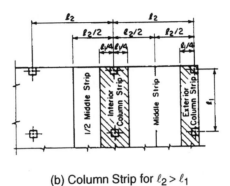

(b) Column Strip for $\ell_2 > \ell_1$

Figure 18-4 Definition of Design Strips

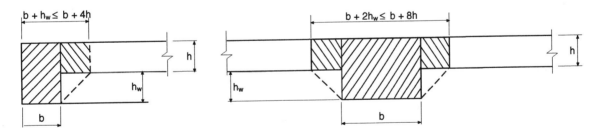

Figure 18-5 Effective Beam Section (13.2.4)

13.3　SLAB REINFORCEMENT

- Minimum area of reinforcement in each direction for two-way slab systems = 0.0018bh (b = slab width, h = total thickness) for Grade 60 bars for either top or bottom steel (13.3.1).

- Maximum bar spacing is 2h, but not more than 18 in. (13.3.2).

- Minimum extensions for reinforcement in slabs without beams (flat plates and flat slabs) are prescribed in Fig. 13.3.8 (13.3.8.1).

Note that the reinforcement details of Fig. 13.3.8 do not apply to two-way slabs with beams between supports or to slabs in non-sway or sway frames resisting lateral loads. For those slabs, a general analysis must be made according to Chapter 12 of the Code to determine bar lengths based on the moment variation but shall not be less than those prescribed in Fig. 13.3.8 (13.3.8.4). Reinforcement details for bent bars were deleted from Fig. 13.3.8 in the '89 code in view of their rare usage in today's construction. Designers who wish to use bent bars in two-way slabs (without beams) should refer to Fig. 13.4.8 of the '83 code, with due consideration of the integrity requirements of 7.13 and 13.3.8 in the current code.

According to 13.3.6, special top and bottom reinforcement must be provided at the exterior corners of a slab with spandrel beams that have a value of α greater than 1.0. The reinforcement must be designed for a moment equal to the largest positive moment per unit width in the panel, and must be placed in a band parallel to the diagonal in the top of the slab and a band perpendicular to the diagonal in the bottom of the slab (Fig. 18-6 (a)); alternatively, it may be placed in two layers parallel to the edges of the slab in both the top and bottom of the slab (Fig. 18-6 (b)). Additionally, the reinforcement must extend at least one-fifth of the longer span in each direction from the corner.

In slabs without beams, all bottom bars in the column strip shall be continuous or spliced with class A splices or with mechanical or welded splices satisying 12.14.3 (13.3.8.5) to provide some capacity for the slab to span to an adjacent support in the event a single support is damaged. Additionally, at least two of these continuous bottom bars shall pass through the column and be anchored at exterior supports. In lift-slab construction and slabs with shearhead reinforcement, clearance may be inadequate and it may not be practical to pass the column strip bottom reinforcing bars through the column. In these cases, two continuous bonded bottom bars in each direction shall pass as close to the column as possible through holes in the shearhead arms or, in the case of lift-slab construction, within the lifting collar (13.3.8.6). This condition was initially addressed in the 1992 Code and was further clarified in 1999.

13.4 OPENINGS IN SLAB SYSTEMS

The code permits openings of any size in any slab system, provided that an analysis is performed that demonstrates that both strength and serviceability requirements are satisfied (13.4.1). For slabs without beams; the analysis of 13.4.1 is waived when the provisions of 13.4.2.1 through 13.4.2.4 are met:

- In the area common to intersecting middle strips, openings of any size are permitted (13.4.2.1).
- In the area common to intersecting column strips, maximum permitted opening size is one-eighth the width of the column strip in either span (13.4.2.2).
- In the area common to one column strip and one middle strip, maximum permitted opening size is limited such that only a maximum of one-quarter of slab reinforcement in either strip may be interrupted (13.4.2.3).

The total amount of reinforcement required for the panel without openings, in both directions, shall be maintained; thus, reinforcement interrupted by the opening must be replaced on each side of the opening. Figure 18-7 illustrates the provisions of 13.4.2 for slabs with $\ell_2 > \ell_1$. Refer to Part 16 for a discussion on the effect of openings in slabs without beams on concrete shear strength (13.4.2.4).

13.5 DESIGN PROCEDURES

Section 13.5.1 permits design (analysis) of two-way slab systems by any method that satisfies code-defined strength requirements (9.2 and 9.3), and all applicable code serviceability requirements, including specified limits on deflections (9.5.3).

13.5.1.1 Gravity Load Analysis—Two methods of analysis of two-way slab systems under gravity loads are addressed in Chapter 13: the simpler Direct Design Method (DDM) of 13.6, and the more complex Equivalent Frame Method (EFM) of 13.7. The Direct Design Method is an approximate method using moment coefficients, while the Equivalent Frame (elastic analysis) Method is more exact. The approximate analysis procedure

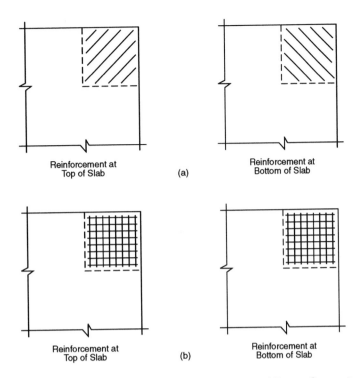

Figure 18-6 Special Reinforcement Required at Corners of Beam-Supported Slabs

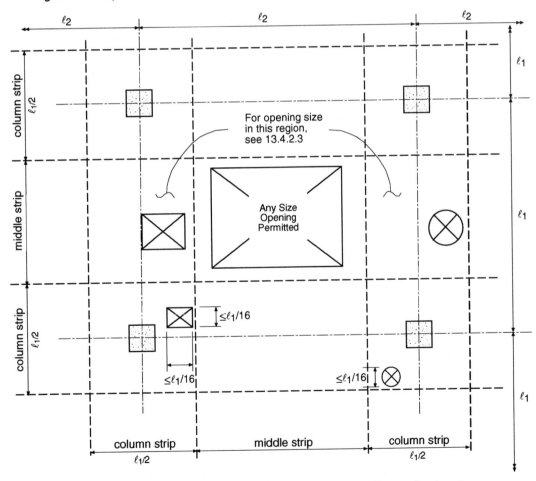

Figure 18-7 Permitted Openings in Slab Systems without Beams for $\ell_2 > \ell_1$

of the Direct Design Method will give reasonably conservative moment values for the stated design conditions for slab systems within the limitations of 13.6.1.

Both methods are for analysis under gravity loads only, and are limited in application to buildings with columns and/or walls laid out on a basically orthogonal grid, i.e., where column lines taken longitudinally and transversely through the building are mutually perpendicular. Both methods are applicable to slabs with or without beams between supports. Note that neither method applies to slab systems with beams spanning between other beams; the beams must be located along column lines and be supported by columns or other essentially nondeflecting supports at the corners of the slab panels.

13.5.1.2 Lateral Load Analysis—For lateral load analysis of frames, the model of the structure may be based upon any approach that is shown to satisfy equilibrium and geometric compatibility and to be in reasonable agreement with test data. Acceptable approaches include plate-bending finite-element models, effective beam width models, and equivalent frame models. The stiffness values for frame members used in the analysis must reflect effects of slab cracking, geometric parameters, and concentration of reinforcement.

During the life of the structure, ordinary occupancy loads and volume changes due to shrinkage and temperature effects will cause cracking of slabs. To ensure that lateral drift caused by wind or earthquakes is not underestimated, cracking of slabs must be considered in stiffness assumptions for lateral drift calculations.

The stiffness of slab members is affected not only by cracking, but also by other parameters such as ℓ_2/ℓ_1, c_1/ℓ_1, c_2/c_1, and on concentration of reinforcement in the slab width defined in 13.5.3.2 for unbalanced moment transfer by flexure. This added concentration of reinforcement increases stiffness by preventing premature yielding and softening in the slab near the column supports. Consideration of the actual stiffness due to these factors is important for lateral load analysis because lateral displacement can significantly affect the moments in the columns, especially in tall moment frame buildings. Also, actual lateral displacement for a single story, or for the total height of a building is an important consideration for building stability and performance.

Cracking reduces stiffness of the slab-beams as compared with that of an uncracked floor. The magnitude of the loss of stiffness due to cracking will depend on the type of slab system and reinforcement details. For example, prestressed slab systems with reduced slab cracking due to prestressing, and slab systems with large beams between columns will lose less stiffness than a conventional reinforced flat plate system.

Prior to the 1999 code, the commentary indicated stiffness values based on Eq. (9-8) were reasonable. However, this was deleted from the commentary in 1999, since factors such as volume change effects and early age loading are not adequately represented in Eq. (9-8). Since it is difficult to evaluate the effect of cracking on stiffness, it is usually sufficient to use a lower bound value. On the assumption of a fully cracked slab with minimum reinforcement at all locations, a stiffness for the slab-beam equal to one-fourth that based on the gross area of concrete ($K_{sb}/4$) should be reasonable. A detailed evaluation of the effect of cracking may also be made. Since slabs normally have more than minimum reinforcement and are not fully cracked, except under very unusual conditions, the one-fourth value should be expected to provide a safe lower bound for stiffness under lateral loads. See R13.5.1.2 for guidance on stiffness assumption for lateral load analysis.

Moments from an Equivalent Frame (or Direct Design) analysis for gravity loading may be combined with moments from a lateral load analysis (13.5.1.3). Alternatively, the Equivalent Frame Analysis can be used for lateral load analysis, if modified to account for reduced stiffness of the slab-beams.

For both vertical and lateral load analyses, moments at critical sections of the slab-beams are transversely distributed in accordance with 13.6.4 (column strips) and 13.6.6 (middle strips).

13.5.4 Shear in Two-Way Slab Systems

If two-way slab systems are supported by beams or walls, the slab shear is seldom a critical factor in design, as the shear force at factored loads is generally well below the shear strength of the concrete.

In contrast, when two-way slabs are supported directly by columns as in flat plates or flat slabs, shear around the columns is of critical importance. Shear strength at an exterior slab-column connection (without edge beams) is especially critical because the total exterior negative slab moment must be transferred directly to the column. This aspect of two-way slab design should not be taken lightly by the designer. Two-way slab systems will normally be found to be quite "forgiving" if an error in the distribution or even in the amount of flexural reinforcement is made, but there will be no forgiveness if the required shear strength is not provided.

For slab systems supported directly by columns, it is advisable at an early stage in design to check the shear strength of the slab in the vicinity of columns as illustrated in Fig. 18-8.

Two types of shear need to be considered in the design of flat plates or flat slabs supported directly on columns. The first is the familiar one-way or beam-type shear, which may be critical in long narrow slabs. Analysis for beam shear considers the slab to act as a wide beam spanning between the columns. The critical section is taken at a distance d from the face of the column. Design against beam shear consists of checking for satisfaction of the requirement indicated in Fig. 18-9(a). Beam shear in slabs is seldom a critical factor in design, as the shear force is usually well below the shear strength of the concrete.

Two-way or "punching" shear is generally the more critical of the two types of shear in slab systems supported directly on columns. Punching shear considers failure along the surface of a truncated cone or pyramid around a column. The critical section is taken perpendicular to the slab at a distance d/2 from the perimeter of a column. The shear force V_u to be resisted can be easily calculated as the total factored load on the area bounded by panel centerlines around the column, less the load applied within the area defined by the critical shear perimeter (see Fig. 18-8).

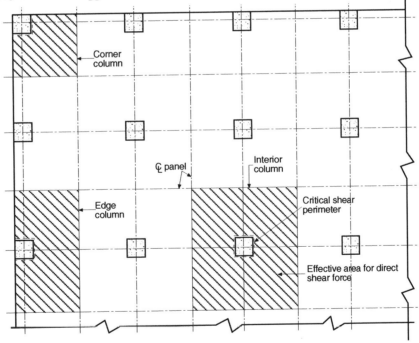

Figure 18-8 Critical Locations for Slab Shear Strength

In the absence of significant moment transfer from the slab to the column, design against punching shear consists of making sure that the requirement of Fig. 18-9(b) is satisfied. For practical design, only direct shear (uniformly distributed around the perimeter b_o) occurs around interior slab-column supports where no (or insignificant) moment is to be transferred from the slab to the column. Significant moments may have to be carried when unbalanced gravity loads on either side of an interior column or horizontal loading due to wind must be transferred from the slab to the column. At exterior slab-column supports, the total exterior slab moment from gravity loads (plus any lateral load moments due to wind or earthquake) must be transferred directly to the column.

18-9

13.5.3 Transfer of Moment in Slab-Column Connections

Transfer of moment between a slab and a column takes place by a combination of flexure (13.5.3) and eccentricity of shear (11.12.6.1). Shear due to moment transfer is assumed to act on a critical section at a distance d/2 from the face of the column (the same critical section around the column as that used for direct shear transfer; see Fig. 18-9(b). The portion of the moment transferred by flexure is assumed to be transferred over a width of slab equal to the transverse column width c_2, plus 1.5 times the slab thickness (1.5h) on either side of the column (13.5.3.2). Concentration of negative reinforcement is to be used to resist moment on this effective slab width. The combined shear stress due to direct shear and moment transfer often governs the design, especially at the exterior slab-column supports.

The portions of the total unbalanced moment M_u to be transferred by eccentricity of shear and by flexure are given by Eqs. (11-39) and (13-1), respectively, where $\gamma_v M_u$ is considered transferred by eccentricity of shear, and $\gamma_f M_u$ is considered transferred by flexure. At an interior square column with $b_1 = b_2$, 40% of the moment is transferred by eccentricity of shear ($\gamma_v M_u = 0.40 M_u$), and 60% by flexure ($\gamma_f M_u = 0.60 M_u$), where M_u is the transfer moment at the centroid of the critical section. The moment M_u at the exterior slab-column support will generally not be computed at the centroid of the critical transfer section. In the Equivalent Frame analysis, moments are computed at the column centerline. In the Direct Design Method, moments are computed at the face of support. Considering the approximate nature of the procedure used to evaluate the stress distribution due to moment transfer, it seems unwarranted to consider a change in moment to the critical section centroid; use of the moment values at column centerline (EFM) or at face of support (DDM) directly would usually be accurate enough.

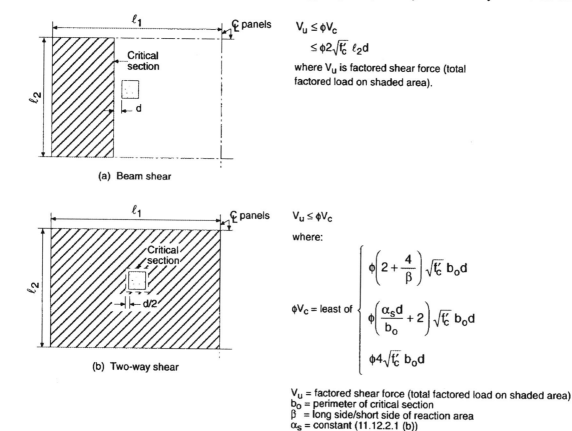

Figure 18-9 Direct Shear at an Interior Slab-Column Support (see Fig. 18-8)

The factored shear stress on the critical transfer section is the sum of the direct shear and the shear caused by moment transfer,

$$v_u = \frac{V_u}{A_c} + \frac{\gamma_v M_u c}{J}$$

For slabs supported on square columns, shear stress v_u must not exceed $\phi 4\sqrt{f'_c}$.

Computation of the combined shear stress involves the following properties of the critical transfer section:

A_c = area of critical section

c = distance from centroid of critical section to face of section where stress is being computed

J = property of critical section analogous to polar moment of inertia

The above properties are given in Part 16. Note that in the case of flat slabs, two different critical sections need to be considered in punching shear calculations, as shown in Fig. 18-10.

Unbalanced moment transfer between slab and an edge column (without spandrel beams) requires special consideration when slabs are analyzed by the Direct Design Method for gravity loads. See discussion on 13.6.3.6 in Part 19.

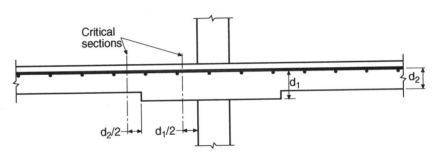

Figure 18-10 Critical Shear-Transfer Sections for Flat Slabs

The provisions of 13.5.3.3 were introduced in the '95 Code. At exterior supports, for unbalanced moments about an axis parallel to the edge, the portion of moment transferred by eccentricity of shear, $\gamma_v M_u$, may be reduced to zero provided that the factored shear at the support (excluding the shear produced by moment transfer) does not exceed 75 percent of the shear strength ϕV_c defined in 11.12.2.1 for edge columns or 50 percent for corner columns. Tests indicate that there is no significant interaction between shear and unbalanced moment at the exterior support in such cases. It should be noted that as $\gamma_v M_u$ is decreased, $\gamma_f M_u$ is increased.

Tests of interior supports have indicated that some flexibility in distributing unbalanced moment by shear and flexure is also possible, but with more severe limitations than for exterior supports. For interior supports, the unbalanced moment transferred by flexure is permitted to be increased up to 25 percent provided that the factored shear (excluding the shear caused by moment transfer) at an interior support does not exceed 40 percent of the shear strength ϕV_c defined in 11.12.2.1.

Note that the above modifications are permitted only when the reinforcement ratio ρ within the effective slab width defined in 13.5.3.2 is less than or equal to $0.375\rho_b$. This provision is intended to improve ductile behavior of the column-slab joint.

SEQUEL

The Direct Design Method and the Equivalent Frame Method for gravity load analysis of two-way slab systems are treated in detail in the following Parts 19 and 20, respectively.

Blank

19

Two-Way Slabs — Direct Design Method

GENERAL CONSIDERATIONS

The Direct Design Method is an approximate procedure for analyzing two-way slab systems subjected to gravity loads only. Since it is approximate, the method is limited to slab systems meeting the limitations specified in 13.6.1. Two-way slab systems not meeting these limitations must be analyzed by more accurate procedures such as the Equivalent Frame Method, as specified in 13.7. See Part 20 for discussion and design examples using the Equivalent Frame Method.

With the publication of ACI 318-83, the Direct Design Method for moment analysis of two-way slab systems was greatly simplified by eliminating all stiffness calculations for determining design moments in an end span. A table of moment coefficients for distribution of the total span moment in an end span (13.6.3.3) replaced the expressions for distribution as a function of the stiffness ratio α_{ec}. As a companion change, the then approximate Eq. (13-4) for unbalanced moment transfer between the slab and an interior column was also simplified through elimination of the α_{ec} term. With these changes, the Direct Design Method became a truly direct design procedure, with all design moments determined directly from moment coefficients. Also, a new 13.6.3.6 was added, addressing a special provision for shear due to moment transfer between a slab without beams and an edge column when the approximate moment coefficients of 13.6.3.3 are used. See discussion on 13.6.3.6 below. Through the 1989 (Revised 1992) edition of the code and commentary, R13.6.3.3 included a "Modified Stiffness Method" reflecting the original distribution, and confirming that design aids and computer programs based on the original distribution as a function of the stiffness ratio α_{ec} were still applicable for usage. The "Modified Stiffness Method" was dropped from R13.6.3.3 in the 1995 edition of the code and commentary.

PRELIMINARY DESIGN

Before proceeding with the Direct Design Method, a preliminary slab thickness h needs to be determined for control of deflections according to the minimum thickness requirements of 9.5.3. Table 18-1 and Fig. 18-3 can be used to simplify minimum thickness computations.

For slab systems without beams, it is advisable at this stage in the design process to check the shear strength of the slab in the vicinity of columns or other support locations in accordance with the special shear provision for slabs (11.12). See discussion on 13.5.4 in Part 18.

Once a slab thickness has been selected, the Direct Design Method, which is essentially a three-step analysis procedure, involves: (1) determining the total factored static moment for each span, (2) dividing the total factored static moment between negative and positive moments within each span, and (3) distributing the negative and the positive moment to the column and the middle strips in the transverse direction.

For analysis, the slab system is divided into design strips consisting of a column strip and two half-middle strip(s) as defined in 13.2.1 and 13.2.2, and as illustrated in Fig. 19-1. Some judgment is required in applying the definitions given in 13.2.1 for slab systems with varying span lengths along the design strip.

13.6.1 Limitations

The Direct Design Method applies within the limitations illustrated in Fig. 19-2:

1. There must be three or more continuous spans in each direction;

2. Slab panels must be rectangular with a ratio of longer to shorter span (centerline-to-centerline of supports) not greater than 2;

3. Successive span lengths (centerline-to-centerline of supports) in each direction must not differ by more than 1/3 of the longer span;

4. Columns must not be offset more than 10% of the span (in direction of offset) from either axis between centerlines of successive columns;

5. Loads must be uniformly distributed, with the unfactored or service live load not more than 2 times the unfactored or service dead load (L/D ≤ 2);

6. For two-way beam-supported slabs, relative stiffness of beams in two perpendicular directions must satisfy the minimum and maximum requirements given in 13.6.1.6; and

7. Redistribution of negative moments by 8.4 is not permitted.

13.6.2 Total Factored Static Moment for a Span

For uniform loading, the total design moment M_o for a span of the design strip is calculated by the simple static moment expression:

$$M_o = \frac{q_u \ell_2 \ell_n^2}{8} \qquad \text{Eq. (13-4)}$$

where q_u is the factored combination of dead and live loads (psf), $q_u = 1.2w_d + 1.6w_\ell$. The clear span ℓ_n (in the direction of analysis) is defined in a straightforward manner for columns or other supporting elements of rectangular cross-section. The clear span starts at the face of support. Face of support is defined as shown in Fig. 19-3. One limitation requires that the clear span not be taken as less than 65% of the span center-to-center of supports (13.6.2.5). The length ℓ_2 is simply the span (centerline-to-centerline) transverse to ℓ_n; however, when the span adjacent and parallel to an edge is being considered, the distance from edge of slab to panel centerline is used for ℓ_2 in calculation of M_o (13.6.2.4).

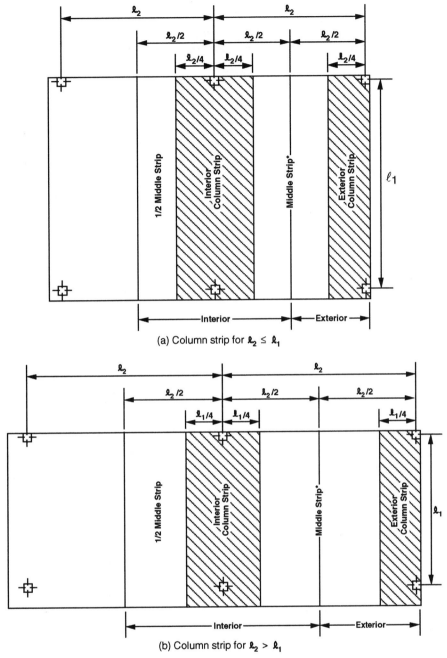

* When edge of exterior design strip is supported by a wall, the factored moment resisted by this middle strip is defined in 13.6.6.3.

Figure 19-1 Definition of Design Strips

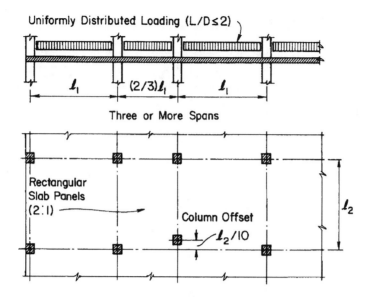

Figure 19-2 Conditions for Analysis by Coefficients

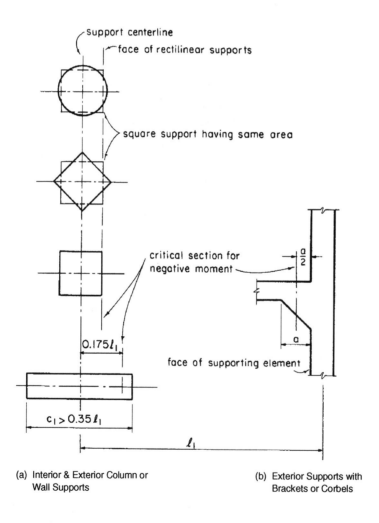

(a) Interior & Exterior Column or Wall Supports

(b) Exterior Supports with Brackets or Corbels

Figure 19-3 Critical Sections for Negative Design Moment

13.6.3 Negative and Positive Factored Moments

The total static moment for a span is divided into negative and positive design moments as shown in Fig. 19-4. End span moments in Fig. 19-4 are shown for a flat plate or flat slab without spandrels (slab system without beams between interior supports and without edge beam). For other end span conditions, the total static moment M_o is distributed as shown in Table 19-1.

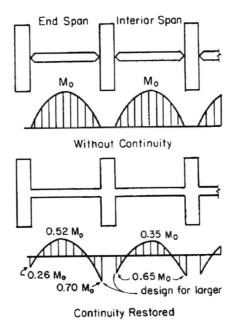

Figure 19-4 Design Strip Moments

Table 19-1 Distribution of Total Static Moment for an End Span

Factored Moment	(1) Slab Simply Supported on Concrete or Masonry Wall	(2) Two-Way Beam-Supported Slabs	Flat Plates and Flat Slabs		(5) Slab Monolithic with Concrete Wall
			(3) Without Edge Beam	(4) With Edge Beam	
Interior Negative	0.75	0.70	0.70	0.70	0.65
Positive	0.63	0.57	0.52	0.50	0.35
Exterior Negative	0	0.16	0.26	0.30	0.65

13.6.3.6 Special Provision for Load Transfer Between Slab and an Edge Column—For columns supporting a slab without beams, load transfer directly between the slab and the supporting columns (without intermediate load transfer through beams) is one of the more critical design conditions for the flat plate or flat slab system. Shear strength of the slab-column connection is critical. This aspect of two-way slab design should not be taken lightly by the designer. Two-way slab systems are fairly "forgiving" of an error in the distribution or even in the amount of flexural reinforcement; however, there is little or no forgiveness if a critical error in the provision of shear strength is made. See Part 16 for special provisions for direct shear and moment transfer at slab-column connections.

Section 13.6.3.6 addresses the potentially critical moment transfer between a beamless slab and an edge column. To ensure adequate shear strength when using the approximate end-span moment coefficients of 13.6.3.3, the 1989 edition of the code required that the full nominal strength M_n provided by the column strip be used in determining the fraction of unbalanced moment transferred by the eccentricity of shear (γ_v) in accordance with 11.12.6 (for end spans without edge beams, the column strip is proportioned to resist the total exterior negative factored moment). This requirement was changed in ACI 318-95. The moment $0.3M_o$ instead of M_n of the column strip must be used in determining the fraction of unbalanced moment transferred by the eccentricity of shear. The total reinforcement provided in the column strip includes the additional reinforcement concentrated over the column to resist the fraction of unbalanced moment transferred by flexure, $\gamma_f M_u = \gamma_f (0.26M_o)$, where the moment coefficient (0.26) is from 13.6.3.3, and γ_f is given by Eq. (13-1).

13.6.4 Factored Moments in Column Strips

The amounts of negative and positive factored moments to be resisted by a column strip, as defined in Fig. 19-1, depends on the relative beam-to-slab stiffness ratio and the panel width-to-length ratio in the direction of analysis. An exception to this is when a support has a large transverse width.

The column strip at the exterior of an end span is required to resist the total factored negative moment in the design strip unless edge beams are provided.

When the transverse width of a support is equal to or greater than three quarters (3/4) of the design strip width, 13.6.4.3 requires that the negative factored moment be uniformly distributed across the design strip.

The percentage of total negative and positive factored moments to be resisted by a column strip may be determined from the tables in 13.6.4.1 (interior negative), 13.6.4.2 (exterior negative) and 13.6.4.4 (positive), or from the following expressions:

Percentage of negative factored moment at interior support to be resisted by column strip

$$= 75 + 30\left(\frac{\alpha_{f1}\ell_2}{\ell_1}\right)\left(1 - \frac{\ell_2}{\ell_1}\right) \qquad (1)$$

Percentage of negative factored moment at exterior support to be resisted by column strip

$$= 100 - 10\beta_t + 12\beta_t\left(\frac{\alpha_{f1}\ell_2}{\ell_1}\right)\left(1 - \frac{\ell_2}{\ell_1}\right) \qquad (2)$$

Percentage of positive factored moment to be resisted by column strip

$$= 60 + 30\left(\frac{\alpha_{f1}\ell_2}{\ell_1}\right)\left(1.5 - \frac{\ell_2}{\ell_1}\right) \qquad (3)$$

Note: When $\alpha_{f1}\ell_2/\ell_1 > 1.0$, use 1.0 in above equations. When $\beta_t > 2.5$, use 2.5 in Eq. (2) above.

For slabs without beams between supports ($\alpha_{f1} = 0$) and without edge beams ($\beta_t = 0$), the distribution of total negative moments to column strips is simply 75 and 100 percent for interior and exterior supports, respectively, and the distribution of total positive moment is 60 percent. For slabs with beams between supports, distribution depends on the beam-to-slab stiffness ratio; when edge beams are present, the ratio of torsional stiffness of edge beam to flexural stiffness of slab also influences distribution. Figs. 19-6, 19-7, and 19-8 simplify evaluation of the beam-to-slab stiffness ratio α_{f1}. To evaluate β_t, stiffness ratio for edge beams, Table 19-2 simplifies calculation of the torsional constant C.

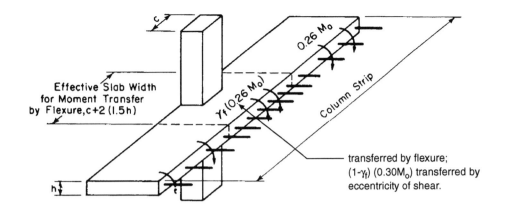

Figure 19-5 Transfer of Negative Moment at Exterior Support Section of Slab without Beams

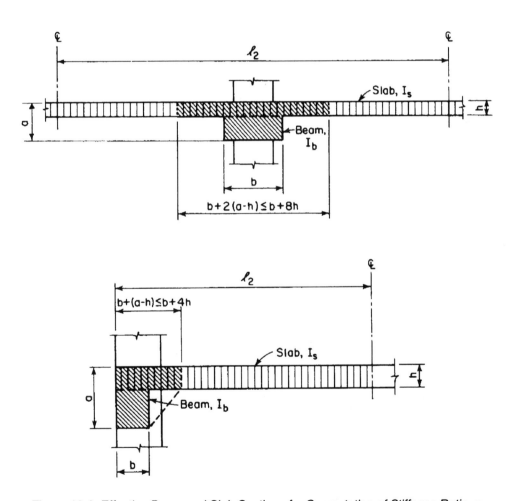

Figure 19-6 Effective Beam and Slab Sections for Computation of Stiffness Ratio α_f

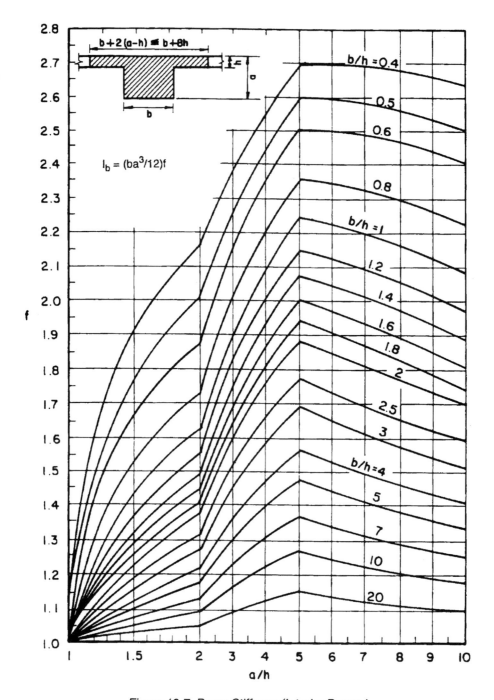

Figure 19-7 Beam Stiffness (Interior Beams)

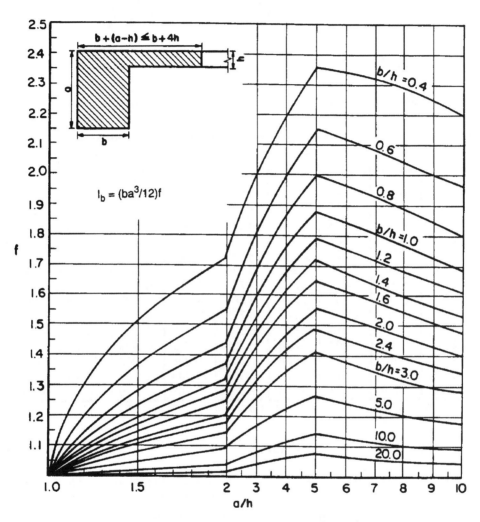

Figure 19-8 Beam Stiffness (Edge Beams)

Table 19-2 Design Aid for Computing C, Cross-Sectional Constant Defining Torsional Properties

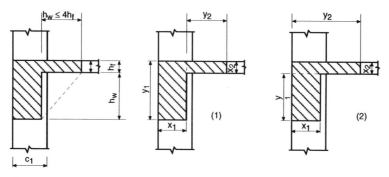

Edge beam (ACI 13.2.4)   Use larger value of C computed from (1) or (2)

y	x*									
	4	5	6	7	8	9	10	12	14	16
12	202	369	592	868	1,118	1,538	1,900	2,557		
14	245	452	736	1,096	1,529	2,024	2,566	3,709	4,738	
16	288	534	880	1,325	1,871	2,510	3,233	4,861	6,567	8,083
18	330	619	1,024	1,554	2,212	2,996	3,900	6,013	8,397	10,813
20	373	702	1,167	1,782	2,553	3,482	4,567	7,165	10,226	13,544
22	416	785	1,312	2,011	2,895	3,968	5,233	8,317	12,055	16,275
24	458	869	1,456	2,240	3,236	4,454	5,900	9,459	13,885	19,005
27	522	994	1,672	2,583	3,748	5,183	6,900	11,197	16,628	23,101
30	586	1,119	1,888	2,926	4,260	5,912	7,900	12,925	19,373	27,197
33	650	1,243	2,104	3,269	4,772	6,641	8,900	14,653	22,117	31,293
36	714	1,369	2,320	3,612	5,284	7,370	9,900	16,381	24,860	35,389
42	842	1,619	2,752	4,298	6,308	8,828	11,900	19,837	30,349	43,581
48	970	1,869	3,183	4,984	7,332	10,286	13,900	23,293	35,836	51,773
54	1,098	2,119	3,616	5,670	8,356	11,744	15,900	26,749	41,325	59,965
60	1,226	2,369	4,048	6,356	9,380	13,202	17,900	30,205	46,813	68,157

* Small side of a rectangular cross-section with dimensions x and y.

13.6.5 Factored Moments in Beams

When a design strip contains beams between columns, the factored moment assigned to the column strip must be distributed between the slab and the beam portions of the column strip. The amount of the column strip factored moment to be resisted by the beam varies linearly between zero and 85 percent as $\alpha_{f1}\ell_2/\ell_1$ varies between zero and 1.0. When $\alpha_{f1}\ell_2/\ell_1$ is equal to or greater than 1.0, 85 percent of the total column strip moment must be resisted by the beam. In addition, the beam section must resist the effects of loads applied directly to the beam, including weight of beam stem projecting above or below the slab.

13.6.6 Factored Moments in Middle Strips

Factored moments not assigned to the column strips must be resisted by the two half-middle strips comprising the design strip. An exception to this is a middle strip adjacent to and parallel with an edge supported by a wall, where the moment to be resisted is twice the factored moment assigned to the half middle strip corresponding to the first row of interior supports (see Fig. 19-1).

13.6.9 Factored Moments in Columns and Walls

Supporting columns and walls must resist any negative moments transferred from the slab system.

For interior columns (or walls), the approximate Eq. (13-7) may be used to determine the unbalanced moment transferred by gravity loading, unless an analysis is made considering the effects of pattern loading and unequal adjacent spans. The transfer moment is computed directly as a function of span length and gravity loading. For the more usual case with equal transverse and adjacent spans, Eq. (13-7) reduces to

$$M_u = 0.07\left(0.5q_{Lu}\ell_2\ell_n^2\right) \qquad (4)$$

where, q_{Lu} = factored live load, psf

ℓ_2 = span length transverse to ℓ_n

ℓ_n = clear span length in the direction of analysis

At exterior column or wall supports, the total exterior negative factored moment from the slab system (13.6.3.3) is transferred directly to the supporting members. Due to the approximate nature of the moment coefficients, it seems unwarranted to consider the change in moment from face of support to centerline of support; use the moment values from 13.6.3.3 directly.

Columns above and below the slab must resist the unbalanced support moment based on the relative column stiffnesses—generally, in proportion to column lengths above and below the slab. Again, due to the approximate nature of the moment coefficients of the Direct Design Method, the refinement of considering the change in moment from centerline of slab-beam to top or bottom of column seems unwarranted.

DESIGN AID — DIRECT DESIGN MOMENT COEFFICIENTS

Distribution of the total free-span moment M_o into negative and positive moments, and then into column and middle strip moments, involves direct application of moment coefficients to the total moment M_o. The moment coefficients are a function of location of span (interior or end), slab support conditions, and type of two-way slab system. For design convenience, moment coefficients for typical two-way slab systems are given in Tables 19-3 through 19-7. Tables 19-3 through 19-6 apply to flat plates or flat slabs with differing end support conditions. Table 19-7 applies to two-way slabs supported on beams on all four sides. Final moments for the column strip and the middle strip are directly tabulated.

Table 19-3 Design Moment Coefficients for Flat Plate or Flat Slab Supported Directly on Columns

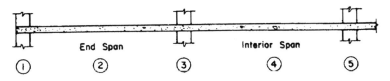

Slab Moments	End Span			Interior Span	
	(1)	(2)	(3)	(4)	(5)
	Exterior Negative	Positive	First Interior Negative	Positive	Interior Negative
Total Moment	0.26M_o	0.52M_o	0.70M_o	0.35M_o	0.65M_o
Column Strip	0.26M_o	0.31M_o	0.53M_o	0.21M_o	0.49M_o
Middle Strip	0	0.21M_o	0.17M_o	0.14M_o	0.16M_o

Note: All negative moments are at face of support.

The moment coefficients of Table 19-4 (flat plate with edge beams) are valid for $\beta_t \geq 2.5$. The coefficients of Table 19-7 (two-way beam-supported slabs) apply for $\alpha_{f1}\ell_2/\ell_1 \geq 1.0$ and $\beta_t \geq 2.5$. Many practical beam sizes will provide beam-to-slab stiffness ratios such that $\alpha_{f1}\ell_2/\ell_1$ and β_t will be greater than these limits, allowing moment coefficients to be taken directly from the tables, without further consideration of stiffnesses and interpolation for moment coefficients. However, if beams are present, the two stiffness parameters α_{f1} and β_t will need to be evaluated. For two-way slabs, and for $E_{cb} = E_{cs}$, the stiffness parameter α_{f1} is simply the ratio of the moments of inertia of the effective beam and slab sections in the direction of analysis, $\alpha_{f1} = I_b/I_s$, as illustrated in Fig. 19-6. Figures 19-7 and 19-8 simplify evaluation of the α_{f1} term.

Table 19-4 Design Moment Coefficients for Flat Plate or Flat Slab with Edge Beams

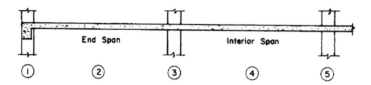

Slab Moments	End Span			Interior Span	
	(1)	(2)	(3)	(4)	(5)
	Exterior Negative	Positive	First Interior Negative	Positive	Interior Negative
Total Moment	0.30M_o	0.50M_o	0.70M_o	0.35M_o	0.65M_o
Column Strip	0.23M_o	0.30M_o	0.53M_o	0.21M_o	0.49M_o
Middle Strip	0.07M_o	0.20M_o	0.17M_o	0.14M_o	0.16M_o

Notes: (1) All negative moments are at face of support.
(2) Torsional stiffness of edge beam is such that $\beta_t \geq 2.5$. For values of β_t less than 2.5, exterior negative column strip moment increases to $(0.30 - 0.03\beta_t)M_o$.

For $E_{cb} = E_{cs}$, relative stiffness provided by an edge beam is reflected by the parameter $\beta_t = C/2I_s$, where I_s is the moment of inertia of the effective slab section spanning in the direction of ℓ_1 and having a width equal to ℓ_2, i.e., $I_s = \ell_2 h^3/12$. The constant C pertains to the torsional stiffness of the effective edge beam cross-section. It is found by dividing the beam section into its component rectangles, each having a smaller dimension x and a larger dimension y, and by summing the contributions of all the parts by means of the equation:

$$C = \Sigma\left(1 - \frac{0.63x}{y}\right)\left(\frac{x^3 y}{3}\right) \qquad (5)$$

The subdivision can be done in such a way as to maximize C. Table 19-2 simplifies calculation of the torsional constant C.

Table 19-5 Design Moment Coefficients for Flat Plate or Flat Slab with End Span Integral with Wall

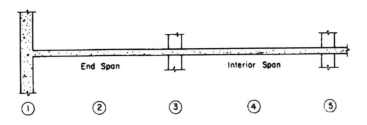

Slab Moments	End Span			Interior Span	
	(1)	(2)	(3)	(4)	(5)
	Exterior Negative	Positive	First Interior Negative	Positive	Interior Negative
Total Moment	$0.65M_o$	$0.35M_o$	$0.65M_o$	$0.35M_o$	$0.65M_o$
Column Strip	$0.49M_o$	$0.21M_o$	$0.49M_o$	$0.21M_o$	$0.49M_o$
Middle Strip	$0.16M_o$	$0.14M_o$	$0.16M_o$	$0.14M_o$	$0.16M_o$

Note: All negative moments are at face of support.

Table 19-6 Design Moment Coefficients for Flat Plate or Flat Slab with End Span Simply Supported on Wall

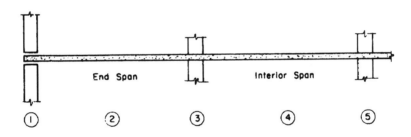

Slab Moments	End Span			Interior Span	
	(1)	(2)	(3)	(4)	(5)
	Exterior Negative	Positive	First Interior Negative	Positive	Interior Negative
Total Moment	0	$0.63M_o$	$0.75M_o$	$0.35M_o$	$0.65M_o$
Column Strip	0	$0.38M_o$	$0.56M_o$	$0.21M_o$	$0.49M_o$
Middle Strip	0	$0.25M_o$	$0.19M_o$	$0.14M_o$	$0.16M_o$

Note: All negative moments are at face of support.

Table 19-7 Design Moment Coefficients for Two-Way Beam-Supported Slab

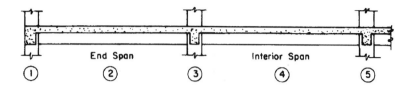

Span Ratio ℓ_2/ℓ_1	Slab and Beam Moments	End Span			Interior Span	
		(1) Exterior Negative	(2) Positive	(3) First Interior Negative	(4) Positive	(5) Interior Negative
	Total Moment	$0.16M_o$	$0.57M_o$	$0.70M_o$	$0.35M_o$	$0.65M_o$
0.5	Column Strip Beam	$0.12M_o$	$0.43M_o$	$0.54M_o$	$0.27M_o$	$0.50M_o$
	Slab	$0.02M_o$	$0.08M_o$	$0.09M_o$	$0.05M_o$	$0.09M_o$
	Middle Strip	$0.02M_o$	$0.06M_o$	$0.07M_o$	$0.03M_o$	$0.06M_o$
1.0	Column Strip Beam	$0.10M_o$	$0.37M_o$	$0.45M_o$	$0.22M_o$	$0.42M_o$
	Slab	$0.02M_o$	$0.06M_o$	$0.08M_o$	$0.04M_o$	$0.07M_o$
	Middle Strip	$0.04M_o$	$0.14M_o$	$0.17M_o$	$0.09M_o$	$0.16M_o$
2.0	Column Strip Beam	$0.06M_o$	$0.22M_o$	$0.27M_o$	$0.14M_o$	$0.25M_o$
	Slab	$0.01M_o$	$0.04M_o$	$0.05M_o$	$0.02M_o$	$0.04M_o$
	Middle Strip	$0.09M_o$	$0.31M_o$	$0.38M_o$	$0.19M_o$	$0.36M_o$

Notes: (1) All negative moments are at face of support.
(2) Torsional stiffness of edge beam is such that $\beta_t \geq 2.5$
(3) $\alpha_{f1}\ell_2/\ell_1 \geq 1.0$

Example 19.1—Two-Way Slab without Beams Analyzed by the Direct Design Method

Use the Direct Design Method to determine design moments for the flat plate slab system in the direction shown, for an intermediate floor.

Story height = 9 ft
Column dimensions = 16 × 16 in.
Lateral loads to be resisted by shear walls
No edge beams
Partition weight = 20 psf
Service live load = 40 psf
f'_c = 4,000 psi, normal weight concrete
f_y = 60,000 psi

Also determine the reinforcement and shear requirements at an exterior column.

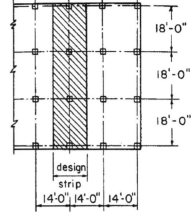

	Code
Calculations and Discussion	Reference

1. Preliminary design for slab thickness h:

    a. Control of deflections.

    For slab systems without beams (flat plate), the minimum overall thickness h with Grade 60 reinforcement is (see Table 18-1):

    9.5.3.2
    Table 9.5(c)

    $$h = \frac{\ell_n}{30} = \frac{200}{30} = 6.67 \text{ in. Use h = 7 in.}$$

    where ℓ_n is the length of clear span in the long direction = 216 - 16 = 200 in.

    This is larger than the 5 in. minimum specified for slabs without drop panels.

    9.5.3.2(a)

    b. Shear strength of slab.

    Use an average effective depth, d ≈ 5.75 in. (3/4-in. cover and No. 4 bar)

    Factored dead load, q_{Du} = 1.2 (87.5 + 20) = 129 psf

    Factored live load, q_{Lu} = 1.6 × 40 = 64 psf

    Total factored load, q_u = 193 psf

    Investigation for wide-beam action is made on a 12-in. wide strip at a distance d from the face of support in the long direction (see Fig.19-9).

    11.12.1.1

    V_u = 0.193 × 7.854 = 1.5 kips

    $V_c = 2\sqrt{f'_c}\, b_w d$

    Eq. (11-3)

19-15

| Example 19.1 (cont'd) | Calculations and Discussion | Code Reference |

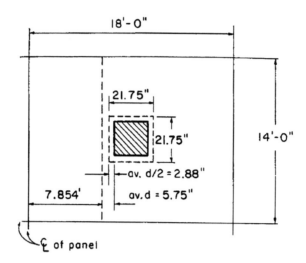

Figure 19-9 Critical Sections for One-Way and Two-Way Shear

$$= \frac{2\sqrt{4,000} \times 12 \times 5.75}{1,000} = 8.73 \text{ kips}$$

$\phi V_c = 0.75 \times 8.73 = 6.6 \text{ kips} > V_u = 1.5 \text{ kips} \quad \text{O.K.}$

Since there are no shear forces at the centerline of adjacent panels (see Fig. 19-9), the shear strength in two-way action at d/2 distance around a support is computed as follows:

$V_u = 0.193 [(18 \times 14) - 1.81^2] = 48.0 \text{ kips}$

$V_c = 4\sqrt{f'_c} \, b_o d$ (for square columns)   Eq. (11-35)

$$= \frac{4\sqrt{4,000} \times (4 \times 21.75) \times 5.75}{1,000} = 126.6 \text{ kips}$$

$V_u = 48.0 \text{ kips} < \phi V_c = 0.75 \times 126.6 \text{ kips} = 95.0 \text{ kips} \quad \text{O.K.}$

Therefore, preliminary design indicates that a 7 in. slab is adequate for control of deflection and shear strength.

2. Check applicability of Direct Design Method: 13.6.1

   There is a minimum of three continuous spans in each direction   13.6.1.1

   Long-to-short span ratio is 1.29 < 2.0   13.6.1.2

		Code
Example 19.1 (cont'd)	**Calculations and Discussion**	**Reference**

Successive span lengths are equal   13.6.1.3

Columns are not offset   13.6.1.4

Loads are uniformly distributed with service live-to-dead load ratio of 0.37 < 2.0   13.6.1.5

Slab system is without beams   13.6.1.6

3. Factored moments in slab:

   a. Total factored moment per span.   13.6.2

   $$M_o = \frac{q_u \ell_2 \ell_n^2}{8}$$ Eq. (13-4)

   $$= \frac{0.193 \times 14 \times 16.67^2}{8} = 93.6 \text{ ft-kips}$$

   b. Distribution of the total factored moment M_o per span into negative and positive   13.6.3
      moments, and then into column and middle strip moments. This distribution involves   13.6.4
      direct application of the moment coefficients to the total moment M_o. Referring to   13.6.6
      Table 19-3 (flat plate without edge beams),

	Total Moment (ft-kips)	Column Strip Moment (ft-kips)	Moment (ft-kips) in Two Half-Middle Strips*
End Span:			
Exterior Negative	$0.26M_o = 24.3$	$0.26M_o = 24.3$	0
Positive	$0.52M_o = 48.7$	$0.31M_o = 29.0$	$0.21M_o = 19.7$
Interior Negative	$0.70M_o = 65.5$	$0.53M_o = 49.6$	$0.17M_o = 15.9$
Interior Span:			
Positive	$0.35M_o = 32.8$	$0.21M_o = 19.7$	$0.14M_o = 13.1$
Negative	$0.65M_o = 60.8$	$0.49M_o = 45.9$	$0.16M_o = 15.0$

*That portion of the total moment M_o not resisted by the column strip is assigned to the two half-middle strips.

Note: The factored moments may be modified by 10 percent, provided the total factored   13.6.7
static moment in any panel is not less than that computed from Eq. (13-4). This modification is omitted here.

4. Factored moments in columns: 13.6.9

   a. Interior columns, with equal spans in the direction of analysis and (different) equal
      spans in the transverse direction.

   $$M_u = 0.07 \left(0.5 q_{Lu} \ell_2 \ell_n^2 \right)$$ Eq. (13-7)

   $$= 0.07 (0.5 \times 1.6 \times 0.04 \times 14 \times 16.67^2) = 8.7 \text{ ft-kips}$$

Example 19.1 (cont'd)   **Calculations and Discussion**   **Code Reference**

With the same column size and length above and below the slab,

$$M_c = \frac{8.7}{2} = 4.35 \text{ ft-kips}$$

This moment is combined with the factored axial load (for each story) for design of the interior columns.

b. Exterior columns.

Total exterior negative moment from slab must be transferred directly to the columns: $M_u = 24.3$ ft-kips. With the same column size and length above and below the slab,

$$M_c = \frac{24.3}{2} = 12.15 \text{ ft-kips}$$

This moment is combined with the factored axial load (for each story) for design of the exterior column.

5. Check slab flexural and shear strength at exterior column

  a. Total flexural reinforcement required for design strip:

  i. Determine reinforcement required for strip moment $M_u = 24.3$ ft-kips

  Assume tension-controlled section ($\phi = 0.9$)   *9.3.2*

  Column strip width $b = \dfrac{14 \times 12}{2} = 84$ in.   *13.2.1*

  $$R_n = \frac{M_u}{\phi b d^2} = \frac{24.3 \times 12,000}{0.9 \times 84 \times 5.75^2} = 117 \text{ psi}$$

  $$\rho = \frac{0.85 f'_c}{f_y}\left(1 - \sqrt{1 - \frac{2R_n}{0.85 f'_c}}\right)$$

  $$= \frac{0.85 \times 4}{60}\left(1 - \sqrt{1 - \frac{2 \times 117}{0.85 \times 4,000}}\right) = 0.0020$$

  $A_s = \rho b d = 0.0020 \times 84 \times 5.75 = 0.96$ in.^{2}

  $\rho_{min} = 0.0018$ *13.3.1*

		Code
Example 19.1 (cont'd)	**Calculations and Discussion**	**Reference**

Min. A_s = 0.0018 × 84 × 7 = 1.06 in.² > 0.96 in.²

Number of No. 4 bars = $\frac{1.06}{0.2}$ = 5.3, say 6 bars

Maximum spacing s_{max} = 2h = 14 in. < 18 in.   13.3.2

Number of No.4 bars based on s_{max} = $\frac{84}{14}$ = 6

Verify tension-controlled section:

$$a = \frac{A_s f_y}{0.85 f'_c b} = \frac{(6 \times 0.2) \times 60}{0.85 \times 4 \times 84} = 0.25 \text{ in.}$$

$$c = \frac{a}{\beta_1} = \frac{0.25}{0.85} = 0.29 \text{ in.}$$

$$\varepsilon_t = \left(\frac{0.003}{c}\right) d_t - 0.003$$

$$= \left(\frac{0.003}{0.29}\right) 5.75 - 0.003 = 0.057 > 0.005$$ 10.3.4

Therefore, section is tension-controlled.

Use 6-No. 4 bars in column strip.

ii. Check slab reinforcement at exterior column for moment transfer between slab and column

Portion of unbalanced moment transferred by flexure = $\gamma_f M_u$ 13.5.3.2

From Fig. 16-13, Case C:

$$b_1 = c_1 + \frac{d}{2} = 16 + \frac{5.75}{2} = 18.88 \text{ in.}$$

$$b_2 = c_2 + d = 16 + 5.75 = 21.75 \text{ in.}$$

$$\gamma_f = \frac{1}{1 + (2/3)\sqrt{b_1/b_2}} = \frac{1}{1 + (2/3)\sqrt{18.88/21.75}} = 0.62$$ Eq. (13-1)

Example 19.1 (cont'd)   **Calculations and Discussion**   **Code Reference**

$\gamma_f M_u = 0.62 \times 24.3 = 15.1$ ft-kips

Note that the provisions of 13.5.3.3 may be utilized; however, they are not in this example.

Assuming tension-controlled behavior, determine required area of reinforcement for $\gamma_f M_u = 15.1$ ft-kips:

Effective slab width $b = c_2 + 3h = 16 + 3(7) = 37$ in.   13.5.3.2

$$R_n = \frac{M_u}{\phi bd^2} = \frac{15.1 \times 12{,}000}{0.9 \times 37 \times 5.75^2} = 165 \text{ psi}$$

$$\rho = \frac{0.85 f'_c}{f_y}\left(1 - \sqrt{1 - \frac{2R_n}{0.85 f'_c}}\right)$$

$$= \frac{0.85 \times 4}{60}\left(1 - \sqrt{1 - \frac{2 \times 165}{0.85 \times 4000}}\right) = 0.0028$$

$A_s = 0.0028 \times 37 \times 5.75 = 0.60$ in.^{2}

Min. $A_s = 0.0018 \times 37 \times 7 = 0.47$ in.^{2} < 0.60 in.^{2}

Number of No. 4 bars $= \dfrac{0.60}{0.2} = 3$

Verify tension-controlled section:

$$a = \frac{A_s f_y}{0.85 f'_c b} = \frac{(3 \times 0.2) \times 60}{0.85 \times 4 \times 37} = 0.29 \text{ in.}$$

$$c = \frac{a}{\beta_1} = \frac{0.29}{0.85} = 0.34 \text{ in.}$$

$$\varepsilon_t = \left(\frac{0.003}{0.34}\right) 5.75 - 0.003 = 0.048 > 0.005$$ 10.3.4

Example 19.1 (cont'd)	Calculations and Discussion	Code Reference

Therefore, section is tension-controlled.

Provide the required 3-No. 4 bars by concentrating 3 of the column strip bars (6-No. 4) within the 37 in. slab width over the column. For symmetry, add one additional No. 4 bar outside of 37-in. width.

Note that the column strip section remains tension-controlled with the addition of 1-No. 4 bar.

iii. Determine reinforcement required for middle strip.

Since all of the moment at exterior columns is transferred to the column strip, provide minimum reinforcement in middle strip:

Min. $A_s = 0.0018 \times 84 \times 7 = 1.06$ in.^{2}

Number of No. 4 bars $= \dfrac{1.06}{0.2} = 5.3$, say 6

Maximum spacing $s_{max} = 2h = 14$ in. < 18 in.   13.3..2

Number of No. 4 bars based on $s_{max} = \dfrac{84}{14} = 6$

Provide No. 4 @ 14 in. in middle strip.

b. Check combined shear stress at inside face of critical transfer section: 11.12.6.1

For shear strength equations, see Part 16.

$$v_u = \dfrac{V_u}{A_c} + \dfrac{\gamma_v M_u c_{AB}}{J_c}$$

Factored shear force at exterior column:

$$V_u = 0.193\left[(14 \times 9.667) - \left(\dfrac{18.88 \times 21.75}{144}\right)\right] = 25.6 \text{ kips}$$

When the end span moments are determined from the Direct Design Method, the fraction of unbalanced moment transferred by eccentricity of shear must be $0.3M_o = 0.3 \times 93.6 = 28.1$ ft-kips.   13.6.3.6

$\gamma_v = 1 - \gamma_f = 1 - 0.62 = 0.38$ Eq. (11-3)

Example 19.1 (cont'd) — Calculations and Discussion

Code Reference

From Fig. 16-13, critical section propeties for edge column bending perpendicular to edge (Case C):

$$A_c = (2b_1 + b_2)d = [(2 \times 18.88) + 21.75] \times 5.75 = 342.2 \text{ in.}^2$$

$$\frac{J_c}{c_{AB}} = \frac{2b_1^2 d(b_1 + 2b_2) + d^3(2b_1 + b_2)}{6b_1}$$

$$= \frac{2(18.88)^2(5.75)[18.88 + (2 \times 21.75)] + 5.75^3[(2 \times 18.88) + 21.75]}{6 \times 18.88}$$

$$= 2{,}357 \text{ in.}^3$$

$$v_u = \frac{25{,}600}{342.2} + \frac{0.38 \times 28.1 \times 12{,}000}{2{,}357}$$

$$= 74.8 + 54.4 = 129.2 \text{ psi}$$

Allowable shear stress $\phi v_n = \phi 4\sqrt{f'_c} = 0.75 \times 4\sqrt{4{,}000} = 189.7 \text{ psi} > v_u$ O.K.   *11.12..6.2*

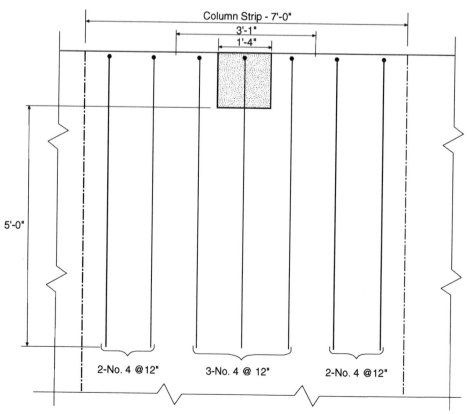

19-22

Example 19.2—Two-Way Slab with Beams Analyzed by the Direct Design Method

Use the Direct Design Method to determine design moments for the slab system in the direction shown, for an intermediate floor.

Story height = 12 ft
Edge beam dimensions = 14 × 27 in.
Interior beam dimensions = 14 × 20 in.
Column dimensions = 18 × 18 in.
Slab thickness = 6 in.
Service live load = 100 psf

f'_c = 4,000 psi (for all members), normal weight concrete
f_y = 60,000 psi

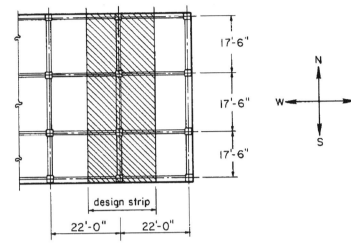

| | Code |
| Calculations and Discussion | Reference |

1. Preliminary design for slab thickness h:

   Control of deflections.

   With the aid of Figs. 19-6, 19-7, and 19-8, beam-to-slab flexural stiffness ratio α_f is computed as follows:

   NS edge beams:

   $\ell_2 = 141$ in.

   $\dfrac{a}{h} = \dfrac{27}{6} = 4.5$

   $\dfrac{b}{h} = \dfrac{14}{6} = 2.33$

   From Fig. 19-8, f = 1.47

   $I_b = \left(\dfrac{ba^3}{12}\right) f$

   $I_s = \dfrac{\ell_2 h^3}{12}$

9.5.3

| Example 19.2 (cont'd) | Calculations and Discussion | Code Reference |

$$\alpha_f = \frac{E_{cb}I_b}{E_{cs}I_s} = \frac{I_b}{I_s}$$

13.0

$$= \left(\frac{b}{\ell_2}\right)\left(\frac{a}{h}\right)^3 f$$

$$= \left(\frac{14}{141}\right)\left(\frac{27}{6}\right)^3 (1.47) = 13.30$$

EW edge beams:

$$\ell_2 = \frac{17.5 \times 12}{2} + \frac{18}{2} = 114 \text{ in.}$$

$$\alpha_f = \left(\frac{14}{114}\right)\left(\frac{27}{6}\right)^3 (1.47) = 16.45$$

NS interior beams:

$$\ell_2 = 22 \text{ ft} = 264 \text{ in.}$$

$$\frac{a}{h} = \frac{20}{6} = 3.33$$

$$\frac{b}{h} = \frac{14}{6} = 2.33$$

From Fig. 19-7, f = 1.61

$$\alpha_f = \left(\frac{14}{114}\right)\left(\frac{27}{6}\right)^3 (1.47) = 16.45$$

EW interior beams:

$$\ell_2 = 17.5 \text{ ft} = 210 \text{ in.}$$

$$\alpha_f = \left(\frac{14}{210}\right)\left(\frac{20}{6}\right)^3 (1.61) = 3.98$$

Since $\alpha_f > 2.0$ for all beams, Eq. (9-13) will control minimum thickness.

9.5.3.3

		Code
Example 19.2 (cont'd)	**Calculations and Discussion**	**Reference**

Therefore,

$$h = \frac{\ell_n \left(0.8 + \frac{f_y}{200{,}000}\right)}{36 + 9\beta}$$

Eq. (9-12)

$$= \frac{246\left(0.8 + \frac{60{,}000}{200{,}000}\right)}{36 + 9(1.28)} = 5.7 \text{ in.}$$

where

$$\beta = \frac{\text{clear span in the long direction}}{\text{clear span in the short direction}} = \frac{20.5}{16} = 1.28$$

ℓ_n = clear span in long direction measured face to face of columns = 20.5 ft = 246 in.

Use 6 in. slab thickness

2. Check applicability of Direct Design Method: 13.6.1

   There is a minimum of three continuous spans in each direction   13.6.1.1
   Long-to-short span ratio is 1.26 < 2.0   13.6.1.2
   Successive span lengths are equal   13.6.1.3
   Columns are not offset   13.6.1.4
   Loads are uniformly distributed with service live-to-dead ratio of 1.33 < 2.0   13.6.1.5

   Check relative stiffness for slab panel: 13.6.1.6

Interior Panel:

$\alpha_{f1} = 3.16 \qquad \ell_2 = 264 \text{ in.}$

$\alpha_{f2} = 3.98 \qquad \ell_1 = 210 \text{ in.}$

$$\frac{\alpha_{f1}\ell_2^2}{\alpha_{f2}\ell_1^2} = \frac{3.16 \times 264^2}{3.98 \times 210^2} = 1.25 \quad 0.2 < 1.25 < 5.0 \quad \text{O.K.}$$

Eq. (13-2)

Exterior Panel:

$\alpha_{f1} = 3.16 \qquad \ell_2 = 264 \text{ in.}$

$\alpha_{f2} = 16.45 \qquad \ell_1 = 210 \text{ in.}$

$$\frac{\alpha_{f1}\ell_2^2}{\alpha_{f2}\ell_1^2} = \frac{3.16 \times 264^2}{16.45 \times 210^2} = 0.3 \quad 0.2 < 0.3 < 5.0 \quad \text{O.K.}$$

Therefore, use of Direct Design Method is permitted.

		Code
Example 19.2 (cont'd)	**Calculations and Discussion**	**Reference**

3. Factored moments in slab:

    Total factored moment per span   *13.6.2*

$$\text{Average weight of beams stem} = \frac{14 \times 14}{144} \times \frac{150}{22} = 9.3 \text{ psf}$$

$$\text{Weight of slab} = \frac{6}{12} \times 150 = 75 \text{ psf}$$

$$w_u = 1.2(75 + 9.3) + 1.6(100) = 261 \text{ psf} \qquad \textit{Eq. (9-2)}$$

$$\ell_n = 17.5 - \frac{18}{12} = 16 \text{ ft}$$

$$M_o = \frac{q_u \ell_2 \ell_n^2}{8} \qquad \textit{Eq. (13-4)}$$

$$= \frac{0.261 \times 22 \times 16^2}{8} = 183.7 \text{ ft-kips}$$

    Distribution of moment into negative and positive moments:

    Interior span: *13.6.3.2*

    Negative moment = $0.65 M_o$ = 0.65×183.7 = 119.4 ft-kips
    Positive moment = $0.35 M_o$ = 0.35×183.7 = 64.3 ft-kips

    End span: *13.6.3.3*

    Exterior negative = $0.16 M_o$ = 0.16×183.7 = 29.4 ft-kips
    Positive   = $0.57 M_o$ = 0.57×183.7 = 104.7 ft-kips
    Interior negative = $0.70 M_o$ = 0.7×183.7 = 128.6 ft-kips

    Note: The factored moments may be modified by 10 percent, provided the total   *13.6.7*
    factored static moment in any panel is not less than that computed from Eq. (13-3).
    This modification is omitted here.

4. Distribution of factored moments to column and middle strips: *13.6.4*

    Percentage of total negative and positive moments to column strip.

Example 19.2 (cont'd) — Calculations and Discussion

At interior support:

$$75 + 30\left(\frac{\alpha_{f1}\ell_2}{\ell_1}\right)\left(1 - \frac{\ell_2}{\ell_1}\right) = 75 + 30(1 - 1.26) = 67\%$$
Eq. (1)

where α_{f1} was computed earlier to be 3.16 (see NS interior beam above)

At exterior support:

$$100 - 10\beta_t + 12\beta_t\left(\frac{\alpha_{f1}\ell_2}{\ell_1}\right)\left(1 - \frac{\ell_2}{\ell_1}\right) = 100 - 10(1.88) + 12(1.88)(1 - 1.26) = 75\%$$
Eq. (2)

where

$$\beta_t = \frac{C}{2I_s} = \frac{17{,}868}{2 \times 4752} = 1.88$$

$$I_s = \frac{\ell_2 h^3}{12} = 4{,}752 \text{ in.}^4$$

C is taken as the larger value computed (with the aid of Table 21-2) for the torsional member shown below.

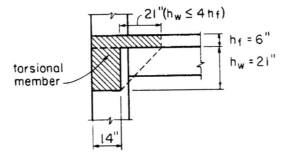

x_1 = 14 in.	x_2 = 6 in.	x_1 = 14 in.	x_2 = 6 in.
y_1 = 21 in.	y_2 = 35 in.	y_1 = 27 in.	y_2 = 21 in.
C_1 = 11,141 in.^{4}	C_2 = 2248 in.^{4}	C_1 = 16,628 in.^{4}	C_2 = 1240 in.^{4}
ΣC = 11,141 + 2248 = 13,389 in.^{4}		ΣC = 16,628 + 1240 = 17,868 in.^{4}	

Positive moment:

$$60 + 30\left(\frac{\alpha_{f1}\ell_2}{\ell_1}\right)\left(1.5 - \frac{\ell_2}{\ell_1}\right) = 60 + 30(1.5 - 1.26) = 67\%$$
Eq. (3)

Example 19.2 (cont'd)   **Calculations and Discussion**   **Code Reference**

Factored moments in column strips and middle strips are summarized as follows:

	Factored Moment (ft-kips)	Column Strip		Moment (ftkips) in Two Half-Middle Strips[2]
		Percent	Moment[1] (ft-kips)	
End Span:				
Exterior Negative	29.4	75	22.1	7.3
Positive	104.7	67	70.1	34.6
Interior Negative	128.6	67	86.2	42.4
Interior Span:				
Negative	119.4	67	80.0	39.4
Positive	64.3	67	43.1	21.2

[1] Since $\alpha_1 \ell_2/\ell_1 > 1.0$, beams must be proportioned to resist 85 percent of column strip moment per 13.6.5.1.

[2] That portion of the factored moment not resisted by the column strip is assigned to the half-middle strips.

5. Factored moments in columns: 13.6.9

   a. Interior columns, with equal spans in the direction of analysis and (different) equal spans in the transverse direction.   13.6.9

   $$M_u = 0.07 \left(0.5 q_{Lu} \ell_2 \ell_n^2\right)$$ Eq. (13-7)

   $$= 0.07 (0.5 \times 1.6 \times 0.1 \times 22 \times 16^2) = 31.5 \text{ ft-kips}$$

   With the same column size and length above and below the slab,

   $$M_c = \frac{31.5}{2} = 15.8 \text{ ft-kips}$$

   This moment is combined with the factored axial load (for each story) for design of the interior columns.

   b. Exterior columns.

   The total exterior negative moment from the slab beam is transferred to the exterior columns; with the same column size and length above and below the slab system:

   $$M_c = \frac{29.4}{2} = 14.7 \text{ ft-kips}$$

	Code
Example 19.2 (cont'd) Calculations and Discussion	**Reference**

6. Shear strength:

   a. Beams.

   Since $\alpha_{f1}\ell_2/\ell_1 > 1$ for all beams, they must resist total shear ($b_w = 14$ in., $d = 17$ in.).   13.6.8.1

   Only interior beams will be checked here, because they carry much higher shear forces than the edge beams.

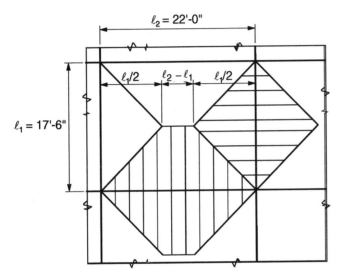

NS beams
$$V_u = \frac{1}{2}q_u\ell_1\frac{\ell_1}{2} = \frac{q_u\ell_1^2}{4}$$

EW beams
$$V_u = \frac{1}{2}q_u\frac{\ell_1}{2}\frac{\ell_1}{2}2 + q_u(\ell_2 - \ell_1)\frac{\ell_1}{2}$$

$$= \frac{q_u\ell_1}{4}(\ell_1 + 2\ell_2 - 2\ell_1) = \frac{q_u\ell_1}{4}(2\ell_2 - \ell_1)$$

NS Beams:

$$V_u = \frac{q_u\ell_1^2}{4} = \frac{0.261\,(17.5)^2}{4} = 20.0 \text{ kips}$$

$$\phi V_c = \phi 2\sqrt{f_c'}\,b_w d \qquad\qquad\qquad\qquad Eq.\ (11\text{-}3)$$

$$= 0.75 \times 2\sqrt{4{,}000} \times 14 \times 17/1{,}000 = 22.6 \text{ kips} > V_u$$

Provide minimum shear reinforcement per 11.5.5.3.   11.5.5.1

EW Beams:

$$V_u = \frac{q_u\ell_1(2\ell_2 - \ell_1)}{4}$$

$$= \frac{0.261 \times 17.5\,[(2 \times 22) - 17.5]}{4} = 30.3 \text{ kips} > \phi V_c = 22.6 \text{ kips N.G.}$$

Example 19.2 (cont'd)	Calculations and Discussion	Code Reference

Required shear strength to be provided by shear reinforcement:

$$V_s = (V_u - \phi V_c)/\phi = (30.3 - 22.6)/0.75 = 10.3 \text{ kips}$$

b. Slabs ($b_w = 12$ in., $d = 5$ in.).   13.6.8.4

$$q_u = (1.2 \times 75) + (1.6 \times 100) = 250 \text{ psf}$$

$$V_u = \frac{q_u \ell_1}{2} = \frac{0.25 \times 17.5}{2} = 2.2 \text{ kips}$$

$$\phi V_c = \phi 2\sqrt{f'_c}\, b_w d$$

$$= 0.75 \times 2\sqrt{4,000} \times 12 \times 5/1,000 = 5.7 \text{ kips} > V_u = 2.2 \text{ kips O.K.}$$

Shear strength of slab is adequate without shear reinforcement.

7. Edge beams must be designed to resist moment not transferred to exterior columns by interior beams, in accordance with 11.6.

20

Two-Way Slabs— Equivalent Frame Method

GENERAL CONSIDERATIONS

The Equivalent Frame Method of analysis converts a three-dimensional frame system with two-way slabs into a series of two-dimensional frames (slab-beams and columns), with each frame extending the full height of the building, as illustrated in Fig. 20-1. The width of each equivalent frame extends to mid-span between column centerlines. The complete analysis of the two-way slab system for a building consists of analyzing a series of equivalent interior and exterior frames spanning longitudinally and transversely through the building. For gravity loading, the slab-beams at each floor or roof (level) may be analyzed separately, with the far ends of attached columns considered fixed (13.7.2.5).

The Equivalent Frame Method of elastic analysis applies to buildings with columns laid out on a basically orthogonal grid, with column lines extending longitudinally and transversely through the building. The analysis method is applicable to slabs with or without beams between supports.

The Equivalent Frame Method may be used for lateral load analysis if the stiffnesses of frame members are modified to account for cracking and other relevant factors. See discussion on 13.5.1.2 in Part 18.

PRELIMINARY DESIGN

Before proceeding with Equivalent Frame analysis, a preliminary slab thickness h needs to be determined for control of deflections, according to the minimum thickness requirements of 9.5.3. Table 18-1 and Fig. 18-3 may be used to simplify minimum thickness computations. For slab systems without beams, it is advisable at this stage of design to check the shear strength of the slab in the vicinity of columns or other support locations, according to the special provisions for slabs of 11.12. See discussion on 13.5.4 in Part 18.

13.7.2 Equivalent Frame

Application of the frame definitions given in 13.7.2, 13.2.1, and 13.2.2 is illustrated in Figs. 20-1 and 20-2. Some judgment is required in applying the definitions given in 13.2.1 for slab systems with varying span lengths along the design strip. Members of the equivalent frame are slab-beams and torsional members (transverse horizontal members) supported by columns (vertical members). The torsional members provide moment transfer between the slab-beams and the columns. The equivalent frame members are illustrated in Fig. 20-3. The initial step in the frame analysis requires that the flexural stiffness of the equivalent frame members be determined.

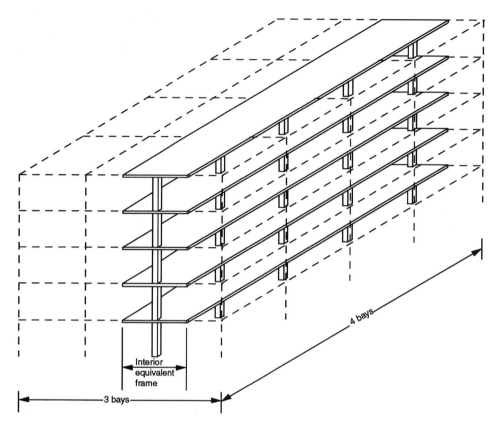

Figure 20-1 Equivalent Frames for 5-Story Building

13.7.3   Slab-Beams

Common types of slab systems with and without beams between supports are illustrated in Figs. 20-4 and 20-5. Cross-sections for determining the stiffness of the slab-beam members K_{sb} between support centerlines are shown for each type. The equivalent slab-beam stiffness diagrams may be used to determine moment distribution constants and fixed-end moments for Equivalent Frame analysis.

Stiffness calculations are based on the following considerations:

a. The moment of inertia of the slab-beam between faces of supports is based on the gross cross-sectional area of the concrete. Variation in the moment of inertia along the axis of the slab-beam between supports must be taken into account (13.7.3.2).

b. A support is defined as a column, capital, bracket or wall. Note that a beam is not considered a supporting member for the equivalent frame (R13.7.3.3).

c. The moment of inertia of the slab-beam from the face of support to the centerline of support is assumed equal to the moment of inertia of the slab-beam at the face of support, divided by the quantity $(1 - c_2/\ell_2)^2$ (13.7.3.3).

The magnification factor $1/(1 - c_2/\ell_2)^2$ applied to the moment of inertia between support face and support centerline, in effect, makes each slab-beam at least a haunched member within its length. Consequently, stiffness and carryover factors and fixed-end moments based on the usual assumptions of uniform prismatic members cannot be applied to the slab-beam members.

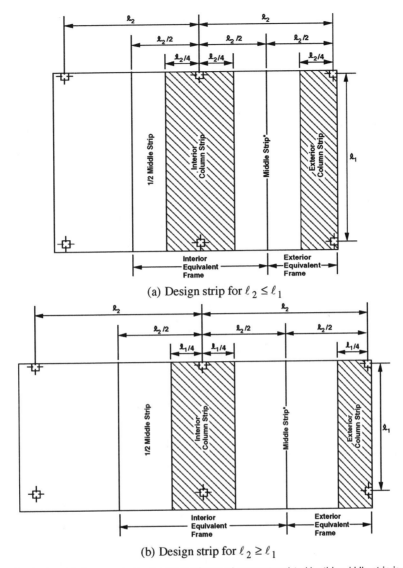

(a) Design strip for $\ell_2 \leq \ell_1$

(b) Design strip for $\ell_2 \geq \ell_1$

*When edge of exterior design strip is supported by a wall, the factored moment resisted by this middle strip is defined in 13.6.6.3.

Figure 20-2 Design Strips of Equivalent Frame

Tables A1 through A6 in Appendix 20A at the end of this chapter give stiffness coefficients, carry-over factors, and fixed-end moment (at left support) coefficients for different geometric and loading configurations. A wide range of column size-to-span ratios in both longitudinal and transverse directions is covered in the tables. Table A1 can be used for flat plates and two-way slabs with beams. Tables A2 through A5 are intended to be used for flat slabs and waffle slabs with various drop (solid head) depths. Table A6 covers the unusual case of a flat plate combined with a flat slab. Fixed-end moment coefficients are provided for both uniform and partially uniform loads. Partial load coefficients were developed for loads distributed over a length of span equal to $0.2\ell_1$. However, loads acting over longer portions of span may be considered by summing the effects of loads acting over each $0.2\ell_1$ interval. For example, if the partial loading extends over $0.6\ell_1$, then the coefficients corresponding to three consecutive $0.2\ell_1$ intervals are to be added. This provides flexibility in the arrangement of loading. For concentrated loads, a high intensity of partial loading may be considered at the appropriate location, and assumed to be distributed over $0.2\ell_1$. For parameter values in between those listed, interpolation may be made. Stiffness diagrams are shown on each table. With appropriate engineering judgment, different span conditions may be considered with the help of information given in these tables.

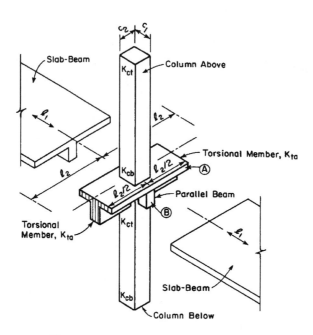

Figure 20-3 Equivalent Frame Members

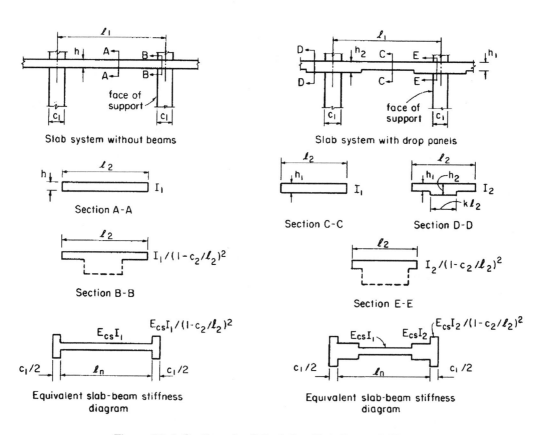

Figure 20-4 Sections for Calculating Slab-Beam Stiffness K_{sb}

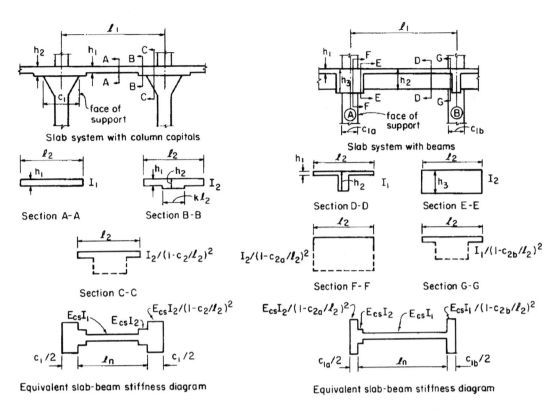

Figure 20-5 Sections for Calculating Slab-Beam Stiffness K_{sb}

13.7.4 Columns

Common types of column end support conditions for slab systems are illustrated in Fig. 20-6. The column stiffness is based on a height of column ℓ_c measured from the mid-depth of the slab above to the mid-depth of the slab below. The column stiffness diagrams may be used to determine column flexural stiffness, K_c. The stiffness diagrams are based on the following considerations:

a. The moment of inertia of the column outside the slab-beam joint is based on the gross cross-sectional area of the concrete. Variation in the moment of inertia along the axis of the column between slab-beam joints is taken into account. For columns with capitals, the moment of inertia is assumed to vary linearly from the base of the capital to the bottom of the slab-beam (13.7.4.1 and 13.7.4.2).

b. The moment of inertia is assumed infinite ($I = \infty$) from the top to the bottom of the slab-beam at the joint. As with the slab-beam members, the stiffness factor K_c for the columns cannot be based on the assumption of uniform prismatic members (13.7.4.3).

Table A7 in Appendix 20A can be used to determine the actual column stiffnesses and carry-over factors.

13.7.5 Torsional Members

Torsional members for common slab-beam joints are illustrated in Fig. 20-7. The cross-section of a torsional member is the largest of those defined by the three conditions given in 13.7.5.1. The governing condition (a), (b), or (c) is indicated below each illustration in Fig. 20-7.

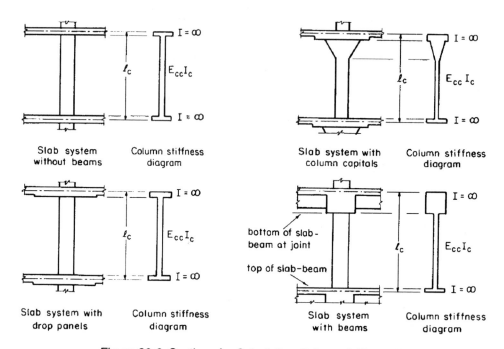

Figure 20-6 Sections for Calculating Column Stiffness K_c

The stiffness K_t of the torsional member is calculated by the following expression:

$$K_t = \Sigma \left[\frac{9E_{cs}C}{\ell_2 \left[1 - (c_2/\ell_2)\right]^3} \right] \qquad (1)$$

where the summation extends over torsional members framing into a joint: two for interior frames, and one for exterior frames.

The term C is a cross-sectional constant that defines the torsional properties of each torsional member framing into a joint:

$$C = \Sigma \left[1 - 0.63 \left(\frac{x}{y}\right) \right] \frac{x^3 y}{3} \qquad (2)$$

where x is the shorter dimension of a rectangular part and y is the longer dimension of a rectangular part.

The value of C is computed by dividing the cross section of a torsional member into separate rectangular parts and summing the C values for the component rectangles. It is appropriate to subdivide the cross section in a manner that results in the largest possible value of C. Application of the C expression is illustrated in Fig. 20-8.

If beams frame into the support in the direction moments are being determined, the torsional stiffness K_t given by Eq. (1) needs to be increased as follows:

$$K_{ta} = \frac{K_t I_{sb}}{I_s}$$

20-6

where K_{ta} = increased torsional stiffness due to the parallel beam (note parallel beam shown in Fig. 20-3)

I_s = moment of inertia of a width of slab equal to the full width between panel centerlines, ℓ_2, excluding that portion of the beam stem extending above and below the slab (note part A in Fig. 20-3).

$$= \frac{\ell_2 h^3}{12}$$

I_{sb} = moment of inertia of the slab section specified for I_s including that portion of the beam stem extending above and below the slab (for the parallel beam illustrated in Fig. 20-3, I_{sb} is for the full tee section shown).

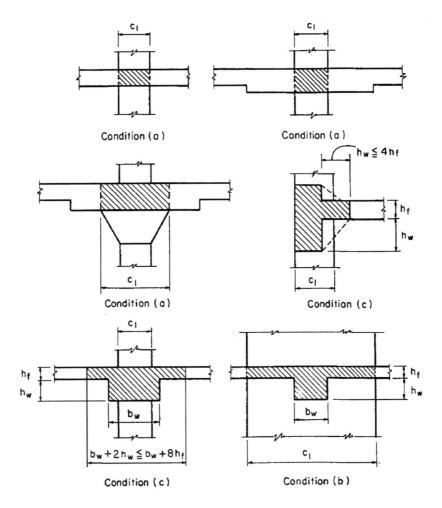

Figure 20-7 Torsional Members

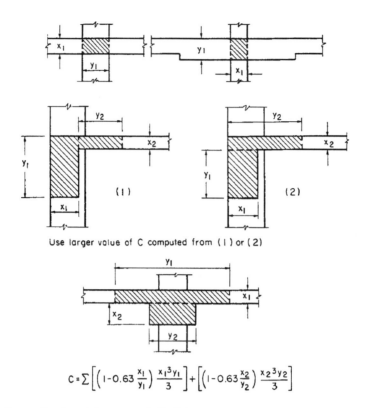

$$C = \sum \left[\left(1-0.63\frac{x_1}{y_1}\right)\frac{x_1^3 y_1}{3}\right] + \left[\left(1-0.63\frac{x_2}{y_2}\right)\frac{x_2^3 y_2}{3}\right]$$

Figure 20-8 Cross-Sectional Constant C, Defining Torsional Properties of a Torsional Member

Equivalent Columns (R13.7.4)

With the publication of ACI 318-83, the equivalent column concept of defining a single-stiffness element consisting of the actual columns above and below the slab-beams plus an attached transverse torsional member was eliminated from the code. With the increasing use of computers for two-way slab analysis by the Equivalent Frame Method, the concept of combining stiffnesses of actual columns and torsional members into a single stiffness has lost much of it attractiveness. The equivalent column was, however, retained in the commentary until the 1989 edition of the code, as an aid to analysis where slab-beams at different floor levels are analyzed separately for gravity loads, especially when using moment distribution or other hand calculation procedures for the analysis. While the equivalent column concept is still recognized by R13.7.4, the detailed procedure contained in the commentary since the '83 edition for calculating the equivalent column stiffness, K_{ec}, was deleted from R13.7.5 of the '95 code.

Both Examples 20.1 and 20.2 utilize the equivalent column concept with moment distribution for gravity load analysis.

The equivalent column concept modifies the column stiffness to account for the torsional flexibility of the slab-to-column connection which reduces its efficiency for transmission of moments. An equivalent column is illustrated in Fig. 20-3. The equivalent column consists of the actual columns above and below the slab-beams, plus "attached" torsional members on both sides of the columns, extending to the centerlines of the adjacent panels. Note that for an edge frame, the attached torsional member is on one side only. The presence of parallel beams will also influence the stiffness of the equivalent column.

The flexural stiffness of the equivalent column K_{ec} is given in terms of its inverse, or flexibility, as follows:

$$\frac{1}{K_{ec}} = \frac{1}{\Sigma K_c} + \frac{1}{\Sigma K_t}$$

For computational purposes, the designer may prefer that the above expression be given directly in terms of stiffness as follows:

$$K_{ec} = \frac{\Sigma K_c \times \Sigma K_t}{\Sigma K_c + \Sigma K_t}$$

Stiffnesses of the actual columns, K_c, and torsional members, K_t must comply with 13.7.4 and 13.7.5.

After the values of K_c and K_t are determined, the equivalent column stiffness K_{ec} is computed. Using Fig. 20-3 for illustration,

$$K_{ec} = \frac{(K_{ct} + K_{cb})(K_{ta} + K_{ta})}{K_{ct} + K_{cb} + K_{ta} + K_{ta}}$$

where K_{ct} = flexural stiffness at top of lower column framing into joint,

K_{cb} = flexural stiffness at bottom of upper column framing into joint,

K_{ta} = torsional stiffness of each torsional member, one on each side of the column, increased due to the parallel beam (if any).

13.7.6 Arrangement of Live Load

In the usual case where the exact loading pattern is not known, the maximum factored moments are developed with loading conditions illustrated by the three-span partial frame in Fig. 20-9, and described as follows:

a. When the service live load does not exceed three-quarters of the service dead load, only loading pattern (1) with full factored live load on all spans need be analyzed for negative and positive factored moments.

b. When the service live-to-dead load ratio exceeds three-quarters, the five loading patterns shown need to be analyzed to determine all factored moments in the slab-beam members. Loading patterns (2) through (5) consider partial factored live loads for determining factored moments. However, with partial live loading, the factored moments cannot be taken less than those occurring with full factored live load on all spans; hence load pattern (1) needs to be included in the analysis.

For slab systems with beams, loads supported directly by the beams (such as the weight of the beam stem or a wall supported directly by the beams) may be inconvenient to include in the frame analysis for the slab loads, $w_d + w_\ell$. An additional frame analysis may be required with the beam section designed to carry these loads in addition to the portion of the slab moments assigned to the beams.

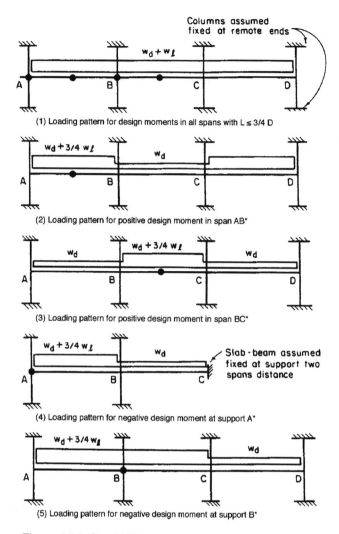

Figure 20-9 Partial Frame Analysis for Vertical Loading

13.7.7   Factored Moments

Moment distribution is probably the most convenient hand calculation method for analyzing partial frames involving several continuous spans with the far ends of upper and lower columns fixed. The mechanics of the method will not be described here, except for a brief discussion of the following two points: (1) the use of the equivalent column concept to determine joint distribution factors and (2) the proper procedure to distribute the equivalent column moment obtained in the frame analysis to the actual columns above and below the slab-beam joint. See Examples 20.1 and 20.2.

A frame joint with stiffness factors K shown for each member framing into the joint is illustrated in Fig. 20-10. Expressions are given below for the moment distribution factors DF at the joint, using the equivalent column stiffness, K_{ec}. These distribution factors are used directly in the moment distribution procedure.

Equivalent column stiffness,

$$K_{ec} = \frac{\Sigma K_c \times \Sigma K_t}{\Sigma K_c + \Sigma K_t}$$

$$= \frac{(K_{ct} + K_{cb})(K_t + K_t)}{K_{ct} + K_{cb} + K_t + K_t}$$

Slab-beam distribution factor,

$$\text{DF (span 2-1)} = \frac{K_{b1}}{K_{b1} + K_{b2} + K_{ec}}$$

$$\text{DF (span 2-3)} = \frac{K_{b2}}{K_{b1} + K_{b2} + K_{ec}}$$

Equivalent column distribution factor (unbalanced moment from slab-beam),

$$\text{DF} = \frac{K_{ec}}{K_{b1} + K_{b2} + K_{ec}}$$

The unbalanced moment determined for the equivalent column in the moment distribution cycles is distributed to the actual columns above and below the slab-beam in proportion to the actual column stiffnesses at the joint. Referring to Fig. 20-10:

$$\text{Portion of unbalanced moment to upper column} = \frac{K_{cb}}{(K_{cb} + K_{ct})}$$

$$\text{Portion of unbalanced moment to lower column} = \frac{K_{ct}}{(K_{cb} + K_{ct})}$$

The "actual" columns are then designed for these moments.

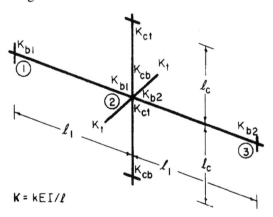

Figure 20-10 Moment Distribution Factors DF

13.7.7.1 - 13.7.7.3 Negative Factored Moments—Negative factored moments for design must be taken at faces of rectilinear supports, but not at a distance greater than $0.175\ell_1$ from the center of a support. This absolute value is a limit on long narrow supports in order to prevent undue reduction in design moment. The support member is defined as a column, capital, bracket or wall. Non-rectangular supports should be treated as square supports having the same cross-sectional area. Note that for slab systems with beams, the faces of beams are not considered face-of-support locations. Locations of the critical section for negative factored moment for various support conditions are illustrated in Fig. 20-11. Note the special requirements illustrated for exterior supports.

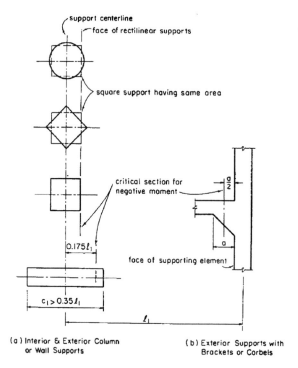

Figure 20-11 Critical Sections for Negative Factored Moment

13.7.7.4 Moment Redistribution—Should a designer choose to use the Equivalent Frame Method to analyze a slab system that meets the limitations of the Direct Design Method, the factored moments may be reduced so that the total static factored moment (sum of the average negative and positive moments) need not exceed M_o computed by Eq. (13-4). This permissible reduction is illustrated in Fig. 20-12.

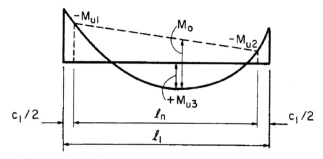

Figure 20-12 Total Static Design Moment for a Span

Since the Equivalent Frame Method of analysis is not an approximate method, the moment redistribution allowed in 8.4 may be used. Excessive cracking may result if these provisions are imprudently applied. The burden of judgment is left to the designer as to what, if any, redistribution is warranted.

13.7.7.5 Factored Moments in Column Strips and Middle Strips—Negative and positive factored moments may be distributed to the column strip and the two half-middle strips of the slab-beam in accordance with 13.6.4, 13.6.5 and 13.6.6, provided that the requirement of 13.6.1.6 is satisfied. See discussion on 13.6.4, 13.6.5, 13.6.6 in Part 19.

APPENDIX 20A   DESIGN AIDS FOR MOMENT DISTRIBUTION CONSTANTS

Table A1 Moment Distribution Constants for Slab-Beam Members

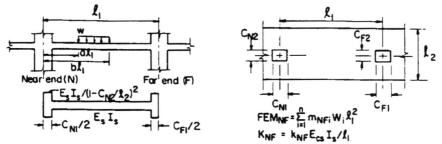

$$FEM_{NF} = \sum_{i=1}^{n} m_{NFi} W_i \ell_1^2$$

$$K_{NF} = k_{NF} E_{cs} I_s / \ell_1$$

C_{N1}/ℓ_1	C_{N2}/ℓ_2	Stiffness Factors k_{NF}	Carry Over Factors C_{NF}	Unif. Load Fixed end M. Coeff. (m_{NF})	Fixed end moment Coeff. (m_{NF}) for (b−a) = 0.2				
					a = 0.0	a = 0.2	a = 0.4	a = 0.6	a = 0.8
$C_{F1} = C_{N1}$; $C_{F2} = C_{N2}$									
0.00	—	4.00	0.50	0.0833	0.0151	0.0287	0.0247	0.0127	0.00226
0.10	0.00	4.00	0.50	0.0833	0.0151	0.0287	0.0247	0.0127	0.00226
	0.10	4.18	0.51	0.0847	0.0154	0.0293	0.0251	0.0126	0.00214
	0.20	4.36	0.52	0.0860	0.0158	0.0300	0.0255	0.0126	0.00201
	0.30	4.53	0.54	0.0872	0.0161	0.0301	0.0259	0.0125	0.00188
	0.40	4.70	0.55	0.0882	0.0165	0.0314	0.0262	0.0124	0.00174
0.20	0.00	4.00	0.50	0.0833	0.0151	0.0287	0.0247	0.0127	0.00226
	0.10	4.35	0.52	0.0857	0.0155	0.0299	0.0254	0.0127	0.00213
	0.20	4.72	0.54	0.0880	0.0161	0.0311	0.0262	0.0126	0.00197
	0.30	5.11	0.56	0.0901	0.0166	0.0324	0.0269	0.0125	0.00178
	0.40	5.51	0.58	0.0921	0.0171	0.0336	0.0276	0.0123	0.00156
0.30	0.00	4.00	0.50	0.0833	0.0151	0.0287	0.0247	0.0127	0.00226
	0.10	4.49	0.53	0.0863	0.0155	0.0301	0.0257	0.0128	0.00219
	0.20	5.05	0.56	0.0893	0.0160	0.0317	0.0267	0.0128	0.00207
	0.30	5.69	0.59	0.0923	0.0165	0.0334	0.0278	0.0127	0.00190
	0.40	6.41	0.61	0.0951	0.0171	0.0352	0.0287	0.0124	0.00167
0.40	0.00	4.00	0.50	0.0833	0.0151	0.0287	0.0247	0.0127	0.00226
	0.10	4.61	0.53	0.0866	0.0154	0.0302	0.0259	0.0129	0.00225
	0.20	5.35	0.56	0.0901	0.0158	0.0318	0.0271	0.0131	0.00221
	0.30	6.25	0.60	0.0936	0.0162	0.0337	0.0284	0.0131	0.00211
	0.40	7.37	0.64	0.0971	0.0168	0.0359	0.0297	0.0128	0.00195
$C_{F1} = 0.5C_{N1}$; $C_{F2} = 0.5C_{N2}$									
0.00	—	4.00	0.50	0.0833	0.0151	0.0287	0.0247	0.0127	0.0023
0.10	0.00	4.00	0.50	0.0833	0.0151	0.0287	0.0247	0.0127	0.0023
	0.10	4.16	0.51	0.0857	0.0155	0.0296	0.0254	0.0130	0.0023
	0.20	4.31	0.52	0.0879	0.0158	0.0304	0.0261	0.0133	0.0023
	0.30	4.45	0.54	0.0900	0.0162	0.0312	0.0267	0.0135	0.0023
	0.40	4.58	0.54	0.0918	0.0165	0.0319	0.0273	0.0138	0.0023
0.20	0.00	4.00	0.50	0.0833	0.0151	0.0287	0.0247	0.0127	0.0023
	0.10	4.30	0.52	0.0872	0.0156	0.0301	0.0259	0.0132	0.0023
	0.20	4.61	0.55	0.0912	0.0161	0.0317	0.0272	0.0138	0.0023
	0.30	4.92	0.57	0.0951	0.0167	0.0332	0.0285	0.0143	0.0024
	0.40	5.23	0.58	0.0989	0.0172	0.0347	0.0298	0.0148	0.0024
0.30	0.00	4.00	0.50	0.0833	0.0151	0.0287	0.0247	0.0127	0.0023
	0.10	4.43	0.53	0.0881	0.0156	0.0305	0.0263	0.0134	0.0023
	0.20	4.89	0.56	0.0932	0.0161	0.0324	0.0281	0.0142	0.0024
	0.30	5.40	0.59	0.0986	0.0167	0.0345	0.0300	0.0150	0.0024
	0.40	5.93	0.62	0.1042	0.0173	0.0367	0.0320	0.0158	0.0025
0.40	0.00	4.00	0.50	0.0833	0.0151	0.0287	0.0247	0.0127	0.0023
	0.10	4.54	0.54	0.0884	0.0155	0.0305	0.0265	0.0135	0.0024
	0.20	5.16	0.57	0.0941	0.0159	0.0326	0.0286	0.0145	0.0025
	0.30	5.87	0.61	0.1005	0.0165	0.0350	0.0310	0.0155	0.0025
	0.40	6.67	0.64	0.1076	0.0170	0.0377	0.0336	0.0166	0.0026
$C_{F1} = 2C_{N1}$; $C_{F2} = 2C_{N2}$									
0.00	—	4.00	0.50	0.0833	0.0151	0.0287	0.0247	0.0127	0.0023
0.10	0.00	4.00	0.50	0.0833	0.0150	0.0287	0.0247	0.0127	0.0023
	0.10	4.27	0.51	0.0817	0.0153	0.0289	0.0241	0.0116	0.0018
	0.20	4.56	0.52	0.0798	0.0156	0.0290	0.0234	0.0103	0.0013
0.20	0.00	4.00	0.50	0.0833	0.0151	0.0287	0.0247	0.0127	0.0023
	0.10	4.49	0.51	0.0819	0.0154	0.0291	0.0240	0.0114	0.0019
	0.20	5.11	0.53	0.0789	0.0158	0.0293	0.0228	0.0096	0.0014

APPENDIX 20A   DESIGN AIDS FOR MOMENT DISTRIBUTION CONSTANTS (cont'd)

Table A2 Moment Distribution Constants for Slab-Beam Members (Drop thickness = 0.25h)

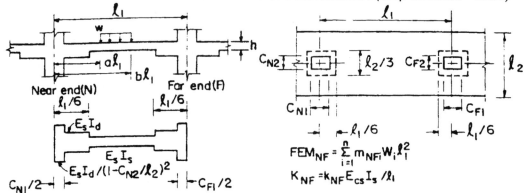

$$FEM_{NF} = \sum_{i=1}^{n} m_{NFi} W_i \ell_1^2$$

$$K_{NF} = k_{NF} E_{cs} I_s / \ell_1$$

C_{N1}/ℓ_1	C_{N2}/ℓ_2	Stiffness Factors k_{NF}	Carry Over Factors C_{NF}	Unif. Load Fixed end M. Coeff. (m_{NF})	Fixed end moment Coeff. (m_{NF}) for (b−a) = 0.2			
					a = 0.0	a = 0.2	a = 0.4	a = 0.6	a = 0.8
$C_{F1} = C_{N1}$; $C_{F2} = C_{N2}$										
0.00	—	4.79	0.54	0.0879	0.0157	0.0309	0.0263	0.0129	0.0022
0.10	0.00	4.79	0.54	0.0879	0.0157	0.0309	0.0263	0.0129	0.0022
	0.10	4.99	0.55	0.0890	0.0160	0.0316	0.0266	0.0128	0.0020
	0.20	5.18	0.56	0.0901	0.0163	0.0322	0.0270	0.0127	0.0019
	0.30	5.37	0.57	0.0911	0.0167	0.0328	0.0273	0.0126	0.0018
0.20	0.00	4.79	0.54	0.0879	0.0157	0.0309	0.0263	0.0129	0.0022
	0.10	5.17	0.56	0.0900	0.0161	0.0320	0.0269	0.0128	0.0020
	0.20	5.56	0.58	0.0918	0.0166	0.0332	0.0276	0.0126	0.0018
	0.30	5.96	0.60	0.0936	0.0171	0.0344	0.0282	0.0124	0.0016
0.30	0.00	4.79	0.54	0.0879	0.0157	0.0309	0.0263	0.0129	0.0022
	0.10	5.32	0.57	0.0905	0.0161	0.0323	0.0272	0.0128	0.0021
	0.20	5.90	0.59	0.0930	0.0166	0.0338	0.0281	0.0127	0.0019
	0.30	6.55	0.62	0.0955	0.0171	0.0354	0.0290	0.0124	0.0017
$C_{F1} = 0.5 C_{N1}$; $C_{F2} = 0.5 C_{N2}$										
0.00	—	4.79	0.54	0.0879	0.0157	0.0309	0.0263	0.0129	0.0022
0.10	0.00	4.79	0.54	0.0879	0.0157	0.0309	0.0263	0.0129	0.0022
	0.10	4.96	0.55	0.0900	0.0160	0.0317	0.0269	0.0131	0.0022
	0.20	5.12	0.56	0.0920	0.0164	0.0325	0.0276	0.0134	0.0022
0.20	0.00	4.79	0.54	0.0879	0.0157	0.0309	0.0263	0.0129	0.0022
	0.10	5.11	0.56	0.0914	0.0162	0.0323	0.0275	0.0133	0.0022
	0.20	5.43	0.58	0.0950	0.0167	0.0337	0.0286	0.0138	0.0022
$C_{F1} = 2 C_{N1}$; $C_{F2} = 2 C_{N2}$										
0.00	—	4.79	0.54	0.0879	0.0157	0.0309	0.0263	0.0129	0.0022
0.10	0.00	4.79	0.54	0.0879	0.0157	0.0309	0.0263	0.0129	0.0022
	0.10	5.10	0.55	0.0860	0.0159	0.0311	0.0256	0.0117	0.0017

APPENDIX 20A DESIGN AIDS FOR MOMENT DISTRIBUTION CONSTANTS (cont'd)

Table A3 Moment Distribution Constants for Slab-Beam Members (Drop thickness = 0.50h)

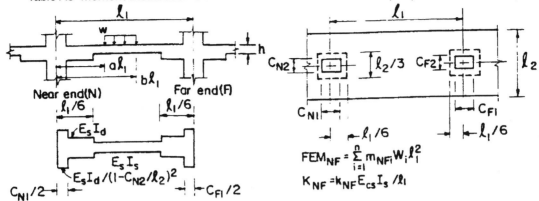

C_{N1}/ℓ_1	C_{N2}/ℓ_2	Stiffness Factors k_{NF}	Carry Over Factors C_{NF}	Unif. Load Fixed end M. Coeff. (m_{NF})	Fixed end moment Coeff. (m_{NF}) for (b−a) = 0.2				
					a = 0.0	a = 0.2	a = 0.4	a = 0.6	a = 0.8
$C_{F1} = C_{N1}$; $C_{F2} = C_{N2}$									
0.00	—	5.84	0.59	0.0926	0.0164	0.0335	0.0279	0.0128	0.0020
0.10	0.00	5.84	0.59	0.0926	0.0164	0.0335	0.0279	0.0128	0.0020
	0.10	6.04	0.60	0.0936	0.0167	0.0341	0.0282	0.0126	0.0018
	0.20	6.24	0.61	0.0940	0.0170	0.0347	0.0285	0.0125	0.0017
	0.30	6.43	0.61	0.0952	0.0173	0.0353	0.0287	0.0123	0.0016
0.20	0.00	5.84	0.59	0.0926	0.0164	0.0335	0.0279	0.0128	0.0020
	0.10	6.22	0.61	0.0942	0.0168	0.0346	0.0285	0.0126	0.0018
	0.20	6.62	0.62	0.0957	0.0172	0.0356	0.0290	0.0123	0.0016
	0.30	7.01	0.64	0.0971	0.0177	0.0366	0.0294	0.0120	0.0014
0.30	0.00	5.84	0.59	0.0926	0.0164	0.0335	0.0279	0.0128	0.0020
	0.10	6.37	0.61	0.0947	0.0168	0.0348	0.0287	0.0126	0.0018
	0.20	6.95	0.63	0.0967	0.0172	0.0362	0.0294	0.0123	0.0016
	0.30	7.57	0.65	0.0986	0.0177	0.0375	0.0300	0.0119	0.0014
$C_{F1} = 0.5C_{N1}$; $C_{F2} = 0.5C_{N2}$									
0.00	—	5.84	0.59	0.0926	0.0164	0.0335	0.0279	0.0128	0.0020
0.10	0.00	5.84	0.59	0.0926	0.0164	0.0335	0.0279	0.0128	0.0020
	0.10	6.00	0.60	0.0945	0.0167	0.0343	0.0285	0.0130	0.0020
	0.20	6.16	0.60	0.0962	0.0170	0.0350	0.0291	0.0132	0.0020
0.20	0.00	5.84	0.59	0.0926	0.0164	0.0335	0.0279	0.0128	0.0020
	0.10	6.15	0.60	0.0957	0.0169	0.0348	0.0290	0.0131	0.0020
	0.20	6.47	0.62	0.0987	0.0173	0.0360	0.0300	0.0134	0.0020
$C_{F1} = 2C_{N1}$; $C_{F2} = 2C_{N2}$									
0.00	—	5.84	0.59	0.0926	0.0164	0.0335	0.0279	0.0128	0.0020
0.10	0.00	5.84	0.59	0.0926	0.0164	0.0335	0.0279	0.0128	0.0020
	0.10	6.17	0.60	0.0907	0.0166	0.0337	0.0273	0.0116	0.0015

20-15

APPENDIX 20A DESIGN AIDS FOR MOMENT DISTRIBUTION CONSTANTS (cont'd)

Table A4 Moment Distribution Constants for Slab-Beam Members (Drop thickness = 0.75h)

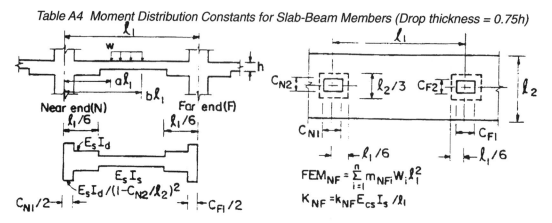

$$FEM_{NF} = \sum_{i=1}^{n} m_{NFi} W_i \ell_1^2$$

$$K_{NF} = k_{NF} E_{cs} I_s / \ell_1$$

C_{N1}/ℓ_1	C_{N2}/ℓ_2	Stiffness Factors k_{NF}	Carry Over Factors C_{NF}	Unif. Load Fixed end M. Coeff. (m_{NF})	Fixed end moment Coeff. (m_{NF}) for (b−a) = 0.2			
					a = 0.0	a = 0.2	a = 0.4	a = 0.6	a = 0.8
$C_{F1} = C_{N1}$; $C_{F2} = C_{N2}$										
0.00	—	6.92	0.63	0.0965	0.0171	0.0360	0.0293	0.0124	0.0017
0.10	0.00	6.92	0.63	0.0965	0.0171	0.0360	0.0293	0.0124	0.0017
	0.10	7.12	0.64	0.0972	0.0174	0.0365	0.0295	0.0122	0.0016
	0.20	7.31	0.64	0.0978	0.0176	0.0370	0.0297	0.0120	0.0014
	0.30	7.48	0.65	0.0984	0.0179	0.0375	0.0299	0.0118	0.0013
0.20	0.00	6.92	0.63	0.0965	0.0171	0.0360	0.0293	0.0124	0.0017
	0.10	7.12	0.64	0.0977	0.0175	0.0369	0.0297	0.0121	0.0015
	0.20	7.31	0.65	0.0988	0.0178	0.0378	0.0301	0.0118	0.0013
	0.30	7.48	0.67	0.0999	0.0182	0.0386	0.0304	0.0115	0.0011
0.30	0.00	6.92	0.63	0.0965	0.0171	0.0360	0.0293	0.0124	0.0017
	0.10	7.29	0.65	0.0981	0.0175	0.0371	0.0299	0.0121	0.0015
	0.20	7.66	0.66	0.0996	0.0179	0.0383	0.0304	0.0117	0.0013
	0.30	8.02	0.68	0.1009	0.0182	0.0394	0.0309	0.0113	0.0011
$C_{F1} = 0.5 C_{N1}$; $C_{F2} = 0.5 C_{N2}$										
0.00	—	6.92	0.63	0.0965	0.0171	0.0360	0.0293	0.0124	0.0017
0.10	0.00	6.92	0.63	0.0965	0.0171	0.0360	0.0293	0.0124	0.0017
	0.10	7.08	0.64	0.0980	0.0174	0.0366	0.0298	0.0125	0.0017
	0.20	7.23	0.64	0.0993	0.0177	0.0372	0.0302	0.0126	0.0016
0.20	0.00	6.92	0.63	0.0965	0.0171	0.0360	0.0293	0.0124	0.0017
	0.10	7.21	0.64	0.0991	0.0175	0.0371	0.0302	0.0126	0.0017
	0.20	7.51	0.65	0.1014	0.0179	0.0381	0.0310	0.0128	0.0016
$C_{F1} = 2 C_{N1}$; $C_{F2} = 2 C_{N2}$										
0.00	—	6.92	0.63	0.0965	0.0171	0.0360	0.0293	0.0124	0.0017
0.10	0.00	6.92	0.63	0.0965	0.0171	0.0360	0.0293	0.0124	0.0017
	0.10	7.26	0.64	0.0946	0.0173	0.0361	0.0287	0.0112	0.0013

APPENDIX 20A   DESIGN AIDS FOR MOMENT DISTRIBUTION CONSTANTS (cont'd)

Table A5 Moment Distribution Constants for Slab-Beam Members (Drop thickness = h)

$$FEM_{NF} = \sum_{i=1}^{n} m_{NFi} W_i \ell_1^2$$

$$K_{NF} = k_{NF} E_{cs} I_s / \ell_1$$

C_{N1}/ℓ_1	C_{N2}/ℓ_2	Stiffness Factors k_{NF}	Carry Over Factors C_{NF}	Unif. Load Fixed end M. Coeff. (m_{NF})	Fixed end moment Coeff. (m_{NF}) for (b−a) = 0.2				
					a = 0.0	a = 0.2	a = 0.4	a = 0.6	a = 0.8
\multicolumn{10}{c}{$C_{F1} = C_{N1}$; $C_{F2} = C_{N2}$}									
0.00	—	7.89	0.66	0.0993	0.0177	0.0380	0.0303	0.0118	0.0014
0.10	0.00	7.89	0.66	0.0993	0.0177	0.0380	0.0303	0.0118	0.0014
	0.10	8.07	0.66	0.0998	0.0180	0.0385	0.0305	0.0116	0.0013
	0.20	8.24	0.67	0.1003	0.0182	0.0389	0.0306	0.0115	0.0012
	0.30	8.40	0.67	0.1007	0.0183	0.0393	0.0307	0.0113	0.0011
0.20	0.00	7.89	0.66	0.0993	0.0177	0.0380	0.0303	0.0118	0.0014
	0.10	8.22	0.67	0.1002	0.0180	0.0388	0.0306	0.0115	0.0012
	0.20	8.55	0.68	0.1010	0.0183	0.0395	0.0309	0.0112	0.0011
	0.30	9.87	0.69	0.1018	0.0186	0.0402	0.0311	0.0109	0.0009
0.30	0.00	7.89	0.66	0.0993	0.0177	0.0380	0.0303	0.0118	0.0014
	0.10	8.35	0.67	0.1005	0.0181	0.0390	0.0307	0.0115	0.0012
	0.20	8.82	0.68	0.1016	0.0184	0.0399	0.0311	0.0111	0.0011
	0.30	9.28	0.70	0.1026	0.0187	0.0409	0.0314	0.0107	0.0009
\multicolumn{10}{c}{$C_{F1} = 0.5C_{N1}$; $C_{F2} = 0.5C_{N2}$}									
0.00	—	7.89	0.66	0.0993	0.0177	0.0380	0.0303	0.0118	0.0014
0.10	0.00	7.89	0.66	0.0993	0.0177	0.0380	0.0303	0.0118	0.0014
	0.10	8.03	0.66	0.1006	0.0180	0.0386	0.0307	0.0119	0.0014
	0.20	8.16	0.67	0.1016	0.0182	0.0390	0.0310	0.0120	0.0014
0.20	0.00	7.89	0.66	0.0993	0.0177	0.0380	0.0303	0.0118	0.0014
	0.10	8.15	0.67	0.1014	0.0181	0.0389	0.0310	0.0120	0.0014
	0.20	8.41	0.68	0.1032	0.0184	0.0398	0.0316	0.0121	0.0013
\multicolumn{10}{c}{$C_{F1} = 2C_{N1}$; $C_{F2} = 0.5C_{N2}$}									
0.00	—	7.89	0.66	0.0993	0.0177	0.0380	0.0303	0.0118	0.0014
0.10	0.00	7.79	0.66	0.0993	0.0177	0.0380	0.0303	0.0118	0.0014
	0.10	8.20	0.67	0.0981	0.0179	0.0382	0.0297	0.0113	0.0010

APPENDIX 20A DESIGN AIDS FOR MOMENT DISTRIBUTION CONSTANTS (cont'd)

Table A6 Moment Distribution Constants for Slab-Beam Members
(Column dimensions assumed equal at near end and far end — $c_{F1} = c_{N1}$, $c_{F2} = c_{N2}$)

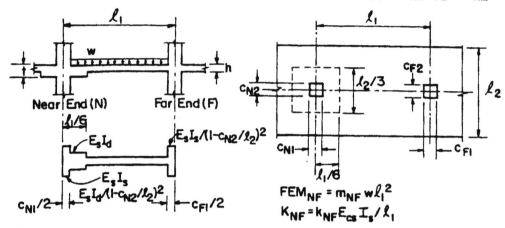

$FEM_{NF} = m_{NF} w \ell_1^2$
$K_{NF} = k_{NF} E_{cs} I_s / \ell_1$

C_1/ℓ_1	C_2/ℓ_2	t = 1.5h						t = 2h					
		k_{NF}	C_{NF}	m_{NF}	k_{FN}	C_{FN}	m_{FN}	k_{NF}	C_{NF}	m_{NF}	k_{FN}	C_{FN}	m_{FN}
0.00	—	5.39	0.49	0.1023	4.26	0.60	0.0749	6.63	0.49	0.1190	4.49	0.65	0.0676
0.10	0.00	5.39	0.49	0.1023	4.26	0.60	0.0749	6.63	0.49	0.1190	4.49	0.65	0.0676
	0.10	5.65	0.52	0.1012	4.65	0.60	0.0794	7.03	0.54	0.1145	5.19	0.66	0.0757
	0.20	5.86	0.54	0.1012	4.91	0.61	0.0818	7.22	0.56	0.1140	5.43	0.67	0.0778
	0.30	6.05	0.55	0.1025	5.10	0.62	0.0838	7.36	0.56	0.1142	5.57	0.67	0.0786
0.20	0.00	5.39	0.49	0.1023	4.26	0.60	0.0749	6.63	0.49	0.1190	4.49	0.65	0.0676
	0.10	5.88	0.54	0.1006	5.04	0.61	0.0826	7.41	0.58	0.1111	5.96	0.66	0.0823
	0.20	6.33	0.58	0.1003	5.63	0.62	0.0874	7.85	0.61	0.1094	6.57	0.67	0.0872
	0.30	6.75	0.60	0.1008	6.10	0.64	0.0903	8.18	0.63	0.1093	6.94	0.68	0.0892
0.30	0.00	5.39	0.49	0.1023	4.26	0.60	0.075	6.63	0.49	0.1190	4.49	0.65	0.0676
	0.10	6.08	0.56	0.1003	5.40	0.61	0.085	7.76	0.62	0.1087	6.77	0.67	0.0873
	0.20	6.78	0.61	0.0996	6.38	0.63	0.092	8.49	0.66	0.1055	7.91	0.68	0.0952
	0.30	7.48	0.64	0.0997	7.25	0.65	0.096	9.06	0.68	0.1047	8.66	0.69	0.0991

APPENDIX 20A DESIGN AIDS FOR MOMENT DISTRIBUTION CONSTANTS (cont'd)

Table A7 Stiffness and Carry-Over Factors for Columns

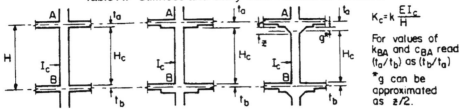

$K_c = k \dfrac{EI_c}{H}$

For values of k_{BA} and c_{BA} read (t_a/t_b) as (t_b/t_a)

*g can be approximated as z/2.

t_a/t_b		H/H_c: 1.05	1.10	1.15	1.20	1.25	1.30	1.35	1.40	1.45	1.50
0.00	k_{AB}	4.20	4.40	4.60	4.80	5.00	5.20	5.40	5.60	5.80	6.00
	C_{AB}	0.57	0.65	0.73	0.80	0.87	0.95	1.03	1.10	1.17	1.25
0.2	k_{AB}	4.31	4.62	4.95	5.30	5.65	6.02	6.40	6.79	7.20	7.62
	C_{AB}	0.56	0.62	0.68	0.74	0.80	0.85	0.91	0.96	1.01	1.07
0.4	k_{AB}	4.38	4.79	5.22	5.67	6.15	6.65	7.18	7.74	8.32	8.94
	C_{AB}	0.55	0.60	0.65	0.70	0.74	0.79	0.83	0.87	0.91	0.94
0.6	k_{AB}	4.44	4.91	5.42	5.96	6.54	7.15	7.81	8.50	9.23	10.01
	C_{AB}	0.55	0.59	0.63	0.67	0.70	0.74	0.77	0.80	0.83	0.85
0.8	k_{AB}	4.49	5.01	5.58	6.19	6.85	7.56	8.31	9.12	9.98	10.89
	C_{AB}	0.54	0.58	0.61	0.64	0.67	0.70	0.72	0.75	0.77	0.79
1.0	k_{AB}	4.52	5.09	5.71	6.38	7.11	7.89	8.73	9.63	10.60	11.62
	C_{AB}	0.54	0.57	0.60	0.62	0.65	0.67	0.69	0.71	0.73	0.74
1.2	k_{AB}	4.55	5.16	5.82	6.54	7.32	8.17	9.08	10.07	11.12	12.25
	C_{AB}	0.53	0.56	0.59	0.61	0.63	0.65	0.66	0.68	0.69	0.70
1.4	k_{AB}	4.58	5.21	5.91	6.68	7.51	8.41	9.38	10.43	11.57	12.78
	C_{AB}	0.53	0.55	0.58	0.60	0.61	0.63	0.64	0.65	0.66	0.67
1.6	k_{AB}	4.60	5.26	5.99	6.79	7.66	8.61	9.64	10.75	11.95	13.24
	C_{AB}	0.53	0.55	0.57	0.59	0.60	0.61	0.62	0.63	0.64	0.65
1.8	k_{AB}	4.62	5.30	6.06	6.89	7.80	8.79	9.87	11.03	12.29	13.65
	C_{AB}	0.52	0.55	0.56	0.58	0.59	0.60	0.61	0.61	0.62	0.63
2.0	k_{AB}	4.63	5.34	6.12	6.98	7.92	8.94	10.06	11.27	12.59	14.00
	C_{AB}	0.52	0.54	0.56	0.57	0.58	0.59	0.59	0.60	0.60	0.61
2.2	k_{AB}	4.65	5.37	6.17	7.05	8.02	9.08	10.24	11.49	12.85	14.31
	C_{AB}	0.52	0.54	0.55	0.56	0.57	0.58	0.58	0.59	0.59	0.59
2.4	k_{AB}	4.66	5.40	6.22	7.12	8.11	9.20	10.39	11.68	13.08	14.60
	C_{AB}	0.52	0.53	0.55	0.56	0.56	0.57	0.57	0.58	0.58	0.58
2.6	k_{AB}	4.67	5.42	6.26	7.18	8.20	9.31	10.53	11.86	13.29	14.85
	C_{AB}	0.52	0.53	0.54	0.55	0.56	0.56	0.56	0.57	0.57	0.57
2.8	k_{AB}	4.68	5.44	6.29	7.23	8.27	9.41	10.66	12.01	13.48	15.07
	C_{AB}	0.52	0.53	0.54	0.55	0.55	0.55	0.56	0.56	0.56	0.56
3.0	k_{AB}	4.69	5.46	6.33	7.28	8.34	9.50	10.77	12.15	13.65	15.28
	C_{AB}	0.52	0.53	0.54	0.54	0.55	0.55	0.55	0.55	0.55	0.55
3.2	k_{AB}	4.70	5.48	6.36	7.33	8.40	9.58	10.87	12.28	13.81	15.47
	C_{AB}	0.52	0.53	0.53	0.54	0.54	0.54	0.54	0.54	0.54	0.54
3.4	k_{AB}	4.71	5.50	6.38	7.37	8.46	9.65	10.97	12.40	13.95	15.64
	C_{AB}	0.51	0.52	0.53	0.53	0.54	0.54	0.54	0.53	0.53	0.53
3.6	k_{AB}	4.71	5.51	6.41	7.41	8.51	9.72	11.05	12.51	14.09	15.80
	C_{AB}	0.51	0.52	0.53	0.53	0.53	0.53	0.53	0.53	0.53	0.52
3.8	k_{AB}	4.72	5.53	6.43	7.44	8.56	9.78	11.13	12.60	14.21	15.95
	C_{AB}	0.51	0.52	0.53	0.53	0.53	0.53	0.53	0.52	0.52	0.52
4.0	k_{AB}	4.72	5.54	6.45	7.47	8.60	9.84	11.21	12.70	14.32	16.08
	C_{AB}	0.51	0.52	0.52	0.53	0.53	0.52	0.52	0.52	0.52	0.51
4.2	k_{AB}	4.73	5.55	6.47	7.50	8.64	9.90	11.27	12.78	14.42	16.20
	C_{AB}	0.51	0.52	0.52	0.52	0.52	0.52	0.52	0.51	0.51	0.51
4.4	k_{AB}	4.73	5.56	6.49	7.53	8.68	9.95	11.34	12.86	14.52	16.32
	C_{AB}	0.51	0.52	0.52	0.52	0.52	0.52	0.51	0.51	0.51	0.50
4.6	k_{AB}	4.74	5.57	6.51	7.55	8.71	9.99	11.40	12.93	14.61	16.43
	C_{AB}	0.51	0.52	0.52	0.52	0.52	0.52	0.51	0.51	0.50	0.50
4.8	k_{AB}	4.74	5.58	6.53	7.58	8.75	10.03	11.45	13.00	14.69	16.53
	C_{AB}	0.51	0.52	0.52	0.52	0.52	0.51	0.51	0.50	0.50	0.49
5.0	k_{AB}	4.75	5.59	6.54	7.60	8.78	10.07	11.50	13.07	14.77	16.62
	C_{AB}	0.51	0.51	0.52	0.52	0.51	0.51	0.51	0.50	0.49	0.49
6.0	k_{AB}	4.76	5.63	6.60	7.69	8.90	10.24	11.72	13.33	15.10	17.02
	C_{AB}	0.51	0.51	0.51	0.51	0.50	0.50	0.49	0.49	0.48	0.47
7.0	k_{AB}	4.78	5.66	6.65	7.76	9.00	10.37	11.88	13.54	15.35	17.32
	C_{AB}	0.51	0.51	0.51	0.50	0.50	0.49	0.48	0.48	0.47	0.46
8.0	k_{AB}	4.78	5.68	6.69	7.82	9.07	10.47	12.01	13.70	15.54	17.56
	C_{AB}	0.51	0.51	0.50	0.50	0.49	0.49	0.48	0.47	0.46	0.45
9.0	k_{AB}	4.79	5.69	6.71	7.86	9.13	10.55	12.11	13.83	15.70	17.74
	C_{AB}	0.50	0.50	0.50	0.50	0.49	0.48	0.47	0.46	0.45	0.45
10.0	k_{AB}	4.80	5.71	6.74	7.89	9.18	10.61	12.19	13.93	15.83	17.90
	C_{AB}	0.50	0.50	0.50	0.49	0.48	0.48	0.47	0.46	0.45	0.44

Example 20.1—Two-Way Slab Without Beams Analyzed by Equivalent Frame Method

Using the Equivalent Frame Method, determine design moments for the slab system in the direction shown, for an intermediate floor.

Story height = 9 ft
Column dimensions = 16 × 16 in.
Lateral loads to be resisted by shear walls
No edge beams
Partition weight = 20 psf
Service live load = 40 psf
f'_c = 4,000 psi (for slabs), normal weight concrete
f'_c = 6,000 psi (for columns), normal weight concrete
f_y = 60,000 psi

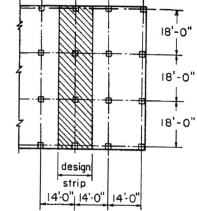

Calculations and Discussion	Code Reference

1. Preliminary design for slab thickness h:

   a. Control of deflections.

   For flat plate slab systems, the minimum overall thickness h with Grade 60 reinforcement is (see Table 18-1):   9.5.3.2

   $$h = \frac{\ell_n}{30} = \frac{200}{30} = 6.67 \text{ in.}$$ Table 9.5 (a)

   but not less than 5 in.   9.5.3.2(a)

   where ℓ_n = length of clear span in the long direction = 216 - 16 = 200 in.

   Try 7 in. slab for all panels (weight = 87.5 psf)

   Note, in addition to ACI 318-05 deflection control requirements, thickness of slab should satisfy the minimum required for fire resistance, as specified in the locally adopted building code.

   b. Shear strength of slab.

   Use average effective depth d = 5.75 in. (3/4 in. cover and No. 4 bar)

   Factored dead load, q_{Du} = 1.2 (87.5 + 20) = 129 psf   9.2.1
   Factored live load, q_{Lu} = 1.6 × 40   = 64 psf
   Total factored load   = 193 psf

   For wide beam action consider a 12-in. wide strip taken at d distance from the face of support in the long direction (see Fig. 20-13).   11.12.1.1

   V_u = 0.193 × 7.854 = 1.5 kips

   $V_c = 2\sqrt{f'_c}\, b_w d$

		Code
Example 20.1 (cont'd)	**Calculations and Discussion**	**Reference**

$\phi V_c = 0.75 \times 2\sqrt{4,000} \times 12 \times 5.75/1,000 = 6.6$ kips $> V_u$ O.K.   *9.3.2.3*

For two-way action, since there are no shear forces at the centerlines of adjacent panels, the shear strength at d/2 distance around the support is computed as follows:

$V_u = 0.193 [(18 \times 14) - 1.81^2] = 48.0$ kips

$V_c = 4\sqrt{f'_c}\, b_o d$ (for square interior column)   *Eq. (11-35)*

$ = 4\sqrt{4,000}\, (4 \times 21.75) \times 5.75/1,000 = 126.6$ kips

$\phi V_c = 0.75 \times 126.6 = 95.0$ kips $> V_u$ O.K.   *9.3.2.3*

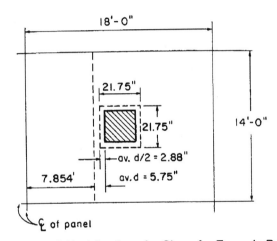

Figure 20-13 Critical Sections for Shear for Example Problem

Preliminary design indicates that a 7 in. overall slab thickness is adequate for control of deflections and shear strength.

2. Frame members of equivalent frame:

   Determine moment distribution factors and fixed-end moments for the equivalent frame members. The moment distribution procedure will be used to analyze the partial frame. Stiffness factors k, carry over factors COF, and fixed-end moment factors FEM for the slab-beams and column members are determined using the tables of Appendix 20-A. These calculations are shown here.

   a. Flexural stiffness of slab-beams at both ends, K_{sb}.

   $$\frac{c_{N1}}{\ell_1} = \frac{16}{(18 \times 12)} = 0.07, \quad \frac{c_{N2}}{\ell_2} = \frac{16}{(14 \times 12)} = 0.1$$

		Code
Example 20.1 (cont'd)	**Calculations and Discussion**	**Reference**

For $c_{F1} = c_{N1}$ and $c_{F2} = c_{N2}$, $k_{NF} = k_{FN} = 4.13$ by interpolation from Table A1 in Appendix 20A.

Thus, $K_{sb} = k_{NF} \dfrac{E_{cs}I_s}{\ell_1} = 4.13 \dfrac{E_{cs}I_s}{\ell_1}$ *Table A1*

$\qquad = 4.13 \times 3.60 \times 10^6 \times 4,802/216 = 331 \times 10^6$ in.-lb

where $I_s = \dfrac{\ell_2 h^3}{12} = \dfrac{168(7)^3}{12} = 4,802$ in.^{4}

$E_{cs} = 57,000 \sqrt{f'_c} = 57,000\sqrt{4,000} = 3.60 \times 10^6$ psi   *8.5.1*

Carry-over factor COF = 0.509, by interpolation from Table A1.

Fixed-end moment FEM = $0.0843 w_u \ell_2 \ell_1^2$, by interpolation from Table A1.

b. Flexural stiffness of column members at both ends, K_c.

Referring to Table A7, Appendix 20A, $t_a = 3.5$ in., $t_b = 3.5$ in.,

$H = 9$ ft = 108 in., $H_c = 101$ in., $t_a/t_b = 1$, $H/H_c = 1.07$

Thus, $k_{AB} = k_{BA} = 4.74$ by interpolation.

$K_c = 4.74 E_{cc} I_c / \ell_c$ *Table A7*

$\qquad = 4.74 \times 4.42 \times 10^6 \times 5461/108 = 1059 \times 10^6$ in.-lb

where $I_c = \dfrac{c^4}{12} = \dfrac{(16)^4}{12} = 5,461$ in.^{4}

$E_{cs} = 57,000\sqrt{f'_c} = 57,000\sqrt{6,000} = 4.42 \times 10^6$ psi   *8.5.1*

$\ell_c = 9$ ft = 108 in.

c. Torsional stiffness of torsional members, K_t.

$K_t = \dfrac{9 E_{cs} C}{\left[\ell_2 (1 - c_2/\ell_2)^3\right]}$ *R.13.7.5*

$\qquad = \dfrac{9 \times 3.60 \times 10^6 \times 13.25}{168(0.905)^3} = 3.45 \times 10^6$ in.-lb

| Example 20.1 (cont'd) | Calculations and Discussion | Code Reference |

where $C = \Sigma (1 - 0.63 \, x/y)(x^3 y/3)$ 13.0

$= (1 - 0.63 \times 7/16)(7^3 \times 16/3) = 1{,}325 \text{ in.}^4$

$c_2 = 16$ in. and $\ell_2 = 14$ ft $= 168$ in.

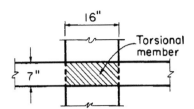

Condition (a) of Fig. 20-7

d. Equivalent column stiffness K_{ec}.

$$K_{ec} = \frac{\Sigma K_c \times \Sigma K_t}{\Sigma K_c + \Sigma K_t}$$

$$= \frac{(2 \times 1{,}059)(2 \times 345)}{[(2 \times 1{,}059) + (2 \times 345)]}$$

$= 520 \times 10^6$ in.-lb

where ΣK_t is for two torsional members, one on each side of column, and ΣK_c is for the upper and lower columns at the slab-beam joint of an intermediate floor.

e. Slab-beam joint distribution factors DF.

At exterior joint,

$$DF = \frac{331}{(331 + 520)} = 0.389$$

At interior joint,

$$DF = \frac{331}{(331 + 331 + 520)} = 0.280$$

COF for slab-beam $= 0.509$

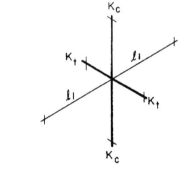

3. Partial frame analysis of equivalent frame:

Determine maximum negative and positive moments for the slab-beams using the moment distribution method. Since the service live load does not exceed three-quarters of the service dead load, design moments are assumed to occur at all critical sections with full factored live load on all spans.   13.7.6.2

$$\frac{L}{D} = \frac{40}{(87.5 + 20)} = 0.37 < \frac{3}{4}$$

a. Factored load and fixed-end moments.

Factored dead load $q_{Du} = 1.2(87.5 + 20) = 129$ psf   Eq. (9-2)

Factored live load $q_{Lu} = 1.6(40) = 64$ psf   Eq. (9-2)

		Code
Example 20.1 (cont'd)	**Calculations and Discussion**	**Reference**

Factored load $q_u = q_{Du} + q_{Lu} = 193$ psf

FEM's for slab-beams $= m_{NF} q_u \ell_2 \ell_1^2$ (Table A1, Appendix 20A)
$= 0.0843 (0.193 \times 14) 18^2 = 73.8$ ft-kips

b. Moment distribution. Computations are shown in Table 20-1. Counterclockwise rotational moments acting on the member ends are taken as positive. Positive span moments are determined from the following equation:

M_u (midspan) $= M_o - (M_{uL} + M_{uR})/2$

where M_o is the moment at midspan for a simple beam.

When the end moments are not equal, the maximum moment in the span does not occur at midspan, but its value is close to that at midspan for this example.

Positive moment in span 1-2:

$+M_u = (0.193 \times 14) 18^2/8 - (46.6 + 84.0)/2 = 44.1$ ft-kips

Positive moment in span 2-3:

$+M_u = (0.193 \times 14) 18^2/8 - (76.2 + 76.2)/2 = 33.2$ ft-kips

Table 20-1 Moment Distribution for Partial Frame

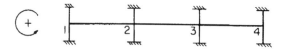

Joint	1	2		3		4
Member	1-2	2-1	2-3	3-2	3-4	4-3
DF	0.389	0.280	0.280	0.280	0.280	0.389
COF	0.509	0.509	0.509	0.509	0.509	0.509
FEM	+73.8	-73.8	+73.8	-73.8	+73.8	-73.8
Dist	-28.7	0.0	0.0	0.0	0.0	28.7
CO	0.0	-14.6	0.0	0.0	14.6	0.0
Dist	0.0	4.1	4.1	-4.1	-4.1	0.0
CO	2.1	0.0	-2.1	2.1	0.0	-2.1
Dist	-0.8	0.6	0.6	-0.6	-0.6	0.8
CO	0.3	-0.4	-0.3	0.3	0.4	-0.3
Dist	-0.1	0.2	0.2	-0.2	-0.2	0.1
CO	0.1	-0.1	-0.1	0.1	0.1	-0.1
Dist	0.0	0.0	0.0	0.0	0.0	0.0
Neg. M	46.6	-84.0	76.2	-76.2	84.0	-46.6
M @ midspan	44.1		33.2		44.1	

| Example 20.1 (cont'd) | Calculations and Discussion | Code Reference |

4. Design moments:

   Positive and negative factored moments for the slab system in the direction of analysis are plotted in Fig. 20-14. The negative design moments are taken at the faces of rectilinear supports but not at distances greater than $0.175\ell_1$ from the centers of supports.   13.7.7.1

   $$\frac{16 \text{ in.}}{2} = 0.67 \text{ ft} < 0.175 \times 18 = 3.2 \text{ ft} \text{ (Use face of support location)}$$

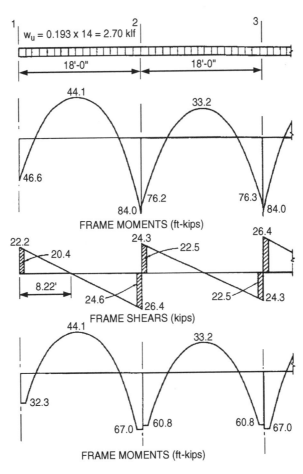

Figure 20-14 Positive and Negative Design Moments for Slab-Beam
(All Spans Loaded with Full Factored Live Load)

5. Total factored moment per span:

   Slab systems within the limitations of 13.6.1 may have the resulting moments reduced in such proportion that the numerical sum of the positive and average negative moments need not be greater than: 13.7.7.4

		Code
Example 20.1 (cont'd)	Calculations and Discussion	Reference

$$M_o = \frac{q_u \ell_2 \ell_n^2}{8} = 0.193 \times 14 \times (16.67)^2 / 8 = 93.9 \text{ ft-kips}$$

End spans: $44.1 + (32.3 + 67.0)/2 = 93.8$ ft-kips

Interior span: $33.2 + (60.8 + 60.8)/2 = 94$ ft-kips

It may be seen that the total design moments from the Equivalent Frame Method yield a static moment equal to that given by the static moment expression used with the Direct Design Method.

6. Distribution of design moments across slab-beam strip: 13.7.7.5

   The negative and positive factored moments at critical sections may be distributed to the column strip and the two half-middle strips of the slab-beam according to the proportions specified in 13.6.4 and 13.6.6. The requirement of 13.6.1.6 does not apply for slab systems without beams, $\alpha = 0$. Distribution of factored moments at critical sections is summarized in Table 20-2.

Table 20-2 Distribution of Factored Moments

	Factored Moment (ft-kips)	Column Strip		Moment (ft-kips) in Two Half-Middle Strips**
		Percent*	Moment (ft-kips)	
End Span:				
Exterior Negative	32.3	100	32.3	0.0
Positive	44.1	60	26.5	17.7
Interior Negative	67.0	75	50.3	16.7
Interior Span:				
Negative	60.8	75	45.6	15.2
Positive	33.2	60	19.9	13.2

* For slab systems without beams
** That portion of the factored moment not resisted by the column strip is assigned to the two half-middle strips.

7. Column moments:

   The unbalanced moment from the slab-beams at the supports of the equivalent frame are distributed to the actual columns above and below the slab-beam in proportion to the relative stiffnesses of the actual columns. Referring to Fig. 20-14, the unbalanced moment at joints 1 and 2 are:

   Joint 1 = +46.6 ft-kips

   Joint 2 = -84.0 + 76.2 = -7.8 ft-kips

   The stiffness and carry-over factors of the actual columns and the distribution of the unbalanced moments to the exterior and interior columns are shown in Fig. 20-15. The design moments for the columns may be taken at the juncture of column and slab.

Example 20.1 (cont'd) — **Calculations and Discussion** — **Code Reference**

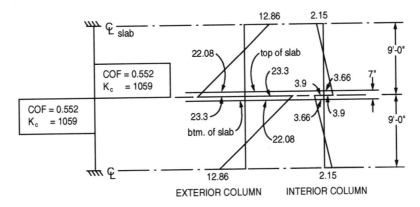

Figure 20-15 Column Moments (Unbalanced Moments from Slab-Beam)

In summary:

Design moment in exterior column = 22.08 ft-kips

Design moment in interior column = 3.66 ft-kips

8. Check slab flexural and shear strength at exterior column

   a. Total flexural reinforcement required for design strip:

      i. Determine reinforcement required for column strip moment M_u = 32.3 ft-kips

         Assume tension-controlled section (ϕ = 0.9)   9.3.2.1

         Column strip width $b = \dfrac{14 \times 12}{2} = 84$ in.   13.2.1

         $R_u = \dfrac{M_u}{\phi b d^2} = \dfrac{32.3 \times 12{,}000}{0.9 \times 84 \times 5.75^2} = 155$ psi

         $\rho = \dfrac{0.85 f'_c}{f_y}\left(1 - \sqrt{1 - \dfrac{2 R_u}{0.85 f'_c}}\right)$

         $= \dfrac{0.85 \times 4}{60}\left(1 - \sqrt{1 - \dfrac{2 \times 155}{0.85 \times 4{,}000}}\right) = 0.0026$

         $A_s = \rho b d = 0.0026 \times 84 \times 5.75 = 1.28$ in.²

         $\rho_{min} = 0.0018$ 13.3.1

         Min $A_s = 0.0018 \times 84 \times 7 = 1.06$ in.² < 1.28 in.²

         Number of No. 4 bars = $\dfrac{1.28}{0.2} = 6.4$, say 7 bars

         Maximum spacing $s_{max} = 2h = 14$ in. < 18 in.   13.3.2

Example 20.1 (cont'd)	Calculations and Discussion	Code Reference

$$c = \frac{a}{\beta_1} = \frac{0.29}{0.85} = 0.34 \text{ in.}$$

$$\epsilon_t = \left(\frac{0.003}{c}\right)d_t - 0.003$$

$$= \left(\frac{0.003}{0.34}\right)5.75 - 0.003 = 0.048 > 0.005$$

Therefore, section is tension-controlled.   10.3.4

Use 7-No. 4 bars in column strip.

ii. Check slab reinforcement at exterior column for moment transfer between slab and column

Portion of unbalanced moment transferred by flexure = $\gamma_f M_u$ 13.5.3.2

From Fig. 16-13, Case C:

$$b_1 = c_1 + \frac{d}{2} = 16 + \frac{5.75}{2} = 18.88 \text{ in.}$$

$$b_2 = c_2 + d = 16 + 5.75 = 21.75 \text{ in.}$$

$$\gamma_f = \frac{1}{1 + (2/3)\sqrt{b_1/b_2}} \quad\quad\quad\text{Eq. (13-1)}$$

$$= \frac{1}{1 + (2/3)\sqrt{18.88/21.75}} = 0.62$$

$$\gamma_f M_u = 0.62 \times 32.3 = 20.0 \text{ ft-kips}$$

Note that the provisions of 13.5.3.3 may be utilized; however, they are not in this example.

Assuming tension-controlled behavior, determine required area of reinforcement for $\gamma_f M_u = 20.0$ ft-kips

Effective slab width $b = c_2 + 3h = 16 + 3(7) = 37$ in.   13.5.3.2

$$R_u = \frac{M_u}{\phi b d^2} = \frac{20 \times 12{,}000}{0.9 \times 37 \times 5.75^2} = 218 \text{ psi}$$

$$\rho = \frac{0.85 f'_c}{f_y}\left(1 - \sqrt{1 - \frac{2R_u}{0.85 f'_c}}\right)$$

Example 20.1 (cont'd)	Calculations and Discussion	Code Reference

$$= \frac{0.85 \times 4}{60}\left(1-\sqrt{1-\frac{2 \times 218}{0.85 \times 4,000}}\right) = 0.0038$$

$A_s = 0.0038 \times 37 \times 5.75 = 0.80$ in.^{2}

Min. $A_s = 0.0018 \times 37 \times 7 = 0.47$ in.^{2} < 0.80 in.^{2} 13.3.1

Number of No. 4 bars = $\frac{0.80}{0.20} = 4$

Verify tension-controlled section:

$$a = \frac{A_s f_y}{0.85 f'_c b} = \frac{(4 \times 0.2) \times 60}{0.85 \times 4 \times 37} = 0.38 \text{ in.}$$

$$c = \frac{a}{\beta_1} = \frac{0.38}{0.85} = 0.45 \text{ in.}$$

$$\epsilon_t = \left(\frac{0.003}{0.45}\right)5.75 - 0.003 = 0.035 > 0.005$$

Therefore, section is tension-controlled.   10.3.4

Provide the required 4-No. 4 bars by concentrating 4 of the column strip bars (7-No. 4) within the 37 in. slab width over the column. For symmetry, add one additional No. 4 bar outside of 37-in. width.

Note that the column strip section remains tension-controlled with the addition of 1-No. 4 bar.

The reinforcement details at the edge column are shown below.

Example 20.1 (cont'd) — Calculations and Discussion

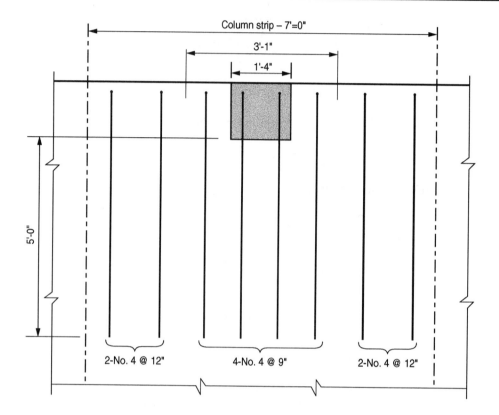

iii. Determine reinforcement required for middle strip.

Provide minimum reinforcement, since $M_u = 0$ (see Table 20-2).

Min. $A_s = 0.0018 \times 84 \times 7 = 1.06$ in.^{2}

Maximum spacing $s_{max} = 2h = 14$ in. < 18 in.   *13.3.2*

Provide No. 4 @ 14 in. in middle strip.

b. Check combined shear stress at inside face of critical transfer section   *11.12.6.1*

For shear strength equations, see Part 16.

$$v_u = \frac{V_u}{A_c} + \frac{\gamma_v M_u}{J/C}$$

From Example 19.1, $V_u = 25.6$ kips

When factored moments are determined by an accurate method of frame analysis, such as the Equivalent Frame Method, unbalanced moment is taken directly from the results of the frame analysis. Also, considering the approximate nature of the moment transfer analysis procedure, assume the unbalanced moment M_u is at the centroid of the critical transfer section.

| Example 20.1 (cont'd) | Calculations and Discussion | Code Reference |

Thus, $M_u = 32.3$ ft-kips (see Table 20-2)

$\gamma_v = 1 - \gamma_f = 1 - 0.62 = 0.38$ Eq. (11-39)

From Example 19.1, critical section properties:

$A_c = 342.2$ in.^{2}

$J/c = 2,357$ in.^{3}

$v_u = \dfrac{25,600}{342.2} + \dfrac{0.38 \times 32.3 \times 12,000}{2,357}$

$\quad = 74.8 + 62.5 = 137.3$ psi

Allowable shear stress $\phi v_n = \phi 4\sqrt{f'_c} = 189.7$ psi $> v_u$ O.K.   11.12.6.2

Example 20.2—Two-Way Slab with Beams Analyzed by Equivalent Frame Method

Using the Equivalent Frame Method, determine design moments for the slab system in the direction shown, for an intermediate floor.

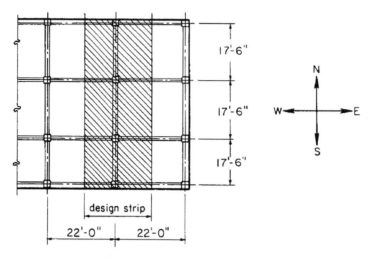

Story height = 12 ft
Edge beam dimensions = 14 × 27 in.
Interior beam dimensions = 14 × 20 in.
Column dimensions = 18 × 18 in.
Service live load = 100 psf

f'_c = 4,000 psi (for all members), normal weight concrete
f_y = 60,000 psi

Calculations and Discussion	Code Reference

1. Preliminary design for slab thickness h.

   Control of deflections: 9.5.3.3

   From Example 19.2, the beam-to-slab flexural stiffness ratios α are:

   α_f = 13.30 (NS edge beam)

       = 16.45 (EW edge beam)

       = 3.16 (NS interior beam)

       = 3.98 (EW interior beam)

   Since all $\alpha_f > 2.0$ (see Fig. 8-2), Eq. (9-13) will control. Therefore,

   $$h = \frac{\ell_n (0.8 + f_y / 200{,}000)}{36 + 9\beta}$$ Eq. (9-12)

Example 20.2 (cont'd)   **Calculations and Discussion**   **Code Reference**

$$= \frac{246\,(0.8 + 60{,}000/200{,}000)}{36 + 9\,(1.28)} = 5.7 \text{ in.}$$

where ℓ_n = clear span in long direction = 20.5 ft = 246 in.

$$\beta = \frac{\text{clear span in long direction}}{\text{clear span in short direction}} = \frac{20.5}{16.0} = 1.28$$

Use 6 in. slab thickness.

2. Frame members of equivalent frame.

   Determine moment distribution constants and fixed-end moment coefficients for the equivalent frame members. The moment distribution procedure will be used to analyze the partial frame for vertical loading. Stiffness factors k, carry-over factors COF, and fixed-end moment factors FEM for the slab-beams and column members are determined using the tables of Appendix 20-A. These calculations are shown here.

   a. Slab-beams, flexural stiffness at both ends K_{sb}:

   $$\frac{c_{N1}}{\ell_1} = \frac{18}{17.5 \times 12} = 0.0857 \approx 0.1$$

   $$\frac{c_{N2}}{\ell_2} = \frac{18}{22 \times 12} = 0.0682$$

   Referring to Table A1, Appendix 20A,

   $$K_{sb} = \frac{4.11 E_c I_{sb}}{\ell_1} = 4.11 \times 25{,}387 E_c / (17.5 \times 12) = 497 E_c$$

   where I_{sb} is the moment of inertia of slab-beam section shown in Fig. 20-16 and computed with the aid of Fig. 20-21 at the end of this Example.

   $I_{sb} = 2.72\,(14 \times 20^3)/12 = 25{,}387$ in.^{4}

   Carry-over factor COF = 0.507

   Fixed-end moment, FEM = $0.0842 q_u \ell_2 \ell_1^2$

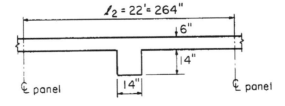

Figure 20-16 Cross-Section of Slab-Beam

		Code
Example 20.2 (cont'd)	**Calculations and Discussion**	**Reference**

b. Column members, flexural stiffness K_c:

t_a = 17 in, t_b = 3 in., t_a/t_b = 5.67

H = 12 ft = 144 in., H_c = 144 - 17 - 3 = 124 in.

H/H_c = 1.16 for interior columns

t_a = 24 in., t_b = 3 in., t_a/t_b = 8.0

H = 12 ft = 144 in., H_c = 144 - 24 - 3 = 117 in.

H/H_c = 1.23 for exterior columns

Referring to Table A7, Appendix 20A,

For interior columns:

$$K_{ct} = \frac{6.82 E_c I_c}{\ell_c} = \frac{6.82 \times 8748 E_c}{144} = 414 E_c$$

$$K_{cb} = \frac{4.99 E_c I_c}{\ell_c} = \frac{4.99 \times 8748 E_c}{144} = 303 E_c$$

For exterior columns:

$$K_{ct} = \frac{8.57 E_c I_c}{\ell_c} = \frac{8.57 \times 8748 E_c}{144} = 512 E_c$$

$$K_{cb} = \frac{5.31 E_c I_c}{\ell_c} = \frac{5.31 \times 8748 E_c}{144} = 323 E_c$$

where $I_c = \frac{(c)^4}{12} = \frac{(18)^4}{12} = 8,748$ in.^{4}

ℓ_c = 12 ft = 144 in.

c. Torsional members, torsional stiffness K_t:

$$K_t = \frac{9 E_c C}{\ell_2 (1 - c_2/\ell_2)^3}$$ R13.7.5

where C = $\Sigma (1 - 0.63 \, x/y)(x^3 y/3)$ 13.0

For interior columns:

K_t = $9 E_c \times 11,698/[264 \, (0.932)^3]$ = $493 E_c$

Example 20.2 (cont'd) — Calculations and Discussion — Code Reference

where $1 - \dfrac{c_2}{\ell_2} = 1 - \dfrac{18}{(22 \times 12)} = 0.932$

C is taken as the larger value computed with the aid of Table 19-2 for the torsional member shown in Fig. 20-17.

x_1 = 14 in.	x_2 = 6 in.	x_1 = 14 in.	x_2 = 6 in.
y_1 = 14 in.	y_2 = 42 in.	y_1 = 20 in.	y_2 = 14 in.
C_1 = 4,738	C_2 = 2,752	C_1 = 10,226	C_2 = 736
ΣC = 4,738 + 2,752 = 7,490 in.^{4}		ΣC = 10,226 + 736 × 2 = 11,698 in.^{4}	

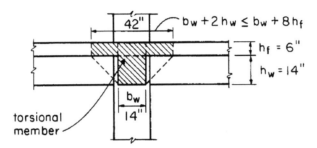

Figure 20-17 Attached Torsional Member at Interior Column

For exterior columns:

$K_t = 9E_c \times 17{,}868/[264\,(0.932)^3] = 752 E_c$

where C is taken as the larger value computed with the aid of Table 19-2 for the torsional member shown in Fig. 20-18.

x_1 = 14 in.	x_2 = 6 in.	x_1 = 14 in.	x_2 = 6 in.
y_1 = 21 in.	y_2 = 35 in.	y_1 = 27 in.	y_2 = 21 in.
C_1 = 11,141	C_2 = 2,248	C_1 = 16,628	C_2 = 1,240
ΣC = 11,141 + 2,248 = 13,389 in.^{4}		ΣC = 16,628 + 1,240 = 17,868 in.^{4}	

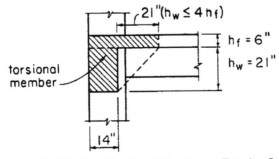

Figure 20-18 Attached Torsional Member at Exterior Column

		Code
Example 20.2 (cont'd)	**Calculations and Discussion**	**Reference**

d. Increased torsional stiffness K_{ta} due to parallel beams:

For interior columns:

$$K_{ta} = \frac{K_t I_{sb}}{I_s} = \frac{493E_c \times 25,387}{4,752} = 2,634E_c$$

For exterior columns:

$$K_{ta} = \frac{752E_c \times 25,387}{4,752} = 4,017E_c$$

where I_s = moment of inertia of slab-section shown in Fig. 20-19.

$$= 264 \, (6)^3/12 = 4,752 \text{ in.}^4$$

I_{sb} = moment of inertia of full T-section shown in Fig. 20-19 and computed with the aid of Fig. 20-21

$$= 2.72 \, (14 \times 20^3/12) = 25,387 \text{ in.}^4$$

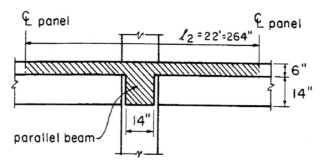

Figure 20-19 Slab-Beam in the Direction of Analysis

e. Equivalent column stiffness, K_{ec}:

$$K_{ec} = \frac{\Sigma K_c \times \Sigma K_{ta}}{\Sigma K_c + \Sigma K_{ta}}$$

where ΣK_{ta} is for two torsional members, one on each side of column, and ΣK_c is for the upper and lower columns at the slab-beam joint of an intermediate floor.

For interior columns:

$$K_{ec} = \frac{(303E_c + 414E_c)(2 \times 2,634E_c)}{(303E_c + 414E_c) + (2 \times 2,634E_c)} = 631E_c$$

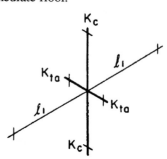

For exterior columns:

		Code
Example 20.2 (cont'd)	**Calculations and Discussion**	**Reference**

$$K_{ec} = \frac{(323E_c + 521E_c)(2 \times 4,017E_c)}{(323E_c + 521E_c) + (2 \times 4,017E_c)} = 764E_c$$

f. Slab-beam joint distribution factors DF:

At exterior joint:

$$DF = \frac{497E_c}{(497E_c + 764E_c)} = 0.394$$

At interior joint:

$$DF = \frac{497E_c}{(497E_c + 497E_c + 631E_c)} = 0.306$$

COF for slab-beam = 0.507

3. Partial frame analysis of equivalent frame.

   Determine maximum negative and positive moments for the slab-beams using the moment distribution method.

   With a service live-to-dead load ratio:

   $$\frac{L}{D} = \frac{100}{75} = 1.33 > \frac{3}{4}$$

   the frame will be analyzed for five loading conditions with pattern loading and partial live load as allowed by 13.7.6.3 (see Fig. 20-9 for an illustration of the five load patterns considered).   13.7.6.3

   a. Factored loads and fixed-end moments:

      Factored dead load, q_{Du} = 1.2 (75 + 9.3) = 101 psf

      $$\left(\frac{14 \times 14}{144} \times \frac{150}{22} = 9.3 \text{ psf is weight of beam stem per foot divided by } \ell_2\right)$$

      Factored live load, q_{Lu} = 1.6 (100) = 160 psf

      Factored load, q_u = q_{Du} + q_{Lu} = 261 psf

      FEM for slab-beams   = $m_{NF} q_u \ell_2 \ell_1^2$ (Table A1, Appendix 20A)

      FEM due to $q_{Du} + q_{Lu}$ = 0.0842 (0.261 × 22) 17.5² = 148.1 ft-kips

		Code
Example 20.2 (cont'd)	**Calculations and Discussion**	**Reference**

FEM due to $q_{Du} + 3/4q_{Lu} = 0.0842 \, (0.221 \times 22) \, 17.5^2 = 125.4$ ft-kips

FEM due to q_{Du} only $= 0.0842 \, (0.101 \times 22) \, 17.5^2 = 57.3$ ft-kips

b. Moment distribution for the five loading conditions is shown in Table 20-3. Counter-clockwise rotational moments acting on member ends are taken as positive. Positive span moments are determined from the equation:

$M_{u(midspan)} = M_o - (M_{uL} + M_{uR})/2$

where M_o is the moment at midspan for a simple beam.

When the end moments are not equal, the maximum moment in the span does not occur at midspan, but its value is close to that at midspan.

Positive moment in span 1-2 for loading (1):

$+M_u = (0.261 \times 22) \, 17.5^2/8 - (93.1 + 167.7)/2 = 89.4$ ft-kips

The following moment values for the slab-beams are obtained from Table 20-3. Note that according to 13.7.6.3, the design moments shall be taken not less than those occurring with full factored live load on all spans.

Maximum positive moment in end span

= the larger of 89.4 or 83.3 = 89.4 ft-kips

Maximum positive moment in interior span*

= the larger of 66.2 or 71.3 = 71.3 ft-kips

Maximum negative moment at end support

= the larger of 93.1 or 86.7 = 93.1 ft-kips

Maximum negative moment at interior support of end span

= the larger of 167.7 or 145.6 = 167.7 ft-kips

Maximum negative moment at interior support of interior span

= the larger of 153.6 or 139.2 = 153.6 ft-kips

4. Design moments.

Positive and negative factored moments for the slab system in the transverse direction are plotted in Fig. 20-20. The negative factored moments are taken at the face of rectilinear supports at distances not greater than $0.175\ell_1$ from the center of supports.

13.7.7.1

$\dfrac{18 \text{ in.}}{2} = 0.75$ ft $< 0.175 \times 17.5 = 3.1$ ft (Use face of support location).

---

* This is the only moment governed by the pattern loading with partial live load. All other maximum moments occur with full factored live load on all spans.

Example 20.2 (cont'd) Calculations and Discussion | Code Reference

Table 20-3 Moment Distribution for Partial Frame (Transverse Direction)

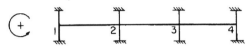

Joint	1	2		3		4
Member	1-2	2-1	2-3	3-2	3-4	4-3
DF	0.394	0.306	0.306	0.306	0.306	0.394
COF	0.507	0.507	0.507	0.507	0.507	0.507

(1) All spans loaded with full factored live load

FEM	148.1	-148.1	148.1	-148.1	148.1	-148.1
Dist	-58.4	0.0	0.0	0.0	0.0	58.4
CO	0.0	-29.6	0.0	0.0	29.6	0.0
Dist	0.0	9.1	9.1	-9.1	-9.1	0.0
CO	4.6	0.0	-4.6	4.6	0.0	-4.6
Dist	-1.8	1.4	1.4	-1.4	-1.4	1.8
CO	0.7	-0.9	-0.7	0.7	0.9	-0.7
Dist	-0.3	0.5	0.5	-0.5	-0.5	0.3
CO	0.3	-0.1	-0.3	0.3	0.1	-0.3
Dist	-0.1	0.1	0.1	-0.1	-0.1	0.1
M	93.1	-167.7	153.6	-153.6	167.7	-93.1

(2) First and third spans loaded with 3/4 factored live load

FEM	125.4	-125.4	57.3	-57.3	125.4	-125.4
Dist	-49.4	20.8	20.8	-20.8	-20.8	49.4
CO	10.6	-25.1	-10.6	10.6	25.1	-10.6
Dist	-4.2	10.9	10.9	-10.9	-10.9	4.2
CO	5.5	-2.1	-5.5	5.5	2.1	-5.5
Dist	-2.2	2.3	2.3	-2.3	-2.3	2.2
CO	1.2	-1.1	-1.2	1.2	1.1	-1.2
Dist	-0.5	0.7	0.7	-0.7	-0.7	0.5
CO	0.4	-0.2	-0.4	0.4	0.2	-0.4
Dist	-0.1	0.2	0.2	-0.2	-0.2	0.1
M	86.7	-119.0	74.6	-74.6	119.0	-86.7
Midspan M	83.3				83.3	

(3) Center span loaded with 3/4 factored live load

FEM	57.3	-57.3	125.4	-125.4	57.3	-57.3
Dist	-22.6	-20.8	-20.8	20.8	20.8	22.6
CO	-10.6	-11.4	10.6	-10.6	11.4	10.6
Dist	4.2	0.3	0.3	-0.3	-0.3	-4.2
CO	0.1	2.1	-0.1	0.1	-2.1	-0.1
Dist	-0.1	-0.6	-0.6	0.6	0.6	0.1
CO	-0.3	0.0	0.3	-0.3	0.0	0.3
Dist	0.1	-0.1	-0.1	0.1	0.1	-0.1
CO	0.0	0.1	0.0	0.0	-0.1	0.0
Dist	0.0	0.0	0.0	0.0	0.0	0.0
M	28.2	-87.9	114.9	-114.9	87.9	-28.2
Midspan M			71.2			

Table cont'd on next page

			Code
Example 20.2 (cont'd)	**Calculations and Discussion**		**Reference**

Table 20-3 Moment Distribution for Partial Frame
(Transverse Direction)
— continued —

(4) First span loaded with 3/4 factored live load and beam-slab assumed fixed st support two spans away

FEM	125.4	-125.4	57.3	-57.3
Dist	-49.4	20.8	20.8	0.0
CO	10.6	-25.0	0.0	10.6
Dist	-4.2	7.7	7.7	0.0
CO	3.9	-2.1	0.0	3.9
Dist	-1.5	0.6	0.6	0.0
CO	0.3	-0.8	0.0	0.3
Dist	-0.1	0.2	0.2	0.0
CO	0.1	-0.1	0.0	0.1
Dist	0.0	0.0	0.0	0.0
M	85.0	-124.0	86.7	-42.4

(5) First and second span loaded with 3/4 factored live load

FEM	125.4	-125.4	125.4	-125.4	57.3	-57.3
Dist	-49.4	0.0	0.0	20.8	20.8	22.6
CO	0.0	-25.1	10.6	0.0	11.4	10.6
Dist	0.0	4.4	4.4	-3.5	-3.5	-4.2
CO	2.2	0.0	-1.8	2.2	-2.1	-1.8
Dist	-0.9	0.5	0.5	0.0	0.0	0.7
CO	0.3	-0.4	0.0	0.3	0.4	0.0
Dist	-0.1	0.1	0.1	-0.2	-0.2	0.0
CO	0.1	-0.1	-0.1	0.1	0.0	-0.1
Dist	0.0	0.0	0.0	0.0	0.0	0.0
M	77.6	-145.8	139.2	-105.7	84.1	-29.5

5. Total factored moment per span.   *13.7.7.4*

Slab systems within the limitations of 13.6.1 may have the resulting moments reduced in such proportion that the numerical sum of the positive and average negative moments are not greater than the total static moment M_o given by Eq. (13-3). Check limitations of 13.6.1.6 for relative stiffness of beams in two perpendicular directions.

For interior panel (see Example 19.2):

$$\frac{\alpha_{f1}\ell_2^2}{\alpha_{f2}\ell_1^2} = \frac{316\,(22)^2}{3.98\,(17.5)^2} = 1.25$$ *13.6.1.6*

$0.2 < 1.25 < 5.0$ O.K.

For exterior panel (see Example 19.2):

$$\frac{3.16\,(22)^2}{16.45\,(17.5)^2} = 0.30$$

$0.2 < 0.30 < 5.0$ O.K.

	Code
Example 20.2 (cont'd) Calculations and Discussion	**Reference**

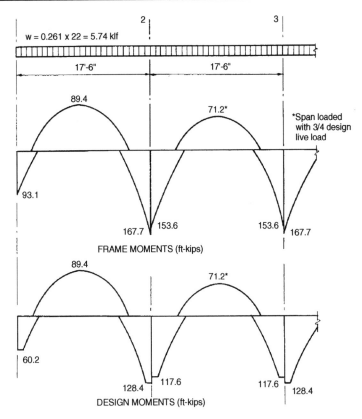

Figure 20-20 Positive and Negative Design Moments for Slab-Beam (All Spans Loaded with Full Factored Live Load Except as Noted)

All limitations of 13.6.1 are satisfied and the provisions of 13.7.7.4 may be applied.

$$M_o = \frac{q_u \ell_2 \ell_n^2}{8} = \frac{0.261 \times 22 \times 16^2}{8} = 183.7 \text{ ft-kips}$$ Eq. (13-3)

End span: $89.4 + (60.2 + 128.4)/2 = 183.7$ ft-kips

Interior span: $71.2 + (117.6 + 117.6)/2 = 188.8$ ft-kips

To illustrate proper procedure, the interior span factored moments may be reduced as follows:

Permissible reduction = $183.7/188.8 = 0.973$

Adjusted negative design moment = $117.6 \times 0.973 = 114.3$ ft-kips
Adjusted positive design moment = $71.2 \times 0.973 = 69.3$ ft-kips
$M_o = 183.7$ ft-kips

Example 20.2 (cont'd)   **Calculations and Discussion**   **Code Reference**

6. Distribution of design moments across slab-beam strip.   13.7.7.5

   Negative and positive factored moments at critical sections may be distributed to the column strip, beam and two-half middle strips of the slab-beam according to the proportions specified in 13.6.4, 13.6.5 and 13.6.6, if requirement of 13.6.1.6 is satisfied.

   a. Since the relative stiffnesses of beams are between 0.2 and 5.0 (see step No. 5), the moments can be distributed across slab-beams as specified in 13.6.4, 13.6.5 and 13.6.6.

   b. Distribution of factored moments at critical section:

   $$\frac{\ell_2}{\ell_1} = \frac{22}{17.5} = 1.257$$

   $$\frac{\alpha_{f1}\ell_2}{\ell_1} = 3.16 \times 1.257 = 3.97$$

   $$\beta_t = \frac{C}{2I_s} = \frac{17{,}868}{(2 \times 4752)} = 1.88$$

   where $I_s = \dfrac{22 \times 12 \times 6^3}{12} = 4{,}752$ in.^{4}

   $C = 17{,}868$ in.^{4} (see Fig. 22-18)

   Factored moments at critical sections are summarized in Table 20-4.

   Table 20-4 Distribution of Design Moments

	Factored Moment (ft-kips)	Column Strip		Moment (ft-kips) in Two Half-Middle Strips**
		Percent*	Moment (ft-kips)	
End Span: Exterior Negative	60.2	75	45.2	15.0
Positive	89.4	67	59.9	29.5
Interior Negative	128.4	67	86.0	42.4
Interior Span: Negative	117.6	67	78.8	38.8
Positive	71.2	67	47.8	23.5

   * Since $\alpha_1\ell_2/\ell_1 > 1.0$ beams must be proportioned to resist 85 percent of column strip moment per 13.6.5.1.
   ** That portion of the factored moment not resisted by the column strip is assigned to the two half-middle strips.

7. Calculations for shear in beams and slab are performed in Example 19.2, Part 19.

| Example 20.2 (cont'd) | Calculations and Discussion | Code Reference |

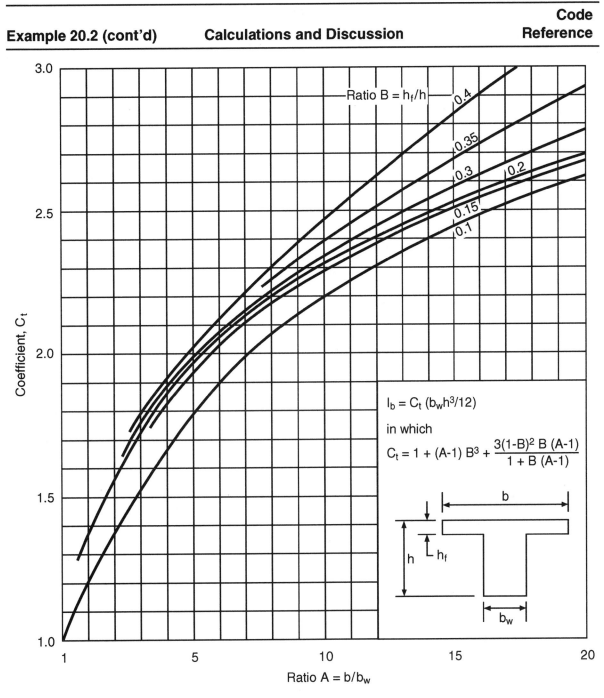

Figure 20-21 Coefficient C_t for Gross Moment of Inertia of Flanged Sections (Flange on One or Two Sides)

Blank

21

Walls

UPDATE FOR THE '05 CODE

Section 14.8.2.3 is updated to reflect the change in design approach that was introduced in 10.3 of the 2002 code. The previous requirement that the reinforcement ratio should not exceed 0.6 ρ_{bal} was replaced by the requirement that the wall be tension-controlled, leading to approximately the same reinforcement ratio.

14.1 SCOPE

Chapter 14 contains the provisions for the design of walls subjected to axial loads, with or without flexure (14.1.1). Cantilever retaining walls with minimum horizontal reinforcement according to 14.3.3 are designed according to the flexural design provisions of Chapter 10 (14.1.2).

14.2 GENERAL

According to 14.2.2, walls shall be designed in accordance with the provisions of 14.2, 14.3, and either 14.4, 14.5, or 14.8. Section 14.4 contains the requirements for walls designed as compression members using the strength design provisions for flexure and axial loads of Chapter 10. Any wall may be designed by this method, and no minimum wall thicknesses are prescribed.

Section 14.5 contains the Empirical Design Method which applies to walls of solid rectangular cross-section with resultant loads for all applicable load combinations falling within the middle third of the wall thickness at all sections along the height of the wall. Minimum thicknesses of walls designed by this method are contained in 14.5.3. Walls of nonrectangular cross-section, such as ribbed wall panels, must be designed by the provisions of 14.4, or if applicable, 14.8.

Section 14.8 contains the provisions of the Alternate Design Method, which are applicable to simply supported, axially loaded members subjected to out-of-plane uniform lateral loads, with maximum moments and deflections occurring at mid-height. Also, the wall cross-section must be constant over the height of the panel. No minimum wall thicknesses are prescribed for walls designed by this method.

All walls must be designed for the effects of shear forces. Section 14.2.3 requires that the design for shear must be in accordance with 11.10, the special shear provisions for walls. The required shear reinforcement may exceed the minimum wall reinforcement prescribed in 14.3.

For rectangular walls containing uniformly distributed vertical reinforcement and subjected to an axial load smaller than that producing balanced failure, the following approximate equation can be used to determine the design moment capacity of the wall (Ref. 21.7 and 21.8):

$$\phi M_n = \phi \left[0.5 A_{st} f_y \ell_w \left(1 + \frac{P_u}{A_{st} f_y}\right)\left(1 - \frac{c}{\ell_w}\right)\right]$$

where

A_{st} = total area of vertical reinforcement, in.²
ℓ_w = horizontal length of wall, in.
P_u = factored axial compressive load, kips
f_y = yield strength of reinforcement, ksi

$$\frac{c}{\ell_w} = \frac{\omega + \alpha}{2\omega + 0.85\beta_1}$$

β_1 = factor relating depth of equivalent rectangular compressive stress block to the neutral axis depth (10.2.7.3)

$$\omega = \left(\frac{A_{st}}{\ell_w h}\right)\frac{f_y}{f_c'}$$

f_c' = compressive strength of concrete, ksi

$$\alpha = \frac{P_u}{\ell_w h f_c'}$$

h = thickness of wall, in.
ϕ = 0.90 (strength primarily controlled by flexure with low axial load)

For a wall subjected to a series of point loads, the horizontal length of the wall that is considered effective for each concentrated load is the least of the center-to-center distance between loads and width of bearing plus four times the wall thickness (14.2.4). Columns built integrally with walls shall conform to 10.8.2 (14.2.5). Walls shall be properly anchored into all intersecting elements, such as floors, columns, other walls, and footings (14.2.6).

Section 15.8 provides the requirements for force transfer between a wall and a footing. Note that for cast-in-place walls, the required area of reinforcement across the interface shall not be less than the minimum vertical reinforcement given in 14.3.2 (15.8.2.2).

14.3   MINIMUM WALL REINFORCEMENT

The minimum wall reinforcement provisions apply to walls designed according to 14.4, 14.5, or 14.8, unless a greater amount is required to resist horizontal shear forces in the plane of the wall according to 11.10.9.

Walls must contain both vertical and horizontal reinforcement. The minimum ratio of vertical reinforcement area to gross concrete area is (1) 0.0012 for deformed bars not larger than No. 5 with $f_y \geq$ 60,000 psi, or for welded wire reinforcement (plain or deformed) not larger than W31 or D31, or (2) 0.0015 for all other deformed bars (14.3.2). The minimum ratio of horizontal reinforcement is (1) 0.0020 for deformed bars not larger than No. 5 with $f_y \geq$ 60,000 psi, or for welded wire reinforcement (plain or deformed) not larger than W31 or D31, or (2) 0.0025 for all other deformed bars (14.3.3).

The minimum wall reinforcement required by 14.3 is provided primarily for control of cracking due to shrinkage and temperature stresses. Also, the minimum vertical wall reinforcement required by 14.3.2 does not substantially increase the strength of a wall above that of a plain concrete wall. It should be noted that the reinforcement and minimum thickness requirements of 14.3 and 14.5.3 may be waived where structural analysis shows adequate strength and wall stability (14.2.7). This required condition may be satisfied by a design using the structural plain concrete provisions in Chapter 22 of the code.

For walls thicker than 10 in., except for basement walls, reinforcement in each direction shall be placed in two layers (14.3.4).

Spacing of vertical and horizontal reinforcement shall not exceed 18 in. nor three times the wall thickness (14.3.5).

According to 14.3.6, lateral ties for vertical reinforcement are not required as long as the vertical reinforcement is not required as compression reinforcement or the area of vertical reinforcement does not exceed 0.01 times the gross concrete area.

A minimum of two No. 5 bars shall be provided around all window and door openings, with minimum bar extension beyond the corner of opening equal to the greater of bar development length or 24 in. (14.3.7).

14.4 WALLS DESIGNED AS COMPRESSION MEMBERS

When the limitations of 14.5 or 14.8 are not satisfied, walls must be designed as compression members by the strength design provisions in Chapter 10 for flexure and axial loads. The minimum reinforcement requirements of 14.3 apply to walls designed by this method. Vertical wall reinforcement need not be enclosed by lateral ties (as for columns) when the conditions of 14.3.6 are satisfied. All other code provisions for compression members apply to walls designed by Chapter 10.

As with columns, the design of walls is usually difficult without the use of design aids. Wall design is further complicated by the fact that slenderness is a consideration in practically all cases. A second-order analysis, which takes into account variable wall stiffness, as well as the effects of member curvature and lateral drift, duration of the loads, shrinkage and creep, and interaction with the supporting foundation, is specified in 10.10.1. In lieu of that procedure, the approximate evaluation of slenderness effects prescribed in 10.11 may be used (10.10.2).

It is important to note that Eqs. (10-11) and (10-12) for EI in the approximate slenderness method were not originally derived for members with a single layer of reinforcement. For members with a single layer of reinforcement, the following expression for EI has been suggested in Ref. 21.2:

$$EI = \frac{E_c I_g}{\beta}\left(0.5 - \frac{e}{h}\right) \geq 0.1 \frac{E_c I_g}{\beta} \qquad \text{Eq. (1)}$$

$$\leq 0.4 \frac{E_c I_g}{\beta}$$

where
E_c = modulus of elasticity of concrete
I_g = moment of inertia of gross concrete section about the centroidal axis, neglecting reinforcement
e = eccentricity of the axial loads and lateral forces for all applicable load combinations
h = overall thickness of wall
$\beta = 0.9 + 0.5\beta_d^2 - 12\rho \geq 1.0$
β_d = ratio of dead load to total load
ρ = ratio of area of vertical reinforcement to gross concrete area

The definition of β_d, included in Eqs. (10-11) and (10-12) for EI, depends on the frame being non-sway or sway. According to 10.0, β_d for non-sway frames is the ratio of the maximum factored axial sustained load to the maximum factored axial load associated with the same load combination. For consistency, the same definition of β_d seems appropriate for the EI expressions for walls in Eq. (1). Note that if it is determined by the provisions of 10.11.4 that a sway condition exists, $\beta_d = 0$ for the case of lateral loads that are not sustained (10.0).

Figure 21-1 shows the comparison of flexural stiffness (EI) by Code Eq. (10-12) and Eq. (1) in terms of E_cI_g. The ratio of EI/E_cI_g is plotted as a function of e/h for several values of β_d, for a constant reinforcement ratio ρ of 0.0015. Note that Code Eq. (10-12) assumes EI to be independent of e/h and appears to overestimate the wall stiffness for larger eccentricities. For walls designed by Chapter 10 with slenderness evaluation by 10.11, Eq. (1) is recommended in lieu of Code Eq. (10-12) for determining wall stiffness. Example 21.1 illustrates this method for a tilt-up wall panel.

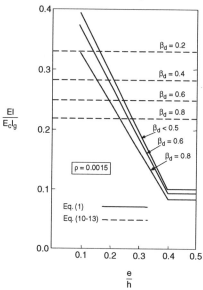

Figure 21-1 Stiffness EI of Walls

When wall slenderness exceeds the limit for application of the approximate slenderness evaluation method of 10.11 ($k\ell_u/r > 100$, i.e. $k\ell_u/h > 30$), 10.10.1 must be used to determine the slenderness effects (10.11.5). The wall panels currently used in some building systems, especially in tilt-up wall construction, usually fall in this high slenderness category. The slenderness analysis must account for the influence of variable wall stiffness, the effects of deflections on the moments and forces, and the effects of load duration.

14.5   EMPIRICAL DESIGN METHOD

The Empirical Design Method may be used for the design of walls if the resultant of all applicable loads falls within the middle one-third of the wall thickness (eccentricity e ≤ h/6), and the thickness is at least the minimum prescribed in 14.5.3 (see Fig. 21-2). Note that in addition to any eccentric axial loads, the effect of any lateral loads on the wall must be included to determine the total eccentricity of the resultant load. The method applies only to walls of solid rectangular cross-section.

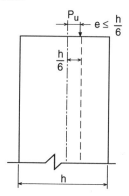

Figure 21-2 Design of Walls by Empirical Design Method (14.5)

Primary application of this method is for relatively short or squat walls subjected to vertical loads only. Application becomes extremely limited when lateral loads need to be considered, because the total load eccentricity must not exceed h/6. Walls not meeting these criteria must be designed as compression members for axial load and flexure by the provisions of Chapter 10 (14.4) or, if applicable, by the Alternate Design Method of 14.8.

When the total eccentricity e does not exceed h/6, the design is performed considering P_u as a concentric axial load. The factored axial load P_u must be less than or equal to the design axial load strength ϕP_n computed by Eq. (14-1):

$$P_u \leq \phi P_n$$

$$\leq 0.55 \phi f'_c A_g \left[1 - \left(\frac{k \ell_c}{32h} \right)^2 \right] \qquad \text{Eq. (14-1)}$$

where
- ϕ = strength reduction factor 0.65(h) corresponding to compression-controlled sections in accordance with 9.3.2.2.
- A_g = gross area of wall section
- k = effective length factor defined in 14.5.2
- ℓ_c = vertical distance between supports

Equation (14-1) takes into consideration both load eccentricity and slenderness effects. The eccentricity factor 0.55 was originally selected to give strengths comparable to those given by Chapter 10 for members with axial load applied at an eccentricity not to exceed h/6.

In order to use Eq. (14-1), the wall thickness h must not be less than 1/25 times the supported length or height, whichever is shorter, nor less than 4 in. (14.5.3.1). Exterior basement walls and foundation walls must be at least 7-1/2 in. thick (14.5.3.2).

With the publication of the 1980 supplement of ACI 318, Eq. (14-1) was modified to reflect the general range of end conditions encountered in wall design, and to allow for a wider range of design applications. The wall strength equation in previous codes was based on the assumption that the top and bottom ends of the wall are restrained against lateral movement, and that rotation restraint exists at one end, so as to have an effective length factor between 0.8 and 0.9. Axial load strength values could be unconservative for pinned-pinned end conditions, which can exist in certain walls, particularly of precast and tilt-up applications. Axial strength could also be overestimated where the top end of the wall is free and not braced against translation. In these cases, it is necessary to reflect the proper effective length in the design equation. Equation (14-1) allows the use of different effective length factors k to address this situation. The values of k have been specified in 14.5.2 for commonly occurring wall end conditions. Equation (14-1) will give the same results as the 1977 Code Eq. (14-1) for walls braced against translation at both ends and with reasonable base restraint against rotation. Reasonable base restraint against rotation implies attachment to a member having a flexural stiffness EI/ℓ at least equal to that of the wall. Selection of the proper k for a particular set of support end conditions is left to the judgment of the engineer.

Figure 21-3 shows typical axial load-moment strength curves for 8-, 10-, and 12-in. walls with f'_c = 4,000 psi and f_y = 60,000 psi.[21.3] The curves yield eccentricity factors (ratios of strength under eccentric loading to that under concentric loading) of 0.562, 0.568, and 0.563 for the 8-, 10-, and 12-in. walls with e = h/6 and ρ = 0.0015.

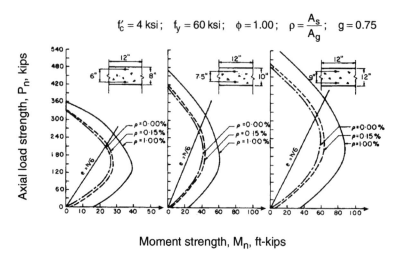

Figure 21-3 Typical Load-Moment Strength Curves for 8-, 10-, and 12-in. Walls

Figure R14.5 in the Commentary shows a comparison of the strengths obtained from the Empirical Design Method and Sect. 14.4 for members loaded at the middle third of the thickness with different end conditions.

Example 21.2 illustrates application of the Empirical Design Method to a bearing wall supporting precast floor beams.

14.8 ALTERNATE DESIGN OF SLENDER WALLS

The alternate design method for walls is based on the experimental research reported in Ref. 21.4. This method has appeared in the Uniform Building Code (UBC) since 1988, and is contained in the 2003 International Building Code (IBC)[21.5]. It is important to note that the provisions of 14.8 differ from those in the UBC and IBC in the following ways: (1) nomenclature and wording has been changed to comply with ACI 318 style, (2) the procedure has been limited to out-of-plane flexural effects on simply supported wall panels with maximum moments and deflections occurring at midspan, and (3) the procedure has been made as compatible as possible with the provisions of 9.5.2.3 for obtaining the cracking moment and the effective moment of inertia.

According to 14.8.1, the provisions of 14.8 are considered to satisfy 10.10 when flexural tension controls the design of a wall. The following limitations apply to the alternate design method (14.8.2):

1. The wall panel shall be simply supported, axially loaded, and subjected to an out-of-plane uniform lateral load. The maximum moments and deflections shall occur at the mid-height of the wall (14.8.2.1).
2. The cross-section is constant over the height of the panel (14.8.2.2).
3. The wall cross sections shall be tension-controlled.
4. Reinforcement shall provide a design moment strength ϕM_n greater than or equal to M_{cr}, where M_{cr} is the moment causing flexural cracking due to the applied lateral and vertical loads. Note that M_{cr} shall be obtained using the modulus of rupture f_r given by Eq. (9-10) (14.8.2.4).
5. Concentrated gravity loads applied to the wall above the design flexural section shall be distributed over a width equal to the lesser of (a) the bearing width plus a width on each side that increases at a slope of 2 vertical to 1 horizontal down to the design flexural section or (b) the spacing of the concentrated loads. Also, the distribution width shall not extend beyond the edges of the wall panel (14.8.2.5) (see Fig. 21-4).
6. The vertical stress P_u/A_g at the mid-height section shall not exceed $0.06 f'_c$ (14.8.2.6).

When one or more of these conditions are not satisfied, the wall must be designed by the provisions of 14.4.

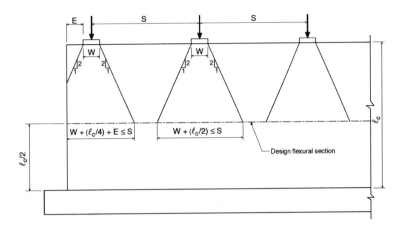

Figure 21-4 Distribution Width of Concentrated Gravity Loads (14.8.2.5)

According to 14.8.3, the design moment strength ϕM_n for combined flexure and axial loads at the mid-height cross-section must be greater than or equal to the total factored moment M_u at this section. The factored moment M_u includes P-Δ effects and is defined as follows:

$$M_u = M_{ua} + P_u \Delta_u \qquad \text{Eq. (14-4)}$$

where M_{ua} = factored moment at the mid-height section of the wall due to factored lateral and eccentric vertical loads

P_u = factored axial load

Δ_u = deflection at the mid-height of the wall due to the factored loads

$$= 5 M_u \ell_c^2 / (0.75) 48 E_c I_{cr} \qquad \text{Eq. (14-5)}$$

ℓ_c = vertical distance between supports

E_c = modulus of elasticity of concrete (8.5)

I_{cr} = moment of inertia of cracked section transformed to concrete

$$= n A_{se}(d - c)^2 + (\ell_w c^3 / 3) \qquad \text{Eq. (14-7)}$$

n = modular ratio of elasticity = $E_s / E_c \geq 6$

E_s = modulus of elasticity of nonprestressed reinforcement

A_{se} = area of effective longitudinal tension reinforcement in the wall segment

$$= (P_u + A_s f_y)/f_y \qquad \text{Eq. (14-8)}$$

A_s = area of longitudinal tension reinforcement in the wall segment

f_y = specified yield stress of nonprestressed reinforcement

d = distance from extreme compression fiber to centroid of longitudinal tension reinforcement

c = distance from extreme compression fiber to neutral axis

ℓ_w = horizontal length of the wall

Note that Eq. (14-4) includes the effects of the factored axial loads and lateral load (M_{ua}), as well as the P-Δ effects ($P_u \Delta_u$).

Substituting Eq. (14-5) for Δ_u into Eq. (14-4) results in the following equation for M_u:

$$M_u = \frac{M_{ua}}{1 - \frac{5P_u \ell_c^2}{(0.75)48E_c I_{cr}}}$$ Eq. (14-6)

Figure 21-5 shows the analysis of the wall according to the provisions of 14.8 for the case of additive lateral and gravity load effects.

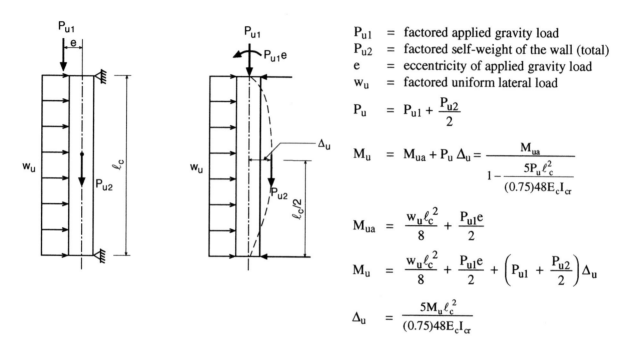

P_{u1} = factored applied gravity load
P_{u2} = factored self-weight of the wall (total)
e = eccentricity of applied gravity load
w_u = factored uniform lateral load

$$P_u = P_{u1} + \frac{P_{u2}}{2}$$

$$M_u = M_{ua} + P_u \Delta_u = \frac{M_{ua}}{1 - \frac{5P_u \ell_c^2}{(0.75)48E_c I_{cr}}}$$

$$M_{ua} = \frac{w_u \ell_c^2}{8} + \frac{P_{u1} e}{2}$$

$$M_u = \frac{w_u \ell_c^2}{8} + \frac{P_{u1} e}{2} + \left(P_{u1} + \frac{P_{u2}}{2}\right)\Delta_u$$

$$\Delta_u = \frac{5M_u \ell_c^2}{(0.75)48E_c I_{cr}}$$

Figure 21-5 Analysis of Wall According to 14.8

The design moment strength ϕM_n of the wall can be determined from the following equation:

$$\phi M_n = \phi A_{se} f_y \left(d - \frac{a}{2}\right)$$ Eq. (2)

where

$$a = \frac{A_{se} f_y}{0.85 f'_c \ell_w}$$

and ϕ is determined in accordance with 9.3.2.

In addition to satisfying the strength requirement of Eq. (14-3), the deflection requirement of 14.8.4 must also be satisfied. In particular, the maximum deflection Δ_s due to service loads, including P-Δ effects, shall not exceed $\ell_c/150$, where Δ_s is:

$$\Delta_s = \frac{5M\ell_c^2}{48 E_c I_e}$$ Eq. (14-9)

where   M = maximum unfactored moment due to service loads, including P-Δ effects

$$= \frac{M_{sa}}{1 - \frac{5P_s \ell_c^2}{48 E_c I_e}} \qquad \text{Eq. (14-10)}$$

and M_{sa} = maximum unfactored applied moment due to service loads, not including P-Δ effects

P_s = unfactored axial load at the design (mid-height) section including effects of self-weight

I_e = effective moment of inertia evaluated using the procedure of 9.5.2.3, substituting M for M_a.

It is important to note that Eq. (14-10) does not provide a closed form solution for M, since I_e is a function of M. Thus, an iterative process is required to determine Δ_s.

Example 21.3 illustrates the design of a nonprestressed precast wall panel by the alternated design method.

11.10   SPECIAL SHEAR PROVISIONS FOR WALLS

For most low-rise buildings, horizontal shear forces acting in the plane of walls are small, and can usually be neglected in design. Such in-plane forces, however, become an important design consideration in high-rise buildings. Design for shear shall be in accordance with the special provisions for walls in 11.10 (14.2.3). Example 21.4 illustrates in-plane shear design of walls, including design for flexure.

DESIGN SUMMARY

A trial procedure for wall design is suggested: first assume a wall thickness h and a reinforcement ratio ρ. Based on these assumptions, check the trial wall for the applied loading conditions.

It is not within the scope of Part 21 to include design aids for a broad range of wall and loading conditions. The intent is to present examples of various design options and aids. The designer can, with reasonable effort, produce design aids to fit the range of conditions usually encountered in practice. For example, strength interaction diagrams such as those plotted in Fig. 21-6(a) (ρ = 0.0015) and Fig. 21-6(b) (ρ = 0.0025) can be helpful design aids for evaluation of wall strength. The lower portions of the strength interaction diagrams are also shown for 6.5-in. thick walls. Design charts, such as the one shown in Fig. 21-7 can also be developed for specific walls. Figure 21-8 may be used to select wall reinforcement.

Prestressed walls are not covered specifically in Part 21. Prestressing of walls is advantageous for handling (precast panels) and for increased buckling resistance. For design of prestressed walls, the designer should consult Ref. 21.6.

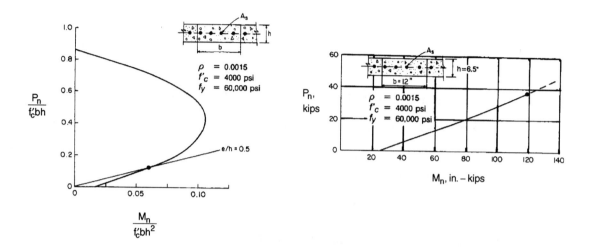

(a) Reinforcement Ratio $\rho = 0.0015$

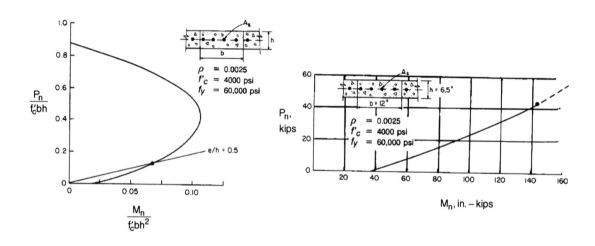

(b) Reinforcement Ratio $\rho = 0.0025$

Figure 21-6 Axial Load-Moment Interaction Diagram for Walls (f'_c = 4000 psi, f_y = 60 ksi)

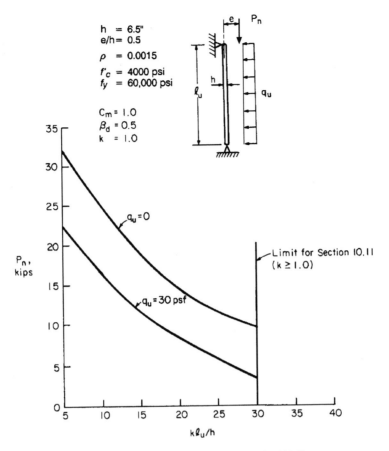

Figure 21-7 Design Chart for 6.5-in. Wall

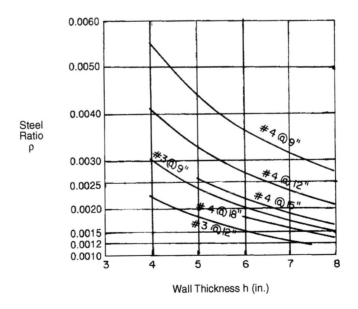

Figure 21-8 Design Aid for Wall Reinforcement

21-11

REFERENCES

21.1 *Uniform Building Code*, Vol. 2, International Conference of Building Officials, Whittier, CA, 1997.

21.2 MacGregor, J.G., "Design and Safety of Reinforced Concrete Compression Members," paper presented at International Association for Bridge and Structural Engineering Symposium, Quebec, 1974.

21.3 Kripanaryanan, K.M., "Interesting Aspects of the Empirical Wall Design Equation," *ACI Journal*, Proceedings Vol. 74, No. 5, May 1977, pp. 204-207.

21.4 Athey, J.W., Ed., "Test Report on Slender Walls," Southern California Chapter of the American Concrete Institute and Structural Engineers Association of Southern California, Los Angeles, CA, 1982.

21.5 *2003 International Building Code*, International Code Council, Falls Church, VA, 2000.

21.6 *PCI Design Handbook - Precast and Prestressed Concrete*, 5th Edition, Prestressed Concrete Institute, Chicago, IL, 1999.

21.7 Iyad M. Alsamsam and Mahmoud E. Kamara. *Simplified Design: Reinforced Concrete Buildings of Moderate Size and Height,* Portland Cement Association, EB104, 2004, pp 6-11.

21.8 Alex E. Cardenas, and Donald D. Magura, *Strength of High-Rise Shear Walls-Rectangular Cross Sections, Response of Multistory Concrete Structures to Lateral Forces,* SP-36, American Concrete Institute, Farmington Hills, MI, 1973, pp 119-150.

Example 21.1—Design of Tilt-up Wall Panel by Chapter 10 (14.4)

Design of the wall shown is required. The wall is restrained at the top edge, and the roof load is supported through 4 in. tee stems spaced at 4 ft on center.

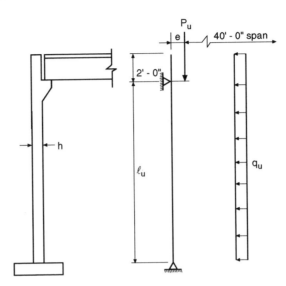

Design data:

Roof dead load = 50 psf
Roof live load = 20 psf
Wind load = 20 psf
Unsupported length of wall ℓ_u = 16 ft
Effective length factor k = 1.0 (pinned-pinned end condition)
Concrete f'_c = 4,000 psi (w_c = 150 pcf)
Reinforcing steel f_y = 60,000 psi
Assume non-sway condition.

Calculations and Discussion	Code Reference

1. Trial wall selection

    Try h = 6.5 in. with assumed e = 6.75 in.

    Try a single layer of No. 4 @ 12 in. vertical reinforcement (A_s = 0.20 in.^{2}/ft) at centerline of wall

    For a 1-ft wide design strip:

    $$\rho_\ell = \frac{A_s}{bh} = \frac{0.20}{(12 \times 6.5)} = 0.0026 > 0.0012 \quad \text{O.K.}$$

    14.3.2 (a)

2. Effective wall length for roof reaction

    Bearing width + 4 (wall thickness) = 4 + 4 (6.5) = 30 in. = 2.5 ft (governs)

    Center-to-center distance between stems = 4 ft

    14.2.4

| Example 21.1 (cont'd) | Calculations and Discussion | Code Reference |

3. Roof loading per foot width of wall

$$\text{Dead load} = \left[50 \times \left(\frac{4}{2.5}\right)\right] \times \frac{40}{2} = 1{,}600 \text{ plf}$$

$$\text{Live load} = \left[20 \times \left(\frac{4}{2.5}\right)\right] \times \frac{40}{2} = 640 \text{ plf}$$

$$\text{Wall dead load at mid-height} = \frac{6.5}{12} \times \left(\frac{16}{2} + 2\right) \times 150 = 813 \text{ plf}$$

4. Factored load combinations

   Load comb. 1: $U = 1.2D + 0.5L_r$ Eq. (9-2)
   $P_u = 1.2(1.6 + 0.81) + 0.5(0.64) = 2.9 + 0.3 = 3.2$ kips
   $M_u = 1.2(1.6 \times 6.75) + 0.5(0.64 \times 6.75) = 15.1$ in.-kips
   $\beta_d = 2.9/3.2 = 0.91$

   Load comb. 2: $U = 1.2D + 1.6L_r + 0.8W$ Eq. (9-3)
   $P_u = 1.2(1.6 + 0.81) + 1.6(0.64) + 0 = 2.9 + 1.0 = 3.9$ kips
   $M_u = 1.2(1.6 \times 6.75) + 1.6(0.64 \times 6.75) + 0.8(0.02 \times 16^2 \times 12/8)$
   $= 26.0$ in.-kips
   $\beta_d = 2.9/3.9 = 0.74$

   Load comb. 3: $U = 1.2D + 1.6W + 0.5L_r$ Eq. (9-4)
   $P_u = 1.2(1.6 + 0.81) + 0 + 0.5(0.64) = 3.2$ kips
   $M_u = 1.2(1.6 \times 6.75) + 1.6(0.02 \times 16^2 \times 12/8) + 0.5(0.64 \times 6.75)$
   $= 27.4$ in.-kips
   $\beta_d = 2.9/3.2 = 0.91$

   Load comb. 4: $U = 0.9D + 1.6W$ Eq. (9-6)
   $P_u = 0.9(1.6 + 0.81) + 0 = 2.2$ kips
   $M_u = 0.9(1.6 \times 6.75) + 1.6(0.02 \times 16^2 \times 12/8)$
   $= 22.0$ in.-kips
   $\beta_d = 2.2/2.2 = 1.0$

5. Check wall slenderness

$$\frac{k\ell_u}{r} = \frac{1.0(16 \times 12)}{(0.3 \times 6.5)} = 98.5 < 100$$ 10.11.5

where $r = 0.3h$ 10.11.12

Therefore, 10.11 may be used to account for slenderness effects.

		Code
Example 21.1 (cont'd)	**Calculations and Discussion**	**Reference**

6. Calculate magnified moments for non-sway case   10.12

$$M_c = \delta_{ns} M_2$$ Eq. (10-8)

$$\delta_{ns} = \frac{C_m}{1 - \left(\dfrac{P_u}{0.75 P_c}\right)} \geq 1$$ Eq. (10-9)

$$P_c = \frac{\pi^2 EI}{(k\ell_u)^2}$$ Eq. (10-10)

$$EI = \frac{E_c I_g}{\beta}\left(0.5 - \frac{e}{h}\right) \geq 0.1 \frac{E_c I_g}{\beta}$$ Eq. (1)

$$\leq 0.4 \frac{E_c I_g}{\beta}$$

$$\frac{e}{h} = \frac{6.75}{6.5} = 1.04 > 0.5$$

Thus, $EI = 0.1 \left(\dfrac{E_c I_g}{\beta}\right)$

$E_c = 57{,}000\sqrt{4000} = 3.605 \times 10^6$ psi   8.5.1

$$I_g = \frac{12 \times 6.5^3}{12} = 274.6 \text{ in.}^4$$

$\beta = 0.9 + 0.5\beta_d^2 - 12\rho \geq 1.0$

$ = 0.9 + 0.5\beta_d^2 - 12(0.0026)$

$ = 0.869 + 0.5\beta_d^2 \geq 1.0$

$$EI = \frac{0.1 \times 3.605 \times 10^6 \times 274.6}{\beta} = \frac{99 \times 10^6}{\beta} \text{ lb-in.}^2$$

$$P_c = \frac{\pi^2 \times 99 \times 10^6}{\beta(16 \times 12)^2 \times 1000} = \frac{26.5}{\beta} \text{ kips}$$

Example 21.1 (cont'd) — Calculations and Discussion

Code Reference

$C_m = 1.0$ for members with transverse loads between supports — 10.12.3.1

Determine magnified moment M_c for each load case.

Load Comb.	P_u (kips)	$M_2 = M_u$ (in.-kips)	β_d	β	EI (lb-in.²)	P_c (kips)	δ_{ns}	M_c (in.-kips)
1	3.2	15.1	0.91	1.28	77 x 10⁶	20.7	1.26	19.0
2	3.9	26.0	0.74	1.14	87 x 10⁶	23.2	1.29	33.5
3	3.2	27.4	0.91	1.28	77 x 10⁶	20.7	1.26	34.5
4	2.2	22.0	1.00	1.37	72 x 10⁶	19.3	1.18	26.0

7. Check design strength vs. required strength

   Assume that the section is tension-controlled for each load combination, i.e., $\varepsilon_t \geq 0.005$ and $\phi = 0.90$. — 10.3.4, 9.3.2

   The following table contains a summary of the strain compatibility analysis for each load combination, based on the assumption above:

Load Comb.	$P_n = P_u/\phi$ (kips)	a (in.)	c (in.)	ε_t (in./in.)
1	3.6	0.38	0.45	0.0187
2	4.3	0.40	0.47	0.0177
3	3.6	0.38	0.45	0.0187
4	2.4	0.35	0.42	0.0205

   For example, the strain in the reinforcement ε_t is computed for load combination No. 2 as follows:

   $P_n = 0.85 f'_c ba - A_s f_y$ — 10.3.1, 10.2.1

   $4.3 = (0.85)(4)(12)a - (0.2)(60) = 40.8a - 12$

   $a = 0.40$ in.

   $c = a/\beta_1 = 0.4/0.85 = 0.47$ in. — 10.2.7.1, 10.2.7.3

   $\varepsilon_t = \dfrac{0.003}{c}(d - c)$ — 10.2.2

   $= \dfrac{0.003}{0.47}(3.25 - 0.47)$

   $= 0.0177 > 0.0050 \rightarrow$ tension-controlled section — 10.3.4

Example 21.1 (cont'd)   **Calculations and Discussion**   **Code Reference**

Note that the strain in the reinforcement for each of the load combinations is greater than 0.0050, so that the assumption of tension-controlled sections ($\phi = 0.90$) is correct.

For each load combination, the required nominal strength will be compared to the computed design strength. The results are tabulated below.

Load Comb.	Required Nominal Strength		Design Strength M_n (in.-kips)
	$P_n = P_u/\phi$ (kips)	$M_n = M_c/\phi$ (in.-kips)	
1	3.6	21.1	47.7
2	4.3	37.2	49.7
3	3.6	38.3	47.7
4	2.4	28.9	44.2

For example, the design strength M_n is computed for load combination No. 2 as follows:

$$M_n = 0.85 f'_c ba \left(\frac{h}{2} - \frac{a}{2}\right) - A_s f_y \left(\frac{h}{2} - d_t\right)$$

$$= 0.85(4)(12)(0.40)\left(\frac{6.5}{2} - \frac{0.40}{2}\right) - 0.2(60)\left(\frac{6.5}{2} - 3.25\right)$$

$$= 49.7 \text{ in.-kips}$$

The wall is adequate with the No. 4 @ 12 in. since the design strength is greater than the required nominal strength for all load combinations.

This conclusion can also be verified by utilizing pcaColumn program. Figure 21-4 shows the interaction diagram for the wall cross section with the applied factored loads

Example 21.1 (cont'd)	Calculations and Discussion	Code Reference

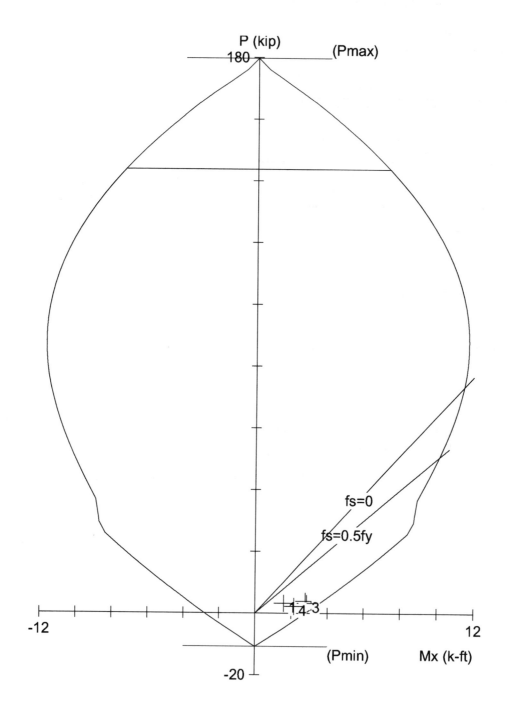

Figure 21-4 Interaction Diagram Generated Using pcaColumn Program

Example 21.2—Design of Bearing Wall by Empirical Design Method (14.5)

A concrete bearing wall supports a floor system of precast single tees spaced at 8 ft on centers. The stem of each tee section is 8 in. wide. The tees have full bearing on the wall. The height of the wall is 15 ft, and the wall is considered laterally restrained at the top.

Design Data:

Floor beam reactions: dead load = 28 kips
live load = 14 kips

$f'_c = 4000$ psi
$f_y = 60,000$ psi

Neglect weight of wall

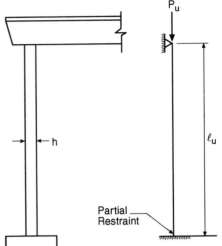

Calculations and Discussion	Code Reference

The general design procedure is to select a trial wall thickness h, then check the trial wall for the applied loading conditions.

1. Select trial wall thickness h

    $h \geq \dfrac{\ell_u}{25}$ but not less than 4 in.   14.5.3.1

    $\geq \dfrac{15 \times 12}{25} = 7.2$ in.

    Try h = 7.5 in.

2. Calculate factored loading

    $P_u = 1.2D + 1.6L$ Eq. (9-2)

    $= 1.2(28) + 1.6(14) = 33.6 + 22.4 = 56.0$ kips

3. Check bearing strength of concrete

    Assume width of stem for bearing equal to 7 in., to allow for beveled bottom edges.

    Loaded area $A_1 = 7 \times 7.5 = 52.5$ in.^{2}

    Bearing capacity $= \phi(0.85 f'_c A_1) = 0.65(0.85 \times 4 \times 52.5) = 116$ kips > 56.0 kips   O.K.   10.17.1

| Example 21.2 (cont'd) | Calculations and Discussion | Code Reference |

4. Calculate design strength of wall

   Effective horizontal length of wall per tee reaction = $\begin{cases} 8 \times 12 = 96 \text{ in.} \\ 7 + 4(7.5) = 37 \text{ in.} \quad (\text{governs}) \end{cases}$ 14.2.4

   $k = 0.8$ 14.5.2

   $$\phi P_n = 0.55 \phi f'_c A_g \left[1 - \left(\frac{k\ell_c}{32h} \right)^2 \right]$$ Eq. (14-1)

   $$= 0.55 \times 0.70 \times 4(37 \times 7.5) \left[1 - \left(\frac{0.8 \times 15 \times 12}{32 \times 7.5} \right)^2 \right]$$

   $= 273 \text{ kips} > 56 \text{ kips} \quad \text{O.K.}$

   The 7.5-in. wall is adequate, with sufficient margin for possible effect of load eccentricity.

5. Determine single layer of reinforcement

   Based on 1-ft width of wall and Grade 60 reinforcement (No. 5 and smaller):

   Vertical $A_s = 0.0012 \times 12 \times 7.5 = 0.108 \text{ in.}^2/\text{ft}$ 14.3.2

   Horizontal $A_s = 0.0020 \times 12 \times 7.5 = 0.180 \text{ in.}^2/\text{ft}$ 14.3.3

   Spacing = $\begin{cases} 3h = 3 \times 7.5 = 22.5 \text{ in.} \\ 18 \text{ in.} \quad (\text{governs}) \end{cases}$ 14.3.5

   Vertical A_s: use No. 4 @ 18 in. on center ($A_s = 0.13 \text{ in.}^2/\text{ft}$)

   Horizontal A_s: use No. 4 @ 12 in. on center ($A_s = 0.20 \text{ in.}^2/\text{ft}$)

   Design aids such as the one in Fig. 21-8 may be used to select reinforcement directly.

Example 21.3—Design of Precast Panel by the Alternate Design Method (14.8)

Determine the required vertical reinforcement for the precast wall panel shown below. The roof loads are supported through the 3.75 in. webs of the 10DT24 which are spaced 5 ft on center.

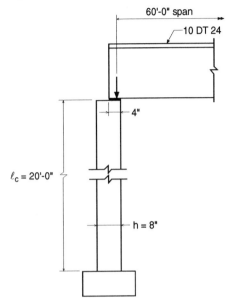

Design data:

  Weight of 10DT24 = 468 plf
  Roof dead load = 20 psf
  Roof live load = 30 psf
  Wind load = 30 psf
  Concrete f'_c = 4000 psi (w_c = 150 pcf)
  Reinforcing steel f_y = 60,000 psi

Calculations and Discussion	Code Reference

1. Trial wall section

   Try h = 8 in.
   Try a single layer of No. 4 @ 9 in. vertical reinforcement (A_s = 0.27 in.^{2}/ft) at centerline of wall.

   For a 1-ft wide design strip: ρ (gross) = $\dfrac{A_s}{\ell_w h} = \dfrac{0.27}{12 \times 8} = 0.0028 > 0.0012$ O.K.   14.3.2

2. Distribution width of interior concentrated loads at mid-height of wall (see Fig. 21-4)

   $W + \dfrac{\ell_c}{2} = \dfrac{3.75}{12} + \dfrac{20}{2} = 10.3$ ft   14.8.2.5

   S = 5.0 ft (governs)

Example 21.3 (cont'd)	Calculations and Discussion	Code Reference

3. Roof loading per foot width of wall

$$\text{Dead load} = \left[\frac{468}{2} + (20 \times 5)\right]\left(\frac{60}{2}\right) = 10{,}020 \text{ lbs/5 ft} = 2{,}004 \text{ plf}$$

$$\text{Live load} = (30 \times 5)\left(\frac{60}{2}\right) = 4{,}500 \text{ lbs/5 ft} = 900 \text{ plf}$$

$$\text{Wall dead load} = \frac{8}{12} \times 20 \times 150 = 2{,}000 \text{ plf}$$

Eccentricity of the roof loads about the panel center line $= \frac{2}{3} \times 4 = 2.7$ in.

4. Factored load combinations at mid-height of wall (see Fig. 21-5)

   a. Load comb. 1: $U = 1.2D + 0.5L_r$ — Eq. (9-2)

   $$P_u = P_{u1} + \frac{P_{u2}}{2}$$

   $$P_{u1} = (1.2 \times 2.0) + (0.5 \times 0.9) = 2.4 + 0.5 = 2.9 \text{ kips}$$

   $$P_{u2} = 1.2 \times 2.0 = 2.4 \text{ kips}$$

   $$P_u = 2.9 + \frac{2.4}{2} = 4.1 \text{ kips}$$

   $$M_u = \frac{M_{ua}}{1 - \frac{5P_u \ell_c^2}{(0.75)\, 48 E_c I_{cr}}}$$ — Eq. (14-6)

   $$M_{ua} = \frac{w_u \ell_c^2}{8} + \frac{P_{u1} e}{2} = 0 + \frac{2.9 \times 2.7}{2} = 3.9 \text{ in.-kips}$$

   $$E_c = 57{,}000 \sqrt{4000} = 3{,}605{,}000 \text{ psi}$$ — 8.5.1

   $$I_{cr} = n A_{se}(d - c)^2 + \frac{\ell_w c^3}{3}$$ — Eq. (14-7)

   $$n = \frac{E_s}{E_c} = \frac{29{,}000}{3605} = 8.0$$

   $$A_{se} = \frac{P_u + A_s f_y}{f_y} = \frac{4.1 + (0.27 \times 60)}{60} = 0.34 \text{ in.}^2/\text{ft}$$ — Eq. (14-8)

21-22

		Code
Example 21.3 (cont'd)	**Calculations and Discussion**	**Reference**

$$a = \frac{A_{se}f_y}{0.85f'_c\ell_w} = \frac{0.34 \times 60}{0.85 \times 4 \times 12} = 0.50 \text{ in.}$$

$$c = \frac{a}{\beta_1} = \frac{0.50}{0.85} = 0.59 \text{ in.}$$

Therefore,

$$I_{cr} = 8.0 \times 0.34 \times (4 - 0.59)^2 + \frac{12 \times 0.59^3}{3} = 32.5 \text{ in.}^4$$

$$\varepsilon_t = \left(\frac{0.003}{c}\right)d_t - 0.003$$

$$= \left(\frac{0.003}{0.59}\right)(4) - 0.003 = 0.0173 > 0.005$$

Therefore, section is tension-controlled   10.3.4

$$\phi = 0.9 \qquad\qquad 9.3.2$$

$$M_u = \frac{3.9}{1 - \frac{5 \times 4.1 \times (20 \times 12)^2}{0.75 \times 48 \times 3,605 \times 32.5}} = 5.4 \text{ in.-kips} \qquad\qquad Eq.\ (14\text{-}6)$$

b. Load comb. 2: $U = 1.2D + 1.6L_r + 0.8W$ *Eq. (9-3)*

$$P_{u1} = (1.2 \times 2) + (1.6 \times 0.9) = 3.8 \text{ kips}$$

$$P_{u2} = 1.2 \times 2.0 = 2.4 \text{ kips}$$

$$P_u = 3.8 + \frac{2.4}{2} = 5.0 \text{ kips}$$

$$M_{ua} = \frac{w_u \ell_c^2}{8} + \frac{P_{u1}e}{2} = \frac{0.8 \times 0.030 \times 20^2}{8} + \frac{3.8 \times (2.7/12)}{2}$$

$$= 1.2 + 0.4 = 1.6 \text{ ft-kips} = 19.2 \text{ in.-kips}$$

$$A_{se} = \frac{5.0 + (0.27 \times 60)}{60} = 0.35 \text{ in.}^2/\text{ft} \qquad\qquad Eq.\ (14\text{-}8)$$

$$a = \frac{0.35 \times 60}{0.85 \times 4 \times 12} = 0.51 \text{ in.}$$

Example 21.3 (cont'd)   **Calculations and Discussion**   **Code Reference**

$$c = \frac{0.51}{0.85} = 0.60 \text{ in.}$$

Therefore,

$$I_{cr} = 8.0 \times 0.35 \times (4 - 0.60)^2 + \frac{12 \times 0.60^3}{3} = 33.2 \text{ in.}^4 \qquad \text{Eq. (14-7)}$$

$$\varepsilon_t = \left(\frac{0.003}{0.60}\right)(4) - 0.003 = 0.0170 > 0.005$$

$$\phi = 0.9 \qquad\qquad 9.3.2$$

$$M_u = \frac{19.2}{1 - \frac{5 \times 5.0 \times (20 \times 12)^2}{0.75 \times 48 \times 3{,}605 \times 33.2}} = 28.8 \text{ in.-kips} \qquad \text{Eq. (14-6)}$$

c.   Load comb. 3: $U = 1.2D + 1.6W + 0.5L_r$ Eq. (9-4)

$$P_{u1} = (1.2 \times 2.0) + (0.5 \times 0.9) = 2.9 \text{ kips}$$

$$P_{u2} = 1.2 \times 2.0 = 2.4 \text{ kips}$$

$$P_u = 2.9 + \frac{2.4}{2} = 4.1 \text{ kips}$$

$$M_{ua} = \frac{1.6 \times 0.03 \times 20^2}{8} + \frac{2.9 \times (2.7/12)}{2}$$

$$= 2.4 + 0.3 = 2.7 \text{ ft-kips} = 32.4 \text{ in.-kips}$$

$$A_{se} = \frac{4.1 + (0.27 \times 60)}{60} = 0.34 \text{ in.}^2/\text{ft} \qquad \text{Eq. (14-8)}$$

$$a = \frac{0.34 \times 60}{0.85 \times 4 \times 12} = 0.5 \text{ in.}$$

$$c = \frac{0.5}{0.85} = 0.59 \text{ in.}$$

Therefore,

$$I_{cr} = 8 \times 0.34 \times (4 - 0.59)^2 + \frac{12 \times 0.59^3}{3} = 32.5 \text{ in.}^4$$

		Code
Example 21.3 (cont'd)	**Calculations and Discussion**	**Reference**

$\phi = 0.9$ as in load combination 1

$$M_u = \frac{32.4}{1 - \frac{5 \times 4.1 \times (20 \times 12)^2}{0.75 \times 48 \times 3605 \times 32.5}} = 45.0 \text{ in.-kips}$$

d. Load comb. 4: $U = 0.9D + 1.6W$ *Eq. (9-6)*

$P_{u1} = 0.9 \times 2.0 = 1.8$ kips

$P_{u2} = 0.9 \times 2.0 = 1.8$ kips

$P_u = 1.8 + \frac{1.8}{2} = 2.7$ kips

$$M_{ua} = \frac{1.6 \times 0.030 \times 20^2}{8} + \frac{1.8 \times (2.7/12)}{2} = 2.6 \text{ ft-kips} = 31.2 \text{ in.-kips}$$

$$A_{se} = \frac{2.7 + (0.27 \times 60)}{60} = 0.32 \text{ in.}^2/\text{ft}$$ *Eq. (14-8)*

$$a = \frac{0.32 \times 60}{0.85 \times 4 \times 12} = 0.47 \text{ in.}$$

$$c = \frac{0.47}{0.85} = 0.55 \text{ in.}$$

Therefore,

$$I_{cr} = 8.0 \times 0.32 \times (4 - 0.55)^2 + \frac{12 \times 0.55^3}{3} = 31.1 \text{ in.}^4$$ *Eq. (14-7)*

$$\varepsilon_t = \left(\frac{0.003}{c}\right) d_t - 0.003 = \left(\frac{0.003}{0.55}\right)(4) - 0.003 = 0.0188 > 0.005$$

$\phi = 0.9$ *9.3.2*

$$M_u = \frac{31.2}{1 - \frac{5 \times 2.7 \times (20 \times 12)^2}{0.75 \times 48 \times 3605 \times 31.1}} = 38.7 \text{ in.-kips}$$ *Eq. (14-6)*

5. Check if section is tenson-controlled.

    Assume section is tension-controlled $\phi = 0.9$ (Fig. R.9.3.2)

    $$P_n = \frac{P_u}{\phi}$$

Example 21.3 (cont'd)   **Calculations and Discussion**   **Code Reference**

Lc1: $U = 1.2D + 0.5\,L_r$
       $P_u = 4.1$ kips

Lc2: $U = 1.2D + 1.6L_r + 0.8W$
       $P_u = 5.0$ kips (controls)

Lc3: $U = 1.2D + 1.6L_r + 0.5W$
       $P_u = 4.1$ kips

Lc4: $U = 0.9D + 1.6W$
       $P_u = 2.7$ kips

$A_s = 0.27\ \text{in.}^2$

$$P_n = \frac{P_u}{\phi} = \frac{5.0}{0.9} = 5.56 \text{ kips}$$

$$a = \frac{P_u + A_s f_y}{0.85 f'_c b} = \frac{5.56 + 0.27 \times 60}{0.85 \times 4 \times 12} = \frac{21.76}{40.8} = 0.533 \text{ in.}$$

$$c = \frac{a}{0.85} = \frac{0.533}{0.85} = 0.627 \text{ in.}$$

$$\varepsilon_t = \frac{0.003}{c}(d-c) = \frac{0.003}{0.627}\left(\frac{8}{2} - 0.627\right)$$

$$= \frac{0.003}{0.627} \times 2.508$$

$$= 0.012 \geq 0.005$$

tension-controlled section.

6. Determine M_{cr}

$$I_g = \frac{1}{12}\ell_w b^3 = \frac{1}{12} \times 12 \times 8^3 = 512 \text{ in.}^4$$

$$y_t = \frac{8}{2} = 4 \text{ in.}$$

$$f_r = 7.5\sqrt{f'_c} = 7.5\sqrt{4000} = 474.3 \text{ psi} \qquad\qquad \textit{Eq. (9-9)}$$

| Example 21.3 (cont'd) | Calculations and Discussion | Code Reference |

$$M_{cr} = \frac{f_r I_g}{y_t} = \frac{474.3 \times 512}{4 \times 1000} = 60.7 \text{ in.-kips}$$

7. Check design moment strength ϕM_n

    a. Load comb. 1

    $$M_n = A_{se}f_y(d - \frac{a}{2}) = 0.34 \times 60 \times \left(4 - \frac{0.5}{2}\right) = 76.5 \text{ in.-kips}$$

    $\phi M_n = 0.9 \times 76.5 = 68.9$ in.-kips $> M_u = 5.4$ in.-kips   O.K.   14.8.3
    $\phantom{\phi M_n = 0.9 \times 76.5 = 68.9 \text{ in.-kips}} > M_{cr} = 60.7$ in.-kips   O.K.   14.8.2.4

    b. Load comb. 2

    $$M_n = 0.35 \times 60 \times \left(4 - \frac{0.51}{2}\right) = 78.7 \text{ in.-kips}$$

    $\phi M_n = 0.9 \times 78.7 = 70.8$ in.-kips $> M_u = 28.8$ in.-kips   O.K.   14.8.3
    $\phantom{\phi M_n = 0.9 \times 78.7 = 70.8 \text{ in.-kips}} > M_{cr} = 60.7$ in.-kips   O.K.   14.8.2.4

    c. Load comb. 3

    $$M_n = 0.34 \times 60 \times \left(4 - \frac{0.5}{2}\right) = 76.5 \text{ in.-kips}$$

    $\phi M_n = 0.9 \times 76.5 = 68.9$ in.-kips $> M_u = 45.0$ in.-kips   O.K.
    $\phantom{\phi M_n = 0.9 \times 76.5 = 68.9 \text{ in.-kips}} > M_{cr} = 60.7$ in.-kips   O.K.

    d. Load comb. 4

    $$M_n = 0.32 \times 60 \times \left(4 - \frac{0.47}{2}\right) = 72.3 \text{ in.-kips}$$

    $\phi M_n = 0.9 \times 72.3 = 65.1$ in.-kips $> M_u = 38.7$ in.-kips   O.K.   14.8.3
    $\phantom{\phi M_n = 0.9 \times 72.3 = 65.1 \text{ in.-kips}} > M_{cr} = 60.7$ in.-kips   O.K.   14.8.2.4

8. Check vertical stress at mid-height section

    Load comb. 2 governs:

    $$\frac{P_u}{A_g} = \frac{5000}{8 \times 12} = 52.1 \text{ psi} < 0.06 f'_c = 0.06 \times 4000 = 240 \text{ psi} \quad \text{O.K.} \qquad 14.8.2.6$$

21-27

Example 21.3 (cont'd)   **Calculations and Discussion**   **Code Reference**

9. Check mid-height deflection Δ_s

$$\Delta_s = \frac{5M\ell_c^2}{48E_c I_e}$$
Eq. (14-9)

$$M = \frac{M_{sa}}{1 - \frac{5P_s \ell_c^2}{48E_c I_e}}$$
Eq. (14-10)

Using Δ_s from Eq. (14-9), Eq. (14-10) can be rewritten as follows:

$M = M_{sa} + P_s \Delta_s$

$$M_{sa} = \frac{w\ell_c^2}{8} + \frac{P_{s1}e}{2} = \frac{0.030 \times 20^2}{8} + \frac{(2.0 + 0.9)(2.7/12)}{2} = 1.8 \text{ ft-kips} = 21.6 \text{ in.-kips}$$

$$P_s = P_{s1} + \frac{P_{s2}}{2} = (2.0 + 0.9) + \frac{2.0}{2} = 3.9 \text{ kips}$$

$$I_e = \left(\frac{M_{cr}}{M}\right)^3 I_g + \left[1 - \left(\frac{M_{cr}}{M}\right)^3\right] I_{cr}$$
Eq. (9-8)

Since I_e is a function of M, no closed form solution for Δ_s is possible. Determine Δ_s by iterative procedure.

Assume $\Delta_s = \frac{\ell_c}{150} = \frac{20 \times 12}{150} = 1.6$ in.

$M = 21.6 + (3.9 \times 1.6) = 27.8$ in.-kips

Since $M_{cr} = 60.7$ in.-kips > M = 27.8 in.-kips, $I_e = I_g = 512$ in.^{4}

$$M = \frac{21.6}{1 - \frac{5 \times 3.9 \times (20 \times 12)^2}{48 \times 3605 \times 512}} = 21.9 \text{ in.-kips}$$
Eq. (14-10)

$$\Delta_s = \frac{5 \times 21.9 \times (20 \times 12)^2}{48 \times 3605 \times 512} = 0.07 \text{ in.}$$
Eq. (14-9)

No further iterations are required since $I_e = I_g$.

Therefore,

$\Delta_s = 0.07$ in. $< \frac{\ell_c}{150} = \frac{20 \times 12}{150} = 1.6$ in.   O.K.

The wall is adequate with No. 4 @ 9 in. vertical reinforcement.

21-28

Example 21.4—Shear Design of Wall

Determine the shear and flexural reinforcement for the wall shown.

h = 8 in.
f'_c = 3000 psi
f_y = 60,000 psi

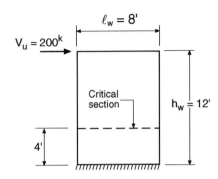

	Calculations and Discussion	Code Reference

1. Check maximum shear strength permitted

   $\phi V_n = \phi 10\sqrt{f'_c}\, hd$ 11.10.3

   where d = $0.8\ell_w$ = 0.8 × 8 × 12 = 76.8 in.   11.10.4

   ϕV_n = 0.75 × 10 $\sqrt{3000}$ × 8 × 76.8/1000 = 252.4 kips > V_u = 200 kips   O.K.

2. Calculate shear strength provided by concrete V_c

   Critical section for shear: 11.10.7

   $\dfrac{\ell_w}{2} = \dfrac{8}{2} = 4$ ft (governs)

   or

   $\dfrac{h_w}{2} = \dfrac{12}{2} = 6$ ft

   $V_c = 3.3\sqrt{f'_c}\, hd + \dfrac{N_u d}{4\ell_w}$ Eq. (11-29)

   $= 3.3\sqrt{3000} \times 8 \times 76.8/1000 + 0 = 111$ kips

   or

Example 21.4 (cont'd)	Calculations and Discussion	Code Reference

$$V_c = \left[0.6\sqrt{f'_c} + \frac{\ell_w\left(1.25\sqrt{f'_c} + \frac{0.2N_u}{\ell_w h}\right)}{\frac{M_u}{V_u} - \frac{\ell_w}{2}}\right] hd$$ Eq. (11-30)

$$= \left[0.6\sqrt{3000} + \frac{96\left(1.25\sqrt{3000} + 0\right)}{96 - 48}\right]\left(\frac{8 \times 76.8}{1000}\right) = 104 \text{ kips (governs)}$$

where $M_u = (12 - 4) V_u = 8V_u$ ft-kips $= 96V_u$ in.-kips

3. Determine required horizontal shear reinforcement

   $V_u = 200$ kips $> \phi V_c /2 = 0.75 (104)/2 = 39.0$ kips   11.10.8

   Shear reinforcement must be provided in accordance with 11.10.9.

   $V_u \leq \phi V_n$ Eq. (11-1)

   $\leq \phi(V_c + V_s)$ Eq. (11-2)

   $\leq \phi V_c + \dfrac{\phi A_v f_y d}{s_2}$ Eq. (11-31)

   $\dfrac{A_v}{s_2} = \dfrac{(V_u - \phi V_c)}{\phi f_y d}$

   $= \dfrac{[200 - (0.75 \times 104)]}{0.75 \times 60 \times 76.8} = 0.0353$

   For   2-No. 3: $s_2 = \dfrac{2 \times 0.11}{0.0353} = 6.2$ in.

   2-No. 4: $s_2 = \dfrac{2 \times 0.20}{0.0353} = 11.3$ in.

   2-No. 5: $s_2 = \dfrac{2 \times 0.31}{0.0353} = 17.6$ in.

   Try 2-No. 4 @ 10 in.

   $\rho_h = \dfrac{A_v}{A_g} = \dfrac{2 \times 0.20}{8 \times 10} = 0.0050 > 0.0025$ O.K.   11.10.9.2

| Example 21.4 (cont'd) | Calculations and Discussion | Code Reference |

$$\text{Maximum spacing} = \begin{cases} \dfrac{\ell_w}{5} = \dfrac{8 \times 12}{5} = 19.2 \text{ in.} \\ 3h = 3 \times 8 = 24.0 \text{ in.} \\ 18.0 \text{ in. (governs)} \end{cases}$$

11.10.9.3

Use 2-No. 4 @ 10 in.

4. Determine vertical shear reinforcement

$$\rho_n = 0.0025 + 0.5\left(2.5 - \dfrac{h_w}{\ell_w}\right)(\rho_h - 0.0025) \geq 0.0025 \qquad \text{Eq. (11-32)}$$

$$= 0.0025 + 0.5\,(2.5 - 1.5)\,(0.0050 - 0.0025)$$

$$= 0.0038$$

$$\text{Maximum spacing} = \begin{cases} \dfrac{\ell_w}{3} = \dfrac{8 \times 12}{3} = 32 \text{ in.} \\ 3h = 3 \times 8 = 24.0 \text{ in.} \\ 18.0 \text{ in. (governs)} \end{cases}$$

11.10.9.5

Use 2-No. 4 @ 13 in. ($\rho_n = 0.0038$)

5. Design for flexure

$M_u = V_u h_w = 200 \times 12 = 2{,}400$ ft-kips

Assume tension-controlled section ($\phi = 0.90$)
with $d = 0.8\ell_w = 0.8 \times 96 = 76.8$ in.
(Note: an exact value of d will be determined by a strain compatibility analysis below)

9.3.2
11.10.4

$$R_n = \dfrac{M_u}{\phi b d^2} = \dfrac{2400 \times 12{,}000}{0.9 \times 8 \times 76.8^2} = 678 \text{ psi}$$

$$\rho = \dfrac{0.85 f'_c}{f_y}\left(1 - \sqrt{1 - \dfrac{2R_n}{0.85 f'_c}}\right)$$

$$= \dfrac{0.85 \times 3}{60}\left(1 - \sqrt{1 - \dfrac{2 \times 678}{0.85 \times 3000}}\right) = 0.0134$$

$A_s = \rho b d = 0.0134 \times 8 \times 76.8 = 8.24$ in.^{2}

	Code
Example 21.4 (cont'd) Calculations and Discussion	**Reference**

Try 9-No. 8 ($A_s = 7.11$ in.²) at each end of wall, which provides less area of steel than that determined based on $d = 0.8\ell_w$.

Check moment strength of wall with 9-No. 8 bars using a strain compatibility analysis (see figure below for reinforcement layout).

From strain compatibility analysis (including No. 4 vertical bars):

$c = 13.1$ in.

$d = 81.0$ in.

$\varepsilon_t = 0.0182 > 0.0050$

Therefore, section is tension-controlled as assumed and $\phi = 0.90$.

$M_n = 3451$ ft-kips

$\phi M_n = 0.9 \times 3451 = 3106$ ft-kips > 2400 ft-kips   O.K.

Use 9-No. 8 bars each side ($A_s = 7.11$ in.²)

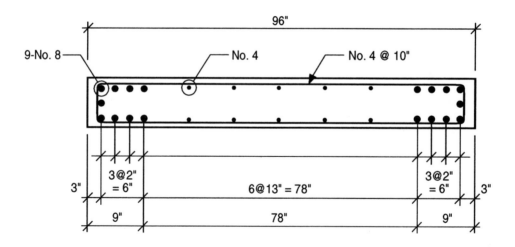

21-32

22

Footings

UPDATE FOR THE '05 CODE

Section 15.5.3 clarifies which proedure and code provisions apply for design of piles where the distance between the axis of the pile and the axis of the column is less then or equal to two times the distance between the top of the pile cap and the top of the pile.

GENERAL CONSIDERATIONS

Provisions of Chapter 15 apply primarily for design of footings supporting a single column (isolated footings) and do not provide specific design provisions for footings supporting more than one column (combined footings). The code states that combined footings shall be proportioned to resist the factored loads and induced reactions in accordance with the appropriate design requirements of the code. Detailed discussion of combined footing design is beyond the scope of Part 22. However, as a general design approach, combined footings may be designed as beams in the longitudinal direction and as an isolated footing in the transverse direction over a defined width on each side of the supported columns. Code references 15.1 and 15.2 are suggested for detailed design recommendations for combined footings.

15.2 LOADS AND REACTIONS

Footings must be designed to safely resist the effects of the applied factored axial loads, shears and moments. The size (base area) of a footing or the arrangement and number of piles is determined based on the allowable soil pressure or allowable pile capacity, respectively. The allowable soil or pile capacity is determined by principles of soil mechanics in accordance with general building codes. The following procedure is specified for footing design:

1. The footing size (plan dimensions) or the number and arrangement of piles is to be determined on the basis of unfactored (service) loads (dead, live, wind, earthquake, etc.) and the allowable soil pressure or pile capacity (15.2.2).

2. After having established the plan dimensions, the depth of the footing and the required amount of reinforcement are determined based on the appropriate design requirements of the code (15.2.1). The service pressures and the resulting shear and moments are multiplied by the appropriate load factors specified in 9.2 and are used to proportion the footing.

For purposes of analysis, an isolated footing may be assumed to be rigid, resulting in a uniform soil pressure for concentric loading, and a triangular or trapezoidal soil pressure distribution for eccentric loading (combined axial and bending effect). Only the computed bending moment that exists at the base of the column or pedestal is to be transferred to the footing. The minimum moment requirement for slenderness considerations in 10.12.3.2 need not be transferred to the footing (R15.2).

15.4 MOMENT IN FOOTINGS

At any section of a footing, the external moment due to the base pressure shall be determined by passing a vertical plane through the footing and computing the moment of the forces acting over the entire area of the footing on one side of the vertical plane. The maximum factored moment in an isolated footing is determined by passing a vertical plane through the footing at the critical sections shown in Fig. 22-1 (15.4.2). This moment is subsequently used to determine the required area of flexural reinforcement in that direction.

In one-way square or rectangular footings and two-way square footings, flexural reinforcement shall be distributed uniformly across the entire width of the footing (15.4.3). For two-way rectangular footings, the reinforcement must be distributed as shown in Table 22-1 (15.4.4).

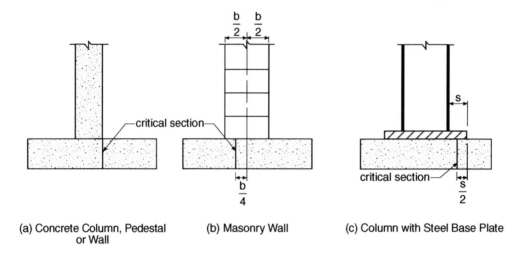

Figure 22-1 Critical Location for Maximum Factored Moment in an Isolated Footing (15.4.2)

Table 22-1 Distribution of Flexural Reinforcement

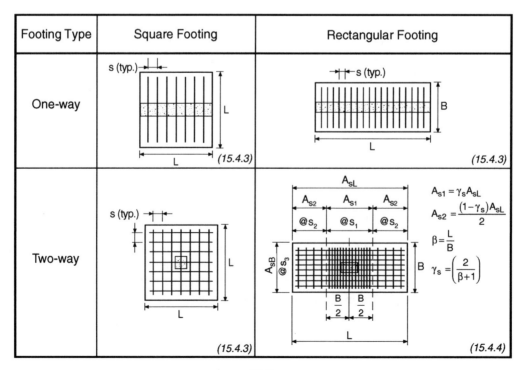

15.5 SHEAR IN FOOTINGS

Shear strength of a footing supported on soil or rock in the vicinity of the supported member (column or wall) must be determined for the more severe of the two conditions stated in 11.12. Both wide-beam action (11.12.1.1) and two-way action (11.12.1.2) must be checked to determine the required footing depth. Beam action assumes that the footing acts as a wide beam with a critical section across its entire width. If this condition is the more severe, design for shear proceeds in accordance with 11.1 through 11.5. Even though wide-beam action rarely controls the shear strength of footings, the designer must ensure that shear strength for beam action is not exceeded. Two-way action for the footing checks "punching" shear strength. The critical section for punching shear is a perimeter b_o around the supported member with the shear strength computed in accordance with 11.12.2.1. Tributary areas and corresponding critical sections for wide-beam action and two-way action for an isolated footing are illustrated in Fig. 22-2. Note that it is permissible to use a critical section with four straight sides for square or rectangular columns (11.12.1.3).

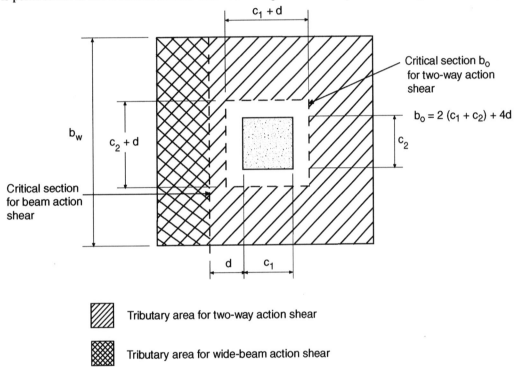

Figure 22-2 Tributary Areas and Critical Sections for Shear

In the design of a footing for two-way action, V_c is the smallest value obtained from Eqs. (11-33), (11-34), and (11-35). Eq. (11-35) established the upper limit of V_c at $4\sqrt{f'_c}\, b_o d$. Eq. (11-33) accounts for the effect of β, which is the ratio of the long side to the short side of the column, concentrated load, or reaction area. As β increases the concrete shear strength decreases (see Fig. 22-3). Eq. (11-34) was developed to account for the effect of b_o/d, and is based on tests that indicated shear strength decreases as b_o/d increases.

If the factored shear force V_u at the critical section exceeds the governing shear strength ϕV_c given by the minimum of Eqs. (11-33), (11-34), or (11-35), shear reinforcement must be provided. For shear reinforcement consisting of bars or wires and single- or multiple-leg stirrups, the shear strength may be increased to a maximum value of $6\sqrt{f'_c}\, b_o d$ (11.12.3.2), provided the footing has an effective depth d greater than or equal to 6 in., but not less than 16 times the shear reinforcement bar diameter (11.12.3). However, shear reinforcement must be designed to carry the shear in excess of $2\sqrt{f'_c}\, b_o d$ (11.12.3.1).

For footing design (without shear reinforcement), the shear strength equations may be summarized as follows:

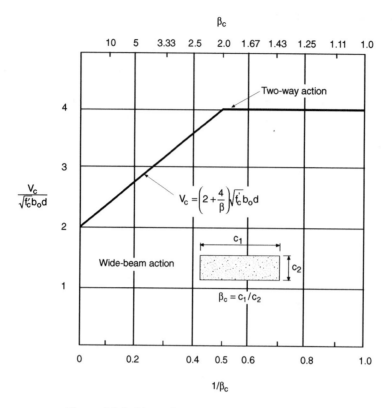

Figure 22-3 Shear Strength of Concrete in Footings

- Wide beam action

$$V_u \leq \phi V_n \qquad \text{Eq. (11-1)}$$

$$\leq \phi\left(2\sqrt{f'_c}\, b_w d\right) \qquad \text{Eq. (11-3)}$$

where b_w and V_u are computed for the critical section defined in 11.12.1.1 (see Fig. 22-2).

- Two-way action

$$V_u \leq \text{minimum of} \begin{cases} \left(2 + \dfrac{4}{\beta}\right)\sqrt{f'_c}\, b_o d & \text{Eq. (11-33)} \\ \left(\dfrac{\alpha_s d}{b_o} + 2\right)\sqrt{f'_c}\, b_o d & \text{Eq. (11-34)} \\ 4\sqrt{f'_c}\, b_o d & \text{Eq. (11-35)} \end{cases}$$

where

β = ratio of long side to short side of the column, concentrated load or reaction area

α_s = 40 for interior columns
= 30 for edge columns
= 20 for corner columns

b_o = perimeter of critical section shown in Fig. 22-2

15.8 TRANSFER OF FORCE AT BASE OF COLUMN, WALL, OR REINFORCED PEDESTAL

With the publication of ACI 318-83, 15.8 addressing transfer of force between a footing and supported member (column, wall, or pedestal) was revised to address both cast-in-place and precast construction. Section 15.8.1 gives general requirements applicable to both cast-in-place and precast construction. Sections 15.8.2 and 15.8.3 give additional rules for cast-in-place and precast construction, respectively. For force transfer between a footing and a precast column or wall, anchor bolts or mechanical connectors are specifically permitted by 15.8.3, with anchor bolts to be designed in accordance with Appendix D. (Prior to the '83 code, connection between a precast member and footing required either longitudinal bars or dowels crossing the interface, contrary to common practice.) Also note that walls are specifically addressed in 15.8 for force transfer to footings.

Section 15.8.3 contains requirements for the connection between precast columns and walls to supporting members. This section refers to 16.5.1.3 for minimum connection strength. Additionally, for precast columns with larger cross-sectional areas than required for loading, it is permitted to use a reduced effective area based on the cross-section required, but not less than one-half the total area when determining the nominal strength in tension.

The minimum tensile strength of a connection between a precast wall panel and its supporting member is required to have a minimum of two ties per panel with a minimum nominal tensile capacity of 10 kips per tie (16.5.1.3(b)).

All forces applied at the base of a column or wall (supported member) must be transferred to the footing (supporting member) by bearing on concrete and/or by reinforcement. Tensile forces must be resisted entirely by reinforcement. Bearing on concrete for both supported and supporting member must not exceed the concrete bearing strength permitted by 10.17 (see discussion on 10.17 in Part 6).

For a supported column, the bearing capacity ϕP_{nb} is

$$\phi P_{nb} = \phi(0.85 f'_c A_1) \qquad \text{10.17.1}$$

where

f'_c = compressive strength of the column concrete

A_1 = loaded area (column area)

$\phi = 0.65$ \hfill 9.3.2.4

For a supporting footing,

$$\phi P_{nb} = \phi(0.85 f'_c A_1) \sqrt{\frac{A_2}{A_1}} \leq 2\phi(0.85 f'_c A_1)$$

where

f'_c = compressive strength of the footing concrete

A_2 = area of the lower base of the largest frustrum of a pyramid, cone, or tapered wedge contained wholly within the footing and having for its upper base the loaded area, and having side slopes of 1 vertical to 2 horizontal (see Fig. R10.17).

Example 22.4 illustrates the design for force transfer at the base of a column.

When bearing strength is exceeded, reinforcement must be provided to transfer the excess load. A minimum area of reinforcement must be provided across the interface of column or wall and footing, even where concrete bearing strength is not exceeded. With the force transfer provisions addressing both cast-in-place and precast construction, including force transfer between a wall and footing, the minimum reinforcement requirements are based on the type of supported member, as shown in Table 22-2.

Table 22-2 Minimum Reinforcement for Force Transfer Between Footing and Supported Member

	Cast-in-Place	Precast
Columns	$0.005A_g$ (15.8.2.1)	$\dfrac{200A_g}{f_y}$ (16.5.1.3 (a))
Walls	see 14.3.2 (15.8.2.2)	see 16.5.1.3(b) and (c)

For cast-in-place construction, reinforcement may consist of extended reinforcing bars or dowels. For precast construction, reinforcement may consist of anchor bolts or mechanical connectors. Reference 22.1 devotes an entire chapter on connection design for precast construction.

The shear-friction design method of 11.7.4 should be used for horizontal force transfer between columns and footings (15.8.1.4; see Example 22.6). Consideration of some of the lateral force being transferred by shear through a formed shear key is questionable. Considerable slip is required to develop a shear key. Shear keys, if provided, should be considered as an added mechanical factor of safety only, with no design shear force assigned to the shear key.

PLAIN CONCRETE PEDESTALS AND FOOTINGS

Plain concrete pedestals and footings are designed in accordance with Chapter 22. See Part 30 for an in-depth discussion and examples.

REFERENCE

22.1 *PCI Design Handbook—Precast and Prestressed Concrete*, MNL-120-04, 6th Edition, Precast/Prestressed Concrete Institute, Chicago, IL, 2004, 750 pp.

Example 22.1—Design for Base Area of Footing

Determine the base area A_f required for a square spread footing with the following design conditions:

Service dead load = 350 kips
Service live load = 275 kips
Service surcharge = 100 psf

Assume average weight of soil and concrete above footing base = 130 pcf

Allowable soil pressure at bottom of footing = 4.5 ksf

Column dimensions = 30 × 12 in.

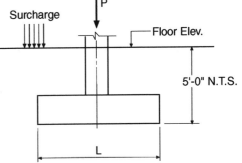

Calculations and Discussion	Code Reference

1. Determination of base area:

    The base area of the footing is determined using service (unfactored) loads with the net permissible soil pressure.

    Weight of surcharge = 0.10 ksf

    Net allowable soil pressure = 4.5 - 0.75 = 3.75 ksf

    Required base area of footing: 15.2.2

    $$A_f = \frac{350 + 275}{3.75} = 167 \text{ ft}^2$$

    Use a 13 × 13 ft square footing (A_f = 169 ft²)

2. Factored loads and soil reaction:

    To proportion the footing for strength (depth and required reinforcement) factored loads are used.   15.2.1

    $P_u = 1.2 (350) + 1.6 (275) = 860$ kips   Eq. (9-2)

    $$q_s = \frac{P_u}{A_f} = \frac{860}{169} = 5.10 \text{ ksf}$$

Example 22.2—Design for Depth of Footing

For the design conditions of Example 22.1, determine the overall thickness of footing required.

$f'_c = 3000$ psi

$P_u = 860$ kips

$q_s = 5.10$ ksf

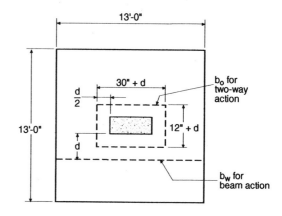

Calculations and Discussion	Code Reference

Determine depth based on shear strength without shear reinforcement. Depth required for shear usually controls the footing thickness. Both wide-beam action and two-way action for strength computation need to be investigated to determine the controlling shear criteria for depth.   11.12

Assume overall footing thickness = 33 in. and average effective thickness d = 28 in. = 2.33 ft

1. Wide-beam action:

   $V_u = q_s \times$ tributary area

   $b_w = 13$ ft = 156 in.

   Tributary area = $13 (6.0 - 2.33) = 47.7$ ft²

   $V_u = 5.10 \times 47.7 = 243$ kips

   $\phi V_n = \phi\left(2\sqrt{f'_c}\, b_w d\right)$ Eq. (11-3)

   $= 0.75\left(2\sqrt{3000} \times 156 \times 28\right)/1000$ 9.3.2.3

   $= 359$ kips $> V_u$ O.K.

2. Two-way action:

   $V_u = q_s \times$ tributary area

   Tributary area = $\left[(13 \times 13) - \dfrac{(30 + 28)(12 + 28)}{144}\right] = 152.9$ ft²

   $V_u = 5.10 \times 152.9 = 780$ kips

Example 22.2 (cont'd)   **Calculations and Discussion**   **Code Reference**

$$\frac{V_c}{\sqrt{f'_c} b_o d} = \text{minimum of} \begin{cases} 2 + \dfrac{4}{\beta} & \text{Eq. (11-35)} \\ \dfrac{\alpha_s d}{b_o} + 2 & \text{Eq. (11-36)} \\ 4 & \text{Eq. (11-37)} \end{cases}$$

$b_o = 2(30 + 28) + 2(12 + 28) = 196$ in.

$\beta = \dfrac{30}{12} = 2.5$

$\dfrac{b_o}{d} = \dfrac{196}{28} = 7$

$\alpha_s = 40$ for interior columns

$$\frac{V_c}{\sqrt{f'_c} b_o d} = \begin{cases} 2 + \dfrac{4}{2.5} = 3.6 \quad \text{(governs)} \\ \dfrac{40}{7} + 2 = 7.7 \\ 4 \end{cases}$$

$\phi V_c = 0.75 \times 3.6\sqrt{3000} \times 196 \times 28/1000$

     $= 812$ kips $> V_u = 780$ kips   O.K.

Example 22.3—Design for Footing Reinforcement

For the design conditions of Example 22.1, determine required footing reinforcement.

$f'_c = 3000$ psi

$f_y = 60,000$ psi

$P_u = 860$ kips

$q_s = 5.10$ ksf

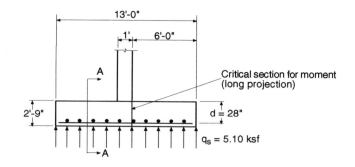

	Code
Calculations and Discussion	**Reference**

1. Critical section for moment is at face of column

    $M_u = 5.10 \times 13 \times 6^2/2 = 1193$ ft-kips

    15.4.2

2. Compute required A_s assuming tension-controlled section ($\phi = 0.9$)

    10.3.4, 9.3.2.1

    Required $R_n = \dfrac{M_u}{\phi b d^2} = \dfrac{1193 \times 12 \times 1000}{0.9 \times 156 \times 28^2} = 130$ psi

    $\rho = \dfrac{0.85 f'_c}{f_y}\left(1 - \sqrt{1 - \dfrac{2R_n}{0.85 f'_c}}\right)$

    $= \dfrac{0.85 \times 3}{60}\left(1 - \sqrt{1 - \dfrac{2 \times 130}{0.85 \times 3000}}\right) = 0.0022$

    $\rho \text{ (gross area)} = \dfrac{d}{h} \times 0.0022 = \dfrac{28}{33} \times 0.0022 = 0.0019$

    Check minimum A_s required for footings of uniform thickness; for Grade 60 reinforcement:

    10.5.4

    $\rho_{min} = 0.0018 < 0.0019$ O.K.

    7.12.2

    Required $A_s = \rho bd$

    $A_s = 0.0022 \times 156 \times 28 = 9.60$ in.^{2}

    Try 13-No. 8 bars ($A_s = 10.27$ in.^{2}) each way

    Note that a lesser amount of reinforcement is required in the perpendicular direction due to lesser M_u, but for ease of placement, the same uniformly distributed reinforcement will be used each way (see Table 22-1). Also note that $d_t = 27$ in. for perpendicular direction.

Example 22.3 (cont'd) — Calculations and Discussion

3. Check net tensile strain (ε_t)

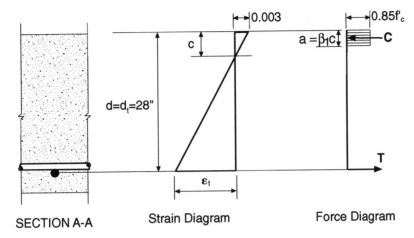

SECTION A-A   Strain Diagram   Force Diagram

$$a = \frac{A_s f_y}{0.85 f'_c b}$$

$$= \frac{10.27 \times 60}{0.85 \times 3 \times 156} = 1.55$$

$$c = \frac{a}{\beta_1} = \frac{1.55}{0.85} = 1.82$$

$$\frac{\varepsilon_t + 0.003}{d_t} = \frac{0.003}{c}$$

$$\varepsilon_t = \left(\frac{0.003}{c}\right) d_t - 0.003$$

$$= \frac{0.003}{1.82} \times 28 - 0.003 = 0.043 > 0.004 \qquad \text{10.3.5}$$

Therefore, section is tension-controlled and initial assumption is valid, O.K.

Thus, use 13-No. 8 bars each way.

4. Check development of reinforcement. **15.6**

   Critical section for development is the same as that for moment (at face of column). **15.6.3**

$$\ell_d = \left[\frac{3}{40} \frac{f_y}{\sqrt{f'_c}} \frac{\psi_t \psi_e \psi_s \lambda}{\left(\frac{c_b + K_{tr}}{d_b}\right)} \right] d_b \qquad \text{Eq. (12-1)}$$

Example 22.3 (cont'd)	Calculations and Discussion	Code Reference

Clear cover (bottom and side) = 3.0 in.

Center-to-center bar spacing = $\dfrac{156 - 2(3) - 2(0.5)}{12}$ = 12.4 in.

c_b = minimum of $\begin{cases} 3.0 + 0.5 = 3.5 \text{ in. (governs)} \\ \dfrac{12.4}{2} = 6.2 \text{ in.} \end{cases}$

12.2.4

K_{tr} = 0 (no transverse reinforcement)

$\dfrac{c_b + K_{tr}}{d_b} = \dfrac{3.5 + 0}{1.0} = 3.5 > 2.5$, use 2.5

12.2.3

ψ_t = 1.0 (less than 12 in. of concrete below bars)

12.2.4

ψ_e = 1.0 (uncoated reinforcement)

$\psi_t \psi_e = 1.0 < 1.7$

ψ_s = 1.0 (larger than No. 7 bars)

λ = 1.0 (normal weight concrete)

$\ell_d = \left[\dfrac{3}{40} \dfrac{60,000}{\sqrt{3000}} \dfrac{1.0 \times 1.0 \times 1.0 \times 1.0}{2.5} \right] \times 1.0 = 32.9$ in. > 12.0 in. O.K.

12.2.1

Since ℓ_d = 32.9 in. is less than the available embedment length in the short direction

$\left(\dfrac{156}{2} - \dfrac{30}{2} - 3 = 60 \text{ in.} \right)$, the No. 8 bars can be fully developed.

Use 13-No. 8 each way.

Example 22.4—Design for Transfer of Force at Base of Column

For the design conditions of Example 22.1, check force transfer at interface of column and footing.

f'_c (column) = 5000 psi

f'_c (footing) = 3000 psi

f_y = 60,000 psi

P_u = 860 kips

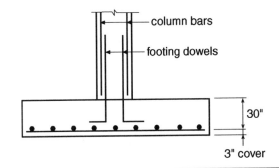

	Code
Calculations and Discussion	**Reference**

1. Bearing strength of column (f'_c = 5000 psi):   15.8.1.1

   $\phi P_{nb} = \phi(0.85 f'_c A_1)$ 10.17.1

   $= 0.65 (0.85 \times 5 \times 12 \times 30) = 995$ kips $> P_u = 860$ kips   O.K.   9.3.2.4

2. Bearing strength of footing (f'_c = 3000 psi):   15.8.1.1

   The bearing strength of the footing is increased by a factor $\sqrt{A_2/A_1} \leq 2$ due to the to the large footing area permitting a greater distribution of the column load.   10.17.1

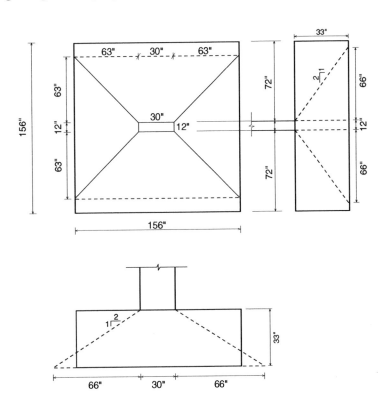

22-13

| Example 22.4 (cont'd) | Calculations and Discussion | Code Reference |

A_1 is the column (loaded) area and A_2 is the plan area of the lower base of the largest frustum of a pyramid, cone, or tapered wedge contained wholly within the support and having for its upper base the loaded area, and having side slopes of 1 vertical to 2 horizontal. For the 30×12 in. column supported on the 13×13 ft square footing, $A_2 = (66 + 12 + 66) \times (63 + 30 + 63)$.

$$\sqrt{\frac{A_2}{A_1}} = \sqrt{\frac{144 \times 156}{30 \times 12}} = 7.9 > 2, \text{ use } 2$$

Note that bearing on the column concrete will always govern until the strength of the column concrete exceeds twice that of the footing concrete.

$$\phi P_{nb} = 2[\phi(0.85 f'_c A_1)]$$

$$= 2[0.65(0.85 \times 3 \times 12 \times 30)] = 1193 \text{ kips} > P_u = 860 \text{ kips} \quad \text{O.K.}$$

3. Required dowel bars between column and footing:

   Even though bearing strength on the column and footing concrete is adequate to transfer the factored loads, a minimum area of reinforcement is required across the interface.
               15.8.2.1

   $A_s \text{ (min)} = 0.005(30 \times 12) = 1.80 \text{ in.}^2$

   Provide 4-No. 7 bars as dowels ($A_s = 2.40$ in.^{2})

4. Development of dowel reinforcement in compression: 12.3.2

   In column:

   $$\ell_{dc} = \left(\frac{0.02 f_y}{\sqrt{f'_c}}\right) d_b \geq (0.0003 f_y) d_b$$

   For No. 7 bars:

   $$\ell_{dc} = \left(\frac{0.02 \times 60,000}{\sqrt{5000}}\right) 0.875 = 14.9 \text{ in.}$$

   $$\ell_{dc(min)} = 0.0003 \times 60,000 \times 0.875 = 15.8 \text{ in.} \quad \text{(governs)}$$

   In footing:

   $$\ell_{dc} = \left(\frac{0.02 \times 60,000}{\sqrt{3000}}\right) 0.875 = 19.2 \text{ in.} \quad \text{(governs)}$$

   $$\ell_{dc(min)} = 0.0003 \times 60,000 \times 0.875 = 15.8 \text{ in.}$$

   Available length for development in footing

       = footing thickness - cover - 2 (footing bar diameter) - dowel bar diameter

       = 33 - 3 - 2(1.0) - 0.875 = 27.1 in. > 19.2 in.

   Therefore, the dowels can be fully developed in the footing.

Example 22.5—Design for Transfer of Force by Reinforcement

For the design conditions given below, provide for transfer of force between the column and footing.

12 × 12 in. tied reinforced column with 4-No. 14 longitudinal bars

f'_c = 4000 psi (column and footing)

f_y = 60,000 psi

P_D = 200 kips

P_L = 100 kips

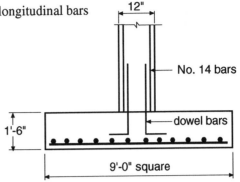

Calculations and Discussion	Code Reference

1. Factored load P_u = (1.2 × 200) + (1.6 × 100) = 400 kips — Eq. (9-2)

2. Bearing strength on column concrete: — 15.8.1.1

    $\phi P_{nb} = \phi(0.85 f'_c A_1) = 0.65(0.85 \times 4 \times 12 \times 12)$ — 10.17.1

    = 318.2 kips < P_u = 400 kips   N.G.

    The column load cannot be transferred by bearing on concrete alone. The excess load (400 - 318.2 = 81.8 kips) must be transferred by reinforcement. — 15.8.1.2

3. Bearing strength on footing concrete: — 15.8.1.1

    $\phi P_{nb} = \sqrt{\dfrac{A_2}{A_1}} \left[\phi(0.85 f'_c A_1)\right]$

    $\sqrt{\dfrac{A_2}{A_1}} = \sqrt{\dfrac{9 \times 9}{1 \times 1}} = 9 > 2$, use 2

    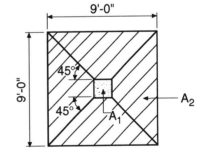

    ϕP_{nb} = 2 (318.2) = 636.4 kips > 400 kips   O.K.

4. Required area of dowel bars: — 15.8.1.2

    $A_s \text{ (required)} = \dfrac{(P_u - \phi P_{nb})}{\phi f_y}$

    $= \dfrac{81.8}{0.65 \times 60} = 2.10 \text{ in.}^2$

    $A_s \text{ (min)} = 0.005 (12 \times 12) = 0.72 \text{ in.}^2$ — 15.8.2.1

    Try 4-No. 8 bars (A_s = 3.16 in.²)

| **Example 22.5 (cont'd)** | **Calculations and Discussion** | **Code Reference** |

5. Development of dowel reinforcement

   a. For development into the column, the No. 14 column bars may be lap spliced with the No. 8 footing dowels. The dowels must extend into the column a distance not less than the development length of the No. 14 column bars or the lap splice length of the No. 8 footing dowels, whichever is greater. — 15.8.2.3

   For No. 14 bars:

   $$\ell_{dc} = \left(\frac{0.02f_y}{\sqrt{f'_c}}\right)d_b = \left(\frac{0.02 \times 60,000}{\sqrt{4000}}\right)1.693 = 32.1 \text{ in. (governs)}$$

   $$\ell_{dc(min)} = (0.0003f_y)d_b = 0.0003 \times 60,000 \times 1.693 = 30.5 \text{ in.}$$

   For No. 8 bars:

   lap length $= 0.0005f_y d_b$ — 12.16.1

   $= 0.0005 \times 60,000 \times 1.0 = 30$ in.

   Development length of No. 14 bars governs.

   The No. 8 dowel bars must extend not less than 33 in. into the column.

   b. For development into the footing, the No. 8 dowels must extend a full development length. — 15.8.2.3

   $$\ell_{dc} = \left(\frac{0.02f_y}{\sqrt{f'_c}}\right)d_b = \left(\frac{0.02 \times 60,000}{\sqrt{4000}}\right) \times 1.0 = 19.0 \text{ in. (governs)}$$

   $$\ell_{dc(min)} = (0.0003f_y)d_b = 0.0003 \times 60,000 \times 1.0 = 18.0 \text{ in.}$$

   This length may be reduced to account for excess reinforcement. — 12.3.3(a)

   $$\frac{A_s \text{ (required)}}{A_s \text{ (provided)}} = \frac{2.10}{3.16} = 0.66$$

   Required $\ell_{dc} = 19 \times 0.66 = 12.5$ in.

   Available length for dowels development $\approx 18 - 5 = 13$ in. > 12.5 in. required, O.K.

   Note: In case the available development length is less than the required development length, either increase footing depth or use larger number of smaller size dowels. Also note that if the footing dowels are bent for placement on top of the footing reinforcement (as shown in the figure), the bent portion cannot be considered effective for developing the bars in compression (12.5.5).

Example 22.6—Design for Transfer of Horizontal Force at Base of Column

For the column and footing of Example 22.5, design for transfer of a horizontal factored force of 85 kips acting at the base of the column.

Design data:

Footing: size = 9 × 9 ft
thickness = 1ft-6 in.

Column: size = 12 × 12 in. (tied)
4-No. 14 longitudinal reinforcement

f'_c = 4000 psi (footing and column)

f_y = 60,000 psi

Calculations and Discussion	Code Reference
1. The shear-friction design method of 11.7 is applicable.	15.8.1.4
Check maximum shear transfer permitted:	11.7.5
$V_u \leq \phi(0.2 f'_c A_c)$ but not greater than $\phi(800 A_c)$	
$\phi V_n = 0.75 (0.2 \times 4 \times 12 \times 12) = 86.4$ kips	
$\phi(800 A_c) = 0.75 \times 800 \times 12 \times 12/1000 = 86.4$ kips	
$V_u = 85$ kips $< \phi(0.2 f'_c A_c)$ and $\phi(800 A_c)$ O.K.	
The shear transfer of 85 kips is permitted at the base of 12 × 12 in. column.	
Strength requirement for shear:	
$V_u \leq \phi V_n$	Eq. (11-1)
$V_n = V_u/\phi = A_{vf} f_y \mu$	Eq. (11-25)
Use $\mu = 0.6$ (concrete not intentionally roughened)	11.7.4.3
and $\phi = 0.75$ (shear)	
Required $A_{vf} = \dfrac{V_u}{\phi f_y \mu} = \dfrac{85}{0.75 \times 60 \times 0.6} = 3.15$ in.²	Eq. (11-25)
A_s (provided) = 3.16 in.² O.K.	
Therefore, use 4-No. 8 dowels (A_s = 3.16 in.²)	

Example 22.6 (cont'd)	Calculations and Discussion	Code Reference

If the 4-No. 8 dowels were not adequate for transfer of horizontal shear, the footing concrete in contact with the column concrete could be roughened to an amplitude of approximately 1/4 in. to take advantage of the higher coefficient of friction of 1.0:

$$\text{Required } A_{vf} = \frac{85}{0.75 \times 60 \times 1.0} = 1.89 \text{ in.}^2$$

2. Tensile development of No. 8 dowels, as required by 11.7.8

   a. Within the column

   $$\ell_d = \left[\frac{3}{40} \frac{f_y}{\sqrt{f'_c}} \frac{\psi_t \psi_e \psi_s \lambda}{\left(\frac{c_b + K_{tr}}{d_b} \right)} \right] d_b \qquad \text{Eq. (12-1)}$$

   Clear cover to No. 8 bar ≈ 3.25 in.

   Center-to-center bar spacing of No. 8 bars ≈ 4.5 in.

   $$c_b = \text{minimum of} \begin{cases} 3.25 + 0.5 = 3.75 \text{ in.} \\ \frac{4.5}{2} = 2.25 \text{ in. (governs)} \end{cases} \qquad 12.2.4$$

   Assume $K_{tr} = 0$ (conservatively consider no transverse reinforcement)

   $$\frac{c_b + K_{tr}}{d_b} = \frac{2.25 + 0}{1.0} = 2.25 < 2.5, \text{ use } 2.25 \qquad 12.2.3$$

   $\psi_t = 1.0$

   $\psi_e = 1.0$

   $\psi_t \psi_e = 1.0 < 1.7$

   $\psi_s = 1.0$

   $\lambda = 1.0$

   $$\ell_d = \left(\frac{3}{40} \frac{60,000}{\sqrt{4000}} \frac{1.0 \times 1.0 \times 1.0}{2.25} \right) \times 1.0 = 31.6 \text{ in.}$$

   Provide at least 32 in. of embedment into the column.

   b. Within the footing

   Use standard hooks at the ends of the No. 8 bars

	Code
Example 22.6 (cont'd) Calculations and Discussion	Reference

$\ell_{dh} = (0.02\psi_e \lambda f_y / \sqrt{f'_c}) d_b$ 12.5.2

$= \left(0.02 \times 1.0 \times 1.0 \times \dfrac{60,000}{\sqrt{4000}} \right) \times 1.0 = 19.0$ in.

Modifications: 12.5.3

cover normal to plane of 90° hook > 2.5 in.

cover on bar extension beyond hook ≥ 2 in.

$\ell_{dh} = 0.7 \times 19 = 13.3$ in.   12.5.3(a)

Min. $\ell_{dh} = 8 \times d_b = 8$ in. < 13.3 in.   12.5.1

Available development length =

= 18 - 5 = 13 in. < 13.3 in. N.G.

Increase footimg depth by 2 in. Total depth = 20 in.

Use 15 in. hook embedment into footing to secure dowels at the footing reinforcement.

Total length of No. 8 dowel = 32 + 15 = 47 in. Use 4 ft-0 in. long dowels.

Note: The top of the footing at the interface between column and footing must be clean and free of laitance before placement of the column concrete.   11.7.9

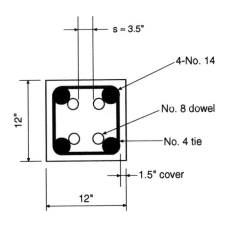
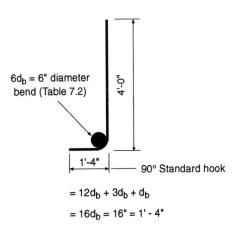

No. 8 dowel detail

Example 22.7—Design for Depth of Footing on Piles

For the footing supported on the piles shown, determine the required thickness of the footing (pile cap).

Footing size = 8.5 × 8.5 ft

Column size = 16 × 16 in.

Pile diameter = 12 in.

f'_c = 4000 psi

Load per pile:

P_D = 20 kips

P_L = 10 kips

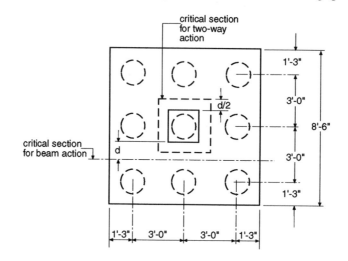

Calculations and Discussion	Code Reference
1. Depth required for shear usually controls footing thickness. Both wide-beam action and two-way action for the footing must be investigated.	11.12
Assume an overall footing thickness of 1 ft-9 in. with an average d ≈ 14 in.	15.7
2. Factored pile loading:	
P_u = 1.2 (20) + 1.6 (10) = 40 kips	Eq. (9-2)
3. Strength requirements for shear	
$V_u \leq \phi V_n$	Eq. (11-1)
a. Wide-beam action for footing:	11.12.1.1
3 piles fall within tributary area	
V_u (neglecting footing wt.) = 3 × 40 = 120 kips	
$\phi V_n = \phi\left(2\sqrt{f'_c} b_w d\right)$	Eq. (11-3)
b_w = 8 ft-6 in. = 102 in.	
$\phi V_n = 0.75 \left(2\sqrt{4000} \times 102 \times 14\right)/1000$ = 135.4 kips > V_u = 120 kips O.K.	

		Code
Example 22.7 (cont'd)	**Calculations and Discussion**	**Reference**

b. Two-way action: 11.12.1.2

  8 piles fall within the tributary area

  $V_u = 8 \times 40 = 320$ kips

$$\frac{V_c}{\sqrt{f'_c}\, b_o d} = \text{smallest value of} \begin{cases} 2 + \dfrac{4}{\beta} & \text{Eq. (11-33)} \\ \dfrac{\alpha_s d}{b_o} + 2 & \text{Eq. (11-34)} \\ 4 & \text{Eq. (11-35)} \end{cases}$$

  $\beta = \dfrac{16}{16} = 1.0$

  $b_o = 4(16 + 14) = 120$ in.

  $\alpha_s = 40$ for interior columns

  $\dfrac{b_o}{d} = \dfrac{120}{14} = 8.6$

$$\frac{V_c}{\sqrt{f'_c}\, b_o d} = \begin{cases} 2 + \dfrac{4}{1} = 6 & \text{Eq. (11-33)} \\ \dfrac{40}{8.6} + 2 = 6.7 & \text{Eq. (11-34)} \\ 4 \ \ (\text{governs}) & \text{Eq. (11-35)} \end{cases}$$

  $\phi V_c = 0.75 \times 4\sqrt{4000} \times 120 \times 14/1000$

    $= 319$ kips $\cong V_u = 320$ kips   O.K.

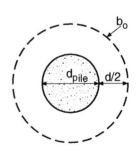

4. Check "punching" shear strength at corner piles. With piles spaced at 3 ft-0 in. on center, critical perimeters do not overlap.

  $V_u = 40$ kips per pile

| **Example 22.7 (cont'd)** | **Calculations and Discussion** | **Code Reference** |

$$\frac{V_c}{\sqrt{f'_c}b_o d} = \text{minimum of} \begin{cases} 2 + \dfrac{4}{\beta} \\ \dfrac{\alpha_s d}{b_o} + 2 \\ 4 \end{cases}$$

Eq. (11-33)
Eq. (11-34)
Eq. (11-35)

β = 1.0 (square reaction area of equal area)

$b_o = \pi(12 + 14) = 81.7$ in.

α_s = 20 (for corner columns) 11.12.2.1

$\dfrac{b_o}{d} = \dfrac{81.7}{14} = 5.8$

$$\frac{V_c}{\sqrt{f'_c}b_o d} = \begin{cases} 2 + \dfrac{4}{1} = 6 \\ \dfrac{20}{5.8} + 2 = 5.4 \\ 4 \text{ (governs)} \end{cases}$$

$\phi V_c = 0.75 \times 4\sqrt{4000} \times 81.7 \times 14/1000 = 217$ kips $> V_u = 40$ kips   O.K.

23

Precast Concrete

GENERAL CONSIDERATIONS

Chapter 16 was completely rewritten for the 1995 code. Previous editions of Chapter 16 were largely performance oriented. The current chapter is more prescriptive, although the word "instructive" may be more appropriate, as the chapter provides much more guidance to the designer of structures which incorporate precast concrete. Not only does the chapter itself provide more requirements and guidelines, but the commentary contains some 25 references, as opposed to 4 in the 1989 code, thus encouraging the designer to make maximum use of the available literature.

The increase in instructive material is most notable in 16.5, Structural Integrity. Requirements for structural integrity were introduced in 7.13 of the 1989 code. For precast construction, this section required only that tension ties be provided in all three orthogonal directions (two horizontal and one vertical) and around the perimeter of the structure, without much further guidance. Reference 23.1 was given for precast bearing wall buildings. The recommendations given in that reference are now codified in 16.5.2. Section 16.5.1 applies primarily to precast structures other than bearing wall buildings and, as is the case with most of the rewritten Chapter 16, is largely a reflection of time-tested industry practice. Note that tilt-up concrete construction is a form of precast concrete. Reference 23.3 addresses all phases of design and construction of tilt-up concrete structures.

16.2   GENERAL

The code requires that precast members and connections be designed for "... loading and restraint conditions from initial fabrication to end use in the structure ..." Often, especially in the case of wall panels, conditions during handling are far more severe than those experienced during service. For this reason, and also because practices and details vary among manufacturers, precast concrete components are most often designed by specialty engineers employed by the manufacturer. Calculations, as well as shop drawings (16.2.4) are then submitted to the engineer-of-record for approval. This procedure is also usually followed in the design of connections. For more information on the relationship between the engineer-of-record and the specialty engineer, see Refs. 23.4 and 23.5.

As stated above, since 1995, the code encourages the use of other publications for the design of precast concrete structures. References 23.2, 23.6, and 23.7 are particularly useful to the designer. Of these, the most widely used is Ref. 23.7, the *PCI Design Handbook*.

Section 16.2.3 states that tolerances must be specified. This is usually done by reference to Industry documents[23.8, 23.9, 23.10], as noted in the commentary. Design of precast concrete members and connections is particularly sensitive to tolerances. Therefore, they should be specified in the contract documents together with required concrete strengths at different stages of construction [16.2.4(b)].

16.3 DISTRIBUTION OF FORCES AMONG MEMBERS

Section 16.3.1 covers distribution of forces perpendicular to the plane of members. Most of the referenced research relates to hollow-core slabs, and is also applicable to solid slabs that are connected by continuous grout keys. Members that are connected together by other means, such as weld plates on double tees, have through extensive use been found to be capable of distributing concentrated loads to adjacent members. It is common to assume for design purposes that up to 25% of a concentrated load can be transferred to each adjacent member; the connections should be designed accordingly. Since load transfer is dependent on compatible deflections, less distribution occurs nearer the support, as shown in Fig. 23-1(a) for hollow core slabs. Flanges of double tees are also designed for transverse load distribution over an effective width as illustrated in Fig. 23-1(b). Other types of decks may not necessarily follow the same pattern, because of different torsional resistance properties, but the same principles are applicable. A typical design is shown in Example 23.1.

Compatibility of the deflections of adjacent units is an important design consideration. For example, in the driving lane of a precast concrete double-tee parking deck, even if each member has adequate strength to carry a full wheel load, it is undesirable for each unit to deflect independently. It is common practice to use more closely spaced connections in those cases, to assure sharing of the load and to eliminate differential deflections between members.

Section 16.3.2 covers distribution of in-plane forces. It requires a continuous load path for such forces. If these forces are tensile, they must be resisted by steel or steel reinforcement. Since these in-plane forces are usually caused by lateral loads such as those due to wind or earthquakes, and since such lateral loads can occur in any direction, nearly all of the continuous load path must be provided by steel or steel reinforcement. In that respect, the reinforcement and connections designed to meet the requirements of this section may also provide the continuous ties required by 16.5.

16.4 MEMBER DESIGN

This section is primarily concerned with minimum reinforcement requirements for precast members. Section 16.4.1 waives transverse steel requirements in prestressed concrete members 12 ft or less in width, except where such steel is required for flexure. The commentary notes that an example of an exception to the waiver is the flanges of single and double tees. This section is intended primarily for hollow core and solid slabs where the transverse connection is typically grouted joints.

Section 16.4.2 reduces the minimum reinforcement in precast, nonprestressed walls, from that required for cast-in-place reinforced concrete walls in Chapter 14, to 0.001 times the gross cross-sectional area of the walls, in accordance with common industry practice. This is in recognition of the fact that much of the shrinkage of precast members occurs prior to attachment to the structure. Spacing of reinforcement in precast walls should not exceed 5 times the wall thickness or 30 in. for interior walls or 18 in. for exterior walls.

When wall panels are load-bearing, they are usually designed as compression members in accordance with Chapter 10. When they are not load-bearing (and often even when they are), the stresses during handling are usually critical. In those cases, it is common to place the lifting and dunnage points so that the stresses during handling do not exceed the modulus of rupture (with a safety factor), especially for architectural precast panels. If cracking is likely, crack control reinforcement in accordance with 10.6.4 is required.

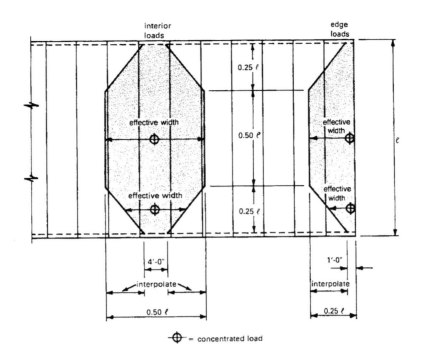

(a) Hollow core slab

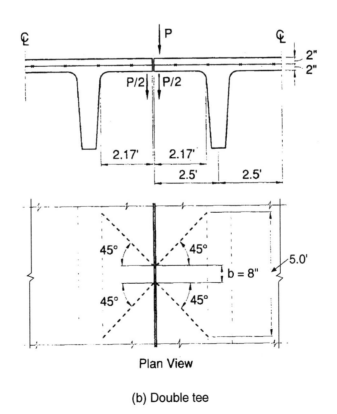

Plan View

(b) Double tee

Figure 23-1 Assumed Load Distribution[23.7]

16.5 STRUCTURAL INTEGRITY

The provisions of 16.5.1.1 are intended to assure that there is a continuous load path from every precast member to the lateral load resisting system. The commentary gives several examples of how this may be accomplished. Section 16.5.1.2 is adopted from a similar requirement in the Uniform Building Code, as explained in the commentary.

Section 16.5.1.3 gives requirements for vertical tension ties. The requirement for columns in 16.5.1.3(a) applies not only to connections of columns to footings, but also to such connections as column splices. The 10,000 lb requirement for each of at least two ties per wall panel in 16.5.1.3(b) is from the *PCI Design Handbook*[23.7], and is the numerical equivalent of a common connection used in the precast concrete industry for many years. This is strictly an empirical value, and is intended to apply only to the dimensions of the hardware items in the connection, without including eccentricities in the design. Section 16.5.1.3(c) permits this connection to be into a reinforced concrete floor slab, as is common with tilt-up construction.

Section 16.5.2 in essence codifies some of the recommendations of Ref. 23.11, which gives numerical values for tension ties in bearing wall buildings. This report is based on a series of tests conducted at the Portland Cement Association's laboratories[23.12] in the late 1970s.

16.6 CONNECTION AND BEARING DESIGN

Section 16.6.1 lists the several ways that precast members can be connected, and then allows design by analysis or test. Special mention is made of 11.7, Shear Friction, as this is a commonly used analysis/design tool. See Part 16 of this document. Examples on application of shear friction design procedure are also given in *PCI Design Handbook*[23.7] and *Design and Typical Details of Connections for Precast and Prestressed Concrete*.[23.6]

Section 16.6.2 describes several important considerations when designing for bearing of precast elements. The minimum bearing lengths of 16.6.2.2(a) are particularly important. It should be emphasized that these are minimum values, and that the structure should be detailed with significantly longer bearing lengths, to allow for tolerance in placement. Section 16.6.2.3 makes it clear that positive moment reinforcement need not extend out of the end of the member, provided it goes at least to the center of the bearing.

16.7 ITEMS EMBEDDED AFTER CONCRETE PLACEMENT

In precasting plants, it has long been common practice to place certain embedded items in the concrete after it has been cast. This practice is recognized in this section and constitutes an exception to the provisions of 7.5.1. Conditions to embed items in the concrete while it is in a plastic state are: (1) embedded item is not required to be hooked or tied to reinforcement within the concrete, (2) embedded item is secured in its position until concrete hardens, and (3) concrete is properly consolidated around each embedment.

16.8 MARKING AND IDENTIFICATION

The purpose of identification marks on precast members is to facilitate construction and avoid placing errors. Each precast member should be marked to show date of manufacture and should be identified according to placing drawings.

16.9 HANDLING

This section re-emphasizes the general requirement of 16.2.1. Handling stresses and deformations must be considered during the design of precast concrete members. Erection steps and hardware required for each step must be shown on contract or erection drawings.

16.10 STRENGTH EVALUATION OF PRECAST CONSTRUCTION

It is always desirable, and certainly safer and more economical, to test a suspect precast concrete member before it is integrated into the structure. This new section describes how Chapter 20 provisions can be applied to this case. The test loads specified in 20.3.2 must be adjusted to simulate the load portion carried by the suspect member when it is in the final, composite mode. The acceptance criteria of 20.5 apply to the isolated precast member.

REFERENCES

23.1 PCI Committee on Building Code and PCI Technical Activities Committee, "Proposed Design Requirements for Precast Concrete," *PCI Journal*, V. 31, No. 6, November-December 1986, pp. 32-47.

23.2 ACI Committee 550, "Design Recommendations for Precast Concrete Structures," (ACI 550R-96, Reapproved 2001). American Concrete Institute, Farmington Hills, MI, 1996. Also in ACI Manual of Concrete Practice, Part 6.

23.3 ACI Committee 551, "Tilt-Up Concrete Structures," (ACI 551R-92, Reapproved 2003), American Concrete Institute, Farmington Hills, MI, 1992. Also in *ACI Manual of Concrete Practice*, Part 6.

23.4 The Case Task Group on Specialty Engineering, "National Practice Guidelines for Specialty Structural Engineers," Council of American Structural Engineers, Washington, DC, 1994, 11 pp.

23.5 ACI Committee on Responsibility in Concrete Construction, "Guidelines for Authorities and Responsibilities in Concrete Design and Construction," *Concrete International*, Vol. 17, No. 9, September 1995, pp. 66-69.

23.6 "Design and Typical Details of Connections for Precast and Prestressed Concrete," MNL-123-88, 2nd Edition, Precast/Prestressed Concrete Institute, Chicago, 1988, 270 pp.

23.7 "PCI Design Handbook—Precast and Prestressed Concrete," MNL-120-04, 6th Edition, Precast/Prestressed Concrete Institute, Chicago, 2004, 750 pp.

23.8 "Manual for Quality Control for Plants and Production of Precast and Prestressed Concrete Products," MNL-116-99, 4th Edition, Precast/Prestressed Concrete Institute, Chicago, 1999, 340 pp.

23.9 "Manual for Quality Control for Plants and Production of Architectural Precast Concrete," MNL-117-96, 3rd Edition, Precast/Prestressed Concrete Institute, Chicago, 1996, 226 pp.

23.10 PCI Committee on Tolerances, "Tolerances for Precast and Prestressed Concrete," *PCI Journal*, V. 30, No. 1, January-February 1985, pp. 26-112.

23.11 PCI Committee on Precast Concrete Bearing Wall Buildings, "Considerations for the Design of Precast Concrete Bearing Wall Buildings to Withstand Abnormal Loads," *PCI Journal*, V. 21, No. 2, March-April 1976, pp. 18-51.

23.12 "Design and Construction of Large-Panel Concrete Structures," six reports, EB100D, 1976-1980, 762 pp.; three studies, EB102D, 1980, 300 pp.; Portland Cement Association, Skokie, IL.

Example 23.1—Load Distribution in Double Tees

Required: Compute the factored moments and shears for each of three double tees of the following roof:

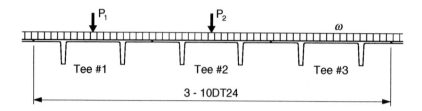

Given:
Double tees = 10DT24 (self weight = 468 plf), h = 24 in.
Span = 60 ft
Superimposed DL = 15 psf, LL = 30 psf
Concentrated dead load on Tee #1, P_1 = 20 kips @ 3 ft from left support
Concentrated dead load on Tee #2, P_2 = 20 kips @ midspan

Calculations and Discussion	**Code Reference**

1. Assume:

    Concentrated dead load P_1 cannot be distributed to adjacent tees since it is near the support.

    Concentrated dead load P_2 is distributed, with 25 percent to adjacent tees and 50 percent to the tee supporting the load, i.e.

    0.25 (20 kips) = 5 kips to Tee #1
    0.50 (20 kips) = 10 kips to Tee #2
    0.25 (20 kips) = 5 kips to Tee #3

2. Factored uniform dead and live loads, for each tee

    DL = 468 + 15 (10 ft width) = 0.618 kip/ft

    LL = 30 (10 ft width) = 0.30 kip/ft

    w_u = 1.2D + 1.6L   Eq. (9-2)

    = 1.2 (0.618) + 1.6 (0.30) = 1.222 kip/ft

3. Factored moments and shears for Tee #1

    Factored concentrated dead load next to support = 1.2 (20) = 24 kips

    Factored concentrated dead load at midspan = 1.2 (5) = 6 kips

    w_u = 1.222 kip/ft

Example 23.1 (cont'd)	Calculations and Discussion	Code Reference

Reaction at left support $= \dfrac{57}{60}(24) + \dfrac{6}{2} + \dfrac{1.222(60)}{2} = 62.46$ kips

For prestressed concrete members, design for shear at distance h/2   11.1.3.2

V_u (left) = 62.46 - 1.222 = 61.24 kips

Reaction at right support $= \dfrac{3}{60}(24) + \dfrac{6}{2} + \dfrac{1.222(60)}{2} = 40.86$ kips

At distance h/2, V_u (right) = 40.86 - 1.222 = 39.64 kips

Maximum moment is at midspan

M_u (max) = 40.86 (30) - 1.222 (30) (15) = 676 ft-kips

4. Factored moments and shears for Tee #2

$$M_u = \dfrac{w_u \ell^2}{8} + \dfrac{P\ell}{4}$$

$$= \dfrac{1.222(60)^2}{8} + \dfrac{1.2(10)(60)}{4} = 730 \text{ ft-kips}$$

Maximum reaction $= \dfrac{w_u \ell}{2} + \dfrac{P}{2}$

$$= \dfrac{1.222(60)}{2} + \dfrac{1.2(10)}{2} = 42.66 \text{ kips}$$

At distance h/2, V_u = 42.66 - 1.222 = 41.44 kips

5. Factored moments and shears for Tee #3

$$M_u = \dfrac{1.222(60)^2}{8} + \dfrac{1.2(5)(60)}{4} = 640 \text{ ft-kips}$$

Maximum reaction $= \dfrac{1.222(60)}{2} + \dfrac{1.2(5)}{2} = 39.66$ kips

At distance h/2, V_u = 39.66 - 1.222 = 38.44 kips

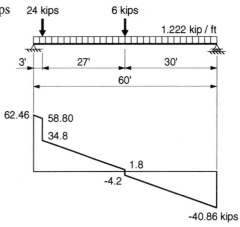

Factored loads on Tee #1

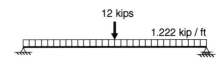

Factored loads on Tee #2

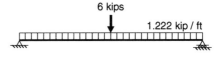

Factored loads on Tee #3

Blank

24

Prestressed Concrete — Flexure

UPDATE FOR THE '05 CODE

Significant changes to the provisions for prestressed concrete design for flexure are:

- Construction joint location limitation of 6.4.4 previously applied to both reinforced and prestressed concrete, they are now waived for prestressed concrete.

- A definition for "Transfer length" was added in 2.2, and was illustrated schematically in Figure R9.3.2.7(a) and R9.3.2.7(b) for pretensioned bonded and debonded strands, respectively. Strength reduction factor" ϕ applicable within transfer and development lengths has been revised (9.3.2.7) to eliminate a discontinuity in computed flexural strength.

- The permissible flexural tensile strength in two-way prestressed slabs of $f_t \leq 7.5\sqrt{f_c'}$ introduced in the 2002 code has been changed to $f_t \leq 6\sqrt{f_c'}$ as prescribed in previous codes (18.3.3) prior to 2002.

- The limiting depth of 36 in. for required skin reinforcement has been changed from the effective depth to the overall member height h (18.4.4.4). The crack control provisions for skin reinforcement of 10.6.7 have been simplified and made consistent with those required for flexural tension reinforcement in 10.6.4.

- An unnecessary sentence describing redistribution of negative moments in R18.10.3 was removed from the commentary to eliminate potential confusion.

GENERAL CONSIDERATIONS

In prestressed members, compressive stresses are introduced into the concrete to reduce tensile stresses resulting from applied loads including the self weight of the member (dead load). Prestressing steel, such as strands, bars, or wires, is used to impart compressive stresses to the concrete. Pretensioning is a method of prestressing in which the tendons are tensioned before concrete is placed and the prestressing force is primarily transferred to the concrete through bond. Post-tensioning is a method of prestressing in which the tendons are tensioned after the concrete has hardened and the prestressing force is primarily transferred to the concrete through the end anchorages.

The act of prestressing a member introduces "prestressing loads" to the member. The induced prestressing loads, acting in conjunction with externally applied loads, must provide serviceability and strength to the member beginning immediately after prestress force transfer and continuing throughout the life of the member. Prestressed structures must be analyzed taking into account prestressing loads, service loads, temperature, creep, shrinkage and the structural properties of all materials involved.

The code states that all provisions of the code apply to prestressed concrete, unless they are in conflict with Chapter 18 or are specifically excluded. The exclusions, listed in 18.1.3, are necessary because some empirical or simplified analytical methods employed elsewhere in the code may not adequately account for the effects of prestressing forces.

Deflections of prestressed members calculated according to 9.5.4 should not exceed the values listed in Table 9.5(b). According to 9.5.4, prestressed concrete members, like any other concrete members, should be designed to have adequate stiffness to prevent deformations which may adversely affect the strength or serviceability of the structure.

PRESTRESSING MATERIALS

The most commonly used prestressing material in the United States is Grade 270 ksi low-relaxation, seven-wire strand, defined by ASTM A 416. The most common size is 1/2-in., although there is increasing use of 0.6-in. strand, especially for post-tensioning. The properties of these strands are as follows:

Nominal Diameter, in.	1/2	0.6
Area, sq. in.	0.153	0.217
Breaking stress f_{pu}, ksi	270	270
Breaking strength, kips	41.3	58.6
Jacking stress, ksi = 0.75 f_{pu}	202.5	202.5

Virtually identical metric strands are used in metric countries.

The Prestressed Concrete Institute's *PCI Design Handbook,* 6th edition, Ref. 24.1, gives a standard stress-strain curve for this material, as shown in Fig. 24-1. This curve is approximated by the two expressions given below the figure.

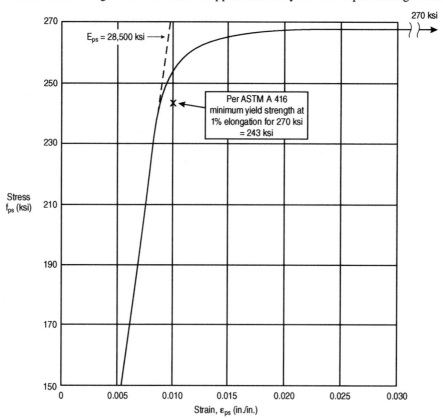

The above curve can be approximated by the following equations: $\varepsilon_{ps} \leq 0.0086$: $f_{ps} = 28{,}500\, \varepsilon_{ps}$ (ksi)

$\varepsilon_{ps} > 0.0086$: $f_{ps} = 270 - \dfrac{0.04}{\varepsilon_{ps} - 0.007}$ (ksi)

Figure 24-1 Stress-Strain Curve for Grade 270, Low Relaxation Strand[24.1]

NOTATION AND TERMINOLOGY

The following symbols are used in 18.4.4, which deals with serviceability requirements for cracked prestressed flexural members.

Δf_{ps} = stress in prestressing steel at service loads less decompression stress, psi. See Fig. 24-2

f_{dc} = decompression stress. Stress in the prestressing steel when stress is zero in the concrete at the same level as the centroid of the tendons, psi. See Fig. 24-2

s = center-to-center spacing of flexural tension steel near the extreme tension face, in. Where there is only one bar or tendon near the extreme tension face, s is the width of extreme tension face

Note, $f_{dc} = f_{se} + f_c \times E_{ps}/E_c$ where f_c is the concrete stress at level of steel under dead load and prestress. f_{dc} may be conservatively taken as f_{se}.

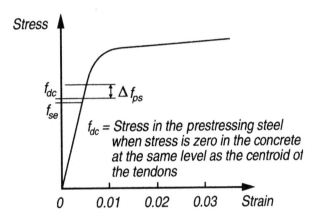

Figure 24-2 Decompression Stress f_{dc}

The following definitions found in 2.2 are consistently used in Chapter 18 and throughout the code. They reflect industry terminology. See Fig. 24-2

Prestressing steel — High-strength steel element, such as wire, bar, or strand, or a bundle of such elements, used to impart prestress forces to concrete.

Tendon — In pretensioned applications, the tendon is the prestressing steel. In post-tensioned applications, the tendon is a complete assembly consisting of anchorages, prestressing steel, and sheathing with coating for unbonded applications or ducts with grout for bonded applications.

Bonded tendon — Tendon in which prestressing steel is bonded to concrete either directly or through grouting.

Unbonded tendon — Tendon in which the prestressing steel is prevented from bonding to the concrete and is free to move relative to the concrete. The prestressing force is permanently transferred to the concrete at the tendon ends by the anchorages only.

Duct — A conduit (plain or corrugated) to accommodate prestressing steel for post-tensioned installation. Requirements for post-tensioning ducts are given in 18.17.

Sheathing — A material encasing prestressing steel to prevent bonding of the prestressing steel with the surrounding concrete, to provide corrosion protection, and to contain the corrosion inhibiting coating.

18.2 GENERAL

The code specifies strength and serviceability requirements for all concrete members, prestressed or nonprestressed. This section requires that, for prestressed members, both strength and behavior at service conditions must be checked. All load stages that may be critical during the life of the structure, beginning with the transfer of the prestressing force to the member and including handling and transportation, must be considered.

This section also calls attention to several structural issues specific to prestressed concrete structures that must be considered in design:

18.2.3...Stress concentrations. See 18.13 for requirements for post-tensioned anchorages.

18.2.4...Compatibility of deformation with adjoining construction. An example of the effect of prestressing on adjoining parts of a structure is the need to include moments caused by axial shortening of prestressed floors in the design of the columns which support the floors.

18.2.5...Buckling of prestressed members. This section addresses the possibility of buckling of any part of a member where prestressing tendons are not in contact with the concrete. This can occur when prestressing steel is in an oversize duct, and with external prestressing described in 18.22.

18.2.6...Section properties. The code requires that the area of open post-tensioning ducts be deducted from section properties prior to bonding of prestressing tendons. For pretensioned members and post-tensioned members after grouting, the commentary allows the use of gross section properties, or effective section properties that may include the transformed area of bonded tendons and nonprestressed reinforcement.

18.3 DESIGN ASSUMPTIONS

In applying fundamental structural principles (equilibrium, stress-strain relations, and geometric compatibility) to prestressed structures, certain simplifying assumptions can be made. For computation of strength (18.3.1), the basic assumptions given for nonprestressed members in 10.2 apply, except that 10.2.4 applies only to nonprestressed reinforcement. For investigation of service load conditions, the "elastic theory" (referring to the linear variation of stress with strain) may be used. Where concrete is cracked, the concrete resists no tension. For analysis at service load conditions, the moduli of elasticity for concrete and nonprestressed reinforcement are given in 8.5. The modulus of elasticity for prestressed reinforcement is not given but can generally be taken as described in Fig. 24-1.

Section 18.3.3 defines three classes of prestressed flexural members, as follows:

Uncracked   Class U: $f_t \leq 7.5\sqrt{f_c'}$

Transition   Class T: $7.5\sqrt{f_c'} < f_t \leq 12\sqrt{f_c'}$

Cracked   Class C: $f_t > 12\sqrt{f_c'}$

Table 24-1 summarizes the applicable requirements for these three classes of prestressed flexural members and, for comparison, for nonprestressed flexural members as well.

Class U and Class T members correspond to those designed by 18.4.2(c) and 18.4.2(d), respectively, of ACI 318-99 and earlier editions of the code. In ACI 318-99, 18.4.2(d) required deflections to be checked by a cracked section analysis if tensile stresses exceeded $6\sqrt{f_c'}$, but the section was not assumed to be cracked unless the stress exceeded $7.5\sqrt{f_c'}$. This inconsistency was eliminated in 2002 Code by setting the dividing tensile stress between Classes U and T at $7.5\sqrt{f_c'}$.

Class C permits design using any combination of prestressing steel and reinforcement. It "fills the gap" between prestressed and nonprestressed concrete. For Class C members, a cracked section analysis or stresses is required by 18.3.4; whereas, for Class T members, an approximate cracked section analysis is required for deflection only. Unfortunately, a cracked section stress analysis for combined flexure and axial load (from the prestress) is complex. Reference 24.2 gives one method of accomplishing this.

Section 18.3.3 requires that prestressed two-way slab systems be designed as Class U with $f_t \leq 6\sqrt{f_c'}$.

Table 24-1 Serviceability Design Requirements

	Prestressed			Nonprestressed
	Class U	Class T	Class C	
Assumed behavior	Uncracked	Transition between uncracked and cracked	Cracked	Cracked
Section properties for stress calculation at service loads	Gross section 18.3.4	Gross section 18.3.4	Cracked section 18.3.4	No requirement
Allowable stress at transfer	18.4.1	18.4.1	18.4.1	No requirement
Allowable compressive stress based on uncracked section properties	18.4.2	18.4.2	No requirement	No requirement
Tensile stress at service loads 18.3.3	$\leq 7.5\sqrt{f_c'}$	$7.5\sqrt{f_c'} < f_t \leq 12\sqrt{f_c'}$	No requirement	No requirement
Deflection calculation basis	9.5.4.1 Gross section	9.5.4.2 Cracked section, bilinear	9.5.4.2 Cracked section, bilinear	9.5.2, 9.5.3 Effective moment of inertia
Crack control	No requirement	No requirement	10.6.4 Modified by 18.4.4.1	10.6.4
Computation of Δf_{ps} or f_s for crack control	—	—	Cracked section analysis	$M/(A_s \times$ lever arm), or $0.6f_y$
Side skin reinforcement	No requirement	No requirement	10.6.7	10.6.7

18.4 SERVICEABILITY REQUIREMENTS — FLEXURAL MEMBERS

Both concrete and prestressing tendon stresses are limited to ensure satisfactory behavior immediately after transfer of prestress and at service loads. The code provides different permissible stresses for conditions immediately after prestress transfer (before time-dependent losses) and for conditions at service loads (after all prestress losses have occurred).

For conditions immediately after prestress transfer, the code allows: extreme fiber compressive stress of $0.60f_{ci}'$; extreme fiber tensile stress of $3\sqrt{f_{ci}'}$, except $6\sqrt{f_{ci}'}$ is permitted at the ends of simply supported members. Where tensile stress exceeds the permissible values, bonded nonprestressed reinforcement shall be provided to resist the total tensile force assuming an uncracked section.

The permissible compressive stress due to prestress plus total service loads is limited to $0.60f_c'$. A permissible stress equal to $0.45f_c'$ has been added for the condition of prestress plus sustained loads. It should be noted that the "sustained loads" mentioned in 18.4.2(a) include any portion of the live load that will be sustained for a sufficient period to cause significant time-dependent deflections.

Concrete tensile stress limitations for Class U and T at service loads apply to the "precompressed" tensile zone which is that portion of the member cross-section in which flexural tension occurs under dead and live loads.

For Class C prestressed members, crack control is accomplished through a steel spacing requirement based on that of 10.6.4 and Eq. (10-4) for nonprestressed concrete. Eq. (10-4) is modified by 18.4.4. The maximum spacing between tendons is reduced to 2/3 of that permitted for bars, to account for lesser bond, compared to deformed bars. The quantity Δf_{ps}, the stress in the prestressing steel at service loads less the decompression stress f_{dc} is the stress in the prestressing steel when the stress is zero in the concrete at the same level as the centroid of the tendons. The code permits f_{dc} to be conservatively taken as the effective prestress f_{se}. The following shows Eq. (10-4), and as modified by 18.4.4.

Eq. (10-4) in 10.6.4:

$$s = 15\left(\frac{40,000}{f_s}\right) - 2.5c_c \le 12\left(\frac{40,000}{f_s}\right)$$

As modified by 18.4.4:

$$s = \frac{2}{3}\left[15\left(\frac{40,000}{\Delta f_{ps}}\right) - 2.5c_c\right]$$

The quantity of Δf_{ps} shall not exceed 36,000 psi. If Δf_{ps} is not greater than 20,000 psi, the above spacing limits need not apply.

The 2/3 modifier is to account for bond characteristics of strands, which are less effective than those of deformed bars. When both reinforcement and bonded tendons are used to meet the spacing requirement, the spacing between a bar and a tendon shall not exceed 5/6 of that given by Eq. (10-4).

Where h of a beam of a Class C exceeds 36 in., skin reinforcement consisting of reinforcement or bonded tendons shall be provided as required by 10.6.7.

18.5 PERMISSIBLE STRESSES IN PRESTRESSING STEEL

The permissible tensile stresses in all types of prestressing steel, in terms of the specified minimum tensile strength f_{pu}, are summarized as follows:

a. Due to tendon jacking force: .... $0.94f_{py}$ but not greater than $0.80f_{pu}$
   low-relaxation wire and strands ($f_{py} = 0.90f_{pu}$) .... $0.80f_{pu}$
   stress-relieved wire and strands, and plain bars (ASTM A722) ($f_{py} = 0.85f_{pu}$) .... $0.80f_{pu}$
   deformed bars (ASTM A722) ($f_{py} = 0.80f_{pu}$) .... $0.75f_{pu}$

b. Immediately after prestress transfer: .... $0.82f_{py}$ but not greater than $0.74f_{pu}$
   low-relaxation wire and strands ($f_{py} = 0.90f_{pu}$) .... $0.74f_{pu}$
   stress-relieved wire and strands, and plain bars ($f_{py} = 0.85f_{pu}$) .... $0.70f_{pu}$
   deformed bars ($f_{py} = 0.80f_{pu}$) .... $0.66f_{pu}$

c. Post-tensioning tendons, at anchorages and couplers, immediately after tendon anchorage .... $0.70f_{pu}$

Note that the permissible stresses given in 18.5.1(a) and (b) apply to both pretensioned and post-tensioned tendons. Pretensioned tendons are often jacked to 75 percent of f_{pu}. This will result in a stress below $0.74 f_{pu}$ after transfer.

18.6 LOSS OF PRESTRESS

A significant factor which must be considered in design of prestressed members is the loss of prestress due to various causes. These losses can dramatically affect the behavior of a member at service loads. Although calculation procedures and certain values of creep strain, friction factors, etc., may be recommended, they are at best only an estimate. For the design of members whose behavior (deflection in particular) is sensitive to prestress losses, the engineer should establish through tests the time-dependent properties of materials to be used in the analysis/design of the structure. Refined analyses should then be performed to estimate the prestress losses. Specific provisions for computing friction loss in post-tensioning tendons are provided in 18.6.2. Allowance for other types of prestress losses are discussed in Ref. 24.1. Note that the designer is required to show on the design drawings the magnitude and location of prestressing forces as required by 1.2.1(g).

ESTIMATING PRESTRESS LOSSES

Lump sum values of prestress losses that were widely used as a design estimate of prestress losses prior to the '83 code edition (35,000 psi for pretensioning and 25,000 psi for post-tensioning) are now considered obsolete. Also, the lump sum values may not be adequate for some design conditions.

Reference 24.3 offers guidance to compute prestress losses and it is adaptable to computer programs. It allows step-by-step computation of losses which is necessary for rational analysis of deformations. The method is too tedious for hand calculations.

Reference 24.4 presents a reasonably accurate and easy procedure for estimating prestress losses due to various causes for pretensioned and post-tensioned members with bonded and unbonded tendons. The procedure, which is intended for practical design applications under normal design conditions, is summarized below. The simple equations enable the designer to estimate the prestress loss from each source rather than using a lump sum value. The reader is referred to Ref. 24.4 for an in-depth discussion of the procedure, including sample computations for typical prestressed concrete beams. Quantities used in loss computations are defined in the summary of notation which follows this section.

COMPUTATION OF LOSSES

Elastic Shortening of Concrete (ES)

For members with bonded tendons:

$$ES = K_{es}E_s \frac{f_{cir}}{E_{ci}} \tag{1}$$

where $K_{es} = 1.0$ for pretensioned members

$K_{es} = 0.5$ for post-tensioned members where tendons are tensioned in sequential order to the same tension. With other post-tensioning procedures, the value for K_{es} may vary from 0 to 0.5.

$$f_{cir} = K_{cir}f_{cpi} - f_g \tag{2}$$

where $K_{cir} = 1.0$ for post-tensioned members

$K_{cir} = 0.9$ for pretensioned members.

For members with unbonded tendons:

$$ES = K_{es}E_s \frac{f_{cpa}}{E_{ci}} \tag{1a}$$

Creep of Concrete (CR)

For members with bonded tendons:

$$CR = K_{cr} \frac{E_s}{E_c}(f_{cir} - f_{cds}) \tag{3}$$

where K_{cr} = 2.0 for pretensioned members

K_{cr} = 1.6 for post-tensioned members

For members made of sand lightweight concrete the foregoing values of K_{cr} should be reduced by 20 percent.

For members with unbonded tendons:

$$CR = K_{cr} \frac{E_s}{E_c} f_{cpa} \tag{3a}$$

Shrinkage of Concrete (SH)

$$SH = 8.2 \times 10^{-6} K_{sh} E_s \left(1 - 0.06 \frac{V}{S}\right)(100 - RH) \tag{4}$$

where K_{sh} = 1.0 for pretensioned members

K_{sh} is taken from Table 24-2 for post-tensioned members.

Table 24-2 Values of K_{sh} for Post-Tensioned Members

Time, days*	1	3	5	7	10	20	30	60
K_{sh}	0.92	0.85	0.80	0.77	0.73	0.64	0.58	0.45

*Time after end of moist curing to application of prestress

Relaxation of Tendons (RE)

$$RE = [K_{re} - J(SH + CR + ES)] \, C \tag{5}$$

where the values of K_{re}, J, and C are taken from Tables 24-3 and 24-4.

Table 24-3 Values of K_{re} and J

Type of Tendon	Kre (psi)	J
270 Grade stress-relieved strand or wire	20,000	0.15
250 Grade stress-relieved strand or wire	18,500	0.14
240 or 235 Grade stress-relieved wire	17,600	0.13
270 Grade low-relaxation strand	5000	0.040
250 Grade low-relaxation wire	4630	0.037
240 or 235 Grade low-relaxation wire	4400	0.035
145 or 160 Grade stress-relieved bar	6000	0.05

Table 24-4 Values of C

f_{pi}/f_{pu}	Stressed relieved strand or wire	Stress-relieved bar or low relaxation strand or wire
0.80		1.28
0.79		1.22
0.78		1.16
0.77		1.11
0.76		1.05
0.75	1.45	1.00
0.74	1.36	0.95
0.73	1.27	0.90
0.72	1.18	0.85
0.71	1.09	0.80
0.70	1.00	0.75
0.69	0.94	0.70
0.68	0.89	0.66
0.67	0.83	0.61
0.66	0.78	0.57
0.65	0.73	0.53
0.64	0.68	0.49
0.63	0.63	0.45
0.62	0.58	0.41
0.61	0.53	0.37
0.60	0.49	0.33

Friction

Computation of friction losses is covered in 18.6.2. When the tendon is tensioned, the friction losses computed can be checked with reasonable accuracy by comparing the measured tendon elongation and the prestressing force applied by the tensioning jack.

SUMMARY OF NOTATION

A_c = area of gross concrete section at the cross-section considered
A_{ps} = total area of prestressing steel
C = a factor used in Eq. (5), see Table 24-4
CR = stress loss due to creep of concrete
e = eccentricity of center of gravity of prestressing steel with respect to center of gravity of concrete at the cross-section considered
E_c = modulus of elasticity of concrete at 28 days
E_{ci} = modulus of elasticity of concrete at time prestress is applied
E_s = modulus of elasticity of prestressing steel. Usually 28,000,000 psi
ES = stress loss due to elastic shortening of concrete
f_{cds} = stress in concrete at center of gravity of prestressing steel due to all superimposed permanent dead loads that are applied to the member after it has been prestressed
f_{cir} = net compressive stress in concrete at center of gravity of prestressing steel immediately after the prestress has been applied to the concrete. See Eq. (2).
f_{cpa} = average compressive stress in the concrete along the member length at the center of gravity of the

		prestressing steel immediately after the prestress has been applied to the concrete
f_{cpi}	=	stress in concrete at center of gravity of prestressing steel due to P_{pi}
f_g	=	stress in concrete at center of gravity of prestressing steel due to weight of structure at time prestress is applied
f_{pi}	=	stress in prestressing steel due to P_{pi}, $= P_{pi}/A_{ps}$
f_{pu}	=	specified tensile strength of prestressing steel, psi
I_c	=	moment of inertia of gross concrete section at the cross-section considered
J	=	a factor used in Eq. (5), see Table 24-3
K_{cir}	=	a factor used in Eq. (2)
K_{cr}	=	a factor used in Eq. (3)
K_{es}	=	a factor used in Eqs. (1) and (1a)
K_{re}	=	a factor used in Eq. (5), see Table 24-3
M_d	=	bending moment due to dead weight of member being prestressed and to any other permanent loads in place at time of prestressing
M_{ds}	=	bending moment due to all superimposed permanent dead loads that are applied to the member after it has been prestressed
P_{pi}	=	prestressing force in tendons at critical location on span after reduction for losses due to friction and seating loss at anchorages but before reduction for ES, CR, SH, and RE
RE	=	stress loss due to relaxation of prestressing steel
RH	=	average relative humidity surrounding the concrete member (see Fig. 24-3)
SH	=	stress loss due to shrinkage of concrete
V/S	=	volume to surface ratio, usually taken as gross cross-sectional area of concrete member divided by its perimeter

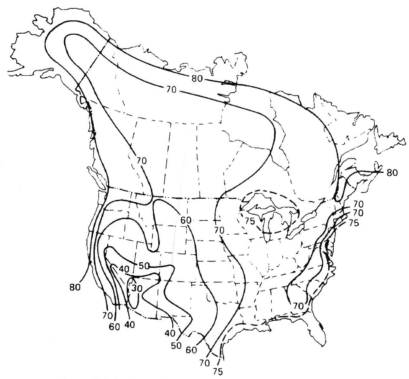

Figure 24-3 Annual Average Ambient Relative Humidity

18.7 FLEXURAL STRENGTH

The flexural strength of prestressed members can be calculated using the same assumptions as for nonprestressed members. Prestressing steel, however, does not have a well defined yield point as does mild reinforcement. As a prestressed cross-section reaches its flexural strength (defined by a maximum compressive concrete strain of 0.003), stress in the prestressed reinforcement at nominal strength, f_{ps}, will vary depending on the amount of prestressing. The value of f_{ps} can be obtained using the conditions of equilibrium, stress-strain relations, and strain compatibility (Design Example 24-4 illustrates the procedure). However, the analysis is quite cumbersome, especially in the case of unbonded tendon. For bonded prestressing, the compatibility of strains can be considered at an individual section, while for unbonded tendon, compatibility relations can be written only at the anchorage points and will depend on the entire cable profile and member loading. To avoid such lengthy calculations, the code allows f_{ps} to be obtained by the approximate Eqs. (18-3), (18-4), and (18-5).

For members with bonded prestressing steel, an approximate value of f_{ps} given by Eq. (18-3) may be used for flexural members reinforced with a combination of prestressed and nonprestressed reinforcement (partially prestressed members), taking into account effects of any nonprestressed tension reinforcement (ω), any compression reinforcement (ω'), the concrete compressive strength f'_c, rectangular stress block factor β_1, and an appropriate factor for type of prestressing material used (γ_p). For a fully prestressed member (without nonprestressed tension or compression reinforcement), Eq. (18-3) reduces to:

$$f_{ps} = f_{pu}\left(1 - \frac{\gamma_p}{\beta_1}\rho_p\frac{f_{pu}}{f'_c}\right)$$

where γ_p = 0.55 for deformed bars $(f_{py}/f_{pu} \geq 0.80)$

   = 0.40 for stress-relieved wire and strands, and plain bars $(f_{py}/f_{pu} \geq 0.85)$

   = 0.28 for low-relaxation wire and strands $(f_{py}/f_{pu} \geq 0.90)$

and β_1, as defined in 10.2.7.3,

β_1 = 0.85 for $f'_c \leq$ 4000 psi
   = 0.80 for f'_c = 5000 psi
   = 0.75 for f'_c = 6000 psi
   = 0.70 for f'_c = 7000 psi
   = 0.65 for $f'_c \geq$ 8000 psi

Eq. (18-3) can be written in nondimensional form as follows:

$$\omega_p = \omega_{pu}\left(1 - \frac{\gamma_p}{\beta_1}\omega_{pu}\right) \tag{6}$$

where $$\omega_p = \frac{A_{ps}f_{ps}}{bd_pf'_c} \tag{7}$$

$$\omega_{pu} = \frac{A_{ps}f_{pu}}{bd_pf'_c} \tag{8}$$

The moment strength of a prestressed member with bonded tendons may be computed using Eq. (18-3) only when all of the prestressed reinforcement is located in the tension zone. When part of the prestressed reinforcement is located in the compression zone of a cross-section, Eq. (18-3), involving d_p, is not valid. Flexural strength for such a condition must be computed by a general analysis based on strain compatibility and equilibrium, using the stress-strain properties of the prestressing steel and the assumptions given in 10.2.

For members with unbonded prestressing steel, an approximate value of f_{ps} given by Eqs. (18-4) and (18-5) may be used. Eq. (18-5) applies to members with high span-to-depth ratios (> 35), such as post-tensioned one-way slabs, flat plates and flat slabs.

With the value of f_{ps} known, the nominal moment strength of a rectangular section, or a flanged section where the stress block is within the compression flange, can be calculated as follows:

$$M_n = A_{ps}f_{ps}\left(d_p - \frac{a}{2}\right) = A_{ps}f_{ps}\left(d_p - 0.59\frac{A_{ps}f_{ps}}{bf'_c}\right) \tag{9}$$

where a = the depth of the equivalent rectangular stress block = $\dfrac{A_{ps}f_{ps}}{0.85bf'_c}$ (10)

or in nondimensional terms:

$$R_n = \omega_p(1 - 0.59\omega_p) \tag{11}$$

where $R_n = \dfrac{M_n}{b(d_p)^2 f'_c}$ (12)

18.8   LIMITS FOR REINFORCEMENT OF FLEXURAL MEMBERS

Prestressed concrete sections are classifed as tension-controlled, transition, or compression-controlled based on net tensile strain. These classifications are defined in 10.3.3 and 10.3.4, with appropriate ϕ-factors in 9.3.2. These requirements are the same as those for nonprestressed concrete.

Figure 24-3 shows the relationship between the coefficient of resistance $\phi M_n/(bd^2)$ and the reinforcement ratio ρ_p for prestressed flexural members. Grade 270 ksi prestressing steel has a useful strength 4.5 times that of Grade 60 reinforcement. Compare Fig. 24-4 to Fig. 7-3. Higher concrete strengths are normally used with prestressed concrete, so Fig. 24-4 shows curves for f'_c from 5000 to 8000 psi; whereas, Fig. 7-3 shows curves for f'_c from 3000 to 6000 psi. The curves for f'_c of 5000 and 6000 psi are almost identical in the two figures.

In both figures, the curves have a break point corresponding to a net tensile strain of 0.005. Beyond that point, the reduction in ϕ in the transition region almost cancels the benefit of increased reinforcement index. For both nonprestressed and prestressed concrete, the best design is to stay in the tension-controlled region, using compression reinforcement, if necessary, to maintain the net tensile strain, ε_t at 0.005 or more.

As in previous ACI 318 codes, there is no absolute limit on the reinforcement index for prestressed members. But it will always be advantageous to design the tension-controlled region at critical sections, as there is little or no gain in design strength in the transition region.

Critical parameters at the tension-controlled limit may be tabulated. The effective prestress f_{se} will normally be at least $0.6 f_{pu}$, or 162 ksi, if a jacking stress of $0.75 f_{pu}$ is used. This amounts to a 20 percent loss. The total steel strain when $\varepsilon_t = 0.005$ is equal to $162/28{,}500 + 0.005 = 0.01068$. Using the stress-strain curve shown in Fig. 24-1, $f_{ps} = 270 - 0.04/(0.01068 - 0.007) = 259$ ksi. A section will be tension-controlled when d_t is taken equal to d_p.

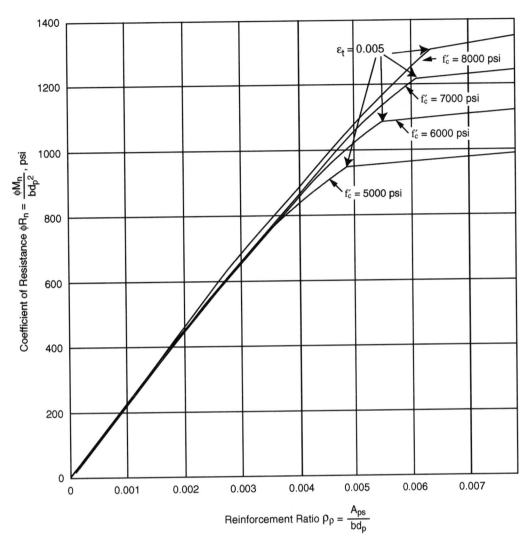

Figure 24-4 Design Strength Curves (ϕR_n vs. ρ_ρ) for Type 270k Low Relaxation Strand

Table 24-5 shows design parameters for prestressed sections at the tension-controlled strain limit, indicated by the added subscript t. The rows for R_{nt}, ϕ_{nt}, and ω_{pt} are identical to those in Table 6-1. The row for ω_{put} shows values slightly higher than ω_{pt}, because ω_{put} is based on f_{pu} of 270 ksi; whereas, ω_{put} is based on f_{ps} of 259 ksi. The final row for ρ_{pt} shows values much smaller than for ρ_t in Table 6-1, because of the much higher strength of the prestressing strand.

The following is a short-cut procedure for finding the flexural strength of sections in which the Grade 270 ksi low-relaxation prestressing steel can reasonably be assumed to be in one layer with $d_p = d_t$, and with $f_{se} \geq 162$ ksi.

1. Assume section is at tension-controlled limit, and $f_{ps} = 259$ ksi.

2. Compute steel tension T and equal compressive force C.

3. Find depth of stress block a and depth to neutral axis c.

4. Is $c/d_p \leq 0.375$? If so, proceed. If not, add compression steel to make $c/d_p \leq 0.375$.

5. Compute provided design strength $\phi M_n = 0.9(T)(d-a/2)$.

Table 24-5 Design Parameters at Strain Limit of 0.005 for Tension-Controlled Sections

	$f'_c = 3000$ $\beta_1 = 0.85$	$f'_c = 4000$ $\beta_1 = 0.85$	$f'_c = 5000$ $\beta_1 = 0.80$	$f'_c = 6000$ $\beta_1 = 0.75$	$f'_c = 8000$ $\beta_1 = 0.65$	$f'_c = 10,000$ $\beta_1 = 0.65$
R_{nt}	683	911	1084	1233	1455	1819
ϕR_{nt}	615	820	975	1109	1310	1637
ω_{pt}	0.2709	0.2709	0.2550	0.2391	0.2072	0.2072
ω_{put}	0.2823	0.2823	0.2657	0.2491	0.2159	0.2159
ρ_{pt}	0.00314	0.00418	0.00492	0.00554	0.00640	0.00800

For $f_{se} \geq 162$ ksi in low-relaxation Grade 270 ksi strand

6. If provided $\phi M_n \geq$ required, stop. Section is adequate. If not proceed.

7. If the deficiency in provided ϕM_n is more than 4 percent, steel must be added. If deficiency is less than 4 percent, strain compatibility may be used in an attempt to find a higher f_{ps} in order to justify adequacy of the section.

Section 18.8.2 requires the total amount of prestressed and nonprestressed reinforcement of flexural members to be adequate to develop a design moment strength at least equal to 1.2 times the cracking moment strength ($\phi M_n = 1.2 M_{cr}$), where M_{cr} is computed by elastic theory using a modulus of rupture equal to $7.5\sqrt{f'_c}$. The provisions of 18.8.2 are analogous to 10.5 for nonprestressed members. They are intended as a precaution against abrupt flexural failure resulting from rupture of the prestressing tendons immediately after cracking. The provision ensures that cracking will occur before flexural strength is reached, and by a large enough margin so that significant deflection will occur to warn that the ultimate capacity is being approached. The typical bonded prestressed member will have a fairly large margin between cracking strength and flexural strength, but the designer must be certain by checking it.

The cracking moment M_{cr} for a prestressed member is determined by summing all the moments that will cause a stress in the bottom fiber equal to the modulus of rupture f_r. Referring to Fig. 24-5 for an unshored prestressed composite member taking compression as negative and tension as positive:

$$-\left(\frac{P_{se}}{A_c}\right) - \left(\frac{P_{se}e}{S_b}\right) + \left(\frac{M_d}{S_b}\right) + \left(\frac{M_a}{S_c}\right) = +f_r$$

Solving for $M_a = \left(f_r + \dfrac{P_{se}}{A_c} + \dfrac{P_{se}e}{S_b}\right) S_c - M_d \left(\dfrac{S_c}{S_b}\right)$

Since $M_{cr} = M_d + M_a$

$$M_{cr} = \left(f_r + \frac{P_{se}}{A_c} + \frac{P_{se}e}{S_b}\right) S_c - M_d \left(\frac{S_c}{S_b} - 1\right) \tag{13}$$

For a prestressed member alone (without composite slab), $S_c = S_b$. Therefore, M_{cr} reduces to

$$M_{cr} = \left(f_r + \frac{P_{se}}{A_c}\right) S_b + P_{se}e \tag{14}$$

Examples 24.6 and 24.7 illustrate computation of the cracking moment strength of prestressed members.

Note that an exception in 18.8.2 waives the $1.2M_{cr}$ requirement for (a) two-way, unbonded post-tensioned slabs, and (b) flexural members with shear and flexural strength at least twice that required by 9-2.

For flexural strength: $\phi M_n \geq 2M_u \geq 2(1.2M_d + 1.6M_\ell)$

For shear strength: $\phi V_n \geq 2V_u \geq 2(1.2V_d + 1.6V_\ell)$

The $1.2M_{cr}$ provision often requires excessive reinforcement for certain prestressed flexural members especially for short span hollow-core members. The exception is intended to limit the amount of additional reinforcement required to amounts that provide for ductility, and is comparable in concept to those for nonprestressed members in 10.5.3.

Introduced in the 1999 edition of the code, the waiver of the $1.2M_{cr}$ provision for two-way, unbonded post-tensioned slabs brings the code in line with current practices which have been shown to be technically sound and safe (Ref. 24.5).

Section 18.8.3 prescribes a qualitative requirement stating that some bonded reinforcement or tendons must be placed as close to the tension face as is practicable.

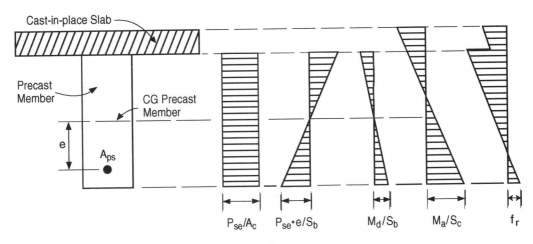

A_{ps} = area of prestressed reinforcement in tensile zone
A_c = area of precast member
S_b = section modulus for bottom of precast member
S_c = section modulus for bottom of composite member
P_{se} = effective prestress force
e = eccentricity of prestress force
M_d = dead load moment of composite member
M_a = additional moment to cause a stress in bottom fiber equal to modulus of rupture f_r

Figure 24-5 Stress Conditions for Evaluating Cracking Moment Strength

18.9   MINIMUM BONDED REINFORCEMENT

A minimum amount of bonded reinforcement is desirable in members with unbonded tendons. Reference to R18.9 is suggested.

For all flexural members with unbonded prestressing tendons, except two-way solid slabs, a minimum area of bonded reinforcement computed by Eq. (18-6) must be uniformly distributed over the precompressed tensile zone as close as practicable to the extreme tension fiber. Figure 26-6 illustrates application of Eq. (18-6).

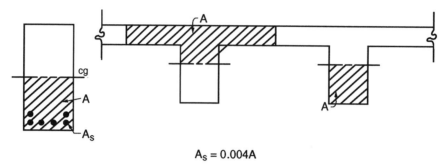

$A_s = 0.004A$

Figure 24-6 Bonded Reinforcement for Flexural Members

For solid slabs, the special provisions of 18.9.3 apply. Depending on the tensile stress in the concrete at service loads, the requirements for positive moment areas of solid slabs are illustrated in Fig. 24-7(a). Formerly, 18.9.3 applied only to flat plates. Starting with ACI 318-02, it also applies to two-way flat slab systems with drop panels.

The requirement for minimum area of bonded reinforcement in two-way flat plates at column supports was revised in the 1999 code edition to reflect the intent of the original research recommendations (Ref. 24.5). This revision increases the minimum reinforcement requirement over interior columns for rectangular panels in one direction, and, for square panels, doubles the minimum reinforcement requirement over exterior columns normal to the slab edge. Figure 24-7(b) illustrates the minimum bonded reinforcement requirements for the negative moment areas at column supports. The bonded reinforcement must be located within the width $c_2 + 2(1.5h)$ as shown, with a minimum of four bars spaced at not more than 12 in. Similarly, minimum bonded reinforcement should be provided parallel to slab edge.

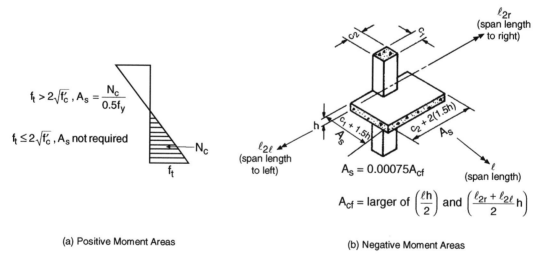

(a) Positive Moment Areas

(b) Negative Moment Areas

Figure 24-7 Bonded Reinforcement for Flat Plates

18.10.4 Redistribution of Negative Moments in Continuous Prestressed Flexural Members

The special provisions for moment redistribution in 8.4, apply equally to prestressed and nonprestressed continuous flexural members. See Part 8 for details.

18.11 COMPRESSION MEMBERS — COMBINED FLEXURE AND AXIAL LOADS

Provisions of the code for calculating the strength of prestressed members are the same as for members without prestressing. Additional considerations include (1) accounting for prestressing strains, and (2) using an appropriate stress-strain relation for the prestressing tendons. Example 24.7 illustrates the calculation procedure.

For compression members with an average concrete stress due to prestressing of less than 225 psi, minimum nonprestressed reinforcement must be provided (18.11.2.1). For compression members with an average concrete stress due to prestressing equal to or greater than 225 psi, 18.11.2.2 requires that all prestressing tendons be enclosed by spirals or lateral ties, except for walls.

REFERENCES

24.1 "PCI Design Handbook – Precast and Prestressed Concrete," MNL 120-04 6th Edition, Precast/Prestressed Concrete Institute, Chicago, 2004, 750 pp.

24.2 Mast, R. F., "Analysis of Cracked Prestressed Sections: A Practical Approach," *PCI Journal,* Vol. 43, No. 4, July-August 1975, pp. 43-75.

24.3 PCI Committee on Prestress Losses, "Recommendations for Estimating Prestress Losses," *PCI Journal*, Vol. 20, No. 4, July-August 1975, pp. 43-75.

24.4 Zia, Paul, et al., "Estimating Prestress Losses," *Concrete International: Design and Construction*, Vol. 1, No. 6, June 1979, pp. 32-38.

24.5 ACI 423.3R-96 Report, ""Recommendations for Concrete Members Prestressed with Unbonded Tendons," American Concrete Institute, Farmington Hills, Michigan.

Example 24.1—Estimating Prestress Losses

For the simply supported double-tee shown below, estimate loss of prestress using the procedures of Ref. 24.4, as outlined earlier under "Computation of Losses." Assume the unit is manufactured in Green Bay, WI.

live load = 40 psf
roof load = 20 psf
dead load = 47 psf = 468 plf
span = 48 ft
f'_{ci} = 3500 psi
f'_c = 5000 psi
8 - 0.5 in. diameter low-relaxation strands
A_{ps} = 8 (0.153 in.^{2}) = 1.224 in.^{2}
e = 9.77 in. (all strands straight)
f_{pu} = 270,000 psi
f_{py} = 0.90f_{pu}
jacking stress = 0.74f_{pu} = 200 ksi

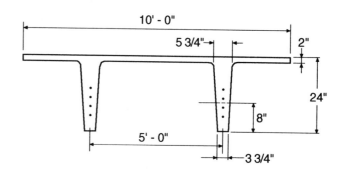

Assume the following for loss computations:
E_{ci} = 3590 ksi
E_c = 4290 ksi
E_s = 28,500 ksi

Section Properties
A_c = 449 in.^{2}
I_c = 22,469 in.^{4}
y_b = 17.77 in.
y_t = 6.23 in.
V/S = 1.35 in.

Calculations and Discussion	**Code Reference**

1. Elastic Shortening of Concrete (ES); using Eq. (1)

$$ES = K_{es}E_s \frac{f_{cir}}{E_{ci}} = 1.0\,(28{,}500)\,\frac{0.725}{3590} = 5.8 \text{ ksi}$$

where

K_{es} = 1.0 for pretensioned members

$$f_{cir} = K_{cir}f_{cpi} - f_g$$

$$= K_{cir}\left(\frac{P_{pi}}{A_c} + \frac{P_{pi}e^2}{I_c}\right) - \frac{M_d e}{I_c}$$

$$= 0.9\left(\frac{245}{449} + \frac{245 \times 9.77^2}{22{,}469}\right) - \frac{1617 \times 9.77}{22{,}469} = 0.725 \text{ ksi}$$

K_{cir} = 0.9 for pretensioned members

		Code
Example 24.1 (cont'd)	**Calculations and Discussion**	**Reference**

$P_{pi} = 0.74 f_{pu} A_{ps} = 0.74 (270)(1.224) = 245$ kips

$M_d = 0.468 \times 48^2 \times \dfrac{12}{8} = 1617$ in.-kips (dead load of unit)

2. Creep of Concrete (CR); using Eq. (3)

$CR = K_{cr} \dfrac{E_s}{E_c} (f_{cir} - f_{cds}) = 2.0 \times \dfrac{28,500}{4290} (0.725 - 0.30) = 5.6$ ksi

where $f_{cds} = M_{ds} \dfrac{e}{I_c} = 691 \times \dfrac{9.77}{22,469} = 0.30$ ksi

$M_{ds} = 0.02 \times 10 \times 48^2 \times \dfrac{12}{8} = 691$ in.-kips (roof load only)

and $K_{cr} = 2.0$ for pretensioned members.

3. Shrinkage of Concrete (SH); using Eq. (4)

$SH = 8.2 \times 10^{-6} K_{sh} E_s \left(1 - 0.06 \dfrac{V}{S}\right)(100 - RH)$

$= 8.2 \times 10^{-6} \times 1.0 \times 28,500 (1 - 0.06 \times 1.35)(100 - 75) = 5.4$ ksi

RH = average relative humidity surrounding the concrete member from Fig. 24-3. For Green Bay, Wisconsin, RH = 75%

and $K_{sh} = 1.0$ for pretensioned members.

4. Relaxation of Tendon Stress (RE); using Eq. (5)

$RE = [K_{re} - J(SH + CR + ES)] C$

$= [5 - 0.04(5.4 + 5.6 + 5.8)] 0.95 = 4.1$ ksi

where, for 270 Grade low-relaxation strand:

$K_{re} = 5$ ksi (Table 24-3)

$J = 0.040$ (Table 24-3)

$C = 0.95$ (Table 24-4 for $\dfrac{f_{pi}}{f_{pu}} = 0.74$)

5. Total allowance for loss of prestress

$ES + CR + SH + RE = 5.8 + 5.6 + 5.4 + 4.1 = 20.9$ ksi   *18.6.1*

		Code
Example 24.1 (cont'd)	**Calculations and Discussion**	**Reference**

6. Stress, f_p, and force, P_p, immediately after transfer.

   Assume that one-fourth of relaxation loss occurs prior to release.

   $f_p = 0.74 f_{pu} - (ES + 1/4\ RE)$

   $\qquad = 0.74\ (270) - [5.8 + 1/4\ (4.1)] = 193.0$ ksi

   $P_p = f_p A_{ps} = 193.0 \times 1.224 = 236$ kips

7. Effective prestress stress f_{se} and effective prestress force P_e after all losses

   $f_{se} = 0.74 f_{pu}$ - allowance for all prestress losses

   $\qquad = 0.74\ (270) - 20.9 = 179$ ksi

   $P_e = f_{se} A_{ps} = 179 \times 1.224 = 219$ kips

Example 24.2—Investigation of Stresses at Prestress Transfer and at Service Load

For the simply supported double-tee considered in Example 24.1, check all permissible concrete stresses immediately after prestress transfer and at service load assuming the unit is used for roof framing. Use losses computed in Example 24.1.

live load = 40 psf
roof load = 20 psf
dead load = 47 psf = 468 plf
span = 48 ft
f'_{ci} = 3500 psi
f'_c = 5000 psi
8 - 0.5 in. diameter low-relaxation strands
A_{ps} = 8 (0.153 in.^{2}) = 1.224 in.^{2}
e = 9.77 in. (all strands straight)
f_{pu} = 270,000 psi
f_{py} = 0.90f_{pu}
jacking stress = 0.74f_{pu} = 200 ksi
stress after transfer = 193 ksi
force after transfer = P_p = 1.224 × 193 = 236 kips

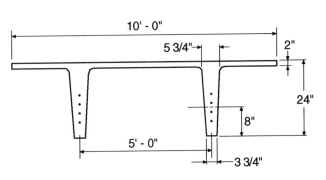

Section Properties
A_c = 449 in.^{2}
I_c = 22,469 in.^{4}
y_b = 17.77 in.
y_t = 6.23 in.
V/S = 1.35 in.

Calculations and Discussion	Code Reference
1. Calculate permissible stresses in concrete.	18.4
At prestress transfer (before time-dependent losses):	18.4.1
Compression: $0.60f'_{ci}$ = 0.60(3500) = 2100 psi	
Tension: $6\sqrt{f'_{ci}}$ = 355 psi (at ends of simply supported members; otherwise $3\sqrt{f'_{ci}}$)	
At service load (after allowance for all prestress losses):	18.4.2
Compression: $0.45f'_c$ = 2250 psi - Due to sustained loads	
Compression: $0.60f'_c$ = 3000 psi - Due to total loads	
Tension: $12\sqrt{f'_c}$ = 849 psi	18.3.3(b)
2. Calculate service load moments at midspan:	

Example 24.2 (cont'd)   **Calculations and Discussion**   **Code Reference**

$$M_d = \frac{w_d \ell^2}{8} = \frac{0.468 \times 48^2}{8} = 134.8 \text{ ft-kips} \quad \text{(beam dead load)}$$

$$M_{ds} = \frac{w_{ds} \ell^2}{8} = \frac{0.02 \times 10 \times 48^2}{8} = 57.6 \text{ ft-kips} \quad \text{(roof dead load)}$$

$$M_{sus} = M_d + M_{ds} = 134.8 + 57.6 = 192.4 \text{ ft-kips} \quad \text{(sustained load)}$$

$$M_\ell = \frac{w_\ell \ell^2}{8} = \frac{0.04 \times 10 \times 48^2}{8} = 115.2 \text{ ft-kips} \quad \text{(live load)}$$

$$M_{tot} = M_d + M_{ds} + M_\ell = 134.8 + 57.6 + 115.2 = 307.6 \text{ ft-kips} \quad \text{(total load)}$$

3. Calculate service load moments at transfer point

   Assume transfer point located at $50d_b = 25$ in. from end of beam. Assume distance from end of beam to center of support is 4 in. Therefore, $x = 25 - 4 = 21$ in. $= 1.75$ ft.   *11.4.3*

   $$M_d = \frac{w_d x}{2}(\ell - x) = \frac{0.468 \times 1.75}{2}(48 - 1.75) = 18.9 \text{ ft-kips} \quad \text{(beam dead load)}$$

   Additional moment calculations at this location are unnecessary because conditions immediately after release govern at this location.

4. Calculate extreme fiber stresses by "linear elastic theory" which leads to the following well known formulas:

   $$f_t = \frac{P}{A} - \frac{Pey_t}{I} + \frac{My_t}{I}$$

   $$f_b = \frac{P}{A} + \frac{Pey_b}{I} - \frac{My_b}{I}$$

   where, from Example 24.1

   $P = P_p = 236$ kips (immediately after transfer)

   $P = P_e = 219$ kips (at service load)

| **Example 24.2 (cont'd)** | **Calculations and Discussion** | **Code Reference** |

Table 24-4 Stresses in Concrete Immediately after Prestress Transfer (psi)

	At Assumed Transfer Point		At Mid Span	
	Top	Bottom	Top	Bottom
P_p/A	+526	+526	+526	+526
$P_p ey/I$	-639	+1824	-639	+1824
$M_d y/I$	+63	-180	+448	-1279
Total	-50 (O.K.)	+2170 (say O.K.)	+335 (O.K.)	+1071 (O.K.)
Permissible	-355	+2100	+2100	+2100

Compression (+)
Tension (-)

Table 24-5 Stresses in Concrete at Service Loads (psi)

	At Midspan – Sustained Loads		At Midspan – Total Loads	
	Top	Bottom	Top	Bottom
P_e/A	+488	+488	+488	+488
$P_e ey/I$	-594	+1695	-594	+1695
My/I	+640	-1826	+1023	-2919
Total	+534 (O.K.)	+357 (O.K.)	+917 (O.K.)	-736 (O.K.)
Permissible	+2250	+2250	+3000	-849

Compression (+)
Tension (-)

Example 24.3—Flexural Strength of Prestressed Member Using Approximate Value for f_{ps}

Calculate the nominal moment strength of the prestressed member shown.

$f'_c = 5000$ psi

$f_{pu} = 270,000$ psi (low-relaxation strands; $f_{py} = 0.90 f_{pu}$)

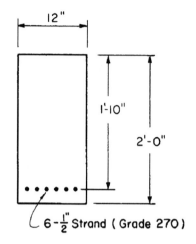

	Calculations and Discussion	Code Reference

1. Calculate stress in prestressed reinforcement at nominal strength using approximate value for f_{ps}. For a fully prestressed member, Eq. (18-3) reduces to:

$$f_{ps} = f_{pu}\left(1 - \frac{\gamma_p}{\beta_1} \rho_p \frac{f_{pu}}{f'_c}\right)$$ Eq. (18-3)

$$= 270\left(1 - \frac{0.28}{0.80} \times 0.00348 \times \frac{270}{5}\right) = 252 \text{ ksi}$$

where

$\gamma_p = 0.28$ for $\dfrac{f_{py}}{f_{pu}} = 0.90$ for low-relaxation strand

$\beta_1 = 0.80$ for $f'_c = 5000$ psi   *10.2.7.3*

$\rho_p = \dfrac{A_{ps}}{bd_p} = \dfrac{6 \times 0.153}{12 \times 22} = 0.00348$

		Code
Example 24.3 (cont'd)	**Calculations and Discussion**	**Reference**

2. Calculate nominal moment strength from Eqs. (9) and (10) of Part 24

    Compute the depth of the compression block:

    $$a = \frac{A_{ps}f_{ps}}{0.85bf'_c} = \frac{0.918 \times 252}{0.85 \times 12 \times 5} = 4.54 \text{ in.} \quad \text{Eq. (10)}$$

    $$M_n = A_{ps}f_{ps}\left(d_p - \frac{a}{2}\right) \quad \text{Eq. (9)}$$

    $$M_n = 0.918 \times 252\left(22 - \frac{4.54}{2}\right) = 4565 \text{ in-kips} = 380 \text{ ft-kips}$$

3. Check to see if tension controlled   *10.3.4*

    $$c/d_p = (a/\beta_1)/d_p = \left(\frac{4.54}{0.80}\right)/22$$

    $c/d_p = 0.258 < 0.375$ *R9.3.2.2*

    Tension controlled $\phi = 0.9$

Example 24.4—Flexural Strength of Prestressed Member Based on Strain Compatibility

The rectangular beam section shown below is reinforced with a combination of prestressed and nonprestressed strands. Calculate the nominal moment strength using the strain compatibility (moment-curvature) method.

f'_c = 5000 psi

f_{pu} = 270,000 psi (low-relaxation strand; f_{py} = 0.9f_{pu})

E_{ps} = 28,500 ksi

jacking stress = 0.74f_{pu}

losses = 31.7 ksi (calculated by method of Ref. 24.4. See 18.6 — Loss of Prestress for procedure.)

	Code
Calculations and Discussion	**Reference**

1. Calculate effective strain in prestressing steel.

    ε = (0.74f_{pu} - losses)/E_{ps} = (0.74 × 270 - 31.7)/28,500 = 0.0059

2. Draw strain diagram at nominal moment strength, defined by the maximum concrete compressive strain of 0.003 and an assumed distance to the neutral axis, c. For f'_c = 5000, β_1 = 0.80.

    18.3.1

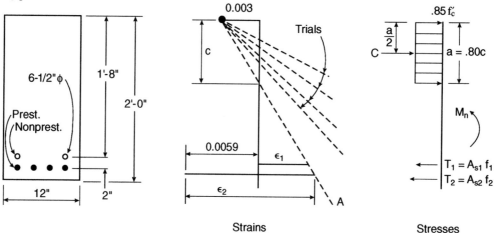

3. Obtain equilibrium of horizontal forces.

    The "strain line" drawn above from point 0 must be located to obtain equilibrium of horizontal forces:

    $C = T_1 + T_2$

    To compute T_1 and T_2, strains ε_1 and ε_2 are used with the stress-strain relation for the strand to determine the corresponding stresses f_1 and f_2. Equilibrium is obtained using the following iterative procedure:

Example 24.4 (cont'd)   **Calculations and Discussion**   **Code Reference**

    a.   assume c (location of neutral axis)

    b.   compute ε_1 and ε_2

    c.   obtain f_1 and f_2 from the equations at the bottom of Fig. 24-1.

    d.   compute $a = \beta_1 c$

    e.   compute $C = 0.85 f'_c ab$

    f.   compute T_1 and T_2

    g.   check equilibrium using $C = T_1 + T_2$

    h.   if $C < T_1 + T_2$, increase c, or vice versa and return to step b of this procedure. Repeat until satisfactory convergence is achieved.

Estimate a neutral axis location for first trial. Estimate stressed strand at 260 ksi, unstressed strand at 200 ksi.

$$T = \Sigma A_{ps} f_s = 0.306 (200) + 0.612 (260) = 220 \text{ kips} = C$$

$$a = C/(0.85 f'_c b) = 220/(0.85 \times 5 \times 12) = 4.32 \text{ in.}$$

$$c = a/\beta_1 = 4.32/0.80 = 5.4 \text{ in. Use } c = 5.4 \text{ in. for first try}$$

The following table summarizes the iterations required to solve this problem:

Trial No.	c in.	ε_1	ε_2	f_1 ksi	f_2 ksi	a in.	C kips	T_1 kips	T_2 kips	$T_1 + T_2$ kips
1	5.4	0.0081	0.0151	231	265	4.32	220	71	162	233
2 O.K.	5.6	0.0077	0.0147	220	265	4.48	228.5	67	162	229

4.   Calculate nominal moment strength.

    Using $C = 228.5$ kips, $T_1 = 67$ kips and $T_2 = 162$ kips, the nominal moment strength can be calculated as follows by taking moments about T_2:

$$M_n = \{[(d_2 - a/2) \times C] - [(d_2 - d_1) \times T_1]\}/12$$

$$= \{[(22 - (4.48/2) \times 228.5] - [(22 - 20) \times 67]\}/12 = 365 \text{ ft-kips}$$

Example 24.5—Tension-Controlled Limit for Prestressed Flexural Member

For the double tee section shown below, check limits for the prestressed reinforcement provided.

f'_c = 5000 psi
22 - 0.5 in. diameter low-relaxation strands
A_{ps} = 22 (0.153 in.^{2}) = 3.366 in.^{2}
f_{pu} = 270,000 psi
f_{py} = 0.90f_{pu}

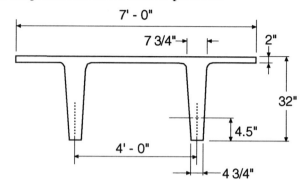

	Calculations and Discussion	Code Reference

Example No. 24.5.1

1. Calculate stress in prestressed reinforcement at nominal strength using Eqs. (6) and (8).

$$\omega_{pu} = \frac{A_{ps}f_{pu}}{bd_p f'_c} = \frac{3.366 \times 270}{84 \times 27.5 \times 5} = 0.079$$

$$f_{ps} = f_{pu}\left(1 - \frac{\gamma_p}{\beta_1}\omega_{pu}\right) = 270\left(1 - \frac{0.28}{0.8} \times 0.079\right) = 263 \text{ ksi} \qquad \text{Eq. (18-3)}$$

where

γ_p = 0.28 for low-relaxation strands

β_1 = 0.80 for f'_c = 5000 psi   10.2.7.3

2. Calculate required depth of concrete stress block.

$$a = \frac{A_{ps}f_{ps}}{0.85bf'_c} = \frac{3.366 \times 263}{0.85 \times 84 \times 5} = 2.48 \text{ in.} > h_f = 2 \text{ in.}$$

3. Calculate area of compression zone.

$$A_c = \frac{A_{ps}f_{ps}}{0.85f'_c} = \frac{3.366 \times 263}{0.85 \times 5} = 208.3 \text{ in.}^2$$

4. Find depth a of stress block, and c.

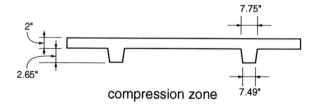

$A = 2 \times 84 = 168 \text{ in.}^2$
$+ 2 \times 2.65 \times \frac{7.75 + 7.49}{2} = 40.4$
                            208.4 in.^{2}

a = 4.65 in.
c = a/β_1 = 4.65/0.8 = 5.81 in.

| Example 24.5 (cont'd) | Calculations and Discussion | Code Reference |

5. Check to see if tension-controlled — *10.3.4*

    $c/d_t = 5.81/30.0 = 0.19 < 0.375$ — *R9.3.2.2*

    (By definition, dimension "d_t" should be measured to the bottom strand)

    Section is tension-controlled

Note: In Step 1, Eq. 18-3 was used to find f_{ps}. But, with the stress block in the web, the value of ω_{pu} used in Eq. (18-5) was not correct, although the error is small in this case. A strain compatibility analysis gives $c = 6.01$ in. and $f_{ps} = 266$ ksi.

Example No. 24.5.2

Check the limits of reinforcement using a 3 in. thick flange on the member in Example 24.5.1. The overall depth remains 32 in.

1. $f_{ps} = 263$ ksi *No change from Example 24.5.1*

2. $a = 2.48$ *No change from Example 24.5.1, Step 2*

    $< h_f = 3$ in.

    Since the stress block is entirely within the flange, the section acts effectively as a rectangular section.

3. Check c/d_p ratio

    $c = a/\beta_1 = 2.48/0.8 = 3.10$ in.

    $c/d_t = 3.10/30.0 = 0.10 < 0.375$ — *R9.3.2.2*

    Section is tension controlled.

Example 24.6—Cracking Moment Strength and Minimum Reinforcement Limit for Non-composite Prestressed Member

For the non-composite prestressed member of Example 24.3, calculate the cracking moment strength and compare it with the design moment strength to check the minimum reinforcement limit.

f'_c = 5000 psi
f_{pu} = 270,000 psi
jacking stress = $0.70 f_{pu}$
Assume 20% losses

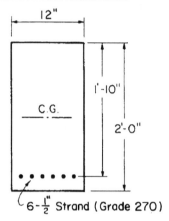

	Code
Calculations and Discussion	Reference

1. Calculate cracking moment strength using Eq. (14) developed in Part 24.

$$M_{cr} = \left(f_r + \frac{P_{se}}{A_c}\right) S_b + (P_{se} \times e) \qquad \text{Eq. (14)}$$

$f_r = 7.5\sqrt{f'_c} = 530$ psi *Eq. (9-10)*

Assuming 20% losses:

$P_{se} = 0.8 \times [6 \times 0.153 \times (0.7 \times 270)] = 139$ kips

$S_b = \dfrac{bh^2}{6} = \dfrac{12 \times 24^2}{6} = 1152$ in.^{3}

$A_c = bh = 12 \times 24 = 288$ in.^{2}

$e = 12 - 2 = 10$ in.

$M_{cr} = \left(0.530 + \dfrac{139}{288}\right) 1152 + (139 \times 10) = 2557$ in.-kips = 213 ft-kips

Note that cracking moment strength needs to be determined for checking minimum reinforcement per 18.8.3.

2. Section 18.8.3 requires that the total reinforcement (prestressed and nonprestressed) must be adequate to develop a design moment strength at least equal to 1.2 times the cracking moment strength. From Example 24.3, M_n = 380 ft-kips.

Example 24.6 (cont'd)	Calculations and Discussion	Code Reference

$\phi M_n \geq 1.2 M_{cr}$ *18.8.3*

$0.9\,(380) > 1.2\,(213)$

$342 > 256$ O.K.

Example 24.7—Cracking Moment Strength and Minimum Reinforcement Limit for Composite Prestressed Member

For the 6 in. precast solid flat slab with 2 in. composite topping, calculate the cracking moment strength. The slab is supported on bearing walls with 15 ft span.

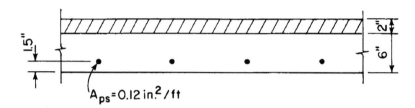

Section properties per foot of width:

A_c = 72 in.^{2} (precast slab)
S_b = 72 in.^{3} (precast slab)
S_c = 132.7 in.^{3} (composite section)

f'_c = 5000 psi (all-lightweight concrete, w_c = 125 pcf)
f_{pu} = 250,000 psi (stress-relieved strand)
jacking stress = $0.70 f_{pu}$
Assume 25% losses

Calculations and Discussion	Code Reference

1. Calculate cracking moment strength using Eq. (13) developed for unshored composite members. All calculations are based on one foot width of slab.   **18.8.3**

$$M_{cr} = \left(f_r + \frac{P_{se}}{A_c} + \frac{P_{se} e}{S_b} \right) S_c - M_d \left(\frac{S_c}{S_b} - 1 \right) \quad (13)$$

$f_r = 0.75 \left(7.5 \sqrt{5000} \right) = 398$ psi reduced for all-lightweight concrete   **9.5.2.3**

Assuming 25% losses:

$P_{se} = 0.75 \, (0.12 \times 0.7 \times 250) = 15.75$ kips

$e = 3 - 1.5 = 1.5$ in.

$w_d = (6 + 2)/12 \times 125 = 83$ psf $= 0.083$ ksf (weight of precast slab + composite topping)

$$M_d = \frac{w_d \ell^2}{8} = \frac{0.083 \times 15^2}{8} = 2.33 \text{ ft-kips} = 28.0 \text{ in.-kips}$$

$$M_{cr} = \left[\left(0.398 + \frac{15.75}{72} + \frac{15.75 \times 1.5}{72} \right) 132.7 \right] - \left[28.0 \left(\frac{132.7}{72} - 1 \right) \right]$$

$= 125.4 - 23.6 = 101.8$ in.-kips

		Code
Example 24.7 (cont'd)	**Calculations and Discussion**	**Reference**

2. Calculate design moment strength and compare with cracking moment strength. All calculations based on one foot width of slab.

    $A_{ps} = 0.12$ in.^{2}, $d_p = 8.0 - 1.5 = 6.5$ in.

    $$\rho_p = \frac{A_{ps}}{bd_p} = \frac{0.12}{12 \times 6.5} = 0.00154$$

    With no additional tension or compression reinforcement, Eq. (18-3) reduces to:

    $$f_{ps} = f_{pu}\left(1 - \frac{\gamma_p}{\beta_1} \rho_p \frac{f_{pu}}{f'_c}\right) = 250\left(1 - \frac{0.4}{0.8} \times 0.00154 \times \frac{250}{5}\right) = 240.4 \text{ ksi}$$

    $$a = \frac{A_{ps}f_{ps}}{0.85f'_c b} = \frac{0.12 \times 240.4}{0.85 \times 5 \times 12} = 0.57 \text{ in.}$$

    $M_n = A_{ps}f_{ps}(d_p - a/2) = 0.12 \times 240.4 (6.5 - 0.57/2) = 179.3$ in.-kips

    $\phi M_n = 0.9 (179.3) = 161.4$ in.-kips

    $\phi M_n \geq 1.2 (M_{cr})$ *18.8.3*

    $161.4 > 1.2 (101.8) = 122.2$ O.K.

Example 24.8—Prestressed Compression Member

For the short column shown, calculate the nominal strength M_n for a nominal axial load $P_n = 30$ kips.

Calculate design strength.

f'_c = 5000 psi
f_{pu} = 270,000 psi (low-relaxation strand)
jacking stress = $0.70 f_{pu}$
Assume 10% losses

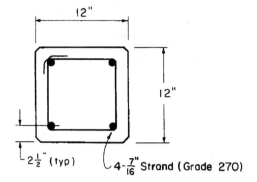

	Calculations and Discussion	Code Reference

Eq. 18-3 should not be used when prestressing steel is in the compression zone. The same "strain compatibility" procedure used for flexure must be used here. The only difference is that for columns the load P_n must be included in the equilibrium of axial forces.

1. Calculate effective prestress.

   $f_{se} = 0.9 \times 0.7 f_{pu} = 0.9 \times 0.7 \times 270 = 170$ ksi

   $P_e = A_{ps} f_{se} = 4 \times 0.115 \times 170 = 78.2$ kips

2. Calculate average prestress on column section.

   $f_{pc} = \dfrac{P_e}{A_g} = \dfrac{78.2}{12^2} = 0.54$ ksi

   Minimum reinforcement as per 10.9.1 not required because $f_{pc} = 0.54$ ksi > 0.225 ksi.   18.11.2.1

   Since $f_{pc} = 0.54$ ksi > 0.225 ksi, lateral ties satisfying the requirements of 18.11.2.2 must enclose all prestressing tendons.

3. Calculate effective strain in prestressing steel.

   $\varepsilon = \dfrac{f_{se}}{E_p} = \dfrac{170}{28,500} = 0.0060$

4. Draw strain diagram at nominal moment strength, defined by the maximum concrete compressive strain of 0.003 and an assumed distance to the neutral axis, c. For $f'_c = 5000$ psi, $\beta_1 = 0.80$.

		Code
Example 24.8 (cont'd)	**Calculations and Discussion**	**Reference**

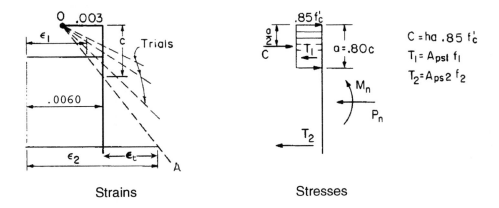

Strains   Stresses

5. Obtain equilibrium of axial forces. The strain line OA drawn above, must be such that equilibrium of axial forces exists.

   $C = T_1 + T_2 + P_n$

   This can be done by trial-and-error as outlined in Example 24.4. Assuming different values of c, the following trial table is obtained:

Trial No.	c in.	ε_1	ε_2	f_1* ksi	f_2* ksi	a in.	C kips	T_1 kips	T_2 kips	$T_1 + T_2 + P_n$ kips
1	3.0	0.0055	0.0125	157	263	2.40	122.4	36.1	60.4	126.5
2	3.2	0.0053	0.0119	152	261	2.56	130.6	35.0	60.2	135.2
3 O.K.	3.1	0.0054	0.0122	154	262	2.48	126.5	35.5	60.3	125.8

*From equation in Fig. 24-1.

6. Calculate nominal moment strength.

   Using C = 126.5 kips (from the sum of the other forces), P_n = 30 kips, T_1 = 35.5 kips, and T_2 = 60.3 kips, the moment strength can be calculated as follows by taking moments about P_n, located at the centroid of the section:

   $M_n = \{[(h/2 - a/2) \times C] - [(h/2 - 2.5) \times T_1] + [(h/2 - 2.5) \times T_2]\}/12$

   $= [(4.76 \times 126.5) - (3.5 \times 35.5) + (3.5 \times 60.3)]/12 = 57.4$ ft-kips

7. Calculate design strength

   $\varepsilon_t = \varepsilon_2 - 0.0060 = 0.0122 - 0.0060 = 0.0062 > 0.005$

   Section is tension-controlled $\phi = 0.9$ 10.3.4

   $\phi P_n = 0.9 \times 30 = 27$ kips   9.3.2.1

   $\phi M_n = 0.9 \times 57.4 = 51.7$ ft-kips

Example 24.9—Cracked Section Design When Tension Exceeds $12\sqrt{f'_c}$

Do the serviceability analysis for the beam shown.

f'_c = 6000 psi
depth d_p = 26 in.
effective prestress f_{se} = 150 ksi
decompression stress f_{dc} = 162 ksi
span = 40 ft

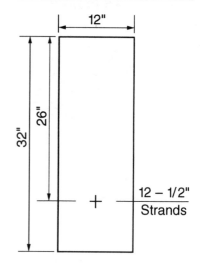

	w k/ft	Midspan moments in.-k
Self-weight	0.413	992
Additional dead load	1.000	2400
Live load	1.250	3000
Sum	2.663	6392

Calculations and Discussion	Code Reference

1. Check tension at service loads, based on gross section.

   $P = A_{ps}f_{se}$ = 1.836 × 150 = 275.4 kip

   P/A = 275.4/384 = 0.717 ksi

   Pe/S = 275.4 × 10/2048   1.345

   $S = bh^2/6 = 12(32)^2/6$ = 2.048 in.^{3}

   $\Sigma M/S$ = 6392/2048   − 3.121
   _____
   − 1.059 ksi tension

   $12\sqrt{f'_c} = 12\sqrt{f'_c}$ = 930 psi = 0.930 ksi

   Tension exceeds $12\sqrt{f'_c}$. Design as a Class C member   18.3.3(c)

2. A cracked section stress analysis is required   18.3.4

   Cracked transformed section properties, similar to those used for working stress analysis of ordinary (nonprestressed) reinforced concrete will be used. The area of steel elements is replaced by a "transformed" area of concrete equal to n times the actual steel area, where n is the ratio of the modulus of elasticity of steel to that of concrete.

   The modular ratio n = E_{ps}/E_c = 28,500/4415 = 6.455

   where $E_c = 57,000\sqrt{f'_c} = 57,000\sqrt{6000}$ = 4415 ksi   8.5.1

   The transformed steel area A_t is:

   $A_t = nA_{ps}$ = 6.455 × 1.836 = 11.85 in.^{2}

24-36

		Code
Example 24.9 (cont'd)	**Calculations and Discussion**	**Reference**

The force P_{dc} at decompression (when the stress in the concrete at the same level as the prestressing steel is zero) is:

$$P_{dc} = A_{ps}f_{dc} = 1.836 \times 162 = 297.4 \text{ kips}$$

3. The stress analysis of a cracked section with axial load (from the prestress) requires, at best, the solution of a cubic equation. A more general approach is to find a neutral axis location that satisfies horizontal force equilibrium and produces the given bending moment. Reference 24.2 gives one way to accomplish this. It is too lengthy to be presented in detail here.

   The results give a neutral axis depth c of 17.26 in., with a concrete stress f_c of 3.048 ksi and a transformed steel stress $\Delta f_{ps}/n$ of 1.545 ksi. The actual Δf_{ps} is $1.545 \times 6.455 = 9.97$ ksi.

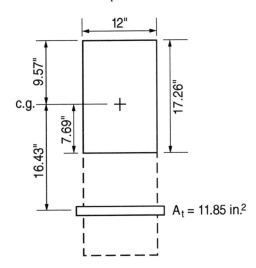
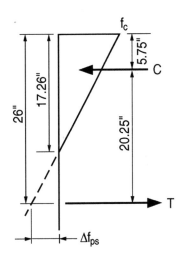

4. The transformed section properties are

   $A = 219$ in.²

   $I = 8524$ in.⁴

   $y_t = 9.57$ in.

5. Equilibrium may be checked manually,

   $C = f_c bc/2 = 3.048\ (12)(17.26)/2$ $= 315.7k$

   C acts at top kern of compression zone

   $\quad = d_c/3$ for rectangular area

   $17.26/3 = 5.75$ in.

   $T = P_{dc} + \Delta f_{ps}\ (A_{ps}) = 297.4 + 9.97\ (1.836)$ $= 315.7k = C$ Check

   $M = C$ or $T \times$ lever arm

   $\quad = 315.7 \times 20.25$ $= 6392$ in.-kips Check

| Example 24.9 (cont'd) | Calculations and Discussion | Code Reference |

6. Check limits on Δf_{ps}

   Δf_{ps} is less than code limit of 36 ksi   O.K.   18.4.3.3

   Δf_{ps} is less than 20 ksi, so the spacing requirements of 18.4.4.1 and 18.4.4.2 need not be applied.   18.4.3.3

7. Check deflection

   Live load deflection calculations based on a cracked section analysis are required for Class C members.   9.5.4.2

   Use the "bilinear moment-deflection relationship," as described in Ref. 24.1   9.5.4.2

8. Find cracking moment M_{cr}, using P_{dc}

   $P/A + Pe/S + M_{cr}/S = f_r$

   modulus of rupture $f_r = 7.5\sqrt{f'_c} = 7.5\sqrt{6000} = 581$ psi   9.5.2.3

   $297/384 + 297 \times 10/2048 + 0.581 = M_{cr}/2048$

   $M_{cr} = 5750$ in.-kips

   $M_d = \underline{3392}$
   $ 2358 =$ live load moment applied to gross section

   balance of M_ℓ $642 =$ live load moment applied to cracked section

9. Compute deflections before and after cracking

   $$\Delta_L = \frac{5}{48} \frac{2358 L^2}{EI_g} + \frac{5}{48} \frac{642 L^2}{EI_{cr}}$$

   $$= \frac{5}{48} \frac{2358 \times 480^2}{4415 \times 32768} + \frac{5}{48} \frac{642 \times 480^2}{4415 \times 8524}$$

   $\Delta_L = 0.39 + 0.41 = 0.80$ in.

   Δ_L is $< (L/360 = 480/365 = 1.33$ in.)   O.K.   9.5.2.6
   Table 9.5(b)

   The live load deflection is shown graphically below.

Example 24.9 (cont'd)	Calculations and Discussion	Code Reference

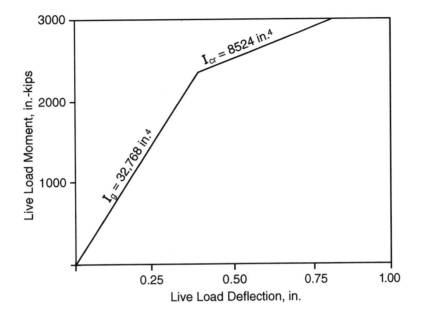

Blank

25

Prestressed Concrete Shear

UPDATE FOR THE '05 CODE

New Section 11.4.1 was added to clarify that the distance from extreme compression fiber to centroid of prestressed and nonprestressed longitudinal tension reinforcement need not be taken less than 80 percent of the overall height of member. Commentary was added in R11.5.6.1 to flag that results of tests on pretensioned concrete hollow core units with overall height greater than 12.5 in. have shown that web shear cracking strengths in end regions can be less than strength V_{cw} computed by Eq. (11-12).

GENERAL CONSIDERATIONS

The basic equations for shear design of prestressed concrete, Eqs. (11-10), (11-11), and (11-12), were introduced in the 1963 code. Although well founded on test results, they have been found difficult to apply in practice. A simplified Eq. (11-9) was introduced in the 1971 code.

In order to understand Eqs. (11-10) and (11-12), it is best to review the principles on which ACI shear design is based. These principles are <u>empirical</u>, based on a large number of tests.

- The shear resisted by concrete and the shear resisted by stirrups are additive.

- The shear resisted by the concrete after shear cracks form is at least equal to the shear existing in the concrete at the location of the shear crack at the time the shear crack forms.

How does one compute the shear resisted by the concrete at the time a shear crack forms? There are two possibilities.

1. Web shear. A diagonal shear crack originates in the web, near the neutral axis, caused by principal tension in the web.

2. Flexure-shear. A crack starts as a flexural crack on the tension face of a flexural member. It then extends up into the web, and develops into a diagonal shear crack. This can happen at a much lower principal tensile stress than that causing a web shear crack, because of the tensile stress concentration at the tip of the crack.

Web Shear

The apparent tensile strength of concrete in direct tension is about $4\sqrt{f'_c}$. When the principal tension at the center of gravity of the cross section reaches $4\sqrt{f'_c}$, a web shear crack will occur. Section 11.4.3.2 states "...V_{cw} shall be computed as the shear...that results in a principal tensile stress of $4\sqrt{f'_c}$ ..."

The compression from the prestress helps to reduce the principal tension. The computation of principal tension due to combined shear and compression can be somewhat tedious. The code gives a simplified procedure.

$$V_{cw} = (3.5\sqrt{f'_c} + 0.3f_{pc})b_w d_p + V_p \qquad \text{Eq. (11-12)}$$

The term V_p in Eq. (11-12) is the vertical component of the tension in the prestressing tendons. This is additive for web shear strength (but not for flexure-shear strength).

A comparison to test results is shown below.

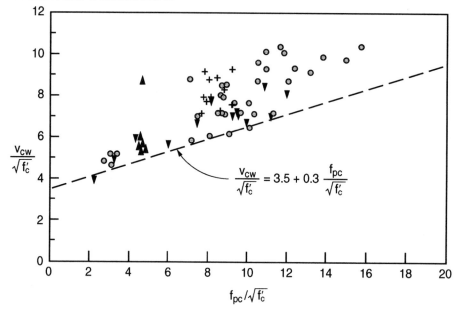

Figure 25-1 Diagonal Cracking in Regions not Previously Cracked

The compression from prestressing increases the shear strength by 30 percent of the P/A level, f_{pc}, of compression.

For nonprestressed beams, the principal tension at the center of gravity of the section is equal to the shear. Why does Eq. (11-3) for shear in nonprestressed members permit only $2\sqrt{f'_c}$ shear resisted by the concrete? Because shear strength is reduced by flexural cracking. In nonprestressed beams, shear is almost always influenced by flexural tension. But, prestressing reduces the flexural cracking.

Flexure-Shear in Prestressed Concrete

In prestressed beams, flexural cracking is delayed by the prestress – usually until loaded beyond service load. It is worthwhile to account for the beneficial effects of prestressing.

In the 1950s, it was thought that draping strands would increase shear strength, by the vertical component V_p of the prestressing force. Tests showed just the opposite. Why? Because draping the strands reduces the flexural cracking strength in the shear span.

The tests were done with concentrated loads; whereas, the dead load of the beam was a uniform load. For this reason, when the shear design method was developed from the test results, the dead load and test load shears were treated separately.

Flexure-Shear

Equation (11-10) is the equation for shear resistance provided by the concrete, as governed by flexural cracks that develop into shear cracks. The shear strength of the concrete at a given cross section is taken equal to the shear at the section at the time a flexural crack occurs, plus a small increment of shear which transforms the flexural crack into an inclined crack. Equation (10-10) may be expressed in words as follows.

V_{ci} = shear existing at the time of flexural cracking plus an added increment to convert it into a shear crack. The added increment is $0.6b_w d_p \sqrt{f'_c}$.

The shear existing at the time of flexural cracking is the dead load shear V_d plus the added shear $V_i M_{cre}/M_{max}$.

What is the origin of the term $V_i M_{cre}/M_{max}$?

The term V_i is the factored ultimate shear at the section, less the dead load shear.

The term M_{cre} is the <u>added</u> moment (over and above stresses due to prestress and dead load) causing $6\sqrt{f'_c}$ tension in the extreme fiber.

The added moment M_{cre} is calculated by finding the bottom fiber stress f_{pe} due to prestress, subtracting the bottom fiber stress f_d due to dead loads, adding $6\sqrt{f'_c}$ tension, and multiplying the result by the section modulus for the section resisting live loads. This is Eq. (11-11) of the code.

$$M_{cr} = (I/y_t)(6\sqrt{f'_c} + f_{pe} - f_d) \qquad \text{Eq. (11-11)}$$

Note: In the above discussion, "bottom" means "tension side" for continuous members.

The term M_{max} is the factored ultimate moment of the section, less the dead load moment.

To better understand the meaning of these terms and their use in Eq. (11-10), refer to Fig. 25-2.

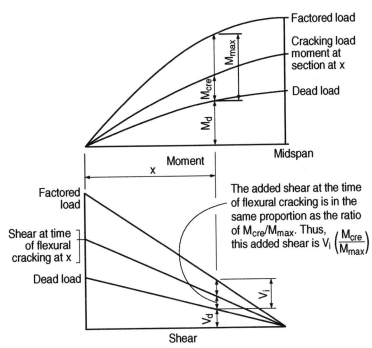

Figure 25-2 Origin of ($V_i M_{cre}/M_{max}$) Term in Eq. (11-10)

The quantity $V_i M_{cre}/M_{max}$ is the shear due <u>to an added</u> load (over and above the dead load) which causes the tensile stress in the extreme fiber to reach $6\sqrt{f'_c}$. The added load is applied to the composite section (if composite).

After a flexural crack forms, a small amount of additional shear is needed to transform the crack into a shear crack. This is determined empirically, as shown in Fig. 25-3.

25-3

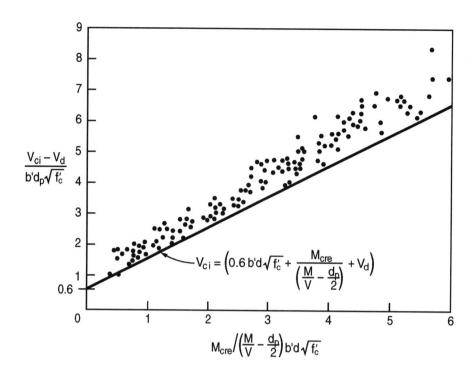

Figure 25-3 Diagonal Cracking in those Regions of Beams Previously Cracked in Flexure

The intercept at 0.6 produces the first term in Eq. (11-10), $0.6 b_w d_p \sqrt{f'_c}$.

Note: The quantity "$-d_p/2$" shown in the expressions of Fig. 25-3 was later dropped, as a conservative simplication.

The notation used in Eqs. (11-10) and (11-11) is follows.

M_{cre} = moment causing flexural cracking at section due to externally applied loads

M_{max} = maximum factored moment at section due to externally applied loads

V_i = factored shear force at section due to externally applied loads occurring simultaneously with M_{max}.

Unfortunately, the subscripts are confusing.

M_{cre} is <u>not</u> the total cracking moment. It is <u>not</u> the same as M_{cr} that is used to check for minimum reinforcement in Example 24.6.

M_{max} is not the total factored moment. It is the total factored moment less the dead load moment.

It would seem that V_i and M_{max} should have the same subscript, because they both relate to the differences between the same two loadings.

To make matters worse, the term "externally applied loads" is ambiguous. Apparently, dead load is not regarded as "externally applied," perhaps because the weight comes from the "internal" mass of the member. But, R11.4.3 says that superimposed dead load on a composite section should be considered an externally applied load. The commentary explains a good reason for this, but the confusion still exists.

The shear strength must be checked at various locations along the shear span, a process that is tedious. For manual shear calculations, the simplified process described in 11.4.2 is adequate for most cases.

11.1 SHEAR STRENGTH FOR PRESTRESSED MEMBERS

The basic requirement for shear design of prestressed concrete members is the same as for reinforced concrete members: the design shear strength ϕV_n must be greater than the factored shear force V_u at all sections (11.1).

$$\phi V_n \geq V_u \qquad \text{Eq. (11-1)}$$

For both reinforced and prestressed concrete members, the nominal shear strength V_n is the sum of two components: the nominal shear strength provided by concrete V_c and the nominal shear strength provided by shear reinforcement V_s.

$$V_n = V_c + V_s \qquad \text{Eq. (11-2)}$$

Therefore,

$$\phi V_c + \phi V_s \geq V_u$$

The nominal shear strength provided by concrete V_c is assumed to be equal to the shear existing at the time an inclined crack forms in the concrete.

Beginning with the 1977 code, shear design provisions have been presented in terms of shear forces V_n, V_c, and V_s, to better clarify application of the material strength reduction factor ϕ for shear design. In force format, the ϕ factor is directly applied to the material strengths, i.e., ϕV_c and ϕV_s.

11.1.2 Concrete Strength

Section 11.1.2 restricts the concrete strength that can be used in computing the concrete contribution because of a lack of shear test data for high strength concrete. The limit does not allow $\sqrt{f'_c}$ to be greater than 100 psi, which corresponds to $f'_c = 10,000$ psi. Note, the limit is expressed in terms of $\sqrt{f'_c}$, as it denotes diagonal tension. The limit can be exceeded if minimum shear reinforcement is provided as specified in 11.1.2.1.

11.1.3 Location for Computing Maximum Factored Shear

Section 11.1.3 allows the maximum factored shear V_u to be computed at a distance from the face of the support when all of the following conditions are satisfied:

a. the support reaction, in the direction of the applied shear, introduces compression into the end regions of the member,

b. loads are applied at or near the top of the member, and

c. no concentrated load occurs between the face of the support and the critical section.

For prestressed concrete sections, 11.1.3.2 states that the critical section for computing the maximum factored shear V_u is located at a distance of h/2 from the face of the support. This differs from the provisions for reinforced (nonprestressed) concrete members, in which the critical section is located at d from the face of the support. For more details concerning maximum factored shear force at supports, see Part 12.

11.2 LIGHTWEIGHT CONCRETE

The adjustments to shear strength for lightweight concrete given in 11.2 apply equally to prestressed and nonprestressed concrete members.

11.4 SHEAR STRENGTH PROVIDED BY CONCRETE FOR PRESTRESSED MEMBERS

Section 11.4 provides two approaches to determining the nominal shear strength provided by concrete V_c. A simplified approach is presented in 11.4.2 with a more detailed approach presented in 11.4.3. In both cases, the shear strength provided by concrete is assumed to be equal to the shear existing at the time an inclined crack forms in the concrete.

11.4.1 NOTATION

For prestressed members, the depth d used in shear calculations is defined as follows.

d = distance from extreme compression fiber to centroid of prestressed and nonprestressed longitudinal tension reinforcement, if any, but need not be less than **0.80*h***.

11.4.2 Simplified Method

The use of this simplified method is limited to prestressed members with an effective prestress force not less than 40 percent of the tensile strength of the flexural reinforcement, which may consist of only prestressed reinforcement or a combination of prestressed and conventional reinforcement.

$$V_c = \left(0.6\sqrt{f'_c} + 700\frac{V_u d_p}{M_u}\right) b_w d \qquad \text{Eq. (11-9)}$$

but need not be less than $2\sqrt{f'_c}\, b_w d$.

V_c must not exceed $5\sqrt{f'_c}\, b_w d$ or V_{cw} (11.4.3.2) computed considering the effects of transfer length (11.4.4) and debonding (11.4.5) which apply in regions near the ends of pretensioned members.

It should be noted that for the term $V_u d_p / M_u$ in Eq. (11-9), d_p must be taken as the actual distance from the extreme compression fiber to the centroid of the prestressed reinforcement rather than the 0.8h allowed elsewhere in the code.

The shear strength must be checked at various locations along the shear span. The commentary notes that for simply supported members subjected to uniform loads, the quantity of $V_u d_p / M_u$ may be expressed as:

$$\frac{V_u d_p}{M_u} = \frac{d_p(\ell - 2x)}{x(\ell - x)}$$

Figure 25-4, useful for a graphical solution, is also given in the commentary.

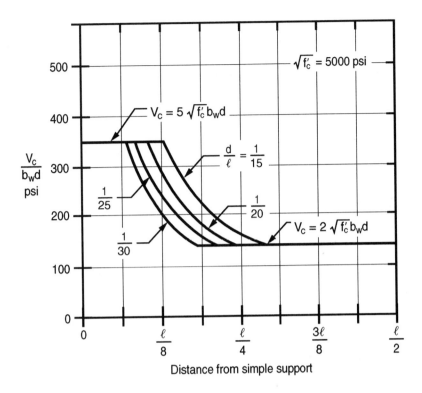

Figure 25-4 Application of Eq. (11-9) to Uniformly Loaded Prestressed Members (Fig. R11.4.2)

The use of this figure is illustrated in Example 25-2. Additional figures for graphical solutions of shear strength are given in Ref. 25.1.

11.4.3 Detailed Method

The origin of this method is discussed under General Considerations, at the beginning of Part 25.

Two types of inclined cracking have been observed in prestressed concrete members: flexure-shear cracking and web-shear cracking. Since the nominal shear strength from concrete is assumed to be equal to the shear causing inclined cracking of the concrete, the detailed method provides equations to determine the nominal shear strength for both types of cracking.

The two types of inclined cracking are illustrated in Fig. 25-5 which is found in R11.4.3. The nominal shear strength provided by concrete V_c is taken as the lesser shear causing the two types of cracking, which are discussed below. The detailed expressions for V_c in 11.4.3 may be difficult to apply without design aids, and should be used only when the simplified expression for V_c in 11.4.2 is not adequate.

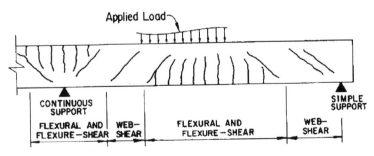

Figure 25-5 Types of Cracking in Concrete Beams (Fig. R11.4.3)

11.4.3.1 Flexure-Shear Cracking, V_{ci} — Flexure-shear cracking occurs when flexural cracks, which are initially vertical, become inclined under the influence of shear. The shear at which this occurs can be taken as

$$V_{ci} = 0.6\sqrt{f'_c}\, b_w d_p + V_d + \frac{V_i M_{cre}}{M_{max}} \qquad \text{Eq. (11-10)}$$

Note that V_{ci} need not be taken less than $1.7\sqrt{f'_c}\, b_w d$.

The added moment M_{cre} to cause flexural cracking is computed using the equation

$$M_{cre} = \left(\frac{I}{y_t}\right)\left(6\sqrt{f'_c} + f_{pe} - f_d\right) \qquad \text{Eq. (11-11)}$$

where f_{pe} is the compressive stress in concrete due to effective prestress force only (after allowance for all prestress losses) at the extreme fiber of the section where tensile stress is caused by externally applied loads.

V_{ci} usually governs for members subject to uniform loading. The total nominal shear strength V_{ci} is assumed to be the sum of three parts:

1. the shear force required to transform a flexural crack into an inclined crack — $0.6\sqrt{f'_c}\, b_w d_p$;

2. the unfactored dead load shear force — V_d; and

3. the portion of the remaining factored shear force that will cause a flexural crack to initially occur — $V_i M_{cre}/M_{max}$.

For non-composite members, V_d is the shear force caused by the unfactored dead load. For composite members, V_d is computed using the unfactored self weight plus unfactored superimposed dead load.

The load combination used to determine V_i and M_{max} is the one that causes maximum moment at the section under consideration. The value V_i is the factored shear force resulting from the externally applied loads occurring simultaneously with M_{max}. For composite members, V_i may be determined by subtracting V_d from the shear force resulting from the total factored loads, V_u. Similarly, $M_{max} = M_n - M_d$. When calculating the cracking moment M_{cre}, the load used to determine f_d is the same unfactored load used to compute V_d.

11.4.3.2 Web-Shear Cracking, V_{cw} — Web-shear cracking occurs when the principal diagonal tension in the web exceeds the tensile strength of the concrete. This shear is approximately equal to

$$V_{cw} = \left(3.5\sqrt{f'_c} + 0.3 f_{pc}\right) b_w d_p + V_p \qquad \text{Eq. (11-12)}$$

where f_{pc} is the compressive stress in concrete (after allowance for all prestress losses) at the centroid of the cross-section resisting externally applied loads or at the junction of the web and flange when the centroid lies within the flange.

V_p is the vertical component of the effective prestress force, which is present only when strands are draped or deflected.

The expression for web shear strength V_{cw} usually governs for heavily prestressed beams with thin webs, especially when the beam is subject to large concentrated loads near simple supports. Eq. (11-12) predicts the shear strength at first web-shear cracking.

An alternate method for determining the web shear strength V_{cw} is to compute the shear force corresponding to dead load plus live load that results in a principal tensile stress of $4\sqrt{f'_c}$ at the centroidal axis of the member, or at the interface of web and flange when the centroidal axis is located in the flange. This alternate method may be advantageous when designing members where shear is critical. Note the limitation on V_{cw} in the end regions of pretensioned members as provided in 11.4.4 and 11.4.5.

11.4.4, 11.4.5 Special Considerations for Pretensioned Members

Section 11.4.4 applies to situations where the critical section located at h/2 from the face of the support is within the transfer length of the prestressing tendons. This means that the full effective prestress force is not available for contributing to the shear strength. A reduced value of effective prestress force must be used using linear interpolation between no stress in the tendons at the end of the member to full effective prestress at the transfer length from the end of the member, which is taken to be 50 diameters (d_b) for strand and $100d_b$ for a single wire.

Section 11.4.5 is provided to ensure that the effect on shear strength of reduced prestress is properly taken into account when bonding of some of the tendons is intentionally prevented (debonding) near the ends of a pretensioned member, as permitted by 12.9.3.

11.5 SHEAR STRENGTH PROVIDED BY SHEAR REINFORCEMENT FOR PRESTRESSED MEMBERS

The design of shear reinforcement for prestressed members is the same as for reinforced nonprestressed concrete members discussed in Part 12, except that V_c is computed differently (as discussed above) and another minimum shear reinforcement requirement applies (11.5.6.4). Therefore, see Part 12 for a complete discussion of design of shear reinforcement.

11.5.6.1 The code permits a slightly wider spacing of (3/4)h (instead of d/2) for prestressed members, because the shear crack inclination is flatter in prestressed members.

As permitted by 11.5.6.2, shear reinforcement may be omitted in any member if shown by physical tests that the required strength can be developed without shear reinforcement. Section 11.5.6.2 clarifies conditions for appropriate tests. Also, commentary discussion gives further guidance on appropriate tests to meet the intent of 11.5.6.2. The commentary also calls attention to the need for sufficient stirrups in all thin-web, post-tensioned members to support the tendons in the design profile, and to provide reinforcement for tensile stresses in the webs resulting from local deviations of the tendons from the design tendon profile.

11.5.6.4 Minimum Reinforcement for Prestressed Members—For prestressed members, minimum shear reinforcement is computed as the smaller of Eqs. (11-13) and (11-14). However, Eq. (11-13) will generally give a higher minimum than Eq. (11-14). Note that Eq. (11-14) may not be used for members with an effective prestress force less than 40 percent of the tensile strength of the prestressing reinforcement.

REFERENCE

25.1 "PCI Design Handbook – Precast and Prestressed Concrete," MNL 120-04 6th Edition, Precast/Prestressed Concrete Institute, Chicago, 2004, 750 pp.

Example 25.1—Design for Shear (11.4.1)

For the prestressed single tee shown, determine shear requirements using V_c by Eq. (11-9).

Precast concrete: f'_c = 5000 psi (sand lightweight, w_c = 120 pcf)
Topping concrete: f'_c = 4000 psi (normal weight, w_c = 150 pcf)
Prestressing steel: Twelve 1/2-in. dia. 270 ksi strands (single depression at midspan)
Span = 60 ft (simple)
Dead load = 725 lb/ft (includes topping)
Live load = 720 lb/ft
f_{se} (after all losses) = 150 ksi

Precast Section:
    A = 570 in.^{2}
    I = 68,917 in.^{4}
    y_b = 26.01 in.
    y_t = 9.99 in.

Composite Section:
    y_{bc} = 29.27 in.

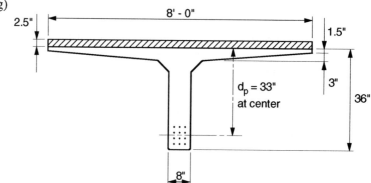

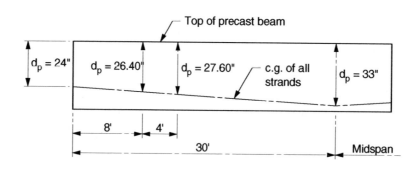

Strand Profile in Precast Girder

Calculations and Discussion	Code Reference

1. Determine factored shear force V_u at various locations along the span. The results are shown in Fig. 25-6.

2. Determine shear strength provided by concrete V_c using Eq. (11-9). The effective prestress f_{se} is greater than 40 percent of f_{pu} (150 ksi > 0.40 × 270 = 108 ksi). Note that the value of d need not be taken less than 0.8h for shear strength computations. Typical computations using Eq. (11-9) for a section 8 ft from support are as follows, assuming the shear is entirely resisted by the web of the precast section: 11.4.1
   11.0

w_u = 1.2 (0.725) + 1.6 (0.720) = 2.022 kips/ft

$$V_u = \left[\left(\frac{60}{2}\right) - 8\right] 2.022 = 44.5 \text{ kips}$$

25-10

| Example 25.1 (cont'd) | Calculations and Discussion | Code Reference |

$M_u = (30 \times 2.022 \times 8) - (2.022 \times 8 \times 4) = 421$ ft-kips

For the non-composite section, at 8 ft from support, determine distance d to centroid of tendons.

d = 26.40 in. (see strand profile)

For composite section, d = 26.4 + 2.5 = 28.9 in. < 0.8h = 30.8 in.   use d = 30.8 in.

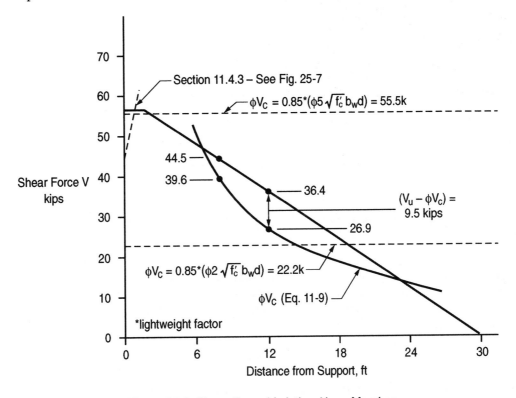

Figure 25-6 Shear Force Variation Along Member

$$V_c = \left(0.6\sqrt{f'_c} + 700\frac{V_u d_p}{M_u}\right) b_w d \qquad \text{Eq. (11-9)}$$

but not less than $2\sqrt{f'_c}\, b_w d$ — 11.4.2

nor greater than $5\sqrt{f'_c}\, b_w d$ — 11.4.2

Since the precast section utilizes sand lightweight concrete, all $\sqrt{f'_c}$ terms must be reduced by the factor 0.85. — 11.2.1.2

Note: Total effective depth, $d_p = 28.9$ in., must be used in $V_u d_p/M_u$ term rather than 0.8h which is used elsewhere. — 11.4.1

Example 25.1 (cont'd) — Calculations and Discussion

$$V_c = \left(0.6 \times 0.85\sqrt{5000} + 700 \times 44.5 \times 28.90 / (421 \times 12)\right) 8 \times 30.8$$

$$= (36 + 178) 8 \times 30.8 = 52.8 \text{ kips} \quad \text{(governs)}$$

$$\geq 2 \times 0.85\sqrt{5000} \times 8 \times 30.8 = 29.6 \text{ kips}$$

$$\leq 5 \times 0.85\sqrt{5000} \times 8 \times 30.8 = 74.0 \text{ kips}$$

$\phi V_c = 0.75 \times 52.8 = 39.6$ (see Fig. 25-6)   *11.2.1.2*

Note: For members simply supported and subject to uniform loading, $V_u d_p / M_u$ in. Eq. (11-9) becomes a simple function of d/ℓ, where ℓ is the span length,

$$V_c = \left[0.6\sqrt{f'_c} + 700 d_p \frac{(\ell - 2x)}{x(\ell - x)}\right] b_w d \qquad Eq.\ (11\text{-}9)$$

where x is the distance from the support to the section being investigated. At 8 ft from the support,

$$V_c = \left[0.6 \times 0.85\sqrt{5000} + 700 \times 28.90 \frac{(60 - 16)}{8(60-8)12}\right] 8 \times 30.8 = 52.8 \text{ kips}$$

3. In the end regions of pretensioned members, the shear strength provided by concrete V_c may be limited by the provisions of 11.4.4. For this design, 11.4.4 does not apply because the section at h/2 is farther out into the span than the bond transfer length (see Fig. 25-7). The following will, however, illustrate typical calculations to satisfy 11.4.4. Compute V_c at face of support, 10 in. from end of member.

Bond transfer length for 1/2-in. diameter strand = 50 (0.5) = 25 in.   *11.4.3*

Prestress force at 10 in. location: $P_{se} = (10/25) \, 150 \times 0.153 \times 12 = 110.2$ kips

Vertical component of prestress force at 10 in. location:

$$\text{slope} = \frac{(d_{CL} - d_{end})}{\dfrac{\ell}{2}} = \frac{(33 - 24)}{30 \times 12} = 0.025$$

$V_p \approx P \times \text{slope} = (110.2)(0.025) = 2.8$ kips

For composite section, d = 28.90 in., use 0.8h = 30.8 in.   *11.4.2.3*

M_d (unfactored weight of precast unit + topping) = 214.4 in.-kips

Distance of composite section centroid above the centroid of precast unit,

	Code
Example 25.1 (cont'd) Calculations and Discussion	**Reference**

$c = y_{bc} - y_b = 29.27 - 26.01 = 3.26$ in.

Tendon eccentricity, $e = d_{end} + 10$ in. $\times$ slope $- y_t = 24 + 10 \times 0.025 - 9.99$

$= 14.26$ in. below the centroid of the precast section

$$f_{pc} \text{ (see notation definition)} = \frac{P}{A_g} - (Pe)\frac{c}{I_g} + M_d\frac{c}{I_g}$$

$$= \frac{110.2}{570} - 110.2(14.26)\left(\frac{3.26}{68,917}\right) + 214.4\left(\frac{3.26}{68,917}\right) = 129 \text{ psi}$$

where A_g and I_g are for the precast section alone.

$V_{cw} = \left(3.5\sqrt{f'_c} + 0.3f_{pc}\right)b_w d_p + V_p$ Eq. (11-12)

$= \left[\left(3.5 \times 0.85\sqrt{5000} + 0.3 \times 129\right)8 \times 28.9\right] + 2800 = 60.4$ kips

$\phi V_{cw} = 0.75 \times 60.4 = 45.3$ kips

The results of this analysis are shown graphically in Fig. 25-7.

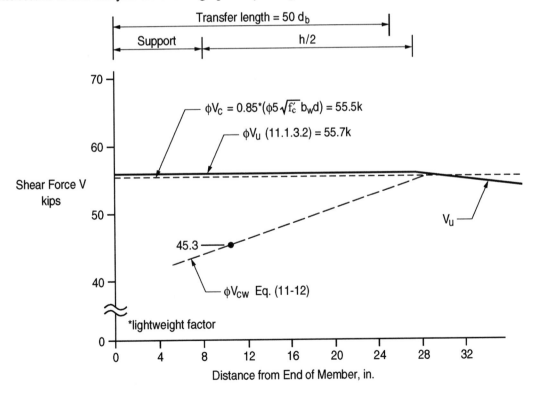

Figure 25-7 Shear Force Variation at End of Member

25-13

Example 25.1 (cont'd)	Calculations and Discussion	Code Reference

4. Compare factored shear V_u with shear strength provided by concrete ϕV_c. Where $V_u > \phi V_c$, shear reinforcement must be provided to carry the excess. Minimum shear reinforcement requirement should also be checked.

Shear reinforcement required at 12 ft from support is calculated as follows:

$d = 30.10$ in. (use in $V_u d_p/M_u$ term)

$M_u = 30 \times 2.24 \times 12 - 2.24 \times 12 \times 6 = 645$ ft-kips

$V_u = \left[\left(\dfrac{60}{2}\right) - 12\right] 2.022 = 36.4$ kips

$V_c = \left[(0.6 \times 0.85\sqrt{5000}) + 700 \times 40.3 \times 30.10/(645 \times 12)\right] 8 \times 30.8 = 35.9$ kips

$\phi V_c = 0.75 \times 35.9 = 26.9$ kips

$A_v = \dfrac{(V_u - \phi V_c) s}{\phi f_y d} = \dfrac{(36.4 - 26.9) 12}{0.75 \times 60 \times 30.8} = 0.082$ in.^{2}/ft

Check minimum required by 11.5.6.3 and 11.5.6.4.

$A_v (\min) = 0.75\sqrt{f'_c}\, \dfrac{b_w s}{f_y} = 0.75\sqrt{5000}\left(\dfrac{8 \times 12}{60,000}\right) = 0.085$ in.^{2}/ft   Eq. (11-13)

but not less than $50\dfrac{b_w s}{f_y}$ (not controlling for $f'_c > 4444$ psi)

$A_v (\min) = \dfrac{A_{ps}}{80}\, \dfrac{f_{pu}}{f_{yt}}\, \dfrac{s}{d} \sqrt{\dfrac{d}{b_w}}$ Eq. (11-14)

$= \dfrac{1.84}{80} \times \dfrac{270}{60} \times \dfrac{12}{30.8} \sqrt{\dfrac{30.8}{8}} = 0.079$ in.^{2}/ft

The lesser A_v (min) from Eqs. (11-13) and (11-14) may be used

The required A_v is very slightly above minimum A_v

Maximum stirrup spacing $= (3/4)d = (3/4) \times 30.8 = 23.1$ in.

Use No. 3 stirrups @ 18 in. for entire member length. ($A_v = 0.147$ in.^{2}/ft)

Example 25.2—Shear Design Using Fig. 25-4

Determine the shear reinforcement for the beam of Example 24.9

f'_c = 6000 psi
depth d_p = 26 in.
effective prestress f_{se} = 150 ksi
decompression stress f_{dc} = 162 ksi
span = 40 ft.

	w k/ft	Midspan moments in.-k
Self-weight	0.413	992
Additional dead load	1.000	2400
Live load	1.250	3000
Sum	2.663	6392

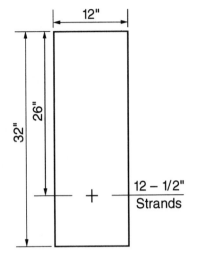

Calculations and Discussion | Code Reference

1. Calculate factored shear at support

   $V_u = 1.2D + 1.6L = [1.2(0.413 + 1.000) + 1.6(1.250)] \times \frac{40}{2}$

   = 73.9 kips

2. Prepare to use Fig. 25-4

   Note: Figure 25-4 is for f'_c = 5000 psi. Its use for f'_c = 6000 psi will be about 10 percent conservative.

   d/ℓ = 26/480 = 1/18.5

   Use curve for ℓ/d = 1/20

   $\dfrac{V_u}{\phi b_w d} = \dfrac{73.9}{0.75 \times 12 \times 26} = 0.316$ ksi = 316 psi

3. Draw line for required nominal shear strength on Fig. 25-4, and find V_s required

| Example 25.2 (cont'd) | Calculations and Discussion | Code Reference |

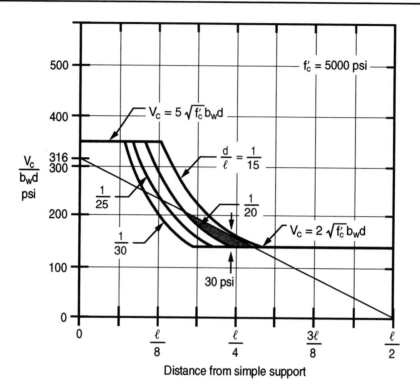

Distance from simple support

The area where shear reinforcement is required is shaded. The maximum nominal shear stress to be resisted by shear reinforcement is 29 psi.

$V_s = 0.03 \text{ ksi} \times b \times d = 0.030 \times 12 \times 26 = 9.4 \text{ kips}$

$A_v = \dfrac{V_s s}{f_{yt} d} = \dfrac{9.4 \times 12}{60 \times 26} = 0.07 \text{ in.}^2/\text{ft}$ *Eq. (11-15)*

4. Check minimum reinforcement.

$A_v = 0.75 \sqrt{f'_c} \dfrac{b_w s}{f_{yt}}$, but not less than *Eq. (11-13)*

$50 \dfrac{b_w s}{f_{yt}}$

$0.75\sqrt{6000} = 58.1$ controls

$A_v = 58.1 \times 12 \times 12/60{,}000 = 0.14 \text{ in.}^2/\text{ft}$

$A_v = \dfrac{A_{ps} f_{pu} s}{80 f_{yt} d} \sqrt{\dfrac{d}{b_w}}$ *Eq. (11-14)*

$A_v = \dfrac{1.836 \times 270 \times 12}{80 \times 60 \times 26} \sqrt{\dfrac{26}{12}} = 0.07 \text{ in.}^2/\text{ft}$

The lesser of A_v by Eqs. (11-13) and (11-14) may be used, but not less than A_v required.

Example 25.2 (cont'd)	**Calculations and Discussion**	**Code Reference**

5. Select stirrups

   $A_v = 0.07$ in.^{2}/ft

   Maximum $s = (3/4)d \leq 24$ in.   *11.5.5.1*

   $s = (3/4)(26) = 19.5$ in.

   Use twin No. 3 @ 18 in.

   $A_v = 0.22/1.5 = 0.15$ in.^{2}/ft   O.K.

   This is required where V_u exceeds $\phi V_c/2$ *11.5.6.1*
   Most designers would provide it for the full length of the member.

Example 25.3—Shear Design Using 11.4.2

For the simple span pretensioned ledger beam shown, determine shear requirements using V_c by Eqs. (11-10) and (11-12).

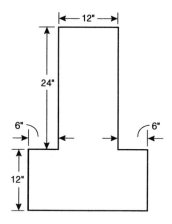

$A = 576$ in.^{2} $w_d = 5.486$ kips/ft
$I = 63,936$ in.^{4} $w_\ell = 5.00$ kips/ft
$h = 36$ in.
$y_b = 15$ in.
$f'_c = 6$ ksi
$\ell = 24$ feet
16 1/2 in. Grade 270 ksi strands, $P = 396.6$ kips
$e_{end} = e_{msp}$ (midspan) $= 10$ in.

Calculations and Discussion	Code Reference

A systematized procedure is needed, to expedite the calculations.

1. Determine midspan moments and end shears

    $M_d = w_d \ell^2/8 = 5.486 \times 24^2/8 = 395$ ft-kips $= 4740$ in.-kips

    $M_\ell = w_\ell \ell^2/8 = 5.00 \times 24^2/8 = 360$ ft-kips $= 4320$ in.-kips

    $M_u = 1.2 M_d + 1.6 M_\ell = 1.2 \times 4740 + 1.6 \times 4320 = 12,600$ in.-kips   Eq. (9-2)

    $M_{max} = M_u - M_d = 12,600 - 4740 = 7860$ in.-kips   11.0

    $V_d = w_d \ell/2 = 5.486 \times 24/2 = 65.8$ kips

    $V_\ell - w_\ell \ell/2 = 5 \times 24/2 = 60.0$ kips

    $V_u = 1.2 V_d + 1.6 V_\ell = 1.2 \times 65.8 + 1.6 \times 60 = 175.0$ kips   Eq. (9-2)

    $V_i = V_u - V_d = 175 - 65.8 = 109.2$ kips   11.0

2. Define factors for converting midspan moments and end shears to moments and shears at a distance x/ℓ from support, for $x/\ell = 0.3$.

    V factor $= 1-2(x/\ell) = 1-2(0.3) = 0.4$

    M factor $= 4(x/\ell -(x/\ell)^2) = 4 \times (0.3-0.3^2) = 0.84$

| Example 25.3 (cont'd) | Calculations and Discussion | Code Reference |

3. Compute V_3, the third term in Eq. (11-10)

   $P/A = 396.6/576 \qquad = 0.689$

   $Pe/S_b = 396.6 \times 10/4262 = 0.930$

   $-M_d/S_b = 0.84\ M_{d\ (msp)}/S_b = -0.934$

   $+6\sqrt{f'_c} = 6\sqrt{6000} = 465 \quad = \underline{0.465}$

   $\phantom{+6\sqrt{f'_c} = 6\sqrt{6000} = 465 \quad = }1.150\ \text{ksi}$

   $M_{cre} = S_b\,(1.150\ \text{ksi}) = 4900\ \text{in.-kips}$ — Eq. (11-11)

   $V_i = 0.4 V_{i\ (end)} = 0.4 \times 109.2 = 43.7\ \text{kips}$

   $M_{max} = 0.84\ M_{max\ (msp)} = 0.84 \times 7860 = 6602\ \text{in.-kips}$

   $V_3 = \dfrac{V_i M_{cre}}{M_{max}} = \dfrac{43.7 \times 4900}{6602} = 32.4\ \text{kips}$ — Eq. (11-10)

4. Compute the remaining terms V_1 and V_2 in Eq. (11-10), and solve for V_{ci} — Eq. (11-10)

   $d = 31$, but not less than $0.8d = 28.8$. Use $d = 31$ — 11.4.1

   $V_1 = 0.6 b_w d_p \sqrt{f'_c} = 0.6 \times 12 \times 31\sqrt{6000} = 17.3\text{k}$

   $V_2 = V_d = 0.4(V_d\,\text{end}) = 0.4 \times 65.8 = 26.3\ \text{kips}$

   $V_{ci} = V_1 + V_2 + V_3 = 17.3 + 26.3 + 32.4 = 76.0\ \text{kips}$ — Eq. (11-10)

5. Compute V_u, and find V_s to be resisted by stirrups

   $V_u = 0.4\ V_u\,(end) = 0.4 \times 175.0 = 70\ \text{kips}$

   ϕ for shear $= 0.75$ — 9.3.2.3

   $V_s = V_n - V_c = V_u/\phi - V_c = 70/0.75 - 76 = 17.3\ \text{kips}$ — Eq. (11-2)

6. Find required stirrups

   $A_v = \dfrac{V_s s}{f_{yt} d} = \dfrac{17.3 \times 12}{60 \times 31} = 0.11\ \text{in.}^2/\text{ft}$

   Minimum requirements

   $A_v = 0.75\sqrt{f'_c}\ b_w s / f_{yt}$ when $f'_c > 4444$ psi — Eq. (11-13)

   $= 0.75\sqrt{6000} \times 12 \times 12/60{,}000 = 0.14\ \text{in.}^2$

   $A_v = \dfrac{A_{ps} f_{pu} s}{80\ f_{yt} d}\sqrt{\dfrac{d}{b_w}} = \dfrac{2.448 \times 270 \times 12}{80 \times 60 \times 31}\sqrt{\dfrac{31}{12}} = 0.086\ \text{in.}^2$ — Eq. (11-14)

   The minimum need only be the lesser of that required by Eqs. (11-13) or (11-14). — 11.5.6.4

   So, the required A_v of $0.09\ \text{in.}^2/\text{ft}$ controls

		Code
Example 25.3 (cont'd)	**Calculations and Discussion**	**Reference**

Maximum spacing = (3/4)d = (3/4)31 = 23.25 in.

Say, twin No. 3 at 18 in., $A_v = 2 \times 0.11/1.5 = 0.15$ in.^{2}/ft.

7. Compute required shear reinforcement at support

   Because the ledger beam is loaded on the ledges, not "near the top," shear must be checked at the support, not at h/2 from the support for prestressed members.   *11.1.3*

   At the support, the prestress force P is assumed to be zero, for simplicity.

   $V_{cw} = (3.5\sqrt{f'_c} + 0.3 f_{pc}) b_w d_p + V_p$ *Eq. (11-12)*
   $= (3.5\sqrt{6000}) \times 12 \times 31 = 100.9$ kips

   $V_s = V_n - V_c = V_u/\phi - V_c = 175/0.75 - 100.9$ *Eq. (11-12)*
   $V_s = 132.4$k

   $A_v = \dfrac{V_s s}{f_{yt} d} = \dfrac{132.4 \times 12}{60 \times 31} = 0.85$ in.^{2} *Eq. (11-15)*

   Say twin No. 4 at 4 in., $A_v = .40/.33 = 1.20$ in.^{2}/ft near end.

   Referring to Step 6, this is above minimum requirements.

8. Repeat the processes described above for various sections along the shear span (not shown). The results are shown below.

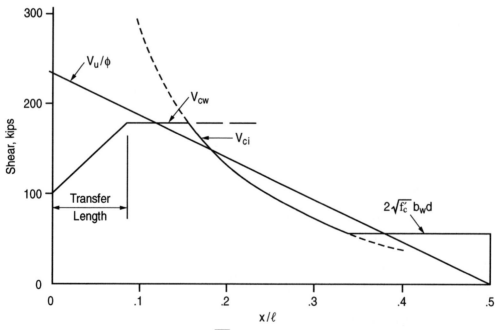

Note: Minimum V_c of $2\sqrt{f'_c} b_w d$ permitted by 11.4.2 was used.

Example 25.3 (cont'd)	Calculations and Discussion	Code Reference

9. Notes:

   1. A spreadsheet can be set up, with each column containing data for various values of x/ℓ and the shear and moment factors in Step 2.

   2. For members with draped tendons, additional factors for varying eccentricity, depth, and tendon slope (for computing V_p) need to be added in Step 2.

   3. For composite members, the portions of dead load applied before and after composite behavior is obtained need to be separated. The dead load applied after the beam becomes composite should not be included in V_d and M_d terms. See R11.4.3.

Blank

26

Prestressed Slab Systems

UPDATE FOR THE '05 CODE

Section 6.4.4, which requires that construction joints in slabs be located in the middle third of the span, is excluded from application to prestressed concrete. Most construction joints in continuous prestressed concrete slabs are located close to the quarter point of the span where the tendon profile is near mid-depth of the member.

Section 18.3.3 defines two-way prestressed slab systems as Class U and reduces the maximum permissible flexural tensile stress f_t from $7.5\sqrt{f'_c}$ to $6\sqrt{f'_c}$.

INTRODUCTION

Six code sections are particularly significant with respect to analysis and design of prestressed slab systems:

Section 11.12.2—Shear strength of prestressed slabs

Section 11.12.6—Shear strength of prestressed slabs with moment transfer

Section 18.3.3—Permissible flexural tensile stresses

Section 18.4.2—Permissible flexural compressive stresses

Section 18.7.2—Determination of f_{ps} for calculation of flexural strength

Section 18.12—Prestressed slab systems

Discussion of each of these code sections is presented below, followed by Example 26.1 of a post-tensioned flat plate. The design example illustrates application of the above code sections as well as general applicability of the code to analysis and design of post-tensioned flat plates.

11.12.2 Shear Strength

Section 11.12.2 contains specific provisions for calculation of shear strength in two-way prestressed concrete systems. At columns of two-way prestressed slabs (and footings) utilizing unbonded tendons and meeting the bonded reinforcement requirements of 18.9.3, the shear strength V_n must not be taken greater than the shear strength V_c computed in accordance with 11.12.2.1 or 11.12.2.2, unless shear reinforcement is provided in accordance with 11.12.3 or 11.12.4. Section 11.12.2.2 gives the following value of the shear strength V_c at columns of two-way prestressed slabs:

$$V_c = \left(\beta_p \sqrt{f'_c} + 0.3 f_{pc}\right) b_o d + V_p \qquad \text{Eq. (11-36)}$$

Equation (11-36) includes the term β_p which is the smaller of 3.5 and $(\alpha_s d/b_o + 1.5)$. The term $\alpha_s d/b_o$ is to account for a decrease in shear strength affected by the perimeter area aspect ratio of the column, where α_s is to be taken as 40 for interior columns, 30 for edge columns, and 20 for corner columns. f_{pc} is the average value of

f_{pc} for the two directions, and V_p is the vertical component of all effective prestress forces crossing the critical section. If the shear strength is computed by Eq. (11-36), the following must be satisfied; otherwise, 11.12.2.1 for nonprestressed slabs applies:

a. no portion of the column cross-section shall be closer to a discontinuous edge than 4 times the slab thickness,

b. f'_c in Eq. (11-36) shall not be taken greater than 5000 psi, and

c. f_{pc} in each direction shall not be less than 125 psi, nor be taken greater than 500 psi.

In accordance with the above limitations, shear strength Eqs. (11-33), (11-34), and (11-35) for nonprestressed slabs are applicable to columns closer to the discontinuous edge than 4 times the slab thickness. The shear strength V_c is the lesser of the values given by these three equations. For usual design conditions (slab thicknesses and column sizes), the controlling shear strength at edge columns will be $4\sqrt{f'_c}b_o d$.

11.12.6 Shear Strength with Moment Transfer

For moment transfer calculations, the controlling shear stress at columns of two-way prestressed slabs with bonded reinforcement in accordance with 18.9.3 is governed by Eq. (11-36), which could be expressed as a shear stress for use in Eq. (11-40) as follows:

$$v_c = \beta_p \sqrt{f'_c} + 0.3 f_{pc} + \frac{V_p}{b_o d} \qquad \text{Eq. (11-36)}$$

If the permissible shear stress is computed by Eq. (11-36), the following must be satisfied:

a. no portion of the column cross-section shall be closer to a discontinuous edge than 4 times the slab thickness,

b. f'_c in Eq. (11-36) shall not be taken greater than 5000 psi, and

c. f_{pc} in each direction shall not be less than 125 psi, nor be taken greater than 500 psi.

For edge columns under moment transfer conditions, the controlling shear stress will be the same as that permitted for nonprestressed slabs. For usual design conditions, the governing shear stress at edge columns will be $4\sqrt{f'_c}$.

18.3.3 Permissible Flexural Tensile Stresses

This section requires that prestressed two-way slab systems be designed as Class U (Uncracked) members but with the permissible flexural tensile stress limited to $6\sqrt{f'_c}$.

18.4.2 Permissible Flexural Compressive Stresses

In 1995, Section 18.4.2 increased the permissible concrete service load flexural compressive stress under total load from $0.45f'_c$ to $0.60f'_c$, but imposed a new limit of $0.45f'_c$ for sustained load. This involves some judgment on the part of designers in determining the appropriate sustained load.

18.7.2 f_{ps} for Unbonded Tendons

In prestressed elements with unbonded tendons having a span/depth ratio greater than 35, the stress in the prestressed reinforcement at nominal strength is given by:

$$f_{ps} = f_{se} + 10{,}000 + \frac{f'_c}{300\rho_p} \qquad \text{Eq. (18-5)}$$

but not greater than f_{py}, nor (f_{se} + 30,000).

Nearly all prestressed one-way slabs and flat plates will have span/depth ratios greater than 35. Equation (18-5) provides values of f_{ps} which are generally 15,000 to 20,000 psi lower than the values of f_{ps} given by Eq. (18-4) which was derived primarily from results of beam tests. These lower values of f_{ps} are more compatible with values of f_{ps} obtained in more recent tests of prestressed one-way slabs and flat plates. Application of Eq. (18-5) is illustrated in Example 26.1.

18.12 SLAB SYSTEMS

Section 18.12 provides analysis and design procedures for two-way prestressed slab systems, including the following requirements:

1. Use of the Equivalent Frame Method of 13.7 (excluding 13.7.7.4 and 13.7.7.5), or more detailed analysis procedures, is required for determination of factored moments and shears in prestressed slab systems. According to References 26.1 and 26.4, for two-way prestressed slabs, the equivalent frame slab-beam strips would not be divided into column and middle strips as for a typical nonprestressed two-way slab, but would be designed as a total beam strip.

2. Spacing of tendons or groups of tendons in one direction shall not exceed 8 times the slab thickness nor 5 ft. Spacing of tendons shall also provide a minimum average prestress, after allowance for all prestress losses, of 125 psi on the slab section tributary to the tendon or tendon group. Special consideration must be given to tendon spacing in slabs with concentrated loads.

3. A minimum of two tendons shall be provided in each direction through the critical shear section over columns. This provision, in conjunction with the limits on tendon spacing outlined in Item 2 above, provides specific guidance for distributing tendons in prestressed flat plates in accordance with the "banded" pattern illustrated in Fig. 26-1. This method of tendon installation is widely used and greatly simplifies detailing and installation procedures.

Calculation of equivalent frame properties is illustrated in Example 26.1. Tendon distribution is also discussed in this example.

References 26.1 and 26.4 illustrate application of ACI 318 requirements for design of one-way and two-way post-tensioned slabs, including detailed design examples.

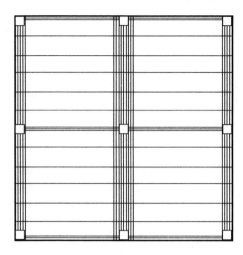

Figure 26-1 Banded Tendon Distribution

REFERENCES

26.1 *Design of Post-Tensioned Slabs Using Unbonded Tendons*, Post-Tensioning Institute, 3rd. ed., Phoenix, AZ, 2004.

26.2 *Continuity in Concrete Building Frames*, Portland Cement Association, Skokie, IL, 1986.

26.3 *Estimating Prestress Losses,* Zia, P., Preston, H. K., Scott, N. L., and Workman, E. B., Concrete International : Design and Construction, V. 1, No. 6, June 1979, pp. 32-38.

26.4 *Design Fundamentals of Post-Tensioned Concrete Floors,* Aalami, B. O., and Bommer, A., Post-Tensioning Institute, Phoenix, AZ, 1999.

Example 26.1—Two-Way Prestressed Slab System

Design a typical transverse equivalent frame strip of the prestressed flat plate with partial plan and section shown in Figure 26-2.

f'_c = 4000 psi; w = 150 pcf (slab and columns)

f_y = 60,000 psi

f_{pu} = 270,000 psi

Live load = 40 psf
Partition load = 15 psf

Reduce live load in accordance with general building code. For this example live load is reduced in accordance with IBC 2003, Section 1607.9.2.

Required minimum concrete cover to tendons 1.5 in. from the bottom of the slab in end spans, 0.75 in. top and bottom elsewhere.

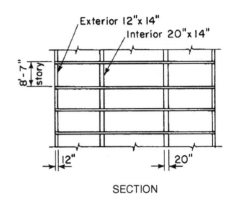

PARTIAL PLAN

SECTION

Figure 26-2 Equivalent Frame

Calculations and Discussion	Code Reference

1. Slab Thickness

   For two-way prestressed slabs, a span/depth ratio of 45 typically results in overall economy and provides satisfactory structural performance.[26.1]

   Slab thickness:

   Longitudinal span: $20 \times 12/45$ = 5.3 in.
   Transverse span: $25 \times 12/45$ = 6.7 in.

   Use 6-1/2 in. slab.

   Slab weight = 81 psf
   Partition load = 15 psf
   Total dead load = 81 + 15 = 96 psf

   Span 2:
   Reduced live load (IBC 1607.9.2)
   Live load = $40[1 - 0.08(500 - 150)/100]$ = 29 psf
   Factored dead load = 1.2×96 = 115 psf
   Factored live load = 1.6×29 = 47 psf

Example 26.1 (cont'd)   **Calculations and Discussion**   **Code Reference**

Total load = 125 psf, unfactored
          = 162 psf, factored

Spans 1 and 3:
Reduced live load (IBC 1607.9.2)
Live load = 40[1 − 0.08(340 − 150)/100] = 34 psf
Factored dead load = 1.2 × 96 = 115 psf
Factored live load = 1.6 × 34 = 55 psf
Total load = 130 psf, unfactored
          = 170 psf, factored

2. Design Procedure

Assume a set of loads to be balanced by parabolic tendons. Analyze an equivalent frame subjected to the net downward loads according to 13.7. Check flexural stresses at critical sections, and revise load balancing tendon forces as required to obtain permissible flexural stresses according to 18.3.3 and 18.4.

When final forces are determined, obtain frame moments for factored dead and live loads. Calculate secondary moments induced in the frame by post-tensioning forces, and combine with factored load moments to obtain design factored moments. Provide minimum bonded reinforcement in accordance with 18.9.

Check design flexural strength and increase nonprestressed reinforcement if required by strength criteria. Investigate shear strength, including shear due to vertical load and due to moment transfer, and compare total to permissible values calculated in accordance with 11.12.2.

3. Load Balancing

Arbitrarily assume the tendons will balance 80% of the slab weight (0.8 × 0.081 = 0.065 ksf) in the controlling span (Span 2), with a parabolic tendon profile of maximum permissible sag, for the initial estimate of the required prestress force F_e:

Maximum tendon sag in Span 2 = 6.5 − 1 − 1 = 4.5 in.

$$F_e = \frac{w_{bal}L^2}{8a} = \frac{0.8(0.081)(25)^2(12)}{8(4.5)} = 13.5 \text{ kips/ft}$$

Assume 1/2 in. diameter (cross-sectional area = 0.153 in.^{2}), 270 ksi seven-wire low relaxation strand tendons with 14 ksi long-term losses (Reference 26.3). Effective force per tendon is 0.153 [(0.7 × 270) − 14] = 26.8 kips, where the tensile stress in the tendons immediately after tendon anchorage = $0.70f_{pu}$.   18.5.1(c)

For a 20-ft bay, 20 × 13.5/26.8 = 10.1 tendons.

	Code
Example 26.1 (cont'd) **Calculations and Discussion**	**Reference**

Use 10-1/2 in. diameter tendons/bay

$F_e = 10 \times 26.8/20 = 13.4$ kips/ft

$f_{pc} = F_e/A = 13.4/(6.5 \times 12) = 0.172$ ksi

Actual balanced load in Span 2:

$$w_{bal} = \frac{8F_e a}{L^2} = \frac{8(13.4)(4.5)}{12 \times 25^2} = 0.064 \text{ ksf}$$

Adjust tendon profile in Spans 1 and 3 to balance same load as in Span 2:

$$a = \frac{w_{bal} L^2}{8F_e} = \frac{0.064(17)^2(12)}{8(13.4)} = 2.1 \text{ in.}$$

Midspan cgs = $(3.25 + 5.5)/2 - 2.1 = 2.275$ in. say 2.25 in.
Actual sag in Spans 1 and 3 = $(3.25 + 5.5)/2 - 2.25 = 2.125$ in.
Actual balanced load in Spans 1 and 3 =

$$w_{bal} = \frac{8(13.4)(2.125)}{17^2(12)} = 0.066 \text{ ksf}$$

4. Tendon Profile

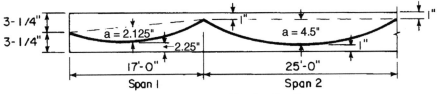

Figure 26-3 Tendon Profile

Net load causing bending:
Span 2:
   $w_{net} = 0.125 - 0.064 = 0.061$ ksf
Spans 1 and 3:
   $w_{net} = 0.130 - 0.066 = 0.064$ ksf

	Code
Example 26.1 (cont'd) Calculations and Discussion	**Reference**

5. Equivalent Frame Properties   *13.7*

   a. Column stiffness.   *13.7.4*

   Column stiffness, including effects of "infinite" stiffness within the slab-column joint (rigid connection), may be calculated by classical methods or by simplified methods which are in close agreement. The following approximate stiffness K_c will give results within five percent of "exact" values.[26.1]

   $K_c = 4EI/(\ell - 2h)$

   where ℓ = center-to-center column height and h = slab thickness.

   For exterior columns (14 × 12 in.):

   $I = 14 \times 12^3/12 = 2016$ in.^{4}

   $E_{col}/E_{slab} = 1.0$

   $K_c = (4 \times 1.0 \times 2016)/[103 - (2 \times 6.5)] = 90$ in.^{3}

   $\Sigma K_c = 2 \times 90 = 180$ in.^{3} (joint total)

   Stiffness of torsional members is calculated as follows: *13.7.5*

   $C = (1 - 0.63 \, x/y) \, x^3 y/3$ *13.0*

   $= [1 - (0.63 \times 6.5/12)] \, (6.5^3 \times 12)/3 = 724$ in.^{4}

   $$K_t = \frac{9CE_{cs}}{\ell_2 (1 - c_2/\ell_2)^3}$$ *R13.7.5*

   $$= \frac{9 \times 724 \times 1.0}{(20 \times 12)(1 - 1.17/20)^3} = 32.5 \text{ in.}^3$$

   $\Sigma K_t = 2 \times 32.5 = 65$ in.^{3} (joint total)

   Exterior equivalent column stiffness (see ACI 318R-89, R13.7.4):

   $1/K_{ec} = 1/\Sigma K_t + 1/\Sigma K_c$

   $K_{ec} = (1/65 + 1/180)^{-1} = 48$ in.^{3}

   For interior columns (14 × 20 in.):

   $I = 14 \times 20^3/12 = 9333$ in.^{4}

	Code
Example 26.1 (cont'd) Calculations and Discussion	**Reference**

$K_c = (4 \times 1.0 \times 9333)/[103 - (2 \times 6.5)] = 415$ in.^{3}

$\Sigma K_c = 2 \times 415 = 830$ in.^{3} (joint total)

$C = [1 - (0.63 \times 6.5/20)] (6.5^3 \times 20)/3 = 1456$ in.^{4}

$$K_t = \frac{9 \times 1456 \times 1.0}{240 (1-1.17/20)^3} = 65 \text{ in.}^3$$

$\Sigma K_t = 2 \times 65 = 130$ in.^{3} (joint total)

$K_{ec} = (1/130 + 1/830)^{-1} = 112$ in.^{3}

   b. Slab-beam stiffness.   13.7.3

Slab stiffness, including effects of infinite stiffness within slab-column joint, can be calculated by the following approximate expression.[26.1]

$K_s = 4EI/(\ell_1 - c_1/2)$

where ℓ_1 = length of span in direction of analysis measured center-to-center of supports and c_1 = column dimension in direction of ℓ_1.

At exterior column:

$K_s = (4 \times 1.0 \times 20 \times 6.5^3)/[(17 \times 12) - 12/2] = 111$ in.^{3}

At interior column (spans 1 & 3):

$K_s = (4 \times 1.0 \times 20 \times 6.5^3)/[(17 \times 12) - 20/2] = 113$ in.^{3}

At interior column (span 2):

$K_s = (4 \times 1.0 \times 20 \times 6.5^3)/[(25 \times 12) - 20/2] = 76$ in.^{3}

   c. Distribution factors for analysis by moment distribution.

Slab distribution factors:

At exterior joints = $111/(111 + 48) = 0.70$

At interior joints for spans 1 and 3 = $113/(113 + 76 + 112) = 0.37$

At interior joints for span 2 = $76/301 = 0.25$

6. Moment Distribution—Net Loads

Since the nonprismatic section causes only very small effects on fixed-end moments and carryover factors, fixed-end moments will be calculated from FEM = $wL^2/12$ and carryover factors will be taken as COF = 1/2.

		Code
Example 26.1 (cont'd)	**Calculations and Discussion**	**Reference**

For Spans 1 and 3, net load FEM $= 0.064 \times 17^2/12 = 1.54$ ft-kips

For Span 2 net load FEM $= 0.061 \times 25^2/12 = 3.18$ ft-kips

Note that since live load is less than three-quarters dead load, patterned or "skipped" live load is not required. Maximum factored moments are based upon full live load on all spans simultaneously. *(13.7.6.2)*

Table 26-1 Moment Distribution—Net Loads
(all moments are in ft-kips)

DF	0.70	0.37	0.25
FEM	-1.54	-1.54	-3.18
Distribution	+1.08	-0.61	+0.41
Carry-over	+0.31	-0.54	-0.21
Distribution	-0.22	+0.12	-0.08
Final	-0.37	-2.57	-3.06

7. Check Net Stresses (tension positive, compression negative)

   a. At interior face of interior column:

   Moment at column face = centerline moment + $Vc_1/3$ (see Ref. 26.2):

   $$-M_{max} = -3.06 + \frac{1}{3}\left(\frac{0.061 \times 25}{2}\right)\left(\frac{20}{12}\right)$$

   $$= -2.64 \text{ ft-kips}$$

   $S = bh^2/6 = 12 \times 6.5^2/6 = 84.5$ in.^{3}

   $$f_{t,b} = -f_{pc} \pm \frac{M_{net}}{S_{t,b}} = -0.172 \pm \frac{12 \times 2.64}{84.5} = 0.172 \pm 0.375 = +0.203, -0.547 \text{ ksi}$$

   Allowable Tension $= 6\sqrt{4000} = 0.379$ ksi *18.3.3*
   At top 0.203 ksi applied < 0.379 allowable OK

   Allowable compression under total load $= 0.60f'_c = 0.6 \times 4000 = 2.4$ ksi *18.4.2(b)*
   At bottom 0.547 ksi applied < 2.4 ksi allowable OK

   Allowable compression under sustained load $= 0.45 \times 4000 = 1.8$ ksi *18.4.2(a)*
   0.547 ksi applied under total load < 1.8 ksi allowable under sustained load OK
   (regardless of value of sustained load).

Example 26.1 (cont'd) — Calculations and Discussion — Code Reference

b. At midspan of Span 2:

$+ M_{max} = (0.061 \times 25^2/8) - 3.18 = +1.59$ ft-kips

$$f_{t,b} = -f_{pc} \mp \frac{M_{net}}{S_{t,b}} = -0.172 \mp \frac{12 \times 1.59}{84.5} = -0.172 \mp 0.226 = -0.398, +0.054 \text{ ksi}$$

Compression at top 0.398 < 1.8 ksi allowable sustained load < 2.4 ksi allowable total load O.K. Tension at bottom 0.054 ksi applied < 0.379 ksi allowable O.K.

When the tensile stress exceeds $2\sqrt{f'_c}$ in positive moment areas, the total tensile force N_c must be carried by bonded reinforcement. For this slab, $2\sqrt{4000} = 0.126$ ksi > 0.054 ksi. Therefore, positive moment bonded reinforcement is not required. When it is, the calculation for the required amount of bonded reinforcement is done as follows (refer to Figure 26-4). 18.9.3.2

$$y = \frac{f_t}{f_t + f_c}(h) \text{ in.}$$

$$N_c = \frac{12(y)(f_t)}{2} \text{ kips/ft}$$

$$A_s = \frac{N_c}{0.5 f_y} \text{ in.}^2 / \text{ft}$$

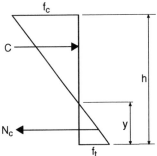

Figure 26-4 Stress Distribution

Determine minimum bar lengths for this reinforcement in accordance with *18.9.4* (Note that conformance to *Chapter 12* is also required.)

Calculate deflections under total loads using usual elastic methods and gross concrete section properties *(9.5.4)*. Limit *computed* deflections to those specified in Table *9.5(b)*.

This completes the service load portion of the design.

8. Flexural Strength

   a. Calculation of design moments.

   Design moments for statically indeterminate post-tensioned members are determined by combining frame moments due to factored dead and live loads with secondary moments induced into the frame by the tendons. The load balancing approach directly includes both primary and secondary effects, so that for service conditions only "net loads" need be considered.

Example 26.1 (cont'd) — Calculations and Discussion

At design flexural strength, the balanced load moments are used to determine secondary moments by subtracting the primary moment, which is simply $F_e \times e$, at each support. For multistory buildings where typical vertical load design is combined with varying moments due to lateral loading, an efficient design approach would be to analyze the equivalent frame under each case of dead, live, balanced, and lateral loads, and combine the cases for each design condition with appropriate load factors. For this example, the balanced load moments are determined by moment distribution as follows:

For spans 1 and 3, balanced load FEM = $0.066 \times 17^2/12$ = 1.59 ft-kips

For span 2, balanced load FEM = $0.064 \times 25^2/12$ = 3.33 ft-kips

Table 26-2 Moment Distribution—Balanced Loads
(all moments are in ft-kips)

DF	0.70	0.37	0.25
FEM	+1.59	+1.59	+3.33
Distribution	-1.11	+0.64	-0.44
Carry-over	-0.32	+0.56	+0.22
Distribution	+0.22	-0.13	+0.09
Final	+0.38	+2.66	+3.20

Since the balanced load moment includes both primary (M_1) and secondary (M_2) moments, secondary moments can be found from the following relationship:

$M_{bal} = M_1 + M_2$, or $M_2 = M_{bal} - M_1$

The primary moment M_1 equals $F_e \times e$ at any point ("e" is the distance between the cgs and the cgc, the "eccentricity" of the prestress force).

Thus, the secondary moments are:

At an exterior column:

$M_2 = 0.38 - (13.4 \times 0/12) = 0.38$ ft-kips

At an interior column:

Spans 1 and 3,

$M_2 = 2.66 - 13.4(3.25 - 1.0)/12 = 0.15$ ft-kips

Span 2,

$M_2 = 3.20 - (13.4 \times 2.25)/12 = 0.69$ ft-kips

Example 26.1 (cont'd)   **Calculations and Discussion**   **Code Reference**

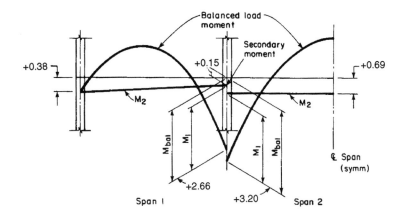

Figure 26-5 Moment Diagram

Factored load moments:
   Spans 1 and 3: w_u = 170 psf
   Span 2: w_u = 162 psf

For spans 1 and 3, factored load FEM = $0.170 \times 17^2/12$ = 4.09 ft-kips

For span 2, factored load FEM = $0.162 \times 25^2/12$ = 8.44 ft-kips

Table 26-3 Moment Distribution—Factored Loads
(all moments are in ft-kips)

DF	0.70	0.37	0.25
FEM	-4.09	-4.09	-8.44
Distribution	+2.86	-1.61	+1.09
Carry-over	+0.81	-1.43	-0.55
Distribution	-0.57	+0.33	-0.22
Final	-0.99	-6.80	-8.12

Combine the factored load and secondary moments to obtain the total negative design moments. The results are given in Table 26-4.

Table 26-4 Design Moments at Face of Column (all moments are in ft-kips)

	Span 1		Span 2
Factored load moments	-0.99	-6.80	-8.12
Secondary moments	+0.38	+0.15	+0.69
Moments at column centerline	-0.61	-6.65	-7.43
Moment reduction to face of column, $Vc_1/3$	+0.48	+0.80	+1.13
Design moments at face of column	-0.13	-5.85	-6.30

		Code
Example 26.1 (cont'd)	**Calculations and Discussion**	**Reference**

Calculate total positive design moments at interior of span:

For span 1,

$$V_{ext} = (0.170 \times 17/2) - (6.65 - 0.61)/17$$

$$= 1.45 - 0.36 = 1.09 \text{ kips/ft}$$

$$V_{int} = 1.45 + 0.36 = 1.81 \text{ kips/ft}$$

Distance x to location of zero shear and maximum positive moment from centerline of exterior column:

$$x = 1.09/0.170 = 6.42 \text{ ft}$$

End span positive moment = $(0.5 \times 1.09 \times 6.42) - 0.61 = 2.89$ ft-kips/ft
(including M_2)

For span 2,

$$V = 0.162 \times 25/2 = 2.03 \text{ kips/ft}$$

Interior span positive moment = $-7.43 + (0.5 \times 2.03 \times 12.5) = 5.26$ ft-kips/ft
(including M_2)

b. Calculation of flexural strength.

Check slab at interior support. Section 18.9.3.3 requires a minimum amount of bonded reinforcement in negative moment areas at column supports regardless of service load stress levels. More than the minimum may be required for flexural strength. The minimum amount is to help ensure flexural continuity and ductility, and to control cracking due to overload, temperature, or shrinkage.

$$A_s = 0.00075 A_{cf} \qquad \textit{Eq. (18-8)}$$

where

A_{cf} = larger cross-sectional area of the slab-beam strips of the two orthogonal equivalent frames intersecting at a column of a two-way slab.

$$A_s = 0.00075 \times 6.5 \times \left(\frac{17 + 25}{2}\right) \times 12 = 1.23 \text{ in.}^2$$

Try 6-No. 4 bars. Space bars at 6 in. on center, so that they are within the column width plus 1.5 times slab thickness on either side of column. *18.9.3.3*

Bar length = $[2 \times (25 - 20/12)/6] + 20/12 = 9$ ft-5 in. *18.9.4.2*

	Code
Example 26.1 (cont'd) **Calculations and Discussion**	**Reference**

For average one-foot strip:

$A_s = 6 \times 0.20/20 = 0.06$ in.^{2}/ft

Initial check of flexural strength will be made considering this reinforcement.

Calculate stress in tendons at nominal strength:

$$f_{ps} = f_{se} + 10{,}000 + \frac{f'_c}{300\rho_p}$$
                                                                                                         Eq. (18-5)

With 10 tendons in 20 ft bay:

$\rho_p = A_{ps}/bd_p = 10 \times 0.153/(20 \times 12 \times 5.5) = 0.00116$

$f_{se} = (0.7 \times 270) - 14 = 175$ ksi   18.5.1, 18.6, Reference 3

$f_{ps} = 175 + 10 + 4/(300 \times 0.00116) = 175 + 10 + 12 = 197$ ksi

f_{ps} shall not be taken greater than $f_{py} = 0.85 f_{pu} = 230$ ksi > 197
or $f_{se} + 30 = 205$ ksi > 197 OK   18.7.2(c)

$A_{ps}f_{ps} = 10 \times 0.153 \times 197/20 = 15.1$ kips/ft

$A_s f_y = 0.06 \times 60 = 3.6$ kips/ft

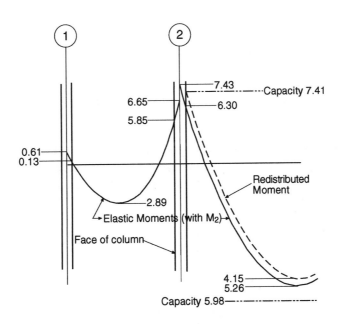

Figure 26-6 Moments in ft-kips

26-15

		Code
Example 26.1 (cont'd)	**Calculations and Discussion**	**Reference**

$$a = \frac{A_{ps}f_{ps} + A_s f_y}{0.85 f'_c b} = \frac{15.1 + 3.6}{0.85 \times 4 \times 12} = 0.46 \text{ in.}$$

$$c = a/\beta_1 = 0.46/0.85 = 0.54 \text{ in.}$$

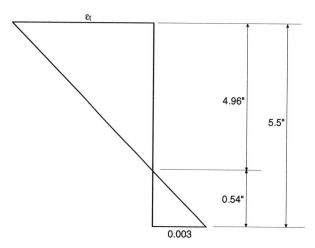

Figure 26-7 Strain Diagram at Interior Support

$\varepsilon_t = (5.5 - 0.54) \times 0.003/0.54 = 0.028$ therefore tension controlled, $\phi = 0.9$ 9.3.2, 10.3.4

Since the bars and tendons are in the same layer:

$$\left(d - \frac{a}{2}\right) = \left(5.5 - \frac{0.46}{2}\right)/12 = 0.44 \text{ ft}$$

$\phi M_n = 0.9 \times (15.1 + 3.6) \times 0.44 = 7.41$ ft-kips/ft > 6.30 ft-kips/ft   OK.   9.3.2.1

Since there is excess negative moment capacity available, use moment redistribution to increase the negative moment and minimize the positive moment demand in Span 2. Note that the actual inelastic moment redistribution occurs at the positive moment section of Span 2.

Permissible change in negative moment = $1000\varepsilon_t = 1000(0.028) = 28\% > 20\%$ max   18.10.4.1
    8.4

Available increase in negative moment = $0.2 \times 6.30 = 1.26$ ft-kips/ft

Actual increase in negative moment = Minimum capacity − Elastic Negative Moment
           = 7.41 − 6.30 = 1.11 ft-kips/ft < 1.26 available   O.K.

	Code
Example 26.1 (cont'd) **Calculations and Discussion**	**Reference**

Minimum design positive moment in Span 2 = 5.26 − 1.11 = 4.15 ft-kips/ft

Capacity at midspan of Span 2 (no bonded reinforcement required):

$A_{ps}f_{ps} = 15.1$ kips/ft

$$a = \frac{15.1}{0.85 \times 4 \times 12} = 0.37 \text{ in.}$$

$$\frac{c}{d_t} = \frac{\frac{0.37}{0.85}}{5.5} = 0.079 < 0.375, \text{ therefore tension controlled.}$$

9.3.2.2
10.3.4

$$\left(d - \frac{a}{2}\right) = \frac{5.5 - \frac{0.37}{2}}{12} = 0.44 \text{ ft}$$

At center of span,

$\phi M_n = 0.9 \times (15.1) \times 0.44 = 5.98$ ft-kips/ft > 4.15 OK at midspan

Check positive moment capacity in Span 1:

$$\left(d - \frac{a}{2}\right) = \frac{(6.5 - 2.25) - \frac{0.37}{2}}{12} = 0.39 \text{ ft}$$

$$\frac{c}{d_t} = \frac{\frac{0.37}{0.85}}{4.25} = 0.102 < 0.375, \text{ therefore, tension controlled}$$

9.3.2.2
10.3.4

$\phi M_n = 0.9 \times (15.1) \times 0.39 = 5.30$ ft-kips/ft > 2.89 OK at midspan

Exterior columns:

A_s minimum = $0.00075 \times 20 \times 12 \times 6.5 = 1.17$ in² use 6-#4 bars

$A_s = 6 \times 0.2/20 = 0.06$ in²/ft

$A_s f_y = 0.06 \times 60 = 3.6$ kips/ft

$\rho_p = 10 \times 0.153/(12 \times 20 \times 3.25) = 0.00196$

$f_{ps} = 175 + 10 + 4/(300 \times 0.00196) = 192$ ksi

		Code
Example 26.1 (cont'd)	**Calculations and Discussion**	**Reference**

$A_s f_{ps} = 10 \times 0.153 \times 192/20 = 14.7$ kips/ft

$a = \dfrac{14.7 + 3.6}{0.85 \times 4 \times 12} = 0.45$ in

$\varepsilon_t = (5.5 - 0.53) \times 0.003/0.53 = 0.028$, therefore, tension controlled, $\phi = 0.9$ 9.3.2
    10.3.4

Tendons:

$\left(d - \dfrac{a}{2}\right) = \dfrac{(3.25) - \dfrac{0.45}{2}}{12} = 0.25$ ft

Rebar:

$\left(d - \dfrac{a}{2}\right) = \dfrac{(5.5) - \dfrac{0.45}{2}}{12} = 0.44$ ft

$\phi M_n = 0.9 \times [(14.7 \times 0.25) + (3.6 \times 0.44)] = 4.73$ ft-kips/ft > 0.13 OK

This completes the design for flexural strength.

9. Shear and Moment Transfer Strength at Exterior Column   11.12.6
    13.5.3

   a. Shear and moment transferred at exterior column.

   $V_u = (0.170 \times 17/2) - (6.65 - 0.61)/17 = 1.09$ kips/ft

   Assume building enclosure is masonry and glass, weighing 0.40 kips/ft.

   Total slab shear at exterior column:

   $V_u = [(1.2 \times 0.40) + 1.09] \times 20 = 31.4$ kips

   Transfer moment $= 20 (0.61) = 12.2$ ft-kips
   (factored moment at exterior column centerline $= 0.61$ ft-kips/ft)

   b. Combined shear stress at inside face of critical transfer section.

   For shear strength equations, see Part 16.

   $v_u = \dfrac{V_u}{A_c} + \dfrac{\gamma_v M_u c}{J}$ R11.12.6.2

		Code
Example 26.1 (cont'd)	**Calculations and Discussion**	**Reference**

where (referring to Table 16-2: edge column-bending perpendicular to edge)

$d \approx 0.8 \times 6.5 = 5.2$ in.

$c_1 = 12$ in.

$c_2 = 14$ in.

$b_1 = c_1 + d/2 = 14.6$ in.

$b_2 = c_2 + d = 19.2$ in.

$c = \dfrac{b_1^2}{(2b_1 + b_2)} = 4.40$ in.

$A_c = (2b_1 + b_2) d = 252$ in.^{2}

$J/c = [2b_1 d (b_1 + 2b_2) + d^3 (2b_1 + b_2)/b_1]/6 = 1419$ in^3

$\gamma_v = 1 - \gamma_f$ *Eq. (11-39)*

$= 1 - \dfrac{1}{1 + \left(\dfrac{2}{3}\right)\sqrt{\dfrac{b_1}{b_2}}} = 0.37$ *13.5.3.2*

$v_u = \dfrac{31400}{252} + \dfrac{0.37 \times 12.2 \times 12000}{1419} = 163$ psi

c. Permissible shear stress (for members without shear reinforcement).   *11.12.6.2*

$\phi v_n = \phi V_c /(b_o d)$ *Eq. (11-20)*
where V_c is defined in 11.12.2.1 or 11.12.2.2

For edge columns:

$\phi v_n = \phi 4\sqrt{f_c'} = 0.85 \times 4\sqrt{4000} = 215$ psi > 163 O.K.   *11.12.2.1*

d. Check moment transfer strength.   *13.5.3*

Although the transfer moment is small, for illustrative purposes, check the moment strength of the effective slab width (width of column plus 1.5 times the slab thickness on each side) for moment transfer. Assume that of the 10 tendons required for the 20 ft bay width, 3 tendons are anchored within the column and are bundled together across   *13.5.3.2*

		Code
Example 26.1 (cont'd)	**Calculations and Discussion**	**Reference**

the building. This amount should be noted on the design drawings. Besides providing flexural strength, this prestress force will act directly on the critical section for shear and improve shear strength. As previously shown, a minimum amount of bonded reinforcement is required at all columns. For the exterior column, the required area is:

$A_s = 0.00075 A_{cf} = 0.00075 \times 6.5 \times 20 \times 12 = 1.17$ in.^{2} *Eq. (18-8)*

Use 6-No. 4 bars, 5 ft in length (including standard end hook).

Calculate stress in tendons:

Effective slab width = $14 + 2(1.5 \times 6.5) = 33.5$ in.

$\rho_p = \dfrac{3 \times 0.153}{33.5 \times 3.25} = 0.0042$

$f_{ps} = 175 + 10 + 4/(300 \times 0.0042) = 188.2$ ksi

Corresponding prestress force = $3 \times 0.153 \times 188.2 = 86.4$ kips

$A_s f_y = 6 \times 0.20 \times 60 = 72.0$ kips

$A_{ps} f_{ps} + A_s f_y = 158.4$ kips

$a = 158.4/(0.85 \times 4 \times 33.5) = 1.39$ in.

tendon $(d_p - a/2) = (3.25 - 1.39/2)/12 = 0.21$ ft

rebar $(d - a/2) = (5.5 - 1.39/2)/12 = 0.40$ ft

$\phi M_n = 0.9 [(86.4 \times 0.21) + (72 \times 0.40)] = 42.25$ ft-kips

$\gamma_f = \dfrac{1}{1 + \dfrac{2}{3}\sqrt{\dfrac{b_1}{b_2}}} = 0.63$ *Eq. (13-1)*

$\gamma_f M_u = 0.63 (12.2) = 7.69$ ft-kips $\ll 42.25$ ft-kips   O.K.

10. **Shear and Moment Transfer Strength at Interior Column**   *11.12.6*
    13.5.3

   a. Shear and moment transferred at interior column.

   Direct shear and moment to the left and right of interior columns is calculated in Step 8 above.

Example 26.1 (cont'd)	Calculations and Discussion	Code Reference

$V_u = (1.81 + 2.03)\, 20 = 76.8$ kips

Transfer moment $= 20\, (7.43 - 6.65) = 15.6$ ft-kips

b. Combined shear stress at face of critical transfer section. For shear strength equations, see Part 16.

$$v_u = \frac{V_u}{A_c} + \frac{\gamma_v M_u c}{J}$$ R11.12.6.2

where (referring to Table 16-1: interior column)

$d \approx 0.8 \times 6.5 = 5.2$ in.

$c_1 = 20$ in.

$c_2 = 14$ in.

$b_1 = c_1 + d = 25.2$ in.

$b_2 = c_2 + d = 19.2$ in.

$A_c = 2\,(b_1 + b_2)\, d = 462$ in.^{2}

$J/c = [b_1 d\,(b_1 + 3b_2) + d^3]/3 = 3664$ in.^{3}

$\gamma_v = 1 - \gamma_f$ Eq. (11-39)

$$= 1 - \frac{1}{1 + \left(\frac{2}{3}\right)\sqrt{\frac{b_1}{b_2}}} = 0.43$$ 13.5.3.2

$$v_u = \frac{76{,}800}{462} + \frac{0.43 \times 15.6 \times 12{,}000}{3664} = 188 \text{ psi}$$

c. Permissible shear stress.

For interior columns, Eq. (11-36) applies: 11.12.2.2

$$\phi v_c = \phi \left(\beta_p \sqrt{f'_c} + 0.3 f_{pc} + \frac{V_p}{b_o d} \right)$$ Eq. (11-36)

26-21

Example 26.1 (cont'd)   **Calculations and Discussion**   **Code Reference**

where $\beta_p = \left(\dfrac{\alpha_s d}{b_o} + 1.5\right)$ but not greater than 3.5

$b_o = 2[(20 + 5.2) + (14 + 5.2)] = 88.8$ in.

$\alpha_s = 40$ for interior columns

$d = 5.2$ in.

$\beta_p = \dfrac{40 \times 5.2}{88.8} + 1.5 = 3.8 > 3.5$, use 3.5

V_p is the shear carried through the critical transfer section by the tendons. For thin slabs, the V_p term must be carefully evaluated, as field placing practices can have a great effect on the profile of the tendons through the critical section. Conservatively, this term may be taken as zero.

R11.12.2.2

$\phi v_c = 0.85\left[3.5\sqrt{4000} + (0.3 \times 172)\right] = 232$ psi > 188 psi O.K.

d.   Check moment transfer strength.

13.5.3

$\gamma_f = \dfrac{1}{1 + \dfrac{2}{3}\sqrt{\dfrac{b_1}{b_2}}} = 0.57$

Eq. (13-1)

Moment transferred by flexure within width of column plus 1.5 times slab thickness on each side = $0.57 (15.6) = 8.89$ ft-kips.

13.5.3.2

Effective slab width = $14 + 2 (1.5 \times 6.5) = 33.5$ in.

Say $A_{ps}f_{ps} = 86.4$ kips (same as exterior column)

$A_s = 0.00075 A_{cf} = 0.00075 \times 6.5 \times (17 + 25)/2 \times 12 = 1.23$ in.^{2}

Eq. (18-8)

Use 6-No. 4 bars ($A_s = 1.20$ in.^{2})

$A_s f_y = 1.20 \times 60 = 72.0$ kips

$A_{ps}f_{ps} + A_s f_y = 86.4 + 72.0 = 158.4$ kips

$a = \dfrac{158.4}{0.85 \times 4 \times 33.5} = 1.39$ in.

		Code
Example 26.1 (cont'd)	**Calculations and Discussion**	**Reference**

$(d - a/2) = (5.5 - 1.39/2)/12 = 0.40$ ft

$\phi M_n = 0.9 \, (158.4 \quad 0.40) = 57.0$ ft-kips $\gg 8.89$ ft-kips   O.K.

This completes the shear design.

11. Distribution of tendons.

   In accordance with 18.12.4, the 10 tendons per 20 ft bay will be distributed in a group of 3 tendons directly through the column with the remaining 7 tendons spaced at 2 ft-6 in. on center (4.6 times slab thickness). Tendons in the perpendicular direction will be placed in a narrow band through and immediately adjacent to the columns.

Blank

27

Shells and Folded Plate Members

INTRODUCTION

Chapter 19, concerning shells and folded plate members, was completely updated for ACI 318-83. Sections 19.2.10 and 19.2.11 were added to the '95 edition. In its present form, Chapter 19 reflects the current state-of-the-knowledge in analysis and design of folded plates and shells. It includes guidance on analysis methods appropriate for different types of shell structures, and provides specific direction as to design and proper placement of shell reinforcement. The Commentary on Chapter 19 should be helpful to designers; its contents reflect current information, including an extended reference listing.

GENERAL CONSIDERATIONS

Code requirements for shells and folded plates must, of necessity, be somewhat general in nature as compared to the provisions for other types of structures where the practice of design has been firmly established. Chapter 19 is specific in only a few critical areas inherent to shell design; otherwise, it refers to standard provisions of the code. It should be noted that strength design is permitted for shell structures, even though most of the shells in this country have been built using working stress design procedures.

The code, the commentary, and the list of references are an excellent source of information and guidance on shell design. The list of references, however, does not exhaust the possible sources of design assistance.

1. Chapter 19 covers the design of a large class of concrete structures that are quite different from the ordinary slab, beam and column construction. Structural action varies from shells with considerable bending in the shell portions (folded plates and barrel shells) to those with very little bending except at the junction of shell and support (hyperbolic paraboloids and domes of revolution). The problems of shell design, therefore, cannot be lumped together, as each type has its own peculiar attributes that must be thoroughly understood by the designer. Even shells classified under one type, such as the hyperbolic paraboloid, vary greatly in their structural action. Studies have shown that gabled hyperbolic paraboloids, for example, are much more complex than the simple membrane theory would indicate. This is one explanation for the lack of a rigid set of rules in the code for the design of shells and folded plate structures.

2. For the reasons given above, design of a shell requires considerable lead time to gain an understanding of the design problems for the particular type of shell. An attempt to design a shell without proper study may invite poor performance. Design of shell structures requires the ability to think in terms of three-dimensional space; this is only gained by study and experience. The conceptual stage is the most critical period in shell design, since this is when vital decisions on form and dimensions must be made.

3. Strength of shell structures is inherent in their shape and is not created by boosting the performance of materials to their limit as in the case of other types of concrete structures such as conventional and prestressed concrete beams. Therefore, the design stresses in the concrete should not be raised to their highest acceptable values, except where required for very large structures. Deflections are normally not a problem if the stresses are low.

4. Shell size is a very important determinant in the analytical precision required for its design. Short spans (up to 60 ft) can be designed using approximate methods such as the beam method for barrel shells, provided the exterior shell elements are properly supported by beams and columns. However, the limits and approximations of any method must be thoroughly understood. Large spans may require much more elaborate analyses. For example, a large hyperbolic paraboloid (150-ft span or more) may require a finite element analysis.

Application of the following code provisions warrants further explanation.

19.2　ANALYSIS AND DESIGN

19.2.6　Prestressed Shells

The components of force produced by prestressing tendons draped in a thin shell must be taken into account in the design. In the case of a barrel shell, it should be noted that the tendon does not lie in one plane, as shown in Fig. 27-1.

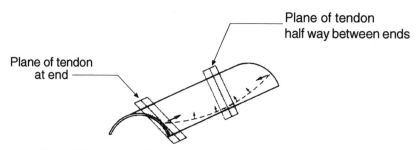

Figure 27-1 Draped Prestressing Tendon in Barrel Shell

19.2.7　Design Method

The Strength Design Method is permitted for the design of shells, but it should be noted that for slab elements intersecting at an angle, and having high tensile stresses at inside corners, the ultimate strength is greatly reduced from that at the center of a concrete slab. Therefore, special attention should be given to the reinforcement used in these areas, and thickness should be greater than the minimum allowed by the strength method.

19.4　SHELL REINFORCEMENT

19.4.6　Membrane Reinforcement

For shells with essentially membrane forces, such as hyperbolic paraboloids and domes of revolution, it is usually convenient to place the reinforcement in the direction of the principal forces. Even though folded plates and barrel shells act essentially as longitudinal beams (traditionally having vertical stirrups as shear reinforcement), an orthogonal pattern of reinforcement (diagonal bars) is much easier to place and also assures end anchorage in the barrel or folded plate. With diagonal bars, five layers of reinforcement may be required at some points.

The direction of principal stresses near the supports is usually about 45 degrees, so that equal areas of reinforcement are needed in each direction to satisfy the requirements of 19.4.4. For illustration, Fig. 27-2 shows a plot of the principal membrane forces in a barrel shell with a span of 60 ft, a rise of 6.3 ft, a thickness of 3.5 in., and a snow load of 25 psf and a roof load of 10 psf. Forces, due to service loads, are shown in kips per linear foot.

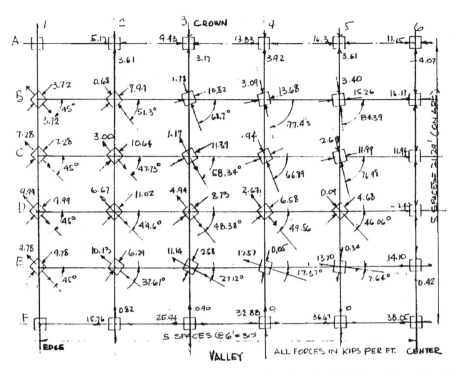

Figure 27-2 Principal Membrane Forces and Direction for 60-ft Span Barrel Shell

19.4.8 Concentration of Reinforcement

In the case of long barrel shells (or domes) it is often desirable to concentrate tensile reinforcement near the edges rather than distribute the reinforcement over the entire tensile zone. When this is done, a minimum amount of reinforcement equal to 0.0035bh must be distributed over the remaining portion of the tensile zone, as shown in Fig. 27-3. This amount in practical terms is twice the minimum steel requirement for shrinkage and temperature stresses.

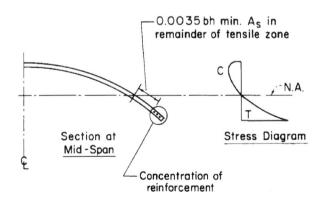

Figure 27-3 Concentration of Shell Reinforcement

19.4.10 Spacing of Reinforcement

Maximum permissible spacing of reinforcement is the smaller of 5 times the shell thickness and 18 in. Therefore, for shells less than 3.6 in. thick, 5 times the thickness controls. For thicker shells, the spacing of bars must not exceed 18 in.

Blank

28

Strength Evaluation of Existing Structures

INTRODUCTION

Chapter 20 was revised in 1995 to flag the need to monitor during load tests not only deflections, but also cracks related to shear and/or bond, along with spalling and crushing of the concrete. In cases involving deterioration of the structure, acceptance of a building should be based on a load test. Further, the acceptance should include a time limit. Periodic inspections and strength reevaluations should be specified depending on the nature of the deterioration. When structure dimensions, size and location of reinforcement, and material properties are known, higher strength reduction factors were introduced in ACI 318-95 for analytical evaluations of the strength of existing structures.

Strength evaluation of an existing structure requires experience and sound engineering judgment. Chapter 20 provides guidance for investigating the safety of a structure when:

1. Materials of a building are considered to be deficient in quality.
2. There is evidence indicating faulty construction.
3. A building has deteriorated.
4. A building will be used for a new function.
5. A building or a portion of it does not appear to satisfy the requirements of the code.

The provisions of Chapter 20 should not be used for approval of special systems of design and construction. Approval of such systems is covered in 1.4.

References 28.1 and 28.2 published by the Concrete Reinforcing Steel Institute (CRSI) are suggested additional guides for strength evaluation of existing structures. Information about reinforcing steel found in old reinforced concrete structures is given in CRSI Engineering Data Report Number 11.[28.3]

20.1 STRENGTH EVALUATION - GENERAL

Strength evaluation of structures can be performed analytically or experimentally. Applicability of the analytical procedure depends on whether the source of deficiency is critical to the structure's strength under: (1) flexural and/or axial load, or (2) shear and/or bond. The behavior and strengths of structural concrete under flexural and/or axial load strengths can be accurately predicted based on Navier's hypothesis of "plane section before loading remains plane after loading." On the other hand, available theories and models are not as reliable to predict the shear and bond behavior and strengths of structural concrete. Code provisions for one- and two-way shear, and for bond are semi-empirical. Shear and bond failures can be brittle.

Analytical strength evaluations suffice for acceptance of buildings if two conditions are met (20.1.2). First, the source of deficiency should be critical to flexural, axial load, or combined flexural and axial load strengths. It cannot be critical to shear or bond strengths. Second, it should be possible to establish the actual building dimensions, size and location of reinforcement, and material properties. If both conditions are not met, strength evaluations should be determined by a load test as prescribed in 20.3. If causes of concern relate to flexure or axial load, but it is not possible or feasible to determine material properties, a physical test may be appropriate.

Analytical evaluations of shear strength are not precluded if they are "well understood." If shear or bond strength is critical to the safety concerns, physical test may be the most efficient solution. Wherever possible and appropriate, it is desirable to support the results of the load tests by analysis (R20.1.3).

If the safety concerns are due to deterioration, strength evaluation may be through a load test. If the building satisfies the acceptance criteria of 20.5, the building should be allowed to remain in service for a specified period of time as a function of the nature of the deterioration. Periodic reevaluations of the building should be conducted.

20.2 DETERMINATION OF REQUIRED DIMENSIONS AND MATERIAL PROPERTIES

If strength evaluation of a building is performed through analysis, actual dimensions, location and size of reinforcement, and material properties should be established. Measurements should be taken at critical sections where calculated stress would reach a maximum value. When shop drawings are available, spot checks should be made to confirm location and size of reinforcing bars shown on the drawings. Nondestructive testing techniques are available to determine location and size of reinforcement, and estimate the strength of concrete. Unless they are already known, actual properties of reinforcing steel or prestressing tendons should be determined from samples extracted from the structure.

An analytical strength evaluation requires the use of the load factors of 9.2 and the strength reduction factors of 20.2.5. One of the purposes of the strength reduction factors ϕ given in R9.3.1 is "to allow for the probability of understrength members due to variations in material strengths and dimensions." When actual member dimensions, size and location of reinforcement, and concrete and reinforcing steel properties are measured, Chapter 20 specifies higher strength reduction factors. A comparison of the strength reduction factors of 20.2.5 to those of 9.3 is given in Table 28-1. The ratios of strength reduction factors of Chapter 20 to those of Chapter 9 are listed in the last column of the table. For analytical evaluation of columns and bearing on concrete, strength reduction factors ϕ of 20.2.5 are about 20 percent higher than those of 9.3. For flexure in beams and axial tension, the increase is 11 percent, while for shear and torsion it is 6 percent.

An increase in strength reduction factors, as specified in Chapter 20, results in an increase in computed member strengths. Nominal axial compressive strength of columns is in great part a function of the product of the column cross sectional area and the concrete compressive strength. As concrete compressive strength is subject to large variability, the strength reduction factors of Chapter 9 are lower for axial compression than for flexure. Because the actual concrete compressive strength is measured for strength evaluation of existing structures (20.1.2), a higher increase in strength reduction factor ϕ is specified for columns in 20.2.5.

Table 28-1 Comparison of Strength Reduction Factors

	Strength reduction factor		
	Ch. 20	Ch. 9	Ch. 20/Ch. 9
Tension-controlled sections, as defined in 10.3.4	1.00	0.90	1.11
Compression-controlled sections, as defined in 10.3.3			
Members with spiral reinforcement conforming to 10.9.3	0.85	0.75	1.21
Other reinforced members	0.80	0.70	1.23
Shear and torsion	0.80	0.75	1.07
Bearing on concrete	0.80	0.65	1.23

20.3 LOAD TEST PROCEDURE

The number and arrangement of spans or panels loaded should be selected to maximize the deflection and stresses in the critical regions of the structural elements of which strength is in doubt (20.3.1). If adjoining elements are expected to contribute to the load carrying capacity, magnitude of the test load or placement should be adjusted to compensate for this contribution. As in earlier editions of the code, the total test load is specified at 0.85 (1.4D + 1.7L), where D is the sum of dead loads or related internal moments and forces, and L is defined as the live loads or related internal moments and forces. The total test load includes the dead load already in place (20.3.2). The portion of the structure being load tested should be at least 56 days old, unless all concerned parties agree to conduct the test at an earlier age (20.3.3).

Note, starting with ACI 318-02, the load factors and strength reduction factors were revised in 9.2 and 9.3, respectively. In spite of these changes, the test load intensity prescribed in 20.3.2 has remained unchanged. Committee 318 felt that it was appropriate to maintain the same factored test load intensity, for designs complying with the new load factors and strength reduction factors of Chapter 9.

20.4 LOADING CRITERIA

Loading criteria are specified in 20.4. Initial values of all response measurements (deflection, strain, crack width, etc.) should be read and recorded not more than one hour before load application. When simulating uniformly distributed loads, arching of the applied loads must be avoided. Figure 28-1 illustrates arching action.[28.1] Sufficient gap should be provided between loading stacks so as to prevent contact, and hence arching, after member deflection, while assuring stability of the test loads.

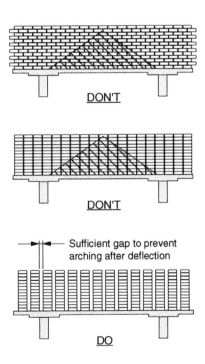

Figure 28-1 Arching Effect Shifts Applied Load to Ends of Span

Test load should be applied in not less than four approximately equal increments. A set of test load and response measurements is to be recorded after each load increment and after the total load has been applied for at least 24 hours. A set of final response measurements is to be recorded 24 hours after the test load is removed.

20.5 ACCEPTANCE CRITERIA

Evidence of failure includes spalling or crushing of concrete (20.5.1), excessive deflections (20.5.2), shear cracks (20.5.3 and 20.5.4), and bond cracks (20.5.5). No simple rules can be developed for application to all types of structures and conditions. However, in members without transverse reinforcement, projection of diagonal (inclined) cracks on an axis parallel to the longitudinal axis of the member should be monitored. If the projection of any diagonal crack is longer than the member depth at mid length of the crack, the member may be deficient in shear. If sufficient damage has occurred so that the structure is considered to have failed that test, retesting is not permitted since it is considered that damaged members should not be put into service even at a lower rating.

Deflection criteria must satisfy the following conditions (20.5.2):

1. When maximum deflection exceeds $\ell_t^2/(20,000h)$, the percentage recovery must be at least 75 percent after 24 hours, where

    h = overall thickness of member, in.
    ℓ_t = span of member under load test, in. (The shorter span for two-way slab systems.) Span is the smaller of (a) distance between centers of supports, and (b) clear distance between supports plus thickness h of member. Span for a cantilever must be taken as twice the distance from support to cantilever end, in.

2. When maximum deflection is less than $\ell_t^2/(20,000h)$, recovery requirement is waived. Figures 28-2 and 28-3 illustrate application of the limiting deflection criteria to the first load test. Figure 28-2 illustrates the limiting deflection versus member thickness for a sample span of 20 ft. Figure 28-3 depicts the limiting deflection versus span for a member 8 in. thick.

3. Members failing to meet the 75 percent recovery criterion may be retested.

4. Before retesting, 72 hours must have elapsed after load removal. On retest, the recovery must be 80 percent.

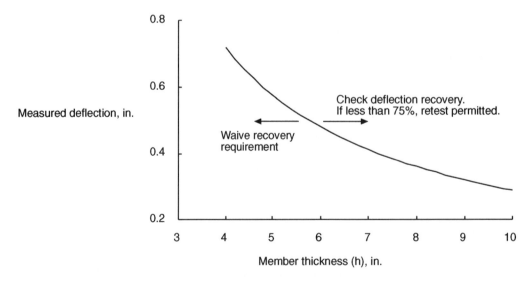

Figure 28-2 Load Testing Acceptance Criteria for Members with Span Length ℓ_t = 20 ft

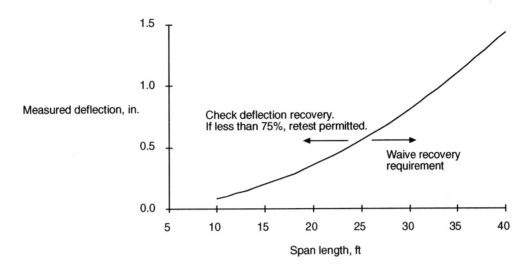

Figure 28-3 Load Testing Criteria for Members with Overall Thickness h = 8 in.

20.6   PROVISION FOR LOWER LOAD RATING

If analytical strength evaluations (20.1.2) indicate that a structure is inadequate, if the deflections of 20.5.2 are exceeded, or if cracks criteria of 20.5.3 are not met, the structure can be used for a lower load rating, if approved by the building official.

20.7   SAFETY

During load testing, shoring normally must be provided under the loaded members to assure safety. The shoring must not interfere with the test procedure or affect the test results. At no time during the load test should the deformed structure touch or bear against the shoring.

REFERENCES

28.1   "Applications of ACI 318 Load Test Requirements," Structural Bulletin No. 16, Concrete Reinforcing Steel Institute, Schaumburg, IL, November 1987.

28.2   "Proper Load Tests Protect the Public," Engineering Data Report Number 27, Concrete Reinforcing Steel Institute, Schaumburg, IL.

28.3   "Evaluation of Reinforcing Steel in Old Reinforced Concrete Structures," Engineering Data Report Number 48, Concrete Reinforcing Steel Institute, Schaumburg, IL, 2001, 4 pp.

Blank

29

Special Provisions for Seismic Design

UPDATE FOR THE '05 CODE

A new term, design story drift ratio, has been defined as the relative difference of design displacements between the top and the bottom of a story, divided by the story height.

Sections 9.4 and 10.9.3 have been modified to allow the use of spiral reinforcement with specified yield strength of up to 100 ksi. A provision added to 21.2.5 specifically prohibits such use in members resisting earthquake-induced forces in structures assigned to Seismic Design Category D, E, or F.

Section 21.5.4 modifies the development length requirements of Chapter 12 for longitudinal beam bars terminating at exterior beam-column joints of structures assigned to high seismic design categories. But then 21.7.2.3 of ACI 318-02 required that all continuous reinforcement in structural walls must be anchored or spliced in accordance with the provisions for reinforcement in tension in 21.5.4. Section 21.9.5.4 of ACI 318-02 further required that all continuous reinforcement in diaphragms, trusses, ties, chords, and collector elements to be anchored or spliced in accordance with the provisions for reinforcement in tension as specified in 21.5.4. Sections 21.7.2.3 and 21.9.5.4 were very confusing to the user, because 21.5.4 is really not applicable to situations covered by those sections. This problem existed with ACI 318 editions prior to 2002 as well.

In a very significant and beneficial change, the requirements of 21.7.2.3 have been modified to remove the reference to beam-column joints in 21.5.4. All reference now is directly to the provisions of Chapter 12. The requirement that mechanical splices of reinforcement conform to 21.2.6, and welded splices to 21.2.7, has now been placed in 21.7.2.3. Consequently, 21.7.6.4(f) of 21.7.6.6 of ACI 318-02 have been deleted.

In a companion change, 21.9.5.4 now requires that all continuous reinforcement in diaphragms, trusses, struts, ties, chords, and collector elements be developed or spliced for f_y in tension.

Structural truss elements, struts, ties, diaphragm chords, and collector elements with compressive stresses exceeding $0.2 f'_c$ at any section are required to be specially confined by 21.9.5.3. The special transverse reinforcement may be discontinued at a section where the calculated compressive stress is less than $0.15 f'_c$. Stresses are calculated for factored forces using a linear elastic model and gross-section properties of the elements considered. In recent seismic codes and standards, collector elements of diaphragms are required to be designed for forces amplified by a factor, to account for the overstrength in the vertical elements of the seismic-force-resisting system. The amplification factor ranges between 2 and 3 for concrete structures, depending upon the document selected and on the type of seismic system. To account for this amplification factor, 21.9.5.3 now additionally states that where design forces have been amplified to account for the overstrength of the vertical elements of the seismic-force-resisting system, the limits of $0.2 f'_c$ and $0.15 f'_c$ shall be increased to $0.5 f'_c$ and $0.4 f'_c$, respectively.

In a very significant change, provisions for shear reinforcement at slab-column joints have been added in a new section 21.11.5, to reduce the likelihood of punching shear failure in two-way slabs without beams. A prescribed amount and detailing of shear reinforcement is required unless either 21.11.5(a) or (b) is satisfied.

BACKGROUND

Special provisions for earthquake resistance were first introduced into the 1971 edition of the ACI code in Appendix A, and were included with minor revisions in ACI 318-77. The original provisions of Appendix A were intended to apply only to reinforced concrete structures located in regions of high seismicity, and designed with a substantial reduction in total lateral seismic forces (as compared with the elastic response forces), in anticipation of inelastic structural behavior. Also, several changes were incorporated into the main body of the 1971 code specifically to improve toughness, in order to increase the resistance of concrete structures to earthquakes or other catastrophic loads. While Appendix A was meant for application to lateral force-resisting frames and walls in regions of high seismicity, the main body of the code was supposed to be sufficient for regions where there is a probability of only moderate or light earthquake damage.

The special provisions of Appendix A were extensively revised for the 1983 code, to reflect current knowledge and practice of the design and detailing of monolithic reinforced concrete structures for earthquake resistance. Appendix A to ACI 318-83 for the first time included special detailing for frames in zones of moderate seismic hazard.

The special provisions for earthquake resistance formed Chapter 21 of the 1989 code edition. This move from an appendix into the main body of the code was made for reasons discussed in Part 1 of this publication.

For buildings located in regions of low seismic risk, or for structures assigned to low seismic performance or design categories, no special design or detailing is required; the general requirements of Chapters 1 through 18 and 22 of the code apply. Concrete structures proportioned by Chapters 1 through 18 and 22 of the code are expected to provide a level of toughness adequate for structures subject to low seismic risk.

For buildings located in regions of moderate seismic hazard, or for structures assigned to intermediate seismic performance or design categories, moment frames proportioned to resist earthquake effects require some additional reinforcement details. The additional detailing requirements apply only to frames (beams, columns, and slabs) to which earthquake-induced forces have been assigned in design. There are no additional requirements for structural walls provided to resist the effects of wind and earthquakes, or for structural components that are not designed to be part of the lateral force-resisting system a of building located in regions of moderate seismic hazard, except for the connection requirements in 21.13 for intermediate precast structural walls. Structural walls proportioned by the non-seismic provisions of the code are considered to have sufficient toughness at drift levels anticipated in structures subject to moderate seismic risk.

For buildings located in regions of high seismic hazard, or for structures assigned to high seismic performance or design categories, where damage to construction has a high probability of occurrence, all structural components, irrespective of whether they are included in the lateral force-resisting system, must satisfy the requirements of 21.2 through 21.10. The special proportioning and detailing provisions of Chapter 21 are intended to produce a concrete structure with adequate toughness to respond inelastically under severe earthquake excitation.

GENERAL CONSIDERATIONS

Economical earthquake-resistant design should aim at providing appropriate dynamic characteristics in structures so that acceptable response levels would result under the design earthquake. The structural properties that can be modified to achieve the desired results are the magnitude and distribution of stiffness and mass and the relative strengths of the structural members.

In some structures, such as slender free-standing towers or smoke stacks, which depend for their stability on the stiffness of the single element making up the structure, or in nuclear containment buildings where a more-than-usual conservatism in design is required, yielding of the principal elements in the structure cannot be tolerated. In such cases, the design needs to be based on an essentially elastic response to moderate-to-strong earthquakes, with the critical stresses limited to the range below yield.

In most buildings, particularly those consisting of frames and other multiply-redundant systems, however, economy is achieved by allowing yielding to take place in some members under moderate-to-strong earthquake motion.

The performance criteria implicit in most earthquake code provisions require that a structure be able to:

1. Resist earthquakes of minor intensity without damage; a structure would be expected to resist such frequent but minor shocks within its elastic range of stresses.

2. Resist moderate earthquakes with negligible structural damage and some nonstructural damage; with proper design and construction, it is expected that structural damage due to the majority of earthquakes will be repairable.

3. Resist major catastrophic earthquakes without collapse; some structural and nonstructural damage is expected.

The above performance criteria allow only for the effects of a typical ground shaking. The effects of landslides, subsidence or active faulting in the immediate vicinity of the structure, which may accompany an earthquake, are not considered.

While no clear quantitative definition of the above earthquake intensity ranges has been given, their use implies the consideration not only of the actual intensity level but also of their associated probability of occurrence with reference to the expected life of a structure.

The principal concern in earthquake-resistant design is the provision of adequate strength and toughness to assure life safety under the most intense earthquake expected at a site during the life of a structure. Observations of building behavior in recent earthquakes, however, have made engineers increasingly aware of the need to ensure that buildings that house facilities essential to post-earthquake operations—such as hospitals, power plants, fire stations and communication centers—not only survive without collapse, but remain operational after an earthquake. This means that such buildings should suffer a minimum amount of damage. Thus, damage control is at times added to life safety as a second design consideration.

Often, damage control becomes desirable from a purely economic point of view. The extra cost of preventing severe damage to the nonstructural components of a building, such as partitions, glazing, ceiling, elevators and other mechanical systems, may be justified by the savings realized in replacement costs and from continued use of a building after a strong earthquake.

The principal steps involved in the earthquake-resistant design of a typical concrete structure according to building code provisions are as follows:

1. Determination of seismic zone, or seismic performance, or design category
   Seismic design category combines the seismic hazard at the site of the structure, the occupancy of the structure, and the soil characteristics at the site of the structure. It's a relatively new concept, for an understanding of which, the reader may consult Ref. 29.1. Seismic performance category is a function only of the seismic hazard at the site of the structure and the occupancy of the structure. Seismic zone considers only the seismic hazard at the site of the structure.

2. Determination of design earthquake forces
   a. calculation of base shear corresponding to computed or estimated fundamental period of vibration of the structure (a preliminary design of the structure is assumed here)
   b. distribution of the base shear over the height of the building

3. Analysis of the structure under the (static) lateral earthquake forces calculated in step 1, as well as under gravity and wind loads, to obtain member design forces.

4. Designing members and joints for the critical combinations of gravity and lateral (wind or seismic) loads.

5. Detailing members for ductile behavior in accordance with the seismic zone, or the seismic performance or design category of the structure.

It is important to note that some buildings are required to be designed by a dynamic, rather than a static, lateral force procedure when one or more criteria of the static procedure are not satisfied.

In the International Building Code (IBC)[29.2], as well as in the model codes that preceeded it, the design base shear represents the total horizontal seismic force that may be assumed acting parallel to the axis of the structure considered. The force in the other horizontal direction is usually assumed to act non-concurrently. Depending on the building and the seismic zone or seismic performance or design category, the seismic forces may need to be applied in the direction that produces the most critical load effect. The requirement that orthogonal effects be considered in the proportioning of a structural element may be satisfied by designing the element for 100 percent of the prescribed seismic forces in one direction plus 30 percent of the prescribed forces in the perpendicular direction. The combination requiring the greater component strength must be used for design. The vertical component of the earthquake ground motion is included in the load combinations involving earthquake forces that are prescribed in the IBC. Special provisions are also required for structural elements that are susceptible to vertical earthquake forces (cantilever beams and slabs; prestressed members).

The code-specified design lateral forces have a general distribution that is compatible with the typical envelope of maximum horizontal shears indicated by elastic dynamic analyses for regular structures. However, the code forces are substantially smaller than those that would be developed in a structure subjected to the anticipated earthquake intensity, if the structure were to respond elastically to such ground excitation. Thus, buildings designed under the present codes would be expected to undergo fairly large deformations when subjected to a major earthquake. These large deformations will be accompanied by yielding in many members of the structure, which is the intent of the codes. The reduced code-specified forces must be coupled with additional requirements for the design and detailing of members and their connections in order to ensure sufficient deformation capacity in the inelastic range.

The capacity of a structure to deform in a ductile manner (i.e., to deform beyond the yield limit without significant loss of strength), allows such a structure to dissipate a major portion of the energy from an earthquake without collapse. Laboratory tests have demonstrated that cast-in-place and precast concrete members and their connections, designed and detailed by the present codes, do possess the necessary ductility to allow a structure to respond inelastically to earthquakes of major intensity without significant loss of strength.

21.2 GENERAL REQUIREMENTS

21.2.1 Scope

Sections 21.2.1.2 through 21.2.1.4 contain the required detailing requirements based on the structural framing system, seismic hazard level at the site, level of energy dissipation planned in the structural design, and the occupancy of the building.

Traditionally, seismic risk levels have been classified as low, moderate, and high. The seismic risk level, or the seismic performance or design category of a building is regulated by the legally adopted building code of the region or is determined by a local authority (1.1.8.3). Table R1.1.8.3 contains a summary of the seismic risk levels, seismic performance categories (SPC), and seismic design categories (SDC) specified in the IBC, the three prior model building codes now called *legacy* codes, as well as other resource documents (see R21.2.1).

The provisions of Chapters 1 through 18 and Chapter 22 of ACI 318 apply to structures in regions of low seismic hazard or to structures assigned to low seismic performance or design categories (21.2.1.2). The design and detailing requirements of these chapters are intended to provide adequate toughness for structures in these regions or assigned to these categories. Ordinary moment frames (cast-in-place or precast) and ordinary structural walls are the structural systems that can be utilized. It is important to note that the requirements of Chapter 21 apply when the design seismic forces are computed using provisions for intermediate or special concrete systems.

In regions of moderate seismic hazard or for structures assigned to satisfy intermediate seismic performance or design categories, intermediate or special moment frames, or ordinary, intermediate, or special structural walls shall be used (21.2.1.3). Provisions for intermediate moment frames and intermediate precast structural walls are contained in 21.12 and 21.13, respectively.

Special moment frames (cast-in-place or precast), special structural walls (cast-in-place or precast), and diaphragms and trusses complying with 21.2.2 through 21.2.8 and 21.3 through 21.10 shall be used in regions of high seismic hazard or for structures assigned to satisfy high seismic performance or design categories (21.2.1.4). Members not proportioned to resist earthquake forces shall comply with 21.11. The provisions of 21.2.2 through 21.2.8 and 21.3 through 21.11 have been developed to provide adequate toughness should the design earthquake occur.

The requirements of Chapter 21 as they apply to various structural components are summarized in Table R21.2.1.

21.2.2   Analysis and Proportioning of Structural Members

The interaction of all structural and nonstructural components affecting linear and nonlinear structural response are to be considered in the analysis (21.2.2.1). Consequences of failure of structural and nonstructural components not forming part of the lateral force-resisting system shall also be considered (21.2.2.2). The intent of 21.2.2.1 and 21.2.2.2 is to draw attention to the influence of nonstructural components on structural response and to hazards from falling objects.

Section 21.2.2.3 alerts the designer to the fact that the base of the structure as defined in analysis may not necessarily correspond to the foundation or ground level. It requires that structural members below base, which transmit forces resulting from earthquake effects to the foundation, shall also comply with the requirements of Chapter 21.

Even though some element(s) of a structure may not be considered part of the lateral force-resisting system, the effect on all elements due to the design displacements must be considered (21.2.2.4).

21.2.3   Strength Reduction Factors

The strength reduction factors of 9.3.2 are not based on the observed behavior of cast-in-place or precast concrete members under load or displacement cycles simulating earthquake effects. Some of those factors have been modified in 9.3.4 in view of the effects on strength due to large displacement reversals into the inelastic range of response.

Section 9.3.4(a) refers to members such as low-rise walls or portions of walls between openings, which are proportioned such as to make it impractical to raise their nominal shear strength above the shear corresponding to nominal flexural strength for the pertinent loading conditions.

21.2.4, 21.2.5   Limitations on Materials

A minimum specified concrete compressive strength f'_c of 3,000 psi and a maximum specified reinforcement yield strength f_y of 60,000 psi are mandated. These limits are imposed as reasonable bounds on the variation of material properties, particularly with respect to their unfavorable effects on the sectional ductilities of members in which they are used. A decrease in the concrete strength and an increase in the yield strength of the tensile reinforcement tend to decrease the ultimate curvature and hence the sectional ductility of a member subjected to flexure.

The statement in 21.2.1.1, referencing 1.1.1, helps to clarify that no maximum specified compressive strength applies. Limitations on the compressive strength of lightweight aggregate concrete is discussed below.

There is evidence suggesting that lightweight concrete ranging in strength up to 12,500 psi can attain adequate ultimate strain capacities. Testing to examine the behavior of high-strength, lightweight concrete under high-intensity, cyclic shear loads, including a critical study of bond characteristics, has not been extensive in the past. However, there are test data showing that properly designed lightweight concrete columns, with concrete strength ranging up to 6,200 psi, maintained ductility and strength when subjected to large inelastic deformations from load reversals. Committee 318 feels that a limit of 5,000 psi on the strength of lightweight concrete is advisable, pending further testing of high-strength lightweight concrete members under reversed cyclic loading. Note that lightweight concrete with a higher design compressive strength is allowed if it can be demonstrated by experimental evidence that structural members made with that lightweight concrete possess strength and toughness equal to or exceeding those of comparable members made with normal weight concrete of the same strength.

Chapter 21 requires that reinforcement for resisting flexure and axial forces in frame members and wall boundary elements be ASTM A 706 Grade 60 low-alloy steel, which is intended for applications where welding or bending, or both, are important. However, ASTM A 615 billet steel bars of Grade 40 or 60 may be used in these members if the following two conditions are satisfied:

$$\text{actual } f_y \leq \text{specified } f_y + 18,000 \text{ psi}$$

$$\frac{\text{actual tensile strength}}{\text{actual } f_y} \geq 1.25$$

The first requirement helps to limit the magnitude of the actual shears that can develop in a flexural member above that computed on the basis of the specified yield strength of the reinforcement when plastic hinges form at the ends of a beam. Note that retests shall not exceed this value by more than an additional 3,000 psi. The second requirement is intended to ensure steel with a sufficiently long yield plateau.

In the "strong column-weak beam" frame intended by the code, the relationship between the moment strengths of columns and beams may be upset if the beams turn out to have much greater moment strengths than intended. Thus, the substitution of Grade 60 steel of the same area for specified Grade 40 steel in beams can be detrimental. The shear strength of beams and columns, which is generally based on the condition of plastic hinges forming at the ends of the members, may become inadequate if the moment strengths of member ends would be greater than intended as a result of the steel having a substantially greater yield strength than specified.

Sections 9.4 and 10.9.3 have been modified to allow the use of spiral reinforcement with specified yield strength of up to 100 ksi. A sentence added to 21.2.5 specifically prohibits such use in members resisting earthquake-induced forces in structures assigned to Seismic Design Category D, E, or F. This is largely the result of some misgiving that high-strength spiral reinforcement may be less ductile than conventional mild reinforcement and that spiral failure has in fact been observed in earthquakes. There are fairly convincing arguments, however, against such specific prohibitions. Spiral failure, primarily observed in bridge columns, have invariably been the result of insufficient spiral reinforcement, rather than the lack of ductility of the spiral reinforcement. Also, prestressing steel, which is the only high-strength steel available on this market, is at least as ductile as welded wire reinforcement which is allowed to be used as transverse reinforcement.

21.2.6   Mechanical Splices

Section 21.2.6 contains provisions for mechanical splices. According to 21.2.6.1, a Type 1 mechanical splice shall conform to 12.14.3.2, i.e., the splice shall develop in tension or compression at least 125 percent of the specified yield strength f_y of the reinforcing bar. A Type 2 mechanical splice shall also conform to 12.14.3.2 and shall develop the specified tensile strength of the spliced bar.

During an earthquake, the tensile stresses in the reinforcement may approach the tensile strength of the reinforcement as the structure undergoes inelastic deformations. Thus, Type 2 mechanical splices can be used at any location in a member (21.2.6.2). The locations of Type 1 mechanical splices are restricted since the tensile stresses in the reinforcement in yielding regions of the member can exceed the strength requirements of 12.14.3.2. Consequently, Type 1 mechanical splices are not permitted within a distance equal to twice the member depth from the face of the column or beam or from sections where yielding of the reinforcement is likely to occur due to inelastic lateral displacements (21.2.6.2).

21.2.7 Welded Splices

The requirements for welded splices are in 21.2.7. Welded splices shall conform to the provisions of 12.14.3.4, i.e., the splice shall develop at least 125 percent of the specified yield strength f_y of the reinforcing bar (21.2.7.1). Similar to Type 1 mechanical splices, welded splices are not permitted within a distance equal to twice the member depth from the face of the column or beam or from sections where yielding of the reinforcement is likely to occur due to inelastic lateral displacements; in yielding regions of the member, the tensile stresses in the reinforcement can exceed the strength requirements of 12.14.3.4 (21.2.7.1).

According to 21.2.7.2, welding of stirrups, ties, inserts or other similar elements to longitudinal reinforcement that is required by design is not permitted. Welding of crossing reinforcing bars can lead to local embrittlement of the steel. If such welding will facilitate fabrication or field installation, it must be done only on bars added expressly for construction. Note that this provision does not apply to bars that are welded with welding operations under continuous competent control, as is the case in the manufacture of welded wire reinforcement.

21.2.8 Anchoring to Concrete

The requirements in this section pertain to anchors resisting earthquake-induced forces in structures located in regions of moderate or high seismic hazard. The design of such anchors must conform to the additional requirements of D.3.3 of Appendix D. See Part 34 for additional information.

21.3 FLEXURAL MEMBERS OF SPECIAL MOMENT FRAMES

The left-hand column of Table 29-1 contains the requirements for flexural members of special moment frames (as noted above, special moment frames, which can be cast-in-place or precast, are required in regions of high seismic hazard or for structures assigned to satisfy high seismic performance or design categories). These requirements typically apply to beams of frames and other flexural members with negligible axial loads (21.3.1). Special precast moment frames must also satisfy the provisions of 21.6, which are discussed below. For comparison purposes, Table 29-1 also contains the corresponding requirements for flexural members of intermediate and ordinary cast-in-place moment frames. See Chapter 16 and Part 23 for additional information on precast systems.

21.3.1 Scope

Flexural members of special moment frames must meet the general requirements of 21.3.1.1 through 21.3.1.4. These limitations have been guided by experimental evidence and observations of reinforced concrete frames that have performed well in past earthquakes. Members must have sufficient ductility and provide efficient moment transfer to the supporting columns. Note that columns subjected to bending and having a factored axial load $P_u \leq A_g f'_c /10$ may be designed as flexural members, where A_g is the gross area of the section.

Table 29-1 Flexural Members of Frames

	Special Moment Frames	**Intermediate and Ordinary CIP Moment Frames**
General	Flexural frame members shall satisfy the following conditions: • Factored axial compressive force $\leq A_g f'_c / 10$ • Clear span $\geq 4 \times$ effective depth • Width to depth ratio ≥ 0.3 • Width ≥ 10 in. • Width $\leq$ width of supporting member + distances on each side of the supporting member not exceeding three-fourths of the depth of the flexural member **21.3.1**	Intermediate — Factored axial compressive force $\leq A_g f'_c / 10$. **21.12.2** Ordinary — No similar requirements.
Flexural Requirements	Minimum reinforcement shall not be less than $$\frac{3\sqrt{f'_c} b_w d}{f_y} \text{ and } \frac{200 b_w d}{f_y}$$ at any section, top and bottom, unless provisions in 10.5.3 are satisfied. **21.3.2.1**	Same requirement, except as provided in 10.5.2, 10.5.3, and 10.5.4, although minimum reinforcement need only be provided at sections where tensile reinforcement is required by analysis. **10.5**
	The reinforcement ratio (ρ) shall not exceed 0.025. **21.3.2.1**	The net tensile strain ε_t at nominal strength shall not be less than 0.004. **10.3.5**
	At least two bars shall be provided continuously at both top and bottom of section. **21.3.2.1**	Provide minimum structural integrity reinforcement. **7.13**
	Positive moment strength at joint face $\geq 1/2$ negative moment strength at that face of the joint. **21.3.2.2**	Intermediate — Positive moment strength at joint face $\geq 1/3$ negative moment strength at that face of the joint. **21.12.4.1** Ordinary — No similar requirement.
	Neither the negative nor the positive moment strength at any section along the member shall be less than 1/4 the maximum moment strength provided at the face of either joint. **21.3.2.2**	Intermediate — Same requirement, except it is needed to provide only 1/5 of the maximum moment strength at the face of either joint at every section along the member. **21.12.4.1** Ordinary — No similar requirement.

A_g = gross area of section
b_w = width of web
d = effective depth of section
f'_c = specified compressive strength of concrete
f_y = specified yield strength of reinforcement

— continued on next page —

Table 29-1 Flexural Members of Frames (cont'd)

	Special Moment Frames	**Intermediate and Ordinary CIP Moment Frames**
Splices	Lap splices of flexural reinforcement are permitted only if hoop or spiral reinforcement is provided over the lap length. Hoop and spiral reinforcment spacing shall not exceed $d/4$ or 4 in. Mechanical splices shall conform to 21.2.6 and welded splices shall conform to 21.2.7. **21.3.2.3, 21.3.2.4**	There is no requirement that splices be enclosed in hoops.
	Lap splices are not to be used: • Within joints. • Within a distance of twice the member depth from the face of the joint. • At locations where analysis indicates flexural yielding caused by inelastic lateral displacements of the frame. **21.3.2.3**	No similar requirement.
Transverse Reinforcement	Hoops are required over a length equal to twice the member depth from the face of the supporting member toward midspan at both ends of the flexural member. **21.3.3.1**	Intermediate — Same requirement. **21.12.4.2** Ordinary — No similar requirement.
	Hoops are required over lengths equal to twice the member depth on both sides of a section where flexural yielding may occur in connection with inelastic lateral displacements of the frame. **21.3.3.1**	Reinforcement for flexural members subject to stress reversals shall consist of closed stirrups extending around flexural reinforcement. Also, provide minimum structural integrity reinforcement. **7.11.2, 7.13**
	Where hoops are required, the spacing shall not exceed: $d/4$ $8 \times$ diameter of smallest longitudinal bar $24 \times$ diameter of hoop bars 12 in. The first hoop shall be located not more than 2 in. from the face of the supporting member. **21.3.3.2**	Intermediate — Same requirement. **21.12.4.2** Ordinary — No similar requirement.
	Where hoops are required, longitudinal bars on the perimeter shall have lateral support conforming to 7.10 5.3. **21.3.3.3**	No similar requirement.
	Where hoops are not required, stirrups with seismic hooks at both ends shall be spaced at a distance not more than $d/2$ throughout the length of the member. **21.3.3.4**	Intermediate — Similar requirement except that seismic hooks are not required. **21.12.4.3**
	Transverse reinforcment must also be proportioned to resist the entire design shear force, neglecting the contribution of concrete to shear strength, if certain conditions are met. **21.3.4**	Intermediate — Transverse reinforcement must also be proportioned to resist the design shear force. **21.12.3** Ordinary — Provide sufficient transverse reinforcement for shear and torsion. **11.5, 11.6**

21.3.2 Flexural Reinforcement

The reinforcement requirements for flexural members of special moment frames are shown in Fig. 29-1. To allow for the possibility of the positive moment at the end of a beam due to earthquake-induced lateral displacements exceeding the negative moment due to gravity loads, 21.3.2.2 requires a minimum positive moment strength at the ends of the beam equal to at least 50 percent of the corresponding negative moment strength. The minimum moment strength at any section of the beam is based on the moment strength at the faces of the supports. These requirements ensure strength and ductility under large lateral displacements. The limiting ratio of 0.025 is based primarily on considerations of steel congestion and also on limiting shear stresses in beams of typical proportions. The requirement that at least two bars be continuous at both the top and the bottom of the beam is for construction purposes.

The flexural requirements for flexural members of intermediate moment frames are similar to those shown in Fig. 29-1 (see Table 29-1).

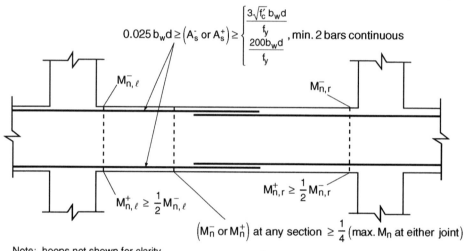

Figure 29-1 Reinforcement Requirements for Flexural Members of Special Moment Frames

Lap splices of flexural reinforcement must be placed at locations away from potential hinge areas subjected to stress reversals under cyclic loading (see Fig. 29-2). Where lap splices are used, they should be designed as tension lap splices and must be properly confined. Mechanical splices and welded splices must conform to 21.2.6 and 21.2.7, respectively.

21.3.3 Transverse Reinforcement

Adequate confinement is required at the ends of flexural members, where plastic hinges are likely to form, in order to ensure sufficient ductility of the members under reversible loads. Transverse reinforcement is also required at these locations to assist the concrete in resisting shear and to maintain lateral support for the reinforcing bars. For flexural members of special moment frames, the transverse reinforcement for confinement must consist of hoops as shown in Fig. 29-3. Hoops must be used for confinement in flexural members of intermediate moment frames as well (21.12.4.2). Shear strength requirements for flexural members are given in 21.3.4 for special moment frames and 21.12.3 for intermediate moment frames.

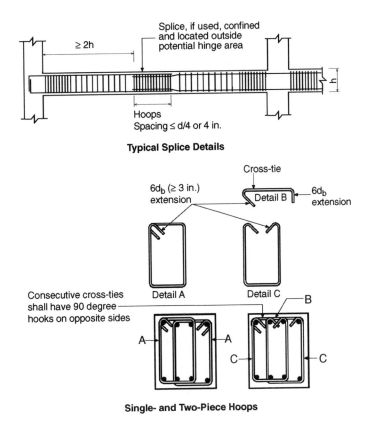

Figure 29-2 Splices and Hoop Reinforcement for Flexural Members of Special Moment Frames

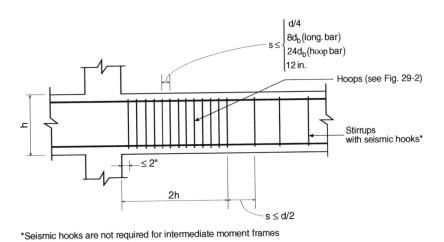

*Seismic hooks are not required for intermediate moment frames

Figure 29-3 Transverse Reinforcement for Flexural Members of Special and Intermediate Moment Frames

21.3.4 Shear Strength Requirements

Typically, larger forces than those prescribed by the governing building code are induced in structural members during an earthquake. Designing for shear forces from a combined gravity and lateral load analysis using the code-prescribed load combinations is not conservative, since in reality the reinforcement may be stressed beyond its yield strength, resulting in larger than anticipated shear forces. Adequate shear reinforcement must be provided so as to preclude shear failure prior to the development of plastic hinges at the ends of the beam. Thus, a flexural member of a special moment frame must be designed for the shear forces associated with probable moment strengths M_{pr} acting at the ends and the factored tributary gravity load along its span (21.3.4.1). The probable moment strength M_{pr} is associated with plastic hinging in the flexural member, and is defined as the strength of the beam with the stress in the reinforcing steel equal to $1.25f_y$ and a strength reduction factor of 1.0:

$$M_{pr} = A_s(1.25f_y)\left(d - \frac{a}{2}\right) \quad \text{(rectangular section with tension reinforcement only)}$$

where $a = \dfrac{A_s(1.25f_y)}{0.85f'_c b}$

Note that sidesway to the right and to the left must both be considered to obtain the maximum shear force (see Fig. 29-4). The use of $1.25f_y$ for the stress in the reinforcing steel reflects the possibility that the actual yield strength may be in excess of the specified value and the likelihood that the deformation in the tensile reinforcement will be in the strain-hardening range. By taking $1.25f_y$ as the stress in the reinforcement and 1.0 as the strength reduction factor, the chance of shear failure preceding flexural yielding is reduced.

In determining the required shear reinforcement over the lengths identified in 21.3.3.1, the contribution of the shear strength of the concrete V_c is taken as zero if the shear force from seismic loading is one-half or more of the required shear strength and the factored axial compressive force including earthquake effects is less than $A_g f'_c /20$ (21.3.4.2). The purpose of this requirement is to provide adequate shear reinforcement to increase the probability of flexural failure. Note that the strength reduction factor ϕ to be used is 0.75 or 0.85, depending on whether Chapter 9 or Appendix C load combinations are used (see 9.3.2.3 or C.3.2.3).

Shear reinforcement shall be in the form of hoops over the lengths specified in 21.3.3.1 (21.3.3.5); at or near regions of flexural yielding, spalling of the concrete shell is very likely to occur. Details of hoop reinforcement are given in 21.3.3.6 (see Fig. 29-2). Where hoops are not required, stirrups with seismic hooks at both ends may be used (21.3.3.4, 21.3.3.5). A minimum amount of transverse reinforcement is required throughout the entire length of flexural members to safeguard against any loading cases that were unaccounted for in design.

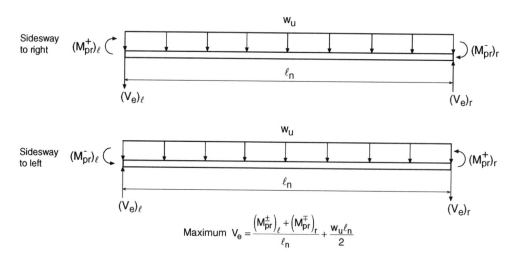

Figure 29-4 Design Shear Forces for Flexural Members of Special Moment Frames

The transverse reinforcement provided within the lengths specified in 21.3.3.1 shall satisfy the requirement for confinement or shear, whichever governs.

A similar analysis is required for frame members of intermediate moment frames except that the nominal moment strength M_n of the member is used instead of the probable moment strength (21.12.3). Also to be found in 21.12.3 is an alternate procedure where the earthquake effects are doubled in lieu of using the nominal moment strength.

21.4   SPECIAL MOMENT FRAME MEMBERS SUBJECTED TO BENDING AND AXIAL LOAD

The left hand column of Table 29-2 contains the requirements for special moment frame members subjected to combined bending and axial loads. These requirements would typically apply to columns of frames and other flexural members that carry a factored axial load $P_u > A_g f'_c/10$. For comparison purposes, Table 29-2 also contains the corresponding requirements for intermediate and ordinary cast-in-place moment frame members subject to combined bending and axial loads.

21.4.1   Scope

Section 21.4.1 is intended primarily for columns of special moment frames. Frame members other than columns that do not satisfy 21.3.1 are proportioned and detailed according to 21.4. The geometric constraints are largely reflective of prior practice. Unlike in the case of flexural members, a column-like member violating the dimensional limitations of 21.4.1 need not be excluded from the lateral force-resisting system, if it is designed as a wall in accordance with 21.7.

21.4.2   Minimum Flexural Strength of Columns

Columns must be provided with sufficient strength so that they will not yield prior to the beam at a beam-column joint. Lateral sway caused by column hinging may result in excessive damage. Yielding of the columns prior to the beams could also result in total collapse of the structure. For these reasons, columns are designed with 20% higher flexural strength as compared to beams meeting at the same joint, as shown in Fig. 29-5. In 21.4.2.2, nominal strengths of the columns and girder are calculated at the joint faces, and those strengths are used in Eq. (21-1). The column flexural strength is calculated for the factored axial load, consistent with the direction of the lateral forces considered, resulting in the lowest flexural strength.

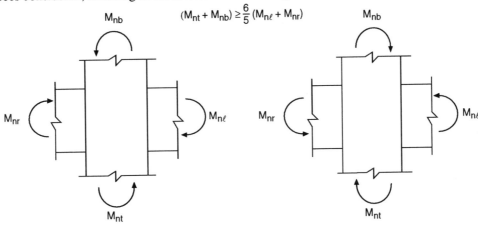

Subscripts ℓ, r, t, and b stand for left support, right support, top of column, and bottom of column, respectively.

Figure 29-5 "Strong Column-Weak Beam" Frame Requirements for Special Moment Frames

When computing the nominal flexural strength of girders in T-beam construction, slab reinforcement within an effective slab width defined in 8.10 shall be considered as contributing to the flexural strength if the slab reinforcement is developed at the critical section for flexure. Research has shown that using the effective flange width in 8.10 gives reasonable estimates of the negative bending strength of girders at interior joints subjected to interstory displacements approaching 2% of the story height.

Table 29-2 Frame Members Subjected to Bending and Axial Loads

	Special Moment Frames	Intermediate and Ordinary CIP Moment Frames
General	Frame members under this classification must meet the following requirements: • Factored axial compressive force > $A_g f'_c / 10$. • Shortest cross-sectional dimension ≥ 12 in. • Ratio of shortest cross-sectional dimension to perpendicular dimension ≥ 0.4. **21.4.1**	Intermediate — Factored axial compressive force > $A_g f'_c / 10$. **21.12.2** Ordinary — No similar requirements.
Flexural Requirements	The flexural strengths of columns shall satisfy the following: $\Sigma M_c \geq (6/5) \Sigma M_g$ where ΣM_c = sum of moments at the faces of the joint, corresponding to the nominal flexural strengths of the columns. ΣM_g = sum of moments at the faces of the joint, corresponding to the nominal flexural strengths of the girders. In T-beam costruction, slab reinforcement within an effective slab width defined in 8.10 shall be considered as contributing to flexural strength. If this requirement is not satisfied, the lateral strength and stiffness of the column shall not be considered when determining the strength and stiffness of the structure, and the column shall conform to 21.11; also, the column must have transverse reinforcement over its full height as specified in 21.4.4.1 through 21.4.4.3. **21.4.2**	No similar requirements.
	The reinforcement ratio (ρ_g) shall not be less than 0.01 and shall not exceed 0.06. **21.4.3.1**	The reinforcement ratio (ρ_g) shall not be less than 0.01 and shall not exceed 0.08. For a compression member with a cross section larger than required by considerations of loading, the reinforcement ratio (ρ_g) can be reduced below 0.01, but never below 0.005 (10.8.4). **10.9**
Splices	Mechanical splices shall conform to 21.2.6 and welded splices shall conform to 21.2.7. Lap splices are permitted only within the center half of the member length, must be tension lap splices, and shall be enclosed within transverse reinforcement conforming to 21.4.4.2 and 21.4.4.3. **21.4.3.2**	There is no restriction on the location of splices which are typically located just above the floor for ease of construction.

Transverse Reinforcement	The transverse reinforcement requirements discussed in the following five items on the next page need only be provided over a length (ℓ_o) from each joint face and on both sides of any section where flexural yielding is likely to occur. The length (ℓ_o) shall not be less than:
	    depth of member
	    1/6 clear span
	    18 in.
	21.4.4.4
		Intermediate — The length (ℓ_o) is the same as for special moment frames.
		21.12.5.2
		
		Ordinary — No similar requirements.

— continued on next page —

A_{ch} = cross-sectional area of member measured out-to-out of transverse reinforcement

A_g = gross area of section

f'_c = specified compressive strength of concrete

f_{yt} = specified yield stress of transverse reinforcement

b_c = cross-sectional dimension of column core measured center-to-center of outer legs of the transverse reinforcement comprising area A_{sh}

h_x = maximum horizontal spacing of hoop or crosstie legs on all faces of the column

s = spacing of transverse reinforcement

s_o = longitudinal spacing of transverse reinforcement within the length ℓ_o.

Table 29-2 Frame Members Subjected to Bending and Axial Loads (cont'd)

	Special Moment Frame	Intermediate and Ordinary CIP Moment Frames
Transverse Reinforcement (continued)	Ratio of spiral reinforcement (ρ_s) shall not be less than the value given by: $$\rho_s = 0.12 \frac{f'_c}{f_{yt}} \geq 0.45 \left(\frac{A_g}{A_{ch}} - 1\right) \frac{f'_c}{f_{yt}}$$ **21.4.4.1**	Ratio of spiral reinforcement (ρ_s) shall not be less than the value given by: $$\rho_s = 0.45 \left(\frac{A_g}{A_{ch}} - 1\right) \frac{f'_c}{f_{yt}}$$ and shall conform to the provisions in 7.10.4. **10.9.3**
	Total cross-sectional area of rectangular hoop reinforcement for confinement (A_{sh}) shall not be less than that given by the following two equations: $$A_{sh} = 0.3 \left(sb_c f'_c / f_{yt}\right)\left[(A_g/A_{ch}) - 1\right]$$ $$A_{sh} = 0.09 \left(sb_c f'_c / f_{yt}\right)$$ **21.4.4.1**	Transverse reinforcement must be provided to satisfy both shear and lateral support requirements for longitudinal bars. **7.10.5, 11.1**
	If the thickness of the concrete outside the confining transverse reinforcement exceeds 4 in., additional transverse reinforcement shall be provided at a spacing ≤ 12 in. Concrete cover on the additional reinforcement shall not exceed 4 in. **21.4.4.1**	No similar requirements.
	Transverse reinforcement shall be spaced at distances not exceeding 1/4 minimum member dimension, 6 × longitudinal bar diameter, 4 in. ≤ s_o = 4 + [(14 - h_x)/3] ≤ 6 in. **21.4.4.2**	Intermediate—Maximum spacing s_o is 8 × smallest longitudinal bar diameter, 24 × hoop bar diameter, 1/2 smallest cross-sectional dimension, or 12 in. First hoop to be located no further than s_o/2 from the joint face. **21.12.5.2** Ordinary — No similar requirement.
	Cross ties or legs of overlapping hoops shall not be spaced more than 14 in. on center in the direction perpendicular to the longitudinal axis of the member. Vertical bars shall not be farther than 6 in. clear from a laterally supported bar. **21.4.4.3, 7.10.5.3**	Vertical bars shall not be farther than 6 in. clear from a laterally supported bar. **7.10.5.3**
	Where the transverse reinforcement as discussed above is no longer required, the remainder of the column shall contain spiral or hoop reinforcement spaced at distances not to exceed 6 × longitudinal bar diameter 6 in. **21.4.4.6**	Intermediate — Outside the length ℓ_o, spacing of transverse reinforcement shall conform to 7.10 and 11.5.4.1. **21.12.5.4** Ordinary — Transverse reinforcement to conform to 7.10 and 11.5.4.1.
	Transverse reinforcement must also be proportioned to resist the design shear force (V_e). **21.4.5**	Intermediate — Transverse reinforcement must also be proportioned to resist the design shear forces specified in 21.12.3. Ordinary — Provide sufficient transverse reinforcement for shear. **11.5.4, 11.5.6**
	Columns supporting reactions from discontinued stiff members, such as walls, shall have transverse reinforcement as specified in 21.4.4.1 through 21.4.4.3 over their full height, if the factored axial compressive force, including earthquake effects, exceeds ($A_g f'_c / 10$). This transverse reinforcement shall extend into the discontinued member for at least the development length of the largest longitudinal reinforcement in the column in accordance with 21.5.4. If the column terminates on a footing or mat, the transverse reinforcement shall extend at least 12 in. into the footing or mat. **21.4.4.5**	No similar requirement.

If Eq. (21-1) is not satisfied at a joint, columns supporting reactions from that joint are to be provided with transverse reinforcement as specified in 21.4.4 over their full height (21.4.2.3), and shall be ignored in determining the calculated strength and stiffness of the structure (21.4.2.1). These columns must also conform to the provisions for frame members not proportioned to resist earthquake motions as given in 21.11 (21.4.2.1).

No similar provisions are included for intermediate or ordinary moment frames.

21.4.3 Longitudinal Reinforcement

The maximum allowable reinforcement ratio is reduced from 8% to 6% for columns in special moment frames (21.4.3.1). This lower ratio prevents congestion of steel, which reduces the chance of improperly placed concrete. It also prevents the development of large shear stresses in the columns. Typically, providing a reinforcement ratio larger than about 3% is not practical or economical.

Mechanical splices shall conform to 21.2.6 and welded splices shall conform to 21.2.7 (21.4.3.2). When lap splices are used, they are permitted only within the center half of the member length and are to be designed as tension lap splices (see Fig. 29-6). Transverse reinforcement conforming to 21.4.4.2 and 21.4.4.3 is required along the length of the lap splice.

There are no restrictions on the location of lap splices in intermediate or ordinary moment frames.

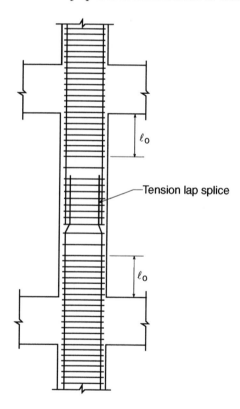

Figure 29-6 Typical Lap Splice Details for Columns in Special Moment Frames

21.4.4 Transverse Reinforcement

Column ends require adequate confinement to ensure column ductility in the event of hinge formation. They also require adequate shear reinforcement in order to prevent shear failure prior to the development of flexural yielding at the column ends. The correct amount, spacing, and location of the transverse reinforcement must be provided so that both the confinement and the shear strength requirements are satisfied. For special moment frames, the transverse reinforcement must be spiral or circular hoop reinforcement or rectangular hoop reinforcement, as shown in Fig. 29-7. Spiral reinforcement is generally the most efficient form of confinement reinforcement; however, the extension of the spirals into the beam-column joint may cause some construction difficulties.

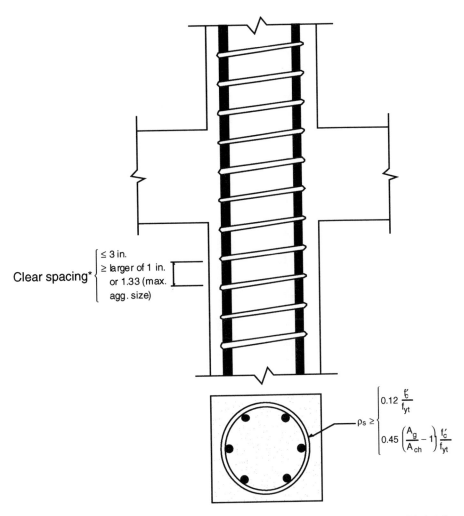

*Clear spacing for spiral reinforcement. Circular hoops to be spaced per 21.4.4.2.

Figure 29-7 Confinement Requirements at Column Ends
(a) spiral or circular hoop reinforcement

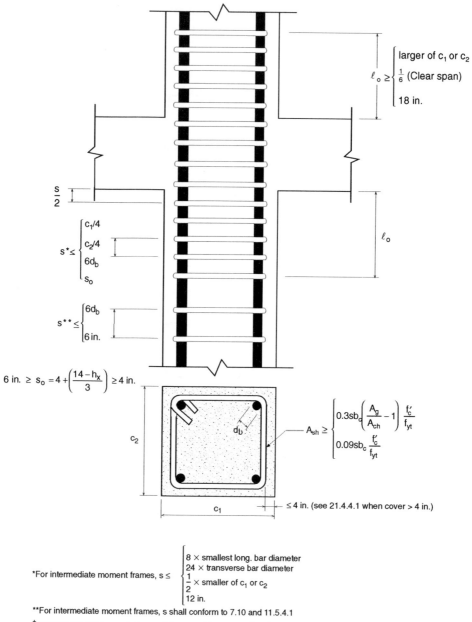

Figure 29-7 Confinement Requirements at Column Ends
(b) rectangular hoop reinforcement

Figure 29-8 shows an example of transverse reinforcement provided by one hoop and three crossties. 90-degree hooks are not as effective as 135-degree hooks. Confinement will be sufficient if crosstie ends with 90-degree hooks are alternated.

The requirements of 21.4.4.2 and 21.4.4.3 must be satisfied for the configuration of rectangular hoop reinforcement. The requirement that spacing not exceed one-quarter of the minimum member dimension is to obtain adequate concrete confinement. Restraining longitudinal reinforcement buckling after spalling is the rationale behind the spacing being limited to 6 bar diameters. Section 21.4.4.2 permits the 4 in. spacing for confinement to be relaxed to a maximum of 6 in. if the horizontal spacing of crossties or legs of overlapping hoops is limited to 8 in.

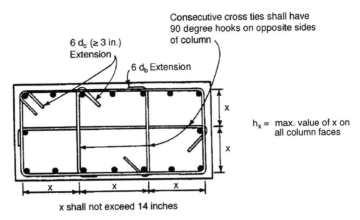

Figure 29-8 Transverse Reinforcement in Columns

Additional transverse reinforcement at a maximum spacing of 12 in. is required when concrete thickness outside the confining transverse reinforcement exceeds 4 in. This additional reinforcement will help reduce the risk of portions of the shell falling away from the column. The required amount of such reinforcement is not specifically indicated; the 1997 UBC [29.3] specifies a minimum amount equal to that required for columns that are not part of the lateral force-resisting system.

For columns supporting discontinued stiff members (such as walls) as shown in Fig. 29-9, transverse reinforcement in compliance with 21.4.4.1 through 21.4.4.3 needs to be provided over the full height of the column and must be extended at least the development length of the largest longitudinal column bars into the discontinued member (wall). The transverse reinforcement must also extend at least 12 in. into the footing or mat, if the column terminates on a footing or mat.

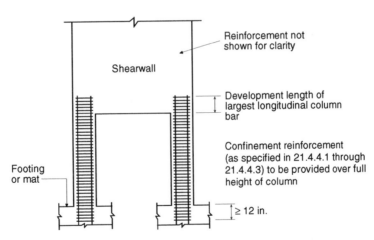

Figure 29-9 Columns Supporting Discontinued Stiff Members

As indicated in Fig. 29-7, there are transverse reinforcement requirements for columns of intermediate moment frames in 21.12.5.

21.4.5 Shear Strength Requirements

In addition to satisfying confinement requirements, the transverse reinforcement in columns must resist the maximum shear forces associated with the formation of plastic hinges in the frame (21.4.5.1). Although the provisions of 21.4.2 are intended to have most of the inelastic deformation occur in the beams, the provisions of 21.4.5.1 recognize that hinging can occur in the column. Thus, as in the case of beams, the shear reinforcement in the columns is based on the probable moment strengths M_{pr} that can be developed at the ends of the column.

The probable moment strength is to be the maximum consistent with the range of factored axial loads on the column; sideway to the right and to the left must both be considered (see Fig. 29- 10). It is obviously conservative to use the probable moment strength corresponding to the balanced point.

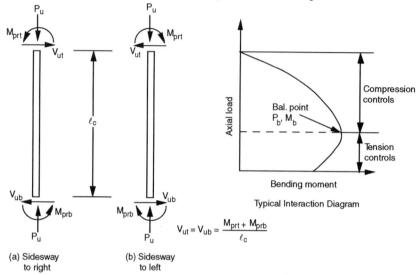

Figure 29-10 Loading Cases for Design of Shear Reinforcement in Columns of Special Moment Frames

Section 21.4.5.1 points out that the column shear forces need not exceed those determined from joint strengths based on the probable moment strengths of the beams framing into the joint. When beams frame on opposite sides of a joint, the combined probable moment strength may be taken as the sum of the negative probable moment strength of the beam on one side of the joint and the positive probable moment strength of the beam on the other side. The combined probable moment strength of the beams is then distributed appropriately to the columns above and below the joint, and the shear forces in the column are computed based on this distributed moment. It is important to note that in no case is the shear force in the column to be taken less than the factored shear force determined from analysis of the structure under the code-prescribed seismic forces (21.4.5.1).

Provisions for proportioning the transverse reinforcement are contained in 21.4.5.2. As in the case of beams, the strength reduction factor ϕ to be used wth the Chapter 9 load combinations is 0.75 (see 9.3.4 and 9.3.2.3).

The shear forces in intermediate frame members subjected to combined bending and axial force are determined in the same manner as for flexural members of intermediate moment frames, i.e., nominal moment strengths at member ends are used to compute the shear forces (21.12.3).

21.5 JOINTS OF SPECIAL MOMENT FRAMES

The overall integrity of a structure is dependent on the behavior of the beam-column joint. Degradation of the joint can result in large lateral deformations which can cause excessive damage or even failure. The left-hand column of Table 29-3 contains the requirements for joints of special moment frames. For intermediate and ordinary cast-in-place frames, the beam-column joints do not require the special design and detailing requirements as for special moment frames. It may be prudent, however, to apply the same line of thinking to intermediate frame joints as to special moment frame joints.

Slippage of the longitudinal reinforcement in a beam-column joint can lead to an increase in the joint rotation. Longitudinal bars must be continued through the joint or must be properly developed for tension (21.5.4) and compression (Chapter 12) in the confined column core. The minimum column size requirement of 21.5.1.4 reduces the possibility of failure from loss of bond during load reversals that take the steel beyond its yield point.

21.5.2 Transverse Reinforcement

The transverse reinforcement in a beam-column joint is intended to provide adequate confinement of the concrete to ensure its ductile behavior and to allow it to maintain its vertical load-carrying capacity even after spalling of the outer shell.

Table 29-3 Joints of Frames

	Special Moment Frames	Intermediate and Ordinary CIP Moment Frames
Longitudinal Beam Reinforcement	Beam longitudinal reinforcement terminated in a column shall be extended to the far face of the confined column core and anchored in tension according to 21.5.4 and in compression according to Chapter 12. **21.5.1.3**	No similar requirement.
	Where longitudinal beam reinforcement extends through a joint, the column dimension parallel to the beam reinforcement shall not be less than 20 times the diameter of the largest longitudinal bar for normal weight concrete. For lightweight aggregate concrete, this dimension shall be not less than 26 times the bar diameter. **21.5.1.4**	No similar requirements.
Transverse Reinforcement	The transverse hoop reinforcement required for column ends (21.4.4) shall be provided within the joint, unless the joint is confined by structural members as specified in 21.5.2.2. If members frame into all four sides of the joint and the member width at the column face is at least 3/4 the column width, the transverse reinforcement can be reduced to 50% of the requirements of 21.4.4.1 within the depth of the shallowest member. The spacing required in 21.4.4.2 shall not exceed 6 in. at these locations. **21.5.2.1, 21.5.2.2**	No similar requirement.

f_c' = specified compressive strength of concrete

f_y = specified yield strength of reinforcement

Table 29-3 Joints of Frames (cont'd)

	Special Moment Frames	Intermediate and Ordinary CIP Moment Frames
Shear Strength	The nominal shear strength of the joint shall not exceed the forces specified below for normal-weight aggregate concrete. For joints confined on all four faces .... $20\sqrt{f'_c}\, A_j$ For joints confined on three faces or on two opposite faces .... $15\sqrt{f'_c}\, A_j$ For other joints: .... $12\sqrt{f'_c}\, A_j$ where: A_j = effective cross-sectional area within a joint in a plane parallel to the plane of the reinforcement generating shear in the joint. The joint depth shall be the overall depth of the column. Where a beam frames into a support of larger width, the effective width of the joint shall not exceed the smaller of: 1. Beam width plus the joint depth. 2. Twice the smaller perpendicular distance from the longitudinal axis of the beam to the column side. A joint is considered to be confined if confining members frame into all faces of the joint. A member that frames into a face is considered to provide confinement at the joint if at least 3/4 of the face of the joint is covered by the framing member. **21.5.3**	Although it is not required, it may be prudent to check the shear strength of the joint in intermediate moment frames. The force in the longitudinal beam reinforcement may be taken as $1.0 f_y$ rather than the $1.25 f_y$ required for special moment frames.
	In determining shear forces in the joints, forces in the longitudinal beam reinforcement at the joint face shall be calculated by assuming that the stress in the flexural tensile reinforcement is $1.25 f_y$. **21.5.1.1**	No similar requirement.
	For lightweight aggregate concrete, the nominal shear strength of the joint shall not exceed 3/4 of the limits given in 21.5.3.1. **21.5.3.2**	No similar requirement.

Minimum confinement reinforcement of the same amount required for potential hinging regions in columns, as specified in 21.4.4, must be provided within a beam-column joint around the column reinforcement, unless the joint is confined by structural members as specified in 21.5.2.2.

For joints confined on all four faces, a 50% reduction in the amount of confinement reinforcement is allowed. A member that frames into a face is considered to provide confinement to the joint if at least three-quarters of the face of the joint is covered by the framing member. The code further allows that where a 50% reduction in the amount of confinement reinforcement is permissible, the spacing specified in 21.4.4.2(b) may be increased to 6 in. (21.5.2.2).

The minimum amount of confinement reinforcement, as noted above, must be provided through the joint regardless of the magnitude of the calculated shear force in the joint. The 50% reduction in the amount of confinement reinforcement allowed for joints having horizontal members framing into all four sides recognizes the beneficial effect provided by these members in resisting the bursting pressures that can be generated within the joint.

21.5.3 Shear Strength

The most significant factor in determining the shear strength of a beam-column joint is the effective area A_j of the joint, as shown in Fig. 29-11. For joints that are confined by beams on all four faces, the shear strength of the joint is equal to $20\sqrt{f'_c}\,A_j$. If the joint is confined only on three faces, or on two opposite faces, the strength must be reduced by 25% to $15\sqrt{f'_c}\,A_j$. For other cases, the shear strength is equal to $12\sqrt{f'_c}\,A_j$. It is important to note that the shear strength is a function of the concrete strength and the cross-sectional area only. Test results show that the shear strength of the joint is not altered significantly with changes in transverse reinforcement, provided a minimum amount of such reinforcement is present. Thus, only the concrete strength or the member size can be modified if the shear strength of the beam-column joint is inadequate. The strength reduction factor ϕ for shear in joints is 0.85 (9.3.4).

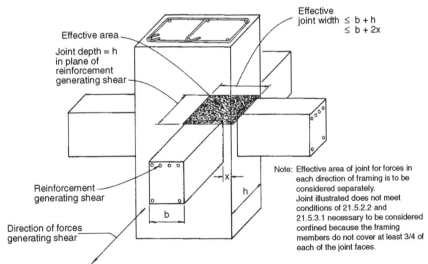

Figure 29-11 Effective Area of Joint (A_j)

The larger the tension force in the steel, the greater the shear in the joint (Fig. 29-12). Thus, the tensile force in the reinforcement is conservatively taken as $1.25 f_y A_s$. The multiplier of 1.25 takes into account the likelihood that due to strain-hardening and actual strengths higher than the specified yield strengths, a larger tensile force may develop in the bars, resulting in a larger shear force.

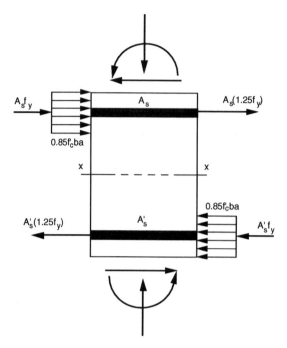

Figure 29-12 Horizontal Shear in Beam-Column Joint

21.5.4   Development Length of Bars in Tension

A standard 90-degree hook located within the confined core of a column or boundary element is depicted in Fig. 29-13. Equation (21-6), based on the requirements of 12.5, includes the factors for hooks enclosed in ties (0.8), satisfaction of minimum cover requirements (0.7), a cyclic load factor (1.1), and a factor of 1.25 for overstrength in the reinforcing steel. The equation for the development length in 12.5.2 $\left[\left(0.02\beta\lambda f_y / \sqrt{f'_c}\right) d_b\right]$ is multiplied by these factors to obtain the equation that is given in 21.5.4.1 for uncoated reinforcing bars with $f_y = 60,000$ psi embedded in normal weight concrete:

$$\ell_{dh} = \frac{0.8 \times 0.7 \times 1.25 \times 1.1 \times 0.02 \times 1.0 \times 1.0 \times 60,000 \times d_b}{\sqrt{f'_c}}$$

$$= \frac{924 d_b}{\sqrt{f'_c}} = \frac{f_y d_b}{65\sqrt{f'_c}}$$

For bar sizes No. 3 through No. 11, the development length ℓ_{dh} for a bar with a standard 90-degree hook in normal-weight aggregate concrete shall not be less than the largest of $8d_b$, 6 in., and the length obtained from Eq. (21-6) shown above. For lightweight aggregate concrete, the development length shall be increased by 25%.

The development length for No. 11 and smaller straight bars is determined by multiplying the development length for hooked bars required by 21.5.4.1 by (a) two-and-a-half (2.5) if the depth of the concrete cast in one lift beneath the bar does not exceed 12 in., and (b) three-and-a-half (3.5) if the depth of the concrete cast in one lift beneath the bar exceeds 12 in. (21.5.4.2). If a portion of a straight bar is not located within the confined core of a column or boundary element, the length of that bar shall be increased by an additional 60%. Provisions for epoxy-coated bars are given in 21.5.4.4.

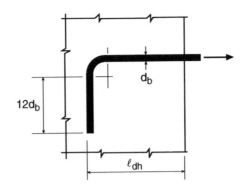

Figure 29-13 Standard 90-Degree Hook

21.6   SPECIAL MOMENT FRAMES CONSTRUCTED USING PRECAST CONCRETE

In addition to the requirements of 21.2 through 21.5, special moment frames constructed using precast concrete must satisfy the requirements of 21.6. The detailing provisions in 21.6.1 for frames with ductile connections and 21.6.2 for frames with strong connections are intended to produce frames that respond to design displacements essentially like cast-in-place special moment frames. Section 21.6.3 provides a design procedure for special moment frames that do not satisfy the appropriate prescriptive requirements of Chapter 21.

21.6.1   Special Moment Frames with Ductile Connections

Special moment frames with ductile connections are designed and detailed so that flexural yielding occurs within the connection regions. Type II mechanical splices or any other technique that provides development in tension and compression of at least the specified tensile strength of the bars and $1.25f_y$, respectively, can be used to make the reinforcement continuous in the connections.

According to 21.6.1(a), the nominal shear strength V_n at the connection must be computed in accordance with the shear-friction design method of 11.7.4. In order to help prevent sliding at the faces of the connection, V_n must be greater than or equal to $2V_e$, where V_e is the design shear force in the beams that is computed according to 21.3.4.1 or the design shear force in the columns that is computed according to 21.4.5.1. Since the ductile connections may be at locations that are not adjacent to the joints, using V_e may be conservative.

Mechanical splices of beam reinforcement must satisfy the requirements of 21.2.6 and must be located at least h/2 from the face of the joint, where h is the overall depth of the beam. This additional requirement is intended to avoid strain concentrations over a short length of reinforcement adjacent to a splice device.

21.6.2   Special Moment Frames with Strong Connections

Special moment frames with strong connections are designed and detailed so that flexural yielding occurs away from the connection regions. Examples of beam-to-beam, beam-to-column, and column-to-footing connections are shown in Fig. R21.6.2.

According to 21.6.2(a), the geometric constraint in 21.3.1.2 related to the clear span to effective depth ratio must be satisfied for any segment between locations where flexural yielding is intended to occur due to the design displacements.

To ensure that strong connections remain elastic and do not slip following the formation of plastic hinges, the design strength of the connection, fS_n, in both flexure and shear must be greater than or equal to the bending moment and shear force, S_e, respectively, corresponding to the development of probable flexural or shear strengths at intended locations of flexural or shear yielding (21.6.2(b)). These provisions are illustrated in Figs. 29-14 and 29-15 for a beam-to-beam and a beam-to-column strong connection, respectively, with sidesway to the right. Sidesway to the left must also be considered.

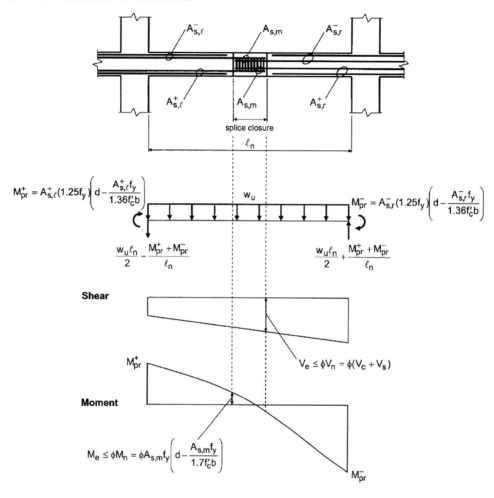

Fig. 29-14 Design Requirements for Beam-to-Beam Strong Connections near Midspan

Section 21.6.2(c) requires that primary longitudinal reinforcement be continuous across connections and be developed outside both the strong connection and the plastic hinge region. Laboratory tests of precast beam-column connections showed that strain concentrations caused brittle fracture of reinforcing bars at the faces of mechanical splices. To avoid this premature fracture, designers should carefully select the locations of strong connections or take other measures, such as using debonded reinforcement in highly stressed regions.

The column-to-column connection requirements of 21.6.2(d) are provided to avoid hinging and strength deterioration of these connections. For columns above the ground floor level, the moments at a joint may be limited by the flexural strengths of the beams framing into that joint (21.4.2.2). Dynamic inelastic analysis and studies of strong ground motion have shown that for a strong column-weak beam deformation mechanism, the beam end moments are not equally divided between the top and bottom columns, even where columns have equal stiffness.[29.4] From an elastic analysis, the moments would be distributed as shown in Fig. 29-16, while the actual distribution is likely to be as shown in Fig. 29-17.

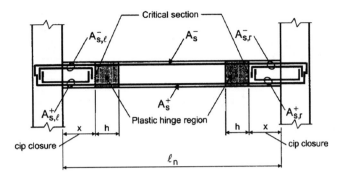

$$M_{pr}^+ = A_s^+(1.25f_y)\left(d - \frac{A_s^+ f_y}{1.36f_c'b}\right) \qquad M_{pr}^- = A_s^-(1.25f_y)\left(d - \frac{A_s^- f_y}{1.36f_c'b}\right)$$

$$V_L = \frac{w_u(\ell_n - 2x)}{2} - \frac{M_{pr}^+ + M_{pr}^-}{\ell_n - 2x}$$

$$V_R = \frac{w_u(\ell_n - 2x)}{2} + \frac{M_{pr}^+ + M_{pr}^-}{\ell_n - 2x}$$

$$M_{e,\ell}^+ = M_{pr}^+ + V_L x$$

$$M_{e,r}^- = M_{pr}^- + V_R x$$

$$\phi M_{n,\ell}^+ = \phi A_{s,\ell}^+ f_y \left(d - \frac{A_{s,\ell}^+ f_y}{1.7f_c'b}\right) \geq M_{e,\ell}^+$$

$$\phi M_{n,r}^- = \phi A_{s,r}^- f_y \left(d - \frac{A_{s,r}^- f_y}{1.7f_c'b}\right) \geq M_{e,r}^-$$

Fig. 29-15 Design Requirements for Beam-to-Column Strong Connections

Fig. 29-16 Bending Moments at Beam-to-Column Connection – Elastic Analysis

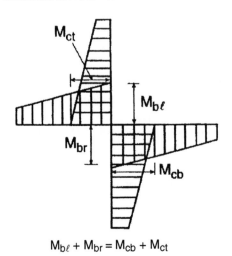

$$M_{b\ell} + M_{br} = M_{cb} + M_{ct}$$

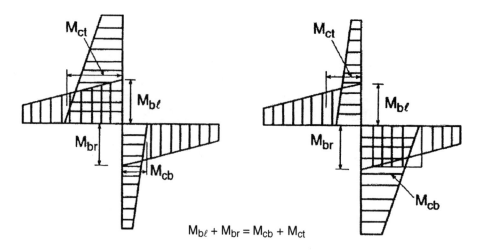

Fig. 29-17 Bending Moments at Beam-to-Column Connection – Inelastic Analysis

Figure 29-18 shows the distribution of the elastic moments M_E (dashed lines) due to the seismic forces and the corresponding envelopes of dynamic moments ωM_E (solid lines) over the full column height, where ω is a dynamic amplification factor. In regions outside of the middle third of the column height, ω is to be taken as 1.4 (21.6.2(d)); thus, connections within these regions must be designed such that $\phi M_n \geq 1.4 M_E$. For connections located within the middle third of the column height, 21.6.2(d) requires that $\phi M_n \geq 0.4 M_{pr}$, where M_{pr} is the maximum probable flexural strength of the column within the story height. Also, the design shear strength ϕV_n of the connection must be greater than or equal to the design shear force V_e computed according to 21.4.5.1.

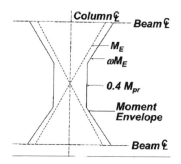

Fig. 29-18 Moment Envelope at Column-to-Column Connection

21.6.3 Non-emulative Design

It has been demonstrated in experimental studies that special moment frames constructed using precast concrete that do not satisfy the provisions of 21.6.1 for frames with ductile connections or 21.6.2 for frames with strong connections can provide satisfactory seismic performance characteristics. For these frames, the requirements of ACI T1.1 Acceptance Criteria for Moment Frames Based on Structural Testing [29.5], as well as the provisions of 21.6.3(a) and 21.6.3(b), must be satisfied.

ACI T1.1 defines minimum acceptance criteria for weak beam/strong column moment frames designed for regions of high seismic risk that do not satisfy the prescriptive requirements of Chapter 21 of ACI 318-99. According to ACI T1.1, acceptance of such frames as special moment frames must be validated by analysis and laboratory tests.

Prior to testing, a design procedure must be developed for prototype moment frames that have the same generic form as those for which acceptance is sought (see 4.0 of ACI T1.1). The design procedure should account for the effects of material nonlinearity (including cracking), deformations of members and connections, and reversed cyclic loading, and must be used to proportion the test modules (see 5.0 for requirements for the test modules). It is also important to note that the overstrength factor (column-to-beam strength ratio) used for the columns of the prototype frame should not be less than 1.2, which is specified in 21.4.2.2 of ACI 318-99.

The test method is described in 7.0. In short, the test modules are to be subjected to a sequence of displacement-controlled cycles that are representative of the drifts expected during the design earthquake for the portion of the frame that is represented by the test module. Figure R5.1 of ACI T1.1 illustrates connection configurations for interior and exterior one-way joints and, if applicable, corner joints that must be tested as a minimum.

The first loading cycle shall be within the linear elastic response range of the module. Subsequent drift ratios are to be between 1.25 and 1.5 times the previous drift ratio, with 3 fully reversed cycles applied at each drift ratio. Testing continues until the drift ratio equals or exceeds 0.035. Cyclic deformation history that satisfies 7.0 is illustrated in Fig. R7.0. Drift ratio is defined in Fig. R2.1.

Section 9.0 provides the detailed acceptance criteria that apply to each module of the test program. The performance of the test module is deemed satisfactory when these criteria are met for both directions of response.

The first criterion is that the test module must attain a lateral resistance greater than or equal to the calculated nominal lateral resistance E_n (see 1.0 for definition of E_n) before the drift ratio exceeds the allowable story drift limitation of the governing building code (see Fig. R9.1). This criterion helps provide adequate initial stiffness.

In order to provide weak beam/strong column behavior, the second criterion requires that the maximum lateral resistance E_{max} recorded in the test must be less than or equal to λE_n where λ is the specified overstrength factor for the test column, which must be greater than or equal to 1.2. Commentary section R9.1.2 provides a detailed discussion on this requirement. Also see Fig. R9.1.

The third criterion requires that the characteristics of the third complete cycle for each test module, at a drift ratio greater than or equal to 0.035, must satisfy 3 criteria regarding peak force value, relative energy dissipation ratio, and drift at zero stiffness. The first of these criteria limits the level of strength degradation, which is inevitable at high drift ratios under revised cyclic loading. A maximum strength degradation of $0.25E_{max}$ is specified (see Fig. R9.1). The second of these criteria sets a minimum level of damping for the frame as a whole by requiring that the relative energy dissipation ratio β be greater than or equal to 1/8. If β is less than 1/8, oscillations may continue for a long time after an earthquake, resulting in low-cycle fatigue effects and possible excessive displacements. The definition of β is illustrated in Fig. R2.4. The third of these criteria helps ensure adequate stiffness around zero drift ratio. The structure would be prone to large displacements following a major earthquake if this stiffness becomes too small. A hysteresis loop for the third cycle between peak drift ratios of 0.035, which has the form shown in Fig. R9.1, is acceptable. An unacceptable hysteresis loop form is shown in Fig. R9.1.3 where the stiffness around the zero drift ratio is unacceptably small for positive, but not for negative, loading.

As noted above, 21.6.3 has additional requirements to those in ACI T1.1. According to 21.6.3(a), the details and materials used in the test specimen shall be representative of those used in the actual structure. Section 21.6.3(b) stipulates additional requirements for the design procedure. Specifically, the design procedure must identify the load path or mechanism by which the frame resists the effects due to gravity and earthquake forces and shall establish acceptable values for sustaining that mechanism. Any portions of the mechanism that deviate from code requirements shall be contained in the test specimens and shall be tested to determine upper bounds for acceptance values. In other words, deviations are acceptable if it can be demonstrated that they do not adversely affect the performance of the framing system.

21.7 SPECIAL REINFORCED CONCRETE STRUCTURAL WALLS AND COUPLING BEAMS

When properly proportioned so that they possess adequate lateral stiffness to reduce interstory distortions due to earthquake-induced motions, structural walls (also called shearwalls) reduce the likelihood of damage to the nonstructural elements of a building. When used with rigid frames, a system is formed that combines the gravity-load-carrying efficiency of the rigid frame with the lateral-load-resisting efficiency of the structural wall.

Observations of the comparative performance of rigid-frame buildings and buildings stiffened by structural walls during earthquakes have pointed to the consistently better performance of the latter. The performance of buildings stiffened by properly designed structural walls has been better with respect to both safety and damage control. The need to ensure that critical facilities remain operational after a major tremor and the need to reduce economic losses from structural and nonstructural damage, in addition to the primary requirement of life safety (i.e., no collapse), has focused attention on the desirability of introducing greater lateral stiffness into earthquake-resistant multistory structures. Structural walls, which have long been used in designing for wind resistance, offer a logical and efficient solution to the problem of lateral stiffening of multistory buildings.

Structural walls are normally much stiffer than regular frame elements and are therefore subjected to correspondingly greater lateral forces due to earthquake motions. Because of their relatively greater depth, the lateral deformation capacities of walls are limited, so that, for a given amount of lateral displacement, structural walls tend to exhibit greater apparent distress than frame members. However, over a broad period range, a structure with structural walls, which is substantially stiffer and hence has a shorter period than a structure with frames, will suffer less lateral displacement than the frame, when subjected to the same ground motion intensity. Structural walls with a height-to-horizontal length ratio, h_w/ℓ_w, in excess of 2 behave essentially as vertical cantilever beams and should therefore be designed as flexural members, with their strength governed by flexure rather than by shear.

Isolated structural walls or individual walls connected to frames will tend to yield first at the base where the bending moment is the greatest. Coupled walls, i.e., two or more walls linked by short, rigidly-connected beams at the floor levels, on the other hand, have the desirable feature that significant energy dissipation through inelastic action in the coupling beams can be made to precede hinging at the bases of the walls.

The left-hand column of Table 29-4 contains the requirements for special reinforced concrete structural walls (recall that special reinforced concrete structural wall are required in regions of high seismic hazard or for structures assigned to high seismic performance or design categories). For comparison purposes, the requirements of ordinary reinforced concrete structural walls are also contained in Table 29-4.

21.7.2 Reinforcement

Special reinforced concrete structural walls are to be provided with reinforcement in two orthogonal directions in the plane of the wall (see Fig. 29-19). The minimum reinforcement ratio for both the longitudinal and the transverse reinforcement is 0.0025, unless the design shear force is less than or equal to $A_{cv}\sqrt{f'_c}$, where A_{cv} is the area of concrete bounded by the web thickness and the length of the wall in the direction of analysis, in which case, the minimum reinforcement must not be less than that given in 14.3. The reinforcement provided for shear strength must be continuous and distributed uniformly across the shear plane with a maximum spacing of 18 in. At least two curtains of reinforcement are required if the in-plane factored shear force assigned to the wall exceeds $2A_{cv}\sqrt{f'_c}$. This serves to reduce fragmentation and premature deterioration of the concrete under load reversals into the inelastic range. Uniform distribution of reinforcement across the height and horizontal length of the wall helps control the width of the inclined (diagonal) cracks.

Section 21.5.4 modifies the development length requirements of Chapter 12 for longitudinal beam bars terminating at exterior beam-column joints of structures assigned to high seismic design categories. But then 21.7.2.3

Table 29-4 Structural Walls

	Special Reinforced Concrete Structural Wall	Ordinary Reinforced Concrete Structural Wall
Reinforcement	The distributed web reinforcement ratios ρ_ℓ and ρ_t shall not be less than 0.0025. If the design shear force $V_u \leq A_{cv}\sqrt{f'_c}$, provide minimum reinforcement per 14.3. **21.7.2.1**	Minimum vertical reinforcement ratio = 0.0012 for No. 5 bars or smaller = 0.0015 for No. 6 bars or larger Minimum horizontal reinforcement ratio = 0.0020 for No. 5 bars or smaller = 0.0025 for No. 6 bars or larger **14.3**
	At least two curtains of reinforcement shall be used in a wall if the in-plane factored shear force (V_u) assigned to the wall exceeds $2 A_{cv}\sqrt{f'_c}$. **21.7.2.2**	Walls more than 10 in. thick require two curtains of reinforcement (except basement walls). **14.3.4**
	Reinforcement spacing each way shall not exceed 18 in. **21.7.2.1**	Reinforcement spacing shall not exceed: 3 × wall thickness 18 in. **14.3.5**
	Reinforcement in structural walls shall be developed or spliced for f_y in tension in accordance with Chapter 12, except: (a) The effective depth shall be permitted to be 0.8 ℓ_w for walls. (b) The requirements of 12.11, 12.12 and 12.13 need not apply. (c) At locations where yielding of longitudinal reinforcement may occur as a result of lateral displacements, development lengths of such reinforcement must be 1.25 times the values calculated for f_y in tension. (d) Mechanical and welded splices of reinforcement must conform to 21.2.6 and 21.2.7, respectively. **21.7.2.3**	The development lengths, spacing, and anchorage of reinforcement shall be as per Chapters 12, 14, and 15.

— continued on next page —

A_{cv} = gross area of concrete section bounded by web thickness and length of section in the direction of shear force considered

A_{cw} = area of concrete section of an individual pier

A_g = gross area of section

b_w = width of web

d = effective depth of section

f'_c = specified compressive strength of concrete

f_{yh} = specified yield strength of transverse reinforcement

h = overall thickness of member

h_w = height of entire wall or segment of wall considered

ℓ_w = length of entire wall or segment of wall in direction of shear force

s = spacing of transverse reinforcement

s_2 = spacing of horizontal reinforcement in walls

ρ_t = ratio of area of distributed transverse reinforcement to gross concrete area perpendicular to that reinforcement

ρ_ℓ = ratio of area of distributed longitudinal reinforcement to gross concrete area perpendicular to that reinforcement

Table 29-4 Structural Walls (cont'd)

	Special Reinforced Concrete Structural Wall	Ordinary Reinforced Concrete Structural Wall
Shear Strength	The nominal shear strength (V_n) for structural walls shall not exceed: $$V_n = A_{cv}\left(\alpha_c\sqrt{f'_c} + \rho_t f_y\right)$$ where α_c is 3.0 for $h_w/\ell_w \leq 1.5$, is 2.0 for $h_w/\ell_w \geq 2.0$, and varies linearly between 3.0 and 2.0 for h_w/ℓ_w between 1.5 and 2.0. 21.7.4.1	The nominal shear strength (V_n) for walls can be calculated using the following methods: $$V_c = 3.3\sqrt{f'_c}\,hd + \frac{N_u d}{4\ell_w}$$ or $$V_c = \left[0.6\sqrt{f'_c} + \frac{\ell_w\left(1.25\sqrt{f'_c} + 0.2\frac{N_u}{\ell_w h}\right)}{\frac{M_u}{V_u} - \frac{\ell_w}{2}}\right]hd$$ where $\frac{M_u}{V_u} - \frac{\ell_w}{2} \geq 0$ $$V_s = \frac{A_v f_y d}{s}$$ $$V_n = V_c + V_s$$ 11.10
	Walls shall have distributed shear reinforcement providing resistance in two orthogonal directions in the plane of the wall. If the ratio (h_w/ℓ_w) does not exceed 2.0, reinforcement ratio (ρ_ℓ) shall not be less than reinforcement ratio (ρ_t). 21.7.4.3	Ratio (ρ_t) of horizontal shear reinforcement area to gross concrete area of vertical section shall not be less than 0.0025. 11.10.9.2 Spacing of horizontal shear reinforcement shall not exceed the smallest of $\ell_w/5$, $3h$, and 18 in. 11.10.9.3 The minimum vertical reinforcement ratio (ρ_ℓ) is a function of (h_w/ℓ_w) and of the horizontal reinforcement as shown below: $$\rho_n = 0.0025 + 0.5\left(2.5 - \frac{h_w}{\ell_w}\right)(\rho_t - 0.0025) \geq 0.0025$$ Also, the vertical shear reinforcement ratio need not exceed the required horizontal shear reinforcement ratio. 11.10.9.4 Spacing of vertical shear reinforcement shall not exceed the smallest of $\ell_w/3$, $3h$, and 18 in. 11.10.9.5
	Nominal shear strength of all wall piers sharing a common lateral force shall not be assumed to exceed $8A_{cv}\sqrt{f'_c}$, and the nominal shear strength of any one of the individual wall piers shall not be assumed to exceed $10A_{cw}\sqrt{f'_c}$. 21.7.4.4	No similar requirement.
	Nominal shear strength of horizontal wall segments and coupling beams shall not be assumed to exceed $10A_{cw}\sqrt{f'_c}$. 21.7.4.5	This limitation also exists for ordinary walls, except A_{cw} is replaced by hd where d may be taken equal to $0.8\ell_w$. 11.10.3

of ACI 318-02 required that all continuous reinforcement in structural walls must be anchored or spliced in accordance with the provisions for reinforcement in tension in 21.5.4. This was very confusing to the user, because 21.5.4 is really not applicable to situations covered by 21.7.2.3. This problem existed with ACI 318 editions prior to 2002 as well.

In a very significant and beneficial change, the requirements of 21.7.2.3 have been modified to remove the reference to beam-column joints in 21.5.4. Because actual forces in longitudinal reinforcement of structural walls may exceed calculated forces, it is now required that reinforcement in structural walls be developed or spliced for f_y in tension in accordance with Chapter 12. The effective depth of member referenced in 12.10.3 is permitted to be taken as $0.8l_w$ for walls. Requirements of 12.11, 12.12, and 12.13 need not be satisfied, because they address issues related to beams and do not apply to walls. At locations where yielding of longitudinal reinforcement is expected, $1.25 f_y$ is required to be developed in tension, to account for the likelihood that the actual yield strength exceeds the specified yield strength, as well as the influence of strain-hardening and cyclic load reversals. Where transverse reinforcement is used, development lengths for straight and hooked bars may be reduced as permitted in 12.2 and 12.5, respectively, because closely spaced transverse reinforcement improves the performance of splices and hooks subjected to repeated cycles of inelastic deformation. The requirement that mechanical splices of reinforcement conform to 21.2.6, and welded splices to 21.2.7, has now been placed in 21.7.2.3. Consequently, 21.7.6.4(f) of 21.7.6.6 of ACI 318-02 have been deleted.

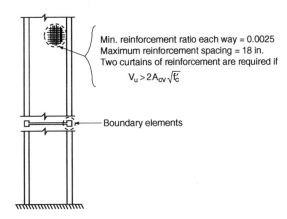

Figure 29-19 Structural Wall Design and Detailing Requirements

21.7.3 Design Forces

A condition similar to that used for the shear design of beams and columns is not as readily established for structural walls, primarily because the shear force at any section is significantly influenced by the forces and deformations at the other sections. Unlike the flexural behavior of beams and columns in a frame, with the forces and deformations determined primarily by the displacements in the end joints, the flexural deformation at any section of a structural wall is substantially influenced by the displacements at locations away from the section under consideration. Thus, for structural walls, the design shear force is determined from the lateral load analysis in accordance with the factored load combinations (21.7.3). The possibility of local yielding, as in the portion of a wall between two window openings, must also be considered; the actual shear forces may be much greater than that indicated by the lateral load analysis based on the factored design forces.

21.7.4 Shear Strength

The nominal shear strength V_n of structural walls is given in 21.7.4.1. The equation for V_n recognizes the higher shear strength of walls with high ratios of shear to moment. Additional requirements for wall segments and wall piers are contained in 21.7.4.2, 21.7.4.4, and 21.7.4.5.

The strength reduction factor f is determined in accordance with 9.3.4. Note that φ for shear must be 0.60 for any structural member that is designed to resist earthquake effects if its nominal shear strength is less than the shear corresponding to the development of the nominal flexural strength of the member. This is applicable to brittle members, such as low-rise walls or portions of walls between openings, which are impractical to reinforce to raise their nominal shear strength above the nominal flexural strength for the pertinent loading conditions.

Walls are to be provided with distributed shear reinforcement in two orthogonal directions in the plane of the wall (21.7.4.3). If the ratio of the height of the wall to the length of the wall is less than or equal to 2.0, the reinforcement ratio ρ_ℓ shall be greater than or equal to the reinforcement ratio ρ_t.

21.7.5 Design for Flexural and Axial Loads

Structural walls subjected to combined flexural and axial loads shall be designed in accordance with 10.2 and 10.3, excluding 10.3.6 and the nonlinear strain requirements of 10.2.2 (21.7.5.1). This procedure is essentially the same as that commonly used for columns. Reinforcement in boundary elements and distributed in flanges and webs must be included in the strain compatibility analysis. Openings in walls must also be considered.

Provisions for the influence of flanges for wall sections forming L-, T-, C-, or other cross-sectional shapes are in 21.7.5.2. Effective flange widths shall be assumed to extend from the face of the web a distance equal to the smaller of one-half the distance to an adjacent wall web and 25 percent of the total wall height.

21.7.6 Boundary Elements of Special Reinforced Concrete Structural Walls

Two approaches for evaluating the need for special boundary elements at the edges of structural walls are provided in 21.7.6. Section 21.7.6.2 allows the use of a displacement-based approach. In this method, the wall is displaced an amount equal to the expected design displacement, and special boundary elements are required to confine the concrete when the calculated neutral axis depth exceeds a certain critical value. Confinement is required over a horizontal length equal to a portion of the neutral axis depth (21.7.6.4). This approach is applicable to walls or wall piers that are essentially continuous in cross-section over the entire height of the wall and designed to have one critical section for flexure and axial loads, i.e., where the inelastic response of the wall is dominated by flexure at a critical, yielding section (21.7.6.2).

According to 21.7.6.2, compression zones must include special boundary elements when

$$c \geq \frac{\ell_w}{600(\delta_u/h_w)}, \quad \delta_u/h_w \geq 0.007 \qquad \text{Eq. (21-8)}$$

where c = distance from the extreme compression fiber to the neutral axis per 10.2.7 calculated for the factored axial force and nominal moment strength, consistent with the design displacement δ_u, resulting in the largest neutral axis depth
ℓ_w = length of the entire wall or segment of wall considered in the direction of the shear force
δ_u = design displacement
h_w = height of entire wall or of the segment of wall considered

The design displacement δ_u is the total lateral displacement expected for the design-basis earthquake, as specified by the governing code for earthquake-resistant design. In the *International Building Code*, ASCE 7 starting with its 1998 edition, and the NEHRP Provisions (1997 edition onwards), the design-basis earthquake is two-thirds of the maximum considered earthquake (MCE), which, in most of the country, has a two percent chance of being exceeded in 50 years. In these documents, the design displacement is computed using a static or dynamic linear-elastic analysis under code-specified actions. Considered in the analysis are the effects of cracking, torsion, P-Δ effects, and modification factors to account for expected inelastic response. In particular, δ_u is determined by multiplying the deflections from an elastic analysis under the prescribed seismic forces by a deflection amplification factor, which is given in the governing code. The deflection amplification factor, which depends on the type of seismic force-resisting system, is used to increase the elastic deflections to levels that would be expected for the design-basis earthquake. The lower limit of 0.007 on the quantity δ_u / h_w is specified to require a moderate wall deformation capacity for stiff buildings.

Typically, the reinforcement for a structural wall section is determined first for the combined effects of bending and axial load, and shear forces in accordance with the provisions outlined above for all applicable load combinations. The distance c can then be obtained from a strain compatibility analysis for each load combination that includes seismic effects, considering sidesway to the left and to the right. The largest c is used in Eq. (21-8) to determine if special boundary elements are required.

When special boundary elements are required, they must extend horizontally from the extreme compression fiber a distance not less than the larger of $c - 0.1\ell_w$ and $c/2$ (21.7.6.4(a); see Fig. 29-20). In the vertical direction, the special boundary elements must extend from the critical section a distance greater than or equal to the larger of ℓ_w or $M_u / 4V_u$ (21.7.6.2). This distance is based on upper bound estimates of plastic hinge lengths, and is beyond the zone over which concrete spalling is likely to occur.

The second approach for evaluating the need for special boundary elements is contained in 21.7.6.3. These provisions have been retained from earlier editions of the code since they are conservative for assessing transverse reinforcement requirements at wall boundaries for many walls. Compression zones shall include special boundary elements where the maximum extreme fiber stress corresponding to the factored forces, including earthquake effects, exceeds $0.2 f'_c$ (see Fig. 29-21). Special boundary elements can be discontinued where the compressive stress is less than $0.15 f'_c$. Note that the stresses are calculated assuming a linear response of the gross concrete section. The extent of the special boundary element is the same as when the approach of 21.7.6.2 is followed.

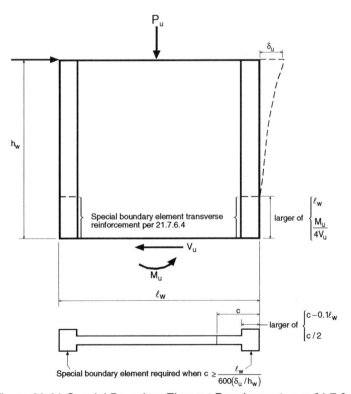

Figure 29-20 Special Boundary Element Requirements per 21.7.6.2

Section 21.7.6.4 contains the details of the reinforcement when special boundary elements are required by 21.7.6.2 or 21.7.6.3. The transverse reinforcement must satisfy the same requirements as for special moment frame members subjected to bending and axial load (21.4.4.1 through 21.4.4.3), excluding Eq. (21-3) (21.7.6.4(c); see Fig. 29-22). Also, the transverse reinforcement shall extend into the support a distance not less than the development length of the largest longitudinal bar in the special boundary element; for footings or mats, the transverse reinforcement shall extend at least 12 in. into the footing or mat (21.7.6.4(d)). Horizontal reinforcement in the wall web shall be anchored within the confined core of the boundary element to develop its specified yield strength (21.7.6.4(e)). To achieve this anchorage, 90-degree hooks or mechanical anchorages are recommended. Mechanical splices and welded splices of the longitudinal reinforcement in the boundary elements shall conform to 21.2.6 and 21.2.7, respectively (21.7.6.4(f)).

When special boundary elements are not required, the provisions of 21.7.6.5 must be satisfied. For the cases when the longitudinal reinforcement ratio at the wall boundary is greater than $400/f_y$, transverse reinforcement, spaced not more than 8 in. on center, shall be provided that satisfies 21.4.4.1(c), 21.4.4.3, and 21.7.6.4(c) (21.7.6.5(a)). This requirement helps in preventing buckling of the longitudinal reinforcement that can be caused by cyclic load reversals. The longitudinal reinforcement ratio to be used includes only the reinforcement at the end of the wall as indicated in Fig. R21.7.6.5. Horizontal reinforcement terminating at the edges of structural walls must be properly anchored per 21.7.6.5(b) in order for the reinforcement to be effective in resisting shear and to help in preventing buckling of the vertical edge reinforcement. The provisions of 21.7.6.5(b) are not required to be satisfied when the factored shear force V_u is less than $A_{cv}\sqrt{f'_c}$.

21.7.7 Coupling Beams

When adequately proportioned and detailed, coupling beams between structural walls can provide an efficient means of energy dissipation under seismic forces, and can provide a higher degree of overall stiffness to the structure. Due to their relatively large depth to clear span ratio, ends of coupling beams are usually subjected to large inelastic rotations. Adequate detailing and shear reinforcement are necessary to prevent shear failure and to ensure ductility and energy dissipation.

Coupling beams with $\ell_n/h \geq 4$ must satisfy the requirement of 21.3 for flexural members of special moment frames, excluding 21.3.1.3 and 21.3.1.4(a) if it can be shown that the beam has adequate lateral stability (21.7.7.1). When $\ell_n/h < 4$, coupling beams with two intersecting groups of diagonally-placed bars symmetrical about the midspan is permitted (21.7.7.2). The diagonal bars are required for deep coupling beams ($\ell_n/h < 2$) with a factored shear force V_u greater than $4\sqrt{f'_c}A_{cw}$, unless it can be shown otherwise that safety and stability are not compromised (21.7.7.3). Experiments have shown that diagonally oriented reinforcement is effectve only if the bars can be placed at a large inclination.

Note that in the 2002 code, h replaces d in the definition of the aspect ratio (clear span/depth) and A_{cw} (formerly A_{cp}) replaced $b_w d$ in the shear equations. The first change simplified the code requirements, since d is not always readily known for beams with multiple layers of reinforcement. The second change removed an inconsistency between 21.6.4.5 and 21.6.7.4 of the 1999 code; A_{cw} is now consistently used in 21.7.4.5 and 21.7.7.4.

Section 21.7.7.4 contains the reinforcement details for the two intersecting groups of diagonally placed bars. Figure 29-23 provides a summary of these requirements. The requirement on side dimensions of the cage and its core is to provide adequate toughness and stability when the bars are stressed beyond yielding. The nominal shear strength of a coupling beam is computed from the following (21.7.7.4(b)):

$$V_n = 2A_{vd}f_y \sin\alpha \leq 10\sqrt{f'_c}A_{cw} \qquad\qquad Eq.\ (21\text{-}9)$$

The additional reinforcement specified in 21.7.7.4(f) is used to confine the concrete outside of the diagonal cores.

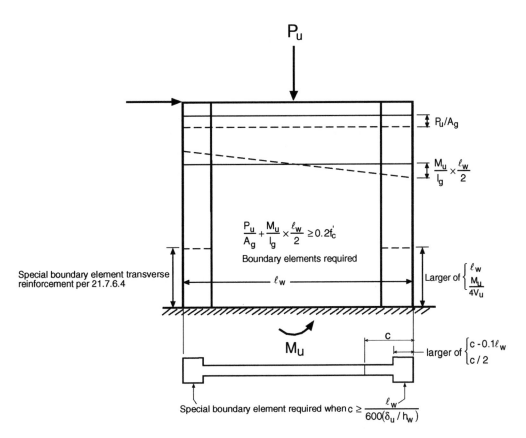

Figure 29-21 Special Boundary Element Requirements per 21.7.6.3

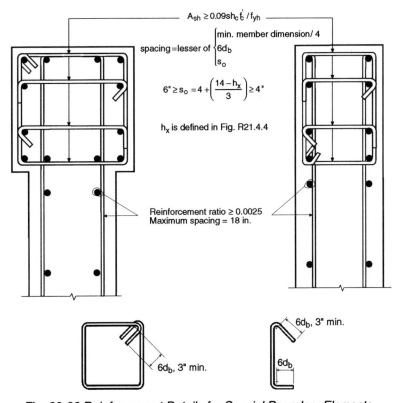

Fig. 29-22 Reinforcement Details for Special Boundary Elements

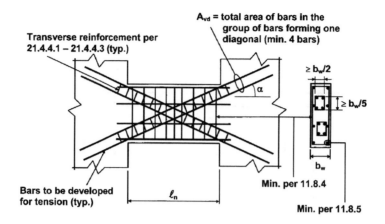

Figure 29-23 Coupling Beam with Diagonally Oriented Reinforcement

21.8 SPECIAL STRUCTURAL WALLS CONSTRUCTED USING PRECAST CONCRETE

According to 21.8.1, special structural walls constructed using precast concrete shall satisfy all requirements of 21.7 for cast-in-place special structural walls and the requirements in 21.13.2 and 21.13.3 for intermediate precast structural walls. Thus, the left-hand column of Table 29-4 may be utilized. The provisions for intermediate precast structural walls are discussed later. Note that the provisions of 21.8 do not apply to tilt-up walls.

21.9 STRUCTURAL DIAPHRAGMS AND TRUSSES

In building construction, diaphragms are structural elements, such as floor or roof slabs, that perform some or all of the following functions:
- Provide support for building elements such as walls, partitions, and cladding, and resist horizontal forces but not act as part of the vertical lateral-force-resisting system
- Transfer lateral forces to the vertical lateral-force-resisting system
- Interconnect various components of the lateral-force-resisting system with appropriate strength, stiffness, and toughness to permit deformation and rotation of the building as a unit

Section 21.9.4 prescribes a minimum thickness of 2 in. for concrete slabs and composite topping slabs serving as diaphragms used to transmit earthquake forces. The minimum thickness is based on what is currently used in joist and waffle slab systems and composite topping slabs on precast floor and roof systems. A minimum of 2.5 in. is required for topping slabs placed over precast floor or roof systems that do not act compositely with the precast system to resist the seismic forces.

Sections 21.9.2 and 21.9.3 provide design criteria for cast-in-place diaphragms. For the case of a cast-in-place composite topping slab on a precast floor or roof system, bonding is required so that the floor or roof system can provide restraint against slab buckling; also, reinforcement is required to ensure shear transfer across the precast joints. Composite action is not required for a cast-in-place topping slab on a precast floor or roof system, provided the topping slab acting alone is designed to resist the seismic forces.

21.9.5 Reinforcement

The minimum reinforcement ratio for structural diaphragms is the same as that required by 7.12 for temperature and shrinkage reinforcement. The maximum reinforcement spacing of 18 in. is intended to control the width of inclined cracks. Sections 21.9.5.1 and 21.9.5.2 contain provisions for welded wire reinforcement used in topping slabs placed over precast floor and roof elements and bonded prestressing tendons used as primary reinforcement in diaphragm chords or collectors, respectively.

According to 21.9.5.3, structural truss elements, struts, ties, diaphragm chords, and collector elements must have transverse reinforcement as specified in 21.4.4.1 through 21.4.4.3 when the compressive stress at any section exceeds $0.2 f'_c$. Note that compressive stress is calculated for the factored forces using a linearly elastic model and gross section properties. The special transverse reinforcement is no longer required where the compressive stress is less than $0.15 f'_c$.

In recent seismic codes and standards, collector elements of diaphragms are required to be designed for forces amplified by a factor Ω_0, to account for the overstrength in the vertical elements of the seismic-force-resisting system. The amplification factor Ω_0, ranges between 2 and 3 for concrete structures, depending upon the document selected and on the type of seismic system. To account for this, 21.9.5.3 now additionally states that where design forces have been amplified to account for the overstrength of the vertical elements of the seismic-force-resisting system. The limits of $0.2 f'_c$ and $0.15 f'_c$ shall be increased to $0.5 f'_c$ and $0.4 f'_c$, respectively.

Section 21.5.4 modifies the development length requirements of Chapter 12 for longitudinal beam bars terminating at exterior beam-column joints of structures assigned to high seismic design categories. But then 21.9.5.4 of ACI 318-02 required that all continuous reinforcement in diaphragms, tresses, ties, chords, and collector elements be anchored or spliced in accordance with the provisions for reinforcement in tension as specified in 21.5.4. This was very confusing to the user, because 21.5.4 is really not applicable to situations covered by 21.9.5.4. This problem existed with ACI 318 editions prior to 2002 as well.

In a very significant and beneficial change, the requirements of 21.9.5.4 have now been modified to remove the reference to beam-column joints in 21.5.4. All reference now is directly to the provisions of Chapter 12. Section 21.9.5.4 now requires that all continuous reinforcement in diaphragms, trusses, struts, ties, chords, and collector elements be developed or spliced for f_y in tension. This is a change similar to that made in 21.7.2.3.

21.9.7 Shear Strength

The shear strength requirements for monolithic structural diaphragms are similar to those for structural walls. In particular, the nominal shear strength V_n is computed from:

$$V_n = A_{cv}\left(2\sqrt{f'_c} + \rho_t f_y\right) \leq 8 A_{cv}\sqrt{f'_c} \qquad \text{Eq. (21-10)}$$

where A_{cv} is the thickness times the width of the diaphragm. Shear reinforcement should be placed perpendicular to the span of the diaphragm.

The nominal shear strength V_n for topping slab diaphragms does not include the contribution from the concrete:

$$V_n = A_{cv}\rho_t f_y \leq 8 A_{cv}\sqrt{f'_c} \qquad \text{Eq. (21-11)}$$

Typically, topping slabs are scored immediately above the boundary between the flanges of adjacent precast floor members to control the location of shrinkage cracks. Thus, these weakened sections of the diaphragm are cracked under service load conditions, and the contribution of the concrete to the overall shear strength of the diaphragm may have already been reduced before the design earthquake occurs. For this reason, the contribution of the concrete is taken as zero. The required web reinforcement must be uniformly distributed in both directions.

21.9.8 Boundary Elements of Structural Diaphragms

According to 21.9.8.1, boundary elements of structural diaphragms are to be proportioned to resist the sum of the factored axial forces acting in the plane of the diaphragm and the force obtained from dividing the factored moment at the section by the distance between the boundary elements of the diaphragm at that section. It is assumed that the factored flexural moments are resisted entirely by chord forces acting at opposite edges of the diaphragm (see Fig. 29-24). It is essential that splices of tensile reinforcement located in the chord and collector elements be fully developed and adequately confined (21.9.8.2). Mechanical and welded splices must conform to 21.2.6 and 21.2.7, respectively. If chord reinforcement is located within a wall, the joint between the diaphragm and the wall should be provided with adequate shear strength to transfer the shear forces.

Reinforcement details for chords and collectors at splices and anchorage zones are given in 21.9.8.3.

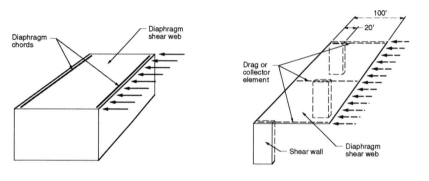

Diaphragm chords resist tension at edges of diaphragm

Drag or collector elements transfer forces from diaphragm to vertical elements

Figure 29-24 Diaphragm Chord and Collector Elements

21.10 FOUNDATIONS

Requirements for foundations supporting buildings in regions of high seismic risk or assigned to high seismic performance or design categories are contained in 21.10. It is important to note that the foundations must also comply with all other applicable provisions of the code. For piles, drilled piers, caissons, and slabs on grade, the provisions of 21.10 supplement other applicable design and construction criteria (see also 1.1.5 and 1.1.6).

21.10.2 Footings, Foundation Mats, and Pile Caps

Detailing requirements are contained in 21.10.2.1 through 21.10.2.4 for footings, mats, and pile caps supporting columns or walls, and are illustrated in Fig. 29-25.

21.10.3 Grade Beams and Slabs on Grade

Grade beams that are designed as ties between pile caps or footings must have continuous reinforcement that is developed within or beyond the supported column, or must be anchored within the pile cap or footing at discontinuities (21.10.3.1).

Section 21.10.3.2 contains geometrical and reinforcement requirements. The smallest cross-sectional dimension of the grade beam shall be greater than or equal to the clear spacing between the connected columns divided by 20; however, this dimension need not be greater than 18 in. Closed ties shall be provided over the length of the beam spaced at a maximum of one-half the smallest orthogonal cross-sectional dimension of the beam or 12 in., whichever is smaller. Both of these provisions are intended to provide reasonable beam proportions.

According to 21.10.3.3, grade beams and beams that are part of a mat foundation that is subjected to flexure from columns that are part of the lateral-force-resisting system shall have reinforcing details conforming to 21.3 for flexural members of special moment frames.

Slabs on grade shall be designed as diaphragms according to the provisions of 21.9 when they are subjected to seismic forces from walls or columns that are part of the lateral-force-resisting system (21.10.3.4). Such slabs shall be designated as structural members on the design drawings for obvious reasons.

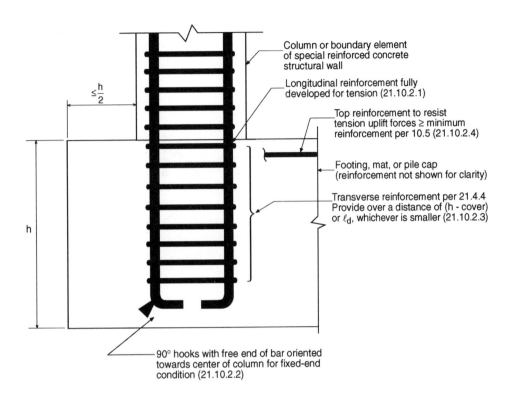

Figure 29-25 Reinforcement Details for Footings, Mats, and Pile Caps per 21.10.2

21.10.4   Piles, Piers, and Caissons

When piles, piers, or caissons are subjected to tension forces from earthquake-induced effects, a proper load path is required to transfer these forces from the longitudinal reinforcement of the column or boundary element through the pile cap to the reinforcement of the pile or caisson. Thus, continuous longitudinal reinforcement is required over the length resisting the tensile forces, and it must be properly detailed to transfer the forces through the elements (21.10.4.2). When grouted or post-installed reinforcing bars are used to transfer tensile forces between the pile cap or mat foundation and a precast pile, a test must be performed to ensure that the grouting system can develop at least 125 percent of the specified yield strength of the reinforcing bar, (21.10.4.3). In lieu of a test, reinforcing bars can be cast in the upper portion of a pile, exposed later by chipping away the concrete, and then mechanically connected or welded to achieve the proper extension.

Transverse reinforcement in accordance with 21.4.4 is required at the top of piles, piers, and caissons over a length equal to at least 5 times the cross-sectional dimension of the member, but not less than 6 ft below the bottom of the pile cap (21.10.4.4(a)). This requirement is based on numerous failures that were observed in earthquakes just below the pile cap, and provides ductility in this region of the pile. Also, for portions of piles in soil that is not capable of providing lateral support, or for piles in air or water, the entire unsupported length plus the length specified in 21.10.4.4(a) must be confined by transverse reinforcement per 21.4.4 (21.10.4.4(b)). Additional requirements for precast concrete driven piles, foundations supporting one- and two-story stud bearing wall construction, and pile caps with batter piles are contained in 21.10.4.5 through 21.10.4.7.

21.11 FRAME MEMBERS NOT PROPORTIONED TO RESIST FORCES INDUCED BY EARTHQUAKE MOTIONS

In regions of high seismic risk or for structures assigned to high seismic performance or design categories, frame members that are assumed not to contribute to lateral resistance shall comply with the requirements of 21.11. Specifically, these members are detailed depending on the magnitude of the moments and shears that are induced when they are subjected to the design displacements. This requirement is intended to enable the gravity load system to maintain its vertical load carrying capacity when subjected to the maximum lateral displacement of the lateral-force-resisting system expected for the design-basis earthquake.

The following summarizes the requirements of 21.11:

(1) Compute moments and shears (E) in all elements that are not part of the lateral-force-resisting system due to the design displacement δ_u. The displacement δ_u is determined based on the provisions of the governing building code. In the IBC, in ASCE 7 starting with its 1998 edition, and in the NEHRP Provisions (1997 and subsequent editions), δ_u is determined from the design-basis earthquake, (two-thirds of the Maximum Considered Earthquake, which for most of the country is an earthquake having a 98% probability of non-exceedance in 50 years) using a static or dynamic linear-elastic analysis, and considering the effects of cracked sections, torsion, and P-Δ effects. δ_u is determined by multiplying the deflections from an elastic analysis under the prescribed seismic forces by a deflection amplification factor, which accounts for expected inelastic response and which is given in the governing code for various seismic-force-resisting systems.

(2) Determine the factored moment M_u and the factored shear V_u in each of the elements that are not part of the lateral-force-resisting system from the more critical of the following load combinations:

$$U = 1.2D + 1.0L + 0.2S + E$$
$$U = 0.9D + E$$

The local factor on L can be reduced to 0.5 except for garages, areas occupied as places of public assembly, and all areas where L > 100 psf.

Note that the E-values (moments and shears) in the above expressions are determined in step 1 above.

(3) If $M_u \leq \phi M_n$ and $V_u \leq \phi V_n$ for an element that is not part of the lateral-force-resisting system, and if such an element is subjected to factored gravity axial forces $P_u \leq A_g f'_c/10$, it must satisfy the longitudinal reinforcement requirements in 21.3.2.1; in addition, stirrups spaced at no more than d/2 must be provided throughout the length of the member. If such an element is subjected to $P_u > A_g f'_c/10$ where $P_u \leq 0.35P_0$ (P_0 is the nominal axial load strength at zero eccentricity), it must conform to 21.4.3, 21.4.4.1(c), 21.4.4.3, and 21.4.5. In addition, ties at a maximum spacing of s_o must be provided throughout the height of the column, where s_o must not exceed the smaller of six times the smallest longitudinal bar diameter and 6 in. If the factored gravity axial force $P_u > 0.35P_0$, the requirements of 21.11.2.3 must be satisfied and the amount of transverse reinforcement provided shall be one-half of that required by 21.4.4.1, with the spacing not exceeding s_o for the full column height.

(4) If M_u or V_u determined in step 2 for an element that is not part of the lateral-force-resisting system exceeds ϕM_n or ϕV_n, or if induced moments and shears due to the design displacements are not calculated, then the structural materials must satisfy 21.2.4 and 21.2.5, and the splices of reinforcement must satisfy 21.2.6 and 21.2.7. If such an element is subjected to $P_u \leq A_g f'_c/10$, it must conform to 21.3.2.1 and 21.3.4; in addition, stirrups spaced at no more than d/2 must be provided throughout the length of the member. If such an element is subjected to $P_u > A_g f'_c/10$, it must be provided with full ductile detailing in conformance with 21.4.3.1, 21.4.4, 21.4.5, and 21.5.2.1. Note that the requirements of 21.4.3.1 used to be requirements of 21.4.3 before 2002. This change was meant to be made in 21.11.2, was inadvertently made in 21.11.3, and needs to be corrected in the future.

Precast concrete frame members assumed not to contribute to lateral resistance must also conform to 21.11.1 through 21.11.3. In addition, the following requirements of 21.11.4 must be satisfied: (a) ties specified in 21.11.2.2 must be provided over the entire column height, including the depth of the beams; (b) structural integrity reinforcement of 16.5 must be provided in all members; and (c) bearing length at the support of a beam must be at least 2 in. longer than the computed bearing length according to 10.17. The 2 in. increase in bearing length is based on an assumed 4% story drift ratio and a 50 in. beam depth, and is considered to be conservative for ground motions expected in high seismic zones.

In a very significant change, provisions for shear reinforcement at slab-column joints have been added in a new section 21.11.5, to reduce the likelihood of punching shear failure in two-way slabs without beams. A prescribed amount and detailing of shear reinforcement is required unless either 21.11.5(a) or (b) is satisfied.

Section 21.11.5(a) requires calculation of shear stress due to the factored shear force and induced moment according to 11.12.6.2. The induced moment is the moment that is calculated to occur at the slab-column joint where subjected to the design displacement defined in 21.1. Section 13.5.1.2 and the accompanying commentary provide guidance on selection of the slab stiffness for the purpose of this calculation.

Section 21.11.5(b) does not require the calculation of induced moments, and is based on research[29.5, 29.6] that identifies the likelihood of punching shear failure considering interstory drift and shear due to gravity loads. The requirement is illustrated in the newly added Fig. R21.11.5. The requirement can be satisfied in several ways: adding slab shear reinforcement, increasing slab thickness, designing a structure with more lateral stiffness to decrease interstory drift, or a combination of two or more of these.

If column capitals, drop panels, or other changes in slab thickness are used, the requirements of 21.1.5 must be evaluated at all potential critical sections.

21.12   REQUIREMENTS FOR INTERMEDIATE MOMENT FRAMES

For comparison purposes with the requirements for special moment frames, the provisions for beams (21.12.4) and columns (21.12.5) in intermediate moment frames have been presented in Table 29-1 and Table 29-2, respectively. The shear provisions of 21.12.3 are also included in those tables.

As was noted above, hoops instead of stirrups are now required at both ends of beams for a distance not less than 2h from the faces of the supports. The likelihood of spalling and loss of shell concrete in some regions of the frame are high. Both observed behavior under actual earthquakes and experimental research have shown that the transverse reinforcement will open at the ends and lose the ability to confine the concrete unless it is bent around the longitudinal reinforcement and its ends project into the core of the element. Similar provisions are now given in 21.12.5 for columns.

Two-way slabs without beams are acceptable lateral-force-resisting systems in regions of low or moderate seismic risk, or for structures assigned to low or moderate seismic performance or design categories. They are not permitted in regions of high seismic risk or for structures assigned to high seismic performance or design categories. Table 29-5, Fig. 29-26, and Fig. 29-27 summarize the detailing requirements for two-way slabs of intermediate moment frames. Provisions for two-way slabs of ordinary moment frames are also presented in Table 29-5.

The provisions of 21.12.6.2 for the band width within which flexural moment transfer reinforcement must be placed at edge and corner slab-column connections are new in the 2002 code. For these connections, flexural moment-transfer reinforcement perpendicular to the edge is not considered fully effective unless it is placed within the specified narrow band width (see Fig. R21.12.6.1).

The shear strength requirements of 21.12.6.8 are also new in the 2002 code. Slab-column frames are susceptible to punching shear failures during earthquakes if the shear stresses due to gravity loads are high. Thus, a limit was inroduced on the allowable shear stress caused by gravity loads, which in turn permits the slab-column connection to have adequate toughness to withstand the anticipated inelastic moment transfer.

21.13   INTERMEDIATE PRECAST STRUCTURAL WALLS

This new section applies to intermediate precast structural walls used to resist forces induced by the design earthquake.

Connections between precast wall panels or between wall panels and the foundation are required to resist forces due to earthquake motions and must provide for yielding that is restricted to steel elements or reinforcement (21.13.2). When Type 2 mechanical splices are used for connecting the primary reinforcement, the strength of the splice should be greater than or equal to 1.5 times f_y of the reinforcement (21.13.3).

Note that the provisions of 21.13, like those of 21.8, do not apply to tilt-up walls.

Table 29-5 Two-Way Slabs Without Beams*

Intermediate — All reinforcement provided to resist M_{slab}, the portion of slab moment balanced by the support moment, must be placed within the column strip defined in 13.2.1. **21.12.6.1** Ordinary — The middle strip is allowed to carry a portion of the unbalanced moment.
Intermediate — The fraction, defined by Eq. (13-1), of the moment M_{slab} shall be resisted by reinforcement placed within the band width specified in 13.5.3.2. Band width for edge and corner connections shall not extend beyond the column face a distance greater than c_t measured perpendicular to the slab span. **21.12.6.2** Ordinary — Similar requirement, except band width restriction for edge and corner connections does not apply.
Intermediate — Not less than one-half of the reinforcement in the column strip at the support shall be placed within the effective slab width specified in 13.5.3.2. **21.12.6.3** Ordinary — No similar requirement.
Intermediate — Not less than one-fourth of the top reinforcement at the support in the column strip shall be continuous throughout the span. **21.12.6.4** Ordinary — No similar requirement.
Intermediate — Continuous bottom reinforcement in the column strip shall not be less than one-third of the top reinforcement at the support in the column strip. **21.12.6.5** Intermediate — Not less than one-half of all middle strip bottom reinforcement and all column strip bottom reinforcement at midspan shall be continuous and shall develop its yield strength at the face of the support as defined in 13.6.2.5. **21.12.6.6** Ordinary — All bottom bars within the column strip shall be continuous or spliced with Class A tension splices or with mechanical or welded splices satisfying 12.14.3. **13.3.8.5**
Intermediate — At discontinuous edges of the slab, all top and bottom reinforcement at the support shall be developed at the face of support as defined in 13.6.2.5. **21.12.6.7** Ordinary — Positive moment reinforcement perpendicular to a discontinuous edge shall extend to the edge of the slab and have embedment of at least 6 in. in spandrel beams, columns, or walls. Negative moment reinforcement perpendicular to a discontinuous edge must be anchored and developed at the face of the support according to provisions in Chapter 12. **13.3.3, 13.3.4**
Intermediate — At the critical sections for columns defined in 11.12.1.2, two-way shear caused by factored gravity loads shall not exceed $0.4\phi V_c$ where V_c is calculated by 11.12.2.1 for nonprestressed slabs and 11.12.2.2 for prestressed slabs. This requirement may be waived if the contribution of the earthquake-induced factored two-way shear stress transferred by eccentricity of shear in accordance with 11.12.6.1 and 11.12.6.2 at the point of maximum stress does not exceed (ϕv_n permitted by 11.12.6.2)/2. **21.12.6.8** Ordinary — No similar requirement.

* *Not permitted as part of the lateral-force-resisting system in regions of high seismic risk or for structures assigned to high seismic performance or design categories.*

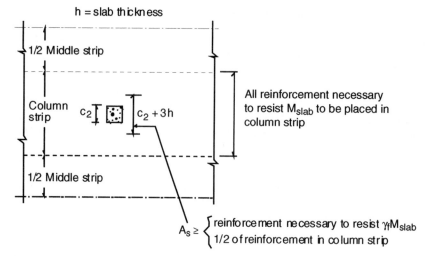

Figure 29-26 Location of Reinforcement in Two-way Slabs without Beams

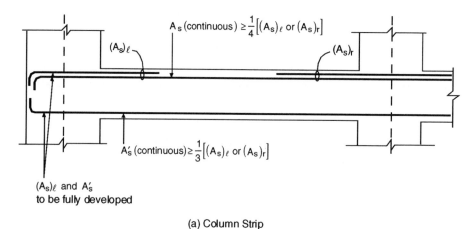

(a) Column Strip

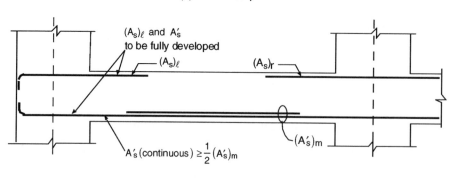

(b) Middle Strip

Figure 29-27 Details of Reinforcement in Two-way Slabs without Beams

REFERENCES

29.1　Ghosh, S.K., "Impact of the Seismic Design Provisions of the International Building Code," Structures and Codes Institute, Northbrook, IL, 2001.

29.2　*International Building Code*, International Code Council, Falls Church, VA, 2000, 2003.

29.3　*Uniform Building Code*, Vol. 2, International Conference of Building Officials (ICBO), Whittier, CA, 1997.

29.4　Ghosh, S. K., Nakaki, S. D., and Krishnan, K., "Precast Structures in Regions of High Seismicity", PCI Journal, Nov.–Dec. 1997, pp. 76-93.

29.5　*Acceptance Criteria for Moment Frames Based on Structural Testing*, ACI T1.1-01, American Concrete Institute, Farmington Hills, MI, 2001.

29.6　Megally, S., and Ghali, A., "Punching Shear Design of Earthquake Resistant Slab-Column Connections," ACI Structural Journal, Vol. 97, No. 5, September-October 2000, pp. 720-730.

29.7　Moehle, J.P., "Seismic Design Considerations for Flat Plate Construction," Mete A. Sozen Symposium, ACI SP-162, J.K. Wight and M.E. Kreger, Editors, American Concrete Institute, Farmington Hills, MI, 1996, pp. 1-35.

Example 29.1—Design of a 12-Story Cast-in-Place Frame-Shearwall Building and its Components

This example, and the 5 examples that follow, illustrate the design and detailing requirements for typical members of a 12-story cast-in-place concrete building.

A typical plan and elevation of the structure are shown in Figs. 29-28(a) and (b) respectively. The columns and structural walls have constant cross-sections throughout the height of the building*, and the bases of the lowest story segments are assumed fixed. The beams and the slabs also have the same dimensions at all floor levels. Although the member dimensions in this example are within the practical range, the structure itself is a hypothetical one, and has been chosen mainly for illustrative purposes. Other pertinent design data are as follows:

Material properties:

    Concrete: $f'_c = 4000$ psi, $w_c = 145$ pcf
    Reinforcement: $f_y = 60,000$ psi

Service loads:

    Live load:
        Floors = 50 psf
        Additional average value to allow for heavier load on corridors = 25 psf
        Total average live load (floors) = 75 psf
        Roof = 20 psf

    Superimposed dead load:
        Average for partitions = 20 psf
        Ceiling and mechanical = 10 psf
        Total average superimposed dead load (floors) = 30 psf
        Roof = 10 psf

Seismic design data:

    The building is located in a region of high seismic risk, and is assigned to a high seismic design or performance category.

    Dual system (special reinforced concrete structural walls with special moment frames) in the N-S direction

    Special moment frames in the E-W direction

---

* The uniformity in member dimensions used in this example has been adopted mainly for simplicity.

Example 29.1 (cont'd)	Calculations and Discussion	Code Reference

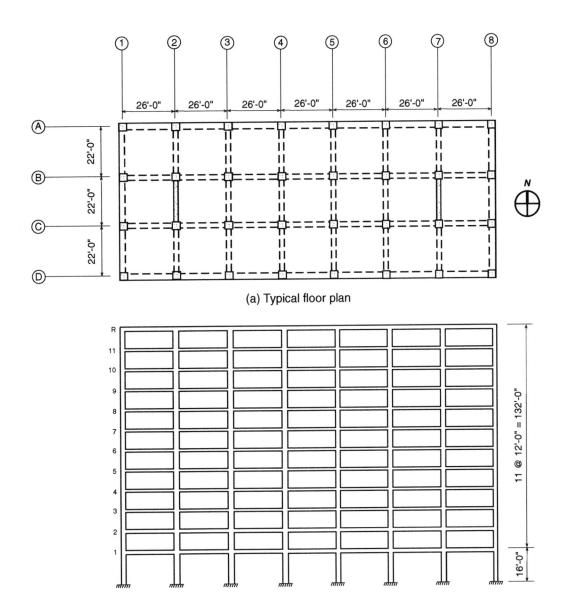

(a) Typical floor plan

Exterior columns: 22 x 22 in.
Interior columns: 26 x 26 in.
Beams: 20 x 24 in.
Slab: 8 in.
Walls: 18 in. web + 32 x 32 in. boundary elements

(b) Longitudinal section

Figure 29-28 Example Building

| **Example 29.1 (cont'd)** | **Calculations and Discussion** | **Code Reference** |

1. Lateral analysis

   The computation of the seismic and wind design forces is beyond the scope of this example.

   A three-dimensional analysis of the building was performed in both the N-S and E-W directions for both seismic and wind load cases. The effects of the seismic forces governed; thus, load combinations containing the effects of wind loads are not considered in the following examples.

2. Gravity analysis

   The Equivalent Frame Method of 13.7 was used to determine the gravity load moments in the members.

   Cumulative service axial loads for the columns and walls were computed considering live load reduction according to ASCE 7.

Example 29.2—Proportioning and Detailing of Flexural Members of Building in Example 29.1

Design a beam on the first floor of a typical interior E-W frame of the example building (Fig. 29-28). The beam has dimensions of b = 20 in. and h = 24 in. (d = 21.5 in.). The slab is 8 in. thick. Use f'_c = 4000 psi and f_y = 60,000 psi.

Calculations and Discussion	Code Reference

1. Check satisfaction of limitations on section dimensions.

    Factored axial compressive force on beams is negligible. O.K.   *21.3.1.1*

    $$\frac{\ell_n}{d} = \frac{(26 \times 12) - 30}{21.5} = 13.1 > 4 \quad \text{O.K.}$$ *21.3.1.2*

    $$\frac{\text{width}}{\text{depth}} = \frac{20}{24} = 0.83 > 0.3 \quad \text{O.K.}$$ *21.3.1.3*

    width = 20 in. > 10 in. O.K.   *21.3.1.4*

    < width of supporting column + (1.5 × depth of beam)

    < 24 + (1.5 × 24) = 60 in. O.K.

2. Determine required flexural reinforcement.

    The required reinforcement for the beams on the first floor level is shown in Table 29-6. The provided areas of steel are within the limits specified in 21.3.2.1. Also given in Table 29-6 are the design moment strengths ϕM_n at each section. The positive moment strength at a joint face must be at least equal to 50% of the negative moment strength provided at that joint. At the exterior negative location, this provision is satisfied since the positive design moment strength of 220.8 ft-kips is greater than 351.2/2 = 175.6 ft-kips. The provision is also satisfied at the interior negative location since 220.8 ft-kips is greater than 414.0/2 = 207.0 ft-kips.   *21.3.2.2*

    Neither the negative nor the positive moment strength at any section along the length of the member shall be less than 25% of the maximum moment strength provided at the face of either joint. In this case, 25% of the maximum design moment strength is equal to 414.0/4 = 103.5 ft-kips. Providing at least 2-No. 8 bars (ϕM_n = 147.9 ft-kips) or 2-No. 7 bars (ϕM_n = 113.2 ft-kips) at any section will satisfy this requirement. However, to satisfy the minimum reinforcement requirement of 21.3.2.1 (i.e., minimum A_s = 1.43 in.^{2}), a minimum of 2-No. 8 bars (A_s = 1.58 in.^{2}) or 3-No. 7 bars (A_s = 1.80 in.^{2}) must be provided at any section. This also automatically satisfies the requirement that 2 bars be continuous at both the top and the bottom of any section.   *21.3.2.2*   *21.3.2.1*

Example 29.2 (cont'd)   **Calculations and Discussion**   **Code Reference**

Table 29-6 Required Reinforcement for Beam of Typical E-W Frame on Floor Level 6

Location		M_u (ft-kips)	Required A_s* (in.²)	Reinforcement*	ϕM_n** (ft-kips)
End Span	Ext. Neg.	-291.9	3.23	5-No. 8	351.2
		138.7	1.48	4-No. 7	220.8
	Positive	145.3	1.55	4-No. 7	220.8
	Int. Neg.	-366.2	4.14	6-No. 8	414.0
		120.1	1.43	4-No. 7	220.8
Interior Span	Positive	125.1	1.43	4-No. 7	220.8
	Negative	-354.3	3.99	5-No. 8	351.2
		135.7	1.45	4-No. 7	220.8

*Max A_s = 0.025 × 20 × 21.5 = 10.75 in.² (21.3.2.1)
Min. A_s = $\sqrt{4{,}000}$ × 20 × 21.5/60,000 = 1.36 in.²
        = 200 × 20 × 21.5/60,000 = 1.43 in.² (governs)
**Does not include slab reinforcement.

3. Calculate required length of anchorage of flexural reinforcement in exterior column.

    Beam longitudinal reinforcement terminated in a column shall be extended to the far face of the confined column core and shall be anchored in tension according to 21.5.4 and in compression according to Chapter 12.   21.5.1.3

    Minimum development length ℓ_{dh} for a bar with a standard 90-degree hook in normal-weight concrete is

    $$\ell_{dh} = \frac{f_y d_b}{65\sqrt{f'_c}}$$ 21.5.4.1

    $$\geq 8d_b$$

    $$\geq 6 \text{ in.}$$

    A standard hook is defined as a 90-degree bend plus a $12d_b$ extension at the free end of the bar.   7.1.2

    For the No. 8 top bars (bend diameter $\geq 6d_b$):   7.2.1

    $$\ell_{dh} = \begin{cases} (60{,}000 \times 1.00)/(65\sqrt{4000}) = 14.6 \text{ in.} & \text{(governs)} \\ 8 \times 1.00 = 8 \text{ in.} \\ 6 \text{ in.} \end{cases}$$

    For the No. 7 bottom bars (bend diamter $\geq 6d_b$):

    $$\ell_{dh} = \begin{cases} (60{,}000 \times 0.875)/(65\sqrt{4000}) = 12.8 \text{ in.} & \text{(governs)} \\ 8 \times 0.875 = 7 \text{ in.} \\ 6 \text{ in.} \end{cases}$$

    Note that the development length ℓ_{dh} is measured from the near face of the column to the far edge of the vertical 12-bar-diameter extension (see Fig. 29-29).

Example 29.2 (cont'd)	Calculations and Discussion	Code Reference

When reinforcing bars extend through a joint, the column dimension must be at least 20 times the diameter of the largest longitudinal bar for normal weight concrete. In this case, the minimum required column dimension is 20 × 1.0 = 20 in., which is less than each of the two column widths that is provided.

21.5.1.4

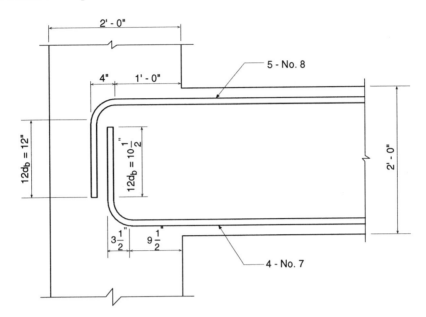

Figure 29-29 Detail of Flexural Reinforcement Anchorage at Exterior Column

4. Determine shear reinforcement requirements.

   Design for shear forces corresponding to end moments that are calculated by assuming the stress in the tensile flexural reinforcement equal to $1.25f_y$ and a strength reduction factor, $\phi = 1.0$ (probable flexural strength), plus shear forces due to factored tributary gravity loads.

   21.3.4.1

   The following equation can be used to compute M_{pr}*:

   $$M_{pr} = A_s(1.25f_y)\left(d - \frac{a}{2}\right)$$

   where $a = \dfrac{A_s(1.25f_y)}{0.85f'_c b}$

---

* The slab reinforcement within the effective slab width defined in 8.10 is not included in the calculation of M_{pr} (note that this reinforcement must be included when computing the flexural strength of the beam when checking the requirements of 21.4.2). It is unlikely that all or even most of the reinforcement within the slab effective width away from the beam will yield when subjected to the forces generated from the design-basis earthquake. Furthermore, including the slab reinforcement in the calculation of M_{pr} would result in a major deviation from how members have been designed in the past. In particular, the magnitude of the negative probable moment strength of the beam would significantly increase if the slab reinforcement were included. This in turn would have a significant impact on the shear strength requirements of the beam (21.3.4) and most likely the columns framing into the joint as well (21.4.5). Such significant increases seem unwarranted when compared to the appropriate provisions in previous editions of the ACI Code and other codes.

	Code
Example 29.2 (cont'd) **Calculations and Discussion**	**Reference**

For example, for sidesway to the right, the interior joint must be subjected to the negative moment M_{pr} which is determined as follows:

For 6-No. 8 top bars, $A_s = 6 \times 0.79 = 4.74$ in.^{2}

$$a = \frac{A_s(1.25f_y)}{0.85f'_c b} = \frac{4.74 \times 1.25 \times 60}{0.85 \times 4 \times 20} = 5.23 \text{ in.}$$

$$M_{pr} = A_s(1.25f_y)\left(d - \frac{a}{2}\right) = 4.74 \times 1.25 \times 60 \times \left(21.5 - \frac{5.23}{2}\right) = 6713.6 \text{ in.-kips} = 559.5 \text{ ft-kips}$$

Similarly, for the exterior joint, the positive moment M_{pr} based on the 4-No. 7 bottom bars is equal to 302.6 ft-kips. The probable flexural strengths for sidesway to the left can be obtained in a similar fashion.

The factored gravity load at midspan is:

$$w_D = \left[\frac{8}{12}(145) + 30\right] \times 22 + \frac{16 \times 20}{144}(145) = 3109 \text{ lbs/ft}$$

$$w_L = 75 \times 22 = 1650 \text{ lbs/ft}$$

$$w_u = 1.2^* w_D + 0.5 w_L = 4.56 \text{ kips/ft} \qquad \text{Eq. (9-5)}$$

Figure 29-30 shows the exterior beam span and the shear forces due to the gravity loads. Also shown are the probable flexural strengths M_{pr} at the joint faces for sidesway to the right and to the left and the corresponding shear forces due to these moments. Note that the maximum combined design shear forces are larger than those obtained from the structural analysis.   *21.3.4.2*

The shear strength of concrete V_c is to be taken as zero when the earthquake-induced shear force calculated in accordance with 21.3.4.1 is greater than or equal to 50% of the total shear force and the factored axial compressive force is less than $A_g f'_c /20$ where A_g is the gross cross-sectioned area of the beam. The beam carries negligible axial forces, and the maximum earthquake-induced shear force, which is equal to 36.3 kips (see Fig. 29-30), is greater than one-half the total design shear force which is equal to $0.5 \times 68.1 = 34.1$ kips. Thus, V_c must be taken equal to zero. The maximum shear force V_s is:

$$\phi V_s = V_u - \phi V_c$$

$$V_s = \frac{V_u}{\phi} - V_c$$

$$= \frac{68.1}{0.75} - 0 = 90.8 \text{ kips}$$

---

* Note that in seismic design complying with the IBC, the factor would be $(1.2 + 0.2S_{DS})$, where S_{DS} is the design spectral response acceleration at short periods at the site of the structure.

	Code
Example 29.2 (cont'd) Calculations and Discussion	Reference

where the strength reduction factor ϕ is 0.75. 9.3.4

Shear strength contributed by shear reinforcement must not exceed $(V_s)_{max}$: 11.5.6.9

$$(V_s)_{max} = 8\sqrt{f'_c}b_w d = 8\sqrt{4000} \times 20 \times 21.5/1000 = 217.6 \text{ kips} > 90.8 \text{ kips} \quad \text{O.K.}$$

Also, V_s is less than $4\sqrt{f'_c}b_w d = 108.8$ kips. 11.5.4.3

Required spacing of No. 3 closed stirrups (hoops) for a factored shear force of 90.8 kips is: 11.5.6.2

$$s = \frac{A_v f_y d}{V_s} = \frac{(4 \times 0.11) \times 60 \times 21.5}{90.8} = 6.3 \text{ in.}$$

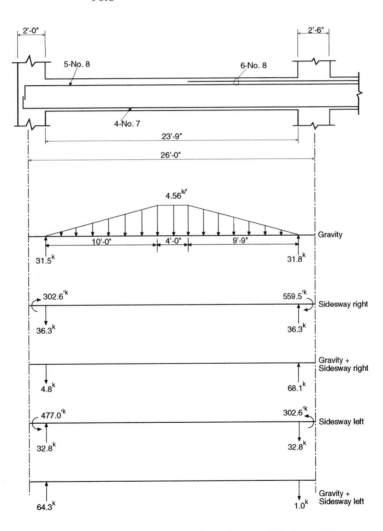

Figure 29-30 Design Shear Forces for Exterior Beam Span of Typical E-N Frame on Floor Level 1

	Code
Example 29.2 (cont'd) **Calculations and Discussion**	**Reference**

Note that 4 legs are required for lateral support of the longitudinal bars.   21.3.3.3

Maximum allowable hoop spacing (s_{max}) within a distance of 2h = 2 × 24 = 48 in.   21.3.3.2
from the face of the support is the smallest of the following:

$$s_{max} = \frac{d}{4} = \frac{21.5}{4} = 5.4 \text{ in. (governs)}$$

$$= 8 \times \text{(diameter of smallest longitudinal bar)} = 8 \times 0.875 = 7.0 \text{ in.}$$

$$= 24 \times \text{(diameter of hoop bar)} = 24 \times 0.375 = 9.0 \text{ in.}$$

$$= 12 \text{ in.}$$

Therefore, hoops must be spaced at 5 in. on center with the first one located at 2 in. from the face of the support. Eleven hoops are to be placed at this spacing.

Where hoops are no longer required, stirrups with seismic hooks at both ends may be used.   21.3.3.4
At a distance of 52 in. from the face of the interior support, $V_u = 63.7$ kips.

With $V_c = 2\sqrt{4000} \times 20 \times 21.5/1000 = 54.4$ kips, the spacing required for
No. 3 stirrups with two legs is 9.3 in. < d/2 = 10.8 in.

A 9 in. spacing, starting at 52 in. from the face of the support will be sufficient
for the remaining portion of the beam.

5. Negative reinforcement cutoff points.

   For the purpose of determining cutoff points for the negative reinforcement at the interior
   support, a moment diagram corresponding to the probable flexural strengths at the beam
   ends and 0.9* times the dead load on the span will be used. The cutoff point for four of   Eq. (9-7)
   the six No. 8 bars at the top will be determined.

   With the design flexural strength of a section with 2-No. 8 top bars = 147.9 ft-kips
   (calculated using $f_s = f_y = 60$ ksi and $\phi = 0.9$, since a section with such light reinforcement will be tension-controlled), the distance from the face of the support to where the
   moment under the loading considered equals 147.9 ft-kips is readily obtained by
   summing moments about section a-a in Fig. 29-31, and equating these to 147.9 ft-kips:

   $$\frac{x}{2}\left(\frac{2.8x}{9.75}\right)\left(\frac{x}{3}\right) - 55.8x + 559.5 = 147.9$$

---

* Note that in seismic design complying with the IBC, the factor would be $(1.2 + 0.2S_{DS})$, where S_{DS} is the design spectral response acceleration at short periods at the site of the structure.

	Code
Example 29.2 (cont'd) **Calculations and Discussion**	**Reference**

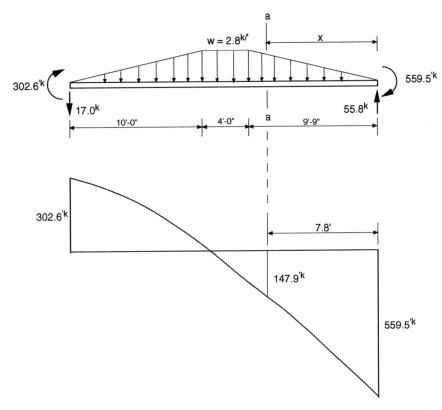

Figure 29-31 Moment Diagram for Cutoff Location of Negative Bars at Interior Support

Solving for x gives a distance of 7.8 ft. The 4-No. 8 bars must extend a distance d = 21.5 in. or $12d_b = 12 \times 1.0 = 12$ in. beyond the distance x. Thus, from the face of the support, the total bar length must be at least equal to 7.8 + (21.5/12) = 9.6 ft. Also, the bars must extend a full development length ℓ_d beyond the face of the support:

12.10.3

12.10.4

$$\ell_d = \left(\frac{3}{40} \frac{f_y}{\sqrt{f'_c}} \frac{\psi_t \psi_e \psi_s \lambda}{\left(\frac{c_b + K_{tr}}{d_b}\right)} \right) d_b \qquad \text{Eq. (12-1)}$$

where ψ_t = reinforcement location factor = 1.3 (top bar)

ψ_e = coating factor = 1.0 (uncoated reinforcement)

ψ_s = reinforcement size factor = 1.0 (No. 8 bar)

λ = lightweight aggregate concrete factor = 1.0 (normal weight concrete)

$$c = \text{spacing or cover dimension} = \begin{cases} 1.5 + 0.375 + \dfrac{1.0}{2} = 2.375 \text{ in.} \\ \dfrac{20 - 2(1.5 + 0.375) - 1.0}{2 \times 5} = 1.525 \text{ in. (governs)} \end{cases}$$

		Code
Example 29.2 (cont'd)	**Calculations and Discussion**	**Reference**

K_{tr} = transverse reinforcement index = 0 (conservative)

$$\frac{c + K_{tr}}{d_b} = \frac{1.525 + 0}{1.0} = 1.525$$

$$\ell_d = \frac{3}{40} \times \frac{60,000}{\sqrt{4000}} \times \frac{1.3 \times 1.0 \times 1.0 \times 1.0}{1.525} \times 1.0 = 60.7 \text{ in.} = 5.1 \text{ ft} < 9.6 \text{ ft}$$

The total required length of the 4-No. 8 bars must be at least 9.6 ft beyond the face of the support.

Flexural reinforcement shall not be terminated in a tension zone unless one or more of the conditions of 12.10.5 are satisfied. In this case, the point of inflection is approximately 11.25 ft from the face of the right support which is greater than 9.6 ft. The 4-No. 8 bars can not be terminated here unless one of the conditions of 12.10.5 is satisfied.

Check if the factored shear force V_u at the cutoff point does not exceed two-thirds of ϕV_n. For No. 3 stirrups spaced at 9 in. on center that are provided in this region: *12.10.5.1*

$$\phi V_n = \phi(V_s + V_c) = 0.75 \times \left(\frac{0.22 \times 60 \times 21.5}{9} + 54.4 \right) = 64.5 \text{ kips}$$

$$\frac{2}{3}\phi V_n = 43.0 \text{ kips} > V_u = 42.7 \text{ kips at 9.6 ft from face of support}$$

Since $2\phi V_n /3 > V_u$, the cutoff point for the 4-No. 8 bars can be 9.6 ft beyond the face of the interior support.

The cutoff point for three of the 5-No. 8 bars at the exterior support can be determined in a similar fashion. These bars can be cut off at 8.4 ft from the face of the exterior support.

6. Flexural reinforcement splices.

    Lap splices of flexural reinforcement must not be placed within a joint, within a distance 2h from faces of supports or within regions of potential plastic hinging. Note that all lap splices have to be confined by hoops or spirals with a maximum spacing or pitch of d/4 or 4 in. over the length of the lap. Lap splices will be determined for the No. 7 bottom bars.   *21.3.2.3*

    Since all of the bars will be spliced within the required length, use a Class B splice.   *12.15.2*

    Required length of splice = $1.3\ell_d \geq 12$ in.   *12.15.1*

    where

$$\ell_d = \left(\frac{3}{40} \frac{f_y}{\sqrt{f'_c}} \frac{\psi_t \psi_e \psi_s \lambda}{\left(\frac{c_b + K_{tr}}{d_b} \right)} \right) d_b$$
     Eq. (12-1)

		Code
Example 29.2 (cont'd)	**Calculations and Discussion**	**Reference**

reinforcement location factor ψ_t = 1.0 (other than top bars) 12.2.4

coating factor ψ_e = 1.0 (uncoated bars)

reinforcement size factor ψ_s = 1.0 (No. 7 bar)

lightweight aggregate concrete factor λ = 1.0 (normal weight concrete)

$$c = 1.5 + 0.375 + \frac{0.875}{2} = 2.31 \text{ in.} \quad \text{(governs)}$$

$$= \frac{1}{2}\left[\frac{20 - 2(1.5 + 0.375) - 0.875}{3}\right] = 2.56 \text{ in.}$$

$$K_{tr} = \frac{A_{tr}f_{yt}}{1500sn} = \frac{(2 \times 0.11)(60,000)}{1500 \times 4.0 \times 4} = 0.55$$

$$\frac{c + K_{tr}}{d_b} = \frac{2.31 + 0.55}{0.875} = 3.3 > 2.5, \text{ use } 2.5$$

Therefore,

$$\ell_d = \frac{3}{40} \times \frac{60,000}{\sqrt{4000}} \times \frac{1.0 \times 1.0 \times 1.0 \times 1.0}{2.5} \times 0.875 = 24.9 \text{ in.}$$

Class B splice length = 1.3 × 24.9 = 32.4 in.

7. Reinforcement details for the beam are shown in Fig. 29-32.

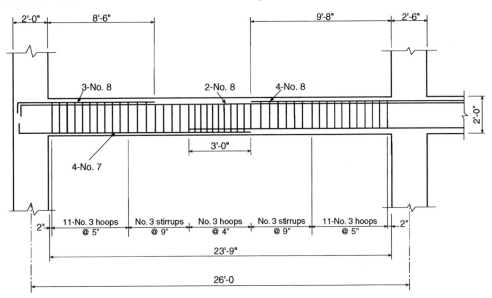

Figure 29-32 Reinforcement Details for Exterior Beam on Floor Level 1

Example 29.3—Proportioning and Detailing of Columns of Building in Example 29.1

Determine the required reinforcement for an edge column supporting the first floor of a typical E-W interior frame. The column dimensions have been established at 24-in. square. Use $f'_c = 4000$ psi and $f_y = 60,000$ psi.

Calculations and Discussion	Code Reference

Table 29-7 contains a summary of the factored axial loads and bending moments for an edge column in the first floor level for seismic forces in the E-W direction.

From Table 29-7, maximum $P_u = 1012$ kips

$P_u = 1012$ kips $> A_g f'_c /10 = (24 \times 24) \times 4/10 = 230$ kips

21.4.1

Thus, the provisions of 21.4 governing special moment frame members subjected to bending and axial load apply.

Table 29-7 Summary of Factored Axial Loads and Bending Moments for an Edge Column in the First Story for Seismic Forces in the E-W Direction

Load Combination	Axial Load, P_u (kips)	Bending Moment, M_u (ft-kips)
1.2D + 1.6L	1002.9	-78.2
1.2D + 0.5L + E	722.8	166.4
1.2D + 0.5L − E	1012.0	-275.6
0.9D + E	459.8	188.1
0.9D − E	749.0	-253.9

1. Check satisfaction of limitations on section dimensions.

    • Shortest cross-sectional dimension = 24 in. > 12 in.   O.K.   21.4.1.1

    • Ratio of shortest cross-sectional dimension to perpendicular dimension = 1.0 > 0.4   O.K.   21.4.1.2

2. Determine required longitudinal reinforcement.

    Based on the load combinations in Table 29-7, a 24×24 in. column with 8-No. 8 bars ($\rho_g = 1.10\%$) is adequate for the column supporting the first floor level.

    Note that $0.01 < \rho_g \leq 0.06$ O.K.   21.4.3.1

	Code
Example 29.3 (cont'd) **Calculations and Discussion**	**Reference**

3. Nominal flexural strength of columns relative to that of beams in E-W direction.

   $$\Sigma M_c \text{ (columns)} \geq \frac{6}{5} \Sigma M_g \text{ (beams)}$$ 21.4.2.2

   The nominal negative flexural strength M_n^- of the beam framing into the column must include the slab reinforcement within an effective slab width equal to: 21.4.2.2

   $(16 \times 8) + 20 = 148$ in.
   $22 \times 12 = 264$ in.   8.10.2
   $(26 \times 12)/4 = 78$ in. (governs)

   The minimum required A_s in the 78-in. effective width is equal to $0.0018 \times 78 \times 8 = 1.12$ in.^{2}, which corresponds to 6-No. 4 bars @ $78/6 = 13$ in. spacing. This spacing is less than the maximum bar spacing ($= 2h = 16$ in.). Provide No. 4 @ 13 in. at both the top and bottom of the slab (according to Fig. 13.3.8, 100 percent of both the top and the bottom reinforcement in the column strip must be continuous or anchored at the support).

   A strain compatibility analysis of the section yields M_n^- of the beam equal to 632 ft-kips.

   For the lower end of the upper column framing into the joint, the minimum nominal flexural strength is 578 ft-kips, which corresponds to $P_u = 922$ kips. Similarly, the minimum M_n is 522 ft-kips for the upper end of the lower column framing into the joint; this corresponds to $P_u = 1012$ kips.

   Therefore,

   $\Sigma M_c = 578 + 522 = 1100$ ft-kips

   $\Sigma M_g = 632$ ft-kips

   $1100 \text{ ft-kips} > \frac{6}{5} \times 632 = 758 \text{ ft-kips}$ O.K.

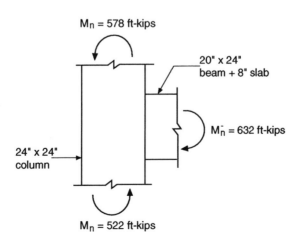

	Code
Example 29.3 (cont'd) Calculations and Discussion	**Reference**

4. Nominal flexural strength of columns relative to that of beams in the N-S direction.

    The beams in the N-S direction framing into columns at the first floor level require 4-No. 7 bars at both the top and the bottom of the section.

    The nominal negative flexural strength M_n^- of the beams framing into the column must include the slab reinforcement within an effective slab width equal to:

    $(22 \times 12)/12 + 20 = 42$ in. (governs)

    $(6 \times 8) + 20 = 68$ in.   8.10.3

    $(23.75 \times 12)/2 + 20 = 162.5$ in.

    The minimum A_s in the 42-in. effective width is equal to $0.0018 \times 42 \times 8 = 0.6$ in.^{2}, which corresponds to 3-No. 4 bars @ $42/3 = 14$ in. spacing. This spacing is less than the maximum bar spacing (= 2h = 16 in.). Provide No. 4 @ 14 in. at both the top and the bottom of the slab (according to Fig. 13.3.8, 100 percent of both the top and the bottom reinforcement in the column strip must be continuous or anchored at the support).

    A strain compatibility analysis of the section yields $M_n^- = 354$ ft-kips and $M_n^+ = 277$ ft-kips.

    For the lower end of the upper column framing into the joint, the minimum nominal flexural strength is 580.4 ft-kips, which corresponds to $P_u = 918$ kips. Similarly, the minimum M_n is 528.6 ft-kips for the upper end of the lower column framing into the joint; this corresponds to $P_u = 1,003$ kips.

    $\Sigma M_g = 354 + 277 = 631$ ft-kips

    $\Sigma M_c = 580 + 529 = 1,109$ ft-kips $> \dfrac{6}{5} \Sigma M_g = \dfrac{6}{5} \times 631 = 757$ ft-kips   O.K.   21.4.4.4
                                                                                                                                                     Eq. (21-1)

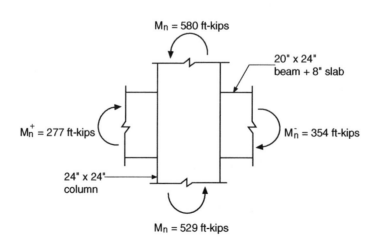

29-62

Example 29.3 (cont'd)	Calculations and Discussion	Code Reference

5. Determine transverse reinforcement requirements.

   a. Confinement reinforcement (see Fig. 29-7(b)).

   Transverse reinforcement for confinement is required over a distance ℓ_o from the column ends where

   $$\ell_o \geq \begin{cases} \text{depth of member} = 24 \text{ in.} \\ 1/6 \text{ (clear height)} = (14 \times 12)/6 = 28 \text{ in. (governs)} \\ 18 \text{ in.} \end{cases}$$

   21.4.4.4

   Maximum allowable spacing of rectangular hoops assuming No. 4 hoops with a crosstie in each direction:

   21.4.4.2

   $s_{max} = 0.25$ (smallest dimension of column) $= 0.25 \times 24 = 6$ in.

   $= 6$ (diameter of longitudinal bar) $= 6 \times 1.0 = 6$ in.

   $= s_o = 4 + \left(\dfrac{14 - h_x}{3}\right) = 4 + \left(\dfrac{14 - 11}{3}\right) = 5$ in. < 6 in. (governs)

   > 4 in.

   where $h_x = \dfrac{24 - 2\left(1.5 + 0.5 + \dfrac{1.0}{2}\right)}{2} + 2\left(\dfrac{1.0}{2} + \dfrac{0.5}{2}\right) = 11$ in.

   Required cross-sectional area of confinement reinforcement in the form of hoops:

   $$A_{sh} \geq \begin{cases} 0.3 s b_c \left[\dfrac{A_g}{A_{ch}} - 1\right] \dfrac{f'_c}{f_{yt}} & \text{Eq. (21-3)} \\ 0.09 s b_c \dfrac{f'_c}{f_{yt}} & \text{Eq. (21-4)} \end{cases}$$

   where

   s = spacing of transverse reinforcement (in.)

   b_c = cross-sectional dimension of column core, measured center-to-center of confining reinforcement (in.) = $24 - 2(1.5 + 0.25) = 20.5$ in.

   A_{ch} = core area of column section, measured outside to-outside of transverse reinforcement (in.^{2}) = $[24 - (2 \times 1.5)]^2 = 441$ in.^{2}

   f_{yt} = specified yield strength of transverse reinforcement (psi)

		Code
Example 29.3 (cont'd)	**Calculations and Discussion**	**Reference**

For a hoop spacing of 5 in. and $f_{yt} = 60,000$ psi, the required cross-sectional area is:

$$A_{sh} \geq \begin{cases} (0.3 \times 5 \times 20.5)\left(\dfrac{576}{441} - 1\right)\dfrac{4000}{60,000} = 0.63 \text{ in.}^2 \quad \text{(governs)} \\ \\ (0.09 \times 5 \times 20.5)\dfrac{4000}{60,000} = 0.62 \text{ in.}^2 \end{cases}$$

No. 4 hoops with one crosstie, as shown in the sketch below, provides 21.4.4.3
$A_{sh} = 3 \times 0.20 = 0.60$ in.^{2} < 0.63 in.^{2} Either accept or reduce hoop spacing to
4 in. so that governing $A_{sh} = 0.50$ in.^{2} < provided $A_{sh} = 0.60$ in.^{2}

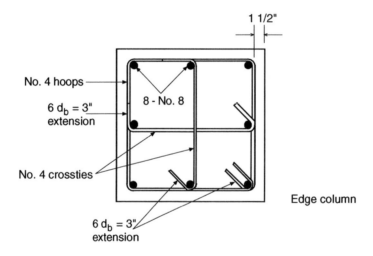

Edge column

b. Transverse reinforcement for shear.

   As in the design of shear reinforcement for beams, the design shear for columns is 21.4.5
   based not on the factored shear forces obtained from a lateral load analysis but rather
   on the nominal flexural strengths provided in the columns. The column design shear
   forces shall be determined from the consideration of the maximum forces that can be
   developed at the faces of the joints, with the probable flexural strengths calculated for
   the factored axial compressive forces resulting in the largest moments acting at the
   joint faces.

   The largest probable flexural strength that may develop in the column can conservatively
   be assumed to correspond to the balanced point of the column interaction diagram.

   With the strength reduction factors equal to 1.0 and $f_y = 1.25 \times 60 = 75$ ksi, the moment
   corresponding to balanced failure is 742 ft-kips. Thus, $V_u = (2 \times 742)/14 = 106$ kips.

| Example 29.3 (cont'd) | Calculations and Discussion | Code Reference |

The shear force need not exceed that determined from joint strengths based on the probable flexural strengths M_{pr} of the members framing into the joint. For seismic forces in the E-W direction, the negative probable flexural strength of the beam framing into the joint at the face of the edge column is 477.0 ft-kips (see Fig. 29-30).

21.4.5.1

Distribution of this moment to the columns is proportional to EI/ℓ of the columns above and below the joint. Since the columns above and below the joint have the same cross-section, reinforcement, and concrete strength, EI is a constant, and the moment is distributed according to $1/\ell$. Therefore, the moment at the top of the first story column is

$$477.0 \left(\frac{12}{12 + 16} \right) = 204.4 \text{ ft-kips}$$

It is possible for the base of the first story column to develop the probable flexural strength of 742.0 ft-kips. Thus, the shear force is

$$V_u = \frac{204.4 + 742.0}{14} = 67.6 \text{ kips}$$

For seismic forces in the N-S direction, the negative probable flexural strength of the beam framing into one side of the column is 302.6 ft-kips (4-No. 7 top bars). The positive probable flexural strength of the beam framing into the other side of the column is also 302.6 ft-kips (4-No. 7 bottom bars). Therefore, at the top of the first story column, the moment is

$$(2 \times 302.6) \left(\frac{12}{12 + 16} \right) = 259.4 \text{ ft-kips}$$

The shear force is

$$V_u = \frac{259.4 + 742.0}{14} = 71.5 \text{ kips}$$

Both of these shear forces are greater than those obtained from analysis.

Since the factored axial forces are greater than $A_g f'_c / 20 = 115$ kips, the shear strength of the concrete may be used:

21.4.5.2

$$V_c = 2\sqrt{f'_c} bd \left(1 + \frac{N_u}{2000 A_g} \right)$$

Eq. (11-4)

		Code
Example 29.3 (cont'd)	**Calculations and Discussion**	**Reference**

Conservatively using the minimum axial load from Table 29-7,

$$V_c = \frac{2\sqrt{4000}\,(24 \times 17.7)}{1000}\left[1 + \frac{459{,}800}{2000 \times (24)^2}\right] = 75.2 \text{ kips}$$

$$V_s = \frac{A_v f_y d}{s} = \frac{(3 \times 0.20) \times 60 \times 17.7}{4.5} = 141.6 \text{ kips}$$

$$\phi(V_c + V_s) = 0.75(75.2 + 141.6) = 162.6 \text{ kips} > V_u = 71.5 \text{ kips O.K.}$$

Thus, the transverse reinforcement spacing over the distance $\ell_o = 28$ in. near the column ends required for confinement is also adequate for shear.

The remainder of the column length must contain hoop reinforcement with center-to-center spacing not to exceed either six times the diameter of the column longitudinal bars ($= 6 \times 1.0 = 6.0$ in.) or 6 in.
21.4.4.6

Use No. 4 hoops and crossties spaced at 4 in. within a distance of 28 in. from the column ends and No. 4 hoops spaced at 6 in. or less over the remainder of the column.

6. Minimum length of lap splices of column vertical bars.

The location of lap splices of column bars must be within the center half of the member length. Also, the splices are to be designed as tension splices. If all the bars are spliced at the same location, the splices need to be Class B. Transverse reinforcement at 4.5 in. is to be provided over the full lap splice length.
21.4.3.2

Required length of Class B splice $= 1.3\ell_d$
12.15.1

where

$$\ell_d = \left(\frac{3}{40}\,\frac{f_y}{\sqrt{f'_c}}\,\frac{\psi_t \psi_e \psi_s \lambda}{\left(\frac{c_b + K_{tr}}{d_b}\right)}\right) d_b$$

Eq. (12-1)

reinforcement location factor $\psi_t = 1.0$ (other than top bars)
12.2.4

coating factor $\psi_e = 1.0$ (uncoated bars)

reinforcement size factor $\psi_s = 1.0$ (No. 7 and larger bars)

lightweight aggregate concrete factor $\lambda = 1.0$ (normal weight concrete)

$$c = 1.5 + 0.5 + \frac{1.0}{2} = 2.5 \text{ in. (governs)}$$

	Code
Example 29.3 (cont'd) **Calculations and Discussion**	**Reference**

$$= \frac{1}{2}\left[\frac{24-2(1.5+0.5)-1.0}{2}\right] = 4.75 \text{ in.}$$

$$K_{tr} = \frac{A_{tr}f_{yt}}{1500sn} = \frac{(3\times 0.20)(60,000)}{1500\times 4.5\times 3} = 1.8$$

where A_{tr} is for 3-No. 4 bars, s (the maximum spacing of transverse reinforcement within ℓ_d) = 4.5 in., and n (number of bars being developed) = 3

$$\frac{c+K_{tr}}{d_b} = \frac{2.5+1.8}{1.0} = 4.3 > 2.5, \text{ use } 2.5$$

Therefore,

$$\ell_d = \frac{3}{40}\times\frac{60,000}{\sqrt{4000}}\times\frac{1.0\times 1.0\times 1.0\times 1.0}{2.5}\times 1.0 = 28.5 \text{ in.}$$

Class B splice length = $1.3 \times 28.5 = 37.1$ in.

Use a 3 ft-2 in. splice length.

7. Reinforcement details for the column are shown in Fig. 29-33. Note that for practical purposes, a 4-in. hoop spacing is used over the entire length of the column.

Example 29.3 (cont'd)	Calculations and Discussion	Code Reference

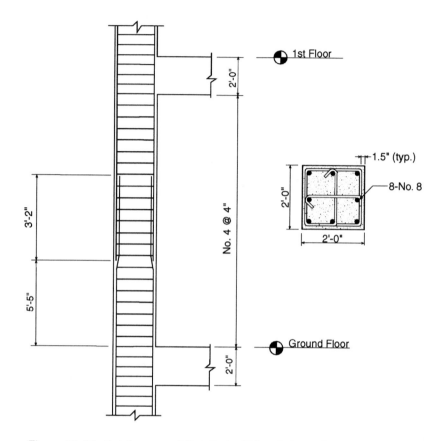

Figure 29-33 Reinforcement Details for Edge Column Supporting Level 1

Example 29.4—Proportioning and Detailing of Exterior Beam-Column Connection of Building in Example 29.1

Determine the transverse reinforcement and shear strength requirements for an exterior beam-column connection between the beam considered in Example 29.2 and the column of Example 29.3. Assume the joint to be located at the first floor level.

	Code
Calculations and Discussion	**Reference**

1. Transverse reinforcement for confinement.

   Section 21.5.2.1 requires the same amount of confinement reinforcement within the joint as for the length ℓ_o at column ends, unless the joint is confined by beams framing into all vertical faces of the column. A member that frames into a face is considered to provide confinement if at least three-quarters of the face of the joint is covered by the framing member.

   In the case of the beam-column joint considered here, beams frame into only three sides of the column. In Example 29.3, confinement requirements at column ends were satisfied by No. 4 hoops with crossties spaced at 4 in.

2. Check shear strength of joint in E-W direction.

   The shear force across section x-x (see Fig. 29-34) of the joint is obtained as the difference between the tensile force from the top flexural reinforcement of the framing beam (stressed to $1.25f_y$) and the horizontal shear from the column above.   21.5.1.1

   $T = A_s(1.25f_y) = (5 \times 0.79)(1.25 \times 60) = 296$ kips

   An estimate of the horizontal shear from the column, V_h, can be obtained by assuming that the beams in the adjoining floors are also deformed so that plastic hinges form at their junctions with the column, with M_{pr} (beam) = 477.0 ft-kips (see Fig. 29-30). By further assuming that the end moments in the beams are resisted by the columns above and below the joint inversely proportional to the column lengths, the average horizontal shear in the column is approximately:

   $$V_h = \frac{2 \times 477.0}{12 + 16} = 34.1 \text{ kips}$$

   Thus, the net shear at section x-x of the joint is $V_u = 296 - 34.1 = 261.9$ kips. Section 21.5.3.1 gives the nominal shear strength of a joint as a function of the area of the joint cross-section, A_j, and the degree of confinement by framing beams. For the joint confined on three faces considered here (note: beam width = 20 in. > 0.75 (column width) = 0.75 × 24 = 18 in.):

   $\phi V_c = \phi 15\sqrt{f'_c}\, A_j$ 21.5.3.1
                                                              9.3.4

         $= 0.85 \times 15 \sqrt{4000} \times 24^2 / 1000 = 464.5$ kips > 261.9 kips   O.K.

Example 29.4 (cont'd)   **Calculations and Discussion**   **Code Reference**

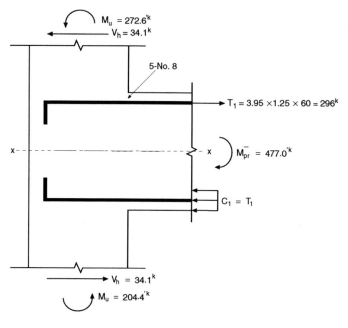

Figure 29-34 Shear Analysis of Exterior Beam-Column Joint in E-W Direction

3. Check shear strength of joint in N-S direction.

    The shear force across section x-x (see Fig. 29-35) of the joint is determined as follows:

    $T_1 = A_s(1.25f_y) = (4 \times 0.60)(1.25 \times 60) = 180$ kips

    The negative and positive probable flexural strengths at the joint are 302.6 ft-kips (4-No. 7 bars top and bottom).

    The average horizontal shear in the column is approximately:

    $$V_u = \frac{2(302.6 + 302.6)}{12 + 16} = 43.2 \text{ kips}$$

    Thus, the net shear at section x-x of the joint is

    $V_u = T_1 + C_2 - V_u = 180 + 180 - 43.2 = 316.8$ kips $< \phi V_c = 464.5$ kips O.K.

29-70

Example 29.4 (cont'd)   **Calculations and Discussion**   **Code Reference**

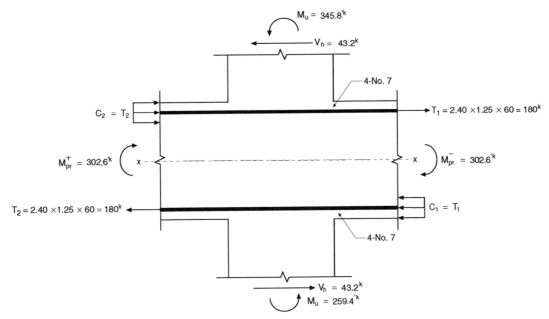

Figure 29-35 Shear Analysis of Exterior Beam-Column Joint in N-S Direction

Note that if the shear strength of the concrete in the joint as calculated above were inadequate, any adjustment would have to take the form of either an increase in the column cross-section (and hence A_j) or an increase in the beam depth (to reduce the amount of flexural reinforcement required and hence the tensile force T) since transverse reinforcement is considered not to have a significant effect on shear strength.

4. Reinforcement details for the exterior joint are shown in Fig. 29-36.

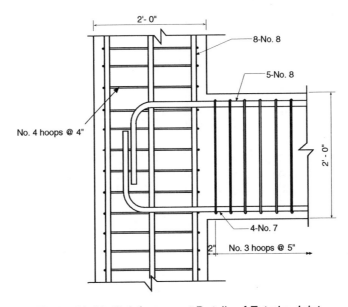

Figure 29-36 Reinforcement Details of Exterior Joint

Example 29.5—Proportioning and Detailing of Interior Beam-Column Connection of Building in Example 29.1

Determine the transverse reinforcement and shear strength requirements for the interior beam-column connection at the first floor of the interior E-W frame considered in the previous examples. The column is 30-in. square and is reinforced with 12-No. 8 bars. The beams have dimensions of b = 20 in. and d = 21.5 in. and are reinforced as noted in Example 29.2 (see Fig. 29-32).

Calculations and Discussion	Code Reference

1. Determine transverse reinforcement requirements.

   a. Confinement reinforcement

   Maximum allowable spacing of rectangular hoops assuming No. 4 hoops with two crossties in both directions:

   s_{max} = 0.25 (least dimension of column) = 0.25 × 30 = 7.5 in.   21.4.4.2

   = 6 (diameter of longitudinal bar) = 6 × 1.00 = 6 in.

   $$= s_o = 4 + \left(\frac{14 - h_x}{3}\right) = 4 + \left(\frac{14 - 9.83}{3}\right) = 5.4 \text{ in.} < 6 \text{ in.} \quad \text{(governs)}$$

   $$> 4 \text{ in.}$$

   where $h_x = \dfrac{30 - 2\left(1.5 + 0.5 + \dfrac{1.00}{2}\right)}{3} + 2\left(\dfrac{1.00}{2} + \dfrac{1}{4}\right) = 9.83$ in.

   With a hoop spacing of 5 in., the required cross-sectional area of confinement reinforcement in the form of hoops is:

   $$A_{sh} \geq \begin{cases} 0.3 sb_c \left(\dfrac{A_g}{A_{ch}} - 1\right)\dfrac{f'_c}{f_{yt}} = (0.3 \times 5 \times 26.5)\left(\dfrac{900}{729} - 1\right)\dfrac{4000}{60,000} = 0.62 \text{ in.}^2 & \text{Eq. (21-3)} \\ \\ 0.09 sb_c \dfrac{f'_c}{f_{yt}} = (0.09 \times 5 \times 26.5)\dfrac{4000}{60,000} = 0.80 \text{ in.}^2 \quad \text{(governs)} & \text{Eq. (21-4)} \end{cases}$$

   Since the joint is framed by beams having widths = 20 in. < 3/4 width of column = 22.5 in.   21.5.2.2
   on all four sides, it is not considered confined and a 50% reduction in the amount of
   confinement reinforcement indicated above is not allowed.

   No. 4 hoops spaced at 5 in. on center provide A_{sh} = 0.20 × 4 = 0.80 in.^{2}

   b. Transverse reinforcement for shear

   Following the same procedure in Example 29.4, the shear forces in the column are obtained for seismic forces in the E-W and N-S directions.

| Example 29.5 (cont'd) | Calculations and Discussion | Code Reference |

The largest probable flexural strength that may develop in the column can conservatively be assumed to correspond to the balanced point of the column interaction diagram.

With the strength reduction factor equal to 1.0 and $f_y = 1.25 \times 60 = 75$ psi, the moment corresponding to balanced failure is 1438 ft-kips. Thus, $V_u = (2 \times 1{,}438)/14 = 205.4$ kips.

The shear force need not exceed that determined from joint strengths based on the M_{pr} of the beams framing into the joint.

21.4.5.1

For seismic forces in the E-W direction, M_{pr}^- of the beam framing into the joint at the face of the interior column is 477.0 ft-kips (5-No. 8 top bars). The M_{pr}^+ is 302.6 ft-kips (4-No. 7 bottom bars) based on the beam framing into the other face of the joint. Distributing the moment to the columns in proportion to $1/\ell$, the moment at the top of the first story column is:

$$(477.0 + 302.6)\left(\frac{12}{12 + 16}\right) = 334.1 \text{ ft-kips}$$

It is possible for the base of the first story column to develop M_{pr} of the column. Thus, the shear force is:

$$V_u = \frac{334.1 + 1438}{14} = 334.1 \text{ ft-kips}$$

For seismic forces in the N-S direction, M_{pr}^- of the beam is 302.6 ft-kips (4-No. 7 top bars) and M_{pr}^+ is 302.6 ft-kips (4-No. 7 bottom bars). Therefore, at the top of the first story column, the moment is approximately:

$$(2 \times 302.6)\left(\frac{12}{12 + 16}\right) = 259.4 \text{ ft-kips}$$

The shear force is:

$$V_u = \frac{259.4 + 1438}{14} = 121.2 \text{ kips}$$

Both of these shear forces are greater than those obtained from analysis.

Since the factored axial forces are greater than $A_g f_c'/20 = 180$ kips, the shear strength of the concrete may be used:

21.4.5.2

$$V_c = 2\sqrt{f_c'}\,bd\left(1 + \frac{N_u}{2000\,A_g}\right)$$

Eq. (11-4)

29-73

Example 29.5 (cont'd) — Calculations and Discussion — Code Reference

Conservatively using the minimum axial force on the column:

$$V_c = \frac{2\sqrt{4000}\,(30 \times 27.5)}{1000}\left[1 + \frac{1{,}192{,}700}{2000 \times (30)^2}\right] = 173.5 \text{ kips}$$

$\phi V_c = 0.75 \times 173.5 = 130.1$ kips $> V_u = 126.6$ kips O.K.

Thus, the transverse reinforcement spacing over the distance $\ell_o = 28$ in. near the column ends required for confinement is also adequate for shear.

Use No. 4 hoops and crossties spaced at 5 in. at the ends of the column.

2. Check shear strength of joint in E-W direction

Following the same procedure as used in Example 29.4, the forces affecting the horizontal shear across a section near mid-depth of the joint shown in Fig. 29-37 are obtained.

Net shear force across section x-x $= T_1 + C_2 - V_h = 296.3 + 180.0 - 55.7 = 420.6$ kips $= V_u$

Shear strength of joint, noting that the joint is not confined on all faces is (i.e., beam width = 20 in. < 0.75 (column width) = 0.75 × 30 = 22.5 in.):

$$\phi V_c = \phi 12\sqrt{f'_c}\, A_j$$

$\qquad = 0.85 \times 12\sqrt{4000} \times 30^2 / 1000 = 580.6$ kips > 420.6 kips   O.K.   21.5.3.1

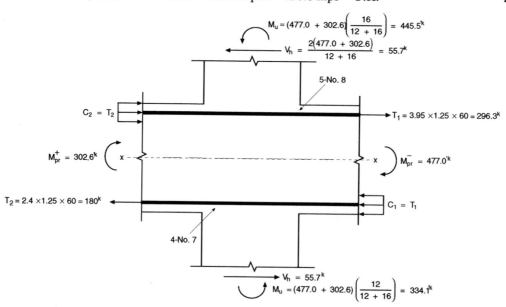

Figure 29-37 Shear Analysis of Interior Beam-Column Joint in E-W Direction

		Code
Example 29.5 (cont'd)	**Calculations and Discussion**	**Reference**

3. Check shear strength of joint in N-S direction

    At both the top and the bottom of the beam, 4-No. 7 bars are required (M_{pr} = 302.6 ft-kips).

    Net shear force across section x-x = $T_1 + C_2 - V_u$ = 180 + 180 - 43.2 = 316.8 kips = V_u

    where $T_1 = C_2 = 2.4 \times 1.25 \times 60 = 180$ kips

    $V_u = 2(302.6 + 302.6)/(12 + 16) = 43.2$ kips

    $\phi V_c = \phi 12\sqrt{f'_c} A_j = 580.6$ kips $> V_u = 316.8$ kips O.K.

Example 29.6—Proportioning and Detailing of Structural Wall of Building in Example 29.1

Design the wall section at the first floor level of the building in Example 29.1. At the base of the wall, $M_u = 49{,}142$ ft-kips and $V_u = 812$ kips.

Calculations and Discussion	Code Reference

1. Determine minimum longitudinal and transverse reinforcement requirements in the wall.

   a. Check if two curtains of reinforcement are required.

   Two curtains of reinforcement shall be provided in a wall if the in-plane factored shear force assigned to the wall exceeds $2A_{cv}\sqrt{f'_c}$, where A_{cv} is the cross-sectional area bounded by the web thickness and the length of section in the direction of the shear force considered. **21.7.2.2**

   $2A_{cv}\sqrt{f'_c} = 2 \times 18 \times 24.5 \times 12 \times \sqrt{4000}/1000 = 669$ kips $< V_u = 812$ kips

   Therefore, two curtains of reinforcement are required.

   Note that $V_u = 812$ kips $<$ upper limit on shear strength $= \phi 8 A_{cv}\sqrt{f'_c} = 2{,}008$ kips   O.K.   **21.7.4.4**

   b. Required longitudinal and transverse reinforcement in wall.

   Minimum distributed web reinforcement ratios $= 0.0025$ with max. spacing $= 18$ in.   **21.7.2.1**

   With A_{cv} (per foot of wall) $= 18 \times 12 = 216$ in.^{2}, minimum required area of reinforcement in each direction per foot of wall $= 0.0025 \times 216 = 0.54$ in.^{2}/ft

   Assuming No. 5 bars in two curtains ($A_s = 2 \times 0.31 = 0.62$ in.^{2}), required spacing is

   $$s = \frac{0.62}{0.54} \times 12 = 13.8 \text{ in.} < 18 \text{ in.}$$

3. Determine reinforcement requirements for shear.   **21.7.4**

   Assume two curtains of No. 5 bars spaced at 12 in. on center. Shear strength of wall:

   $$\phi V_n = \phi A_{cv}\left(\alpha_c \sqrt{f'_c} + \rho_t f_y\right)$$ Eq. (21-7)

   where $\phi = 0.75$ and $\alpha_c = 2.0$ for $h_w/\ell_w = 148/24.5 = 6 > 2$

   $A_{cv} = 18 \times 24.5 \times 12 = 5292$ in.^{2}

   $$\rho_t = \frac{0.62}{18 \times 12} = 0.0029$$

   $\phi V_n = (0.75 \times 5292)[2\sqrt{4000} + (0.0029 \times 60{,}000)]/1000 = 1193$ kips > 812 kips   O.K.

		Code
Example 29.6 (cont'd)	**Calculations and Discussion**	**Reference**

Therefore, use two curtains of No. 5 bars spaced at 12 in. on center in the horizontal direction.

The reinforcement ratio ρ_ℓ shall not be less than the ratio ρ_t when the ratio h_w / ℓ_w is less than 2.0. Since h_w / ℓ_w is equal to 6.0, the minimum reinforcement ratio will be used.   21.7.4.3

Use 2 curtains of No. 5 bars spaced at 12 in. on center in the vertical direction.

3. Determine reinforcement requirements for combined flexural and axial loads.

   Structural walls subjected to combined flexural and axial loads shall be designed in accordance with 10.2 and 10.3 except that 10.3.6 and the nonlinear strain requirements of 10.2.2 do not apply.   21.7.5.1

   Assume that each 30 × 30 in. column at the end of the wall is reinforced with 24-No. 11 bars. It was determined above that 2-No. 5 bars at a spacing of 12 in. are required as vertical reinforcement in the web. With this reinforcement, the wall is adequate to carry the factored load combinations per 9.2.

4. Determine if special boundary elements are required.

   The need for special boundary elements at the edges of structural walls shall be evaluated in accordance with 21.7.6.2 or 21.7.6.3. The provisions of 21.7.6.2 are used in this example.   21.7.6.1

   Compression zones shall be reinforced with special boundary elements where

   $$c \geq \frac{\ell_w}{600(\delta_u / h_w)}, \quad \delta_u / h_w \geq 0.007$$ Eq. (21-8)

   In this case, ℓ_w = 24.5 ft = 294 in., h_w = 148 ft = 1776 in., δ_u = 13.5 in. and δ_u / h_w = 0.0076 > 0.007. Therefore, special boundary elements are required if c is greater than or equal to 294/(600 × 0.0076) = 64.5 in.

   The distance c to be used in Eq. (21-8) is the largest neutral axis depth calculated for the factored axial force and nominal moment strength consistent with the design displacement δ_u. From a strain compatibility analysis, the largest c is equal to 68.1 in. corresponding to an axial load of 3649 kips and nominal moment strength of 97,302 ft-kips, which is greater than 64.5 in. Thus, special boundary elements are required.

   The special boundary element shall extend horizontally from the extreme compression fiber a distance not less than $c - 0.1\ell_w$ = 68.1 – (0.1 × 294) = 38.7 in. (governs) or c/2 = 68.1/2 = 34.1 in. Considering the placement of the vertical bars in the web, confine 45 in. at both ends of the wall.   21.7.6.4(a)

5. Determine special boundary element transverse reinforcement.

   Transverse reinforcement shall satisfy the requirements of 21.4.4.1 through 21.4.4.3 except Eq. (21-3) need not be satisfied.   21.7.6.4(c)

Example 29.6 (cont'd)	Calculations and Discussion	Code Reference

- Confinement of 30 × 30 in. boundary elements

  Maximum allowable spacing of rectangular hoops assuming No. 4 hoops and crossties around every longitudinal bar in both directions of the 30 × 30 in. boundary elements:

  s_{max} = 0.25 (minimum member dimension) = 0.25 × 30 = 7.5 in.
  = 6 (diameter of longitudinal bar) = 6 × 1.41 = 8.5 in.
  = $s_o = 4 + \left(\frac{14 - h_x}{3}\right) = 4 + \left(\frac{14 - 6.0}{3}\right)$ = 6.7 in. ⩾ 6.0 in.; use 6 in. (governs)

  where h_x = maximum horizontal spacing of hoop or crosstie legs on all faces of the 30 × 30 in. boundary element.

  Required cross-sectional area of transverse reinforcement in the 30 × 30 in. boundary elements, assuming s = 6.0 in.:

  $$A_{sh} = \frac{0.09 s h_c f'_c}{f_y} = \frac{0.09 \times 6.0 \times [30 - (2 \times 1.5) - 0.5] \times 4}{60} = 0.95 \text{ in.}^2$$

  No. 4 hoops with crossties around every longitudinal bar in the 30 × 30 in. boundary elements provide A_{sh} = 7 × 0.2 = 1.40 in.^{2} > 0.95 in.^{2}

- Confinement of web

  Maximum allowable spacing of No. 5 transverse reinforcement:

  s_{max} = 0.25 (minimum member dimension) = 0.25 × (45 - 30) = 3.75 in. (governs)
  = 6 (diameter of longitudinal bar) = 6 × 0.625 = 3.75 in.
  = $s_o = 4 + \left(\frac{14 - 13.25}{3}\right)$ = 4.25 in.

  For confinement in the direction parallel to the wall, assuming s = 3.0 in.:

  b_c = 18 - (2 × 1.5) - 0.625 = 14.375 in.

  $$A_{sh} = \frac{0.09 \times 3.0 \times 14.375 \times 4}{60} = 0.26 \text{ in.}^2$$

  Using 2-No. 5 horizontal bars, A_{sh} = 2 × 0.31 = 0.62 in.^{2} > 0.26 in.^{2}

  For confinement in the direction perpendicular to the wall:

  b_c = 45 - 30 = 15 in.

  $$A_{sh} = \frac{0.09 \times 3.0 \times 15 \times 4}{60} = 0.27 \text{ in.}^2$$

  With a No. 5 hoop and crosstie, A_{sh} = 2 × 0.31 = 0.62 in.^{2} > 0.27 in.^{2}

	Code
Example 29.6 (cont'd) **Calculations and Discussion**	**Reference**

The transverse reinforcement of the boundary element shall extend vertically a distance of $\ell_w = 24.5$ ft (governs) or $M_u/4V_u = 49{,}142/(4 \times 812) = 15.1$ ft from the critical section.   21.7.6.2(b)

6. Determine required development and splice lengths.

    Reinforcement in structural walls shall be developed or spliced for f_y in tension in accordance with Chapter 12, except that at locations where yielding of longitudinal reinforcement is likely to occur as a result of lateral displacements, development of longitudinal reinforcement shall be 1.25 times the values calculated for f_y in tension.   21.7.2.3

    a. Lap splice for No. 11 vertical bars in boundary elements.*

        Class B splices are designed for the No. 11 vertical bars.

        Required length of Class B splice = $1.3\ell_d$ 12.15.1

        where

        $$\ell_d = \left(\frac{3}{40} \frac{f_y}{\sqrt{f'_c}} \frac{\psi_t \psi_e \psi_s \lambda}{\left(\frac{c_b + K_{tr}}{d_b}\right)} \right) d_b$$ Eq. (12-1)

        $\psi_t = 1.3$ for top bars; a = 1.0 for other bars   12.2.4

        $\psi_e = 1.0$ for uncoated bars

        $\psi_s = 1.0$ for No. 7 and larger bars

        $\lambda = 1.0$ for normal weight concrete (1.3 for lightweight concrete)

        Assume no more than 50% of the bars spliced at any one location.

        $c = 1.5 + 0.5 + \dfrac{1.41}{2} = 2.7$ in. (governs)

        $= \dfrac{1}{2}\left[\dfrac{30 - 2(1.5 + 0.5) - 1.41}{3}\right] = 4.1$ in.

        $K_{tr} = \dfrac{A_{tr} f_{yt}}{1500 sn} = \dfrac{(4 \times 0.20)(60{,}000)}{1500 \times 6.0 \times 4} = 1.3$

        where A_{tr} is for 4-No. 5 bars, s = 6.0 in., and n (number of bars being developed) = 4 in one layer at one location.

        $\dfrac{(c_b + K_{tr})}{d_b} = \dfrac{(2.7 + 1.3)}{1.41} = 2.8 > 2.5$, use 2.5

---

* The use of mechanical connectors may be considered as an alternative to lap splices for these large bars.

Example 29.6 (cont'd)	Calculations and Discussion	Code Reference

Therefore,

$$\ell_d = \frac{3}{40} \times \frac{1.25 \times 60{,}000}{\sqrt{4000}} \times \frac{1.0}{2.5} \times 1.41 = 50.2 \text{ in.}$$

Class B splice length $= 1.3 \times 50.2 = 65.3$ in.

Use a 5 ft-6 in. splice length.

Note that splices beyond the first story can be 25% shorter, or 4 ft-6 in. long, as long as the same reinforcement continues.

b. Lap splice for No. 5 vertical bars in wall web.

Again assuming no more than 50% of bars spliced at any one location, the length of the Class B splice is determined as follows:

$$\ell_d = \left(\frac{3}{40} \frac{f_y}{\sqrt{f'_c}} \frac{\psi_t \psi_e \psi_s \lambda}{\left(\frac{c_b + K_{tr}}{d_b} \right)} \right) d_b \qquad \text{Eq. (12-1)}$$

reinforcement location factor $\psi_t = 1.0$ (other than top bars) 12.2.4

coating factor $\psi_e = 1.0$ (uncoated bars)

reinforcement size factor $\psi_s = 0.8$ (No. 6 and smaller bars)

lightweight aggregate concrete factor $\lambda = 1.0$ (normal weight concrete)

$c = 0.75 + 0.625 + \dfrac{0.625}{2} = 1.7$ in. (governs)

$= \dfrac{1}{2} \times 12 = 6$ in.

$K_{tr} = 0$

$\dfrac{(c_b + K_{tr})}{d_b} = \dfrac{1.7}{0.625} = 2.7 > 2.5$, use 2.5

Therefore,

$$\ell_d = \frac{3}{40} \times \frac{1.25 \times 60{,}000}{\sqrt{4000}} \times \frac{0.8}{2.5} \times 0.625 = 17.8 \text{ in.}$$

Example 29.6 (cont'd)   **Calculations and Discussion**   **Code Reference**

Class B splice length = $1.3 \times 17.8 = 23.1$ in.

Use a 2 ft-0 in. splice length.

Although all the No. 5 bars will not yield at the base, it is simpler to base the splice lengths of all No. 5 bars on possible yielding. Beyond the first story, the splice lengths may be reduced to 1 ft-8 in.

c. Development length for No. 5 horizontal bars in wall assuming no hooks are used within boundary element.   21.5.4

$$\ell_d = \left[\frac{3}{40} \frac{1.25 f_y}{\sqrt{f_c'}} \frac{\psi_t \psi_e \psi_s \lambda}{\left(\frac{c+K_{tr}}{d_b}\right)} \right] d_b \qquad \text{Eq. (12-1)}$$

Since it is reasonable to assume that the depth of concrete cast in one lift beneath a horizontal bar will be greater than 12 in.,

reinforcement factor $\psi_t = 1.0$

coating factor $\psi_e = 1.0$ (uncoated bars)

reinforcement size factor $\psi_s = 0.8$ (No. 6 and smaller bars)

lightweight aggregate concrete factor $\lambda = 1.0$ (normal weight concrete)

$c = 0.75 + \dfrac{0.625}{2} = 1.06$ in. (governs)

$ = \dfrac{1}{2} \times 12 = 6$ in.

$K_{tr} = 0$

$\dfrac{(c_b + K_{tr})}{d_b} = \dfrac{1.06}{0.625} = 1.7 < 2.5$

Therefore,

$\ell_d = \dfrac{3}{40} \times \dfrac{1.25 \times 60{,}000}{\sqrt{4000}} \times \dfrac{1.3 \times 0.8}{1.7} \times 0.625 = 34.0$ in.

This length cannot be accommodated within the confined core of the boundary element, thus hooks are needed.

Anchor horizontal bars to longitudinal reinforcement in boundary element.   21.7.6.4(e)

No lap splices would be required for the No. 5 horizontal bars (full length bars weigh approximately 25 lbs. and are easily installed).

Example 29.6 (cont'd)   **Calculations and Discussion**   **Code Reference**

7. Reinforcement details for structural wall are shown in Fig. 29-38.

   Note that the No. 5 bars at 3 in. that are required for confinement in the direction parallel to the web are developed into the boundary element and into the web beyond the face of the 2 ft-6 in. boundary element [see Fig. 29-38(b)].

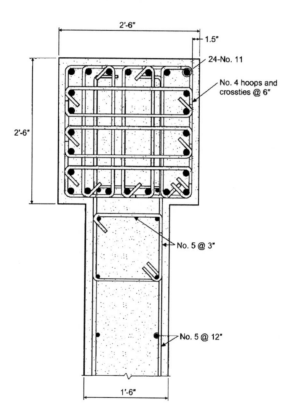

Figure 29-38(a) Reinforcement Details for the Structural Wall

Example 29.7—Design of 12-Story Precast Frame Building using Strong Connections*

This example illustrates the design and detailing requirements for typical beam-to-beam, column-to-column, and beam-to-column connections for the precast building shown in Fig. 29-39. In particular, details are developed for: (1) a strong connection near midspan of an interior beam that is part of an interior frame on the third floor level, (2) a column-to-column connection at mid-height between levels 2 and 3 of an interior column stack that is part of an interior frame, and (3) a strong connection at the interface between an exterior beam at the second floor level of an exterior frame and the continuous corner column to which it is connected. Pertinent design data are as follows:

Material Properties:

Concrete (w_c = 150 pcf): f'_c = 6000 psi for columns in the bottom six stories
    = 4000 psi elsewhere

Reinforcement: f_y = 60,000 psi

Service Loads:

Live load = 50 psf
Superimposed dead load = 42.5 psf

Member Dimensions:

Beams in N-S direction: 24 × 26 in.
Beams in E-W direction: 24 × 20 in.
Columns: 24 × 24 in.
Slab: 7 in.

Calculations and Discussion	Code Reference

1. Seismic design forces

   The computation of the seismic design forces is beyond the scope of this example. Traditional analysis methods can be used for precast frames, although care should be taken to approximate the component stiffness in a way that is appropriate for the precast components being used. For emulation design (as illustrated in this example), it is reasonable to model the beams and columns as if they were monolithic concrete.

2. Strong connection near beam midspan

   a. Required flexural reinforcement

   Special moment frames with strong connections constructed using precast concrete shall satisfy all requirements for special moment frames constructed with cast-in-place concrete, in addition to the provisions of 21.6.2.   21.6.2

   The required reinforcement for the beams on the third floor level is shown in Table 29-8. The design moments account for all possible load combinations per 9.2.1, and the provided areas of steel are within the limits specified in 21.3.2.1. Also given in Table 29-8 are the flexural moment strengths ϕM_n at each section. Note that at each location, the section is tension-controlled, so that ϕ = 0.9.   9.3.2.1

---

*This example has been adapted from: Ghosh, S.K., Nakaki, S.D., and Krishnan, K., "Precast Structures in Regions of High Seismicity: 1997 UBC Design Provisions", PCI Journal, Vol. 42, No. 6, November-December 1997, pp. 76-93.

| Example 29.7 (cont'd) | Calculations and Discussion | Code Reference |

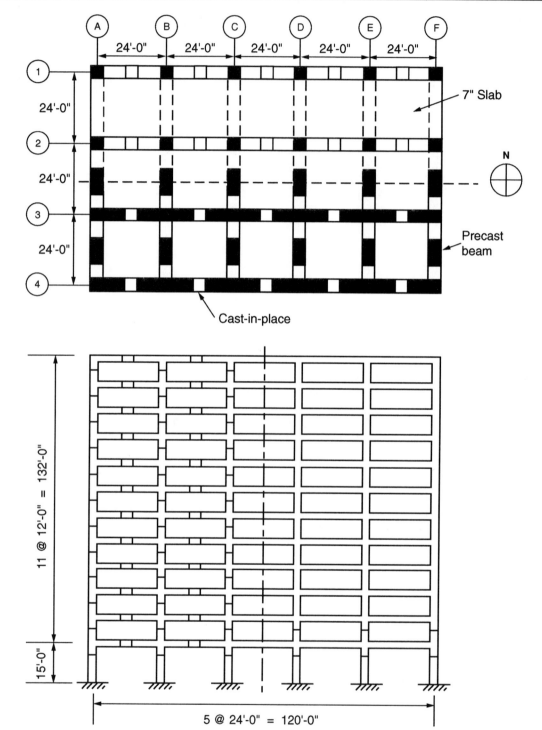

Fig 29-39 Example Building

	Code
Example 29.7 (cont'd) **Calculations and Discussion**	**Reference**

Table 29-8 Required Reinforcement for E-W Third Floor Beam

Location	M_u (ft-kips)	A_s* (in.^{2})	Reinforcement	ϕM_n (ft-kips)
Supports	-510.8	7.79	8-No. 9	-522.0
	+311.7	4.38	5-No. 9	+351.0
Interior	+63.0	1.40	2-No. 9	+150.3

*Max. A_s = 0.025 × 24 × 17.44 = 10.46 in.^{2} (21.3.2.1)
Min. A_s = 3$\sqrt{4000}$ × 24 × 17.44/60,000 = 1.32 in.^{2} (10.5.1)
        = 200 × 24 × 17.44/60,000 = 1.40 in.^{2} (governs)

Three No. 9 bars are made continuous at the top and bottom throughout the spans, providing negative and positive design moment strengths of 220.6 ft-kips.

Check provisions of 21.3.2.2 for moment strength along span of beam:

At the supports, ϕM_n^+ (5-No. 9) = 351.0 ft-kips > $\phi M_n^-/2$ = 261.0 ft-kips   O.K.   21.3.2.2

At other sections, ϕM_n (3-No. 9) = 220.6 ft-kips > $\phi M_n^-/4$ = 130.5 ft-kips   O.K.

b. Lap splice length   12.2.1

Lap splices of flexural reinforcement must not be placed within a joint, within a distance   21.3.2.3
2h from faces of supports or within regions of potential plastic hinging. Note that all lap splices must be confined by hoops or spirals with a maximum spacing or pitch of d/4 = 4.4 in. or 4 in. (governs) over the length of the lap. Lap splice lengths will be determined for the No. 9 top and bottom bars.

$$\ell_d = \left(\frac{3}{40} \frac{f_y}{\sqrt{f'_c}} \frac{\psi_t \psi_e \psi_s \lambda}{\frac{c_b + K_{tr}}{d_b}} \right) d_b \qquad \text{Eq. (12-1)}$$

where $\dfrac{c + K_{tr}}{d_b} \leq 2.5$

ψ_t = 1.3 for top bars; α = 1.0 for other bars   12.2.4

ψ_e = 1.0 for uncoated bars

ψ_s = 1.0 for No. 7 and larger bars

λ = 1.0 for normal weight concrete (1.3 for lightweight concrete)

Example 29.7 (cont'd) — Calculations and Discussion — Code Reference

$$c = 1.5 + 0.5 + \frac{1.128}{2} = 2.56 \text{ in. (governs)}$$

$$= \frac{24 - 2(1.5 + 0.5) - 1.128}{2 \times 2} = 4.72 \text{ in.}$$

$$\frac{c}{d_b} = \frac{2.56}{1.128} = 2.27, \text{ which makes it reasonable to take } \frac{c_b + K_{tr}}{d_b} = 2.5$$

Thus, for top bars:

$$\ell_d = \frac{3}{40} \frac{60,000}{\sqrt{4000}} \frac{1.3}{2.5} d_b = 37 d_b = 37 \times 1.128 = 41.7 \text{ in.}$$

For bottom bars:

$$\ell_d = \frac{3}{40} \frac{60,000}{\sqrt{4000}} \frac{1.0}{2.5} d_b = 28.5 d_b = 28.5 \times 1.128 = 32.2 \text{ in.}$$

Note that 2-No. 9 top bars are adequate in the interior of the span, i.e., ϕM_n(2-No. 9) = 150.3 ft-kips > $\phi M_n^- / 4$ = 130.5 ft-kips. Thus, the top bar development length can be reduced by an excess reinforcement factor of (A_s required/A_s provided) = 2/3:
12.2.5

$$\ell_d = 2/3 \times 41.7 = 27.8 \text{ in.}$$

Since all of the reinforcement is spliced at the same location, Type B splices are to be used for both the top and bottom bars.
12.15.2

Type B splice length = 1.3 × 32.2 = 41.9 in. > 12 in. for the bottom bar
12.15.1

Provide 3 ft-6 in. splice length for both the top and the bottom bars.

c. Reinforcing bar cutoff points

For the purpose of determining the cutoff points for the reinforcement, a moment diagram corresponding to the probable moment strengths at the beam ends and 0.9 times the dead load on the span will be used, since this will result in the longest bar lengths. The cutoff point for 5 of the 8-No. 9 bars at the top will be determined.

Determine probable moment strengths M_{pr}^+ and M_{pr}^- with $f_s = 1.25 f_y = 75$ ksi and $\phi = 1.0$, ignoring compression steel.
21.0

For 5-No.9 bottom bars:

$$a = \frac{A_s f_s}{0.85 f_c' b} = \frac{5 \times 75}{0.85 \times 4 \times 24} = 4.6 \text{ in.}$$

$$M_{pr}^+ = A_s f_s \left(d - \frac{a}{2} \right) = (5 \times 75)\left(17.44 - \frac{4.6}{2} \right) / 12 = 473.1 \text{ ft-kips}$$

where d = 20 – 1.5 (clear cover) – 0.5 (diameter of No. 4 stirrup) – 0.564 (diameter of No. 9 bar/2) = 17.44 in.

Similarly, for 8-No. 9 top bars: $M_{pr}^- = 688.2$ ft-kips

| Example 29.7 (cont'd) | Calculations and Discussion | Code Reference |

Dead load on beam:

$$w_D = \left(\frac{7}{12} \times 0.150 \times 24\right) + (0.0425 \times 24) + \left(\frac{24 \times 13 \times 0.150}{144}\right) = 3.45 \text{ kips/ft at midspan}$$

$0.9 w_D = 0.9 \times 3.45 = 3.11$ kips/ft

The distance from the face of the interior support to where the moment under the loading considered equals ϕM_n(3-No. 9) = 220.6 ft-kips is readily obtained by summing moments about section A-A (see Fig. 29-40):

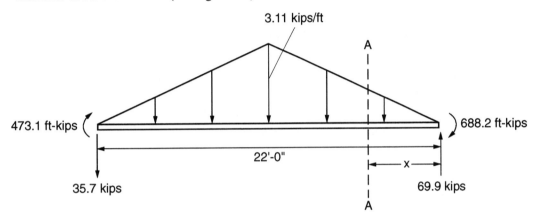

Fig. 29-40 Cutoff Location of Negative Bars

$$\frac{x}{2}\left(\frac{3.11x}{11}\right)\left(\frac{x}{3}\right) + 688.2 - 220.6 - 69.9x = 0$$

Solving for x gives a distance of 6.91 ft from the face of the support.

The 5-No. 9 bars must extend a distance d = 17.44 in. or $12d_b$ = 13.54 in. beyond the distance x. Thus, from the face of the support, the total bar length must be at least equal to 6.91 + (17.44/12) = 8.4 ft. Also, the bars must extend a full development length ℓ_d beyond the face of the support:

12.10.3

12.10.4

$$\ell_d = \left(\frac{3}{40} \frac{f_y}{\sqrt{f'_c}} \frac{\psi_t \psi_e \psi_s \lambda}{\frac{c_b + K_{tr}}{d_b}}\right) d_b$$

Eq. (12-1)

where $\dfrac{c + K_{tr}}{d_b} \leq 2.5$

$\psi_t = 1.3$ for top bars

12.2.4

$\psi_e = 1.0$ for uncoated bars

$\psi_s = 1.0$ for No. 7 and larger bars

$\lambda = 1.0$ for normal weight concrete

	Code
Example 29.7 (cont'd) **Calculations and Discussion**	**Reference**

$$c = 2.56 \text{ in. or } \frac{24 - 2(1.5 + 0.5) - 1.128}{2 \times 7} = 1.35 \text{ in. (governs)}$$

$K_{tr} = 0$ (conservative)

$$\frac{c + K_{tr}}{d_b} = \frac{1.35 + 0}{1.128} = 1.2$$

$$\ell_d = \frac{3}{40} \frac{60,000}{\sqrt{4000}} \frac{1.3}{1.2} d_b = 77 d_b = 77 \times 1.128 = 86.9 \text{ in.} = 7.2 \text{ ft} < 8.4 \text{ ft}$$

The total required length of the 5-No. 9 bars must be at least 8.4 ft beyond the face of the support.

Flexural reinforcement shall not be terminated in a tension zone unless one or more of the conditions of 12.10.5 are satisfied. In this case, the point of inflection is approximately 10.7 ft from the face of the right support, which is greater than 8.4 ft. The 5-No. 9 bars can not be terminated here unless one of the conditions of 12.10.5 is satisfied.

Check if the factored shear force V_u at the cutoff point does not exceed two-thirds of ϕV_n. In this region of the beam, it can be shown that No. 4 stirrups @ 8 in. are required. However, No. 4 stirrups @ 6 in. will be provided to satisfy 12.10.5.1.

$$\phi V_n = \phi(V_c + V_s) = 0.75 \times \left(2\sqrt{4000} \times 24 \times 17.44 + \frac{0.4 \times 60,000 \times 17.44}{6}\right)/1000 = 92.0 \text{ kips}$$

$$\frac{2}{3}\phi V_n = 61.3 \text{ kips} > V_u = 60.0 \text{ kips at 8.4 ft from face of support}$$

Since $2\phi V_n/3 > V_u$, the cutoff point for the 5-No. 9 bars can be 8.4 ft beyond the face of the interior support.

The cutoff point for 2 of the 5-No. 9 bottom bars can be determined in a similar fashion. These bars can be cut off at 8.4 ft from the face of the exterior support as well, which is short of the splice closure.

d. Check connection strength

    For strong connections: $\phi S_n \geq S_e$ 21.6.2(b)

    where S_n = nominal flexural or shear strength of the connection   21.0

    S_e = moment or shear at connection corresponding to development of probable strength at intended yield locations, based on the governing mechanism of inelastic lateral deformation, considering both gravity and earthquake load effects

    At the connection,

$$\phi V_n = \phi(V_c + V_s) = 0.75 \times \left(2\sqrt{4000} \times 24 \times 17.44 + \frac{0.4 \times 60,000 \times 17.44}{4}\right)/1000 = 118.2 \text{ kips}$$

		Code
Example 29.7 (cont'd)	**Calculations and Discussion**	**Reference**

Gravity load on beam:

$1.2w_D + 0.5w_L = (1.2 \times 3.45) + (0.5 \times 0.05 \times 24) = 4.74$ kips/ft Eq. (9-5)

Maximum shear force at connection due to gravity and earthquake load effects occurs at 9.125 ft from face of right support (see Fig. 29-41):

$$V_e = 78.9 - \left(\frac{1}{2} \times 9.125 \times \frac{4.74 \times 9.125}{11}\right) = 61.0 \text{ kips} < \phi V_n = 118.2 \text{ kips} \quad \text{O.K.}$$

At the connection, ϕM_n (3-No. 9) = 220.6 ft-kips

Maximum moment at connection due to gravity and earthquake load effects occurs at 9.125 ft from face of left support (see Fig. 29-41):

$$M_e = 473.1 - (26.8 \times 9.125) - \left(\frac{1}{2} \times 9.125\right)\left(\frac{4.74 \times 9.125}{11}\right)\left(\frac{1}{3} \times 9.125\right) = 174.0 \text{ ft-kips}$$

$< \phi M_n = 220.6$ ft-kips O.K.

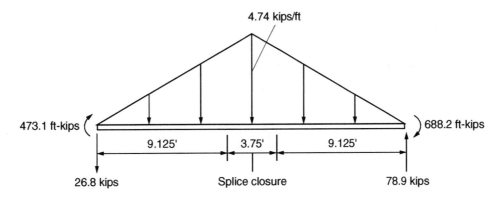

Fig. 29-41 Connection Strength

e. Reinforcement details

The reinforcement details for the beam are shown in Fig. 29-42.

Example 29.7 (cont'd)	Calculations and Discussion	Code Reference

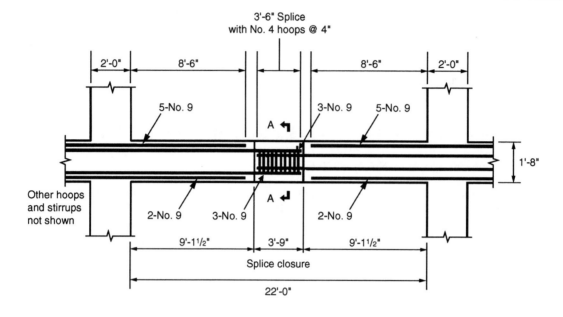

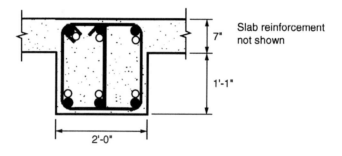

Section A-A

Fig. 29-42 Reinforcement Details for Beam-to-Beam Connection

3. Column-to-column connection at mid-height

   a. Determine required longitudinal reinforcement

   A summary of the design forces for the interior column between levels 2 and 3, which is part of an interior longitudinal frame, is contained in Table 29-9. The design forces account for all possible load combinations per 9.2.1.

Example 29.7 (cont'd)   **Calculations and Discussion**   **Code Reference**

Table 29-9 Design Forces for Interior Column between the Second and Third Floors

Load Combination	Axial load, P_u (kips)	Moment, M_u (ft-kips) Top	Moment, M_u (ft-kips) Bottom	Shear, V_u (kips)
1.2D + 1.6L	1402.6	-8.0	5.5	1.2
1.2D + 0.5L + E	1609.8	-408.3	467.5	70.7
1.2D + 0.5L - E	1195.4	392.3	-456.5	73.1
0.9D + E	1125.8	-405.5	466.3	71.1
0.9D - E	711.4	395.1	-457.8	72.7

It can be shown that 12-No. 10 bars are adequate for all load combinations.

Check longitudinal reinforcement ratio:

$$\rho_g = \frac{A_{st}}{bh} = \frac{12 \times 1.27}{24 \times 24} = 0.0265$$

$\rho_{min} = 0.01 < \rho_g = 0.0265 < \rho_{max} = 0.06$ O.K.   21.4.3.1

b. Nominal flexural strength of columns relative to that of beams

$$\Sigma M_c \text{ (columns)} \geq \frac{6}{5} \Sigma M_g \text{ (beams)}$$ 21.4.2.2

For the top end of the lower column framing into the joint between the second and the third floor levels, $M_n = 1182.3$ ft-kips, which corresponds to $P_u = 711.4$ kips. Similarly, for the bottom end of the upper column framing into the same joint, $M_n = 1168.4$ ft-kips, which corresponds to $P_u = 655.5$ kips.

Thus,

$\Sigma M_c = 1182.3 + 1168.4 = 2350.7$ ft-kips

The nominal negative flexural strength M_n^- of the beam framing into the column must include the slab reinforcement within an effective slab width equal to:

16 (slab thickness) + beam width = $(16 \times 7) + 24 = 136$ in.   8.10.2
Center-to-center beam spacing = $24 \times 12 = 288$ in.
Span/4 = $(24 \times 12)/4 = 72$ in. (governs)

The minimum required A_s in the 72-in. effective width = $0.0018 \times 72 \times 7 = 0.91$ in.^{2} which corresponds to 5-No. 4 bars @ 72/5 = 14.4 in. Since maximum bar spacing = 2h = 14 in., provide No. 4 @ 14 in. at both the top and the bottom of the slab (according to ACI 318 Fig. 13.3.8, 100 percent of both the top and the bottom reinforcement in the column strip must be continuous or anchored at the support).

Example 29.7 (cont'd)   Calculations and Discussion   Code Reference

From a strain compatibility analysis, $M_n^- = 736.0$ ft-kips and $M_n^+ = 459.0$ ft-kips

Thus,

$\Sigma M_g = 736.0 + 459.0 = 1195.0$ ft-kips

2350.7 ft-kips $> \dfrac{6}{5} \times 1195.0 = 1434.0$ ft-kips   O.K.   Eq. (21-1)

The intent of 21.4.2.2 is to prevent a story mechanism, rather than prevent local yielding in a column. The 6/5 factor is clearly insufficient to prevent column yielding if the adjacent beams both hinge. Therefore, confinement reinforcement is required in the potential hinge regions of a frame column.

c. Minimum connection strength

At column-to-column connections, $\phi M_n \geq 0.4 M_{pr}$ when bars are spliced within the middle third of the clear column height.   21.6.2(d)

For the column between the second and the third floor levels with $P_u = 711.4$ kips, it can be shown from a strain compatibility analysis that $M_{pr} = 1244.1$ ft-kips.

Also, as indicated above, $M_n = 1182.3$ ft-kips for $P_u = 711.4$ kips. From a strain compatibility analysis, $\varepsilon_t = 0.00223$, so that $\phi = 0.48 + (83 \times 0.00223) = 0.67$.   9.3.2.2

Therefore,

$\phi M_n = 0.67 \times 1182.3 = 792.1$ ft-kips $> 0.4 M_{pr} = 0.4 \times 1244.1 = 497.6$ ft-kips   O.K.

Splice all twelve bars at mid-height, as shown in Fig. 29-43.

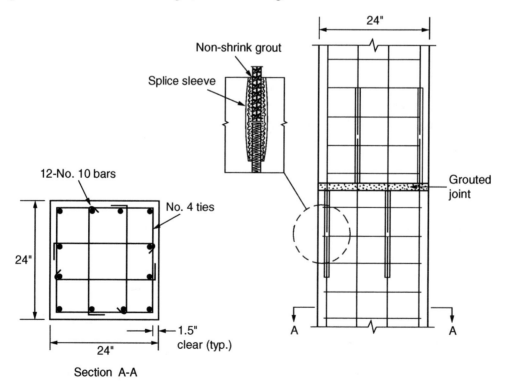

Fig. 29-43 Reinforcement Details for Column-to-Column Connection

		Code
Example 29.7 (cont'd)	**Calculations and Discussion**	**Reference**

4. Column-face strong connection in beam

   A strong connection is to be designed at the interface between a precast beam at the second floor level of the building that forms the exterior span of an exterior transverse frame and the continuous corner column to which it is connected.

   a. Required flexural reinforcement

   From the combined effects of gravity and earthquake forces, the required flexural reinforcement at the top of the beam is 5-No. 9 bars and is 4-No. 9 bars at the bottom. All possible load combinations of 9.2.1 were considered.

   b. Strength design of connection

   The beam-to-column connection similar to the one depicted in ACI 318 Fig. R21.6.2(c) will be provided.

   The strong connection must be designed for the probable moment strength of the beam plus the moment at the face of the column due the shear force at the critical section.   *21.6.2*

   Determine probable moment strengths M_{pr}^+ and M_{pr}^- with $f_s = 1.25 f_y = 75$ ksi and $\phi = 1.0$, ignoring compression steel.   *21.0*

   For 4-No. 9 bottom bars:

   $$a = \frac{A_s f_s}{0.85 f'_c b} = \frac{4 \times 75}{0.85 \times 4 \times 24} = 3.7 \text{ in.}$$

   $$M_{pr}^+ = A_s f_s \left(d - \frac{a}{2} \right) = (4 \times 75)\left(23.44 - \frac{3.7}{2} \right)/12 = 539.8 \text{ ft-kips}$$

   where d = 26 − 1.5 (clear cover) − 0.5 (diameter of No. 4 stirrup) − 0.564 (diameter of No. 9 bar/2) = 23.44 in.

   Similarly, for 5-No. 9 top bars: $M_{pr}^- = 660.6$ ft-kips

   Assuming a 2 ft-6 in. cast-in-place closure, the shear forces at the critical sections, and the moments at the connections can be determined for the two governing load combinations as follows (see Fig. 29-44).

   Load combination 1: U = 1.2D + 0.5L + E   *Eq. (9-5)*

   $$w_D = \left(\frac{7}{12} \times 0.150 \times 13 \right) + (0.0425 \times 13) + \left(\frac{24 \times 19 \times 0.150}{144} \right) = 2.17 \text{ kips/ft at midspan}$$

   $w_L = 0.05 \times 13 = 0.65$ kips/ft at midspan

   $w_{u, mid} = (1.2 \times 2.17) + (0.5 \times 0.65) = 2.93$ kips/ft

   and

   $w_{u, end} = 2.93 \times \frac{2.5}{11} = 0.67$ kips/ft

| Example 29.7 (cont'd) | Calculations and Discussion | Code Reference |

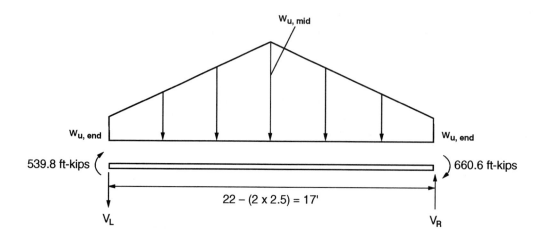

Fig. 29-44 Shear Forces at Critical Sections

From Fig. 29-44:

$$V_R(17) = \left(0.67 \times 17 \times \frac{17}{2}\right) + \left[\frac{1}{2} \times 17 \times (2.93 - 0.67) \times \frac{17}{2}\right] + 539.8 + 660.6$$

or, $V_R = 85.9$ kips

$$V_L = 85.9 - (0.67 \times 17) - \frac{1}{2}[(2.93 - 0.67) \times 17] = 55.3 \text{ kips}$$

$$M_{e,\ell}^+ = 539.8 + (55.3 \times 2.5) = 678.1 \text{ ft-kips}$$

$$M_{e,r}^- = 660.6 + (85.9 \times 2.5) = 875.4 \text{ ft-kips}$$

Load combination 2: U = 0.9D + E   Eq. (9-7)

$w_{u,mid} = 0.9 w_D = 0.9 \times 2.17 = 1.95$ kips/ft

and

$w_{u,end} = 1.95 \times \frac{2.5}{11} = 0.44$ kips/ft

From Fig. 29-44:

$$V_R(17) = \left(0.44 \times 17 \times \frac{17}{2}\right) + \left[\frac{1}{2} \times 17 \times (1.95 - 0.44) \times \frac{17}{2}\right] + 539.8 + 660.6$$

		Code
Example 29.7 (cont'd)	**Calculations and Discussion**	**Reference**

or, $V_R = 80.8$ kips

$$V_L = 80.8 - (0.44 \times 17) - \frac{1}{2}[(1.95 - 0.44) \times 17] = 60.5 \text{ kips}$$

$$M^+_{e,\ell} = 539.8 + (60.5 \times 2.5) = 691.1 \text{ ft-kips}$$

$$M^-_{e,r} = 660.6 + (80.8 \times 2.5) = 862.6 \text{ ft-kips}$$

Thus, the governing moments at the connections are

$$M^+_{e,\ell} = 691.1 \text{ ft-kips and } M^-_{e,r} = 875.4 \text{ ft-kips}$$

At the bottom of the connection, provide an additional 4-No. 9 bars to the 4-No. 9 bars (2 layers) and at the top of the section, provide an additional 5-No. 9 bars to the 5-No. 9 bars (2 layers). From a strain compatibility analysis considering all of the reinforcement in the section:

$$\phi M^+_n = 729.3 \text{ ft-kips} > M^+_{e,\ell} = 691.1 \text{ ft-kips} \quad \text{O.K.}$$

$$\phi M^-_n = 888.2 \text{ ft-kips} > M^-_{e,r} = 875.4 \text{ ft-kips} \quad \text{O.K.}$$

For both the positive and negative moment capacities, the strain in the extreme tension steel was determined from the strain compatibility analysis to be greater than 0.005 so that the section is tension-controlled.

Maximum reinforcement ratio = $\dfrac{10 \times 1.0}{24 \times 22.33} = 0.019 < 0.025$ O.K.   *21.3.2.1*

where the effective depth d was determined from the strain compatibility analysis.

c. Anchorage and splices

Per 21.5.4.1, the minimum development length for a bar with a standard 90-degree hook in normal weight aggregate concrete is:

$$\ell_{dh} = \frac{f_y d_b}{65\sqrt{f'_c}} = \frac{(60,000)(1.128)}{65\sqrt{4000}} = 16.5 \text{ in.} \quad \quad \quad \textit{Eq. (21-6)}$$

Figure 29-45 shows the reinforcement details for the connection.

| Example 29.7 (cont'd) | Calculations and Discussion | Code Reference |

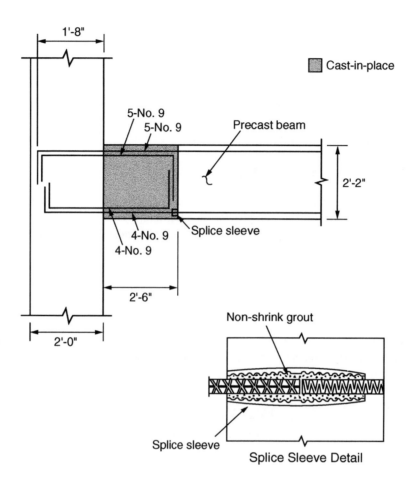

Fig. 29-45 Reinforcement Details for Connection

Example 29.8—Design of Slab Column Connections According to 21.11.5

Figure 29.46 shows the partial plan of a 5-story building assigned to Seismic Design Category D (i.e., high seismic design category). The seismic-force-resisting system consists of a building frame, where shear walls (not shown in figure) resist the seismic forces. Check the slab-column connections at columns B1 and B2 for the provisions of 21.11.5 assuming that the induced moments transferred between the slab and column under the design displacement are not computed.

Material Properties:

    Concrete (w_c = 150 pcf): f'_c = 4000 psi

    Reinforcement: f_y = 60 ksi

Service Loads:

    Live load = 50 psf

    Superimposed dead load = 30 psf

Member Dimensions:

    Slab thickness = 9 in.

    Columns = 24 × 24 in.

Additional Data:

    Story height = 10 ft

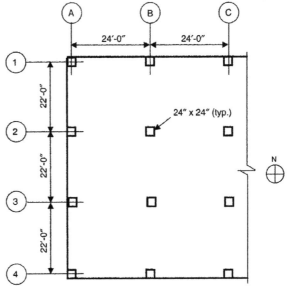

Figure 29-46 Partial Floor Plan

Design displacements and story drifts in the N-S direction:

Story	Design Displacement (in.)	Story Drift (in.)*
5	1.5	0.3
4	1.2	0.4
3	0.8	0.3
2	0.5	0.3
1	0.2	0.2

* Story drift = design displacement at top of story — design displacement at bottom of story

Calculations and Discussion	Code Reference

1. Column B1

    a. Determine factored shear force V_u due to gravity loads on slab critical section for two-way action

        w_D = (9/12) × 0.15 + 0.03 = 0.143 ksf

| Example 29.8 (cont'd) | Calculations and Discussion | Code Reference |

$w_L = 0.05$ ksf

$w_u = 1.2w_D + 0.5w_L = 1.2 \times 0.143 + 0.5 \times 0.05 = 0.2$ ksf   21.11.5, 9.2.1(a)

Critical section dimensions: 11.12.1.2

Use average d = 9 − 1.25 = 7.75 in.

$b_1 = 24 + (7.75/2) = 27.875$ in.

$b_2 = 24 + 7.75 = 31.75$ in.

$V_u = 0.2[(24 \times 12) - (27.875 \times 31.75/144)] = 56$ kips

b. Determine two-way shear design strength ϕV_c

For square columns, Eq. (11-35) governs: 11.12.21

$$V_c = 4\sqrt{f'_c}b_o d$$

where $b_o = b_2 + 2b_1 = 31.75 + 2 \times 27.875 = 87.5$ in.

Thus,

$$V_c = 4\sqrt{4000} \times 87.5 \times 7.75/1000 = 172 \text{ kips}$$

$\phi V_c = 0.75 \times 172 = 129$ kips   9.3.2.3

c. Check criterion in 21.11.5(b)

Since induced moments are not computed, the requirements of 21.11.5(b) must be satisfied.

Maximum story drift at 4th floor level = 0.4 in.

Design story drift ratio = story drift/story height = $0.4/(10 \times 12) = 0.003$

Limiting design story drift ratio:

$0.035 - 0.05(V_u/\phi V_c) = 0.035 - 0.05(56/129) = 0.013 > 0.005$

Since the design story drift ratio = 0.003 < 0.013, slab shear reinforcement satisfying the requirements of 21.11.5 need not be provided.

2. Column B2

a. Determine factored shear force V_u due to gravity loads on slab critical section for two-way action

Example 29.8 (cont'd)	Calculations and Discussion	Code Reference

$w_u = 1.2w_D + 0.5w_L = 1.2 \times 0.143 + 0.5 \times 0.05 = 0.2$ ksf   21.11.5, 9.2.1(a)

Critical section dimensions: 11.12.1.2

$b_1 = b_2 = 24 + 7.75 = 31.75$ in.

$V_u = 0.2[(24 \times 22) - (31.75^2/144)] = 104$ kips

b. Determine two-way shear design strength ϕV_c

For square columns, Eq. (11-35) governs: 11.12.21

$$V_c = 4\sqrt{f'_c}\,b_o d$$

where $b_o = 4 \times 31.75 = 127.0$ in.

Thus,

$$V_c = 4\sqrt{4000} \times 127.0 \times 7.75/1000 = 249 \text{ kips}$$

$\phi V_c = 0.75 \times 249 = 187$ kips   9.3.2.3

c. Check criterion in Section 21.11.5(b)

Maximum story drift at 4th floor level = 0.4 in.

Design story drift ratio = story drift/story height = $0.4/(10 \times 12) = 0.003$

Limiting design story drift ratio:

$0.035 - 0.05(V_u/\phi V_c) = 0.035 - 0.05(104/187) = 0.007 > 0.005$

Since the design story drift ratio = 0.003 < 0.007, slab shear reinforcement satisfying the requirements of 21.11.5 need not be provided.

Blank

30

Structural Plain Concrete

BACKGROUND

With publication of the 1983 edition of ACI 318, provisions for structural plain concrete were incorporated into the code by reference. The document referenced was ACI 318.1, *Building Code Requirements for Structural Plain Concrete*. This method of regulating plain concrete continued with the 1989 edition of ACI 318. For the 1995 edition, the provisions formally contained in the ACI 318.1 standard were incorporated into Chapter 22 of the code and publication of ACI 318.1 was discontinued. While the presentation of some provisions is different, few technical changes have been made since the 1989 edition of ACI 318.1. Technical changes that were made are discussed at the appropriate location in this part.

22.1, 22.2 SCOPE AND LIMITATIONS

By definition, structural plain concrete is concrete in members that either contains no reinforcement or contains less reinforcement than the minimum amount specified for reinforced concrete in other chapters of ACI 318 and Appendices A through C (22.2.1). The designer should take special note of 22.2.2. Since the structural integrity of structural plain concrete members depends solely on the properties of the concrete, it limits the use of plain concrete to: members that are continuously supported by soil or by other structural members capable of providing vertical support continuous throughout the length of the plain concrete member; members in which arch action assures compression under all conditions of loading; and walls and pedestals. Chapter 22 of ACI 318 contains specific design provisions for structural plain concrete walls, footings and pedestals.

Section 22.1.1.2 indicates that sidewalks and other slabs-on-grade are not regulated by the code unless they transmit vertical loads or lateral forces from other parts of the structure to the soil. Section 1.1.6 also stipulates that if a slab transmits vertical loads or lateral forces from the structure to the soil, it also must comply with the code. In addition, 22.2.3 points out that the design and construction of portions of structural plain concrete foundation piers and cast-in-place piles embedded in ground capable of providing adequate lateral support are not governed by Chapter 22. Provisions for these elements are typically found in the general building code.

22.3 JOINTS

Structural plain concrete members must be small enough or provided with contraction (control) joints so as to create elements that are flexurally discontinuous (22.3.1). This requires that the build-up of tensile stresses due to external loads and internal loads, such as from drying shrinkage, temperature and moisture changes, and creep, must be limited to permissible values. Section 22.3.2 emphasizes several items that will influence the size of elements and, consequently, the spacing of contraction joints. These include: climatic conditions; selection and proportioning of materials; mixing, placing and curing of concrete; degree of restraint to movement; stresses due to external and internal loads to which the element is subjected; and construction techniques. Where contraction joints are provided, the member thickness must be reduced a minimum of 25% if the joint is to be effective. For additional information on drying shrinkage of concrete, other causes of volume changes of concrete, and the use of contraction joints to relieve stress build-up, see Refs. 30.1 through 30.3.

While not a part of the provisions, R22.3 gives an exception to the above requirement for contraction joints. It indicates that where random cracking due to creep, shrinkage and temperature effects will not affect the structural integrity, and is otherwise acceptable, such as transverse cracks in a continuous wall footing, contraction joints are not necessary.

22.4   DESIGN METHOD

As for reinforced concrete designed in accordance with Chapters 1 through 21, the provisions of Chapter 22 are based on the strength design methodology. Load combinations and load factors are found in 9.2, and are the same as those used for the design of reinforced concrete. The load combinations and load factors in the 1999 and earlier ACI codes were replaced with load combinations and load factors from ASCE 7-98 into the 2002 code. The strength reduction factor, ϕ, is found in 9.3.5. It was reduced from 0.65 (found in the 1999 and earlier editions of the ACI code) to 0.55 in the 2002 code, and applies for all stress conditions (i.e., flexure, compression, shear and bearing). Everything else remaining the same, the reduction in ϕ results in a 15.4% decrease in the design strength. Although some load factors in 9.2 are less than those in C.2, they have not been reduced enough to completely compensate for the lower design strength. While each case needs to be investigated, generally speaking a more economical design will be obtained by using the load and strength reduction factors of Appendix C. If snow loads or roof live loads are included in the controlling gravity load, depending on the magnitude of these loads with respect to floor live load, use of the load and strength reduction factors of Chapter 9 may be more economical.

To quickly determine which one of the two sets of load and strength reduction factors should be used, compute the governing load/load effect (e.g., P_u or M_u) using the load factors in 9.2 and C.2. These values can then be divided by the corresponding strength reduction factors from 9.3.5 and C.3.5, respectively, to determine the nominal loads/load effects. Satisfying the lower nominal load/load effect may be more economical.

Numerous figures and tables are provided in the main body of this part to assist the user in designing structural members of plain concrete. They are based on the load factors and strength reduction factor (0.55) of Chapter 9. An appendix to this part contains similar figures and tables based on the load factors and strength reduction factor (0.65) found in Appendix C. To facilitate comparing companion figures and tables, their assigned numbers are the same except those corresponding to the appendix, are prefaced with the letter "C."

A linear stress-strain relationship in both tension and compression is assumed for members subject to flexure and axial loads. The allowable stress design procedures contained in Appendix A - Alternate Design Method of the 1999 and earlier editions of the code do not apply to structural members of plain concrete. That Appendix was removed from the 2002 code.

Where the provisions for contraction joints and/or size of members have been observed in accordance with 22.3, tensile strength of plain concrete is permitted to be considered (22.4.5). Tension is not to be considered beyond the outside edges of the panel, contraction joints or construction joints, nor is flexural tension allowed to be assumed between adjacent structural plain concrete elements (22.4.7).

Section 22.4.8 permits the entire cross-section to be considered effective in resisting flexure, combined flexure and axial load, and shear; **except that for concrete cast on the ground, such as a footing, the overall thickness, h, shall be assumed to be 2 in. less than actual**. The commentary indicates that this provision is necessary to allow for unevenness of the excavation and for some contamination of the concrete adjacent to the soil. No strength shall be assigned to any steel reinforcement that may be present (22.4.6).

As in the past, 22.2.4, through its reference to 1.1.1, requires that the minimum specified compressive strength of concrete, f'_c, used in design of structural plain concrete elements shall not be less than 2500 psi. This provision is considered necessary due to the fact that safety and load-carrying capability is based solely on the strength and quality of the concrete.

22.5 STRENGTH DESIGN

Permissible stresses of ACI 318.1-89 were replaced with formulas for calculating nominal strengths for flexure, compression, shear and bearing. The nominal moment strength, M_n, is given by:

$$M_n = 5\sqrt{f'_c}\, S_m \qquad \text{Eq. (22-2)}$$

for flexural tension controlled sections, and

$$M_n = 0.85 f'_c S_m \qquad \text{Eq. (22-3)}$$

for flexural compression controlled sections.

The nominal axial compression strength, P_n, is given by:

$$P_n = 0.60 f'_c \left[1 - \left(\frac{\ell_c}{32h}\right)^2\right] A_1 \qquad \text{Eq. (22-5)}$$

Note that the effective length factor, k, is missing from the numerator of the ratio $\ell_c/32h$. This change from ACI 318.1 was made because it was felt that it is always conservative to assume k = 1, which is based on both ends being fixed against translation. Also, it was recognized that it is difficult to obtain fixed connections in typical types of construction utilizing structural plain concrete walls. If a connection fixed against rotation is provided at one or both ends, the engineer can always assume k = 0.8 as in the past. However, before doing so the engineer should verify that the member providing rotational restraint has a flexural stiffness EI/ℓ at least equal to that of the wall.

For members subject to combined flexural and axial compression, two interaction equations are given and both must be satisfied. For the compression face:

$$\frac{P_u}{\phi P_n} + \frac{M_u}{\phi M_n} \leq 1 \qquad \text{Eq. (22-6)}$$

where $M_n = 0.85 f'_c S_m$

and for the tension face:

$$\frac{M_u}{S_m} - \frac{P_u}{A_g} \leq 5\phi\sqrt{f'_c} \qquad \text{Eq. (22-7)}$$

The nominal moment strength, M_n, for use in Eq. (22-6) (i.e., $0.85\, f'_c S_m$) is more conservative than in the 1989 edition of ACI 318.1 in which it was $f'_c S_m$. Other nominal strengths of Chapter 22 are consistent with those calculated using permissible stresses of ACI 318.1-89.

The nominal shear strength, V_n, is given by:

$$V_n = \frac{4}{3}\sqrt{f'_c}\, b_w h \qquad \text{Eq. (22-9)}$$

for beam action, and by

$$V_n = \left[\frac{4}{3} + \frac{8}{3\beta}\right]\sqrt{f'_c}\, b_o h \leq 2.66\sqrt{f'_c}\, b_o h \qquad \text{Eq. (22-10)}$$

for two-way action, or punching shear.

In Eq. (22-10), the expression [4/3 + 8/(3β)] reduces the nominal shear strength for concentrated loads with long-to-short-side ratios β greater than 2. Where the ratio is equal to or less than 2, the expression takes on the maximum permitted value of 2.66.

The equations for computing nominal flexural and shear strengths apply to normal weight aggregate concrete. If lightweight aggregate concrete is used, the strengths may need to be reduced in accordance with 22.5.6. Where the average splitting tensile strength of lightweight concrete, f_{ct}, is specified <u>and</u> concrete is proportioned in accordance with 5.2 of ACI 318, $\sqrt{f'_c}$ shall be replaced with $f_{ct}/6.7$, but the value of $f_{ct}/6.7$ shall not exceed $\sqrt{f'_c}$. Where f_{ct} is not specified, the value of $\sqrt{f'_c}$ shall be multiplied by 0.75 for all lightweight aggregate concrete and by 0.85 for sand-lightweight aggregate concrete, with linear interpolation permitted for mix designs using partial sand replacement.

Nominal bearing strength, B_n, is given by:

$$B_n = 0.85 f'_c A_1 \qquad \text{Eq. (22-12)}$$

where A_1 is the loaded area. If the supporting surface is wider on all sides than A_1, the bearing strength may be increased by $\sqrt{A_2/A_1}$, but by not more than 2. A_2 is the area of the lower base of the largest frustum of a pyramid, cone, or tapered wedge contained wholly within the support and having for its upper base the loaded area, A_1, and having side slopes of 1 vertical to 2 horizontal. See Part 6 for determination of A_2.

22.6 WALLS

22.6.5 Empirical Design Method

The provisions offer two alternatives for designing plain concrete walls. The simpler of the two is referred to as the *empirical design method*. It is only permitted for walls of solid rectangular cross-section where the resultant of <u>all</u> factored loads falls within the middle one-third of the overall thickness of the wall. In determining the effective eccentricity, the moment induced by lateral loads must be considered in addition to any moment induced by the eccentricity of the axial load. Limiting the eccentricity to one-sixth the wall thickness assures that all portions of the wall remain under compression. Under the empirical design method, the nominal axial load strength, P_n, is determined from:

$$P_n = 0.45 f'_c A_g \left[1 - \left(\frac{\ell_c}{32h} \right)^2 \right] \qquad \text{Eq. (22-14)}$$

This is a single-strength equation considering only the axial load. Moments due to eccentricity of the applied axial load and/or lateral loads can be ignored since an eccentricity not exceeding h/6 is assumed.

To assist the code user in the design of plain concrete walls using the empirical design method, Fig. 30-1 has been provided. By entering the figure with the required axial load strength, one can select the wall thickness that will yield a design axial load strength, ϕP_n, that is equal to or greater than required. For intermediate values of f'_c, the required wall thickness can be determined by interpolation.

22.6.3 Combined Flexure and Axial Load

The second method, which may be used for all loading conditions, must be used where the resultant of all factored loads falls outside of the middle one-third of the wall thickness (i.e., e > h/6). In this procedure the wall must be proportioned to satisfy the provisions for combined flexure and axial loads of interaction Eqs. (22-6) and (22-7). Where the effective eccentricity is less than 10% of the wall thickness, h, an assumed eccentricity of not less than 0.10h is required.

To utilize this method, one must generally proceed on a trial and error basis by assuming a wall thickness and specified compressive strength of concrete, f'_c, and determine if the two interaction equations are satisfied. This

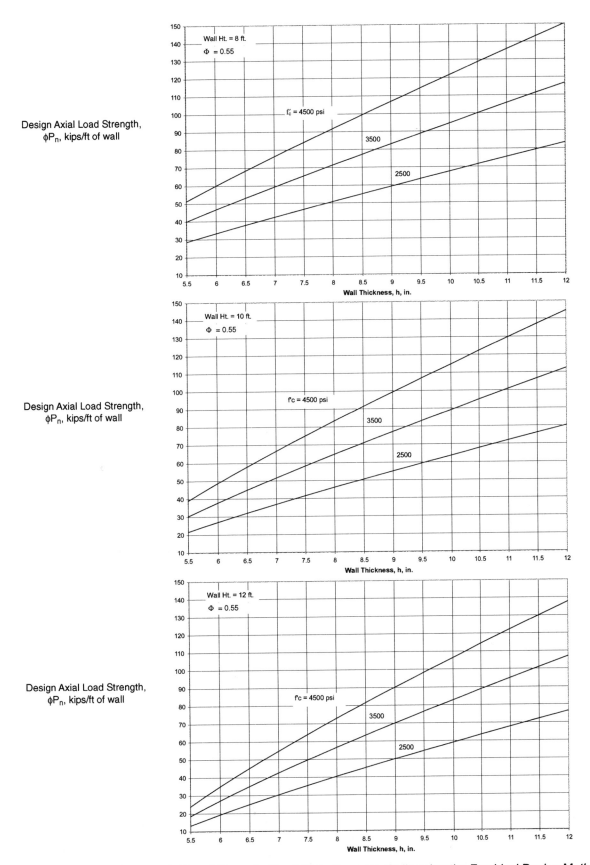

Figure 30-1 Design Axial Load Strength, P_n, of Plain Concrete Walls using the Empirical Design Method

process can proceed on a more structured basis if it is first determined whether Eq. (22-6) or (22-7) will control. Each equation can be rearranged to solve for M_u. Then, by setting the two equations equal to one another, rearranging to get both terms with P_u on the left side, and introducing constants to make units consistent, the resulting equation (1), which is shown below, can be solved for the axial load, P_u. The computed value of P_u is the axial load at which the moment strength is greatest and is the same regardless of whether Eq. (22-6) or (22-7) is used. If the required axial load strength is less than the computed value of P_u, the design is governed by Eq. (22-7). Conversely, if the required axial load strength is greater than the computed valued of P_u, the design is governed by Eq. (22-6). The equation for solving for P_u is:

$$\frac{S_m P_u}{12 A_g} + \frac{M_n P_u}{P_n} = \phi M_n - \frac{5\phi\sqrt{f'_c} S_m}{12,000} \qquad (1)$$

where M_n is determined from Eq. (22-3), axial loads are in kips, moments are in ft-kips, section modulus is in in.^{3}, area is in in.^{2}, and $\sqrt{f'_c}$ is in psi.

If it is determined that Eq. (22-6) governs, equation (1) can be rearranged, with A_g and S_m expressed in terms of h. Quadratic equation (2) can be solved for the required wall thickness h:

If Eq. (22-6) governs, the required wall thickness is best determined by trial and error. Several iterations may be necessary before the most economical design solution is achieved.

$$0.06\phi\sqrt{f'_c}\, h^2 + P_u h - 72 M_u = 0 \qquad (2)$$

where the axial load is in kips, the moment is in ft-kips, $\sqrt{f'_c}$ is in psi, and the thickness is in inches. If the required wall thickness is more than that assumed, another iteration is necessary. If it is significantly less than assumed, it may be advisable to repeat the process to determine if a more economical thickness and/or concrete strength can be justified.

The design process can be greatly simplified with the use of axial load-moment strength curves such as those shown in Figs. 30-2 and 30-3. To use the curves, enter with the known factored axial load, P_u, and determine if the design moment strength, ϕM_n, equals or exceeds the required factored moment strength, M_u. Of course, the curves can also be used by entering with the required factored moment strength, M_u, and determining if the design axial load strength, ϕP_n, equals or exceeds the required factored axial load strength, P_u.

If the effective eccentricity due to all factored loads is less than 0.10h, the design axial load strength, ϕP_n, is determined by projecting horizontally to the left from the intersection of the line labeled "e = h/10" and the curve representing the specified compressive strength of concrete, f'_c. For example, Fig. 30-2 shows that for an 8-in. wall, 8 ft in height constructed of concrete with a specified compressive strength, f'_c, of 2500 psi, the design axial load strength, ϕP_n, is approximately 68 kips/ft of wall. This assumes that the wall is loaded concentrically and there are no lateral loads to induce moments (i.e., $\phi M_n = 0$). However, when the axial load is applied at the required minimum eccentricity of 0.10h, the design axial load strength is reduced to approximately 50 kips/ft of wall. The moment corresponding to the 50-kip load being applied at the minimum eccentricity of 0.10h is approximately 3.3 ft-kips/ft of wall.

A line labeled "e = h/6" has also been included on Figs. 30-2 and 30-3 to assist the user in identifying when the effective eccentricity exceeds this value. If the intersection of the axial load, P_u, and moment, M_u, lies to the right of the line, a portion of the wall is under tension due to the induced moment.

Walls of plain concrete are typically used as basement walls and above grade walls in residential and small commercial buildings. In most cases the axial loads are small compared to the design axial load compressive strength, ϕP_n, of the wall. Therefore, Figs. 30-4 through 30-6 have been developed which include only the lower range of values of axial loads from Figs. 30-2 and 30-3. Where small axial loads are acting in conjunction with moments, the design is governed by flexural tension [Eq. (22-7)] rather than by combined axial and flexural compression [Eq. (22-6)]. An

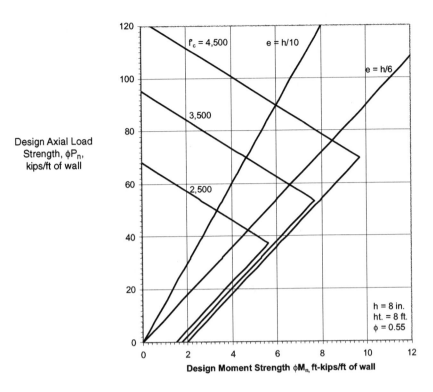

Figure 30-2 Design Strength Interaction Diagrams for 8.0-in. Wall, 8 ft in Height

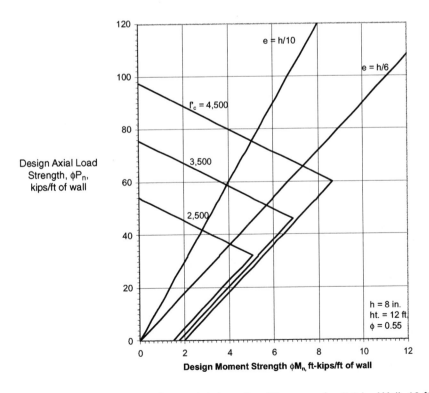

Figure 30-3 Design Strength Interaction Diagrams for 8.0-in. Wall, 12 ft in Height

examination of Eq. (22-7) will reveal that the design moment strength for lightly-loaded walls is not a function of the wall's height; therefore, the format of Figs. 30-4 through 30-6 is somewhat different than that of Figs. 30-2 and 30-3. To assist the user in verifying that the wall being designed is controlled by Eq. (22-7) instead of Eq. (22-6), Figs. 30-7 through 30-9 have been provided. These figures show the value of the design axial load strength, ϕP_n, that corresponds to the maximum value of the design moment strength, ϕM_n. For example, Fig. 30-7 shows an 8-in. wall 8 ft high has a design axial load strength, ϕP_n, of approximately 37.4 kips/ft of wall when a moment equal to the maximum design moment strength, ϕM_n, is applied. From Fig. 30-2 the maximum design moment strength, ϕM_n, is approximately 5.6 ft-kips/ft of wall when a factored load of approximately 37 kips/ft of wall is applied. When using Figs. 30-4 through 30-6, the user should always verify that the required axial load strength, P_u, is less than the value determined from Figs. 30-7 through 30-9. Also, the provisions of 22.6.6.2 should not be overlooked. They require that the thickness of the wall be not less than the larger of 1/24 the unsupported height or length of the wall and 5-1/2 in. A close examination of Figs. 30-7 through 30-9 will show that in almost every case covered, the design axial load strength exceeds 15 kips/ft of wall, which is significantly greater than the factored load on typical walls in low-rise residential buildings.

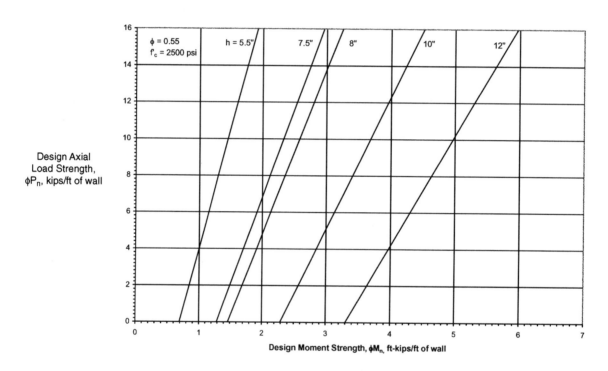

Figure 30-4 Design Strength Interaction Diagrams for Lightly-Loaded Plain Concrete Walls (f'_c = 2500 psi)

In typical construction of basement or foundation walls retaining unbalanced backfill, some walls of the building are generally nonload bearing walls. In this case, especially where the wall may be backfilled before all the dead load that will eventually be on the wall is in place, it is prudent to design the wall assuming no axial load is acting in conjunction with the lateral soil load. For this condition, equation (2) simplifies to:

$$0.06\phi \sqrt{f'_c}\, h^2 - 72M_u = 0 \qquad (3)$$

In this form, the equation can be rearranged to solve for required wall thickness:

$$h = (72M_u/0.06\phi \sqrt{f'_c})^{1/2} \qquad (4)$$

or to solve for required specified compressive strength of concrete:

$$f'_c = (72M_u/0.06\phi h^2)^2 \qquad (5)$$

In equations (3), (4), and (5), f'_c is in psi, h is in inches, and M_u is in ft-kips/ft of wall

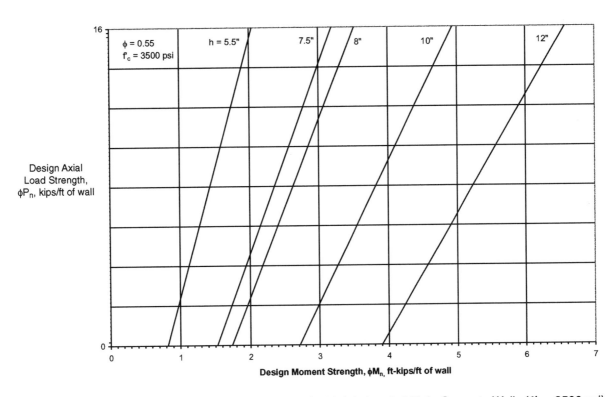

Figure 30-5 Design Strength Interaction Diagrams for Lightly-Loaded Plain Concrete Walls (f'_c = 3500 psi)

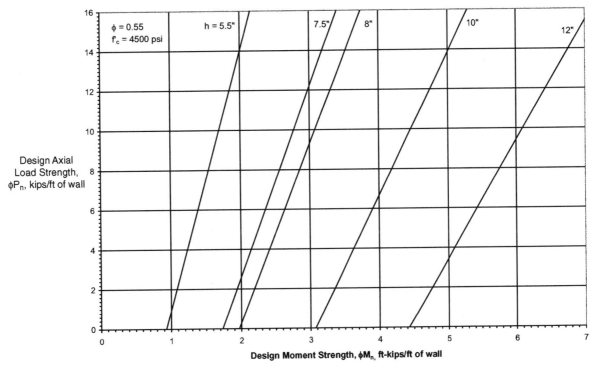

Figure 30-6 Design Strength Interaction Diagrams for Lightly-Loaded Plain Concrete Walls (f'_c = 4500 psi)

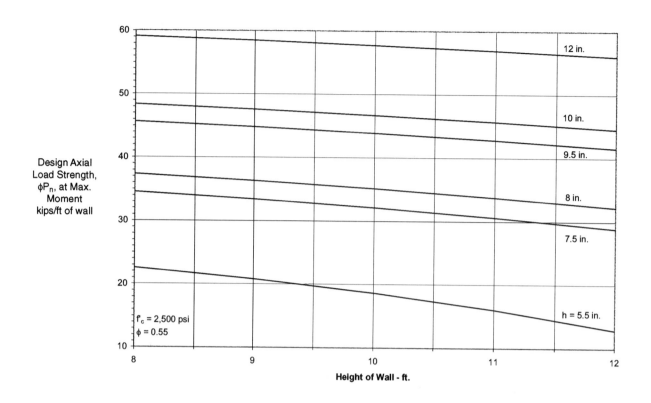

Figure 30-7 Design Axial Load Strength of Plain Concrete Walls at Maximum Design Moment Strength (f'_c = 2500 psi)

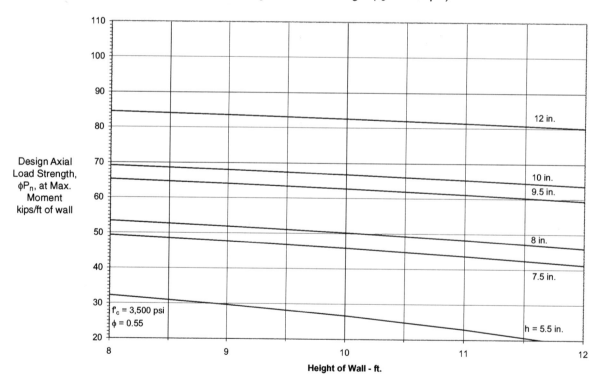

Figure 30-8 Design Axial Load Strength of Plain Concrete Walls at Maximum Design Moment Strength (f'_c = 3500 psi)

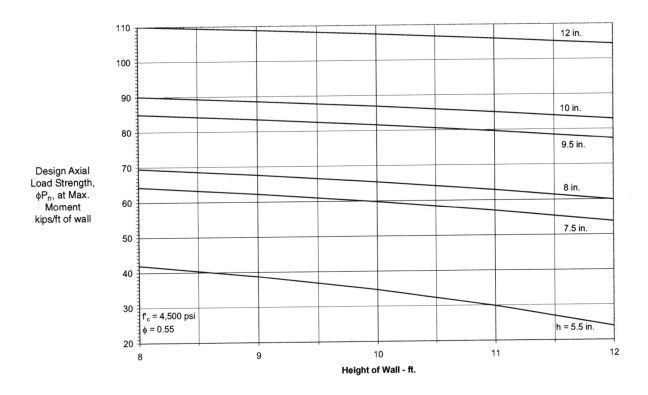

Figure 30-9 Design Axial Load Strength of Plain Concrete Walls at Maximum Design Moment Strength (f'_c = 4500 psi)

Comparison of the Two Methods

Since 22.6.5.1 limits use of the empirical method to cases where the resultant of all factored loads falls in the middle one-third of the wall, one would assume that this method is more conservative for effective eccentricities less that one-sixth the wall thickness. However, this is not the case. The empirical method has an implied eccentricity factor [ratio of strength given by Eq. (22-14) to that given by Eq. (22-5)] of 0.75 (i.e., 0.45/0.60). However, the curves of Figs. 30-2 and 30-3 have an eccentricity factor (ratio of strength where load is applied at eccentricity of 0.10h to that of load applied concentrically) of 0.725 for 2500 psi concrete. Therefore, for effective eccentricities of less than one-sixth the wall thickness, the empirical method will yield a greater nominal axial load strength.

Section 22.6.6.3 requires that exterior basement and foundation walls must be not less than 7-1/2 in. thick. Note though, some building codes permit these walls to be 5-1/2 in. in thickness. Section 22.6.6.2 requires the thickness of other walls to be not less than 5-1/2 in., but not less than 1/24 the unsupported height or length of the wall, whichever is shorter.

Other limitations of 22.6.6 must also be observed. They include: the wall must be braced against lateral translation (22.6.6.4); and not less than 2-No. 5 bars, extending at least 24 in. beyond the corners, shall be provided around all door and window openings (22.6.6.5).

To facilitate the design of simply-supported walls subject to lateral loads from wind and/or soil, Tables 30-1 and 30-2 have been provided. The tables give factored moments due to various combinations of wind and soil lateral loads, and varying backfill heights. The tables also accommodate exterior walls completely above grade (no lateral load due to backfill), as well as walls that are not subject to lateral loading from wind. Tables 30-1 and 30-2 are to be used with the load factors of 9.2, and Tables C30-1 and C30-2 are to be used with the load factors of C.2. Note that the only difference between the two tables in each set is that the first table was developed using a load factor on wind of 1.6; whereas, the second table utilizes a load factor of 1.3. In each set of tables the load factor on the lateral load due to the soil is the same; either 1.6 or 1.7. Table 30-1 or C30-1 is to be used where load combinations in 9.2. or C.2, respectively, are being investigated where the wind load has been reduced by a directionality factor as in the IBC and ASCE 7-98 [9.2.1(b)]. Table 30-2 or C30-2 is to be used where load combinations in 9.2 or C.2, respectively, are being investigated where the wind load has not been reduced by a directionality factor, such as in the NBC, SBC, and UBC, and in editions of ASCE 7 prior to 1998.

For exterior walls partially above and below grade, the moments in the table assume that the wind load is acting in the same direction as the lateral load due to the soil (i.e., inward). Most contemporary wind design standards require that exterior walls be designed for both inward and outward acting pressures due to wind. Generally, the higher absolute value of wind loading occurs where the wall is under negative pressure (i.e., the force is acting outward on the wall). Section 9.2.1(d) stipulates that where earth pressure counteracts wind load, which is the case with wind acting outward, the load factor on H must be set equal to zero in load combination Eq. (9-6). Except for situations where the backfill height is small compared to the overall wall height, and depending upon the relative magnitude of the design wind pressure and lateral soil pressure, the moment due to lateral soil loads and inward-acting wind of Tables 30-1 and 30-2 will generally apply. For situations where outward-acting wind controls, it is simpler to design the wall as though the full height of the wall is exposed to the outward-acting wind pressure. Tables 30-1, 30-2, C30-1, and C30-2 can be used in this manner by assuming the backfill height is zero.

Before designing a structural plain concrete wall that will be resisting wind uplift and/or overturning forces, the appropriate load combinations of 9.2 or C.2 need to be investigated. If the entire wall cross-section will be in tension due to the factored axial and lateral forces, the wall must be designed as a reinforced concrete wall, or other means must be employed to transfer the uplift forces to the foundation. This condition can occur frequently in the design of walls supporting lightweight roof systems subject to net uplift forces from wind loads.

22.7 FOOTINGS

It is common practice throughout the United States, including high seismic risk areas, to use structural plain concrete footings for the support of walls for all types of structures. In addition, plain concrete is frequently used for footings supporting columns and pedestals, particularly in residential construction. These uses of structural plain concrete are permitted by Chapter 22; however, 22.7.3 prohibits the use of structural plain concrete footings which are supported on piles. In addition, contemporary building codes also have limitations on the use of structural plain concrete footings for structures in regions of moderate seismic risk, or for structures assigned to intermediate seismic performance or design categories. See section below on 22.10 for limitations on the use of structural plain concrete footings for structures in regions of high seismic risk, or for structures assigned to high seismic performance or design categories.

Many architects and engineers specify that two No. 4 or No. 5 longitudinal bars be included in footings supporting walls. However, typically these footings have no reinforcement in the transverse direction, or the amount provided is less than that required by the code to consider the footing reinforced. Such footings must be designed as structural plain concrete, since in the transverse direction the footing is subjected to flexural and possibly shear stresses due to the projection of the footing beyond the face of the supported member.

The base area of footings must be determined from unfactored loads and moments, if any, using permissible soil bearing pressures. Once the base area of the footing is selected, factored loads and moments are used to proportion the thickness of the footing to satisfy moment and, where applicable, shear strength requirements. Sections 22.7.5 and 22.7.6 define the critical sections for computing factored moments and shears. The locations are summarized in Table 30-3. Figure 22-2 illustrates the location of the critical sections for beam action and two-way action shear for a footing supporting a column or pedestal.

Footings must be proportioned to satisfy the requirements for moment in accordance with Eq. (22-2). For footings supporting columns, pedestals or concrete walls, if the projection of the footing beyond the face of the supported member does not exceed the footing thickness, h, it is not necessary to check for beam action shear since the location of the critical section for calculating shear falls outside the footing. Where beam action shear must be considered, the requirements of Eq. (22-9) must be satisfied. In addition, for footings supporting columns, pedestals or other concentrated loads, if the projection of the footing beyond the critical section exceeds h/2, it is necessary to determine if the requirements of Eq. (22-10) are satisfied for two-way action (punching) shear. Generally, flexural strength will govern the thickness design of plain concrete footings; however, the engineer should not overlook the possibility that beam action shear or two-way action shear may control. It must be remembered that the provisions of 22.4.8 require that for plain concrete members cast on soil, the thickness, h, used to compute flexural and shear strengths is the overall thickness minus 2 in. Thus, for a footing with an overall thickness of 8 in. (which is the minimum overall thickness permitted by 22.7.4), the thickness, h, used to compute strengths is 6 in. Some building codes permit 6-in. thick footings for residential and other small buildings. In this case the thickness, h, for strength computation purposes is 4 in.

Table 30-1 Factored Moments Induced in Walls by Lateral Soil and/or Wind Load (ft-kips/linear ft)
(For Use with Chapter 9 Load Factors: Soil 1.6, Wind 1.6)

Wall Ht. (ft.)	Back fill Ht. (ft.)	Unfactored Design Lateral Soil Load (psf per foot of depth)																		
		30	30	30	30	30	45	45	45	45	45	60	60	60	60	60	100	100	100	100	100
		Unfactored Design Wind Pressure (psf)																			
		0	10	20	40	80	0	10	20	40	80	0	10	20	40	80	0	10	20	40	80
8	0	0.00	0.13	0.26	0.51	1.02	0.00	0.13	0.26	0.51	1.02	0.00	0.13	0.26	0.51	1.02	0.00	0.13	0.26	0.51	1.02
8	1	0.01	0.13	0.25	0.50	1.00	0.01	0.13	0.25	0.50	1.00	0.01	0.13	0.26	0.50	1.00	0.02	0.14	0.26	0.51	1.01
8	2	0.05	0.14	0.26	0.48	0.93	0.08	0.16	0.27	0.50	0.95	0.10	0.18	0.29	0.51	0.96	0.17	0.23	0.34	0.56	1.00
8	3	0.15	0.21	0.29	0.48	0.85	0.23	0.28	0.35	0.53	0.90	0.31	0.36	0.42	0.59	0.95	0.51	0.56	0.62	0.75	1.10
8	4	0.33	0.37	0.41	0.51	0.78	0.49	0.53	0.57	0.66	0.90	0.65	0.69	0.73	0.82	1.02	1.09	1.12	1.16	1.25	1.42
8	5	0.57	0.59	0.62	0.67	0.79	0.85	0.87	0.90	0.95	1.06	1.13	1.16	1.18	1.23	1.34	1.88	1.91	1.93	1.98	2.09
8	6	0.86	0.88	0.89	0.91	0.96	1.30	1.31	1.32	1.34	1.39	1.73	1.74	1.75	1.78	1.83	2.88	2.89	2.90	2.93	2.98
8	7	1.21	1.21	1.21	1.22	1.23	1.81	1.81	1.82	1.82	1.84	2.41	2.42	2.42	2.43	2.44	4.02	4.03	4.03	4.04	4.05
8	8	1.58	1.58	1.58	1.58	1.58	2.36	2.36	2.36	2.36	2.36	3.15	3.15	3.15	3.15	3.15	5.26	5.26	5.26	5.26	5.26
9	0	0.00	0.16	0.32	0.65	1.30	0.00	0.16	0.32	0.65	1.30	0.00	0.16	0.32	0.65	1.30	0.00	0.16	0.32	0.65	1.30
9	1	0.01	0.16	0.32	0.64	1.27	0.01	0.16	0.32	0.64	1.27	0.01	0.17	0.32	0.64	1.27	0.02	0.17	0.33	0.65	1.28
9	2	0.05	0.18	0.32	0.62	1.20	0.08	0.20	0.34	0.63	1.22	0.10	0.21	0.36	0.65	1.23	0.17	0.27	0.40	0.69	1.27
9	3	0.16	0.24	0.36	0.61	1.12	0.24	0.31	0.42	0.67	1.17	0.32	0.39	0.48	0.72	1.23	0.53	0.60	0.68	0.88	1.37
9	4	0.34	0.40	0.47	0.65	1.05	0.51	0.57	0.63	0.78	1.17	0.69	0.74	0.80	0.94	1.30	1.14	1.20	1.26	1.38	1.66
9	5	0.60	0.65	0.69	0.78	1.01	0.91	0.95	0.99	1.08	1.28	1.21	1.25	1.29	1.38	1.57	2.01	2.05	2.09	2.18	2.36
9	6	0.94	0.96	0.99	1.04	1.16	1.41	1.43	1.46	1.51	1.62	1.88	1.90	1.93	1.98	2.09	3.13	3.15	3.18	3.23	3.33
9	7	1.33	1.35	1.36	1.38	1.43	2.00	2.01	2.03	2.05	2.10	2.67	2.68	2.69	2.72	2.77	4.45	4.46	4.47	4.50	4.55
9	8	1.78	1.78	1.78	1.79	1.80	2.66	2.67	2.67	2.68	2.69	3.55	3.56	3.56	3.57	3.58	5.92	5.92	5.93	5.93	5.95
9	9	2.24	2.24	2.24	2.24	2.24	3.37	3.37	3.37	3.37	3.37	4.49	4.49	4.49	4.49	4.49	7.48	7.48	7.48	7.48	7.48
10	0	0.00	0.20	0.40	0.80	1.60	0.00	0.20	0.40	0.80	1.60	0.00	0.20	0.40	0.80	1.60	0.00	0.20	0.40	0.80	1.60
10	1	0.01	0.20	0.40	0.79	1.57	0.01	0.20	0.40	0.79	1.57	0.01	0.20	0.40	0.79	1.58	0.02	0.21	0.41	0.80	1.58
10	2	0.05	0.22	0.40	0.77	1.51	0.08	0.23	0.42	0.78	1.52	0.11	0.25	0.43	0.80	1.54	0.18	0.30	0.48	0.84	1.58
10	3	0.16	0.28	0.44	0.76	1.43	0.25	0.35	0.50	0.82	1.48	0.33	0.42	0.56	0.87	1.53	0.55	0.64	0.74	1.03	1.67
10	4	0.36	0.44	0.54	0.80	1.35	0.54	0.61	0.70	0.93	1.47	0.71	0.79	0.87	1.08	1.60	1.19	1.27	1.34	1.52	1.96
10	5	0.64	0.70	0.76	0.91	1.31	0.95	1.01	1.08	1.22	1.55	1.27	1.33	1.39	1.53	1.82	2.12	2.18	2.24	2.37	2.64
10	6	1.00	1.04	1.09	1.18	1.39	1.50	1.54	1.59	1.68	1.87	2.00	2.04	2.09	2.18	2.36	3.33	3.38	3.42	3.51	3.69
10	7	1.44	1.47	1.49	1.55	1.66	2.16	2.19	2.22	2.27	2.38	2.88	2.91	2.94	2.99	3.10	4.81	4.83	4.86	4.91	5.02
10	8	1.95	1.96	1.97	2.00	2.05	2.92	2.93	2.95	2.97	3.02	3.89	3.91	3.92	3.94	3.99	6.49	6.50	6.52	6.54	6.59
10	9	2.50	2.50	2.51	2.51	2.53	3.75	3.75	3.76	3.76	3.78	5.00	5.00	5.01	5.01	5.03	8.33	8.34	8.34	8.35	8.36
10	10	3.08	3.08	3.08	3.08	3.08	4.62	4.62	4.62	4.62	4.62	6.16	6.16	6.16	6.16	6.16	10.26	10.26	10.26	10.26	10.26
12	0	0.00	0.29	0.58	1.15	2.30	0.00	0.29	0.58	1.15	2.30	0.00	0.29	0.58	1.15	2.30	0.00	0.29	0.58	1.15	2.30
12	1	0.01	0.29	0.57	1.14	2.28	0.01	0.29	0.57	1.14	2.28	0.01	0.29	0.58	1.14	2.28	0.02	0.30	0.58	1.15	2.29
12	2	0.06	0.30	0.58	1.12	2.21	0.08	0.32	0.59	1.14	2.22	0.11	0.34	0.61	1.15	2.24	0.18	0.39	0.65	1.20	2.28
12	3	0.17	0.36	0.61	1.12	2.13	0.26	0.43	0.67	1.17	2.18	0.34	0.50	0.73	1.23	2.23	0.57	0.70	0.90	1.38	2.38
12	4	0.38	0.51	0.71	1.15	2.06	0.57	0.69	0.86	1.28	2.18	0.76	0.88	1.02	1.42	2.30	1.26	1.38	1.51	1.83	2.66
12	5	0.69	0.80	0.92	1.25	2.01	1.03	1.14	1.25	1.53	2.25	1.37	1.48	1.59	1.84	2.51	2.29	2.39	2.50	2.73	3.26
12	6	1.10	1.19	1.28	1.49	2.03	1.65	1.74	1.83	2.02	2.46	2.20	2.28	2.37	2.56	2.97	3.66	3.75	3.84	4.02	4.40
12	7	1.61	1.68	1.75	1.89	2.20	2.42	2.49	2.55	2.69	2.99	3.23	3.29	3.36	3.50	3.78	5.38	5.45	5.51	5.65	5.92
12	8	2.22	2.27	2.31	2.41	2.61	3.34	3.38	3.43	3.52	3.71	4.45	4.49	4.54	4.63	4.82	7.41	7.46	7.50	7.59	7.78
12	9	2.92	2.94	2.97	3.03	3.14	4.37	4.40	4.43	4.48	4.59	5.83	5.86	5.89	5.94	6.05	9.72	9.75	9.77	9.83	9.94
12	10	3.68	3.69	3.70	3.73	3.78	5.51	5.53	5.54	5.56	5.62	7.35	7.36	7.38	7.40	7.45	12.25	12.27	12.28	12.30	12.35
12	11	4.48	4.49	4.49	4.50	4.51	6.73	6.73	6.73	6.74	6.75	8.97	8.97	8.98	8.98	8.99	14.95	14.95	14.95	14.96	14.97
12	12	5.32	5.32	5.32	5.32	5.32	7.98	7.98	7.98	7.98	7.98	10.64	10.64	10.64	10.64	10.64	17.74	17.74	17.74	17.74	17.74

Table 30-2 Factored Moments Induced in Walls by Lateral Soil and/or Wind Load (ft-kips/linear ft)
(For Use with Chapter 9 Load Factors: Soil 1.6, Wind 1.3)

Wall Ht. (ft.)	Backfill Ht. (ft.)	Unfactored Design Lateral Soil Load (psf per foot of depth)																			
		30	30	30	30	30	45	45	45	45	45	60	60	60	60	60	100	100	100	100	100
		Unfactored Design Wind Pressure (psf)																			
		0	10	20	40	80	0	10	20	40	80	0	10	20	40	80	0	10	20	40	80
8	0	0.00	0.10	0.21	0.42	0.83	0.00	0.10	0.21	0.42	0.83	0.00	0.10	0.21	0.42	0.83	0.00	0.10	0.21	0.42	0.83
8	1	0.01	0.10	0.21	0.41	0.81	0.01	0.11	0.21	0.41	0.81	0.01	0.11	0.21	0.41	0.81	0.02	0.11	0.21	0.42	0.82
8	2	0.05	0.12	0.21	0.40	0.76	0.08	0.14	0.23	0.41	0.78	0.10	0.16	0.25	0.43	0.79	0.17	0.22	0.30	0.47	0.83
8	3	0.15	0.20	0.26	0.41	0.71	0.23	0.27	0.32	0.46	0.76	0.31	0.35	0.40	0.52	0.81	0.51	0.55	0.60	0.70	0.96
8	4	0.33	0.36	0.39	0.47	0.68	0.49	0.52	0.55	0.62	0.80	0.65	0.68	0.72	0.78	0.94	1.09	1.12	1.15	1.21	1.35
8	5	0.57	0.59	0.61	0.65	0.74	0.85	0.87	0.89	0.93	1.02	1.13	1.15	1.17	1.21	1.30	1.88	1.90	1.92	1.96	2.05
8	6	0.86	0.87	0.88	0.90	0.94	1.30	1.31	1.32	1.34	1.38	1.73	1.74	1.75	1.77	1.81	2.88	2.89	2.90	2.92	2.96
8	7	1.21	1.21	1.21	1.22	1.23	1.81	1.81	1.82	1.82	1.83	2.41	2.42	2.42	2.43	2.44	4.02	4.03	4.03	4.04	4.05
8	8	1.58	1.58	1.58	1.58	1.58	2.36	2.36	2.36	2.36	2.36	3.15	3.15	3.15	3.15	3.15	5.26	5.26	5.26	5.26	5.26
9	0	0.00	0.13	0.26	0.53	1.05	0.00	0.13	0.26	0.53	1.05	0.00	0.13	0.26	0.53	1.05	0.00	0.13	0.26	0.53	1.05
9	1	0.01	0.13	0.26	0.52	1.03	0.01	0.13	0.26	0.52	1.03	0.01	0.14	0.26	0.52	1.04	0.02	0.14	0.27	0.53	1.04
9	2	0.05	0.15	0.27	0.51	0.98	0.08	0.17	0.29	0.52	1.00	0.10	0.19	0.30	0.54	1.01	0.17	0.24	0.35	0.58	1.06
9	3	0.16	0.22	0.32	0.52	0.93	0.24	0.30	0.38	0.57	0.98	0.32	0.38	0.44	0.63	1.04	0.53	0.59	0.65	0.80	1.18
9	4	0.34	0.39	0.44	0.58	0.90	0.51	0.56	0.61	0.72	1.02	0.69	0.73	0.78	0.88	1.15	1.14	1.19	1.23	1.33	1.55
9	5	0.60	0.64	0.67	0.75	0.92	0.91	0.94	0.97	1.04	1.20	1.21	1.24	1.27	1.34	1.49	2.01	2.05	2.08	2.15	2.29
9	6	0.94	0.96	0.98	1.02	1.11	1.41	1.43	1.45	1.49	1.58	1.88	1.90	1.92	1.96	2.05	3.13	3.15	3.17	3.21	3.29
9	7	1.33	1.34	1.35	1.37	1.42	2.00	2.01	2.02	2.04	2.08	2.67	2.68	2.69	2.71	2.75	4.45	4.46	4.47	4.49	4.53
9	8	1.78	1.78	1.78	1.79	1.80	2.66	2.67	2.67	2.68	2.69	3.55	3.56	3.56	3.56	3.57	5.92	5.92	5.93	5.93	5.94
9	9	2.24	2.24	2.24	2.24	2.24	3.37	3.37	3.37	3.37	3.37	4.49	4.49	4.49	4.49	4.49	7.48	7.48	7.48	7.48	7.48
10	0	0.00	0.16	0.33	0.65	1.30	0.00	0.16	0.33	0.65	1.30	0.00	0.16	0.33	0.65	1.30	0.00	0.16	0.33	0.65	1.30
10	1	0.01	0.16	0.32	0.64	1.28	0.01	0.17	0.32	0.64	1.28	0.01	0.17	0.33	0.65	1.28	0.02	0.17	0.33	0.65	1.29
10	2	0.05	0.18	0.33	0.63	1.23	0.08	0.20	0.35	0.65	1.24	0.11	0.22	0.36	0.66	1.26	0.18	0.27	0.41	0.71	1.30
10	3	0.16	0.25	0.38	0.64	1.18	0.25	0.32	0.44	0.70	1.23	0.33	0.40	0.50	0.75	1.28	0.55	0.62	0.70	0.92	1.43
10	4	0.36	0.42	0.50	0.70	1.14	0.54	0.60	0.67	0.84	1.27	0.71	0.78	0.84	0.99	1.40	1.19	1.25	1.31	1.45	1.77
10	5	0.64	0.69	0.74	0.85	1.15	0.95	1.00	1.05	1.16	1.41	1.27	1.32	1.37	1.48	1.71	2.12	2.17	2.22	2.32	2.53
10	6	1.00	1.04	1.07	1.15	1.31	1.50	1.54	1.57	1.64	1.80	2.00	2.04	2.07	2.14	2.29	3.33	3.37	3.40	3.47	3.62
10	7	1.44	1.46	1.48	1.53	1.62	2.16	2.18	2.21	2.25	2.34	2.88	2.90	2.93	2.97	3.06	4.81	4.83	4.85	4.89	4.98
10	8	1.95	1.96	1.97	1.99	2.03	2.92	2.93	2.94	2.96	3.00	3.89	3.90	3.91	3.93	3.98	6.49	6.50	6.51	6.53	6.57
10	9	2.50	2.50	2.51	2.51	2.52	3.75	3.75	3.75	3.76	3.77	5.00	5.00	5.00	5.01	5.02	8.33	8.34	8.34	8.34	8.35
10	10	3.08	3.08	3.08	3.08	3.08	4.62	4.62	4.62	4.62	4.62	6.16	6.16	6.16	6.16	6.16	10.26	10.26	10.26	10.26	10.26
12	0	0.00	0.23	0.47	0.94	1.87	0.00	0.23	0.47	0.94	1.87	0.00	0.23	0.47	0.94	1.87	0.00	0.23	0.47	0.94	1.87
12	1	0.01	0.23	0.47	0.93	1.85	0.01	0.24	0.47	0.93	1.85	0.01	0.24	0.47	0.93	1.85	0.02	0.24	0.47	0.94	1.86
12	2	0.06	0.25	0.47	0.92	1.80	0.08	0.27	0.49	0.93	1.82	0.11	0.29	0.51	0.95	1.83	0.18	0.34	0.55	0.99	1.87
12	3	0.17	0.32	0.52	0.93	1.75	0.26	0.39	0.58	0.98	1.80	0.34	0.46	0.64	1.04	1.85	0.57	0.68	0.82	1.19	2.00
12	4	0.38	0.48	0.63	0.98	1.72	0.57	0.67	0.79	1.12	1.84	0.76	0.86	0.97	1.26	1.97	1.26	1.36	1.46	1.69	2.33
12	5	0.69	0.77	0.87	1.12	1.73	1.03	1.12	1.21	1.41	1.97	1.37	1.46	1.55	1.74	2.24	2.29	2.37	2.46	2.64	3.05
12	6	1.10	1.17	1.24	1.40	1.80	1.65	1.72	1.79	1.94	2.28	2.20	2.27	2.34	2.49	2.81	3.66	3.73	3.80	3.95	4.25
12	7	1.61	1.67	1.72	1.84	2.08	2.42	2.47	2.53	2.64	2.87	3.23	3.28	3.34	3.44	3.67	5.38	5.43	5.49	5.59	5.82
12	8	2.22	2.26	2.30	2.37	2.53	3.34	3.37	3.41	3.48	3.64	4.45	4.48	4.52	4.60	4.75	7.41	7.45	7.49	7.56	7.71
12	9	2.92	2.94	2.96	3.00	3.10	4.37	4.40	4.42	4.46	4.55	5.83	5.85	5.88	5.92	6.01	9.72	9.74	9.76	9.81	9.90
12	10	3.68	3.69	3.70	3.72	3.76	5.51	5.52	5.53	5.55	5.60	7.35	7.36	7.37	7.39	7.43	12.25	12.26	12.27	12.29	12.33
12	11	4.48	4.49	4.49	4.49	4.51	6.73	6.73	6.73	6.74	6.75	8.97	8.97	8.97	8.98	8.99	14.95	14.95	14.95	14.96	14.97
12	12	5.32	5.32	5.32	5.32	5.32	7.98	7.98	7.98	7.98	7.98	10.64	10.64	10.64	10.64	10.64	17.74	17.74	17.74	17.74	17.74

Figure 30-10 has been provided to aid in the selection of footing thickness to satisfy flexural strength requirements. The figure is entered with the *factored* soil bearing pressure. Project vertically upward to the curve that represents the length that the footing projects beyond the critical section at which the moment must be calculated (see Table 30-3). Read horizontally to the left to determine the minimum required footing thickness. Two (2) in. must be added to this value to satisfy 22.4.8. The thicknesses in the figure are based on a specified compressive strength of concrete, f'_c of 2500 psi. For higher strength concrete, the thickness can be reduced by multiplying by the factor:

$$(2500/\text{specified compressive strength of concrete})^{0.25}$$

As the exponent in the equation suggests, a large increase in concrete strength results in only a small decrease in footing thickness. For example, doubling the concrete strength only reduces the thickness 16 percent.

Table 30-3 Locations for Computing Moments and Shears in Footings*

Supported Member	Moment	Shear – Beam Action	Shear – Two-Way (punching)
Concrete Wall	at face of wall	h from face of wall	Not Applicable
Masonry Wall	1/2 way between center of wall and face of wall	h from face of wall	Not Applicable
Column or Pedestal	at face of column or pedestal	h from face of column or pedestal	h/2 from face of column or pedestal
Column with Steel Base Plate	1/2 way between face of column and edge of steel base plate	h from 1/2 way between face of column and edge of steel base plate	h/2 from 1/2 way between face of column edge of steel base plate

* h = thickness of footing for moment and shear computation purposes.

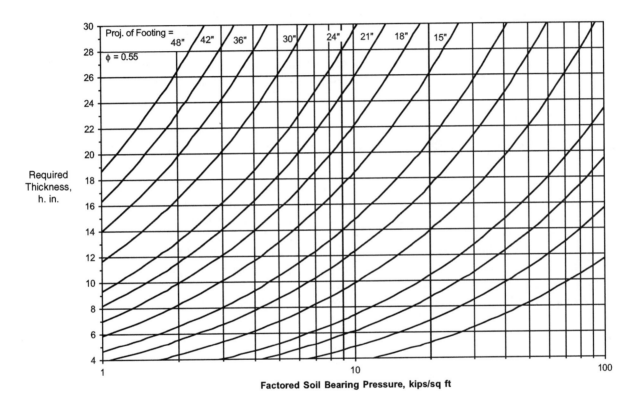

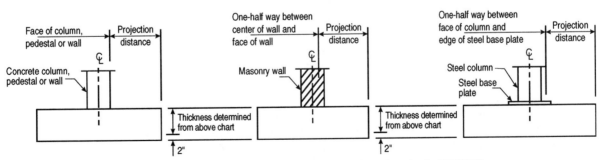

* For f'_c greater than 2500 psi, multiple thickness determined from above chart by $(2500/f'_c)^{0.25}$

Figure 30-10 Thickness of Footing Required to Satisfy Flexural Strength for Various Projection Distances, in. ($f'_c = 2500$ psi*)

22.8 PEDESTALS

Pedestals of plain concrete are permitted by 22.8.2 provided the unsupported height does not exceed three times the average least plan dimension. The design must consider all vertical and lateral loads to which the pedestal will be subjected. The nominal bearing strength, B_n, must be determined from Eq. (22-12). Where moments are induced due to eccentricity of the axial load and/or lateral loads, the pedestal shall be designed for both flexural and axial loads and satisfy interaction Eqs. (22-6) and (22-7). In Eq. (22-6), P_n is replaced with B_n, nominal bearing strength.

Pedestal-like members with heights exceeding three times the least lateral dimension are defined as *columns* by the code and must be designed as reinforced concrete members. Columns of structural plain concrete are prohibited by Chapter 22.

Some contemporary building codes prohibit the use of structural plain concrete pedestals to resist seismic lateral forces in structures at moderate seismic risk or assigned to intermediate seismic performance or design categories. See Table 1-3 and section below on 22.10.

22.10 PLAIN CONCRETE IN EARTHQUAKE-RESISTING STRUCTURES

The '99 edition of the code included a new section 22.10 to address a seismic design issue not covered previously. This concerns the use of plain concrete elements in structures subject to earthquake ground motions intense enough to cause significant structural damage to the elements or partial or total collapse of the structure. By default, the model building codes in use in the U.S. had assumed responsibility for this subject. The requirements, based on similar provisions in *The BOCA National Building Code*[30.4] and *Standard Building Code*[30.5], prohibit the use of structural plain concrete foundation elements for structures in regions at high seismic risk, or for structures assigned to high seismic performance or design categories, except for three specific cases cited in the provisions. See Table 1-3 for an explanation of how seismic risk assigned by the model building codes can be correlated to the requirements of ACI 318.

The provisions prohibit the use of structural plain concrete foundation elements for structures assigned a "high" seismic risk in accordance with Table 1-3, except for the following specific cases. They are:

1. In detached one-and two-family dwellings not exceeding three stories in height and constructed with wood or steel stud bearing walls, the following are permitted:
    a. plain concrete footings supporting walls, columns or pedestals; and
    b. plain concrete foundation or basement walls provided
        i. the wall is not less than 7-1/2 in. thick, and
        ii. it retains no more than 4 ft of unbalanced fill.

2. In structures other than covered by 1 above, plain concrete footings supporting cast-in-place reinforced concrete walls or reinforced masonry walls are permitted provided the footing has at least two continuous No. 4 longitudinal reinforcing bars that provide an area of steel of not less than 0.002 times the gross transverse cross-sectional area of the footing. Continuity of reinforcement must be provided at corners and intersections. In the 2002 ACI code, the requirement was added to limit the use of this provision to situations where the supported wall is either of cast-in-place reinforced concrete or reinforced masonry.

Although Chapter 22 of the code has no limitations on the use of structural plain concrete elements for structures in areas of moderate seismic risk, or for structures assigned to intermediate seismic performance or design categories in accordance with Table 1-3, model building codes in use in the U.S. either prohibit their use or generally require that some reinforcement be included to provide some ductility and tie the elements together. Where construction is contemplated in these areas, the legally adopted building code should be consulted to determine the specific limitations.

REFERENCES

30.1 *Joints in Walls Below Grade*, CR059 Portland Cement Association, Skokie, IL, 1982.

30.2 *Building Movements and Joints*, Portland Cement Association, Skokie, IL, 1982.

30.3 Kosmatka, Steven H., Kerkhoff, Beatrix, and Panarese, William C.; *Design and Control of Concrete Mixtures*, EB001, Fourteenth Edition, Portland Cement Association, Skokie, IL, 2002.

30.4 *The BOCA National Building Code*, Building Officials and Code Administrators International, Country Club Hills, IL, 1999.

30.5 *Standard Building Code*, Southern Building Code Congress International, Birmingham, AL 1999.

APPENDIX 30A

This Appendix includes figures and tables that parallel those of Part 30. The figures of this Appendix are compatible with the load factors and strength reduction factor ($\phi = 0.65$) of ACI 318-02 and ACI 318-05, Appendix C. (In the body of Part 30, figures and tables are compatible with load factors and strength reduction factor ($\phi = 0.55$) of ACI 318-02 and ACI 318-05, Chapter 9.)

Included are the following:

Table C30-1	*Table C30-1 Factored Moments Induced in Walls by Lateral Soil and/or Wind Loads (ft/kips/linear ft) (For Use with Appendix C Load Factors: Soil 1.7, Wind 1.6)*
Table C30-2	*Table C30-1 Factored Moments Induced in Walls by Lateral Soil and/or Wind Loads (ft/kips/linear ft) (For Use with Appendix C Load Factors: Soil 1.7, Wind 1.3)*
Figure C30-1(a-c)	*Figure C30-1 Design Axial Load Strength, P_{nw}, of Plain Concrete Walls using the Empirical Design Method*
Figure C30-2	*Figure C30-2 Design Strength Interaction Diagrams for 8.0-in Wall, 8 ft in Height*
Figure C30-3	*Figure C30-3 Design Strength Interaction Diagrams for 8.0-in Wall, 12 ft in Height*
Figure C30-4	*Figure C30-4 Design Strength Interaction Diagrams for Lightly-Loaded Plain Concrete Walls ($f'_c = 2500$ psi)*
Figure C30-5	*Figure C30-5 Design Strength Interaction Diagrams for Lightly-Loaded Plain Concrete Walls ($f'_c = 3500$ psi)*
Figure C30-6	*Figure C30-6 Design Strength Interaction Diagrams for Lightly-Loaded Plain Concrete Walls ($f'_c = 4500$ psi)*
Figure C30-7	*Figure C30-7 Design Axial Load Strength of Plain Concrete Walls at Maximum Design Moment Strength ($f'_c = 2500$ psi)*
Figure C30-8	*Figure C30-8 Design Axial Load Strength of Plain Concrete Walls at Maximum Design Moment Strength ($f'_c = 3500$ psi)*
Figure C30-9	*Figure C30-9 Design Axial Load Strength of Plain Concrete Walls at Maximum Design Moment Strength ($f'_c = 4500$ psi)*
Figure C30-10	*Figure C30-10 Thickness of Footing Required to Satisfy Flexural Strength for Various Projection Distances, in. ($f'_c = 2500$ psi*)*

Table C30-1 Factored Moments Induced in Walls by Lateral Soil and/or Wind Loads (ft/kips/linear ft)
(For Use with Appendix C Load Factors: Soil 1.7, Wind 1.6)

Wall Ht. (ft.)	Back fill Ht. (ft.)	\multicolumn{20}{c}{Unfactored Design Lateral Soil Load (psf per foot of depth)}
		30	30	30	30	30	45	45	45	45	45	60	60	60	60	60	100	100	100	100	100
		\multicolumn{20}{c}{Unfactored Design Wind Pressure (psf)}
		0	10	20	40	80	0	10	20	40	80	0	10	20	40	80	0	10	20	40	80
8	0	0.00	0.13	0.26	0.51	1.02	0.00	0.13	0.26	0.51	1.02	0.00	0.13	0.26	0.51	1.02	0.00	0.13	0.26	0.51	1.02
8	1	0.01	0.13	0.25	0.50	1.00	0.01	0.13	0.25	0.50	1.00	0.02	0.13	0.26	0.50	1.00	0.03	0.14	0.26	0.51	1.01
8	2	0.05	0.15	0.26	0.48	0.93	0.08	0.17	0.28	0.50	0.95	0.11	0.19	0.29	0.52	0.96	0.18	0.24	0.34	0.56	1.01
8	3	0.16	0.22	0.30	0.48	0.86	0.25	0.30	0.37	0.54	0.91	0.33	0.38	0.44	0.60	0.97	0.55	0.60	0.65	0.78	1.12
8	4	0.35	0.39	0.43	0.53	0.80	0.52	0.56	0.60	0.69	0.92	0.69	0.73	0.77	0.86	1.06	1.15	1.19	1.23	1.31	1.49
8	5	0.60	0.63	0.65	0.70	0.82	0.90	0.93	0.95	1.00	1.11	1.20	1.23	1.25	1.30	1.41	2.00	2.03	2.05	2.10	2.20
8	6	0.92	0.93	0.94	0.97	1.02	1.38	1.39	1.40	1.43	1.48	1.84	1.85	1.86	1.88	1.93	3.06	3.07	3.08	3.11	3.16
8	7	1.28	1.29	1.29	1.30	1.31	1.92	1.93	1.93	1.94	1.95	2.57	2.57	2.57	2.58	2.59	4.28	4.28	4.28	4.29	4.30
8	8	1.68	1.68	1.68	1.68	1.68	2.51	2.51	2.51	2.51	2.51	3.35	3.35	3.35	3.35	3.35	5.58	5.58	5.58	5.58	5.58
9	0	0.00	0.16	0.32	0.65	1.30	0.00	0.16	0.32	0.65	1.30	0.00	0.16	0.32	0.65	1.30	0.00	0.32	0.65	1.30
9	1	0.01	0.16	0.32	0.64	1.27	0.01	0.16	0.32	0.64	1.27	0.02	0.17	0.32	0.64	1.27	0.03	0.17	0.33	0.65	1.28
9	2	0.06	0.18	0.33	0.62	1.20	0.08	0.20	0.34	0.64	1.22	0.11	0.22	0.36	0.65	1.24	0.19	0.27	0.41	0.70	1.28
9	3	0.17	0.25	0.37	0.62	1.13	0.26	0.33	0.43	0.68	1.18	0.34	0.41	0.50	0.74	1.24	0.57	0.63	0.71	0.91	1.39
9	4	0.36	0.42	0.49	0.66	1.07	0.55	0.60	0.66	0.81	1.19	0.73	0.78	0.84	0.98	1.33	1.21	1.27	1.33	1.45	1.73
9	5	0.64	0.68	0.73	0.82	1.04	0.96	1.00	1.05	1.13	1.33	1.28	1.32	1.37	1.45	1.64	2.14	2.18	2.22	2.31	2.48
9	6	1.00	1.02	1.05	1.10	1.21	1.49	1.52	1.55	1.60	1.71	1.99	2.02	2.04	2.10	2.20	3.32	3.35	3.37	3.42	3.53
9	7	1.42	1.43	1.44	1.47	1.52	2.13	2.14	2.15	2.18	2.23	2.84	2.85	2.86	2.88	2.93	4.73	4.74	4.75	4.77	4.82
9	8	1.89	1.89	1.89	1.90	1.91	2.83	2.83	2.84	2.84	2.86	3.77	3.78	3.78	3.79	3.80	6.29	6.29	6.30	6.30	6.32
9	9	2.39	2.39	2.39	2.39	2.39	3.58	3.58	3.58	3.58	3.58	4.77	4.77	4.77	4.77	4.77	7.95	7.95	7.95	7.95	7.95
10	0	0.00	0.20	0.40	0.80	1.60	0.00	0.20	0.40	0.80	1.60	0.00	0.20	0.40	0.80	1.60	0.00	0.20	0.40	0.80	1.60
10	1	0.01	0.20	0.40	0.79	1.57	0.01	0.20	0.40	0.79	1.57	0.02	0.20	0.40	0.79	1.58	0.03	0.21	0.41	0.80	1.58
10	2	0.06	0.22	0.40	0.77	1.51	0.09	0.24	0.42	0.79	1.52	0.11	0.26	0.44	0.80	1.54	0.19	0.31	0.49	0.85	1.59
10	3	0.18	0.29	0.44	0.77	1.43	0.26	0.36	0.51	0.83	1.49	0.35	0.44	0.57	0.89	1.54	0.58	0.67	0.77	1.06	1.70
10	4	0.38	0.46	0.56	0.82	1.37	0.57	0.65	0.73	0.96	1.50	0.76	0.83	0.92	1.11	1.63	1.26	1.34	1.42	1.59	2.02
10	5	0.68	0.74	0.80	0.95	1.34	1.01	1.07	1.14	1.27	1.60	1.35	1.41	1.47	1.60	1.90	2.25	2.31	2.37	2.50	2.77
10	6	1.06	1.11	1.15	1.24	1.45	1.59	1.64	1.68	1.77	1.96	2.13	2.17	2.21	2.30	2.49	3.54	3.59	3.63	3.72	3.89
10	7	1.53	1.56	1.58	1.64	1.75	2.30	2.32	2.35	2.40	2.51	3.06	3.09	3.12	3.17	3.28	5.11	5.13	5.16	5.21	5.32
10	8	2.07	2.08	2.09	2.12	2.17	3.10	3.12	3.13	3.15	3.20	4.14	4.15	4.16	4.19	4.24	6.90	6.91	6.92	6.95	7.00
10	9	2.66	2.66	2.66	2.67	2.68	3.98	3.99	3.99	4.00	4.01	5.31	5.32	5.32	5.33	5.34	8.85	8.86	8.86	8.87	8.88
10	10	3.27	3.27	3.27	3.27	3.27	4.91	4.91	4.91	4.91	4.91	6.54	6.54	6.54	6.54	6.54	10.91	10.91	10.91	10.91	10.91
12	0	0.00	0.29	0.58	1.15	2.30	0.00	0.29	0.58	1.15	2.30	0.00	0.29	0.58	1.15	2.30	0.00	0.29	0.58	1.15	2.30
12	1	0.01	0.29	0.57	1.14	2.28	0.01	0.29	0.57	1.14	2.28	0.02	0.29	0.58	1.14	2.28	0.03	0.30	0.58	1.15	2.29
12	2	0.06	0.31	0.58	1.12	2.21	0.09	0.32	0.60	1.14	2.23	0.12	0.34	0.61	1.16	2.24	0.19	0.39	0.66	1.20	2.29
12	3	0.18	0.37	0.62	1.12	2.13	0.27	0.44	0.68	1.18	2.19	0.37	0.51	0.74	1.24	2.25	0.61	0.74	0.93	1.40	2.40
12	4	0.40	0.53	0.73	1.17	2.07	0.60	0.73	0.89	1.31	2.20	0.81	0.93	1.07	1.46	2.34	1.34	1.46	1.59	1.89	2.72
12	5	0.73	0.84	0.96	1.29	2.04	1.09	1.20	1.32	1.58	2.30	1.46	1.56	1.68	1.92	2.57	2.43	2.54	2.64	2.87	3.39
12	6	1.17	1.26	1.35	1.55	2.08	1.75	1.84	1.93	2.12	2.55	2.34	2.42	2.51	2.70	3.10	3.89	3.98	4.07	4.25	4.62
12	7	1.71	1.78	1.85	1.99	2.30	2.57	2.64	2.70	2.84	3.13	3.43	3.50	3.56	3.70	3.98	5.72	5.78	5.85	5.98	6.26
12	8	2.36	2.41	2.45	2.55	2.74	3.54	3.59	3.63	3.73	3.92	4.72	4.77	4.82	4.91	5.10	7.87	7.92	7.97	8.06	8.24
12	9	3.10	3.13	3.15	3.21	3.32	4.65	4.67	4.70	4.76	4.87	6.20	6.22	6.25	6.31	6.42	10.33	10.35	10.38	10.44	10.55
12	10	3.91	3.92	3.93	3.96	4.01	5.86	5.87	5.88	5.91	5.96	7.81	7.82	7.84	7.86	7.91	13.02	13.03	13.04	13.07	13.12
12	11	4.76	4.77	4.77	4.78	4.79	7.15	7.15	7.15	7.16	7.17	9.53	9.53	9.54	9.54	9.56	15.88	15.89	15.89	15.89	15.91
12	12	5.65	5.65	5.65	5.65	5.65	8.48	8.48	8.48	8.48	8.48	11.31	11.31	11.31	11.31	11.31	18.84	18.84	18.84	18.84	18.84

Table C30-2 Factored Moments Induced in Walls by Lateral Soil and/or Wind Loads (ft-kips/linear ft)
(For Use with Appendix C Load Factors: Soil 1.7, Wind 1.3)

Wall Ht. (ft.)	Back fill Ht. (ft.)	30	30	30	30	30	45	45	45	45	45	60	60	60	60	60	100	100	100	100	100
		\multicolumn{20}{c}{Unfactored Design Lateral Soil Load (psf per foot of depth)}
		0	10	20	40	80	0	10	20	40	80	0	10	20	40	80	0	10	20	40	80
8	0	0.00	0.10	0.21	0.42	0.83	0.00	0.10	0.21	0.42	0.83	0.00	0.10	0.21	0.42	0.83	0.00	0.10	0.21	0.42	0.83
8	1	0.01	0.11	0.21	0.41	0.81	0.01	0.11	0.21	0.41	0.81	0.02	0.11	0.21	0.41	0.81	0.03	0.12	0.22	0.42	0.82
8	2	0.05	0.13	0.22	0.40	0.76	0.08	0.15	0.23	0.42	0.78	0.11	0.17	0.25	0.43	0.80	0.18	0.23	0.30	0.48	0.84
8	3	0.16	0.21	0.27	0.41	0.72	0.25	0.29	0.34	0.47	0.77	0.33	0.37	0.42	0.54	0.83	0.55	0.59	0.63	0.73	0.99
8	4	0.35	0.38	0.41	0.49	0.69	0.52	0.55	0.58	0.65	0.82	0.69	0.72	0.76	0.82	0.98	1.15	1.18	1.22	1.28	1.42
8	5	0.60	0.62	0.64	0.68	0.78	0.90	0.92	0.94	0.98	1.07	1.20	1.22	1.24	1.28	1.37	2.00	2.02	2.04	2.08	2.17
8	6	0.92	0.93	0.94	0.96	1.00	1.38	1.39	1.40	1.42	1.46	1.84	1.85	1.86	1.88	1.92	3.06	3.07	3.08	3.10	3.14
8	7	1.28	1.29	1.29	1.29	1.30	1.92	1.93	1.93	1.93	1.95	2.57	2.57	2.57	2.58	2.59	4.28	4.28	4.28	4.29	4.30
8	8	1.68	1.68	1.68	1.68	1.68	2.51	2.51	2.51	2.51	2.51	3.35	3.35	3.35	3.35	3.35	5.58	5.58	5.58	5.58	5.58
9	0	0.00	0.13	0.26	0.53	1.05	0.00	0.13	0.26	0.53	1.05	0.00	0.13	0.26	0.53	1.05	0.00	0.13	0.26	0.53	1.05
9	1	0.01	0.13	0.26	0.52	1.03	0.01	0.13	0.26	0.52	1.03	0.02	0.14	0.27	0.52	1.04	0.03	0.14	0.27	0.53	1.04
9	2	0.06	0.15	0.27	0.51	0.98	0.08	0.17	0.29	0.53	1.00	0.11	0.19	0.31	0.54	1.02	0.19	0.25	0.36	0.59	1.06
9	3	0.17	0.23	0.32	0.52	0.94	0.26	0.31	0.39	0.58	0.99	0.34	0.40	0.46	0.65	1.05	0.57	0.62	0.68	0.83	1.21
9	4	0.36	0.41	0.46	0.59	0.91	0.55	0.59	0.64	0.75	1.05	0.73	0.77	0.82	0.93	1.18	1.21	1.26	1.31	1.40	1.62
9	5	0.64	0.68	0.71	0.78	0.95	0.96	1.00	1.03	1.10	1.25	1.28	1.32	1.35	1.42	1.57	2.14	2.17	2.21	2.27	2.41
9	6	1.00	1.02	1.04	1.08	1.17	1.49	1.52	1.54	1.58	1.67	1.99	2.01	2.03	2.08	2.16	3.32	3.34	3.36	3.41	3.49
9	7	1.42	1.43	1.44	1.46	1.50	2.13	2.14	2.15	2.17	2.21	2.84	2.85	2.86	2.88	2.92	4.73	4.74	4.75	4.77	4.81
9	8	1.89	1.89	1.89	1.90	1.91	2.83	2.83	2.84	2.84	2.85	3.77	3.78	3.78	3.79	3.80	6.29	6.29	6.30	6.30	6.31
9	9	2.39	2.39	2.39	2.39	2.39	3.58	3.58	3.58	3.58	3.58	4.77	4.77	4.77	4.77	4.77	7.95	7.95	7.95	7.95	7.95
10	0	0.00	0.16	0.33	0.65	1.30	0.00	0.16	0.33	0.65	1.30	0.00	0.16	0.33	0.65	1.30	0.00	0.16	0.33	0.65	1.30
10	1	0.01	0.16	0.32	0.64	1.28	0.01	0.17	0.32	0.64	1.28	0.02	0.17	0.33	0.65	1.28	0.03	0.17	0.33	0.65	1.29
10	2	0.06	0.18	0.33	0.63	1.23	0.09	0.20	0.35	0.65	1.25	0.11	0.22	0.37	0.67	1.26	0.19	0.28	0.42	0.71	1.31
10	3	0.18	0.26	0.38	0.65	1.18	0.26	0.34	0.45	0.71	1.24	0.35	0.42	0.52	0.77	1.30	0.58	0.65	0.73	0.94	1.45
10	4	0.38	0.44	0.52	0.72	1.16	0.57	0.63	0.70	0.87	1.29	0.76	0.82	0.89	1.03	1.43	1.26	1.33	1.39	1.52	1.84
10	5	0.68	0.73	0.78	0.89	1.18	1.01	1.06	1.11	1.22	1.46	1.35	1.40	1.45	1.55	1.78	2.25	2.30	2.35	2.45	2.66
10	6	1.06	1.10	1.13	1.21	1.37	1.59	1.63	1.66	1.74	1.89	2.13	2.16	2.20	2.27	2.41	3.54	3.58	3.61	3.68	3.83
10	7	1.53	1.55	1.57	1.62	1.71	2.30	2.32	2.34	2.38	2.47	3.06	3.08	3.11	3.15	3.24	5.11	5.13	5.15	5.19	5.28
10	8	2.07	2.08	2.09	2.11	2.15	3.10	3.11	3.12	3.14	3.18	4.14	4.15	4.16	4.18	4.22	6.90	6.91	6.92	6.94	6.98
10	9	2.66	2.66	2.66	2.67	2.68	3.98	3.99	3.99	3.99	4.01	5.31	5.31	5.32	5.32	5.33	8.85	8.86	8.86	8.86	8.87
10	10	3.27	3.27	3.27	3.27	3.27	4.91	4.91	4.91	4.91	4.91	6.54	6.54	6.54	6.54	6.54	10.91	10.91	10.91	10.91	10.91
12	0	0.00	0.23	0.47	0.94	1.87	0.00	0.23	0.47	0.94	1.87	0.00	0.23	0.47	0.94	1.87	0.00	0.23	0.47	0.94	1.87
12	1	0.01	0.24	0.47	0.93	1.85	0.01	0.24	0.47	0.93	1.85	0.02	0.24	0.47	0.93	1.85	0.03	0.25	0.48	0.94	1.86
12	2	0.06	0.26	0.48	0.92	1.80	0.09	0.27	0.49	0.94	1.82	0.12	0.29	0.51	0.95	1.84	0.19	0.35	0.56	1.00	1.88
12	3	0.18	0.33	0.53	0.93	1.75	0.27	0.40	0.59	0.99	1.81	0.37	0.48	0.65	1.05	1.87	0.61	0.71	0.85	1.22	2.02
12	4	0.40	0.51	0.65	1.00	1.73	0.60	0.70	0.82	1.15	1.86	0.81	0.90	1.01	1.30	2.00	1.34	1.44	1.54	1.77	2.39
12	5	0.73	0.82	0.91	1.15	1.76	1.09	1.18	1.27	1.48	2.02	1.46	1.54	1.63	1.83	2.31	2.43	2.52	2.60	2.78	3.18
12	6	1.17	1.24	1.31	1.47	1.85	1.75	1.82	1.89	2.05	2.38	2.34	2.41	2.48	2.63	2.94	3.89	3.96	4.03	4.18	4.48
12	7	1.71	1.77	1.82	1.94	2.18	2.57	2.63	2.68	2.79	3.02	3.43	3.48	3.54	3.65	3.87	5.72	5.77	5.82	5.93	6.15
12	8	2.36	2.40	2.44	2.51	2.67	3.54	3.58	3.62	3.69	3.85	4.72	4.76	4.80	4.87	5.02	7.87	7.91	7.95	8.02	8.17
12	9	3.10	3.12	3.14	3.19	3.28	4.65	4.67	4.69	4.74	4.83	6.20	6.22	6.24	6.28	6.37	10.33	10.35	10.37	10.42	10.50
12	10	3.91	3.92	3.93	3.95	3.99	5.86	5.87	5.88	5.90	5.94	7.81	7.82	7.83	7.85	7.89	13.02	13.03	13.04	13.06	13.10
12	11	4.76	4.77	4.77	4.78	4.79	7.15	7.15	7.15	7.16	7.17	9.53	9.53	9.53	9.54	9.55	15.88	15.88	15.89	15.89	15.90
12	12	5.65	5.65	5.65	5.65	5.65	8.48	8.48	8.48	8.48	8.48	11.31	11.31	11.31	11.31	11.31	18.84	18.84	18.84	18.84	18.84

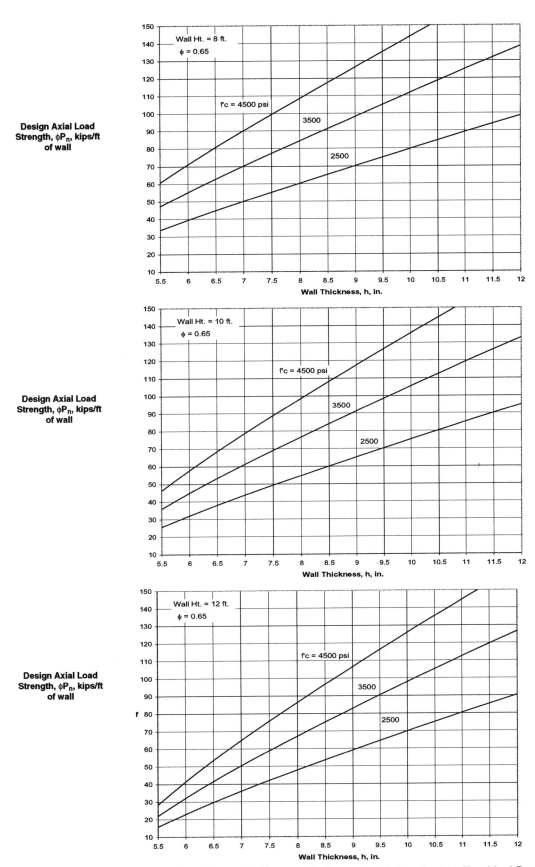

Figure C30-1 Design Axial Load Strength, P_n, of Plain Concrete Walls using the Empirical Design Method

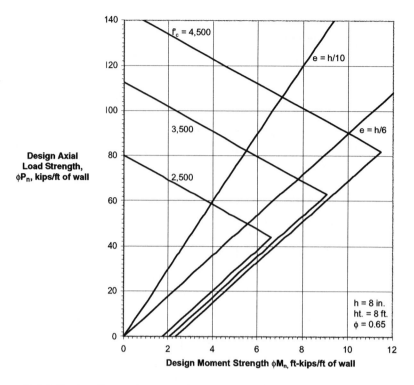

Figure C30-2 Design Strength Interaction Diagrams for 8.0-in. Wall, 8 ft in Height

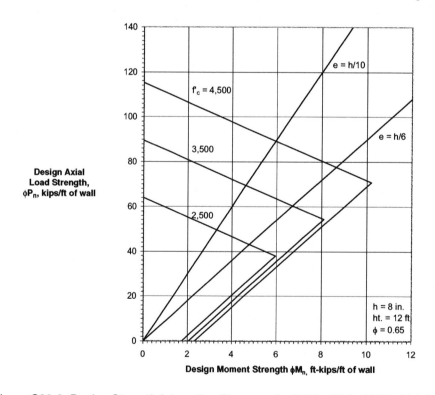

Figure C30-3 Design Strength Interaction Diagrams for 8.0-in. Wall, 12 ft in Height

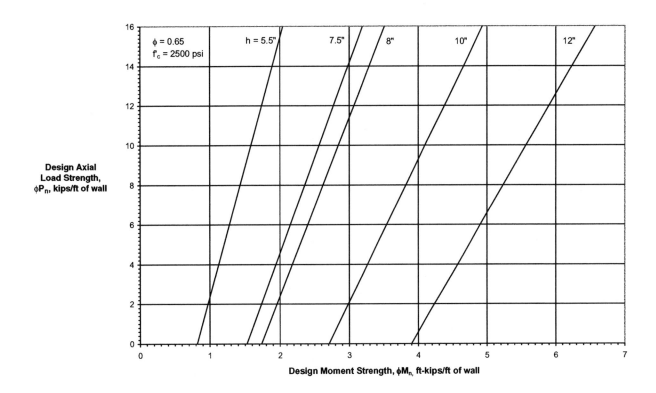

Figure C30-4 Design Strength Interaction Diagrams for Lightly-Loaded Plain Concrete Walls (f'_c = 2500 psi)

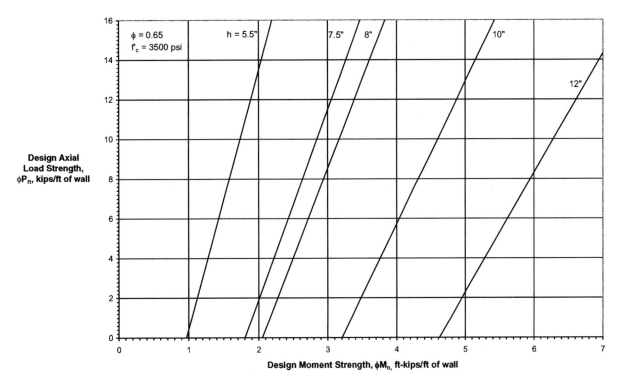

Figure C30-5 Design Strength Interaction Diagrams for Lightly-Loaded Plain Concrete Walls (f'_c = 3500 psi)

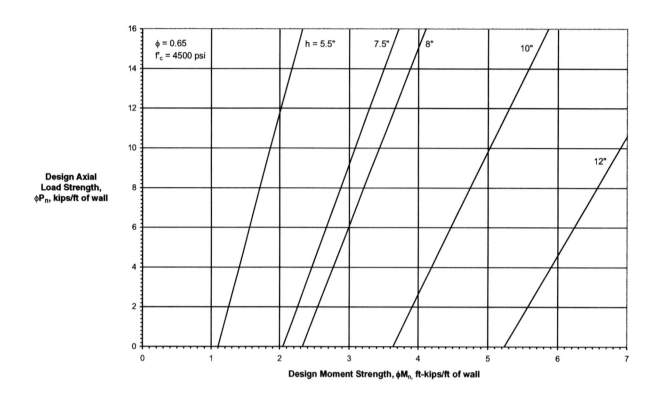

Figure C30-6 Design Strength Interaction Diagrams for Lightly-Loaded Plain Concrete Walls (f'_c = 4500 psi)

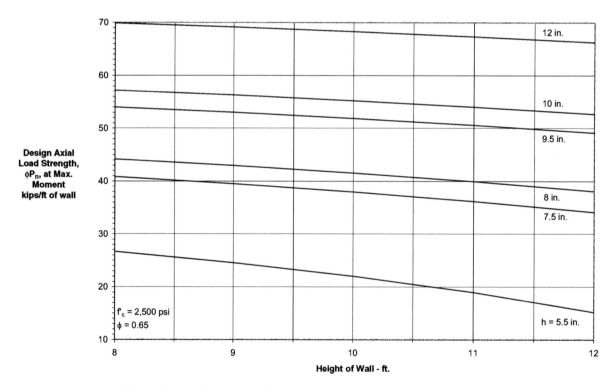

Figure C30-7 Design Axial Load Strength of Plain Concrete Walls at Maximum Design Moment Strength (f'_c = 2500 psi)

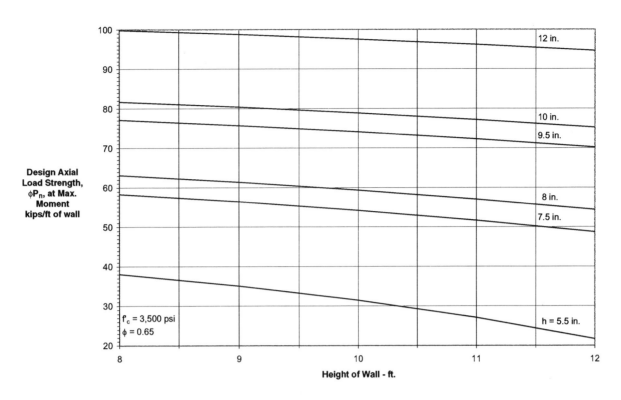

Figure C30-8 Design Axial Load Strength of Plain Concrete Walls at Maximum Design Moment Strength (f'_c = 3500 psi)

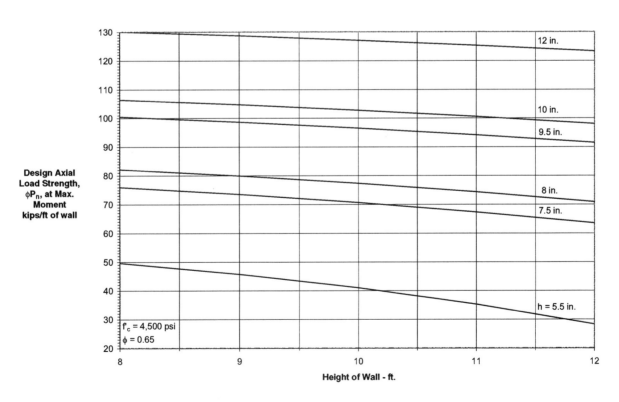

Figure C30-9 Design Axial Load Strength of Plain Concrete Walls at Maximum Design Moment Strength (f'_c = 4500 psi)

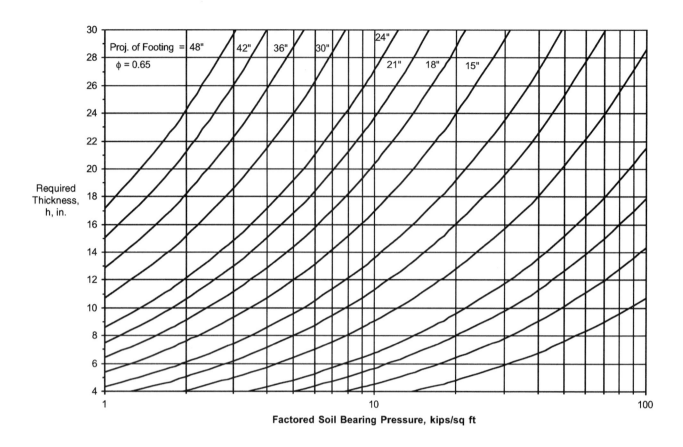

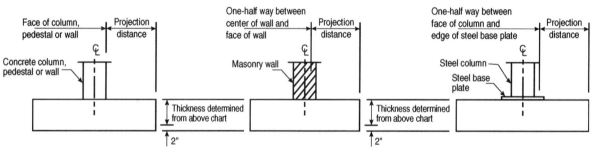

*For f'_c greater than 2500 psi, multiple thickness determined from above chart by $(2500/f'_c)^{0.25}$

Figure C30-10 Thickness of Footing Required to Satisfy Flexural Strength for Various Projection Distances, in.
($f'_c = 2500$ psi*)

Example 30.1—Design of Plain Concrete Footing and Pedestal

Proportion a plain concrete square footing with pedestal for a residential occupancy building. Design in conformance with Chapter 22 using the load and strength reduction factors of Chapter 9. Perform a second design using the alternate load and strength reduction factors of Appendix C to determine which results in a more economical design.

Design data:

Service dead load = 40 kips
Service floor live load = 40 kips
Service roof live load = 7.5 kips
Service roof snow load = 10 kips
Service surcharge = 0
Pedestal dimensions = 12 × 12 in.
Permissible soil bearing pressure = 2.5 ksf
f'_c = 2500 psi

	Calculations and Discussion	Code Reference

1. Determine base area of footing:

   The base area is determined by using unfactored service gravity loads and the permissible soil bearing pressure.

   22.7.2

   $$A_f = \frac{40 + 40 + 10}{2.5} = 36 \text{ ft}^2$$

   Use 6 × 6 ft square footing (A_f = 36 ft²)

2. Determine the applicable load combinations that must be considered.

   To proportion the footing for strength, factored loads must be used. Two sets of load factors are provided; one in 9.2 and the other in C.2. Because of the significant difference between strength reduction factor, ϕ, to be used with each set of load factors (0.55 in 9.3.5 versus 0.65 in C.3.5), a design in accordance with each set of factors should be evaluated to determine which alternate will provide the more economical solution.

   22.7.1
   9.2
   9.3.5
   C.2
   C.3.5

   The required strength must at least equal the largest factored load determined from applicable load combinations. One of the following load combinations will govern:

   1. U = 1.2D + 1.6L + 0.5S   *Eq. (9-2)*
   2. U = 1.2D + 0.5L + 1.6S   *Eq. (9-3)*
   3. U = 1.2D + 0.5L + 0.5S   *Eq. (9-4)*

   Note that in Combination 1, T is being neglected. In Combinations 2 and 3 the factor on L is 0.5 in accordance with 9.2.1(a).

   9.2.1(a)

Example 30.1 (cont'd)	Calculations and Discussion	Code Reference

3. Calculate the factored axial load, P_u, for each load combination.

   By observation it can be seen that either Combination 1 or 2 will yield the largest factored axial load.

   1. $P_u = 1.2D + 1.6L + 0.5S = 1.2(40) + 1.6(40) + 0.5(10) = 117$ kips — Eq. (9-2)

   2. $P_u = 1.2D + 0.5L + 1.6S = 1.2(40) + 0.5(40) + 1.6(10) = 64.8$ kips — Eq. (9-3)

   Use $P_u = 117$ kips

   Upon reviewing the applicable load combination of C.2, it is obvious that Eq. (C-1) will control:

   $P_u = 1.4D + 1.7L = 1.4(40) + 1.7(50) = 141$ kips — Eq. (C-1)

   To quickly determine which one of the two sets of load and strength reduction factors to use, compare the nominal axial load strength, P_n, required by the factored loads of 9.2 to that required by C.2. — 9.3.5, C.3.5

   Chapter 9 $P_n = P_u/\phi = 117/0.55 = 212.7$ kips

   Appendix C $P_n = P_u/\phi = 141/0.65 = 216.9$ kips

   Since the nominal axial load strength required by the load and strength reduction factors of Chapter 9 is less than the corresponding load required by Appendix C, design according to Chapter 9 will be more economical. Regardless of the load and strength reduction factors used, the design procedures will be the same.

4. Calculate the factored soil bearing pressure.

   Since the footing must be proportioned for strength by using factored loads and induced reactions, the factored soil bearing pressure must be used. — 22.7.1

   $q_s = \dfrac{P_u}{A_f} = \dfrac{117}{36} = 3.25$ ksf

5. Determine the footing thickness required to satisfy moment strength.

   For plain concrete, flexural strength will usually control thickness. The critical section for calculating moment is at the face of the concrete pedestal (see figure above). — 22.7.5(a)

   $M_u = q_s (b) \left(\dfrac{b-c}{2}\right)\left(\dfrac{b-c}{4}\right)$

   $= 3.25 (3) (2.5)^2 = 60.9$ ft-kips

   $\phi M_n \geq M_u$ — Eq. (22-1)

| Example 30.1 (cont'd) | Calculations and Discussion | Code Reference |

$\phi = 0.55$ for all stress conditions   *9.3.5*

$\phi M_n = 5\phi\sqrt{f'_c}\, S_m$ *Eq. (22-2)*

$$= \frac{5\,(0.55)\left(\sqrt{2500}\right)(6)(12)h^2}{(1000)(6)} \geq 60.9 \text{ ft-kips}$$

Solving for h:

$$h \geq \left[\frac{60.9\,(12)(1000)(6)}{5\,(0.55)\left(\sqrt{2500}\right)(6)(12)}\right]^{0.5} = 21.0 \text{ in.}$$

Alternate solution using Fig. 30-10:

Enter figure with factored soil bearing pressure (3.25 kips/sq. ft). Project upward to the distance that the footing projects beyond the face of the pedestal, which is 30 in. (36 - 6). Then project horizontally and read approximately 21 in. as the required footing thickness.

For concrete cast on the soil, the bottom 2 in. of concrete cannot be considered for strength computations (the reduced overall thickness is to allow for unevenness of the excavation and for some contamination of the concrete adjacent to the soil).   *22.4.8*

Use overall footing thickness of 24 in.

6. Check for beam action shear. Use effective thickness of h = 22 in. = 1.83 ft.

   The critical section for beam action shear is located a distance equal to the thickness, h, away from the face of the pedestal, or 0.67 ft (3 - 0.5 - 1.83) from the edge of the footing.   *22.7.6.1*
   22.7.6.2(a)

   $V_u = q_s b\left[\left(\dfrac{b}{2}\right) - \left(\dfrac{c}{2}\right) - h\right] = 3.25\,(6)(0.67) = 13.07 \text{ kips}$

   $\phi V_n \geq V_u$ *Eq. (22-8)*

   $V_n = \left(\dfrac{4}{3}\right)\sqrt{f'_c}\, b_w h$ *Eq. (22-9)*

   $\phi V_n = \dfrac{4\,(0.55)\left(\sqrt{2500}\right)(72)(22)}{(3)(1000)} = 58.08 \text{ kips} > 13.07 \text{ kips} \quad \text{O.K.}$

7. Check for two-way action (punching) shear.

   The critical section for two-way action shear is located a distance equal to one-half the footing thickness, h, away from the face of the pedestal.   *22.7.6.1*
   22.7.6.2(b)

| Example 30.1 (cont'd) | Calculations and Discussion | Code Reference |

$$V_u = q_s [b^2 - (c+h)^2] = 3.25 \left[6^2 - \left(1 + \frac{22}{12}\right)^2 \right] = 90.91 \text{ kips}$$

$$\phi V_n \geq V_u \qquad \qquad Eq.\ (22\text{-}8)$$

$$V_n = \left[\frac{4}{3} + \frac{8}{3\beta} \right] \sqrt{f'_c} b_o h \leq 2.66 \sqrt{f'_c} b_o h \qquad \qquad Eq.\ (22\text{-}10)$$

where β_c is the ratio of the long-to-short side of the supported load. In this case, $\beta_c = 1$.

Since $\left[\left(\frac{4}{3}\right) + \left(\frac{8}{3}\right) \right] = 4.0 > 2.66$,

$$V_n = 2.66 \sqrt{f'_c} b_o h$$

$$\phi V_n = \frac{2.66 (0.55) \left(\sqrt{2500}\right)(34)(4)(22)}{1000} = 218.87 \text{ kips} > 90.91 \text{ kips} \quad \text{O.K.}$$

8. Check bearing strength of pedestal.   *22.8.3*

   $P_u = 141$ kips (from Step 3)

   $\phi B_n \geq P_u$ *Eq. (22-11)*

   $B_n = 0.85 f'_c A_1$ *Eq. (22-12)*

   $$\phi B_n = \frac{0.85 (0.55)(2500)(12 \times 12)}{1000} = 168.3 \text{ kips} > 117 \text{ kips} \quad \text{O.K.}$$

Example 30.2—Design of Plain Concrete Basement Wall

A plain concrete basement wall is to be used to support a 2-story residential occupancy building of wood frame construction with masonry veneer. The height of the wall is 10 ft (distance between the top of the concrete slab and the wood-framed floor, both of which provide lateral support of the wall). The backfill height is 7 ft and the wall is laterally restrained at the top. Design of the wall is required in accordance with Chapter 22 using the load and strength reduction factors of Chapter 9. Perform a second design using the alternate load and strength reduction factors of Appendix C to determine which results in a more economical design.

Design data:

Service dead load = 1.6 kips per linear foot
Service floor live load = 0.8 kips per linear foot
Service roof live load = 0.4 kips per linear foot
Service roof snow load = 0.3 kips per linear foot
Service lateral earth pressure = 60 psf/ft of depth
Service wind pressure = 20 psf inward, 25 psf outward
Assume factored roof dead load plus factored wind uplift load on roof (Eq. 9-6) = 0
Eccentricity of axial loads = 0

Calculations and Discussion	Code Reference

Design using load and strength reduction factors of Chapter 9.

1. The wall must be designed for vertical, lateral, and other loads to which it will be subjected. Therefore, determine the applicable load combinations that must be considered.   *22.6.2 / 9.2*

    1. $U = 1.2D + 1.6L + 0.5L_r + 1.6H$ *Eq. (9-2)*
    2. $U = 1.2D + 0.5L + 1.6L_r + 1.6H$ *Eq. (9-3)*
    3. $U = 1.2D + 0.5L + 0.5L_r\ 1.6W + 1.6H$ *Eq. (9-4)*
    4. $U = 0.9D + 1.6W + 1.6H$ *Eq. (9-6)*
    5. $U = 1.6H$ *Eq. (9-6)*

    Note that in Combination 1, T is being neglected. In Combinations 2 and 3 the factor on L is 0.5 in accordance with 9.2.1(a). In Combinations 2 and 4, 1.6H has been included since the lateral soil load will always be acting. In Combination 5, D has been omitted since this condition may occur during construction.

2. Calculate the axial load, P_u, for each load combination.

    1. $P_u = 1.2D + 1.6L + 0.5L_r = 1.2(1.6) + 1.6(0.8) + 0.5(0.4) = 3.40$ kips/ft   *Eq. (9-2)*
    2. $P_u = 1.2D + 0.5L + 1.6L_r = 1.2(1.6) + 0.5(0.8) + 1.6(0.4) = 2.96$ kips/ft   *Eq. (9-3)*
    3. $P_u = 1.2D + 0.5L + 0.5L_r = 1.2(1.6) + 0.5(0.8) + 0.5(0.4) = 2.52$ kips/ft   *Eq. (9-4)*

| Example 30.2 (cont'd) | Calculations and Discussion | Code Reference |

    4. $P_u = 0.9D = 0.9(1.6) = 1.44$ kips/ft   *Eq. (9-6)*

    5. $P_u = 0$ *Eq. (9-6)*

3. Calculate the moment, M_u, for load combination.

    1. M_u due to 1.6H   *Eq. (9-2)*

    2. M_u due to 1.6H   *Eq. (9-3)*

    3. M_u due to 1.6W + 1.6H   *Eq. (9-4)*

    4. M_u due to 1.6W + 1.6H   *Eq. (9-6)*

    5. M_u due to 1.6H   *Eq. (9-6)*

The maximum moment occurs at the location of zero shear. To determine this location with respect to the top of the wall, first calculate the reaction at the top of the wall. If wind is acting and in the same direction as the lateral soil load, and the resultant of the wind load, W, is greater than the reaction at the top of the wall, the location of zero shear is some distance "X" below the top of the wall (above the top of the backfill). Otherwise, the zero shear location will be some distance "X" below the top of the backfill. Next, sum the horizontal forces above "X" (the location of zero shear). Finally, solve for "X." See figure below.

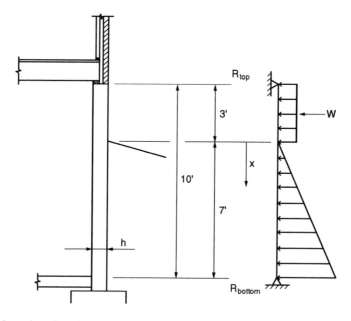

For load Combinations 3 and 4, the reaction at the top of the wall is:

$$R_{top} = \frac{1.6\,(20)\,(3)\,(8.5) + [1.6\,(60)7^3]/6}{10} = 630.4 \text{ plf}$$

Example 30.2 (cont'd)	Calculations and Discussion	Code Reference

W = 1.6(20)(3) = 96 plf

W < R_{top}

Therefore, location of zero shear is below top of backfill.

Sum horizontal forces at "X":

630.4 - 1.6 (20) (3) - [1.6 (60) (X^2)]/2 = 0

630.4 - 96 - 48X^2 = 0

Solving for distance "X":

$$X = \left(\frac{630.4 - 96}{48}\right)^{0.5} = 3.37 \text{ ft}$$

The point of zero shear is 3.0 + 3.37 = 6.37 ft from the top of the wall.

Note: It is generally simpler to compute the location of zero shear with respect to the top of the wall since doing so with respect to the bottom will involve solving a quadratic equation.

Compute the maximum moment, M_u, due to the wind and lateral soil loads:

$$M_u = \frac{630.4 \ (6.37) - 1.6 \ (20) \ (3) \ (6.37 - 1.5) - [1.6 \ (60) \ (3.37)^3]/6}{1000} = 2.94 \text{ ft-kips/ft}$$

Alternately, the maximum moment can be determined from Table 30-1 since the load factor on wind is 1.6.

From Table 30-1, for a 10-ft high wall with 7 ft of backfill, for unfactored wind and soil loads of 20 psf and 60 psf/ft, respectively, the moment, M_u, is 2.94 ft-kips/ft, which is the same as calculated above.

Next determine from the moment Table 30-1 for load Combinatios 1, 2 and 5 (with no wind acting). From Table 30-1 for a 10-ft high wall with 7 ft of backfill, for unfactored wind and soil loads of 0 psf and 60 psf/ft, respectively, the moment, M_u, is 2.88 ft-kips/ft. Note that either Table 30-1 or 30-2 can be used since this load combination does not include wind. The moment is read from the zero column for "unfactored design wind pressure."

4. Calculate the effective eccentricities for load Combinations 1, 3, and 4 to determine if the wall can be designed by the empirical method (i.e., e ≤ h/6). Assume a conservative wall thickness of 12 in.

    22.6.3
    22.6.5

    Allowable e = 12/6 = 2 in = 0.167 ft

    For load Combination 1:

    $$e = \frac{2.88}{3.40} = 0.85 \text{ ft} > 0.167 \text{ ft}$$

30-33

		Code
Example 30.2 (cont'd)	**Calculations and Discussion**	**Reference**

For load Combination 3:

$$e = \frac{2.94}{2.52} = 1.17 \text{ ft} > 0.167 \text{ ft}$$

For load Combination 4:

$$e = \frac{2.94}{1.44} = 2.04 \text{ ft} > 0.167 \text{ ft}$$

Since the effective eccentricity exceeds h/6, the wall cannot be designed by the empirical method. The wall must be designed taking into consideration both flexure and axial compression.

22.5.3
22.6.3

5. Determine the required wall thickness to satisfy the axial load and induced moments by using the appropriate interaction equation. The factored axial loads and moments for the various load combinations are summarized in the following table.

Eq. (22-6)
Eq. (22-7)

Load Combination	Axial Load, P_u, kips/ft	Moment, M_u, ft-kips/ft
1	3.40	2.88
2	2.96	2.88
3	2.52	2.94
4	1.44	2.94
5	0	2.88

By observation, the combination of very low axial load and relatively high moment will be governed by interaction Eq. (22-7).

22.5.3

$$\frac{M_u}{S} - \frac{P_u}{A_g} \leq 5\phi\sqrt{f'_c}$$

Eq. (22-6)

By rearranging the equation and substituting for S and A_g in terms of "h," the required thickness can be determined by solving the following quadratic equation (3):

$$0.06\phi\sqrt{f'_c}\, h^2 + P_u h - 72 M_u = 0 \qquad (3)$$

Determine the required wall thickness, h, to satisfy load Combination 5 ($\phi = 0.55$, $P_u = 0$ and $M_u = 2.88$ ft-kips/ft), since this combination has the highest moment and lowest axial load. Since $P_u = 0$, equation (3) simplifies to equation (4). Use equation (4) to solve for required, "h," assuming $f'_c = 4000$ psi.

$$h = (72 M_u / 0.06\phi \sqrt{f'_c})^{1/2} = [((72)(2.88))/((0.6)(0.55)(\sqrt{4000}))]^{1/2} = 9.97 \text{ in.} \qquad (4)$$

Assume a 10-inch wall with $f'_c = 4000$ psi.

Check preliminary wall selection for all load combinations by interpolating between values from Figs. 30-5 and 30-6 for $f'_c = 3500$ and 4500 psi, respectively. The following table summarizes the required axial load, P_u, and moment, M_u, strengths for the various

	Code Reference
Example 30.2 (cont'd) **Calculations and Discussion**	

load combinations, and indicates the approximate design moment strength, ϕM_n, determined from the figures based on a wall thickness of 10 in.

Load Combination	Axial Load, P_u kips/ft	Moment, M_u ft-kips/ft	Approximate Design Moment Strength,[1] ϕM_n ft-kips/ft	$\phi M_n / M_u$
1	3.40	2.88	Note 2	—
2	2.96	2.88	(3.1 + 3.5)/2 = 3.3	1.15
3	2.52	2.94	Note 2	—
4	1.44	2.94	(2.9 + 3.3)/2 = 3.1	1.05
5	0	2.88	(2.7 + 3.1)/2 = 2.9	1.01

[1] Values interpolated from Figs. 30-5 and 30-6.
[2] There is no need to evaluate this combination since another combination has the same moment and a lower axial load. Given equal moments, the combination with the lower axial load will govern.

Since the ratio of design moment strength, ϕM_n, to required moment strength, M_u, exceeds 1 in all cases, the 10-in. wall is adequate for axial loading and induced moments.

Tentatively, use 10-in. wall with concrete $f'_c = 4000$ psi and check for shear.

6. Check for shear strength.

    Shear strength will rarely govern the design of a wall; nevertheless, it should not be overlooked. The shear will be greatest at the bottom of the wall. The critical section for calculating shear is located at wall thickness, h, above the top of the floor slab.   *22.5.4*   *22.6.4*

    For shear, load Combinations 3 and 4, which are the same, will control. Calculate the reaction at the bottom of the wall.

    $$R_{bottom} = \frac{[(1.6)(20)(3)^2/2] + [((1.6)(60)(7)^2/2)(3 + (2/3)(7))]}{10} = 1818 \text{ plf}$$

    $$V_u = 1818 - 1.6(60)\{[(7 - 10/12)(10/12)] + [(10/12)^2/2]\}$$

    $$= 1291 \text{ plf} = 1.29 \text{ klf}$$

    $\phi V_n \geq V_u$ *Eq. (22-8)*

    $\phi V_n = \dfrac{4\phi\sqrt{f'_c}\, b_w h}{3}$ *Eq. (22-9)*

    $= \dfrac{4(0.55)(\sqrt{4000})(12)(10)}{3(1000)} = 5.57 \text{ klf} > 1.29 \text{ klf} \quad \text{O.K.}$

7. Use 10-in. wall with specified compressive strength of concrete, $f'_c = 4000$ psi.

Example 30.2 (cont'd)	Calculations and Discussion	Code Reference

Design using load and strength reduction factors of Appendix C.

C1. The wall must be designed for vertical, lateral, and other loads to which it will be subjected. Therefore, determine the applicable load combinations that must be considered.

    1.   U = 0.75(1.4D + 1.7L) + 1.6W + 1.7H   *Eq. (C-2)*

    2.   U = 0.9D + 1.6W + 1.7H   *Eq. (C-3)*

    3.   U = 1.4D + 1.7L + 1.7H   *Eq. (C-4)*

    4.   U = 0.9D + 1.7H   *Eq. (C-3)*

    5.   U = 1.7H   *Eq. (C-3)*

Code Reference: 22.6.2, C.2

In Combinations 1 and 2, 1.7H has been added since lateral earth pressure will always be acting and for consistency with Eq. (9-2) and (9-3). For consistency with Eq. (9-6) and since dead load, D, may not be acting during early phases of construction, it has been omitted.

C2. Calculate the factored axial load, P_u, for each load combinations.

    1. P_u = 0.75 (1.4D + 1.7L) = 0.75((1.4)(1.6) + (1.7)(1.2)) = 3.21 kips/ft   *Eq. (C-2)*

    2. P_u = 0.9D = 0.9(1.6) = 1.44 kips/ft   *Eq. (C-3)*

    3. P_u = 1.4D + 1.7L = 1.4(1.6) + 1.7(1.2) = 4.28 kips/ft   *Eq. (C-4)*

    4. P_u = 0.9D = 0.9(1.6) = 1.44 kips/ft   *Eq. (C-3)*

    5. P_u = 0   *Eq. (C-3)*

C3. Calculate the factored moment, M_u, for each load combinations using Table 30-3.

    1. M_u = 1.6W + 1.7H = 3.12 ft-kips/ft   *Eq. (C-2)*

    2. M_u = 1.6W + 1.7H = 3.12 ft-kips/ft   *Eq. (C-3)*

    3. M_u = 1.7H = 3.06 ft-kips/ft   *Eq. (C-4)*

    4. M_u = 1.7H = 3.06 ft-kips/ft   *Eq. (C-3)*

    5. M_u = 1.7H = 3.06 ft-kips/ft   *Eq. (C-3)*

C4. By observation, the effective eccentricities exceed one-sixth the wall thickness; therefore, the empirical design procedure cannot be used.   *22.6.5.1*

C5. Determine the required wall thickness by using the appropriate interaction equation. Since lightly-loaded walls are governed by Eq. (22-7), Figures 30-4 through C30-6 will be used to design the wall for axial loads and flexure. The following table shows for each load combination the factored axial load and moment, and corresponding approximate design moment strength, ϕM_n, and ratio of over- or under-strength based on the assumed wall thickness and concrete strength. Assume a wall thickness of 10 in. and f'_c = 3000 psi.   *22.5.3, Eq. (22-6), Eq. (22-7)*

Example 30.2 (cont'd) — Calculations and Discussion — Code Reference

Load Combination	Axial Load, P_u kips/ft	Moment, M_u ft-kips/ft	Approximate Design Moment Strength,[1] ϕM_n ft-kips/ft	$\phi M_n / M_u$
1	3.21	3.12	Note 2	—
2	1.44	3.12	$(2.9 + 3.4)/2 = 3.15$	1.031
3	4.28	3.06	$(3.3 + 3.8)/2 = 3.55$[3]	1.16
4	1.44	3.06	Note 2	—
5	0	3.06	$(2.7 + 3.2)/2 = 2.95$	0.96

[1] Values interpolated from Figs. C30-4 and C30-5 based on 10-inch wall with 3000 psi concrete.
[2] There is no need to evaluate this combination since another combination has the same moment and a lower axial load. Given equal moments, the combination with the lower axial load will govern.
[3] While it was not necessary to check this load combination, it was done to illustrate that given equal induced moments, the combination with the higher axial, P_u, load yields a greater design moment strength, ϕM_n.

Since the 10-in. wall of 3000 psi concrete exceeds the requirements for all combinations except where no dead load is present (Combination 5), and in that case it is less than 4% under-strength, tentatively use these parameters.

C6. Check for shear strength.

As indicated for the first part of this design example in which the load and strength reduction factors of Chapter 9 were used, shear strength will rarely govern the design of a wall. Since in that example the wall was greatly over-designed for shear, there is no need to check for shear in this example.

22.5.4
22.6.4

C7. Use 10-in. wall with concrete specified compressive strength, $f'_c = 3000$ psi.

This example showed that use of the load and strength reduction factors of Appendix C resulted in a more economical design. To quickly determine which one of the two sets of load and strength reduction factors should be used, compare the nominal moment strength required by the factored loads of 9.2 to that required by C.2. The set requiring the lower nominal strength will generally be more economical.

Chapter 9 $M_n = M_u/\phi = 2.88/0.55 = 5.24$ ft-kips/ft

Appendix C $M_n = M_u/\phi = 3.06/0.65 = 4.71$ kips

Since the nominal moment strength required by the load and strength reduction factors of Appendix C is less than the corresponding strength required by Chapter 9, a design according to Appendix C will be more economical as the example shows. Regardless of the load and strength reduction factors used, the design procedures are the same.

Blank

Alternate (Working Stress) Design Method

INTRODUCTION

Although the Working Stress Design (WSD) was deleted from the code in the 2002 edition, the current Commentary Section R1.1 states, "The Alternate Design Method of the 1999 code may be used in place of applicable sections of this code." Note that the Commentary is not mandatory language and, thus, does not bear legal status. Therefore, in jurisdictions that adopt the current code, designers that intend to design by the 1999 WSD are cautioned to first seek approval of the local building official of the jurisdiction where the structure will be built.

GENERAL CONSIDERATIONS

Prior to the 1956 edition of the code, the working stress design method, which was very similar to the alternate design method of Appendix A, was the only method available for design of reinforced concrete members. The (ultimate) strength design method was introduced as an appendix to the 1956 code. In the next edition of the code (1963), strength design was moved to the body of the code as an alternative to working stress design. Because of the widespread acceptance of the strength design method, the 1971 code covered the working stress method in less than one page. The working stress method was moved out of the body of the code and into an appendix with the 1983 edition of the code. The method then became referred to as the "alternate design method." It remained an appendix through the 1999 code.

The alternate design method presented in Appendix A of the 1999 code is a method that seeks to provide adequate structural safety and serviceability by limiting stresses at service loads to certain prescribed limits. These "allowable stresses" are well within the range of elastic material behavior for concrete in compression and steel in tension (and compression). Concrete is assumed to be cracked and provide no resistance in tension. The stress in the concrete is represented by a linear elastic stress distribution. The steel is generally transformed into an equivalent area of concrete for design.

The alternate design method is identical to the "working stress design method" used prior to 1963 for members subject to flexure without axial loads. The procedures for the design of compression members with flexure, shear design, and bond stress and development of reinforcement follow the procedures of the strength design method of the body of the code with factors applied to reflect design at service loads. The design procedures of the alternate design method have not been updated as thoroughly as the remainder of the code.

The replacement of the working stress design method and alternate design method by the strength design method can be attributed to several factors including:

- the uniform treatment of all types of loads, i.e., all load factors are equal to unity. The different variability of different types of loads (dead and live load) is not acknowledged.

- the unknown factor of safety against failure (as discussed below)

- and the typically more conservative designs, which generally require more reinforcement or larger member sizes for the same design moments when compared to the strength design method.

It should be noted that in general, reinforced concrete members designed using working stresses, or the alternate design method, are less likely to have cracking and deflection problems than members designed using strength methods with Grade 60 reinforcement. This is due to the fact that with strength design using Grade 60 reinforcement, the stresses at service loads have increased significantly from what they were with working stress design.

Therefore, crack widths and deflection control are more critical in members designed using strength design methods because these factors are directly related to the stress in the reinforcement.

Today, the alternate design method is rarely used, except for a few special types of structures or by designers who are not familiar with strength design. Footings seem to be the members most often designed using the alternate design method. Note that ACI 350, Environmental Engineering Concrete Structures, governs the design of water retaining structures.

COMPARISON OF WORKING STRESS DESIGN WITH STRENGTH DESIGN

To illustrate the variability of the factor of safety against failure by the working stress design versus the strength design method, a rectangular and a T-section with dimensions shown in Figs. 31-1 and 31-2, respectively, were analyzed. In both cases, f'_c = 4000 psi, f_y = 60 ksi, and amount of reinforcement was varied between minimum flexural reinforcement per 10.5.1 and a maximum of $0.75\rho_b$ per Appendix B. Flexural strengths were computed using three procedures:

1. Nominal flexural strength, M_n, using the rectangular stress block of 10.2.7. Results are depicted by the solid lines.

2. Nominal flexural strength based on equilibrium and compatibility. This detailed analysis was performed using program Response 2000[31.1] assuming representative stress-strain relationships for concrete and reinforcing steel are shown in Fig. 31.1. Results are depicted by symbol "+."

3. Working stress analysis using linear elastic stress-strain relationships for concrete and reinforcement, and permissible service load stresses of Appendix A of the 1999 code, as noted below. The results are depicted by the dashed lines for M_s.

Obervations:

(a) Flexural strength based on the rectangular stress block, M_n, is very similar to the prediction based on detailed analysis using strain compatibility and equilibrium.

(b) The factor of safety, represented by the ratio $\phi M_n/M_s$ is highly variable. For the rectangular sections, that ratio ranges between 2.3 and 2.6 while for the T-section it ranges between 2.3 and 2.4. In comparison, for flexural design using Chapter 9 load and strength reduction factors, the factor of safety (L.F./ϕ) ranges between 1.2/0.9 = 1.33 where dead load dominates, and 1.6/0.9 = 1.78 where live load dominates. For Appendix C load and strength reduction factors, those ratios are 1.4/0.9 = 1.56 and 1.7/0.9 = 1.89, respectively.

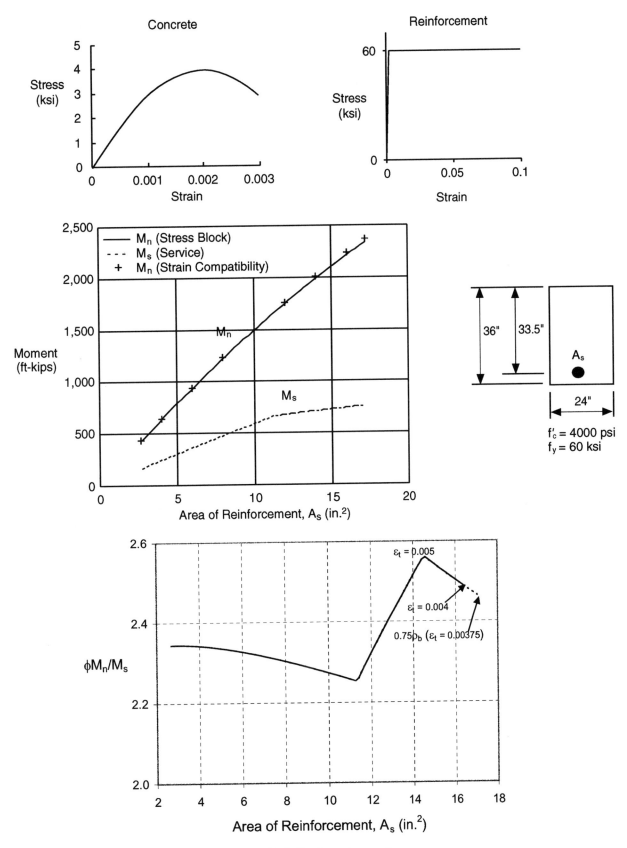

Figure 31-1 Rectangular section.

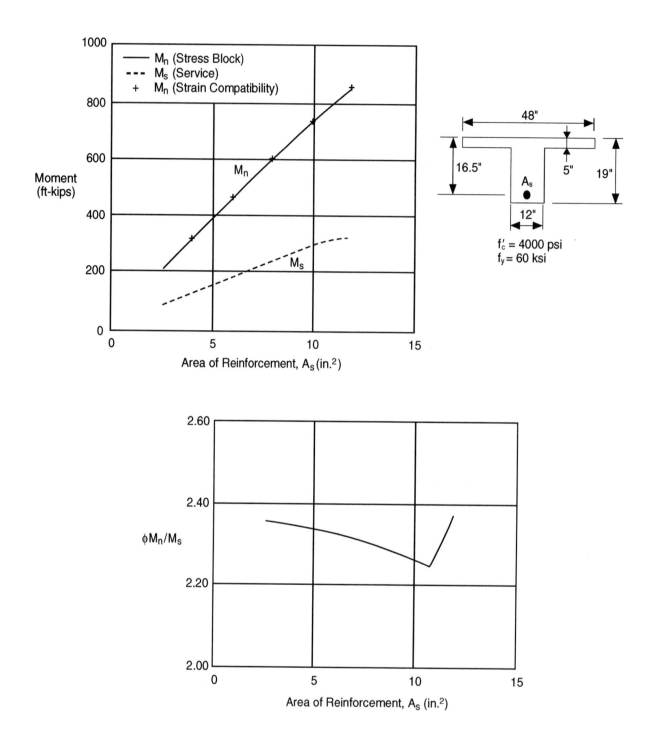

The following sections highlight provisions of Appendix A of the 1999 code.

SCOPE (A.1 OF '99 CODE)

The code specifies that any nonprestressed reinforced concrete member may be designed using the alternate design method of Appendix A. Prestressed concrete members are designed using a similar approach that is contained in code Chapter 18.

All other requirements of the code shall apply to members designed using the alternate design method, except the moment redistribution provisions of 8.4. This includes such items as distribution of flexural reinforcement and slenderness of compression members, as well as serviceability items such as control of deflections and crack control.

GENERAL (A.2 OF '99 CODE)

Load factors for all types of loads are taken to be unity for this design method. When wind and earthquake loads are combined with other loads, the member shall be designed to resist 75% of the total combined effect. This is similar to the provisions of the original working stress design method which allowed an overstress of one-third for load combinations including wind and earthquake.

When dead loads act to reduce the effects of other loads, 85% of the dead load may be used in computing load effects.

PERMISSIBLE SERVICE LOAD STRESSES (A.3 OF '99 CODE)

Concrete stresses at service loads must not exceed the following:

Flexure	Extreme fiber stress in compression	$0.45 f'_c$
Bearing	On loaded area	$0.3 f'_c$

Permissible concrete stresses for shear are also given in this section (A.3 of '99 Code) and in greater detail in A.7 of '99 Code.

Tensile stresses in reinforcement at service loads must not exceed the following:

Grade 40 and 50 reinforcement	20,000 psi
Grade 60 reinforcement or greater and welded wire fabric (plain or deformed)	24,000 psi

Permissible tensile stresses for a special case are also given in A.3.2(c).

FLEXURE (A.5 OF '99 CODE)

Members are designed for flexure using the following assumptions:

- Strains vary linearly as the distance from the neutral axis. A non-linear distribution of strain must be used for deep members (see 10.7).
- Under service load conditions, the stress-strain relationship of concrete in compression is linear for stresses not exceeding the permissible stress.
- In reinforced concrete members, concrete resists no tension.
- The modular ratio, $n = E_s/E_c$, may be taken as the nearest whole number, but not less than 6. Additional provisions are given for lightweight concrete.
- In members with compression reinforcement, an effective modular ratio of $2E_s/E_c$ must be used to transform the compression reinforcement for stress computations. The stress in the compression reinforcement must not exceed the permissible tensile stress.

DESIGN PROCEDURE FOR FLEXURE

The following equations are used in the alternate design method for the flexural design of a member with a <u>rectangular cross section</u>, reinforced with only tension reinforcement. They are based on the assumptions stated above and the notation defined in Fig. 31-3. See Refs. 33.2 and 33.3 or other texts on reinforced concrete design for derivation of these equations. Equations can also be developed for other cross sections, such as members with flanges or compression reinforcement.

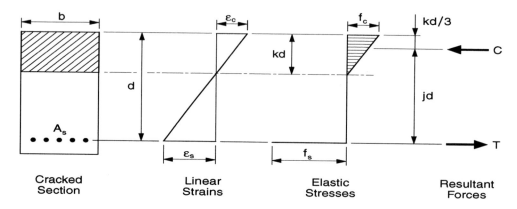

Figure 31-3 Assumptions for Alternate Design Method for Flexure

$$k = \sqrt{2\rho n + (\rho n)^2} - \rho n$$

where

$$\rho = \frac{A_s}{bd}$$

$$n = \frac{E_s}{E_c} \geq 6$$

$$j = 1 - \left(\frac{k}{3}\right)$$

$$f_s = \frac{M_s}{A_s jd}$$

$$f_c = \frac{2M_s}{kjbd^2}$$

SHEAR AND TORSION (A.7 OF '99 CODE)

Shear and torsion design in Appendix A of ACI 318-99 is based on the strength design methods of code Chapter 11 ('99 code) with modified coefficients that allow use of the equations for unfactored loads at service load conditions.

A complete set of the modified equations is presented for shear design for the convenience of the user (A.7 of the '99 Code). Since the equations appear in the same form as in code Chapter 11, they will not be discussed here.

REFERENCES

33.1 Bentz, Evans C. and Collins, Michael P., "Response 2000 – Reinforced Concrete Sectional Analysis using the Modified Compression Field Theory." Downloadable at http://www.ecf.utoronto.ca/~bentz/r2k.htm

33.2 MacGregor, J.G., *Reinforced Concrete: Mechanics and Design*, 2nd Edition, Prentice Hall, Englewood Cliffs, NJ, 1997, 939 pp.

33.3 Leet, Kenneth, *Reinforced Concrete Design*, McGraw-Hill, New York, 1984, 544 pp.

Example 31.1—Design of Rectangular Beam with Tension Reinforcement Only

Given the rectangular beam of Example 7.1, modify the beam depth and/or required reinforcement to satisfy the permissible stresses of the alternate design method. The service load moments are: M_d = 56 ft-kips and $M\ell$ = 35 ft-kips.

f'_c = 4000 psi
f_y = 60,000 psi
A_s = 2.40 in.^{2}
b = 10 in.
h = 16 in.
d = 13.5 in.

Calculations and Discussion	Code Reference

1. To compare a design using the alternate design method to the load factor method of the code, check the service load stresses in concrete and steel in the design given in Example 10.1.

    $M_s = M_d + M_\ell = (56 + 35)(12) = 1092$ in.-kips

    $E_c = 57,000\sqrt{f'_c} = 57,000\sqrt{4000} = 3,605,000$ psi   8.5.1

    $n = \dfrac{E_s}{E_c} = \dfrac{29,000,000}{3,605,000} = 8.04$ Use n = 8   A.5.4

    $\rho = \dfrac{A_s}{bd} = \dfrac{2.40}{(10 \times 13.5)} = 0.0178$

    $\rho n = (0.0178)(8) = 0.142$

    $k = \sqrt{2\rho n + (\rho n)^2} - \rho n = \sqrt{2(0.142) + (0.142)^2} - 0.142 = 0.41$

    $j = 1 - \dfrac{k}{3} = 1 - \dfrac{0.41}{3} = 0.863$

    $f_s = \dfrac{M_s}{A_s j d} = \dfrac{1092}{[(2.40)(0.863)(13.5)]} = 39.05$ ksi > 24.0 ksi allowed   N.G.   A.3.2

    $f_c = \dfrac{2M_s}{kjbd^2} = \dfrac{2(1092)}{[(0.41)(0.863)(10)(13.5)^2]} = 3.39$ ksi > 0.45(4.00) = 1.80 ksi allowed   N.G.   A.3.1

    Note: The above calculations are based on the assumption of linear-elastic material behavior. Since both f_c and f_s exceed the permissible stresses, increase the beam depth.

2. Check stresses in concrete and reinforcement with an increased member depth, with the same area of reinforcement.

    h = 24 in.   d = 21.5 in.

    $\rho = \dfrac{A_s}{bd} = \dfrac{2.40}{(10 \times 21.5)} = 0.0112$

| Example 31.1 (cont'd) | Calculations and Discussion | Code Reference |

$\rho n = (0.0112)(8) = 0.0893$

$k = \sqrt{2\rho n + (\rho n)^2} - \rho n = \sqrt{2(0.0893) + (0.0893)^2} - 0.0893 = 0.343$

$j = 1 - \dfrac{k}{3} = 1 - \dfrac{0.343}{3} = 0.886$

$f_s = \dfrac{M_s}{A_s j d} = \dfrac{1092}{[(2.40)(0.886)(21.5)]} = 23.89 \text{ ksi} < 24.0 \text{ ksi allowed} \quad \text{O.K.}$

$f_c = \dfrac{2M_s}{kjbd^2} = \dfrac{2(1092)}{[(0.343)(0.886)(10)(21.5)^2]} = 1.55 \text{ ksi} < 0.45(4.0) = 1.80 \text{ ksi allowed} \quad \text{O.K.}$

Note: It was necessary to increase the effective depth by nearly 60% in order to satisfy allowable stresses using the same quantity of reinforcement.

3. Compute the design moment strength, ϕM_n, of the modified member to determine the factor of safety (FS).

$a = \dfrac{A_s f_y}{0.85 b f_c'} = \dfrac{(2.40)(60)}{[(0.85)(10)(4.00)]} = 4.24 \text{ in.}$

$M_n = A_s f_y \left(d - \dfrac{a}{2}\right) = (2.40)(60)\left[21.5 - \left(\dfrac{4.24}{2}\right)\right] = 2791 \text{ in.-kips}$

$\phi M_n = 0.9(2791) = 2512 \text{ in.-kips}$

$FS = \dfrac{\phi M_n}{M_s} = \dfrac{2512}{1092} = 2.30$

Blank

32

Alternative Provisions for Reinforced and Prestressed Concrete Flexural and Compression Members

B.1 SCOPE

Section 8.1.2 allows the use of Appendix B to design reinforced and prestressed concrete flexural and compression members. The Appendix contains the provisions that were displaced from the main body of the code when the Unified Design Provisions (formerly Appendix B to the 1999 Code) were incorporated in the Code in 2002. Since it may be judged that an appendix is not an official part of a legal document unless it is specifically adopted, reference is made to Appendix B in the main body of the code in order to make it a legal part of the code.

Appendix B contains provisions for moment redistribution, design of flexural and compression members, and prestressed concrete that were in the main body of the code for many years prior to 2002. The use of these provisions is equally acceptable to those in the corresponding sections of the main body of the code.

Section B.1 contains the sections in Appendix B that replace those in the main body of the code when Appendix B is used in design. It must be emphasized that when any section of Appendix B is used, all sections of this appendix must be substituted in the main body of the code. All other sections in the body of the code are applicable.

According to RB.1, load factors and strength reduction factors of either Chapter 9 or new Appendix C (see Part 33) may be used. It is the intent that strength reduction factors given in Chapter 9 or Appendix C for tension-controlled sections be utilized for members subjected to bending only. Similarly, strength reduction factors for compression-controlled sections should be used for members subjected to flexure and axial load with ϕP_n greater than or equal to $0.10 f'_c A_g$ or the balanced axial load ϕP_b, whichever is smaller (see 9.3.2.2 and C3.2.2). For other cases, ϕ can be increased linearly to 0.90 as ϕP_n decreases from $0.10 f'_c A_g$ or ϕP_b, whichever is smaller, to zero (9.3.2.2 and C3.2.2).

B.8.4 REDISTRIBUTION OF NEGATIVE MOMENTS IN CONTINUOUS NONPRESTRESSED FLEXURAL MEMBERS

Section B.8.4 permits a redistribution of negative moments in continuous flexural members if reinforcement percentages do not exceed a specified amount.

A maximum 10 percent adjustment of negative moments was first permitted in the 1963 ACI Code (see Fig. RB.8.4). Experience with the use of that provision, though satisfactory, was still conservative. The 1971 code increased the maximum adjustment percentage to that shown in Fig. 32-1. The increase was justified by additional knowledge of ultimate and service load behavior obtained from tests and analytical studies. Appendix B retains the same adjustment percentage criteria.

A comparison between the permitted amount of redistribution according to 8.4 and B.8.4 as a function of the strain in the extreme tension steel ε_t is depicted in Figure 32-2.

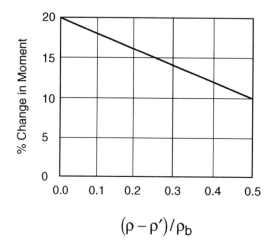

Figure 32-1 Permissible Moment Redistribution for Nonprestressed Members

Application of B.8.4 will permit, in many cases, substantial reduction in total reinforcement required without reducing safety, and reduce reinforcement congestion in negative moment regions.

According to 8.9, continuous members must be designed to resist more than one configuration of live loads. An elastic analysis is performed for each loading configuration, and an envelope moment value is obtained for the design of each section. Thus, for any of the loading conditions considered, certain sections in a given span will reach the ultimate moment, while others will have reserve capacity. Tests have shown that a structure can continue to carry additional loads if the sections that reached their moment capacities continue to rotate as plastic hinges and redistribute the moments to other sections until a collapse mechanism forms.

Recognition of this additional load capacity beyond the intended original design suggests the possibility of redesign with resulting savings in material. Section B.8.4 allows a redesign by decreasing or increasing the elastic negative moments for each loading condition (with the corresponding changes in positive moment required by statics). These moment changes may be such as to reduce both the maximum positive and negative moments in the final moment envelope. In order to ensure proper rotation capacity, the percentage of steel in the sections must conform to B.8.4, which is shown in Fig. 32-1.

In certain cases, the primary benefit to be derived from B.8.4 will be simply a reduction of negative moment at the supports, to avoid reinforcement congestion or reduce concrete dimensions. In this case, the steel percentage must still conform to Fig. 32-1.

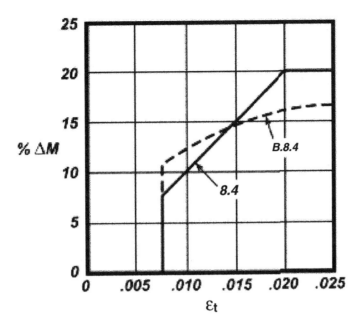

Figure 32-2 Comparison of Permissible Moment Redistribution for Nonprestressed Members

Limits of applicability of B.8.4 may be summarized as follows:

1. Provisions apply to continuous nonprestressed flexural members. Moment redistribution for prestressed members is addressed in B.18.10.4.

2. Provisions do not apply to members designed by the approximate moments of 8.3.3, or to slab systems designed by the Direct Design Method (see 13.6.1.7 and RB.8.4).

3. Bending moments must be determined by analytical methods, such as moment distribution, slope deflection, etc. Redistribution is not allowed for moments determined through approximate methods.

4. The reinforcement ratios ρ or $(\rho - \rho')$ at a cross-section where moment is to be adjusted must not exceed one-half of the balanced steel ratio, ρ_b, as defined by Eq. (B.8-1).

5. Maximum allowable percentage increase or decrease of negative moment is given by Eq. (B-1):

$$20\left(1 - \frac{\rho - \rho'}{\rho_b}\right)$$

6. Adjustment of negative moments is made for each loading configuration considered. Members are then proportioned for the maximum adjusted moments resulting from all loading conditions.

7. Adjustment of negative support moments for any span requires adjustment of positive moments in the same span (B.8.4.2). A decrease of a negative support moment requires a corresponding increase in the positive span moment for equilibrium.

8. Static equilibrium must be maintained at all joints before and after moment redistribution.

9. In the case of unequal negative moments on the two sides of a fixed support (i.e., where adjacent spans are unequal), the difference between these two moments is taken into the support. Should either or both of these negative moments be adjusted, the resulting difference between the adjusted moments is taken into the support.

10. Moment redistribution may be carried out for as many cycles as deemed practical, provided that after each cycle of redistribution, a new allowable percentage increase or decrease in negative moment is calculated, based on the final steel ratios provided for the adjusted support moments from the previous cycle.

11. After the design is completed and the reinforcement is selected, the actual steel ratios provided must comply with Fig. 32-1 for the percent moment redistribution taken, to ensure that the requirements of B.8.4 are met.

Examples that illustrate these requirements can be found in Part 9 of Notes on ACI 318-99.

B.10.3 GENERAL PRINCIPLES AND REQUIREMENTS – NONPRESTRESSED MEMBERS

The flexural strength of a member is ultimately reached when the strain in the extreme compression fiber reaches the ultimate (crushing) strain of the concrete, ε_u. At that stage, the strain in the tension reinforcement could just reach the strain at first yield ($\varepsilon_s = \varepsilon_y = f_y/\varepsilon_u$), be less than the yield strain, or exceed the yield strain. Which steel strain condition exists at ultimate concrete strain depends on the relative proportion of reinforcement to concrete. If the steel amount is low enough, the strain in the tension steel will greatly exceed the yield strain ($\varepsilon_s \gg \varepsilon_y$) when the concrete strain reaches ε_u, with large deflection and ample warning of impending failure (ductile failure condition). With a larger quantity of steel, the strain in the tension steel may not reach the yield strain ($\varepsilon_s < \varepsilon_y$) when the concrete strain reaches ε_u, which would mean small deflection and little warning of impending failure (brittle failure condition). For design it is desirable to restrict the ultimate strength condition so that a ductile failure mode would be expected.

The provisions of B.10.3.3 are intended to ensure a ductile mode of failure by limiting the amount of tension reinforcement to 75% of the balanced steel to ensure yielding of steel before crushing of concrete. The balanced steel will cause the strain in the tension steel to just reach yield strain when concrete reaches the crushing strain.

The maximum amount of reinforcement permitted in a rectangular section with tension reinforcement only is

$$\rho_{max} = 0.75\rho_b = 0.75\left[0.85\beta_1\frac{f'_c}{f_y} \times \frac{87,000}{87,000+f_y}\right]$$

where ρ_b is the balanced reinforcement ratio for a rectangular section with tension reinforcement only.

The maximum amount of reinforcement permitted in a flanged section with tension reinforcement only is

$$\rho_{max} = 0.75\left[\frac{b_w}{b}(\rho_b+\rho_f)\right]$$

where b_w = width of the web
 b = width of the effective flange (see 8.10)

$$\rho_f = A_{sf} / b_w d$$

h_f = thickness of the flange

A_{sf} = area of reinforcement required to equilibrate compressive strength of overhanging flanges (see Part 6)

The maximum amount of reinforcement permitted in a rectangular section with compression reinforcement is (B10.3.3)

$$\rho_{max} = 0.75\rho_b + \rho' \frac{f'_{sb}}{f_y}$$

where $\rho' = A'_s / bd$

A'_s = area of compression reinforcement

f'_{sb} = stress in compression reinforcement at balanced strain condition

$$= 87{,}000 - \frac{d'}{d}(87{,}000 + f_y) \leq f_y$$

d' = distance from extreme compression fiber to centroid of compression reinforcement

Note that with compression reinforcement, the portion of ρ_b contributed by the compression reinforcement ($\rho' f'_{sb} / f_y$) need not be reduced by the 0.75 factor. For ductile behavior of beams with compression reinforcement, only that portion of the total tension steel balanced by compression in the concrete (ρ_b) need be limited.

It should be realized that the limit on the amount of tension reinforcement for flexural members is a limitation for ductile behavior. Tests have shown that beams reinforced with the computed amount of balanced reinforcement actually behave in a ductile manner with gradually increasing deflections and cracking up to failure. Sudden compression failures do not occur unless the amount of reinforcement is considerably higher than the computed balanced amount.

One reason for the above is the limit on the ultimate concrete strain assumed at $\varepsilon_u = 0.003$ for design. The actual maximum strain based on physical testing may be much higher than this value. The 0.003 value serves as a lower bound on limiting strain. Unless unusual amounts of ductility are required, the $0.75\rho_b$ limitation will provide ample ductile behavior for most designs.

Comparison of the design using the unified design method and the provisions of B.10.3 for a rectangular beam with tension reinforcement only and for a rectangular beam with compression reinforcement can be found in Part 7 of Notes on ACI 318-99 (Examples 7.1 and 7.3).

B18.1   SCOPE–PRESTRESSED CONCRETE

This section contains a list of the provisions in the code that do not apply to prestressed concrete. Section RB.18.1.3 provides detailed commentary and specific reasons on why some sections are excluded.

B.18.8 LIMITS FOR REINFORCEMENT OF PRESTRESSED FLEXURAL MEMBERS

The requirements of B.18.8 for percentage of reinforcement are illustrated in Fig. 32-3. Note that reinforcement can be added to provide a reinforcement index higher than $0.36\beta_1$; however, this added reinforcement cannot be assumed to contribute to the moment strength.

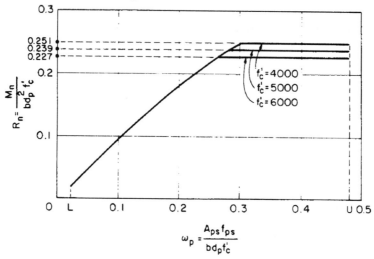

Figure 32-3 Permissible Limits of Prestressed Reinforcement and Influence on Moment Strength

Section B.18.8.3 requires the total amount of prestressed and nonprestressed reinforcement of flexural members to be adequate to develop a design moment strength at least equal to 1.2 times the cracking moment strength ($\phi M_n \geq 1.2 M_{cr}$), where M_{cr} is computed by elastic theory using a modulus of rupture equal to $7.5\sqrt{f_c'}$ (see 9.5.2.3). The provisions of B.18.8.3 are analogous to 10.5 for nonprestressed members and are intended as a precaution against abrupt flexural failure resulting from rupture of the prestressing tendons immediately after cracking. The provision ensures that cracking will occur before flexural strength is reached, and by a large enough margin so that significant deflection will occur to warn that the ultimate capacity is being approached. The typical prestressed member will have a fairly large margin between cracking strength and flexural strength, but the designer must be certain by checking it.

The cracking moment M_{cr} for a prestressed member is determined by summing all the moments that will cause a stress in the bottom fiber equal to the modulus of rupture f_r. Refer to Part 24 for detailed equations to compute M_{cr} for prestressed members.

Note that an exception in B.18.8.3 waives the $1.2 M_{cr}$ requirement for (a) two-way unbonded post-tensioned slabs, and (b) flexural members with shear and flexural strength at least twice that required by 9.2. See Part 24 for more information.

B.18.10.4 REDISTRIBUTION OF NEGATIVE MOMENTS IN CONTINUOUS PRESTRESSED FLEXURAL MEMBERS

Inelastic behavior at some sections of prestressed concrete beams and slabs can result in a redistribution of moments when member strength is approached. Recognition of this behavior can be advantageous in design under certain circumstances. Although a rigorous design method for moment redistribution is complex, a rational method can be realized by permitting a reasonable adjustment of the sum of the elastically calculated factored gravity load moments and the unfactored secondary moments due to prestress. The amount of adjustment should be kept within predetermined safe limits.

According to B.18.10.4.1, the maximum allowable percentage increase or decrease of negative moment in a continuous prestressed flexural member is

$$20\left[1 - \frac{\omega_p + \frac{d}{d_p}(\omega - \omega')}{0.36\beta_1}\right]$$

Note that redistribution of negative moments is allowed only when bonded reinforcement is provided at the supports in accordance with 18.9. The bonded reinforcement ensures that beams and slabs with unbonded tendons act as flexural members after cracking and not as a series of tied arches.

Similar to nonprestressed members, adjustment of negative support moments for any span requires adjustment of positive moments in the same span (B.18.10.4.2). A decrease of a negative support moment requires a corresponding increase in the positive span moment for equilibrium.

The amount of allowable redistribution depends on the ability of the critical sections to deform inelastically by a sufficient amount. Sections with larger amounts of reinforcement will not be able to undergo sufficient amounts of inelastic deformations. Thus, redistribution of negative moments is allowed only when the section is designed so that the appropriate reinforcement index is less than $0.24\beta_1$ (see B.18.10.4.3). This requirement is in agreement with the requirements of B.8.4 for nonprestressed members. Note that each of the expressions in B.18.10.4.3 is equal to $0.85a/d_p$ where a is the depth of the equivalent rectangular stress distribution for the section under consideration (see 10.2.7.1).

A comparison between the permitted amount of redistribution according to 18.10.4 and B.18.10.4 of the 2005 code as a function of the strain in the extreme tension steel ε_t is depicted in Figure 32-4.

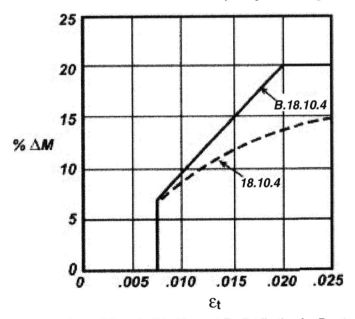

Figure 32-4 Comparison of Permissible Moment Redistribution for Prestressed Members

Blank

33

Alternative Load and Strength Reduction Factors

C.1   GENERAL

Section 9.1.3 allows the use of load factor combinations and strength reduction factors of Appendix C to design structural members. Since it may be judged that an appendix is not an official part of a legal document unless it is specifically adopted, reference is made in 9.1.3 to Appendix C in the main body of the code in order to make it a legal part of the code. Appendix C contains revised versions of the load and strength reduction factors that were formerly in Chapter 9 of the 1999 Code and earlier editions.

The load and strength reduction factors in new Appendix C have evolved since the early 1960s when the strength design method was originally introduced in the code. Some of the factors have been changed from the values in the 1999 code for reasons stated below. In any case, these sets of factors are still considered to be reliable for the design of concrete structural members.

It is important to note that a consistent set of load and strength reduction factors must be utilized when designing members. It is not permissible to use the load factors of Chapter 9 in conjunction with the strength reduction factors of Appendix C.

C.2   REQUIRED STRENGTH

In general,

> Design Strength ≥ Required Strength

or

> Strength Reduction Factor × Nominal Strength ≥ Load Factor × Service Load Effects

Part 5 contains a comprehensive discussion on the philosophy of the strength design method, including the reasons why load factors and strength reduction factors are required.

Section C.2 prescribes load factors for specific combinations of loads. A list of these combinations is given in Table 33-1. The numerical value of the load factor assigned to each type of load is influenced by the degree of accuracy with which the load can usually be assessed, the variation which may be expected in the load during the lifetime of a structure, and the probability of simultaneous occurrence of different load types. Hence, dead loads, because they can usually be more accurately determined and are less variable, are assigned a lower load factor (1.4) as compared to live loads (1.7). Also, weight and pressure of liquids with well-defined densities and controllable maximum heights are assigned a reduced load factor of 1.4 due the lesser probability of overloading (see C.2.4). A higher load factor of 1.7 is required for earth and groundwater pressures due to considerable uncertainty of their magnitude and recurrence (see C.2.3). Note that while most usual combinations of loads are included, it should not be assumed that all cases are covered

Table 33-1 Required Strength for Different Load Combinations

Code Section	Loads†	Required Strength	Code Eq. No.
C.2.1	Dead (D) & Live (L)	$U = 1.4D + 1.7L$	C-1
C.2.2	Dead, Live & Wind (W)††	(i) $U = 1.4D + 1.7L$ (ii) $U = 0.75(1.4D + 1.7L + 1.6W)$ (iii) $U = 0.9D + 1.6W$	C-1 C-2 C-3
C.2.2	Dead, Live & Earthquake (E)††	(i) $U = 1.4D + 1.7L$ (ii) $U = 0.75(1.4D + 1.7L + 1.0E)$ (iii) $U = 0.9D + 1.0E$	C-1 C-2 C-3
C.2.3	Dead, Live & Earth and Groundwater Pressure (H)*	(i) $U = 1.4D + 1.7L$ (ii) $U = 1.4D + 1.7L + 1.7H$ (iii) $U = 0.9D + 1.7H$ where D or L reduces H	C-1 C-4
C.2.4	Dead, Live & Fluid Pressure (F)**	(i) $U = 1.4D + 1.7L$ (ii) $U = 1.4D + 1.7L + 1.4F$ (iii) $U = 0.9D + 1.4F$ where D or L reduces F	C-1
C.2.5	Impact (I)***	In all of the above equations, substitute (L+I) for L when impact must be considered.	
C.2.6	Dead, Live and Effects from Differential Settlement, Creep, Shrinkage, Expansion of Shrinkage-Compensating Concrete, or Temperature (T)	(i) $U = 1.4D + 1.7L$ (ii) $U = 0.75(1.4D + 1.4T + 1.7L)$ (iii) $U = 1.4(D + T)$	C-1 C-5 C-6

† D, L, W, E, H, F, and T represent the designated service loads or their corresponding effects such as moments, shears, axial forces, torsion, etc. Note: E is a service-level earthquake force.

†† Where wind load W has not been reduced by a directionality factor, it is permitted to use 1.3W in place of 1.6W in Eq. (C-2) and (C-3). Where earthquake load E is based on service-load seismic forces, 1.4E shall be used in place of 1.0E in Eq. (C-2) and (C-3).

* Weight and pressure of soil and water in soil. (Groundwater pressure is to be considered part of earth pressure with a 1.7 load factor.)

** Weight and pressure of fluids with well-defined densities and controllable maximum heights

*** Impact factor is required for design of parking structures, loading docks, warehouse floors, elevator shafts, etc.

The load factors for wind and earthquake forces have changed from those in Chapter 9 of the 1999 code. Since the wind load equation in ASCE 7-02 and IBC 2003 includes a factor for wind directionality that is equal to 0.85 for buildings, the corresponding load factor for wind in the load combination equations was increased accordingly (1.3/0.85 = 1.53, rounded up to 1.6). The code allows use of the previous wind load factor of 1.3 when the design wind load is obtained from other sources that do not include the wind directionality factor.

The most recent legacy model building codes and the 2003 IBC specify strength-level earthquake forces; thus, the earthquake load factor was reduced to 1.0. The code requires use of the previous load factor for earthquake loads, which is 1.4, when service-level earthquake forces from earlier editions of the model codes or other resource documents are used.

C.3 DESIGN STRENGTH

As noted above, the design strength of a member is the nominal strength of the member, which is determined in accordance with code requirements, multiplied by the appropriate strength reduction factor, ϕ. The purposes of the strength reduction factors are given in Part 5 and RC.3.

The ϕ-factors prescribed in C.3, which have changed from those given in Chapter 9 of the 1999 code, are contained in Table 33-2. Prior to the 2002 code, ϕ-factors were given in terms of the type of loading for members subjected to axial load, flexure, or combined flexure and axial load. Now, for these cases, the ϕ-factor is determined by the strain conditions at a cross-section at nominal strength. Figure RC3.2 shows the variation of ϕ with the net tensile strain ε_t for Grade 60 reinforcement and prestressing steel. The Unified Design Provisions are described in detail in Parts 5 and 6. As noted above, the ϕ-factors given in C.3 are consistent with the load factors given in C.2.

Table 33-2 Strength Reduction Factors ϕ in the Strength Design Method

Tension-controlled sections	0.90
Compression-controlled sections Members with spiral reinforcement conforming to 10.9.3 Other reinforced members	0.75 0.70
Shear and torsion	0.85
Bearing on concrete (except for post-tensioned anchorage zones and strut-and-tie models)	0.70
Post-tensioned anchorage zones	0.85
Strut-and-tie models (Appendix A)	0.85

Blank

34

Anchoring to Concrete

UPDATE FOR '05 CODE

ACI 318-05 incorporates the second edition of Appendix D, Anchoring to Concrete. Significant revisions were made primarily to clarify application of the provisions in special conditions. Several subscripts to the notations were revised to make each parameter used in the equations more meaningful and to reflect the intent of that parameter. Noteworthy changes include:

- Clarify how to determine the concrete breakout strength of post-installed mechanical anchors when the results of ACI 355.2 product evaluation indicate that values of the coefficient for basic concrete breakout strength, k_c, and the factor used to modify tensile strength of anchors based on presence or absence of cracks, $\psi_{c,N}$, are different than the default values provided in Appendix D (D.5.2.2, RD.5.2.2, D.5.2.6, and RD.6.2.6)
- Determine the value of h_{ef} to be used in computing the concrete breakout strength for anchors loaded in tension and located close to three or four edges, i.e. in narrow members (D.5.2.3 and RD.5.2.3)
- Clarify computation of eccentricities for tension, e'_N (D.5.2.4 and RD.5.2.4) and for shear e'_V (D.6.2.5 and RD.6.2.5)
- Introduce a new modifier $\psi_{cp,N}$ to the basic concrete breakout equations to account for uncracked concrete, use of post-installed anchors, and a free edge near the anchors (new D.5.2.7), and provide conservative default values for the critical edge distance c_{ac} used in determining $\psi_{cp,N}$ (new D.8.6)
- Require testing according to ACI 355.2 product evaluation report if the contribution of post-installed anchor sleeves to shear strength is considered [D.6.1.2(c)]
- Provide guidance for computing the nominal shear concrete breakout strength for anchor groups. [RD.6.2.1 and Fig. RD.6.2.1(b)] For anchors, excluding welded studs, the concrete breakout failure for shear must be determined for all potential concrete breakout failure surfaces. For welded studs, only the breakout failure surface originating from the anchors located farthest from the free edge needs to be considered.
- Clarify computation of shear breakout strength close to three or four edges, i.e. in narrow, thin members (D.6.2.4 and RD.6.2.4)
- Provide an equation to compute the nominal concrete pryout strength of a group of anchors (D.4.1.2 and D.6.3)

INTRODUCTION

Appendix D, Anchoring to Concrete, was introduced in ACI 318-02. It provides requirements for the design of anchorages to concrete using both cast-in-place and post-installed mechanical anchors. The following presents an overview regarding the development and publication of ACI 318 Appendix D. As of the late 1990's, ACI 318 and the American Institute of Steel Construction LRFD and ASD Specifications were silent regarding the design of anchorage to concrete. ACI 349-85 Appendix B and the Fifth Edition of PCI Design Handbook provided the primary sources of design information for connections to concrete using cast-in-place anchors. The design of connections to concrete using post-installed anchors has typically been based on information provided by individual anchor manufacturers.

During the 1990's, ACI Committee 318 took the lead in developing building code provisions for the design of anchorages to concrete using both cast-in-place and post-installed mechanical anchors. Committee 318 received support from ACI Committee 355 (ACI 355), Anchorage to Concrete, and ACI Committee 349, Concrete Nuclear Structures. Concurrent with the ACI 318 effort to develop design provisions, ACI 355 was involved with developing a test method for evaluating the performance of post-installed mechanical anchors in concrete. During the code cycle leading to ACI 318-99, a proposed Appendix D to ACI 318 dealing with the design of anchorages to concrete using both cast-in-place and post-installed mechanical anchors was approved by ACI 318. Final adoption of the proposed appendix awaited ACI 355 approval of a test method for evaluating the performance of post-installed mechanical anchors in concrete under the ACI consensus process.

Since ACI 355 was not able to complete the test method for evaluating post-installed mechanical anchors on time to meet the publication deadlines for the ACI 318-99 code, an attempt was made to process an ACI 318 Appendix D reduced in scope to only cast-in-place anchors (i.e., without post-installed mechanical anchors). However, there was not sufficient time to meet the deadlines established by the International Code Council for submittal of the published ACI 318-99 standard to be referenced in the International Building Code (IBC 2000). As a result, the anchorage to concrete provisions originally intended for ACI 318-99 Appendix D (excluding provisions for post-installed mechanical anchors) were submitted and approved for incorporation into Section 1913 of IBC 2000.

At the end of 2001, ACI Committee 355 completed ACI 355.2-01 titled "Evaluating the Performance of Post-Installed Mechanical Anchors." Availability of ACI 355.2 led the way to incorporating into ACI 318-02 a new Appendix D, Anchoring to Concrete, which provided design requirements for both cast-in-place and post-installed mechanical anchors. As a result, Section 1913 of IBC 2003 references ACI 318 Appendix D. It is anticipated that IBC 2006 Section 1913 will reference ACI 318-05 Appendix D, which in turn adopts ACI 355.2-04 "Qualification of Post-Installed Mechanical Anchors in Concrete" by reference.

It should be noted that ACI 318-05 Appendix D does not address adhesive and grouted anchors. It is anticipated that design provisions for adhesive and grouted anchors will be incorporated into ACI 318-08.

HISTORICAL BACKGROUND OF DESIGN METHODS

The 45-degree cone method used in ACI 349-85 Appendix B and the PCI Design Handbook, Fifth Edition, was developed in the mid 1970's. In the 1980's, comprehensive tests of different types of anchors with various embedment lengths, edge distances, and group effects were performed at the University of Stuttgart on both uncracked and cracked concrete. The Stuttgart test results led to the development of the Kappa (K) method that was introduced in ACI 349 and ACI 355 in the late 1980's. In the early 1990's, the K method was improved, and made user-friendlier at the University of Texas at Austin. This effort resulted in the Concrete Capacity Design (CCD) method. During this same period, an international database was assembled. During the mid 1990's, the majority of the work of ACI Committees 349 and 355 was to evaluate both the CCD method and the 45-degree cone method using the international database of test results. As a result of this evaluation, ACI Committees 318, 349, and 355 proceeded with implementation of the CCD method. The design provisions of ACI 318 Appendix D and ACI 349-01 Appendix B are based on the CCD method. Differences between the CCD method and the 45-degree cone method are discussed below.

GENERAL CONSIDERATIONS

The design of anchorages to concrete must address both strength of the anchor steel and that associated with the embedded portion of the anchors. The lesser of these two strengths will control the design.

The strength of the anchor steel depends on the steel properties and size of the anchor. The strength of the embedded portion of the anchorage depends on its embedment length, strength of the concrete, proximity to other anchors, distance to free edges, and the characteristics of the embedded end of the anchor (headed, hooked, expansion, undercut, etc.).

The primary difference between the ACI 318 Appendix D provisions and those of the 45-degree cone method lies in the calculation of the embedment capacity for concrete breakout (i.e., a concrete cone failure). In the 45-degree cone method, the calculation of breakout capacity is based on a 45-degree concrete cone failure model that results in an equation based on the embedment length squared (h_{ef}^2). The ACI 318 Appendix D provisions account for fracture mechanics and result in an equation for concrete breakout that is based on the embedment length to the 1.5 power ($h_{ef}^{1.5}$). Although the 45-degree concrete cone failure model gives conservative results for anchors with $h_{ef} \leq 6$ in., the ACI 318 Appendix D provisions have been shown to give a better prediction of embedment strength for both single anchors and for anchors influenced by edge and group effects.

In addition to better prediction of concrete breakout strength, the ACI 318 Appendix D provisions simplify the calculation of the effects of anchor groups and edges by using a rectangular area bounded by $1.5h_{ef}$ from each anchor and free edges rather than the overlapping circular cone areas typically used in the 45-degree cone method.

DISCUSSION OF DESIGN PROVISIONS

The following provides a section-by-section discussion of the design provisions of ACI 318-05 Appendix D. Section, equation, and figure numbers in the following discussion and examples refer to those used in ACI 318-05 Appendix D. Note that notation for Appendix D is presented in 2.1 of ACI 318.

D.1   DEFINITIONS

The definitions presented are generally self-explanatory and are further explained in the text and figures of Appendix D. The following tables are provided as an aid to the designer in determining values for many of the variables:

Table 34-1: This table provides information on the types of materials typically specified for cast-in-place anchor applications. The table provides values for specified tensile strength f_{uta} and specified yield strength f_{ya} as well as the elongation and reduction in area requirements necessary to determine if a material should be considered as a brittle or ductile steel element. As shown in Table 34-1, all typical anchor materials satisfy the ductile steel element requirements of D.1. When using cast-in-place anchor materials not given in Table 34-1, the designer should refer to the appropriate material specification to be sure the material falls within the ductile steel element definition. Some high strength materials may not meet these requirements and must be considered as brittle steel elements.

Table 34-2: This table provides information on the effective cross-sectional area A_{se} and bearing area A_{brg} for threaded cast-in-place anchors up to 2 in. diameter.

Table 34-3: This table provides a fictitious sample information table for post-installed mechanical anchors that have been tested in accordance with ACI 355.2. This type of table will be available from manufactures that have tested their products in accordance with ACI 355.2. The table provides all of the values necessary for design of a particular post-installed mechanical anchor. The design of post-installed mechanical anchors must be based on this type of table unless values assumed in the design are specified in the project specifications (e.g., the pullout strength N_p).

As a further commentary on the five percent fractile in D.1 – Definitions, the five percent fractile is used to determine the nominal embedment strength of the anchor. It represents a value such that if 100 anchors are tested there is a 90% confidence that 95 of the anchors will exhibit strengths higher than the five percent fractile value. The five percent fractile is analogous to the use of f'_c for concrete strength and f_{ya} for steel strength in the nominal strength calculations in other parts of the ACI 318 code. For example, ACI 318 Section 5.3 requires that the required average compressive strength of the concrete f'_{cr} be statistically greater than the specified value of f'_c used in design calculations. For steel, f_{ya} represents the specified yield strength of the material. Since ASTM specifications give the minimum specified yield strength, the value of f_{ya} used in design is in effect a zero percent fractile (i.e., the designer is ensured that the actual steel used will have a yield value higher than the

Table 34-1 Properties of Cast-in-Place Anchor Materials

Material specification[1]	Grade or type	Diameter (in.)	Tensile strength, for design f_{uta} (ksi)	Tensile strength, min. (ksi)	Yield strength, min. ksi	Yield strength, min. method	Elongation, min. %	Elongation, min. length	Reduction of area, min., (%)
AWS D1.1[2]	B	1/2 – 1	65	65	51	0.2%	20	2"	50
ASTM A 307[3]	A	≤ 4	60	60	—	—	18	2"	—
	C	≤ 4	58	58-80	36	—	23	2"	—
ASTM A 354[4]	BC	≤ 4	125	125	109	0.2%	16	2"	50
	BD	≤ 4	125	150	130	0.2%	14	2"	40
ASTM A 449[5]	1	≤ 1	120	120	92	0.2%	14	4D	35
		1 – 1-1/2	105	105	81	0.2%	14	4D	35
		> 1-1/2	90	90	58	0.2%	14	4D	35
ASTM F 1554[6]	36	≤ 2	58	58-80	36	0.2%	23	2"	40
	55	≤ 2	75	75-95	55	0.2%	21	2"	30
	105	≤ 2	125	125-150	105	0.2%	15	2"	45

Notes:
1. The materials listed are commonly used for concrete anchors. Although other materials may be used (e.g., ASTM A 193 for high temperature applications, ASTM A 320 for low temperature applications), those listed are preferred for normal use. Structural steel bolting materials such as ASTM A 325 and ASTM A 490 are not typically available in the lengths needed for concrete anchorage applications.
2. *AWS D1.1-04 Structural Welding Code - Steel -* This specification covers welded headed studs or welded hooked studs (unthreaded). None of the other listed specifications cover welded studs.
3. *ASTM A 307-04 Standard Specification for Carbon Steel Bolts and Studs, 60,000 psi Tensile Strength -* This material is commonly used for concrete anchorage applications. Grade A is headed bolts and studs. Grade C is nonheaded bolts (studs), either straight or bent, and is equivalent to ASTM A 36 steel. Note that although a reduction in area requirement is not provided, A 307 may be considered a ductile steel element. Under the definition of "Ductile steel element" in D.1, the code states: "A steel element meeting the requirements of ASTM A 307 shall be considered ductile."
4. *ASTM A 354-04 Standard Specification for Quenched and Tempered Alloy Steel Bolts, Studs, and Other Externally Threaded Fasteners -* The strength of Grade BD is equivalent to ASTM A 490.
5. *ASTM A 449-04b Standard Specification for Quenched and Tempered Steel Bolts and Studs -* This specification is referenced by ASTM A 325 for "equivalent" anchor bolts.
6. *ASTM F 1554-04 Standard Specification for Anchor Bolts -* This specification covers straight and bent, headed and headless, anchor bolts in three strength grades. Anchors are available in diameters ≤ 4 in. but reduction in area requirements vary for anchors > 2 in.

Table 34-2 Dimensional Properties of Threaded Cast-in-Place Anchors

Anchor Diameter (d_o) (in.)	Gross Area of Anchor (in.²)	Effective Area of Anchor (A_{se}) (in.²)	Bearing Area of Heads and Nuts (A_{brg}) (in.²) Square	Heavy Square	Hex	Heavy Hex
0.250	0.049	0.032	0.142	0.201	0.117	0.167
0.375	0.110	0.078	0.280	0.362	0.164	0.299
0.500	0.196	0.142	0.464	0.569	0.291	0.467
0.625	0.307	0.226	0.693	0.822	0.454	0.671
0.750	0.442	0.334	0.824	1.121	0.654	0.911
0.875	0.601	0.462	1.121	1.465	0.891	1.188
1.000	0.785	0.606	1.465	1.855	1.163	1.501
1.125	0.994	0.763	1.854	2.291	1.472	1.851
1.250	1.227	0.969	2.228	2.773	1.817	2.237
1.375	1.485	1.160	2.769	3.300	2.199	2.659
1.500	1.767	1.410	3.295	3.873	2.617	3.118
1.750	2.405	1.900	—	—	—	4.144
2.000	3.142	2.500	—	—	—	5.316

Table 34-3 Sample Table of Anchor Data for a Fictitious Post-Installed Torque-Controlled Mechanical Expansion Anchor as Presumed Developed from Qualification Testing in Accordance with ACI 355.2-04.

(Note: Fictitious data for example purposes only – data are not from a real anchor)

Anchor system is qualified for use in both cracked and uncracked concrete in accordance with test program of Table 4.2 of *ACI 355.2-04*. The material, ASTM F1554 grade 55, meets the ductile steel element requirements of *ACI 318-05 Appendix D* (tensile test elongation of at least 14 percent and reduction in area of at least 30 percent).

Characteristic	Symbol	Units	Nominal anchor diameter			
Installation information						
Outside diameter	d_o	in.	3/8	–	5/8	–
Effective embedment depth	h_{ef}	in.	1.75 / 2.75 / 4.5	2.5 / 3.5 / 5.5	3 / 4.5 / 6.5	3.5 / 5 / 8
Installation torque	T_{inst}	ft-lb	30	65	100	175
Minimum edge distance	c_{min}	in.	1.75	2.5	3	3.5
Minimum spacing	s_{min}	in.	1.75	2.5	3	3.5
Minimum concrete thickness	h_{min}	in.	$1.5h_{ef}$	$1.5h_{ef}$	$1.5h_{ef}$	$1.5h_{ef}$
Critical edge distance @ h_{min}	c_{ac}	in.	2.1	3.0	3.6	4.0
Anchor data						
Anchor material			ASTM F 1554 Grade 55 (meets ductile steel element requirements)			
Category number	1, 2, or 3	–	2	2	1	1
Yield strength of anchor steel	f_{ya}	psi	55,000	55,000	55,000	55,000
Ultimate strength of anchor steel	f_{uta}	psi	75,000	75,000	75,000	75,000
Effective tensile stress area	A_{se}	in.²	0.0775	0.142	0.226	0.334
Effective shear stress area	A_{se}	in.²	0.0775	0.142	0.226	0.334
Effectiveness factor for uncracked concrete	k_{uncr}	–	24	24	24	24
Effectiveness factor for cracked concrete used for ACI 318 design	k_c*	–	17	17	17	17
$\psi_{c,N}$ for ACI 318 design in cracked concrete	$\psi_{c,N}$*	–	1.0	1.0	1.0	1.0
$\psi_{c,N} = k_{uncr}/k_{cr}$ for ACI 318 design in uncracked concrete	$\psi_{c,N}$*	–	1.4	1.4	1.4	1.4
Pullout or pull-through resistance from tests	N_p	lb	h_{ef}: 1.75 / 2.75 / 4.5; N_p: 1,354 / 2,667 / 5,583	h_{ef}: 2.5 / 3.5 / 5.5; N_p: 2,312 / 3,830 / 7,544	h_{ef}: 3 / 4.5 / 6.5; N_p: 4,469 / 8,211 / 14,254	h_{ef}: 3.5 / 5 / 8; N_p: 5,632 / 9,617 / 19,463
Tension resistance of single anchor for seismic loads	N_{eq}	lb	1.75: 903; 4.5: 3,722	2.5: 1,541; 5.5: 5,029	3: 2,979; 6.5: 9,503	3.5: 3,755; 8: 12,975
Shear resistance of single anchor for seismic loads	V_{eq}	lb	2,906	5,321	8,475	12,543
Axial stiffness in service load range	β	lb/in.	55,000	57,600	59,200	62,000
Coefficient of variation for axial stiffness in service load range.	ν	%	12	11	10	9

*These are values used for k_c and $\psi_{c,N}$ in ACI 318 for anchors qualified for use only in both cracked and uncracked concrete.

minimum specified value). All embedment strength calculations in Appendix D are based on a nominal strength calculated using 5 percent fractile values (e.g., the k_c values used in calculating basic concrete breakout strength are based on the 5 percent fractile).

D.2 Scope

These provisions apply to cast-in-place and post-installed mechanical anchors (such as those illustrated in Fig. RD.1) that are used to transmit structural loads between structural elements and safety related attachments to structural elements. The type of anchors included are cast-in-place headed studs, headed bolts, hooked rods (J and L bolts), and post-installed mechanical anchors that have met the anchor assessment requirements of ACI 355.2. Other types of cast-in-place anchors (e.g., specialty inserts) and post-installed anchors (e.g., adhesive, grouted, and pneumatically actuated nails or bolts) are currently excluded from the scope of Appendix D as well as post-installed mechanical anchors that have not met the anchor assessment requirements of ACI 355.2. As noted in D.2.4, these design provisions do not apply to anchorages loaded with high cycle fatigue and impact loads.

D.3 GENERAL REQUIREMENTS

The analysis methods prescribed in D.3 to determine loads on individual anchors in multiple anchor applications depend on the type of loading, rigidity of the attachment base plate, and the embedment of the anchors.

For multiple-anchor connections loaded concentrically in pure tension, the applied tensile load may be assumed to be evenly distributed among the anchors if the base plate has been designed so as not to yield. Prevention of yielding in the base plate will ensure that prying action does not develop in the connection.

For multiple-anchor connections loaded with an eccentric tension load or moment, distribution of loads to individual anchors should be determined by elastic analysis unless calculations indicate that sufficient ductility exists in the embedment of the anchors to permit a redistribution of load among individual anchors. If sufficient ductility is provided, a plastic design approach may be used. The plastic design approach requires ductile steel anchors sufficiently embedded so that embedment failure will not occur prior to a ductile steel failure. The plastic design approach assumes that the tension load (either from eccentric tension or moment) is equally distributed among the tension anchors. For connections subjected to moment, the plastic design approach is analogous to multiple layers of flexural reinforcement in a reinforced concrete beam. If the multiple layers of steel are adequately embedded and are a sufficient distance from the neutral axis of the member, they may be considered to have reached yield.

For both the elastic and plastic analysis methods of multiple-anchor connections subjected to moment, the exact location of the compressive resultant cannot be accurately determined by traditional concrete beam analysis methods. This is true for both the elastic linear stress-strain method (i.e., the transformed area method) and the ACI 318 stress block method since plane sections do not remain plane. For design purposes, the compression resultant from applied moment may be assumed to be located at the leading edge of the compression element of the attached member unless base plate stiffeners are provided. If base plate stiffeners are provided, the compressive resultant may be assumed to be located at the leading edge of the base plate.

Sections D.3.3.1 to D.3.3.5 provide special requirements for anchor design that includes seismic loads. Appendix D should not be used for the design of anchors in plastic hinge zones where high levels of cracking and spalling may be expected due to a seismic event. The Appendix D design provisions and the anchor evaluation criteria of ACI 355.2 are based on cracks that might occur normally in concrete (the cracked concrete tests and simulated seismic tests in ACI 355.2 are based on anchor performance in cracks from 0.012 in. to 0.020 in.). In regions of moderate or high seismic risk, or for structures assigned to intermediate or high seismic performance or design categories (see Table 1-3 in Part 1 for equivalent terminology used in building codes) all values for ϕN_n and ϕV_n must be reduced by multiplying by an additional factor of 0.75. Further, the strength of the connection must be controlled by the strength of ductile steel elements and not the embedment strength or the strength of brittle steel elements unless the structural attachment has been designed to yield at a load no greater than the design strength of the anchors, reduced by the factor of 0.75. Section RD.3.3 provides a detailed discussion of these requirements.

D.4 GENERAL REQUIREMENTS FOR STRENGTH OF ANCHORS

This section provides a general discussion of the failure modes that must be considered in the design of anchorages to concrete. The section also provides strength reduction factors, ϕ, for each type of failure mode. The failure modes that must be considered include those related to the steel strength and those related to the strength of the embedment.

Failure modes related to steel strength are simply tensile failure [Fig. RD.4.1(a)(i)] and shear failure [Fig. RD.4.1(b)(i)] of the anchor steel. Anchor steel strength is relatively easy to compute but typically does not control the design of the connection unless there is a specific requirement that the steel strength of a ductile steel element must control the design.

Embedment failure modes that must be considered are illustrated in Appendix D Fig. RD.4.1. They include:

- concrete breakout - a concrete cone failure emanating from the embedded end of tension anchors [Fig. RD.4.1(a)(iii)] or from the entry point of shear anchors located near an edge [Fig. RD.4.1(b)(iii)]
- pullout - a straight pullout of the anchor such as might occur for an anchor with a small head [Fig. RD.4.1(a)(ii)]
- side-face blowout - a spalling at the embedded head of anchors located near a free edge [Fig. RD.4.1(a)(iv)]
- concrete pryout - a shear failure mode that can occur with a short anchor popping out a wedge of concrete on the back side of the anchor [Fig. RD.4.1(b)(ii)]
- splitting - a tensile failure mode related to anchors placed in relatively thin concrete members [Fig. RD.4.1(a)(v)]

As noted in D.4.2, the use of any design model that results in predictions of strength that are in substantial agreement with test results is also permitted by the general requirements section. If the designer feels that the 45-degree cone method, or any other method satisfy this requirement he or she is permitted to use them. If not, the design provisions of the remaining sections of Appendix D should be used provided the anchor diameter does not exceed 2 in. and the embedment length does not exceed 25 in. These restrictions represent the upper limits of the database that the Appendix D design provisions are based on.

In the selection of the appropriate ϕ related to embedment failure modes, the presence of supplementary reinforcement designed to tie a potential failure prism to the structural member determines whether the ϕ for Condition A or Condition B applies. For the case of cast-in-place anchors loaded in shear directed toward a free edge, the supplementary reinforcement required for Condition A might be achieved by the use of hairpin reinforcement. It should be noted that for determining pullout strength for a single anchor, N_{pn}, and pryout strengths for a single anchor in shear, V_{cp}, or a group V_{cpg}, D.4.4(c) indicates that Condition B applies in all cases regardless of whether supplementary reinforcement is provided or not. In the case of post-installed anchors it is doubtful that this type of reinforcement will have been provided and Condition B will normally apply. The selection of ϕ for post-installed anchors also depends on the anchor category determined from the ACI 355.2 product evaluation tests. As part of the ACI 355.2 product evaluation tests, product reliability tests (i.e., sensitivity to installation variables) are performed and the results used to establish the appropriate category for the anchor. Since each post-installed mechanical anchor may be assigned a different category, product data tables resulting from ACI 355.2 testing should be referred to. Example data are shown in Table 34-3.

Table 34-4 summarizes the strength reduction factors, ϕ, to be used with the various governing conditions depending upon whether the load combinations of 9.2 or Appendix C are used.

D.5 DESIGN REQUIREMENTS FOR TENSILE LOADING

Methods to determine the nominal tensile strength as controlled by steel strength and embedment strength are presented in the section on tensile loading. The nominal tensile strength of the steel is based on the specified tensile strength of the steel Eq. (D-3). The nominal tensile strength of the embedment is based on (1) concrete breakout strength, Eq. (D-4) for single anchors or Eq. (D-5) for groups of anchors, (2) pullout strength, Eq. (D-14), or (3) side-face blowout strength, Eq. (D-17) for single anchors or Eq. (D-18) for groups. When combined with the appropriate strength reduction factors from D.4.4 or D.4.5, the smallest of these strengths will control the design tensile strength of the anchorage.

D.5.1 Steel Strength of Anchor in Tension

The tensile strength of the steel, N_{sa}, is determined from Eq. (D-3) using the effective cross-sectional area of the anchor A_{se} and the specified tensile strength of the anchor steel f_{uta}.

For cast-in-place anchors (i.e., threaded anchors, headed studs and hooked bars), the effective cross-sectional area of the anchor A_{se} is the net tensile stress area for threaded anchors and the gross area for headed studs that are welded to a base plate. These areas are provided in Table 34-2. For anchors of unusual geometry, the nominal steel strength may be taken as the lower 5% fractile of test results. For post-installed mechanical anchors the effective cross-sectional area of the anchor A_{se} must be determined from the results of the ACI 355.2 product evaluation tests. Example data are shown in Table 34-3.

The value of f_{uta} used in Eq. (D-3) is limited to $1.9f_{ya}$ or 125,000 psi. The limit of $1.9f_{ya}$ is intended to ensure that the anchor does not yield under service loads and typically is applicable only to stainless steel materials. The limit of 125,000 psi is based on the database used in developing the Appendix D provisions. Table 34-1 provides values for f_{ya} and f_{uta} for typical anchor materials. Note that neither of the limits applies to the typical anchor materials given in Table 34-1. For anchors manufactured according to specifications having a range for specified tensile strength, f_{uta} (e.g., ASTM F 1554), the lower limit value should be used to calculate the design strength. For post-installed mechanical anchors, both f_{ya} and f_{uta} must be determined from the results of the ACI 355.2 product evaluation tests. Example data are shown in Table 34-3.

D.5.2 Concrete Breakout Strength of Anchor in Tension

Figure RD.4.1(a)(iii) shows a typical concrete breakout failure (i.e., concrete cone failure) for a single headed cast-in-place anchor loaded in tension. Eq. (D-4) gives the concrete breakout strength for a single anchor, N_{cb}, while Eq. (D-5) gives the concrete breakout strength for a group of anchors in tension, N_{cbg}.

The individual terms in Eq. (D-4) and Eq. (D-5) are discussed below:

N_b: The basic concrete breakout strength for a single anchor located away from edges and other anchors (N_b) is given by Eq. (D-7) or Eq. (D-8). As previously noted, the primary difference between these equations and those of the 45-degree concrete cone method is the use of $h_{ef}^{1.5}$ in Eq. (D-7) [or alternatively $h_{ef}^{5/3}$ for anchors with $h_{ef} \geq 11$ in. in Eq. (D-8)] rather than h_{ef}^2. The use of $h_{ef}^{1.5}$ accounts for fracture mechanics principles and can be thought of as follows:

$$N_b = \frac{k\sqrt{f_c'}h_{ef}^2}{h_{ef}^{0.5}} \left[\frac{\text{general 45° concrete cone equation}}{\text{modification factor for fracture mechanics}} \right]$$

Resulting in:

$$N_b = k_c\sqrt{f_c'}h_{ef}^{1.5} \qquad \qquad Eq.\ (D\text{-}7)$$

The fracture mechanics approach accounts for the high tensile stresses that exist at the embedded head of the anchor while other approaches (such as the 45-degree concrete cone method) assume a uniform distribution of stresses over the assumed failure surface.

Table 34-4 Strength Reduction Factors for Use with Appendix D

Strength Governed by	Strength Reduction Factor, ϕ, for use with Load Combinations in			
	Section 9.2		Appendix C	
Ductile steel element				
Tension, N_{sa}	0.75		0.80	
Shear, V_{sa}	0.65		0.75	
Brittle steel element				
Tension, N_{sa}	0.65		0.70	
Shear, V_{sa}	0.60		0.65	
Concrete	Condition		Condition	
	A	B	A	B
Shear				
Breakout, V_{cb} and V_{cbg}	0.75	0.70	0.85	0.75
Pryout, V_{cp} and V_{cpg}	0.70	0.70	0.75	0.75
Tension				
Cast-in headed studs, headed bolts, or hooked bolts				
Breakout and side face blowout, N_{cb}, N_{cbg}, N_{sb} and N_{sbg}	0.75	0.70	0.85	0.75
Pullout, N_{pn}	0.70	0.70	0.75	0.75
Post-installed anchors with category determined per ACI 355.2				
Category 1 (low sensitivity to installation and high reliability)				
Breakout and side face blowout, N_{cb}, N_{cbg}, N_{sb} and N_{sbg}	0.75	0.65	0.85	0.75
Pullout, N_{pn}	0.65	0.65	0.75	0.75
Category 2 (med. sensitivity to installation and med. reliability)				
Breakout and side face blowout, N_{cb}, N_{cbg}, N_{sb} and N_{sbg}	0.65	0.55	0.75	0.65
Pullout, N_{pn}	0.55	0.55	0.65	0.65
Category 3 (high sensitivity to installation and low reliability)				
Breakout and side face blowout, N_{cb}, N_{cbg}, N_{sb} and N_{sbg}	0.55	0.45	0.65	0.55
Pullout, N_{pn}	0.45	0.45	0.55	0.55

The numeric constant k_c of 24 in Eq. (D-7) [or k_c of 16 in Eq. (D-8) if $h_{ef} \geq 11$ in.] is based on the 5% fractile of test results on headed cast-in-place anchors in cracked concrete. These k_c values must be used unless higher values of k_c are justified by ACI 355.2 product-specific tests. The value of k_c must not exceed 24. Note that the crack width used in tests to establish these k_c values was 0.012 in. If larger crack widths are anticipated, confining reinforcement to control crack width to about 0.012 in. should be provided or special testing in larger cracks should be performed.

$\dfrac{A_{Nc}}{A_{Nco}}$: This factor accounts for adjacent anchors and/or free edges. For a single anchor located away from free edges, the A_{Nco} term is the projected area of a 35-degree failure plane, measured relative to the surface of the concrete, and defined by a square with the sides $1.5h_{ef}$ from the centerline of the anchor [Fig. RD.5.2.1(a)]. The A_{Nc} term is a rectilinear projected area of the 35-degree failure plane at the surface of the concrete with sides $1.5h_{ef}$ from the centerline of the anchor(s) as limited by adjacent anchors and/or free edges. The definition of A_{Nc} is shown in Fig. RD.5.2.1(b). For a single anchor located at least $1.5h_{ef}$ from the closest free edge and $3h_{ef}$ from other anchors, A_{Nc} equals A_{Nco}.

Where a plate or washer is used to increase the bearing area of the head of an anchor, $1.5h_{ef}$ can be measured from the effective perimeter of the plate or washer where the effective perimeter is defined in D.5.2.8. Where a plate or washer is used, the projected area A_{Nc} can be based on $1.5h_{ef}$ measured from the effective perimeter of the plate or washer where the effective perimeter is defined in D.5.2.8 and shown in Fig. 34-1.

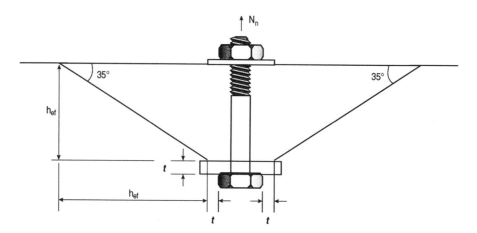

Figure 34-1 Effect of Washer Plate on Projected Area of Concrete Breakout

$\psi_{ec,N}$: This factor is applicable when multiple rows of tension anchors are present and the elastic design approach is used. In this case, the individual rows of tension anchors are assumed to carry different levels of load with the centerline of action of the applied tension load at an eccentricity (e'_N) from the centroid of the tension anchors. If the plastic design approach is used, all tension anchors are assumed to carry the same load and the eccentricity factor, $\psi_{ec,N}$, is taken as 1.0.

$\psi_{ed,N}$: This factor accounts for the non-uniform distribution of stresses when an anchor is located near a free edge of the concrete that are not accounted for by the $\dfrac{A_{Nc}}{A_{Nco}}$ term.

$\psi_{c,N}$: This factor is taken as 1.0 if cracks in the concrete are likely to occur at the location of the anchor(s). If calculations indicate that concrete cracking is not likely to occur under service loads (e.g., $f_t < f_r$), then $\psi_{c,N}$ may be taken as 1.25 for cast-in-place anchors or 1.4 for post-installed anchors.

$\psi_{cp,N}$: This factor is taken as 1.0 except when the design assumes uncracked concrete, uses post-installed anchors, and has a free edge near the anchors.

D.5.3 Pullout Strength of Anchor in Tension

A schematic of the pullout failure mode is shown in Fig. RD.4.1(a)(ii). The pullout strength of cast-in-place anchors is related to the bearing area at the embedded end of headed anchors, A_{brg}, and the properties of embedded hooks (e_h and d_o) for J-bolts and L-bolts. Obviously, if an anchor has no head or hook it will simply pull out of the concrete and not be able to achieve the concrete breakout strength associated with a full concrete cone failure (D.5.2). With an adequate head or hook size, pullout will not occur and the concrete breakout strength can be achieved. Eq. (D-14) provides the general requirement for pullout while Eq. (D-15) and Eq. (D-16) provide the specific requirements for headed and hooked anchors, respectively.

For headed anchors, the bearing area of the embedded head (A_{brg}) is the gross area of the head less the gross area of the anchor shaft (i.e., not the area of the embedded head). Washers or plates with an area larger than the head of an anchor can be used to increase the bearing area, A_{brg}, thus increasing the pullout strength (see D.5.2.8). In regions of moderate or high seismic risk, or for structures assigned to immediate or high seismic performance or design categories, where a headed bolt is being designed as a ductile steel element according to D.3.3.4, it may be necessary to use a bolt with a larger head or a washer in order to increase the design pullout strength, ϕN_{pn}, to assure that yielding of the steel takes place prior failure of the embedded portion of the anchor. Table 34-2 provides values for A_{brg} for standard bolt heads and nuts. Tables 34-5A, B and C can be used to quickly determine scenarios where the head of a bolt will not provide adequate pullout strength and will need to be increased in size.

For J-bolts and L-bolts, the minimum length of the hook measured from the inside surface of the shaft of the anchor is $3d_o$ while the maximum length for calculating pullout strength by Eq. (D-16) is $4.5d_o$. For other than high strength concrete, it is difficult to achieve design pullout strength of a hooked bolt that is equal to or greater than the design tensile strength of the steel. For example, a 1/2 in. diameter hooked bolt with the maximum hook length of $4.5d_o$ permitted in evaluating pullout strength in Eq. (D-16) requires that f'_c be at least 8700 psi to develop the design tensile strength of an ASTM A 307, Grade C, or ASTM F 1554, Grade 36 anchor (f_{uta} = 58,000 psi). This essentially prohibits the use of hooked bolts in many applications subject to seismic tensile loading due to the limitations of D.3.3.4 that the anchor strength must be governed by the ductile anchor steel.

For post-installed mechanical anchors, the value for the pullout strength, N_p, must be determined from the results of the ACI 355.2 product evaluation tests. Example data are shown in Table 34-3.

D.5.4 Concrete Side-Face Blowout Strength of Headed Anchor in Tension

The side-face blowout strength is associated with the lateral pressure that develops around the embedded end of headed anchors under load. Where the minimum edge distance for a single headed anchor is less than $0.4 h_{ef}$, side-face blowout must be considered using Eq. (D-17). If an orthogonal free edge (i.e., an anchor in a corner) is located less than three times the distance from the anchor to the nearest edge) then an additional reduction factor of $[(1+ c_{a2}/c_{a1})/4]$, where c_{a1} is the distance to the nearest edge and c_{a2} is the distance to the orthogonal edge, must be applied to Eq. (D-17).

For multiple anchor groups, the side-face blowout strength is the given by Eq. (D-18) provided the spacing between individual anchors parallel to a free edge is greater than or equal to six times the distance to the free edge. If the spacing of the anchors in the group is less than six times the distance to the free edge, Eq. (D-18) must be used.

D.6 DESIGN REQUIREMENTS FOR SHEAR LOADING

Methods to determine the nominal shear strength as controlled by steel strength and embedment strength are specified in D.6. The nominal shear strength of the steel is based on the specified tensile strength of the steel using Eq. (D-19) for headed studs, Eq. (D-20) for headed and hooked bolts, and for post-installed anchors. The nominal shear strength of the embedment is based on concrete breakout strength Eq. (D-21) for single anchors or Eq. (D-22) for groups of anchors, or pryout strength Eq. (D-29) for single anchors or Eq. (D-30) for groups. When combined with the appropriate strength reduction factors from D.4.4, the smaller of these strengths will control the design shear strength of the anchorage.

D.6.1 Steel Strength of Anchor in Shear

For cast-in-place anchors, the shear strength of the steel is determined from Eq. (D-19) for headed studs and Eq. (D-20) for headed and hooked bolts using the effective cross-sectional area of the anchor, A_{se}, and the specified tensile strength of the anchor steel, f_{uta}. For post-installed mechanical anchors, the shear strength of the steel is determined from Eq. (D-20) using the effective cross-sectional area of the anchor, A_{se}, and the specified tensile strength of the anchor steel, f_{uta}, unless the ACI 355.2 anchor qualification report provides a value for V_{sa}.

For cast-in-place anchors (i.e., headed anchors, headed studs and hooked bars), the effective cross-sectional area of the anchor (A_{se}) is the net tensile stress area for threaded anchors and the gross area for headed studs that are welded to a base plate. These areas are provided in Table 34-2. If the threads of headed anchors, L-, or J-bolts are located well above the shear plane (at least two diameters) the gross area of the anchor may be used for shear. For anchors of unusual geometry, the nominal steel strength may be taken as the lower 5% fractile of test results. For post-installed mechanical anchors the effective cross-sectional area of the anchor, A_{se}, or the nominal shear strength, V_{sa}, must be determined from the results of the ACI 355.2 product evaluation tests. Example data are shown in Table 34-3.

The value of f_{uta} used in Eq. (D-19) and Eq. (D-20) is limited to $1.9f_{ya}$ or 125,000 psi. The limit of $1.9f_{ya}$ is intended to ensure that the anchor does not yield under service loads and typically is applicable only to stainless steel materials. The limit of 125,000 psi is based on the database used in developing the Appendix D provisions. Table 34-1 provides values for f_{ya} and f_{uta} for typical anchor materials. Note that neither of the limits applies to the typical anchor materials given in Table 34-1. For anchors manufactured according to specifications having a range for specified tensile strength, f_{uta} (e.g., ASTM F 1554), the lower limit value should be used to calculate the design strength. For post-installed mechanical anchors, f_{ya} and f_{uta} must be determined from the results of the ACI 355.2 product evaluation tests. Example data are shown in Table 34-3.

When built-up grout pads are present, the nominal shear strength values given by Eq. (D-19) and Eq. (D-20) must be reduced by 20% to account for the flexural stresses developed in the anchor if the grout pad fractures upon application of the shear load.

D.6.2 Concrete Breakout Strength of Anchor in Shear

Fig. RD.4.1(b)(iii) shows typical concrete breakout failures for anchors loaded in shear directed toward a free edge. Eq. (D-21) gives the concrete breakout strength for a single anchor while Eq. (D-22) gives the concrete breakout strength for groups of anchors in shear. In cases where the shear is directed away from the free edge, the concrete breakout strength in shear need not be considered.

The individual terms in Eq. (D-21) and Eq. (D-22) are discussed below:

V_b : The basic concrete breakout strength for a single anchor in cracked concrete loaded in shear, directed toward a free edge (V_b) without any other adjacent free edges or limited concrete thickness is given by Eq. (D-24) for typical bolted connections and Eq. (D-25) for connections with welded studs or other anchors welded to the attached base plate. The primary difference between these equations and those using the 45-degree concrete cone method is the use of $c_{a1}^{1.5}$ rather than c_{a1}^2. The use of $c_{a1}^{1.5}$ accounts for fracture mechanics principles in the same way that $h_{ef}^{1.5}$ does for tension anchors. The fracture mechanics approach accounts for the high tensile stresses that exist in the concrete at the point where the anchor first enters the concrete.

ℓ_e, d_o : The terms involving ℓ_e and d_o in Eq. (D-24) and Eq. (D-25) relate to the shear stiffness of the anchor. A stiff anchor is able to distribute the applied shear load further into the concrete than a flexible anchor.

$\dfrac{A_{Vc}}{A_{Vco}}$: This factor accounts for adjacent anchors, concrete thickness, and free edges. For a single anchor in a thick concrete member with shear directed toward a free edge, the A_{Vco} term is the projected area on the side of the free edge of a 35-degree failure plane radiating from the point where the anchor first enters the concrete and directed toward the free edge [see Fig. RD.6.2.1(a)]. The A_{Vc} term is a rectilinear projected area of the 35-degree failure plane on the side of the free edge with sides $1.5 h_{ef}$ from the point where the anchor first enters the concrete as limited by adjacent anchors, concrete thickness and free edges. The definition of A_{Vc} is shown in Fig. RD.6.2.1(b).

$\psi_{ec,V}$: This factor applies when the applied shear load does not act through the centroid of the anchors loaded in shear [see Fig. RD.6.2.5]

$\psi_{ed,V}$: This factor accounts for the non-uniform distribution of stresses when an anchor is located in a corner that is not accounted for by the $\dfrac{A_{Vc}}{A_{Vco}}$ term [see Fig. RD.6.2.1(d)].

$\psi_{c,V}$: This factor is taken as 1.0 if cracks in the concrete are likely to occur at the location of the anchor(s) and no supplemental reinforcement has been provided. If calculations indicate that concrete cracking is not likely to occur (e.g., $f_t < f_r$ at service loads), then $\psi_{c,V}$ may be taken as 1.4. Values of $\psi_{c,V} > 1.0$ may be used if cracking at service loads is likely, provided No. 4 bar or greater edge reinforcement is provided (see D.6.2.7).

D.6.3 Concrete Pryout Strength of Anchor in Shear

The concrete pryout strength of an anchor in shear may control when an anchor is both short and relatively stiff. Fig. RD.4.1(b)(ii) shows this failure mode. As a mental exercise, this failure mode may be envisioned by thinking of a No. 8 bar embedded 2 in. in concrete with 3 ft. of the bar sticking out. A small push at the top of the bar will cause the bar to "pryout" of the concrete.

D.7 INTERACTION OF TENSILE AND SHEAR FORCES

The interaction requirements for tension and shear are based on a trilinear approximation to the following interaction equation (see Fig. RD.7):

$$\left[\frac{N_{ua}}{\phi N_n}\right]^{\frac{5}{3}} + \left[\frac{V_{ua}}{\phi V_n}\right]^{\frac{5}{3}} = 1$$

In the trilinear simplification, D.7.1 permits the full value of ϕN_n if $V_{ua} \leq 0.2\, \phi V_n$ and Section D.7.2 permits the full value of ϕV_n if $N_{ua} \leq 0.2\phi N_n$. If both of these conditions are not satisfied, the linear interaction of Eq. (D-31) must be used.

The most important aspect of the interaction provisions is that both ϕN_n and ϕV_n are the smaller of the anchor strengths as controlled by the anchor steel or the embedment. Tests have shown that the interaction relationship is valid whether steel strength or embedment strength controls for ϕN_n or ϕV_n.

D.8 REQUIRED EDGE DISTANCES, SPACINGS, AND THICKNESSES TO PRECLUDE SPLITTING FAILURE

Section D.8 provides minimum edge distance, spacing, and member thickness requirements to preclude a possible splitting failure of the structural member. For untorqued cast-in-place anchors (e.g., headed studs or headed bolts that are not highly preloaded after the attachment is installed), the minimum edge distance and member thickness is controlled by the cover requirements of 7.7 and the minimum anchor spacing is $4d_o$. For torqued cast-in-place anchors (e.g., headed bolts that are highly pre-loaded after the attachment is installed), the minimum edge distance and spacing is $6d_o$ and the member thickness is controlled by the cover requirements of 7.7.

Post-installed mechanical anchors can exert large lateral pressures at the embedded expansion device during installation that can lead to a splitting failure. Minimum spacing, edge distance, and member thickness requirements for post-installed anchors should be determined from the product-specific test results developed in the ACI 355.2 product evaluation testing. Example data are shown in Table 34-3. In the absence of the product-specific test results, the following should be used: a minimum anchor spacing of $6d_o$; a minimum edge distance of $6d_o$ for undercut anchors, $8d_o$ for torque-controlled anchors, and $10d_o$ for displacement controlled anchors; and a minimum member thickness of $1.5h_{ef}$ but need not exceed h_{ef} plus 4 in. Examples of each of these types of anchors are shown in ACI 355.2. In all cases, the minimum edge distance and member thickness should meet the minimum cover requirements of 7.7.

For untorqued anchors, D.8.4 provides a method to use a large diameter anchor nearer to an edge or with closer spacing than that required by D.8.1 to D.8.3. In this case, a fictitious anchor diameter d'_o is used in evaluating the strength of the anchor and in determining the minimum edge and spacing requirements.

For post-installed mechanical anchors, D.8.6 provides conservative default values for the critical edge distance c_{ac} used to determine $\psi_{cp,N}$. ACI 355.2 anchor qualification reports will provide values of c_{ac} associated with individual products (see sample Table 34-3.)

D.9　INSTALLATION OF ANCHORS

Cast-in-place anchors should be installed in accordance with construction documents. For threaded anchors, a metal or plywood template mounted above the surface of the concrete with nuts on each side of the template should be used to hold the anchors in a fixed position while the concrete is placed, consolidated, and hardens. Project specifications should require that post-installed anchors be installed in accordance with the manufacturer's installation instructions. As noted in RD.9, ACI 355.2 product evaluation testing is based on the manufacturer's installation instructions. As part of the ACI 355.2 product evaluation tests, product reliability tests (i.e., sensitivity to installation variables) are performed and the results are used to determine the category of the anchor to be used in the selection of the appropriate ϕ in D.4.4.

DESIGN TABLES FOR SINGLE CAST-IN ANCHORS

Tables have been provided to assist in the design of single anchors subject to tensile or shear loads. Tables 34-5A, B, and C provide design tensile strengths, ϕN_n, of single anchors in concrete with f'_c of 2500, 4000, and 6000 psi, respectively. Tables 34-6A, B, and C provide design shear strengths, ϕV_n, of single anchors in concrete with f'_c of 2500, 4000, and 6000 psi, respectively. A number of specified tensile strengths of steel, f_{uta}, are included to accommodate most anchor materials in use today. Notes accompany each group of tables that explain the assumptions used to develop the tables and how to adjust values for conditions that differ from those assumed.

According to D.8.2, minimum edge distances for cast-in headed anchors that will not be torqued must be based on minimum cover prescribed in 7.7. Thus, technically, concrete cover as low as 3/4 in. is permitted. If such a small cover is provided to the anchor shaft, the head of the anchor would end up having a cover smaller than 3/4 in. For corrosion protection, and in consideration of tolerances on placement (location and alignment) of anchors, it is recommended to provide a minimum concrete cover on cast-in anchors of 1-1/2 in. Tables 34-5 and 34-6 include design strengths for cast-in anchors with a minimum cover of 1-1/2 in.

NOTES FOR TENSION TABLES 34-5A, B AND C

NP – Not practical. Resulting edge distance, c_{a1}, yields less than 3/4 in. cover.

All Notation are identical to those used in 2.1 starting with ACI 318-05.

1. Design strengths in table are for single cast-in anchors near one edge only. The values do not apply where the distance between adjacent anchors is less than $3h_{ef}$, or where the perpendicular distance, c_{a2}, to the edge distance being considered, c_{a1}, is less than $1.5h_{ef}$.

2. In regions of moderate or high seismic risk (UBC Zone 2, 3 or 4), or in structures assigned to intermediate or high seismic performance or design categories (IBC Seismic Design Category C, D, E or F), the design strengths in the table must be reduced by 25%. In addition, the anchor must be designed so strength is governed by a ductile steel element, unless D.3.3.5 is satisfied. Therefore, the design strengths based on the three concrete failure modes, ϕN_{cb}, ϕN_{pn}, and ϕN_{sb}, must exceed the design strength of the steel in tension, ϕN_{sa}. This requirement effectively precludes the use of hooked anchor bolts in the seismic zones noted above.

3. For design purposes the tensile strength of the anchor steel, f_{uta}, must not exceed $1.9f_{ya}$ or 125,000 psi.

4. Design strengths in table are based on strength reduction factor, ϕ, of Section D.4.4. Factored tensile load N_{ua} must be computed from the load combinations of 9.2. Design strengths for concrete breakout, ϕN_{cb}, pullout, ϕN_{pn}, and sideface blowout, ϕN_{sb}, are based on Condition B. Where supplementary reinforcement is provided to satisfy Condition A, design strengths for ϕN_{cb} and ϕN_{sb} may be increased 7.1% to account for the increase in strength reduction factor from 0.70 to 0.75. This increase does not apply to pullout strength, ϕN_{pn}.

5. Design strengths for concrete breakout in tension, ϕN_{cb}, are based on N_b determined in accordance with Eq. (D-7) and apply to headed and hooked anchors. To determine the design strength of headed bolts with embedment depth, h_{ef}, greater than 11 in. in accordance with Eq. (D-8), multiply the table value by $[2(h_{ef}^{5/3})]/[3(h_{ef}^{1.5})]$.

6. Where analysis indicates that there will be no cracking at service load levels ($f_t < f_r$) in the region of the anchor, the design strengths for concrete breakout in tension, ϕN_{cb}, may be increased 25%.

7. The design strengths for pullout in tension, ϕN_{pn}, for headed bolts with diameter, d_o, less than 1-3/4 in. are based on bolts with regular hex heads. The design strengths for 1-3/4 and 2-in. bolts are based on heavy hex heads. For bolts with d_o less than 1-3/4 in. having heads with a larger bearing area, A_{brg}, than assumed, the design strengths may be increased by multiplying by the bearing area of the larger head and dividing by the bearing area of the regular hex head.

8. The design strengths for pullout in tension, ϕN_{pn}, for hooked bolts with hook-length, e_h, between 3 and 4.5 times diameter, d_o, may be determined by interpolation.

9. Where analysis indicates there will be no cracking at service load levels ($f_t < f_r$) in the region of the anchor, the design strengths for pullout in tension, ϕN_{pn}, may be increased 40%.

10. The design strengths for side-face blowout in tension, ϕN_{sb}, are applicable to headed bolts only and where edge distance, c_{a1}, is less than $0.4h_{ef}$. The values for $0.4h_{ef}$ are shown for interpolation purposes only. The design strengths for bolts with diameter, d_o, less than 1-3/4 in. are based on bolts with regular hex heads. The design strengths for 1-3/4 and 2 in. bolts are based on bolts with heavy hex heads. For bolts with d_o less than 1-3/4 in. having heads with a larger bearing area, A_{brg}, than assumed, the design strengths may be increased by multiplying by the square root of the quotient resulting from dividing the bearing area of the larger head by the bearing area of the regular hex head ($\sqrt{A_{brg2}/A_{brg1}}$).

NOTES FOR SHEAR TABLES 34-6A, B AND C

NP – Not practical. Resulting edge distance, c_{a1}, yields less than 3/4 in. cover.

All Notation are identical to those used in 2.1 starting with ACI 318-05.

1. Design strengths in table are for single cast-in anchors near one edge only. The values do not apply where the distance to an edge measured perpendicular to c_{a1} is less than $1.5c_{a1}$. See Note 9.

    The values do not apply where the distance between adjacent anchors is less than $3c_{a1}$, where c_{a1} is the distance from the center of the anchor to the edge in the direction of shear application.

2. In regions of moderate or high seismic risk (UBC Zone 2, 3 or 4), or in structures assigned to intermediate or high seismic performance or design categories (IBC Seismic Design Category C, D, E or F), the design strengths in the table must be reduced by 25%. In addition, the anchor must be designed so failure is initiated by a ductile steel element, unless D.3.3.5 is satisfied. This means that all the design strengths based on the two concrete failure modes, ϕV_{cb} and ϕV_{cp}, must equal or exceed the design strength of the steel in shear, ϕV_{sa}.

3. Concrete pryout strength, ϕV_{cp}, is to be taken equal to tension breakout strength, ϕN_{cb}, where h_{ef} is less than 2.5 in., and to be taken as twice ϕN_{cb} where h_{ef} is equal to or greater than 2.5 in. Condition B (see D.4.4) must be assumed even where supplementary reinforcement qualifying for Condition A is present (i.e., strength reduction factor, ϕ, must be taken equal to 0.70).

4. For design purposes the tensile strength of the anchor steel, f_{uta}, must not exceed $1.9 f_{ya}$ or 125,000 psi.

5. Design strengths in table are based on strength reduction factor, ϕ, of Section D.4.4. Factored shear load V_{ua} must be computed from the load combinations of 9.2. Design strengths for concrete breakout, ϕV_{cb}, are based on Condition B. Where supplementary reinforcement is provided to satisfy Condition A, design strengths may be increased 7.1% to account for the increase in strength reduction factor from 0.70 to 0.75.

6. Where analysis indicates that there will be no cracking at service load levels ($f_t < f_r$) in the region of the anchor, the design strengths for concrete breakout in shear, ϕV_{cb}, may be increased 40%.

7. In regions of members where analysis indicates cracking at service level loads, the strengths in the table for concrete breakout, ϕV_{cb}, may be increased in accordance with the factors in D.6.2.7 if edge reinforcement is provided in accordance with that section.

8. The design strengths for concrete breakout, ϕV_{cb}, are based on the shear load being applied perpendicular to the edge. If the load is applied parallel to the edge, the strengths may be increased 100%.

9. Where the anchor is located near a corner with an edge distance perpendicular to direction of shear, c_{a2}, less than $1.5c_{a1}$, design strengths for concrete breakout, ϕV_{cb}, shall be reduced by multiplying by modification factor, $\psi_{ed,V}$, determined from Eq. (D-28). The calculated values in the table do not apply where two edge distances perpendicular to direction of shear, c_{a2}, are less than $1.5c_{a1}$. See D.6.2.4.

10. This value of thickness, h, is not practical since the head or hook would project below the bottom surface of the concrete. It was chosen to facilitate mental calculation of the actual edge distance, c_{a1}, since the variable used in the calculation c_{a1} is a function of embedment depth, h_{ef}.

11. Linear interpolation for intermediate values of edge distance, c_{a1}, is permissible. Linear interpolation for intermediate values of embedment depth, h_{ef}, is unconservative.

12. For 3/4 in. cover and for $c_{a1} = 0.25 h_{ef}$ and $0.50 h_{ef}$, see portion of table for $h = h_{ef}$.

13. For 3/4 in. cover and for $c_{a1} = 0.25 h_{ef}$ and $0.50 h_{ef}$, see portion of table for $h = h_{ef}$. For $c_{a1} = h_{ef}$, see portion of table for $h = 1.5 h_{ef}$.

Table 34-5A. Design Strengths for Single Cast-In Anchors Subject to Tensile Loads ($f'_c = 2500$ psi)[1,2,4]
Notes pertaining to this table are given on Page 34-14

d_o in.	h_{ef} in.	ϕN_{sa} - Tension Strength of Anchor									ϕN_{cb} - Tension Breakout[4,5,6]					ϕN_{pn} - Pullout[9]			ϕN_{sb} - Sideface Blowout[4,10]		
		f_{uta} - for design purposes[3] - psi									c_{a1} - edge distance in.				head[7]	"J" or "L" hook[8]		1-1/2-in. cover	c_{a1} - edge distance in.		
		58,000	60,000	75,000	90,000	105,000	120,000	125,000	1-1/2-in. cover	0.25h_{ef}	0.5h_{ef}	h_{ef}	≥1.5h_{ef}		$e_h = 3d_o$	$e_h = 4.5d_o$		0.25h_{ef}	0.4h_{ef}
1/4	2	1,392	1,440	1,800	2,160	2,520	2,880	3,000	1,580	NP	NP	1,782	2,376	1,638	295	443	3,113	NP	NP
	3	1,392	1,440	1,800	2,160	2,520	2,880	3,000	2,401	NP	NP	3,274	4,365	1,638	295	443	3,113	NP	NP
	4	1,392	1,440	1,800	2,160	2,520	2,880	3,000	3,336	NP	3,584	5,040	6,720	1,638	295	443	3,113	NP	NP
	5	1,392	1,440	1,800	2,160	2,520	2,880	3,000	4,371	NP	5,009	7,044	9,391	1,638	295	443	3,113	NP	3,831
	6	1,392	1,440	1,800	2,160	2,520	2,880	3,000	5,496	NP	6,584	9,259	12,345	1,638	295	443	3,113	NP	4,597
3/8	2	3,393	3,510	4,388	5,265	6,143	7,020	7,313	1,613	NP	NP	1,782	2,376	2,296	664	997	3,827	NP	NP
	3	3,393	3,510	4,388	5,265	6,143	7,020	7,313	2,438	NP	NP	3,274	4,365	2,296	664	997	3,827	NP	NP
	4	3,393	3,510	4,388	5,265	6,143	7,020	7,313	3,377	NP	3,584	5,040	6,720	2,296	664	997	3,827	NP	NP
	5	3,393	3,510	4,388	5,265	6,143	7,020	7,313	4,415	NP	5,009	7,044	9,391	2,296	664	997	3,827	NP	4,536
	6	3,393	3,510	4,388	5,265	6,143	7,020	7,313	5,543	NP	6,584	9,259	12,345	2,296	664	997	3,827	NP	5,443
1/2	2	6,177	6,390	7,988	9,585	11,183	12,780	13,313	1,646	NP	NP	1,782	2,376	4,074	1,181	1,772	5,287	NP	NP
	3	6,177	6,390	7,988	9,585	11,183	12,780	13,313	2,475	NP	NP	3,274	4,365	4,074	1,181	1,772	5,287	NP	NP
	4	6,177	6,390	7,988	9,585	11,183	12,780	13,313	3,418	NP	3,584	5,040	6,720	4,074	1,181	1,772	5,287	NP	NP
	5	6,177	6,390	7,988	9,585	11,183	12,780	13,313	4,459	NP	5,009	7,044	9,391	4,074	1,181	1,772	5,287	NP	6,042
	6	6,177	6,390	7,988	9,585	11,183	12,780	13,313	5,591	NP	6,584	9,259	12,345	4,074	1,181	1,772	5,287	NP	7,250
	7	6,177	6,390	7,988	9,585	11,183	12,780	13,313	6,806	6,806	8,297	11,668	15,557	4,074	1,181	1,772	5,287	5,287	8,458
	8	6,177	6,390	7,988	9,585	11,183	12,780	13,313	8,099	8,316	10,137	14,255	19,007	4,074	1,181	1,772	5,287	6,042	9,667
5/8	3	9,831	10,170	12,713	15,255	17,798	20,340	21,188	2,513	NP	NP	3,274	4,365	6,356	1,846	2,769	6,839	NP	NP
	4	9,831	10,170	12,713	15,255	17,798	20,340	21,188	3,459	NP	3,584	5,040	6,720	6,356	1,846	2,769	6,839	NP	NP
	5	9,831	10,170	12,713	15,255	17,798	20,340	21,188	4,504	NP	5,009	7,044	9,391	6,356	1,846	2,769	6,839	NP	7,547
	6	9,831	10,170	12,713	15,255	17,798	20,340	21,188	5,639	NP	6,584	9,259	12,345	6,356	1,846	2,769	6,839	NP	9,056
	7	9,831	10,170	12,713	15,255	17,798	20,340	21,188	6,857	NP	8,297	11,668	15,557	6,356	1,846	2,769	6,839	NP	10,565
	8	9,831	10,170	12,713	15,255	17,798	20,340	21,188	8,153	8,316	10,137	14,255	19,007	6,356	1,846	2,769	6,839	7,547	12,074
	9	9,831	10,170	12,713	15,255	17,798	20,340	21,188	9,522	9,923	12,096	17,010	22,680	6,356	1,846	2,769	6,839	8,490	13,584
	10	9,831	10,170	12,713	15,255	17,798	20,340	21,188	10,960	11,621	14,167	19,922	26,563	6,356	1,846	2,769	6,839	9,433	15,093
3/4	4	14,529	15,030	18,788	22,545	26,303	30,060	31,313	3,500	NP	3,584	5,040	6,720	9,156	2,658	3,987	8,491	NP	NP
	5	14,529	15,030	18,788	22,545	26,303	30,060	31,313	4,549	NP	5,009	7,044	9,391	9,156	2,658	3,987	8,491	NP	9,057
	6	14,529	15,030	18,788	22,545	26,303	30,060	31,313	5,687	NP	6,584	9,259	12,345	9,156	2,658	3,987	8,491	NP	10,869
	7	14,529	15,030	18,788	22,545	26,303	30,060	31,313	6,908	NP	8,297	11,668	15,557	9,156	2,658	3,987	8,491	NP	12,680
	8	14,529	15,030	18,788	22,545	26,303	30,060	31,313	8,207	8,316	10,137	14,255	19,007	9,156	2,658	3,987	8,491	9,057	14,492
	9	14,529	15,030	18,788	22,545	26,303	30,060	31,313	9,579	9,923	12,096	17,010	22,680	9,156	2,658	3,987	8,491	10,190	16,303
	10	14,529	15,030	18,788	22,545	26,303	30,060	31,313	11,020	11,621	14,167	19,922	26,563	9,156	2,658	3,987	8,491	11,322	18,115
	12	14,529	15,030	18,788	22,545	26,303	30,060	31,313	14,097	15,277	18,623	26,189	34,918	9,156	2,658	3,987	8,491	13,586	21,738
7/8	4	14,529	15,030	18,788	22,545	26,303	30,060	31,313	3,500	NP	3,584	5,040	6,720	12,474	3,618	5,426	10,242	NP	NP
	5	20,097	20,790	25,988	31,185	36,383	41,580	43,313	5,736	NP	6,584	9,259	12,345	12,474	3,618	5,426	10,242	NP	12,686
	6	20,097	20,790	25,988	31,185	36,383	41,580	43,313	8,261	8,316	10,137	14,255	19,007	12,474	3,618	5,426	10,242	NP	16,915
	8	20,097	20,790	25,988	31,185	36,383	41,580	43,313	14,161	15,277	18,623	26,189	34,918	12,474	3,618	5,426	10,242	10,572	25,373
	12	20,097	20,790	25,988	31,185	36,383	41,580	43,313	19,235	21,350	26,026	36,600	48,800	12,474	3,618	5,426	10,242	15,858	31,716
	15	20,097	20,790	25,988	31,185	36,383	41,580	43,313	24,803	28,065	34,213	48,112	64,149	12,474	3,618	5,426	10,242	19,822	38,059
	18	20,097	20,790	25,988	31,185	36,383	41,580	43,313	29,505	35,938	56,000	78,750	105,000	12,474	3,618	5,426	10,242	23,787	52,860
	25	20,097	20,790	25,988	31,185	36,383	41,580	43,313	39,505	45,938	56,000	78,750	105,000	12,474	3,618	5,426	10,242	33,037	52,860

Table 34-5A. Design Strengths for Single Cast-In Anchors Subject to Tensile Loads (f'_c = 2500 psi)[1, 2, 4] (cont'd.)
Notes pertaining to this table are given on Page 34-14

d_o in.	h_{ef} in.	ϕN_{sa} - Tension Strength of Anchor									ϕN_{cb} - Tension Breakout[4, 5, 6]					ϕN_{pn} - Pullout[9]				ϕN_{sb} - Sideface Blowout[4, 10]		
		f_{uta} - for design purposes[3] - psi									c_{a1} - edge distance in.						"J" or "L" hook[8]		head[7]	c_{a1} - edge distance in.		
		58,000	60,000	75,000	90,000	105,000	120,000	125,000	1-1/2-in. cover	0.25h_{ef}	0.5h_{ef}	h_{ef}	≥1.5h_{ef}		e_h = 3d_o	e_h = 4.5d_o		1-1/2-in. cover	0.25h_{ef}	0.4h_{ef}
1	6	26,361	27,270	34,088	40,905	47,723	54,540	56,813	5,784	NP	6,584	9,259	12,345	16,282	4,725	7,088	12,078	NP	14,494
	9	26,361	27,270	34,088	40,905	47,723	54,540	56,813	9,693	9,923	12,096	17,010	22,680	16,282	4,725	7,088	12,078	13,588	21,741
	12	26,361	27,270	34,088	40,905	47,723	54,540	56,813	14,226	15,277	18,623	26,189	34,918	16,282	4,725	7,088	12,078	18,118	28,988
	15	26,361	27,270	34,088	40,905	47,723	54,540	56,813	19,307	21,350	26,026	36,600	48,800	16,282	4,725	7,088	12,078	22,647	36,235
	18	26,361	27,270	34,088	40,905	47,723	54,540	56,813	24,881	28,065	34,213	48,112	64,149	16,282	4,725	7,088	12,078	27,176	43,482
	21	26,361	27,270	34,088	40,905	47,723	54,540	56,813	30,908	35,366	43,113	60,627	80,837	16,282	4,725	7,088	12,078	31,706	50,729
	25	26,361	27,270	34,088	40,905	47,723	54,540	56,813	39,595	45,938	56,000	78,750	105,000	16,282	4,725	7,088	12,078	37,745	60,392
1-1/8	6	33,191	34,335	42,919	51,503	60,086	68,670	71,531	5,833	NP	6,584	9,259	12,345	20,608	5,980	8,970	14,013	NP	16,306
	9	33,191	34,335	42,919	51,503	60,086	68,670	71,531	9,750	9,923	12,096	17,010	22,680	20,608	5,980	8,970	14,013	15,287	24,459
	12	33,191	34,335	42,919	51,503	60,086	68,670	71,531	14,291	15,277	18,623	26,189	34,918	20,608	5,980	8,970	14,013	20,383	32,612
	15	33,191	34,335	42,919	51,503	60,086	68,670	71,531	19,378	21,350	26,026	36,600	48,800	20,608	5,980	8,970	14,013	25,478	40,766
	18	33,191	34,335	42,919	51,503	60,086	68,670	71,531	24,958	28,065	34,213	48,112	64,149	20,608	5,980	8,970	14,013	30,574	48,919
	21	33,191	34,335	42,919	51,503	60,086	68,670	71,531	30,991	35,366	43,113	60,627	80,837	20,608	5,980	8,970	14,013	35,670	57,072
	25	33,191	34,335	42,919	51,503	60,086	68,670	71,531	39,685	45,938	56,000	78,750	105,000	20,608	5,980	8,970	14,013	42,464	67,943
1-1/4	6	42,152	43,605	54,506	65,408	76,309	87,210	90,844	5,882	NP	6,584	9,259	12,345	25,438	7,383	11,074	16,041	NP	18,117
	9	42,152	43,605	54,506	65,408	76,309	87,210	90,844	9,807	9,923	12,096	17,010	22,680	25,438	7,383	11,074	16,041	16,984	27,175
	12	42,152	43,605	54,506	65,408	76,309	87,210	90,844	14,355	15,277	18,623	26,189	34,918	25,438	7,383	11,074	16,041	22,646	36,233
	15	42,152	43,605	54,506	65,408	76,309	87,210	90,844	19,450	21,350	26,026	36,600	48,800	25,438	7,383	11,074	16,041	28,307	45,292
	18	42,152	43,605	54,506	65,408	76,309	87,210	90,844	25,036	28,065	34,213	48,112	64,149	25,438	7,383	11,074	16,041	33,969	54,350
	21	42,152	43,605	54,506	65,408	76,309	87,210	90,844	31,075	35,366	43,113	60,627	80,837	25,438	7,383	11,074	16,041	39,630	63,408
	25	42,152	43,605	54,506	65,408	76,309	87,210	90,844	39,776	45,938	56,000	78,750	105,000	25,438	7,383	11,074	16,041	47,179	75,486
1-3/8	6	50,460	52,200	65,250	78,300	91,350	104,400	108,750	5,931	NP	6,584	9,259	12,345	30,786	8,933	13,400	18,166	NP	19,930
	9	50,460	52,200	65,250	78,300	91,350	104,400	108,750	9,865	9,923	12,096	17,010	22,680	30,786	8,933	13,400	18,166	18,685	29,895
	12	50,460	52,200	65,250	78,300	91,350	104,400	108,750	14,420	15,277	18,623	26,189	34,918	30,786	8,933	13,400	18,166	24,913	39,860
	15	50,460	52,200	65,250	78,300	91,350	104,400	108,750	19,521	21,350	26,026	36,600	48,800	30,786	8,933	13,400	18,166	31,141	49,826
	18	50,460	52,200	65,250	78,300	91,350	104,400	108,750	25,114	28,065	34,213	48,112	64,149	30,786	8,933	13,400	18,166	37,369	59,791
	21	50,460	52,200	65,250	78,300	91,350	104,400	108,750	31,158	35,366	43,113	60,627	80,837	30,786	8,933	13,400	18,166	43,597	69,756
	25	50,460	52,200	65,250	78,300	91,350	104,400	108,750	39,866	45,938	56,000	78,750	105,000	30,786	8,933	13,400	18,166	51,902	83,043
1-1/2	12	61,335	63,450	79,313	95,175	111,038	126,900	132,188	14,486	15,277	18,623	26,189	34,918	36,638	10,631	15,947	20,383	27,178	43,484
	15	61,335	63,450	79,313	95,175	111,038	126,900	132,188	19,593	21,350	26,026	36,600	48,800	36,638	10,631	15,947	20,383	33,972	54,355
	18	61,335	63,450	79,313	95,175	111,038	126,900	132,188	25,192	28,065	34,213	48,112	64,149	36,638	10,631	15,947	20,383	40,766	65,226
	21	61,335	63,450	79,313	95,175	111,038	126,900	132,188	31,242	35,366	43,113	60,627	80,837	36,638	10,631	15,947	20,383	47,561	76,097
	25	61,335	63,450	79,313	95,175	111,038	126,900	132,188	39,957	45,938	56,000	78,750	105,000	36,638	10,631	15,947	20,383	56,620	90,592
1-3/4	12	82,650	85,500	106,875	128,250	149,625	171,000	178,125	14,616	15,277	18,623	26,189	34,918	58,016	14,470	21,705	27,075	34,199	54,719
	15	82,650	85,500	106,875	128,250	149,625	171,000	178,125	19,737	21,350	26,026	36,600	48,800	58,016	14,470	21,705	27,075	42,749	68,399
	18	82,650	85,500	106,875	128,250	149,625	171,000	178,125	25,348	28,065	34,213	48,112	64,149	58,016	14,470	21,705	27,075	51,299	82,079
	21	82,650	85,500	106,875	128,250	149,625	171,000	178,125	31,409	35,366	43,113	60,627	80,837	58,016	14,470	21,705	27,075	59,849	95,758
	25	82,650	85,500	106,875	128,250	149,625	171,000	178,125	40,138	45,938	56,000	78,750	105,000	58,016	14,470	21,705	27,075	71,249	113,998
2	12	108,750	112,500	140,625	168,750	196,875	225,000	234,375	14,747	15,277	18,623	26,189	34,918	74,424	18,900	28,350	32,279	38,735	61,976
	15	108,750	112,500	140,625	168,750	196,875	225,000	234,375	19,881	21,350	26,026	36,600	48,800	74,424	18,900	28,350	32,279	48,419	77,470
	18	108,750	112,500	140,625	168,750	196,875	225,000	234,375	25,504	28,065	34,213	48,112	64,149	74,424	18,900	28,350	32,279	58,102	92,964
	21	108,750	112,500	140,625	168,750	196,875	225,000	234,375	31,577	35,366	43,113	60,627	80,837	74,424	18,900	28,350	32,279	67,786	108,458
	25	108,750	112,500	140,625	168,750	196,875	225,000	234,375	40,320	45,938	56,000	78,750	105,000	74,424	18,900	28,350	32,279	80,698	129,116

Table 34-5B. Design Strengths for Single Cast-In Anchors Subject to Tensile Loads ($f'_c = 4000$ psi)[1,2,4]

Notes pertaining to this table are given on Page 34-14

d_o in.	h_{ef} in.	ϕN_{sa} - Tension Strength of Anchor f_{uta} - for design purposes[3] - psi									ϕN_{cb} - Tension Breakout[4,5,6] c_{a1} - edge distance in.						ϕN_{pn} - Pullout[9]			ϕN_{sb} - Sideface Blowout[4,10] c_{a1} - edge distance in.		
		58,000	60,000	75,000	90,000	105,000	120,000	125,000	1-1/2-in. cover	0.25h_{ef}	0.5h_{ef}	h_{ef}	$\geq 1.5h_{ef}$	head[7]	$e_h = 3d_o$	$e_h = 4.5d_o$	1-1/2-in. cover	0.25h_{ef}	0.4h_{ef}
1/4	2	1,392	1,440	1,800	2,160	2,520	2,880	3,000	1,998	NP	NP	2,254	3,005	2,621	473	709	3,937	NP	NP
	3	1,392	1,440	1,800	2,160	2,520	2,880	3,000	3,037	NP	NP	4,141	5,521	2,621	473	709	3,937	NP	NP
	4	1,392	1,440	1,800	2,160	2,520	2,880	3,000	4,220	NP	4,533	6,375	8,500	2,621	473	709	3,937	NP	NP
	5	1,392	1,440	1,800	2,160	2,520	2,880	3,000	5,528	NP	6,336	8,910	11,879	2,621	473	709	3,937	NP	4,846
	6	1,392	1,440	1,800	2,160	2,520	2,880	3,000	6,952	NP	8,328	11,712	15,616	2,621	473	709	3,937	NP	5,815
3/8	2	3,393	3,510	4,388	5,265	6,143	7,020	7,313	2,040	NP	NP	2,254	3,005	3,674	1,063	1,595	4,841	NP	NP
	3	3,393	3,510	4,388	5,265	6,143	7,020	7,313	3,084	NP	NP	4,141	5,521	3,674	1,063	1,595	4,841	NP	NP
	4	3,393	3,510	4,388	5,265	6,143	7,020	7,313	4,271	NP	4,533	6,375	8,500	3,674	1,063	1,595	4,841	NP	NP
	5	3,393	3,510	4,388	5,265	6,143	7,020	7,313	5,584	NP	6,336	8,910	11,879	3,674	1,063	1,595	4,841	NP	5,737
	6	3,393	3,510	4,388	5,265	6,143	7,020	7,313	7,012	NP	8,328	11,712	15,616	3,674	1,063	1,595	4,841	NP	6,885
1/2	2	6,177	6,390	7,988	9,585	11,183	12,780	13,313	2,082	NP	NP	2,254	3,005	6,518	1,890	2,835	6,687	NP	NP
	3	6,177	6,390	7,988	9,585	11,183	12,780	13,313	3,131	NP	NP	4,141	5,521	6,518	1,890	2,835	6,687	NP	NP
	4	6,177	6,390	7,988	9,585	11,183	12,780	13,313	4,323	NP	4,533	6,375	8,500	6,518	1,890	2,835	6,687	NP	NP
	5	6,177	6,390	7,988	9,585	11,183	12,780	13,313	5,641	NP	6,336	8,910	11,879	6,518	1,890	2,835	6,687	NP	7,642
	6	6,177	6,390	7,988	9,585	11,183	12,780	13,313	7,072	NP	8,328	11,712	15,616	6,518	1,890	2,835	6,687	NP	9,171
	7	6,177	6,390	7,988	9,585	11,183	12,780	13,313	8,609	8,609	10,495	14,759	19,678	6,518	1,890	2,835	6,687	6,687	10,699
	8	6,177	6,390	7,988	9,585	11,183	12,780	13,313	10,245	10,518	12,823	18,032	24,042	6,518	1,890	2,835	6,687	7,642	12,228
5/8	3	9,831	10,170	12,713	15,255	17,798	20,340	21,188	3,179	NP	NP	4,141	5,521	10,170	2,953	4,430	8,651	NP	NP
	4	9,831	10,170	12,713	15,255	17,798	20,340	21,188	4,375	NP	4,533	6,375	8,500	10,170	2,953	4,430	8,651	NP	NP
	5	9,831	10,170	12,713	15,255	17,798	20,340	21,188	5,697	NP	6,336	8,910	11,879	10,170	2,953	4,430	8,651	NP	9,546
	6	9,831	10,170	12,713	15,255	17,798	20,340	21,188	7,133	NP	8,328	11,712	15,616	10,170	2,953	4,430	8,651	NP	11,455
	7	9,831	10,170	12,713	15,255	17,798	20,340	21,188	8,674	NP	10,495	14,759	19,678	10,170	2,953	4,430	8,651	NP	13,364
	8	9,831	10,170	12,713	15,255	17,798	20,340	21,188	10,313	10,518	12,823	18,032	24,042	10,170	2,953	4,430	8,651	9,546	15,273
	9	9,831	10,170	12,713	15,255	17,798	20,340	21,188	12,044	12,551	15,300	21,516	28,688	10,170	2,953	4,430	8,651	10,739	17,182
	10	9,831	10,170	12,713	15,255	17,798	20,340	21,188	13,864	14,700	17,920	25,200	33,600	10,170	2,953	4,430	8,651	11,932	19,091
3/4	4	14,529	15,030	18,788	22,545	26,303	30,060	31,313	4,428	NP	4,533	6,375	8,500	14,650	4,253	6,379	10,741	NP	NP
	5	14,529	15,030	18,788	22,545	26,303	30,060	31,313	5,754	NP	6,336	8,910	11,879	14,650	4,253	6,379	10,741	NP	11,457
	6	14,529	15,030	18,788	22,545	26,303	30,060	31,313	7,194	NP	8,328	11,712	15,616	14,650	4,253	6,379	10,741	NP	13,748
	7	14,529	15,030	18,788	22,545	26,303	30,060	31,313	8,738	NP	10,495	14,759	19,678	14,650	4,253	6,379	10,741	NP	16,040
	8	14,529	15,030	18,788	22,545	26,303	30,060	31,313	10,381	10,518	12,823	18,032	24,042	14,650	4,253	6,379	10,741	11,457	18,331
	9	14,529	15,030	18,788	22,545	26,303	30,060	31,313	12,116	12,551	15,300	21,516	28,688	14,650	4,253	6,379	10,741	12,889	20,622
	10	14,529	15,030	18,788	22,545	26,303	30,060	31,313	13,939	14,700	17,920	25,200	33,600	14,650	4,253	6,379	10,741	14,321	22,914
	12	14,529	15,030	18,788	22,545	26,303	30,060	31,313	17,831	19,324	23,556	33,126	44,168	14,650	4,253	6,379	10,741	17,185	27,497
7/8	4	20,097	20,790	25,988	31,185	36,383	41,580	43,313	4,428	NP	4,533	6,375	8,500	19,958	5,788	8,682	12,955	NP	NP
	6	20,097	20,790	25,988	31,185	36,383	41,580	43,313	7,255	NP	8,328	11,712	15,616	19,958	5,788	8,682	12,955	NP	16,047
	8	20,097	20,790	25,988	31,185	36,383	41,580	43,313	10,450	10,518	12,823	18,032	24,042	19,958	5,788	8,682	12,955	13,373	21,396
	12	20,097	20,790	25,988	31,185	36,383	41,580	43,313	17,913	19,324	23,556	33,126	44,168	19,958	5,788	8,682	12,955	20,059	32,094
	15	20,097	20,790	25,988	31,185	36,383	41,580	43,313	24,331	27,006	32,921	46,295	61,727	19,958	5,788	8,682	12,955	25,074	40,118
	18	20,097	20,790	25,988	31,185	36,383	41,580	43,313	31,374	35,500	43,276	60,857	81,142	19,958	5,788	8,682	12,955	30,088	48,141
	25	20,097	20,790	25,988	31,185	36,383	41,580	43,313	49,970	58,107	70,835	99,612	132,816	19,958	5,788	8,682	12,955	41,789	66,863

Table 34-5B. Design Strengths for Single Cast-In Anchors Subject to Tensile Loads ($f'_c = 4000$ psi)[1,2,4] (cont'd.)
Notes pertaining to this table are given on Page 34-14

d_o in.	h_{ef} in.	ϕN_{sa} - Tension Strength of Anchor f_{uta} - for design purposes[3] - psi									ϕN_{cb} - Tension Breakout[4,5,6]					ϕN_{pn} - Pullout[9]			ϕN_{sb} - Sideface Blowout[4,10]		
		58,000	60,000	75,000	90,000	105,000	120,000	125,000			1-1/2-in. cover	c_{a1} - edge distance in.					"J" or "L" hook[8]		c_{a1} - edge distance in.		
												$0.25h_{ef}$	$0.5h_{ef}$	h_{ef}	$\geq 1.5h_{ef}$	head[7]	$e_h = 3d_o$	$e_h = 4.5d_o$	1-1/2-in. cover	$0.25h_{ef}$	$0.4h_{ef}$
1	6	26,361	27,270	34,088	40,905	47,723	54,540	56,813	7,316	NP	8,328	11,712	15,616	26,051	7,560	11,340	15,278	NP	18,334
	9	26,361	27,270	34,088	40,905	47,723	54,540	56,813	12,260	12,551	15,300	21,516	28,688	26,051	7,560	11,340	15,278	17,188	27,500
	12	26,361	27,270	34,088	40,905	47,723	54,540	56,813	17,995	19,324	23,556	33,126	44,168	26,051	7,560	11,340	15,278	22,917	36,667
	15	26,361	27,270	34,088	40,905	47,723	54,540	56,813	24,421	27,006	32,921	46,295	61,727	26,051	7,560	11,340	15,278	28,646	45,834
	18	26,361	27,270	34,088	40,905	47,723	54,540	56,813	31,472	35,500	43,276	60,857	81,142	26,051	7,560	11,340	15,278	34,376	55,001
	21	26,361	27,270	34,088	40,905	47,723	54,540	56,813	39,096	44,735	54,534	76,688	102,251	26,051	7,560	11,340	15,278	40,105	64,168
	25	26,361	27,270	34,088	40,905	47,723	54,540	56,813	50,084	58,107	70,835	99,612	132,816	26,051	7,560	11,340	15,278	47,744	76,390
1-1/8	6	33,191	34,335	42,919	51,503	60,086	68,670	71,531	7,378	NP	8,328	11,712	15,616	32,973	9,568	14,352	17,725	NP	20,626
	9	33,191	34,335	42,919	51,503	60,086	68,670	71,531	12,333	12,551	15,300	21,516	28,688	32,973	9,568	14,352	17,725	19,337	30,939
	12	33,191	34,335	42,919	51,503	60,086	68,670	71,531	18,076	19,324	23,556	33,126	44,168	32,973	9,568	14,352	17,725	25,782	41,252
	15	33,191	34,335	42,919	51,503	60,086	68,670	71,531	24,511	27,006	32,921	46,295	61,727	32,973	9,568	14,352	17,725	32,228	51,565
	18	33,191	34,335	42,919	51,503	60,086	68,670	71,531	31,570	35,500	43,276	60,857	81,142	32,973	9,568	14,352	17,725	38,674	61,878
	21	33,191	34,335	42,919	51,503	60,086	68,670	71,531	39,201	44,735	54,534	76,688	102,251	32,973	9,568	14,352	17,725	45,119	72,191
	25	33,191	34,335	42,919	51,503	60,086	68,670	71,531	50,198	58,107	70,835	99,612	132,816	32,973	9,568	14,352	17,725	53,713	85,941
1-1/4	6	42,152	43,605	54,506	65,408	76,309	87,210	90,844	7,440	NP	8,328	11,712	15,616	40,701	11,813	17,719	20,290	NP	22,916
	9	42,152	43,605	54,506	65,408	76,309	87,210	90,844	12,405	12,551	15,300	21,516	28,688	40,701	11,813	17,719	20,290	21,484	34,374
	12	42,152	43,605	54,506	65,408	76,309	87,210	90,844	18,158	19,324	23,556	33,126	44,168	40,701	11,813	17,719	20,290	28,645	45,832
	15	42,152	43,605	54,506	65,408	76,309	87,210	90,844	24,602	27,006	32,921	46,295	61,727	40,701	11,813	17,719	20,290	35,806	57,290
	18	42,152	43,605	54,506	65,408	76,309	87,210	90,844	31,668	35,500	43,276	60,857	81,142	40,701	11,813	17,719	20,290	42,967	68,748
	21	42,152	43,605	54,506	65,408	76,309	87,210	90,844	39,307	44,735	54,534	76,688	102,251	40,701	11,813	17,719	20,290	50,129	80,206
	25	42,152	43,605	54,506	65,408	76,309	87,210	90,844	50,313	58,107	70,835	99,612	132,816	40,701	11,813	17,719	20,290	59,677	95,483
1-3/8	6	50,460	52,200	65,250	78,300	91,350	104,400	108,750	7,502	NP	8,328	11,712	15,616	49,258	14,293	21,440	22,978	NP	25,210
	9	50,460	52,200	65,250	78,300	91,350	104,400	108,750	12,478	12,551	15,300	21,516	28,688	49,258	14,293	21,440	22,978	23,634	37,815
	12	50,460	52,200	65,250	78,300	91,350	104,400	108,750	18,241	19,324	23,556	33,126	44,168	49,258	14,293	21,440	22,978	31,512	50,420
	15	50,460	52,200	65,250	78,300	91,350	104,400	108,750	24,693	27,006	32,921	46,295	61,727	49,258	14,293	21,440	22,978	39,391	63,025
	18	50,460	52,200	65,250	78,300	91,350	104,400	108,750	31,767	35,500	43,276	60,857	81,142	49,258	14,293	21,440	22,978	47,269	75,630
	21	50,460	52,200	65,250	78,300	91,350	104,400	108,750	39,412	44,735	54,534	76,688	102,251	49,258	14,293	21,440	22,978	55,147	88,235
	25	50,460	52,200	65,250	78,300	91,350	104,400	108,750	50,427	58,107	70,835	99,612	132,816	49,258	14,293	21,440	22,978	65,651	105,041
1-1/2	12	61,335	63,450	79,313	95,175	111,038	126,900	132,188	18,323	19,324	23,556	33,126	44,168	58,621	17,010	25,515	25,783	34,377	55,004
	15	61,335	63,450	79,313	95,175	111,038	126,900	132,188	24,783	27,006	32,921	46,295	61,727	58,621	17,010	25,515	25,783	42,972	68,755
	18	61,335	63,450	79,313	95,175	111,038	126,900	132,188	31,865	35,500	43,276	60,857	81,142	58,621	17,010	25,515	25,783	51,566	82,505
	21	61,335	63,450	79,313	95,175	111,038	126,900	132,188	39,518	44,735	54,534	76,688	102,251	58,621	17,010	25,515	25,783	60,160	96,256
	25	61,335	63,450	79,313	95,175	111,038	126,900	132,188	50,542	58,107	70,835	99,612	132,816	58,621	17,010	25,515	25,783	71,619	114,591
1-3/4	12	82,650	85,500	106,875	128,250	149,625	171,000	178,125	18,488	19,324	23,556	33,126	44,168	92,826	23,153	34,729	34,247	43,259	69,215
	15	82,650	85,500	106,875	128,250	149,625	171,000	178,125	24,965	27,006	32,921	46,295	61,727	92,826	23,153	34,729	34,247	54,074	86,519
	18	82,650	85,500	106,875	128,250	149,625	171,000	178,125	32,063	35,500	43,276	60,857	81,142	92,826	23,153	34,729	34,247	64,889	103,822
	21	82,650	85,500	106,875	128,250	149,625	171,000	178,125	39,730	44,735	54,534	76,688	102,251	92,826	23,153	34,729	34,247	75,704	121,126
	25	82,650	85,500	106,875	128,250	149,625	171,000	178,125	50,771	58,107	70,835	99,612	132,816	92,826	23,153	34,729	34,247	90,123	144,198
2	12	108,750	112,500	140,625	168,750	196,875	225,000	234,375	18,654	19,324	23,556	33,126	44,168	119,078	30,240	45,360	40,830	48,996	78,394
	15	108,750	112,500	140,625	168,750	196,875	225,000	234,375	25,148	27,006	32,921	46,295	61,727	119,078	30,240	45,360	40,830	61,245	97,992
	18	108,750	112,500	140,625	168,750	196,875	225,000	234,375	32,261	35,500	43,276	60,857	81,142	119,078	30,240	45,360	40,830	73,494	117,591
	21	108,750	112,500	140,625	168,750	196,875	225,000	234,375	39,942	44,735	54,534	76,688	102,251	119,078	30,240	45,360	40,830	85,743	137,189
	25	108,750	112,500	140,625	168,750	196,875	225,000	234,375	51,001	58,107	70,835	99,612	132,816	119,078	30,240	45,360	40,830	102,075	163,320

Table 34-5C. Design Strengths for Single Cast-In Anchors Subject to Tensile Loads (f'_c = 6000 psi)[1,2,4]
Notes pertaining to this table are given on Page 34-14

d_o in.	h_{ef} in.	ϕN_{sa} - Tension Strength of Anchor f_{uta} - for design purposes[3] - psi										ϕN_{cb} - Tension Breakout[4,5,6] c_{a1} - edge distance in.				h_{ef}	$\geq 1.5h_{ef}$	head[7]	ϕN_{pn} - Pullout[9] "J" or "L" hook[8]			ϕN_{sb} - Sideface Blowout[4,10] c_{a1} - edge distance in.		
		58,000	60,000	75,000	90,000	105,000	120,000	125,000				1-1/2-in. cover	0.25h_{ef}	0.5h_{ef}					$e_h = 3d_o$	$e_h = 4.5d_o$	1-1/2-in. cover	0.25h_{ef}	0.4h_{ef}
1/4	2	1,392	1,440	1,800	2,160	2,520	2,880	3,000				2,447	NP	NP	2,761	3,681	3,931	709	1,063	4,822	NP	NP
	3	1,392	1,440	1,800	2,160	2,520	2,880	3,000				3,720	NP	NP	5,071	6,762	3,931	709	1,063	4,822	NP	NP
	4	1,392	1,440	1,800	2,160	2,520	2,880	3,000				5,168	NP	5,552	7,808	10,411	3,931	709	1,063	4,822	NP	NP
	5	1,392	1,440	1,800	2,160	2,520	2,880	3,000				6,771	NP	7,760	10,912	14,549	3,931	709	1,063	4,822	NP	5,935
	6	1,392	1,440	1,800	2,160	2,520	2,880	3,000				8,514	NP	10,200	14,344	19,125	3,931	709	1,063	4,822	NP	7,122
3/8	2	3,393	3,510	4,388	5,265	6,143	7,020	7,313				2,498	NP	NP	2,761	3,681	5,510	1,595	2,392	5,929	NP	NP
	3	3,393	3,510	4,388	5,265	6,143	7,020	7,313				3,777	NP	NP	5,071	6,762	5,510	1,595	2,392	5,929	NP	NP
	4	3,393	3,510	4,388	5,265	6,143	7,020	7,313				5,231	NP	5,552	7,808	10,411	5,510	1,595	2,392	5,929	NP	NP
	5	3,393	3,510	4,388	5,265	6,143	7,020	7,313				6,840	NP	7,760	10,912	14,549	5,510	1,595	2,392	5,929	NP	7,027
	6	3,393	3,510	4,388	5,265	6,143	7,020	7,313				8,588	NP	10,200	14,344	19,125	5,510	1,595	2,392	5,929	NP	8,432
1/2	2	6,177	6,390	7,988	9,585	11,183	12,780	13,313				2,550	NP	NP	2,761	3,681	9,778	2,835	4,253	8,190	NP	NP
	3	6,177	6,390	7,988	9,585	11,183	12,780	13,313				3,835	NP	NP	5,071	6,762	9,778	2,835	4,253	8,190	NP	NP
	4	6,177	6,390	7,988	9,585	11,183	12,780	13,313				5,295	NP	5,552	7,808	10,411	9,778	2,835	4,253	8,190	NP	NP
	5	6,177	6,390	7,988	9,585	11,183	12,780	13,313				6,908	NP	7,760	10,912	14,549	9,778	2,835	4,253	8,190	NP	9,360
	6	6,177	6,390	7,988	9,585	11,183	12,780	13,313				8,662	NP	10,200	14,344	19,125	9,778	2,835	4,253	8,190	NP	11,232
	7	6,177	6,390	7,988	9,585	11,183	12,780	13,313				10,544	10,544	12,854	18,076	24,101	9,778	2,835	4,253	8,190	8,190	13,104
	8	6,177	6,390	7,988	9,585	11,183	12,780	13,313				12,547	12,882	15,704	22,084	29,446	9,778	2,835	4,253	8,190	9,360	14,976
5/8	3	9,831	10,170	12,713	15,255	17,798	20,340	21,188				3,893	NP	NP	5,071	6,762	15,254	4,430	6,645	10,595	NP	NP
	4	9,831	10,170	12,713	15,255	17,798	20,340	21,188				5,359	NP	5,552	7,808	10,411	15,254	4,430	6,645	10,595	NP	NP
	5	9,831	10,170	12,713	15,255	17,798	20,340	21,188				6,978	NP	7,760	10,912	14,549	15,254	4,430	6,645	10,595	NP	11,691
	6	9,831	10,170	12,713	15,255	17,798	20,340	21,188				8,736	NP	10,200	14,344	19,125	15,254	4,430	6,645	10,595	NP	14,029
	7	9,831	10,170	12,713	15,255	17,798	20,340	21,188				10,623	NP	12,854	18,076	24,101	15,254	4,430	6,645	10,595	NP	16,367
	8	9,831	10,170	12,713	15,255	17,798	20,340	21,188				12,630	12,882	15,704	22,084	29,446	15,254	4,430	6,645	10,595	11,691	18,706
	9	9,831	10,170	12,713	15,255	17,798	20,340	21,188				14,751	15,372	18,739	26,352	35,136	15,254	4,430	6,645	10,595	13,152	21,044
	10	9,831	10,170	12,713	15,255	17,798	20,340	21,188				16,979	18,004	21,947	30,864	41,151	15,254	4,430	6,645	10,595	14,614	23,382
3/4	4	14,529	15,030	18,788	22,545	26,303	30,060	31,313				5,423	NP	5,552	7,808	10,411	21,974	6,379	9,568	13,155	NP	NP
	5	14,529	15,030	18,788	22,545	26,303	30,060	31,313				7,047	NP	7,760	10,912	14,549	21,974	6,379	9,568	13,155	NP	14,032
	6	14,529	15,030	18,788	22,545	26,303	30,060	31,313				8,811	NP	10,200	14,344	19,125	21,974	6,379	9,568	13,155	NP	16,838
	7	14,529	15,030	18,788	22,545	26,303	30,060	31,313				10,702	NP	12,854	18,076	24,101	21,974	6,379	9,568	13,155	NP	19,644
	8	14,529	15,030	18,788	22,545	26,303	30,060	31,313				12,714	12,882	15,704	22,084	29,446	21,974	6,379	9,568	13,155	14,032	22,451
	9	14,529	15,030	18,788	22,545	26,303	30,060	31,313				14,839	15,372	18,739	26,352	35,136	21,974	6,379	9,568	13,155	15,786	25,257
	10	14,529	15,030	18,788	22,545	26,303	30,060	31,313				17,071	18,004	21,947	30,864	41,151	21,974	6,379	9,568	13,155	17,540	28,064
	12	14,529	15,030	18,788	22,545	26,303	30,060	31,313				21,839	23,667	28,851	40,571	54,095	21,974	6,379	9,568	13,155	21,048	33,676
7/8	4	20,097	20,790	25,988	31,185	36,383	41,580	43,313				5,423	NP	5,552	7,808	10,411	29,938	8,682	13,023	15,866	NP	NP
	6	20,097	20,790	25,988	31,185	36,383	41,580	43,313				8,886	NP	10,200	14,344	19,125	29,938	8,682	13,023	15,866	NP	19,654
	8	20,097	20,790	25,988	31,185	36,383	41,580	43,313				12,798	12,882	15,704	22,084	29,446	29,938	8,682	13,023	15,866	16,378	26,205
	12	20,097	20,790	25,988	31,185	36,383	41,580	43,313				21,939	23,667	28,851	40,571	54,095	29,938	8,682	13,023	15,866	24,567	39,307
	15	20,097	20,790	25,988	31,185	36,383	41,580	43,313				29,799	33,075	40,320	56,700	75,600	29,938	8,682	13,023	15,866	30,709	49,134
	18	20,097	20,790	25,988	31,185	36,383	41,580	43,313				38,425	43,478	53,002	74,534	99,379	29,938	8,682	13,023	15,866	36,851	58,961
	25	20,097	20,790	25,988	31,185	36,383	41,580	43,313				61,200	71,166	86,755	121,999	162,665	29,938	8,682	13,023	15,866	51,181	81,890

Table 34-5C. Design Strengths for Single Cast-In Anchors Subject to Tensile Loads (f'_c = 6000 psi)[1,2,4] (cont'd.)

Notes pertaining to this table are given on Page 34-14

d_o in.	h_{ef} in.	ϕN_{sa} - Tension Strength of Anchor f_{uta} - for design purposes[3] - psi										ϕN_{cb} - Tension Breakout[4,5,6] c_{a1} - edge distance in.					ϕN_{pn} - Pullout[9]			ϕN_{sb} - Sideface Blowout[4,10] c_{a1} - edge distance in.		
		58,000	60,000	75,000	90,000	105,000	120,000	125,000	1-1/2-in. cover	0.25h_{ef}	0.5h_{ef}	h_{ef}	≥1.5h_{ef}	head[7]	"J" or "L" hook[8] e_h = 3d_o	e_h = 4.5d_o	1-1/2-in. cover	0.25h_{ef}	0.4h_{ef}
1	6	26,361	27,270	34,088	40,905	47,723	54,540	56,813	8,961	NP	10,200	14,344	19,125	39,077	11,340	17,010	18,712	NP	22,454
	9	26,361	27,270	34,088	40,905	47,723	54,540	56,813	15,016	15,372	18,739	26,352	35,136	39,077	11,340	17,010	18,712	21,051	33,681
	12	26,361	27,270	34,088	40,905	47,723	54,540	56,813	22,039	23,667	28,851	40,571	54,095	39,077	11,340	17,010	18,712	28,068	44,908
	15	26,361	27,270	34,088	40,905	47,723	54,540	56,813	29,910	33,075	40,320	56,700	75,600	39,077	11,340	17,010	18,712	35,084	56,135
	18	26,361	27,270	34,088	40,905	47,723	54,540	56,813	38,545	43,478	53,002	74,534	99,379	39,077	11,340	17,010	18,712	42,101	67,362
	21	26,361	27,270	34,088	40,905	47,723	54,540	56,813	47,882	54,789	66,790	93,924	125,232	39,077	11,340	17,010	18,712	49,118	78,589
	25	26,361	27,270	34,088	40,905	47,723	54,540	56,813	61,340	71,166	86,755	121,999	162,665	39,077	11,340	17,010	18,712	58,474	93,559
1-1/8	6	33,191	34,335	42,919	51,503	60,086	68,670	71,531	9,036	NP	10,200	14,344	19,125	49,459	14,352	21,528	21,709	NP	25,261
	9	33,191	34,335	42,919	51,503	60,086	68,670	71,531	15,104	15,372	18,739	26,352	35,136	49,459	14,352	21,528	21,709	23,683	37,892
	12	33,191	34,335	42,919	51,503	60,086	68,670	71,531	22,139	23,667	28,851	40,571	54,095	49,459	14,352	21,528	21,709	31,577	50,523
	15	33,191	34,335	42,919	51,503	60,086	68,670	71,531	30,020	33,075	40,320	56,700	75,600	49,459	14,352	21,528	21,709	39,471	63,154
	18	33,191	34,335	42,919	51,503	60,086	68,670	71,531	38,665	43,478	53,002	74,534	99,379	49,459	14,352	21,528	21,709	47,365	75,784
	21	33,191	34,335	42,919	51,503	60,086	68,670	71,531	48,011	54,789	66,790	93,924	125,232	49,459	14,352	21,528	21,709	55,259	88,415
	25	33,191	34,335	42,919	51,503	60,086	68,670	71,531	61,480	71,166	86,755	121,999	162,665	49,459	14,352	21,528	21,709	65,785	105,256
1-1/4	6	42,152	43,605	54,506	65,408	76,309	87,210	90,844	9,112	NP	10,200	14,344	19,125	61,051	17,719	26,578	24,850	NP	28,066
	9	42,152	43,605	54,506	65,408	76,309	87,210	90,844	15,193	15,372	18,739	26,352	35,136	61,051	17,719	26,578	24,850	26,312	42,099
	12	42,152	43,605	54,506	65,408	76,309	87,210	90,844	22,239	23,667	28,851	40,571	54,095	61,051	17,719	26,578	24,850	35,083	56,132
	15	42,152	43,605	54,506	65,408	76,309	87,210	90,844	30,131	33,075	40,320	56,700	75,600	61,051	17,719	26,578	24,850	43,853	70,165
	18	42,152	43,605	54,506	65,408	76,309	87,210	90,844	38,786	43,478	53,002	74,534	99,379	61,051	17,719	26,578	24,850	52,624	84,198
	21	42,152	43,605	54,506	65,408	76,309	87,210	90,844	48,141	54,789	66,790	93,924	125,232	61,051	17,719	26,578	24,850	61,395	98,231
	25	42,152	43,605	54,506	65,408	76,309	87,210	90,844	61,620	71,166	86,755	121,999	162,665	61,051	17,719	26,578	24,850	73,089	116,942
1-3/8	6	50,460	52,200	65,250	78,300	91,350	104,400	108,750	9,188	NP	10,200	14,344	19,125	73,886	21,440	32,160	28,142	NP	30,876
	9	50,460	52,200	65,250	78,300	91,350	104,400	108,750	15,283	15,372	18,739	26,352	35,136	73,886	21,440	32,160	28,142	28,946	46,314
	12	50,460	52,200	65,250	78,300	91,350	104,400	108,750	22,340	23,667	28,851	40,571	54,095	73,886	21,440	32,160	28,142	38,595	61,751
	15	50,460	52,200	65,250	78,300	91,350	104,400	108,750	30,242	33,075	40,320	56,700	75,600	73,886	21,440	32,160	28,142	48,243	77,189
	18	50,460	52,200	65,250	78,300	91,350	104,400	108,750	38,906	43,478	53,002	74,534	99,379	73,886	21,440	32,160	28,142	57,892	92,627
	21	50,460	52,200	65,250	78,300	91,350	104,400	108,750	48,270	54,789	66,790	93,924	125,232	73,886	21,440	32,160	28,142	67,541	108,065
	25	50,460	52,200	65,250	78,300	91,350	104,400	108,750	61,760	71,166	86,755	121,999	162,665	73,886	21,440	32,160	28,142	80,406	128,649
1-1/2	12	61,335	63,450	79,313	95,175	111,038	126,900	132,188	22,441	23,667	28,851	40,571	54,095	87,931	25,515	38,273	31,578	42,103	67,365
	15	61,335	63,450	79,313	95,175	111,038	126,900	132,188	30,353	33,075	40,320	56,700	75,600	87,931	25,515	38,273	31,578	52,629	84,207
	18	61,335	63,450	79,313	95,175	111,038	126,900	132,188	39,027	43,478	53,002	74,534	99,379	87,931	25,515	38,273	31,578	63,155	101,048
	21	61,335	63,450	79,313	95,175	111,038	126,900	132,188	48,399	54,789	66,790	93,924	125,232	87,931	25,515	38,273	31,578	73,681	117,889
	25	61,335	63,450	79,313	95,175	111,038	126,900	132,188	61,901	71,166	86,755	121,999	162,665	87,931	25,515	38,273	31,578	87,715	140,345
1-3/4	12	82,650	85,500	106,875	128,250	149,625	171,000	178,125	22,643	23,667	28,851	40,571	54,095	139,238	34,729	52,093	41,944	52,982	84,771
	15	82,650	85,500	106,875	128,250	149,625	171,000	178,125	30,576	33,075	40,320	56,700	75,600	139,238	34,729	52,093	41,944	66,227	105,963
	18	82,650	85,500	106,875	128,250	149,625	171,000	178,125	39,269	43,478	53,002	74,534	99,379	139,238	34,729	52,093	41,944	79,472	127,156
	21	82,650	85,500	106,875	128,250	149,625	171,000	178,125	48,659	54,789	66,790	93,924	125,232	139,238	34,729	52,093	41,944	92,718	148,348
	25	82,650	85,500	106,875	128,250	149,625	171,000	178,125	62,182	71,166	86,755	121,999	162,665	139,238	34,729	52,093	41,944	110,378	176,605
2	12	108,750	112,500	140,625	168,750	196,875	225,000	234,375	22,846	23,667	28,851	40,571	54,095	178,618	45,360	68,040	50,006	60,008	96,012
	15	108,750	112,500	140,625	168,750	196,875	225,000	234,375	30,800	33,075	40,320	56,700	75,600	178,618	45,360	68,040	50,006	75,010	120,016
	18	108,750	112,500	140,625	168,750	196,875	225,000	234,375	39,511	43,478	53,002	74,534	99,379	178,618	45,360	68,040	50,006	90,012	144,019
	21	108,750	112,500	140,625	168,750	196,875	225,000	234,375	48,919	54,789	66,790	93,924	125,232	178,618	45,360	68,040	50,006	105,014	168,022
	25	108,750	112,500	140,625	168,750	196,875	225,000	234,375	62,463	71,166	86,755	121,999	162,665	178,618	45,360	68,040	50,006	125,016	200,026

Table 34-6A. Design Strengths for Single Cast-In Anchors Subject to Shear Loads (f'_c = 2500 psi)[1,2,3,5]

Notes pertaining to this table are given on Page 34-15

d_o in.	h_{ef} in.	ϕV_{sa} - Shear Strength of Anchor f_{uta} - for design purposes[4] - psi										1-1/2-in. cover	ϕV_{cb} - Shear Breakout[5,6,7,8,9] $h = h_{ef}$[10] and c_{a1}=[11]				$h = 1.5 h_{ef}$ and c_{a1}=[11,12]				$h = 2.25 h_{ef}$ and c_{a1}=[11,13]			
		58,000	60,000	75,000	90,000	105,000	120,000	125,000		0.25h_{ef}	0.5h_{ef}	h_{ef}	1.5h_{ef}	3h_{ef}	h_{ef}	1.5h_{ef}	3h_{ef}	1.5h_{ef}	2h_{ef}	3h_{ef}
1/4	2	724	749	936	1,123	1,310	1,498	1,560	316	NP	NP	350	429	606	525	643	910	965	1,114	1,364
	3	724	749	936	1,123	1,310	1,498	1,560	385	NP	NP	643	788	1,114	965	1,182	1,671	1,772	2,047	2,507
	4	724	749	936	1,123	1,310	1,498	1,560	385	NP	525	990	1,213	1,715	1,485	1,819	2,573	2,729	3,151	3,859
	5	724	749	936	1,123	1,310	1,498	1,560	385	NP	734	1,384	1,695	2,397	2,076	2,542	3,596	3,814	4,404	5,393
	6	724	749	936	1,123	1,310	1,498	1,560	385	NP	965	1,819	2,228	3,151	2,729	3,342	4,727	5,013	5,789	7,090
3/8	2	1,764	1,825	2,282	2,738	3,194	3,650	3,803	363	NP	NP	395	484	685	593	726	1,027	1,090	1,258	1,541
	3	1,764	1,825	2,282	2,738	3,194	3,650	3,803	499	NP	NP	788	965	1,364	1,182	1,447	2,047	2,171	2,507	3,070
	4	1,764	1,825	2,282	2,738	3,194	3,650	3,803	499	NP	643	1,213	1,485	2,101	1,819	2,228	3,151	3,342	3,859	4,727
	5	1,764	1,825	2,282	2,738	3,194	3,650	3,803	499	NP	899	1,695	2,076	2,936	2,542	3,114	4,404	4,671	5,393	6,606
	6	1,764	1,825	2,282	2,738	3,194	3,650	3,803	499	NP	1,182	2,228	2,729	3,859	3,342	4,093	5,789	6,140	7,090	8,683
1/2	2	3,212	3,323	4,154	4,984	5,815	6,646	6,923	403	NP	NP	431	528	747	647	792	1,120	1,188	1,372	1,680
	3	3,212	3,323	4,154	4,984	5,815	6,646	6,923	574	NP	NP	859	1,052	1,487	1,288	1,578	2,231	2,366	2,733	3,347
	4	3,212	3,323	4,154	4,984	5,815	6,646	6,923	608	NP	743	1,400	1,715	2,426	2,101	2,573	3,638	3,859	4,456	5,458
	5	3,212	3,323	4,154	4,984	5,815	6,646	6,923	608	NP	1,038	1,957	2,397	3,390	2,936	3,596	5,085	5,393	6,228	7,627
	6	3,212	3,323	4,154	4,984	5,815	6,646	6,923	608	NP	1,364	2,573	3,151	4,456	3,859	4,727	6,684	7,090	8,187	10,026
	7	3,212	3,323	4,154	4,984	5,815	6,646	6,923	608	608	1,719	3,242	3,971	5,615	4,863	5,956	8,423	8,934	10,316	12,635
	8	3,212	3,323	4,154	4,984	5,815	6,646	6,923	608	743	2,101	3,961	4,851	6,861	5,942	7,277	10,291	10,915	12,604	15,437
5/8	3	5,112	5,288	6,611	7,933	9,255	10,577	11,018	647	NP	NP	918	1,125	1,590	1,377	1,687	2,386	2,530	2,922	3,578
	4	5,112	5,288	6,611	7,933	9,255	10,577	11,018	685	NP	794	1,497	1,834	2,594	2,246	2,751	3,890	4,126	4,765	5,836
	5	5,112	5,288	6,611	7,933	9,255	10,577	11,018	716	NP	1,160	2,188	2,680	3,790	3,282	4,020	5,685	6,030	6,963	8,528
	6	5,112	5,288	6,611	7,933	9,255	10,577	11,018	716	NP	1,525	2,876	3,523	4,982	4,315	5,284	7,473	7,927	9,153	11,210
	7	5,112	5,288	6,611	7,933	9,255	10,577	11,018	716	NP	1,922	3,625	4,439	6,278	5,437	6,659	9,417	9,989	11,534	14,126
	8	5,112	5,288	6,611	7,933	9,255	10,577	11,018	716	830	2,349	4,429	5,424	7,671	6,643	8,136	11,506	12,204	14,092	17,259
	9	5,112	5,288	6,611	7,933	9,255	10,577	11,018	716	991	2,802	5,284	6,472	9,153	7,927	9,708	13,729	14,562	16,815	20,594
	10	5,112	5,288	6,611	7,933	9,255	10,577	11,018	716	1,160	3,282	6,189	7,580	10,720	9,284	11,370	16,080	17,055	19,694	24,120
3/4	4	7,555	7,816	9,770	11,723	13,677	15,631	16,283	761	NP	839	1,582	1,937	2,739	2,372	2,906	4,109	4,358	5,033	6,164
	5	7,555	7,816	9,770	11,723	13,677	15,631	16,283	796	NP	1,226	2,311	2,831	4,003	3,467	4,246	6,005	6,369	7,354	9,007
	6	7,555	7,816	9,770	11,723	13,677	15,631	16,283	826	NP	1,671	3,151	3,859	5,458	4,727	5,789	8,187	8,683	10,026	12,280
	7	7,555	7,816	9,770	11,723	13,677	15,631	16,283	826	NP	2,106	3,971	4,863	6,878	5,956	7,295	10,316	10,942	12,635	15,474
	8	7,555	7,816	9,770	11,723	13,677	15,631	16,283	826	910	2,573	4,851	5,942	8,403	7,277	8,912	12,604	13,369	15,437	18,906
	9	7,555	7,816	9,770	11,723	13,677	15,631	16,283	826	1,085	3,070	5,789	7,090	10,026	8,683	10,635	15,040	15,952	18,420	22,560
	10	7,555	7,816	9,770	11,723	13,677	15,631	16,283	826	1,271	3,596	6,780	8,304	11,743	10,170	12,455	17,615	18,683	21,574	26,422
	12	7,555	7,816	9,770	11,723	13,677	15,631	16,283	826	1,671	4,727	8,912	10,915	15,437	13,369	16,373	23,155	24,560	28,359	34,733
7/8	4	10,450	10,811	13,514	16,216	18,919	21,622	22,523	838	NP	878	1,656	2,029	2,869	2,485	3,043	4,304	4,565	5,271	6,455
	6	10,450	10,811	13,514	16,216	18,919	21,622	22,523	908	NP	1,750	3,300	4,042	5,716	4,950	6,063	8,574	9,094	10,501	12,861
	8	10,450	10,811	13,514	16,216	18,919	21,622	22,523	937	983	2,779	5,240	6,418	9,076	7,860	9,627	13,614	14,440	16,674	20,421
	12	10,450	10,811	13,514	16,216	18,919	21,622	22,523	937	1,805	5,105	9,627	11,790	16,674	14,440	17,685	25,010	26,528	30,631	37,516
	15	10,450	10,811	13,514	16,216	18,919	21,622	22,523	937	2,523	7,135	13,453	16,477	23,302	20,180	24,716	34,953	37,073	42,809	52,430
	18	10,450	10,811	13,514	16,216	18,919	21,622	22,523	937	3,316	9,379	17,685	21,660	30,631	26,528	32,489	45,947	48,734	56,273	68,921
	25	10,450	10,811	13,514	16,216	18,919	21,622	22,523	937	5,428	15,352	28,947	35,453	50,138	43,421	53,179	75,207	79,769	92,109	112,811

Table 34-6A. Design Strengths for Single Cast-In Anchors Subject to Shear Loads (f'_c = 2500 psi)[1,2,3,5] (cont'd.)
Notes pertaining to this table are given on Page 34-15

d_o in.	h_{ef} in.	ϕV_{sa} - Shear Strength of Anchor										ϕV_{cb} - Shear Breakout [5,6,7,8,9]											
		f_{uta} - for design purposes [4] - psi										1-1/2-in. cover	0.25h_{ef}	$h = h_{ef}$[10] and $c_{a1} \geq$[1]			$h = 1.5h_{ef}$ and $c_{a1} \geq$[11,12]				$h = 2.25h_{ef}$ and $c_{a1} \geq$[11,13]		
		58,000	60,000	75,000	90,000	105,000	120,000	125,000						0.5h_{ef}	h_{ef}	1.5h_{ef}	3h_{ef}	h_{ef}	1.5h_{ef}	3h_{ef}	1.5h_{ef}	2h_{ef}	3h_{ef}
1	6	13,708	14,180	17,726	21,271	24,816	28,361	29,543				992	NP	1,822	3,435	4,207	5,950	5,153	6,311	8,924	9,466	10,930	13,387
	9	13,708	14,180	17,726	21,271	24,816	28,361	29,543				1,050	1,253	3,545	6,684	8,187	11,578	10,026	12,280	17,366	18,420	21,269	26,050
	12	13,708	14,180	17,726	21,271	24,816	28,361	29,543				1,050	1,930	5,458	10,291	12,604	17,825	15,437	18,906	26,737	28,359	32,746	40,106
	15	13,708	14,180	17,726	21,271	24,816	28,361	29,543				1,050	2,697	7,627	14,382	17,615	24,911	21,574	26,422	37,366	39,633	45,764	56,050
	18	13,708	14,180	17,726	21,271	24,816	28,361	29,543				1,050	3,545	10,026	18,906	23,155	32,746	28,359	34,733	49,119	52,099	60,159	73,679
	21	13,708	14,180	17,726	21,271	24,816	28,361	29,543				1,050	4,467	12,635	23,824	29,179	41,265	35,737	43,768	61,898	65,652	75,809	92,846
	25	13,708	14,180	17,726	21,271	24,816	28,361	29,543				1,050	5,802	16,412	30,946	37,901	53,600	46,419	56,851	80,400	85,277	98,469	120,600
1-1/8	6	17,259	17,854	22,318	26,781	31,245	35,708	37,196				1,076	NP	1,887	3,559	4,358	6,164	5,338	6,538	9,245	9,806	11,323	13,868
	9	17,259	17,854	22,318	26,781	31,245	35,708	37,196				1,167	1,329	3,760	7,090	8,683	12,280	10,635	13,025	18,420	19,537	22,560	27,630
	12	17,259	17,854	22,318	26,781	31,245	35,708	37,196				1,167	2,047	5,789	10,915	13,369	18,906	16,373	20,053	28,359	30,079	34,733	42,539
	15	17,259	17,854	22,318	26,781	31,245	35,708	37,196				1,167	2,860	8,090	15,255	18,683	26,422	22,882	28,025	39,633	42,037	48,540	59,450
	18	17,259	17,854	22,318	26,781	31,245	35,708	37,196				1,167	3,760	10,635	20,053	24,560	34,733	30,079	36,840	52,099	55,259	63,808	78,149
	21	17,259	17,854	22,318	26,781	31,245	35,708	37,196				1,167	4,738	13,401	25,270	30,949	43,768	37,904	46,423	65,652	69,635	80,407	98,478
	25	17,259	17,854	22,318	26,781	31,245	35,708	37,196				1,167	6,154	17,407	32,823	40,200	56,851	49,235	60,300	85,277	90,450	104,442	127,915
1-1/4	6	21,919	22,675	28,343	34,012	39,681	45,349	47,239				1,161	NP	1,948	3,673	4,498	6,362	5,509	6,747	9,542	10,121	11,687	14,314
	9	21,919	22,675	28,343	34,012	39,681	45,349	47,239				1,259	1,372	3,881	7,317	8,962	12,674	10,976	13,443	19,011	20,165	23,284	28,517
	12	21,919	22,675	28,343	34,012	39,681	45,349	47,239				1,286	2,157	6,102	11,506	14,092	19,929	17,259	21,138	29,893	31,706	36,611	44,840
	15	21,919	22,675	28,343	34,012	39,681	45,349	47,239				1,286	3,015	8,528	16,080	19,694	27,851	24,120	29,541	41,777	44,311	51,166	62,665
	18	21,919	22,675	28,343	34,012	39,681	45,349	47,239				1,286	3,963	11,210	21,138	25,888	36,611	31,706	38,832	54,917	58,248	67,260	82,376
	21	21,919	22,675	28,343	34,012	39,681	45,349	47,239				1,286	4,994	14,126	26,636	32,623	46,136	39,955	48,934	69,204	73,401	84,757	103,805
	25	21,919	22,675	28,343	34,012	39,681	45,349	47,239				1,286	6,487	18,349	34,599	42,374	59,926	51,898	63,562	89,890	95,342	110,092	134,834
1-3/8	6	26,239	27,144	33,930	40,716	47,502	54,288	56,550				1,248	NP	2,004	3,779	4,629	6,546	5,669	6,943	9,819	10,415	12,026	14,729
	9	26,239	27,144	33,930	40,716	47,502	54,288	56,550				1,353	1,412	3,993	7,530	9,222	13,042	11,295	13,833	19,563	20,749	23,959	29,344
	12	26,239	27,144	33,930	40,716	47,502	54,288	56,550				1,409	2,263	6,400	12,067	14,780	20,901	18,101	22,169	31,352	33,254	38,398	47,028
	15	26,239	27,144	33,930	40,716	47,502	54,288	56,550				1,409	3,162	8,944	16,865	20,655	29,211	25,297	30,983	43,816	46,474	53,663	65,724
	18	26,239	27,144	33,930	40,716	47,502	54,288	56,550				1,409	4,157	11,757	22,169	27,152	38,398	33,254	40,728	57,598	61,092	70,542	86,396
	21	26,239	27,144	33,930	40,716	47,502	54,288	56,550				1,409	5,238	14,816	27,937	34,215	48,388	41,905	51,323	72,581	76,984	88,894	108,872
	25	26,239	27,144	33,930	40,716	47,502	54,288	56,550				1,409	6,804	19,244	36,287	44,443	62,851	54,431	66,664	94,277	99,996	115,465	141,416
1-1/2	12	31,894	32,994	41,243	49,491	57,740	65,988	68,738				1,535	2,363	6,684	12,604	15,437	21,831	18,906	23,155	32,746	34,733	40,106	49,119
	15	31,894	32,994	41,243	49,491	57,740	65,988	68,738				1,535	3,303	9,342	17,615	21,574	30,510	26,422	32,360	45,764	48,540	56,050	68,647
	18	31,894	32,994	41,243	49,491	57,740	65,988	68,738				1,535	4,342	12,280	23,155	28,359	40,106	34,733	42,539	60,159	63,808	73,679	90,238
	21	31,894	32,994	41,243	49,491	57,740	65,988	68,738				1,535	5,471	15,474	29,179	35,737	50,539	43,768	53,605	75,800	80,407	92,846	113,713
	25	31,894	32,994	41,243	49,491	57,740	65,988	68,738				1,535	7,106	20,100	37,901	46,419	65,646	56,851	69,628	98,469	104,442	120,600	147,704
1-3/4	12	42,978	44,460	55,575	66,690	77,805	88,920	92,625				1,743	2,475	7,001	13,201	16,167	22,864	19,801	24,251	34,296	36,377	42,004	51,444
	15	42,978	44,460	55,575	66,690	77,805	88,920	92,625				1,798	3,567	10,090	19,026	23,302	32,954	28,539	34,953	49,431	52,430	60,541	74,147
	18	42,978	44,460	55,575	66,690	77,805	88,920	92,625				1,798	4,689	13,264	25,010	30,631	43,319	37,516	45,947	64,979	68,921	79,583	97,468
	21	42,978	44,460	55,575	66,690	77,805	88,920	92,625				1,798	5,909	16,714	31,517	38,600	54,589	47,275	57,900	81,883	86,850	100,286	122,824
	25	42,978	44,460	55,575	66,690	77,805	88,920	92,625				1,798	7,676	21,710	40,938	50,138	70,906	61,406	75,207	106,359	112,811	130,262	159,538
2	12	56,550	58,500	73,125	87,750	102,375	117,000	121,875				1,960	2,576	7,287	13,740	16,828	23,799	20,610	25,242	35,698	37,863	43,721	53,547
	15	56,550	58,500	73,125	87,750	102,375	117,000	121,875				2,049	3,765	10,648	20,079	24,591	34,778	30,118	36,887	52,166	55,331	63,891	78,250
	18	56,550	58,500	73,125	87,750	102,375	117,000	121,875				2,076	5,013	14,180	26,737	32,746	46,310	40,106	49,119	69,465	73,679	85,077	104,198
	21	56,550	58,500	73,125	87,750	102,375	117,000	121,875				2,076	6,317	17,868	33,693	41,265	58,358	50,539	61,898	87,536	92,846	107,210	131,305
	25	56,550	58,500	73,125	87,750	102,375	117,000	121,875				2,076	8,206	23,209	43,764	53,600	75,802	65,646	80,400	113,702	120,600	139,256	170,554

Table 34-6B. Design Strengths for Single Cast-In Anchors Subject to Shear Loads ($f'_c = 4000$ psi)[1, 2, 3, 5]
Notes pertaining to this table are given on Page 34-15

d_o in.	h_{ef} in.	ϕV_{sa} - Shear Strength of Anchor										1-1/2-in. cover	ϕV_{cb} - Shear Breakout[5, 6, 7, 8, 9]									
		f_{uta} - for design purposes[4] - psi											$h = h_{ef}$[10] and $c_{a1} =$[11]			$h = 1.5h_{ef}$ and $c_{a1} =$[11, 12]			$h = 2.25h_{ef}$ and $c_{a1} =$[11, 13]		
		58,000	60,000	75,000	90,000	105,000	120,000	125,000					0.25h_{ef}	0.5h_{ef}	h_{ef}	1.5h_{ef}	3h_{ef}	h_{ef}	1.5h_{ef}	3h_{ef}	1.5h_{ef}	2h_{ef}	3h_{ef}
1/4	2	724	749	936	1,123	1,310	1,498	1,560	399	NP	NP	443	542	767	664	814	1,151	1,220	1,409	1,726
	3	724	749	936	1,123	1,310	1,498	1,560	487	NP	NP	814	996	1,409	1,220	1,495	2,114	2,242	2,589	3,171
	4	724	749	936	1,123	1,310	1,498	1,560	487	NP	664	1,253	1,534	2,170	1,879	2,301	3,254	3,452	3,986	4,882
	5	724	749	936	1,123	1,310	1,498	1,560	487	NP	928	1,751	2,144	3,032	2,626	3,216	4,548	4,824	5,570	6,822
	6	724	749	936	1,123	1,310	1,498	1,560	487	NP	1,220	2,301	2,818	3,986	3,452	4,228	5,979	6,341	7,322	8,968
3/8	2	1,764	1,825	2,282	2,738	3,194	3,650	3,803	459	NP	NP	500	613	866	750	919	1,299	1,378	1,591	1,949
	3	1,764	1,825	2,282	2,738	3,194	3,650	3,803	631	NP	NP	996	1,220	1,726	1,495	1,831	2,589	2,746	3,171	3,883
	4	1,764	1,825	2,282	2,738	3,194	3,650	3,803	631	NP	814	1,534	1,879	2,657	2,301	2,818	3,986	4,228	4,882	5,979
	5	1,764	1,825	2,282	2,738	3,194	3,650	3,803	631	NP	1,137	2,144	2,626	3,714	3,216	3,939	5,570	5,908	6,822	8,355
	6	1,764	1,825	2,282	2,738	3,194	3,650	3,803	631	NP	1,495	2,818	3,452	4,882	4,228	5,178	7,322	7,766	8,968	10,983
1/2	2	3,212	3,323	4,154	4,984	5,815	6,646	6,923	510	NP	NP	545	668	944	818	1,002	1,417	1,502	1,735	2,125
	3	3,212	3,323	4,154	4,984	5,815	6,646	6,923	726	NP	NP	1,086	1,330	1,881	1,629	1,996	2,822	2,993	3,456	4,233
	4	3,212	3,323	4,154	4,984	5,815	6,646	6,923	769	NP	939	1,771	2,170	3,068	2,657	3,254	4,602	4,882	5,637	6,904
	5	3,212	3,323	4,154	4,984	5,815	6,646	6,923	769	NP	1,313	2,476	3,032	4,288	3,714	4,548	6,432	6,822	7,878	9,648
	6	3,212	3,323	4,154	4,984	5,815	6,646	6,923	769	NP	1,726	3,254	3,986	5,637	4,882	5,979	8,455	8,968	10,355	12,683
	7	3,212	3,323	4,154	4,984	5,815	6,646	6,923	769	769	2,175	4,101	5,023	7,103	6,151	7,534	10,655	11,301	13,049	15,982
	8	3,212	3,323	4,154	4,984	5,815	6,646	6,923	769	939	2,657	5,010	6,136	8,678	7,516	9,205	13,017	13,807	15,943	19,526
5/8	3	5,112	5,288	6,611	7,933	9,255	10,577	11,018	818	NP	NP	1,161	1,422	2,012	1,742	2,134	3,018	3,201	3,696	4,526
	4	5,112	5,288	6,611	7,933	9,255	10,577	11,018	867	NP	1,004	1,894	2,320	3,281	2,841	3,480	4,921	5,220	6,027	7,381
	5	5,112	5,288	6,611	7,933	9,255	10,577	11,018	906	NP	1,468	2,768	3,390	4,794	4,152	5,085	7,191	7,627	8,807	10,787
	6	5,112	5,288	6,611	7,933	9,255	10,577	11,018	906	NP	1,930	3,638	4,456	6,302	5,458	6,684	9,453	10,026	11,578	14,180
	7	5,112	5,288	6,611	7,933	9,255	10,577	11,018	906	NP	2,432	4,585	5,615	7,941	6,878	8,423	11,912	12,635	14,589	17,868
	8	5,112	5,288	6,611	7,933	9,255	10,577	11,018	906	1,050	2,971	5,602	6,861	9,703	8,403	10,291	14,554	15,437	17,825	21,831
	9	5,112	5,288	6,611	7,933	9,255	10,577	11,018	906	1,253	3,545	6,684	8,187	11,578	10,026	12,280	17,366	18,420	21,269	26,050
	10	5,112	5,288	6,611	7,933	9,255	10,577	11,018	906	1,468	4,152	7,829	9,588	13,560	11,743	14,382	20,340	21,574	24,911	30,510
3/4	4	7,555	7,816	9,770	11,723	13,677	15,631	16,283	963	NP	1,061	2,001	2,450	3,465	3,001	3,675	5,198	5,513	6,366	7,796
	5	7,555	7,816	9,770	11,723	13,677	15,631	16,283	1,007	NP	1,550	2,923	3,581	5,064	4,385	5,371	7,595	8,056	9,303	11,393
	6	7,555	7,816	9,770	11,723	13,677	15,631	16,283	1,044	NP	2,114	3,986	4,882	6,904	5,979	7,322	10,355	10,983	12,683	15,533
	7	7,555	7,816	9,770	11,723	13,677	15,631	16,283	1,044	NP	2,664	5,023	6,151	8,699	7,534	9,227	13,049	13,841	15,982	19,574
	8	7,555	7,816	9,770	11,723	13,677	15,631	16,283	1,044	1,151	3,254	6,136	7,516	10,629	9,205	11,273	15,943	16,910	19,526	23,915
	9	7,555	7,816	9,770	11,723	13,677	15,631	16,283	1,044	1,373	3,883	7,322	8,968	12,683	10,983	13,452	19,024	20,178	23,299	28,536
	10	7,555	7,816	9,770	11,723	13,677	15,631	16,283	1,044	1,608	4,548	8,576	10,503	14,854	12,864	15,755	22,281	23,633	27,289	33,422
	12	7,555	7,816	9,770	11,723	13,677	15,631	16,283	1,044	2,114	5,979	11,273	13,807	19,526	16,910	20,711	29,289	31,066	35,872	43,934
7/8	4	10,450	10,811	13,514	16,216	18,919	21,622	22,523	1,060	NP	1,111	2,095	2,566	3,629	3,143	3,849	5,444	5,774	6,667	8,165
	6	10,450	10,811	13,514	16,216	18,919	21,622	22,523	1,149	NP	2,214	4,174	5,113	7,230	6,262	7,669	10,845	11,503	13,283	16,268
	8	10,450	10,811	13,514	16,216	18,919	21,622	22,523	1,185	1,243	3,515	6,628	8,118	11,480	9,942	12,177	17,220	18,265	21,091	25,831
	12	10,450	10,811	13,514	16,216	18,919	21,622	22,523	1,185	2,283	6,458	12,177	14,913	21,091	18,265	22,370	31,636	33,555	38,746	47,454
	15	10,450	10,811	13,514	16,216	18,919	21,622	22,523	1,185	3,191	9,025	17,017	20,842	29,475	25,526	31,263	44,213	46,894	54,149	66,319
	18	10,450	10,811	13,514	16,216	18,919	21,622	22,523	1,185	4,194	11,863	22,370	27,398	38,746	33,555	41,096	58,119	61,644	71,181	87,178
	25	10,450	10,811	13,514	16,216	18,919	21,622	22,523	1,185	6,865	19,418	36,616	44,845	63,420	54,923	67,267	95,130	100,901	116,510	142,695

Table 34-6B. Design Strengths for Single Cast-In Anchors Subject to Shear Loads ($f'_c = 4000$ psi)[1,2,3,5] (cont'd.)
Notes pertaining to this table are given on Page 34-15

d_o in.	h_{ef} in.	ϕV_{sa} - Shear Strength of Anchor											ϕV_{cb} - Shear Breakout [5,6,7,8,9]											
		f_{uta} - for design purposes[4] - psi										1-1/2-in. cover	0.25h_{ef}	$h = h_{ef}$[10]				$h = 1.5h_{ef}$ and c_{a1} =[11]			$h = 1.5h_{ef}$ and c_{a1} =[11,12]			$h = 2.25h_{ef}$ and c_{a1} =[11,13]		
		58,000	60,000	75,000	90,000	105,000	120,000	125,000						0.5h_{ef}	h_{ef} and c_{a1} =[1]	1.5h_{ef}	3h_{ef}	h_{ef}	1.5h_{ef}	3h_{ef}	1.5h_{ef}	2h_{ef}	3h_{ef}			
1	6	13,708	14,180	17,726	21,271	24,816	28,361	29,543	1,254	NP	2,304	4,345	5,322	7,526	6,518	7,982	11,289	11,973	13,826	16,933			
	9	13,708	14,180	17,726	21,271	24,816	28,361	29,543	1,329	1,585	4,484	8,455	10,355	14,645	12,683	15,533	21,967	23,299	26,904	32,950			
	12	13,708	14,180	17,726	21,271	24,816	28,361	29,543	1,329	2,441	6,904	13,017	15,943	22,547	19,526	23,915	33,820	35,872	41,421	50,730			
	15	13,708	14,180	17,726	21,271	24,816	28,361	29,543	1,329	3,411	9,648	18,192	22,281	31,510	27,289	33,422	47,265	50,132	57,888	70,898			
	18	13,708	14,180	17,726	21,271	24,816	28,361	29,543	1,329	4,484	12,683	23,915	29,289	41,421	35,872	43,934	62,132	65,901	76,096	93,198			
	21	13,708	14,180	17,726	21,271	24,816	28,361	29,543	1,329	5,650	15,982	30,136	36,909	52,197	45,204	55,363	78,295	83,044	95,891	117,442			
	25	13,708	14,180	17,726	21,271	24,816	28,361	29,543	1,329	7,339	20,759	39,144	47,941	67,799	58,716	71,912	101,699	107,868	124,555	152,548			
1-1/8	6	17,259	17,854	22,318	26,781	31,245	35,708	37,196	1,361	NP	2,387	4,501	5,513	7,796	6,752	8,269	11,695	12,404	14,323	17,542			
	9	17,259	17,854	22,318	26,781	31,245	35,708	37,196	1,476	1,681	4,756	8,968	10,983	15,533	13,452	16,475	23,299	24,713	28,536	34,949			
	12	17,259	17,854	22,318	26,781	31,245	35,708	37,196	1,476	2,589	7,322	13,807	16,910	23,915	20,711	25,365	35,872	38,048	43,934	53,808			
	15	17,259	17,854	22,318	26,781	31,245	35,708	37,196	1,476	3,618	10,233	19,296	23,633	33,422	28,944	35,449	50,132	53,173	61,399	75,198			
	18	17,259	17,854	22,318	26,781	31,245	35,708	37,196	1,476	4,756	13,452	25,365	31,066	43,934	38,048	46,599	65,901	69,898	80,711	98,851			
	21	17,259	17,854	22,318	26,781	31,245	35,708	37,196	1,476	5,993	16,951	31,964	39,147	55,363	47,946	58,721	83,044	88,082	101,708	124,566			
	25	17,259	17,854	22,318	26,781	31,245	35,708	37,196	1,476	7,785	22,018	41,518	50,849	71,912	62,277	76,274	107,868	114,411	132,110	161,801			
1-1/4	6	21,919	22,675	28,343	34,012	39,681	45,349	47,239	1,469	NP	2,464	4,646	5,690	8,047	6,969	8,535	12,070	12,802	14,783	18,105			
	9	21,919	22,675	28,343	34,012	39,681	45,349	47,239	1,593	1,735	4,909	9,256	11,336	16,032	13,884	17,004	24,048	25,506	29,452	36,071			
	12	21,919	22,675	28,343	34,012	39,681	45,349	47,239	1,627	2,729	7,718	14,554	17,825	25,208	21,831	26,737	37,812	40,106	46,310	56,718			
	15	21,919	22,675	28,343	34,012	39,681	45,349	47,239	1,627	3,814	10,787	20,340	24,911	35,229	30,510	37,366	52,844	56,050	64,721	79,266			
	18	21,919	22,675	28,343	34,012	39,681	45,349	47,239	1,627	5,013	14,180	26,737	32,746	46,310	40,106	49,119	69,465	73,679	85,077	104,198			
	21	21,919	22,675	28,343	34,012	39,681	45,349	47,239	1,627	6,317	17,868	33,693	41,265	58,358	50,539	61,898	87,536	92,846	107,210	131,305			
	25	21,919	22,675	28,343	34,012	39,681	45,349	47,239	1,627	8,206	23,209	43,764	53,600	75,802	65,646	80,400	113,702	120,600	139,256	170,554			
1-3/8	6	26,239	27,144	33,930	40,716	47,502	54,288	56,550	1,579	NP	2,535	4,781	5,855	8,280	7,171	8,783	12,420	13,174	15,212	18,631			
	9	26,239	27,144	33,930	40,716	47,502	54,288	56,550	1,712	1,786	5,051	9,524	11,665	16,497	14,287	17,497	24,745	26,246	30,306	37,118			
	12	26,239	27,144	33,930	40,716	47,502	54,288	56,550	1,782	2,862	8,095	15,264	18,695	26,438	22,896	28,042	39,658	42,063	48,571	59,487			
	15	26,239	27,144	33,930	40,716	47,502	54,288	56,550	1,782	4,000	11,313	21,332	26,127	36,949	31,999	39,190	55,423	58,785	67,879	83,135			
	18	26,239	27,144	33,930	40,716	47,502	54,288	56,550	1,782	5,258	14,872	28,042	34,345	48,571	42,063	51,517	72,856	77,275	89,230	109,284			
	21	26,239	27,144	33,930	40,716	47,502	54,288	56,550	1,782	6,626	18,740	35,337	43,279	61,206	53,006	64,919	91,809	97,378	112,443	137,713			
	25	26,239	27,144	33,930	40,716	47,502	54,288	56,550	1,782	8,606	24,342	45,900	56,216	79,501	68,850	84,324	119,252	126,486	146,053	178,878			
1-1/2	6	31,894	32,994	41,243	49,491	57,740	65,988	68,738	1,942	NP	2,989	4,781	5,855	8,280	7,171	8,783	12,420	13,174	15,212	18,631			
	9	31,894	32,994	41,243	49,491	57,740	65,988	68,738	1,942	2,989	8,455	15,943	19,526	27,614	23,915	29,289	41,421	43,934	50,730	62,132			
	12	31,894	32,994	41,243	49,491	57,740	65,988	68,738	1,942	4,178	11,816	22,281	27,289	38,592	33,422	40,933	57,888	61,399	70,898	86,832			
	15	31,894	32,994	41,243	49,491	57,740	65,988	68,738	1,942	5,492	15,533	29,289	35,872	50,730	43,934	53,808	76,096	80,711	93,198	114,143			
	18	31,894	32,994	41,243	49,491	57,740	65,988	68,738	1,942	6,920	19,574	36,909	45,204	63,928	55,363	67,805	95,891	101,708	117,442	143,837			
	21	31,894	32,994	41,243	49,491	57,740	65,988	68,738	1,942	8,989	25,425	47,941	58,716	83,037	71,912	88,074	124,555	132,110	152,548	186,832			
1-3/4	12	42,978	44,460	55,575	66,690	77,805	88,920	92,625	2,205	3,131	8,855	16,698	20,450	28,921	25,046	30,675	43,382	46,013	53,131	65,073			
	15	42,978	44,460	55,575	66,690	77,805	88,920	92,625	2,274	4,512	12,763	24,066	29,475	41,684	36,099	44,213	62,526	66,319	76,578	93,789			
	18	42,978	44,460	55,575	66,690	77,805	88,920	92,625	2,274	5,932	16,777	31,636	38,746	54,795	47,454	58,119	82,193	87,178	100,665	123,289			
	21	42,978	44,460	55,575	66,690	77,805	88,920	92,625	2,274	7,475	21,142	39,866	48,825	69,050	59,799	73,238	103,574	109,857	126,852	155,362			
	25	42,978	44,460	55,575	66,690	77,805	88,920	92,625	2,274	9,709	27,462	51,782	63,420	89,680	77,674	95,130	134,535	142,695	164,770	201,802			
2	12	56,550	58,500	73,125	87,750	102,375	117,000	121,875	2,479	3,259	9,217	17,380	21,286	30,103	26,070	31,929	45,155	47,894	55,303	67,732			
	15	56,550	58,500	73,125	87,750	102,375	117,000	121,875	2,592	4,762	13,469	25,398	31,106	43,991	38,097	46,659	65,986	69,989	80,816	98,979			
	18	56,550	58,500	73,125	87,750	102,375	117,000	121,875	2,626	6,341	17,936	33,820	41,421	58,578	50,730	62,132	87,868	93,198	107,615	131,801			
	21	56,550	58,500	73,125	87,750	102,375	117,000	121,875	2,626	7,991	22,602	42,618	52,197	73,817	63,928	78,295	110,726	117,442	135,611	166,089			
	25	56,550	58,500	73,125	87,750	102,375	117,000	121,875	2,626	10,380	29,358	55,358	67,799	95,882	83,037	101,699	143,823	152,548	176,147	215,735			

Table 34-6C. Design Strengths for Single Cast-In Anchors Subject to Shear Loads (f'_c = 6000 psi)[1,2,3,5]

Notes pertaining to this table are given on Page 34-15

d_o in.	h_{ef} in.	ϕV_{sa} - Shear Strength of Anchor											ϕV_{cb} - Shear Breakout[5,6,7,8,9]										
		f_{uta} - for design purposes[4] - psi											1-1/2-in. cover	$h = h_{ef}$[10] and $c_{a1} =$[11]			$h = 1.5 h_{ef}$ and $c_{a1} =$[11,12]				$h = 2.25 h_{ef}$ and $c_{a1} =$[11,13]		
		58,000	60,000	75,000	90,000	105,000	120,000	125,000						0.25h_{ef}	0.5h_{ef}	h_{ef}	1.5h_{ef}	3h_{ef}	h_{ef}	1.5h_{ef}	3h_{ef}	1.5h_{ef}	2h_{ef}	3h_{ef}
1/4	2	724	749	936	1,123	1,310	1,498	1,560	489	NP	NP	542	664	939	814	996	1,409	1,495	1,726	2,114
	3	724	749	936	1,123	1,310	1,498	1,560	596	NP	NP	996	1,220	1,726	1,495	1,831	2,589	2,746	3,171	3,883
	4	724	749	936	1,123	1,310	1,498	1,560	596	NP	814	1,534	1,879	2,657	2,301	2,818	3,986	4,228	4,882	5,979
	5	724	749	936	1,123	1,310	1,498	1,560	596	NP	1,137	2,144	2,626	3,714	3,216	3,939	5,570	5,908	6,822	8,355
	6	724	749	936	1,123	1,310	1,498	1,560	596	NP	1,495	2,818	3,452	4,882	4,228	5,178	7,322	7,766	8,968	10,983
3/8	2	1,764	1,825	2,282	2,738	3,194	3,650	3,803	563	NP	NP	613	750	1,061	919	1,125	1,591	1,688	1,949	2,387
	3	1,764	1,825	2,282	2,738	3,194	3,650	3,803	772	NP	NP	1,220	1,495	2,114	1,831	2,242	3,171	3,363	3,883	4,756
	4	1,764	1,825	2,282	2,738	3,194	3,650	3,803	772	NP	996	1,879	2,301	3,254	2,818	3,452	4,882	5,178	5,979	7,322
	5	1,764	1,825	2,282	2,738	3,194	3,650	3,803	772	NP	1,393	2,626	3,216	4,548	3,939	4,824	6,822	7,236	8,355	10,233
	6	1,764	1,825	2,282	2,738	3,194	3,650	3,803	772	NP	1,831	3,452	4,228	5,979	5,178	6,341	8,968	9,512	10,983	13,452
1/2	2	3,212	3,323	4,154	4,984	5,815	6,646	6,923	625	NP	NP	668	818	1,157	1,002	1,227	1,735	1,840	2,125	2,602
	3	3,212	3,323	4,154	4,984	5,815	6,646	6,923	889	NP	NP	1,330	1,629	2,304	1,996	2,444	3,456	3,666	4,233	5,185
	4	3,212	3,323	4,154	4,984	5,815	6,646	6,923	942	NP	1,151	2,170	2,657	3,758	3,254	3,986	5,637	5,979	6,904	8,455
	5	3,212	3,323	4,154	4,984	5,815	6,646	6,923	942	NP	1,608	3,032	3,714	5,252	4,548	5,570	7,878	8,355	9,648	11,816
	6	3,212	3,323	4,154	4,984	5,815	6,646	6,923	942	NP	2,114	3,986	4,882	6,904	5,979	7,322	10,355	10,983	12,683	15,533
	7	3,212	3,323	4,154	4,984	5,815	6,646	6,923	942	NP	2,664	5,023	6,151	8,699	7,534	9,227	13,049	13,841	15,982	19,574
	8	3,212	3,323	4,154	4,984	5,815	6,646	6,923	942	1,151	3,254	6,136	7,516	10,629	9,205	11,273	15,943	16,910	19,526	23,915
5/8	3	5,112	5,288	6,611	7,933	9,255	10,577	11,018	1,002	NP	NP	1,422	1,742	2,464	2,134	2,613	3,696	3,920	4,526	5,544
	4	5,112	5,288	6,611	7,933	9,255	10,577	11,018	1,061	NP	1,230	2,320	2,841	4,018	3,480	4,262	6,027	6,393	7,381	9,040
	5	5,112	5,288	6,611	7,933	9,255	10,577	11,018	1,110	NP	1,798	3,390	4,152	5,872	5,085	6,228	8,807	9,342	10,787	13,211
	6	5,112	5,288	6,611	7,933	9,255	10,577	11,018	1,110	NP	2,363	4,456	5,458	7,718	6,684	8,187	11,578	12,280	14,180	17,366
	7	5,112	5,288	6,611	7,933	9,255	10,577	11,018	1,110	NP	2,978	5,615	6,878	9,726	8,423	10,316	14,589	15,474	17,868	21,884
	8	5,112	5,288	6,611	7,933	9,255	10,577	11,018	1,110	NP	3,638	6,861	8,403	11,883	10,291	12,604	17,825	18,906	21,831	26,737
	9	5,112	5,288	6,611	7,933	9,255	10,577	11,018	1,110	1,286	4,342	8,187	10,026	14,180	12,280	15,040	21,269	22,560	26,050	31,904
	10	5,112	5,288	6,611	7,933	9,255	10,577	11,018	1,110	1,535	5,085	9,588	11,743	16,607	14,382	17,615	24,911	26,422	30,510	37,366
3/4	4	7,555	7,816	9,770	11,723	13,677	15,631	16,283	1,180	NP	1,299	2,450	3,001	4,244	3,675	4,501	6,366	6,752	7,796	9,549
	5	7,555	7,816	9,770	11,723	13,677	15,631	16,283	1,233	NP	1,899	3,581	4,385	6,202	5,371	6,578	9,303	9,867	11,393	13,954
	6	7,555	7,816	9,770	11,723	13,677	15,631	16,283	1,279	NP	2,589	4,882	5,979	8,455	7,322	8,968	12,683	13,452	15,533	19,024
	7	7,555	7,816	9,770	11,723	13,677	15,631	16,283	1,279	NP	3,262	6,151	7,534	10,655	9,227	11,301	15,982	16,951	19,574	23,973
	8	7,555	7,816	9,770	11,723	13,677	15,631	16,283	1,279	1,409	3,986	7,516	9,205	13,017	11,273	13,807	19,526	20,711	23,915	29,289
	9	7,555	7,816	9,770	11,723	13,677	15,631	16,283	1,279	1,681	4,756	8,968	10,983	15,533	13,452	16,475	23,299	24,713	28,536	34,949
	10	7,555	7,816	9,770	11,723	13,677	15,631	16,283	1,279	1,969	5,570	10,503	12,864	18,192	15,755	19,296	27,289	28,944	33,422	40,933
	12	7,555	7,816	9,770	11,723	13,677	15,631	16,283	1,298	2,589	7,322	13,807	16,910	23,915	20,711	25,365	35,872	38,048	43,934	53,808
7/8	4	10,450	10,811	13,514	16,216	18,919	21,622	22,523	1,361	NP	1,522	2,566	3,143	4,445	3,849	4,714	6,667	7,072	8,165	10,001
	6	10,450	10,811	13,514	16,216	18,919	21,622	22,523	1,407	NP	2,711	5,113	6,262	8,855	7,669	9,392	13,283	14,089	16,268	19,924
	8	10,450	10,811	13,514	16,216	18,919	21,622	22,523	1,451	1,522	4,305	8,118	9,942	14,060	12,177	14,913	21,091	22,370	25,831	31,636
	12	10,450	10,811	13,514	16,216	18,919	21,622	22,523	1,451	2,796	7,909	14,913	18,265	25,831	22,370	27,398	38,746	41,096	47,454	58,119
	15	10,450	10,811	13,514	16,216	18,919	21,622	22,523	1,451	3,908	11,053	20,842	25,526	36,099	31,263	38,289	54,149	57,434	66,319	81,224
	18	10,450	10,811	13,514	16,216	18,919	21,622	22,523	1,451	5,137	14,530	27,398	33,555	47,454	41,096	50,332	71,181	75,499	87,178	106,771
	25	10,450	10,811	13,514	16,216	18,919	21,622	22,523	1,451	8,408	23,783	44,845	54,923	77,674	67,267	82,385	116,510	123,578	142,695	174,765

Table 34-6C. Design Strengths for Single Cast-In Anchors Subject to Shear Loads (f'_c = 6000 psi)[1,2,3,5] (cont'd.)
Notes pertaining to this table are given on Page 34-15

d_o in.	h_{ef} in.	ϕV_{sa} - Shear Strength of Anchor											ϕV_{cb} - Shear Breakout[5,6,7,8,9]										
		f_{uta} - for design purposes[4] - psi											$h = h_{ef}$[10] and $c_{a1} =$[11]				$h = 1.5h_{ef}$ and $c_{a1} =$[11,12]				$h = 2.25h_{ef}$ and $c_{a1} =$[11,13]		
		58,000	60,000	75,000	90,000	105,000	120,000	125,000	1-1/2-in. cover	0.25h_{ef}	0.5h_{ef}	h_{ef}	1.5h_{ef}	3h_{ef}	h_{ef}	1.5h_{ef}	3h_{ef}	1.5h_{ef}	2h_{ef}	3h_{ef}
1	6	13,708	14,180	17,726	21,271	24,816	28,361	29,543	1,536	NP	2,822	5,322	6,518	9,217	7,982	9,776	13,826	14,664	16,933	20,739
	9	13,708	14,180	17,726	21,271	24,816	28,361	29,543	1,627	1,942	5,492	10,355	12,683	17,936	15,533	19,024	26,904	28,536	32,950	40,356
	12	13,708	14,180	17,726	21,271	24,816	28,361	29,543	1,627	2,989	8,455	15,943	19,526	27,614	23,915	29,289	41,421	43,934	50,730	62,132
	15	13,708	14,180	17,726	21,271	24,816	28,361	29,543	1,627	4,178	11,816	22,281	27,289	38,592	33,422	40,933	57,888	61,399	70,898	86,832
	18	13,708	14,180	17,726	21,271	24,816	28,361	29,543	1,627	5,492	15,533	29,289	35,872	50,730	43,934	53,808	76,096	80,711	93,198	114,143
	21	13,708	14,180	17,726	21,271	24,816	28,361	29,543	1,627	6,920	19,574	36,909	45,204	63,928	55,363	67,805	95,891	101,708	117,442	143,837
	25	13,708	14,180	17,726	21,271	24,816	28,361	29,543	1,627	8,989	25,425	47,941	58,716	83,037	71,912	88,074	124,555	132,110	152,548	186,832
1-1/8	6	17,259	17,854	22,318	26,781	31,245	35,708	37,196	1,667	NP	2,924	5,513	6,752	9,549	8,269	10,128	14,323	15,192	17,542	21,485
	9	17,259	17,854	22,318	26,781	31,245	35,708	37,196	1,807	2,059	5,825	10,983	13,452	19,024	16,475	20,178	28,536	30,267	34,949	42,804
	12	17,259	17,854	22,318	26,781	31,245	35,708	37,196	1,807	3,171	8,968	16,910	20,711	29,289	25,365	31,066	43,934	46,599	53,808	65,901
	15	17,259	17,854	22,318	26,781	31,245	35,708	37,196	1,807	4,431	12,533	23,633	28,944	40,933	35,449	43,416	61,399	65,124	75,198	92,099
	18	17,259	17,854	22,318	26,781	31,245	35,708	37,196	1,807	5,825	16,475	31,066	38,048	53,808	46,599	57,072	80,711	85,607	98,851	121,067
	21	17,259	17,854	22,318	26,781	31,245	35,708	37,196	1,807	7,340	20,761	39,147	47,946	67,805	58,721	71,918	101,708	107,878	124,566	152,562
	25	17,259	17,854	22,318	26,781	31,245	35,708	37,196	1,807	9,534	26,967	50,849	62,277	88,074	76,274	93,416	132,110	140,124	161,801	198,165
1-1/4	6	21,919	22,675	28,343	34,012	39,681	45,349	47,239	1,799	NP	3,018	5,690	6,969	9,855	8,535	10,453	14,783	15,680	18,105	22,174
	9	21,919	22,675	28,343	34,012	39,681	45,349	47,239	1,951	2,126	6,012	11,336	13,884	19,635	17,004	20,826	29,452	31,239	36,071	44,178
	12	21,919	22,675	28,343	34,012	39,681	45,349	47,239	1,992	3,342	9,453	17,825	21,831	30,874	26,737	32,746	46,310	49,119	56,718	69,465
	15	21,919	22,675	28,343	34,012	39,681	45,349	47,239	1,992	4,671	13,211	24,911	30,510	43,147	37,366	45,764	64,721	68,647	79,266	97,081
	18	21,919	22,675	28,343	34,012	39,681	45,349	47,239	1,992	6,140	17,366	32,746	40,106	56,718	49,119	60,159	85,077	90,238	104,198	127,616
	21	21,919	22,675	28,343	34,012	39,681	45,349	47,239	1,992	7,737	21,884	41,265	50,539	71,473	61,898	75,809	107,210	113,713	131,305	160,815
	25	21,919	22,675	28,343	34,012	39,681	45,349	47,239	1,992	10,050	28,426	53,600	65,646	92,838	80,400	98,469	139,256	147,704	170,554	208,885
1-3/8	6	26,239	27,144	33,930	40,716	47,502	54,288	56,550	1,933	NP	3,105	5,855	7,171	10,141	8,783	10,756	15,212	16,135	18,631	22,818
	9	26,239	27,144	33,930	40,716	47,502	54,288	56,550	2,097	2,187	6,186	11,665	14,287	20,204	17,497	21,430	30,306	32,145	37,118	45,460
	12	26,239	27,144	33,930	40,716	47,502	54,288	56,550	2,183	3,505	9,914	18,695	22,896	32,380	28,042	34,345	48,571	51,517	59,487	72,856
	15	26,239	27,144	33,930	40,716	47,502	54,288	56,550	2,183	4,899	13,856	26,127	31,999	45,253	39,190	47,998	67,879	71,997	83,135	101,819
	18	26,239	27,144	33,930	40,716	47,502	54,288	56,550	2,183	6,440	18,214	34,345	42,063	59,487	51,517	63,095	89,230	94,643	109,284	133,845
	21	26,239	27,144	33,930	40,716	47,502	54,288	56,550	2,183	8,115	22,952	43,279	53,006	74,962	64,919	79,509	112,443	119,263	137,713	168,664
	25	26,239	27,144	33,930	40,716	47,502	54,288	56,550	2,183	10,540	29,813	56,216	68,850	97,369	84,324	103,275	146,053	154,913	178,878	219,080
1-1/2	12	31,894	32,994	41,243	49,491	57,740	65,988	68,738	2,378	3,661	10,355	19,526	23,915	33,820	29,289	35,872	50,730	53,808	62,132	76,096
	15	31,894	32,994	41,243	49,491	57,740	65,988	68,738	2,378	5,117	14,472	27,289	33,422	47,265	40,933	50,132	70,898	75,198	86,832	106,347
	18	31,894	32,994	41,243	49,491	57,740	65,988	68,738	2,378	6,726	19,024	35,872	43,934	62,132	53,808	65,901	93,198	98,851	114,143	139,796
	21	31,894	32,994	41,243	49,491	57,740	65,988	68,738	2,378	8,476	23,973	45,204	55,363	78,295	67,805	83,044	117,442	124,566	143,837	176,164
	25	31,894	32,994	41,243	49,491	57,740	65,988	68,738	2,378	11,009	31,139	58,716	71,912	101,699	88,074	107,868	152,548	161,801	186,832	228,822
1-3/4	12	42,978	44,460	55,575	66,690	77,805	88,920	92,625	2,701	3,834	10,845	20,450	25,046	35,421	30,675	37,570	53,131	56,354	65,073	79,697
	15	42,978	44,460	55,575	66,690	77,805	88,920	92,625	2,786	5,527	15,631	29,475	36,099	51,052	44,213	54,149	76,578	81,224	93,789	114,868
	18	42,978	44,460	55,575	66,690	77,805	88,920	92,625	2,786	7,265	20,548	38,746	47,454	67,110	58,119	71,181	100,665	106,771	123,289	150,997
	21	42,978	44,460	55,575	66,690	77,805	88,920	92,625	2,786	9,155	25,894	48,825	59,799	84,568	73,238	89,698	126,852	134,547	155,362	190,278
	25	42,978	44,460	55,575	66,690	77,805	88,920	92,625	2,786	11,891	33,634	63,420	77,674	109,847	95,130	116,510	164,770	174,765	201,802	247,156
2	12	56,550	58,500	73,125	87,750	102,375	117,000	121,875	3,036	3,991	11,289	21,286	26,070	36,869	31,929	39,105	55,303	58,658	67,732	82,955
	15	56,550	58,500	73,125	87,750	102,375	117,000	121,875	3,175	5,832	16,496	31,106	38,097	53,877	46,659	57,145	80,816	85,718	98,979	121,224
	18	56,550	58,500	73,125	87,750	102,375	117,000	121,875	3,216	7,766	21,967	41,421	50,730	71,744	62,132	76,096	107,615	114,143	131,801	161,423
	21	56,550	58,500	73,125	87,750	102,375	117,000	121,875	3,216	9,787	27,681	52,197	63,928	90,407	78,295	95,891	135,611	143,837	166,089	203,416
	25	56,550	58,500	73,125	87,750	102,375	117,000	121,875	3,216	12,712	35,956	67,799	83,037	117,431	101,699	124,555	176,147	186,832	215,735	264,221

Example 34.1—Single Headed Bolt in Tension Away from Edges

Design a single headed bolt installed in the bottom of a 6 in. slab to support a 5000 lb service dead load.

$f'_c = 4000$ psi

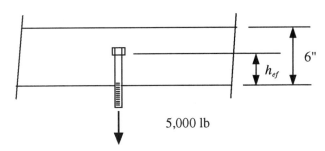

	Code
Calculations and Discussion	**Reference**

1. Determine factored design load (only dead load is present)   *9.2*

   $N_u = 1.4 (5000) = 7000$ lb   *Eq. (9-1)*

2. Determine anchor diameter and material   *D.5.1*

   The strength of most anchors is likely to be controlled by the embedment strength rather than the steel strength. As a result, it is usually economical to design the anchor using a mild steel rather than a high strength steel. ASTM F 1554 "Standard Specification for Anchor Bolts, Steel, 36, 55, and 105-ksi Yield Strength," covers straight and bent, headed and headless, anchors in three strength grades.

   Assume an ASTM F 1554 Grade 36 headed anchor for this example.

   The basic requirement for the anchor steel is:

   $\phi N_{sa} \geq N_{ua}$ *Eq. (D-1)*
        D.4.1.2

   where:

   $\phi = 0.75$ *D.4.4(a)*

   Per the Ductile Steel Element definition in D.1, ASTM F 1554 Grade 36 steel qualifies as a ductile steel element (23% minimum elongation in 2 in. which is greater than the 14% required and a minimum reduction in area of 40% that is greater than the 30% required, see Table 34.1). This results in $\phi = 0.75$ rather than $\phi = 0.65$ if the steel had not met the ductile steel element requirements.

   $N_{sa} = n A_{se} f_{uta}$ *Eq. (D-3)*

   For design purposes, Eq. (D-1) with Eq. (D-3) may be rearranged as:

   $$A_{se} \geq \frac{N_{ua}}{\phi \, n \, f_{uta}}$$

Example 34.1 (cont'd)	Calculations and Discussion	Code Reference

where:

N_{ua} = 7000 lbs
ϕ = 0.75
n = 1
f_{uta} = 58,000 psi

Per ASTM F 1554, Grade 36 has a specified minimum yield strength of 36 ksi and a specified tensile strength of 58-80 ksi (see Table 34-1). For design purposes, the minimum tensile strength of 58 ksi should be used.

Note: Per D.5.1.2, f_{uta} shall not be taken greater than $1.9f_{ya}$ or 125,000 psi. For ASTM F 1554 Grade 36, $1.9f_{ya}$ = 1.9(36,000) = 68,400 psi, therefore use the specified minimum f_{uta} of 58,000 psi.

D.5.1.2

Substituting:

$$A_{se} = \frac{7000}{0.75(1)(58,000)} = 0.161 \text{ in.}^2$$

Per Table 34-2, a 5/8 in. diameter threaded anchor will satisfy this requirement (A_{se} = 0.226 in.^{2}).

3. Determine the required embedment length (h_{ef}) based on concrete breakout

D.5.2

The basic requirement for the single anchor embedment is:

$\phi N_{cb} \geq N_{ua}$

Eq. (D-1)
D.4.1.2

where:

ϕ = 0.70

D.4.4(c)

Condition B applies, no supplementary reinforcement has been provided to tie the failure prism associated with the concrete breakout failure mode of the anchor to the supporting structural member. This is likely to be the case for anchors loaded in tension attached to a slab. Condition A (with ϕ = 0.75) may apply when the anchoring is attached to a deeper member (such as a pedestal or beam) where there is space available to install supplementary reinforcement across the failure prism.

$$N_{cb} = \frac{A_{Nc}}{A_{Nco}} \psi_{ed,N} \psi_{c,N} \psi_{cp,N} N_b$$

Eq. (D-4)

where:

$\frac{A_{Nc}}{A_{Nco}}$ and $\psi_{ed,N}$ terms are 1.0 for single anchors away from edges

For cast-in anchors $\psi_{cp,N}$ = 1.0

$$N_b = 24\sqrt{f'_c}\, h_{ef}^{1.5}$$

Eq. (D-7)

		Code
Example 34.1 (cont'd)	**Calculations and Discussion**	**Reference**

For design of a single anchor away from edges, Eq. (D-1) with Eq. (D-4) and Eq. (D-7) may be rearranged as:

$$h_{ef} = \left(\frac{N_{ua}}{\phi \, \psi_{c,N} \, 24\sqrt{f'_c}} \right)^{\frac{2}{3}}$$

where:

$\psi_{c,N}$ = 1.0 for locations where concrete cracking is likely to occur (i.e., the bottom of the slab)   *D.5.2.6*

Substituting:

$$h_{ef} = \left(\frac{7000}{0.70(1.0)24\sqrt{4000}} \right)^{\frac{2}{3}} = 3.51 \text{ in.}$$

Select 4 in. embedment for this anchor.

Note: The case of a single anchor away from an edge is essentially the only case where h_{ef} can be solved for directly. Whenever edges or adjacent anchors are present, the solution for h_{ef} is iterative.

4. Determine the required head size for the anchor   *D.5.3*

   The basic requirement for pullout strength (i.e., the strength of the anchor related to the embedded anchor having insufficient bearing area so that the anchor pulls out without a concrete breakout failure) is:

   $\phi N_{pn} \geq N_{ua}$ *Eq. (D-1)*   *D.4.1.2*

   where:

   $\phi = 0.70$ *D.4.4(c)*

   Condition B applies for pullout strength in all cases

   $N_{pn} = \psi_{c,P} \, N_p$ *Eq. (D-14)*

   where:

   $N_p = A_{brg} \, 8 \, f'_c$ *Eq. (D-15)*

   $\psi_{c,P}$ = 1.0 for locations where concrete cracking is likely to occur (i.e., the bottom of the slab)   *D.5.3.6*

   For design purposes Eq. (D-1) with Eq. (D-14) and Eq. (D-15) may be rearranged as:

   $$A_{brg} = \frac{N_{ua}}{\phi \psi_{c,P} \, 8 \, f'_c}$$

		Code
Example 34.1 (cont'd)	Calculations and Discussion	Reference

Substituting:

$$A_{brg} = \frac{7000}{0.70(1.0)(8)(4000)} = 0.313 \text{ in}^2$$

As shown in Table 34-2, any type of standard head (square, heavy square, hex, or heavy hex) is acceptable for this 5/8 in. diameter anchor. ASTM F 1554 specifies a hex head for Grade 36 bolts less than 1-1/2 in. in diameter.

5. Evaluate side-face blowout   *D.5.4*

Since this anchor is located far from a free edge of concrete ($c_{a1} \geq 0.4\, h_{ef}$) this type of failure mode is not applicable.

6. Required edge distances, spacings, and thicknesses to preclude splitting failure   *D.8*

Since this is a cast-in-place anchor and is located far from a free edge of concrete, the only requirement is that the minimum cover requirements of Section 7.7 should be met. Assuming this is an interior slab, the requirements of Section 7.7 will be met with the 4 in. embedment length plus the head thickness. The head thickness for square, hex, and heavy hex heads and nuts are at most equal to the anchor diameter (refer to ANSI B.18.2.1 and ANSI B.18.2.2 for exact dimensions). This results in ~1-3/8 in. cover from the top of the anchor head to the top of the slab.

7. Summary:

Use an ASTM F 1554 Grade 36, 5/8 in. diameter headed anchor with a 4 in. embedment.

Alternate design using Table 34-5B

Note: Step numbers correspond to those in the main example above, but prefaced with "A".

Table 34-5B has been selected because it contains design tension strength values based on concrete with f'_c = 4000 psi. Table Note 4 indicates that the values in the table are based on Condition B (no supplementary reinforcement), and Notes 6 and 10 indicate that cracked concrete was assumed.

A2. Determine anchor diameter and material.   *D.5.1*
         Eq. (D-3)

Tentatively try a bolt complying with ASTM F 1554 Grade 36 with a f_{uta} of 58,000 psi. Using the factored tensile load from Step 1 above (7000 lb), and 58,000 psi as the value of f_{uta}, go down the column for 58,000 until an anchor size has a design tensile strength, ϕN_{sa}, equal to or greater than 7000 lb. Table 34-6B shows that a 5/8 in. diameter bolt has a design tensile strength equal to

ϕN_{sa} = 9831 lb > N_u = 7000 lb   O.K.

Since this is greater than the required strength, tentatively use a 5/8 in. headed bolt.

| Example 34.1 (cont'd) | Calculations and Discussion | Code Reference |

A3. Determine the required embedment length (h_{ef}) based on concrete breakout strength (ϕN_{cb})

D.5.2
Eq. (D-4)

Since the anchor will be far from an edge, use the column labeled ">1.5h_{ef}." In this case "far from an edge" means that the edge distance, c_{a1}, must equal or exceed 1.5h_{ef}. A 5/8 in. bolt with 3 in. embedment has a design tension breakout strength,

$\phi N_{cb} = 5521$ lb $< N_{ua} = 7000$ lb

A 5/8-in. bolt with 4 in. embedment has

$\phi N_{cb} = 8500$ lb $> N_{ua} = 7000$ lb   O.K.

Tentatively use 5/8 in. bolt with embedment depth of 4 in.

A4. Determine if the bearing area of the head of the 5/8-in. bolt, A_{brg}, is large enough to prevent anchor pullout (ϕN_{pn}).

D.5.3
Eq. (D-14)

Values for design tension pullout strength, ϕN_{pn}, in Table 34-5B for headed bolts with a diameter of less than 1-3/4 in. are based on a regular hex head (Table Note 7). Under the column labeled "head w/o washer," a 5/8 in. bolt has a design pullout strength,

$\phi N_{pn} = 10,170$ lb $> N_{ua} = 7000$ lb   O.K.

A5. Determine if the anchor has enough edge distance, c_{a1}, to prevent side-face blow-out (ϕN_{sb}).

D.5.4.1

Since the anchor is farther from an edge than 0.4h_{ef} (0.4 x 4 in. = 1.6 in.), side-face blowout does not need to be considered.

A6. Required edge distances, spacings, and thicknesses to preclude splitting failure.

See Step 6 above.

A7. Summary:

Use an ASTM F 1554 Grade 36, 5/8 in. diameter headed bolt with a 4 in. embedment.

Example 34.2—Group of Headed Studs in Tension Near an Edge

Design a group of four welded, headed studs spaced 6 in. on center each way and concentrically loaded with a 10,000 lb service dead load. The anchor group is to be installed in the bottom of an 8 in. thick slab with the centerline of the connection 6 in. from a free edge of the slab.

$f'_c = 4000$ psi

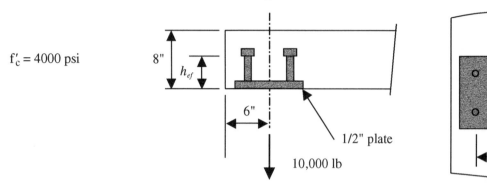

Calculations and Discussion	Code Reference
1. Determine factored design load	9.2
$N_{ua} = 1.4\,(10,000) = 14,000$ lb	Eq. (9-1)
2. Determine anchor diameter	D.5.1

Assume AWS D1.1 Type B welded, headed studs.

The basic requirement for the anchor steel is:

$\phi N_{sa} \geq N_{ua}$ Eq. (D-1)
   D.4.1.2

where:

$\phi = 0.75$ D.4.4(a)i

Per the Ductile Steel Element definition in D.1, AWS D1.1 Type B studs qualify as a ductile steel element (20% minimum elongation in 2 in. which is greater than the 14% required and a minimum reduction in area of 50% that is greater than the 30% required, see Table 34-1).

$N_{sa} = n\,A_{se}\,f_{uta}$ Eq. (D-3)

For design purposes, Eq. (D-1) with Eq. (D-3) may be rearranged as:

$A_{se} = \dfrac{N_{ua}}{\phi\,n\,f_{uta}}$

| Example 34.2 (cont'd) | Calculations and Discussion | Code Reference |

where:

N_{ua} = 14,000 lbs
ϕ = 0.75
n = 4
f_{uta} = 60,000 psi

Note: Per D5.1.2, f_{uta} shall not be taken greater than $1.9f_{ya}$ or 125,000 psi. For AWS D1.1 headed studs, $1.9f_{ya}$ = 1.9(50,000) = 95,000 psi, therefore use the specified minimum f_{uta} of 60,000 psi.

D.5.1.2

Substituting:

$$A_{se} = \frac{14,000}{0.75(4)(60,000)} = 0.078 \text{ in.}^2$$

Per Table 34-2, 1/2 in. diameter welded, headed studs will satisfy this requirement (A_{se} = 0.196 in.²).

Note: Per AWS D1.1 Table 7.1, Type B welded studs are 1/2 in., 5/8 in., 3/4 in., 7/8 in., and 1 in. diameters. Although individual manufacturers may list smaller diameters they are not explicitly covered by AWS D1.1

3. Determine the required embedment length (h_{ef}) based on concrete breakout

D.5.2

Two different equations are given for calculating concrete breakout strength; for single anchors Eq. (D-4) applies, and for anchor groups Eq. (D-5) applies. An "anchor group" is defined as:

"A number of anchors of approximately equal effective embedment depth with each anchor spaced at less than three times its embedment depth from one or more adjacent anchors."

D.1

Since the spacing between anchors is 6 in., they must be treated as a group if the embedment depth exceeds 2 in. Although the embedment depth is unknown, at this point it will be assumed that the provisions for an anchor group will apply.

The basic requirement for embedment of a group of anchors is:

$\phi N_{cbg} \geq N_{ua}$

Eq. (D-1)
D.4.1.2

where:

ϕ = 0.70

D.4.4(c)ii

Condition B applies since no supplementary reinforcement has been provided (e.g., hairpin type reinforcement surrounding the anchors and anchored into the concrete).

| Example 34.2 (cont'd) | Calculations and Discussion | Code Reference |

$$N_{cbg} = \frac{A_{Nc}}{A_{Nco}} \psi_{ec,N} \psi_{ed,N} \psi_{c,N} \psi_{cp,N} N_b$$

Eq. (D-5)

Since this connection is likely to be affected by both group effects and edge effects, the embedment length h_{ef} cannot be determined from a closed form solution. Therefore, an embedment length must be assumed. The strength of the connection is then determined and compared with the required strength.

Note: Welded studs are generally available in fixed lengths. Available lengths may be determined from manufacturers' catalogs. For example, the Nelson Stud http://www.nelsonstudwelding,com/ has an effective embedment of 4 in. for a standard 1/2 in. concrete anchor stud.

Assume an effective embedment length of h_{ef} = 4.5 in.

Note: The effective embedment length h_{ef} for the welded stud anchor is the effective embedment length of the stud plus the thickness of the embedded plate.

Evaluate the terms in Eq. (D-5) with h_{ef} = 4.5 in.

Determine A_{Nc} and A_{Nco} for the anchorage:

D.5.2.1

A_{Nc} is the projected area of the failure surface as approximated by a rectangle with edges bounded by 1.5 h_{ef} (1.5 x 4.5 = 6.75 in. in this case) and free edges of the concrete from the centerlines of the anchors.

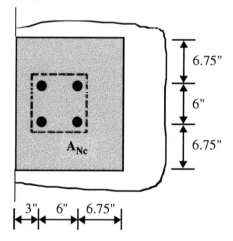

A_{Nc} = (3+6+6.75)(6.75+6+6.75) = 307 in.^{2}

A_{Nco} = 9 h_{ef}^2 = 9 (4.5)^2 = 182 in.^{2}

Eq. (D-6)
D.5.2.1

Check: $A_{Nc} \leq nA_{Nco}$ 307 < 4(182)   O.K.

Determine $\psi_{ec,N}$:

D.5.2.4

$\psi_{ec,N}$ = 1.0 (no eccentricity in the connection)

| Example 34.2 (cont'd) | Calculations and Discussion | Code Reference |

Determine $\psi_{ed,N}$ since $c_{a1} < 1.5\, h_{ef}$ — *D.5.2.5*

$$\psi_{ed,N} = 0.7 + 0.3\frac{c_{a,min}}{1.5\, h_{ef}}$$ — *Eq. (D-11)*

$$\psi_{ed,N} = 0.7 + 0.3\frac{3.0}{1.5(4.5)} = 0.83$$

Determine $\psi_{c,N}$: — *D.5.2.6*

$\psi_{c,N} = 1.0$ for locations where concrete cracking is likely to occur (i.e., the bottom of the slab)

For cast-in anchors $\psi_{cp,N} = 1.0$ — *D.5.2.6*

Determine N_b: — *D.5.2.2*

$$N_b = 24\sqrt{f'_c}\, h_{ef}^{1.5} = 24\sqrt{4000}\,(4.5)^{1.5} = 14{,}490 \text{ lb}$$ — *Eq. (D-7)*

Substituting into Eq. (D-5):

$$N_{cbg} = \left[\frac{307}{182}\right](1.0)(0.83)(1.0)(14{,}490) = 20{,}287 \text{ lb}$$

The final check on the assumption of $h_{ef} = 4.5$ in. is satisfied by meeting the requirements of Eq. (D-1):

$(0.70)(20{,}287) \geq 14{,}000$

$14{,}201 > 14{,}000$ O.K.

Specify a 4 in. length for the welded, headed studs with the 1/2 in.-thick base plate.

4. Determine if welded stud head size is adequate for pullout — *D.5.3*

   $\phi N_{pn} \geq N_{ua}$ — *Eq. (D-1)* / *D.4.1.2*

   where:

   $\phi = 0.70$ — *D.4.4(c)ii*

   Condition B applies for pullout strength in all cases.

   $N_{pn} = \psi_{c,P} N_p$ — *Eq. (D-14)*

   where:

   $N_p = A_{brg}\, 8\, f'_c$ — *Eq. (D-15)*

Example 34.2 (cont'd)	Calculations and Discussion	Code Reference

$\psi_{c,P}$ = 1.0 for locations where concrete cracking is likely to occur (i.e., the bottom of the slab) — D.5.3.6

For design purposes Eq. (D-1) with Eq. (D-14) and Eq. (D-15) may be rearranged as:

$$A_{brg} = \frac{N_{ua}}{\phi\, \psi_{c,P}\, 8\, f'_c}$$

For the group of four studs the individual factored tension load N_u on each stud is:

$$N_{ua} = \frac{14,000}{4} = 3500 \text{ lb}$$

Substituting:

$$A_{brg} = \frac{3500}{0.70\,(1.0)\,(8)\,(4000)} = 0.156 \text{ in}^2$$

The bearing area of welded, headed studs should be determined from manufacturers' catalogs. As shown on the Nelson Stud web page the diameter of the head for a 1/2 in. diameter stud is 1 in.

$$A_{brg,\,provided} = \frac{\pi}{4}(1.0^2 - 0.5^2) = 0.589 \text{ in}^2 > 0.156 \text{ in}^2 \quad \text{O.K.}$$

5. Evaluate side-face blowout — D.5.4

   Side-face blowout needs to be considered when the edge distance from the centerline of the anchor to the nearest free edge is less than $0.4 h_{ef}$. For this example:

   $0.4 h_{ef} = 0.4\,(4.5) = 1.8$ in. < 3 in. actual edge distance O.K.

   The side-face blowout failure mode is not applicable.

6. Required edge distances, spacings, and thickness to preclude splitting failure — D.8

   Since a welded, headed anchor is not torqued the minimum cover requirements of 7.7 apply.

   Per 7.7 the minimum clear cover for a 1/2 in. bar not exposed to earth or weather is 3/4 in. which is less than the 2-3/4 in. provided (3 - 1/4 = 2-3/4 in.) O.K.

7. Summary:

   Use ASW D1.1 Type B 1/2 in. diameter welded studs with an effective embedment of 4.5 in. (4 in. from the stud plus 1/2 in. from the embedded plate).

Example 34.3—Group of Headed Studs in Tension Near an Edge with Eccentricity

Determine the factored tension load capacity (N_{ua}) for a group of four 1/2 in.×4 in. AWS D1.1 Type B headed studs spaced 6 in. on center each way and welded to a 1/2-in.-thick base plate. The centerline of the structural attachment to the base plate is located 2 in. off of the centerline of the base plate resulting in an eccentricity of the tension load of 2 in. The fastener group is installed in the bottom of an 8 in.-thick slab with the centerline of the connection 6 in. from a free edge of the slab.

Note: This is the configuration chosen as a solution for Example 34.2 to support a 14,000 lb factored tension load centered on the connection. The only difference is the eccentricity of the tension load. From Example 34.2, the spacing between anchors dictates that they be designed as an anchor group.

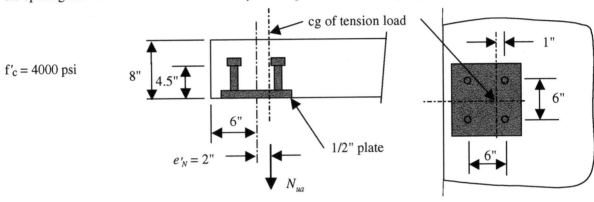

	Code
Calculations and Discussion	Reference

1. Determine distribution of loads to the anchors

   D.3.1

   Assuming an elastic distribution of loads to the anchors, the eccentricity of the tension load will result in a higher force on the interior row of fasteners. Although the studs are welded to the base plate, their flexural stiffness at the joint with the base plate is minimal compared to that of the base plate. Therefore, assume a simple support condition for the base plate:

   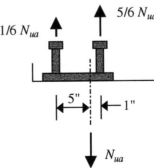

   The two interior studs will control the strength related to the steel ϕN_{sa} and the pullout strength ϕN_{pn} (i.e., 5/6 N_{ua} must be less than or equal to $\phi N_{sa,\ 2\ studs}$ and $\phi N_{pn,\ 2\ studs}$). Rearranged:

   $N_{ua} \leq 6/5\ \phi N_{sa,\ 2\ studs}$ and $N_{ua} \leq 6/5\ \phi N_{pn,\ 2\ studs}$)

Example 34.3 (cont'd)	Calculations and Discussion	Code Reference

2. Determine the design steel strength as controlled by the two anchors with the highest tensile load (ϕN_{sa}) — D.5.1

$$\phi N_{sa} = \phi\, n\, A_{se}\, f_{uta}$$ Eq. (D-3)

where:

$\phi = 0.75$ — D.4.4(a)i

Per the Ductile Steel Element definition in D.1, AWS D1.1 Type B studs qualify as a ductile steel element [20% minimum elongation in 2 in. which is greater than the 14% required and a minimum reduction in area of 50% that is greater than the 30% required, see Table 34-1)].

$n = 2$ (for the two inner studs with the highest tension load)

$A_{se} = 0.196$ in.^{2} (see Table 34-2)

$f_{uta} = 65,000$ (see Table 34-1)

Substituting:

$\phi N_{sa,\, 2studs} = 0.75\, (2)\, (0.196)\, (65,000) = 19,110$ lb

Therefore, the maximum N_{ua} as controlled by the anchor steel is:

$\phi N_{sa} = 6/5\, \phi N_{sa, 2studs} = 6/5\, (19,110) = 22,932$ lb

3. Determine design breakout strength (ϕN_{cbg}) — D.5.2

The only difference between concrete breakout strength in this example and Example 34.2 is the introduction of the eccentricity factor $\psi_{ec,N}$.

From Example 34.2 with $\psi_{ec,N} = 1.0$:

$$N_{cbg} = \frac{A_{Nc}}{A_{Nco}}\, \psi_{ec,N}\, \psi_{ed,N}\, \psi_{c,N}\, \psi_{cp,N}\, N_b = 14,201 \text{ lb}$$ Eq. (D-5)

Determine $\psi_{ec,N}$ for this example ($e_N = e'_N = 2$ in $< s/2 = 3$ in.): D.5.2.4

$$\psi_{ec,N} = \frac{1}{\left(1 + \dfrac{2 e'_N}{3 h_{ef}}\right)}$$ Eq. (D-9)

where:

$e'_N = 2$ in. (distance between centroid of anchor group and tension force)

$h_{ef} = 4.5$ in. (1/2 in. plate plus 4 in. embedment of headed stud)

Example 34.3 (cont'd)	Calculations and Discussion	Code Reference

Substituting:

$$\psi_{ec,N} = \frac{1}{\left(1+\frac{2(2)}{3(4.5)}\right)} = 0.77$$

Therefore:

$\phi N_{cbg} = (0.77)(14{,}201) = 10{,}935$ lb

4. Determine the design pullout strength as controlled by the two anchors with the highest tensile load (ϕN_{pn})

    D.5.3

    $\phi N_{pn,\,1\,stud} = \phi\,\psi_{c,P}\,N_p = \phi\,\psi_{c,P}\,A_{brg}\,8\,f'_c$

    Eq. (D-14)
    Eq. (D-15)

    where:

    $\phi = 0.70$ – Condition B applies for pullout.

    D.4.4(c)ii

    $\psi_{c,P} = 1.0$ for locations where concrete cracking is likely to occur (i.e., the bottom of the slab)

    D.5.3.6

    $A_{brg} = 0.589$ in.^{2} (see Step 4 of Example 35.2)

    Substituting:

    $\phi N_{pn,\,1\,stud} = (0.70)(1.0)(0.589)(8.0)(4000) = 13{,}194$ lb

    For the two equally loaded inner studs:

    $\phi N_{pn,2\,studs} = 2\,(13{,}194) = 26{,}387$ lb

    Therefore, the maximum N_{ua} as controlled by pullout is:

    $\phi N_{pn} = 6/5\,\phi N_{pn,\,2\,studs} = 6/5\,(26{,}387) = 31{,}664$ lb

5. Evaluate side-face blowout

    D.5.4

    Side-face blowout needs to be considered when the edge distance from the centerline of the anchor to the nearest free edge is less than 0.4 h_{ef}. For this example:

    0.4 h_{ef} = 0.4 (4.5) = 1.8 in. < 3 in. actual edge distance   O.K.

    The side-face blowout failure mode is not applicable.

Example 34.3 (cont'd)	Calculations and Discussion	Code Reference

6. Required edge distances, spacings, and thickness to preclude splitting failure — *D.8*

   Since a welded, headed anchor is not torqued the minimum cover requirements of 7.7 apply.

   Per 7.7 the minimum clear cover for a 1/2 in. bar not exposed to earth or weather is 3/4 in. which is less than the 2-3/4 in. provided (3 in. - 1/4 in. = 2-3/4 in.) O.K. — *7.7.1(c)*

7. Summary:

Steel strength, (ϕN_{sa}):	22,932 lb
Embedment strength – concrete breakout, (ϕN_{cbg}):	10,935 lb ← controls
Embedment strength – pullout, (ϕN_{pn}):	31,664 lb
Embedment strength – side-face blowout, (ϕN_{sb}):	N/A

   The maximum factored tension load N_{ua} for this anchorage is 10,935 lb

   Note: Example 34.2 with the same connection but without an eccentricity was also controlled by concrete breakout strength but had a factored load capacity of 14,201 lb (see Step 3 of Example 34.2).

Example 34.4—Single Headed Bolt in Shear Near an Edge

Determine the reversible service wind load shear capacity for a single 1/2 in. diameter headed anchor with a 7 in. embedment installed with its centerline 1-3/4 in. from the edge of a concrete foundation.

Note: This is the minimum anchorage requirement at the foundation required by IBC 2000 Section 2308.6 for conventional light-frame wood construction. The 1-3/4 in. edge distance represents a typical connection at the base of wood framed walls using 2×4 members.

$f'_c = 4000$ psi

ASTM F 1554 Grade 36

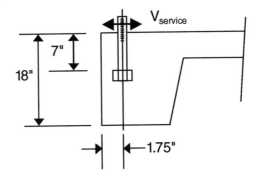

	Code
Calculations and Discussion	**Reference**

1. This problem provides the anchor diameter, embedment length, and material properties, and requires computing the maximum service shear load capacity to resist wind load. In this case, it is best to first determine the controlling factored shear load, V_{ua}, based on the smaller of the steel strength and embedment strength then as a last step determine the maximum service load. Step 6 of this example provides the conversion of the controlling factored shear load V_{ua} to a service load due to wind.

2. Determine V_{ua} as controlled by the anchor steel   *D.6.1*

   $\phi V_{sa} \geq V_{ua}$ *Eq. (D-2)*
                                                                                         D.4.1.2

   where:

   $\phi = 0.65$ *D.4.4(a)i*
   Per the Ductile Steel Element definition in D.1, ASTM F 1554 Grade 36 steel qualifies as a ductile steel element.

   $V_{sa} = n\, 0.6\, A_{se}\, f_{uta}$ *Eq. (D-20)*

   To determine V_{ua} for the steel strength Eq. (D-2) can be combined with Eq. (D-20) to give:

   $V_{ua} = \phi V_{sa} = \phi\, n\, 0.6\, A_{se}\, f_{uta}$

Example 34.4 (cont'd)	Calculations and Discussion	Code Reference

where:

$\phi = 0.65$
$n = 1$
$A_{se} = 0.142$ in.^{2} for the 1/2 in. threaded bolt (Table 34-2)
$f_{uta} = 58,000$ psi

> Per ASTM F 1554 Grade 36 has a specified minimum yield strength of 36 ksi and a specified tensile strength of 58-80 ksi (see Table 34-1). For design purposes, the minimum tensile strength of 58 ksi should be used.

> Note: Per D6.1.2, f_{uta} shall not be taken greater than $1.9f_{ya}$ or 125,000 psi. For ASTM F 1554 Grade 36, $1.9f_{ya} = 1.9(36,000) = 68,400$ psi. Therefore, use the specified minimum f_{uta} of 58,000 psi.

Substituting, V_{ua} as controlled by steel strength is:

$V_{ua} = \phi V_{sa} = 0.65\,(1)(0.6)(0.142)(58,000) = 3212$ lb

3. Determine V_{ua} for embedment strength governed by concrete breakout strength with shear directed toward a free edge

    $\phi V_{cb} \geq V_{ua}$

D.6.2

Eq. (D-2)
D.4.1.2

   where:

   $\phi = 0.70$

D.4.4(c)i

   No supplementary reinforcement has been provided (i.e., hairpin type reinforcement surrounding the anchor and anchored into the concrete).

   $V_{cb} = \dfrac{A_{Vc}}{A_{Vco}}\,\psi_{ed,V}\,\psi_{c,V}\,V_b$

Eq. (D-21)

   where:

   $\dfrac{A_{Vc}}{A_{Vco}}$ and $\psi_{ed,V}$ terms are 1.0 for single shear anchors not influenced by more than one free edge (i.e., the member thickness is greater than $1.5c_{a1}$ and the distance to an orthogonal edge c_{a2} is greater than $1.5c_{a1}$)

   $\psi_{c,V} = 1.0$ for locations where concrete cracking is likely to occur (i.e., the edge of the foundation is susceptible to cracks)

D.6.2.7

| Example 34.4 (cont'd) | Calculations and Discussion | Code Reference |

$$V_b = 7\left(\frac{\ell_e}{d_o}\right)^{0.2} \sqrt{d_o} \sqrt{f'_c} \, c_{a1}^{1.5}$$

Eq. (D-24)

where:

ℓ_e = load bearing length of the anchor for shear, not to exceed $8d_o$

2.1
D.6.2.2

For this problem $8d_o$ will control since the embedment depth h_{ef} is 7 in.

$\ell_e = 8d_o = 8\,(0.5) = 4.0$ in.

To determine V_{ua} for the embedment strength governed by concrete breakout strength Eq. (D-2) can be combined with Eq. (D-21) and Eq. (D-24) to give:

$$V_{ua} = \phi V_{cb} = \phi \frac{A_{Vc}}{A_{Vco}} \psi_{ed,V} \, \psi_{c,V} \, 7\left(\frac{\ell_e}{d_o}\right)^{0.2} \sqrt{d_o} \sqrt{f'_c} \, c_{a1}^{1.5}$$

Eq. (D-21)
Eq. (D-24)

Substituting, V_u for the embedment strength as controlled by concrete breakout strength is:

$$V_{ua} = \phi V_{cb} = 0.70\,(1.0)(1.0)(1.0)(7)\left(\frac{8\,(0.5)}{(0.5)}\right)^{0.2} \sqrt{0.5} \sqrt{4000}\,(1.75)^{1.5} = 769 \text{ lb}$$

4. Determine V_{ua} for embedment strength governed by concrete pryout strength

D.6.3

Note: The pryout failure mode is normally only a concern for shallow, stiff anchors. Since this example problem addresses both shear directed toward the free edge and shear directed inward from the free edge, the pryout strength will be evaluated.

$\phi V_{cp} \geq V_{ua}$

Eq. (D-2)
D.4.1.2

where:

$\phi = 0.70$ – Condition B applies for pryout strength in all cases

D.4.4(c)i

$V_{cp} = k_{cp} N_{cb}$

Eq. (D-29)

where:

$k_{cp} = 2.0$ for $h_{ef} \geq 2.5$ in.

$$N_{cb} = \frac{A_{Nc}}{A_{Nco}} \psi_{ed,N} \, \psi_{c,N} \, \psi_{cp,N} N_b$$

Eq. (D-4)

| Example 34.4 (cont'd) | Calculations and Discussion | Code Reference |

Evaluate the terms of Eq. (D-4) for this problem:

A_{Nc} is the projected area of the tensile failure surface as approximated by a rectangle with edges bounded by 1.5 h_{ef} (1.5 × 7 = 10.5 in.) and free edges of the concrete from the centerline of the anchor.

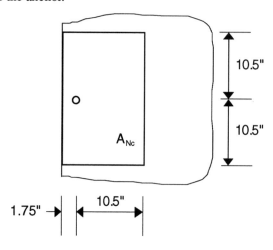

$A_{Nc} = (1.75 + 10.5)(10.5 + 10.5) = 257$ in.^{2}

$A_{Nco} = 9\, h_{ef}^2 = 9\,(7.0)^2 = 441$ in.^{2} *Eq. (D-6)*

Determine $\psi_{ed,N}$:   *D.5.2.5*

$$\psi_{ed,N} = 0.7 + 0.3 \frac{c_{a,min}}{1.5 h_{ef}}$$ *Eq. (D-11)*

$$\psi_{ed,N} = 0.7 + 0.3 \frac{1.75}{1.5(7.0)} = 0.75$$

Determine $\psi_{c,N}$:   *D.5.2.6*

$\psi_{c,N} = 1.0$ for locations where concrete cracking is likely to occur (i.e., the edge of the foundation is susceptible to cracks)

Determine N_b for the fastening: *D.5.2.2*

$$N_b = 24\sqrt{f'_c}\, h_{ef}^{1.5} = 24\sqrt{4000}\,(7.0)^{1.5} = 28{,}112 \text{ lb}$$ *Eq. (D-7)*

Substituting into Eq. (D-4):

$$N_{cb} = \left[\frac{257}{441}\right](0.75)(1.0)(28{,}112) = 12{,}287 \text{ lb}$$

		Code
Example 34.4 (cont'd)	**Calculations and Discussion**	**Reference**

To determine V_{ua} for the embedment strength governed by pryout strength Eq. (D-2) can be combined with Eq. (D-29) to give:

$V_{ua} = \phi V_{cp} = \phi k_{cp} N_{cb}$

Substituting, V_{ua} for the embedment strength governed by pryout is:

$V_{ua} = \phi V_{cp} = 0.70\,(2.0)\,(12{,}287) = 17{,}202$ lb

5. Required edge distances, spacings, and thickness to preclude splitting failure *D.8*

   Since a headed anchor used to attach wood frame construction is not likely to be torqued significantly, the minimum cover requirements of 7.7 apply.

   Per 7.7 the minimum clear cover for a 1/2 in. bar is 1-1/2 in. when exposed to earth or weather. The clear cover provided for the bolt is exactly 1-1/2 in. (1-3/4 in. to bolt centerline less one half bolt diameter). Note that the bolt head will have slightly less cover (1-3/16 in. for a hex head) say O.K. (note that this is within the minus 3/8 in. tolerance allowed for cover) *7.7*
7.5.2.1

6. Summary:

   The factored shear load ($V_{ua} = \phi V_n$) based on steel strength and embedment strength (concrete breakout and pryout) can be summarized as:

   Steel strength, (ϕV_{sa}):   3212 lb
   Embedment strength – concrete breakout, (ϕV_{cb}):   769 lb ← controls
   Embedment strength – pryout, (ϕV_{cp}):   17,202 lb

   In accordance with 9.2 the load factor for wind load is 1.6:

   $$V_{service} = \frac{V_{ua}}{1.6} = \frac{769}{1.6} = 481 \text{ lb}$$ *9.2*

   The reversible service load shear strength from wind load of the IBC 2000 Section 2308.6 minimum foundation connection for conventional wood-frame construction (1/2 in. diameter bolt embedded 7 in.) is 481 lb per bolt. The strength of the attached member (i.e., the 2×4 sill plate) also needs to be evaluated.

   Note that this embedment strength is only related to the anchor being installed in concrete with a specified compressive strength of 4000 psi. In many cases, concrete used in foundations such as this is specified at 2500 psi, the minimum strength permitted by the code. Since the concrete breakout strength controlled the strength of the connection, a revised strength based on using 2500 psi concrete rather than the 4000 psi concrete used in the example can be determined as follows:

   $$V_{service@2500} = 481\frac{\sqrt{2500}}{\sqrt{4000}} = 380 \text{ lb}$$

		Code
Example 34.4 (cont'd)	**Calculations and Discussion**	**Reference**

Alternate design using Table 34-6B.

Note: Step numbers correspond to those in the main example above, but prefaced with "A".

Table 34-6B has been selected because it contains design shear strength values based on concrete with f'_c = 4000 psi. Table Note 5 indicates that the values in the table are based on Condition B (no supplementary reinforcement), and Note 6 indicates that cracked concrete was assumed.

A2. Determine V_{ua} as controlled by the anchor steel

D.6.1
Eq. (D-20)

From Step 2, use ASTM F 1554, Grade 36 headed bolt with f_{uta} = 58,000 psi. From Table 34-6B, for specified compressive strength of concrete, f'_c, = 4000 psi, determine the design shear strength, ϕV_{sa}, for a 1/2-in. bolt.

ϕV_{sa} = 3212 lb

A3. Determine V_{ua} for embedment strength governed by concrete breakout strength with shear directed toward a free edge

D.6.2
Eq. (D-21)

Determine the design concrete breakout strength in shear, ϕV_{cb}, based on 7-in. embedment, and an edge distance, c_{a1}, of 1-3/4 in. In the table c_{a1}, is a function of embedment depth, h_{ef}. Therefore, the edge distance is:

$c_{a1} = c_{a1}/h_{ef} = 1.75/7 = 0.25 h_{ef}$

From table, the design concrete breakout strength in shear is,

ϕV_{cb} = 769 lb

A4. Determine V_{ua} for embedment strength governed by concrete pryout strength

D.6.3

Determine the design concrete pryout strength in shear, ϕV_{cp}, based on 7-in. embedment, and an edge distance of 1-3/4 in. This cannot be determined from Table 34-6B; however, since

$\phi V_{cp} = \phi k_{cp} N_{cb}$

Eq. (D-29)

where k_{cp} = 2, since h_{ef} > 2.5 in., and N_{cb} can be determined from Table 34-5B.

Note that the values in Tables 34-5 and 34-6 are design strengths, thus they include the strength reduction factor, ϕ. Since Table 34-5B is based on Condition B (with no supplementary reinforcement), the ϕ-value used for the concrete tensile strength calculations was 0.70, which is the same as to be used to determine the concrete pryout strength in shear. Therefore, the design concrete breakout strength, ϕN_{cb}, value from Table 34-5B can be used above without adjustment. From Table 34-5B, for an edge distance, c, equal to $0.25 h_{ef}$

		Code
Example 34.4 (cont'd)	**Calculations and Discussion**	**Reference**

$\phi N_{cb} = 8609$ lb

Substituting in Equation (D-29)

$\phi V_{cp} = k_{cp} \phi N_{cb} = (2)(8609) = 17{,}218$ lb

Note that the above value differs slightly from that obtained in Step #4 above. The table values are more precise due to rounding that occurred in the long-hand calculations.

A5. Required edge distances, spacings, and thickness to preclude splitting failure

See Step 5 above.

A6. Determine service wind shear load:

The factored shear load ($V_{ua} = \phi V_n$) based on steel strength and embedment strength (concrete breakout and pryout) can be summarized as:

Steel strength, (ϕV_{sa}): 3212 lb
Embedment strength - concrete breakout, (ϕV_{cb}): 769 lb ← controls
Embedment strength - pryout, (ϕV_{cp}): 17,218 lb

From this point, the allowable service wind load shear capacity of the 1/2 in. anchor is determined as in Step 6 above.

Example 34.5—Single Headed Bolt in Tension and Shear Near an Edge

Determine if a single 1/2 in. diameter hex headed anchor with a 7 in. embedment installed with its centerline 1-3/4 in. from the edge of a concrete foundation is adequate for a service tension load from wind of 1000 lb and reversible service shear load from wind of 400 lb.

Note: This is an extension of Example 34.4 that includes a tension load on the fastener as well as a shear load.

$f'_c = 4000$ psi

ASTM F 1554 Grade 36 hex head anchor

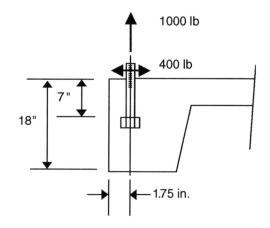

	Code
Calculations and Discussion	Reference

1. Determine the factored design loads

    $N_{ua} = 1.6 (1000) = 1600$ lb

    $V_{ua} = 1.6 (400) = 640$ lb

    9.2

2. This is a tension/shear interaction problem where values for both the design tensile strength (ϕN_n) and design shear strength (ϕV_n) will need to be determined. ϕN_n is the smallest of the design tensile strengths as controlled by steel (ϕN_{sa}), concrete breakout (ϕN_{cb}), pullout (ϕN_{pn}), and side-face blowout (ϕN_{sb}). ϕV_n is the smallest of the design shear strengths as controlled by steel (ϕV_{sa}), concrete breakout (ϕV_{cb}), and pryout (ϕV_{cp}).

    D.7
    D.4.1.2

3. Determine the design tensile strength (ϕN_n)

    D.5

    a. Steel strength, (ϕN_{sa}):

    D.5.1

    $\phi N_{sa} = \phi \, n \, A_{se} \, f_{uta}$

    Eq. (D-3)

    where:

    $\phi = 0.75$

    D.4.4(a)i

    Per the Ductile Steel Element definition in D.1, ASTM F 1554 Grade 36 steel qualifies as a ductile steel element.

Example 34.5 (cont'd)	Calculations and Discussion	Code Reference

$A_{se} = 0.142$ in.^{2} (see Table 34-2)

$f_{uta} = 58,000$ psi (see Table 34-1)

Substituting:

$\phi N_{sa} = 0.75 \ (1) \ (0.142) \ (58,000)) = 6177$ lb

b. Concrete breakout strength (ϕN_{cb}):   *D.5.2*

Since no supplementary reinforcement has been provided, $\phi = 0.70$ *D.4.4(c)ii*

In the process of calculating the pryout strength for this fastener in Example 34.4 Step 4, N_{cb} for this fastener was found to be 12,287 lb

$\phi N_{cb} = 0.70 \ (12,287) = 8601$ lb

c. Pullout strength (ϕN_{pn})   *D.5.3*

$\phi N_{pn} = \phi \ \psi_{c,P} \ N_p$ *Eq. (D-14)*

where:

$\phi = 0.70$ – Condition B applies for pullout strength in all cases   *D.4.4(c)ii*

$\psi_{c,P} = 1.0$, cracking may occur at the edges of the foundation   *D.5.3.6*

$N_p = A_{brg} \ 8 \ f'_c$ *Eq. (D-15)*

$A_{brg} = 0.291$ in.^{2}, for 1/2 in. hex head bolt (see Table 34-2)

Pullout Strength (ϕN_{pn})

$\phi N_{pn} = 0.70 \ (1.0) \ (0.291) \ (8) \ (4000) = 6518$ lb

d. Concrete side-face blowout strength (ϕN_{sb})   *D.5.4*

The side-face blowout failure mode must be investigated when the edge distance (c) is less than $0.4 \ h_{ef}$ *D.5.4.1*

$0.4 \ h_{ef} = 0.4 \ (7) = 2.80$ in. > 1.75 in.

Therefore, the side-face blowout strength must be determined

$\phi N_{sb} = \phi \left(160 \ c_{a1} \sqrt{A_{brg}} \ \sqrt{f'_c} \right)$ *Eq. (D-17)*

Example 34.5 (cont'd)	Calculations and Discussion	Code Reference

where:

$\phi = 0.70$, no supplementary reinforcement has been provided

D.4.4(c)ii

$c_{a1} = 1.75$ in.

$A_{brg} = 0.291$ in.^{2}, for 1/2 in. hex head bolt (see Table 34-2)

Substituting:

$$\phi N_{sb} = 0.70 \left(160 \, (1.75) \sqrt{0.291} \sqrt{4000} \right) = 6687 \text{ lb}$$

Summary of steel strength, concrete breakout strength, pullout strength, and side-face blowout strength for tension:

Steel strength, (ϕN_{sa}):	6177 lb ← controls	D.5.1
Embedment strength – concrete breakout, (ϕN_{cb}):	8601 lb	D.5.2
Embedment strength – pullout, (ϕN_{pn}):	6518 lb	D.5.3
Embedment strength – side-face blowout, (ϕN_{sb}):	6687 lb	D.5.4

Check $\phi N_n \geq N_{ua}$

6177 lb > 1600 lb   O.K.

Eq. (D-1)

Therefore:

$\phi N_n = 6177$ lb

4. Determine the design shear strength (ϕV_n)

D.6

Summary of steel strength, concrete breakout strength, and pryout strength for shear from Example 35.4, Step 6:

Steel strength, (ϕV_{sa}):	3212 lb	D.6.1
Embedment strength – concrete breakout, (ϕV_{cb}):	769 lb ← controls	D.6.2
Embedment strength – pryout, (ϕV_{cp}):	17,202 lb	D.6.3

Check $\phi V_n \geq V_{ua}$

769 lb > 640 lb   O.K

Eq. (D-2)

Therefore:

$\phi V_n = 769$ lb

5. Check tension and shear interaction

D.7

If $V_{ua} \leq 0.2 \phi V_n$ then the full tension design strength is permitted

D.7.1

$V_{ua} = 640$ lb

$0.2 \phi V_n = 0.2 \, (769) = 154$ lb < 640 lb

V_{ua} exceeds $0.2 \phi V_n$, the full tension design strength is not permitted

Example 34.5 (cont'd)	Calculations and Discussion	Code Reference

If $N_{ua} \leq 0.2 \phi N_n$ then the full shear design strength is permitted — D.7.2

$N_{ua} = 1600$ lb

$0.2 \phi N_n = 0.2 (6177) = 1235$ lb < 1600 lb

N_{ua} exceeds $0.2\phi N_n$, the full shear design strength is not permitted

The interaction equation must be used — D.7.3

$$\frac{N_{ua}}{\phi N_n} + \frac{V_{ua}}{\phi V_n} \leq 1.2$$ Eq. (D-29)

$$\frac{1600}{6177} + \frac{640}{769} = 0.26 + 0.83 = 1.09 < 1.2 \quad \text{O.K.}$$

6. Required edge distances, spacings, and thickness to preclude splitting failure — D.8

   Since a headed anchor used to attach wood frame construction is not likely to be torqued significantly, the minimum cover requirements of 7.7 apply.

   Per 7.7 the minimum clear cover for a 1/2 in. bar is 1-1/2 in. when exposed to earth or weather. The clear cover provided for the bolt is exactly 1-1/2 in. (1-3/4 in. to bolt centerline less one half bolt diameter). Note that the bolt head will have slightly less cover (1-3/16 in. for a hex head) say O.K. (note that this is within the minus 3/8 in. tolerance allowed for cover) — 7.7, 7.5.2.1, D.5

7. Summary

   Use a 1/2 in. diameter ASTM F 1554 Grade 36 hex headed anchor embedded 7 in.

Alternate design using Tables 34-5B and 34-6B

   Note: Step numbers correspond to those in the main example above, but prefaced with "A".

   Tables 34-5 and 34-6 have been selected because they contain design tension and shear values, respectively, based on concrete with $f'_c = 4000$ psi. Table Notes 4 and 5, respectively, indicate that the values in the tables are based on Condition B (no supplementary reinforcement). Cracked concrete is assumed in both tables (Table 34-5 Notes 6 and 10, and Table 34-6 Note 6).

   A3. Determine the design tensile strength (ϕN_n): — D.5.1, Eq. (D-3)

   A3a. Determine the design tensile strength of steel (ϕN_{sa}):

   Based on Step 3a, assume an ASTM F 1554, Grade 36 bolt, with a $f_{uta} = 58,000$ psi.

Example 34.5 (cont'd)	Calculations and Discussion	Code Reference

Using Table 34-5B, under the column for 58,000 a 1/2-in. diameter bolt has a design tensile strength,

ϕN_{sa} = 6177 lb.

A3b. Determine design concrete breakout strength (ϕN_{cb}):
    D.5.2
    Eq. (D-4)

Since breakout strength varies with edge distance for anchors close to an edge (c_{a1} < 1.5h_{ef}), determine the edge distance as a function of embedment depth. Since c = 1-3/4 in.

$c_{a1} = c_{a1}/h_{ef} = 1.75/7 = 0.25 h_{ef}$

Under column labeled "0.25h_{ef}" for a 1/2 in. bolt with 7 in. embedment depth,

ϕN_{cb} = 8609 lb

Note that the above value differs slightly from that obtained in Step 3b above. The table values are more precise due to rounding that occurred in the long-hand calculations.

A3c. Determine design concrete pullout strength (ϕN_{pn})
    D.5.3
    Eq. (D-14)

From the table under the column labeled "head" for a 1/2 in. bolt

ϕN_{pn} = 6518 lb

A3d. Determine design concrete side-face blowout strength (ϕN_{sb})
    D.5.4

Side face blowout is not applicable where the edge distance is equal to or greater than 0.4h_{ef}. In this case edge distance, c_{a1}, as calculated above is 0.25h_{ef}; therefore, it must be evaluated. From the table under the column labeled "0.25h_{ef}" for a 1/2 in. bolt with 7 in. embedment,
    D.5.4.1

ϕN_{sb} = 6687 lb

Summary of steel strength, concrete breakout strength, pullout strength, and side-face blowout strength for tension:

Steel strength, (ϕN_{sa}):   6177 lb ← controls
Embedment strength – concrete breakout, (ϕN_{cb}):   8609 lb
Embedment strength – pullout, (ϕN_{pn}):   6518 lb
Embedment strength – side-face blowout, (ϕN_{sb}):   6687 lb

Therefore:

ϕN_n = 6177 lb

| **Example 34.5 (cont'd)** | **Calculations and Discussion** | **Code Reference** |

A4. Determine the design shear strength (ϕV_n)

Summary of steel strength, concrete breakout strength, and pryout strength for shear from Step A6 of Example 34.4, alternate solution using Table 34-6B

Steel strength, (ϕV_{sa}):   3212 lb
Embedment strength - concrete breakout, (ϕV_{cb}):   769 lb ← controls
Embedment strength - pryout, (ϕV_{cp}):   17,218 lb

Therefore:

$\phi V_n = 769$ lb

A5. Check tension and shear interaction.

See Step 5 above.

A6. Required edge distances, spacings, and thickness to preclude splitting failure

See Step 6 above.

A7. Summary

Use a 1/2 in. diameter ASTM F 1554 Grade 36 hex headed bolt embedded 7 in.

Example 34.6—Group of L-Bolts in Tension and Shear Near Two Edges

Design a group of four L-bolts spaced as shown to support a 10,000 lb factored tension load and 5000 lb reversible factored shear load resulting from wind load. The connection is located at the base of a column in a corner of the building foundation.

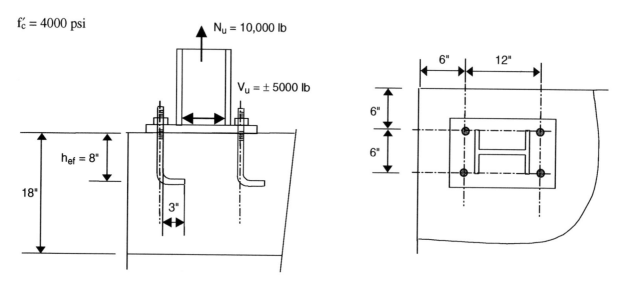

$f'_c = 4000$ psi

Note: OSHA Standard 29 CFR Part 1926.755 requires that the column anchorage use a least four anchors and be able to sustain a minimum eccentric gravity load of 300 lb located 18 in. from the face of the extreme outer face of the column in each direction. The load is to be applied at the top of the column. The intent is that the column be able to sustain an iron worker hanging off the side of the top of the column.

	Calculations and Discussion	Code Reference
1.	The solution to this example is found by assuming the size of the anchors, then checking compliance with the design provisions. Try four 5/8 in. ASTM F 1554 Grade 36 L-bolts with $h_{ef} = 8$ in. and a 3 in. extension, e_h, as shown in the figure.	
2.	This is a tension/shear interaction problem where values for both the design tensile strength (ϕN_n) and design shear strength (ϕV_n) will need to be determined. ϕN_n is the smallest of the design tensile strengths as controlled by steel (ϕN_{sa}), concrete breakout (ϕN_{cb}), pullout (ϕN_{pn}), and side-face blowout (ϕN_{sb}). ϕV_n is the smallest of the design shear strengths as controlled by steel (ϕV_{sa}), concrete breakout (ϕV_{cb}), and pryout (ϕV_{cp}).	D.7 D.4.1.2
3.	Determine the design tensile strength (ϕN_n)	D.5
	a. Steel strength, (ϕN_{sa}):	D.5.1

		Code
Example 34.6 (cont'd)	**Calculations and Discussion**	**Reference**

$\phi N_{sa} = \phi\, n\, A_{se}\, f_{uta}$ Eq. (D-3)

where:

$\phi = 0.75$ D.4.4(a)i

Per Table 34-1, the ASTM F 1554 Grade 36 L-bolt meets the Ductile Steel Element definition of D.1.

$A_{se} = 0.226$ in.^{2} (see Table 34-2)

$f_{uta} = 58,000$ psi (see Table 34-1)

Substituting:

$\phi N_{sa} = 0.75\,(4)\,(0.226)\,(58,000) = 39,324$ lb

b. Concrete breakout strength (ϕN_{cbg}):   D.5.2

Since the spacing of the anchors is less than 3 times the effective embedment depth h_{ef} (3×8 = 24), the anchors must be treated as an anchor group.   D.1

$\phi N_{cbg} = \phi\, \dfrac{A_{Nc}}{A_{Nco}}\, \Psi_{ec,N}\, \Psi_{ed,N}\, \Psi_{c,N}\, \Psi_{cp,N}\, N_b$ Eq. (D-5)

Since no supplementary reinforcement has been provided, $\phi = 0.70$ D.4.4(c)ii

Determine A_{Nc} and A_{Nco}:   D.5.2.1

A_{Nc} is the projected area of the failure surface as approximated by a rectangle with edges bounded by 1.5 h_{ef} (1.5 × 8.0 = 12.0 in. in this case) and free edges of the concrete from the centerlines of the anchors.

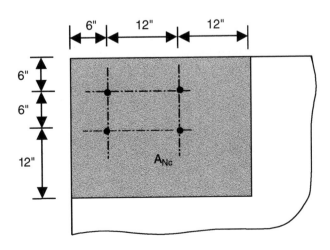

34-57

		Code
Example 34.6 (cont'd)	**Calculations and Discussion**	**Reference**

$A_{Nc} = (6 + 12 + 12)(6 + 6 + 12) = 720$ in.^{2}

$A_{Nco} = 9\, h_{ef}^2 = 9\,(8)^2 = 576$ in.^{2} Eq. (D-6)

Check: $A_{Nc} \leq n A_{Nco}$ 720 < 4(576)   O.K.

Determine $\psi_{ec,N}$:   D.5.2.4

$\psi_{ec,N} = 1.0$ (no eccentricity in the connection)

Determine $\psi_{ed,N}$ [$c_{a,min} < 1.5\, h_{ef}$, $6 < 1.5\,(8)$]:   D.5.2.5

$\psi_{ed,N} = 0.7 + 0.3 \dfrac{c_{a,min}}{1.5\, h_{ef}}$ Eq. (D-11)

$\psi_{ed,N} = 0.7 + 0.3 \dfrac{6.0}{1.5\,(8.0)} = 0.85$

Determine $\psi_{c,N}$:   D.5.2.6

$\psi_{c,N} = 1.0$ for locations where concrete cracking is likely to occur (i.e., the edge of the foundation)

Determine $\psi_{cp,N}$:   D.5.2.7

For cast-in-place anchors, $\psi_{cp,N} = 1.0$

Determine N_b: D.5.2.2

$N_b = 24 \sqrt{f'_c}\, h_{ef}^{1.5} = 24 \sqrt{4000}\,(8.0)^{1.5} = 34{,}346$ lb   Eq. (D-7)

Substituting into Eq. (D-5):

$\phi N_{cbg} = 0.70 \left[\dfrac{720}{576}\right](1.0)\,(0.85)\,(1.0)(1.0)\,(34{,}346) = 25{,}545$ lb

c.   Pullout strength (ϕN_{pn})   D.5.3

$\phi N_{pn} = \phi\, \psi_{c,P}\, N_p$ Eq. (D-14)

where:

$\phi = 0.70$, Condition B always applies for pullout strength   D.4.4(c)ii

$\psi_{c,P} = 1.0$, cracking may occur at the edges of the foundation   D.5.3.6

N_p for the L-bolts:

$N_p = 0.9\, f'_c\, e_h\, d_o$ Eq. (D-16)

Example 34.6 (cont'd)	Calculations and Discussion	Code Reference

e_h = maximum effective value of $4.5d_o$ = 4.5 (0.625) = 2.81 in.

$e_{h,provided}$ = 3 in. > 2.81 in., therefore use $e_h = 4.5d_o$ = 2.81 in.   D.5.3.5

Substituting into Eq. (D-14) and Eq. (D-16) with 4 L-bolts (ϕN_{pn})

ϕN_{pn} = 4 (0.70) (1.0) [(0.9) (4000) (2.81) (0.625)] = 17,703 lb

Note: If 5/8 in. hex head bolts where used ϕN_{pn} would be significantly increased as shown below:

N_p for the hex head bolts:

$N_p = A_{brg}\, 8\, f'_c$ Eq. (D-15)

A_{brg} = 0.454 in.², for 5/8 in. hex head bolt (see Table 34-2)

Substituting into Eq. (D-12) and Eq. (D-13) with 4 bolts (ϕN_{pn})

ϕN_{pn} = 4 (0.70) (1.0) (0.454) (8) (4000) = 40,678 lb

The use of hex head bolts would increase the pullout capacity by a factor of 2.3 over that of the L-bolts.

d. Concrete side-face blowout strength (ϕN_{sb})   D.5.4

The side-face blowout failure mode must be investigated for headed anchors where the edge distance (c_{a1}) is less than 0.4 h_{ef}. Since L-bolts are used here the side face blowout failure is not applicable. The calculation below is simply to show that if headed anchors were used the anchors are far enough from the edge that the side-face blowout strength is not applicable.   D.5.4.1

0.4 h_{ef} = 0.4 (8) = 3.2 in. < 6.0 in.

Therefore, the side-face blowout strength is not applicable (N/A).

Summary of design strengths based on steel strength, concrete breakout strength, pullout strength, and side-face blowout strength for tension:

Steel strength, (ϕN_{sa}):	39,324 lb	D.5.1
Embedment strength - concrete breakout, (ϕN_{cbg}):	25,545 lb	D.5.2
Embedment strength - pullout, (ϕN_{pn}):	17,703 lb ← controls	D.5.3
Embedment strength - side-face blowout, (ϕN_{sb}):	N/A	D.5.4

Therefore:

ϕN_n = 17,703 lb

Example 34.6 (cont'd)	Calculations and Discussion	Code Reference

Note: If hex head bolts were used the concrete breakout strength of 25,545 lb would control rather than the L-bolt pullout strength of 17,703 lb (i.e., 44% higher tensile capacity if hex head bolts were used).

4. Determine the design shear strength (ϕV_n)   D.6

   a. Steel strength, (ϕV_{sa}):   D.6.1

   $$\phi V_{sa} = \phi \, n \, 0.6 \, A_{se} \, f_{uta} \qquad \text{Eq. (D-19)}$$

   where:

   $\phi = 0.65$ D.4.4(a)ii

   Per Table 34-1, the ASTM F 1554 Grade 36 meets the Ductile Steel Element definition of Section D.1.

   $A_{se} = 0.226 \text{ in.}^2$ (see Table 34-2)

   $f_{uta} = 58,000$ psi (see Table 34-1)

   Substituting:

   $\phi V_{sa} = 0.65 \, (4) \, (0.6) \, (0.226) \, (58,000) = 20,448$ lb

   b. Concrete breakout strength (ϕV_{cbg}):   D.6.2

   Two potential concrete breakout failures need to be considered. The first is for the two anchors located near the free edge toward which the shear is directed (when the shear acts from right to left). For this potential breakout failure, these two anchors are assumed to carry one-half of the shear (see Fig. RD.6.2.1(b) upper right). For this condition, the total breakout strength for shear will be taken as twice the value calculated for these two anchors. The reason for this is that although the four-anchor group may be able to develop a higher breakout strength, the group will not have the opportunity to develop this strength if the two anchors nearest the edge fail first. The second potential concrete breakout failure is for the entire group transferring the total shear load. This condition also needs to be considered and may control when anchors are closely spaced or where the concrete member thickness is limited. For the case of welded studs, only the breakout strength of entire group for the total shear force needs to be considered (see Fig. RD.6.2.1(b) lower right), however this is not permitted for cast-in-place anchors that are installed through holes in the attached base plate.

   $$\phi V_{cbg} = \phi \, \frac{A_{Vc}}{A_{Vco}} \, \psi_{ec,V} \, \psi_{ed,V} \, \psi_{c,V} \, V_b \qquad \text{Eq. (D-22)}$$

   Determine the values of ϕ, $\psi_{ec,V}$, and $\psi_{c,V}$ (these are the same for both potential concrete breakout failures):

		Code
Example 34.6 (cont'd)	**Calculations and Discussion**	**Reference**

No supplementary reinforcement has been provided, $\phi = 0.70$ — D.4.4(c)i

There is no eccentricity in the connection, $\psi_{ec,V} = 1.0$ — D.6.2.5

For locations where concrete cracking is likely to occur (i.e., the edge of the foundation), $\psi_{c,V} = 1.0$ — D.6.2.6

For concrete breakout failure of the two anchors located nearest the edge:

Determine A_{Vc} and A_{Vco}:

A_{Vc} is the projected area of the shear failure surface on the free edge toward which shear is directed. The projected area is determined by a rectangle with edges bounded by $1.5\, c_{a1}$ ($1.5 \times 6.0 = 9.0$ in. in this case) and free edges of the concrete from the centerlines of the anchors and the surface of the concrete. Although the $1.5\, c_{a1}$ distance is not specified in D.6.2.1, it is shown in Fig. RD.6.2.1(b).

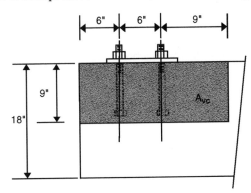

$A_{Vc} = (6+6+9)(9) = 189$ in.^{2}

$A_{Vco} = 4.5\, c_{a1}^2 = 4.5\, (6)^2 = 162$ in.^{2} — Eq. (D-23)

Check: $A_{Vc} \leq n A_{Vco}$ 189 < 2(162) O.K. — D.6.2.1

Determine $\psi_{ed,V}$ [$c_{a2} < 1.5\, c_{a1}$, $6 < (1.5 \times 6)$]: — D.6.2.6

$$\psi_{ed,V} = 0.7 + 0.3 \frac{c_{a2}}{1.5\, c_{a1}}$$ — Eq. (D-28)

$$\psi_{ed,V} = 0.7 + 0.3 \frac{6.0}{1.5\,(6.0)} = 0.90$$

The single anchor shear strength, V_b:

$$V_b = 7 \left(\frac{\ell_e}{d_o}\right)^{0.2} \sqrt{d_o}\, \sqrt{f'_c}\, c_{a1}^{1.5}$$ — Eq. (D-24)

Example 34.6 (cont'd) — Calculations and Discussion

where:

ℓ_e = load bearing length of the anchor for shear, not to exceed $8d_o$

For this problem $8d_o$ will control:

Substituting into Eq. (D-24):

$$V_b = (7)\left(\frac{5.0}{0.625}\right)^{0.2}\sqrt{0.625}\sqrt{4000}\,6.0^{1.5} = 7797 \text{ lb}$$

Substituting into Eq. (D-22) the design breakout strength of the two anchors nearest the edge toward which the shear is directed is:

$$\phi V_{cbg} = 0.70\left(\frac{189}{162}\right)(1.0)(0.90)(1.0)(7797) = 5731 \text{ lb}$$

The total breakout shear strength of the four anchor group related to an initial concrete breakout failure of the two anchors located nearest the free egde is:

$$\phi V_{cbg} = 2(5731) = 11{,}462 \text{ lb}$$

For concrete breakout failure of the entire four anchor group:

Determine A_{Vc} and A_{Vco}:

A_{Vc} is the projected area of the shear failure surface on the free edge toward which shear is directed. The projected area is determined by a rectangle with edges bounded by 1.5 c_{a1} (1.5 × 18.0 = 27.0 in. in this case) and free edges (side and bottom) of the concrete from the centerlines of the anchors and the surface of the concrete. Although the 1.5 c_{a1} distance is not specified in Section D.6.2.1, it is shown in Commentary Figure RD.6.2.1(b).

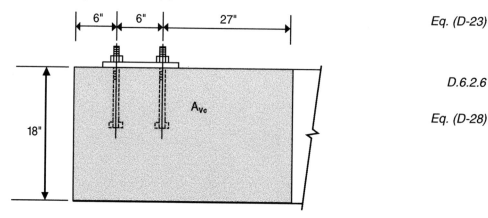

Code Reference:

Eq. (D-23)

D.6.2.6

Eq. (D-28)

Example 34.6 (cont'd)	**Calculations and Discussion**	**Code Reference**

$A_{Vc} = (6+6+27)(18) = 702$ in.^{2} *Eq. (D-24)*

$A_{Vco} = 4.5 \, c_{a1}^2 = 4.5 \, (18)^2 = 1458$ in.^{2}

Check: $A_{Vc} \leq nA_{Vco}$ $702 < 2 \, (1458)$ O.K.

Determine $\psi_{ed,V}$ [$c_{a2} < 1.5 \, c_{a1}$, $6 < (1.5 \times 18)$]:

$$\psi_{ed,V} = 0.7 + 0.3 \frac{c_{a2}}{1.5 \, c_{a1}}$$

$$\psi_{ed,V} = 0.7 + 0.3 \frac{6.0}{1.5 \, (18.0)} = 0.77$$

The single anchor shear strength, V_b:

$$V_b = 7 \left(\frac{\ell_e}{d_o}\right)^{0.2} \sqrt{d_o} \sqrt{f_c'} \, c_{a1}^{1.5}$$

where:

$\ell_e = 5.0$ in. (no change)

Substituting into Eq. (D-24):

$$V_b = (7) \left(\frac{5.0}{0.625}\right)^{0.2} \sqrt{0.625} \sqrt{4000} \, 18.0^{1.5} = 40,513 \text{ lb}$$

Substituting into Eq. (D-22) the design breakout strength of the four anchor group is:

$$\phi V_{cbg} = 0.70 \left(\frac{702}{1458}\right) (1.0) \, (0.77) \, (1.0) \, (40,513) = 10,514 \text{ lb}$$

The concrete breakout shear strength of the four anchor group is controlled by the breakout of the full group.

$\phi V_{cbg} = 10,514$ lb

Example 34.6 (cont'd)	Calculations and Discussion	Code Reference

c. Pryout strength (ϕV_{cp})   *D.6.3*

Note: The pryout failure mode is normally only a concern for shallow, stiff anchors. Since this example problem addresses both shear directed toward the free edge and shear directed away from the free edge, the pryout strength will be evaluated.

$$\phi V_{cpg} = \phi\, k_{cp}\, N_{cbg} \qquad \text{Eq. (D-30)}$$

where:

$\phi = 0.70$, Condition B always applies for pryout strength   *D.4.4(c)i*

$k_{cp} = 2.0$ for $h_{ef} \geq 2.5$ in.

From Step 3(b) above

$$N_{cbg} = \left[\frac{720}{576}\right](1.0)(0.85)(1.0)(1.0)(34{,}346) = 36{,}493\ \text{lb}$$

Substituting into Eq. (D-30):

$$\phi V_{cpg} = 0.70\,(2.0)(36{,}493) = 51{,}090\ \text{lb}$$

Summary of design strengths based on steel strength, concrete breakout strength, and pryout strength for shear:

Steel strength, (ϕV_{sa}):	20,448 lb	*D.6.1*
Embedment strength - concrete breakout, (ϕV_{cbg}):	10,514 lb ← controls	*D.6.2*
Embedment strength - pryout, (ϕV_{cp}):	51,090 lb	*D.6.3*

Therefore:

$\phi V_n = 10{,}514$ lb

5. Check tension and shear interaction   *D.7*

If $V_{ua} \leq 0.2\phi V_n$ then the full tension design strength is permitted   *D.7.1*

$V_{ua} = 5000$ lb

$0.2\phi V_n = 0.2\,(10{,}514) = 2103$ lb < 5000 lb

V_{ua} exceeds $0.2\phi V_n$, the full tension design strength is not permitted

If $N_{ua} \leq 0.2\phi N_n$ then the full shear design strength is permitted   *D.7.2*

$N_{ua} = 10{,}000$ lb

Example 34.6 (cont'd) — Calculations and Discussion

$0.2\phi N_n = 0.2\,(17{,}703) = 3541\text{ lb} < 10{,}000\text{ lb}$

N_{ua} exceeds $0.2\phi N_n$, the full shear design strength is not permitted

The interaction equation must be used. *D.7.3*

$$\frac{N_{ua}}{\phi N_n} + \frac{V_{ua}}{\phi V_n} \le 1.2 \quad\quad\text{Eq. (D-29)}$$

$$\frac{10{,}000}{17{,}703} + \frac{5000}{10{,}514} = 0.56 + 0.48 = 1.04 < 1.2 \quad\text{O.K.}$$

6. Required edge distances, spacings, an thicknesses to preclude splitting failure *D.8*

   Since cast-in-place L-bolts are not likely to be highly torqued, the minimum cover requirements of 7.7 apply.

   Per 7.7 the minimum clear cover for a 5/8 in. bar is 1-1/2 in. when exposed to earth or weather. The clear cover provided for the bolt exceeds this requirement with the 6 in. edge distance to the bolt centerline – O.K.

7. Summary

   Use 5/8 in. diameter ASTM F 1554 Grade 36 L-bolts with an embedment of 8 in. (measured to the upper surface of the L) and a 3 in. extension, e_h, as shown in the figure.

   Note: The use of hex head bolts rather than L-bolts would significantly increase the tensile strength of the connection. If hex head bolts were used, the design tensile strength would increase from 17,719 lb as controlled by the pullout strength of the L-bolts to 25,545 lb as controlled by concrete breakout for hex head bolts.

Example 34.7—Group of Headed Bolts in Moment and Shear Near an Edge in a Region of Moderate or High Seismic Risk

Design a group of four headed anchors spaced as shown for a reversible 18.0 k-ft factored moment and a 5.0 kip factored shear resulting from lateral seismic load in a region of moderate or high seismic risk. The connection is located at the base of an 8 in. steel column. $f'_c = 4000$ psi

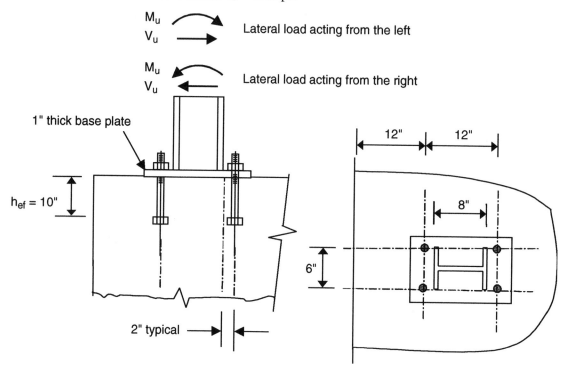

Calculations and Discussion	Code Reference

1. The solution to this example is found by assuming the size of the anchors, then checking for compliance with the design provisions for seismic loadings in regions of moderate or high seismic risk. For this example, assume four 3/4 in. ASTM F 1554 Grade 36 hex head anchors with $h_{ef} = 10$ in.

2. Since this connection is subjected to seismic load in a region of moderate or high seismic risk, the design tensile strength is $0.75\phi N_n$ and design shear strength is $0.75\phi V_n$. Unless the attachment has been designed to yield at a load lower than the design strength of the anchors (including the 0.75 factor), the strength of the anchors must be controlled by the tensile and shear strengths of ductile steel elements (D.3.3.3). To ensure ductile behavior, $0.75\phi N_{sa}$ must be larger than the concrete breakout (ϕN_{cb}), pullout (ϕN_{pn}), and side-face blowout (ϕN_{sb}). Further, $0.75\phi V_{sa}$ must be larger than concrete breakout (ϕV_{cb}), and pryout (ϕV_{cp}).   D.3.3

| Example 34.7 (cont'd) | Calculations and Discussion | Code Reference |

3. This problem involves the design of the connection of the steel column to the foundation for lateral loads coming from either the left or the right of the structure as shown below:

   Lateral load acting from the left:

   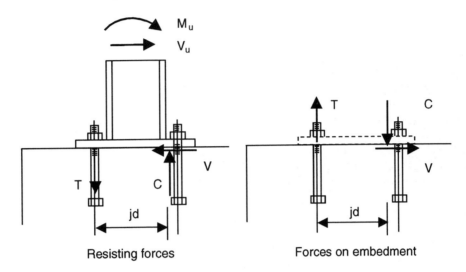

   Resisting forces   Forces on embedment

   Lateral load acting from the right:

   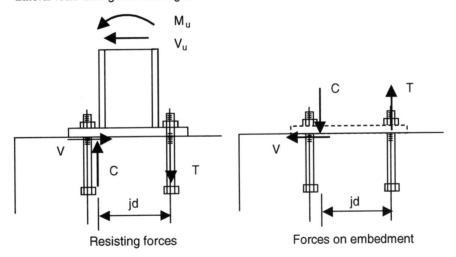

   Resisting forces   Forces on embedment

   As shown in the figures above, due to the free edge on the left, the critical case for tension on the anchors occurs when the lateral load is acting from the left while the critical case for shear occurs when the lateral load is acting from the right.

4. Distribution of the applied moment and shear loads to the anchors

   Tension in the anchors resulting from the applied moment - The exact location of the compressive resultant from the applied moment cannot be accurately determined by traditional concrete beam methods. This is true for both the elastic linear stress-strain

Example 34.7 (cont'd)	Calculations and Discussion	Code Reference

method (i.e., the transformed area method) and the ACI 318 stress block method since plane sections do not remain plane and different cross-sections and materials are utilized on each side of the connection. These methods require additional work that is simply not justified and in many cases can yield unconservative results for the location of the compressive resultant. The actual location of the compressive resultant is dependent on the stiffness of the base plate.

If the base plate rotates as a rigid body the compressive resultant will be at the leading edge of the base plate. For example, take a book, lay it on your desk and lift one end. The end opposite of the one being lifted is where the compressive resultant is located; this is rigid base plate behavior where the compressive resultant is located at the leading edge of the base plate. The assumption of rigid base plate behavior is conservative for determining base plate thickness but is unconservative for determining the tension force in the anchors since it provides a maximum distance (lever arm) between the tensile and compressive resultants from the applied moment.

If the base plate is flexible, the compressive resultant will be very near the edge of the attached structural member that is in compression from the applied moment. For example; take a piece of paper, lay it on your desk and lift one end. A portion of the paper opposite of the one being lifted will remain flat on the desktop. Since this portion of the paper remains flat, it has no curvature and therefore carries no moment. For this case, the compressive resultant must be located at the point where the piece of paper with one end lifted first contacts the desktop. References D.4 and D.5 of the ACI 318 Commentary show that the minimum distance between the edge of the attached structural member that is in compression from the applied moment and the compressive resultant from the applied moment is equal to the yield moment of the base plate divided by the compressive resultant from the applied moment. Since the determination of this distance adds unwarranted difficulty to the calculations, it is conservative to assume that the compressive resultant is located at the edge of the attached structural member that is in compression from the applied moment when determining the tensile resultant in the anchors from the applied moment.

For this example, the internal moment arm jd will be conservatively determined by assuming flexible base plate behavior with the compressive resultant located at the edge of the compression element of the attached member.

$jd = 2 + 8 = 10$ in.

By summing moments about the location of the compressive resultant (see figures in Step 3):

$M_u = T \, (jd)$

Example 34.7 (cont'd)	Calculations and Discussion	Code Reference

where:

$M_u = 18.0$ k-ft $= 216,000$ in.-lb
$T = N_{ua}$ (i.e., the factored tensile load acting on the anchors in tension)
$jd = 2+8 = 10$ in.

Rearranging and substituting:

$$N_{ua} = \frac{M_u}{jd} = \frac{216,000}{10} = 21,600 \text{ lb}$$

Shear – Although the compressive resultant from the applied moment will allow for the development of a frictional shear resistance between the base plate and the concrete, the frictional resistance will be neglected for this example and the anchors on the compression side will be designed to transfer the entire shear. The assumption of the anchors on the compression side transferring the entire shear is supported by test results reported in Ref. D.4, D.5, and D.6. This assumption is permitted by D.3.1 which allows for plastic analysis where the nominal strength is controlled by ductile steel elements (as required by D.3.3.4).

References D.4, D.5, D.6 and ACI 349-01 *Code Requirements for Nuclear Safety Related Concrete Structures* B.6.1.4 provide information regarding the contribution of friction to the shear strength. As noted in these references, the coefficient of friction between the steel base plate and concrete may be assumed to be 0.40. For this example, the frictional shear resistance is likely to have the potential to transfer 8640 lbs ($0.40 \times 21,600$). Although the potential frictional resistance between the base plate and the concrete will be neglected in this example, it does exist and will be located at the compressive reaction (i.e., near the anchors in the compression zone).

To summarize, the assumption of the entire shear being transferred by the anchors in the compression zone is permitted by D.3.1, represents a conservative condition for shear design, is supported by test results, and best represents where the shear will actually be transferred to the concrete if the friction force were considered.

$V_u = 5000$ lb on the two anchors on the compression side

5. Determine the design tensile strength for seismic load ($0.75\phi N_n$)   D.5

    a. Steel strength, (ϕN_{sa}):   D.5.1

        $\phi N_{sa} = \phi \, n \, A_{se} \, f_{uta}$ Eq. (D-3)

        where:

        $\phi = 0.75$ D.4.4(a)i

| Example 34.7 (cont'd) | Calculations and Discussion | Code Reference |

Per Table 34-1, the ASTM F 1554 Grade 36 bolt meets the Ductile Steel Element definition of Section D.1.

$A_{se} = 0.334$ in.^{2} (see Table 34-2)

$f_{uta} = 58,000$ psi (see Table 34-1)

Substituting:

$\phi N_{sa} = 0.75\ (2)\ (0.334)\ (58,000) = 29,058$ lb

b. Concrete breakout strength (ϕN_{cbg}):   D.5.2

Since the spacing of the anchors is less than 3 times the effective embedment depth h_{ef} (3 × 10 in. = 30 in.), the anchors must be treated as an anchor group.   D.1

$$\phi N_{cbg} = \phi \frac{A_{Nc}}{A_{Nco}} \psi_{ec,N}\ \psi_{ed,N}\ \psi_{c,N}\ \psi_{cp,N}\ N_b$$ Eq. (D-5)

Since no supplementary reinforcement has been provided, $\phi = 0.70$ D.4.4(c)ii

Determine A_{Nc} and A_{Nco}:   D.5.2.1

A_{Nc} is the projected area of the failure surface as approximated by a rectangle with edges bounded by 1.5 h_{ef} (1.5 × 10.0 = 15.0 in.) and free edges of the concrete from the centerlines of the anchors.

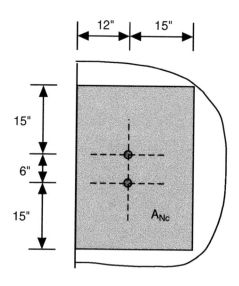

34-70

		Code
Example 34.7 (cont'd)	**Calculations and Discussion**	**Reference**

$A_{Nc} = (12 + 15)(15 + 6 + 15) = 972$ in.^{2}

$A_{Nco} = 9 h_{ef}^2 = 9(10)^2 = 900$ in.^{2} *Eq. (D-6)*

Check: $A_{Nc} \leq n A_{Nco}$ $972 < 2(900)$ O.K.

Determine $\psi_{ec,N}$:   *D.5.2.4*

$\psi_{ec,N} = 1.0$ (no eccentricity in the connection)

Determine $\psi_{ed,N}$:   *D.5.2.5*

$$\psi_{ed,N} = 0.7 + 0.3 \frac{c_{a,min}}{1.5 h_{ef}}$$ *Eq. (D-11)*

$$\psi_{ed,N} = 0.7 + 0.3 \frac{12.0}{1.5(10.0)} = 0.94$$

Determine $\psi_{c,N}$:   *D.5.2.6*

$\psi_{c,N} = 1.0$ for locations where concrete cracking is likely to occur (i.e., the edge of the foundation)

Determine $\psi_{cp,N}$ *D.5.2.7*

For cast-in-place anchors, $\psi_{cp,N} = 1.0$

Determine N_b: *D.5.2.2*

$$N_b = 24 \sqrt{f_c'} \ h_{ef}^{1.5} = 24 \sqrt{4000} \ (10.0)^{1.5} = 48,000 \text{ lb}$$ *Eq. (D-7)*

Substituting into Eq. (D-5):

$$\phi N_{cbg} = 0.70 \left[\frac{972}{900} \right] (1.0)(0.94)(1.0)(1.0)(48,000) = 34,111 \text{ lb}$$

c.   Pullout strength (ϕN_{pn})   *D.5.3*

$\phi N_{pn} = \phi \ \psi_{c,P} \ N_p$ *Eq. (D-14)*

where:

$\phi = 0.70$, Condition B always applies for pullout strength   *D.4.4(c)ii*

Example 34.7 (cont'd)	Calculations and Discussion	Code Reference

$\psi_{c,P} = 1.0$, cracking may occur at the edges of the foundation

D.5.3.6

N_p for the hex head bolts:

$N_p = A_{brg} \, 8 \, f'_c$

Eq. (D-15)

$A_{brg} = 0.654$ in.^{2}, for 3/4 in. hex head bolt (see Table 34-2)

Substituting into Eq. (D-14) and Eq. (D-15) with 2 bolts (ϕN_{pn})

$\phi N_{pn} = 2 \, (0.70) \, (1.0) \, (0.654) \, (8) \, (4000) = 29{,}299$ lb

   d. Concrete side-face blowout strength (ϕN_{sb})

D.5.4

      The side-face blowout failure mode must be investigated when the edge distance (c) is less than $0.4 \, h_{ef}$

D.5.4.1

      $0.4 \, h_{ef} = 0.4 \, (10) = 4.0$ in. < 12.0 in.

      Therefore, the side-face blowout strength is not applicable (N/A)

Summary of design strengths based on steel strength, concrete breakout strength, pullout strength, and side-face blowout strength for tension:

Steel strength, (ϕN_{sa}):	29,058 lb ← controls	D.5.1
Embedment strength - concrete breakout, (ϕN_{cbg}):	34,111 lb	D.5.2
Embedment strength - pullout, (ϕN_{pn}):	29,299 lb	D.5.3
Embedment strength - side-face blowout, (ϕN_{sb}):	N/A	D.5.4

Therefore:

$\phi N_n = 29{,}058$ lb and is controlled by a ductile steel element as required in D.3.3.4

For seismic load in a region of moderate or high seismic risk, the design tensile strength is $0.75 \phi N_n$:

D.3.3.3

$0.75 \, \phi N_n = 0.75 \, (29{,}058) = 21{,}794$ lb and is controlled by a ductile steel element

Check if $N_{ua} \leq 0.75 \phi N_n$

21,600 lb $<$ 21,794 lb   O.K. for tension

6. Determine the design shear strength (ϕV_n)

D.6

   a. Steel strength, (ϕV_{sa}):

D.6.1

      $\phi V_{sa} = \phi \, n \, 0.6 \, A_{se} \, f_{uta}$

Eq. (D-20)

| Example 34.7 (cont'd) | Calculations and Discussion | Code Reference |

where:

$\phi = 0.65$ *D.4.4(a)ii*

Per Table 34-1, the ASTM F 1554 Grade 36 meets the Ductile Steel Element definition of Section D.1.

$A_{se} = 0.334$ in.^{2} (see Table 34-2)

$f_{uta} = 58,000$ psi (see Table 34-1)

Substituting:

$\phi V_{sa} = 0.65\ (2)\ (0.6)\ (0.334)\ (58,000) = 15,110$ lb

b. Concrete breakout strength (ϕV_{cbg}):   *D.6.2*

$$\phi V_{cbg} = \phi \frac{A_{Vc}}{A_{Vco}}\ \psi_{ec,V}\ \psi_{ed,V}\ \psi_{c,V}\ V_b$$ *Eq. (D-22)*

Since no supplementary reinforcement has been provided, $\phi = 0.70$ *D.4.4(c)i*

Determine A_{Vc} and A_{Vco}:   *D.6.2.1*

A_{Vc} is the projected area of the shear failure surface on the free edge toward which shear is directed. The projected area is determined by a rectangle with edges bounded by 1.5 c_{a1} (1.5 × 12.0 = 18.0 in.) and free edges of the concrete from the centerlines of the anchors and surface of the concrete.

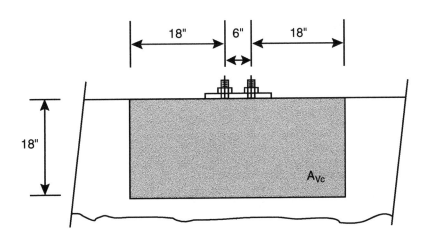

Example 34.7 (cont'd)	Calculations and Discussion	Code Reference

$A_{Vc} = (18 + 6 + 18)(18) = 756$ in.^{2}

$A_{Vco} = 4.5\ c_{a1}^2 = 4.5\ (12)^2 = 648$ in.^{2} *Eq. (D-23)*

Check: $A_{Vc} \leq nA_{Vco}$ $756 < 2(648)$ O.K.

Determine $\psi_{ec,V}$:   *D.6.2.5*

$\psi_{ec,V} = 1.0$ (no eccentricity in the connection)

Determine $\psi_{ed,V}$:   *D.6.2.6*

$\psi_{ed,V} = 1.0$ (no orthogonal free edge)   *Eq. (D-27)*

Determine $\psi_{c,V}$:

$\psi_{c,V} = 1.0$ for locations where concrete cracking is likely to occur (i.e., the edge of the foundation)

Determine V_b for an anchor:

$$V_b = 7\left(\frac{\ell_e}{d_o}\right)^{0.2} \sqrt{d_o}\ \sqrt{f'_c}\ c_{a1}^{1.5}$$ *Eq. (D-24)*

where:

ℓ_e = load bearing length of the anchor for shear, not to exceed $8d_o$ *2.1*

For this problem $8d_o$ will control:

$\ell_e = 8d_o = 8\ (0.75) = 6.0$ in. < 10 in.   therefore, use $8d_o$

Substituting into Eq. (D-24):

$$V_b = (7)\left(\frac{8\ (0.75)}{0.75}\right)^{0.2} \sqrt{0.75}\ \sqrt{4000}\ (12.0)^{1.5} = 24{,}157\ \text{lb}$$

Substituting into Eq. (D-22):

$$\phi V_{cbg} = 0.70\left(\frac{756}{648}\right)(1.0)\ (1.0)\ (1.0)\ (24{,}157) = 19{,}728\ \text{lb}$$

Example 34.7 (cont'd)	Calculations and Discussion	Code Reference

c. Pryout strength (ϕV_{cpg})   *D.6.3*

Note: The pryout failure mode is normally only a concern for shallow, stiff anchors. Since this example problem addresses both shear directed toward the free edge and shear directed inward from the free edge, the pryout strength will be evaluated.

$\phi V_{cpg} = \phi\, k_{cp}\, N_{cbg}$ *Eq. (D-30)*

where:

$\phi = 0.70$, Condition B always applies for pryout strength   *D.4.4(c)i*

$k_{cp} = 2.0$ for $h_{ef} > 2.5$ in.

From Step 5(b) above

$\phi N_{cbg} = \phi \dfrac{A_{Nc}}{A_{Nco}}\, \psi_{ec,N}\, \psi_{ed,N}\, \psi_{c,N}\, \psi_{cp,N}\, N_b$ *Eq. (D-5)*

$N_{cbg} = \left[\dfrac{972}{900}\right](1.0)(0.94)(1.0)(1.0)(48,000) = 48,730$ lb

Substituting into Eq. (D-30):

$\phi V_{cpg} = 0.70\,(2.0)\,(48,730) = 68,222$ lb

Summary of design strengths based on steel strength, concrete breakout strength, and pryout strength for shear:

Steel strength, (ϕV_{sa}):	15,110 lb ← controls	*D.6.1*
Embedment strength - concrete breakout, (ϕV_{cbg}):	19,728 lb	*D.6.2*
Embedment strength - pryout, (ϕV_{cp}):	68,222 lb	*D.6.3*

Therefore:

$\phi V_n = 15,110$ lb and is controlled by a ductile steel element as required in D.3.3.4

For seismic load in a region of moderate or high seismic risk, the design shear strength is $0.75\phi V_n$:   *D.3.3.3*

$0.75\, \phi V_n = 0.75\,(15,110) = 11,333$ lb and is controlled by a ductile steel element

Check if $V_{ua} \le 0.75\,\phi V_n$

5000 lb < 11,333 lb   O.K. for shear

Example 34.7 (cont'd)	Calculations and Discussion	Code Reference

7. Required edge distances, spacings, and thicknesses to preclude splitting failure

   D.8

   Since cast-in-place anchors are not likely to be highly torqued, the minimum cover requirements of 7.7 apply.

   Per 7.7 the minimum clear cover for a 3/4 in. bar is 1-1/2 in. when exposed to earth or weather. The clear cover provided for the bolt exceeds this requirement with the 12 in. edge distance to the bolt centerline   O.K.

8. Summary

   Use 3/4 in. diameter ASTM F 1554 Grade 36 hex head anchors with h_{ef} = 10 in.

Note: OSHA Standard 29 CFR Part 1926.755 requires that column anchorages use at least four anchors and be able to sustain a minimum eccentric gravity load of 300 pounds located 18 in. from the face of the extreme outer face of the column in each direction. The load is to be applied at the top of the column. The intent is that the column be able to sustain an iron worker hanging off the side of the top of the column. This connection will satisfy the OSHA requirement but calculations are not included in the example.

Example 34.8—Single Post-Installed Anchor in Tension and Shear Away from Edges

Design a single post-installed mechanical anchor installed in the bottom of an 8 in. slab to support a factored 4500 lb tension load and a factored 2000 lb shear load (seismic loads from regions of moderate to high seismic risk are not included).

$f'_c = 4000$ psi

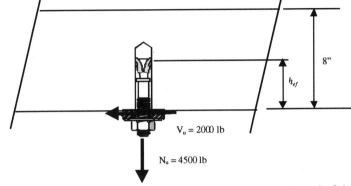

Note: This example for a single post-installed mechanical anchor is provided at the end of the design examples of Part 34 since additional calculations to account for group effects, edge conditions, eccentricity, and tension/shear interaction covered in the previous examples for cast-in-place anchors are essentially the same as for post-installed mechanical anchors.

Similarities between post-installed mechanical anchors and cast-in-place anchors:

- For group and edge conditions, A_{Nc}, A_{Nco}, A_{Vc}, and A_{Vco} are determined in the same manner.
- For eccentric loads, $\psi_{ec,N}$ and $\psi_{ec,V}$ are determined in the same manner.
- For edge effects, $\psi_{ed,N}$ and $\psi_{ed,V}$ are determined in the same manner.
- For anchors used in areas where concrete cracking may occur, $\psi_{c,N}$ and $\psi_{c,V} = 1.0$.

The unique properties of post-installed mechanical anchors are provided by the ACI 355.2 product evaluation report (refer to the sample in Table 34-3 for anchor data for a fictitious post-installed torque-controlled mechanical expansion anchor). The unique properties associated with each post-installed mechanical anchor product are:

- effective embedment length h_{ef}
- effective cross sectional area A_{se} in tension and shear
- specified yield strength f_{ya} and specified ultimate strength f_{uta}
- minimum edge distance $c_{a,min}$ for the anchor
- minimum member thickness h_{min} for the anchor
- minimum spacing s for the anchor
- critical edge distance c_{ac} for $\psi_{ep,N}$ with uncracked concrete design (D.5.2.7)
- category of the anchor for determination of the appropriate ϕ factor for embedment strength
- coefficient for basic concrete breakout strength k_c for use in Eq. (D-7)
- factor $\psi_{c,N}$ for uncracked concrete design
- pullout strength N_p of the anchor

Calculations and Discussion	Code Reference

1. The solution to this example is found by assuming the size of the anchor, then checking compliance with the design provisions. Try the fictitious 5/8 in. post-installed torque-controlled mechanical expansion anchor with a 4.5 in. effective embedment depth, shown in Table 34-3.

| Example 34.8 (cont'd) | Calculations and Discussion | Code Reference |

2. This is a tension/shear interaction problem where values for both the design tensile strength (ϕN_n) and design shear strength (ϕV_n) will need to be determined. ϕN_n is the smallest of the design tensile strengths as controlled by steel (ϕN_{sa}), concrete breakout (ϕN_{cb}), pullout (ϕN_{pn}), and side-face blowout (ϕN_{sb}). ϕV_n is the smallest of the design shear strengths as controlled by steel (ϕV_{sa}), concrete breakout (ϕV_{cb}), and pryout (ϕV_{cp}).

D.7
D.4.1.2

3. Determine the design tensile strength (ϕN_n)

D.5

    a. Steel strength, (ϕN_{sa}):

D.5.1

$$\phi N_{sa} = \phi \, n \, A_{se} \, f_{uta}$$

Eq. (D-3)

where:

$\phi = 0.75$

D.4.4(a)i

As shown in Table 34-3, this anchor meets ductile steel requirements.

$A_{se} = 0.226$ in.^{2} (see Table 34-3)

$f_{uta} = 75,000$ psi (see Table 34-3)

Note: Per D.5.1.2, f_{uta} shall not be taken greater than $1.9 f_{ya}$ or 125,000 psi. From Table 34-3, $f_{ya} = 55,000$ psi and $1.9 f_{ya} = 1.9(55,000) = 104,500$ psi, therefore use the specified minimum f_{uta} of 75,000 psi.

D.5.1.2

Substituting:

$\phi N_{sa} = 0.75 \, (1) \, (0.226) \, (75,000) = 12,712$ lb

    b. Concrete breakout strength (ϕN_{cb}):

D.5.2

$$\phi N_{cb} = \phi \, \frac{A_{Nc}}{A_{Nco}} \, \psi_{ed,N} \, \psi_{c,N} \, \psi_{cp,N} \, N_b$$

Eq. (D-4)

where:

$\phi = 0.65$

D.4.4

From Table 34-3, this post-installed anchor is Category 1 and no supplementary reinforcement has been provided.

		Code
Example 34.8 (cont'd)	**Calculations and Discussion**	**Reference**

$\dfrac{A_{Nc}}{A_{Nco}}$ and $\psi_{ed,N}$ terms are 1.0 for single anchors away from edges

$\psi_{c,N} = 1.0$ and $\psi_{cp,N} = 1.0$ for locations where concrete cracking is likely to occur (i.e., the bottom of the slab)

$$N_b = k_c \sqrt{f'_c}\, h_{ef}^{1.5}$$ Eq. (D-7)

where:

$k_c = 17$

Note: $k_c = 17$ for post-installed anchors unless the ACI 355.2 product evaluation report indicates a higher value may be used. For the case of this torque-controlled mechanical expansion anchor, $k_c = 17$ per Table 34-3.   RD.5.2.2

$h_{ef} = 4.5$ in. (Table 34-3)

Therefore,

$$N_b = 17\sqrt{4000}\; 4.5^{1.5} = 10{,}264 \text{ lb}$$

Substituting:

$\phi N_{cb} = 0.65\,(1.0)\,(1.0)\,(1.0)\,(10{,}264) = 6672$ lb

c. Pullout strength (ϕN_{pn})   D.5.3

$\phi N_{pn} = \phi\, \psi_{c,P}\, N_p$ Eq. (D-14)

where:

$\phi = 0.65$, Category 1 and no supplementary reinforcement has been provided   D.4.4

$\psi_{c,P} = 1.0$, cracking may occur at the edges of the foundation   D.5.3.6

$N_p = 8211$ lb (see Table 34-3)

Substituting:

$\phi N_{pn} = 0.65\,(1.0)\,(8211) = 5337$ lb

d. Concrete side-face blowout strength (ϕN_{sb})   D.5.4

This anchor is not located near any free edges therefore the side-face blowout strength is not applicable.   D.5.4.1

Summary of steel strength, concrete breakout strength, pullout strength, and side-face blowout strength for tension:

Example 34.8 (cont'd)	Calculations and Discussion	Code Reference

Steel strength, (ϕN_{sa}): 12,712 lb — D.5.1
Embedment strength - concrete breakout, (ϕN_{cb}): 6672 lb — D.5.2
Embedment strength - pullout, (ϕN_{pn}): 5337 lb ← controls — D.5.3
Embedment strength - side-face blowout, (ϕN_{sb}): N/A — D.5.4

Therefore:

$\phi N_n = 5337$ lb

4. Determine the design shear strength (ϕV_n) — D.6

   a. Steel strength, (ϕV_{sa}): — D.6.1

   $\phi V_{sa} = \phi\, n\, (0.6 A_{se}\, f_{uta})$ — Eq. (D-20)

   where:

   $\phi = 0.65$

   As shown in Table 34-3, this anchor meets ductile steel requirements.

   $A_{se} = 0.226$ in.^{2} (see Table 34-3)

   $f_{uta} = 75,000$ psi (see Table 34-3)

   Note: Per D.5.1.2, f_{uta} shall not be taken greater than $1.9 f_{ya}$ or 125,000 psi. From Table 3, $f_{ya} = 55,000$ psi and $1.9 f_{ya} = 1.9(55,000) = 104,500$ psi. Therefore, use the specified minimum f_{uta} of 75,000 psi.

   Substituting:

   $\phi V_{sa} = 0.65\,(1)\,(0.6)\,(0.226)\,(75,000) = 6610$ lb

   b. Concrete breakout strength (ϕV_{cb}): — D.6.2

   This anchor is not located near any free edges therefore the concrete breakout for shear is not applicable.

   c. Pryout strength (ϕV_{cp}) — D.6.3

   $\phi V_{cp} = \phi\, k_{cp}\, N_{cb}$ — Eq. (D-29)

   where:

   $\phi = 0.65$, Category 1 and no supplementary reinforcement has been provided — D.4.4

   $k_{cp} = 2.0$ for $h_{ef} > 2.5$ in.

		Code
Example 34.8 (cont'd)	**Calculations and Discussion**	**Reference**

$$N_{cb} = \frac{A_{Nc}}{A_{Nco}} \psi_{ed,N} \psi_{c,N} \psi_{cp,N} N_b$$

Eq. (D-4)

From Step 3 above:

$N_{cb} = (1.0)(1.0)(1.0)(10,264) = 10,264$ lb

Substituting into Eq. (D-29):

$\phi V_{cp} = 0.65(2.0)(10,264) = 13,343$ lb

Summary of steel strength, concrete breakout strength, and pryout strength for shear:

Steel strength, (ϕV_{sa}):	6610 lb ← controls	D.6.1
Embedment strength - concrete breakout, (ϕV_{cb}):	N/A	D.6.2
Embedment strength - pryout, (ϕV_{cp}):	13,343 lb	D.6.3

Therefore:

$\phi V_n = 6610$ lb

5. Check tension and shear interaction

    D.7

    If $V_{ua} \leq 0.2 \phi V_n$ then the full tension design strength is permitted

    D.7.1

    $V_{ua} = 2000$ lb

    $0.2 \phi V_n = 0.2(6610) = 1322$ lb

    V_{ua} exceeds $0.2 \phi V_n$, the full tension design strength is not permitted

    If $N_{ua} \leq 0.2 \phi N_n$ then the full shear design strength is permitted

    D.7.2

    $N_{ua} = 4500$ lb

    $0.2 \phi N_n = 0.2(5337) = 1067$ lb

    N_{ua} exceeds $0.2 \phi N_n$, the full shear design strength is not permitted

    The interaction equation must be used

    D.7.3

    $$\frac{N_{ua}}{\phi N_n} + \frac{V_{ua}}{\phi V_n} \leq 1.2$$

    Eq. (D-29)

Example 34.8 (cont'd)	Calculations and Discussion	Code Reference

$$\frac{4500}{5337}+\frac{2000}{6610}=0.84+0.30=1.14<1.2 \quad \text{O.K.}$$

6. Required edge distances, spacings, and thickness to preclude splitting failure   D.8

    Since this anchor is located away from edges, only the limits on embedment length h_{ef} related to member thickness are applicable. Per D.8.5, h_{ef} shall not exceed 2/3 of the member thickness or the member thickness less 4 in. D.8 does permit the use of larger values of h_{ef} provided product-specific tests have been performed in accordance with ACI 355.2.

    As shown in Table 34-3, the ACI 355.2 product evaluation report for this anchor provides the minimum thickness as 1.5 h_{ef} = 1.5(4.5) = 6.75 in. which is less than the 8 in. provided   O.K.

7. Summary

    The fictitious 5/8 in. diameter post-installed torque-controlled mechanical expansion anchor with 4.5 in. effective embedment depth shown in Table 34-3 is O.K. for the factored tension and shear loads.